International encyclopedia of statistics

International Encyclopedia of STATISTICS

International Encyclopedia of STATISTICS

Edited by

WILLIAM H. KRUSKAL
University of Chicago

and

JUDITH M. TANUR
State University of New York at Stony Brook

VOLUME 1

THE FREE PRESS
A Division of Macmillan Publishing Co., Inc.
NEW YORK

Collier Macmillan Publishers
LONDON

THE *International Encyclopedia of Statistics* includes articles on statistics and articles relevant to statistics that were first published in the *International Encyclopedia of the Social Sciences* (1968). These articles have been brought up to date by the addition of postscripts or by revision in whole or in part and have been supplemented by new articles and biographies listed in the Introduction.

THE FREE PRESS
A Division of Macmillan Publishing Co., Inc.
866 Third Avenue, New York, N.Y. 10022

COLLIER MACMILLAN CANADA, LTD.

Library of Congress Catalog Card Number: 78-17324

PRINTED IN THE UNITED STATES OF AMERICA

printing number

3 4 5 6 7 8 9 10

Library of Congress Cataloging in Publication Data
Main entry under title:

International encyclopedia of statistics.

Includes index.
1. Statistics—Dictionaries. 2. Social sciences—Statistical methods—Dictionaries. I. Kruskal, William H. II. Tanur, Judith M.
HA17.I63 001.4'22'03 78-17324
ISBN 0-02-917960-2 (set)
ISBN 0-02-917970-X (Vol. I)
ISBN 0-02-917980-7 (Vol. II)

Contents

List of Articles

Foreword

DURING THE YEARS when my fellow editors and I were working on the *International Encyclopedia of the Social Sciences*—from 1962 to 1968—we discussed and debated from time to time its anticipated audience. We knew it would be a diverse audience, but that did not help much in setting editorial policy and in making editorial decisions. Accordingly, we created in our imagination three ideal types of reader among social scientists: the graduate student, reading in a particular field of specialization; the professor, reading in an adjacent field; and the student or professor at an Asian, Latin American, or African university, out of touch with recent developments in the social sciences.

As far as we know, the *International Encyclopedia of the Social Sciences* (*IESS*) reaches all these audiences—and other audiences as well. The intervening years have provided us with many opportunities for hindsight in our choices of articles and contributors, and we feel that we accurately predicted (or perhaps determined) the core audience.

The articles on statistics in *IESS*, collected in the present volumes, were special from the beginning. In the first place, statistics has an intricate historical and contemporary relationship to the social science disciplines, a relationship described by William H. Kruskal in his article on the field of statistics. In the second place, statistics (along with econometrics) is especially difficult to write about for an audience with limited technical training.

Accordingly, the articles on statistics are relevant not only to the three ideal types of reader indicated above but also to professional statisticians and to research scientists in the natural and social sciences. This, it seems to me, is the major rationale for the present volumes.

W. Allen Wallis, the distinguished statistician–economist–university chancellor who served as chairman of the *IESS* Editorial Advisory Board, made the following remark at the publication-day press luncheon (as quoted in the *New York Times*): "Statistics is a field in which you can recover your amateur status very fast. But our articles here will stand up for a long time. If they were brought together under one cover, it would be the single best book in the field."

Now, a decade later, this "single best book in the field" is a reality, and I am delighted, both because of its value to all the sciences and because of the prescient glow it casts upon Allen Wallis's 1968 remark. Credit for this important accomplishment goes primarily to my friends and colleagues William H. Kruskal and Judith M. Tanur. Their vision of the relevance of statistics to scientific research of high quality made the statistics articles in *IESS* possible; their discernment of the importance of an updated and supplemented edition made this new work a reality.

DAVID L. SILLS
Social Science Research Council
New York City

Preface

THE ARTICLES ON statistics in the *International Encyclopedia of the Social Sciences* (*IESS*) have a separate potential usefulness that could not be fully exploited when they were embedded in a multivolume encyclopedia, especially in a specialized encyclopedia. The present pair of volumes makes it possible to exploit that previously incompletely realized potential.

The statistics articles of *IESS* constitute, for example, an excellent reference work for the professional statistician. But scattered through a set of 17 large and specialized volumes, they form an inconvenient statistics reference work, and they do not readily come to the attention of statistical professionals working outside the social sciences.

Similarly, the *IESS* statistics articles together might serve as an excellent textbook for a statistics course, but such a use would be cumbersome when the articles are scattered through 17 volumes.

The present volumes thus may be used as an invaluable reference work and as a stimulating textbook.

The statistics articles in *IESS* have a valuable characteristic that is difficult to find in other expositions of statistics: each is consciously constructed for readers varying widely in knowledge and sophistication, and each is comparatively self-contained. Articles of this character can be widely useful to others besides social scientists, and these volumes bring them to everyone's attention.

In the decade since *IESS* appeared, it has increasingly been recognized that statistical techniques and statistical principles of reasoning are appropriate in a wide range of activities, not only in science but also in practical affairs. The editors of these volumes have acknowledged this by bringing in material from outside the social sciences and even outside science in the ordinary sense.

The scope and quality of these articles from *IESS* well justify the title *International Encyclopedia of Statistics*. To those of us who participated in the preparation of *IESS*, it is a real satisfaction to see the usefulness of the work extended in this way.

W. ALLEN WALLIS
Chancellor
University of Rochester

Introduction

THE *International Encyclopedia of Statistics* draws together, expands, and brings up to date the statistics articles of the *International Encyclopedia of the Social Sciences* (*IESS*), edited by David L. Sills and published by The Macmillan Company and The Free Press in 1968. At the time, those articles formed by far the most extensive treatment of statistics in any encyclopedic reference work: some 70 articles on statistics proper, numerous articles on social science topics with strong statistical flavor, and about 45 biographies of statisticians and others important in the development of statistics.

The present work reproduces and amends the *IESS* statistics articles, the statistical biographies, and many of the social science articles with statistical import. To these have been added 5 new articles and 12 new biographies.

In planning this encyclopedia, we had two editorial goals: we wished to present the *IESS* statistics articles, whole and intact, in a convenient format, yet we wished also to amend and update them in order to record and explain changes in the discipline since publication of *IESS* ten years ago. Thus, each author of an *IESS* article relevant to statistics was asked to prepare a postscript covering recent advances and second thoughts, to supply emendations, and to add a fresh bibliography. Most authors responded gallantly on all counts. The postscripts make excellent, informative reading; the emendations correct the few errors that slipped in and give current data on recent editions of classic works; the new bibliographies reflect recent scholarship.

Primarily because some contributors to *IESS* are dead, postscripts and new bibliographies for a few articles were kindly prepared by others. Their willing help is specifically acknowledged in the final section of this Introduction.

We have been flexible in arrangements for updating. A few *IESS* authors asked for a chance to rewrite their articles *ab ovo* because of major scientific advances; we happily agreed to such rewritings (and to the addition of joint authors in two instances). Conversely, a few *IESS* authors believe their articles unaffected by the passage of time; we reprint those articles as first published. Other authors saw no need for substantive postscripts but supplied emendations and new bibliographies. Still others argued that use of a postscript alone would be awkward;

when appropriate, whole paragraphs of their original articles were replaced or new paragraphs inserted.

To mark this variety of amendment to *IESS* material, we set up simple systems of notation. Details are given later in this Introduction.

Articles specially written or completely revised for this encyclopedia are

DATA ANALYSIS, EXPLORATORY	*David F. Andrews*
EPIDEMIOLOGY	*Mervyn Susser*
ERRORS, *article on* EFFECTS OF ERRORS IN STATISTICAL ASSUMPTIONS	*Janet D. Elashoff and Robert M. Elashoff*
FACTOR ANALYSIS AND PRINCIPAL COMPONENTS, *article on* BILINEAR METHODS	*Joseph B. Kruskal*
MULTIVARIATE ANALYSIS, *article on* CORRELATION METHODS	*Robert F. Tate*
PUBLIC POLICY AND STATISTICS	*William B. Fairley*
SCALING, MULTIDIMENSIONAL	*J. Douglas Carroll and Joseph B. Kruskal*
STATISTICS AS LEGAL EVIDENCE	*Hans Zeisel*
SURVEY ANALYSIS, *article on* METHODS OF SURVEY ANALYSIS	*Hanan C. Selvin and Stephen J. Finch*
TIME SERIES, *general article*	*Christopher Bingham*

A list of new biographies appears later in this Introduction.

In the *IESS* statistics articles, motivating discussions and examples are mainly of a social science character, with "social science" interpreted broadly. For the present work, we encouraged authors of postscripts and fresh articles to draw on all scientific fields for illustrations. Our hope was to make this work useful to anyone who wishes to gain, or to regain, an understanding of statistical topics in the widest practicable sense. In particular, we include a considerable number of articles about fields in which statistical methods are significantly applied, for example, epidemiology, public policy, industrial quality control, demography, geography, econometrics, and so on.

Indeed, we envisioned from the start a heterogeneous readership, and we tried to keep the articles as self-contained as possible; at the same time, we introduced extensive cross-references and bibliographies as aids to systematic study. One aim was to begin each article with a readily accessible, nontechnical summary of the article's topic and coverage.

We hope that this encyclopedia will be of use to

Students (of all fields) in schools, colleges, and universities—as background reading and as a way of rapidly filling gaps in background and of surveying topics

Teachers—for study assignments, as a source of expository suggestions, and for guidance of students into fields new to them

Statisticians—as readings to use in advisory relationships and in teaching

Research scientists—as a guide to statistical understanding and to the literature, a guide organized for reference rather than as a cumulative textbook

We might add to this list journalists, librarians, philosophers, and poets. Our constant attempt was to obtain expositions that combine accuracy and clarity, breadth and grace, invitation and objectivity.

The reader who seeks an introduction to the discipline of statistics—what it is in a general way, why it is important, what criticisms have been aimed at it—should turn to STATISTICS, *article on* THE FIELD. That article provides an extended substantive introduction as well as an organized tour through the articles.

ADVANCES IN STATISTICS

This encyclopedia reflects, of course, its editors' views of the most important parts of statistics in the 1960s, together with changes or advances that have taken place since then. We hope that these views are not parochial or idiosyncratic; indeed, we tried conscientiously to obtain expositions of some topics that are regarded as far more important by others than by ourselves. Of the major statistical changes and advances in the past ten years, we include discussions of the following:

Exploratory data analysis. There has been a reaction against excessive preoccupation in statistical theory and practice with modes of inference based on probabilistic models and criteria. This reaction, together with the wide availability of high-speed computation, has introduced a flood of fresh ideas for inspection, analysis, and compression of data in ways less formally anchored than had become usual. These ideas—set forth in DATA ANALYSIS, EXPLORATORY—include new graphical and tabular methods; more important, these ideas legitimate new flexibility and freedom.

Yet we predict that the pendulum will swing partly back. One original motivation for probabilistic ideas in inference was, after all, the moderation of premature conviction by the enthusiastic investigator. Exploratory data analysis surely widens our horizons, yet, for some of us, it may result in unnecessary confusion or contradiction. A synthesis of the data-analytic approach with that of probabilistic inference seems to us likely.

One important aspect of recent stress on exploratory data analysis has been intensive work on methods—old and new—for exploring data sets of forms that appear frequently. Two-way and three-way arrays of data are often appropriate for analysis and simplification by methods described in FACTOR ANALYSIS AND PRINCIPAL COMPONENTS and in SCALING, MULTIDIMENSIONAL.

Robustness. Along with the enthusiasm for exploratory data analysis has come increased concern for the stability—called robustness—of existing inferential procedures under deviations from the usual assumptions on which they are based. Statisticians have, of course, long investigated the effects of nonnormality when normality was assumed, dependence when independence was assumed, and so on; the last decade has seen fresh attacks on such questions.

These attacks, furthermore, have gone beyond description of effects alone. There has been extensive discussion of alternative procedures that are—or are hoped to be—more robust than conventional procedures. One rarely gets something for nothing, however, and the costs of some of the new procedures are (1) less robustness to deviations other than those designed against, (2) loss of effectiveness when the usual assumptions in fact hold, (3) greater complexity in computation and in behavior, and (4) greater difficulty in understanding and exposition. These matters are discussed in DATA ANALYSIS, EXPLORATORY; ERRORS; and ESTIMATION.

Categorical data. Major steps have been taken in developing new methods for analysis of categorical, or counted, data. These include analogues to linear hypothesis methods, although one must be careful not to push that analogy too hard. See the postscripts to COUNTED DATA and to SURVEY ANALYSIS.

Computation. The role of computation in both statistical practice and statistical theory has continued to grow. Computational hardware becomes less and less expensive, and software more and more accessible. At the same time, the dangers of thoughtless computation, confused computation, and misinterpretations of computation are unhappily more severe. Far too widely held is the concept of throwing data onto the mercy of the program in the computer—the *deus in machina.* See the postscript to COMPUTATION and the articles on TIME SERIES.

Public policy. There has been an increasing concern for the statistical aspects of wide national programs or decisions, both governmental and private. Some of the issues have concerned the accuracy of government statistics; problems of integrity and ethics for the working statistician; statistics in cost–benefit policy calculations; confidentiality and privacy in censuses and surveys; clear, honest communication of statistical studies; the use of what have come to be called social indicators; and evaluation of social programs, especially by controlled, randomized field trials (that is, by proper experiments). For relevant discussions, see the postscript to GOVERNMENT STATISTICS, the new PUBLIC POLICY AND STATISTICS, and the revised STATISTICS AS LEGAL EVIDENCE.

Bibliographical advances are listed after the postscript to STATISTICS, *article on* THE FIELD.

BIOGRAPHIES

Part of *IESS* editorial policy for selecting biographees was to exclude scientists alive in the 1960s unless they were born before 1891 (and were deemed of sufficient stature and influence). After much discussion, we decided to continue that policy, except that we modified "alive in the 1960s" to "alive in the 1970s." Besides biographies of scholars who died between 1967 and 1977, we have added two that might have appeared in *IESS*, articles on Kepler and Price. The new biographies are

BIRNBAUM, ALLAN	*Dennis V. Lindley*
HOTELLING, HAROLD	*Wassily Hoeffding*
KEPLER, JOHANNES	*O. B. Sheynin*
LAZARSFELD, PAUL F.	*James S. Coleman*
MAHALANOBIS, P. C.	*C. Radhakrishna Rao*
PRICE, RICHARD	*William H. Kruskal*
RÉNYI, ALFRÉD	*István Vincze*
SAVAGE, LEONARD JIMMIE	*William H. Kruskal*
SHEWHART, WALTER A.	*W. Edwards Deming*
VAN DANTZIG, DAVID	*J. Hemelrijk*
WILCOXON, FRANK	*Ralph A. Bradley and Myles Hollander*
YOUDEN, W. J.	*H. H. Ku*

All *IESS* biographies of statisticians are, of course, reproduced and updated here. We also include some *IESS* biographies of social scientists who were not primarily statisticians but whose statistical influence has been substantial, for example, J. M. Keynes and C. E. Spearman.

Since *IESS* appeared, there has been published the *Dictionary of Scientific Biography*, edited by Charles C. Gillispie (New York: Scribner's, 1970–1976). This excellent 14-volume reference work includes biographies of many of the figures for whom biographies appear in the present work. Naturally, our biographies stress the statistical sides of the work of polymathous scientists. Because we have not always provided additional bibliographic references to this dictionary, we here commend it to the reader's attention.

EDITORIAL PRACTICES

We have carried along the *IESS* organization by alphabetical ordering and its use of composite entries to group articles on closely related topics. Headnotes have been added to some articles to clarify the consequences of our narrowing the broad context of *IESS* to focus on statistics, to explain, for example, minor modifications of titles and to note companion articles in *IESS* less relevant to statistics.

Text. In editorial style, we tried to follow the conventions established by the staff of *IESS*, and we believe that a high degree of consistency has been achieved.

Bibliographic citations. We have continued the admirable *IESS* practice of giving parenthetical dates of first publication of works cited in a chronological context, where development of ideas is important. If pages of a subsequent edition are cited, the date of first publication is given in brackets, followed by the date of the edition used for page references. Occasionally, albeit rarely, a work that remained in manuscript for many years is cited parenthetically by the date when it was written rather than when it was first published.

Cross-references. In all material reprinted from *IESS*, cross-references to articles not reprinted here were transformed into bibliographic citations. In this way, each contributor's original intention of directing readers to topics in *IESS* less relevant to statistics than those included here remains undisturbed.

Bibliographies. Matching the excellence of the *IESS* bibliographies presented a special challenge, for the *IESS* staff succeeded in producing bibliographies that are more than usually accurate and informative. We scrupulously followed the *IESS* system of verification of all bibliographic data by a library search team, and we did, of course, endeavor to equal *IESS* in both content and style.

Thus, the selected bibliographies that follow each article do not merely document the article. They also contain explicit suggestions for further reading, sources for further bibliography, sources for historical and current data, and the titles of journals concerned more or less exclusively with the topic of the article.

The bibliographic staff consistently attempted to simplify the reader's task in locating cited works in libraries. Besides paying unusually close attention to the accuracy of citations, and to determining the most widely available editions in English, we followed, as a rule, the practices of the U.S. Library of Congress in deciding whether to list a work by name of author or editor, by title, or by name of the organization that published or sponsored the work.

As in *IESS*, dates of first publication of works now in subsequent editions are given in parentheses, and other apposite details of publishing history are given in annotations following a small arrow.

Index. One notable difference from *IESS* editorial practice is our indexing of all scholars' names mentioned in bibliographic citations. Thus our Index lists not only the pages on which a scholar's work is discussed but also those on which any work is cited. Although the editors of *IESS* had hoped to follow this useful practice, time constraints made such exhaustive indexing impossible.

NOTATION OF AMENDMENTS

To mark inserted, revised, and emended paragraphs or bibliography entries in *IESS* material, we have used three easily remembered symbols,

► = inserted
● = revised
○ = emended

Postscripts and additional bibliographies, whose newness is self-evident from their headings, are not marked with these symbols, although new and revised articles are signified by both a symbol and a headnote.

To make specific cross-references between reprinted *IESS* material and new material we have used boldface superscript numbers in paragraph indentations and before bibliography entries. The presence of a superscript so placed in a main article or original bibliography signals a pertinent amendment in the postscript or additional bibliography. For example, in STATISTICS, *article on* THE FIELD, the reader will find a paragraph that begins

> [3]An extreme case of the breadth-of-inference problem is represented by the case study, . . .

and in the postscript a corresponding paragraph that begins

> [3]In connection with case studies, broadly viewed, I cite Margaret Mead's presidential address to the American Association for the Advancement of Science.

As a further example, in the original bibliography of DISTRIBUTIONS, STATISTICAL, *article on* APPROXIMATIONS TO DISTRIBUTIONS, the following entry appears:

> [2]ELDERTON, W. PALIN (1906) 1953 *Frequency Curves and Correlation.* 4th ed. Washington: Harren.

and in the additional bibliography this corresponding item:

> [2]ELDERTON, W. PALIN; and JOHNSON, NORMAN L. 1969 *Systems of Frequency Curves.* New York: Cambridge Univ. Press. → A drastic revision of Elderton (1906), which is now out of print.

These simple systems of symbols and key numbers have allowed us to call attention to advances and changes since the 1968 publication of *IESS* with minimum editorial disturbance of the original material.

ACKNOWLEDGMENTS

David L. Sills, editor of *IESS*, has been a constant fount of ideas, encouragement, and moral support. We thank him for all. We also thank W. Allen Wallis, chairman of the *IESS* Editorial Advisory Board, and Jeremiah Kaplan, president of Macmillan Publishing Co., for their encouragement and faith. We express further appreciation to Bert F. Hoselitz and J. Arthur Greenwood, who made special contributions in the early preparation of *IESS;* to Elinor G. Barber, for her devoted attention to the *IESS* biographies; and to George Lowy, for his meticulous direction of our bibliographical search.

We are especially grateful to the contributors of postscripts and emendations to articles not originally their own,

Oskar A. Anderson	OSKAR N. ANDERSON
Albert Jacquard	DEMOGRAPHY, *article on* DEMOGRAPHY AND POPULATION GENETICS
Edward B. Perrin	LIFE TABLES
Thomas W. Pullum	SOCIAL MOBILITY
Leopold K. Schmetterer	VON MISES, RICHARD
Glenn Shafer	BERNOULLI FAMILY
O. B. Sheynin	BORTKIEWICZ, LADISLAUS VON
Stephen M. Stigler	LAPLACE, PIERRE SIMON DE; PEIRCE, CHARLES SANDERS

Edward Barry and Charles Smith, senior officers of The Free Press, and we have been mutually patient; their understanding is appreciated. Claude Conyers, who has served as our redactor in chief, has been an intelligent and indefatigable colleague. Valerie Klima, production manager, has paid careful attention to unusually complicated details of bookmaking.

Besides these, so many people were helpful in connection with specific articles that we cannot name them individually, and we apologize for this generalized but heartfelt expression of thanks. Specific acknowledgments by contributors appear at the back of volume 2. We also thank the members of the *IESS* Editorial Advisory Board, many of whom were concerned one way or another with statistics.

Finally, we thank the contributors of the articles and postscripts that follow. Their rewards must primarily be the gratitude that we hope is yours.

WILLIAM H. KRUSKAL

Department of Statistics
University of Chicago

JUDITH M. TANUR

Department of Sociology
State University of New York
at Stony Brook

A

ACCEPTANCE SAMPLING

See under QUALITY CONTROL, STATISTICAL.

ANALYSIS, CROSS-SECTION

See CROSS-SECTION ANALYSIS.

ANALYSIS OF VARIANCE

See under LINEAR HYPOTHESES.

ANDERSON, OSKAR N.

Oskar Nikolayevich Anderson (1887–1960), a pioneer of applied sampling-survey techniques, and remembered also for his contributions to the variate-difference method, has been described as "perhaps the most widely known statistician in Central Europe. . . . He provided a link between the Russian school of statistics . . . ([A.A.] Markoff [Sr.], Tschuprow) and the Anglo-American school. . . . Through his origin in the flourishing Russian school of 'probabilistes,' . . . Anderson belongs to the so-called 'continental' school of statistics, and worked in the tradition of the well known German statisticians [W.] Lexis and [L.] von Bortkiewicz. He might be the last representative of this approach . . ." (Tintner 1961, p. 273).

Anderson was born in Minsk, Byelorussia. His father was a professor of Finno–Ugric languages at the University of Kazan. Although the Andersons were Russian subjects, they were ethnically German. In 1906 Anderson was graduated from the gymnasium in Kazan with a gold medal. After studying mathematics at the University of Kazan for a year, he entered the economics department of the Polytechnical Institute in St. Petersburg. There he became an outstanding pupil of Aleksandr A. Chuprov (or Tschuprow), whose strong permanent influence on Anderson is evident, even in such detailed matters as taxonomic conventions. As B. I. Karpenko wrote: "The ideas of A. A. Chuprov penetrate into foreign science not only directly but also through the writings of his students . . . e.g., O. N. Anderson, who wrote a series of valuable works, in particular a serious book on the theory of statistics, which (in German) expounds A. A. Chuprov's ideas . . ." (Karpenko 1957, p. 317).

From 1912 to 1917 Anderson taught in a commercial gymnasium in St. Petersburg. While there he also obtained a law degree. In 1915 he took part in an expedition to Turkestan to make a survey of agricultural production under irrigation in the Syr Darya River area. There, as chief scientific consultant, Anderson made one of the earliest applications of sampling methods. It apparently had only one Russian precedent: on January 6, 1910, Chuprov had presented a paper on the application of sampling techniques to data from the 1898–1900 rural census, but this application had had an "experimental, rather than a practical, character" (Volkov 1961, p. 159).

In 1917 Anderson served as research economist for a large cooperative society in southern Russia and also underwent further training in statistics at the Commercial Institute in Kiev, where he became a docent in 1918. Concurrently, he held an executive position in the demographic institute of the Academy of Sciences in Kiev. While in Kiev, he came to know and was perhaps influenced by Eugen Slutsky.

Like many Russian students of his time, Anderson had leftist sympathies. In 1920, however, he and his family left Russia as a result of the political upheavals. The following year he became a high school principal in Budapest; from 1924 to 1933 he was a professor at the Commercial Institute in Varna, Bulgaria; and from 1935 to 1942 he held a similar position at the University of Sofia. From the mid-1920s on, he was a member of the Supreme Statistical Council of the Bulgarian government. He successfully advocated the use of sampling techniques—in addition to a complete enumeration—in the 1926 census of population and manufacture. Another large-scale sample survey instigated by Anderson covered Bulgarian agricultural production and producers in 1931–1932. In 1936 he began a complete redesigning of the acreage and crop statistics, basing them on purposive sampling.

In 1933 Anderson went to England and Germany on a Rockefeller stipend and, as a result of this trip, he published his first textbook (1935). He was a charter member of the Econometric Society. He contributed the article "Statistical Method" to the *Encyclopaedia of the Social Sciences* (1934). From the middle 1930s on, he served as an adviser for the League of Nations. In 1940 the Bulgarian government sent him to Germany, then at war, to study rationing. In 1942 Anderson accepted an appointment at the University of Kiel, Germany, and from 1947 to his death he was a professor of statistics in the economics department of the University of Munich. He was coeditor of the *Mitteilungsblatt für mathematische Statistik* (later *Metrika*) from its inception until he died. At the time of his death, his authority in German statistical circles was unrivaled. It was mainly through his efforts that the statistical training for economists at German universities was improved or was maintained at a reasonable level, despite various adverse influences.

Major contributions. One author has summed up Anderson's lifework in the following manner:

> The course of outer events in Oskar Anderson's life reflects the turbulence and agonies of a Europe torn by wars and revolutions. His scientific work, always marked by personal involvement, is of sufficient stature to be of lasting interest. . . . Some of Anderson's endeavours were ahead of his time, along lines that have not yet received adequate attention. Thus his emphasis on causal analysis of nonexperimental data is a reminder that this important sector of applied statistics is far less developed than descriptive statistics and experimental analysis. . . . The main strength of Anderson's scientific *œuvre* lies, I think, in the systematic coordination of theory and application. Only to a relatively small extent does his importance derive from specific contributions. . . . (Wold 1961, pp. 651–653)

Despite the cogent appraisal by Wold, it seems desirable to take up some of Anderson's particular contributions.

Sample surveys. The Turkestan sample survey of 1915 and a demographic sampling study of 1916–1917 are contained in manuscripts that were lost; it seems safe to assume that the lost papers were valuable ones. We do have access to the first Bulgarian sample survey, which Anderson designed and whose implementation he supervised (1929*a*).

Variate-difference method in time series. Utilizing, in essence, the theorem that the nth difference of a polynomial of degree n is a constant and the $(n + 1)$st zero, the variate-difference method attempts to analyze time series on the basis of few assumptions, but including the sensitive one that random errors are not autocorrelated. Anderson developed the method concurrently with William S. Gosset (for details, see Tintner 1940, pp. 10–15). The method has not met with general favor, although it is still taken as a point of departure for various theoretical studies. Anderson himself was aware of its limitations (1954, pp. 178–180 in 1957 edition). In a different context, his incisive critique of the Harvard method of time-series analysis (1929*b*) is recognized as having definitively discredited mechanical procedures of that kind.

Quantitative economics. Anderson's study of the "verifiability" of the quantity theory of money (1931) is an econometric classic because Anderson took advantage of the then new awareness of the importance of random residuals. Several of his papers analyze the causes of divergent movements of agricultural and industrial prices. A critique of N. D. Kondratieff's work on long waves (in business cycles) is among Anderson's lost papers. His contributions to index-number theory were both constructive and critical. For reasons of error accumulation, he was specifically against chain indexing (1952).

Probability theory and nonparametric methods. Anderson's was essentially an eclectic, and somewhat modified, frequency point of view (e.g. 1954, pp. 98–100 in 1957 edition). In particular, he felt that finite, rather than infinite, urn models were more appropriate in most social contexts; correspondingly, he was against too facile an invocation of the central-limit theorem. In his papers on nonparametric methods (1955*a*; 1955*b*), he intended to make correlation and regression models applicable to a wider range of socioeconomic phenomena.

Textbooks. In his first textbook (1935), Anderson tried to expound twentieth-century statistical methods using preuniversity-level mathematics. Because of the time of publication and the then predominant doctrines, its influence seems to have been stronger outside of Germany than within. His second textbook (1954), however, went through three editions in three years; it was unusual for its highly personal anecdotal style and its abundance of historical, biographical, institutional, and mathematical asides.

Students. Anderson's best-known students, all professors in Germany, are Hans Kellerer, at Munich; Heinrich Strecker, at Tübingen; and Anderson's son Oskar, at Mannheim. What clearly characterizes them as Andersonians, especially the former two, is a strong interest and activity in sampling-survey design, a concern for the implemental side of statistics, and avowed reservations against abstractions unrelated to practice. Strecker has also worked on variants of the variate-difference method.

EBERHARD M. FELS

[*For discussion of the history and subsequent development of the fields in which Anderson worked, see* NONPARAMETRIC STATISTICS; PROBABILITY; SAMPLE SURVEYS; TIME SERIES.]

WORKS BY ANDERSON

(1929*a*) 1949 *Über die repräsentative Methode und deren Anwendung auf die Aufarbeitung der Ergebnisse der bulgarischen landwirtschaftlichen Betriebszählung vom 31. XII. 1926.* Munich: Bayerisches Statistisches Landesamt. → First published in Bulgarian.

1929*b* *Zur Problematik der empirisch-statistischen Konjunkturforschung: Kritische Betrachtung der Harvard-Methoden.* Bonn: Schroeder.

1931 Ist die Quantitätstheorie statistisch nachweisbar? *Zeitschrift für Nationalökonomie* 2:523–578.

(1934) 1959 Statistical Method. Volume 14, pages 366–371 in *Encyclopaedia of the Social Sciences.* New York: Macmillan.

1935 *Einführung in die mathematische Statistik.* Vienna: Springer.

1952 Wieder eine Indexverkettung? *Mitteilungsblatt für mathematische Statistik* 4:32–47.

(1954) 1962 *Probleme der statistischen Methodenlehre in den Sozialwissenschaften.* 4th ed. Würzburg (Germany): Physica-Verlag. → Pages cited in text refer to the 1957 edition.

1955*a* Eine "nicht-parametrische" (verteilungsfreie) Ableitung der Streuung (Variance) des multiplen ($R_{z.xy}$) und partiellen ($R_{xy.z}$) Korrelationskoeffizienten im Falle der sogenannten Null-Hypothese, sowie der dieser Hypothese entsprechenden mittleren quadratischen Abweichungen (Standard Deviations) der Regressionskoeffizienten. *Mitteilungsblatt für mathematische Statistik* 7:85–112.

1955*b* Wann ist der Korrelationsindex von Fechner "Gesichert" (Significant)? *Mitteilungsblatt für mathematische Statistik* 7:166–167.

1963 *Ausgewählte Schriften.* 2 vols. Edited by H. Kellerer et al. Tübingen (Germany): Mohr. → Contains most of Anderson's extant papers—translated into German if originals are in other languages—and a biography.

SUPPLEMENTARY BIBLIOGRAPHY

KARPENKO, B. I. 1957 Zhizn' i deiatel'nost' A. A. Chuprova (The Life and Activity of A. A. Chuprov). Akademiia nauk SSSR, Otdelenie ekonomicheskikh, filosofskikh i pravovykh nauk, *Uchenye zapiski po statistike* 3:282–317.

TINTNER, GERHARD 1940 *The Variate Difference Method.* Bloomington, Ind.: Principia Press.

TINTNER, GERHARD 1961 The Statistical Work of Oskar Anderson. *Journal of the American Statistical Association* 56:273–280. → Contains a bibliography.

VOLKOV, A. G. 1961 Vyborochnoe nabliudenie naseleniia (The Sample Survey of Population). Akademiia nauk SSSR, Otdelenie ekonomicheskikh, filosofskikh i pravovykh nauk, *Uchenye zapiski po statistike* 6:157–184.

WOLD, HERMAN 1961 Oskar Anderson: 1887–1960. *Annals of Mathematical Statistics* 32:651–660. → Contains a bibliography.

APPROXIMATIONS TO DISTRIBUTIONS

See under DISTRIBUTIONS, STATISTICAL.

ASSOCIATION, STATISTICS OF

See under STATISTICS, DESCRIPTIVE.

ATTRIBUTES, STATISTICS OF

See COUNTED DATA; STATISTICS, DESCRIPTIVE; SURVEY ANALYSIS; *and the biographies of* LAZARSFELD *and* YULE.

AUTOCORRELATION

See TIME SERIES.

AUTOREGRESSIVE SERIES

See TIME SERIES.

AVERAGES

See STATISTICS, DESCRIPTIVE, *article on* LOCATION AND DISPERSION.

AVERAGES, LAW OF

See PROBABILITY.

B

BABBAGE, CHARLES

Charles Babbage (1792–1871), English mathematician, did pioneering work on calculating machines and in operations research and was active in winning public support for science. A man far ahead of his time, he was generally recognized only long after his death. However, his work strikingly anticipated certain key developments in modern thought. The great electronic computers, whose uses have multiplied enormously since they were developed in the mid-twentieth century, are based on principles first stated by Babbage.

His dream was to mechanize the abstract operations of mathematics for use in industry. His first idea was that of the "difference engine," a machine for integrating difference equations, which formed mathematical tables by interpolation and set them directly into type. Babbage pointed out the advantages such a machine would have for the government in preparing its lengthy tables for navigation and astronomy. With the enthusiastic approval of the Royal and Astronomical societies, the government of England agreed to grant funds for the construction of such a machine. Work proceeded for about eight years but stopped abruptly after a dispute between Babbage and his chief engineer. Shortly thereafter Babbage thought of another machine, the "analytical engine," built on an entirely new principle—internal programming—and wholly superseding and transcending the difference engine. Babbage explained his new idea to the first lord of the Treasury and asked for an official decision on whether to continue and complete the original difference engine or to suspend work on it until the analytical engine was further developed. The government had already spent £17,000 on the difference engine, and Babbage had contributed a large amount from his private fortune. After years of correspondence with various government officials, Babbage was advised that the prime minister, Sir Robert Peel, had decided the government must abandon the project because of the expense involved.

Babbage continued to work on his analytical engine. The machine he envisioned (which he called "the Engine eating its own tail") was one that could change its operations in accordance with the results of its own calculations. The machine could make judgments by comparing numbers and then, acting on the result of its comparisons, could proceed along lines not specified in advance by its instructions. These notions are acknowledged as the backbone of modern digital computers. Bound by the technology of his time, Babbage had to translate his great idea into wholly mechanical form, using a mass of intricate clockwork in pewter, brass, and steel, with punched cards modeled on those of the Jacquard loom.

After some years of work on his analytical engine, Babbage decided to design a second difference engine, which would incorporate the improvements suggested by his work on the analytical engine. He again asked for government support but was again refused. Babbage completed only small bits of a working engine and did not publish any detailed descriptions of them other than the informal ones in the autobiography he wrote as a disappointed old man, *Passages From the Life of a Philosopher* (1864). After his death one of his sons, Major Henry P. Babbage, compiled and published a book including papers both by Babbage himself and by his contemporaries, entitled *Babbage's Calculating Engines.*

While working on his engines, Babbage became deeply involved in the problems of establishing and maintaining in his machine shop and drafting room the new standards of precision that his designs demanded. Under his direction the machinists he employed developed tools and methods far ahead of contemporary practice; these developments alone might have justified the government's expenditures. Babbage also invented a scheme of mechanical symbols that could make clear the action of all the complicated moving parts of his machinery. The detailed drawings of his engines were models for their day.

The son of a banker in Devon, who later left him a considerable fortune, Charles Babbage was educated mostly at home, with mathematics his favorite subject. At Cambridge University his closest friends were John Herschel (later the astronomer royal) and George Peacock (later the dean of Ely); with them Babbage solemnly entered into a compact to "do their best to leave the world wiser than they found it." They began their mission by translating Sylvestre Lacroix's *An Elementary Treatise on the Differential and Integral Calculus* and founding the Analytical Society, whose purpose was to put "English mathematicians on an equal basis with their continental rivals." Babbage published a variety of mathematical papers after receiving his M.A. from Cambridge in 1817. His interest in mathematics led directly to a concern for accurate and readable mathematical tables. A chance conversation with Herschel, while the two were checking a table of calculations done for the Astronomical Society (which they had recently helped to found), led Babbage to his dream of a machine for calculating mathematical tables, a dream that was to become the obsession of his life.

Although he never abandoned the pursuit of his engines, his great curiosity and enthusiasm led him onto many other paths. The problems he encountered in the construction of his own machines aroused his interest in the general problems of manufacturing. After a tour of factories throughout England and the Continent, Babbage wrote his most popular book, *On the Economy of Machinery and Manufactures* (1832). The book included a detailed description and classification of the tools and machinery he had observed, together with a discussion of the "economical processes of manufacturing." A pioneer work in the field that, one hundred years later, we call operations research, the *Economy* is still good reading.

In addition to pure and applied mathematics, Babbage wrote papers on physics and geology, astronomy and biology. He even ventured into the fields of archeology and apologetics and wrote one of the first clear popular accounts of the theory of life insurance. He also enjoyed making suggestions for practical inventions of all kinds, ranging from the cowcatcher on a railway locomotive to a system of flashing signals for lighthouses.

An enthusiastic conference man, Babbage was an active member of learned societies all over the world. He was instrumental, with Herschel, in founding the Royal Astronomical Society in 1820, the British Association for the Advancement of Science (BAAS) in 1831, and the Statistical Society in 1834. For years Babbage led an assault on the decline of science in England, attacked the neglect of science in the universities, and urged government support of scientists. He pointed out that only men with private fortunes could pursue abstract science and that "scientific knowledge today hardly exists among the higher classes." The chief target of his book *Reflections on the Decline of Science in England* (1830) was the Royal Society, to which he had been elected while still at Cambridge. He attacked the autocratic misrule of the society by a social clique and pointed out that only a small proportion of the society's members ever contributed papers to its *Transactions.* His book received a good deal of support from other members, and within the next twenty years the Royal Society did succeed in reorganizing itself in response to their criticisms. In an appendix to the *Decline of Science,* Babbage reprinted without comment an account of "an annual Congress of German naturalists meeting in each successive year in some great town." This account probably inspired the first meeting of the BAAS in 1831, with Babbage taking a leading part in shaping its constitution.

Babbage was deeply committed to the belief that careful analysis, mathematical procedures, and statistical calculations—using high-speed computation—could be reliable guides in practical and productive life. This conviction, combined with the wide range of his organizational and scientific interests, gives him still a wonderful modernity.

PHILIP MORRISON AND EMILY MORRISON

[*For discussion of the subsequent development of Babbage's ideas, see* COMPUTATION. *See also* Maron 1968.]

WORKS BY BABBAGE

1830 *Reflections on the Decline of Science in England, and on Some of Its Causes.* London: Fellowes.

(1832) 1841 *On the Economy of Machinery and Manufactures.* 4th ed. enl. London: Knight.

1864 *Passages From the Life of a Philosopher.* London: Longmans.

Babbage's Calculating Engines: Being a Collection of Papers Relating to Them; Their History and Construction. Edited by Henry P. Babbage. London: Spon. 1889.

Charles Babbage and His Calculating Engines: Selected Writings by Charles Babbage and Others. Edited with an introduction by Philip and Emily Morrison. New York: Dover. 1961.

SUPPLEMENTARY BIBLIOGRAPHY

BOWDEN, BERTRAM V. (editor) (1953) 1957 *Faster Than Thought: A Symposium on Digital Computing Machines.* London: Pitman.

►MARON, M. E. 1968 Cybernetics. Volume 4, pages 3–6 in *International Encyclopedia of the Social Sciences.* Edited by David L. Sills. New York: Macmillan and Free Press.

MULLETT, CHARLES F. 1948 Charles Babbage: A Scientific Gadfly. *Scientific Monthly* 67:361–371.

BARGAINING

See GAME THEORY, *especially the article on* ECONOMIC APPLICATIONS.

BAYES, THOMAS

Thomas Bayes (1702–1761) was the eldest son of the Reverend Joshua Bayes, one of the first nonconformist ministers to be publicly ordained in England. The younger Bayes spent the last thirty years of his comfortable, celibate life as Presbyterian minister of the meeting house, Mount Sion, in the fashionable town of Tunbridge Wells, Kent. Little is known about his personal history, and there is no record that he communicated with the well-known scientists of his day. Circumstantial evidence suggests that he was educated in literature, languages, and science at Coward's dissenting academy in London (Holland 1962). He was elected a fellow of the Royal Society in 1742, presumably on the basis of two metaphysical tracts he published (one of them anonymously) in 1731 and 1736 (Barnard 1958). The only mathematical work from his pen consists of two articles published posthumously in 1764 by his friend Richard Price, one of the pioneers of social security (Ogborn 1962). The first is a short note, written in the form of an undated letter, on the divergence of the Stirling (de Moivre) series ln(z!). It has been suggested that Bayes' remark that the use of "a proper number of the first terms of the . . . series" will produce an accurate result constitutes the first recognition of the asymptotic behavior of a series expansion (see Deming's remarks in Bayes [1764] 1963). The second article is the famous "An Essay Towards Solving a Problem in the Doctrine of Chances," with Price's preface, footnotes, and appendix (followed, a year later, by a continuation and further development of some of Bayes' results).

The "Problem" posed in the Essay is: "*Given* the number of times in which an unknown event has happened and failed: *Required* the chance that the probability of its happening in a single trial lies somewhere between any two degrees of probability that can be named." A few sentences later Bayes writes: "By *chance* I mean the same as probability" ([1764] 1963, p. 376).

If the number of successful happenings of the event is p and the failures q, and if the two named "degrees" of probability are b and f, respectively, Proposition 9 of the Essay provides the following answer expressed in terms of areas under the curve $x^p(1-x)^q$:

$$(1) \qquad \int_b^f x^p(1-x)^q\,dx \Big/ \int_0^1 x^p(1-x)^q\,dx.$$

This is based on the assumption (Bayes' "Postulate 1") that all values of the unknown probability are equally likely before the observations are made. Bayes indicated the applicability of this postulate in his famous "Scholium": "that the . . . rule is the proper one to be used in the case of an event concerning the probability of which we absolutely know nothing antecedently to any trials made concerning it, seems to appear from the following consideration; viz. that concerning such an event I have no reason to think that, in a certain number of trials, it should rather happen any one possible number of times than another" (*ibid.*, pp. 392–393).

The remainder of Bayes' Essay and the supplement (half of which was written by Price) consists of attempts to evaluate (1) numerically, (a) by expansion of the integrand and (b) by integration by parts. The results are satisfactory for p and q small but the approximations for large p, q are only of historical interest (Wishart 1927).

Opinions about the intellectual and mathematical ability evidenced by the letter and the essay are extraordinarily diverse. Netto (1908), after outlining Bayes' geometrical proof, agreed with Laplace ([1812] 1820) that it is *ein wenig verwickelt* ("somewhat involved"). Todhunter (1865) thought that the résumé of probability theory that precedes Proposition 9 was "excessively obscure." Molina (in Bayes [1764] 1963, p. xi) said that "Bayes and Price . . . can hardly be classed with the great mathematicians that immediately preceded or followed them," and Hogben (1957, p. 133) stated that "the ideas commonly identified with the name of Bayes are largely [Laplace's]."

On the other hand von Wright (1951, p. 292) found Bayes' Essay "a masterpiece of mathematical elegance and free from . . . obscure philosophical pretentions." Barnard (1958, p. 295) wrote that Bayes' "mathematical work . . . is of the very highest quality." Fisher ([1956] 1959, p. 8) concurred with these views when he said Bayes' "mathematical contributions . . . show him to have been in the first rank of independent thinkers. . . ."

The subsequent history of mathematicians' and philosophers' extensions and criticisms of Proposition 9—the only statement that can properly be called Bayes' theorem (or rule)—is entertaining and instructive. In his first published article on probability theory, Laplace (1774), without mentioning Bayes, introduced the principle that if p_j is the probability of an observable event resulting from "cause" j ($j = 1, 2, 3, \cdots, n$) then the probability that "cause" j is operative to produce the observed event is

$$p_j \Big/ \sum_{j=1}^{n} p_j . \tag{2}$$

This is Principle III of the first (1812) edition of Laplace's probability text, and it implies that the prior (antecedent, initial) probabilities of each of the "causes" is the same. However, in the second (1814) edition Laplace added a few lines saying that if the "causes" are not equally probable a priori (2) would become

$$\omega_j p_j \Big/ \sum_{j=1}^{n} \omega_j p_j , \tag{3}$$

where ω_j is the prior probability of cause j and p_j is now the probability of the event, given that "cause" j is operative. He gave no illustrations of this more general formula.

Laplace (1774) applied his new principle (2) to find the probability of drawing m white and n black tickets in a specified order from an urn containing an infinite number of white and black tickets in an unknown ratio and from which p white and q black tickets have already been drawn. His solution, namely,

$$\int_0^1 x^{p+m}(1-x)^{q+n}\,dx \Big/ \int_0^1 x^p (1-x)^q\,dx = \frac{(p+m)!\,(q+n)!\,(p+q+1)!}{p!\,q!\,(p+q+m+1)!}, \tag{4}$$

was later (1778–1781; 1812, chapter 6) generalized by the bare statement that if all values of x are not equally probable a factor $z(x)$ representing the a priori probability density (*facilité*) of x must appear in both integrands. However, Laplace's own views on the applicability of expressions like (4) were stated in 1778 (1778–1781, p. 264) and agree with those of Bayes' Scholium: "Lorsqu'on n'a aucune donnée *a priori* sur la possibilité d'un événement, il faut supposer toutes les possibilités, depuis zéro jusqu'à l'unité, également probables. . . ." ("When nothing is given a priori as to the probability of an event, one must suppose all probabilities, from zero to one, to be equally likely. . . .") Much later Karl Pearson (1924, p. 191) pointed out that Bayes was "considering excess of one variate . . . over a second . . . as the determining factor of occurrence" and this led naturally to a generalization of the measure in the integrals of (1). Fisher (1956) has even suggested that Bayes himself had this possibility in mind.

Laplace's views about prior probability distributions found qualified acceptance on the Continent (von Kries 1886) but were subjected to strong criticism in England (Boole 1854; Venn 1866; Chrystal 1891; Fisher 1922), where a relative frequency definition of probability was proposed and found incompatible with the uniform prior distribution (for example, E. S. Pearson 1925). However, developments in the theory of inference (Keynes 1921; Ramsey 1923–1928; Jeffreys 1931; de Finetti 1937; Savage 1954; Good 1965) suggest that there are advantages to be gained from a "subjective" or a "logical" definition of probability and this approach gives Bayes' theorem, in its more general form, a central place in inductive procedures (Jeffreys 1939; Raiffa & Schlaifer 1961; Lindley 1965).

HILARY L. SEAL

[*For the historical context of Bayes' work, see* STATISTICS, *article on* THE HISTORY OF STATISTICAL METHOD; *and the biography of* LAPLACE. *For discussion of the subsequent development of his ideas, see* BAYESIAN INFERENCE; PROBABILITY; *and the biographies of* FISHER *and* PEARSON.]

BIBLIOGRAPHY

BARNARD, G. A. 1958 Thomas Bayes: A Biographical Note. *Biometrika* 45:293–295.

BAYES, THOMAS (1764) 1963 *Facsimiles of Two Papers by Bayes.* New York: Hafner. → Contains "An Essay Towards Solving a Problem in the Doctrine of Chances, With Richard Price's Foreword and Discussion," with a commentary by Edward C. Molina; and "A Letter on Asymptotic Series From Bayes to John Canton," with a commentary by W. Edwards Deming. Both essays first appeared in Volume 53 of the *Philosophical Transactions*, Royal Society of London, and retain the original pagination.

BOOLE, GEORGE (1854) 1951 *An Investigation of the Laws of Thought, on Which Are Founded the Mathematical Theories of Logic and Probabilities.* New York: Dover.

CHRYSTAL, GEORGE 1891 On Some Fundamental Principles in the Theory of Probability. Actuarial Society of Edinburgh, *Transactions* 2:419–439.

DE FINETTI, BRUNO 1937 La prévision: Ses lois logiques, ses sources subjectives. Paris, Université de, Institut Henri Poincaré, *Annales* 7:1–68.

FISHER, R. A. (1922) 1950 On the Mathematical Foundations of Theoretical Statistics. Pages 10.307a–10.368 in R. A. Fisher, *Contributions to Mathematical Statistics*. New York: Wiley. → First published in Volume 222 of the *Philosophical Transactions*, Series A, Royal Society of London.

FISHER, R. A. (1956) 1959 *Statistical Methods and Scientific Inference*. 2d ed., rev. New York: Hafner; London: Oliver & Boyd.

GOOD, IRVING J. 1965 *The Estimation of Probabilities: An Essay on Modern Bayesian Methods*. Cambridge, Mass.: M.I.T. Press.

HOGBEN, LANCELOT T. 1957 *Statistical Theory; the Relationship of Probability, Credibility and Error: An Examination of the Contemporary Crisis in Statistical Theory From a Behaviourist Viewpoint*. London: Allen & Unwin.

HOLLAND, J. D. 1962 The Reverend Thomas Bayes, F.R.S. (1702–1761). *Journal of the Royal Statistical Society* Series A 125:451–461.

JEFFREYS, HAROLD (1931) 1957 *Scientific Inference*. 2d ed. Cambridge Univ. Press.

JEFFREYS, HAROLD (1939) 1961 *Theory of Probability*. 3d ed. Oxford: Clarendon.

KEYNES, J. M. (1921) 1952 *A Treatise on Probability*. London: Macmillan. → A paperback edition was published in 1962 by Harper.

KRIES, JOHANNES VON (1886) 1927 *Die Principien der Wahrscheinlichkeitsrechnung: Eine logische Untersuchung*. 2d ed. Tübingen (Germany): Mohr.

LAPLACE, PIERRE S. (1774) 1891 Mémoire sur la probabilité des causes par les événements. Volume 8, pages 27–65 in Pierre S. Laplace, *Oeuvres complètes de Laplace*. Paris: Gauthier-Villars.

LAPLACE, PIERRE S. (1778–1781) 1893 Mémoire sur les probabilités. Volume 9, pages 383–485 in Pierre S. Laplace, *Oeuvres complètes de Laplace*. Paris: Gauthier-Villars.

LAPLACE, PIERRE S. (1812) 1820 *Théorie analytique des probabilités*. 3d ed., rev. Paris: Courcier.

LINDLEY, DENNIS V. 1965 *Introduction to Probability and Statistics From a Bayesian Viewpoint*. 2 vols. Cambridge Univ. Press.

NETTO, E. 1908 Kombinatorik, Wahrscheinlichkeitsrechnung, Reihen-Imaginäres. Volume 4, pages 199–318, in Moritz Cantor (editor), *Vorlesungen über Geschichte der Mathematik*. Leipzig: Teubner.

OGBORN, MAURICE E. 1962 *Equitable Assurances: The Story of Life Assurance in the Experience of The Equitable Life Assurance Society, 1762–1962*. London: Allen & Unwin.

PEARSON, EGON S. 1925 Bayes' Theorem, Examined in the Light of Experimental Sampling. *Biometrika* 17: 388–442.

PEARSON, KARL 1924 Note on Bayes' Theorem. *Biometrika* 16:190–193.

PRICE, RICHARD 1765 A Demonstration of the Second Rule in the Essay Towards a Solution of a Problem in the Doctrine of Chances. Royal Society of London, *Philosophical Transactions* 54:296–325. → Reprinted by Johnson in 1965.

RAIFFA, HOWARD; and SCHLAIFER, ROBERT 1961 *Applied Statistical Decision Theory*. Harvard University Graduate School of Business Administration, Studies in Managerial Economics. Boston: The School.

RAMSEY, FRANK P. (1923–1928) 1950 *The Foundations of Mathematics and Other Logical Essays*. New York: Humanities.

○SAVAGE, LEONARD J. (1954) 1972 *The Foundations of Statistics*. Rev. ed. New York: Dover. → Includes a new preface.

TODHUNTER, ISAAC (1865) 1949 *A History of the Mathematical Theory of Probability From the Time of Pascal to That of Laplace*. New York: Chelsea.

VENN, JOHN (1866) 1888 *The Logic of Chance: An Essay on the Foundations and Province of the Theory of Probability, With Special Reference to Its Logical Bearings and Its Application to Moral and Social Science*. 3d ed. London: Macmillan.

WISHART, JOHN 1927 On the Approximate Quadrature of Certain Skew Curves, With an Account of the Researches of Thomas Bayes. *Biometrika* 19:1–38.

WRIGHT, GEORG H. VON 1951 *A Treatise on Induction and Probability*. London: Routledge.

Postscript

It is difficult to stay current with the extensive literature dealing with methods flowing from Thomas Bayes' original suggestion. It is also difficult to maintain a clear mind in the profusion of discussions about whether to use so-called Bayesian techniques. One may cite the proceedings of the 1970 Symposium on the Foundations of Statistical Inference (1971) and the philosophical treatise by Stegmüller (1973). Two pairs of eminent authors, with long experience in both theory and application of statistics, have adopted very different approaches toward the Bayesian viewpoint: Box and Tiao (1973) and Kempthorne and Folks (1971).

HILARY L. SEAL

ADDITIONAL BIBLIOGRAPHY

BOX, GEORGE E. P.; and TIAO, GEORGE C. 1973 *Bayesian Inference in Statistical Analysis*. Reading, Mass.: Addison-Wesley.

¹DE FINETTI, BRUNO (1937) 1964 Foresight: Its Logical Laws, Its Subjective Sources. Pages 93–158 in Henry E. Kyberg, Jr. and Howard E. Smokler, *Studies in Subjective Probability*. New York: Wiley. → First published in French.

KEMPTHORNE, OSCAR; and FOLKS, LEROY 1971 *Probability, Statistics and Data Analysis*. Ames: Iowa State Univ. Press.

STEGMÜLLER, WOLFGANG 1973 *Personelle und statistische Wahrscheinlichkeit*. Volume 2: *Statistisches Schliessen, statistische Begründung, statistische Analyse*. Berlin: Springer.

SYMPOSIUM ON THE FOUNDATIONS OF STATISTICAL INFERENCE, UNIVERSITY OF WATERLOO, 1970 1971 *Foundations of Statistical Inference: Proceedings*. Edited by V. P. Godambe and D. A. Sprott. Toronto: Holt.

BAYESIAN INFERENCE

Bayesian inference or Bayesian statistics is an approach to statistical inference based on the theory of subjective probability. A formal Bayesian analysis leads to probabilistic assessments of the object of uncertainty. For example, a Bayesian inference might be, "The probability is .95 that the mean of a normal distribution lies between 12.1 and 23.7." The number .95 represents a degree of belief, either in the sense of *subjective probability coherent* or *subjective probability rational* [*see* PROBABILITY, *article on* INTERPRETATIONS, *which should be read in conjunction with the present article*]; .95 need not correspond to any "objective" long-run relative frequency. Very roughly, a degree of belief of .95 can be interpreted as betting odds of 95 to 5 or 19 to 1. A degree of belief is always potentially a basis for action; for example, it may be combined with utilities by the principle of maximization of expected utility [*see* DECISION THEORY; *see also* Georgescu-Roegen 1968].

By contrast, the sampling theory or classical approach to inference leads to probabilistic statements about the method by which a particular inference is obtained. Thus a classical inference might be, "A .95 confidence interval for the mean of a normal distribution extends from 12.1 to 23.7" [*see* ESTIMATION, *article on* CONFIDENCE INTERVALS AND REGIONS]. The number .95 here represents a long-run relative frequency, namely the frequency with which intervals obtained by the method that resulted in the present interval would in fact include the unknown mean. (It is not to be inferred from the fact that we used the same numbers, .95, 12.1, and 23.7, in both illustrations that there will necessarily be a numerical coincidence between the two approaches.)

The term Bayesian arises from an elementary theorem of probability theory named after the Rev. Thomas Bayes, an English clergyman of the eighteenth century, who first enunciated a special case of it and proposed its use in inference. Bayes' theorem is used in the process of making Bayesian inferences, as will be explained below. For a number of historical reasons, however, current interest in Bayesian inference is quite recent, dating, say, from the 1950s. Hence the term "neo-Bayesian" is sometimes used instead of "Bayesian."

An illustration of Bayesian inference. For a simple illustration of the Bayesian approach, consider the problem of making inferences about a Bernoulli process with parameter p. A Bernoulli process can be visualized in terms of repeated independent tosses of a not necessarily fair coin. It generates heads and tails in such a way that the probability of heads on a single trial is always equal to a parameter p regardless of the previous history of heads and tails. The subjectivistic counterpart of this description of a Bernoulli process is given by de Finetti's concept of exchangeable events [*see* PROBABILITY, *article on* INTERPRETATIONS].

Suppose first that we have no direct sample evidence from the process. Based on experience with similar processes, introspection, general knowledge, etc., we may be willing to translate our judgments about the process into probabilistic terms. For example, we might assess a (subjective) probability distribution for $\tilde{p}$. The tilde ($\sim$) indicates that we are now thinking of the parameter p as a random variable. Such a distribution is called a *prior* (or *a priori*) distribution because it is usually assessed prior to sample evidence. Purely for illustration, suppose that the prior distribution of $\tilde{p}$ is uniform on the interval from 0 to 1: the probability that $\tilde{p}$ lies in any subinterval is that subinterval's length, no matter where the subinterval is located between 0 and 1. Now suppose that on three tosses of a coin we observe heads, heads, and tails. The probability of observing this sample, conditional on $\tilde{p} = p$, is $p^2(1-p)$. If we regard this expression as a function of p, it is called the *likelihood function* of the sample. Bayes' theorem shows how to use the likelihood function in conjunction with the prior distribution to obtain a revised or *posterior* distribution of $\tilde{p}$. Posterior means after the sample evidence, and the posterior distribution represents a reconciliation of sample evidence and prior judgment. In terms of inferences about $\tilde{p}$, we may write Bayes' theorem in words as follows: Posterior probability (density) at p, given the observed sample, equals

$$\frac{\text{Prior probability (density) at } p \times \text{likelihood function}}{\text{Prior probability of obtaining the observed sample}}.$$

Expressed mathematically,

$$(1) \qquad f''(p|r,n) = \frac{f'(p)\, p^r(1-p)^{n-r}}{\int_0^1 f'(p)\, p^r(1-p)^{n-r}\, dp},$$

where $f'(p)$ denotes the prior density of $\tilde{p}$, $p^r(1-p)^{n-r}$ denotes the likelihood if r heads are observed in n trials, and $f''(p|r,n)$ denotes the posterior density of $\tilde{p}$ given the sample evidence. In our example, $f'(p) = 1$ for $0 \leqslant p \leqslant 1$ and 0 otherwise; $r = 2$; $n = 3$; and

$$\int_0^1 f'(p)\, p^r(1-p)^{n-r}\, dp = \int_0^1 p^2(1-p)\, dp = 1/12,$$

so that

$$f''(p|r=2, n=3) = \begin{cases} 12\, p^2(1-p), & 0 \leqslant p \leqslant 1, \\ 0, & \text{otherwise.} \end{cases}$$

Thus we emerge from the analysis with an explicit posterior probability distribution for $\tilde{p}$. This distribution characterizes fully our judgments about $\tilde{p}$. It could be applied in a formal decision-theoretic analysis in which utilities of alternative acts are functions of p. For example, we might make a Bayesian point estimate of p (each possible point estimate is regarded as an act), and the seriousness of an estimation error (loss) might be proportional to the square of the error. The best point estimate can then be shown to be the mean of the posterior distribution; in our example this would be .6. Or we might wish to describe certain aspects of the posterior distribution for summary purposes; it can be shown, for example, that, where P refers to the posterior distribution,

$$P(\tilde{p} < .194) = .025 \quad \text{and} \quad P(\tilde{p} > .932) = .025,$$

so that a .95 *credible interval* for $\tilde{p}$ extends from .194 to .932. Again, it can easily be shown that $P(\tilde{p} > .5) = .688$: the posterior probability that the coin is "biased" in favor of heads is a little over $\frac{2}{3}$.

The likelihood principle. In our example, the effect of the sample evidence was wholly transmitted by the likelihood function. All we needed to know from the sample was $p^r(1-p)^{n-r}$; the actual sequence of individual observations was irrelevant *so long as we believed the assumption of a Bernoulli process.* In general, a full Bayesian analysis requires as inputs for Bayes' theorem only the likelihood function and the prior distribution. Thus the import of the sample evidence is fully reflected in the likelihood function, a principle known as the likelihood principle [*see* LIKELIHOOD]. Alternatively, given that the sample is drawn from a Bernoulli process, the import of the sample is fully reflected in the numbers r and n, which are called *sufficient statistics* [*see* SUFFICIENCY]. (If the sample size, n, is fixed in advance of sampling, it is said that r alone is sufficient.)

The likelihood principle implies certain consequences that do not accord with traditional ideas. Here are examples: (1) Once the data are in, there is no distinction between sequential analysis and analysis for fixed sample size. In the Bernoulli example, successive samples of n_1 and n_2 with r_1 and r_2 successes could be analyzed as one pooled sample of $n_1 + n_2$ trials with $r_1 + r_2$ successes. Alternatively, a posterior distribution could be computed after the first sample of n_1; this distribution could then serve as a prior distribution for the second sample; finally, a second posterior distribution could be computed after the second sample of n_2. By either route the posterior distribution after $n_1 + n_2$ observations would be the same. Under almost any situation that is likely to arise in practice, the "stopping rule" by which sampling is terminated is irrelevant to the analysis of the sample. For example, it would not matter whether r successes in n trials were obtained by fixing r in advance and observing the rth success on the nth trial, or by fixing n in advance and counting r successes in the n trials. (2) For the purpose of statistical reporting, the likelihood function is the important information to be conveyed. If a reader wants to perform his own Bayesian analysis, he needs the likelihood function, not a posterior distribution based on someone else's prior nor traditional analyses such as significance tests, from which it may be difficult or impossible to recover the likelihood function.

Vagueness about prior probabilities. In our example we assessed the prior distribution of $\tilde{p}$ as a uniform distribution from 0 to 1. It is sometimes thought that such an assessment means that we "know" $\tilde{p}$ is so distributed and that our claim to knowledge might be verified or refuted in some way. It is indeed possible to imagine situations in which the distribution of $\tilde{p}$ might be known, as when one coin is to be drawn at random from a number of coins, each of which has a known p determined by a very large number of tosses. The frequency distribution of these p's would then serve as a prior distribution, and all statisticians would apply Bayes' theorem in analyzing sample evidence. But such an example would be unusual. Typically, in making an inference about $\tilde{p}$ for a *particular* coin, the prior distribution of $\tilde{p}$ is not a description of some distribution of p's but rather a tool for expressing judgments about $\tilde{p}$ based on evidence other than the evidence of the particular sample to be analyzed.

Not only do we rarely know the prior distribution of $\tilde{p}$, but we are typically more or less vague when we try to assess it. This vagueness is comparable to the vagueness that surrounds many decisions in everyday life. For example, a person may decide to offer $21,250 for a house he wishes to buy, even though he may be quite vague about what amount he "should" offer. Similarly, in statistical inference we may assess a prior distribution in the face of a certain amount of vagueness. If we are not willing to do so, we cannot pursue a *formal* Bayesian analysis and must evaluate sample evidence intuitively, perhaps aided by the tools of descriptive statistics and classical inference.

Vagueness about prior probabilities is not the only kind of vagueness to be faced in statistical analysis, and the other kinds of vagueness are equally troublesome for approaches to statistics that do not use prior probabilities. Vagueness about the likelihood function, that is, the process generating the data, is typically substantial and hard to deal with. Moreover, both classical and Bayesian decision theory bring in the idea of utility, and utilities often are vague.

In assessing prior probabilities, skillful self-interrogation is needed in order to mitigate vagueness. Self-interrogation may be made more systematic and illuminating in several ways. (1) *Direct judgmental assessment.* In assessing the prior distribution of $\tilde{p}$, for example, we might ask: For what p would we be indifferent to an even money bet that $\tilde{p}$ is above or below this value? (Answer: the .50-quantile or median.) If we were told that $\tilde{p}$ is above the .50-quantile just assessed, but nothing more, for what value of p would we now be indifferent in such a bet? (Answer: the .75-quantile.) Similarly we might locate other key quantiles or key relative heights on the density function. (2) *Translation to equivalent but hypothetical prior sample evidence.* For example, we might feel that our prior opinion about $\tilde{p}$ is roughly what it would have been if we had initially held a uniform prior, and then seen r heads in n hypothetical trials from the process. The implied posterior distribution would serve as the prior. (3) *Contemplation of possible sample outcomes.* Sometimes we may find it easy to decide directly what our posterior distribution *would be* if a certain hypothetical sample outcome were to materialize. We can then work backward to see the prior distribution thereby implied. Of course, this approach is likely to be helpful only if the hypothetical sample outcomes are easy to assimilate. For example, if we make a certain technical assumption about the general shape of the prior (beta) distribution [*see* DISTRIBUTIONS, STATISTICAL, *article on* SPECIAL CONTINUOUS DISTRIBUTIONS], the answers to the following two simply stated questions imply a prior distribution of $\tilde{p}$: (1) How do we assess the probability of heads *on a single trial*? (2) If we were to observe a head on a single trial (this is the hypothetical future outcome), how would we assess the probability of heads on a second trial?

These approaches are intended only to be suggestive. If several approaches to self-interrogation lead to substantially different prior distributions, we must either try to remove the internal inconsistency or be content with an intuitive analysis. Actually, from the point of view of subjective probability coherent, the discovery of internal inconsistency in one's judgments is the only route toward more rational decisions. The danger is not that internal inconsistencies will be revealed but that they will be suppressed by self-deception or glossed over by lethargy.

It may happen that vagueness affects only unimportant aspects of the prior distribution: theoretical or empirical analysis may show that the posterior distribution is insensitive to these aspects of the distribution. For example, we may be vague about many aspects of the prior distribution, yet feel that it is nearly uniform over all values of the parameter for which the likelihood function is not essentially zero. This has been called a diffuse, informationless, or locally uniform prior distribution. These terms are to be interpreted relative to the spread of the likelihood function, which depends on the sample size; a prior that is diffuse relative to a large sample may not be diffuse relative to a small one. If the prior distribution is diffuse, the posterior distribution can be easily approximated from the assumption of a strictly uniform prior distribution. The latter assumption, known historically as Bayes' postulate (not to be confused with Bayes' theorem), is regarded mainly as a device that leads to good approximations in certain circumstances, although supporters of subjective probability rational sometimes regard it as more than that in their approach to Bayesian inference. The uniform prior is also useful for statistical reporting, since it leads to posterior distributions from which the likelihood is easily recovered and presents the results in a readily usable form to any reader whose prior distribution is diffuse.

Probabilistic prediction. A distribution, prior or posterior, of the parameter $\tilde{p}$ of a Bernoulli process implies a probabilistic prediction for any future sample to be drawn from the process, assuming that the stopping rule is given. For example, the denominator in the right-hand side of Bayes' formula for Bernoulli sampling (equation 1) can be interpreted as the probability of obtaining the particular sample actually observed, given the prior distribution of $\tilde{p}$. If Mr. A and Mr. B each has a distribution for $\tilde{p}$, and a new sample is then observed, we can calculate the probability of the sample in the light of each prior distribution. The ratio of these probabilities, technically a marginal likelihood ratio, measures the extent to which the data favor Mr. A over Mr. B or vice versa. This idea has important consequences for evaluating judgments, selecting statistical models, and performing Bayesian tests of significance.

In connection with the previous paragraph a separate point is worth making. The posterior distributions of Mr. A and Mr. B are bound to grow closer together as sample evidence piles up, so long as neither of the priors was dogmatic. An example of a dogmatic prior would be the opinion that $\tilde{p}$ is exactly .5.

In an important sense the predictive distribution of future observations, which is derived from the posterior distribution, is more fundamental to Bayesian inference than the posterior distribution itself.

Multivariate inference and nuisance parameters. Thus far we have used one basic example, inferences about a Bernoulli process. To introduce some additional concepts, we now turn to inferences about the mean μ of a normal distribution with unknown variance σ^2. In this case we begin with a *joint* prior distribution for $\tilde{\mu}$ and $\tilde{\sigma}^2$. The likelihood function is now a function of two variables, μ and σ^2. An inspection of the likelihood function will show not only that the *sequence* of observations is irrelevant to inference but also that the magnitudes are irrelevant except insofar as they help determine the sample mean $\bar{x}$ and variance s^2, which, along with the sample size n, are the sufficient statistics of this example. The prior distribution combines with the likelihood essentially as before except that a double integration (or double summation) is needed instead of a single integration (or summation). The result is a joint posterior distribution of $\tilde{\mu}$ and $\tilde{\sigma}^2$.

If we are interested only in $\tilde{\mu}$, then $\tilde{\sigma}^2$ is said to be a *nuisance parameter*. In principle it is simple to deal with a nuisance parameter: we integrate it out of the posterior distribution. In our example this means that we must find the marginal distribution of $\tilde{\mu}$ from the joint posterior distribution of $\tilde{\mu}$ and $\tilde{\sigma}^2$.

Multivariate problems and nuisance parameters can always be dealt with by the approach just described. The integrations required may demand heavy computation, but the task is straightforward. A more difficult problem is that of assessing multivariate prior distributions, especially when the number of parameters is large, and research is needed to find better techniques for avoiding self-contradictions and meeting the problems posed by vagueness in such assessments.

Design of experiments and surveys. So far we have talked only about problems of analysis of samples, without saying anything about what kind of sample evidence, and how much, should be sought. This kind of problem is known as a problem of *design*. A formal Bayesian solution of a design problem requires that we look beyond the posterior distribution to the ultimate decisions that will be made in the light of this distribution. What is the best design depends on the purposes to be served by collecting the data. Given the specific purpose and the principle of maximization of expected utility, it is possible to calculate the expected utility of the best act for any particular sample outcome. We can repeat this for each possible sample outcome for a given sample design. Next, we can weight each such utility by the probability of the corresponding outcome in the light of the prior distribution. This gives an over-all expected utility for any proposed design. Finally, we pick the sample design with the highest expected utility. For two-action problems—for example, deciding whether a new medical treatment is better or worse than a standard treatment—this procedure is in no conflict with the traditional approach of selecting designs by comparing operating characteristics, although it formalizes certain things—prior probabilities and utilities—that often are treated intuitively in the traditional approach.

Comparison of Bayesian and classical inference. Certain common statistical practices are subject to criticism, either from the point of view of Bayesian or of classical theory: for example, estimation problems are frequently regarded as tests of null hypotheses [*see* HYPOTHESIS TESTING], and .05 or .01 significance levels are used inflexibly. Bayesian and classical theory are in many respects closer to each other than either is to everyday practice. In comparing the two approaches, therefore, we shall confine the discussion to the level of underlying theory. In one sense the basic difference is the acceptance of subjective probability judgment as a *formal* component of Bayesian inference. This does not mean that classical theorists would disavow judgment, only that they would apply it informally after the purely statistical analysis is finished: judgment is the "second span in the bridge of inference." Building on subjective probability, Bayesian theory is a unified theory, whereas classical theory is diverse and *ad hoc*. In this sense Bayesian theory is simpler. In another sense, however, Bayesian theory is more complex, for it incorporates more into the formal analysis. Consider a famous controversy of classical statistics, the problem of comparing the means of two normal distributions with possibly unequal and unknown variances, the so-called Behrens–Fisher problem [*see* LINEAR HYPOTHESES]. Conceptually this problem poses major difficulties for some classical theories (not Fisher's fiducial inference; see Fisher 1939) but none for Bayesian theory. In application, however, the Bayesian approach faces the problem of assessing a prior distribution involving four random variables. Moreover, there may be messy computational work after the prior distribution has been assessed.

In many applications, however, a credible interval emerging from the assumption of a diffuse prior distribution is identical, or nearly identical, to the corresponding confidence interval. There is a difference of interpretation, illustrated in the opening two paragraphs of this article, but in practice many people interpret the classical result in the Bayesian way. There often are numerical similarities between the results of Bayesian and classical analyses of the same data; but there can also be substantial differences, for example, when the prior distribution is nondiffuse and when a genuine null hypothesis is to be tested.

Often it may happen that the problem of vagueness, discussed at some length above, makes a formal Bayesian analysis seem unwise. In this event Bayesian theory may still be of some value in selecting a descriptive analysis or a classical technique that conforms well to the general Bayesian approach, and perhaps in modifying the classical technique. For example, many of the classical developments in sample surveys and analysis of experiments can be given rough Bayesian interpretations when vagueness about the likelihood (as opposed to prior probabilities) prevents a full Bayesian analysis. Moreover, even an abortive Bayesian analysis may contribute insight into a problem.

Bayesian inference has as yet received much less theoretical study than has classical inference. Such commonplace and fundamental ideas of classical statistics as randomization and nonparametric methods require re-examination from the Bayesian view, and this re-examination has scarcely begun. It is hard at this writing to predict how far Bayesian theory will lead in modification and reinterpretation of classical theory. Before a fully Bayesian replacement is available, there is certainly no need to discard those classical techniques that seem roughly compatible with the Bayesian approach; indeed, many classical techniques are, under certain conditions, good approximations to fully Bayesian ones, and useful Bayesian interpretations are now known for almost all classical techniques. From the classical viewpoint, the Bayesian approach often leads to procedures with desirable sampling properties and acts as a stimulus to further theoretical development. In the meanwhile, the interaction between the two approaches promises to lead to fruitful developments in statistical inference; and the Bayesian approach promises to illuminate a number of problems—such as allowance for selectivity—that are otherwise hard to handle.

HARRY V. ROBERTS

[*See also the biography of* BAYES.]

BIBLIOGRAPHY

The first book-length development of Bayesian inference, which emphasizes heavily the decision-theoretic foundations of the subject, is Schlaifer 1959. *A more technical development of the subject is given by* Raiffa & Schlaifer 1961. *An excellent short introduction with an extensive bibliography is* Savage 1962. *A somewhat longer introduction is given by Savage and other contributors in* Joint Statistics Seminar 1959. *This volume also discusses advantages and disadvantages of the Bayesian approach. An interesting application of Bayesian inference, along with a penetrating discussion of underlying philosophy and a comparison with the corresponding classical analysis, is given in* Mosteller & Wallace 1964. *This study gives a specific example of how one might cope with vagueness about the likelihood function. Another example is to be found in* Box & Tiao 1962. *A thorough development of Bayesian inference from the viewpoint of "subjective probability rational" is to be found in* Jeffreys 1939. *A basic paper on fiducial inference is* Fisher 1939.

BAYES, THOMAS (1764) 1963 *Facsimiles of Two Papers by Bayes.* New York: Hafner. → Contains "An Essay Toward Solving a Problem in the Doctrine of Chances, With Richard Price's Foreword and Discussion," with a commentary by Edward C. Molina, and "A Letter on Asymptotic Series From Bayes to John Canton," with a commentary by W. Edwards Deming. Both essays first appeared in Volume 53 of the Royal Society of London's *Philosophical Transactions* and retain the original pagination.

BOX, GEORGE E. P.; and TIAO, GEORGE C. 1962 A Further Look at Robustness Via Bayes's Theorem. *Biometrika* 49:419–432.

EDWARDS, WARD; LINDMAN, HAROLD; and SAVAGE, LEONARD J. 1963 Bayesian Statistical Inference for Psychological Research. *Psychological Review* 70:193–242.

FISHER, R. A. (1939) 1950 The Comparison of Samples With Possibly Unequal Variances. Pages 35.173a–35.180 in R. A. Fisher, *Contributions to Mathematical Statistics.* New York: Wiley. → First published in Volume 9 of the *Annals of Eugenics.*

►GEORGESCU-ROEGEN, NICHOLAS 1968 Utility. Volume 16, pages 236–267 in *International Encyclopedia of the Social Sciences.* Edited by David L. Sills. New York: Macmillan and Free Press.

JEFFREYS, HAROLD (1939) 1961 *Theory of Probability.* 3d ed. Oxford: Clarendon.

JOINT STATISTICS SEMINAR, UNIVERSITY OF LONDON 1959 *The Foundations of Statistical Inference.* A discussion opened by Leonard J. Savage at a meeting of the Seminar. London: Methuen; New York: Wiley.

LINDLEY, DENNIS V. 1965 *Introduction to Probability and Statistics From a Bayesian Viewpoint.* 2 vols. Cambridge Univ. Press.

MOSTELLER, FREDERICK; and WALLACE, DAVID L. 1963 Inference in an Authorship Problem: A Comparative Study of Discrimination Methods Applied to the Authorship of the Disputed Federalist Papers. *Journal of the American Statistical Association* 58:275–309.

MOSTELLER, FREDERICK; and WALLACE, DAVID L. 1964 *Inference and Disputed Authorship: The Federalist.* Reading, Mass.: Addison-Wesley.

PRATT, JOHN W.; RAIFFA, HOWARD; and SCHLAIFER, ROBERT 1964 The Foundations of Decision Under Uncertainty: An Elementary Exposition. *Journal of the American Statistical Association* 59:353–375.

RAIFFA, HOWARD; and SCHLAIFER, ROBERT 1961 *Applied Statistical Decision Theory.* Graduate School of Business Administration, Studies in Managerial Economics. Boston: Harvard Univ., Division of Research.

SAVAGE, LEONARD J. 1962 Bayesian Statistics. Pages 161–194 in Symposium on Information and Decision Processes, Third, Purdue University, 1961, *Recent Developments in Information and Decision Processes.*

Edited by Robert E. Machol and Paul Gray. New York: Macmillan.

SCHLAIFER, ROBERT 1959 *Probability and Statistics for Business Decisions: An Introduction to Managerial Economics Under Uncertainty.* New York: McGraw-Hill.

Postscript

Since 1968 Bayesian ideas have been greatly developed in the literature on statistical methodology. Perhaps the most technically impressive achievements have come in the area of econometrics, where highly intricate models have received a thorough and systematic treatment from the Bayesian viewpoint. The work of Arnold Zellner has been especially influential. His *Introduction to Bayesian Inference in Econometrics,* published in 1971, covers virtually the entire field of econometrics, and frequently gives in parallel the classical ("sampling-theory" now seems to me a better term) and Bayesian treatments of the same problems. A sampling of subsequent work in Bayesian econometrics is given in a volume edited by Fienberg and Zellner (1975). Dedicated to Leonard J. Savage, it includes his last technical paper on Bayesian ideas, "Elicitation of Personal Probabilities and Expectations." (Inexplicably, my original list of references failed to include Savage's major work, *The Foundations of Statistics,* 1954.)

The development of Bayesian methods in more traditional areas of statistical practice is illustrated by Box and Tiao (1973).

On the whole, the tendency toward philosophical disputation between Bayesians and non-Bayesians during the early 1960s appears to have abated, and many statisticians would probably now describe themselves as eclectic, drawing on Bayesian concepts (or non-Bayesian ones) as they appear to fulfill a need. Thus much of the recent published work gives little attention to questions about "foundations." (One explanation for this change of emphasis is the increasing realization by statisticians of the importance of exploratory data analysis, which seems to lie in part outside formal theoretical frameworks.) But Lindley's *Bayesian Statistics* (1971) provided a review of deeper issues, and de Finetti's *Probability, Induction and Statistics* (also in memory of Savage) appeared in 1972.

Also, Bayesian inference seems gradually to be invading statistics curricula, although often as an advanced course or a module at the end of a traditional course. The effect on practice is harder to generalize about, in part because much practical work draws on theory only in a relatively mechanical way and in part because almost all Bayesian techniques based upon diffuse prior distributions have exact or approximate counterparts in sampling-theory methodology, although, as explained in the main article, the interpretation of numerical results is very different, even though the numbers are the same or approximately so. Nonetheless, it is safe to say that for many statisticians (including me) the Bayesian ideas are extremely helpful in approaching applications.

What appears not to have happened is extensive reliance on nondiffuse prior distributions, a reliance much feared by some early critics of Bayesian methods, who thought that posterior distributions would be manipulated at will by the insertion of an appropriate nondiffuse prior. (I regard this fear as groundless, and give reasons in an article in the volume edited by Fienberg and Zellner, 1975.) Clearly, jointly diffuse prior distributions are often appropriate, especially for reporting, but at the same time the full potential of Bayesian methodology cannot be attained if nondiffuse distributions are never assessed. A part of the explanation may reside in practical difficulties, mathematical and psychological, of making the desired assessments. The psychological problems of making judgmental assessments for Bayesian inference (and decision theory as well) have been explored at great length by a school of psychologists and others interested in human behavior under uncertainty. A recent review article by Hogarth (1975) summarizes much of this work and gives an extensive bibliography. In practice, of course, some statistical procedures such as those used in backward stepwise regression can be interpreted as an indirect way of imposing nondiffuse prior judgments on the data. Further, progress has been made in dealing with certain difficulties inherent in the assessment of jointly diffuse prior distributions in situations in which many parameters are involved. Bayesian work in this area, which owes much to the work of Charles Stein in the sampling-theory tradition, is summarized in the monograph by Lindley (1971, pp. 49–59).

HARRY V. ROBERTS

ADDITIONAL BIBLIOGRAPHY

BOX, GEORGE E. P.; and TIAO, GEORGE C. 1973 *Bayesian Inference in Statistical Analysis.* Reading, Mass.: Addison-Wesley.

DE FINETTI, BRUNO 1972 *Probability, Induction and Statistics: The Art of Guessing.* New York: Wiley.

FIENBERG, STEPHEN E.; and ZELLNER, ARNOLD 1975 *Studies in Bayesian Econometrics and Statistics.* Amsterdam: North-Holland.

HOGARTH, ROBIN M. 1975 Cognitive Processes and the Assessment of Subjective Probability Distributions. *Journal of the American Statistical Association* 70: 271–294. → Includes comments by Robert L. Winkler and rejoinder by Robin M. Hogarth.

LINDLEY, DENNIS V. 1971 *Bayesian Statistics: A Review.* Regional Conference Series in Applied Mathematics No. 2. Philadelphia: Society for Industrial and Applied Mathematics.

SAVAGE, LEONARD J. (1954) 1972 *The Foundations of Statistics*. Rev. ed. New York: Dover. → Includes a new preface.

ZELLNER, ARNOLD 1971 *An Introduction to Bayesian Inference in Econometrics*. New York: Wiley.

BENINI, RODOLFO

Rodolfo Benini (1862–1956) was born in Cremona. At the very young age of 27 he was appointed to the chair of history of economics at Bari. His academic life led him from Bari to Perugia in 1896 and then to Pavia, where from 1897 to 1907 he taught political economics and statistics. In 1908 Benini went to Rome and there held the chair of statistics until 1928; from then until 1956 he taught economics at the same university. It is difficult to name any single predecessor or teacher who had a particularly strong influence on Benini, since he worked in a great variety of social science fields and showed in each a great degree of independent thought and creativity. As immediate predecessors of Benini we may certainly list Angelo Messedaglia, Ridolfo Livi, Maffeo Pantaleoni, and especially Vilfredo Pareto. In a broader sense we might well cite the whole system of Italian economic and social thought, which up to this century had developed quite independently and which was ahead of the rest of European thought in many instances. It is the connection of statistical knowledge and economic and social theory in Benini which led him to very constructive results in all the fields he engaged in.

Together with Süssmilch, Quetelet, and Achille Guillard, Benini can be regarded as one of the founders of demography as a separate science (1896; 1901*a*). In his *Principii di demografia* he distinguished between a qualitative and a quantitative theory of population. Both his concepts differ from the standard use in the literature. In the "qualitative" theory Benini elaborated the concept of descriptive statistics to include rates of birth and mortality; life expectancies; the fertility of women as a decreasing function of their age; and normal distributions of physiological characteristics of men and women. In the analysis of the cohesion of social groups Benini developed an attraction–repulsion index to measure the association between dichotomous characteristics of husbands and wives (1898; 1901*a*; 1928*a*). Classifying a population by any given characteristic (e.g., literate vs. illiterate males—m_1, m_2—and females—f_1, f_2), Benini arranged the relative frequencies ρ_{ij} $(i, j = 1, 2)$ of the possible combinations in 2×2 tables such as

Table 1

		MEN		
		m_1	m_2	Σ
WOMEN	f_1	ρ_{11}	ρ_{12}	$\rho_{1\cdot}$
	f_2	ρ_{21}	ρ_{22}	$\rho_{2\cdot}$
	Σ	$\rho_{\cdot 1}$	$\rho_{\cdot 2}$	1

Table 1. From this classification Benini then derived a first measure of attraction (repulsion) with

$$\gamma_{ij} = \rho_{ij} - \rho_{i\cdot}\,\rho_{\cdot j}, \qquad i, j = 1, 2$$

where $\gamma_{11} = \gamma_{22} = -\gamma_{21} = -\gamma_{12}$. The index of attraction (when $\gamma_{ij} > 0$) or repulsion (when $\gamma_{ij} < 0$), a later measure he put forth, is given by Benini as

$$\frac{\gamma_{11}}{\min \rho_{1\cdot}, \rho_{\cdot 1} - \rho_{1\cdot}\rho_{\cdot 1}} \equiv \frac{\rho_{11} - \rho_{\cdot 1}\,\rho_{1\cdot}}{\min \rho_{1\cdot}, \rho_{\cdot 1} - \rho_{1\cdot}\rho_{\cdot 1}},$$

the absolute value of the index ranging from 0 to 1. This statistic was further elaborated on by Benini (1928*a*) and later by many others in Italy and elsewhere (Goodman & Kruskal 1954–1963).

In the "quantitative" theory Benini tried to discover evolutionary laws of aggregate societies and their latent structures. He denied the general validity of Malthus' theorem. Uniform predictions cannot be made from empirical evidence, said Benini. He expected, however, that the continuing process of urbanization of societies would lead to an adjustment of the growth rates of social aggregates, not so much by delayed marriages as by advanced education and new methods of birth control. Moreover, because of an instinct of imitation, the lower classes of urban centers would aspire to the behavior and social status of the next higher classes. This would cause a continual process of assimilation—a narrowing and adjustment of the average age of marriages within and between social groups; an adjustment of the proportion of unmarried people; and an adjustment of birth and death rates. Thus eventually, the evolution of societies will lead to a stationary state of social aggregates and a gradual elimination of structural differences.

Benini engaged in statistical research in economics similar to his statistical work in demography. He hoped to reduce economic science by systematization and by empirically verified formulas or laws to the concise system of expressions characteristic of physics and chemistry (1907, p. 1053; 1908, p. 17). In 1894 he tried to estimate the distribution of property among social classes (1894). After Pareto's work on the distribution of incomes appeared (1896; 1897), Benini extended this work and modified the Pareto distribution in his analysis of the distribution of property. The function proposed by Benini for the distribution of property values was

$$\log F(x) = \log k - a(\log x)^2, \qquad \text{for } x \geqslant x_0,$$

where $F(x)$ is the proportion of property values greater than or equal to x, that is, the distribution function cumulated to the right. There is a truncation point, x_0, forming a lower limit. Another way of expressing Benini's distribution is to say that it is that of a random variable, X, such that $[\log (X - x_0)]^2$ has a negative exponential distribution. [*See* DISTRIBUTIONS, STATISTICAL, *article on* SPECIAL CONTINUOUS DISTRIBUTIONS.] In contrast, Pareto's distribution is such that $\log (X - x_0)$ has a negative exponential distribution. The attraction–repulsion index and the modification of the Pareto distribution are Benini's most original contributions to statistics.

As early as 1907 Benini established empirical estimates of price elasticities, demand curves, and Engel curves (see 1907; 1908). The demand curve underlying his estimates was of the general hyperbolic form $\log y = a + b \log x$. In the case of prices and demand for coffee in Italy, Benini came to estimates of $a = 3.63161$ and $b = -0.384$. In an extension of this work (1908) Benini estimated, in addition to other demand curves, income-induced increases in expenditures for housing. The underlying model is similar to that shown above, with b now positive and ranging from 0 to 1 (for Dresden and Breslau $b = 0.617$). In this Benini preceded A. C. Pigou, Lenoir, Lehfeld, R. Frisch, and H. Schultz. Benini, however, had no means of establishing confidence intervals for his estimates, although the signs of his estimates agreed with what one would expect. (See Fox 1968.)

Benini's other economic ideas are spread over a long series of articles, and it is hard to consider them all. They can be reduced, however, to some few central results which underlie his writings. Most important among them is Benini's notion that in any exchange transaction there exist a minimum price and a maximum price (p min and p max) at which the transaction still can take place (1928*b*). According to Benini, the most likely price between two equally endowed parties (equality of bargaining power, in whichever way this is defined) would be the point equidistant from the boundary points p min and p max. Given, however, unequal endowment of the contracting parties, there would result a shift in the price and an exploitation of the weaker partner. On this basis Benini explained how profits arise at the expense of labor income.

Benini believed that protectionism in international trade is justified on three grounds: first, by the existence of the same kind of exploitation of a weaker party described above for the general market place (in this case the parties are foreign enterprises or foreign states); second, by the cumulative effect of that exploitation associated with the power of states in international relations; and third, by the vulnerability of infant industries. For all these reasons the state has to fulfill special functions and is therefore introduced by Benini as an additional factor of production.

In connection with the distribution of property and income Benini observed that in some countries, among them Italy, a doubling of property was associated with a threefold increase in total income (per person or household), at least as long as the income derived from labor constituted a significant part of total income derived from labor and property. From this Benini then derived his fiscal axiom that a proportional taxation of incomes would lead in those countries to a less than proportional taxation of property and that a proportional taxation of property would lead to a progressive taxation of incomes.

His empirical results also led him to extend Galton's law concerning the progressive elimination of economic and social divergencies (structures). This process occurs as extreme points are continuously eliminated. Their elimination induces an asymptotic approach to stationary states. This process was previously noted in Benini's quantitative theory of the growth of social aggregates.

In addition to his work in demography, sociology, and economics, Benini developed a unique interest in the works of Dante. He undertook to reveal a "second beauty" in Dante by uncovering the quantitative consistency of the structure of *The Divine Comedy*, which up to then had generally been neglected (there had been some exceptions—Busnelli, Angelitti, and Moore). (It is apparent that Dante did incorporate quantitative relations and symbols in the structure of *The Divine Comedy*, as is immediately apparent from the fact that each of the three parts of *The Divine Comedy* contains 33 canti and *The Inferno* an additional introductory one, bringing the total to exactly 100. Moreover, each of the three parts ends with the word "stelle," and so on.) Many enigmas and allegorical elements in *The Divine Comedy* have yet to be interpreted. To do so it is important to appreciate Dante's knowledge of astronomy and the way he incorporated this knowledge into *The Divine Comedy*. Similarly, by quantitative analysis the structure and dimensions of the Inferno and Purgatory may be ascertained. In addition, this kind of analysis may explain the relationship of the calendar used by Dante and the dates he attributed to events in *The Divine Comedy*. The exact dates of Dante's poetic voyage with Vergil and, later, with Beatrice through the regions of the Inferno, Purgatory, and Paradise can thus be established. The main assumption on which Benini based his investigation is that the rigorous structure of the poem will not allow for obvious contradictions or omissions. Explanations may be found by applying medieval concepts to these enigmas. In addition to explanations of the above kind, Benini showed that Dante believed Purgatory to be located on Mount Sinai and established the date of birth of Cacciaguida, an ancestor of Dante, and the date when Cacciaguida's son Alighiero died. Benini also thought he had discovered that Dante's poetic technique differed, depending on the seriousness of the moral crimes he was describing. The new perspectives contained in Benini's contribution to the knowledge of Dante's poem lead to the very margin where one might discover something in Dante that the poet himself was unaware of. Benini knew of this danger and tried to avoid such pitfalls. It took some time for Benini's work on Dante to find support, just as Benini's achievements in combining statistics and social science were not immediately appreciated.

KLAUS-PETER HEISS

[*For discussion of the subsequent development of Benini's ideas, see* STATISTICS, DESCRIPTIVE, *article on* ASSOCIATION. *See also* Allais 1968; Frey 1968; Sauvy 1968.]

WORKS BY BENINI

1892 Sulle dottrine economiche di Antonio Serra: Appunti critici. *Giornale degli economisti* Series 2 5:222–248.

1894 Distribuzione probabile della ricchezza privata in Italia per classi di popolazione. *Riforma sociale* 1: 862–869.

1896 Di alcuni punti oscuri della demografia. *Giornale degli economisti* Series 2 13:97–128, 297–327, 509–534.

1897 Di alcune curve descritte da fenomeni economici aventi relazione colla curva del reddito o con quella del patrimonio. *Giornale degli economisti* Series 2 14:177–214.

1898 Le combinazioni simpatiche in demografia. *Rivista italiana di sociologia* 2:152–171.

1899 Gerarchie sociali: Contributo alla teoria qualitativa della popolazione. *Rivista italiana di sociologia* 3: 17–49.

1901*a* *Principii di demografia.* Florence: Barbèra.

1901*b* Tecnica e logica dei rapporti statistici. *Giornale degli economisti* Series 2 23:503–516.

1905*a* I diagrammi a scala logaritmica (a proposito della graduazione per valore delle successioni ereditarie in Italia, Francia e Inghilterra). *Giornale degli economisti* Series 2 30:222–231.

(1905*b*) 1923 *Principii di statistica metodologica.* Turin: Unione Tipografico–Editrice Torinese.

1907 Sull' uso delle formole empiriche nell' economia applicata. *Giornale degli economisti* Series 2 35: 1053–1063.

1908 Una possibile creazione del metodo statistico: "L'economia politica induttiva." *Giornale degli economisti* Series 2 36:11–34.

1912 L'azione recente dell' oro sui prezzi generali delle merci. Società Italiana per il Progresso delle Scienze, Rome, *Atti* 6:97–123.

1928*a* Gruppi chiusi e gruppi aperti in alcuni fatti collettivi di combinazioni. International Statistical Institute, *Bulletin* 23, no. 2:362–383.

1928*b* Un ritorno ai preliminari dell' economia politica. *Economia* New Series 1:411–428.

1952 *Dante tra gli splendori de' suoi enigmi risolti, ed altri saggi.* Rome: Edizioni dell' Ateneo.

SUPPLEMENTARY BIBLIOGRAPHY

►ALLAIS, MAURICE 1968 Pareto, Vilfredo: I. Contributions to Economics. Volume 11, pages 399–411 in *International Encyclopedia of the Social Sciences.* Edited by David L. Sills. New York: Macmillan and Free Press.

BACHI, ROBERTO 1929 I principali scritti di Rodolfo Benini. *Giornale degli economisti* Series 4 69:1068–1076.

BARI (CITY), UNIVERSITÀ, FACOLTÀ DI ECONOMIA E COMMERCIO 1956 *Studi in memoria di Rodolfo Benini.* Bari: The University.

►FOX, KARL A. 1968 Demand and Supply: Econometric Studies. Volume 4, pages 104–111 in *International Encyclopedia of the Social Sciences.* Edited by David L. Sills. New York: Macmillan and Free Press.

►FREY, LUIGI 1968 Pantaleoni, Maffeo. Volume 11, pages 379–380 in *International Encyclopedia of the Social Sciences.* Edited by David L. Sills. New York: Macmillan and Free Press.

GOODMAN, LEO A.; and KRUSKAL, WILLIAM H. 1954–1963 Measures of Association for Cross-classifications. Parts 1–3. *Journal of the American Statistical Association* 49:732–764; 54:123–163; 58:310–364.

PARETO, VILFREDO 1896 La curva delle entrate e le osservazioni del Prof. Edgeworth. *Giornale degli economisti* Series 2 13:439–448.

PARETO, VILFREDO 1897 Aggiunta allo studio sulla curva delle entrate. *Giornale degli economisti* Series 2 14: 15–26.

[Rodolfo Benini]. 1929 *Giornale degli economisti* Series 4 69:837–966. → Contains articles on Benini by Corrado Gini and others.

►SAUVY, ALFRED 1968 Population: II. Population Theories. Volume 12, pages 349–358 in *International Encyclopedia of the Social Sciences.* Edited by David L. Sills. New York: Macmillan and Free Press.

BERNOULLI FAMILY

The Bernoullis, a Swiss family, acquired its fame in the history of science by producing eight or nine mathematicians of the first rank within three generations. They were all descendants of Niklaus Bernoulli, a prominent merchant in the city of Basel. Each of these mathematicians was compelled by his parents to study for one of the established professions before being permitted to embark upon his real interest, mathematics. Within the group there were four in particular who contributed to the theory of probability and mathematical statistics: Jakob (Jacques) I, Johann (Jean) I, Niklaus (Nicolas) I, and Daniel I.

Jakob I and Johann I

The first in the line of the Bernoulli mathematicians, Jakob I (1654–1705), was the son of the merchant Niklaus. He completed theological studies and then spent six years traveling in England, France, and Holland. Returning to Basel, he lectured on physics at the university until he was appointed professor of mathematics in 1687. His younger brother, Johann I (1667–1748), studied for a medical degree, at the same time receiving instruction in mathematics from Jakob. Later Johann, too, became a professor of mathematics, teaching at the University of Groningen in Holland until he returned to Basel as Jakob's successor.

The brothers were inspired by the works of Leibniz on the infinitesimal calculus, and they became his chief protagonists on the Continent. The new methods enabled them to solve an abundance of mathematical problems, many with applications to mechanics and physics. They applied differentiation and integration to find the properties of many important curves: they determined the form of the catenary curve (hanging chain) and the isochrone, or tautochrone (cycloid), and the form of a sail subject to wind pressure. Jakob was particularly fascinated by the logarithmic spiral, which he requested be engraved on his tombstone. Both used infinite series as a tool; the Bernoulli numbers were introduced by Jakob.

Johann was perhaps even more productive as a scientist than was Jakob. He studied the theory of differential equations, discovered fundamental principles of mechanics, and wrote on the laws of optics. Although the first textbook on the calculus, *Analyse des infiniment petits* (1696), was written by Antoine de L'Hospital, it was largely based upon the author's correspondence with Johann.

The personal relations between the brothers were marred by violent public strife, mainly disputes about priority in the discovery of scientific results. Particularly bitter was their controversy over the brachystochrone, the curve of most rapid descent of a particle sliding from one point to another under the influence of gravity. The problem was of great theoretical interest, since it raised for the first time a question whose solution required the use of the principles of the calculus of variation (the solution is a cycloid).

During his stay in Holland, Jakob became interested in the theory of probability. In his lifetime he published very little on the subject—only a few scattered notes in the *Acta eruditorum*. His main work on probability, the *Ars conjectandi* ("The Art of Conjecturing"), was printed posthumously in Basel in 1713. It is divided into four books. The first is an extensive commentary upon Huygens' pioneer treatise: "De ratiociniis in ludo aleae" (1657; "On Calculations in Games of Chance"). The second gives a systematic presentation of the theory of permutations and combinations, and in the third this is applied to a series of contemporary games, some quite involved. For each, Jakob computed the mathematical expectations of the participants.

The fourth book shows the greatest depth. Here Jakob tried to analyze the events to which probability theory is applicable; in other words, he dealt with the basic question of mathematical statistics: when is it possible to determine an unknown probability from experience? He emphasized that a great number of observations are necessary. Furthermore, he pointed out "something which perhaps no one has thought of before," namely, that (in modern terminology) it is necessary to prove mathematically that as the number of observations increases, the relative number of successes must be within an arbitrarily small (but fixed) interval around the theoretical probability with a probability that tends to 1. This he did, with complete rigor and without the use of calculus, by examining the binomial probabilities and estimating their sums. Illustrations with numerical computations for small intervals are given. The author concluded with some philosophical observations which show the importance he attached to his theorem.

To the *Ars conjectandi* Jakob added a supplement on the *jeu de paume* (similar to the game of tennis), in the form of a letter to a friend. Here he computed the chances of winning for a player at any stage of the game, given players with equal skill and players with differing skill, and in the latter cases he determined how great an advantage the more skilled one can allow the other.

Niklaus I

Niklaus I (1687–1759) was a nephew of Jakob I and Johann I; his father was a portrait painter. True to the family tradition, Niklaus studied for one of the older professions, jurisprudence, while on the side he attended the lectures in mathematics of his two uncles. His law thesis straddled both fields: ". . . de usu artis conjectandi in jure" (1709). He accepted a professorship in mathematics at Padua in 1716 but disliked the university there and returned to Basel in 1719. In 1722 he was appointed professor of logic; in 1731 he changed to a chair of jurisprudence.

When Jakob I died, his *Ars conjectandi* was not in finished form and the publisher asked Johann I to serve as editor. When Johann refused, Niklaus was suggested. He refused also, doubting his competence, but he was finally prevailed upon to accept the undertaking. Niklaus published little in the field of mathematics, probably because of his excessive modesty. But, as editor of the *Ars conjectandi*, he entered into extensive correspondence with the two other pioneers in probability, Rémond de Montmort and Abraham de Moivre. Both appealed for his support in the priority feud that arose between them.

Niklaus also corresponded with the Dutch physicist van s'Gravesande on a curious statistical phenomenon that had first been pointed out by Arbuthnot. It was generally accepted that births of boys as compared to girls correspond to a game of chance, with the same probability, $p = \frac{1}{2}$, for each. Nevertheless, the birth records in London showed that for 82 successive years there had been more males born than females, a most unlikely occurrence under the assumption of equal probabilities. Van s'Gravesande and Arbuthnot were inclined to see this as an example of divine intervention in the laws of nature, while Niklaus took the view that it was more rational to assume that the probability for the birth of a male child is slightly greater than one-half.

Daniel I

Johann I had three sons who were mathematicians: Niklaus II (1695–1726), Daniel I (1700–1782), and Johann II (1710–1790). He compelled each one to acquire a professional degree. Niklaus II studied law and began his career in Berne, as a professor in this subject, in 1723. In 1725 he was appointed to a professorship of mathematics at the Imperial Academy in St. Petersburg, but he died shortly after his arrival there. Johann II also studied jurisprudence; eventually he became his father's successor as professor of mathematics in Basel. He continued the Bernoulli dynasty, having three sons who were mathematicians: Johann III, Daniel II, and Jakob II.

Daniel I studied medicine, but his first mathematical book had already appeared when he was 24 years old and the next year he was called to a mathematical professorship at the Imperial Academy in St. Petersburg, remaining there from 1725 to 1733. Upon his return to Basel he became professor of medicine and botany; in 1750 he was appointed to a professorship in physics, which suited him better.

Daniel I was a prolific writer, even by the standards of the Bernoulli family; no less than ten times were his works awarded prizes by the French Academy of Sciences. His main interests centered in theoretical physics, the foundation of mechanics and, later, probability. Some of his best-known papers deal with celestial mechanics, the tides, and the laws governing a vibrating string.

Most important among the papers on probability by Daniel I is the study *Specimen theoriae novae de mensura sortis* (1738). The basis for this work is the well-known Petersburg paradox, which at that time was a much discussed topic in connection with the concept of expected value. The paradox was first mentioned by Niklaus I in his correspondence with Montmort and is reproduced in the 1713 edition of Montmort's book *Essay d'analyse sur les jeux de hazard*. The fact that the expectation is infinite led Daniel to introduce a moral expectation or marginal utility, now fundamental in economic investigations. He assumed that for a person with a fortune of size x, the utility of an increase Δx is proportional to Δx and inversely proportional to x, giving an expression

$$u = a \log x + b$$

for the utility. In the same paper Daniel also pointed out that a similar idea had already been proposed by the Swiss mathematician G. Cramer in a letter of 1728 to Niklaus I. Daniel also wrote a few other papers on probability, but they are of lesser importance; a number of them are concerned with questions arising from mortality statistics.

ØYSTEIN ORE

[*For the historical context of the Bernoullis' work, see* STATISTICS, *article on* THE HISTORY OF STATISTICAL METHOD; *and the biography of* MOIVRE; *for discussion of the subsequent development of their ideas, see* PROBABILITY.]

WORKS BY DANIEL I

1724 *Exercitationes quaedam mathematicae.* Venice (Italy): Apud Dominicum Lovisam.

(1738) 1954 Exposition of a New Theory on the Measurement of Risk. *Econometrica* 22:23–36. → First published as "Specimen theoriae novae de mensura sortis."

WORKS BY JAKOB I

(1713) 1899 *Wahrscheinlichkeitsrechnung (Ars conjectandi)*. 2 vols. Leipzig: Engelmann. → First published posthumously in Latin.

1744 *Jacobi Bernoulli . . . Opera.* 2 vols. Geneva: Cramer & Fratrum Philibert. → Published posthumously.

WORKS BY JOHANN I

1742 *Johannis Bernoulli . . . Opera omnia.* 4 vols. Geneva: Bousquet.

Der Briefwechsel von Johann Bernoulli. Volume 1. Basel: Birkhauser, 1955.

WORKS BY NIKLAUS I

1709 *Dissertatio inauguralis mathematico-juridica de usu artis conjectandi in jure.* Basel: Mechel.

SUPPLEMENTARY BIBLIOGRAPHY

HUYGENS, CHRISTIAAN 1657 De ratiociniis in ludo aleae. Pages 521–534 in Frans van Schooten, *Exercitationum mathematicarum libri quinque.* Leiden (Netherlands): Elsevier.

L'HOSPITAL, GUILLAUME FRANÇOIS ANTOINE DE 1696 *Analyse des infiniment petits, pour l'intelligence des lignes courbes.* Paris: Imprimerie Royale.

[MONTMORT, PIERRE RÉMOND DE] (1708) 1713 *Essay d'analyse sur les jeux de hazard.* 2d ed. Paris: Quillau. → Published anonymously.

Postscript

The philosophical ideas on probability that Jakob I expressed in Book IV of his *Ars conjectandi* continue to inspire discussion and controversy, especially among students of subjective probability. See, for example, Hacking (1975, chapter 16).

Excellent biographies of the Bernoullis can now be found in the *Dictionary of Scientific Biography* (vol. 2, 1970).

GLENN SHAFER

[*See also* PROBABILITY, *article on* FORMAL PROBABILITY.]

ADDITIONAL BIBLIOGRAPHY

The 1913 edition of Jakob I's Ars conjectandi *was reprinted in Latin in 1968 by the Belgian publishing house Culture et Civilisation. A translation of Book IV into English by Bing Sung was issued as* Translations From James Bernoulli, *Technical Report No. 2 of the Department of Statistics of Harvard University, in 1966, and is available in microfiche from the Clearinghouse for Scientific and Technical Information, Washington. In 1968 Dover published, in one volume, Daniel's* Hydrodynamics *and Johann's* Hydraulics *in English translation by Thomas Carmody and Helmut Kobus.*

FELLMAN, E. A.; and FLECKENSTEIN, J. O. 1970 Johann (Jean) I Bernoulli. Volume 2, pages 51–55 in *Dictionary of Scientific Biography.* Edited by Charles C. Gillispie. New York: Scribner's.

FLECKENSTEIN, J. O. 1970 Johann (Jean) II and Johann (Jean) III Bernoulli; Nikolaus I Bernoulli. Volume 2, pages 56–57 in *Dictionary of Scientific Biography.* Edited by Charles C. Gillispie. New York: Scribner's.

HACKING, IAN 1975 *The Emergence of Probability: A Philosophical Study of Early Ideas About Personality, Induction and Statistical Inference.* Cambridge Univ. Press.

HOFMANN, J. E. 1970 Jakob (Jacques) I Bernoulli. Volume 2, pages 46–51 in *Dictionary of Scientific Biography.* Edited by Charles C. Gillispie. New York: Scribner's.

STRAUB, HANS 1970 Daniel Bernoulli. Volume 2, pages 36–46 in *Dictionary of Scientific Biography.* Edited by Charles C. Gillispie. New York: Scribner's.

BIAS

See ERRORS, *article on* NONSAMPLING ERRORS; ESTIMATION; FALLACIES, STATISTICAL; INTERVIEWING IN SOCIAL RESEARCH.

BIENAYMÉ, JULES

Jules Bienaymé, statistician and mathematician, was born in Paris in 1796 and died there in 1878. He received his secondary education in Bruges and later at the Lycée Louis le Grand in Paris. His studies at the École Polytechnique, where he enrolled in 1815, ended the following year, because that institution was dissolved when its students persisted in their loyalty to the Napoleonic regime. In 1818 Bienaymé became lecturer in mathematics at Saint-Cyr, the French equivalent of West Point. In the end, he joined the civil service as a general inspector of finance.

After Bienaymé became a civil servant, he began his studies of actuarial science, statistics, and probability. Baron Louis, France's able minister of finance during the Bourbon restoration, was inclined to make use of technical advice, and Bienaymé became closely associated with Louis's work. Bienaymé's career in the civil service was not interrupted by the revolution of 1830, but after the revolution of 1848 he retired and devoted all his time to scientific work.

Bienaymé's retirement made possible his active participation in the affairs of various scientific societies. He became a member of the Société Philomatique (an association for the advancement of science) and, on July 5, 1852, he was elected a member of the Institut de France (Académie des Sciences). At the time of his death, he was a corresponding member of the Science Academy of St Petersburg and of the Central Commission of Statistics of Belgium, and an honorary member of the Chemical Conference Association of Naples. As a member of the Académie des Sciences he acted for 23 years as a referee for the Montyon Prize, the highest French award for achievement in statistics, and his interesting judgments of the candidates for this distinction can be found in the records of the academy.

Bienaymé published many papers in the proceedings of the academy. Among these is an important one on runs, giving a theorem for the probable number of maxima and minima of a sequence of observed numbers. In 1853 Bienaymé discovered a very important inequality: the probability that the inequality $|X| \geqslant t\sigma$ is true is less than or equal to $1/t^2$, X being a random variable with mean zero and standard deviation σ (1853*a*). The Russian mathematician Chebyshev independently published the same discovery some twelve years later.

Scientific controversies had considerable appeal for Bienaymé. He debated with Cauchy about the relative merits of the least squares method and of an interpolation procedure proposed by the latter (1853*b*). He also criticized the extension by Poisson of a theorem of Jacques Bernoulli, the so-called law of large numbers (1855). In addition to the criticism of Poisson, this paper contains a keen analysis of meteorological data, especially those having to do with rainfall.

In spite of his retirement, Bienaymé had considerable influence, as a statistical expert, in the government of Napoleon III. In 1864 Napoleon's minister, Dumas, praised Bienaymé in the French Senate for the help he had given to the administration in connection with the actuarial work required for the creation of a retirement fund.

DANIEL DUGUÉ

[*For the historical context of Bienaymé's work, see the biographies of the* BERNOULLI FAMILY; POISSON. *For discussion of the subsequent development of his ideas, see* NONPARAMETRIC STATISTICS, *article on* RUNS; PROBABILITY.]

WORKS BY BIENAYMÉ

1837 De la durée de la vie en France depuis le commencement du XIXe siècle. *Annales d'hygiène publique et de médicine légale* 18:177–218.

1838*a* Mémoire sur la probabilité des resultats moyens des observations: Démonstration directe de la règle de Laplace. Académie des Sciences, Paris, *Mémoires présentés par divers savants; sciences mathématiques et physiques* 2d Series 5:513–558.

1838*b* Probabilité des jugements et des témoinages. Société Philomatique de Paris, *Extraits des procès-verbaux des séances* 5th Series 3:93–96.

1853*a* Considérations à l'appui de la découverte de Laplace sur la loi de probabilité dans la méthode des moindres carrés. Académie des Sciences, Paris, *Comptes-rendus hebdomadaires des séances* 37:309–324.

1853*b* Remarques sur les différences qui distinguent l'interpolation de M. Cauchy de la méthode des moindres carrés, et qui assurent la supériorité de cette méthode. Académie des Sciences, Paris, *Comptes-rendus hebdomadaires des séances* 37:5–13.

1855 Communication sur un principe que M. Poisson avait cru découvrir et qu'il avait appelé loi des grands nombres. Académie des Sciences Morales et Politiques, *Séances et travaux* 31:379–389.

1875 Application d'un théorème nouveau du calcul des probabilités. Académie des Sciences, Paris, *Comptes-rendus hebdomadaires des séances* 81:417–423.

SUPPLEMENTARY BIBLIOGRAPHY

M. de la Gournerie donne lecture de la note suivante, sur les travaux de M. Bienaymé. 1878 Académie des Sciences, Paris, *Comptes-rendus hebdomadaires des séances* 87:617–619.

Notice sur les travaux scientifiques de M. I. J. Bienaymé . . . inspecteur général des finances. 1852 Paris: Bachelier.

Postscript

Bienaymé's work is being restudied in its historical context by modern scholars such as Heyde and Seneta ([1972] 1977*a*; 1977*b*).

DANIEL DUGUÉ

[*See also the biographies of* COURNOT; QUETELET.]

ADDITIONAL BIBLIOGRAPHY

HEYDE, C. C.; and SENETA, E. (1972) 1977*a* The Simple Branching Process, a Turning Point Test and a Fundamental Inequality: A Historical Note on I. J. Bienaymé. Volume 2, pages 406–409 in Maurice G. Kendall and R. L. Plackett (editors), *Studies in the History of Statistics and Probability.* London: Griffin; New York: Macmillan. → First published in *Biometrika* 59:680–683.

HEYDE, C. C.; and SENETA, E. 1977*b* *I. J. Bienaymé: Statistical Theory Anticipated.* New York: Springer.

BILINEAR METHODS

See under FACTOR ANALYSIS AND PRINCIPAL COMPONENTS.

BINOMIAL DISTRIBUTION

See DISTRIBUTIONS, STATISTICAL, *article on* SPECIAL DISCRETE DISTRIBUTIONS.

BIOASSAY

See QUANTAL RESPONSE; SCREENING AND SELECTION.

BIRNBAUM, ALLAN

► *This article was specially written for this volume.*

Allan Birnbaum (1923–1976) took his undergraduate training in mathematics at the University of California, Berkeley, where he stayed to do graduate work not only in mathematics but in science and philosophy. This led to a lifelong interest in the philosophy of science, greatly influenced by the works of Ernest Nagel and Hans Reichenbach. In 1947 he moved from the Bay area, where he was born, to Columbia University, where he obtained his PH.D. in mathematical statistics in 1954. He stayed at Columbia until 1959, then moved to the Courant Institute of Mathematical Sciences at New York University. He remained there until 1972, when he left for an extended visit to Britain. He resigned his post at the Courant Institute in 1974 upon his appointment to the Chair of Statistics at City University, London. He had one son.

His early published research was in the Neyman–Pearson tradition; his 1955 paper is a typical example. By the mid-1950s, however, his main interest had shifted to the foundations of statistical inference. He wrote a number of technical reports on that subject, which he summarized and extended in "On the Foundations of Statistical Inference" (1962). In this important paper, Birnbaum studied some properties that a notion of statistical evidence might possess and, in particular, explored connections among the notions of likelihood, conditionality, and sufficiency. His conclusions were in a sense obvious to Bayesian statisticians. To those, however, who relied primarily on a sampling theory approach, Birnbaum's demonstration that the acceptable notions of conditionality and sufficiency led to an apparently unacceptable strong likelihood notion caused considerable discussion. The paper should be read in conjunction with "Concepts of Statistical Evidence" (1969), wherein Birnbaum concluded that the requirements put on statistical evidence in his earlier paper could not all be satisfied and that, for most statistical purposes, some "sensible" version of the Neyman–Pearson approach should be adopted. Birnbaum strongly resented suggestions that this was a retreat from his earlier paper, which he regarded as an exploration of properties that might be required of statistical evidence, rather than an assertion of what *should* be done [*see* LIKELIHOOD, *written by Birnbaum*]. His latest views are contained in "The Neyman–Pearson Theory as Decision Theory, and as Inference Theory" (1978).

Birnbaum had several other professional interests. An important one was in methodological problems of testing abilities and aptitudes, and he contributed substantially (1968*a*) to a book on this topic written primarily by Lord and Novick. Later fields of interest were statistical genetics, where he worked on statistical methods of data analysis, and genetic theory, including quantitative inheritance and evolutionary theory; see, for example, his 1972 paper "The Random Phenotype Concept." He was widely read in subjects ranging from the history of probability and statistics to statistical theory and the philosophy of science.

An important factor in the later years of Birnbaum's life was his concern for statistical scholarship, which he expressed in an article published in 1971. He was keen that mathematicians in particular should appreciate the role that statistics should play in education. His resignation from the Courant Institute was, in part, due to philosophical differences between him and his colleagues there. He spent a period in England, first at Cambridge and then in London, before settling permanently in that city. Unfortunately, his relations with some of his immediate colleagues at the City University were, again, difficult.

Both in his published works and especially in papers presented at scientific meetings, Birnbaum could be difficult to understand, and the careless listener might think his statements diffuse. In fact, however, he had a scrupulous regard for fairness and accuracy, and this could give a misleading impression of prolixity. On less formal occasions, he could present his views concisely and forcibly. Personally he was quiet and unassuming and was widely regarded with affection and respect.

DENNIS V. LINDLEY

[*See* LIKELIHOOD *for a discussion of some of the issues raised by Birnbaum and his work; related material appears in* PROBABILITY, *article on* INTERPRETATIONS; BAYESIAN INFERENCE; SUFFICIENCY.]

WORKS BY BIRNBAUM

1955 Characterizations of Complete Classes of Tests of Some Multiparametric Hypotheses, With Applications to Likelihood Ratio Tests. *Annals of Mathematical Statistics* 26:21–36.

1962 On the Foundations of Statistical Inference. *Journal of the American Statistical Association* 57:269–306. → Discussion on pages 307–326.

1968*a* Some Latent Trait Models and Their Use in Inferring an Examinee's Ability. Part 5, chapters 17–20, pages 397–479 in Frederic M. Lord and Melvin R. Novick, with contributions by Allan Birnbaum, *Statistical Theories of Mental Test Scores*. Reading, Mass.: Addison-Wesley.

1968*b* Likelihood. Volume 9, pages 299–301 in *International Encyclopedia of the Social Sciences*. Edited by David L. Sills. New York: Macmillan and Free Press.

1969 Concepts of Statistical Evidence. Pages 112–143 in Sidney Morgenbesser, Patrick Suppes, and Morton White (editors), *Philosophy, Science, and Method: Essays in Honor of Ernest Nagel*. New York: St. Martin's.

1971 A Perspective for Strengthening Scholarship in Statistics. *American Statistician* 25, no. 3:14–17. → A briefer, earlier version appeared in *New York Statistician* 22 (1970):1–2.

1972 The Random Phenotype Concept, With Applications. *Genetics* 72:739–758.

1978 The Neyman–Pearson Theory as Decision Theory, and as Inference Theory, With a Criticism of the Lindley–Savage Argument for Bayesian Theory. *Synthese* 36, no. 1:19–49.

SUPPLEMENTARY BIBLIOGRAPHY

Medical Research: Statistics and Ethics. 1977 *Science* 198:677–705. → Comprises articles adapted from lectures and discussions presented at the Birnbaum Memorial Symposium held at the Memorial Sloan–Kettering Cancer Center, New York, May 27: Valerie Miké and Robert A. Good, "Old Problems, New Challenges"; John W. Tukey, "Some Thoughts on Clinical Trials, Especially Problems of Multiplicity"; John P. Gilbert, Bucknam McPeek, and Frederick Mosteller, "Statistics and Ethics in Surgery and Anesthesia";

Victor Herbert, "Acquiring New Information While Retaining Old Ethics"; Jerome Cornfield, "Carcinogenic Risk Assessment"; and André Cournand, "The Code of the Scientist and Its Relationship to Ethics."

Synthese 36 (1978), no. 1. → A special issue in "Foundations of Probability and Statistics" dedicated to Birnbaum.

BIVARIATE DISTRIBUTIONS

See DISTRIBUTIONS, STATISTICAL.

BORTKIEWICZ, LADISLAUS VON

● Ladislaus von Bortkiewicz (first spelled Bortkewitsch as in the Russian transcription) was born of Polish descent in 1868 in St. Petersburg and graduated from the Faculty of Law at the university there. His first papers (1890 and 1891) were published in Russian; in all, seven of his papers were in Russian. He continued his studies in Göttingen under Lexis, where he wrote his doctoral thesis (1893). In 1895 he became *Privatdozent* in Strassburg and subject to the influence of Knapp, but he returned to Russia in 1899. Besides serving as clerk in the general office of the Railway Pension Committee in St. Petersburg from 1897 to 1901, Bortkiewicz taught at the Alexandrowsky Lyceum there during the period 1899–1900. He became associate professor at the University of Berlin in 1901, and finally in 1920 he became full professor *ad personam* of economics and statistics. From 1906 to 1923 he also served as *Dozent* at the Berlin Handelshochschule. He remained in Berlin for thirty years, until his death in 1931. With rare exceptions (as noted above), he wrote in German. He was one of the few representatives of mathematical statistics in Germany and as such a lonely figure, highly respected but rarely understood.

Besides classical economics, the work of Bortkiewicz covered population statistics and theory, actuarial science, mathematical statistics, probability theory, mathematical economics, and physical statistics—fields separate in content but analogous in methodology. He contributed to the process of consolidating each of these disciplines and did classic work in mathematical statistics.

Many of his investigations dealt with mortality tables. In a stationary population, the birth rate equals the death rate and the expectation of life of a newborn equals the reciprocal of the common value of the two rates. For increasing populations it was believed that the expectation of life could be obtained from the observed birth and death rates. Bortkiewicz showed (1893), however, that a correct answer can be obtained only by the construction of a mortality table. He returned to this problem when dealing with different methods of comparing mortality rates (1904*b*; 1911). In an increasing population there are more infants and fewer old people than in a stationary one. The first influence raises, the second lowers, the general mortality. Bortkiewicz showed that the second influence prevails in general so that the growth of the population tends to decrease the mortality rate. The study of life tables led him to actuarial science (see 1903; 1929).

The work that made his name widely known was a brochure (1898) of sixty pages, *Das Gesetz der kleinen Zahlen* ("The Law of Small Numbers"). Poisson had shown in 1837 that besides the usual normal limit for Bernoulli's distribution there is a second limit, requiring that the number (n) of observations increase and the probability (p) decrease so that the product (np) has a limiting value. In this distribution n and p enter only through their product $\lambda = np$, which is the expected number of happenings. (The Poisson limit is primarily useful as an approximation when λ is small.) Poisson's important derivation remained practically unknown for sixty years; at least, its importance was not recognized. Bortkiewicz was the first to note the fact that events in a large population, with low frequency, can be fitted by a Poisson distribution even when the probability of an event varies somewhat between the strata of the population. This is what he called the law of small numbers—the name refers to small numbers of events (see also 1915*a*).

A striking example was the number of soldiers killed by horse kicks per year per Prussian army corps. Fourteen corps were examined, each for twenty years. For over half the corps–year combinations there were no deaths from horse kicks; for other combinations the number of deaths ranged up to four. Presumably the risk of lethal horse kicks varied over years and corps, yet the over-all distribution was remarkably well fitted by a Poisson distribution.

In this distribution the variance, that is, the square of the standard deviation, is equal to the expectation. The corresponding observed quotient should therefore be near unity. This is called "normal dispersion" in the Lexis theory. The law of small numbers says that rare events usually show normal dispersion; for a mathematical explanation of this fact consistent with Lexis' theory, see Gosset (1919). Bortkiewicz computed tables of Poisson's distribution and discussed estimation of its expectation by the sample mean. In addition he discussed errors of estimation in quantitative terms and used them as criteria for the validity of the theory. Thus Bortkiewicz created an important instrument for mathematical statistics and probability theory. However, the name he gave it was unfortunate because it implied a nonexistent contrast to the law of large numbers and led to much confusion and unnecessary argument. [*For a discussion of the law of large numbers see* PROBABILITY, *article on* FORMAL PROBABILITY; *see also the biography of* POISSON.] It would have been better to speak of "rare events."

Many recent studies on the meaning of the different derivations and uses of Poisson's formula are linked to Bortkiewicz's discovery. The Poisson distribution has become the subject of important work that extends to statistically dependent events and varying probabilities. Large parts of operational research and queueing theory are based on the Poisson distribution.

Bortkiewicz also contributed to the theory of runs with the publication of his book *Die Iterationen* (1917). This work was motivated by an attack made by the psychologist Karl Marbe (1916–1919) on the easy assumption of independence in applications of probability theory, for example, to successive flips of a coin or to sequences of male and female births. Marbe believed that a run of male births leads to a heightened probability of a female birth, as nature tries to equalize or make uniform the sex ratio. Bortkiewicz showed, however, that Marbe's mathematics was wrong and that a mathematically correct approach gives agreement between theoretical independence and observed sequences in cases of the kind discussed.

Bortkiewicz devoted a book (1913) to the statistical interpretation of radioactivity. He showed that regularities considered as physical laws could be expressed by existing theorems on mean values of stochastic processes.

He also showed (1922*a*) that the extreme values, which had been considered unsuitable for the analysis and the characterization of a distribution, are statistical variables depending upon initial distribution and sample size. He gave the exact distribution of the normal range and computed its mean for sample sizes up to 20. With primitive equipment, he reached good numerical results (Tippett 1925) and checked them by many observations. The statistical importance of this work is obscured by mathematical complexities incident to the normal distribution.

The study of dispersion was central to the thought of Bortkiewicz. He confirmed and extended (1904*b*) the ideas of his teacher, Lexis, and strengthened them by his derivation of the standard error of the coefficient of dispersion (1918). He defended the importance and originality of Lexis (1930). The generalization of these methods led to the modern analysis of variance.

From the start Bortkiewicz worked on political economy and, like Lexis, he shared none of the usual vulgar prejudices to Marx. According to Schumpeter ([1932] 1960, p. 303) "By far his most important achievement is his analysis of the theoretical framework of the Marxian system [(1906–1907; 1907)], much the best thing ever written on it and, incidentally, on its other critics. A similar masterpiece is his paper on the theories of rent of Rodbertus and Marx [(1910–1911)]."

Bortkiewicz succeeded in embedding in a mathematical form both Marx's determination of the average profit rate for simple reproduction and Marx's transformation, implicit therein, of value into price. According to Marx (*Das Kapital*, vol. 3), to yield the average profit rate the total surplus value is divided by the total capital, that is, the sum of constant and variable capital. In this solution, however, the input is measured in values and the output in prices. Bortkiewicz was the first of Marx's many critics to see this inconsistency. He made the necessary modifications that rendered the Marxian scheme of surplus values and prices consistent. However, his dry presentation prevented the Marxists (except for Klimpt) from accepting his method.

His investigations on price index numbers (1923–1924) are noteworthy contributions to mathematical economics. Irving Fisher (1922) had developed many such numbers. Bortkiewicz brought clarity and order into this system of index numbers by stating the requirements that such a number must satisfy in order to fulfill its purpose.

Bortkiewicz argued vigorously for his views. In 1910 he wrote an article attacking Alfred Weber's geometrical representation of the location of industries. His polemical article (1915*b*) against the Pearson school clarified his fundamental attitude —namely, that it is worthless to construct formulas to reproduce observations if these formulas have no theoretical meaning. In another polemic (1922*b*) against Pearson he insisted on Helmert's priority in discovering the distribution of the mean square residual when individual errors are normal. Yet Bortkiewicz's answer (1923) was quite mild when Keynes (1921, p. 403, note 2) wrote, ". . . Bortkiewicz does not get any less obscure as he goes on. The mathematical argument is right enough and often brilliant. But what it is all really about, and what it really amounts to, and what the premises are, it becomes increasingly perplexing to decide."

In his article (1931*a*) on the disparity of income distributions Bortkiewicz used Pareto's law. While Pareto had not been very clear about the role of his basic parameter α, Bortkiewicz established different measures of income concentration and showed that α is such a measure. Bortkiewicz's work on concentration was published in ignorance of the prior work of Gini.

Bortkiewicz had a characteristic way of working. He presented each problem from all sides with extreme thoroughness and patience after an extensive study of the literature. This multiple foundation makes the solution unassailable, but the reader can trace no single line from premises to conclusion: the central line of thought is entwined with numerous sidelines and extensive polemics, especially on matters of scientific priority. He criticized with equal zeal and profundity important and insignificant mistakes, printing errors, and numerical miscalculations. A large part of his work appeared as reviews and critical analyses in remote journals. His writings stimulated numerous scientists in Germany, in the northern European countries and in Italy, but not in England. He did not create a school, perhaps because of his austere character and his poor teaching. He underestimated his own work and even doubted, wrongly, its practical significance. His cautious nature forbade him to strive for external honors. He was, from 1903, a member of both the International Statistical Institute, which then consisted mainly of administrative statisticians, and the Swedish Academy of Sciences. He maintained objectivity in the face of popular slogans as well as "untimely opinions." He was a true scholar of the old school and his life was passed in enviable quietness.

Four of his contributions are decisive: the proof that the Poisson distribution corresponds to a statistical reality; the introduction of mathematical statistics into the study of radioactivity; the inception of the statistical theory of extreme values; and the lonely effort to construct a Marxian econometry.

E. J. GUMBEL

[*Other relevant material may be found in* DISTRIBUTIONS, STATISTICAL, *article on* APPROXIMATIONS TO DISTRIBUTIONS; NONPARAMETRIC STATISTICS, *articles on* ORDER STATISTICS *and* RUNS; QUEUES; *and in the biography of* LEXIS. *See also* Allais 1968; Dobb 1968.]

WORKS BY BORTKIEWICZ

1890 Smertnost' i dolgovechnost' muzhskago pravoslavnago naseleniia evropeiskoi Rossii (Mortality and Lifespan of the Male Russian Orthodox Population of European Russia). Imp. Akademiia Nauk, *Zapiski* 63: Supplement no. 8.

1891 Smertnost' i dolgovechnost' zhenskago pravoslavnogo naseleniia evropeiskoi Rossii (Mortality and Lifespan of the Female Russian Orthodox Population of European Russia). Imp. Akademiia Nauk, *Zapiski* 66: Supplement no. 3.

1893 *Die mittlere Lebensdauer: Die Methoden ihrer Bestimmung und ihr Verhältnis zur Sterblichkeitsmessung.* Jena (Germany): Fischer.

1898 *Das Gesetz der kleinen Zahlen.* Leipzig: Teubner.

1901 Anwendungen der Wahrscheinlichkeitsrechnung auf Statistik. Volume 1, pages 821–851 in *Encyklopädie der mathematischen Wissenschaften.* Leipzig: Teubner.

1903 Risicoprämie und Sparprämie bei Lebensversicherungen auf eine Person. *Assekuranz-Jahrbuch* 24, no. 2:3–16.

1904*a* Über die Methode der "Standard Population." International Statistical Institute, *Bulletin* 14, no. 2: 417–437.

1904*b* Die Theorie der Bevölkerungs- und Moralstatistik nach Lexis. *Jahrbücher für Nationalökonomie und Statistik* 82:230–254.

(1906–1907) 1952 Value and Price in the Marxian System. *International Economic Papers* 2:5–60. → First published in German.

(1907) 1949 On the Correction of Marx's Fundamental Theoretical Construction in the Third Volume of *Capital.* Pages 197–221 in Eugen von Böhm-Bawerk, *Karl Marx and the Close of His System.* New York: Kelley. → First published in German.

1909–1911 Statistique. Part 1, volume 4, pages 453–490 in *Encyclopédie des sciences mathématiques.* Paris: Gauthier-Villars. → Substantially the same as Bortkiewicz 1901, with changes by the translator, F. Oltramare.

1910 Eine geometrische Fundierung der Lehre vom Standort der Industrien. *Archiv für Sozialwissenschaft und Sozialpolitik* 30:759–785.

1910–1911 Die Rodbertus'sche Grundrententheorie und die Marx'sche Lehre von der absoluten Grundrente. *Archiv für die Geschichte des Sozialismus und der Arbeiterbewegung* 1:1–40, 391–434.

1911 Die Sterbeziffer und der Frauenüberschuss in der stationären und der progressiven Bevölkerung. International Statistical Institute, *Bulletin* 19:63–141.

1913 *Die radioaktive Strahlung als Gegenstand wahrscheinlichkeitstheoretischer Untersuchungen.* Berlin: Springer.

1915*a* Über die Zeitfolge zufälliger Ereignisse. International Statistical Institute, *Bulletin* 20, no. 2:30–111.

1915*b* Realismus und Formalismus in der mathematischen Statistik. *Allgemeines statistisches Archiv* 9: 225–256.

1917 *Die Iterationen: Ein Beitrag zur Wahrscheinlichkeitstheorie.* Berlin: Springer.

1918 Der mittlere Fehler des zum Quadrat erhobenen Divergenzkoeffizienten. Deutsche Mathematiker-Vereinigung, *Jahresbericht* 27:71–126.

1919 *Bevölkerungswesen.* Leipzig: Teubner.

1922*a* Die Variationsbreite beim Gaussschen Fehlergesetz. *Nordisk statistisk tidskrift* 1:11–38, 193–220.

1922*b* Das Helmertsche Verteilungsgesetz für die Quadratsumme zufälliger Beobachtungsfehler. *Zeitschrift für angewandte Mathematik und Mechanik* 2:358–375.

1923 Wahrscheinlichkeit und statistische Forschung nach Keynes. *Nordisk statistisk tidskrift* 2:1–23.

1923–1924 Zweck und Struktur einer Preisindexzahl. *Nordisk statistisk tidskrift* 2:369–408; 3:208–251, 494–516.

1929 Korrelationskoeffizient und Sterblichkeitsindex. *Blät-*

ter für Versicherungs-Mathematik und verwandte Gebiete 1:87–117.

1930 Lexis und Dormoy. *Nordic Statistical Journal* 2: 37–54.

1931*a* Die Disparitätsmasse der Einkommensstatistik. International Statistical Institute, *Bulletin* 25, no. 3: 189–291.

1931*b* The Relations Between Stability and Homogeneity. *Annals of Mathematical Statistics* 2:1–22.

SUPPLEMENTARY BIBLIOGRAPHY

►ALLAIS, MAURICE 1968 Pareto, Vilfredo: I. Contributions to Economics. Volume 11, pages 399–411 in *International Encyclopedia of the Social Sciences*. Edited by David L. Sills. New York: Macmillan and Free Press.

ANDERSON, OSKAR 1931 Ladislaus von Bortkiewicz. *Zeitschrift für Nationalökonomie* 3:242–250.

○ANDERSSON, THOR 1931 Ladislaus von Bortkiewicz: 1868–1931. *Nordic Statistical Journal* 3:9–26. → Includes a bibliography. Reprinted, with supplementary materials, in *Nordisk statistisk tidskrift* 10 (1931): 1–16.

CRATHORNE, A. R. 1928 The Law of Small Numbers. *American Mathematical Monthly* 35:169–175.

►DOBB, MAURICE 1968 Economic Thought: IV. Socialist Thought. Volume 4, pages 446–454 in *International Encyclopedia of the Social Sciences*. Edited by David L. Sills. New York: Macmillan and Free Press.

FISHER, IRVING (1922) 1927 *The Making of Index Numbers: A Study of Their Varieties, Tests, and Reliability*. 3d ed., rev. Boston: Houghton Mifflin.

FREUDENBERG, KARL 1951 Die Grenzen für die Anwendbarkeit des Gesetzes der kleinen Zahlen. *Metron* 16: 285–310.

GINI, C. 1931 Observations . . . à la communication . . . du M. L. von Bortkiewicz. International Statistical Institute, *Bulletin* 25, no. 3:299–306.

GOSSET, WILLIAM S. (1919) 1943 An Explanation of Deviations From Poisson's Law in Practice. Pages 65–69 in [William S. Gosset], *"Student's" Collected Papers*. Cambridge Univ. Press.

GUMBEL, E. J. 1931 L. von Bortkiewicz. *Deutsches statistisches Zentralblatt* 23, cols. 231–236.

GUMBEL, E. J. 1937 Les centenaires. *Aktuárske védy* (Prague) 7:10–17.

KEYNES, JOHN M. (1921) 1952 *A Treatise on Probability*. London: Macmillan. → A paperback edition was published in 1962 by Harper.

KÜHNE, OTTO 1922 *Untersuchungen über die Wert und Preisrechnung des Marxschen Systems: Eine dogmenkritische Auseinandersetzung mit L. von Bortkiewicz*. Greifswald (Germany): Bamberg.

LORENZ, CHARLOTTE 1951 *Forschungslehre der Sozialstatistik*. Volume 1: Allgemeine Grundlegung und Anleitung. Berlin: Duncker & Humblot.

MARBE, KARL 1916–1919 *Die Gleichförmigkeit in der Welt: Untersuchungen zur Philosophie und positiven Wissenschaft*. 2 vols. Munich: Beck.

NEWBOLD, ETHEL M. 1927 Practical Applications of the Statistics of Repeated Events, Particularly to Industrial Accidents. *Journal of the Royal Statistical Society* 90:487–535.

SCHUMACHER, HERMANN 1931 Ladislaus von Bortkiewicz *Allgemeines statistisches Archiv* 21:573–576.

SCHUMPETER, JOSEPH A. (1932) 1960 Ladislaus von Bortkiewicz: 1868–1931. Pages 302–305 in Joseph A. Schumpeter, *Ten Great Economists From Marx to Keynes*. New York: Oxford Univ. Press. → First published in Volume 42 of the *Economic Journal*.

TIPPETT, L. H. C. 1925 On the Extreme Individuals and the Range of Samples Taken From a Normal Population. *Biometrika* 17:364–387.

WEBER, ERNA 1935 *Einführung in die Variations- und Erblichkeits-statistik*. Munich: Lehmann.

WINDSOR, CHARLES P. 1947 Quotations: *Das Gesetz der kleinen Zahlen*. *Human Biology* 19:154–161.

WOYTINSKY, WLADIMIR S. 1961 *Stormy Passage; A Personal History Through Two Russian Revolutions to Democracy and Freedom: 1905–1960*. New York: Vanguard.

Postscript

Transliteration and citation practices have varied widely in treating Bortkiewicz's name. Although I prefer to transliterate the original Cyrillic as "Vladislav Bortkiewitch," the title of my biography (Sheynin 1970) makes explicit the ambiguity of conventional spellings. The particle "von" does not, of course, appear in Bortkiewicz's early writings in Russian, and even though it does appear in his later works in German, some modes of alphabetical arrangement (including that of this very volume) disregard it.

[1]Sheynin (1966) has shown that the distribution of the mean square residual when individual errors are normal had been derived still earlier by the German scholar of optics Ernst Abbe. (See also Kendall 1971.)

O. B. SHEYNIN

ADDITIONAL BIBLIOGRAPHY

KENDALL, MAURICE G. 1971 The Work of Ernst Abbe. *Biometrika* 58:369–373.

LOREY, W. 1932 Ladislav von Bortkiewicz. *Versicherungsarchiv* 3:199–206.

SHEYNIN, O. B. 1966 Origin of the Theory of Errors. *Nature* 211:1003–1004.

SHEYNIN, O. B. 1970 Bortkiewicz (or Bortkewitsch), Ladislaus (or Vladislav) Josephowitsch. Volume 2, pages 318–319 in *Dictionary of Scientific Biography*. Edited by Charles C. Gillispie. New York: Scribner's.

BOWLEY, ARTHUR LYON

Arthur Lyon Bowley (1869–1957), British statistician, was born at Bristol and brought up in a conventional and religious family. Before he went to Trinity College, Cambridge, in 1888 with a major scholarship in mathematics, he spent nine years at Christ's Hospital, a boarding school of a strictly religious foundation where pupils wore a traditional costume and were subject to spartan conditions. The school left a lasting impression on his character; later in life he was governor of the school for some years.

Bowley followed a conventional course in mathematics at Cambridge, graduating as a wrangler in 1891, but an interest in economics and social problems, which was to be the mainspring of his life's work, was already evident in his undergraduate days. He was in contact with Alfred Marshall and others active in the developing social sciences, and he was deeply affected, not so much by refinements of economic analysis as by the problems of social reform in Britain at the end of the century.

In studying these problems Bowley used both contemporary and historical material. His first work was *A Short Account of England's Foreign Trade in the Nineteenth Century* (1893), for which he received the Cobden Prize at Cambridge in 1892. Foreign trade was a subject on which he continued to write throughout his life, but a more important early interest was the relation between movements of wages and prices, the subject of the first paper he read to the Royal Statistical Society in March 1895 (1895*a*). From 1895 to 1906 the *Journal* of the society and the more recently established *Economic Journal* published many papers by him on this topic, sometimes by Bowley alone and sometimes with G. H. Wood as coauthor. Bowley approached the subject with statistical and historical care verging on the pedantic, yet at the same time with a deep and sympathetic appreciation of the human problems involved.

It was on Marshall's recommendation that Bowley was invited to join the small and mainly part-time staff of the London School of Economics when the first session began in 1895. Thus was laid the main path of his professional career. Over a period of more than forty years, until his retirement in 1936, Bowley taught statistics to successive generations of students of the social sciences at the school. He developed an intimate friendship with Edwin Cannan and remained to take his place as an elder statesman in a large and distinguished faculty of economists, historians, and social and political theorists. He was never a socialist in the sense of Webb and other founding fathers of the school, but as a good liberal he found the senior common room a congenial and stimulating background to his activities in teaching and research.

The London School of Economics, although increasingly the locus of his work, did not for many years provide Bowley with his livelihood. In 1895, when living and teaching mathematics at a school in Leatherhead, he bicycled from there on the Wednesday half holidays to lecture at the London School in the early evening. Later, from 1900 to 1913, he was a member of the mathematics staff at University Extension College, Reading, and remained as a lecturer in economics there until 1919. Meanwhile he became a part-time reader in statistics at the London School in 1908, receiving the title of professor in 1915. It was only in 1919, with the establishment of a chair in statistics in the University of London, of which he was the first holder, that he became a full-time member of the school's faculty.

As a mathematician Bowley was competent without being very original, and he became increasingly old-fashioned in his mathematical formulations. He published relatively little, and nothing of real substance, either in mathematical statistics or in mathematical economics, although both fields developed rapidly in very exciting directions in his middle and later life. Much of his work was in mathematical form as a matter of convenience, but the mathematics itself was incidental to his main purposes. First and foremost Bowley was a practitioner in applied statistics, with the whole of the social sciences as his field, and for most of his career he had to make bricks with very little straw. Always a severe critic of British official statistics, and highly respected in official quarters, he was called upon far too seldom to advise on the development of government statistics. British economic and social statistics in the 1920s and 1930s would undoubtedly have been improved, particularly by the use of sampling techniques, if he had had more to do with them. His main influence was through his private researches and in discussions at the international level.

There can be little doubt that Bowley's major contribution was to the development of sampling techniques and their application to economic and social studies. While he was forming his ideas in the 1890s, the great debate on the "representative method" was taking place among official statisticians in Europe and the United States. It was from these discussions against the rather narrow background of official statistics that the modern corpus of sampling techniques developed, with applications in all fields of scientific inquiry. Anders N. Kiaer (1838–1919), the distinguished chief of the Norwegian Bureau of Statistics for 46 years, led the case for sampling at a series of sessions of the International Statistical Institute from 1895 (Bern) to 1901 (Budapest). He was at first opposed by a majority of leading official statisticians, but his ideas rapidly gained ground, being greatly supported by the report of Carroll D. Wright, at Budapest, on sampling experience in the U.S. Department of Labor. Bowley, elected to the institute in

1903, was immediately attracted by the possibilities of the "representative method." With characteristic care, he explored for himself both the appropriate mathematical formulation of sampling precision and the best ways of interpreting the results of sample surveys to laymen.

Between 1912 and 1914 Bowley directed sample surveys of working-class households in five English towns, and in presenting his results in *Livelihood and Poverty* (Bowley & Burnett-Hurst 1915), he was far ahead of his time in explaining both the method and the errors of sampling. He devoted a chapter to the four sources of error: incorrect information, loose definitions, bias in selection of sample, and the calculable errors of sampling. It is true that he did not distinguish the method of cluster or systematic sampling he adopted (selection of 1 in *n* down a listing of the frame) from simple random sampling. Even so, his exposition of 1915 would have been readily accepted two generations later.

It was only appropriate, therefore, that Bowley became a member of the committee set up by the International Statistical Institute in 1924, presenting their "Report on the Representative Method in Statistics" (Jensen 1926) at the Rome session in 1925. Bowley's hand is clearly visible in the major recommendation "that the investigation should be so arranged wherever possible, as to allow of a mathematical statement of the precision of the results, and that with these results should be given an indication of the extent of the error to which they are liable" (Jensen 1926, p. 378), as well as in Annex A to the report "Measurement of the Precision Attained in Sampling" (1926*a*). Bowley himself continued to practice what he preached, notably in his resurvey of the five English towns published in *Has Poverty Diminished?* (1925) and in *The New Survey of London Life and Labour,* Volume 3 (1932*a*; 1932*b*) and Volume 6 (1934).

Another pioneering work undertaken by Bowley was the estimation of the distribution of national income, a task in which he was in the end rather less successful than might have been expected. His first essays were "The Division of the Product of Industry" and "The Change in the Distribution of the National Income: 1880–1913." Later he worked with Josiah Stamp on the more elaborate "The National Income: 1924" (see Bowley 1919–1927) and with the National Institute of Economic and Social Research on *Studies in the National Income: 1924–1938* (1942), a series that was curtailed by the outbreak of World War II. This work was a natural development of his early interest in wages and of his continuing concern with the redistribution of income as a tool in social reform. In the years between the two world wars he found British data quite inadequate for his purpose; economists did not agree on even the concept of national income. His work was influential when the first official estimates of the British national income were made, under the inspiration of J. M. Keynes, during World War II. But it was not in his careful and precise nature to take undue risks in handling scattered data. This he left to others, notably to the more adventurous, almost buccaneering, spirit of Colin Clark.

A third area in which Bowley's pioneering was influential was his regular reporting on and analysis of the current economic position for the London and Cambridge Economic Service. This service began publication as a private venture, dependent on subscriptions for its bulletins, in January 1923. At first it was a cooperative project with the Harvard University Committee on Economic Research, which had issued its Harvard Economic Service bulletins for some time, and the aims were set out by William H. Beveridge in an introductory article on the study of business cycles. Bowley was the first editor of the service, serving in this capacity for more than twenty years, until 1945, and continuing as a regular contributor until 1953. Under his guidance the service was soon set on a very profitable course of its own, independent of its Harvard parent (which failed to survive the crisis of 1929) and of various schools of business cycle research.

Two characteristics of Bowley as an editor were outstanding. One was the skill with which he listened at editorial meetings to the diverse and outspoken views of economists before writing his own pithy assessment of the economic position in order to represent just as much as, and no more than, the majority of economists could agree upon at that moment. The other was his conviction that any analysis of the present or forecast of the future was dependent on long runs of carefully prepared statistical series covering the whole range of economic and social matters. He and his research associates were indefatigable in designing and improving index numbers and in devising ways of presenting them most effectively. From the beginning he showed his series in graphical form, often by the use of ratio scales and often after adjustment for seasonal variations. He was one of the earliest champions of these now well-recognized devices. In these and other aspects he never shirked the task of explaining highly technical matters to a lay public.

Bowley was an effective if rather dour committeeman, and he held many offices, in the British Association, in the Royal Statistical Society, and in the International Statistical Institute, among others. He received many honors, which culminated in his appointment as knight bachelor soon after his eightieth birthday.

R. G. D. ALLEN

[*For discussion of the subsequent development of Bowley's work in sampling, see* INDEX NUMBERS; SAMPLE SURVEYS. *See also* Ruggles 1968.]

WORKS BY BOWLEY

(1893) 1922 *A Short Account of England's Foreign Trade in the Nineteenth Century: Its Economic and Social Results.* 3d ed. London: Allen & Unwin.

1895*a* Changes in Average Wages (Nominal and Real) in the United Kingdom Between 1860 and 1891. *Journal of the Royal Statistical Society* 58:223–278.

1895*b* Comparison of the Rates of Increase of Wages in the United States and in Great Britain: 1860–1891. *Economic Journal* 5:369–383.

1897 Relations Between the Accuracy of an Average and That of Its Constituent Parts. *Journal of the Royal Statistical Society* 60:855–866.

●1900 *Wages in the United Kingdom in the Nineteenth Century.* Cambridge Univ. Press.

(1901) 1937 *Elements of Statistics.* 6th ed. New York: Scribner; London: King.

(1910) 1951 *An Elementary Manual of Statistics.* 7th ed. London: Macdonald & Evans.

1911 The Measurement of the Accuracy of an Average. *Journal of the Royal Statistical Society* 75:77–88.

1915 BOWLEY, ARTHUR L.; and BURNETT-HURST, A. R. *Livelihood and Poverty: A Study in the Economic Conditions of Working-class Households in Northhampton, Warrington, Stanley and Reading.* London: Bell.

(1919–1927) 1938 *Three Studies on the National Income.* London School of Economics and Political Science. → Contains "The Division of the Product of Industry"; "The Change in the Distribution of National Income: 1880–1913," by Arthur L. Bowley; and "The National Income: 1924," by Arthur L. Bowley and Josiah Stamp.

1924 *The Mathematical Groundwork of Economics: An Introductory Treatise.* Oxford: Clarendon.

1925 BOWLEY, ARTHUR L.; and HOGG, MARGARET H. *Has Poverty Diminished? A Sequel to* Livelihood and Poverty. London: King.

1926*a* Measurement of the Precision Attained in Sampling. International Statistical Institute, *Bulletin* 22, part 1:6–62.

1926*b* The Influence on the Precision of Index-numbers of Correlation Between the Prices of Commodities. *Journal of the Royal Statistical Society* 89:300–319.

1928 Notes on Index Numbers. *Economic Journal* 38: 216–237.

1930*a* Area and Population. Pages 58–83 in London School of Economics and Political Science, *The New Survey of London Life and Labour.* Volume 1: Forty Years of Change. London: King.

1930*b* London Occupations and Industries. Pages 315–340 in London School of Economics and Political Science, *The New Survey of London Life and Labour.* Volume 1: Forty Years of Change. London: King.

1932*a* The House Sample Analysis. Pages 29–96 in London School of Economics and Political Science, *The New Survey of London Life and Labour.* Volume 3: Survey of Social Conditions: 1. The Eastern Area. London: King.

1932*b* BOWLEY, ARTHUR L.; and SMITH, H. LLEWELLYN. Overcrowding. Pages 216–253 in London School of Economics and Political Science, *The New Survey of London Life and Labour.* Volume 3: Survey of Social Conditions: 1. The Eastern Area. London: King.

1934 The House Sample Analysis. Pages 29–117 in London School of Economics and Political Science, *The New Survey of London Life and Labour.* Volume 6: Survey of Social Conditions: 2. The Western Area. London: King.

1935 ALLEN, ROY G. D.; and BOWLEY, ARTHUR L. *Family Expenditure: A Study of Its Variation.* London School of Economics and Political Science, Studies in Statistics and Scientific Method, No. 2. London: King.

►1937 *Wages and Income in the United Kingdom Since 1860.* Cambridge Univ. Press.

1942 BOWLEY, ARTHUR L. (editor). *Studies in the National Income: 1924–1938.* Cambridge Univ. Press.

SUPPLEMENTARY BIBLIOGRAPHY

JENSEN, ADOLPH 1926 Report on the Representative Method in Statistics. International Statistical Institute, *Bulletin* 22, part 1:359–380.

►RUGGLES, RICHARD 1968 Economic Data: I. General. Volume 4, pages 365–369 in *International Encyclopedia of the Social Sciences.* Edited by David L. Sills. New York: Macmillan and Free Press.

BUSINESS CYCLES: MATHEMATICAL MODELS

This article was first published in IESS *with a companion article less relevant to statistics.*

A mathematical model of business cycles is not necessarily a special kind of business cycle theory as far as economic content is concerned. The mathematical formulation is an instrument for organizing our factual knowledge and our hypotheses. For this purpose, mathematical tools may be not only useful, but indispensable. Use of these tools may produce fruitful theories that could not have been discovered by verbal reasoning, and a precise mathematical formulation may serve to verify or reject previous theories set forth in a loose, verbal form and to clear the way for more systematic empirical studies.

It is not easy to date the origin of mathematical business cycle models. Fragments of such models may be found in even classical economic theory. However, it is probably fair to say that the development of explicit and complete mathematical business cycle models does not date further back than the early 1930s (see Frisch 1933; Kalecki 1935; Tinbergen 1935). These first models were of a highly macroeconomic type, involving only a few key variables to characterize the economic system. Subsequently, a very large number and variety of such models have been developed (see, for example, Samuelson 1939; Metzler 1941; Hicks 1950; Goodwin 1951). A good survey of some of these models is found in Allen's textbook on mathematical economics (1957, pp. 209–280).

More detailed models, involving a large number of economic variables, have also been developed by Tinbergen (1938–1939), Klein (1950), and others. The purpose of these detailed models has been not only to furnish a more detailed theoretical explanation of business cycles but also to pave the way for verification and measurement by means of principles of statistical inference. Electronic computers play an increasingly important role in this kind of business cycle research.

General features. Facts and data concerning the ups and downs of business activity constitute a bewildering mass of information. Any attempt to write "the whole story of what happens" during booms and depressions is not only hopeless but also rather unrewarding as far as gaining real understanding is concerned. Somehow one has to look for principles of systematic classification and for simplifying ideas of *simulation* that can help to reduce the number of things to be taken into account. With this in mind, what are the general features of the dynamic process we call business cycles?

Apart from some relatively crude theories that explain business cycles as something "coming from the outside" (sunspot theories or the like), all theories of business cycles focus attention on the idea that what we observe is a result of human decision and action. The *driving force* is the prospect of profit or economic advantage, of one kind or another. The release, strength, and direction of such forces can be regarded as reactions to a system of *signals* that guide the economic activities of the various individuals or groups. These signals are prices of goods and services or other data that enter into the calculations of economic gains or losses for each decision unit.

The forces thus released are counteracted by various elements of inertia and friction due, in part, to human hesitation and slowness and to constraints set by nature or by rigid institutions.

There is also an intricate network of "feedbacks," with the characteristic property that the feedback line from the activity of *one* decision group usually connects with the signal system guiding some *other* decision unit. Clearly, if elements of inertia and delay are present in such a system, a continued process of adjustment of some kind is almost unavoidable.

If we view the process of booms and depressions through the framework just described, the analogy with models of force and motion in mechanical engineering and related fields becomes striking. To utilize the idea of such analogies while at the same time being on guard against stretching the analogy too far is one of the main principles of mathematical model building in the field of business cycle analysis.

The notion of dynamic equilibrium. Economists have long had a deeply rooted feeling that a "normal" situation in business activity is a state of affairs where *motion is absent* (except, perhaps, for some kind of trend). Strangely enough, the use of mathematics to set up systems of general market equilibrium may have strengthened the hold that this notion of normality has on so many economists. Given the wide acceptance of this notion, it is understandable that ups and downs in business activity are often looked upon as "deviations from normal," as imperfections of the market, as unforeseeable and unwanted exceptions to the rule. But such ideas are not particularly fruitful as a basis for understanding business cycles.

The point is, of course, that in a stationary situation, such as the equilibrium of the Walrasian system, the forces in operation are not zero. They are in fact very strong, but they happen to *balance at zero motion*. This is, however, a very special case of a balance of forces. The more general case is a balance brought about by sustained *motion* of certain elements of the economy. This explains why it is indeed possible to represent the process of change in business activity by means of mathematical equations based on the principle of forces in balance. The economic forces may be in balance at various rates of motion of the economic magnitudes involved, and this general idea of dynamic equilibrium is fundamental in mathematical business cycle theories.

Explanation of turning points. A central problem in business cycle theory has, of course, been to explain the turning points, i.e., to explain why expansion should turn into contraction, and vice versa. One of the major contributions of the mathematical approach to business cycle theory has been to demonstrate that the explanation of turning points is no more difficult than the explanation of any other phase of cyclical movements. Employing the notion of dynamic equilibrium, one can simply say that the relative strength of the economic forces in operation at any time will determine whether the motion necessary for balance will be up or down.

The reason why the mathematical approach is

superior to a verbal analysis is obvious. By verbal reasoning it is simple enough to enumerate the various economic forces involved in a process of development, but it is often difficult, if not impossible, to determine the direction of the motion resulting from the relative strengths of the various forces.

Effects of learning—irreversibility. One objection to business cycle theories in the form of rigid mathematical models has been that they lead to a monotonous recurrence of booms and depressions of the same kind, while in fact "history never repeats itself." Certain mathematical models are indeed open to this criticism, but others are not.

One of the remedies for this deficiency is the explicit introduction of elements of learning into the model. For example, the pattern of consumers' demand may gradually change as a consequence of accumulating experience, or the way in which producers form their expectations (their basis for action) may gradually change as a result of their comparing past expectations with realizations. In recent years, more and more attention has been given to such elements as necessary parts of mathematical business cycle models (cf. Goodwin 1951).

While it is possible to introduce elements of irreversibility into a model in this way, it should be realized that a model must be based on the assumption that there are *some* aspects of economic development that repeat themselves. Otherwise, no theory, mathematical or verbal, is feasible.

Main types of models. By classifying the various business cycle models according to the principles involved rather than according to the particular economic variables dealt with, it is possible to group the models in a two-by-two table. First, we consider whether the principal active forces responsible for motion are assumed to come from the *outside* or are assumed to be *endogenous* parts of the economic system itself. The first type of model is sometimes called an *open* model, and the second type is called *closed.* Second, for each of these two types of models, we consider whether the cycles are produced because *the driving force is itself cyclical* ("forced oscillations") or because of the particular way in which the *economic system responds* to the stimulating forces ("free oscillations").

These principles of classification are helpful, even though a really comprehensive business cycle model may contain elements that would place it in all four categories simultaneously (cf. Samuelson 1947, pp. 335–349).

Some explicit examples will illustrate many of the points discussed above.

The cobweb model with external forces. Let $x^d(t)$ be the demand for product x at time t, and let $p(t)$ be the price of x at time t. Assume that

$$x^d(t) = f[p(t)]. \tag{1}$$

Let $x^s(t)$ be the supply of x at time t, and assume that

$$x^s(t) = g[p(t-\theta)] + v(t), \tag{2}$$

where θ is positive and $v(t)$ is some external force that independently influences $x^s(t)$. For example, $v(t)$ may be some weather factor or perhaps some influence from another economic sector that is independent of the one considered here. For market clearance at time t, $x^d(t)$ must equal $x^s(t)$. Let $x(t)$ be the quantity of x at which the market clears at time t. From (1) and (2) and the market-clearance condition, it will generally be possible to derive

$$x(t) = G[x(t-\theta)] + v(t). \tag{3}$$

The usual shape of supply and demand curves would imply that the first derivative of the function G is negative.

If $v(t)$ were a constant, independent of t, this model would be the usual textbook case of the "cobweb" (Allen 1957, pp. 2–6). In such a model, there could be business cycles of period 2θ, which would either eventually die out or go on vigorously forever.

Now let us consider the effect of changes in $v(t)$. If $x(t)$ oscillated but tended toward some constant when $v(t)$ was constant, a change in $v(t)$ to a new level would generally set the variable $x(t)$ in motion again; and $x(t)$ would go on oscillating for some time, even if $v(t)$ were to remain constant at the new level. In other words, the driving force $v(t)$ need not itself oscillate systematically in order to generate oscillations in $x(t)$. It is sufficient that $v(t)$ change occasionally, perhaps in a quite irregular manner.

If $v(t)$ should have a cycle of its own, this would, of course, have certain consequences for the resulting time shape of $x(t)$. But $x(t)$ would, in addition, have cyclical properties that are not present in $v(t)$, but are a consequence of the functional form G and of the lag θ. In this case, the model is one of "free" oscillations with an external driving force.

Consider now the special case where θ is equal to zero. From (3) it can be seen that in this case $x(t)$ can be expressed directly as a function of $v(t)$, assuming (3) permits such a solution. Thus, $x(t)$ could not move except when $v(t)$ is in motion. If $v(t)$ had a cyclical nature, these cycles would, in some manner or other, be reflected in $x(t)$ as "forced" oscillations.

Investment cycles—a closed model. Let $C(t)$ denote consumption, $I(t)$ net investment, and $Y(t)$ income at time t in a closed economy. Then we have

$$(4) \qquad Y(t) = C(t) + I(t).$$

Consider a simple "Keynesian" consumption function:

$$(5a) \qquad C(t) = f[Y(t)],$$

or, as an alternative, a dynamic version:

$$(5b) \qquad C(t) = F[Y(t), Y(t-1)].$$

Suppose that for some reason there are outside forces causing independent oscillations in the rate of investment $I(t)$. Then, in the case of the consumption function (5a), we should have *forced oscillations* in $C(t)$ and $Y(t)$. If the consumption function were (5b) instead of (5a), consumption and income could be subject to both free oscillations and forced oscillations.

The model above would be called *open*, because it does not "explain" the behavior of investment. The idea of the acceleration principle can be used to close the model. Let us first consider a very simple version of this idea.

Let $K(t)$ be the (physical) amount of capital present in the economy at time t, and let $K^*(t)$ denote the amount of capital that producers *would like to have* at that time. If these two amounts of capital are equal, producers would be satisfied. If, on the other hand, $K^*(t)$ is larger than $K(t)$, the demand for new capital *per unit of time* would be unlimited, i.e., producers would be willing to buy any rate of investment that could be supplied. Assuming that there is a capacity limit on total production in the economy, there would be an upper limit, a "ceiling," on the amount of output of capital goods. In other words, the rate of investment would be restricted on the *supply* side. If, instead, $K^*(t)$ is below $K(t)$, there would be no demand for new capital goods, not even for replacement purposes. In this situation it is, therefore, *demand* that determines investment, and the rate of depreciation establishes a (negative) *floor* under which demand cannot fall (unless capital is purposely destroyed).

The model now "explains" investment, provided we know the determinants of the desired capital stock, $K^*(t)$, and how existing capital depreciates. But if this knowledge is lacking, the model is still an *open* model and the question is how to close it. The simple assumptions that have been introduced for this purpose (cf. Allen 1957, pp. 242–247) are that the desired capital stock is a function of total net output; more specifically, that $K^*(t)$ and $Y(t)$ are proportional and that there is a constant rate of depreciation.

Under these assumptions, it is easy to indicate the characteristic properties that the model would have. Suppose that consumption is given by (5a) and that the capacity to produce investment goods is sufficiently high for the amount of capital to *reach and to exceed* the amount of desired capital. After the desired amount of capital has been reached, output obviously must fall below capacity. But as output falls, so does the desired amount of capital, and this decline cannot stop until gross output of investment goods is zero. This would then lead to a minimum level of net output and thus to a minimum level of desired capital. Only after the existing amount of capital has been worn down below the minimum desired level could there be any demand for new capital goods. But *when* that situation eventually occurs, output must increase again. The amount of desired capital must then increase, and output will again reach full capacity. Thus, we have a *closed model, with free and maintained oscillations.*

This version of the model is, however, unsatisfactory in several respects. First, an explanation of why the desired amount of capital should be a function of output is needed. Second, there is some question as to whether it is safe to assume that the level of desired capital will ever actually be reached. And third, there is some question as to whether the closed model is not a somewhat artificial product, obtained by neglecting such things as wage policy and monetary policy.

It may be of interest to indicate briefly how some of these defects could be remedied. Let $X(t)$ denote total gross output, and let depreciation be equal to $\delta K(t)$, where δ is a constant. We then have

$$(6) \qquad X(t) = Y(t) + \delta K(t).$$

Assume that $X(t)$ is the output of a "classical" production function:

$$(7) \qquad X(t) = \phi[N(t), K(t)],$$

where $N(t)$ is employment and where complementarity is assumed to exist between inputs N and K. Let the (real) wage rate, $w(t)$, be an increasing function of employment:

$$(8) \qquad w(t) = W[N(t)].$$

Finally, let $r(t)$ be the rate of interest.

Consider two different situations:

Situation 1. Suppose, tentatively, that there is

no limit on $X(t)$ from the demand side. Then we may assume that employment is determined by setting the marginal productivity of labor equal to the wage rate, provided total wages are below current revenues. If the marginal productivity of capital is greater than or equal to $r + \delta$, the demand for X will in fact be unlimited, because of the demand for an increased amount of capital.

Situation 2. Suppose, tentatively, that output, $X(t)$, is limited by effective demand, the role of producers being simply that of producing to order. Then employment follows from (7), with $K(t)$ given, provided total wages according to (8) are below current revenues. If the corresponding marginal productivity of capital is less than or equal to $r + \delta$, demand for $X(t)$ will in fact be limited and will be equal to consumers' demand, because producers will not want any new capital.

Whether in this model there will be a switching back and forth between the two situations, similar to that in the simpler model previously discussed, depends in an essential way on policy concerning the rate of interest. If situation 1 exists but is about to break down, lowering the rate of interest could prolong the situation. It should be noted that if situation 2 is allowed to occur, the reduction of the rate of interest that could then get us back to situation 1 would generally be much greater than the reduction that would be sufficient to maintain a situation 1 already in existence.

This model has a great deal of flexibility and can be extended to include technical progress, rachet effects in consumers' demand, and so on.

TRYGVE HAAVELMO

[*Directly related is the entry* ECONOMETRIC MODELS, AGGREGATE. *See also* Baumol 1968.]

BIBLIOGRAPHY

ALLEN, R. G. D. (1957) 1963 *Mathematical Economics.* 2d ed. New York: St. Martins; London: Macmillan.

►BAUMOL, WILLIAM J. 1968 Statics and Dynamics in Economics. Volume 15, pages 169–177 in *International Encyclopedia of the Social Sciences.* Edited by David L. Sills. New York: Macmillan and Free Press.

FRISCH, RAGNAR (1933) 1965 Propagation Problems and Impulse Problems in Dynamic Economics. Pages 155–185 in American Economic Association, *Readings in Business Cycles.* Edited by R. A. Gordon and L. R. Klein. Homewood, Ill.: Irwin.

GOODWIN, R. M. 1951 The Nonlinear Accelerator and the Persistence of Business Cycles. *Econometrica* 19:1–17.

HICKS, JOHN R. 1950 *A Contribution to the Theory of the Trade Cycle.* Oxford: Clarendon.

KALECKI, M. 1935 A Macrodynamic Theory of Business Cycles. *Econometrica* 3:327–344.

KLEIN, LAWRENCE R. 1950 *Economic Fluctuations in the United States: 1921–1941.* New York: Wiley.

METZLER, LLOYD A. (1941) 1965 The Nature and Stability of Inventory Cycles. Pages 100–129 in American Economic Association, *Readings in Business Cycles.* Edited by R. A. Gordon and L. R. Klein. Homewood, Ill.: Irwin.

SAMUELSON, PAUL A. (1939) 1944 Interactions Between the Multiplier Analysis and the Principle of Acceleration. Pages 261–269 in American Economic Association, *Readings in Business Cycle Theory.* Edited by Gottfried Haberler. Philadelphia: Blakiston. → First published in Volume 21 of the *Review of Economic Statistics.*

SAMUELSON, PAUL A. (1947) 1958 *Foundations of Economic Analysis.* Harvard Economic Studies, Vol. 80. Cambridge, Mass.: Harvard Univ. Press. → A paperback edition was published in 1965 by Atheneum.

TINBERGEN, JAN 1935 Annual Survey: Quantitative Business Cycle Theory. *Econometrica* 3:241–308.

TINBERGEN, JAN 1938–1939 *Statistical Testing of Business-cycle Theories.* 2 vols. Geneva: League of Nations, Economic Intelligence Service. → Volume 1: *A Method and Its Application to Investment Activity.* Volume 2: *Business Cycles in the United States of America: 1919–1932.*

C

CANONICAL CORRELATION

See MULTIVARIATE ANALYSIS, *article on* CORRELATION METHODS.

CAUSATION

A cause is something that occasions or effects a result (the usual lexical definition) or a uniform antecedent of a phenomenon (J. S. Mill's definition). When a question is asked in the form "Why . . .?" it can usually be answered appropriately by a statement in the form "Because . . . ." Thus, to state the causes of a phenomenon is at least one way to explain the phenomenon; and a careful explication of the concept of causation in science must rest on a prior analysis of the notions of scientific explanation and scientific law.

Explanations in terms of causation are sought both for particular events and for classes of events or phenomena. Thus, a statement of the causes of World War II might include references to German economic difficulties during the 1930s or to the failure of the League of Nations to halt the Ethiopian conquest. On the other hand, a statement of the causes of war might include references to the outward displacement of aggression arising from internal frustrations, the absence of legitimized institutions for legal settlements of disputes between nations, and so on.

Causal explanations generally involve a combination of particular and general statements. In classical price theory, for example, a drop in the price of a commodity can be caused by an increase in its supply and/or a decrease in demand. If an explanation is desired for a drop in the price of wheat in a particular economy in a given year, it may be sought in the unusually large wheat crop of that year. (The large wheat crop may be explained, in turn, by a combination of general laws asserting that the size of the wheat crop is a function of acreage, rainfall, fertilization, and specific facts relevant to these laws—the actual acreage planted, rainfall, and amount of fertilizer applied that year.)

A general paradigm can be given for this kind of causal explanation. Let a be a particular situation (for example, the wheat market in 1965); let $A(x)$ be a statement about situation x (for example, the supply in market x increases); and let $B(x)$ be another such statement about x (for example, the price in market x declines). Suppose there is an accepted scientific law of the form

$$(x)(A(x) \rightarrow B(x))$$

(for example, in any market, if the supply of the commodity increases, the price will decline). Upon substitution of a for x, this becomes $(A(a) \rightarrow B(a))$; if the supply of the commodity in market a increases, its price will decline (for example, if the supply of wheat increased in 1965, its price declined). Then $A(a)$ and $(x)(A(x) \rightarrow B(x))$ provide, conjointly, a causal explanation for $B(a)$. That is to say, $A(a)$ occasions or effects $B(a)$, while $A(x)$ is the uniform antecedent of $B(x)$. Thus, the paradigm incorporates both the lexical and Mill's definitions of cause.

Explication of causation along these lines gives rise to three sets of problems that have been discussed extensively by philosophers of science. The first of these may be called the "problem of Hume" because all treatments of it in modern times—both those that agree with Hume and those that oppose him—take Hume's analysis as their starting point (see Hume 1777). It is the logical and epistemological problem of the nature of the connection between the "if" and the "then" in a scientific law. Is it a "necessary" connection, or a connection in fact, and how is the existence of the connection verified?

The second problem may be called the problem of causal ordering or causal asymmetry. If $A(a)$ is the cause of $B(a)$, we do not ordinarily think that $B(a)$, or its absence, can cause $A(a)$ or its absence. But in the standard predicate calculus of formal logic, $(x)(A(x) \rightarrow B(x))$ implies $(x)(\sim B(x) \rightarrow \sim A(x))$, where "$\sim$" stands for "not." ("If much rainfall, other things equal, then a large wheat crop" implies "If a small wheat crop, other things equal, then not much rainfall.") While we accept the inverse inference, we do not regard it as causal. (The size of the wheat crop does not retrospectively affect the amount of rain; *knowledge* of the size of the wheat crop may, however, affect our *inference* of how much rain there had been.) Thus two statements corresponding to the same truth-function (that is, either both are true or both are false) need not express the same causal ordering. The asymmetry between $A(x)$ and $B(x)$ cannot, therefore, rest solely on observation of the situations in which either or both of these predicates hold. How can the causal relation be defined to preserve the asymmetry between cause and effect?

The third set of problems surrounding causation are the psychological problems. Michotte has explored the circumstances under which one event will in fact be perceived as causing another. Piaget and his associates have investigated the meanings that "cause," "why," "because," and other terms of explanation have for children at various ages. In succeeding sections we shall examine these three sets of problems: the problem of Hume, the problem of causal ordering, and the psychological problems of causal perception and inference.

The problem of Hume

David Hume pointed out that even though empirical observation could establish that events of type *B* had in each case followed events of type *A*, observation could never establish that the connection between *A* and *B* was necessary or that it would continue to hold for new observations. Thus, general statements like "If *A*, then *B*" can serve as convenient summaries of numbers of particular facts but cannot guarantee their own validity beyond the limits of the particular facts they summarize.

Hume did not deny, of course, that people commonly make general inductions of the form "If *A*, then *B*" from particular facts; nor that they use these generalizations to predict new events; nor that such predictions are often confirmed. He did deny that the inductive step from particulars to generalization could be provided with a deductive justification, or that it could establish a "necessary" connection between antecedent and consequent, that is, a connection that could not fail in subsequent observations. Modern philosophers of science in the empiricist tradition hold positions very close to Hume's. Extensive discussions and references to the literature can be found in Braithwaite (1953, chapter 10), Popper ([1934] 1959, chapter 3), and Nagel (1961, pp. 52–56).

In everyday usage, however, a distinction is often made between generalizations that denote "lawful" regularities and those that denote "accidental" regularities. "If there is a large wheat crop, the price of wheat will fall" states a lawful regularity. "If X is a man, then X's skin color is not bright greenish blue" states an accidental regularity. But since our basis for accepting both generalizations is the same—we have observed confirming instances and no exceptions—the distinction between laws, on the one hand, and generalizations that are only factually or accidentally true, on the other, must be sought not in the empirical facts these generalizations denote but in their relations to other generalizations, that is, in the structure of scientific theories.

Within some given body of scientific theory, a general statement may be called a *law* if it can be deduced from other statements in that theory. The connection between the quantity of a commodity offered on a market and the price is lawful, relative to general economic theory, because it can be deduced from other general statements in economic theory: that price reaches equilibrium at the level where quantity offered equals quantity demanded; that the quantity demanded is smaller (usually) at the higher price. These latter statements may be derived, in turn, from still others: statements about the characteristics of buyers' utility functions, postulates that buyers act so as to maximize utility, and so on.

Although lawfulness does not exempt a generalization from the need for empirical verification, a scientific law's logical connections with others may subject it to indirect disconfirmation; and its direct disconfirmation may affect the validity of other generalizations in the system. A scientific theory may be viewed as a system of simultaneous relations from which the values of particular observations can be deduced. When a reliable observation conflicts with the prediction, some change must be made in the theory; but there is no simple or general way to determine where in the system the change must be made.

Nothing has been said here about the issue of determinism versus indeterminism, which is often discussed in relation to the problem of Hume (see Braithwaite 1953, chapter 10; Popper [1934] 1959, chapter 3; Nagel 1961, chapter 10). The concept of causal ordering is entirely compatible with either deterministic or probabilistic scientific theories. In a probabilistic theory, events are only incompletely determined by their causes, and formalization of such theories shows that the causal relations implicit in them hold between probability distributions of events rather than between the individual events. Thus, in a system described by a so-called Markov process, the probability distribution of the system among its possible states at time t is causally (and not probabilistically) determined by the probability distribution at time $t-1$. [*See* MARKOV CHAINS.] In this situation, disconfirming a theory requires not just a simple disconfirming observation but rather a sufficiently large set of observations to show that a predicted distribution does not hold.

Causal ordering

It was pointed out above that we cannot replace "*A* causes *B*" with the simple truth-functional $(x)(A(x) \rightarrow B(x))$ without creating difficulties of interpretation. For $(x)(A(x) \rightarrow B(x))$ implies $(x)(\sim B(x) \rightarrow \sim A(x))$, while we do not ordinarily infer "*not-B* causes *not-A*" from "*A* causes *B*." On the contrary, if *A* causes *B*, it will usually also be the case that *not-A* causes *not-B*. Thus, if a large wheat crop causes a low price, we will expect a small wheat crop to cause a high price. The appropriate asymmetrical statement is that "the size of the wheat crop is a cause of the price of wheat," and the example shows that the asymmetry here is different from that of "if–then" statements. Attempts (Burks 1951) to base a definition of the causal relation on the logical relation of implication have been unsuccessful for this reason.

The concept of scientific law, introduced in the last section, provides an alternative approach to explicating causal ordering. This approach makes the causal connection between two variables depend on the context provided by a scientific theory —a whole set of laws containing these variables (Simon [1947–1956] 1957, chapters 1, 3). This approach views the causal ordering as holding between *variables* rather than between particular *values* of those variables. As noted in the last paragraph, the variable "size of the wheat crop" (x) is to be taken as the cause for "price of wheat" (p), a large crop causing a low price and a small crop causing a high price ($p = f(x)$, with $dp/dx < 0$).

Linear structures. For concreteness, consider the important special situation where a scientific theory takes the form of a set of n simultaneous linear algebraic equations in n variables. Then, apart from certain exceptional cases, these equations can be solved for the unique values of the variables. (In the general case the system is called a *linear structure.*) In solving the equations, algebraic manipulations are performed that do not change the solutions. Equations are combined until one equation is derived that contains only a single variable. This equation is then solved for that variable, the value is inserted in the remaining equations, and the process is repeated until the values of all variables have been found. In general, there is no single, set order in which the variables must be evaluated. Hence, from an algebraic viewpoint, there is no distinction between "independent" and "dependent" variables in the system. The variables are all interdependent, and the linear structure expresses that interdependence.

However, it may be found in particular systems of this kind that certain subsets of equations containing corresponding subsets of variables can be solved independently of the remaining equations. Such subsets are called *self-contained subsets.* In the extreme case, a particular equation may contain only one variable, which can then be evaluated as the dependent variable of that equation. Substituting its value as an independent variable in another equation, we may find that only one dependent variable remains, which can now be evaluated.

A causal ordering among variables of a linear structure that has one or more self-contained subsets can now be defined as follows: Consider the minimal self-contained subsets (those that do not themselves contain smaller self-contained subsets) of the system. With each such subset, associate the variables that can be evaluated from that subset alone. These are the *endogenous* or dependent variables of that subset and are exogenous to the rest of the system. Call them *variables of order zero.* Next, substitute the values of these variables in the remaining equations of the system, and repeat the whole process for these remaining equations, obtaining the variables of order one, two, and so on, and the corresponding subsets of equations in which they are the dependent variables. Now if a variable of some order occurs with nonzero coefficient in an equation of the linear structure belonging to a subset of higher order, the former variable has a *direct causal connection* to the endogenous variables of the latter subset.

An example. The wheat price example will help make the above notions more concrete. Suppose a theory in the following form: (1) The amount of rain in a given year is taken as exogenous to the remainder of the system; that is, it is set equal to a constant. (2) The wheat crop is assumed to increase linearly with the amount of rain (within some range of values). (3) The price of wheat is assumed to move inversely, but linearly, with the size of the crop. The system thus contains three equations in three variables. The first equation, determining the amount of rain, is the only self-contained subset. Hence, amount of rain is a variable of order zero. Given its value, the second equation can be solved for the size of the wheat crop, a variable of order one. Finally, the third equation can be solved for the price of wheat, a variable of order two. Thus, there is a direct causal connection from the amount of rain to the size of the crop and from the size of the crop to the price.

Operational meaning—mechanisms. The causal ordering would be altered, of course, if before solving the equations the system is modified by taking linear combinations of them. This process is algebraically admissible, for it does not change the solutions, and, indeed, is employed as an essential means for solving simultaneous equations. If the causal ordering is not invariant under such transformations, can it be said to have operational meaning?

Operational meaning is assigned to the equations of the initial, untransformed system by associating with each equation a *mechanism* (meaning an identifiable locus of intervention or alteration in the system). "Intervention" may be human (for example, experimentation) or natural (for example, change in initial conditions). Just as the operational identity of individual variables in a system depends on means for measuring each independently of the others, so the operational identity of mechanisms depends on means for intervention in each independently of the others. Thus, in the wheat price model, the nature of the mechanism determining rainfall is unspecified, but the mechanism can be "modified" by taking a sample of years with different amounts of rain. Coefficients in the mechanism relating rainfall to the size of the wheat crop can be modified by irrigation or by growing drought-resistant strains of wheat. The mechanism relating the size of the crop to price can be modified by changing buyers' incomes.

When particular mechanisms can be identified and causal ordering inferred, this knowledge permits predictions to be made of the effects on the variables of a system of specific modifications of the mechanisms—whether these be produced by policy intervention, experimental manipulation, or the impact of exogenous variables. Thus, although there is algebraic equivalence between a system in which each equation corresponds to a separate mechanism and systems obtained by taking linear combinations of the original equations, the derived systems are not operationally equivalent to the original one from the standpoint of control, experiment, or prediction. In the statistical literature, the equations that represent mechanisms and causal ordering are called *structural equations;* certain equivalent equations derived for purposes of statistical estimation are called *reduced-form equations.*

Temporal sequence. The method described here for defining causal ordering accounts for the asymmetry of cause and effect but does not base the asymmetry on temporal precedence. It imposes no requirement that the cause precede the effect. We are free to limit our scientific theories to those in which all causes *do* precede their effects, but the definition of causal ordering does not require us to do so. Thus, in experimental situations, average values of the independent variables over the period of the experiment may be interpreted as the causal determinants of the values of the dependent variables over the same period.

Many scientific theories do, however, involve temporal sequence. In dynamic theories, the state of the system at one point in time is (causally) determined by the state of the system at an earlier point in time. Generally, a set of initial conditions is given, specifying the state of the system at a point in time taken as origin. The initial conditions, together wth the differential or difference equations of the system (the general laws), induce a causal ordering like that defined above.

Interdependence. If almost all the variables in a dynamic system are directly interdependent, so that the value of each variable at a given time depends significantly on the values of almost all other variables at a slightly earlier time, the causal ordering provides little information and has little usefulness. When the interrelations are sparse, however, so that relatively few variables are directly dependent on each other, a description of the causal ordering provides important information about the structure of the system and about the qualitative characteristics of its dynamic behavior (for example, the presence or absence of closed feedback loops). For this reason, the language of causation is used more commonly in relation to highly organized and sparsely connected structures —man-made mechanisms and organisms with their systems of organs—than in relation to some of the common systems described by the partial differential equations of chemistry and physics, where the interactions are multitudinous and relatively uniform.

Psychology of causal inference

The considerations of the preceding sections are purely logical and say nothing of the circumstances under which persons will infer causal connections between phenomena. Extensive studies of the psychology of causal inference have been made by Michotte and Piaget.

Michotte (1946) has shown that a perception of causal connection can be induced, for example, by two spots of light, the first of which moves toward the second and stops on reaching it, while the second then continues the motion in the same direction. Using numerous variants of this scheme, he has explored the circumstances under which subjects will or will not interpret the events causally. His evidence tends to show that the process by which the subject arrives at a causal interpretation is "direct," subconscious, and perceptual—that causal attribution is not a conscious act of inference or induction from a sequence of events perceived independently. While the general distinction intended by Michotte is clear, its detailed interpretation must depend on a more precise understanding of the neural mechanisms of visual perception. In particular, little is known as yet about the respective roles of peripheral and central mechanisms or of innate and learned processes in the perception of causality.

Piaget (1923; 1927) has brought together a sizable body of data on children's uses of causal language. A principal generalization from these data is that the earliest uses of "Why?" are directed toward motivation of actions and justification of rules; demands for naturalistic causal explanations appear only as the child's egocentrism begins to wane.

Spurious correlation

Previous sections have dealt with the principal problems surrounding the concept of causation. The remaining sections treat some significant applications of causal language to the social sciences.

It has often been pointed out that a statistical correlation between two variables is not sufficient grounds for asserting a causal relation between them. A correlation is called *spurious* if it holds between two variables that are not causally related. The definition of causal ordering provides a means of distinguishing genuine from spurious correlation [*see* FALLACIES, STATISTICAL; MULTIVARIATE ANALYSIS, *article on* CORRELATION METHODS].

Since the causal orderings among variables can be determined only within the context of a scientific theory—a complete structure—it is only within such a context that spurious correlation can be distinguished from genuine correlation. Thus, to interpret the correlation between variables x and y in causal terms, either there must be added to the system the other variables that are most closely connected with x and y, or sufficient assumptions of independence of x and/or y from other variables must be introduced to produce a self-contained system. When this has been done, the simple correlation between x and y can be replaced with their partial correlation (other variables being held constant) in the larger self-contained system. The partial correlation provides a basis for estimating the coefficients of the self-contained system and hence can be given a causal interpretation in the manner outlined earlier. It can be shown (Simon [1947–1956] 1957, chapter 2) that all causal inference from correlation coefficients involves, explicitly or implicitly, this procedure.

Suppose, for example, that per capita candy consumption is found to be (negatively) correlated with marital status. Can it be concluded that marriage causes people to stop eating candy or that candy eating inhibits marriage? The question can be answered only in the context of a more complete theory of behavior (Zeisel [1947] 1957, p. 198). If age is introduced as a third variable, it is found to have a high negative correlation with candy consumption but a high positive correlation with marital status. When age is held constant, the partial correlation of candy consumption with marital status is almost zero. If age is taken as the exogenous variable, these facts permit the inference that age causally influences both candy consumption and marital status but that there is no causal connection between the latter two variables —their correlation was spurious.

Practical techniques for interpreting correlations and distinguishing spurious from genuine relations have been discussed by Blalock (1964), Hyman (1955, chapters 6, 7), Kendall and Lazarsfeld

(1950, pp. 135–167), Simon ([1947–1956] 1957, chapter 2), and Zeisel ([1947] 1957, chapter 9).

Purpose and motivation

Since the social sciences are much concerned with purposeful and goal-oriented behavior, it is important to ascertain how causal concepts are to be applied to systems exhibiting such behavior. Purposeful behavior is oriented toward achieving some desired future state of affairs. It is not the future state of affairs, of course, that produced the behavior but the intention or motive to realize this state of affairs. An intention, if it is to be causally efficacious for behavior, must reside in the central nervous system of the actor prior to or at the time of action. Hence, present intention, and not the future goal, provides the causal explanation for the behavior. The influence of expectations and predictions on behavior can be handled in the same way: the expectations are *about* the future but exist at present in the mind (and brain) of the actor (Rosenblueth et al. 1943).

The simplest teleological system that illustrates these points is a house thermostat. The desired state of affairs is a specified air temperature. The thermostat setting is the thermostat's (present) representation of that goal—its intention. Measurements of the difference between actual temperature and setting are the causal agents that produce corrective action. Thus, the causal chain runs from the setting and the temperature-measuring device, to the action of a heat source, to the temperature of the room.

The term *function* (in its sociological, not mathematical, sense) can be analyzed similarly. To say that the family has the function of nurturing children is to say (*a*) that it is causally efficacious to that end; (*b*) that it operates in a goal-oriented fashion toward that end; and, possibly, (*c*) that it contributes causally to the survival of the society.

As Piaget's studies show, teleological explanation —explanation in terms of the motives for and justifications of action—is probably the earliest kind of causal analysis observable in children. Similarly, children tend to interpret causation anthropomorphically—to treat the cause as an active, living agent rather than simply a set of antecedent circumstances. Thus teleological explanation, far from being distinct from causal explanation in science, is probably the prototype for all causal analysis.

Influence and power relations. An influence or power mechanism, in the terms of the present discussion, is simply a particular kind of causal mechanism—the cause and effect both being forms of human behavior. The asymmetry of the causal relation is reflected in the asymmetry of these mechanisms, considered singly. This does not mean that there cannot be reciprocal relations and feedback loops but simply that the influence of A on B can be analyzed (conceptually and sometimes empirically) independently of the influence of B on A (Simon [1947–1956] 1957, chapter 4).

In sum, causal language is useful language for talking about a scientific theory, especially when the variables the theory handles are interconnected, but sparsely so, and especially when there is interest in intervention (for reasons of policy or experiment) in particular mechanisms of the system. In a formalized theory, a formal analysis can be made of the causal relations asserted by the theory. When causal language is used in this way—and most everyday use fits this description—it carries no particular philosophical implications for the problem of Hume or the issue of determinism. Causal concepts are as readily applied to living teleological systems as to inanimate systems, and influence and power relations are special cases of causal relations.

HERBERT A. SIMON

[*See also* PREDICTION; SCIENTIFIC EXPLANATION; Dahl 1968.]

BIBLIOGRAPHY

BLALOCK, HUBERT M. JR. 1964 *Causal Inferences in Nonexperimental Research.* Chapel Hill: Univ. of North Carolina Press.

BRAITHWAITE, RICHARD B. 1953 *Scientific Explanation.* Cambridge Univ. Press.

BREDEMEIER, HARRY C. 1966 [Review of] *Cause and Effect,* edited by Daniel Lerner. *American Sociological Review* 31:280–281.

BROWN, ROBERT R. 1963 *Explanation in Social Science.* London: Routledge; Chicago: Aldine.

BURKS, ARTHUR W. 1951 The Logic of Causal Propositions. *Mind* 60:363–382.

Cause and Effect. Edited by Daniel Lerner. 1965 New York: Free Press.

►DAHL, ROBERT A. 1968 Power. Volume 12, pages 405–415 in *International Encyclopedia of the Social Sciences.* Edited by David L. Sills. New York: Macmillan and Free Press.

HUME, DAVID (1777) 1900 *An Enquiry Concerning Human Understanding.* Chicago: Open Court. → A paperback edition was published in 1955 by Bobbs-Merrill.

HYMAN, HERBERT H. 1955 *Survey Design and Analysis: Principles, Cases, and Procedures.* Glencoe, Ill.: Free Press.

KENDALL, PATRICIA L.; and LAZARSFELD, PAUL F. 1950 Problems of Survey Analysis. Pages 133–196 in Robert K. Merton and Paul F. Lazarsfeld (editors), *Continuities in Social Research: Studies in the Scope and Method of* The American Soldier. Glencoe, Ill.: Free Press.

MICHOTTE, ALBERT (1946) 1963 *The Perception of Causality*. Paterson, N.J.: Littlefield. → First published in French.

NAGEL, ERNEST 1961 *The Structure of Science: Problems in the Logic of Scientific Explanation*. New York: Harcourt.

PIAGET, JEAN (1923) 1959 *The Language and Thought of the Child*. 3d ed., rev. New York: Humanities Press. → First published as *Le langage et la pensée chez l'enfant*.

PIAGET, JEAN (1927) 1930 *The Child's Conception of Physical Causality*. New York: Harcourt; London: Routledge. → First published as *La causalité physique chez l'enfant*. A paperback edition was published in 1960 by Littlefield.

POPPER, KARL R. (1935) 1959 *The Logic of Scientific Discovery*. New York: Basic Books; London: Hutchinson. → First published as *Logik der Forschung*.

ROSENBLUETH, A.; WIENER, NORBERT; and BIGELOW, J. 1943 Behavior, Purpose and Teleology. *Philosophy of Science* 10, no. 1:18–24.

SIMON, HERBERT A. (1947–1956) 1957 *Models of Man, Social and Rational: Mathematical Essays on Rational Human Behavior in a Social Setting*. New York: Wiley.

WOLD, HERMAN (editor) 1964 *Econometric Model Building: Essays on the Causal Chain Approach*. Amsterdam: North-Holland Publishing.

ZEISEL, HANS (1947) 1957 *Say It With Figures*. 4th ed., rev. New York: Harper.

Postscript

There has been a rapid spread, especially in sociology and political science, of use of the techniques for estimating structural (causal) relations among the variables in multi-equation systems (see Blalock 1964; Duncan 1966; Goldberger & Duncan 1973). Such techniques had of course been used earlier on a large scale in economics, and to a more limited extent in psychology. [*See* STATISTICAL IDENTIFIABILITY; SIMULTANEOUS EQUATION ESTIMATION.]

Following Sewall Wright (1934), the term "path coefficients" is now commonly used to denote the regression coefficients (often in normalized form) of the structural equations. [*See* SURVEY ANALYSIS, *article on* METHODS OF SURVEY ANALYSIS; MULTIVARIATE ANALYSIS, *article on* CORRELATION METHODS.] The important choices lie not in deciding between the terminology of "causal analysis" and "path coefficients" but (1) in deciding whether the statistical parameters are to be stated as regression coefficients (usually preferable) or as normalized beta coefficients and (2) in specifying correctly the structural equations on which the estimates are based. Useful discussions of these fundamental conceptual issues, as well as examples of applications of the techniques of causal analysis, can be found in Blalock (1964; 1967), Duncan (1966), Wright (1960), Goldberger (1970), Goldberger and Duncan (1973), and Campbell and Stanley (1966).

HERBERT A. SIMON

ADDITIONAL BIBLIOGRAPHY

BLALOCK, HUBERT M. JR. 1967 Path Coefficients Versus Regression Coefficients. *American Journal of Sociology* 72:675–676. → A letter to the editor.

CAMPBELL, DONALD T.; and STANLEY, JULIAN C. 1966 *Experimental and Quasi-experimental Designs for Research*. Chicago: Rand McNally.

DUNCAN, OTIS DUDLEY 1966 Path Analysis: Sociological Examples. *American Journal of Sociology* 72:1–16.

GOLDBERGER, ARTHUR S. 1970 On Boudon's Method of Linear Causal Analysis. *American Sociological Review* 35:97–100.

GOLDBERGER, ARTHUR S.; and DUNCAN, OTIS DUDLEY 1973 *Structural Equation Models in the Social Sciences*. New York: Academic Press.

GRANGER, C. W. J. 1969 Investigating Causal Relations by Econometric Models and Cross-spectral Methods. *Econometrica* 37:424–438.

SIMS, CHRISTOPHER A. 1972 Money, Income, and Causality. *American Economic Review* 62:540–552.

WOLD, HERMAN 1966 On the Definition and Meaning of Causal Concepts. Pages 265–295 in R. Peltier and H. Wold (editors), *Technique des modeles dans les sciences humaines*. Entretiens de Monaco en Sciences Humaines, Session 1964. Monaco: Union Européenne d'Editions.

WRIGHT, SEWALL 1934 The Method of Path Coefficients. *Annals of Mathematical Statistics* 5:161–215.

WRIGHT, SEWALL 1960 Path Coefficients and Path Regressions: Alternative or Complementary Concepts? *Biometrics* 16:189–202.

CENSORED DATA

See STATISTICAL ANALYSIS, SPECIAL PROBLEMS OF, *article on* TRUNCATION AND CENSORSHIP.

CENSUS

A census of the population—that is, a counting of the people within the boundaries of a country—has become indispensable to any modern government. How many people are there? What are their basic socioeconomic characteristics? Where do they live, and how are they affected by the processes of social and biological change? These questions arise daily in the governments of all industrially developed countries, not to mention the governments of those that are still developing.

Censuses have come to include many topics other than population. Censuses of manufactures, agriculture, mineral industries, housing, and business establishments are taken by many countries, often independently of the census of population.

The following discussion is concerned primarily with censuses of population, but many of the comments—especially the comments on methods, tabulation, and quality of results—apply with equal force to other kinds of censuses.

Some early censuses. Counting the people, or some portion of them, is a practice that is probably as old as government itself. No one knows which ruler first enumerated the men for military purposes, or drew up a list of households with a view to taxing them. Figures obtained from censuses have long served as items of political propaganda, particularly in order to justify territorial expansion.

Population counts were reported in ancient Japan and were taken by the ancient Egyptians, Greeks, Hebrews, Persians, and Romans. Many of these early censuses appear to have covered only part of the population, often the men of military age, and the results were generally treated as state secrets. In Europe censuses on a city-wide basis or, as in Switzerland, on a canton-wide basis, were reported in the fifteenth and sixteenth centuries. A 1449 census of Nuremberg presumably was taken to determine the needed food supplies when a siege was threatened. The city of Madras in India reportedly took a census in 1687.

Various censuses have been claimed as the first held in modern times with purposes and methods resembling those of today. Among these are a census in New France (the early possessions of France in North America) taken at intervals between 1665 and 1754 and a census in Sweden in 1749. Census taking began early in the North American and South American colonies. The British Board of Trade ordered 27 censuses in the North American colonies between 1635 and 1776, and censuses were taken by these former colonies between independence in 1776 and the establishment of the United States.

The oldest continuous periodic census is that of the United States, which has been conducted every ten years since 1790. The census of the United Kingdom dates back to 1801, and a census has been taken there every ten years except in 1941, during World War II.

If one were to define a modern census as one in which information is collected separately about each individual instead of each household, then the beginning of modern censuses would have to be dated about the middle of the nineteenth century. Censuses along these lines were taken in Brussels in 1842, in all Belgium in 1846, in Boston in 1845, and in the entire United States in 1850. This is the procedure that is now generally in use.

International activities. Toward the end of the nineteenth century, the International Statistical Institute recommended international publication of the results of all censuses. Around the turn of the century the institute repeated its recommendation, pointing out that the results of some 68 different censuses were available, covering about 43 per cent of the world's population. The institute had already adopted, in 1897, a set of rules for conducting censuses and presenting their results, while the whole topic had been discussed by the International Demographic Congress as early as 1878.

One of the early proposals considered by the United Nations was that it develop plans for a 1950 world census of population. Although it found that conditions were not ripe for such an effort, the United Nations took steps to foster census taking and made recommendations to improve the comparability of results. The Inter American Statistical Institute meantime had undertaken a program for the 1950 Census of the Americas, and 18 of the 21 countries in the Americas took censuses between 1945 and 1954. Throughout the world at least 150 areas took censuses in this period, collecting individual data on more than 2 billion people. For the decade centered on 1960, the number of censuses was about 180, including 2.2 billion people.

The use of censuses. The reasons for taking a census vary with the needs of the countries involved. The current concern with social and economic development is one of the prime reasons. Much information is needed in order to institute programs that will improve health, literacy, education, income, levels of living, supplies of food and other consumer goods, agricultural production, and industrial output. Census data are also collected in order to determine the representativeness of legislative bodies, the number of persons eligible to vote, and the areas or groups that have a claim to benefits deriving from the state.

The census provides a basis for much demographic, economic, and social research. It makes possible the identification and description of such groups as the labor force, economically dependent persons, recent migrants to cities, rural and urban populations, racial or religious minorities, refugees, scientific and technical workers, and others. Comparisons of successive censuses show changes in the numbers, characteristics, and location of the population. The census returns are also used as a frame from which samples are selected for subsequent inquiries.

The modern census

The United Nations (1958, p. 4) gives the following definition of a modern population census:

"A census of population may be defined as the total process of collecting, compiling and publishing demographic, economic and social data pertaining, at a specified time or times, to all persons in a country or delimited territory."

It also listed six essential features of a census, as follows. (1) A census must have *national sponsorship*. Only a national government can provide the necessary resources and enact suitable legislation, although provincial and local governments may share a part of the responsibility and sometimes a part of the cost. (2) A census must cover a *precisely defined territory;* boundary changes that affect comparisons between successive censuses should be clearly and explicitly stated. (3) *All persons* in the scope of the census must be included without duplication or omission. (4) The people must be counted as of a *fixed time*. Persons born after the census date are to be excluded, and persons who die after the census date are to be included. Some information, such as that relating to labor force participation or migration, may relate not to the census date but to another period, which must be clearly defined. (5) Census data must be obtained separately for *each individual*. This does not preclude making some entries for the entire household—and in exceptional circumstances summarized information for a group of persons may be acceptable—but the objective of a modern census, insofar as possible, is to collect data separately for each individual. (6) The data from a census must be *published*. Although at one time census reports were treated as state secrets, it is now recognized that a census is not complete until the data are compiled and published.

What is included. The content of a census, no less than its purpose, is determined by the country's needs at the time. Questions of high importance in one country may be of relatively little importance in another, and questions of great significance at one time may be of little significance later. Every national census has changed over the years. Conditions in the country, alternative sources of information, and the ability of the census organization to provide the proposed information chiefly determine the census content.

The United Nations has approved a set of recommendations for national population censuses that includes the following list of question topics: location at time of census and/or place of usual residence, relation to head of household or family, sex, age, marital status, place of birth, citizenship, whether economically active or not, occupation, industry, status (as employer, employee, etc.), language, ethnic or nationality characteristics, literacy, level of education, school attendance, and the number of children born to each woman (Principles and Recommendations . . . 1958). For countries that could not include all items in the list, the following were suggested as a minimum: sex, age, marital status, and some indication of economic activity. Each item listed, of course, requires specific definition. Some items, such as marital status, type of economic activity, or level of education, apply to only part of the population.

Collection of data. Discussions of census methods generally distinguish between a count based on the actual location of the population on the census date and one which relates each person to the place of which he is a resident. Frequently residence is interpreted as the place where the person usually lives, although "usual residence" has no statutory meaning. On either basis national totals are usually about equal; community totals, however, may differ substantially.

To facilitate international comparison, the United Nations (Principles and Recommendations . . . 1958) has suggested the adoption of an "international conventional total." This would include all persons present in the country on the census date except foreign military, naval, and diplomatic personnel: it would also include the country's own military, naval and diplomatic personnel and their families located abroad, and merchant seamen normally resident in the country but at sea on the census date. The same document recommends that counts or estimates of the following groups should be given where feasible: indigenous inhabitants and nomadic tribes, civilian national residents temporarily abroad on the census date, and civilian aliens temporarily in the country on the census date. This is because whether or not these groups are included in a national census is usually determined by the laws and needs of each country; thus detailed comparisons between national totals can be made only if separate enumerations of these groups are made available.

Customarily two basic methods of collecting census data are recognized—direct enumeration and self-enumeration. Under the direct enumeration system, an enumerator collects information directly from the individual concerned, the head of the household, or some other member of the household who may be authorized to report for him. Under the self-enumeration system, the questionnaire is given to the individual, or the head of the household, who is expected to enter the required information and return the questionnaire to the census office. In self-enumeration the enumerator often delivers and collects the questionnaire. He may assist persons in making the entries, and he is responsible for assuring the accuracy and completeness of the completed questionnaires.

In a few countries the people are required to present themselves at a designated place to be enumerated. In a few others the population is immobilized on the census date, and no one is permitted on the streets unless he has been enumerated. In the great majority of cases the census enumerator seeks out the potential respondents and delivers questionnaires to them or asks them directly for the required information.

Census legislation normally provides that respondents must give full and correct information. This information must be given confidential treatment by all census employees. For persons who do not wish to have other members of their households or local officials see the information about themselves, census offices have provided special forms on which the person can insert the information and send it directly to the local, regional, or national office.

Population registers and lists of households, where available, are often used to help make the census complete. In some censuses the first step in the field work is to establish a list of housing units. The numbers assigned to these units are then used to control the completeness of the enumeration.

The questionnaire must be designed to permit ready entry of the required information and also be adapted to the tabulation procedure. In self-enumeration there is customarily a separate form for each household, with space for entering information for each household member. When information is collected directly by enumerators they may carry a separate form for each household or a form on which information is recorded for a number of households. Sometimes a separate document is used for each person.

Some countries have a long-established tradition of taking a census once every ten years, a few countries require one every five years, but most countries take a census when it is needed and do not have a fixed date. In a few unusual situations censuses have been taken more often than once in five years. Whether by law or by custom, the ten-year term appears to be most common.

In an increasing number of countries a pilot census is taken before the full census to test the inquiries, the procedures, the development of the field organization, and sometimes the tabulation program.

Use of sampling. Sampling has been extensively used in connection with censuses, although a census in principle requires that information be collected for each person. The term "sample census," which is sometimes used, is a misnomer; a collection of information relating to only a specified part of a population should be termed a sample survey. In some censuses a part of the questions are asked of only a sample of all individuals; in this way more information can be collected without a comparable increase in respondent burden or tabulation work load. Many countries base some tabulations on a sample of the population instead of the total to provide preliminary totals for early release or to reduce the cost of some final tabulations. Sampling has been used to control the quality of the work at several stages of the processing of the data. In some countries a sample has been selected for a pretest and the resulting data have been used to validate the processing operations.

Tabulation of results. To be useful, individual census returns must be converted to statistical summaries. Tabulation methods vary from simple hand counts to processing on high-speed electronic computers. Processing may be done in the field, in provincial offices, or at the central office.

Questionnaires as received from the enumerators must be reviewed to locate incomplete or inconsistent entries. Procedures have been developed for correcting such errors by utilizing other information on the questionnaires or by using probability distributions based on data from other sources. In many instances the entries on the schedule must be converted to a numerical or other suitable code to facilitate counting and grouping into appropriate categories.

The most common procedure for grouping the entries is to punch cards and tabulate them by mechanical means. More recently, electronic computers have been adapted to this work, and devices have been developed to transfer the original questionnaire entries directly to the magnetic tape used in these computers. The capacity and speed of the computers have led to important advances in the amount of material tabulated and in the timeliness of the publications.

Completeness and accuracy. In view of the many public and private uses for census data, there is increasing concern with the quality of the results in terms both of completeness of coverage and accuracy of returns.

Completeness of coverage has received major attention. Unless special precautions are taken, some individuals may not be counted and some areas may not be enumerated. Undercounts of the population are more frequent than overcounts, although a few zealous census officials have been known to overcount the population of an area. Omission of individuals may arise from incomplete enumeration of a household or an area. Underenumeration has been particularly marked in the case of infants and young adults who are highly mobile or who lack firm occupational attachments. In cultures where men are considered more valuable than women, the number of women may be understated. In a few instances a fear that the census may be used for military conscription has led to some omission of men or to misreporting them as women.

Reporting residence may present problems. Some persons, such as migratory workers, have no usual place of residence. Others, such as college students, members of the armed forces, and persons engaged in long-distance transportation, may live away from their families for short or extended periods. Unless precise instructions are given for reporting these persons, they may be counted twice or not at all.

Age is often misreported either because people see some advantage in giving the wrong age or because they do not know their correct age. In many countries births are not recorded, and age is not precisely known. Some reasons for misreporting are that a person believes that despite pledges of confidentialness, the data will be used to his disadvantage; a parent believes that overstating a young child's age may gain him entrance to school or that overstating an older child's age may free him from compulsory school attendance; an older person may overstate his age to get social security benefits or to acquire the higher status sometimes accorded to the very old. Often people report a number ending in zero or five—even when asked for year of birth.

Nationality, citizenship, and mother tongue are particularly subject to misstatement when the correct answer may be considered detrimental in the context of the political situation of the group controlling the area.

Improvement of quality. Public confidence in the census greatly affects accuracy of returns. Public confidence can often be increased through an informational campaign, which explains that the data about any individual cannot be used to his detriment. Police officers, who are used as enumerators in some countries because of their availability and authority, are specifically excluded in others because they may give the impression that the information is not collected solely for statistical purposes.

To improve the quality of results, research has been devoted to reliability of response, role of enumerators, question wording, and attitudes of respondents in different situations. Increased attention is also being given to training the temporary field forces in census methods and concepts and in training the office staff in principles of editing and coding.

The quality of census statistics can often be determined by analyzing them after the complete census tabulations have been published. Formerly this work was performed mainly by scholars or by census officials willing to do scholarly work on their own time; but modern census offices increasingly are undertaking such analysis as part of their regular work. Checks are made for internal consistency of the census, for consistency with previous censuses, and also with other statistics, including estimates, that are entirely independent of the census. Both underenumeration and overenumeration of an age group can be determined by comparing the census results with the comparable age cohort for the previous census, adjusted for deaths and migration [*see* COHORT ANALYSIS]. Inaccuracies in age reporting that result from the preference of respondents for certain ages can be identified by internal analysis of the data. Sometimes the accuracy of census information for individuals is tested by comparison with information given for these same individuals in administrative records. Sample surveys may be taken to check on completeness of coverage and accuracy of census returns.

These evaluations aid in increasing the accuracy of future censuses, but they also provide users of the statistics with information about the reliability of the data. Although there has been some hesitation about revealing errors in the census, there is a growing recognition that full and frank disclosure leads not only to improvement in future censuses but also to increased public confidence.

CONRAD TAEUBER

[*See also* DEMOGRAPHY; GOVERNMENT STATISTICS; VITAL STATISTICS; Grauman 1968; Mayer 1968; Whitney 1968.]

BIBLIOGRAPHY

BRUNSMAN, HOWARD G. 1963 Significance of Electronic Computers for Users of Census Data. Pages 269–277 in Milbank Memorial Fund, *Emerging Techniques in Population Research.* New York: The Fund.

CANADA, BUREAU OF STATISTICS 1955 *Ninth Census of Canada, 1951: Administrative Report.* Volume 11. Ottawa: Cloutier.

COALE, ANSLEY J. 1955 The Population of the United States in 1950 Classified by Age, Sex, and Color: A Revision of Census Figures. *Journal of the American Statistical Association* 50:16–54.

ECKLER, A. ROSS; and HURWITZ, WILLIAM N. 1958 Response Variances and Biases in Censuses and Surveys. International Statistical Institute, *Bulletin* 36, part 2: 12–35.

►GRAUMAN, JOHN V. 1968 Population: VI. Population Growth. Volume 12, pages 376–381 in *International Encyclopedia of the Social Sciences*. Edited by David L. Sills. New York: Macmillan and Free Press.

JAFFE, A. J. 1947 A Review of the Censuses and Demographic Statistics of China. *Population Studies* 1:308–337.

►MAYER, KURT B. 1968 Population: IV. Population Composition. Volume 12, pages 362–370 in *International Encyclopedia of the Social Sciences*. Edited by David L. Sills. New York: Macmillan and Free Press.

MAYR, GEORG VON (1895–1917) 1926 *Statistik und Gesellschaftslehre*. 3 vols. 2d ed., enl. Tübingen (Germany): Mohr.

Principles and Recommendations for National Population Censuses. 1958 United Nations Statistical Office, *Statistical Papers* Series M, No. 27.

STEINBERG, JOSEPH; and WAKSBERG, JOSEPH 1956 *Sampling in the 1950 Census of Population and Housing*. Bureau of the Census Working Paper, No. 4. Washington: U.S. Department of Commerce.

TAEUBER, CONRAD; and HANSEN, MORRIS H. 1964 A Preliminary Evaluation of the 1960 Census of Population. *Demography* 1, no. 1:1–14.

TAEUBER, CONRAD; and TAEUBER, IRENE B. 1958 *The Changing Population of the United States*. U.S. Bureau of the Census, Census Monograph Series, 1950. New York: Wiley.

UNITED NATIONS, STATISTICAL OFFICE 1958 *Handbook of Population Census Methods*. Volume 1: General Aspects of a Population Census. New York: United Nations.

U.S. BUREAU OF LABOR 1900 *The History and Growth of the United States Census*. Prepared for the Senate Committee on the Census. Washington: Government Printing Office.

U.S. BUREAU OF THE CENSUS *Bureau of the Census Catalog*. → Published since 1947.

U.S. BUREAU OF THE CENSUS 1909 *A Century of Population Growth From the First Census of the United States to the Twelfth: 1790–1900*. Washington: Government Printing Office.

U.S. BUREAU OF THE CENSUS 1955 *The 1950 Censuses: How They Were Taken*. Procedural Studies of the 1950 Census, No. 2. Washington: Government Printing Office.

U.S. BUREAU OF THE CENSUS 1960*a* *Inquiries Included in Each Population Census, 1790 to 1960*. Washington: Government Printing Office.

U.S. BUREAU OF THE CENSUS 1960*b* *The Post-enumeration Survey, 1950: An Evaluation Study of the 1950 Censuses of Population and Housing*. Bureau of the Census Technical Paper, No. 4. Washington: Government Printing Office.

U.S. BUREAU OF THE CENSUS 1964 *Evaluation and Research Program of the U.S. Censuses of Population and Housing, 1960: Accuracy of Data on Population Characteristics as Measured by CPS–Census Match*. Series ER 60, No. 5. Washington: Government Printing Office.

U.S. BUREAU OF THE CENSUS 1966 *1960 Censuses of Population and Housing: Procedural History*. Washington: Government Printing Office.

U.S. LIBRARY OF CONGRESS, CENSUS LIBRARY PROJECT 1950 *Catalog of United States Census Publications, 1790–1945*. Washington: Government Printing Office.

►WHITNEY, VINCENT H. 1968 Population: V. Population Distribution. Volume 12, pages 370–376 in *International Encyclopedia of the Social Sciences*. Edited by David L. Sills. New York: Macmillan and Free Press.

WILLCOX, WALTER F. 1930 Census. Volume 2, pages 295–300 in *Encyclopaedia of the Social Sciences*. New York: Macmillan.

ZELNIK, MELVIN 1961 Age Heaping in the United States Census: 1880–1950. *Milbank Memorial Fund Quarterly* 39, no. 3:540–573.

ZELNIK, MELVIN 1964 Errors in the 1960 Census Enumeration of Native Whites. *Journal of the American Statistical Association* 59:437–459.

Postscript

Census taking has expanded to cover nearly the entire population of the world. The United Nations reports that during the decade centering on 1970 more than 170 censuses were taken. Efforts are being made to enable a few countries that have never had a census to take one before 1980, and international assistance for this purpose is being provided. Most of the countries affected are in Africa.

National governments and the international organizations have been giving increased attention to methodological issues involved in census taking and to developing measures of accuracy of the census results. The use of electronic computers for census tabulations has become almost universal. Many of the census offices have found means of releasing microdata, without any individual identification, thereby providing invaluable assistance to social and economic research workers.

CONRAD TAEUBER

ADDITIONAL BIBLIOGRAPHY

ECKLER, A. ROSS 1971 *The Bureau of the Census*. New York: Praeger.

KAHN, E. J. JR. 1974 *The American People: The Findings of the 1970 Census*. New York: Weybright & Talley. → A paperback edition was published in 1975 by Penguin.

SCOTT, ANN HERBERT 1968 *Census U.S.A.: Fact Finding for the American People, 1790–1970*. New York: Seabury.

TAEUBER, IRENE B.; and TAEUBER, CONRAD 1971 *People of the United States in the Twentieth Century*. Washington: U.S. Bureau of the Census; Government Printing Office.

U.S. BUREAU OF THE CENSUS 1972–1975 *1970 Census of Population and Housing: Evaluation and Research Program*. Series PHC(E). Washington: Government Printing Office. → Fifteen reports.

U.S. BUREAU OF THE CENSUS 1973 *1970 Census of Population and Housing: Procedural History*. Washington: Government Printing Office. → Nine reports.

U.S. BUREAU OF THE CENSUS 1974 *Catalog of United States Census Publications, 1790–1972*. Washington: Government Printing Office.

CENTRAL TENDENCY
See STATISTICS, DESCRIPTIVE, *article on* LOCATION AND DISPERSION.

CHARTS
See GRAPHIC PRESENTATION.

CHI-SQUARE DISTRIBUTIONS
See DISTRIBUTIONS, STATISTICAL, *article on* SPECIAL CONTINUOUS DISTRIBUTIONS.

CHI-SQUARE STATISTICS
See COUNTED DATA; GOODNESS OF FIT; SIGNIFICANCE, TESTS OF.

CLASS, SOCIAL
See SOCIAL MOBILITY.

CLASSIFICATION
See CLUSTERING; MULTIVARIATE ANALYSIS, *article on* CLASSIFICATION AND DISCRIMINATION; SCALING, MULTIDIMENSIONAL.

CLIQUES
See COALITIONS, THE STUDY OF; SOCIOMETRY.

CLUSTERING

A battery of psychological tests provides a profile of test scores for each subject. Can the subjects be grouped sensibly into a moderate number of classes or clusters so that the subjects in each class have similar profiles? Can the clustering process be made automatic and feasible while producing subjectively meaningful results?

These types of data and questions arise throughout the social and natural sciences. The "subjects" or units may be such things as census tracts, nations, tribes, legislators, plants, fossils, microorganisms, documents, languages, or corporations. The data consist of a vector or profile of measurements or observations for each unit, one measurement on each of a number of variables or characters. Both quantitative variables and binary characters, scored only "present" or "absent," are frequent and important.

Often, the profile of measurements on a unit may consist of some measure of interaction with each other unit. For example, the units might be importing nations and the variables the relative value of imports from each other (exporting) nation. Special attention must be given to any use of the interaction of the unit with itself. Social choices between individuals give other examples, and special techniques for analyzing such data have been developed [*see* SOCIOMETRY].

The role of units and variables may be interchanged. In the preceding example, if clustering on the basis of export pattern were desired, the old "variables" would become new "units." (The measurements would change, since relative value would be computed with respect to a different base.) Or, again, from the voting record of legislators on each of many issues, one might want to cluster legislators or to cluster issues.

Clustering methods form a loosely organized body of techniques for the analysis of such data. As with most methods of data analysis, the aim is to find, to describe, and, hopefully, to lead to an explanation of some simple structure in a complex mass of data. Clustering methods are distinguished by the type of structure that is sought.

Examples in two dimensions. Geometric notions motivate the terminology and methods of clustering, although only for one or two variables can the geometric representation be used directly. For example, suppose the units are essays, putatively written by a single author, and that the variables are two quantitative indicators of style, say mean sentence length and the rate of occurrence of the word *of* in the essay.

In Figure 1, each essay is represented by a point whose coordinates are the two measurements on the essay. Units are similar if their corresponding points are close together; clusters of units correspond to geometric clusters of points. In Figure 1, three clusters are visually evident and so clear-cut that a search for a theoretical explanation is natural. Might the essays have been written by three different authors?

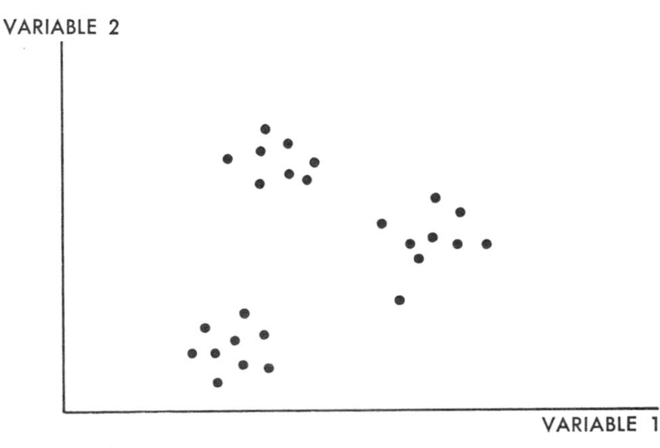

Figure 1 — Data exhibiting evident clustering

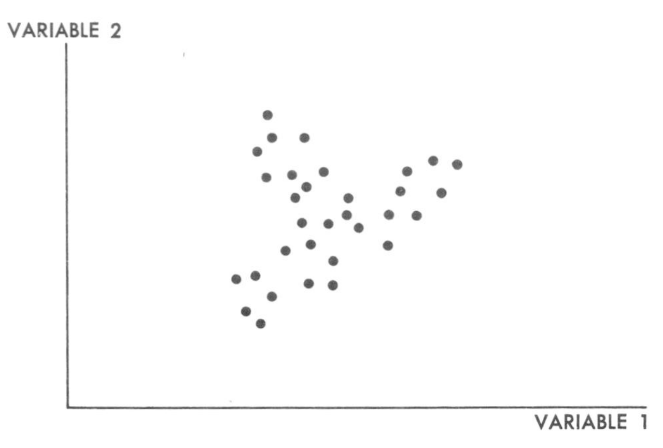

Figure 2 — Data exhibiting less clear-cut clustering

Figure 2 illustrates data in which the existence of two overlapping clusters is suggested, but the identification is incomplete. Further variables might permit better definition of clusters, but even the imperfect clustering may aid the description and understanding of the data.

Extensions to more than two dimensions. The aim of most clustering methods is to imitate and automate the process that can be done well visually in two dimensions and to extend it to any number of dimensions. The gain is the extension to many dimensions; that the process is in some ways inferior to visual inspection in two dimensions should be neither surprising nor disturbing.

If it is possible to group units into a moderate number of classes, within which the units have similar profiles, then a reduction of the data has been achieved that is easily described, that facilitates the analysis of further data, and that may be suggestive of theoretical structures underlying the data.

Each class can be described by a typical or average profile and, secondarily, by some measure of variability (shape and magnitude) within the class. Once the classes are well determined, future units can be placed in one of the classes through techniques of discriminant analysis [*see* MULTIVARIATE ANALYSIS, *article on* CLASSIFICATION AND DISCRIMINATION].

In some instances the clusters may correspond to underlying types explained by theory. In others they may represent only a convenient empirical reduction of the data. Interest may lie exclusively with the units included in the analysis or may extend to additional units.

The clustering process may be carried out on a finer scale, dividing a class into subclasses. More generally, an entire hierarchical classification or taxonomy can be obtained. It is in this latter form that clustering methods have been developed and applied in biology.

The technology of clustering is in an early and rapidly changing state. With a few exceptions, the development of clustering methods began only in the late 1950s, as large, high-speed computers became widely available. Many methods have been proposed and some have been tried and proved useful, but none is thoroughly understood or firmly established. The next few years should bring dramatic advances in the formulation of new methods, in techniques of computer execution, and in theoretical foundations. Most important, understanding of how and why procedures work or fail will be gained through experience and comparative experimentation.

Methods of clustering

Methods of clustering can usually be broken down into these steps: selection of units and of variables for inclusion in the data, determination of measures of similarity between pairs of units, grouping of units into clusters, interpretation and description of the results of grouping, possible iterations, and further analysis. Not all methods fit this decomposition, but each step has some counterpart in most methods.

Selection of units and variables. Specification of the data requires selection both of units and of variables and a choice of scale for each variable. The importance of selection of units is not yet fully understood. One effect is clear: including many units of one type will ensure that that type shows up as a cluster. Overrepresentation of a type has no strong effect on the clustering of the rest. Selection may be easy. In a study of voting behavior of a legislature, all legislators would be used. In a socioeconomic study of counties in the United States, a population of units is evident and a random or stratified sample might be included. Mostly, though, there is no clear universe of units, and those easily available are not a probability sample from any population. Understanding must await more experience with the results of clustering, especially in applications that go beyond a description of the original data.

Selection of variables and characters is more critical than selection of units and determines what aspects of behavior of units will be represented or emphasized in the clustering. Overrepresentation of variables in some area can strongly affect the entire outcome. Not only is there no natural universe of variables, but the notion of what constitutes an individual variable is not well defined. Variables can be refined, subdivided, and combined. Two variables that are functionally dependent or that are highly associated statistically can usually be recognized and avoided, but variables will inevitably have some statistical dependence over various groups of units. The choice of variables is a highly subjective, crucial element of the clustering process.

Scaling of variables. The data in any clustering process may be arranged in a rectangular array whose rows correspond to the units being clustered and whose columns correspond to the observed variables or characters. The entries in a row form the vector or profile of measurements of a unit on the respective variables. Most work has used variables that are measured on a numerical scale or else binary characters that can be so represented by denoting presence by a 1 and absence by a 0. Although the latter may be treated as a quantitative variable, there are substantial gains, primarily in computer technique, to be achieved by taking advantage of the 0-1 nature of binary characters. Much development has been restricted to binary variables. While any variable may be represented, at least approximately, by several binary variables, the representation introduces new arbitrariness and other difficulties.

With general quantitative variables, the problem of scaling or weighting is critical. If, for example, the first variable in Figure 2 were the median years of schooling (in a county, say) and the second were the median family income measured in thousands of dollars, to change to income measured in hundreds of dollars would expand the vertical scale by a factor of 10 and greatly distort the picture. While the eye may adjust for a bad choice, the numerical measures used in two and more dimensions are severely affected by a bad choice. Imposing an arbitrary, standard scaling by requiring each variable to have a standard deviation of 1 over the units included is a frequent choice. It is less subjective but no less arbitrary and often poorer than imposing a choice based on an external standard or on subjective judgment.

Choice of *nonlinear* scaling is even more difficult but offers potentially great benefits. [*See* STATISTICAL ANALYSIS, SPECIAL PROBLEMS OF, *article on* TRANSFORMATIONS OF DATA.]

Superficially, there is no scaling problem for binary variables that all take the same values, 0 and 1; but the difference between two variables, one about equally often 0 or 1 and another that takes the value 1 for about 2 per cent of the units, may indicate a need for some scaling of binary variables also.

Measures of similarity. The geometric representation of figures 1 and 2 can be extended conceptually to more than two variables. Generally, if the measurements of the first unit on k variables are denoted by $(x_{11}, \cdots, x_{1k})$, then the unit can be represented as a point in k-dimensional space with the measurements as coordinates. If the second unit has measurements $(x_{21}, \cdots, x_{2k})$, similarity of the two units corresponds to "closeness of the points" representing them. One natural measure of dissimilarity between points 1 and 2 is Euclidean squared distance,

$$(x_{11} - x_{21})^2 + (x_{12} - x_{22})^2 + \cdots + (x_{1k} - x_{2k})^2.$$

Unless the variables have been carefully scaled, a weighted distance,

$$w_1(x_{11} - x_{21})^2 + w_2(x_{12} - x_{22})^2 + \cdots + w_k(x_{1k} - x_{2k})^2,$$

is needed to make sense of the analysis. In order to allow for patterns of statistical dependence among the variables, a more complex weighting is required, such as

$$\sum_{i=1}^{k} \sum_{j=1}^{k} w_{ij} (x_{1i} - x_{2i})(x_{1j} - x_{2j}).$$

There is no uniquely correct choice of weights, but a careful subjective choice based on external knowledge of the variables, observed pattern of variability, and computational feasibility should be workable and will be preferable to an arbitrary but objective choice such as using the equally weighted Euclidean distance. New theory and methods—likely requiring lengthy and iterative computation—to make more effective use of internal patterns of variability to guide the choice and adjustment of weights will gradually be developed (see Ihm 1965).

For quantitative variables, the measures of similarity commonly used are equivalent to one of the weighted or unweighted squared distances. For binary variables, a greater variety of measures are in use. The equally weighted squared distance, when applied to variables taking only the values 0 and 1, yields the number of variables for which the items fail to match, a measure equivalent to the simple matching coefficient—the proportion of characters for which the two units have the same state. If a 1 represents possession of an attribute, a positive match, 1-1, may be more indicative of similarity than a negative match, 0-0, and numerous ways of taking account of the difference have been proposed (see Sokal & Sneath 1963, section 6.2).

Computation in grouping into clusters. Clustering methods require vast amounts of computation. Two examples illustrate some magnitudes and their relevance to methodology.

Suppose 100 units are to be clustered, so that there are 4,950 pairs of units. To compute unweighted distances for all pairs of units on the basis of, say, 50 variables requires sums of about 250,000 products. This is easily handled by currently available computers, although attention to computational technique, especially in the storage of data, will be critical as the number of units increases.

Consider next an operation that in principle is natural and attractive: Divide the 100 units into two groups in all possible ways, evaluate a criterion measuring the homogeneity of the groups for each partition, and choose the best one. With 100 units, there are $\frac{1}{2}(2^{100} - 2) \cong 6.3 \cdot 10^{29}$ groupings. Even if successive groupings could be generated and the criterion evaluated in a spectacularly efficient ten machine cycles, and using the fastest currently available machine cycle of 10^{-6} second, the process would require about 10^{17} years. Division into three and more groups is worse. Hence, many conceptually useful processes can never be realized in practice. Some modification or restriction is required to reduce drastically the possibilities to be considered, and even then results that only approximate conceptually optimum procedures must be accepted.

As an example of a possible extreme simplification, suppose that the units were dated and that time was an important variable. If the units in any cluster were required to be adjacent in the time sequence, then the number of partitions of 100 units into two groups is reduced to 99, into three groups to 4,851 (Fisher 1958).

A second possibility, based on the preceding one, is to require the grouping to be ordered on some variable, not prespecified but chosen to make the grouping best. When all variables are dichotomous, this procedure becomes remarkably simple and has been highly developed and used in plant classification with large numbers of units and variables. Lance and Williams (1965) survey these methods.

Hierarchical methods. A third possibility is to reverse directions, building up clusters by combining one or more units at a time. Most of the procedures now used fall in this category. One commonly used version is built on repeated application of the following rule: Find the closest pair of clusters and combine them into a new, larger cluster and compute the distance between the new cluster and each of the remaining clusters. Initially, the clusters are the N units, treated as single-unit clusters. The first new cluster will be a two-unit cluster; the next may be a second two-unit cluster or it may be a three-unit cluster.

Since the process starts with N clusters and each stage reduces the number by one, the process terminates after exactly $N - 1$ steps, with all N items combined in one huge cluster. Of chief interest are the intermediate results, especially the last few, where the units are sorted into a moderate number of clusters. An interpretative phase is needed.

The output can be well presented in a linkage tree or dendrogram, as illustrated in Figure 3. The

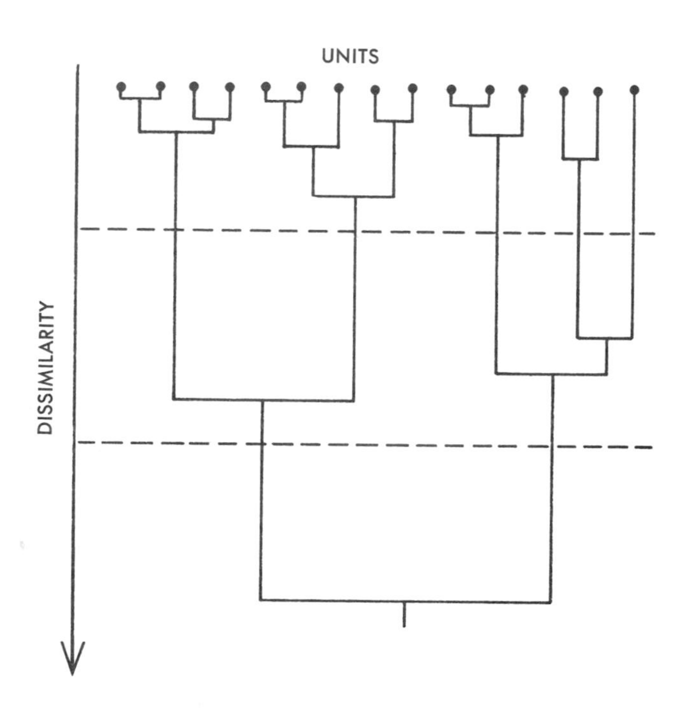

Figure 3 — Linkage tree of a stepwise clustering, showing two plausible slices

initial units are placed on a line in an order that is easily determined from the process. Vertical lines are dropped from each unit until it joins with another. If the distance measure is used as the vertical scale, then the vertical location of each joint shows how dissimilar the two groups joined were. Thus clusters formed near the top represent homogeneous groups, clusters formed farther down represent groups in which some units differ in larger ways, and clusters formed far down represent clusters formed only because the process goes to its logical end of one big group. The linkage tree generated by the process is a salient feature and potential advantage. A hierarchical nested structure is obtained upon choice of several horizontal slices of the tree. Slices may be made at standard vertical positions, or to give desired numbers of groups, or at natural breaks in the particular tree—where there are long vertical stretches without joints.

Another salient feature of this version is how little must be specified as part of the process. Once a distance measure is specified—between two clusters of any size, not just between two units—the process is completely determined. All judgment of how many clusters or how many levels of clustering is postponed to the next phase.

A major variation that gives up both salient features begins in the same way but proceeds by building up the first cluster until adding more units would make it too inhomogeneous. Then another cluster is constructed, etc. The process requires a rule for judging when to stop adding to a cluster (as well as the distance measure to determine which unit to add). As described, the process gives no hierarchy of clusters, but it could be reapplied to the clusters of the first round.

How should a measure of similarity between units be extended to a measure of similarity between clusters? Should it be based on the most similar units in the clusters? the least similar? an average? Should all units in a cluster be weighted equally, or should two joining stems be weighted equally? Methods based on these and many other possibilities have been tried. The choices matter—some favor large clusters, others smaller, equal-sized clusters—but no adequate basis for choice yet exists.

Where to stop clustering or how to cut the tree are open questions. The objectives of the analysis are essential to the decision, but perhaps some statistical models may be developed that would aid the choice.

Iteration. The possibility of treating clustering as an iterative process with the results of one clustering used to improve the next has just begun to be explored and will surely play an essential part in any thorough clustering procedure. Ihm (1965) illustrates one use of this idea to deal with the choice of weights. Ball (1965) describes a composite procedure that is highly iterative.

Relation to factor analysis. Factor analysis has frequently been used in lieu of direct clustering procedures. Both methods of analysis are attempts to discover and describe structure in an unstructured rectangular array of data. The methods seek different types of structure, although there is much overlap. Described in conventional statistical terms, factor analysis is an attempt to find a few "independent" variables such that regression on those variables fits the observed data. Geometrically, it is an attempt to locate a linear subspace that nearly contains the data [*see* FACTOR ANALYSIS].

In contrast, cluster analysis is an attempt to find an analysis of variance classification of the units (a one-way or nested design) that fits the observed data, that is, that reduces unexplained variation [*see* LINEAR HYPOTHESES, *article on* ANALYSIS OF VARIANCE].

Any cluster structure can be explained by a factor structure, although generally as many factors are needed as there are clusters. However, if two or three factors explain most of the variation, then by estimating factor scores for each unit, the units can be represented as points in two or three dimensions and clusters determined visually. This is one of the oldest routes for finding clusters.

Interchanging the role of units and variables in a factor analysis is common when clusters are sought. Then the process of rotation to "simple structure" may show directly the presence of clusters of units. These approaches permit the use of the older, better developed techniques of factor analysis.

Role of statistics. Clustering methodology has not yet advanced to the stage where sources of error and variation are considered formally; consequently, formal statistical methods play almost no role at present.

Unfortunately, even informal consideration of statistical variation has been neglected, in part because, typically, neither units nor variables are simple random samples, so conventional statistical theory is not *directly* applicable. Thus, although measures of similarity may be algebraically equivalent to correlation coefficients, conventional sampling theory and tests of hypotheses are not applicable or relevant.

Statistical variation. Several levels of statistical variation require consideration. At the lowest level are variation and error in the measurements. At this level there may be measurement error, as in determining the median income in a census tract, or there may be intraunit variability, as when units are Indian castes and tribes being clustered on the basis of physical measurements on individual members. Measurement error and intraunit variability, even on binary attributes, are present more often than is hoped or admitted, and they degrade the quality of clustering. Sometimes the variation must be accepted, but often it can be reduced by improved measurement technique or by inclusion of more than one instance of each unit.

A second level is variation of units within a cluster. This variation cannot be estimated until clusters are at least tentatively determined, but this is the important variability for determining statistical scales and weights, as would be used, for example, in classifying new units into established clusters.

Variation between clusters is the largest source and is the variation explainable by the clustering. Its magnitude depends heavily on the selection of units. Unfortunately, scaling variables according to over-all variability reflects largely this cluster-to-cluster variability.

Statistical models for variability in clustering are not yet highly developed. The fundamental paper by Rao (1948) and current work by Ihm (1965) are of note. One statistical model for clustering is very old. The distributions of observed data like those illustrated in figures 1 and 2 are naturally represented as mixtures whose component distributions correspond to the clusters. This formulation has greater conceptual than practical value; direct numerical use is troublesome even for a single variable [*see* DISTRIBUTIONS, STATISTICAL, *article on* MIXTURES OF DISTRIBUTIONS].

Development and applications. After a long, slow history in social science, notably in psychology and anthropology, numerical clustering methods underwent a rebirth of rapid, serious development —first through work in numerical taxonomy, both in conventional biology and in microbiology, and later in automatic classification and pattern recognition. These developments have spread throughout the social and natural sciences.

The book *Principles of Numerical Taxonomy* by Sokal and Sneath (1963) provides the most comprehensive presentation of methods and principles and a thorough bibliography. Later developments in the biological area are represented in such periodicals as the *Journal of Ecology*, *Journal of Bacteriology*, *Journal of General Microbiology*, and the newsletter *Taxometrics*. The symposium of the Systematics Association (Heywood & McNeill 1964) is valuable.

The history of clustering in anthropology and linguistics has been surveyed by Driver (1965) in a detailed discussion of principles and results. [*Applications in geography are described in* GEOGRAPHY, STATISTICAL.]

Clustering methods in psychology, often as an adjunct of factor analysis and often under the name "pattern analysis," have developed from Zubin (1938) and Tryon (1939). Many grouping ideas were introduced by McQuitty (1954; 1964) in a long series of papers.

Work on political districting is exemplified by Weaver and Hess (1963) and Kaiser (1966). Developments and applications in pattern recognition, information retrieval, and automatic classification are in large part not yet in the public literature (Needham 1965). Ball (1965) surveys approaches taken in many areas.

DAVID L. WALLACE

[*Other relevant material may be found in* MULTIVARIATE ANALYSIS, *article on* CLASSIFICATION AND DISCRIMINATION. *See also* Tiryakian 1968.]

BIBLIOGRAPHY

BALL, GEOFFREY H. 1965 Data Analysis in the Social Sciences. Volume 27, part 1, pages 533–560 in American Federation of Information Processing Societies Conference, *Proceedings*. Fall Joint Computer Conference. Washington: Spartan Books; London: Macmillan.

DRIVER, HAROLD E. 1965 Survey of Numerical Classification in Anthropology. Pages 301–344 in Dell Hymes (editor), *The Use of Computers in Anthropology*. The Hague: Mouton.

FISHER, WALTER D. 1958 On Grouping for Maximum Homogeneity. *Journal of the American Statistical Association* 53:789–798.

HEYWOOD, VERNON H.; and MCNEILL, J. (editors) 1964 *Phenetic and Phylogenetic Classification*. London: Systematics Association.

IHM, PETER 1965 Automatic Classification in Anthropology. Pages 357–376 in Dell Hymes (editor), *The Use of Computers in Anthropology*. The Hague: Mouton.

KAISER, HENRY F. 1966 An Objective Method for Establishing Legislative Districts. *Midwest Journal of Political Science* 10, no. 2:200–213.

LANCE, G. N.; and WILLIAMS, W. T. 1965 Computer Programs for Monothetic Classification ("Association Analysis"). *Computer Journal* 8:246–249.

MCQUITTY, LOUIS L. 1954 Pattern Analysis Illustrated in Classifying Patients and Normals. *Educational and Psychological Measurement* 14:598–604.

MCQUITTY, LOUIS L. 1964 Capabilities and Improvements of Linkage Analysis as a Clustering Method. *Educational and Psychological Measurement* 24:441–456.

NEEDHAM, R. M. 1965 Computer Methods for Classification and Grouping. Pages 345–356 in Dell Hymes (editor), *The Use of Computers in Anthropology*. The Hague: Mouton.

RAO, C. RADHAKRISHNA 1948 The Utilization of Multiple Measurements in Problems of Biological Classification. *Journal of the Royal Statistical Society* Series B 10: 159–193. → Pages 194–203 contain an especially interesting discussion of this paper.

SOKAL, ROBERT R.; and SNEATH, PETER H. A. 1963 *Principles of Numerical Taxonomy*. San Francisco: Freeman.

►TIRYAKIAN, EDWARD A. 1968 Typologies. Volume 16, pages 177–186 in *International Encyclopedia of the Social Sciences*. Edited by David L. Sills. New York: Macmillan and Free Press.

TRYON, ROBERT C. 1939 *Cluster Analysis: Correlation Profile and Orthometric (Factor) Analysis for the Isolation of Unities in Mind and Personality*. Ann Arbor, Mich.: Edwards.

WARD, JOE H. JR. 1963 Hierarchical Grouping to Optimize an Objective Function. *Journal of the American Statistical Association* 58:236–244.

WEAVER, JAMES B.; and HESS, SIDNEY W. 1963 A Procedure for Nonpartisan Districting: Development of Computer Techniques. *Yale Law Journal* 73:288–308.
ZUBIN, JOSEPH 1938 A Technique for Measuring Likemindedness. *Journal of Abnormal and Social Psychology* 33:508–516.

Postscript

The applications of clustering techniques have grown at a rapid rate over the past decade, in large part because of the increased capability and accessibility of computers. Programs for carrying out cluster analyses are included in many of the widely used systems for statistical computation, and special-purpose programs are easily implemented from sample programs provided in several books (for example, Anderberg 1973; Hartigan 1975). The advances in the development and exploration of clustering algorithms have not been matched by advances in theoretical foundations or in the development of models for handling uncertainty or error. For better or worse, clustering remains exclusively a part of exploratory data analysis. [*See* DATA ANALYSIS, EXPLORATORY.]

Biological systematics and taxonomy account for the greatest number and variety of applications. The influential book by Sokal and Sneath (1963) has been extended and revised in Sneath and Sokal (1973); their discussion of principles and guide to applications are complemented by the mathematical treatment in Jardine and Sibson (1971). Hartigan (1973) gives an accessible introduction and review of general clustering ideas and methods. Cormack (1971) provides a critical review and extensive bibliography. The proceedings of the 1970 Anglo–Romanian Conference on Mathematics in the Archeological and Historical Sciences (1971) are rich in applications and methodology of clustering and of related techniques for the exposure of multivariate relationships [*see* SCALING, MULTIDIMENSIONAL].

Clustering is represented by a chapter in almost every textbook on multivariate analysis or on pattern recognition published since 1970. Books devoted to clustering are becoming more common: Tryon and Bailey (1970), Anderberg (1973), and Hartigan (1975) are notable.

DAVID L. WALLACE

ADDITIONAL BIBLIOGRAPHY

ANDERBERG, MICHAEL R. 1973 *Cluster Analysis for Applications*. New York: Academic Press.
ANGLO–ROMANIAN CONFERENCE ON MATHEMATICS IN THE ARCHEOLOGICAL AND HISTORICAL SCIENCES, MAMAIA, RUMANIA, *1970* 1971 *Mathematics in the Archaeological and Historical Sciences*. Edited by F. R. Hodson, D. G. Kendall, and P. Tautu. Edinburgh Univ. Press.
CORMACK, R. M. 1971 A Review of Classification. *Journal of the Royal Statistical Society* Series A 134: 321–367. → Includes discussion.
HARTIGAN, JOHN A. 1973 Clustering. *Annual Review of Biophysics and Bioengineering* 2:81–101.
HARTIGAN, JOHN A. 1975 *Clustering Algorithms*. New York: Wiley.
JARDINE, NICHOLAS; and SIBSON, ROBIN 1971 *Mathematical Taxonomy*. New York: Wiley.
SNEATH, PETER H. A.; and SOKAL, ROBERT R. 1973 *Numerical Taxonomy: The Principles and Practice of Numerical Classification*. San Francisco: Freeman.
TRYON, ROBERT C.; and BAILEY, DANIEL D. 1970 *Cluster Analysis*. New York: McGraw-Hill.

COALITIONS, THE STUDY OF

This article was first published in IESS *as "Coalitions: I. The Study of Coalitions" with a companion article less relevant to statistics.*

The word "coalition" has long been used in ordinary English to refer to a group of people who come together (usually on a temporary basis) to obtain some end. Typically, a coalition has been regarded as a parliamentary or political grouping less permanent than a party or a faction or an interest group (see Epstein 1968). Recently, however, the word has acquired a technical significance in social science theories with the elaboration (in the last two decades) of the theory of n-person games. The notion of coalition formation is central to this theory, since coalitions are the characteristic form of social organization by which the outcomes of such games are determined. To the degree that the theory provides a model for the study of national decision making in elections, parliaments, committees, cabinets, etc., or of international decision making in wars, diplomatic maneuvers, and international organizations—to that degree, coalitions are the characteristic form of social organization for political decision making generally.

The originators of the theory of n-person games, John von Neumann and Oskar Morgenstern, observed a fundamental difference, with respect to discovering the best way to win, between two-person games and games involving more than two persons: In two-person games, the problem for each player is to select the best strategy against his opponent; but in three-person or larger games, the problem for each player is to select partner(s) who can collectively win. They called the artifact resulting from the mutual selection of partners a *coalition*, and they constructed the whole theory of n-person games about the process of forming coalitions. Since politics is often defined as the authoritative allocation of values and since, in all but dictatorial or duopolistic situations, allocation is a process of coalition formation (Riker 1962, chapter 1), it is apparent that a theory of coalitions is a central part of a theory of politics.

Main problems

Three main questions have been dealt with in the theory: (1) How should winnings be divided to ensure victory for a player or a coalition? (2) Given a particular set of rules, what chance does a particular player have to be in a crucial position in a winning coalition? (3) Which potential partners should come together in particular play? The theories relating to each of these questions will be summarized and then attempts at verification and use of the theories will be discussed.

The division of winnings. The von Neumann–Morgenstern notion of a *solution* is the main contribution here. The substance of this notion is that, while it is not possible to specify a uniquely preferable coalition, it is possible to specify a set of preferable imputations, that is, a set of preferable ways to distribute gains and losses. To explain this notion the following vocabulary is required.

Let there be a set of players, I, where I is given by $\{1, 2, \cdots, n\}$, and let the subsets of I, which are *coalitions*, be designated by $S, T, \cdots$. Let the payment to each player at the end of the game be designated by x_i, where $i = 1, 2, \cdots, n$, so that the totality of payments, which is an *imputation* is a vector, $\boldsymbol{x} = (x_1, x_2, \cdots, x_n)$. Let the payment to a coalition, S, be designated by a function, v, which is a real valued set function with at least the following properties:

(1) $v(\phi) = 0$, where ϕ is the empty coalition; and (2) $v(S \cup T) \geqslant v(S) + v(T)$, where S and T are disjoint subsets of I.

The second property, superadditivity, records the fact that for at least some players in some games there is an increment in payoff from the very act of joining together. A coalition, S, is said to be *effective* for an imputation, $\boldsymbol{x}$, if

$$\sum_{i \text{ in } S} x_i \leqslant v(S).$$

That is, a coalition is effective if it can win as much or more than the sum of the payoffs to its members. An imputation, $\boldsymbol{x}$, is said to dominate an imputation, $\boldsymbol{y}$, if

(1) S is not empty
(2) S is effective for $\boldsymbol{x}$
(3) $x_i > y_i$, for all i in S.

A solution, $\boldsymbol{V}$, is a set of imputations such that (1) no $\boldsymbol{y}$ in $\boldsymbol{V}$ is dominated by an $\boldsymbol{x}$ in $\boldsymbol{V}$; and (2) every $\boldsymbol{y}$ not in $\boldsymbol{V}$ is dominated by an $\boldsymbol{x}$ in $\boldsymbol{V}$. To illustrate, for the zero-sum three-person game in normal form (for definitions of "zero-sum" and "normal form" see, e.g., Riker & Niemi 1964) the set $\boldsymbol{V}$ is $\{(\frac{1}{2}, \frac{1}{2}, -1), (\frac{1}{2}, -1, \frac{1}{2}), (-1, \frac{1}{2}, \frac{1}{2})\}$. The essence of this definition is that, regardless of which winning coalition forms [i.e., (1, 2), (1, 3), or (2, 3)], still any one that does form ought to adopt a split-the-winnings-equally kind of imputation. The heuristic rationale leading to this definition is the observation that a player, i, who seeks to obtain $x_i > \frac{1}{2}$ is creating a situation in which he may receive the worst possible payoff, i.e., $x_i = -1$. If, in the course of negotiations, a coalition of players 1 and 2 has arrived at a tentative imputation $\boldsymbol{y}$, where $\boldsymbol{y} = (\frac{3}{4}, \frac{1}{4}, -1)$, player 2 is likely to be especially receptive to offers from player 3 to form (2, 3) with imputation $\boldsymbol{x} = (-1, \frac{1}{2}, \frac{1}{2})$ and where, of course, $\boldsymbol{y}$ is now dominated by $\boldsymbol{x}$. Thus player 1 is undone by his greed. On the other hand, none of the imputations in the solution are susceptible to that kind of renegotiation of coalitions, for no partner in a winning coalition is likely to feel disadvantaged by the imputation adopted. Hence, if an imputation in the solution is adopted, it is likely to be stable as are also the coalitions associated with it.

In short, a solution specifies a set of preferable imputations, but not preferable coalitions. Since, however, von Neumann and Morgenstern had heuristically stated the problem of n-person games to be that of finding partners and since, in contrast, their mathematical "solution" specified not partners but rather the division of spoils among unspecified partners, much dissatisfaction has been expressed with the notion of a solution, and a number of alternatives have been offered (von Neumann & Morgenstern 1944). The spirit of these alternatives, like the spirit of the von Neumann–Morgenstern solution, is to pick out a limited number of acceptable imputations from the (usually) infinite number of imputations possible.

Ψ-stability: The most interesting of these alternatives is the notion of Ψ-stability, which can be explained with the following vocabulary: Let a partition of I into disjoint subsets be defined as a coalition structure, $\boldsymbol{\tau}$. For example, if $n = 3$, the partitions $(\{1, 2\}, \{3\})$ and $(\{1, 3\}, \{2\})$ are possible coalition structures. Let Ψ denote a rule of admissible changes from $\boldsymbol{\tau}$, and let $\Psi(\boldsymbol{\tau})$ denote the set of coalition structures resulting from the application of Ψ to $\boldsymbol{\tau}$. For example, if for $n = 3$, $\boldsymbol{\tau}$ is the partition $(\{1, 2\}, \{3\})$, and if Ψ is a rule that per-

mits any coalition formed by the addition of a single player to an existent coalition, then $\Psi(\tau)$ is the following set of alternative partitions:

$$[(\{1, 2, 3\}), (\{1, 3\}, \{2\}), (\{2, 3\}, \{1\})].$$

Since the point of this argument is to set some limit on possible imputations and coalitions, one wishes to find pairs, $(\boldsymbol{x}, \tau)$, of an imputation and a coalition structure such that players would not wish to depart from the pair (once they have reached it) by means of the changes permitted by Ψ. Such pairs, $(\boldsymbol{x}, \tau)$, are in a kind of equilibrium, which Luce called *Ψ-stability* (1954). Formally defined, a pair, $(\boldsymbol{x}, \tau)$, is Ψ-stable if:

(1) $v(S) \leqslant \sum_{i \text{ in } S} x_i$, for all S in $\Psi(\tau)$, and

(2) if $x_i = v(\{i\})$, then $\{i\}$ is in τ.

The first condition states that, for any coalition, S, that might be formed by the application of Ψ to τ, the players in this potential coalition cannot better themselves. The second condition states that if a player receives no more than he can get in the worst possible circumstances (i.e., when he is in a single-member coalition), then he must in fact be in such a condition in τ, that is to say for a member, i, of a multimember coalition in τ, $x_i > v(\{i\})$.

The advantage of this definition is that it permits some discussion of actual partners along with a discussion of imputations. Its disadvantage is that it requires a precise specification of Ψ for a particular game—and in social situations this is usually difficult to specify. Furthermore, there is an embarrassingly *ad hoc* flavor to the whole definition inasmuch as the specification of Ψ must be in terms of standards of behavior prevailing among particular players of a game.

Other alternatives to solution theory have also been offered, but they are no more satisfactory than what they purport to supplant; so instead of summarizing them here, the reader is referred to Luce and Raiffa, where Ψ-stability is also admirably discussed (1957, pp. 220–245).

The chance of obtaining a crucial position. Here the main contribution is Shapley's notion of a value for n-person games, which is an a priori method for estimating whether or not, for a particular player, a game is worth playing (Shapley 1953; Shapley & Shubik 1954). Suppose, in a game of n players, a coalition of k players, where $k \leqslant n$, is necessary to win. Let coalitions of k players be constructed by permuting the n players so that the first k players in any permutation are the minimally winning coalition, which is defined as a winning coalition that ceases to be winning if the kth player is subtracted. Let the kth position be designated as the *pivot* and let the number of times a player, i, occupies the pivot position be designated by p_i. Since there are $n!$ permutations of n, the chance, v, that a player, i, occupies the pivotal position is $v(i) = p_i/n!$.

Underlying this measure of value are two crucial sociological assumptions: (1) Since the notion of pivoting is defined with respect to minimal winning coalitions, one infers that membership of a coalition in excess of the minimally winning size is irrelevant. This is a version of the size principle to be discussed below. (2) The expectation about imputations is quite different from imputations prescribed in the von Neumann–Morgenstern solution. In a solution, the division of winnings among equally weighted members is, regardless of their position in the coalition, equal. But in the sociological theory underlying Shapley's value, there is a time dimension to membership in a coalition so that the player in the pivotal position can expect to receive more than others. Over time, these advantages are expected to average out; but still the single imputation in the specific play is not an imputation in the solution. The difference is, of course, the assumption of the existence of a time dimension (and perhaps of a differentiation of roles in the coalition-formation process).

[2]Which players should become partners. The main contributions here are those by Gamson (1961*a*, 1961*b*) and Riker (1962). Underlying these contributions is the *size principle*, which is the assertion that, with perfect and complete information, players should prefer minimally winning coalitions to larger winning ones. Using this principle, it is possible to show that if players have unequal weights then some possible coalitions are preferable to others, and indeed in some distributions of weights one possible coalition is uniquely preferable to others. Similarly, some players have unique advantages over others in the sense that the advantaged players can expect to be included in *any* preferable coalition. Since these two kinds of advantage are a function of the kind of ways in which the several weights can add up to k, which is the minimally winning size, it is not possible to specify in general these advantages for players and coalitions. Tables specifying these advantages for relevant variations in partitions of the total weight in the set of players are set forth in Gamson and in Riker for $n = 3$, 4, or 5. The most interesting result of such specifications is that *usually* the least weighty players ought to combine with each other, which leads to the somewhat paradoxical assertion that the weightiest player is *usually* the weakest in terms of combinatorial advantages. Note, however, that this conclusion follows from an argument in which perfect and complete information is assumed. Weakening the assumptions about information also weakens the force of this conclusion. As information is rendered less perfect and less complete, players may be expected to attempt to increase the size of winning coalitions above the minimum in order to guarantee victory.

Verification

Much more energy has been expended on the elaboration of the theory of coalitions than on the verification of it. The paucity of attempts at verification is explicable in terms of a theoretical difficulty: The whole theory is normative in the sense that it specifies what rational players should do to obtain the best possible payoff. It does not specify what real players will in fact do. To render the normative theory into a descriptive one, it must be assumed (1) that some players (that is, at least the winners) are rational, in the sense that they prefer to win rather than lose, and (2) that some political situations are sufficiently analogous to abstract games so that the theory of games is also a theory of political coalitions. Many social scientists are suspicious of both assumptions, and, given their reluctance to adopt them, the theory remains unverifiable just as all normative theories are unverifiable as truth-functional sentences. Nevertheless, some social scientists have been willing to assume that institutions that favor winners encourage the existence of rational political men and that political situations in which participants perceive their problem as one of winning are quite analogous to abstract games. Hence they have regarded the theory of coalitions as descriptive and have sought to verify it, hoping thereby to verify as well the assumption about rationality.

Verification and the "solution." Several experiments, most of which are well summarized by Rapoport and Orwant (1962) and Riker and Niemi (1964), have been conducted to test some features of the notion of a solution. In general, the results of these have not verified the Von Neumann–Morgenstern theory, except when special precautions have been taken in the experimental design to encourage subjects to behave highly competitively. Usually the imputations arrived at have been of the sort $(1/n, 1/n, \cdots, 1/n)$, instead of the prescribed $\alpha_1, \alpha_2, \cdots, \alpha_k, \beta_1, \beta_2, \cdots, \beta_{n-k}$, where $\alpha_i = \alpha_j$ and $\alpha_i > 0 > \beta_i$. The latter imputation has been achieved with fair consistency only when, for example, stooges have been inserted among sophisticated subjects to insist upon imputations of the prescribed kind. Several explanations for the failure have been advanced, to wit, that the experimental design (1) failed to protect the outcome from the influence of variables in subjects' personalities (e.g., their attitudes toward gambling); (2) failed to provide stakes large enough to induce rational behavior; and (3) permitted subjects to perceive the experimenter as another player in an $(n+1)$-player game, in which the imputation $(1/n, 1/n, \cdots, 1/n, -n)$ may be in the solution (provided the game is regarded as discriminatory). Further experiments with improved designs are necessary before any conclusion can be drawn on the truth or falsity of solution theory.

As for the notion of Ψ-stability, the one attempt to use it in analyzing a real situation (which attempt amounts indirectly to an attempt at verification) has produced interesting and intuitively plausible, but far from definitive, results. Luce and Rogow (1956) analyzed a simplified version of the three houses of the national legislature in the United States (i.e., the president, in his legislative capacity, the Senate, and the House of Representatives) in which there were assumed to exist two parties each with some members who always voted with the majority of the party (die-hards) and some who might on occasion vote with a majority of the other party (defectors). They examined the legislative branch according to variations in numbers of party members and numbers of die-hards and defectors in each party and produced two conclusions that do not otherwise seem immediately obvious: (1) a president is weak when *either* party has a more than two-thirds majority in Congress and (2) a president is strong when there is no party that can obtain a two-thirds majority even with the help of defectors from the other party. These nonobvious, but intuitively satisfying, conclusions suggest that much can be learned by further application of Ψ-stability theory to other situations.

Verification and use of value. It is hard to imagine how the notion of value might be verified, for it is a method to evaluate rather than predict or prescribe. Riker tried to determine by statistical analyses of roll calls in an assembly whether or not legislators seek to improve their chances of occupying the pivotal position; but his data and procedures were too crude to lead to any sure conclusions (1959). The notion of value has, however, been used to evaluate several real constitutions, treating them as games to be played: the Congress in the United States; the Security Council of the United Nations (Shapley & Shubik 1954); the Electoral College in the United States (Mann &

Shapley 1960; 1962); and the federal relationship in the United States, Canada, and Australia (Riker & Schaps 1957).

Verification of prescriptions on partners. Vinacke and Arkoff (1957) and Gamson (1961*b*) have attempted to verify Caplow's predictions in experimental situations. Their results, with some reservations, tend to verify his specifications of preferable coalitions. Riker has offered historical evidence for the acceptance of the size principle by politicians in national and international political situations (1962, chapters 3, 7). Barth, assuming the size principle, has shown that Afghan chieftains conduct their intertribal diplomacy in terms of it (1959). Although far from satisfactorily verified, this feature of the theory of coalitions is closer to verification than any other.

Further developments

Considering the previous development of the theory of coalitions, it seems likely that future expansion of the theory can be expected only if additional sociological assumptions can be incorporated in the mathematical theory. The original theory of solutions to *n*-person games contained a bare minimum of assumptions about behavior, viz., (1) rational motives, i.e., participants' desire to win rather than to lose; (2) sets and subsets or the notion of coalitions itself; and (3) super additivity. Out of these minimal assumptions, von Neumann and Morgenstern were able to arrive at the theory of solutions, which, however, seemed inadequate because it specified only the division of winnings and not the choice of partners. Luce's step toward the discussion of partners (i.e., Ψ-stability) involved the introduction of an additional sociological assumption, namely, the existence of a standard of behavior, Ψ, that admitted some kinds of bargaining about membership in coalitions but not others. Caplow's discussion of partners and Riker's elaboration of the size principle required the introduction of both the notion of differentials in weights (which is substantially equivalent to the introduction of the notion of "power") and the notion of a majority. Since these elaborations of the theory of coalitions depended upon the addition of sociological premises, it can be expected that future elaborations will depend on future additions.

Dynamics of growth. One promising additional sociological assumption is the notion that coalitions go through a process of growth. Let the situation at the beginning of any decision-making or allocative process be such that participants are partitioned into n single-member subsets. Define *coalitions* as winning or losing subsets when a winning subset exists, and define *protocoalitions* as subsets when no winning coalition exists. Then the process of decision making is the transformation of some protocoalition into a winning coalition. If a winning coalition consists of k members, where $\frac{1}{2}n < k \leqslant n$, then the process of decision making or allocation is the development of a protocoalition from 1 member to 2 members, from 2 to 3 members, $\cdots$, from $(k-2)$ members to $(k-1)$ members, and from $(k-1)$ to k members. There are several crucial stages in this process, especially the movement from one-member to two-member protocoalitions and the movement from $(k-1)$ to k members (which latter may be called the "end-play"). Riker has set forth some of the strategic considerations of the end-play, but these have as yet to be developed in a general statement (1962, chapter 6). His conclusions can be summarized, however, as the assertion that the strategy of end-play is to find, for a given structure of protocoalitions, that coalition that most nearly approaches the minimally winning size. Doubtless many other strategic considerations enter into the growth process, although these are as yet unspecified.

Roles. Defining roles as positions specified by the rules of any particular process of coalition formation, von Neumann and Morgenstern assumed that while roles of players may differ with respect to advantages in the rules, still the roles are identical with respect to the kind of behavior in bargaining. Furthermore, they assumed that players could rotate indiscriminately among roles. There is no reason to suppose that either assumption is appropriate for the description of the natural world. Indeed, it is likely that in reality there exists a variety of roles in the coalition-formation process: for example, *leader,* the role of initiating that bargaining by which protocoalitions are enlarged; *follower,* the role of moving, at the instigation of a leader, from a single-member protocoalition to a larger one; *pivot,* already defined as that follower who becomes the kth member of a minimally winning coalition; *reliable follower,* the role of irrevocably accepting membership in a multimember protocoalition; *defecting follower,* the role of accepting and subsequently rejecting membership in a multimember protocoalition; *wallflower,* a member of a single-member coalition who is not sought out for the role of follower; etc. Doubtless the theory of coalition formation may be rendered more appropriate for the study of nature by mathematizing the essential features of behavior in each of these roles. For example, with respect to the role of leader, it is likely, as Riker has argued, that his payments are different in kind from the payments to followers, that the leader (who quite possibly starts out as a wallflower) is willing to forgo material reward for the sake of obtaining the psychic reward of leadership and that the leader may, for the sake of retaining his leadership, pay out more material rewards than his prospective coalition can win. Assuming these possibilities do in fact prevail, then the notion of a solution must be modified. Solution theory requires a symmetric kind of imputation (although it allows for unequal rewards for roles with unequal advantages in the rules). But if differentials in behavior may affect imputations, then the notion of "equal rewards for equal advantages in the rules" must be abandoned for a notion of unequal rewards according tc whether the participant plays the role of leader or follower, with the leader accepting the lesser material reward. (The sociological, but not the mathematical, consequences of accepting the notion of these differentials in roles is set forth in Riker 1962.)

In the beginning of this article, it was suggested that a theory of coalitions amounted to a theory of politics. The subsequent considerations, however, suggest that a theory of coalitions adequate to serve as a theory of politics has not yet been developed. The hope for the next generation is that it will be.

WILLIAM H. RIKER

[*Also relevant are the entries* DECISION MAKING; GAME THEORY; SIMULATION.]

BIBLIOGRAPHY

BARTH, FREDRIK 1959 Segmentary Opposition and the Theory of Games: A Study of Pathan Organization. *Journal of the Royal Anthropological Institute of Great Britain and Ireland* 89:5–21.

BUCHANAN, JAMES M.; and TULLOCK, GORDON 1962 *The Calculus of Consent: Logical Foundations of Constitutional Democracy.* Ann Arbor: Univ. of Michigan Press.

CAPLOW, THEODORE 1956 A Theory of Coalitions in the Triad. *American Sociological Review* 21:489–493.

►EPSTEIN, LEON D. 1968 Parliamentary Government. Volume 11, pages 419–426 in *International Encyclopedia of the Social Sciences.* Edited by David L. Sills. New York: Macmillan and Free Press.

GAMSON, WILLIAM A. 1961*a* A Theory of Coalition Formation. *American Sociological Review* 26:373–382.

GAMSON, WILLIAM A. 1961*b* An Experimental Test of a Theory of Coalition Formation. *American Sociological Review* 26:565–573.

LUCE, R. DUNCAN 1954 A Definition of Stability for *n*-Person Games. *Annals of Mathematics* 59:357–366.

LUCE, R. DUNCAN; and RAIFFA, HOWARD 1957 *Games and Decisions: Introduction and Critical Survey.* A Study of the Behavioral Models Project, Bureau of Applied Social Research, Columbia University. New York: Wiley. → First issued in 1954 as *A Survey of the Theory of Games,* Columbia University, Bureau of Applied Social Research, Technical Report No. 5.

LUCE, R. DUNCAN; and ROGOW, ARNOLD A. 1956 A Game Theoretic Analysis of Congressional Power Distributions for a Stable Two-party System. *Behavioral Science* 1:83–95.

MANN, IRWIN; and SHAPLEY, L. S. 1960 *Values of Large Games, IV: Evaluating the Electoral College by Montecarlo Techniques.* U.S. Air Force Project, Memorandum RM-2651. Santa Monica, Calif.: RAND Corp.

MANN, IRWIN; and SHAPLEY, L. S. 1962 *Values of Large Games, VI: Evaluating the Electoral College Exactly.* Memorandum RM-3158-PR. Santa Monica, Calif.: RAND Corp.

RAPOPORT, ANATOL 1960 *Fights, Games, and Debates.* Ann Arbor: Univ. of Michigan Press.

RAPOPORT, ANATOL; and ORWANT, CAROL 1962 Experimental Games: A Review. *Behavioral Science* 7:1–37.

RIKER, WILLIAM H. 1959 A Test of the Adequacy of the Power Index. *Behavioral Science* 4:120–131.

RIKER, WILLIAM H. 1962 *The Theory of Political Coalitions.* New Haven: Yale Univ. Press.

RIKER, WILLIAM H.; and NIEMI, DONALD 1962 The Stability of Coalitions on Roll Calls in the House of Representatives. *American Political Science Review* 56: 58–65.

RIKER, WILLIAM H.; and NIEMI, RICHARD G. 1964 Anonymity and Rationality in the Essential Three-person Game. *Human Relations* 17:131–141.

RIKER, WILLIAM H.; and SCHAPS, RONALD 1957 Disharmony in Federal Government. *Behavioral Science* 2: 276–290.

SHAPLEY, L. S. 1953 A Value for *n*-Person Games. Volume 2, pages 307–317 in H. W. Kuhn and A. W. Tucker (editors), *Contributions to the Theory of Games.* Princeton Univ. Press.

SHAPLEY, L. S.; and SHUBIK, MARTIN 1954 A Method for Evaluating the Distribution of Power in a Committee System. *American Political Science Review* 48: 787–792.

SHUBIK, MARTIN (editor) 1964 *Game Theory and Related Approaches to Social Behavior: Selections.* New York: Wiley.

VINACKE, W. E.; and ARKOFF, ABE 1957 An Experimental Study of Coalitions in the Triad. *American Sociological Review* 22:406–414.

○VON NEUMANN, JOHN; and MORGENSTERN, OSKAR (1944) 1953 *Theory of Games and Economic Behavior.* 3d ed., rev. Princeton Univ. Press. → A paperback edition was published in 1964 by Wiley.

Postscript

[1]In addition to the solution notions mentioned in the main article, there should also have been a consideration of the bargaining set (Auman & Maschler 1964). The main social scientific developments on solution theory have been attempts at verification of the von Neumann and Morgenstern theory and the bargaining set theory (Riker 1971; Riker & Zavoina 1970; Buckley & Weston 1973).

[2]With respect to the size principle, there have been works of theoretical and empirical signifi-

cance. Inspired by the criticism of Butterworth (1971), who, relying on an idiosyncratic notion of the characteristic function, charged that the size principle was not generally true, Shepsle (1974) attempted to infer the size principle from solution theory. Empirically, a number of studies have investigated the appearance or nonappearance of minimum winning coalitions in places where they might be expected according to the size principle: Congress (Koehler 1975; Hinckley 1972); state legislatures (Lutz & Murray 1974); and cabinet formation in European governments (Browne 1971; Damgaard 1969; DeSwaan 1973).

WILLIAM H. RIKER

ADDITIONAL BIBLIOGRAPHY

AUMANN, ROBERT J.; and MASCHLER, MICHAEL 1964 The Bargaining Set for Cooperative Games. Pages 443–476 in Melvin Dresher, Lloyd S. Shapley, and A. W. Tucker (editors), *Advances in Game Theory*. Princeton Univ. Press.

BROWNE, ERIC C. 1971 Testing Theories of Coalition Formation in the European Context. *Comparative Political Studies* 3:391–408.

BUCKLEY, JONES J.; and T. EDWARD WESTEN 1973 The Summetric Solution to a Five-person Constant-sum Game as a Description of Experimental Game Outcomes. *Journal of Conflict Resolution* 17:703–719.

BUTTERWORTH, ROBERT L. 1971 A Research Note on the Size of Winning Coalitions. *American Political Science Review* 65:741–748.

DAMGAARD, ERICK 1969 The Parliamentary Basis of Danish Governments: The Patterns of Coalition Formation. Volume 4, pages 30–57 in *Scandinavian Political Studies: A Yearbook*. → Published by the political science associations in Denmark, Finland, Norway, and Sweden.

DESWAAN, ABRAM 1973 *Coalition Theories and Cabinet Formations: A Study of Formal Theories of Coalition Formation Applied to Nine European Parliaments After 1918*. New York: American Elsevier.

HINCKLEY, BARBARA 1972 Coalitions in Congress: Size and Ideological Distance. *Midwest Journal of Political Science* 16:197–207.

KOEHLER, DAVID H. 1975 Legislative Coalition Formation: The Meaning of Minimal Winning Size With Uncertain Participation. *American Journal of Political Science* 19:27–39.

LUTZ, DONALD S.; and MURRAY, RICHARD W. 1974 Redistricting Decisions in the American States: A Test of the Minimal Winning Coalition Hypothesis. *American Journal of Political Science* 18:233–255.

RIKER, WILLIAM H. 1971 An Experimental Examination of Formal and Informal Rules of a Three-person Game. Pages 115–140 in Bernhardt Lieberman (editor), *Social Choice*. New York: Gordon & Breach.

RIKER, WILLIAM H.; and ZAVOINA, WILLIAM J. 1970 Rational Behavior in Politics: Evidence From a Three-person Game. *American Political Science Review* 64: 48–60.

SHEPSLE, KENNETH A. 1974 On the Size of Winning Coalitions. *American Political Science Review* 68:505–518.

COMBINATIONS AND PERMUTATIONS

See PROBABILITY, *article on* FORMAL PROBABILITY.

COMPUTATION

In recent years there has occurred enormous technological development and widespread application of automatic information-handling techniques. These are relevant to three broad classes of tasks with which the scientist frequently finds himself faced: (*a*) *data processing*, the sorting and summarization of large files of information, usually representing the results of empirical observation; (*b*) *mathematical computation*, the derivation of numerical values associated with mathematically defined constructs (for example, matrix and differential equation systems); and (*c*) *simulation*, the generation of data supposed to represent the "behavior" of some temporal process (for example, a physical or economic system). Of course, a given task may have more than one of these components; the widely used technique of regression analysis, for example, may be thought of as consisting of a data-processing phase, for the selection and reduction of input data, followed by a computational phase, for the calculation of regression coefficients and related quantities.

Historically, various special-purpose mechanical aids have been used to facilitate operations in one or another of the areas cited. Most recently, the general-purpose automatic digital computer has been developed as an effective and flexible tool for handling all three types of application. Accordingly, after a brief survey of older methods, the main body of the present discussion will be concerned with this device.

Older methods

Methods for data processing are strongly conditioned by the characteristics of the medium in which the data are recorded. A file in which each data entry is recorded on a physically separate carrier, such as a card, has obvious manipulatory advantages. Thus traditional techniques for data processing have centered around the concept of a *unit record*. Various schemes for aiding the manual manipulation of files of cards have been devised; a useful one depends on notching the edges of cards in coded patterns to permit mechanization of selection and sorting. The most significant line of development here, however, has been that of the punched card system, in which special-purpose machines are used to perform sorting, collating, reproducing, and tabulating operations on decks of cards carrying information entirely in the form of coded patterns of punched holes. Such systems, the basic principles of which were developed by Herman Hollerith in the closing years of the last century, are in widespread use today. Their efficient use depends on the fact that file manipulation procedures can often be specified as a sequence of elementary operations repetitively performed. Processing of any complexity, however, may require many runs of the data cards through the same or different machines, and this is a factor that limits the magnitude of the tasks that can be attempted by the use of these techniques.

Mathematical computation can be mechanized in two basic ways, *digital* and *analog*. In digital computation numbers are represented and manipulated as symbolic entities; the abacus, in which the configuration of beads on a rod is taken as a coded representation of a decimal digit, is an elementary example of a computational aid based on the digital principle. In analog computation, by contrast, numbers are represented by the measure of physical aspects of some system; thus the slide rule, on which numerical magnitudes are marked off as lengths on a series of scales, is a basic example of a device for computing by the analog method.

The desk calculator is a special-purpose digital device that serves to mechanize the most onerous part of pencil-and-paper computing; since it performs only the fundamental arithmetic operations on numbers manually entered into its keyboard, however, elaborate computations still require a great deal of routine human labor. This sort of process can be mechanized one step further by incorporating elementary arithmetic capability into punched card equipment, but the inability of such machines to do more than a repetitive sequence of quite simple operations on a given run is still a limitation.

Manual analog methods include graphic techniques, such as the plotting of lines representing equations in order to determine their solution as an intersection point, and the use of nomograms, which may depend on predrawn forms suited to particular problems. In addition, a number of mechanical devices more advanced than the slide rule have been developed, such as the mechanical integrator. The evolutionary end product of this development is the modern analog computer, in which numerical magnitudes are represented as electrical voltages and circuit units interconnected by means of plug wires correspond to mathematical systems with various parameters. Analog techniques are subject to fundamental limitations regarding accuracy, since they depend on the precision with which physical quantities can be measured, and are restricted as to the types of problems that can be handled, since they must be capable of being formulated in physical terms.

Analog techniques also can be applied successfully in the area of simulation; in fact, here the distinction between mathematical computation and simulation becomes blurred. For, if an analog computer is set up to "solve" the differential equation that governs the behavior of a physical system through time, it can be thought of in a very real sense as a simulator of that system. This has led to the exploitation of such techniques in experimental work, where there is to be interaction, say, between an actual physical system and a simulated one.

Simulation by analog techniques is often useful but is limited; by contrast, any system with a mathematical description can be simulated by digital techniques. In fact, however, before the advent of the general-purpose computer, digital simulation of systems of any complexity was so inefficient as to be of little practical utility.

Further reading on the material of this section may be found in Brooks and Iverson (1963, pp. 52–144), Calingaert (1965, pp. 87–126), and Borko (1962, pp. 22–57).

General-purpose digital computer

The modern automatic digital computer, at the center of a system incorporating appropriate input, output, and auxiliary information storage devices, can be applied to data processing, mathematical computation, and simulation tasks in an integrated and flexible manner. This versatility is partly due to the physical speed and capacity made possible by contemporary technology, and partly due to a feature of the computer that was largely missing in earlier mechanisms—the internally stored program. Once a program or schedule of operations has been prepared, it can be introduced into the computer and used to control a long complicated sequence of steps, which are thus executed automatically at an extremely rapid rate without human intervention. The possibility of preparing programs of a general nature, which can then be tailored to particular cases by the specification of parameter values, and the fact that programs, once prepared, may be run by different users on different computers make the digital computer a tool of such power as to permit consideration of projects that would not formerly have been contemplated at all; this in turn often leads to the development of more advanced models and hence acts to influence the substance of the discipline to which the computer is applied. Thus, for example, the availability of programs incorporating provision for preliminary data editing, extensive computation of residuals, and graphic display of results has markedly influenced the manner in which statistical problems may be attacked.

This potential, however, is limited by the capacity of people to plan and implement effective programs; furthermore, the nature of computer operation is such as to make it most desirable for the user to have some comprehension of its inner workings rather than to depend completely on a professional programmer to interpret his needs. The following sections therefore contain a compact introduction to the subject, which, it is hoped, will at least enable the reader to become aware of some of the more obvious pitfalls that the computer user must anticipate.

Computer structure. In order to avoid too narrow a conception of function, it is appropriate to think of the general task for which an automatic digital computer is designed to be "symbol processing," rather than "computation" in the strictest (numerical) sense. The organization and operation of a computer may be basically described in terms of five functional units, as depicted schematically in the block diagram of Figure 1.

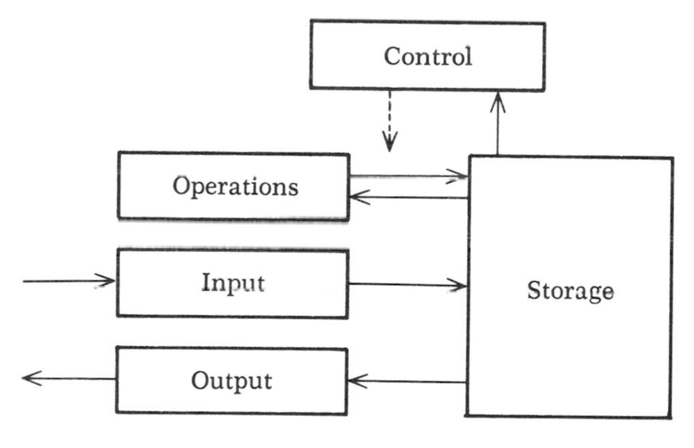

Figure 1 — Computer block diagram

The *storage* unit (usually picturesquely termed the "memory") acts as a repository for symbolic information and also as an intermediary for all interaction among other units. Internal processing of information is implemented by the transmission of symbolic operands from storage to the *operations* unit, where they are subjected to arithmetic or other transformations, the results of which are then transmitted back to storage. In order to set this process up initially an *input* unit is provided to introduce information from the external world into the storage unit, and in order that final results may be provided for the user there must also be an *output* unit for the opposite function.

Thus far, the functions described are roughly analogous to those performed by a man doing a hand calculation: he maintains a worksheet (storage) from which he derives numbers to be entered into the keyboard of his desk calculator (operations) and on which he copies the results; before the calculation proper starts some initial values may have to be entered (input), and when it is finished a separate sheet containing only the relevant results is prepared (output). The word "automatic" applied to a computer, however, implies mechanization not only of these physical processes but of the directing function supplied by the man, who causes a sequence of actions to be performed on the basis (presumably) of a predetermined schedule. In the computer the actuating function is performed by the *control* unit, and what distinguishes the computer from earlier devices is that the sequence of actions called for by this control can be in accordance with a pattern of arbitrary complexity, as specified by a symbolic *program* recorded in storage. Before processing can begin, therefore, a program must be introduced (via the input unit) into storage; the fact that the computer can be switched from one task to another by merely introducing a different program is what gives it its "general purpose" designation. Note also that the program, because it consists of symbolic information recorded in storage, can be altered by internal operations; this gives rise to powerful and flexible procedural techniques in which programs systematically modify themselves in order to achieve maximum processing efficiency.

Technological considerations dictate the use of bistable, or two-state, devices as the basic components of digital systems; this leads to the conception of all symbolic information ultimately coded in terms of just two symbols, the *bits* 0, 1. Particular patterns of several bits may then be interpreted as coded *characters* (letters, digits, etc.), which play the role of the elementary units of symbol processing. This notion is familiar from consideration of punched card or teletype tape codes, in which specified patterns of no-punch, punch represent characters drawn from a given set. In a computer, as in these contexts, the basic character set may include in addition to numeric digits the letters of the alphabet and selected special symbols such as mathematical signs and punctuation marks.

The unit of transmission of information within the computer may be the character, but on the more powerful computers efficiency dictates that characters be transmitted and operated upon in larger aggregates called for convenience *words*. Data-processing applications generally involve much manipulation of *alphanumeric* words, considered simply as arbitrary strings of characters; arithmetic operations, however, are meaningful only when applied to *numeric* words, which represent numbers in a more or less conventional manner. Finally, words formed according to a particular format are interpretable by the control unit as *instruction* words, and it is in terms of these that programs are specified for automatic execution.

Such special interpretations of computer words are applied only by the operations and control units; as far as storage goes, there is no distinction. Storage is organized into a series of subunits called *locations*, each of which is designated uniquely by an integer label, its *address*. The capacity of each addressed location is generally the same, being a character or a word of some standard length. Computer instructions are then typically designed to specify an operation to be performed and one or more storage addresses giving the locations at which operands are to be found or results stored.

Computers with storage capacities of the order of hundreds of thousands of characters and which carry out internal operations at the rate of hundreds of thousands a second have become available in the last few years, and in the more advanced systems both capacity and rate may effectively be in the millions. Although basic input/output operations (card or paper tape reading/punching, or printing) are generally much slower, delays can be reduced to a considerable extent by designing these units so that their operation can proceed concurrently with the main internal sequencing. Finally, the capabilities of the computer can be considerably enhanced by imbedding it in a system containing additional bulk storage (perhaps millions of characters stored in magnetic form on drums or disks) and intermediate input/output devices (magnetic tape units, for example), all communicating via the internal storage unit.

Further discussion may be found in Gregory and Van Horn ([1960] 1963, pp. 58–125), Green (1963, pp. 12–29), and Borko (1962, pp. 60–91).

Computer procedures. The popular conception of computer function seems to be that reflected in statements such as "Now they have a computer that can check everybody's tax return" or "Now they've come up with a computer that can simulate world politics." Although the area of intended application in the broad sense may influence design, as the previous discussion suggests, computers are not built to do specific tasks but are programmed to do them. This is important to the user because it shows the need for focusing attention not on computer availability but on program availability. And although the uninitiated but reasonably knowledgeable scientist may be aware of this fact, he often fails to anticipate the extent to which unqualified descriptions such as "regression analysis" fail to characterize a task definitively; details of matters such as input/output formats and computational methods are what determine whether or not a program is appropriate to the needs of a given user, and here information may be poorly communicated or not provided at all. Even worse, it may be erroneously communicated, partly due to lack of insight on the part of the program designer and partly because the tools for precise communication in the area of symbolic processing are as yet only imperfectly developed.

For these reasons it is desirable that the user have at least a rudimentary notion of the nature of computer instructions and programming techniques. This section and the one following outline briefly some of the concepts involved.

The first problem to be faced when planning to use a computer is that of precise specification of the task to be carried out, in a form sufficiently perspicuous so that the procedure can be communicated in human terms. Some form of *flow chart* can generally fulfill this function on several levels, from over-all to detailed. The basic idea here will be illustrated by an example.

Imagine that it is required to design an automatic procedure for computing the sample mean and standard deviation,

$$\bar{x} = \frac{1}{n}\sum_{i=1}^{n} x_i, \qquad s = \sqrt{\frac{1}{n-1}\sum_{i=1}^{n}(x_i - \bar{x})^2},$$

for several sets of x_i, not all of which necessarily contain the same number n of items (in all cases it must be that $n \geqslant 2$, however). If these are to be processed in a single input file (perhaps a deck of punched cards) some means must be provided for distinguishing the data sets from each other, and one practical way to do this is to suppose each file entry (card) contains two values x, m, the first of which is normally data (an x_i value) and the second of which is a "tag" to distinguish legitimate data items ($m = 0$) from "dummy" items used to signal the end of a data set ($m = 1$) or the end of the entire file ($m = 2$).

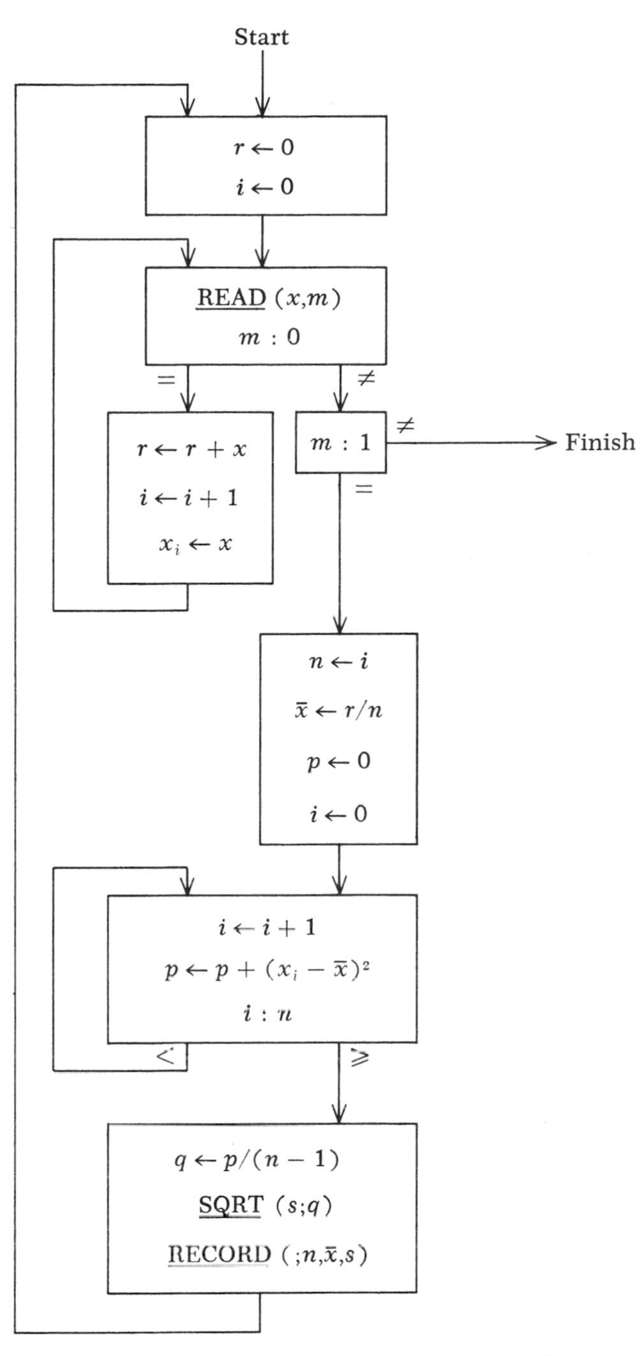

Figure 2 — Flow chart for mean and standard deviation calculations

A complete flow chart for this task is given in Figure 2; it covers the repetitive reading of data sets and the computation and recording (for example, printing) of n, $\bar{x}$, and s for each. The flow chart is a formal specification of a procedure that is described as follows:

Initially, set the sum r and the index i to 0, in anticipation of a data run. Then read a pair of values x, m and test (in two steps) whether m is 0 or 1. If $m = 0$ add x to the sum r, and also increment the value i and enter x as x_i in a data list; then return to read the next entry. If $m = 1$ the data set is complete, and the current value of i is then n for the set and $\bar{x}$ is determined as r/n. Now compute a second sum p by adding values $(x_i - \bar{x})^2$ for $i = 1, 2, \cdots, n$; following this compute s as the square root of $q = p/(n-1)$ and record the values n, $\bar{x}$, s. Repeat this entire process until the test gives $m \neq 0$ and $m \neq 1$ (so presumably $m = 2$); this indicates that there are no more input files and the procedure is finished.

The stylized formalism of the flow chart employs mathematical notation of the conventional sort but also makes use of special symbols signifying procedural action. These are "$\leftarrow$", meaning "set" (to a value); ":", meaning "compare" (for branching); and "____", meaning "execute" (a standard subprocedure). Thus "$r \leftarrow r + x$" means "set r to its former value plus x"; "$i{:}n$" means "compare i and n" (which causes the procedure to repeat as indicated if $i < n$ and go ahead if $i \geqslant n$); and "SQRT$(s;q)$" means "execute a subprocedure SQRT$(s;q)$" (which is designed to have the effect $s \leftarrow \sqrt{q}$).

The flow chart shows the structure of the computation in terms of procedural loops and re-entry points, without concern for details of printing formats, etc., which of course must be supplied if the procedure is to be actually carried out by computer. It is important to realize that these flow chart conventions are independent of the characteristics (and the idiosyncrasies) of any particular computer but that they do reflect general aspects of computer operation; thus both the notion that indices such as i can be manipulated independently in dealing with sets of values x_i and that parametrized subroutines such as SQRT(;) can be called upon and executed by a main program correspond to important features of actual computer operation.

The flow chart symbolism given here, for which detailed rules can be stated, is suitable primarily for expressing procedures of mathematical computation. The same basic idea, however, can be used to specify data-processing procedures; the main additional concept needed is a more versatile way of representing external files of given format and organization. Also, simulation and other internal processing applications that may involve nonnumeric elements can be handled by an extension of the flow charting technique.

Although no standardization of flow chart convention exists, and the choice depends to a certain extent on the application at hand, it nevertheless seems reasonable to ask that a flow chart represent an actual computer program accurately enough so that the filling in of details is routine and is unlikely to lead to serious discrepancies between what the user thought was to be done and what the programmer in fact specified was to be done.

For additional discussion of flow charting see Calingaert (1965, pp. 19–23).

Computer programs. To get some insight into the way in which a flow chart specification translates into a computer program, assume a computer with 1,000 storage locations 000, 001, $\cdots$, 999, each of which holds a 12-character word interpretable as an alphanumeric datum, a number, or an instruction in the format "$\omega\alpha\beta\gamma$", where ω is a 3-character alphabetic operation code and each of α, β, γ is a 3-digit storage address.

Then the instruction

ADD 500 501 500

might then have the significance "Add the contents of storage locations 500 and 501, and record the result in storage location 500." This would be appropriate only in case the 12-character words initially contained in storage locations 500 and 501 were in numeric format (say, 11 decimal digits preceded by algebraic sign), so that the addition operation has meaning. The execution of this instruction by the control unit results in the destruction of the original number in storage location 500, since it is replaced by the sum; the number in storage location 501, however, is not altered by the operation.

In order to be executed by the control unit, the ADD instruction itself must exist some place in storage, say at location 099; thus one must distinguish between the locations (here 500, 501) to which the instruction explicitly *refers* and the location (here 099) at which the instruction *resides*. Concern with the location of the instruction itself is necessary because of the sequential nature of computer operation; somehow there must be specified the order in which instructions are obtained and executed by the control unit, and the most straightforward way to do this is to design the computer so that instructions are executed in sequential order by resident address unless there is an explicit indication (in an instruction) to the contrary. To see the manner in which such an indication is typically set forth, consider a second hypothetical instruction

COM 500 550 099

which is to be given the interpretation "Compare the numbers in storage locations 500 and 550, and if the former is less than the latter, take the next instruction from storage location 099." This effects a "conditional jump" in the instruction sequence; if the stated condition is met, the next instruction executed is that residing in location 099, but otherwise it is the instruction whose address follows in sequence. If the COM instruction itself resides in location 100, then the pair of instructions introduced form a "program loop" schematized as follows:

(099) ADD 500 501 500
(100) COM 500 550 099.

The effect of this piece of program is to add the contents of 501 repeatedly to the contents of 500, which therefore acts as an "accumulator" for the sum; this addition process is continued (by virtue of the "jump back" to 099) until the accumulated sum exceeds the number residing in location 550, at which time the program proceeds in the normal sequence with the instruction following that numbered 100.

If one imagines that locations 500, 501, and 550 contain values, i, 1, and n, respectively, then the two-word loop is a computer implementation of the flow chart component shown in Figure 3 (which happens to perform a rather meaningless function

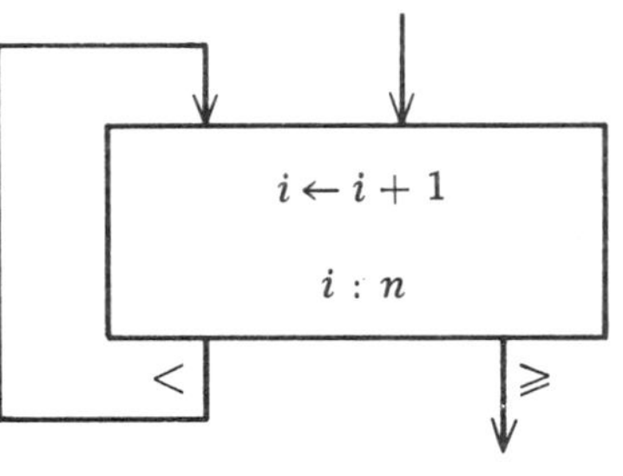

Figure 3 — Loop pattern

in itself but with appropriate instructions inserted between the ADD and COM becomes one of the loops of Figure 2). This would serve to suggest the manner in which flow charts may be translated into computer programs. Two important techniques in this connection, *indexing* of a set of values with a computed value (so as to be able to reference a general element x_i) and *linking* of subprocedures into a program (so as to be able to execute a subroutine such as SQRT and jump back to the main program at the proper point when finished) may be accomplished in a computer by programmed modification of instruction addresses.

Although the example illustrates some facets of computer control that are typical, the details of coding, format, etc. vary widely in practice. For example, it is common to have several different coding systems employed in the same computer for various purposes, perhaps a character code used for general alphanumeric information, a different code for number representation (often based on the binary radix 2 instead of the decimal radix 10), and a special format, not interpretable in either alphanumeric or pure numeric terms, for instructions. Also, the basic "three address" pattern of instructions given here is not typical; usually instructions are "one address" and require explicit recognition of registers for retaining intermediate results in the operations and other units. Thus on most computers the addition operation of the example requires three instructions: one that clears an accumulator register in the operations unit and records the contents of location 500 in it, one that adds the contents of 501 to this and retains the sum in the accumulator register, and one that records the contents of the accumulator in location 500. One or more special index registers are also customarily provided to facilitate indexing and linking operations.

Besides instructions for internal operations such as those described, the computer must have instructions for controlling the operation of input/output and other auxiliary devices with which it communicates; these would of course be used in procedures such as the READ and RECORD subroutines of the earlier example. The repertoire of a large computer may number instructions in the hundreds, many of these representing minor variants of each other to permit processing efficiency.

Two characteristic features of computer programs may be perceived from the discussion: a procedure is formulated for execution in terms of minutely small steps, and the symbolic representation of these steps is not conducive to quick apprehension of their effect in the aggregate. Not only does the preparation of a program for a given task require a great amount of routine labor, but the bookkeeping necessitated by the numerical addressing scheme tends to produce errors. Since computers are generally capable of performing routine tasks more reliably than humans, the question naturally arises of whether some of the burden of program preparation cannot be passed along to the very devices intended to execute the programs. This leads to the concept of program-preparation programs, which translate programs written in some language more or less "procedure-oriented" into direct computer code.

A basic example is the *assembly program*, which permits the programmer to write instructions in a form such as

```
GO:  ADD, I, ONE, I
     COM, I, N, GO.
```

These instructions may be thought of as assembly language versions of those given earlier. GO is a label corresponding to the instruction address 099, and I, ONE, and N are labels corresponding to the data addresses 500, 501, and 550, respectively. The programmer writes in terms of such mnemonic labels, and the assembly program translates these symbolic instructions (made up of letters, commas, spaces, etc.) into corresponding direct computer instructions in appropriate format; generally, the programmer does not specify at all the actual numerical addresses that will be assigned to the various labels by the assembly program. There are also other features incorporated into assembly programs to simplify the specification of numeric constants, etc.

Assembly language makes a program more perspicuous but does not compress its level of detail. Higher order program-preparation programs exist for this purpose; for example, an *algebraic translator*, which permits specification of the computer action already discussed in a form such as

```
GO:  I = I + 1
     IF I > N, GO.
```

This describes the procedure in terms much closer to a flow chart specification and also in general achieves an over-all symbolic compression. The program to translate such a language into computer code is correspondingly more complicated, however, and the resulting computer version may be considerably more inefficient than a program prepared directly in assembly language. There is an added disadvantage, which can only be appreciated through experience: the programmer is a level removed from the actual computer process, and this raises problems in the area of precise specification of computer procedures.

Further discussion of computer programming is given in Borko (1962, pp. 92–133), Green (1963, pp. 30–64, 75–99), and McCracken and Dorn (1964, pp. 1–42).

Program systems. From the previous section it may be seen that the user does not in general deal with the computer directly but through the mediation of some program system that provides for translation from procedure-oriented language into actual computer code. On the larger computers, where efficiency is at a premium, this function is expanded to include the batching of programs for sequential execution with a minimum of human intervention and the keeping of the computing log and accounting records. Such a system makes for highly efficient computer utilization but removes the user one level more from direct access to the computer; "debugging" under such circumstances, for example, requires a succession of runs, with the user progressively detecting and correcting the errors in his program.

This implies that in addition to a programming language the user must learn (or have access to someone who knows) the operating techniques associated with the particular computer installation with which he deals. The knowledge of what translators and diagnostic programs are available, and of details concerning the formats in which control information must be submitted with programs in order that they be usable, assumes a perhaps disproportionate but nevertheless critical importance. Although details vary widely from installation to installation, the type of support one may look for is indicated by the following categorization of "system routines":

loader—an elementary "job initiation" program that controls the reading of a program to be executed, perhaps supplying address modification based on the locations the program is to occupy in storage, etc.

assembler—a program that translates a program in assembly language (which may consist of several sections, each programmed using symbolic addresses and mnemonic operation codes) into a single coherent computer code (in a form that can be processed by a loader).

translator—(often called a "compiler") a program that translates a program in some algebraic or other problem-oriented language into some more elementary (perhaps assembly language) form.

monitor—a program that synchronizes translating, assembling, and loading functions and, further, has a certain batching, scheduling, and accounting capability permitting a whole set of jobs to be run with minimal human intervention.

multiprocessor—a supermonitor based on an ability to carry out several functions concurrently (for example, those of a central processor and several input/output devices) and dynamically schedule these for maximal efficiency.

The order of listing here corresponds to both historical evolution and increasing level of complexity; monitor and multiprocessing facilities, in particular, are appropriate only to the more powerful computer installations. At the time of this writing there is evolving an even more advanced program system concept, growing out of the idea of multiprocessing: an executive system that integrates one or more central processors, several auxiliary input/output processors, and a large number of remote user-operated consoles in a way so as to make efficient use of the centralized facilities but at the same time permit each user to interact directly, in an "on-line" fashion, with the computer complex. In order to achieve this, the system must be able to run programs piecemeal and respond, in what is to a human a short time, to a multiplicity of external demands. The requirements of this sort of "time-sharing" operation have led to radical new concepts in both the computer-design and programming areas.

Translators are generally identified by the problem-oriented language that they accept as input, such as Fortran, Algol (two general algebraic languages), Cobol (a standard file-processing language), Simscript (a simulation language), etc. A similar function is accomplished less efficiently but often more straightforwardly by an *interpreter* program, which both translates and executes simultaneously, thereby carrying out the whole computing task in one rather than several phases.

These system programs, along with various diagnostic programs for analyzing both computer operation and program operation, may be categorized as *service routines*; they serve the user by facilitating the execution of the particular procedure whose results he desires. This function may be contrasted with that of *standard routines* and *standard subroutines*, generally available on a "library" basis, which actually carry out all or a part of the desired task. One may anticipate the availability of subroutines for at least some of the following:

radix conversion—if the computer uses a number base other than 10, which is often the case. (This function will usually be incorporated at the assembly level.)

special arithmetic—"floating-point" (flexible scale) and "double-precision" (extended length) numeric manipulation if not directly available on the computer instruction level; also complex arithmetic, etc.

function evaluation—calculation of square root, exponential and logarithm, sine and cosine, etc. for arbitrary arguments. (A call for standard functions may be incorporated into translator specifications.)

polynomial and matrix operations—addition, multiplication by scalar, multiplication; evaluation, calculation of zeros (for polynomials); transposition, inversion, diagonalization (for matrices).

input/output editing—formating and sequencing card and paper tape reading and punching, magnetic tape, drum and disk file operations, printing and other forms of graphic display. (This function will sometimes be incorporated at the monitor level.)

By definition, a subroutine is supposed to serve as part of a larger program; its usefulness therefore depends not only on its conception with regard to generality and comprehensiveness but on its being accompanied by an appropriate set of specifications characterizing its action and the directions for "connecting" it into the program that uses it. Standard routines, by contrast, are programs for tasks that can be regarded as independent in their own right; the appropriate form of these depends very much on the applications environment. Programs for regression, factor analysis, linear programming calculations, and solution of differential equations are generally in this category, as well as those for standard file-processing operations such as sorting and input/output media conversion (card-to-tape, tape-to-card, etc.).

The importance to successful computer usage of a good available set of service routines, standard routines, and subroutines cannot be underestimated, and it is this that leads to "software" costs equaling or exceeding "hardware" costs for an installation, a fact often unanticipated by the administration that set up the budget. Some alleviation of the difficulty is provided by the existence of various "user organizations," generally associated with a particular class or type of computer, which maintain a pool of programs available to their members. These are only guaranteed up to minimum standards of documentation, however, and the problem of making effective use of work already done by others remains one of the most formidable and often exasperating problems facing the user who wishes to reap the benefits afforded by computers without investing the time to understand fully the device that he is entrusting with a major role in his research or operations.

The topics in this section are further developed in Borko (1962, pp. 140–170), Green (1963, pp. 65–74), and Gregory and Van Horn ([1960] 1963, pp. 439–478).

Error analysis

Although computer applications in general are subject to the problem that imperfect understanding of procedures by the user may lead to unsatisfactory results, computation related to familiar mathematical concepts is especially dangerous in this regard. This is because the user often starts out with some preconceptions based on his mathematical understanding and is thus likely to neglect or misjudge the effect of computational error, which may be considerable. It is customary to measure error in either absolute or relative units, depending on circumstances; thus if x is a computed value and it is assumed that it represents some "true" value x^T, then $\delta = x - x^T$, or possibly its magnitude $|\delta|$, is defined as the *absolute error* in x, and δ/x^T (which is approximated by δ/x) is defined as the *relative error*. Computational error may also be classified according to source as follows: *inherent error,* reflecting inaccuracies in initially given data; *generated error*, reflecting inaccuracies due to the necessity of rounding or otherwise truncating the numeric results of arithmetic operations; and *analytic error*, reflecting inaccuracies due to the use of a computing procedure that represents only an approximation to the theoretical result desired.

In the example of the mean and standard deviation calculation given earlier, one might wish to analyze the effect of inherent error in the input x values on the computed $\bar{x}$ and s values, that is, to analyze the manner in which errors δ_i attributed to the x_i induce errors $\delta_{\bar{x}}$ and δ_s in $\bar{x}$ and s, respectively. In the case of simple functions the effect can be estimated using the traditional techniques of calculus; in a calculation of complicated structure, however, it may be difficult even to get a feeling for the manner in which errors in input values propagate. Of course, in this type of situation, inherent errors in the data may be just what the calculation itself is concerned with, and hence it might be more appropriate to consider the input values x as exact as far as the computation is concerned. Even then it would be important, however, to have some idea of the "sensitivity" of the computed results as a function of input; one might want to know, for example, that the closer to zero $\bar{x}$ gets, the more relative effect it will show from given relative changes in x values.

In any actual implementation of the procedure diagrammed in Figure 2 there will be generated error introduced at each arithmetic step, the exact nature of which depends on the rounding rules followed by the computer arithmetic unit, which are often obscure to one who communicates only via a procedure-oriented language. Furthermore, analytic error is necessarily introduced at the stage where the square root is called for, since functions such as this can be computed only approximately. The closeness of the approximation, and hence the level at which analytic error is introduced, depends on properties of the square root procedure, which also are frequently not very well documented for the user.

When it is considered that error effects of all three types may be interwoven in a calculation, and may propagate through all the calculational steps in a complicated manner, it is seen that the problem of error assessment is nontrivial. Furthermore, it is not generally taken into account how catastrophic relative error propagation effects can be; the subtraction of two numbers each with a relative error of order 10^{-8} which happen to have a relative difference of order 10^{-6} will yield a result with relative error of order 10^{-2}, and the arithmetic procedures used on most contemporary computers are such that this loss of significance is not brought to the user's attention unless special monitoring techniques are programmed into the calculation. And while it is generally true that both generated and analytic error can be minimized by computational techniques (carrying more precision and using more refined approximations), an assessment is required to know when such measures should be resorted to. Poor computational technique can magnify the effect of generated and analytic error unnecessarily; a typical case is use of the formula $\sum x_i^2 - n\bar{x}^2$ instead of $\sum(x_i - \bar{x})^2$ for calculating the standard deviation, since then the result tends to depend on subtraction of two large and nearly equal numbers. Finally, inherent error, as its name implies, is dependent only on the level of input error and the functional connection between input and output values, neither of which are subject to remedial action in the computational context. When small relative errors in input induce large relative errors in output, a process is said to be "ill-conditioned"; if this obtains, the most one can hope for is a means of detecting the situation.

The subject of error analysis, in its several facets, has been extensively studied; although no comprehensive set of directives can be given, the user will be well repaid for becoming familiar with at least some of the work already done in this area, as indicated in McCracken and Dorn (1964, pp. 43–64) and Hamming (1962, pp. 24–40).

ROBERT L. ASHENHURST

[*See also* SIMULATION.]

BIBLIOGRAPHY

BORKO, HAROLD (editor) 1962 *Computer Applications in the Behavioral Sciences.* Englewood Cliffs, N.J.: Prentice-Hall.

BROOKS, FREDERICK P. JR.; and IVERSON, KENNETH E. 1963 *Automatic Data Processing.* New York: Wiley.

CALINGAERT, PETER 1965 *Principles of Computation.* Reading, Mass.: Addison-Wesley.

GREEN, BERT F. JR. 1963 *Digital Computers in Research: An Introduction for Behavioral and Social Scientists.* New York: McGraw-Hill.

GREGORY, ROBERT H.; and VAN HORN, RICHARD L. (1960) 1963 *Automatic Data Processing Systems.* 2d ed. Belmont, Calif.: Wadsworth.

HAMMING, R. W. 1962 *Numerical Methods for Scientists and Engineers.* New York: McGraw-Hill.

MCCRACKEN, DANIEL D.; and DORN, WILLIAM S. 1964 *Numerical Methods and FORTRAN Programming.* New York: Wiley.

YOUDEN, W. J. 1965 *Computer Literature Bibliography: 1946–1963.* National Bureau of Standards, Miscellaneous Publication No. 266. Washington: U.S. National Bureau of Standards.

Postscript

Advances in computing methodology have been many and varied. Describing even those developments that affect only the research user or the general social milieu would require a great deal of space. Hence the present treatment must be only a sketch. Developments will be considered under four headings: "Computer systems," "Operating systems," "Applications systems," and "Computers and society."

Computer systems. Developments in computer system hardware over the past few years have improved technological effectiveness in ways far transcending computation in the narrow sense. In addition to improvements in components at all levels (processors are cheaper and faster, memories are cheaper and larger, etc.), a significant degree of what may be called configuration flexibility has been achieved. It is now less appropriate to think of "a computer" than a *computer system* consisting of interconnected units of functional types: *processor/memory, input/output, mass storage, remote transmission,* with various *interface* units performing mediating control functions among them (see "General-purpose digital computer" in the main article). A simple processor/memory with an input/output unit directly attached is a computer in the traditional sense (see the block diagram, Figure 1, in the main article). Mass storage includes the drum, disk, and magnetic tape units, along with some more recently developed specialized devices of truly massive storage capacity (for example the trillion-bit laser store). One speaks of the *central processing unit* (CPU) surrounded by *peripheral units.*

Remote transmission devices permit peripheral units to be interconnected over substantial distances so that, for example, the input/output for a computer system may be dispersed and located far from its main processor/memory and mass storage units. Thus one or more *remote job entry stations,* each with a card reader and printer, may be associated with a general-purpose computer system at a central location. This increased modularization permits a computer system to be tailored to a particular kind of processing application (for example, heavy computing versus heavy input/output). If this were the only function of modularization, the result would be merely to improve computational efficiency and user convenience. A difference in kind emerges, however, with the connection of *remote terminals* that permit manual or device interaction with the central computer system, rather than mere job submission. New modes of computation must be defined to take advantage of this capability, as discussed in the next section.

Further possibilities, only beginning to be exploited, are offered by the formation of *computer networks*—the interconnection of complete computer systems (each with a distinct processor/memory) located at dispersed points. Networks improve computing for the user in a variety of ways. For example, a set of independent general-purpose computer systems may be connected to permit program and data sharing among users at the different sites; "national networks" are of this variety. Another configuration is a set of less powerful, but still general-purpose, systems that communicate with a more powerful central system; these "regional networks" are implemented by such organizations as state university systems. Such interconnection modes can of course be realized on a more restricted geographical basis also, but more often a "local network" is not of such a general character. Instead it consists of a single general-purpose system connected to a number of special-purpose input/output devices, such as versatile terminals, with a typewriter unit and perhaps a graphic display; or data acquisition systems interfaced with laboratory or other devices monitoring realworld phenomena.

Coupled with this configuration flexibility, the development of smaller and less expensive processor/memory devices and (to some extent) peripheral units, have made the range of configuration options enormous. These smaller devices have also rendered decentralized computing (networks or individual computer systems) once again competitive with centralized computing (large CPU, with perhaps remote peripheral units). The *minicomputer system* of today is as powerful as the large general-purpose system of a few years ago, at a small fraction of the price. Thus one sees a proliferation of minicomputer systems (perhaps in networks) even within a single organization not dispersed geographically. The *microprocessor*, a self-contained processor/memory of limited capacity, can function as a versatile, programmed control device in circumstances that only extreme miniaturization makes possible, such as traffic and building control devices.

These developments have not only changed the cost-effectiveness of computing and control in a more traditional sense, they have enormously increased the feasibility of interactive or real-time computing, where a human or device interacts with the computing process through a remote terminal or an interfacing device. This has inevitably affected the use of computing equipment and in particular has changed the character of operating systems, the software through which the computer systems are made responsive to applications requirements. This is considered in the next section.

An overview and perspective on computers and computer systems is given by Davis (1977), in a special issue of *Science* devoted to the electronics revolution. Hardware configuration possibilities are discussed in Booth (1973, chapter 5) and the particular attributes of interactive communication devices in Meadow (1970, chapter 2). Many aspects of institutional networking are covered in Greenberger et al. (1974). Minicomputers and microprocessors, in the context of scientific applications, are considered in Robinson (1975), Dessy (1976), and Arnold (1976).

Operating systems. The main article refers briefly to the evolution of *operating systems,* from the simple program loader through the complex multiprocessor, the latter being based on the capability of computer systems to carry on a multiplicity of functions concurrently (see "Program systems" in the main article). The terms "on-line" and "time-sharing" there serve to describe the resulting system from the user's point of view. Since that writing such systems have become commonplace, and their technology highly developed. It should suffice to distinguish two main types of computing, by *job* and by *demand,* and further to subdivide the demand category into that furnishing *real-time* response and that furnishing *minimal-delay* response. Job-oriented computing refers to the classical operation where jobs are submitted in entirety, batched for sequential execution, and the results returned some time later. The term "batch" is still used to describe this mode, although the execution is no longer strictly sequential, since in fact many jobs are processed concurrently. Thus the computer system now functions more efficiently, but to the user waiting for results it seems much the same (except that programs and data that are used repeatedly may be stored on disk and so invoked as part of the job control, and jobs may be submitted from remote job entry stations instead of being handed over an input counter). Demand-oriented computing, however, presents a new dimension. Programs may be constructed, debugged, and run interactively; programs and data may be edited in a similar fashion; and programs themselves may be designed to have interaction points in them as they run, where data may be introduced and/or produced at an input/output device. When this interaction involves input/output devices mediating the functioning of some physical process running on its own time scale, or some human process where a time scale is arbitrarily impressed, the demand response is termed real-time. When the time scale is not fixed, and the interaction is with a human user who can wait for a response (but will not wait forever) the response is termed minimal-delay.

Other terms used to describe demand computing are "interactive" and "on-line," both of which suggest the response from the user (human or non-human) point of view. The term "time-sharing," which is also used, really expresses how a computer and operating system manage to respond to many users with the appearance of concurrency. In fact, the basic speed of the hardware is much faster than that of the demand-response user, and hence the operating system can be programmed to handle multiple demands by switching attention rapidly among them. Thus it is simple for a computer system to interact with a single user on a demand basis, but efficiency considerations usually dictate that the interaction be with many users. Terms such as *multiplexing*, *multiaccessing*, and *spooling* (from SPOOL— Simultaneous Peripheral Operations On Line) refer to particular mechanisms for achieving concurrent operation, while the more generic terms *multiprogramming* (one processor running a multiplicity of programs by switching among them intermittently) and *multiprocessing* (more than one processor running in some sort of coordinated fashion with intercommunication) describe the underlying techniques.

These technical developments, however, are useful only insofar as they can be employed to advantage in applications, and to see how this can be done required some attention to the environment as well as to the content of applications. This is the subject of the following section.

A perspective on operating systems and software is given in three articles in the special issue of *Science* previously mentioned: Madnick (1977), Mills (1977), and Holt and Stevenson (1977). Aspects of software configurations are discussed in Booth (1973, chapters 6–12), and for the particular case of conversational time-sharing in Meadow (1970, chapters 3–4).

Applications systems. Work in applications, in the context of academic research and instruction, of course still depends on the availability of appropriate application programs (see "Computer procedures" and "Computer programs" in the main article). Improvement in computer and operating systems, as well as in high-level language capabilities, has continued to make easier the task of programming applications, or getting them programmed.

The term "applications systems" is used here to suggest that computation may be increasingly thought of in terms of systems or packages that do not require the user to write programs, even in a higher-level language. Rather, the user interacts (often on a demand basis) with a set of standard routines that carry out well-defined procedures (for example, statistical calculations) upon being supplied the necessary data and parameters (the latter includes all procedural options, as well as auxiliary quantities, that need specification). Even when the user must write some program statements, improvements in libraries of standard subroutines (for example, for matrix manipulation) render the task much simpler than before.

Of particular interest to researchers in the social sciences are two varieties of applications systems, *statistical processing systems* and *research database systems*. The former are generalized collections of routines that perform the standard calculations of statistics, presumably using methods that attend to computational efficiency and error control (see "Error analysis" in the main article), and also permit the user to generate and refer to files of data on which the calculations are run. Research database systems are large collections of data with associated routines for retrieving individual records or generating summary reports. These kinds of systems might seem similar (summary reports are obtained by statistical calculations, and a database is a standardized file of data), but the emphases are quite different. The crucial distinction is that the data content forms an intrinsic part of the database system (hence its name), and the responsibility for maintaining the body of information is an explicit and ongoing task associated with the system. By contrast, developers of a statistical processing system seek to introduce good computational and file-handling routines, but the contents of the data files are supplied by, and are the responsibility of, each individual user.

These kinds of systems form an example of more

general information systems that have become an integral part of many nonacademic organizational activities. Business and government have come to depend on such systems, and they account for much of the social impact of the computer, as discussed in the concluding section.

Before turning to a more general survey of such orthodox applications, however, developments should be briefly mentioned in the area of *cognitive simulation* and *artificial intelligence*. These terms refer to attempts to imitate human thought processes, or perhaps only produce similar results by different means. Work on systems with formalized rules, such as game playing and theorem proving, and more general systems involving natural language processing, has taken place under this rubric. Although advances in understanding certain kinds of symbolic processes have been made, and even in some cases incorporated into systems designed for applications (for example, algebraic formula manipulation and document retrieval by abstract), the general goals of artificial intelligence research still seem elusive.

Various aspects of general application systems are considered in three *Science* articles: Balderston, Carman, and Hoggatt (1977), Evans (1977), and Potter (1977). General applications in research are covered in Baker et al. (1977). Statistical applications are covered in Sterling and Pollack (1968). Slysz (1974) surveys types of statistical processing systems available, while the University of California Health Sciences Computing Facility (1964), Nie, Bent, and Hull (1970), and Roberts (1974) describe particular systems. A committee on evaluation of statistical software was established by the Statistical Computing Section of the American Statistical Association in 1974 (see Francis et al. 1975). Since then it has encouraged evaluation of statistical packages; these are beginning to appear in the annual *Symposium on the Interface of Computer Science and Statistics* and in the *Proceedings* of the Statistical Computing Section of the American Statistical Association. (See Berk & Francis 1978; Muller 1978; and the discussion following these two companion articles.) Similar evaluative activities have been undertaken by professional organizations in other countries, and these have prompted the formation of the International Association for Statistical Computing as a branch of the International Statistical Institute; the *Proceedings* of this organization will also publish evaluations. Research database systems for the social sciences is the subject of Bisco (1970). Artificial intelligence is covered in Feigenbaum and Feldman (1963) and Minsky (1969), and a penetrating critique of its progress and prospects is given by Dreyfus (1972).

Computers and society. A great deal has been written about the supposed role of the computer as either a boon or a threat to society. Much of this has been poorly considered or speculative in terms of actual social impact, and only recently have more reasoned studies begun to appear, including some scholarly contributions. Meanwhile, tremendous advances have occurred in the development and integration of information systems into organizational practice. There has also been relevant activity on the legislative and judicial fronts. The present treatment can give only the barest outline.

The most clearly perceived computer-related threats to the individual arise from the way information systems are used by either government agencies or private organizations in dealing with the public they supposedly serve. Two basic informational activities involved in almost all such interactions are *record keeping* and *transaction monitoring*. For some government agencies, a recording function (for example, disease registration) is a basic mission; others combine recording with service functions of a transactional nature (for example, welfare payments). The widespread consolidation of government information into computer-based "databanks," and possible interagency exchange of information, have been prominent areas of public concern. This concern has been accentuated by the fact that the use of databanks has been most noticeable in the areas of health, education, and welfare, where information tends to be personal. Many commercial enterprises combine the two functions (for example, subscription fulfillment and credit extension). In these situations individuals, as clients or customers, may feel powerless in the face of the information system; and if they prefer not to be a part of it, their options are limited. Airline reservation systems furnish a good, and in fact rather noncontroversial, example.

The financial industries—banking, brokerage, and insurance—have been among the first to develop extensive data-processing facilities. The initial systems were internal, but there has since been an increase in computerizing the interaction with the public. Repeated predictions of a "checkless, cashless society" are now possible based on the concept of an EFT system—*E*lectronic *F*unds *T*ransfer. Under such a system the exchange of tangible instruments such as checks or cash is replaced by computer-systematized debits or credits in a comprehensive information system maintained (presumably) by the banks. Thus "paychecks" may be automatically deposited and "purchases" automatically deducted from an individual's "bank account." The prospect of EFT is seen as a boon by the banks and the retail industry, and as a threat by those who regard it as a further encroachment on individual privacy. In particular, information in the system allows inferences to be made regarding an individual's activities in a wide sphere, not just financial. Such information—where a person has been and with whom dealing—could be adversely used in the wrong hands.

Another three-letter concept that affects the individual as consumer is the POS system—Point Of Sale. Here the computer terminal replaces not only the cash register (as it has done already in the larger chain stores) but the whole inventory maintenance and price control system as well. A POS system requires that coded machine-readable labels be placed on purchase items, a process that renders the usual price-marking superfluous from the store's operational point of view. Thus, POS systems are attractive to the retailers (who agreed to introduce the machine-readable labeling with surprisingly little extended discussion), but they put difficulties in the path of the rational shopper.

These issues have sometimes been framed in terms of the need for humanizing information systems, but it is of course complicated to assess the facets of *humanity* as a quality.

One set of such questions involves the notion of *privacy*, a notion that generally applies to persons independent of the functioning of organizational information systems. Privacy guidelines translate into questions of information system design via the two related concepts of *confidentiality* (who is or is not authorized access to information, according to privacy guidelines), and *security* (the actual incorporation into an information system of procedures, both computer-based and manual, for enforcing the confidentiality dictates). Privacy, confidentiality, and security are primarily associated with the record-keeping aspects of information systems.

Concepts associated with transaction monitoring are less well delineated, but one could take *propriety* as a basic quality relating to the treatment of individuals who are (more or less) forced to deal with an organization. Propriety guidelines serve to avoid the host of behaviors attributed to "the computer" in jokes and cartoons (unresponsiveness to requests for clarification, etc.). Propriety guidelines get translated into questions of system design via the two related concepts of *congruity* (what should be the specific response of the system to various situations that may arise) and *integrity* (the actual incorporation into an information system of procedures, both computer-based and manual, for enforcing the congruity dictates).

Since the consequences of the information-processing revolution were first perceived, with varying degrees of clarity, there has been a call for action in the legislative and judicial arenas to protect the public from one or another evil consequence (for the most part, protection of business and other organizational interests was thought to be adequately provided by extension and interpretation of existing legal doctrine). Although the initial thrust was in the area of privacy, propriety is also beginning to be addressed.

The landmark of privacy legislation in the United States is the Privacy Act of 1974, which sets forth rules governing practices in maintaining government databanks on citizens. The act prohibits databanks whose very existence is secret, and declares that the individual should have access to personal information in government databanks except those maintained for security or police purposes, or in certain other exceptional cases. Procedures for redress of grievances resulting from inaccurate information are also set down. Finally, guidelines for confidentiality are given, governing release of data to individuals and between agencies. In this respect the act apparently counters the disclosure provisions of the earlier Freedom of Information Act, which was designed to make proceedings within bureaucratic units open to the public. Indeed, many agencies opposed both acts without inconsistency. At this writing, similar legislation is under consideration for the private sector, and related legislation on the state and local levels is beginning to appear. The federal Privacy Protection Study Commission has also been established. Similar trends are noted in other countries, and the Swedish National Data Act was the first such legislation passed.

The regulation of criminal justice and security intelligence information systems poses a problem since these systems are by nature exceptions to the general rules. In addition to the obvious possibilities of privacy infringement, there have been instances of insufficient attention to propriety—for example, police pursuit and detainment based on erroneous records. In the private sector, no such blanket legislation has been attempted (which, in view of the conceptual difficulties, is perhaps just as well). But certain areas have been singled out for legislative treatment. An early landmark here is the Fair Credit Reporting Act of 1972, which includes privacy provisions and spells out the beginnings of a consumer-credit protection code. A much greater step is represented by the Fair Credit Billing Act of 1975, which sets forth procedures by which credit card companies and retail firms extending credit must interact with customers concerning complaints and disputes.

Such legislation is so new that there has been little opportunity for judicial test. The court cases have been mainly concerned with the business and commercial area, and as mentioned above, are based on more traditional law. The most spectacular court case is the IBM antitrust case of 1974, but it is computer-related only incidentally, by virtue of the principal product of IBM. There have been reported cases where a company sues a computer vendor for failure of information systems to live up to promise, and questions of patent and copyright law applied to computer systems in nontraditional ways are under scrutiny. New and ingenious ways to perpetrate fraud and other types of crime, always a specter raised by the doomsayers, are also appearing.

Computers and public policy can now be dealt with substantially as a legitimate area for scholarly study, as an increasing amount of literature shows.

General perspectives on computers in relation to man and society are given in Weizenbaum (1972) and Simon (1977). The influence of computer technology on organizations is studied in Whisler (1970) and Laudon (1974). The general question of the computer threat to society is discussed in Wessel (1974), and the more particular aspects of databanks and privacy are comprehensively considered in Miller (1971) and Westin and Baker (1972). The report of the DHEW Secretary's Advisory Committee on Automated Personal Data Systems (1973) lays the groundwork for the federal Privacy Act of 1974, and the work of the Privacy Protection Study Commission. Guidelines for humanizing information systems are set forth in Sterling (1975), and a framework for the consideration of information and public policy is outlined in McCracken (1974). Two discussions of a particular current public policy program, the relation between computer and communication systems, are discussed in Farber and Baran (1977) and Irwin and Johnson (1977). Finally, four books that should be useful for general perspectives on issues of computers and society are Sackman and Borko (1972), Gotlieb and Borodin (1973), Weizenbaum (1976), and Mowshowitz (1976).

ROBERT L. ASHENHURST

ADDITIONAL BIBLIOGRAPHY

ARNOLD, JAMES T. 1976 Microprocessor Application: A Less Sophisticated Approach. *Science* 192:519–523.

BAKER, W. O. et al. 1977 Computers and Research. *Science* 195:1134–1139.

BALDERSTON, F. E.; CARMAN, JAMES M.; and HOGGATT, AUSTIN C. 1977 Computers in Banking and Marketing. *Science* 195:1115–1119.

BERK, KENNETH N.; and FRANCIS, IVOR S. 1978 A Review of the Manuals for BMDP and SPSS. *Journal of the American Statistical Association* 73:65–71. → A companion article to Muller (1978), which follows, as do several comments and rejoinders.

BISCO, RALPH L. (editor) 1970 *Data Bases, Computers and the Social Sciences.* New York: Wiley. → Based on papers presented at the fourth annual conference of the Council of Social-Science Data Archives held at the University of California, Los Angeles, June 1967.

BOOTH, GRAYCE M. 1973 *Functional Analysis of Information Processing: A Structured Approach for Simplifying Systems Design.* New York: Wiley.

CALIFORNIA, UNIVERSITY OF, LOS ANGELES, HEALTH SCIENCES COMPUTING FACILITY (1964) 1975 *BMD Biomedical Computer Programs.* 2d ed. Edited by Wilfrid J. Dixon. Berkeley: Univ. of California Press.

DAVIS, RUTH M. 1977 Evolution of Computers and Computing. *Science* 195:1096–1102.

DESSY, RAYMOND E. 1976 Microprocessors? An End User's View. *Science* 192:511–518.

DREYFUS, HUBERT L. 1972 *What Computers Can't Do: A Critique of Artificial Reason.* New York: Harper & Row.

EVANS, LAWRENCE B. 1977 Impact of the Electronics Revolution on Industrial Process Control. *Science* 195:1146–1151.

FARBER, DAVID; and BARAN, PAUL 1977 The Convergence of Computing and Telecommunications Systems. *Science* 195:1166–1170.

FEIGENBAUM, EDWARD A.; and FELDMAN, JULIAN (editors) 1963 *Computers and Thought.* New York: McGraw-Hill.

FRANCIS, I.; HEIBERGER, R. M.; and VELLEMAN, P. F. 1975 Criteria and Considerations in the Evaluation of Statistical Program Packages. *American Statistician* 29:52–56.

GOTLIEB, CALVIN C.; and BORODIN, ALLAN 1973 *Social Issues in Computing.* New York: Academic Press.

GREENBERGER, MARTIN et al. (editors) 1974 *Networks for Research and Education: Sharing Computer and Information Resources Nationwide.* Cambridge, Mass.: M.I.T. Press.

HOLT, H. O.; and STEVENSON, F. L. 1977 Human Performance Considerations in Complex Systems. *Science* 195:1205–1209.

IRWIN, MANLEY R.; and JOHNSON, STEVEN C. 1977 The Information Economy and Public Policy. *Science* 195:1170–1174.

LAUDON, KENNETH C. 1974 *Computers and Bureaucratic Reform: The Political Functions of Urban Information Systems.* New York: Wiley.

MCCRACKEN, DANIEL D. (editor) 1974 A Problem-list of Issues Concerning Computers and Public Policy. Association for Computing Machinery, *Communications of the ACM* 17:495–503. → A report of the ACM Committee on Computers and Public Policy.

MADNICK, STUART E. 1977 Trends in Computers and Computing: The Information Utility. *Science* 195: 1191–1199.

MEADOW, CHARLES T. 1970 *Man–Machine Communication.* New York: Wiley.

MILLER, ARTHUR R. 1971 *The Assault on Privacy: Computers, Data Banks, and Dossiers.* Ann Arbor: Univ. of Michigan Press.

MILLS, HARLAN D. 1977 Software Engineering. *Science* 195:1199–1205.

MINSKY, MARVIN L. (editor) 1968 *Semantic Information Processing*. Cambridge, Mass.: M.I.T. Press.

MOWSHOWITZ, ABBE 1976 *The Conquest of Will: Information Processing in Human Affairs*. Reading, Mass.: Addison-Wesley.

MULLER, MERVIN E. 1978 A Review of the Manuals for BMDP and SPSS. *Journal of the American Statistical Association* 73:71–80. → A companion article to Berk and Francis (1978), followed by several comments and rejoinders.

NIE, NORMAN H.; BENT, DALE H.; and HULL, C. HADLAI (1970) 1975 *Statistical Package for the Social Sciences*. 2d ed. New York: McGraw-Hill.

POTTER, ROBERT J. 1977 Electronic Mail. *Science* 195: 1160–1164.

ROBERTS, HARRY V. 1974 *Conversational Statistics*. Edited by Christine Doerr. Cupertino, Calif.: Hewlett-Packard.

ROBINSON, ARTHUR L. 1975 Multiple Minicomputers: Inexpensive and Reliable Computing. *Science* 187: 337–338.

SACKMAN, HAROLD; and BORKO, H. (editors) 1972 *Computers and the Problems of Society*. Montvale, N.J.: AFIPS Press.

SIMON, HERBERT A. 1977 What Computers Mean for Man and Society. *Science* 195:1186–1191.

SLYSZ, WILLIAM D. 1974 An Evaluation of Statistical Software in the Social Sciences. Association for Computing Machinery, *Communications of the ACM* 17:326–332.

STERLING, THEODOR D. 1975 Humanizing Computerized Information Systems. *Science* 190:1168–1172.

STERLING, THEODOR D.; and POLLACK, SEYMOUR V. 1968 *Introduction to Statistical Data Processing*. Englewood Cliffs, N.J.: Prentice-Hall.

U.S. DEPARTMENT OF HEALTH, EDUCATION, AND WELFARE, SECRETARY'S ADVISORY COMMITTEE ON AUTOMATED PERSONAL DATA SYSTEMS 1973 *Records, Computers, and the Rights of Citizens: Report*. DHEW Publication No. (05) 73-94. Washington: Government Printing Office; Cambridge, Mass.: M.I.T. Press.

WEIZENBAUM, JOSEPH 1972 On the Impact of the Computer on Society. *Science* 176:609–614.

WEIZENBAUM, JOSEPH 1976 *Computer Power and Human Reason: From Judgment to Calculation*. San Francisco: Freeman.

WESSEL, MILTON R. 1974 *Freedom's Edge: The Computer Threat to Society*. Reading, Mass.: Addison-Wesley.

WESTIN, ALAN F.; and BAKER, MICHAEL A. 1972 *Databanks in a Free Society: Computers, Record-keeping and Privacy*. New York: Quadrangle.

WHISLER, THOMAS L. 1970 *Information Technology and Organizational Change*. Belmont, Calif.: Wadsworth.

COMPUTERS

See COMPUTATION; OPERATIONS RESEARCH; SIMULATION; *and the biographies of* BABBAGE; VON NEUMANN; WIENER.

CONCENTRATION CURVES

See GRAPHIC PRESENTATION; STATISTICS, DESCRIPTIVE, *article on* LOCATION AND DISPERSION; *and the biography of* MAHALANOBIS.

CONCEPTS AND INDICES

See SURVEY ANALYSIS.

CONDORCET

Marie Jean Antoine Nicolas Caritat, marquis de Condorcet (1743–1794), was a French mathematician, philosopher, and politician, the author of a philosophy of progress, of a program for educational reform, and one of the first to apply the calculus of probabilities to the analysis of voting and to social phenomena in general. He was among the most original thinkers of the revolutionary age.

Born at Ribemont in Picardy, Condorcet had a brilliant scholastic career with the Jesuits at Rheims and at the Collège de Navarre in Paris. He attracted the attention of such mathematicians as d'Alembert, Clairaut, Fontaine, and Lagrange, although his mathematical publications had no permanent importance. At the same time he attended the salon of Mlle. de Lespinasse, became a friend of Turgot's, and, with d'Alembert, made a pilgrimage in 1770 to Fernet to see Voltaire. He remained on excellent terms with Voltaire. He married Sophie de Grouchy in 1787; their salon at the Hôtel des Monnaies became one of the most brilliant in Paris.

Condorcet pursued three careers simultaneously and with varying success—an academic one, an administrative one, and a political one. As assistant to the secretary of the Académie des Sciences from 1769 on, he wrote a large number of *éloges* and recorded proceedings; he was elected a member of the Académie Française in 1782. As an administrator, he was appointed *inspecteur des monnaies* in 1774 and entrusted with various scientific missions by Turgot, then the minister of finance. Finally, in the context of politics, he drew up, in 1789, the petition of the nobility of Nantes (one of the *cahiers de doléances*). As a deputy to the Legislative Assembly, he was an active member of the Committee on Public Education. After being elected to the Convention, he was chosen to prepare the Girondist draft for the constitution, but although his proposals were almost always passed on the floor, they were very rarely put into effect. In 1793 he shared the fate of the Girondins: his arrest was ordered in July 1793, but he managed to remain hidden in Paris until March of the following year; then he was arrested, and he died under mysterious circumstances in the prison of Bourg la Reine.

An enlightened nobleman. Prominent in scientific, political, and worldly circles, Condorcet is a typical example of the enlightened segment of the nobility that supported the Revolution. He was one of the first to be converted to the idea of republicanism. His ideology of social progress, his economic liberalism, and his faith in the omnipotence of rational knowledge allied him with the rising bourgeoisie rather than with his own class.

Condorcet might well be called the last of the Encyclopedists. He took an active part in the publication of the *Supplément à l'encyclopédie*, and more particularly in recasting its mathematical portion, the *Encyclopédie méthodique*, in 1784–1785. His curiosity was universal, and his most characteristic effort was his attempt to join and coordinate mathematics and philosophy and thus "satisfy two passions at once," as he wrote to Frederick the Great. He considered philosophy to include everything relating to the knowledge of man: "the metaphysical and social sciences, those that have man himself as their object . . ." (1847–1849, vol. 6, p. 494). His epistemological principles were borrowed from Condillac, but Condorcet differed greatly from Condillac in his conception of man's destiny and the fate of human societies.

Condorcet's best-known work, the *Sketch for a Historical Picture of the Progress of the Human Mind* (1795), was written while he was in hiding in Paris in 1793–1794. In it he presented a history of the errors and the advances of humanity, in order to predict, direct, and accelerate its forward march. In Condorcet's view, historical development coincides with the spread and triumph of the light of reason.

Condorcet believed that education, more than anything else, produced the triumph of the Enlightenment. He regarded inequality of education as one of the main sources of tyranny and advocated public education that would offer to all who might benefit from it the "aid hitherto confined to the children of the rich." Nonetheless, his interests remained those of the bourgeoisie and the enlightened nobility, and his educational recommendations consist, in effect, of two separate programs based on class distinctions; one is for the lower classes and is essentially technical; the other is for the ruling classes and develops the critical faculties of future citizens.

"Social mathematics." An essentially new feature of Condorcet's educational program was the importance he gave to science at every level of instruction, in particular to applied science. In addition, he believed that the social sciences should be taught in institutes and *lycées*. One of his central ideas is the importance to the progress of humanity of developing an "art of society" (*art social*).

In essence, Condorcet conceived of this *art social* as an "application of mathematics to the moral sciences," a discipline at once empirical and deductive, making use of genuine mathematical models of human phenomena. Condorcet was convinced that "the truths of the moral and political sciences can be as certain as those that make up the system of the physical sciences" (1785, p. 1). However, he believed that this certainty generally does not apply to causal relationships but to probable connections. The primary concern of what he called "social mathematics" is the application of the calculus of probability to the description and prediction of human phenomena.

This mathematics, as it emerges in Condorcet's published writings and unpublished manuscripts, comprises a statistical description of societies, an economic science (physiocratic in inspiration, but oriented toward the more recent idea of collective welfare), and a combinatory and probabilistic theory of intellectual operations.

This last theory appears in connection with the theory of voting, which is by far the best-developed part of social mathematics (1785). Voting was viewed by Condorcet as making manifest not so much a compromise among a number of conflicting forces as a *true* opinion. He assumed, therefore, that the question being voted on has a true solution that is independent of the wishes of those voting and that these voters express in their individual choices their greater or lesser understanding of that truth. The problem of structuring the process of voting is that of producing the maximum probability of a collective choice of the true solution; this may be done by varying the size of the voting body and the kind of majority required, as well as by considering the likelihood that each voter will make a correct decision. Condorcet was thus led to the construction of statistical models of voting bodies that permit the appraisal of the probability that collective decisions will correspond to the true answers to particular problems. This is, then, a real problem in operational research—how to establish voting procedures such that the chances of the emergence of correct decisions are maximized.

Since he conceived of voting as a collective search for the truth, Condorcet had to deal with the problem of defining precisely what a collective decision or judgment is. Given a series of dichotomous questions, the "yes" or "no" responses to them may be called "judgments"; how, then, may a single collective judgment or a coherent hierarchy of judgments be established on the basis of a set of individual judgments? Charles de Borda, 1733–1799, had shown that in the choice of the "best" among three candidates the one so designated by a simple majority might not be the same as the one arrived at when each is compared with the two others. Condorcet showed that such a weighing of preferences may reveal a circular order of preferences among the candidates—$A > B > C > A$—and, consequently, the possibility of an inconsistent collective choice even when individual choices are consistent. The mathematical apparatus available to Condorcet was too crude for him to obtain results that could be applied empirically, but he did open the way to a highly original and important social scientific conception.

○ **Influence.** The ideology of progress of Condorcet's *Esquisse* directly influenced Auguste Comte. His influence on the application of mathematics to human affairs can be found in such nineteenth-century works as those of Poisson and Cournot. The questions involved in the formation of a collective opinion have been taken up again by, among others, Georges Th. Guilbaud, Kenneth J. Arrow, and Duncan Black, although they have been reformulated in the light of welfare economics and the theory of games.

GILLES-GASTON GRANGER

[*For discussion of the subsequent development of Condorcet's ideas, see* GAME THEORY; MODELS, MATHEMATICAL; *and the biographies of* COURNOT; POISSON. *See also* Arrow 1968; König 1968.]

WORKS BY CONDORCET

1785 *Essai sur l'application de l'analyse à la probabilité des décisions rendues à la pluralité des voix.* Paris: Imprimerie Royale.

(1795) 1955 *Sketch for a Historical Picture of the Progress of the Human Mind.* New York: Noonday. → First published as *Esquisse d'un tableau historique des progrès de l'esprit humain.*

►(1795–1849) 1974 *Mathématique et société.* Edited by Rohsoli Rashed. Paris: Hermann.

1847–1849 *Oeuvres.* 12 vols. Edited by A. Condorcet O'Connor and M. F. Arago. Paris: Firmin-Didot.

SUPPLEMENTARY BIBLIOGRAPHY

ARAGO FRANÇOIS 1879 Condorcet: A Biography. Pages 180–235 in Smithsonian Institution, *Annual Report of the Board of Regents, 1878.* Washington: The Institution.

►ARROW, KENNETH J. 1968 Economic Equilibrium. Volume 4, pages 376–389 in *International Encyclopedia of the Social Sciences.* Edited by David L. Sills. New York: Macmillan and Free Press.

►BAKER, KEITH M. 1967 Scientism, Elitism and Liberalism: The Case of Condorcet. Pages 129–165 in *Studies on Voltaire and the Eighteenth Century.* Geneva: Institut et Musée Voltaire, Les Délices.

►BAKER, KEITH M. 1975 *Condorcet: From Natural Philosophy to Social Mathematics.* Univ. of Chicago Press.

►BLACK, DUNCAN 1948 The Decision of Committees Using a Special Majority. *Econometrica* 16:245–261.

CAHEN, LÉON 1904 *Condorcet et la révolution française.* Paris: Alcan.

GRANGER, GILLES-GASTON 1956 *La mathématique sociale du marquis de Condorcet.* Paris: Presses Universitaires de France. → Contains a comprehensive bibliography of Condorcet's mathematical works.

GUILBAUD, GEORGES TH. 1952 La théorie de l'intérêt général. *Économie appliquée* 4:501–584.

►KÖNIG, RENÉ 1968 Comte, Auguste. Volume 3, pages 201–206 in *International Encyclopedia of the Social Sciences.* Edited by David L. Sills. New York: Macmillan and Free Press.

TODHUNTER, ISAAC (1865) 1949 *A History of the Mathematical Theory of Probability From the Time of Pascal to That of Laplace.* New York: Chelsea.

CONFIDENCE INTERVALS AND REGIONS

See under ESTIMATION; *see also* LINEAR HYPOTHESES, *article on* MULTIPLE COMPARISONS.

CONFIDENTIALITY

See ETHICAL ISSUES IN THE SOCIAL SCIENCES; PUBLIC POLICY AND STATISTICS.

CONFLICT RESOLUTION

See GAME THEORY.

CONFOUNDING

See EXPERIMENTAL DESIGN.

CONGESTION

See QUEUES.

CONTINGENCY TABLES

See COUNTED DATA; STATISTICS, DESCRIPTIVE, *article on* ASSOCIATION; SURVEY ANALYSIS.

CONTINUOUS DISTRIBUTIONS

See under DISTRIBUTIONS, STATISTICAL.

CONTROL

See QUALITY CONTROL, STATISTICAL, *and the biography of* SHEWHART.

CONTROL CHARTS

See QUALITY CONTROL, STATISTICAL, *article on* PROCESS CONTROL.

CORRELATION

See under MULTIVARIATE ANALYSIS; *see also* FACTOR ANALYSIS AND PRINCIPAL COMPONENTS; NONPARAMETRIC STATISTICS, *article on* RANKING METHODS; STATISTICS; STATISTICS, DESCRIPTIVE, *article on* ASSOCIATION.

CORRELOGRAM

See TIME SERIES.

COUNTED DATA

Counted data subject to sampling variability arise in demographic sampling, in survey research, in learning experiments, and in almost every other branch of social science. The counted data may relate to a relatively simple investigation, for example, estimating sex ratio at birth in some specified human population, or to a complex problem, investigating the interaction among qualitative responses of animals to stimuli in a physiological experiment. Further, a counted data approach is sometimes useful even when the actual data are inherently not counted; for example, a classical approach to so-called goodness of fit uses the counts of numbers of continuous observations in cells or intervals. Again, some nonparametric tests are based on a related device. [*See* GOODNESS OF FIT; NONPARAMETRIC STATISTICS.]

Investigations leading to counted data are often described by giving percentages of individuals falling in the various categories. It is essential that the total numbers of individuals also be reported; otherwise reliability and sampling error cannot be estimated.

The structure of this article is as follows. First, simple procedures relating to one or two sample percentages are considered. These procedures exemplify the basic chi-square approach; they may be regarded as methods for treating particular contingency tables in a way falling in the domain of the basic chi-square theorem. Second, special aspects of contingency tables are considered in some detail: power, single degrees of freedom, ordered alternatives, dependent samples, measures of association, multidimensional contingency tables. Under the last topic is considered the important topic of three-factor interactions. Third, some alternatives to chi-square are briefly mentioned.

Binomial model

Consider an experiment in which animals of a group are independently subjected to a stimulus. Assume two, and only two, responses are possible (A and $\bar{A}$). Of 20 animals exposed independently to the stimulus, responses of type A are exhibited by 16. Such a count, or the corresponding percentage, 80 per cent, may be the basis of an estimate of the probability of an A response in all animals of this kind; or it may be the basis of a test of the hypothesis that responses A and $\bar{A}$ are equally likely. The evaluation of either this estimate or the test is dependent upon the assumptions underlying the data collection.

One of the basic models associated with such experiments is the binomial. The binomial model is associated with a series of independent trials in each of which an event A may or may not occur and for which it is assumed that the probability of occurrence of A, denoted p, is constant from trial to trial. If the number of occurrences of A among n such trials is v, then v/n is the maximum likelihood and also the minimum variance unbiased estimator of p. [*For further discussion of the binomial distribution, see* DISTRIBUTIONS, STATISTICAL, *article on* SPECIAL DISCRETE DISTRIBUTIONS.]

Additional insight as to the reliability of the estimator is obtained from a confidence interval for p. Tables and graphs have been prepared to provide such confidence intervals for appropriate levels of confidence. The best known of these is the graph, by Clopper and Pearson (1934). This graph or the tables that have been computed for the same purpose (for example, Owen 1962) determine so-called central confidence limits; that is, the intervals that are "false," in the sense that they do not include the true parameter value, are equally divided between those that are too low and those that are too high. [*See* ESTIMATION, *article on* CONFIDENCE INTERVALS AND REGIONS.]

Confidence intervals may also be used to test a null hypothesis that p has the value p_0. If p_0 is not included in the $1-\alpha$ level confidence interval, then the null hypothesis $p = p_0$ is rejected at level α.

Equivalently, a direct test may be made of this hypothesis by utilizing the extensive tables of the binomial distribution. Two of the best known are those of Harvard University (1955) and the U.S. National Bureau of Standards (1950).

More usually both confidence intervals and test procedures are based upon an approximation to the distribution, that is, on the fact that $(v-np)\cdot[np(1-p)]^{-\frac{1}{2}}$ has a limiting standard normal distribution. [*See* DISTRIBUTIONS, STATISTICAL, *article on* APPROXIMATIONS TO DISTRIBUTIONS.]

Denote by $Z_{1-\alpha}$ the $100(1-\alpha)$ percentile of the standard normal distribution. The null hypothesis $p = p_0$ tested against the alternative $p \neq p_0$ is rejected at level α on the basis of an observation of

v successes in n trials if

$$|v - np_0| - \tfrac{1}{2} > Z_{1-\frac{1}{2}\alpha}\,[np_0(1-p_0)]^{\frac{1}{2}};$$

in case the alternatives of interest are limited to one side of p_0, say $p > p_0$, the test procedure at level α is to reject H_0 if

$$v - np_0 - \tfrac{1}{2} > Z_{1-\alpha}\,[np_0\,(1-p_0)]^{\frac{1}{2}}.$$

The subtracted $\frac{1}{2}$ is the so-called continuity correction—useful when a discrete distribution is being approximated by a continuous one.

Thus, in the experiment described above, the experimenter might be testing whether the choice is made at random between A and $\bar{A}$ against the possibility that A is the preferred response. This is a test of the hypothesis $p = \frac{1}{2}$ against the alternative $p > \frac{1}{2}$. Corresponding to the conventional 5 per cent significance level, $Z_{0.95} = 1.64$; then if $v = 16$ (16 A responses are observed), the hypothesis is rejected at the 5 per cent level since

$$16 - 10 - \tfrac{1}{2} > 1.64\,[(20)(\tfrac{1}{2})(\tfrac{1}{2})]^{\frac{1}{2}}.$$

The normal approximation to the binomial is thought to be quite satisfactory if $np_0(1-p_0)$ is 5 or more. However, for many practical situations the normal approximation provides an adequate test (in the sense that the type I error is sufficiently close to the specified level) for values of $np_0(1-p_0)$ well below the bound of 5 mentioned above.

The simplest confidence limits for p based on the normal approximation are

$$\frac{v}{n} \pm Z_{1-\frac{1}{2}\alpha}\sqrt{\frac{1}{n}\frac{v}{n}\left(1-\frac{v}{n}\right)}.$$

The binomial model requires independence of the successive trials. Much sampling, especially of human populations, is, however, done without replacement so that successive observations are in fact dependent and the correct model is not the binomial but the hypergeometric. In sampling theory this is taken into account by the finite population correction, which modifies the variance. Thus, where the binomial variance is $np(1-p)$, the hypergeometric variance for a sample of size n from a population of size N is $np(1-p)(1-n/N)$. If n is a small fraction of N, the finite population correction is negligible; thus, the binomial model is often used as an acceptable approximation.

Chi-square tests

For one or two proportions. The statistic $(|v - np_0| - \frac{1}{2})[np_0(1-p_0)]^{-\frac{1}{2}}$, which for sufficiently large n may be used to test the hypothesis $p = p_0$ against $p \neq p_0$, yields, when squared, an equivalent test procedure based on the chi-square distribution with one degree of freedom; this follows from the fact that the square of a standard normal variable has a chi-square distribution with one degree of freedom. [*See* DISTRIBUTIONS, STATISTICAL, *article on* SPECIAL CONTINUOUS DISTRIBUTIONS.]

Following recent practice, "X^2" is written for the test statistic, and the symbol "χ^2" is reserved for the distributional form.

$$X^2 = \frac{(|v - np_0| - \frac{1}{2})^2}{np_0(1-p_0)}$$
$$= \frac{(|v - np_0| - \frac{1}{2})^2}{np_0} + \frac{(|n - v - n(1-p_0)| - \frac{1}{2})^2}{n(1-p_0)}.$$

This algebraic identity shows that the statistic X^2 may be written (neglecting the continuity correction term, $\frac{1}{2}$) as (observed − expected)^2/expected, summed over the two categories A and $\bar{A}$. Such a measure of deviation of observations from their expected values under a null hypothesis is of wide application.

For example, consider the counts of individuals with characteristic A that occur in two independent random samples and suppose that the null hypothesis at test is that the probability of occurrence of A is the same in both populations; call the common (but unspecified) probability p. The observations may be tabulated as in Table 1.

Table 1 — Observations in two samples

		NUMBERS OF OBSERVED		
		A's	$\bar{A}$'s	Totals
SAMPLE	1	v_{11}	v_{12}	n_1
	2	v_{21}	v_{22}	n_2
	Totals	$v_{\cdot 1}$	$v_{\cdot 2}$	n

If p were known, then under the null hypothesis the expectation of the number of A's in sample 1 would be n_1p and in sample 2 the expectation would be n_2p, where p is the probability of occurrence of A. Since p is unknown, however, it must be estimated from the data [*see* ESTIMATION, *article on* POINT ESTIMATION].

If the hypothesis were true, the two samples could be pooled and the usual (minimum variance unbiased) estimator of p would be $v_{\cdot 1}/n$. With this estimator the estimated expected number of A's in sample 1 is $n_1(v_{\cdot 1}/n)$ and in sample 2 is $n_2(v_{\cdot 1}/n)$. Similarly the estimated expected numbers of $\bar{A}$'s are $n_1(v_{\cdot 2}/n)$ and $n_2(v_{\cdot 2}/n)$ in the two samples. These estimated expectations are tabulated in Table 2.

Table 2 — Estimated expected number of observations in two samples

		ESTIMATED EXPECTATIONS FOR NUMBER OF OBSERVED	
		A's	$\bar{A}$'s
SAMPLE	1	$n_1\left(\frac{v_{\cdot 1}}{n}\right)$	$n_1\left(\frac{v_{\cdot 2}}{n}\right)$
	2	$n_2\left(\frac{v_{\cdot 1}}{n}\right)$	$n_2\left(\frac{v_{\cdot 2}}{n}\right)$

An expression similar to X^2 can be calculated for each sample where now, however, p_0 is replaced by the estimator $v_{\cdot 1}/n$. These expressions are

$$\frac{[|v_{11} - n_1(v_{\cdot 1}/n)| - \frac{1}{2}]^2}{n_1(v_{\cdot 1}/n)(1 - v_{\cdot 1}/n)}$$

and

$$\frac{[|v_{21} - n_2(v_{\cdot 1}/n)| - \frac{1}{2}]^2}{n_2(v_{\cdot 1}/n)(1 - v_{\cdot 1}/n)}.$$

Since the estimator of p will tend to be close to the true value for large sample sizes, it is intuitive to conjecture that each of these are squares of normal variables (at least approximately for large samples). The sum does have a limiting chi-square distribution but with one degree of freedom, not two. The "loss" of the degree of freedom comes from estimating the unknown parameter, p. The test statistic, which more formally written is

$$X^2 = \sum_{i=1}^{2} \sum_{j=1}^{2} \frac{(|v_{ij} - n_i v_{\cdot j}/n| - \frac{1}{2})^2}{n_i v_{\cdot j}/n},$$

may be simplified to

$$\frac{n(|v_{11}v_{22} - v_{12}v_{21}| - n/2)^2}{n_1 n_2 v_{\cdot 1} v_{\cdot 2}}.$$

If $|v_{11}v_{22} - v_{12}v_{21}|$ is less than or equal to $n/2$, the correction term is inappropriate and possibly misleading. In practice this problem rarely arises.

Basic chi-square theorem. The above chi-square test statistics for one or two proportions may, as was seen, be written as sums of terms whose numerators are squared deviations of the observed counts from those "expected" under the null hypothesis. (*Expected* is placed in quotation marks to emphasize that the "expectations" are often estimated expectations obtained via estimation of unknown parameters.) The denominators may be regarded as weights to standardize the ratios. This pattern may be widely extended.

For example, consider a questionnaire with respondents placing themselves in five categories: strongly favor, mildly favor, neutral, mildly oppose, strongly oppose. The n independent responses might furnish data for a test of the hypothesis that each of the responses is equally likely. If the probabilities of the five responses are denoted p_1 through p_5 this null hypothesis specifies $p_1 = p_2 = p_3 = p_4 = p_5 = \frac{1}{5}$ and under the null hypothesis the expected number of responses in each category is $n/5$. The appropriate weights in the denominator of the chi-square test statistic are suggested by the expanded form of X^2 given above; each term (observed − expected)^2 is divided by its expected value. Thus in this example,

$$X^2 = \sum_{j=1}^{5} \frac{(v_j - n/5)^2}{n/5}.$$

That these weights lead to the usual kind of null distribution can be shown by considering the multinomial distribution, the extension of the binomial distribution to a series of independent trials with several outcomes rather than just two. If the null hypothesis is true, this X^2 has approximately a chi-square distribution with four degrees of freedom.

More generally, suppose that on each of n independent trials of an experiment exactly one of the events $E_1, \cdots, E_J$ occurs. Let p_j, depending in a given way (under the null hypothesis under test) on unknown parameters $\theta_1, \cdots, \theta_m$, be the probability that E_j occurs and suppose there are asymptotically efficient estimators of the θ's, from which are obtained asymptotically efficient estimators of the p_j, denoted $\hat{p}_j$; thus $n\hat{p}_j$ estimates the expected frequency of occurrence of E_j under the null hypothesis. Let the random variable v_j be the number of times E_j actually occurs in the n trials. Then

$$X^2 = \sum_{j=1}^{J} \frac{(v_j - n\hat{p}_j)^2}{n\hat{p}_j}$$

has, under the null hypothesis for large n and under mathematical regularity conditions, approximately the chi-square distribution with $J - m - 1$ degrees of freedom. When the null hypothesis is false, X^2 tends to be larger on the average than when it is true, so that a right-hand tail critical region is appropriate, that is, the null hypothesis is rejected for large values of X^2.

Note that the above "chi-square" statistic is of form

$$\text{Sum}\left[\frac{(\text{observed number} - \text{“expected” number})^2}{\text{“expected” number}}\right].$$

(The quotation marks around *expected* indicate that this is actually an asymptotically efficient *estimator* of the expectation under the null hypothesis.)

The above development can readily be extended

to I independent sequences of trials, with n_i trials in the ith sequence, p_{ij} denoting the probability under the null hypothesis of event j for sequence i, and v_{ij} denoting the number of times E_j occurs in sequence i. As before,

$$X^2 = \sum_{i=1}^{I} \sum_{j=1}^{J} \frac{(v_{ij} - n_i \hat{p}_{ij})^2}{n_i \hat{p}_{ij}}$$

is, for large n_i, approximately chi-square with $I(J-1) - m$ degrees of freedom, under the null hypothesis, and with appropriate regularity conditions. Note that when $I = 1$, $J - m - 1$ degrees of freedom are obtained, as before.

The primary problem in such tests is the derivation of asymptotically efficient estimators. For example, such estimators may be maximum likelihood estimators or minimum chi-square estimators. The latter are the θ's that minimize X^2, the test statistic, subject to whatever functional restraints are imposed upon the p_{ij}'s. Neyman (1949) has given a method of determining modified minimum chi-square estimators, a method that reduces to solving only linear equations, as many as there are unknown parameters to estimate. A review of the methods of generating such minimum chi-square estimators for this model, and for a more general one, is given by Ferguson (1958).

It is easily seen that the comparison of two percentages is a special case of the general theorem. Here $I = J = 2$ and the null hypothesis can be put in the form $p_{11} = p_{12} = \theta$; $p_{21} = p_{22} = 1 - \theta$. Here p_{11} is the probability of A occurring on a trial in the first series, p_{12} is the probability of A occurring on a trial in the second series; p_{21}, p_{22} are defined similarly with respect to $\bar{A}$. The maximum likelihood estimator of θ is $v_{.1}/n$ and the degrees of freedom are seen to be one from insertion in the general formula.

Proofs of the basic chi-square theorem and statements of the mathematical regularity conditions may be found in Cramér (1946) or Neyman (1949).

Power of the chi-square test. The chi-square test is extensively used as an omnibus test without particular alternatives in view. Frequently such applications are almost useless in the sense that their sensitivity (that is, power) is very low. It is therefore important not only to make such tests but also to specify the alternatives of interest and to determine the power, that is, the probability that the null hypothesis is rejected when in fact such alternatives are true. A fairly complete theory of the power of chi-square tests has been given recently by Mitra (1958) and Diamond (1963).

Because chi-square tests are based upon a limiting distribution theorem it is necessary to express the alternative in a special form, depending on the sample size, n, in order to obtain meaningful results. Consider first the case where $I = 1$ and the null hypothesis completely specifies the p_j^0 as numerical constants. (In the questionnaire experiment above, since there are five responses the null hypothesis that the responses are equally likely specifies $p_j^0 = \frac{1}{5}$.) Write an alternative p_j^A in the form

$$p_j^A = p_j^0 + \frac{C_j}{\sqrt{n}}, \qquad \text{where } \sum_{j=1}^{J} C_j = 0.$$

If in fact $p_j = p_j^A$ then the test statistic X^2 has a limiting *noncentral* chi-square distribution with noncentrality parameter

$$\lambda = \sum_{j=1}^{J} C_j^2/p_j^0$$

and with $J - 1$ degrees of freedom [*see* DISTRIBUTIONS, STATISTICAL, *article on* SPECIAL CONTINUOUS DISTRIBUTIONS].

The λ required to obtain a specified probability of rejection of an alternative p_i^A for tests at significance levels 0.01 and 0.05 has been tabulated; such a table is given, for example, by Owen (1962, pp. 61–62). These tables are useful not only in calculating the power function but also in specifying sample size in advance. For the example where there are five responses and the null hypothesis is $p_1^0 = p_2^0 = p_3^0 = p_4^0 = p_5^0 = 0.20$, consider the alternative $p_1^A = p_2^A = p_3^A = p_4^A = 0.15$, $p_5^A = 0.40$. Then $C_j = \sqrt{n}(0.05)$, $j = 1, \cdots, 4$; $C_5 = -\sqrt{n}(0.20)$, so that $\lambda = n(.25)$. To achieve a probability of 0.80 of rejecting the null hypothesis for this alternative, it is found from the tables that λ must be 11.94 (four degrees of freedom and 0.05 significance level). This requires a sample size of 11.94/.25 or, to the nearest whole number, 48.

For the comparison of two samples, a similar power theory is available. Consider two sequences of n_i trials each of which results in an outcome $E_1, E_2, \cdots, E_J$. Here p_{ij} $(i = 1, 2, j = 1, \cdots, J)$ is the probability of outcome j on sequence i and the null hypothesis of homogeneity is $p_{11} = p_{21}$, $p_{12} = p_{22}$, $\cdots$, $p_{1J} = p_{2J}$. Now consider a sequence of alternatives $p_{ij}^A(n)$ which for some p_j $(j = 1, 2, \cdots, J)$ satisfy the equations $p_{ij}^A(n) = p_j + C_{ij}/\sqrt{n}$, where $n = n_1 + n_2$ and $\sum_j C_{1j} = \sum_j C_{2j} = 0$. Then, for the sequence of alternatives, X^2 has, in the limit as $n \to \infty$, a noncentral chi-square distribution with $J - 1$ degrees of freedom and noncentrality parameter λ, where

$$\lambda = \frac{n_1 n_2}{n^2} \sum_{j=1}^{J} \frac{(C_{1j} - C_{2j})^2}{p_j}.$$

In actual practice, when the statistician considers

a specified alternative for finite n, p_j is not uniquely defined; it is convenient to define $p_j = \frac{1}{2}[p^A_{1j}(n) + p^A_{2j}(n)]$ but whether other choices of p_j might improve the goodness of the asymptotic approximation to the actual power appears not to have been investigated. For the case $J = 2$, and with $n_1 = n_2$, a nomogram is available showing the sample size required to obtain a specified level of power for one-sided hypotheses, that is, for comparison of an experimental and a standard group (Columbia University 1947, chapter 7). In the general case the formulation of λ is more difficult.

Contingency tables. In the example of comparing two percentages, the observations were conveniently set out in a 2×2 array. Similarly, in the more general comparative experiment (the power of which was just discussed), it would be convenient to set out the observations in a $2 \times J$ array. These are special cases of *contingency tables*, which, in general, have r rows and c columns; counted data that may be so represented arise, for example, in many experiments and surveys.

Such arrays or contingency tables may arise in at least three different situations, which may be illustrated by specific examples:

(1) Double polytomy: A sample of n voters is taken from an electoral list and each voter is classified into one of r party affiliations $P_1, \cdots, P_r$ and into one of c educational levels $E_1 \cdots, E_c$. Denote by p_{ij} the probability that a voter belongs to party i and educational level j, so that $\sum_i \sum_j p_{ij} = 1$. The usual null hypothesis of interest is that the classifications are independent, that is,

$$p_{ij} = p_i . p._j ,$$

where $p_i. =$ probability a voter is in party i (regardless of educational level) and $p._j =$ probability a voter is in educational level j, again regardless of the other classification variable (that is, $p_i. = \sum_j p_{ij}$ and $p._j = \sum_i p_{ij}$). In this case both the vertical and horizontal marginal totals of the $r \times c$ sample array are random.

(2) Comparative trials: Consider instead of a single sample from the general electoral roll, r samples of sizes n_i from the r different party rolls. The voters in each sample are classified as to educational level (levels $E_1, \cdots, E_c$). Denote as before by p_{ij} the probability that a voter drawn from party i belongs to educational level j (so that $\sum_j p_{ij} = 1$). The hypothesis of homogeneity specifies that $p_{1j} = p_{2j} = \cdots = p_{rj}$ for each j. In this case the row totals are fixed $(n_1, \cdots, n_r)$, while the column totals are random. Into this category falls the two-sample experiment discussed earlier; in that case $r = 2$.

(3) Independence trials (fixed marginal totals): Consider a group of n manufactured articles, of which fixed proportions are in each of the quality categories $C_1, \cdots, C_c$. The articles have been divided into r groups of fixed size $n_1, \cdots, n_r$ for further processing or for shipment to customers. The question arises whether the partitioning into the r groups can reasonably be considered to have been done randomly, that is, independently of how the articles fall into the quality categories. Since the number of articles in each of the categories $C_1, \cdots, C_c$ as well as the $n_1, \cdots, n_r$ are fixed, both marginal totals are fixed in this situation.

For these three cases let v_{ij} denote the number of individuals falling into row i, column j, and denote by $v_i.$ and $v._j$ the row and column totals, whether fixed or random. While different probability models are associated with the three cases, the approximate or large sample chi-square test is identical. The test statistic is

$$X^2 = \sum_{i=1}^{r} \sum_{j=1}^{c} \frac{[v_{ij} - (v_i.v._j)/n]^2}{(v_i.v._j)/n},$$

which has, if the null hypothesis is true, an approximate chi-square distribution with $(r-1) \cdot (c-1)$ degrees of freedom.

For the comparative trials case, this is an extension of the comparison of two percentages. The maximum likelihood estimator of the common value of $p_{1j}, p_{2j}, \cdots, p_{rj}$ is $v._j/n$ under the null hypothesis. The comparative trials model consists of r sequences of trials, each of which may result in one of c events; $c - 1$ parameters are estimated. Because $\sum_{j=1}^{c} p_{ij} = 1$, as soon as $c - 1$ of the probabilities are estimated the final one is determined. Hence the degrees of freedom are $r(c - 1) - (c-1) = (r-1)(c-1)$.

In the double polytomy case there are $(r-1) + (c-1)$ independent parameters to be estimated under the null hypothesis: $p_1., p_2., \cdots, p_{r-1}., p._1, p._2, \cdots, p._{c-1}$, since again the restrictions $\sum_i p_i. = \sum_j p._j = 1$ provide the last two needed values. The maximum likelihood estimators of the $p_i.$ are the $v_i./n$ and of the $p._j$ are the $v._j/n$, so that the estimated expected values are $n(v_i./n)(v._j/n)$ or $v_i.v._j/n$. The degrees of freedom in this case are $(rc-1) - (r-1) - (c-1)$ or $(r-1)(c-1)$, since there is only one sequence of trials with rc outcomes.

Like all chi-square tests, these are based upon asymptotic distribution theory and are satisfactory in practice for "large" sample sizes. A number of rules of thumb have been established in regard to the acceptable lower limit of sample size so that the actual type I error, or probability of rejecting the null hypothesis when true, does not depart too far from the prescribed significance level. For a careful discussion of this problem, and of procedures to adopt when the samples are too small, see Cochran (1952; 1954).

2×2 *tables.* The special case of contingency tables with $r = c = 2$ has been extensively studied, and the so-called Fisher exact test is available. Given $v_{1\cdot}, v_{2\cdot}, v_{\cdot 1}, v_{\cdot 2}$, under any of the null hypotheses, v_{11} has a specific hypergeometric distribution; hence probabilities of deviations as numerically large as, or larger than, the observed deviation can be calculated and a test can be made. The application of the test is now greatly facilitated by use of tables by Finney et al. (1963). For the comparative trials model and the double dichotomy model this exact test is a conditional test, given the marginal counts.

While the hypotheses associated with the three different models in $r \times c$ tables, in general, and 2×2 tables, in particular, can be tested by the same chi-square procedure, the power of the test varies according to the model. For the 2×2 case, approximations and tables have been given for each of the three models. The most recent of these are by Bennett and Hsu (1960) for comparative and independence trials and Harkness and Katz (1964) for the double dichotomy model. Earlier approximations are discussed and compared by these authors.

Single degrees of freedom. The statistic X^2 used to test the several null hypotheses possible for $r \times c$ contingency tables can be partitioned into $(r-1)(c-1)$ uncorrelated X^2 terms, each of which has a limiting chi-square distribution with one degree of freedom when the null hypothesis is true.

Planned comparisons. Planned subcomparisons, however, can be treated most easily by forming new contingency tables and calculating the approximate X^2 statistic. For example, in the comparison of three experimental learning methods with a standard method the observations might be recorded for each pupil as successful or unsuccessful and tabulated in a 4×2 table. These are four comparative trials; and X^2, the statistic to test homogeneity, has, under the null hypothesis, an approximate chi-square distribution with three degrees of freedom.

In this situation, two subcomparisons might be indicated: the standard method versus the combined experimental groups and in the experimental groups among themselves. Tables 3*a* and 3*b* show the two new contingency tables. The X^2 statistics calculated from these two subtables may be

Table 3a — Comparison between standard method and combined experimental methods

	Successful	Unsuccessful
Standard (method 1)	v_{11}	v_{12}
Experimental methods combined	$v_{21} + v_{31} + v_{41}$	$v_{22} + v_{32} + v_{42}$

Table 3b — Comparison among experimental methods

	Successful	Unsuccessful
Experimental (method 2)	v_{21}	v_{22}
Experimental (method 3)	v_{31}	v_{32}
Experimental (method 4)	v_{41}	v_{42}

used to make the indicated secondary tests. The two X^2 values (with one and two degrees of freedom respectively) will not sum to the X^2 calculated for the whole 4×2 array. Short-cut formulas for a partition that is additive and references to other papers on this subject are given by Kimball (1954).

Unplanned comparisons. As in the analysis of variance of linear models, distinction should be made between such planned comparisons and unplanned comparisons. Goodman (1964*a*) has given a procedure to find confidence intervals for a family of "contrasts" among multinomial probabilities for the $r \times c$ contingency table in the comparative-trials model. A "contrast" is any linear function of the probabilities p_{ij}, with coefficients summing to zero, that is,

$$\sum_{i=1}^{r}\sum_{j=1}^{c} b_{ij}p_{ij}, \qquad \sum_{i=1}^{r} b_{ij} = 0 \text{ for each } j.$$

[*See* LINEAR HYPOTHESES, *article on* MULTIPLE COMPARISONS.]

Thus in the comparison of teaching methods experiment referred to above, where p_{i1} is the probability of a pupil being successful when taught by method i, the unplanned comparisons or contrasts might be $p_{21} - p_{31}$, $p_{21} - p_{41}$, $p_{31} - p_{41}$. These represent pairwise comparisons of the three experimental methods.

Denote a contrast by θ; an estimator of p_{ij} is $\hat{p}_{ij} = v_{ij}/v_{i\cdot}$ and an estimator of θ is

$$\hat{\theta} = \sum_{i=1}^{r}\sum_{j=1}^{c} b_{ij}\hat{p}_{ij}.$$

An estimator of the variance of $\hat{\theta}$ is

$$S^2(\hat{\theta}) = \sum_{i=1}^{r}\left\{(v_{i\cdot}^{-1}\left[\sum_{j=1}^{c} b_{ij}^2\hat{p}_{ij} - \left(\sum_{j=1}^{c} b_{ij}\hat{p}_{ij}\right)^2\right]\right\}.$$

The large-sample joint confidence intervals for θ with confidence coefficient $1-\alpha$ have the form $\hat{\theta} - S(\hat{\theta})L, \hat{\theta} + S(\hat{\theta})L$, where L is the square root of the upper $100(1-\alpha)$th percentage point of the chi-square distribution with $(r-1)(c-1)$ degrees of freedom. An experiment in which one or more of the totality of all such possible intervals fail to include the true θ may be called a violation. The probability of such a violation is α.

If instead of all contrasts, only a few, say G, are of interest, then L in the last formula may be replaced by $Z_{1-\alpha/2G}$ (the $100\,[1-\alpha/2G]$ percentile of the standard normal distribution), which often will be smaller than L and hence yield shorter confidence intervals while the probability of a violation is still less than or at most equal to α.

Comparative trials; ordered alternatives. In the comparative trials model, with $r \times 2$ contingency tables, frequently the only alternative of interest is an ordered set of p_{i1}'s. For example in a 2×2 comparative trial involving a control and a test group, the question may be to decide whether the groups are the same or whether the test group yields "better" results than the control group. In the 2×2 case, this situation is handled simply by working with the signed square root of X^2, which has a standard normal distribution if the null hypothesis is true. A one-sided alternative is then treated in the same manner as a test for a percentage referred to earlier.

For the more general $r \times 2$ table, the most complete treatment is that of Bartholomew (1959); his test, however, requires special tables. If the experimenter believes that the p_{i1} have a functional relationship to a known associated variable, x_i, then a specific test can be derived from the basic theorem. Such a test would be a particular example of a planned comparison. Many authors have given short-cut formulas and worked out examples of this type of problem (cf. Cochran 1954; Armitage 1955).

Comparative trials; dependent samples. A sample is taken of n voters who have voted in the last two national elections for one of the major parties. Denote the parties by L and C, and suppose that in the sample 45 per cent voted L in the first election and 55 per cent voted L in the second. Does this indicate a significant change in voter behavior in the subpopulation of which this is a sample? To make such a comparison in matched or dependent samples, it is necessary to obtain information on the actual changes in party preference [*see* PANEL STUDIES]. These can be read from a 2×2 table such as Table 4.

Table 4 — Voter preference in two elections

		ELECTION 1	
		L	C
ELECTION 2	L	v_{11}	v_{12}
	C	v_{21}	v_{22}

Such a 2×2 table with random marginals appears to fall into the double-dichotomy model, but the hypothesis of independence is not of interest here. The changes are indicated by the off-diagonal elements v_{12}, v_{21}, and the hypothesis of no net change is equivalent to the hypothesis that, given $v_{12} + v_{21}$, v_{12} is binomially distributed with probability $\frac{1}{2}$. Thus, the test of comparison of two percentages in identical or matched samples reduces to the test for a percentage. If the normal approximation is adequate, the square of the normal deviate, with the continuity correction, is

$$X^2 = \frac{(|v_{12} - v_{21}| - 1)^2}{v_{12} + v_{21}},$$

which has a limiting chi-square distribution with one degree of freedom under the null hypothesis that the probabilities are the same in the two matched groups. Cochran (1950) has extended this test to the $r \times c$ case. The test described above does not at all depend on v_{11} and v_{22}, but of course these quantities would enter into procedures pointed toward issues other than testing the null hypothesis of no net change.

Chi-square tests of goodness of fit. Chi-square tests have been used extensively to test whether sample observations might have arisen from a population with a specified form, such as binomial, Poisson, or normal. Such chi-square tests are again special cases of the general theory outlined above, although there are many other types of tests for goodness of fit [*see* GOODNESS OF FIT].

There are some special problems in connection with some nonstandard chi-square tests of goodness of fit for the binomial and Poisson distributions. The standard chi-square test of goodness of fit for these two discrete distributions requires (in most cases) estimation of the mean. The sample mean is an efficient estimator of the population mean; it appears to make little difference whether the sample mean is computed from the raw or grouped data.

There is evidence that two simpler tests are more powerful, at least for some alternatives, for testing whether a set of counts does come from one of the distributions. These test statistics are the so-called indices of dispersion, studied by Lexis and Bortkiewicz, which in fact compare two estimators of the variance—the usual sample estimator and the estimator derivable from the fact that for these distributions the variance is a function of the mean. Alternatively they may be viewed as chi-square tests, conditional on the total count and placed in the framework of the basic theorem. Thus if the observations are $v_1, \cdots, v_n$, which according to the null hypothesis come from a Poisson distribution, the appropriate index of dispersion test statistic is

$$X^2 = \frac{\sum_{i=1}^{n} (v_i - \bar{v})^2}{\bar{v}},$$

where $\bar{v}$ is the sample mean of the v_i. For large n and if the null hypothesis is true, X^2 is approximately distributed as chi-square with $n-1$ degrees of freedom [*see* BORTKIEWICZ; LEXIS].

The corresponding test for the binomial can be expressed similarly, but it is also useful to set out the n observations in a $2 \times c$ contingency table such as Table 5.

Table 5 — Arrangement of data to test for binomial

	Sample 1	Sample 2	...	Sample c
Successes	x_1	x_2	...	x_c
Failures	$n_1 - x_1$	$n_2 - x_2$	...	$n_c - x_c$
Total	n_1	n_2	...	n_c

The variance test is equivalent to the chi-square test of homogeneity in this $2 \times c$ array, which has, of course, $c-1$ degrees of freedom.

Whereas in general the chi-square test is a one-tailed test, that is, the null hypothesis is rejected for large values of the statistics, the dispersion tests are often two-tailed tests, not necessarily with equal probability in the tails. The reason for this is that a too small value of X^2 reflects a pattern that is *more* regular than that expected by chance, and such patterns may correspond to important alternatives to the null hypothesis of homogeneous randomness.

Contingency table association measures. If in the double-dichotomy model the hypothesis of independence is rejected, it is logical to seek a measure of association between the classifications. Distinction must be made between purely descriptive measures and sampling estimators of such measures. [*See* STATISTICS, DESCRIPTIVE, *article on* ASSOCIATION.]

A large number of such measures have been presented, usually related to the X^2 statistic used to test the null hypothesis of independence. Goodman and Kruskal (1954–1963) have emphasized the need to choose measures of association that have contextual meaning in the light of some probability model with predictive or explanatory value. They distinguish between two cases—no ordering among the categories and directed ordering among them.

Multidimensional contingency tables. The analysis of data that have been categorized into three or more classifications involves not only a considerable increase in the variety of possibilities but also introduces some new conceptual problems. The basic test of mutual independence is, however, a straightforward extension of the two-dimensional one and a simple application of the main theorem. The test will be discussed for three classifications.

This is a test of the hypothesis that p_{ijk}, the probability of an observation falling in row i, column j, and layer k, can be factored into $p_{i..}p_{.j.}p_{..k}$. Under this null hypothesis the estimated expected value in cell ijk is $n^{-2}(v_{i..})(v_{.j.})(v_{..k})$, where the dots indicate summation over the corresponding subscripts of the observed counts, v_{ijk}. The X^2 statistic has the usual form, sum of (observed − expected)^2/expected, and has $rcl - r - c - l + 2$ degrees of freedom if there are r rows, c columns, and l layers.

Tests for partial independence, for example that $p_{ijk} = (p_{i..})(p_{.jk})$, or for homogeneity (between layers, for example) may be derived similarly. New concepts and new tests are introduced by the idea of *interaction* between the different classifications.

In linear models, interactions are measures of nonadditivity of the effects due to different classifications. With contingency models several definitions of interaction have been given; the present treatment follows Goodman (1964*b*). Consider, for example, samples drawn from rural and urban populations and classified by sex and age, with age treated dichotomously (see Table 6).

Table 6 — Classification of rural and urban samples by sex and age

	URBAN			RURAL		
	Young	Old	Totals	Young	Old	Totals
Male	20	14	34	34	11	45
Female	42	24	66	36	19	55
Totals	62	38	100	70	30	100

For the urban group there is a sex ratio 20/42 among the "young" and 14/24 among the "old." The ratio of these may be regarded as a measure of the interaction of age and sex in the urban population. Similarly the same ratio of sex ratios is a measure of the interaction in the rural population. These are, of course, sample values; and the population interactions must be defined in terms of the probabilities p_{ijk}. It is useful to define

$$\Delta_k = \frac{p_{11k}}{p_{21k}} \Big/ \frac{p_{12k}}{p_{22k}}$$

and to write the three-factor no interaction hypothesis as $\Delta_1 = \Delta_2$ for this $2 \times 2 \times 2$ contingency table. The maximum likelihood estimator of Δ_k is $d_k = v_{11k}v_{22k}/v_{12k}v_{21k}$, and its variance can be estimated consistently by $s_k^2 = d_k^2 u_k$ where $u_k = \sum_i \sum_j v_{ijk}^{-1}$.

A simple statistic to test the hypothesis $\Delta_1 = \Delta_2$ is $X^2 = (d_1 - d_2)^2/(s_1^2 + s_2^2)$, which, if the null hypothesis is true, has a large sample chi-square distribution with one degree of freedom.

For the data given $d_1 = 0.816$, $d_2 = 1.631$, $s_1^2 = 0.125$, $s_2^2 = 0.534$, and $X^2 = 1.01$ so that the three-factor no interaction hypothesis is not rejected at the usual significance levels. Goodman has extended this test in an obvious way to the $2 \times 2 \times l$ contingency table. Here the test statistic is

$$X^2 = \sum_{k=1}^{l} \frac{(d_k - \bar{d})^2}{s_k^2}, \qquad \text{where } \bar{d} = \frac{\sum_{k=1}^{l} d_k/s_k^2}{\sum_{k=1}^{l} s_k^2},$$

which has $l - 1$ degrees of freedom. The extension to $r \times c \times l$ tables is based on logarithms of frequencies rather than the actual frequencies. Goodman also provides confidence intervals for the interactions Δ_k and indicates a number of equivalent tests. A bibliography of the very extensive literature on this topic is given in this paper.

Alternatives to chi-square

While the chi-square tests are classical for the analysis of counted data, with the original simple tests going back to Karl Pearson, they are not the likelihood ratio tests. The latter are based upon statistics of the form

$$\sum_{j=1}^{J} v_j \log \frac{v_j}{np_j} = \sum_{j=1}^{J} v_j \log [1 + (v_j - np_j)/np_j]$$

in the general case with one sequence of trials. It is easy to show, by expanding minus twice the logarithm of the likelihood ratio statistic in a power series, that its leading term is X^2, and that further terms are of a smaller order than the leading term so that the tests are equivalent in the limit. However, they are not equivalent for small samples.

Another test for contingency tables (comparative trials model) is that of C. A. B. Smith (1951). Further work appears to be necessary before any of these alternatives is accepted as preferable to the chi-square tests. The most widely used alternative analysis is that indicated in the next section.

ANOVA of transformed counted data. Contingency tables have an obvious analogy to similar arrays of measured data that are often treated by analysis of variance techniques (ANOVA). The analysis of variance models are more satisfactory if the data are such that (1) effects are additive, (2) error variability is constant, (3) the error distribution is symmetrical and nearly normal, and (4) the errors are statistically independent. [*See* LINEAR HYPOTHESES, *article on* ANALYSIS OF VARIANCE.]

Counted data that arise from a binomial or multinomial model fail most obviously on the second property since the variances of such data vary with the mean. However, some function or transformation of the observations may have approximately constant variance. Transformations that have been derived for counted data to make variances nearly constant have been found empirically often to improve the degree of approximation to which properties (1) and (3) hold also. These transformations include: (*a*) Arc sine transformation for proportions

$$y = \text{arc sin } \sqrt{v/n},$$

which is applicable to dichotomous data with an equal number of trials in each sequence. If the number of trials (n_i) varies from sequence to sequence the problem is more complicated (see Cochran 1943). (*b*) Square root transformation: $y = \sqrt{v+1}$ for Poisson data. (*c*) Logarithmic transformation: $y = \log (v + 1)$ for data such that the standard deviation is proportional to the mean. Use of the arc sine transformation, and subsequent analysis, is facilitated by the use of binomial probability paper; graphic techniques are simple and usually adequate. The basic reference for such procedures is Mosteller and Tukey (1949).

Refinements of these transformations and discussion of the choice of transformations is given a thorough treatment by Tukey (1957). If a suitable transformation has been made, the whole battery of tests that have been developed in analysis of variance (including covariance techniques) is applicable. Estimation problems may be more subtle; in some situations estimates may be given in the transformed variable but in others it may be desirable to transform back to the original variable. [*See* STATISTICAL ANALYSIS, SPECIAL PROBLEMS OF, *article on* TRANSFORMATIONS OF DATA.]

Precautions in the analysis of counted data

The transformations discussed above were derived to apply to data that conform to such models as the binomial or Poisson and that could be analyzed by chi-square methods. However, counted data often arise from models that do not conform to the basic assumption; in particular independence may be lacking, so that the chi-square tests are not valid. Such data are often transformed and treated by analysis of variance procedures; the justification for this is largely empirical. Examples of situations where this is necessary are experimental responses of animals in a group where dependence may be present, eye estimates of the numbers in a group of people, and comparisons of proportions in heterogeneous and unequal-sized groups. In such situations care is necessary that the proper transformation is selected to achieve the properties listed above and in the interpretation of the results of the analysis.

The lack of independence and the presence of extraneous sources of variation are frequent sources of error in the analysis of counted data because the chi-square tests are invalidated by such factors. A discussion of these errors and others is found in Lewis and Burke (1949). The two careful expository papers by Cochran (1952; 1954) represent an excellent source of further reading on this topic. See also the monograph by Maxwell (1961).

DOUGLAS G. CHAPMAN

[*See also* QUANTAL RESPONSE.]

BIBLIOGRAPHY

ARMITAGE, P. 1955 Tests for Linear Trends in Proportions and Frequencies. *Biometrics* 11:375–386.

BARTHOLOMEW, D. J. 1959 A Test of Homogeneity for Ordered Alternatives. Parts 1–2. *Biometrika* 46:36–48, 328–335.

BENNETT, B. M.; and HSU, P. 1960 On the Power Function of the Exact Test for the 2 × 2 Contingency Table. *Biometrika* 47:393–398.

CLOPPER, C. J.; and PEARSON, E. S. 1934 The Use of Confidence or Fiducial Limits Illustrated in the Case of the Binomial. *Biometrika* 26:404–413.

COCHRAN, WILLIAM G. 1943 Analysis of Variance for Percentages Based on Unequal Numbers. *Journal of the American Statistical Association* 38:287–301.

COCHRAN, WILLIAM G. 1950 The Comparison of Percentages in Matched Samples. *Biometrika* 37:256–266.

COCHRAN, WILLIAM G. 1952 The χ^2 Test of Goodness of Fit. *Annals of Mathematical Statistics* 23:315–345.

COCHRAN, WILLIAM G. 1954 Some Methods for Strengthening the Common χ^2 Tests. *Biometrics* 10:417–451.

COLUMBIA UNIVERSITY, STATISTICAL RESEARCH GROUP 1947 *Techniques of Statistical Analysis for Scientific and Industrial Research and Production and Management Engineering.* Edited by Churchill Eisenhart, Millard W. Hastay, and W. Allen Wallis. New York: McGraw-Hill.

CRAMÉR, H. 1946 *Mathematical Methods of Statistics.* Princeton Univ. Press. → See especially Chapter 30.

DIAMOND, EARL L. 1963 The Limiting Power of Categorical Data Chi-square Tests Analogous to Normal Analysis of Variance. *Annals of Mathematical Statistics* 34:1432–1441.

FERGUSON, THOMAS S. 1958 A Method of Generating Best Asymptotically Normal Estimates With Application to the Estimation of Bacterial Densities. *Annals of Mathematical Statistics* 29:1046–1062.

FINNEY, DAVID J. et al. 1963 *Tables for Testing Significance in a 2 × 2 Contingency Table.* Cambridge Univ. Press.

GOODMAN, LEO A. 1964*a* Simultaneous Confidence Intervals for Contrasts Among Multinomial Population. *Annals of Mathematical Statistics* 35:716–725.

GOODMAN, LEO A. 1964*b* Simple Methods for Analyzing Three-factor Interaction in Contingency Tables. *Journal of the American Statistical Association* 59:319–352.

GOODMAN, LEO A.; and KRUSKAL, WILLIAM H. 1954–1963 Measures of Association for Cross-classifications. Parts 1–3. *Journal of the American Statistical Association* 49:732–764; 54:123–163; 58:310–364.

HARKNESS, W. L.; and KATZ, LEO 1964 Comparison of the Power Functions for the Test of Independence in 2 × 2 Contingency Tables. *Annals of Mathematical Statistics* 35:1115–1127.

HARVARD UNIVERSITY, COMPUTATION LABORATORY 1955 *Tables of the Cumulative Binomial Probability Distribution.* Cambridge, Mass.: Harvard Univ. Press.

KIMBALL, A. W. 1954 Short Cut Formulas for the Exact Partition of χ^2 in Contingency Tables. *Biometrics* 10:452–458.

LEWIS, D.; and BURKE, C. J. 1949 The Use and Misuse of the Chi-square Test. *Psychological Bulletin* 46:433–489. → Discussion of the article may be found in subsequent issues of this bulletin: 47:331–337, 338–340, 341–346, 347–355; 48:81–82.

MAXWELL, ALBERT E. 1961 *Analyzing Qualitative Data.* New York: Wiley.

MITRA, SUJIT KUMAR 1958 On the Limiting Power Function of the Frequency Chi-square Test. *Annals of Mathematical Statistics* 29:1221–1233.

MOSTELLER, FREDERICK; and TUKEY, JOHN W. 1949 The Uses and Usefulness of Binomial Probability Paper. *Journal of the American Statistical Association* 44:174–212.

NEYMAN, JERZY 1949 Contribution to the Theory of the χ^2 Test. Pages 239–273 in Berkeley Symposium on Mathematical Statistics and Probability, *Proceedings.* Edited by Jerzy Neyman. Berkeley: Univ. of California Press.

OWEN, DONALD B. 1962 *Handbook of Statistical Tables.* Reading, Mass.: Addison-Wesley. A list of addenda and errata is available from the author.

SMITH, C. A. B 1951 A Test for Heterogeneity of Proportions. *Annals of Eugenics* 16:15–25.

TUKEY, JOHN W. 1957 On the Comparative Anatomy of Transformations. *Annals of Mathematical Statistics* 28:602–632.

U.S. NATIONAL BUREAU OF STANDARDS 1950 *Tables of the Binomial Probability Distribution.* Applied Mathematics Series, No. 6. Washington: Government Printing Office.

Postscript

The chi-square statistic to test hypotheses about one or two proportions has long been defined with the $\frac{1}{2}$ continuity correction. A recent discussion by Conover (1974) suggests that the use of the correction may be inappropriate, although the matter is still controversial. It is striking that this simple question, discussed in detail at least since Barnard's treatment in 1947, remains unsettled.

The work of Meng and Chapman (1966) leads to easier evaluation of the noncentrality parameter λ so that the power of tests for contingency tables may now be calculated routinely.

There has been extensive development in the analysis of multidimensional contingency tables. This has followed several different paths. Goodman has greatly extended the work that is briefly described in the main article. Goodman (1973) includes references to his numerous earlier papers; also useful is a survey of Goodman's work through 1972 in Davis (1974). Goodman's work is concerned mainly with the development of log-linear models, a development that has now been summarized in Bishop et al. (1975). Another important book in this area is Haberman (1974).

To grasp the essential idea, consider a log-linear model applied to the example of Table 6. We begin by writing $\log p_{ijk} = \mu + \mu_{1(i)} + \mu_{2(j)} + \mu_{3(k)} + \mu_{12(ij)} + \mu_{13(ik)} + \mu_{23(jk)} + \mu_{123(ijk)}$ where μ is the grand mean; $\mu_{1(i)}$, $\mu_{2(j)}$, $\mu_{3(k)}$ are main effects; and the other terms represent interactions. The analogy of this with analysis of variance procedures is clearly seen, but there are some important differences. The aim of the procedure is to simplify the model as much as possible, that is, to find which of the parameters in the linear representation of $\log p_{ijk}$ can be set equal to zero and still retain a good fit to the data. Further, one wishes to obtain estimates, often maximum likelihood estimates, of the parameters that remain in the simplified model. These yield estimates of the cell probabilities. Not only are complete tables treated, but also more difficult cases of incomplete tables, that is, tables having cells with zero frequency. Such zeros may come from sampling variation, but they may also be part of the inherent structure of the problem. For example, if the cross-tabulation is of surgical operation by gender, the cell for male hysterectomy is necessarily empty—a structural zero. [*Further discussion of these log-linear models can be found in* SURVEY ANALYSIS, *article on* METHODS OF SURVEY ANALYSIS.]

Alternative approaches use information theory (for example, Ku et al. 1971) and a modification of analysis of variance (Margolin & Light 1974).

Other books that deal with counted data are Cox (1970), Fleiss (1973), Mosteller and Rourke (1973), and Plackett (1974). The Lewis and Burke material cited in the main bibliography (Lewis & Burke 1949) has been reprinted by Steger (1971) and by Lieberman (1971). A follow-up note was published in 1962 by Lana and Lubin.

DOUGLAS G. CHAPMAN

ADDITIONAL BIBLIOGRAPHY

BARNARD, G. A. 1947 Significance Tests for 2 × 2 Tables. *Biometrika* 34:123–138.

BISHOP, YVONNE M. M.; FIENBERG, STEPHEN E.; and HOLLAND, PAUL W. 1975 *Discrete Multivariate Analysis.* Cambridge, Mass.: M.I.T. Press.

CONOVER, W. J. 1974 Some Reasons for Not Using the Yates Continuity Correction. *Journal of the American Statistical Association* 69:374–382. → Includes comments and rejoinder.

COX, D. R. 1970 *The Analysis of Binary Data.* London: Methuen; New York: Wiley.

DAVIS, JAMES A. 1974 Hierarchical Models for Significance Tests in Multivariate Contingency Tables: An Exegesis of Goodman's Recent Papers. Pages 189–231 in Herbert L. Costner (editor), *Sociological Methodology, 1973–1974.* San Francisco: Jossey-Bass.

FLEISS, JOSEPH L. 1973 *Statistical Methods for Rates and Proportions.* New York: Wiley.

GOODMAN, LEO A. 1973 Guided and Unguided Methods for the Selection of Models for a Set of *T* Multi-dimensional Contingency Tables. *Journal of the American Statistical Association* 68:165–175.

HABERMAN, SHELBY J. 1974 *The Analysis of Frequency Data.* Univ. of Chicago Press.

KU, HARRY W. et al. 1971 On the Analysis of Multidimensional Contingency Tables. *Journal of the American Statistical Association* 66:55–64.

LANA, ROBERT E.; and LUBIN, ARDIE 1962 Chi Square Revisited. *American Psychologist* 17:793 only.

LIEBERMAN, BERNHARDT (editor) 1971 *Contemporary Problems in Statistics: A Book of Readings for the Behavioral Sciences.* New York: Oxford Univ. Press.

MARGOLIN, BARRY H.; and LIGHT, RICHARD J. 1974 An Analysis of Variance for Categorical Data: II. Small Sample Comparisons With Chi Square and Other Competitors. *Journal of the American Statistical Association* 69:755–764.

MENG, ROSA C.; and CHAPMAN, DOUGLAS G. 1966 The Power of Chi Square Tests for Contingency Tables. *Journal of the American Statistical Association* 61:965–975.

MOSTELLER, FREDERICK; and ROURKE, ROBERT E. K. 1973 *Sturdy Statistics: Nonparametrics and Order Statistics.* Reading, Mass.: Addison-Wesley.

PLACKETT, R. L. 1974 *The Analysis of Categorical Data.* London: Griffin; New York: Hafner.

STEGER, JOSEPH A. (editor) 1971 *Readings in Statistics for the Behavioral Scientist.* New York: Holt.

COURNOT, ANTOINE AUGUSTIN

Antoine Augustin Cournot (1801–1877), French mathematician, economist, and philosopher, was born at Gray (Haute-Saône), the son of a notary. He came from a family of farmers who had lived in Franche-Comté since at least the middle of the sixteenth century. Until age 15, Cournot attended the *collège* at Gray; then, for the next four years, he read a great deal on his own, especially works by scientists and philosophers, such as Laplace and Leibniz. After preparing at Besançon, Cournot was admitted in 1821 to the scientific section of the École Normale Supérieure, in Paris. The school was closed by the reactionary regime the following year, but Cournot stayed in Paris and in 1823 received the licentiate in sciences. He attended sessions at the Académie des Sciences and associated with the principal scholars of the day; it was through one of them that he met Proudhon. In

October 1823, Cournot entered the household of Marshal Gouvion Saint-Cyr in the double capacity of literary adviser to the marshal, who wanted to complete some unfinished manuscripts, and tutor to the marshal's son.

Cournot spent ten years with the marshal, all the while continuing his scientific studies. In 1829 he received a doctorate in science, writing a main thesis in mechanics and a supplementary one in astronomy. He also studied law. A series of articles that he published on scientific questions attracted the attention of the great mathematician Poisson, who was then professor at the École Polytechnique and later at the University of Paris. Poisson was then in charge of the instruction in mathematics throughout France and arranged Cournot's appointment to the chair of mathematical analysis in the faculty of sciences at Lyon. Cournot taught for only one year, however; subsequently, most of his life was spent in university administration, in which he was very successful. He became, successively, rector of the Académie de Grenoble in 1835, inspector general of the University of Dijon, and rector of the Académie de Dijon from 1854 to 1862. Cournot then accepted no further public positions and returned to Paris, where he died just as he was about to apply for membership in the Institut de France. Toward the end of his life he was nearly blind.

Cournot's administrative work left him ample time for a great deal of scientific writing. Unfortunately, his books suffer from a lack of the stimulation that contact with an audience gives to teachers. Ten of his works appeared between 1838 and 1877. These books have three major themes: (1) algebra, infinitesimal analysis, and calculus of probabilities; (2) the theory of wealth; and (3) the philosophy of science, the philosophy of history, and even general philosophy. These themes were interdependent in Cournot's work, and although the economic side will be stressed here, the profound unity of his thought must be remembered. Cournot started with mathematics, was then attracted to economics, and ended with a general interpretation of the world that was infused with his profound understanding of probabilities. It is the concept of probability that integrates the three parts of his work.

Cournot's rather melancholic and solitary temperament considerably delayed the influence he was ultimately to have. Modest and self-effacing, he did nothing to make his books attractive. They tend to be austere, crowded with facts and proofs. Even the titles he chose reflect his modesty: *Researches*, *Essay*, and *Considerations*. It is not surprising, then, that he is less well known in the field of probability than Laplace and Poisson, less appreciated among economists than Bastiat, Say, or Proudhon (to mention only French economists), and less quoted in philosophy than Comte or Spencer. And yet Cournot, having the rare knowledge required to use the language of all of these authors, combined many of their qualities and made a quasi-prophetic synthesis of their ideas.

Mathematics and probability theory. Cournot's mathematical works, econometrics aside, appeared between 1840 and 1850. In his *Exposition de la théorie des chances et des probabilités* (1843), he put forward a definition of statistics as that science which deals with collecting and coordinating numerous facts of every kind, in such a way as to obtain numerical relationships that are markedly independent of the anomalies of chance, and that manifest the existence of uniformly operating causes whose effects have been confounded, however, with other, accidental effects.

Cournot, living in the first half of the nineteenth century, stood at the crossroads of two ways of mathematical endeavor: the one originated with Pascal and Fermat and led to the work of Jacques Bernoulli, Gauss, Laplace, and Poisson on the doctrine of chances; the other, renouncing the mathematical study of chance and uncertainty, focused on the mathematics of rigorous determinations by algorithms that admit no margin of uncertainty—as if science were perfectly deterministic. This concept that the perfect is the determined was to play a major role in the development of the early science of economics, as one can see from the still current term "perfect competition." Cournot foresaw that science could not be intrinsically and definitively tied to such determinism. He believed that a science of margins and chances is not only viable but perhaps better suited to the needs of economics than a science of absolute, exact equilibria. Nonetheless, Cournot is often regarded as having introduced into economic theory the use of deterministic mathematics.

Economics. Cournot wrote three works in economics, in 1838, 1863, and 1877; thus, his career began and ended with writings on economics. When *Recherches sur les principes mathématiques de la théorie des richesses* (1838*a*) first appeared, it was such a fiasco that Cournot remained silent on the subject of economics for 25 years. He then wrote *Principes de la théorie des richesses* (1863), putting into "literary" language what he had previously said in the language of mathematics. Despite the concessions Cournot made in the form of presentation, the *Principes* was no better received than the *Recherches*; indeed, if Cournot had produced only his nonmathematical work, he would never have been recognized as anything but a minor figure.

Strictly speaking, to be sure, Cournot was not the first to have used mathematical language to express economic problems. There was Nicolas Canard, in France, whom Cournot did quote, but Canard did not have Cournot's scope and erudition. Great erudition is needed to combine economics and mathematics, and if mathematical economics has had its setbacks, the reason is that it has not always been guided by such clear minds as Cournot's.

Cournot's great merit is that, without saying so explicitly, he was the first to construct a true theory of prices and markets. Chapter 4 of *Recherches*, entitled "Of the Law of Demand," is the first model of its kind. Cournot was interested exclusively in the demand that is followed by an actual sale and that is, therefore, observable and measurable. He is recognized for having revealed the concept of function and for having made available to economists the immensely useful language of functional concepts. Sales are, in general, a decreasing function of price. This function is continuous, at least when the number of consumers is not limited. Moreover, the sales function is not purely abstract; it may be constructed on the basis of mean annually observed data. And since it is not merely an a priori function but an empirical, experimental one, Cournot may be considered to have made the initial step in developing econometrics.

In this econometric mode of analysis, a related idea is that of imperfection, hence of uncertainty and, in turn, of chance in measurement. Here again, Cournot's knowledge of probabilities saved him from incorrectly associating mathematics with the idea of rigorous precision. He asserted that even if the object of numerical calculation "were unattainable, it would be nevertheless not improper to introduce the unknown law of demand into the analytical combinations, by means of an indeterminate symbol; for it is well known that one of the most important functions of analysis consists precisely in assigning determinate relations between quantities to which numerical values and even algebraic forms are absolutely unassignable" ([1838*a*] 1960, p. 48). We must give up trying to grasp what cannot be grasped rigorously. It is for this reason, perhaps, that we must reason mathematically ". . . by showing what determinate relations exist between unknown quantities, analysis reduces these unknown quantities to the smallest possible number, and guides the observer to the best observations for discovering their values. It reduces and coordinates statistical documents; and it diminishes the labour of statisticians . . ." (*ibid.*, pp. 48–49).

Cournot's first construction is worked out in this spirit: it consists in drawing up, within suitable limits, tables of correspondence between the values of the demand, $D = f(p)$, and the price, p. In a second stage, the function $pf(p)$, the total value of the amount sold, is considered. This becomes the crux of the theory of markets. Long before Alfred Marshall, Cournot presented a theory of elasticity: depending on whether $\Delta D/\Delta p < D/p$ or $\Delta D/\Delta p > D/p$, the price increase will make the product $pf(p)$ larger or smaller. Hence commercial statistics should begin by dividing merchandise into two categories, depending on whether their current prices are lower or higher than the value making $pf(p)$ a maximum. In this formulation, Cournot was truly a great innovator.

Instead of attacking the general price equilibrium directly, as Léon Walras and Vilfredo Pareto were later to do, Cournot proceeded by gradual steps. He started with monopoly, then considered competition limited to a few participants, and in the end took up the case of indefinite or unlimited competition, using the intercommunication of markets to complete his theory. This procedure has been criticized, for economists in the classical tradition follow the inverse procedure, starting with unlimited competition or, as they call it, perfect competition and ending up with monopoly. However, Cournot's point of view has again been adopted in the modern theory of games and of the rational search for decisions. Although his theory of duopoly was criticized in 1883 by the mathematician Joseph Bertrand, the theory of bilateral monopoly was later to be erected on a similar base. The model with two parties was changed to a three-sided model (triopoly), and as the number of parties increased, the system was called oligopoly, pliopoly, and finally polypoly; the larger the number, the more closely the model approached that of classical competition. Today, makers of models no longer believe that there is an irreducible difference between models with many elements and those with few, but instead they relate the theory of markets to the general theory of economic interaction; hence their work is in the tradition of Cournot's model, at first so poorly understood. There is still much to be learned from this 1838 model. Although Cournot has been rehabilitated, his contribution to economics has not been exhausted.

Philosophy. In Cournot's philosophical works there emerge a philosophy of order and one of history. Cournot firmly maintained that, appearances to the contrary, chance does not imply disorder. By virtue of the theory of probabilities it is possible to see regularities: it is, as it were, the point of intersection of multiple independent causal series. And it permits the joining of the sciences with philosophy. The meaning of history, for all its incoherence, and the meaning of the future, for all its unpredictability, are profoundly related. To discover the course of ideas and events is the function of knowledge and the vocation of the human mind.

Cournot was a pioneer. He did nothing to court his contemporaries, and they, in turn, not only failed to appreciate him but ignored him. By a fitting reversal, his triumph came 80 years after his death. The most advanced of the econometric school recognize him as their ancestor. The theory of probabilities, whose full import Cournot realized, is a vital component of the structure of recent science. The problems that did not interest the men of his time are those that guide the building of tomorrow's world; Cournot had anticipated the concern with predictions and decisions that now preoccupies economists.

HENRI GUITTON

[*For the historical context of Cournot's work, see biographies of the* BERNOULLI FAMILY; GAUSS; LAPLACE; POISSON. *For discussion of the subsequent development of his ideas, see* PROBABILITY. *See also* Boulding 1968; Fox 1968; Markham 1968.]

WORKS BY COURNOT

1829 *Mémoire sur le mouvement d'un corps rigide soutenu par un plan fixe.* Paris: Hachette.

(1838*a*) 1960 *Researches Into the Mathematical Principles of the Theory of Wealth.* New York: Kelley. → First published in French.

1838*b* Mémoire sur les applications du calcul des chances à la statistique judiciaire. *Journal de mathématiques pures et appliquées* 3:257–334. → This is an early example of work on what is now called "latent structure."

(1841) 1857 *Traité élémentaire de la théorie des fonctions et du calcul infinitésimal.* 2d ed. Paris: Hachette.

1843 *Exposition de la théorie des chances et des probabilités.* Paris: Hachette.

1847 *De l'origine et des limites de la correspondance entre l'algèbre et la géométrie.* Paris: Hachette.

(1851) 1956 *An Essay on the Foundations of Our Knowledge.* New York: Liberal Arts Press. → First published in French.

(1861) 1911 *Traité de l'enchaînement des idées fondamentales dans les sciences et dans l'histoire.* New ed. Paris: Hachette.

1863 *Principes de la théorie des richesses.* Paris: Hachette.

(1872) 1934 *Considérations sur la marche des idées et des évènements dans les temps modernes.* 2 vols. Paris: Boivin.

(1875) 1923 *Matérialisme, vitalisme, rationalisme: Études des données de la science en philosophie.* Paris: Hachette.

1877 *Revue sommaire des doctrines économiques.* Paris: Hachette.

1913 *Souvenirs (1760–1860).* With an introduction by E. P. Bottinelli. Paris: Hachette. → Published posthumously.

SUPPLEMENTARY BIBLIOGRAPHY

[A. A. Cournot.] 1905 *Revue de métaphysique et de morale* 13:291–343. → The entire issue is devoted to Cournot.

BERTRAND, J. 1883 [Book Reviews of] *Théories mathématiques de la richesse sociale*, par Léon Walras; *Recherches sur les principes mathématiques de la théorie de la richesse*, par Augustin Cournot. *Journal des savants* [1883]:499–508.

BOMPAIRE, FRANÇOIS 1931 *Du principe de liberté économique dans l'oeuvre de Cournot et dans celle de l'école de Lausanne (Walras, Pareto).* Paris: Sirey.

BOTTINELLI, E. P. 1913 *A. Cournot: Métaphysicien de la connaissance.* Paris: Hachette.

►BOULDING, KENNETH E. 1968 Demand and Supply: I. General. Volume 4, pages 96–104 in *International Encyclopedia of the Social Sciences.* Edited by David L. Sills. New York: Macmillan and Free Press.

EDGEWORTH, F. Y. (1894) 1963 Antoine Augustin Cournot. Volume 1, pages 445–447 in Robert H. I. Palgrave, *Palgrave's Dictionary of Political Economy.* New York: Kelley.

►FOX, KARL A. 1968 Demand and Supply: II. Econometric Studies. Volume 4, pages 104–111 in *International Encyclopedia of the Social Sciences.* Edited by David L. Sills. New York: Macmillan and Free Press.

LIEFMANN-KIEL, ELISABETH 1937 Die wissenschaftliche Methode und das Gesamtwerk Cournots. *Archiv für mathematische Wirtschafts- und Sozialforschung* 3: 238–251.

LOISEAU, GEORGES 1913 *Les doctrines économiques de Cournot.* Paris: Rousseau.

►MARKHAM, JESSE W. 1968 Oligopoly. Volume 11, pages 281–290 in *International Encyclopedia of the Social Sciences.* Edited by David L. Sills. New York: Macmillan and Free Press.

MENTRÉ, FRANÇOIS 1908 *Cournot et la renaissance du probabilisme au XIXe siècle.* Paris: Rivière.

MENTRÉ, FRANÇOIS 1927 *Pour qu'on lise Cournot.* Paris: Beauchesne.

MILHAUD, GASTON S. (1902–1911) 1927 *Études sur Cournot.* Paris: Vrin.

ROY, RENÉ 1933 Cournot et l'école mathématique. *Econometrica* 1:13–22.

SEGOND, J. 1911 *Cournot et la psychologie vitaliste.* Paris: Alcan.

COVARIANCE, ANALYSIS OF

See LINEAR HYPOTHESES, *article on* ANALYSIS OF VARIANCE.

CROSS-SECTION ANALYSIS

Empirical analysis is concerned with the establishment of quantitative or qualitative relations between observable variables. From a temporal point of view two kinds of data are used in empirical analysis—cross-section data and time series data. Cross-section data are observations on variables at a point of time, whereas time series data are observations covering several time periods. Sometimes the two kinds of data are used together to overcome some specific difficulties.

The unit of observation in a cross section is generally, although not necessarily, elementary, such as a firm or a consumer; the unit of observation in a time series is generally an aggregate. The empirical association between the type of data and the level of aggregation contributes to differences in the results obtained with the two kinds of data.

While cross-section analysis is used in many of the social sciences, this article focuses on applications of cross-section analysis in economics. [*For other applications, see* SURVEY ANALYSIS.]

Some of the more important areas of economics in which cross-section data have been used are the estimation of Engel curves (Liviatan 1964; Prais & Houthakker 1955), the estimation of consumption functions (Friedman 1957), the estimation of production functions (Bronfenbrenner & Douglas 1939), and the estimation of investment functions (Kuh 1963).

Reasons for using cross-section data. There are several reasons why cross-section data are often superior to time series data for estimating economic relations.

(1) Cross-section data contain large variations in some variables whose variations over time are only moderate and often subject to trend. For instance, there is a much larger variation in income among consumers at a particular point of time than in average per capita income over time. Likewise, there are wide variations in productive capacity and sales among firms in a cross section, but only relatively small variations in these variables in a time series.

(2) The size of a cross-section sample of data can usually be increased enough to make sampling variance relatively negligible (Kuh 1963).

(3) Multicollinearity among variables in a cross section is often less acute than among corresponding variables in a time series.

(4) The problem of interdependent disturbances, which frequently arises in the analysis of time series because of trends in time series data, usually does not arise in the analysis of cross-section data, simply because the order of the observations has no meaning.

(5) In some cases cross-section data are more reliable; and the variables that can be measured in a cross section often correspond more closely to the variables defined and studied in economic theory. For instance, cross-section data on consumer budgets furnish a precise account of consumption by commodity, and such data can be more useful than time series data for estimating demand functions. On the other hand, some variables are subject to larger errors in a cross section, although the errors are of a different nature than those encountered in a time series. For example, the discrepancy between observed income and permanent income is larger in a cross section than in an aggregate time series (Friedman 1957). This problem is discussed below.

(6) The distinction between the individual economic unit and the market, which is basic in economic analysis, leads to different classifications of variables at the microeconomic and macroeconomic levels of analysis. For instance, prices can be assumed to be given (exogenous) for a consumer, but they should be considered as dependent (endogenous) variables when markets are analyzed. Cross-section analysis usually deals with the behavior of microeconomic units; therefore it can often proceed on the basis of a simpler economic model than time series analysis of macroeconomic data.

It should be noted that cross-section analyses had been undertaken long before the statistical advantages of cross-section data were recognized; and in some cases, such as the study of Engel curves, cross-section analyses also preceded the rationale that was to be provided for them by economic theory (Staehle 1934–1935; Stigler 1954).

The scope of cross-section analysis. The explanatory variables appearing in economic relations can be divided into three groups: (1) variables that vary in a cross section and over time; (2) variables that are stable in a cross section and vary over time; (3) variables that vary in a cross section and are stable over time.

It is the existence of the first group that makes cross-section analysis valuable in economics. For instance, income is an important variable in consumption functions, and its impact on consumption can be learned from cross-section analysis. If income did not vary in a cross section, cross-section analysis of consumption functions would be impossible; if income did not vary over time, the results of cross-section analysis would be of little interest.

The existence of the second group implies that cross-section analysis by itself is not sufficient to explain the variations of many economic variables. For example, variations in consumption resulting from changes in income can be determined by cross-section analysis, since income varies in a cross section; but variations in consumption resulting from changes in prices cannot be determined by cross-section analysis, since prices generally dc not vary in a cross section. Thus, to obtain complete economic relations, variables in the second group must be included in the relations, and the coefficients of these variables must be estimated by time series analysis.

The variables in the third group may be very important in explaining variations of economic variables in a cross section. For instance, age and sex and their interaction may explain more of the variations in consumption of certain commodities among individuals than does income. Yet they may be relatively unimportant variables to consider in most economic decisions or predictions, since neither the average age nor the sex composition of the population varies much over time.

The variables in the third group are largely noneconomic, that is, they are not endogenous variables in current economic theory. Among the important variables in this group are those measuring the uncertainty faced by decision-making units, particularly firms. According to the theory of the firm facing no uncertainty, the amount of a good supplied by a firm and the amounts of productive factors demanded by a firm depend on the price of the good and on the prices of the productive factors. Given input and output prices, the firm should choose the input–output configuration that maximizes its profits. But in many empirical analyses of cross-section data, it is found that prices explain only a small proportion of variations in input demands and output supplies among firms. The unexplained variations may be attributable to differences in the degree of uncertainty among firms or to differences in the response to uncertainty among firms. Firms may be certain of the prices that will prevail when they execute their decisions, but they may be uncertain of the quantity of output that can be sold or the amounts of inputs that can be purchased at those prices.

Given uncertainty of this kind, a firm may deviate from its profit maximizing input–output configuration, so as to avoid partially the costs that would be incurred if plans cannot be realized, e.g., the costs of undesired inventory accumulations. In such cases a firm might consider all input–output configurations that yield profits greater than a preassigned fraction, say 95 per cent, of maximum profits and select one of these configurations for final execution. The range of acceptable configurations may be very large. Hence, firms may differ considerably in their decisions, the differences resulting from variations in uncertainty and in the response to uncertainty among firms.

From these considerations it is evident that economic variables may well explain only a small fraction of the variations of a dependent variable in a cross-section relation and that other variables that appear or should appear in the relation will perform the major explanatory role. Thus, there is no a priori requirement that cross-section relations produce high degrees of explanation or that, if they do, their explanatory powers result from the economic variables included (Grunfeld & Griliches 1960). Yet when all individuals are taken together, the noneconomic variables generally offset each other, and economic variables, such as prices, may turn out to be the important explanatory variables.

The problem of multiperiod relations. In general, the time horizon of economic decisions extends beyond a single period. Consequently, "true" economic relations contain variables of several periods; but in cross-section analysis one can usually observe only the current variables, and the problem of inferring a complete relation from such partial information arises. To illustrate, according to the permanent income hypothesis the consumption function is a relation between permanent consumption and permanent income. While permanent consumption may be approximately equal to observed consumption, permanent income is an average of incomes of several periods. Hence, consumption in any given year will depend, apart from errors, on a stream of income which, except for its current component, is unobserved. Similarly, in deciding whether to invest in durable assets, a firm takes into account the profits that the assets will yield not only in the year of the investment but also in future years. Thus, in order to ascertain empirically the determinants of business investment, it is necessary to take into account firms' expectations of the future, which are unobservable.

The nature of the problem can also be seen by expressing an economic relation as

$$f(\boldsymbol{x}_0, \boldsymbol{x}_1, \cdots, \boldsymbol{x}_t, \cdots) = 0,$$

where $\boldsymbol{x}_t$ is the vector of variables from period t that enter the relation. The problem discussed above exists when observations are available only for a particular $\boldsymbol{x}_t$ (as is the case in cross-section analysis) and when f is not separable with respect to that $\boldsymbol{x}_t$. In such cases the estimated cross-section relation is subject to bias resulting from the omission of the unobserved variables. This is all that can be said at this level of generality.

More specific conclusions can be arrived at when the underlying theory is more specific with respect to the economic relation to be estimated and with respect to the relationships between the observed and the unobserved variables. For instance, in the case of the permanent income hypothesis the stream of income over time in the true relation is replaced by one unobserved variable—permanent income. For purposes of estimation it is assumed that observed income measures permanent income with an unsystematic error. With specifications of the properties of the error term, the estimation problem reduces to a regression problem with errors in the variables, and the appropriate statistical methods are applied [*see* LINEAR HYPOTHESES, *article on* REGRESSION].

It should be noted that if the measurement errors are unsystematic, then aggregating the observations will reduce the measurement errors. Thus, while measurement errors are a serious problem in estimating most multiperiod relations with cross-section data, they usually do not present much of a problem in estimating such relations with aggregate time series data.

In studying multiperiod relations, considerable information can be gained by taking repeated observations on microeconomic units over time. Liviatan (1963) used such information to perform a rich variety of tests of the permanent income hypothesis.

The study of multiperiod relations is further complicated by the existence of uncertainty in the decisions of microeconomic units. Consumers and firms do not have complete information on the future values of the variables exogenous to them. In such cases observed values of variables cannot be identified with expectations, and the utilization of repeated observations does not solve this problem. Assumptions must be made regarding the formation of expectations and regarding the behavior of individuals under uncertainty. Some of the consequences of uncertainty were noted in the preceding section.

Estimation. An initial step in estimating cross-section relations is choosing the explanatory variables to be included. Returning to the classifications noted above, variables in the second group obviously cannot be included, since they do not vary in a cross section. Variables in the first group should be included for two reasons. First, their variations in the cross section may contribute to explaining the variations in the dependent variable of the relation. Second, since they vary over time, cross-section estimates of their coefficients will be useful in making intertemporal forecasts with the estimated relation. Variables in the third group, which are specific to the cross section, may be of little interest for forecasting intertemporal changes in the dependent variables. However, if they are correlated with variables in the first group, they should also be included in the relation, to avoid bias in the estimates of the coefficients of the variables in the first group. For example, the size of the family is included as a variable in cross-section studies of Engel curves, even though it may have little to contribute in making intertemporal forecasts (Liviatan 1963; Prais & Houthakker 1955).

Sometimes the variables in the third group are not quantifiable but their attributes can be specified. They are then introduced into the analysis by grouping individuals according to the attributes and estimating separate relations for each group. For example, in studying consumption functions one of the variables in the third group might be "place of residence." Individuals in a cross-section sample might then be grouped according to geographical areas. After fitting the relation to each group, one may test the hypothesis that individuals in all groups behave in the same manner, i.e., that the same relation holds for all the groups and that the so-called group effects are insignificant. Covariance analysis provides the statistical framework for testing the equality of intercepts and the equalities of some or all of the slopes in the group relations. In such analysis, within-group variations of the variables (deviations of observations in the group from the group mean) are utilized, so at least two observations per group are required.

It may happen that the appropriate groups are identical to the units of the observations. For example, managerial ability is not quantifiable but must be allowed for in a cross-section study of production functions. However, the unit of observation may be the firm, and managerial ability probably differs for each firm. In that case, covariance analysis is impossible with just one cross section, since there is only one observation per group. If managerial ability is to be handled by the use of covariance analysis, repeated observations over time must be made on each firm. Since this calls for a combination of time series and cross-section data, some of the variables in the second group must also be included in the relation. If some of these variables are not directly quantifiable, their effects may be allowed for by introducing different intercepts and slopes for the various years in the sample (Mundlak 1963).

Analysis that is based merely on within-group variations of the variables ignores the between-group variations (deviations of the group means from the mean of all the observations), which are often much larger. The between-group variations can be utilized if the explanatory variables in the estimated relations are not correlated with the group effects. For example, if income is uncorrelated with "place of residence," then the mean consumptions of the groups can be regressed on the mean incomes of the groups. This is particularly desirable when the variables are subject to unsystematic measurement errors (as is the case when testing the permanent income hypothesis), because averaging observations for each group will eliminate most of the measurement errors. [*See* LINEAR HYPOTHESES, *article on* ANALYSIS OF VARIANCE.]

The problem of measurement errors is also handled in cross-section analysis by the use of instrumental variables [*see* SIMULTANEOUS EQUATION ESTIMATION]. For instance, in estimating the consumption function it is assumed that observed income measures permanent income with error. If this error is not serially correlated, the income of one year may be used as an instrumental variable for estimating the consumption function in another year. Note that this again calls for repeated observations over time on the incomes of the microeconomic units.

When variables in the relation to be estimated are jointly determined, the explanatory variables in the relation may not be independent of the disturbance term. For instance, factor inputs may not be independent of the disturbance term in a production function (Mundlak 1963; Walters 1963). In such cases estimating the relation by direct least squares will result in biased estimates of the coefficients. Various multiequation estimation procedures are available to overcome least-squares bias. However, they depend fundamentally on restrictions that may be satisfied only over time and not in a cross section. An exception is the instrumental variables method which, for example, uses lagged (or lead) factor inputs as instrumental variables in estimating a production function.

Special problems exist in estimating dynamic cross-section relations that involve adjustment processes [*see* DISTRIBUTED LAGS]. The empirical implications of many of the currently used adjustment models may be more applicable to group behavior than to individual behavior. While individuals facing uncertainty may not react instantaneously to changes in the variables on which they base their decisions, changes in their decisions may be discrete rather than continuous. It may be advantageous for them to make larger adjustments less often. However, the adjustment models employed in empirical work generally assume continuous adjustment. Since the frequency and size of adjustments may vary among individuals, continuous adjustment might be the result for the group. Here again, repeated observations over time on individuals can be utilized to surmount this difficulty.

Problems of application. The application of cross-section estimates of economic relations to intertemporal predictions of aggregates is subject to several difficulties. The cross-section estimates may depend on the values of the variables that are constant in the cross section but vary with time. Presumably this problem should be solved by a correct specification of the estimated relations, so that the variables that are stable in the cross section will not affect the estimates of the coefficients of the cross-section variables. For instance, income coefficients should be independent of prices, so that an estimated income coefficient will be applicable to periods with different prices. While such independence may nearly exist for income coefficients and prices, it may not exist in relations such as investment functions, where less regularity is the rule.

Furthermore, estimation of the coefficients of variables that are constant in a cross section but vary over time requires time series data. In estimating their coefficients by time series analysis, it is possible to use cross-section estimates of the coefficients of the variables in the first group in the time series relation. This of course can be done only if the cross-section estimates do not vary much from year to year. Income elasticities estimated from cross-section analysis are often grafted onto time series demand equations (Tobin 1950).

Finally, the transformation of estimates obtained for individuals to estimates applicable to markets is somewhat problematic. Aggregation over individuals is sensitive to the distribution of the explanatory variables among individuals and may lead to aggregate relations that differ in form from the individual relations (Houthakker 1955; Tobin 1950).

YAIR MUNDLAK

BIBLIOGRAPHY

BRONFENBRENNER, MARTIN; and DOUGLAS, P. H. 1939 Cross-section Studies in the Cobb–Douglas Function. *Journal of Political Economy* 47:761–785.

FRIEDMAN, MILTON 1957 *A Theory of the Consumption Function.* National Bureau of Economic Research, General Series, No. 63. Princeton Univ. Press.

GRUNFELD, YEHUDA; and GRILICHES, Z. 1960 Is Aggregation Necessarily Bad? *Review of Economics and Statistics* 42:1–13.

HOUTHAKKER, HENDRIK S. 1955 The Pareto Distribution and the Cobb–Douglas Production Function in Activity Analysis. *Review of Economic Studies* 23, no. 1:27–31.

KLEIN, LAWRENCE R. 1953 *Textbook of Econometrics.* Evanston, Ill.: Row, Peterson. → See especially pages 211–241.

KUH, EDWIN 1963 *Capital Stock Growth: A Micro-econometric Approach.* Contributions to Economic Analysis, 32. Amsterdam: North-Holland Publishing.

LIVIATAN, NISSAN 1963 Tests of the Permanent-income Hypothesis Based on a Reinterview Savings Survey.

Pages 29–59 in *Measurement in Economics: Studies in Mathematical Economics and Econometrics in Memory of Yehuda Grunfeld.* Stanford Univ. Press. → A "Note" by Milton Friedman and a reply by Liviatan appear on pages 59–66.

LIVIATAN, NISSAN 1964 *Consumption Patterns in Israel.* Jerusalem: Falk Project for Economic Research in Israel.

MICHIGAN, UNIVERSITY OF, SURVEY RESEARCH CENTER 1954 *Contributions of Survey Methods to Economics.* Edited by Lawrence R. Klein. New York: Columbia Univ. Press.

MUNDLAK, YAIR 1963 Estimation of Production and Behavioral Functions From a Combination of Cross-section and Time-series Data. Pages 138–166 in *Measurement in Economics: Studies in Mathematical Economics and Econometrics in Memory of Yehuda Grunfeld.* Stanford Univ. Press.

PRAIS, S. J.; and HOUTHAKKER, H. S. 1955 *The Analysis of Family Budgets With an Application to Two British Surveys Conducted in 1937–1939 and Their Detailed Results.* Cambridge Univ. Press.

STAEHLE, HANS 1934–1935 Annual Survey of Statistical Information: Family Budgets. *Econometrica* 2:349–362; 3:106–118.

STIGLER, GEORGE J. 1954 The Early History of Empirical Studies of Consumer Behavior. *Journal of Political Economy* 62:95–113.

TOBIN, JAMES 1950 A Statistical Demand Function for Food in the U.S.A. *Journal of the Royal Statistical Society* Series A 113, part 2:113–141.

WALTERS, ALAN A. 1963 Production and Cost Functions: An Econometric Survey. *Econometrica* 31:1–66.

CUMULATIVE DISTRIBUTION FUNCTION

See PROBABILITY.

CURVE FITTING

See DATA ANALYSIS, EXPLORATORY; GOODNESS OF FIT; LINEAR HYPOTHESES, *article on* REGRESSION.

CYBERNETICS

See the biography of WIENER.

CYCLES

See under TIME SERIES; *see also* BUSINESS CYCLES: MATHEMATICAL MODELS.

D

DANTZIG, DAVID VAN

See VAN DANTZIG, DAVID.

DATA ANALYSIS, EXPLORATORY

► *This article was specially written for this volume.*

Exploratory data analysis is the manipulation, summarization, and display of data to make them more comprehensible to human minds, thus uncovering underlying structure in the data and detecting important departures from that structure.

These goals have always been central to statistics and indeed to all scientific inquiry. In the 1960s and 1970s exploratory data analysis has seen a renaissance, and powerful statistical imaginations have widened and deepened its scope. A central figure in this renaissance has been John Wilder Tukey, whose *Exploratory Data Analysis* ([1970] 1977) provides a detailed introduction. The development of exploratory data analysis has also been aided by the increasing availability of electronic computers for calculation and, perhaps more important, for efficient graphical display. These machines have also made possible the collection and organization of very large amounts of data, thus often presenting the analyst with a problem of sensible selection from an overbundance of data.

Exploratory data analysis has much in common with the field called descriptive statistics. [*See* STATISTICS, DESCRIPTIVE.] To the aims of descriptive statistics, namely, the summarization and presentation of data, data analysis adds the goals of discovering structure and anomalies. Both exploratory data analysis and descriptive statistics differ from much of formal statistics in flexibility of application and in relative indifference toward probabilistic criteria for inference. Tukey (1977) has developed an analogy in which exploratory data analysis is the activity of a detective gathering evidence and questioning assumptions in an attempt to make a case that later may be formally tested in the court of statistical inference. This detectivelike inquiry has long been an important, though rarely discussed, component of the statistician's art. In many current problems it is the critical component.

Exploratory analysis of a body of data begins not with assumptions or a statistical model but with an examination of the data. At first this may be just a rough overview; graphs or plots of the data are particularly useful. The dominant patterns in the data are then described by a statistical model. Typically this is a simple model whose parameters may be readily estimated. Since at this stage the form of the model may only be guessed at, strong assumptions (such as normality, additivity of effects, independence of observations) should not be made. Robust estimation procedures, which work reasonably well under a variety of conditions are required. [*See* ERRORS, *article on* EFFECTS OF ERRORS IN STATISTICAL ASSUMPTIONS; *see also* ESTIMATION.]

The estimated model, describing the gross features of the data, may then be removed, so to speak, from the data. This removal typically involves a subtraction or other transformation based on the estimated model. With the first large component of the structure removed, the reduced data may be examined for finer structure. Again this structure may be modeled, estimated, and removed. The process may be repeated until all of the apparent structure has been removed and the residual, reduced data appears as patternless, "random" noise. At each stage, examination of data is aided by graphs, plots, tabulations, or other displays.

The success of this activity depends on the knowledge and persistence of the investigator. The modeling requires understanding of the basic problem, and of the data acquisition and measurement processes. One must anticipate what variables are likely to be important and how they are likely to interact. Persistence is required to continue the repeated peeling away, layer after layer, from the data. The investigator must continue to ask "Is anything else apparent in the data?"

This step-by-step procedure does not usually lead directly to an appropriate analysis. Often an investigator must backtrack or start anew. The discovery of fundamental relations may require substantial modifications in the model. Variables that at one point seem important may turn out to be otherwise.

The iterative procedure of examination, estimation, and removal is different from the procedures of formal statistical inference. In formal analysis, literally taken, rigorous inferences are drawn only from procedures selected when the experiment or survey was designed. In the formal framework post hoc procedures are suspect, not only because they change probabilistic characteristics in complex ways but, more fundamentally, because of concern with subjective bias or prior prejudice. Before a formal inference may be drawn, however, we must first discover what the inference is to be about and how data should be acquired and represented to facilitate the inference procedure. The activity of exploratory analysis precedes the procedures of formal statistical inference. Formal inference procedures can be applied only when there is sufficient understanding to transform the problem into one approximately satisfying rather specific assumptions.

Exploratory analysis has different goals and is used at a different stage of an inquiry than are formal inference procedures. Exploratory analysis requires different techniques. Graphical displays are used to discover patterns and structure *and* to display departures from an estimated pattern. Estimation or fitting procedures must be simple, flexible, and well understood so that they may be applied in many situations. The tools are typically more like jackknives than razors. Statistical power and efficiency are relatively unimportant properties of the techniques used at this stage, compared with robustness and generality of application.

The following sections describe some of the statistical tools useful in exploratory data analysis.

Exploratory graphical procedures

Graphical tools have always played an important role in statistics. [*See* GRAPHIC PRESENTATION.] These descriptive procedures are used primarily to exhibit the over-all structure or pattern of the data, that is, to summarize a statistical model. Exploratory graphical procedures, in contrast, are most often used to show *departures* from a contemplated structure. This general difference in purpose leads to the predominant use of different types of display.

Appropriate displays can quickly show how well a statistical model fits, where the fit is poor, and even how the model might be improved. To do this, the displays should have known behavior under "ideal" situations. For example, if the ideal data form a horizontal straight line, departures from the ideal may be easily detected. Different departures will register in different ways. The most important departures should be the most apparent.

Stem-and-leaf displays. A stem-and-leaf display is a simple kind of data representation, requiring only pencil and paper. It also has advantages as a computer-generated display. For a full description of the method and its applications see Tukey (1977); the following is only an introductory example.

In a psychological experiment to compare the effectiveness of several proposals for computer language design, subjects were given programming tasks and then scored on their performance. (See Weissman 1974.) The scores for 48 subjects were obtained; they were 69, 9, 37, 18, 44, 38, 26, 20, 34, ... The scores are to be displayed to assist in the possible reexpression prior to further manipulation. [*See* STATISTICAL ANALYSIS, SPECIAL PROBLEMS OF, *article on* TRANSFORMATIONS OF DATA.] Most of the observations fall in the range from 0 to 100 or from 0 to 10 in the "tens" digit. If these tens digits are written in a column (stem), the units digits (leaves) may be appended as illustrated in Table 1.

One score is recorded, and hence one digit is added at each step in constructing the final display. The largest score, 136, is displayed. There is no need to squeeze the rest of the display. The break in the display indicates that this score is well separated from the rest. Many variations of this scheme will work.

The display is useful in sorting the data; further, many standard descriptive statistics may be found directly from the display. For example, the median is 26, the quartiles are 19 and 38. The positive

Table 1 – Steps in constructing a stem-and-leaf display*

Step 0	Step 1	Step 2	Step 6	Step 48
0	0	0 9	0 9	0 9
1	1	1	1 8	1 84459945986
2	2	2	2	2 60680585952815 0669
3	3	3	3 78	3 7848086495
4	4	4	4 4	4 4130
5	5	5	5	5 2
6	6 9	6 9	6 9	6 92
7	7	7	7	7
8	8	8	8	8
9	9	9	9	9
10	10	10	10	10
				13 6

* Numbers are programming scores for 48 subjects.

Source: Data from Weissman 1974.

skewness of the data is evident. Attention is directed to the unusually large values 136, 69, and 62. [*See* NONPARAMETRIC STATISTICS, *article on* ORDER STATISTICS; STATISTICAL ANALYSIS, SPECIAL PROBLEMS OF, *article on* OUTLIERS *for techniques that deal with unusually large or small values.*]

A stem-and-leaf display is itself a crude histogram. A standard histogram is a more formal, although for small data sets a less informative, display. Histograms are more practical, however, when there are more than about 200 observations.

Displays based on distributions. Sometimes a display is used to show how far a batch of numbers departs from a particular distribution or kind of distribution. Such a display might indicate what transformation would produce numbers for which the assumed distribution is more reasonable. Histograms and stem-and-leaf displays would be considered for this purpose if more sensitive methods were not available. Difficulties arise when comparing a histogram with a curved approximation as in Figure 1, where the comparison is between the data in Table 1 and a (truncated) normal distribution. [*See* GOODNESS OF FIT.]

A more effective display is obtained by subtracting the fitted curved distribution from the histogram. Figure 2 illustrates the suspended histogram resulting from this procedure. This display directs attention to the pattern of departures rather than to the known shape of the contemplated distribution.

There are more sensitive displays. If sorted, ordered data (order statistics) $x_{(1)} < x_{(2)} < \cdots < x_{(n)}$ are to be compared with a cumulative distribution $F(x)$, a plot of i against $F(x_{(i)})$ in the ideal case yields a straight line. [*See* NONPARAMETRIC STATISTICS, *article on* ORDER STATISTICS.] This plot has

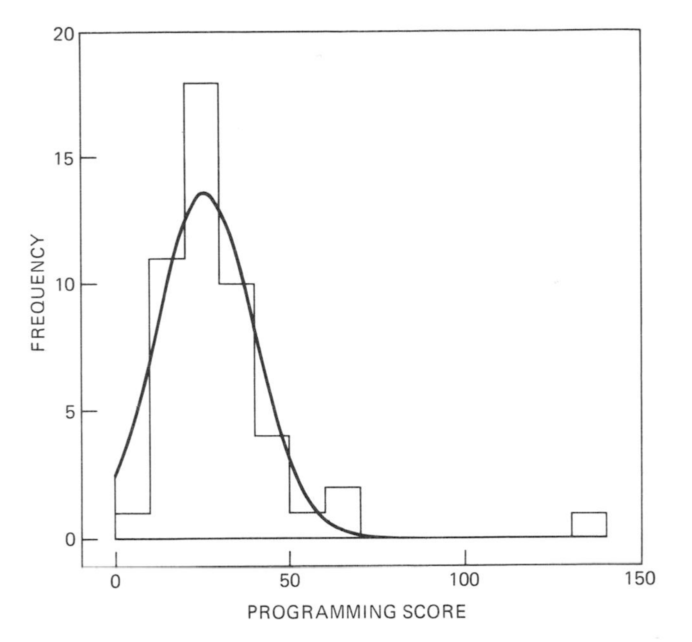

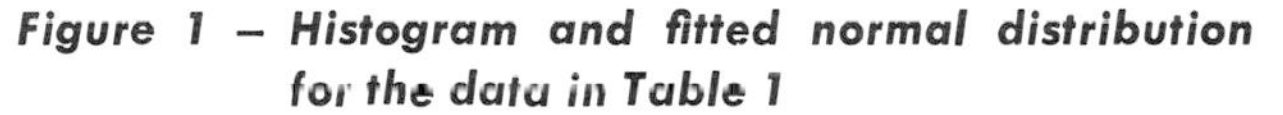

Figure 1 – Histogram and fitted normal distribution for the data in Table 1

Source: Data from Weissman 1974.

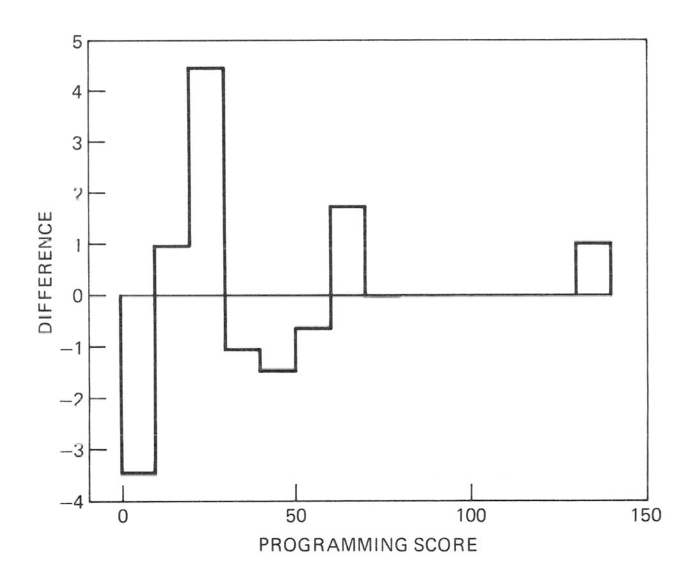

Figure 2 – Suspended histogram displaying the differences in Figure 1

Source: Data from Weissman 1974.

the advantage that each datum is represented without grouping. It has the disadvantage of shrinking the length of the tails of the sample relative to the length of the middle. Most common distributions for data are bell-shaped and resemble the normal distributions in the middle. The important differences, which will be emphasized by a good display, often occur in the tails.

Quantile plots achieve this goal. A quantile plot is constructed by plotting $x_{(i)}$ against $z_{(i)}$, a corresponding quantile point in the comparison distribution. One possibility is to choose $z_{(i)}$ so that

$$F(z_{(i)}) = i/(n+1).$$

(The addition of 1 in the denominator above yields a set of probabilities symmetric about 0.5.) This relation between quantiles, z, and proportions, $i/(n+1)$, is used in some settings where F is a standard normal distribution and $z+5$ is called the probit of $i/(n+1)$. Some special graph papers are available, labeled in terms of probability, so that one can plot $x_{(i)}$ against $i/(n+1)$ with ease. These are called probability graph papers and are available for a variety of distribution families. Figure 3 plots the data of Table 1 using normal probability paper.

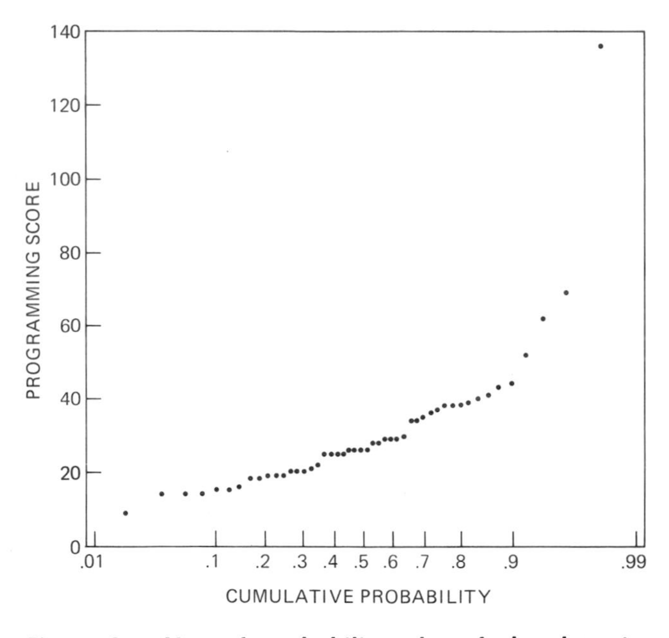

Figure 3 – Normal probability plot of the data in Table 1

Source: Data from Weissman 1974.

The ideal configuration of this plot, if the distribution assumption (here normality) holds, is a straight line with positive slope. The unusual observations 136, 69, and 62 are clearly visible. The upward curve suggests skewness in the data. Comparisons with horizontal lines are even more striking. For examples of flat probability plots, see Tukey (1977) and Andrews and Tukey (1972). Daniel and Wood (1972, chapter 3) give many examples of quantile plots.

Plots of residuals. The displays described above present a single batch of numbers. In most problems we are interested also in relations between variables or batches. Often these relations are summarized in a statistical model that is fitted to the data. The discrepancies or differences between the data and the fitted values are called residuals. Plots of residuals are useful in assessing the adequacy of the fitted model, in uncovering a dependence of the data on factors missing in the model, and in pointing to other improvements in the model. The examination of residual plots is an important part of all regression analyses where the form of the regression and the coefficients form the model, and we can think of the dependent variable or the response as the data being modeled. Some plots and their uses are described below. In general these plots should eventually appear as a scatter of points with no pattern or structure. Patterns are characteristic of residuals from incomplete models, that is, models that can be improved.

Plots of residuals against fitted values. If a good model has been appropriately fitted to the data, the residuals tend to resemble random noise and exhibit no marked dependence on other variables. Thus a plot of residuals against the fitted values should appear as patternless random scatter. Patterns or trends in such a plot suggest an inadequacy of the model or of the fitting procedure. For example, the spread or variability of the residuals may seem to change with the fitted values. This is an indication that the variables might be more appropriately expressed after some transformation. It is often an indication that important factors, related to the ones included in the model, have been neglected. Box and Cox (1964) discuss an experiment in which the survival times of animals subject to treatment combinations involving three poisons and four antidotes are analyzed. Figure 4 is a plot of residuals against fitted values for this experiment. The plot suggests that the variability of the survival times increases with the average or expected survival time. This awkward dependence is often removed by first transforming the response. Figure 5 is a similar plot based on the residuals calculated using survival rate = 1/time. The inverse transformation has removed much of the apparent trend in variability.

Trends in any plot of residuals thus indicate that the response is not linear in the factors considered. Such trends suggest that perhaps other factors should be considered, or perhaps another transformation of these factors will help.

Observations departing unusually from the pattern of the rest can be readily detected from a residual plot. Such observations require special attention. If they cannot be related to the other observations, they are usually set aside and treated separately. The bulk of the data may be fitted first and then attention focused on the interesting anomalies.

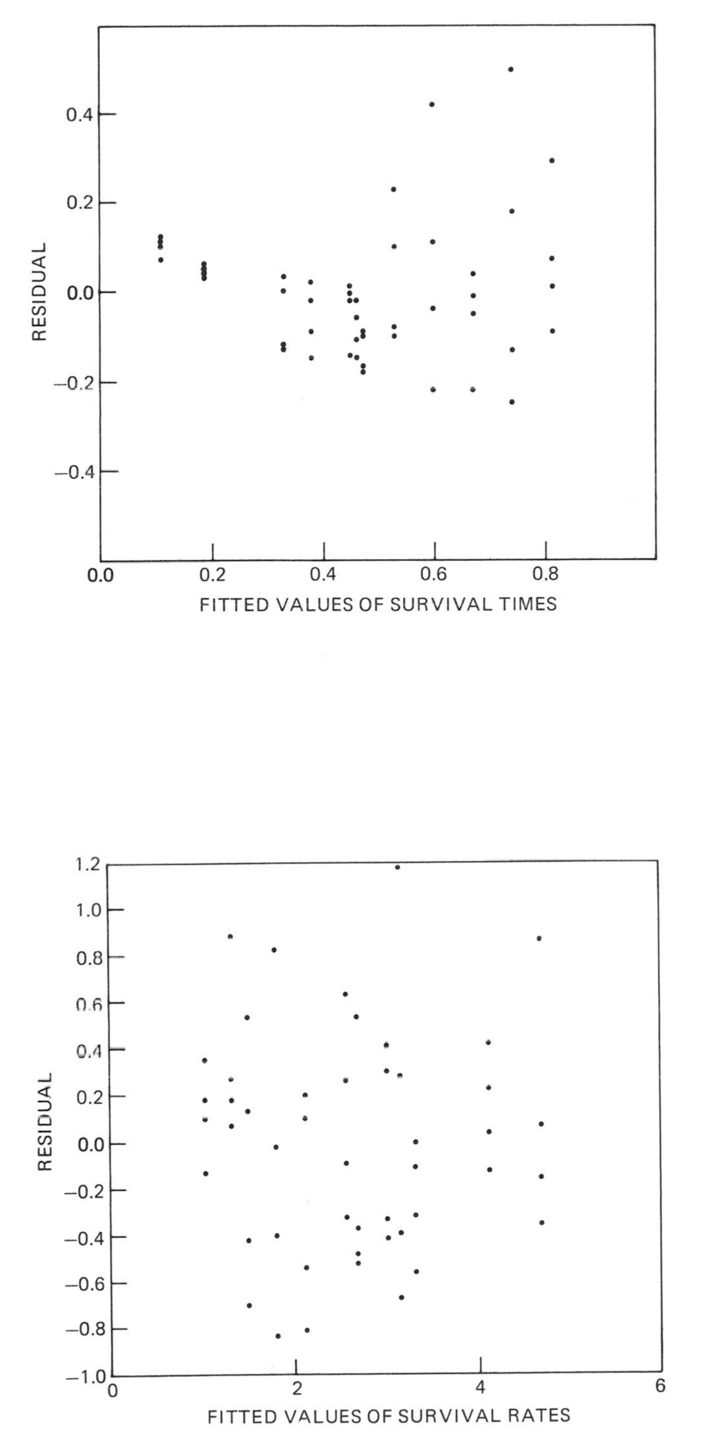

Figure 5 – Same plot as Figure 4 but based on residuals calculated after first transforming the response

Source: Data from Box & Cox 1964.

Plots of residuals against variables included in the model. Plotting residuals in turn against each of the variables included in the model gives useful information about the dependence of the response on these variables. When this dependence has been appropriately formulated, these plots will appear as a random scatter. Trends or patterns suggest a more complex dependence (for example, quadratic).

The plot also indicates where the information about the parameters comes from. If the plot is well distributed along the "variable" axis, the estimates of the parameters of the model are influenced by most of the observations. However, if the variable points are clumped except for a few stragglers, the estimated values of the parameters will be strongly influenced by whether these stragglers are included or excluded. [*Some ways of dealing with outliers are discussed in* NONPARAMETRIC STATISTICS, *article on* ORDER STATISTICS; ERRORS, *article on* EFFECTS OF ERRORS IN STATISTICAL ASSUMPTIONS; *and* STATISTICAL ANALYSIS, SPECIAL PROBLEMS OF, *article on* OUTLIERS.]

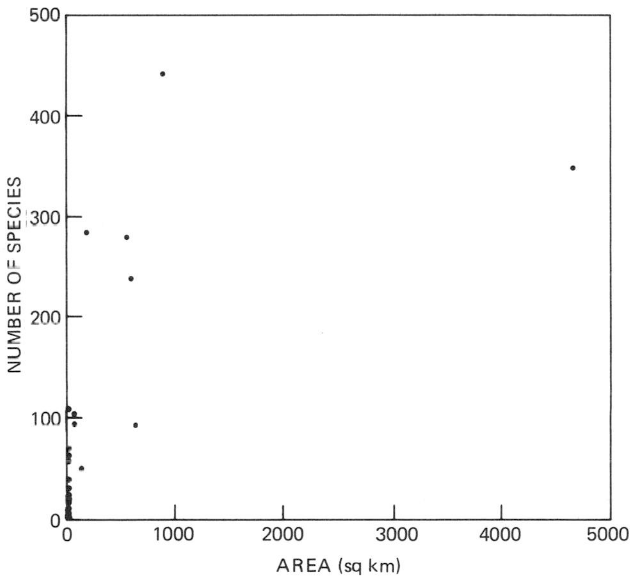

Figure 6 – Number of species plotted against area of 23 of the Galápagos Islands

Source: Data from Johnson & Raven 1973.

Figure 6 presents an extreme example of the effect of stragglers. The data come from a study of the relation between species diversity and geographic characteristics of the Galápagos Islands (Johnson & Raven 1973). For this plot the number of species and area (sq km) are considered. Large islands tend to have more species. A straight line

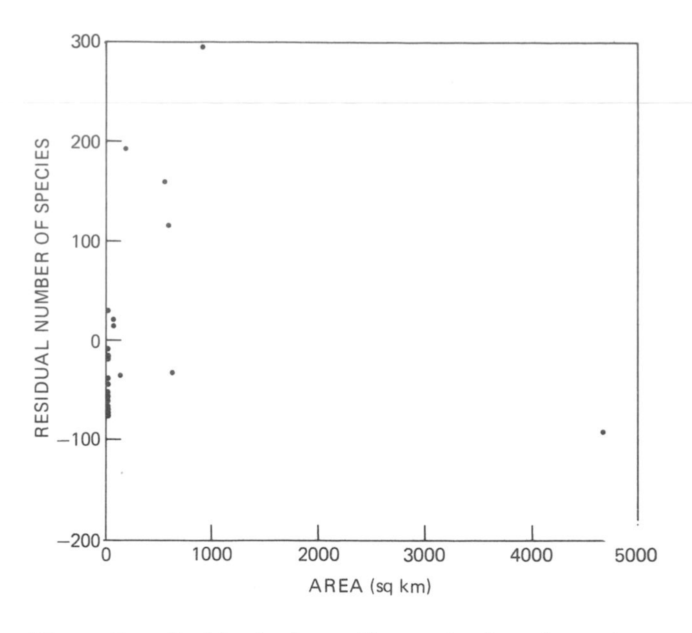

Figure 7 – Residuals from Figure 6 plotted against area, Galápagos Islands*

* Note the residual in the lower right corner.

Source: Data from Johnson & Raven 1973.

was fitted to the number of species as a function of area, and the residuals from this fit are plotted against area in Figure 7.

This plot suggests modifications to the simple straight-line fit. One island is so large (more than 4,500 sq km) that it alone has largely determined the fitted line. The residuals of the remaining islands show a marked upward trend. If the large island were set aside, the line might better fit the remaining points. The fit would still not be satisfactory, however, because a few islands very much larger than the rest would still remain and a good fit could not be reached without transforming the variables. Taking logarithms of *both* variables, number of species and area, should spread the points across the page and improve the fit. (Note that in the logarithmic model the relation between number and area is no longer a straight line.) The residuals calculated after excluding the largest island and taking logarithms are shown in Figure 8. These indicate a much more reasonable fit.

Plots of residuals against variables not included in the model. If a complete model has been appropriately fitted, the residuals tend to appear independent of other variables. Thus, a plot of residuals against some other variable not included in the model should appear as a random scatter unless the variable under consideration would add to the completeness of the model. If there are trends or patterns in this plot, they often suggest the form in which this variable should enter an improved model. Figure 9 is a plot of the residuals (from the fit of log species on log area) in the Galápagos example versus log elevation (measured at the highest point of each island). The upward trend here suggests that (aside from some small islands

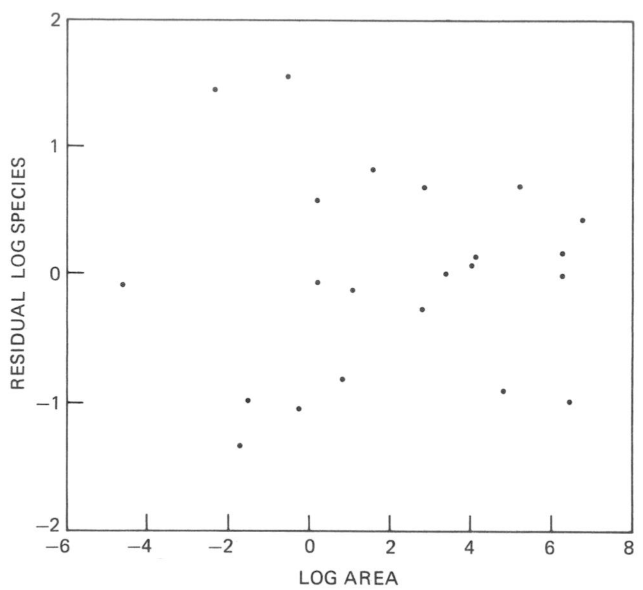

Figure 8 – Residuals from logarithmic model* plotted against logarithm of area, Galápagos Islands

* Excluding the largest island.

Source: Data from Johnson & Raven 1973.

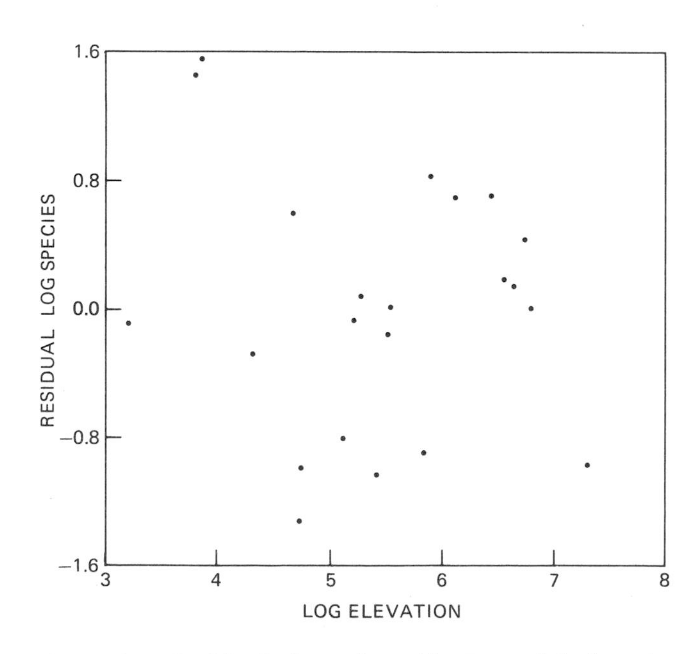

Figure 9 – Residuals from logarithmic model plotted against logarithm of elevation, Galápagos Islands

with large residuals—located at upper left in the figure) elevation is related to species and may be useful in the model.

Hazards of graphical procedures. Graphical procedures are not without their hazards. The importance of effects seen in plots is difficult to assess. Good displays draw attention to anomalies in the data. Some of these will be pure artifacts of no importance, but looking at them is a small price to pay for the opportunity to make significant discoveries.

Transformations, reexpression

Only by rare good fortune does an experimenter begin measuring on a scale well suited to understanding and making simple models. Typically, data are collected in the form most convenient for measurement or in a purely conventional form. The data may be just counts or measurements on an arbitrary scale of five or seven points. They may be scores on a test. Frequently a simple transformation of the data is useful. [*See, for example,* STATISTICAL ANALYSIS, SPECIAL PROBLEMS OF, *article on* TRANSFORMATIONS OF DATA.] In some cases a transformation may be suggested by the *type* of data. For example, if the data are in the form of counts it is usually easier to model the square roots of the counts than to study the counts themselves. (There is some theory to suggest that this transformation usually stabilizes variances, and, in addition, empirical experience has shown that square roots of counts lend themselves to simpler models than do counts themselves.) If the data are obtained by measuring the time taken until something happens, it is usually easier to work with rate, in units of 1/time, than it is to work with units of time. In most instances, however, several transformations should be tried. Box and Cox (1964) proposed methods for selecting transformations on the basis of the data themselves.

In some situations the appropriate expression for the data involves the combination of several variables. In the study of elections the raw data are the votes for candidates in an electoral unit. It is almost always more informative to study fractions or percentages. [*See* COUNTED DATA.] It helps even more to consider swings—changes in the percentages from previous years, for example. Bartholomew and Basset (1971) give an introduction to voting statistics. See also Tufte (1974).

In the study of human fertility, the basic data are frequently the number and birthdates of children in a family. The time intervals between the births of surviving children are, however, more useful, particularly when the age of the mother and family size are also considered. These intervals are based on typically reliable data, are relatively uncorrelated, and are strongly related to other explanatory variables such as age and current family size.

In studies of the learning process, the basic data might include time to learn, extent of learning (test scores), and retention (later test scores). The relation of these to other variables is of interest. Different combinations of these variables are useful for different purposes. Rate of learning (the first test score divided by learning time) may be useful for some purposes, as may the forgotten proportion (the difference between the first and second scores divided by the first score). A variety of such measures can and should be investigated.

In exploratory analysis, flexibility, guided by understanding, is essential in choosing the appropriate expression for the data.

Robust summaries, resistant fits

A previous section discussed plots of residuals from fitted values. In the exploratory stages, residuals that clearly indicate the inadequacies of the model developed to date are required. This section deals with the fitting of models to produce useful residuals. In formal inferential analysis, robust procedures may be advocated on the basis of efficiency (small variance). [*See* ERRORS, *article on* EFFECTS OF ERRORS IN STATISTICAL ASSUMPTIONS.]

Estimates of location. Location is perhaps the simplest and most studied problem. The arithmetic mean or average is commonly used to estimate location because it is easily calculated and because the sample average is analogous to the population total. Further, under ideal conditions (when the population distribution is normal) the arithmetic mean has optimal properties. Under other (more realistic) conditions, however, the arithmetic mean has undesirable properties. Among the most serious of these is the contamination of residuals. If one number in a batch is greatly altered, *all* the residuals from the mean are substantially changed as shown in Table 2. One or more gross outliers tend to make most or all the residuals numerically large. This complicates the detection of outliers and the estimation of the variability of the remaining data.

Table 2 – The effect of moving one observation on the residuals from the arithmetic mean

Set	Data	Mean	Residuals
1	715, 698, 690, 705, 707	703	12, −5, −13, 2, 4
2	715, 698, 740, 705, 707	713	2, −15, 27, −8, −6

The residuals from the *median* are much less sensitive to gross perturbations in a few data values. Altering one number will usually make a large change in only *one* residual. Table 3 illustrates this on the same data as Table 2. Outliers are more readily detected, and the remaining residuals may still be used to assess the variability of the remaining data. There are other estimates of location that are, like the median, insensitive or resistant to gross deviations in a small number of observations. See Andrews et al. (1972) for a fuller discussion of this problem. [*See also* ESTIMATION, *article on* POINT ESTIMATION.]

Table 3 – The effect of moving one observation on the residuals from the median

Set	*Data*	*Median*	*Residuals*
1	715, 698, 690, 705, 707	705	10, −7, −15, 0, 2
2	715, 698, <u>740</u>, 705, 707	707	8, −9, 3, −2, 0

Robust estimation in two-way tables. Responses are often measured in patterns including all combinations of the levels of two factors. In such cases it is helpful if the data or their transforms fit a simple additive model,

$$\begin{bmatrix}\text{fitted}\\\text{values}\end{bmatrix} = \begin{bmatrix}\text{average}\\\text{effect}\end{bmatrix} + \begin{bmatrix}\text{row}\\\text{effect}\end{bmatrix} + \begin{bmatrix}\text{column}\\\text{effect}\end{bmatrix}.$$

In the example discussed in Box and Cox (1964) and mentioned above, survival times were measured for animals subject to one of three poisons and one of four antidotes. The data can be set out in four columns and three rows. The results of the experiment are relatively easy to understand and communicate if the above additive model holds. If this model applies, the residuals (observations – fitted values) should be relatively small and without structure.

One way of calculating the residuals from the least-squares (normal maximum-likelihood) fit of the parameters of such a model is to cast the data into a two-way layout, where each row represents a level of the first factor (for example, a poison) and each column a level of the second (for example, an antidote). Then

(1) calculate the grand average and subtract this from the data;

(2) calculate each row average of what is left and subtract this from the corresponding row;

(3) calculate each column average of what is left and subtract this from the corresponding column.

The result of these operations is a set of residuals.

Table 4 – Calculation of residuals by means*

Display survival times for antidotes and poisons to get

		Antidotes a	b	c	
Poisons	A	79	63	68	
	B	66	66	69	grand average: 72.
	C	77	78	82	

Subtract grand average to get

	a	b	c	
A	7	−9	−4	−2
B	−6	−6	−3	row averages: −5.
C	5	6	10	7

Subtract row averages to get

	a	b	c	
A	9	−7	−2	
B	−1	−1	2	column averages: 2 −3 1.
C	−2	−1	3	

Subtract column averages to get

	a	b	c	
A	7	−4	−3	
B	−3	2	1	residuals.
C	−4	2	2	

* Data are hypothetical survival times; row averages are poison effects; column averages are antidote effects.

These are illustrated in Table 4. Table 5 presents the analogous treatment by medians. The median calculation is nonlinear and should be iterated until all row and column medians are near or equal to zero. (The corresponding result for row and column means occurs at the first iteration when using averages.) This iteration accounts for the additional step in Table 5. These two analyses present different views of the data. Table 4 is typical of a relatively poor fit, in which the residuals are about as large as the row and column averages themselves. Perhaps there is a large interaction between two factors. Table 5 presents the residuals of a better fit with one unusual observation. One or both of these explanations may be plausible. Further information is required to distinguish between them. Is one treatment situation inherently different? Was an error made in recording the data? The median analysis (Table 5) indicates which number is suspect more clearly than the fit by means (Table 4).

2^k factorial experiments. Factorial experiments with two levels for each factor present another opportunity for analysis by medians. In a 2^k factorial experiment each of k factors A, B, . . . is assigned two levels, + and −, and all possible

Table 5 – Calculation of residuals by medians*

Display survival times for antidotes and poisons to get

		Antidotes a	b	c	
Poisons	A	79	63	68	
	B	66	66	69	grand median: 69.
	C	77	78	82	

Subtract grand median to get

	a	b	c	
A	10	−6	−1	−1
B	−3	−3	0	row medians: −3.
C	8	9	13	9

Subtract row medians to get

	a	b	c	
A	11	−5	0	
B	0	0	3	column medians: 0 0 3.
C	−1	0	3	

Subtract column medians to get

	a	b	c	
A	11	−5	−3	−3
B	0	0	0	row medians: 0.
C	−1	0	0	0

Subtract row medians to get

	a	b	c	
A	14	−2	0	
B	0	0	0	residuals.
C	−1	0	0	

* Data are hypothetical survival times.

combinations of these treatments are considered. (The resulting number of combinations may be very large. In these instances only a selection of the combinations is investigated. Appropriate selections form so-called fractional factorial experiments.) [*See* EXPERIMENTAL DESIGN.]

In an experiment with three factors there are eight possible treatment combinations. These may be paired to give four estimates of the effect of the first factor. Let $(A^+B^+C^+)$ denote the response when all three factors are at the upper level, with similar notation for the other combinations. The four estimates of the effect of A, $d_1, \cdots, d_4$ may be written as in Table 6. The effect of factor A may be estimated by the median (average of the middle two in this case) of the d_i.

Table 6 – Four estimates of the effect of A in the 2^3 factorial experiment

$$(A^+B^+C^+) - (A^-B^+C^+) = d_1$$
$$(A^+B^+C^-) - (A^-B^+C^-) = d_2$$
$$(A^+B^-C^+) - (A^-B^-C^+) = d_3$$
$$(A^+B^-C^-) - (A^-B^-C^-) = d_4$$

Afifi and Azen (1972, p. 184) describe an experiment to investigate the effects of three factors on the adrenal weight (mgm/100 gm) of young animals. The factors were whether or not nurslings were separated from their mother (A), and whether the animals were kept with their siblings from weaning to maturity (B), and in their adult life (C). Four weights were found for each condition. The data for part of the experiment are recorded in Table 7, with the over-all median already subtracted. The large residual, 13.5, deserves special attention. Large unexplained deviations cast doubt on the measurement process and hence on all the data and on the results derived therefrom. If some explanation for the 13.5 can be found, faith in the experiment may be restored. The medians for the treatment combinations may be used to estimate the effects of the three factors as suggested in Table 6. The effect of A can be estimated as in Table 8 from differences involving changes only in the level of factor A. Note that the unusual observation has not greatly affected this estimate.

Table 7 – A 2^3 factorial experiment with four replicates, using medians to represent treatment combinations and to uncover outliers*

	Data				*Median*	*Residuals*			
$(A^+B^+C^+)$	1	−5	−9	−5	−5	6	0	−4	0
$(A^+B^+C^-)$	−12	−8	−9	−11	−10	−2	2	1	−1
$(A^+B^-C^+)$	1	−6	−1	2	0	1	−6	−1	2
$(A^+B^-C^-)$	−4	−3	−7	−4	−4	0	1	−3	0
$(A^-B^+C^+)$	15	17	15	13	15	0	2	0	−2
$(A^-B^+C^-)$	2	9	19	1	5.5	−3.5	3.5	13.5	−4.5
$(A^-B^-C^+)$	14	11	13	12	12.5	1.5	−1.5	.5	−.5
$(A^-B^-C^-)$	−7	5	−1	−1	−1	−6	6	0	0

* Data are adrenal weights in mg/100 gm, residualized on the grand median. See text for description of factors *A*, *B*, and *C*.

Source: Data from Afifi & Azen 1972.

Table 8 – A 2^3 factorial experiment continued, using medians to estimate the main effect of A

Median

$$\left.\begin{array}{l}(A^+B^+C^+) - (A^-B^+C^+) = -5 - 15 = -20 \\ (A^+B^+C^-) - (A^-B^+C^-) = -10 - 5.5 = -15.5 \\ (A^+B^-C^+) - (A^-B^-C^+) = 0 - 12.5 = -12.5 \\ (A^+B^-C^-) - (A^-B^-C^-) = -4 + 1 = -3\end{array}\right\} \quad \textit{Median} = -14$$

The effect of *A* may be removed from the treatment medians by subtracting or adding −7 according to whether the level of factor *A* is + or −. This yields new medians: 2, −3, 7, 3, 8, −1.5, 5.5, −8, which may be corrected by subtracting their median (2.5) to have median equal zero, giving −.5, −5.5, 4.5, .5, 5.5, −4, 3, −10.5. The effects of *B* (−1) and *C* (3.5) may be similarly estimated and removed to yield residual treatment medians, corrected to have zero median: −2.5, −.5, .5, 3.5, 3.5, 1.0, −1, −7.5. The range of these medians is now about one-half what it was in Table 7. Removing the effects of the three factors separately has markedly reduced the magnitude of the residuals from the treatment medians. It is unlikely that large *systematic* interactions can be removed. The treatment residual of −7.5, however, corresponding to treatment ($A^-B^-C^-$) indicates a much larger (negative) effect than the other combinations. This analysis by medians suggests that while factors *A*, *B*, and *C* separately have effects −7, −1, and 3.5, respectively, the combined effect is noticeably lower when all factors are at the lower level. The corresponding analysis by averages describes this same set of data by main effects and several two- and three-factor interactions.

This example illustrates the successive estimation and removal of structure. First the over-all median was subtracted. Then the effects of the three factors was removed. This removal was necessary in order to examine the treatment residuals to see the unusual nature of the ($A^-B^-C^-$) treatment.

Robust regression estimates. Linear regression calculations play a large and central role in statistical analysis. These models are commonly fitted by least squares, and the estimates have the same normal optimality and nonnormal lack of robustness as the arithmetic mean. Before the least-squares method was widely used, regression models were sometimes fitted by minimizing the sum of absolute deviations. Alternatives along lines suggested by Huber (1973) are currently being explored.

DAVID F. ANDREWS

[*See also* GRAPHIC PRESENTATION; STATISTICS, DESCRIPTIVE.]

BIBLIOGRAPHY

AFIFI, A. A.; and AZEN, S. P. 1972 *Statistical Analysis: A Computer Oriented Approach.* New York: Academic Press.

ANDREWS, DAVID F. et al. 1972 *Robust Estimates of Location: Survey and Advances.* Princeton Univ. Press.

ANDREWS, DAVID F.; and TUKEY, JOHN W. 1972 Teletypewriter Plots Can Be Fast: Six-line Plots, Including Probability Plots. *Applied Statistics* 22:192–202.

ANSCOMBE, F. J. 1973 Graphs in Statistical Analysis. *American Statistician* 27, no. 1:17–21.

ANSCOMBE, F. J.; and TUKEY, JOHN W. 1963 The Examination and Analysis of Residuals. *Technometrics* 5:141–160.

BARTHOLOMEW, DAVID J.; and BASSET, E. E. 1971 *Let's Look at the Figures: A Quantitative Approach to Human Affairs.* Harmondsworth (England): Penguin.

BEATON, ALBERT E.; and TUKEY, JOHN W. 1974 The Fitting of Power Series, Meaning Polynomials, Illustrated on Band-Spectroscopic Data. *Technometrics* 16:147–192.

BOX, GEORGE E. P.; and COX, D. R. 1964 An Analysis of Transformations. *Journal of the Royal Statistical Society* Series B 26:211–243. → Discussion on pages 244–252.

DANIEL, CUTHBERT 1959 Use of Half-normal Plots in Interpreting Factorial Two Level Experiments. *Technometrics* 1:311–341.

DANIEL, CUTHBERT; and WOOD, FRED S. 1971 *Fitting Equations to Data: A Computer Analysis of Multifactor Data for Scientists and Engineers.* New York: Wiley.

GNANADESIKAN, R. 1973 Graphical Methods for Informal Inference in Multivariate Data Analysis. Volume 4, pages 195–206 in International Statistical Institute, 39th Session, *Proceedings.* Vienna: Österreichisches Statistiches Zentralamt.

GNANADESIKAN, R. 1977 *Methods for Statistical Data Analysis of Multivariate Observations.* New York: Wiley.

GNANADESIKAN, R.; and KETTENRING, J. J. 1972 Robust Estimates, Residuals, and Outlier Detection With Multiresponse Data. *Biometrics* 28:81–124.

GNANADESIKAN, R.; and WILK, M. B. 1969 Data Analytic Methods in Multivariate Statistical Analysis. Pages 593–638 in International Symposium on Multivariate Analysis, Second, Wright State University, 1968, *Multivariate Analysis: Proceedings.* Edited by Paruchuri R. Krishnaiah. New York: Academic Press.

HUBER, PETER J. 1973 Robust Regression: Asymptotics, Conjectures and Monte Carlo. *Annals of Statistics* 1:799–821.

JOHNSON, MICHAEL P.; and RAVEN, PETER H. 1973 Species Number and Endemism: The Galápagos Archipelago Revisited. *Science* 179:893–895.

MOSTELLER, FREDERICK; and TUKEY, JOHN W. 1968 Data Analysis Including Statistics. Volume 2, pages 80–203 in Gardner Lindzey and Elliot Aronson (editors), *Handbook of Social Psychology.* 2d ed. Volume 2: *Research Methods.* Reading, Mass.: Addison-Wesley.

TUFTE, EDWARD R. 1974 *Data Analysis for Politics and Policy.* Englewood Cliffs, N.J.: Prentice-Hall.

TUKEY, JOHN W. (1970) 1977 *Exploratory Data Analysis.* Reading, Mass.: Addison-Wesley. → First published in a "limited preliminary edition."

TUKEY, JOHN W.; and WILK, M. B. (1965) 1970 Data Analysis and Statistics: Techniques and Approaches. Pages 370–390 in Edward R. Tufte (editor), *The Quantitative Analysis of Social Problems.* Reading, Mass.: Addison-Wesley.

WEISSMAN, L. M. 1974 A Methodology for Studying the Psychological Complexity of Computer Programs. Ph.D. dissertation, Univ. of Toronto.

WILK, M. B.; and GNANADESIKAN, R. 1968 Probability Plotting Methods for the Analysis of Data. *Biometrika* 55:1–17.

DATA DREDGING

See FALLACIES, STATISTICAL; SIGNIFICANCE, TESTS OF; SURVEY ANALYSIS, *article on* METHODS OF SURVEY ANALYSIS.

DE MOIVRE, ABRAHAM

See MOIVRE, ABRAHAM DE.

DEATH RATES

See LIFE TABLES *and the biographies of* KŐRÖSY *and* WILLCOX.

DECISION MAKING

The articles under this heading were first published in IESS with a companion article less relevant to statistics.

I
PSYCHOLOGICAL ASPECTS

Men must choose what to do. Often, choices must be made in the absence of certain knowledge of their consequences. However, an abundance of fallible, peripheral, and perhaps irrelevant information is usually available at the time of an important choice; the effectiveness with which this information is processed may control the appropriateness of the resulting decision. This article is concerned with laboratory studies of human choices and of certain kinds of human information processing leading up to these choices. It is organized around two concepts and two principles. The two concepts are utility, or the subjective value of an outcome, and probability, or how likely it seems to the decision maker that a particular outcome will occur if he makes a particular decision. Both of the principles are normative or prescriptive; they specify what an ideal decision maker would do and thus invite comparison between performances of ideal and of real decision makers. One, the principle of maximizing expected utility, in essence asserts that you should choose the action that on the average will leave you best off. The other, a principle of probability theory called Bayes' theorem, is a formally optimal rule for transforming opinions in the light of new information, and so specifies how you should process information. The basic conclusions reached as a result of comparison of actual human performance with these two principles is that men do remarkably well at conforming intuitively to ideal rules, except for a consistent inefficiency in information processing.

Utility

Utility measurement and expected utility. The concepts of utility and probability have been with us since at least the eighteenth century. But serious psychological interest in any version of them did not begin until the 1930s, when Kurt Lewin wrote about valence (utility) and several probability-like concepts. Lewin had apparently been influenced by some lectures on decision theory that the mathematician John von Neumann had given in Berlin in 1928. But the Lewinian formulations were not very quantitative, and the resulting research did not lead to explicit psychological concern with decision processes. However, in 1944 von Neumann and Morgenstern published their epochal book *Theory of Games and Economic Behavior*. The theory of games as such has been remarkably unfruitful in psychological research, mostly because of its dependence on the absurdly conservative minimax principle that in effect instructs you to deal with your opponent as though he were going to play optimally, no matter how inept you may know him to be. But von Neumann and Morgenstern rather incidentally proposed an idea that made utility measurable; that proposal is the historical origin of most psychological research on decision processes since then. Their proposal amounts to assuming that men are rational, in a rather specific sense, and to designing a set of procedures exploiting that assumption to measure the basic subjective quantities that enter into a decision.

Since the origin of probability theory, the idea has been obvious that bets (and risky acts more generally) can be compared in attractiveness. Formally, every bet has an expectation, or expected value (EV), which is simply the average gain or loss or money per bet that you might expect to accrue if you played the bet many times. To calculate the EV, you multiply each possible dollar outcome of the bet by the probability of that outcome, and sum the products. In symbols, the EV of the ith bet is calculated as follows, where V_{ij} is the payoff for the jth outcome of the ith bet and p_j is the probability of obtaining that payoff:

$$EV_i = \sum_j p_j V_{ij}. \qquad (1)$$

Bets can be ordered in terms of their *EV*, and it seems plausible to suppose that men should prefer a bet with a higher *EV* to a bet with a lower one. But a little thought shows that men buy insurance in spite of the fact that the insurance companies pay their employees and build buildings, and thus must take in more money in premiums than they pay out in benefits. Thus insurance companies are in the business of selling bets that are favorable to themselves and unfavorable to their customers. Nevertheless, it is doubtful that anyone would call buying insurance irrational. This and other considerations led to a reformulation of the notion that men should order bets in terms of *EV*. The seventeenth-century British utilitarian philosophers had distinguished between objective value, or price, and subjective value, or utility. If the utility of some object to you is different from its price, then surely your behavior should attempt to maximize not expected value in dollars but expected utility. That is, you should substitute $u(V_{ij})$, the *utility* of the payoff, for the *j*th outcome of bet *i*, for the payoff itself, V_{ij} in equation (1). Since it is utility, not payoff, that you attempt to maximize in this model, it is called the expected utility maximization model, or *EU* model. In symbols,

$$EU_i = \sum_j p_j u(V_{ij}). \tag{2}$$

Von Neumann and Morgenstern proposed simply that one should use equation (2) to measure utility, by assuming that men make choices rationally. Several specific implementations of this idea will be examined below.

Freud and psychiatry have taught us, perhaps too stridently, to look for irrational motivations behind human acts, and introspection confirms this lesson of thousands of years of human folly. Why bother, then, with measurement based on the assumption that men are rational? Three kinds of answers seem clear. First, rationality, as decision theorists think of it, has nothing to do with what you want, but only with how you go about implementing your wants. If you would rather commit rape than get married and rather get drunk than commit rape, the decision theorist tells you only that, to be rational, you should get drunk rather than get married. The compatibility of your tastes with your, or society's, survival or welfare is not his concern. So it is easy to be irrational in Freud's sense, and yet rational from a standpoint of decision theory. Second, men often want to implement their tastes in a consistent (which means rational) way, and when large issues are at stake, they often manage to do so. In fact, knowledge of the rules of rational behavior can help one make rational decisions, that is, knowledge of the theory helps make the theory true. Third, the most important practical justification of these or any other scientific procedures is that they work. Methods based on von Neumann and Morgenstern's ideas do produce measurements, and those measurements provide predictors of behavior. The following review of experiments supports this statement.

The Mosteller and Nogee experiment. The von Neumann–Morgenstern proposal was elaborated by Friedman and Savage (1948), a mathematical economist and a statistician writing for economists, and then was implemented experimentally by Mosteller and Nogee (1951), a statistician and a graduate student in social psychology. (No discipline within the social or mathematical sciences has failed to contribute to, or use, decision theory.) Mosteller and Nogee asked subjects to accept or reject bets of the form "If you beat the following poker-dice hand, you will win $X; otherwise, you will lose $0.05." A value of X was found such that the subject was indifferent between accepting and rejecting the bet. Arbitrary assignment of 0 utiles (the name for the unit of utility, as gram is the name for a unit of weight) to no transaction (rejection of the bet) and of −1 utile to losing a nickel fixed the origin and unit of measurement of the utility scale, and the calculated probability *p* of beating the poker-dice hand was substituted into the following equation:

$$pu(\$X) + (1-p)u(-\$0.05) = u(\$0).$$

Since $u(\$X)$, the utility of $X, is the only unknown in this equation, it is directly solvable. Mosteller and Nogee used two groups of subjects in this experiment: Harvard undergraduates and National Guardsmen. For the Harvard undergraduates, they found the expected decreasing marginal utility; that is, the utility function rose less and less rapidly as the amount of money increased. For the National Guardsmen, they found the opposite; the larger the amount of money, the steeper the slope of the utility function.

Perhaps the most important of the many criticisms of the Mosteller–Nogee experiment is about the role that probabilities display in it. The probability of beating the specified poker-dice hand is not at all easy to calculate; still, it is substituted into equation (2), which is supposed to represent the decision-making processes of the subject. Actually, Mosteller and Nogee did display these probabilities as numbers to their subjects. But they also displayed as numbers the values of the amounts of money for which the subjects were gambling. Why should we assume that the subjects make some subjective transformation that changes those numbers called dollars into subjective quantities called utilities, while the numbers called probabilities remain unchanged? Mosteller and Nogee made the point themselves, and reanalyzed their data using equation (2), but treating p rather than $u(\$X)$ as the unknown quantity. But this is no more satisfactory. The fundamental fact is that equation (2) has at least two unknowns, p and $u(\$X)$. This fact has been recognized by renaming the *EU* model; it is now called the *SEU* (subjectively expected utility) model. The addition of the *S* means only that the probabilities which enter into equation (2) must be inferred from behavior of the person making the decision, rather than calculated from some formal mathematical model. Of course, the person making the decision may not have had a very orderly set of probabilities. In particular, his probabilities for an exhaustive set of mutually exclusive events may not add up to 1. Thus there are really two different *SEU* models, depending on whether or not the probabilities are assumed to add up to 1. (In the latter case, the utilities must be measured on a ratio, not an interval, scale.) This article will examine the topic of subjective probabilities at length later.

In the *SEU* model every equation like equation (2) has at least two unknowns. A single equation with two unknowns is ordinarily insoluble. All subsequent work on utility and probability measurement has been addressed in one way or another to solution of this problem of too many unknowns. Edwards (1953; 1954*a*; 1954*b*) gave impetus to further analysis of the problem by exhibiting sets of preferences among bets that could not be easily accounted for by any plausible utility function for money but that seemed to require instead the notion that subjects simply prefer to gamble at some probabilities rather than others, and indeed will accept less favorable bets embodying preferred probabilities over more favorable bets embodying less preferred probabilities.

The Davidson, Suppes, and Siegel experiment. The next major experiment in the utility measurement literature was performed by Davidson, Suppes, and Siegel (see Davidson & Suppes 1957), two philosophers and a graduate student in psychology. The key idea in their experiment—the pair of subjectively equally likely events—was taken from a neglected paper by the philosopher Ramsey (1926). Suppose that you find yourself committed to a bet in which you stand to win some large sum if event *A* occurs and to lose some other large sum if it does not occur. Now suppose it is found that by paying you a penny I could induce you to substitute for the original bet another one exactly like it except that now you are betting on *A* not occurring. Now, after the substitution, suppose that I could induce you to switch back to the original bet by offering you yet another penny. Clearly, for you, the probability that *A* will occur is equal to the probability that it will not, that is, the two events are subjectively equally likely.

If we assume that either *A* or not-*A* must happen, and if we assume that for you (as for any probability theorist) the probabilities of any set of events no more than one of which can happen and some one of which must happen—that is, an exhaustive set of mutually exclusive events—add up to 1, then the sum of the probabilities of *A* and not-*A* must be 1. Now, if those two numbers are equal, then each must be 0.5.

Davidson, Suppes, and Siegel hunted for subjectively equally likely events, and finally used a die with one nonsense syllable (e.g., *ZEJ*) on three of its faces and another (*ZOJ*) on the other three. (They were very lucky that the same event turned out equally likely for all subjects.) They fixed two amounts of money, −4 cents and 6 cents. They found an amount of money, *X* cents, such that the subject was indifferent to receiving 6 cents if, say, *ZOJ* occurred and *X* cents if *ZEJ* occurred, and receiving −4 cents for sure. It follows from the *SEU* model that

$$p(ZOJ)\,u(6¢) + p(ZEJ)\,u(X¢) = u(-4¢).$$

A little algebra shows that

$$u(6¢) - u(-4¢) = u(-4¢) - u(X¢).$$

That is, the distance on the utility-of-money scale from 6 cents to −4 cents is equal to the distance from −4 cents to *X* cents (of course *X* cents is a larger loss than −4 cents). Once two equal intervals on the utility scale have been determined, it is no longer necessary to use a sure thing as one of the options. If *A*, *B*, *C*, and *D* are decreasing amounts of money, if the distance from *B* to *C* is equal to the distance from *C* to *D*, and if the subject is indifferent between a subjectively equally likely bet in which he wins *A* for *ZOJ* and *D* otherwise, and another in which he wins *B* for *ZOJ* and *C* otherwise, then the distance from *A* to *B* is equal to the other two distances. Davidson, Suppes, and Siegel used this procedure to construct a set of equally spaced points on their subjects' utility-for-money functions. Thereafter, they used the resulting utility functions as a basis for measuring the probability that one face of a symmetrical four-faced die would come up. (If the four faces were considered equally likely and if their probabilities added to 1, then that probability would be 0.25.)

The most important finding of the Davidson, Suppes, and Siegel experiment was that a good many internal consistency checks on their utility functions worked out well. Once a number of equal intervals on the utility function have been determined, many predictions can be made about preferences among new bets; some of these were tested, and in general they were successful. The utility functions were typically more complicated than those of the Mosteller–Nogee experiment and differed from subject to subject; they were seldom linearly related to money. The subjective probability of the face of the four-faced die was typically found to be in the region of 0.20.

The Davidson–Suppes–Siegel procedure remains intellectually valid, though criticisms of details are easy to make. However, it is unlikely that future utility measurement experiments will use it. The prior determination of the subjectively equally likely event is less attractive than procedures now available for determination of both utilities and probabilities from the same set of choices.

Among a substantial set of utility-measurement experiments, only two others are reviewed here. They both embody sophisticated ideas taken from recent developments in measurement theory, and they both use what amounts to a simultaneous-equations approach to the solution of a system of equations like equation (2) for utilities and probabilities, treating both as unknowns.

The Lindman experiment. Harold Lindman (1965) began by giving a subject a two-outcome bet of the form that if a spinner stopped in a specified region, the subject would win $X, while if it did not, he would win $Y. Then he invited the subject to state the minimum price for which he would sell the bet back to the experimenter. After the subject had stated that amount, $Z, Lindman operated a random device that specified an experimenter's price. If the experimenter's price was at least as favorable to the subject as the subject's price, then the sale took place, the experimenter paid his price to the subject, and the bet was not played. Otherwise, the sale did not take place, and the subject played the bet. Since the sale, if it took place, always took place at the experimenter's price, it was to the subject's advantage to name the actual minimum amount of money that he considered just as valuable as the bet. Thus Lindman could write

$$pu(\$X) + (1-p)u(\$Y) = u(\$Z).$$

This equation has at least four unknowns, three utilities and a probability. If we question whether subjects unsophisticated about probability theory make their probabilities add up to 1, it may have five unknowns, since then both the probability of the event and the probability of its complement can be treated as unknowns. Thus, in a system of such equations there will always be many more unknowns than equations, even though the same probabilities and amounts of money are used in many different bets. However, the system can be rendered soluble, and even overdetermined, by taking advantage of the fact that if $Z is between $X and $Y (as it will be in the example), then $u(\$Z)$ will be between $u(\$X)$ and $u(\$Y)$. In more formal language, the relation between $u(\$X)$ and $X is monotonic. Lindman exploited this fact by fitting a series of line segments to his utility functions; by controlling the number of line segments, he controlled both the number of unknowns and the amount of curvilinearity (more precisely, changes in slope) he could introduce into the utility function.

Lindman's results are complex, orderly, and pretty. He obtained a variety of shapes of utility functions for money from his different subjects. When he analyzed the data without assuming additivity of probabilities, he found that the actual sums were very close to 1; the data strongly support the idea that his subjects do in fact make probability judgments that add to 1. The probability and utility functions that were found predicted choices among bets very well indeed. There were some interactions between probabilities and utilities of a kind not appropriate to the *EU* model, but they were not major ones.

The Tversky experiment. Deviating from the nearly universal use of college student subjects, Amos Tversky used prisoners at Jackson Prison, the largest state prison in Michigan, as subjects (1964). He used cigarettes, candy bars, and money, rather than money alone, as the valuable objects whose utilities were to be measured. Tversky's research was based on an application of simultaneous conjoint measurement, a new approach to fundamental measurement which emphasizes the idea of additive structures. Consider a bet in which with probability p you win $X and with probability $1-p$ no money changes hands. Suppose such a bet is worth just $Z to you. A matrix with different values of p on one axis and different values of $X on the other defines a family of such bets. If the utility of no money changing hands is taken to be 0, then the *EU* model can be written for any such bet in logarithmic form as follows:

$$\log p_j + \log u(\$X_{ij}) = \log u(\$Z_{ij}).$$

That is, in logarithmic form this is an additive model. If all the values of $\$Z_{ij}$ for such a matrix of bets are known, then the rules for additive representations of two-dimensional matrices permit solution (by complex computer methods) of a system of inequalities that give close values of p_j and $u(\$X_{ij})$. Tversky did just this, both for gambles and for commodity bundles consisting of so many packs of cigarettes and so many candy bars.

The main finding of Tversky's study was a consistent discrepancy between behavior under risky and riskless conditions. If probabilities are not forced to add to 1 in the data analysis, then the form that this discrepancy takes is that the probabilities of winning are consistently overestimated relative to the probabilities of losing. If probabilities are forced to add to 1, then the utilities measured under risky conditions are consistently higher than those measured under riskless conditions. The latter finding would normally be interpreted as reflecting the attractiveness or utility of gambling as an activity, independently of the attractiveness of the stakes and prizes. The discrepancy between Tversky's finding and Lindman's remains unexplained.

Summary. In all of these studies, the choices among bets made under well-defined experimental conditions turn out to be linked via some form or other of the *SEU* model to choices among other bets made by the same subjects under more or less the same conditions. That is, the *SEU* model permits observation of coherence among aspects of subjects' gambling behavior. Of course it would be attractive to find that such coherence would hold over a larger range of risk-taking activity. A substantial disappointment of these studies is that the individual utility and probability functions vary so much from one person to another. It would be scientifically convenient if different people had the same tastes—but experience offers no reason for hoping that they will, and the data clearly say that they do not.

At any rate, these studies offer no support for those who reject a priori the idea that men make "rational" decisions. An a priori model of such decision making turns out to predict very well the behavior of a variety of subjects in a variety of experiments. Nor is this finding surprising. A very general and intuitively appealing model of almost any kind of human behavior is contained in the following dialogue:

Question: What is he doing?
Answer: He's doing the best he can.

Probability

Identification rules for probability. The previous discussion has probably been somewhat confusing to many readers not familiar with what is now going on in statistical theory. It has consistently treated probability as a quantity to be inferred from the behavior of subjects, rather than calculated from such observations as the ratio of heads to total flips of a coin. It is intuitively reasonable to think that men make probability judgments just as they make value judgments and that these judgments can be discovered from men's behavior. But how might such subjective probabilities relate to the more familiar quantities that we estimate by means of relative frequencies?

This question is a controversial one in contemporary statistics. Considered as a mathematical quantity, a probability is a value that obeys three quite simple rules: it remains between 0 and 1; 0 means impossibility and 1 means certainty; and the probability of an event's taking place plus the probability of its not taking place add up to 1. These three properties are basic to all of the elaborate formal structure of probability theory considered as a topic in mathematics. Nor are they, or their consequences, at all controversial. What are controversial are the identification rules linking these abstract numbers with observations made or makable in the real world. The usual relative-frequency rules suffer from a number of intellectual difficulties. They require an act of faith that a sequence of relative frequencies will in fact approach a limit as the number of observations increases without limit. They are very vague and subjective while pretending to be otherwise; this fact is most conspicuous in the specification that relative frequencies are supposed to be observed under "substantially similar conditions," which means that the conditions should be similar enough but not too similar. (A coin always tossed in *exactly* the same way would presumably fall with the same face up every time.) Perhaps most important, the frequentistic set of rules is just not applicable to many, perhaps most, of the questions about which men might be uncertain. What is the probability that your son will be a straight-A student in his senior year in high school? While an estimate might be made by counting the fraction of senior boys in the high school he is likely to attend who have straight-A records, a much better estimate would be based primarily on his own personal characteristics, past grade record, family background, and the like.

The Bayesian approach. Dissatisfaction with the frequentistic set of identification rules, for these and other more technical reasons, has caused a set of statisticians, probability theorists, and philosophers led by Leonard J. Savage, author of *The Foundations of Statistics* (1954), to adopt a different, personalistic, set of such rules. According to the personalistic set of identification rules, a probability is an opinion about how likely an event is. If it is an opinion, it must be someone's opinion; I will remind you of this from time to time by reference to your probability for something, and by calling such probabilities personal. Not any old opinion can be a probability; probabilities are orderly, which mostly means that they add up to 1. This requirement of orderliness is extremely constraining, so much so that no real man is likely to be able to conform to it in his spontaneous opinions. Thus the "you" whose opinions I shall refer to is a slightly idealized you, the person you would presumably like to be rather than the person you are.

Those who use the personalistic identification rules for probabilities are usually called Bayesians, for the rather unsatisfactory reason that they make heavier use of a mathematical triviality named Bayes' theorem than do nonpersonalists. Bayes' theorem, an elementary consequence of the fact that probabilities add up to 1, is important to Bayesians because it is a formally optimal rule for revising opinions on the basis of evidence. Consider some hypothesis H. Your opinion that it is true at some given time is expressed by a number $p(H)$, called the prior probability of H. Now you observe some datum D, with unconditional probability $p(D)$ and with conditional probability $p(D|H)$ of occurring if H is true. After that, your former opinion about H, $p(H)$, is revised into a new opinion about H, $p(H|D)$, called the posterior probability of H on the basis of D. Bayes' theorem says these quantities are related by the equation

$$p(H|D) = \frac{p(D|H)\,p(H)}{p(D)}. \tag{3}$$

An especially useful form of Bayes' theorem is obtained by writing it for two different hypotheses, H_A and H_B, on the basis of the same datum D, and then dividing one equation by the other:

$$p(H_A|D) = \frac{p(D|H_A)}{p(D)}\,p(H_A),$$

$$p(H_B|D) = \frac{p(D|H_B)}{p(D)}\,p(H_B),$$

$$\frac{p(H_A|D)}{p(H_B|D)} = \frac{p(D|H_A)}{p(D|H_B)}\,\frac{p(H_A)}{p(H_B)},$$

or, in simpler notation,

$$\Omega_1 = L\Omega_0. \tag{4}$$

In equation (4), Ω_0, the ratio of $p(H_A)$ to $p(H_B)$, is called the prior odds, Ω_1 is called the posterior odds, and L is the likelihood ratio. Equation (4) is perhaps the most widely useful form of Bayes' theorem.

Conservatism in information processing. Bayes' theorem is of importance to psychologists as well as to statisticians. Psychologists are very much interested in the revision of opinion in the light of information. If equation (3) or equation (4) is an optimal rule for how such revisions should be made, it is appropriate to compare its prescriptions with actual human behavior.

Consider a very simple experiment by Phillips and Edwards (1966). Subjects were presented with a bookbag full of poker chips. They were told that it had been chosen at random, with 0.5 probability, from two bookbags, one containing 700 red and 300 blue chips, while the other contained 700 blue and 300 red. The question of interest is which bag this one is. Subjects were to answer the question by estimating the probability that this was the predominantly red bookbag. On the basis of the information so far available, that probability is 0.5, as all subjects agreed.

Now, the experimenter samples randomly with replacement from the bookbag. At this point let me invite you to be a subject in this experiment. Suppose that in 12 samples, with replacement, the experimenter gets red, red, red, blue, red, blue, red, blue, red, red, blue, red—that is, 8 reds and 4 blues. Now, on the basis of all the evidence you have, what is the probability that this is the predominantly red bookbag? Write down an intuitive guess before starting to read the next paragraph.

Let us apply equation (4) to the problem. The prior odds are 1:1, so all we need is the likelihood ratio. To derive that is straightforward; for the general binomial case of r reds in n samples, where H_A says that the probability of a red is p_A (and of a blue is q_A) and H_B says the probability of a red is p_B (and of a blue is q_B),

$$L = \frac{p_A^r q_A^{n-r}}{p_B^r q_B^{n-r}}.$$

In this particular case, made much simpler by the fact that $p_A = q_B$ and vice versa, it is simply

$$L = \left(\frac{p_A}{q_A}\right)^{2r-n}$$

Of course $2r - n = r - (n - r)$ and is the difference between the number of reds and the number of blues in the sample—in this case, 4. So the likelihood ratio is $(7/3)^4 = 29.64$. Since the prior odds are 1:1, the posterior odds are then 29.64:1. And so the posterior probability that this is the predominantly red bookbag is 0.97.

If you are like Phillips and Edwards' subjects, the number you wrote down wasn't nearly so high as 0.97. It was probably about 0.70 or 0.80. Phillips and Edwards' subjects, and indeed all subjects who have been studied in experiments of this general variety, are conservative information processors, unable to extract from data anything like as much certainty as the data justify. A variety of experiments has been devoted to this conservatism phenomenon. It is a function of response mode; Phillips and Edwards have shown that people are a bit less conservative when estimating odds than when estimating probability. It is a function of the diagnosticity of the information; Peterson, Schneider, and Miller (1965) have shown that the larger the number of poker chips presented at one time, the greater is the conservatism, and a number of studies by various investigators have shown that the more diagnostic each individual poker chip, the greater the conservatism.

Conservatism could be attributed to either or both of two possible failures in the course of human information processing. First, the subjects might be unable to perceive the data-generating process accurately; they might attribute to data less diagnostic value than the data in fact have. Or the subjects might be unable to combine information properly, as Bayes' theorem prescribes, even though they may perceive the diagnostic value of any individual datum correctly. Data collected by Beach (1966) favor the former hypothesis; data collected by Edwards (1966) and by Phillips (1965) favor the latter. It seems clear by now that this formulation of the possible causes of conservatism is too simple to account for the known facts, but no one has yet proposed a better one.

Probabilistic information processing. If men are conservative information processors, then it seems reasonable to expect that this fact has practical consequences. One practical consequence is familiar to anyone who has ever grumbled over a hospital bill: human conservatism in processing available diagnostic information may lead to collection of too much information, where the collection process is costly. An even more serious consequence may arise in situations in which speed of a response is crucial: a conservative information processor may wait too long to respond because he is too uncertain what response is appropriate. This consequence of conservatism is especially important in the design of large military information-processing systems. In the North American Air Defense System, for example, the speed with which the system declares we are under attack, if we are, may make a difference of millions of lives.

Edwards (1962; 1965*b*) has proposed a design for diagnostic systems that overcomes the deficiency of human conservatism. A probabilistic information processing system (PIP) is designed in terms of Bayes' theorem. For vague, verbal data and vague, verbal hypotheses, experts must estimate $p(D|H)$ or L as appropriate (usually L rather than $p(D|H)$ will be appropriate). They make these estimates separately for each datum and each hypothesis or pair of hypotheses of interest to the system. Then a computer uses equation (3) or equation (4) to synthesize these separate judgments into a posterior distribution that reflects how all the hypotheses stand in the light of all the data. This distribution is of course revised each time a new datum becomes available.

A number of studies of PIP have been performed (see, e.g., Schum, Goldstein, & Southard 1966; Kaplan & Newman 1966; Edwards 1966). The studies have generally found it more efficient than competitive systems faced with the same information. In a large unpublished simulation study, Edwards and his associates found that data that would lead PIP to give 99:1 odds in favor of some hypothesis would lead its next-best competitor to give less than 5:1 odds in favor of that hypothesis. Even larger discrepancies in favor of PIP appear in Phillips' study. Thus a combination of human judgment and Bayes' theorem seems to be capable of doing a better job of information processing than either alone.

Are men rational?

All in all, the evidence favors rationality. Men seem to be able to maximize expected utility rather well, in a too-restricted range of laboratory tasks. There are, of course, a number of well-known counterexamples to the idea that men consistently do what is best for them. More detailed analysis of such experiments (e.g., probability learning experiments) indicates that substantial deviations from rationality seldom occur unless they cost little; when a lot is at stake and the task isn't too complex for comprehension, men behave in such a way as to maximize expected utility.

The comparison of men with Bayes' theorem is less favorable to men. The conservatism phenomenon is a large, consistent deviation from optimal information-processing performance. Nevertheless, it is surprising to those newly looking at this area that men can do as well as they do at probability or odds estimation.

The topic of the articulation between diagnosis and action selection will receive much more study in the next few years than it has so far. What little evidence is available suggests that men do remarkably well at doing what they should, given the information on hand. But the surface of this topic has scarcely been scratched.

Total rejection of the notion that men behave rationally is as inappropriate as total acceptance would be. On the whole, men do well; exactly how well they do, depends in detail on the situation they are in, what's at stake, how much information they have, and so on. The main thrust of psychological theory in this area is likely to be a detailed spelling out of just how nearly rational men can be expected to be under given circumstances.

WARD EDWARDS

[*Directly related is the entry* DECISION THEORY; *see also* Johnson 1968; Taylor 1968. *Other relevant material may be found in* BAYESIAN INFERENCE; GAME THEORY; MODELS, MATHEMATICAL; PROBABILITY; *and in the biographies of* BAYES; VON NEUMANN; *see also* Devereux 1968; Georgescu-Roegen 1968; Mack 1968; Parsons 1968; Pollack 1968; Urmson 1968.]

BIBLIOGRAPHY

○ Edwards 1954c; 1961 *provide two reviews of decision theory from a psychological point of view; these constitute a good starting point for more intensive study. For those who find these papers too difficult,* Edwards, Lindman, & Phillips 1965 *is an easier and more up-to-date but less thorough introduction to the topic.* Luce & Raiffa 1957, *though somewhat out-of-date, remains unique for its clear and coherent exposition of the mathematical content of decision theory.* Luce & Suppes 1965 *performs a similar job at chapter rather than book length and much more recently; its emphasis is on the probabilistic models of choice and decision. The easiest introduction to Bayesian statistics is* Edwards, Lindman, & Savage 1963 *or* Schlaifer 1961. *By far the most authoritative treatment of the topic is* Raffa & Schlaifer 1961, *but* Lindley 1965 *and* Good 1965 *are also important.* Beach 1966; Kaplan & Newman 1966; Phillips, Hays, & Edwards 1966; Schum, Goldstein, & Southard 1966; Slovic 1966; Peterson & Phillips 1966 *provide a good sample of that work, comparing men with Bayes' theorem as information processors.* Wasserman & Silander 1958 *is an annotated bibliography that emphasizes the sorts of interpersonal topics and social applications here ignored; it is a good guide to older parts of that literature. There is a much more up-to-date supplement, but it is not widely available.* Rapoport & Orwant 1962 *reviews the literature on experimental games.* Kogan & Wallach 1964 *is on personality variables in decision making.*

BEACH, LEE ROY 1966 Accuracy and Consistency in the Revision of Subjective Probabilities. Institute of Electrical and Electronics Engineers, *Transactions on Human Factors in Electronics* 7:29–37.

BRIGGS, GEORGE E.; and SCHUM, DAVID A. 1965 Automated Bayesian Hypothesis-selection in a Simulated Threat-diagnosis System. Pages 169–176 in Congress on the Information Systems Sciences, Second, *Proceedings.* Washington: Spartan.

○DAVIDSON, DONALD; and SUPPES, PATRICK 1957 *Decision Making: An Experimental Approach.* Stanford Univ. Press. → In collaboration with Sidney Siegel. A paperback edition was published in 1977 by the Univ. of Chicago Press.

►DEVEREUX, EDWARD C. JR. 1968 Gambling. Volume 6, pages 53–62 in *International Encyclopedia of the Social Sciences.* Edited by David L. Sills. New York: Macmillan and Free Press.

EDWARDS, WARD 1953 Probability-preferences in Gambling. *American Journal of Psychology* 66:349–364.

EDWARDS, WARD 1954*a* Probability-preferences Among Bets With Differing Expected Values. *American Journal of Psychology* 67:56–67.

EDWARDS, WARD 1954*b* The Reliability of Probability-preferences. *American Journal of Psychology* 67:68–95.

EDWARDS, WARD 1954*c* The Theory of Decision Making. *Psychological Bulletin* 51:380–417.

EDWARDS, WARD 1961 Behavioral Decision Theory. *Annual Review of Psychology* 12:473–498.

EDWARDS, WARD 1962 Dynamic Decision Theory and Probabilistic Information Processing. *Human Factors* 4:59–73.

EDWARDS, WARD 1965*a* Optimal Strategies for Seeking Information: Models for Statistics, Choice Reaction Times, and Human Information Processing. *Journal of Mathematical Psychology* 2:312–329.

EDWARDS, WARD 1965*b* Probabilistic Information Processing Systems for Diagnosis and Action Selection. Pages 141–155 in Congress on the Information System Sciences, Second, *Proceedings.* Washington: Spartan.

EDWARDS, WARD 1966 Non-conservative Probabilistic Information Processing Systems Final Report. ESD Final Report No. 05893-22-F. Unpublished manuscript.

EDWARDS, WARD; LINDMAN, HAROLD; and PHILLIPS, LAWRENCE D. 1965 Emerging Technologies for Making Decisions. Volume 2, pages 261–325 in Frank Barron et al. (editors), *New Directions in Psychology.* New York: Holt.

EDWARDS, WARD; LINDMAN, HAROLD; and SAVAGE, LEONARD J. 1963 Bayesian Statistical Inference for Psychological Research. *Psychological Review* 70:193–242.

EDWARDS, WARD; and PHILLIPS, LAWRENCE D. 1964 Man as Transducer for Probabilities in Bayesian Command and Control Systems. Pages 360–401 in Maynard W. Shelly and Glenn L. Bryan (editors), *Human Judgments and Optimality.* New York: Wiley.

FRIEDMAN, MILTON; and SAVAGE, LEONARD J. 1948 The Utility Analysis of Choices Involving Risk. *Journal of Political Economy* 56:279–304.

►GEORGESCU-ROEGEN, NICHOLAS 1968 Utility. Volume 16, pages 236–267 in *International Encyclopedia of the Social Sciences.* Edited by David L. Sills. New York: Macmillan and Free Press.

GOOD, I. J. 1965 *The Estimation of Probabilities: An Essay on Modern Bayesian Methods.* Cambridge, Mass.: M.I.T. Press.

►JOHNSON, DONALD M. 1968 Reasoning and Logic. Volume 13, pages 344–350 in *International Encyclopedia of the Social Sciences.* Edited by David L. Sills. New York: Macmillan and Free Press.

KAPLAN, R. J.; and NEWMAN, J. R. 1966 Studies in Probabilistic Information Processing. Institute of Electrical and Electronics Engineers, *Transactions on Human Factors in Electronics* 7:49–63.

KOGAN, NATHAN; and WALLACH, MICHAEL A. 1964 *Risk Taking: A Study in Cognition and Personality.* New York: Holt.

LINDLEY, DENNIS V. 1965 *Introduction to Probability and Statistics From a Bayesian Viewpoint.* 2 vols. New York: Cambridge Univ. Press. → Volume 1: *Probability.* Volume 2: *Inference.*

LINDMAN, HAROLD R. 1965 The Simultaneous Measurement of Utilities and Subjective Probabilities. Ph.D. dissertation, Univ. of Michigan.

LUCE, R. DUNCAN; and RAIFFA, HOWARD 1957 *Games and Decisions: Introduction and Critical Survey.* A study of the Behavioral Models Project, Bureau of Applied Social Research, Columbia University. New York: Wiley.

LUCE, R. DUNCAN; and SUPPES, PATRICK 1965 Preference, Utility, and Subjective Probability. Volume 3, pages 249–410 in R. Duncan Luce, Robert R. Bush, and Eugene Galanter (editors), *Handbook of Mathematical Psychology.* New York: Wiley.

►MACK, MARY PETER 1968 Bentham, Jeremy. Volume 2, pages 55–58 in *International Encyclopedia of the Social Sciences.* Edited by David L. Sills. New York: Macmillan and Free Press.

MOSTELLER, FREDERICK; and NOGEE, PHILIP 1951 An Experimental Measurement of Utility. *Journal of Political Economy* 59:371–409.

►PARSONS, TALCOTT 1968 Utilitarianism: II. Sociological Thought. Volume 16, pages 229–236 in *International Encyclopedia of the Social Sciences.* Edited by David L. Sills. New York: Macmillan and Free Press.

PETERSON, CAMERON R.; and MILLER, ALAN J. 1965 Sensitivity of Subjective Probability Revision. *Journal of Experimental Psychology* 70:117–121.

PETERSON, CAMERON R.; and PHILLIPS, LAWRENCE D. 1966 Revision of Continuous Subjective Probability Distributions. Institute of Electrical and Electronics Engineers, *Transactions on Human Factors in Electronics* 7:19–22.

PETERSON, CAMERON R.; SCHNEIDER, ROBERT J.; and MILLER, ALAN J. 1965 Sample Size and the Revision of Subjective Probabilities. *Journal of Experimental Psychology* 70:522–527.

PETERSON, CAMERON R. et al. 1965 Internal Consistency of Subjective Probabilities. *Journal of Experimental Psychology* 70:526–533.

PHILLIPS, LAWRENCE D. 1965 Some Components of Probabilistic Inference. Ph.D. dissertation, Univ. of Michigan.

PHILLIPS, LAWRENCE D.; and EDWARDS, WARD 1966 Conservatism in a Simple Probability Inference Task. Unpublished manuscript. Univ. of Michigan, Institute of Science and Technology.

PHILLIPS, LAWRENCE D.; HAYS, WILLIAM L.; and EDWARDS, WARD 1966 Conservatism in Complex Probabilistic Inference. Institute of Electrical and Electronics Engineers, *Transactions on Human Factors in Electronics* 7:7–18.

►POLLACK, IRWIN 1968 Information Theory. Volume 7, pages 331–337 in *International Encyclopedia of the Social Sciences.* Edited by David L. Sills. New York: Macmillan and Free Press.

RAIFFA, HOWARD; and SCHLAIFER, ROBERT 1961 *Applied Statistical Decision Theory.* Graduate School of Business Administration, Studies in Managerial Economics. Boston: Harvard Univ., Division of Research.

RAMSEY, FRANK P. (1926) 1964 Truth and Probability. Pages 61–92 in Henry E. Kyburg, Jr. and Howard E. Smokler (editors), *Studies in Subjective Probabilities.* New York: Wiley.

RAPOPORT, ANATOL; and ORWANT, CAROL 1962 Experimental Games: A Review. *Behavioral Science* 7:1–37.

○SAVAGE, LEONARD J. (1954) 1972 *The Foundations of Statistics.* Rev. ed. New York: Dover.

SCHLAIFER, ROBERT 1961 *Introduction to Statistics for Business Decisions.* New York: McGraw-Hill.

SCHUM, D. A.; GOLDSTEIN, I. L.; and SOUTHARD, J. T. 1966 Research on a Simulated Bayesian Information-processing System. Institute of Electrical and Electronics Engineers, *Transactions on Human Factors in Electronics* 7:37–48.

SLOVIC, PAUL 1966 Value as a Determiner of Subjective Probability. Institute of Electrical and Electronics Engineers, *Transactions on Human Factors in Electronics* 7:22–28.

►TAYLOR, DONALD W. 1968 Problem Solving. Volume 12, pages 505–511 in *International Encyclopedia of the Social Sciences.* Edited by David L. Sills. New York: Macmillan and Free Press.

TVERSKY, AMOS 1964 Additive Choice Structures. Ph.D. dissertation, Univ. of Michigan.

►URMSON, J. O. 1968 Utilitarianism: I. The Philosophy. Volume 16, pages 224–229 in *International Encyclopedia of the Social Sciences.* Edited by David L. Sills. New York: Macmillan and Free Press.

○VON NEUMANN, JOHN; and MORGENSTERN, OSKAR (1944) 1953 *Theory of Games and Economic Behavior.* 3d ed., rev. Princeton Univ. Press. → A paperback edition was published in 1964 by Wiley.

WASSERMAN, PAUL S.; and SILANDER, FRED S. 1958 *Decision Making: An Annotated Bibliography.* Ithaca, N.Y.: Cornell Univ., Graduate School of Business and Public Administration.

Postscript

The literature on decision making has grown substantially and shifted its orientation. Descriptive and static theories have been joined by theories of a more specifically psychological nature.

A major study in a Las Vegas casino setting indicates that, to a good first approximation, subjects simply choose among bets so as to maximize expected value (Goodman et al. 1977). Coombs and his collaborators (Coombs & Bowen 1971*a;* 1971*b;* Coombs & Huang 1970*a;* 1970*b*) are developing a theory of risky choice that treats expected value and risk (an undefined quantity) as the determiners of risky choices.

There has been declining interest in research that compares human decision making with presumably rational approaches via Bayes' theorem. Conservatism, clearly established as an experimental fact, seems to result from a problem of aggregating evidence; men can judge diagnosticity, meaning likelihood ratios, for individual data quite well, but they aggregate them poorly (Wheeler 1972). Yet some writers doubt the generality of these laboratory findings in the real world (Winkler & Murphy 1973), and some criticize the artificiality of the Bayesian experimental paradigms (for example, Wallsten 1972).

Major current research about decision processes deals with the heuristics that people in fact use to make decisions. See Kahneman and Tversky (1972; 1973); Tversky and Kahneman (1974); Slovic and Lichenstein (1971).

Bayesian ideas, in a variety of technological versions, are in practical use in various real information-processing settings, ranging from medical diagnosis to social, industrial, and military contexts. See, for example, Gustafson et al. (1971). The frontier of decision technology now lies in the development and validation of practical methods for measuring the utility of complex objects such as social programs, budgets, and so on. Research on this topic is rapidly increasing in volume and cogency.

A framework for both practical and theoretical work is the emerging discipline of decision analysis, which combines all these ideas and tools into practical techniques for making decisions. For an early exposition, see Raiffa (1968); for a later one, see Brown, Kahr, and Peterson (1974).

For a point of view skeptical of the applicability of decision analysis to public policy problems, see Boruch (1973).

WARD EDWARDS

ADDITIONAL BIBLIOGRAPHY

BORUCH, ROBERT F. 1973 Problems in Research Utilization: Use of Social Experiments, Experimental Results, and Auxiliary Data in Experiments. New York Academy of Sciences, *Annals* 218:56–77.

BROWN, REX V.; KAHR, ANDREW S.; and PETERSON, CAMERON R. 1974 *Decision Analysis for the Manager.* New York: Holt.

COOMBS, CLYDE H.; and BOWEN, JAMES N. 1971*a* A Test of VE-theories of Risk and the Effect of the Central Limit Theorem. *Acta Psychologica* 35:15–28.

COOMBS, CLYDE H.; and BOWEN, JAMES N. 1971*b* Additivity of Risk in Portfolios. *Perception and Psychophysics* 10:43–46.

COOMBS, CLYDE H.; and HUANG, LILY C. 1970*a* Polynomial Psychophysics of Risk. *Journal of Mathematical Psychology* 7:317–338.

COOMBS, CLYDE H.; and HUANG, LILY C. 1970*b* Tests of a Portfolio Theory of Risk Preference. *Journal of Experimental Psychology* 85:23–29.

GOODMAN, BARBARA C. et al. 1977 When SEU Theory Is Irrelevant. Unpublished manuscript, Univ. of Michigan.

GUSTAFSON, D. H. et al. 1971 Initial Evaluation of a Subjective Bayesian Diagnostic System. *Health Services Research* 6:204–213.

KAHNEMAN, DANIEL; and TVERSKY, AMOS 1972 Subjective Probability: A Judgment of Representativeness. *Cognitive Psychology* 3:430–454.

KAHNEMAN, DANIEL; and TVERSKY, AMOS 1973 On the Psychology of Prediction. *Psychological Review* 80: 237–251.

RAIFFA, HOWARD 1968 *Decision Analysis: Introductory Lectures on Making Choices Under Uncertainty.* Reading, Mass.: Addison-Wesley.

SLOVIC, PAUL; and LICHTENSTEIN, SARAH C. 1971 Comparison of Bayesian and Regression Approaches to the Study of Information Processing in Judgment. *Organizational Behavior and Human Performance* 6:649–744.

TVERSKY, AMOS; and KAHNEMAN, DANIEL 1974 Judgment Under Uncertainty: Heuristics and Biases. *Science* 185:1124–1131.

WALLSTEN, T. S. 1972 Conjoint Measurement Framework for the Study of Probabilistic Information Processing. *Psychological Review* 79:245–260.

WHEELER, GLORIA E. 1972 Misaggregation Versus Response Bias as Explanations for Conservative Inference. Ph.D. dissertation, Univ. of Michigan.

WINKLER, ROBERT L.; and MURPHY, ALLAN H. 1973 Experiments in the Laboratory and the Real World. *Organizational Behavior and Human Performance* 10:252–270.

II
ECONOMIC ASPECTS

The distinction between prescriptive and descriptive theories of decision is similar to that between logic—a system of formally consistent rules of thought—and the psychology of thinking. A logical rule prescribes, for example, that if you believe all *X* to be *Y*, you should also believe that all non-*Y* are non-*X*, but not that all *Y* are *X*; and if you believe, in addition, that all *Y* are *Z*, you should believe all *X* to be *Z* (a logical rule known as transitivity of inclusion). Here, *X*, *Y*, and *Z* are objects or propositions. Such prescriptive rules do not state that all people of a given culture, social position, age, and so forth, always comply with them. If, for example, the links in a chain of reasoning are numerous, the rule of transitivity will probably be broken, at least by children or by unschooled or impatient people.

Descriptions, and consequently predictions, of "illogical" behavior, as indeed of all human behavior, are presumably a task for psychologists or anthropologists. Such predictions are obviously important to practicing lawyers, politicians, salesmen, organizers, teachers, and others who work with people, just as predictions of the behavior of metals and animals are important to engineers and dairy farmers. These practitioners are well advised to know the frequencies of the various types of illogical behavior of their clients, adversaries, or students. But they are also well advised to avoid logical errors in their own behavior, as when they apply their knowledge of men (and of all nature, for that matter) to win lawsuits or elections or to be successful in organizing or teaching. For example, their propositions *X*, *Y*, and *Z* may be about another's madness, yet should obey the rule of transitivity of inclusion. Moreover, if you want to train future lawyers, statesmen, or businessmen, you are well advised to learn, from your own or other people's experiences, the techniques needed to make your pupil not only knowledgeable about other men's logical frailties but also strong in his own thinking and able to solve the brain twisters his future will offer. These pedagogical techniques are, of course, objects of descriptive study.

To illustrate the prescriptive–descriptive distinction in the domain of decisions, let us consider one of the rules proposed by prescriptive decision theory. We choose the rule called transitivity of preferences (other proposed rules will be discussed below) because of its special similarity with transitivity of inclusion. If you prefer a to b and b to c, you should prefer a to c. Here, a, b, and c are, in general, your actions with uncertain outcomes, although in the special case of "sure actions," the outcome, e.g., gaining or losing an object, is unique, in which case preferring action a is, in effect, the same as preferring the outcome of action a. If an individual disobeys the transitivity rule and prefers a to b, b to c, and c to a, we would say that he does not know what he wants. Surely we would advise or train a practitioner to develop the ability for concentrated deliberation, weighing advantages and disadvantages in some appropriate way. This should result in a consistent ranking, from the most to the least preferable, with ties (i.e., indifference between some alternatives) not excluded, of alternative actions from which the decision maker may be forced to choose.

► That human "rationality" is "bounded" is shown by Slovic et al. (1974) by many examples; our "perceptual and cognitive capabilities" are limited (see section 8 and, in a more general vein, Marschak & Radner 1972, chapter 9, sec. 6, "Uncertainty About the Outcome and Cost of Logical Operations"). Indeed, how does, or should, the experienced engineer or accountant decide how many digits to drop? He exercises wise "reluctance to bother about trifles," to quote Ramsey (1923–1928), a founder of normative decision theory. It can be relaxed in many respects. For example, Sen (1969) and Fishburn (1970) pointed out that the indifference (as distinguished from preference) relation need not be transitive, even in a normative context. This is analogous to the point made by Hamming (1965) and Rothstein (1965) that the inevitable rounding-off by computer makes the equality relation between numbers intransitive.

► Thus, the boundary between prescriptive decision logic and the descriptive psychology of decision making is fluid. As pointed out by another founder, Savage (1954, sec. 5.6), the discrepancy between a prescribed principle and a person's actual preferances, maintained "after thorough deliberation, . . . seems intolerable in a normative theory. Analogous circumstances forced D. Bernoulli [see section 6] to abandon the theory of mathematical expectation for that of utility." (See also Marschak 1976.)

► To train the decision maker in habits of "thorough deliberation" is itself a task for psychologists. This was well stated by leading British statisticians in the discussion following addresses by Tversky (1974) and Suppes (1974).

The statement that "given a set of feasible alternatives, the 'reasonable' (rational) man should choose the best" and the parallel statement that "an actual man does choose the best (and if the best is not feasible, the second best, etc.)" are both empty if the ranking order is not supposed to be stable over a period long enough to make the statements of practical relevance. If the ranking order is stable, economics as the study of the best allocation of available resources becomes possible.

The boundary between descriptive and prescriptive economics did not worry early writers. Gresham's law, Ricardo's explanations of rent and of the high postwar price of bullion, and Böhm-Bawerk's theory of interest are deductions from the assumption of consistent ranking of alternatives by reasonable men. But those authors also practiced induction and looked for historical facts that confirmed their deductions. The two approaches are not inconsistent if one assumes that in the cultures considered, housewives and adult, gainfully occupied men have by and large been "reasonable," at least when deciding not on wars and divorces but on matters of major interest to the economists of the time—for example, quantities such as prices, demands, and outputs, as well as less quantifiable choices such as location and internal organization (division of labor) of a plant. These matters were studied under the assumption of a tolerably well-known future.

The descriptive and prescriptive approaches became more clearly distinguished as more attention was directed to economic decisions under uncertainty (Fisher 1906; Hicks 1939; Hart 1940) and, stimulated by the *Theory of Games and Economic Behavior* (von Neumann & Morgenstern 1944), to nonquantifiable decisions traditionally assigned to political and military science or to sociology. In search of tools for their descriptive work, economists were also influenced by modern statistical decision theory. [*See* DECISION THEORY.] It then appeared that the reasonable man was not sufficiently approached even by the *ideal type* of an industrial entrepreneur, who evidently must be aided by a sophisticated hypothesis tester, a statistician, or, more recently and generally, an expert in operations research. [See OPERATIONS RESEARCH.] As such experts, armed with computers, do indeed progressively penetrate the economic world, predictive and descriptive economics (e.g., the prediction of aggregate inventories) may again become closer to what can be deductively derived from rational decision rules.

To return to the analogy between theories of thinking and of decision, prescriptive theories of thinking and decision are concerned with formal consistency between sentences or decisions, not with their content. If one believes that Los Angeles is on the Moon and that the Moon is in Africa, it is logical for him to believe that Los Angeles is in Africa, although none of the three statements is true. Similarly, if a man prefers killing to stealing and stealing to drinking strong liquor, he should prefer killing to drinking, although none of these actions is laudable or hygienic. Hicks (1939) used Milton's "reason is also choice" as a motto to the theory of consumption. Ramsey (1923–1928) called his theory of decision an extension of logic, an attitude adopted more recently by other logicians, for example, Carnap (1962), von Wright (1963), and Jeffrey (1965). The mutual penetration of logic and decision theory will appear even more founded and urgent when we come to discuss the cost of thinking and deciding (section 8).

In sections 1, 3, 4, and 5 we shall discuss some prescriptive decision rules and outline some simple experiments whose outcomes determine whether the subject did or did not comply with these rules. Whenever such experiments or similar ones have actually been performed, we shall refer to available descriptive evidence, namely, the frequency of subjects' obeying these rules or obeying some other stated behavior patterns, such as the probabilistic decision models of section 2. In section 6 the expected utility theorem implied by the prescriptive rules is discussed. Sections 7 and 8 point to unsolved problems on the frontier of decision theory.

1. Complete ordering of actions

Before we consider the notion of a complete ordering of alternative actions or decisions (we need not distinguish between the two here), it is convenient to restate the *transitivity* of preferences by defining the relation "not preferred to" or "not better than," written symbolically as $\leqslant$. If $a \leqslant b$ and $b \leqslant c$, then $a \leqslant c$. It is a *reflexive* relation, i.e., $a \leqslant a$. A relation that is transitive and reflexive is called an *ordering* relation. When $a \leqslant b$ and $b \leqslant a$, we write $a \sim b$ and say that the decision maker is indifferent between a and b, even though a and b are not identical. For example, an investor may be indifferent between two different portfolios of securities. Thus "$\leqslant$" (unlike the relation "not larger than" applied to numbers) is a *weak ordering* relation. Further, when $a \leqslant b$ and not $b \leqslant a$, we write $a < b$ and say that b is preferred to, or is better than, a.

○ Shall we say that a reasonable man always either prefers b to a or a to b or is indifferent? Or can he, in addition, refuse to compare them? If he can, the ordering of the actions by the relation "not better than" is only a *partial* one. This position has been taken, for example, by Aumann (1962), Chipman (in Interdisciplinary Research Conference 1960), and Wolfowitz (1962). But can one escape choosing? As in Pascal's (1670) famous immortality bet, "Il faut parier!" The avoiding of choice itself is a decision. Like other decisions, it is made more or less desirable by the prizes and penalties it entails. If this is granted, the transitive relation "$\leqslant$" is also *connective* and induces a *complete ordering*. Then any subset of actions can be arranged according to their desirabilities (with ties not excluded). Rank numbers, also called *ordinal utilities* (priorities), can then be assigned to the alternative actions provided that certain weak conditions are satisfied (Debreu 1959, sec. 4, n. 2, and sec. 4.6).

Tests of transitivity of preferences. As a matter of descriptive theory we ask whether people (of a given culture, etc.) do act as if there exist complete orderings of their actions by preferences. In actual experiments the penalty for refusing to choose was assumed to be strong enough to force subjects to make genuine choices in the form of verbal statements of preferences or, in some cases, actually to stake money on one wager rather than another. With connectivity of the preference relation thus assured and its reflexivity assumed by definition, the transitivity of the relation remained to be tested. The alternative actions were represented by multidimensional objects—a bundle of tickets to various shows (Papandreou 1957), marriage partners classified by three three-valued characteristics (May 1954; Davis 1958), a monetary wager (Edwards 1953; Davidson & Marschak 1959), a price policy affecting both the firm's rate of return and its share of the market, an air trip varying in both cost and duration (MacCrimmon 1965), and so forth. The subject responded by a triad of binary choices—a or b? b or c? c or a?—but in most experiments these choices were separated in time by choices from other triads, that is, "a or b?" was followed by "a' or b'?" rather than by "b or c?" Moreover, neither of the two actions "dominated" the other (see section 3), and no subject could use paper and pencil.

In all experiments, the transitivity rule was violated with varying frequencies. The proportion of the number of violations by all subjects to the maximum theoretical number of intransitive triads that were possible in the given experimental designs ranged from .04 in MacCrimmon's tests of business executives to .27 in May's tests of students. Since the experiments record actual choices and not preferences, intransitivity might be exhibited by a person who, although consistent, is strictly indifferent among three alternatives. (To approach, more generally, the case of "near indifference," the hypothesis tested must be rephrased in probabilistic terms. This will be discussed in section 2.) Thus the outcome of an experiment depends on the nature of the alternatives offered, making the experiments mentioned not quite comparable.

It would also be of great interest to study the effects of the passage of time and the effects of learning, especially if different methods of training are applied, such as "post-mortem" discussion (a domain opened by MacCrimmon), sequential modifications of the list of choices, etc. In fact, what would be the effect of supplying the subjects with paper and pencil and training them to use these prerequisites of our culture to tabulate the decision problem in an "outcome matrix" as in Table 1 below?

It has been suggested (Quandt 1956; Simon 1957) that intransitivities of decisions occur when the subject is unable to pay simultaneous attention to the several dimensions of each object of choice. This is somewhat analogous to the psychophysical finding that discrimination is greater between sounds differing in pitch but of equal loudness than between sounds differing in both (Pollack 1955). This suggests that the probabilistic approach common in psychophysics might be tried in a description of decision behavior. As to prescriptive theory, one may recall Benjamin Franklin's suggestion to score the pros and cons of each decision or the practice of adding scores when grading students or judging young ladies in beauty contests. Again, these or any other ways of overcoming the multidimensionality of decisions through deliberate and simultaneous marshaling of all the dimensions will in general require paper and pencil.

►A way to make a decision maker regret, and abandon, his intransitive ways was suggested by Pratt et al. (1965, secs. 2, 3, 4): if a person prefers *c* to *a*, *a* to *b*, *b* to *c* and if he owns *a*, he would be willing to pay a premium to exchange *a* for *c*; he would pay another premium to exchange *c* for *b*, then pay still another premium to exchange *b* for *a*—and end up where he began, *minus* three premiums.

2. Probabilistic ordering of actions

Since the prescriptive rule of transitivity is often violated, it is necessary to recast descriptive decision theory in different terms. The hypothesis that transitivity of decisions may be achieved by learning or training has already been mentioned. It is also traditional in experimental psychology, especially since Fechner's (1860) studies of perception, to describe behavior by probability distributions. Economists, by contrast, have succeeded in the past in making useful predictions of the aggregate behavior of large numbers of individuals based on the assumption that individuals are, by and large, "reasonable." The probabilistic approach is rather new in economics, perhaps as young as the econometricians' attempts to test aggregate economic models statistically and to relate these models to observable behavior of sampled individual households and firms.

○ Luce and Suppes (1965) have comprehensively surveyed the literature and the experimental evidence relevant to probabilistic orderings. For a more recent survey of relevant axioms and theorems, see Tversky (1976). One model that implies all the others suggested so far is that presented by Luce (1959). Fixed positive numbers $v_a, v_b, v_c, \cdots$, called *strict utilities*, are attached to the alternative actions $a, b, c, \cdots$. The set of actions includes all alternatives that may ever be offered to a given subject. The strict utilities are assumed to have the following property: if the subject must choose from a particular subset of alternatives, he will choose each with a probability proportional to its strict utility. For example, suppose $v_a:v_b:v_c:v_d = 1:2:3:4$. Then, if asked to choose from the subset (a, b, c), the subject will choose a, b, or c with respective probabilities $1/6$, $2/6$, $3/6$; if he must choose from the pair (a, b), he will choose a or b with respective probabilities $1/3$, $2/3$. The hypothesis can be tested at least "in principle" (i.e., assuming that representative samples from the universe of all choice situations can be obtained). One would observe the relative frequencies of choices from various subsets and infer, for example, that the ratio of the probability of choosing a to the probability of choosing b from the pair (a, b) does or does not change if a third alternative is also offered.

In another model, constants $u_a, u_b, u_c, \cdots$, called *strong utilities*, are associated with the alternatives $a, b, c, \cdots$. The strong utilities are assumed to have the following property: in the binary choice between a and b, the probability $p(a, b)$ of choosing a is an increasing function of the difference

$u_a - u_b$ (except when $p(a, b)$ is equal to 1 or 0). It follows that $p(a,b) \gtreqless \frac{1}{2}$ according as $u_a \gtreqless u_b$. Note that the case $p(a, b) = \frac{1}{2}$ provides an operational definition of "indifference" based on actual choices rather than on the verbal statement "I am indifferent." This model was used, in effect, in Thurstone's discussion of public opinion and of soldiers' food preferences (1927–1955). It is analogous to Fechner's perception model, which associates a physical stimulus a with a subjective "sensation" u_a; the subject perceives a to be heavier (or louder or brighter) than b with a probability $p(a, b)$ that increases with the difference $u_a - u_b$ (provided $p(a, b)$ is neither 0 nor 1).

It can be shown that if strict utilities exist, then strong utilities also exist (putting $u_a = \log v_a$, etc.), but not the converse. The strong utility model implies, in turn, the still less restrictive *weak utility* model, which assigns the ordinal numbers $w_a, w_b, w_c, \cdots$ to the alternatives $a, b, c, \cdots$, such that $p(a,b) \gtreqless \frac{1}{2}$ according as $w_a \gtreqless w_b$. Clearly, if the strong utilities exist, so do the weak utilities (putting $w_a = u_a$, etc.), but not the converse since differences between ordinal numbers are undefined. Both the weak utility model and the strong utility model can be tested in principle for possible rejection, for they have implications that involve choice probabilities only, avoiding the intervening utility concept. In particular, the models imply, respectively, the *weak stochastic transitivity*,

$$\text{if } p(a,b) \geqslant \tfrac{1}{2} \text{ and } p(b,c) \geqslant \tfrac{1}{2}, \text{ then } p(a,c) \geqslant \tfrac{1}{2},$$

and the *strong stochastic transitivity*,

$$\text{if } p(a,b) \geqslant \tfrac{1}{2} \text{ and } p(b,c) \geqslant \tfrac{1}{2}, \text{ then } p(a,c) \geqslant \max\,[p(a,b), p(b,c)].$$

Of these two forms of stochastic transitivity, the strong implies the weak one; it is, in turn, implied by the exact transitivity rule treated in section 1 if the choice probabilities are assumed to take on only three values: 0, $\frac{1}{2}$, and 1, corresponding to the preference relations $a < b$, $a \sim b$, and $b < a$.

Another probabilistic model also implied by, and less restrictive than, Luce's strict utility model is the *random utility* model. It does not imply, nor is it implied by, the strong or the weak utility model. It asserts, for a given subject, the existence of a fixed probability distribution on the set of all preference orderings of all alternatives, so that for a set of N alternatives, a probability is attached to each of their $N!$ permutations. In the nonprobabilistic economics of consumer's choice the (ordinal) utility function that attaches a constant rank to each commodity bundle is a fixed one. In the random utility model this function is, instead, visualized as drawn at random from a set of such functions, according to a probability distribution that is fixed for the given subject. Thus the word "utility" designates here a random number, not a constant as in the other probabilistic models described.

► Tversky (1972) designed experiments that permit separation of the chooser's indifference between two alternatives from his perceptual failure to discriminate between them when they are similar.

What relevance do these (and related) models have for a prescriptive theory of decision? Just as not all men are always consistent, not all men are always in good health. Doctors and nurses are busy measuring the temperatures and blood pressures of the sick ones. The probability distributions supposed to characterize a "stochastic decision maker" may vary in the degree of their closeness to the ideal limit—consistency. For example, consistency is approached in the strong utility model as the values of $p(a, b)$ concentrate near 1 or 0 or $\frac{1}{2}$. We can thus trace the path of progress of learning or training for consistency.

3. Inadmissibility of dominated actions

An action's result (outcome) depends, in general, not only on the action but also on the "state of the world (nature or environment)," which is not in the decision maker's control. For our purposes an action (called an act by Savage [1954]) can be defined by the results it yields at various states of the world. Thus an action is a function from the set of states to the set of results. (This function has constant value if the action is a "sure action.") In a decision situation, only some such functions are available; their "feasible set" (the "offered set of alternatives" of section 2) depends, for example, on the decision maker's resources, technology, and market as he views them.

An *event* is a subset of the states of nature. In particular, the subset consisting of all states that, for any given feasible action, result in the same outcome is called an *outcome-relevant event*. In a decision situation the set of states is partitioned into such events, with all irrelevant details omitted.

Even if the well-disciplined decision maker can consistently rank multidimensional results according to his preferences, action under uncertainty remains multidimensional because of the multiplicity of events. Actions are bets, or wagers. Yet it seems reasonable to prescribe (as in section 1) that bets be ordered; indeed, some authors prescribe that bets be completely ordered (Ramsey 1923–

1928; Savage 1954; de Finetti 1937). It will be shown in section 6 that a complete ordering corresponds to that of numerical "expected utilities" of actions *provided* that the decision maker is consistent in the sense of obeying both the traditional rules of logic and certain postulates that are plausible enough to qualify as extensions of logic. These postulates will be discussed in this and subsequent sections, essentially following Savage.

Action a is said to dominate action b if the results of a are sometimes (i.e., when some events take place) better than the results of b and are never worse. Is it not reasonable that you should prefer the dominating action? Any action that is dominated by some feasible action is thus inadmissible. Consider, for example, the actions listed in Table 1.

Table 1 — An outcome matrix (in dollars)

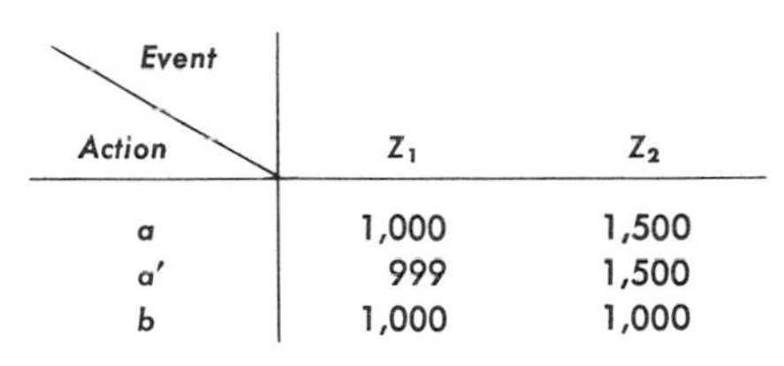

Action \ Event	Z_1	Z_2
a	1,000	1,500
a'	999	1,500
b	1,000	1,000

If action a is taken and the state of the world is Z_1, a gain of \$1,000 will be realized; if action a is taken and the state of the world is Z_2, a gain of \$1,500 will be realized; and so on. Such a table is called an *outcome matrix*.

The inadmissibility postulate would order these actions as follows: $a' < a, b < a$. It does not, by itself, induce complete ordering of the actions. Thus, it is silent about the preference relation between actions a' and b in Table 1. The "expected utility" rule of section 6 will determine this relation roughly in the following manner: depending on the "probabilities" of events and on the "utilities" of results, the \$500 advantage of a' over b if Z_2 happens may or may not outweigh its \$1 disadvantage if Z_1 happens. This appeals to common sense but remains vague until, with help of other postulates, we define the concepts of "probability" and "utility."

Does the inadmissibility rule describe people's actual behavior? It is difficult to make an individual violate the rule when the decision situation is presented with clarity, for example, by an outcome matrix whose entries have an obvious ordering (sums of money, as in Table 1, or good health, light illness, and death). But consider the following experiment. A subject is given an envelope containing v dollars; he is permitted to convince himself that the envelope does contain v dollars. He writes, but does not tell the experimenter, his asking price, a, for the envelope and its contents. He will then receive a bid of x dollars. If x exceeds a, he will receive x dollars; otherwise, he will keep the v dollars. There will be no further negotiations. Subjects often ask more than the true value, setting $a = a_1 > v$. In Figure 1, the stipulated result r of such action is plotted against possible levels of the bid x (state of nature). Comparison with the corresponding plot for the "honest" asking price, $a = a_0 = v$, shows that a_1 is dominated by a_0. In fact, an asking price $a_2 < v$ (not shown in Figure 1) is also dominated by a_0. Hence the honest asking price is the only admissible one. It seems that some subjects expect further negotiations despite the explicit warning. We surmise that they would waive uncalled-for associations and habits if trained to plot or tabulate payoffs.

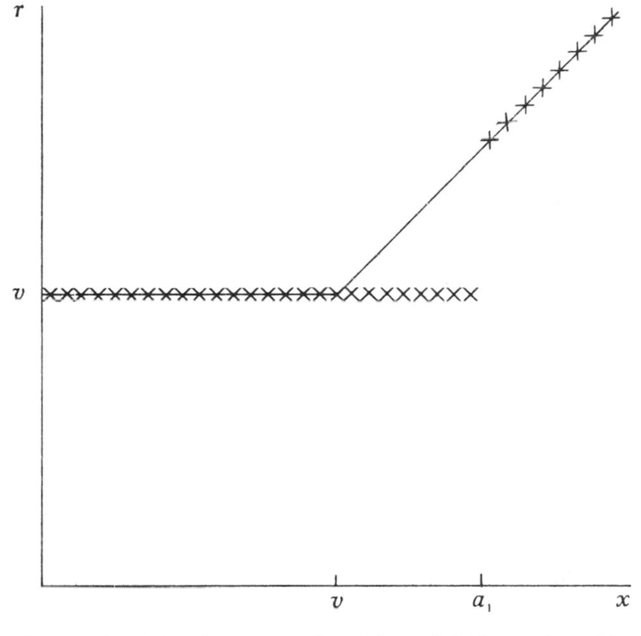

Figure 1 — Result r as a function of bid x when the asking price is equal to v (solid line) and to $a_1 > v$ (crosses)

4. Irrelevance of nonaffected outcomes

Table 2 presents a matrix of outcomes measured in per cent of return on a firm's investment. We may interpret a, b, a', and b' as the firm's investing in the development of alternative products. The outcomes of these actions depend on the mutually exclusive events (say, business conditions) Z_1, Z_2, and Z_3. Suppose the firm prefers a to b. This preference cannot be due to any difference in the outcomes if the event Z_1 happens, for these outcomes are identically 5. Therefore, the firm's preference of a to b must be due to its preferring the wager "-200 if Z_2, 100 if Z_3" to the return of 5 with certainty. But then the firm should also prefer a' to b', for, again, if Z_1 happens, the outcome (-200) is

Table 2 — Outcome matrix for the return on a firm's investment (per cent)

Action \ Event	Z_1	Z_2	Z_3
a	5	−200	100
b	5	5	5
a'	−200	−200	100
b'	−200	5	5

Table 3 — Outcome matrix for the funnel experiment (in dollars)

Bet \ Event	RB	BR	RR	BB
1R	100	0	100	0
2R	0	100	100	0
2B	100	0	0	100
1B	0	100	0	100

not affected by the firm's choice. Its preference as between a' and b' must depend on its preference as between the wager "−200 if Z_2, 100 if Z_3" and the certainty of 5, just as in the previous case. By the same reasoning, if the firm is indifferent between a and b, it should be indifferent between a' and b'. The rule enunciated here can be regarded as a generalization of the admissibility rule of the previous section, with wagers admitted as a form of outcomes so that, for example, action a is described as follows: if Z_1 happens, you get 5; otherwise you get a lottery ticket, losing 200 if Z_2 happens and gaining 100 if Z_3 happens.

○ **Tests of the irrelevance rule.** Most business executives violated the irrelevance rule when MacCrimmon (1965) *verbally* gave them the choices in Table 2—first a versus b, then a' versus b'. However, a large majority of my students, who also never heard of the irrelevance principle but knew how to draw up a matrix of outcomes, complied with the rule. Allais (1953) performed a similar experiment, but instead of describing the three events as alternative business conditions, he gave them numerical probabilities (an extraneous notion in the present context), the probability of one of the events being only 0.01—thus possibly introducing perceptual failure to discriminate (see section 3). Most of Allais's respondents, including some decision theorists, violated the principle (but see Savage's introspective discussion of his own second thought [1954, sec. 5.6g]).

The following type of experiment has been discussed widely. Funnel 1 contains an equal number of red and black balls, and the subject is invited to convince himself of this; but he is not permitted to look into funnel 2, which he only knows to contain one or more balls whose colors are either black or red. When a handle is pulled, each funnel will release only the bottom ball. In the following bets, he will win either $100 or nothing:

Bet $1R$: bottom ball in funnel 1 is red.
Bet $1B$: bottom ball in funnel 1 is black.
Bet $2R$: bottom ball in funnel 2 is red.
Bet $2B$: bottom ball in funnel 2 is black.

The outcomes are as indicated in Table 3, where, for example, event RB is "bottom ball in funnel 1 is red, bottom ball in funnel 2 is black." Suppose the subject always prefers to use funnel 1 rather than funnel 2, that is, he prefers bet $1R$ to $2R$ and also bet $1B$ to $2B$. He will thus treat the results unaffected by his choice—namely, those earned when events RR and BB occur—as *relevant* to his choice! Yet similar experiments performed by Chipman (see Interdisciplinary Research Conference 1960) and Ellsberg (1961) suggest that many people do indeed prefer funnel 1 to funnel 2. Some of my student subjects who did so motivated their choice by stating the sharp distinction between "risk" (funnel 1) and "uncertainty" (funnel 2), a distinction taught to economists since Knight (1921). Others stated that they could base their bets on more information when using funnel 1 than when using funnel 2. But information, although never harmful, can be useless. In experiments of Raiffa (1961) and Fellner (1965), subjects were, in effect, willing to pay for such useless information.

5. Definitions of probabilities

Having defined events $Z_1, Z_2, Z_3, \cdots$ as subsets of the set X of all states of nature, it is consistent with current mathematical language and current English to require that any numbers, $P(Z_1)$, $P(Z_2)$, $P(Z_3)$, $\cdots$, claimed to be the probabilities of these events, satisfy the following conditions: (1) they should be nonnegative; (2) for any two mutually exclusive events, say Z_1 and Z_2, the number assigned to the event "Z_1 or Z_2" (meaning the occurrence of either Z_1 or Z_2) should equal the sum of $P(Z_1)$ and $P(Z_2)$; (3) $P(X)$, i.e., $P(Z_1$ or Z_2 or Z_3 or $\cdots)$, should be equal to one.

Personal probabilities. The numbers $P(Z_1)$, $P(Z_2)$, $P(Z_3)$, $\cdots$ are an individual's personal probabilities of these events if, in addition to having the mathematical properties just stated, they describe his behavior in the following sense: whenever $P(Z_1) > P(Z_2)$, for example, he will prefer betting on Z_1 to betting on Z_2, assuming he wants to win the bet. This, too, is consistent with ordinary English—he will prefer betting on the victory of the Democrats in the next election to betting on the truth of the proposition that California is longer than Norway if and only if he considers winning the former bet "more probable" than winning the latter one. Here, "betting on Z_1" means taking an action that yields a result s (for success) if Z_1 occurs, a result that is more desirable than the result f (for failure) if Z_1 does not occur. The probabilities of events must depend on the events only, i.e., the individual's preferences between bets must be the same for all pairs of results s, f ($100, $0; *status quo*, loss of prestige; etc.) provided only that s is better than f. This postulate of *independence of beliefs on rewards* must be added to those of sections 1, 3, and 4. Will the subject reverse his judgment about the comparative chances of any two events—as revealed by his choices between two bets—if the prizes are changed? If not, he satisfies the postulate. It seems to be satisfied by practically all subjects asked to rank several bets according to their preferences, the rewards being first a pair s, f, then a different pair s', f'.

Suppose a subject is indifferent to bets on any of the eight horses running a race. His preferences would thus imply that $P(Z_1) = P(Z_2) = \cdots = P(Z_8)$, where Z_i is the event that the ith horse wins (ties are excluded for simplicity). The nonnegativity property is satisfied if we put $P(Z_i) = \frac{1}{8}$ for all i. What about the other two mathematical properties required? Suppose the subject considers double bets with the same prizes s, f as for single bets. He will prefer every double bet, e.g., the bet on the event "Z_1 or Z_2," to any single bet (here he has, in effect, applied the inadmissibility postulate), and he will be indifferent between any two double bets (here the postulates of irrelevance and of intransitivity apply). Similarly, he will prefer triple bets to double bets and will be indifferent between any two triple bets, and so on. Consistent with these preferences, we can put

$$P(Z_i \text{ or } Z_j) = 2/8 = P(Z_i) + P(Z_j)$$

for any two horses i and j,

$$P(Z_i \text{ or } Z_j \text{ or } Z_k) = 3/8 = P(Z_i \text{ or } Z_j) + P(Z_k)$$

for any three horses i, j, and k, and so on. In general, we can assign probability $P(Z) = m/8$ to the event Z that one of the m specified horses will win (where $1 \leqslant m \leqslant 8$). Then the numbers $P(Z)$ are the subject's personal probabilities, for they agree with his preferences between bets and also satisfy the three mathematical requirements stated at the beginning of this section.

○ Instead of a horse race, a subject is asked to imagine a dial divided into n equal sectors. Suppose the hand of the dial is spun and comes to rest in the ith sector. We define this occurrence as event Z_i. If the subject is convinced that the dial's mechanism is "fair," i.e., that the events $Z_1, Z_2, \cdots, Z_n$ are symmetrical (exchangeable), he will be indifferent among bets on any one of these n events. [*See* PROBABILITY, *article on* INTERPRETATIONS.] Following Borel (1939, chapter 5), he can then assess his personal probability of any event T, say "rain tomorrow," by finding a number m ($1 \leqslant m \leqslant n$) such that betting on T is not more desirable than betting on "Z_1 or Z_2 or $\cdots$ or Z_m" and is not less desirable than betting on "Z_1 or Z_2 or $\cdots$ or Z_{m-1}." Then his personal probability of T, $P(T)$, satisfies the inequality

$$\frac{m-1}{n} \leqslant P(T) \leqslant \frac{m}{n}.$$

By making n arbitrarily large, one can assess $P(T)$ arbitrarily closely, and by using dial arcs which represent any fraction, rational or irrational, of the dial's circumference, one can define personal probabilities ranging continuously from 0 to 1. Note that such assessments of personal probabilities, when determined in an experiment, are based not on the subject's verbal statement of numbers he calls probabilities but on his actual choices. They may therefore be useful in predicting actions provided that the subject is consistent. Such elicitation of personal probabilities has been carried out by Staël von Holstein (1970), and Savage (1971) suggested further experimental designs. Pratt et al. (1965, chapter 2, sec. 6) provided methods to check (and train) for consistent evaluation of probabilities.

● If the subject is not consistent, the violations committed may or may not be similar in principle to those incurred in any instrument readings—a theme of probabilistic psychology touched upon in section 2. Some systematically occurring inconsistencies in assessing probabilities were observed by Tversky and Kahneman (1974).

○ **Objective probabilities.** As a special case, personal probabilities of some real-world events may be "objective," i.e., the same for different people. This is particularly the case when there is agreement that the events come sufficiently close, for all practical purposes of those involved, to fulfilling certain symmetry requirements. Approximate symmetry is assumed for the positions of a roulette dial like Borel's and for the occurrences of death among many similar males of age 20. Such requirements are strictly satisfied only by idealized, mathematically defined events—events that are never observed empirically. A "fair" coin, a "fair" roulette dial, a "homogeneous" population of males aged 20 (or a "random" sample from such a population) are all mathematical constructs. The mathematical theory of probability applies rules of logic to situations in which strict symmetry and the three properties stated at the beginning of this section hold (refining property 2 in order to accommodate the case of an infinite X). If decision makers agree that certain events are approximately symmetric, and if they apply logical rules, then their choices between betting on (predicting) any two events will agree; their personal (in this case also objective) probabilities will coincide with those given by mathematical theory. Clarity requires us, however, to distinguish between mathematical probabilities and objective—or, better, *inter-subjective*—probabilities assigned by decision makers to empirical events, just as we distinguish between a geometric rectangle and the shape of an actual sheet of paper. For details, see Marschak (1973; 1974*a*, essay 17).

6. Expected utility

The four postulates discussed thus far—complete ordering of actions, inadmissibility of dominated actions, irrelevance of nonaffected outcomes, and independence of beliefs on rewards—appear about as convincing as the rules of logic (and about as subject to transgression by people not trained in untwisting brain twisters). Together with a "continuity" postulate (to be introduced presently), they imply the following rule, which is more complicated and less immediately convincing: The consistent man behaves as if he (1) assigned personal probabilities $P(Z)$ to events Z, (2) assigned numerical utilities $u(r)$ to the results r of his actions, and (3) chose the action with the highest "expected utility." The expected utility $\omega(a)$ of action a is the weighted average

$$\omega(a) = \sum_r u(r) \cdot P(Z_{ra}),$$

where the event Z_{ra} is the set of all states for which action a yields result r.

The rule is trivially true when the choice is among sure actions; if action a always yields result r, then $P(Z_{ra}) = 1$, so that $\omega(a) = u(r)$.

Consider now actions with two possible results—success s and failure f. This is the case, for example, when actions are two-prize bets (as in section 5) or when the decision maker is a "satisficer" (Simon 1957) for whom all outcomes below his "aspiration level" are equally bad and all others equally good. In section 5 we saw that of two two-prize bets, a consistent decision maker prefers the bet that has the higher probability of success. Since s is better than f, we can assign numerical utilities

$$u(s) = 1 > 0 = u(f),$$

and we see that the expected utility $\omega(b)$ of a two-prize bet b coincides with its probability of success $P(Z_{sb})$, since

$$\omega(b) = 1 \cdot P(Z_{sb}) + 0 \cdot P(Z_{fb}) = P(Z_{sb}).$$

Thus the satisficer maximizes the probability of reaching his aspiration level.

As the next step, we compute the probability of success and hence the expected utility $\omega(c)$ of a bet c compounded of n simple two-prize bets or lottery tickets $b_1, \cdots, b_n$ on n different (but not necessarily mutually exclusive) events $T_1, \cdots, T_n$. Lottery ticket b_i is a bet on the event T_i, and the subject will receive ticket b_i if Z_i happens. The events $Z_1, \cdots, Z_n$ are mutually exclusive events one of which must happen, and the events "Z_i and T_i" (the occurrence of Z_i and T_i) are pairwise independent in the sense that

$$P(Z_i \text{ and } T_i) = P(Z_i) \cdot P(T_i).$$

We can thus regard ticket b_i as the result yielded by action c when Z_i happens. Hence $P(Z_i) = P(Z_{b_ic})$ in the present notation. Moreover, we have just shown that the expected utility of a simple two-prize bet can be measured by its probability of success, so that $P(T_i) = \omega(b_i)$. Clearly, the probability of success of the compound bet c is the probability of the event "(Z_1 and T_1) or (Z_2 and T_2) or $\cdots$ or (Z_n and T_n)"; by mathematical property 2 of probabilities (section 5), this is equal to $\sum_i P(T_i) \cdot P(Z_i)$. Hence,

$$\omega(c) = \sum_i \omega(b_i) \cdot P(Z_{b_ic}),$$

i.e., the expected utility rule is valid for the special case where each result of an action is a two-prize bet.

To extend this in a final step to the general case, let s be the best and f the worst of all results of an action. In the preference notation of section 1, $f \leqslant r \leqslant s$ for any result r. Consider the continuous range of all bets b whose two prizes are s and f and whose success probabilities take all the values between (and including) 1 and 0. Then, for any b, $f \leqslant b \leqslant s$, and for a given r, $r \leqslant b$ or $b \leqslant r$ depending on the bet's success probability. A plausible *continuity postulate* asserts, for each r, the existence of a bet, say b_r, such that $r \sim b_r$. We can therefore assign to r a utility $u(r) = \omega(b_r)$. A decision maker should then be indifferent between an action a that yields various results r with respective probabilities $P(Z_{ra})$ and a bet c compounded of the corresponding two-prize bets b_r just described entering with the same probabilities $P(Z_{ra})$. That is to say, $P(Z_{b_r c}) = P(Z_{ra})$. The expected utility rule follows, since

$$\omega(a) = \omega(c) = \sum_r \omega(b_r) \cdot P(Z_{b_r c})$$
$$= \sum_r u(r) \cdot P(Z_{ra}).$$

Some insight into this derivation of the expected utility rule is provided to the trainee in decision making by letting him rank his preferences among the tickets to four lotteries. Each ticket is described by prizes contingent on two alternative events, one of which must occur. An example of such a decision problem is presented in Table 4, where p is written for $P(Z)$ for brevity. If the event Z is "a coin is tossed and comes up heads" (we refer below to this event simply as "heads") and the subject regards the coin as "fair," then $p = \frac{1}{2}$. But Z may also be, for example, "the next sentence spoken in this room will contain the pronoun 'I.' " In any case, when the inadmissibility postulate is applied, it is evident from the last column of Table 4 that ticket a is better than c and worse than b. Furthermore, the decision maker should be indifferent between tickets a and d.

Table 4 — Decision problem involving lottery tickets

Lottery ticket	Prize if event Z happens	Prize if event Z does not happen	Probability of gaining $100
a	$100	$0	p
b	Lottery ticket a	$100	$p^2 + 1 - p > p$
c	Lottery ticket a	$0	$p^2 < p$
d	Lottery ticket b	Lottery ticket c	$p(p^2 + 1 - p) + (1 - p)p^2 - p$

Cash equivalents and numerical utilities. Let us define the *cash equivalent* of ticket a, denoted $k(a)$, as the highest price the decision maker would offer for ticket a; $k(b)$, $k(c)$, and $k(d)$ are defined similarly. If asked to name his cash equivalent for each lottery ticket in Table 4, the decision maker should name amounts such that $k(b) > k(a) = k(d) > k(c)$. If he fails to do this, he is inconsistent, and no scale of numerical utilities can describe his behavior. If he is consistent, and if the event Z is "heads," the following utilities for some money gains can be ascribed to him:

$$\omega(\$100) = 1;$$
$$\omega[k(b)] = \tfrac{3}{4};$$
$$\omega[k(a)] = \omega[k(d)] = \tfrac{1}{2};$$
$$\omega[k(c)] = \tfrac{1}{4};$$
$$\omega(\$0) = 0.$$

Some but not all subjects conform with the required ranking of the lottery tickets. Therefore, in any empirical estimation of a subject's utilities and personal probabilities, one must check whether the subject is consistent, at least in some approximate sense. As pointed out in the simpler context of section 2, probabilistic models of decision and of learning to decide are needed for any descriptive theory, and they too may fail.

○ **Behavior toward risk.** Again suppose that the event Z in Table 4 is "heads." If the subject has named cash equivalents $k(a) = k(d) = \$50$, $k(c) = \$25$, and $k(b) = \$75$ (and similarly for further, easily conceived compound lotteries with utilities $\frac{1}{8}, \frac{3}{8}, \cdots, \frac{1}{16}, \cdots$), we would infer that over the observed range he is indifferent between a "fair bet" and the certainty of getting its expected gain. We would say that he is "indifferent to risk." His utility function of money gain is a straight line. On the other hand, if he has named cash equivalents $k(a) = k(d) < \$50$, $k(c) < \frac{1}{2}k(a)$, and $k(b) < \frac{1}{2}[100 + k(a)]$, his utility function of money gain will be a concave curve (any of its chords will lie below the corresponding arc), and we would say that he is "averse to risk." If the inequality signs in the preceding sentence are reversed, the utility function is convex (chords are above arcs), and we would say that over the observed range the subject "loves risk." When the utility function is either concave or convex, the decision maker maximizes the expected value, not of money gain, but of some nonlinear function of money gain. Relevant measurements based on observed choices between wagers were performed by Mosteller and Nogee (1951), Becker et al. (1964), and Swalm (1966).

Daniel Bernoulli (1738) pointed out that the utility function is concave in the case of the "Petersburg paradox." Marshall (1890) also assumed risk aversion as an economic fact (or perhaps as a prescription from Victorian morals) equivalent to that of "decreasing marginal utility of money." Some economic implications of risk aversion were given by Pratt (1964) and Arrow (1965). It should be noted that this assumption is inconsistent with the behavior of a satisficer for whom utility is a step-function of money gain and the behavior of a merchant for whom the disutility of bankruptcy is the same regardless of the amount owed.

► Having rejected the irrelevance rule (see section 4) and hence the expected utility theorem, Allais (1953) proposed replacing the maximization of expected utility by the maximization of a function of the mean and the variance of a numerical "psychological value" of money gain, as communicated verbally by the decision maker to the experimenter.

The maxmin rule. In competition with the expected utility rule (and the postulates underlying it) is the "conservative" or "maxmin" rule, which states that the decision maker should maximize the minimum payoff; that is to say, the maxmin rule proposes that preferences among actions be based on each action's worst result only. Thus in the case of Table 1 the maxmin rule would prescribe that $a' < b$ even if Z_1 has the probability of, say, an earthquake, and presumably $b \sim a$, contradicting the inadmissibility postulate. However, because of the strong appeal of the inadmissibility postulate, it is usually proposed that the maxmin rule be applied only when domination is absent; this leads to the ordering $a' < b < a$ rather than $a' < b \sim a$. This proposal is not too satisfactory since it violates a plausible continuity principle (Milnor 1954). A small modification in outcomes —by $1 in Table 1, or by 1 cent for that matter— can make or break domination and thus reverse the preference ordering. Indeed, should we advise people not to live in San Francisco or Tokyo because of possible but not very probable earthquakes? Should we advise pedestrians never to cross streets because of the possibility of being struck by an auto? Such considerations compel us to balance the advantages against the disadvantages of competing decisions, weighing them by appropriately defined probabilities—the expected utility principle.

►**Multi-criterion decisions.** Sometimes the outcome of a feasible action a can be represented as a vector of n (say) numerical criteria, $f(a) = [f_1(a), \cdots, f_n(a)]$. The case is familiar in consumption economics (see, for example, Debreu 1959): under certainty, the consumer maximizes the utility of a "bundle" of commodity amounts over the set A of feasible actions, or choices (for example, A may be the budget constraint, given the n market prices):

$$\max_{a \epsilon A} U[f(a)],$$

where the utility, U, is "ordinal," that is, determined up to a monotone transformation. Lancaster (1971) has further proposed considering each commodity itself as a vector of criteria (for example, an automobile's cost, comfort, fuel consumption). This opens up the problem of a consumer's soul searching for trade-offs (marginal rates of substitution) between criteria. Most important are applications to decisions of governmental or other nonprofit organizations: the "cost–benefit analysis" (see, for example, Kendall 1971), the search for, and use of, the nation's "welfare function" (Johansen 1974). For the case of certainty, Geoffrion et al. (1972) developed a soul-searching dialogue between the decision maker (a person in charge of a university department) and a computer, provided it is sufficient to determine the trade-offs in the neighborhood of a specified feasible vector, such as the last year's budget allocation. The man–machine dialogue allows for checks on the decision maker's consistency and thus for his "training."

► Under uncertainty, a (cardinal) utility additive in the criterion values presents great convenience: for then, and only then, is the expected utility, $E[(U(f(a))]$, monotone in the expectations of the criterion-values, $E[f_i(a)], i = 1, \cdots, n$. See Fishburn (1970) and Marschak (1971, appendix 1).

► To be sure, additivity is sometimes excluded a priori. For example, an airline's revenue depends on the number of passenger-miles per year, and thus on the product, not the weighted sum, of the criteria "speed" and "occupancy." Or, to illustrate the above theorem, consider the not unusual case of two "goals" ($n = 2$: two examinations to be passed, for example), such that, without loss of generality, utility equals 1, if both goals are attained, or zero otherwise. Then, putting $f_i(a) = 1$ or 0, $i = 1,2$ (without loss of generality) we have $U[f(a)] = f_1(a) \cdot f_2(a)$. Now suppose the joint probabilities of the two criterion-values, given two alternative actions, a and b, are

$$\begin{array}{cc|cc} & & \multicolumn{2}{c}{f_2(a)} \\ & & 1 & 0 \\ \hline f_1(a) & 1 & 0 & \frac{1}{2} \\ & 0 & \frac{1}{2} & 0 \end{array} \qquad \text{and} \qquad \begin{array}{cc|cc} & & \multicolumn{2}{c}{f_2(b)} \\ & & 1 & 0 \\ \hline f_1(b) & 1 & \frac{1}{4} & 0 \\ & 0 & 0 & \frac{3}{4} \end{array}$$

Then

$$E[U(f(a))] = 0 < \tfrac{1}{4} = E[U(f(b))];$$

and yet

$$E[f_1(a)] = E[f_2(a)] = \tfrac{1}{2} > \tfrac{1}{4} = E[f_1(b)] = E[f_2(b)].$$

Still, assuming additivity (for example, assuming commensurability of benefit and cost) may well serve as a first approximation. Alternatively, its convenience may be sacrificed and, for example, products of criterion-values admitted as components of utility (Keeney 1974). Raiffa (1968, chapter 9) proposed assessing the decision maker's criterion-values $f_i(a)$ by letting him choose between wagers, on the lines of Mosteller and Nogee (1951); and applications to medical treatment were made. Appropriate man–computer interaction programs have been worked out, for example, by Sarin (1975). Many relevant contributions will be found in Cochrane and Zeleny (1973). On the negative side, the criticism by Brewer (1973) deserves attention.

A historical note. Early theories prescribing maximization of expected utility were confined to special cases. Two types of restrictions were imposed. First, the decision maker was advised to maximize the expected value of his money wealth or gain, i.e., utility was in effect identified with a money amount or some linear function of a money amount. Some other quantifiable good, such as the number of prisoners taken or the number of patients cured, might play the same role as money, but nonquantifiable rewards and penalties of actions were unnecessarily excluded from the set of results. Second, probabilities were restricted to objective ones—a special case of personal probabilities.

It is difficult to imagine that an experienced Bronze Age player who used dice that had "tooled edges and threw absolutely true" (David 1962) would bet much more than 1:5 on the coming of the ace. Cardano's efforts in computing the gambler's odds (Ore 1953) suggest at any rate that by the sixteenth century the rule of maximizing average money gains computed on the basis of objective probabilities was taken for granted.

We have already cited Bernoulli and Marshall as having proclaimed utility a nonlinear function of money, thus lifting the first of the above restrictions. Marshall also applied the utility concept to commodity bundles. Von Neumann and Morgenstern (1944) extended it further to all possible results of actions and derived the expected utility rule from simple consistency postulates. This was simplified by others, especially Herstein and Milnor (1953). All these writers dealt only with objective probabilities, leaving out the important cases in which symmetries between relevant events are not agreed upon.

Bayes (1764) can be credited with the idea of personal probabilities. He thought of them as being revealed by an individual's choices among wagers—not just on cards and coins, but on horses and fighting cocks as well! Thus Bayes removed the second restriction, but he retained the first in assuming that utility was in effect identified with money gains, so that betting 9 guineas against 1 implies a corresponding ratio of probabilities. In 1937, de Finetti provided mathematical rigor for this approach.

○ In 1926, Ramsey (1923–1928) stated, perhaps for the first time, simple consistency postulates that imply the existence of both personal probabilities (of any events, regardless of symmetries) and utilities (of any results of actions, quantifiable or not). Savage (1954) restated the consistency postulates and, partly following de Finetti and von Neumann and Morgenstern, proved that they imply the expected utility rule. For original expositions, see Pratt, Raiffa, and Schlaifer (1964; 1965).

○ Reviews of the works of other contemporary authors can be found in Arrow (1951), Luce and Suppes (1965), Ożga (1965), Borch (1968), Fishburn (1964; 1970), and Shackle (1968). Surveys and bibliographies as well as much original material by 21 leading authors on both the prescriptive and descriptive aspects of decision theory can be found in Shelly and Bryan (1964).

7. Strategies

The existence of numerical utilities describing a decision maker's "tastes" and of probabilities describing his "beliefs" has been shown to follow from rules of consistency. To formulate and solve the problem of choosing a good decision, both tastes and beliefs must be assumed fixed over some given period of time (but see Koopmans [1964] on "flexibility"). In general, the probabilities the decision maker assigns at any time to the various states of nature and thus to the results of his actions depend on his information at that time. New information may also uncover new feasible actions. We shall generalize the decision concept accordingly in several steps.

A *strategy* (also called a decision function or response rule) is the assignment, in advance of information, of specific actions to respond to the different messages that the decision maker may receive from an information source. If, more generally, messages, actions, and results form time sequences (the case of "earning while learning"), we have a *sequential strategy* (also called an extensive or dynamic strategy). To each possible sequence of future messages, a sequential strategy assigns a sequence of actions (see, for example, Theil 1964). An optimal sequential strategy maximizes the weighted average of utilities assigned to all possible sequences of results, possibly taking into account "impatience" by means of a discount rate (Fisher 1930; Koopmans 1960). The weights are personal *joint* probabilities of sequences of events *and* messages. This amounts to the same thing as saying that the probabilities of events are revised each time a message is received.

It is still more general to redefine action in order to include in it the choice of information sources to be used and thus of "questions to be asked" at a given time. The resulting problem of finding an optimal *informational strategy* is a task of economic theories of information and organization and also of statistical decision theory (e.g.,

Raiffa & Schlaifer 1961; DeGroot 1970) where events and information sources correspond to hypotheses and experiments, respectively.

Finally, if we allow the decision maker to receive messages about the feasibility and outcomes of actions he has not formerly considered, we obtain the still more general concept of *exploratory strategy*. True, it has been almost proverbially tenuous to assign probabilities to the results of industrial or scientific research; yet these undertakings are not different in principle from many other ventures and bets, such as those discussed in section 5. ○ The complex strategies noted here are hardly maximized by the actual entrepreneur, although the penetration of industry by professionals may again bring descriptive and prescriptive economics to their pristine closeness. Today many descriptive hypotheses in this field use the concept of aspiration level—the boundary between success and failure. The actual decision maker—the "satisficer" (Simon 1957)—is said to revise his aspiration level upward or downward depending on whether he has or has not reached it by previous action; exploration for actions not previously considered is triggered by failure. The sequence of actions generated by the dynamic aspiration-level model will, in general, differ from that prescribed by dynamic programming. [*See* PROGRAMMING.] Yet, with utilities assigned to results of actions in a particular way, it is possible that the aspiration-level mechanism is indeed optimal. It has been inspired, in fact, by adaptive feedbacks observed in live organisms; such feedbacks presumably have maximized the probability of the survival of the species. Light is also thrown on the process of scientific research if the scientist is regarded as a "satisficer," or even under somewhat more general conditions (Marschak 1974*b*).

8. Cost of decision making

○ One action or strategy may appear better than another as long as we disregard the toil and trouble of decision making itself, i.e., the efforts of gathering information and of processing it into an optimal decision. The ranking of actions may be reversed when we take these efforts into account and deal, in this sense, with "net" rather than "gross" expected utilities of actions. A small increase in expected profit may not be worth a good night's sleep. In statistics, we stop sequential sampling earlier the higher the cost is of obtaining each observation. As an approximation to some logically required, but very complicated, decision rule, we may use a linear decision rule (e.g., prescribe inventories to be proportional to turnover) in order to lessen computational costs. The "incrementalism" observed and recommended by Lindblom (1965) in the field of political decisions corresponds to the common mathematical practice of searching for a global optimum in the neighborhood of a local optimum or possibly in the neighborhood of the *status quo*. How many local search steps one should undertake and how often he should jump (hopefully) toward a global optimum will presumably depend on the costs of searching (see, for example, Gel'fand & Tsetlin 1962). Indeed, some strategies may be too complex to be computable in a finite amount of time, even though they respond to each message by a feasible action. There is, after all, a limit on the capacity of computers and on the brain capacity of decision makers. Hence, some strategies may have infinite costs. The effects of human capability limitations on the structure of organization have been studied by Beckmann (1960) and by Drenick and Levis (1974).

The net utility of an action is often represented as the difference betwen "gross utility" and "decision cost." A prescriptive theory would presumably require that personal probabilities be assigned not only to the outcomes of the actions but also to the efforts of estimating the outcomes and of searching for the action with the highest expected utility.

On a more general level, it is not strictly permissible to represent net utility as a difference between gross utility and decision cost. Even if the two were measurable in dollars, say, utility may be a nonlinear function of money gains. Yet the assumption that net utility is separable into these components simplifies the theory—it reduces the cost of thinking! Almost all prescriptive theory to date deals with gross utility only. Little attention has been given to decision costs that might be subtractable and thus definable in arriving at a net utility concept. Still less attention has been given to a net utility concept that cannot be decomposed into gross-utility and decision-cost components.

The elements for a theory of the expected cost of using inanimate computers, given the (statistical) population of future problems, are probably available. However, the current classification of computation problems as "scientific" and "business" (differing in their comparative needs for speed and memory) is certainly much too rough. A theory of mechanical computation costs would resemble the theory of the cost of manufacturing with complex equipment and known technology when capacities and operations are scheduled optimally. On the other hand, so little is known about the "technology" of human brains! If economists would join forces with students of the psychology of problem solving, insights would undoubtedly be gained into both the descriptive and prescriptive aspects of decision making.

JACOB MARSCHAK

[*Directly related are the entries* DECISION THEORY; GAME THEORY, *article on* THEORETICAL ASPECTS. *See also* Georgescu-Roegen 1968.]

BIBLIOGRAPHY

ALLAIS, M. 1953 Le comportement de l'homme rationnel devant le risque: Critique des postulats et axiomes de l'école américaine. *Econometrica* 21:503–546.

ARROW, KENNETH J. 1951 Alternative Approaches to the Theory of Choice in Risk-taking Situations. *Econometrica* 19:404–437.

ARROW, KENNETH J. 1958 Utilities, Attitudes, Choices: A Review Note. *Econometrica* 26:1–23.

ARROW, KENNETH J. 1963 Utility and Expectation in Economic Behavior. Pages 724–752 in Sigmund Koch (editor), *Psychology: A Study of a Science.* Volume 6: Investigations of Man as Socius: Their Place in Psychology and the Social Sciences. New York: McGraw-Hill.

ARROW, KENNETH J. 1965 *Aspects of the Theory of Risk-bearing.* Helsinki: Academic Bookstore.

AUMANN, ROBERT J. 1962 Utility Theory Without the Completeness Axiom. *Econometrica* 30:445–462. → Corrections in Volume 32 of *Econometrica.*

BAYES, THOMAS (1764) 1958 An Essay Towards Solving a Problem in the Doctrine of Chances. *Biometrika* 45:296–315.

BERNOULLI, DANIEL (1738) 1954 Exposition of a New Theory on the Measurement of Risk. *Econometrica* 22:23–36. → First published as "Specimen theoriae novae de mensura sortis."

BOREL, ÉMILE 1939 *Valeur pratique et philosophie des probabilités.* Paris: Gauthier-Villars.

CARNAP, R. 1962 The Aim of Inductive Logic. Pages 303–318 in International Congress for Logic, Methodology, and Philosophy of Science, Stanford, California, 1960, *Logic, Methodology, and Philosophy of Science: Proceedings.* Edited by Ernest Nagel, Patrick Suppes, and Alfred Tarski. Stanford Univ. Press.

CHERNOFF, HERMAN; and MOSES, LINCOLN E. 1959 *Elementary Decision Theory.* New York: Wiley.

DAVID, FLORENCE N. 1962 *Games, Gods and Gambling: The Origins and History of Probability and Statistical Ideas From the Earliest Times to the Newtonian Era.* New York: Hafner.

DAVIDSON, DONALD; and MARSCHAK, JACOB 1959 Experimental Tests of a Stochastic Decision Theory. Pages 233–269 in Charles W. Churchman and Philburn Ratoosh (editors), *Measurement: Definitions and Theories.* New York: Wiley.

DAVIS, JOHN M. 1958 The Transitivity of Preferences. *Behavioral Science* 3:26–33.

DEBREU, GERARD 1959 *Theory of Value.* New York: Wiley.

DE FINETTI, BRUNO (1937) 1964 Foresight: Its Logical Laws, Its Subjective Sources. Pages 93–158 in Henry E. Kyburg and Howard E. Smokler (editors), *Studies in Subjective Probabilities.* New York: Wiley. → First published in French.

EDWARDS, WARD 1953 Probability-preferences in Gambling. *American Journal of Psychology* 66:349–364.

ELLSBERG, DANIEL 1961 Risk, Ambiguity, and the Savage Axioms. *Quarterly Journal of Economics* 75:643–669.

FECHNER, GUSTAV T. (1860) 1907 *Elemente der Psychophysik.* 2 vols. 3d ed. Leipzig: Breitkopf & Härtel. → An English translation of Volume 1 was published by Holt in 1966.

FELLNER, WILLIAM 1965 *Probability and Profit.* Homewood, Ill.: Irwin.

FISHBURN, PETER C. 1964 *Decision and Value Theory.* New York: Wiley.

FISHER, IRVING 1906 The Risk Element. Pages 265–300 in Irving Fisher, *The Nature of Capital and Income.* New York: Macmillan.

FISHER, IRVING (1930) 1961 *The Theory of Interest.* New York: Kelley. → Revision of the author's *The Rate of Interest* (1907).

FRANKLIN, BENJAMIN (1772) 1945 How to Make a Decision. Page 786 in *A Benjamin Franklin Reader.* Edited by Nathan G. Goodman. New York: Crowell. → A letter to J. Priestly.

GEL'FAND, I. M.; and TSETLIN, M. L. 1962 Some Methods of Control for Complex Systems. *Russian Mathematical Surveys* 17, no. 1:95–117.

►GEORGESCU-ROEGEN, NICHOLAS 1968 Utility. Volume 16, pages 236–267 in *International Encyclopedia of the Social Sciences.* Edited by David L. Sills. New York: Macmillan and Free Press.

HART, ALBERT G. (1940) 1951 *Anticipations, Uncertainty and Dynamic Planning.* New York: Kelley.

HERSTEIN, I. N.; and MILNOR, JOHN 1953 An Axiomatic Approach to Measurable Utility. *Econometrica* 21:291–297.

HICKS, JOHN R. (1939) 1946 *Value and Capital: An Inquiry Into Some Fundamental Principles of Economic Theory.* 2d ed. Oxford: Clarendon.

INTERDISCIPLINARY RESEARCH CONFERENCE, UNIVERSITY OF NEW MEXICO 1960 *Decisions, Values and Groups: Proceedings.* Edited by D. Willner. New York: Pergamon. → See especially the article by J. S. Chipman, "Stochastic Choice and Subjective Probability."

JEFFREY, RICHARD C. 1965 *The Logic of Decision.* New York: McGraw-Hill.

KNIGHT, FRANK H. (1921) 1933 *Risk, Uncertainty and Profit.* London School of Economics and Political Science Series of Reprints of Scarce Tracts in Economic and Political Science, No. 16. London School of Economics.

KOOPMANS, TJALLING 1960 Stationary Ordinal Utility and Impatience. *Econometrica* 28:287–309.

KOOPMANS, TJALLING 1964 On Flexibility of Future Preference. Pages 243–254 in Maynard W. Shelly and Glenn L. Bryan (editors), *Human Judgments and Optimality.* New York: Wiley.

LINDBLOM, CHARLES E. 1965 *The Intelligence of Democracy: Decision Making Through Mutual Adjustment.* New York: Free Press.

LUCE, R. DUNCAN 1959 *Individual Choice Behavior: A Theoretical Analysis.* New York: Wiley.

LUCE, R. DUNCAN; and RAIFFA, HOWARD 1957 *Games and Decisions: Introduction and Critical Survey.* A study of the Behavioral Models Project, Bureau of Applied Social Research, Columbia University. New York: Wiley. → First issued in 1954 as *A Survey of the Theory of Games,* Columbia University, Bureau of Applied Social Research, Technical Report No. 5.

LUCE, R. DUNCAN; and SUPPES, PATRICK 1965 Preference, Utility, and Subjective Probability. Volume 3, pages 249–410 in R. Duncan Luce, Robert R. Bush, and Eugene Galanter (editors), *Handbook of Mathematical Psychology*. New York: Wiley.

MACCRIMMON, K. R. 1965 An Experimental Study of the Decision-making Behavior of Business Executives. Ph.D. dissertation, Univ. of California at Los Angeles.

MARSCHAK, JACOB 1950 Rational Behavior, Uncertain Prospects, and Measurable Utility. *Econometrica* 18:111–141.

MARSCHAK, JACOB (1954) 1964 Scaling of Utility and Probability. Pages 95–109 in Martin Shubik (editor), *Game Theory and Related Approaches to Social Behavior: Selections*. New York: Wiley.

MARSHALL, ALFRED (1890) 1961 *Principles of Economics*. 9th ed. New York: Macmillan. → See especially the Mathematical Appendix, note 9.

MAY, KENNETH O. 1954 Intransitivity, Utility and the Aggregation of Preference Patterns. *Econometrica* 22:1–13.

MILNOR, JOHN 1954 Games Against Nature. Pages 49–60 in Robert M. Thrall, C. H. Combs, and R. L. Davis (editor), *Decision Processes*. New York: Wiley.

NEYMAN, JERZY 1950 *First Course in Probability and Statistics*. New York: Holt.

ORE, ØYSTEIN 1953 *Cardano, the Gambling Scholar*. Princeton Univ. Press; Oxford Univ. Press. → Includes a translation by Sidney Henry Gould from the Latin of Cardano's *Book on Games of Chance*.

OŻGA, S. ANDREW 1965 *Expectations in Economic Theory*. London: Weidenfeld & Nicolson.

PAPANDREOU, ANDREAS G. 1957 *A Test of a Stochastic Theory of Choice*. University of California Publications in Economics, Vol. 16, No. 1. Univ. of California Press. → In collaboration with O. H. Sauerlender, O. H. Brownlee, L. Hurwicz, and W. Franklin.

PASCAL, BLAISE (1670) 1961 *Pensées: Notes on Religion and Other Subjects*. New York: Doubleday. → See especially Section 3: De la nécessité du pari.

POLLACK, IRWIN 1955 Sound Level Discrimination and Variation of Reference Testing Conditions. *Journal of the Acoustical Society of America* 27:474–480.

PRATT, JOHN W. 1964 Risk Aversion in the Small and in the Large. *Econometrica* 32:122–136.

PRATT, JOHN W.; RAIFFA, HOWARD; and SCHLAIFER, ROBERT 1964 The Foundations of Decision Under Uncertainty: An Elementary Exposition. *Journal of the American Statistical Association* 59:353–375.

QUANDT, RICHARD 1956 A Probabilistic Theory of Consumer Behavior. *Quarterly Journal of Economics* 70:507–536.

RAIFFA, HOWARD 1961 Risk, Ambiguity, and the Savage Axioms: Comment. *Quarterly Journal of Economics* 75:690–694.

RAIFFA, HOWARD; and SCHLAIFER, ROBERT 1961 *Applied Statistical Decision Theory*. Graduate School of Business Administration, Studies in Managerial Economics. Boston: Harvard Univ., Division of Research.

RAMSEY, FRANK P. (1923–1928) 1950 *The Foundations of Mathematics and Other Logical Essays*. New York: Humanities. → See especially Chapter 7, "Truth and Probabilities," and Chapter 8, "Further Considerations."

○SAVAGE, LEONARD J. (1954) 1972 *The Foundations of Statistics*. Rev. ed. New York: Dover.

SHACKLE, L. S. (1949) 1952 *Expectation in Economics*. 2d ed. Cambridge Univ. Press.

SHACKLE, L. S. 1955 *Uncertainty in Economics and Other Reflections*. Cambridge Univ. Press.

SHACKLE, L. S. 1961 *Decision, Order and Time in Human Affairs*. Cambridge Univ. Press.

►SHACKLE, L. S. 1968 Economic Expectations. Volume 4, pages 389–395 in *International Encyclopedia of the Social Sciences*. Edited by David L. Sills. New York: Macmillan and Free Press.

SHELLY, MAYNARD W.; and BRYAN, GLENN L. (editors) 1964 *Human Judgments and Optimality*. New York: Wiley.

SIMON, HERBERT A. 1957 A Behavioral Model of Rational Choice. Pages 241–260 in Herbert A. Simon, *Models of Man*. New York: Wiley.

SIMON, HERBERT A. 1959 Theories of Decision-making in Economics and Behavioral Science. *American Economic Review* 49:253–283.

THEIL, HENRI 1964 *Optimal Decision Rules for Government and Industry*. Amsterdam: North Holland Publishing; Chicago: Rand McNally.

THURSTONE, LOUIS L. (1927–1955) 1959 *The Measurement of Values*. Univ. of Chicago Press. → Selections from previously published papers.

VON NEUMANN, JOHN; and MORGENSTERN, OSKAR (1944) 1964 *Theory of Games and Economic Behavior*. 3d ed. New York: Wiley.

WALD, ABRAHAM (1950) 1964 *Statistical Decision Functions*. New York: Wiley.

WOLFOWITZ, J. 1962 Bayesian Inference and Axioms of Consistent Decision. *Econometrica* 3:471–480.

WRIGHT, GEORGE H. VON 1963 *The Logic of Preference*. Edinburgh Univ. Press.

ADDITIONAL BIBLIOGRAPHY

A collection of articles edited by Maurice Allais and Ole Hagen is planned for publication by Reidel as Rational Decisions Under Uncertainty. *Of special interest are "The Foundations of a Positive Theory of Choice Involving Risk" by Allais and "Utilities, Psychological Values, and the Training of Decision-makers" by Jacob Marschak.*

ARROW, KENNETH J. 1971 *Essays in the Theory of Risk-bearing*. Chicago: Markham.

BECKER, GORDON; DEGROOT, MORRIS H.; and MARSCHAK, JACOB (1964) 1974 Measuring Utility by a Single-response Sequential Method. Volume 1, pages 317–328 in Jacob Marschak, *Economic Information, Decision, and Prediction*. Volume 1: *Economics of Decision*. Dordrecht (Netherlands): Reidel.

BECKMANN, MARTIN J. 1960 Some Aspects of Returns to Scale in Business Administration. *Quarterly Journal of Economics* 74:464–471.

BORCH, KARL H. 1968 *The Economics of Uncertainty*. Princeton Studies in Mathematical Economics No. 2. Princeton Univ. Press.

BREWER, GARRY D. 1973 *Politicians, Bureaucrats and the Consultants: A Critique of Urban Problem Solving*. New York: Basic Books.

COCHRANE, JAMES L.; and ZELENY, MILAN (editors) 1973 *Multiple Criteria Decision-making*. Columbia: Univ. of South Carolina Press.

DEGROOT, MORRIS H. 1970 *Optimal Statistical Decisions*. New York: McGraw-Hill.

DRENICK, R. F.; and LEVIS, A. H. 1974 *A Mathematical Theory of Organization*. Part 1. Technical Report. Brooklyn: Polytechnic Institute of New York.

FISHBURN, PETER C. 1970 *Utility Theory for Decision Making*. New York: Wiley.

GEOFFRION, A. M.; DYER, J. S.; and FEINBERG, A. 1972 An Interactive Approach for Multi-criterion Optimization, With an Application to the Operation of an

Academic Department. *Management Science* 19:357–368.

HAMMING, R. W. 1965 Numerical Analysis vs. Mathematics. *Science* 148:473–475.

JOHANSEN, LEIF 1974 Establishing Preference Functions for Macroeconomic Decision Models: Some Observations on Ragnar Frisch's Contributions. *European Economic Review* 5:41–66.

KEENEY, R. L. 1974 Multiplicative Utility Functions. *Operations Research* 22:22–34.

KENDALL, MAURICE G. (editor) 1971 *Cost–Benefit Analysis: A Symposium.* New York: American Elsevier.

KOOPMANS, TJALLING C. 1970 *Scientific Papers of Tjalling C. Koopmans.* Berlin and New York: Springer. → See especially pages 387–410, "Stationary Ordinal Utility and Impatience" (1960), and pages 469–481, "On Flexibility of Future Research" (1964).

LANCASTER, KELVIN J. 1971 *Consumer Demand: A New Approach.* New York: Columbia Univ. Press.

MARSCHAK, JACOB 1971 Economics of Information Systems. Pages 32–106 in Michael D. Intriligator (editor), *Frontiers of Quantitative Economics.* Amsterdam: North-Holland.

MARSCHAK, JACOB 1973 Intersubjective Wahrscheinlichkeit. *Heidelberger Jahrbuecher* 17:14–26.

MARSCHAK, JACOB 1974*a* *Economic Information, Decision, and Prediction.* Volume 1: *Economics of Decision.* Dordrecht (Netherlands): Reidel. → Contains eighteen essays published between 1950 and 1974, some written jointly with Gordon M. Becker, H. D. Block, Donald Davidson, and Morris H. DeGroot.

MARSCHAK, JACOB 1974*b* Information, Decision, and the Scientist. Pages 145–178 in Colin Cherry (editor), *Pragmatic Aspects of Human Communication.* Dordrecht (Netherlands): Reidel.

MARSCHAK, JACOB 1976 Guided Soul-searching for Multi-criterion Decisions. Pages 1–16 in Milan Zeleny (editor), *Multiple Criteria Decision Making, Kyoto, 1975.* Berlin and New York: Springer.

MARSCHAK, JACOB; and RADNER, ROY 1972 *Economic Theory of Teams.* Cowles Foundation for Research in Economics at Yale University, Monograph 22. New Haven: Yale Univ. Press.

MOSTELLER, FREDERICK; and NOGEE, PHILIP 1951 An Experimental Measurement of Utility. *Journal of Political Economy* 59:371–409.

PRATT, JOHN W.; RAIFFA, HOWARD; and SCHLAIFER, ROBERT 1965 *Introduction to Statistical Decision Theory.* New York: McGraw-Hill.

RAIFFA, HOWARD 1968 *Decision Analysis: Introductory Lectures on Making Choices Under Uncertainty.* Reading, Mass.: Addison-Wesley.

ROTHSTEIN, J. 1965 Numerical Analysis: Pure or Applied Mathematics? *Science* 149:1049–1050.

SARIN, R. R. 1975 Interactive Procedures for Evaluation of Multi-attributed Alternatives. Working Paper No. 232. Los Angeles: Western Management Science Institute.

SAVAGE, LEONARD J. 1971 Elicitation of Personal Probabilities and Expectations. *Journal of the American Statistical Association* 66:783–801.

SEN, AMARTYA K. 1969 Quasi-transitivity, Rational Choice and Collective Decisions. *Review of Economic Studies* 36:381–394.

SLOVIC, PAUL; KUNREUTHER, H.; and WHITE, GILBERT F. 1974 Decision Processes, Rationality, and Adjustment to Natural Hazards. Pages 187–205 in Gilbert F. White (editor), *Natural Hazards: Local, National, Global.* New York: Oxford Univ. Press.

STAËL VON HOLSTEIN, CARL-AXEL S. 1970 *Assessment and Evaluation of Subjective Probability Distributions.* Stockholm: Economic Research Institute, School of Economics.

SUPPES, PATRICK 1974 The Measurement of Belief. *Journal of the Royal Statistical Society* Series B 36:160–175. → Discussion on pages 175–191.

SWALM, RALPH O. 1966 Utility Theory: Insights Into Risk Taking. *Harvard Business Review* 44:123–136.

TVERSKY, AMOS 1972 Elimination by Aspects: A Theory of Choice. *Psychological Review* 79:281–299.

TVERSKY, AMOS 1974 Assessing Uncertainty. *Journal of the Royal Statistical Society* Series B 36:148–159. → Discussion on pages 175–191.

TVERSKY, AMOS 1976 Representations of Choice Probabilities. Volume 2, chapter 16 in David H. Krantz et al. (editors), *Foundations of Measurement.* New York: Academic Press.

TVERSKY, AMOS; and KAHNEMAN, DANIEL 1974 Judgment Under Uncertainty: Heuristics and Biases. *Science* 185:1124–1131.

DECISION THEORY

Statistical decision theory was introduced by Abraham Wald (1939) as a generalization of the classic statistical theories of hypothesis testing and estimation. Ostensibly designed to put these theories in a common mathematical framework, it accomplished much more. It eliminated fundamental gaps in the previous theories and widened the scope of statistics to embrace the science of decision making under uncertainty. Although statistical decision theory represented a novel synthesis, many of its individual elements of novelty had been anticipated in spirit by Gauss and by the Neyman–Pearson theory of testing hypotheses. Furthermore, in formal structure it has much in common with game theory [*see* GAME THEORY].

In a decision problem the statistician must select one action from a set of available actions. A strategy (also called a decision function) is a plan that tells how to use the data to select an action and is evaluated by reference to the cost of its expected consequences. The difficulty in selecting the best strategy derives from the fact that the consequences of an action depend on the unknown state of nature.

The proper role of decision theory is the subject of considerable controversy. Because it makes essential use of costs, many statisticians feel that decision theory may be suitable for problems of the marketplace but not those of pure science. Another issue is the difficulty in assigning costs or values to the consequences.

Crosscutting this controversy is another, which concerns criteria for selecting a strategy; such criteria may be regarded as principles of statistical inference.

An example. To illustrate the decision-theory formulation, consider a simplified and artificial example. An archeologist has recently uncovered a skull that may belong to one of two populations, A or B. A skull positively identified as A would set off an extensive and costly research effort; a skull positively identified as B would not justify any research effort. The archeologist proposes to take one of three actions, the extensive research effort (action a_1), a lesser research effort (action a_2), or dropping the inquiry (action a_3). The possibilities that the skull belongs to either population A or to population B are called states of nature and labeled θ_1 and θ_2 respectively. Assume that after considerable introspection the archeologist has attached values to the consequences. By tradition, statisticians think in terms of losses, so the best situation (action a_1 and state θ_1) will be treated as having zero loss. The losses are represented in Table 1.

Table 1 — Losses

		ACTION		
		a_1	a_2	a_3
STATE OF NATURE	θ_1	0	2	5
	θ_2	4	3	2

One may ask why the loss corresponding to θ_2 and a_3 is not zero since a_3 is the best action under state θ_2. If the situation (θ_2, a_3) is compared with the conceivable alternative (θ_1, a_1), the latter seems preferable and it is reasonable to assign a positive loss to (θ_2, a_3). To compare (θ_2, a_3) with the alternatives under θ_2, one uses *regret*. The regret for (θ_2, a_3) would be zero, while the regrets for (θ_2, a_1) and (θ_2, a_2) are 2 and 1 respectively.

Thus far three of the basic elements of decision theory have been presented: the available actions, the possible states of nature, and the consequences, or loss table. Because the state of nature is not known, the selection of a best action is not trivial. Ordinarily a scientist will perform some experiment to diminish the costly effect of ignorance. Suppose that the archeologist decides to measure the cranial capacity of the skull and classify it as either large (z_1), medium (z_2), or small (z_3).

There would be no difficulty in using the data if all members of population A had large skulls and all members of population B had small ones. Unfortunately the experiment is not so informative: while skulls from A tend to be large and skulls from B tend to be small, there is appreciable overlap. Table 2 gives the fraction of skulls of population A and of population B that are large, medium, and small; this information, supposedly based on the results of extensive prior work, is not always readily available.

Table 2 — Probability of observing z given θ

	z_1	z_2	z_3
θ_1	.6	.3	.1
θ_2	.2	.3	.5

There are many ways in which the data can be used. Any rule that prescribes an action for each possible result of the experiment is called a strategy; Table 3 lists the 27 such rules for this problem. A discussion of the concept of strategy is given by Girshick (1954).

A typical reasonable strategy is s_6, which may be abbreviated (a_1, a_2, a_3). This strategy calls for action a_1 if z_1 is observed, a_2 if z_2 is observed, and a_3 if z_3 is observed. The strategies s_1, s_{14}, and s_{27} ignore the data; s_{22} seems to be unreasonable.

To determine how good a strategy is, it must be evaluated. How to evaluate the expected costs of $s_6 = (a_1, a_2, a_3)$ will be illustrated. If θ_1 is the true state of nature, z_1 is observed and a_1 taken with loss 0 and probability .6, z_2 is observed and a_2 taken with loss 2 and probability .3, and z_3 is observed and a_3 taken with loss 5 and probability .1. The expected loss is

$$(.6)0 + (.3)2 + (.1)5 = 1.1.$$

Similarly, if θ_2 is the true state, the expected loss is

$$(.2)4 + (.3)3 + (.5)2 = 2.7.$$

The expected losses for the 27 strategies are given in Table 4. Justification for the relevance of expected loss derives from the theory of utility (Chernoff & Moses 1959, chapter 4; Luce & Raiffa 1957, chapter 2; Georgescu-Roegen 1968).

Table 3 — Actions for the 27 strategies

		STRATEGIES																										
		s_1	s_2	s_3	s_4	s_5	s_6	s_7	s_8	s_9	s_{10}	s_{11}	s_{12}	s_{13}	s_{14}	s_{15}	s_{16}	s_{17}	s_{18}	s_{19}	s_{20}	s_{21}	s_{22}	s_{23}	s_{24}	s_{25}	s_{26}	s_{27}
DATA	z_1	a_1	a_1	a_1	a_1	a_1	a_1	a_1	a_1	a_1	a_2	a_2	a_2	a_2	a_2	a_2	a_2	a_2	a_2	a_3	a_3	a_3	a_3	a_3	a_3	a_3	a_3	a_3
	z_2	a_1	a_1	a_1	a_2	a_2	a_2	a_3	a_3	a_3	a_1	a_1	a_1	a_2	a_2	a_2	a_3	a_3	a_3	a_1	a_1	a_1	a_2	a_2	a_2	a_3	a_3	a_3
	z_3	a_1	a_2	a_3	a_1	a_2	a_3	a_1	a_2	a_3	a_1	a_2	a_3	a_1	a_2	a_3	a_1	a_2	a_3	a_1	a_2	a_3	a_1	a_2	a_3	a_1	a_2	a_3

Table 4 — Expected losses

		STRATEGIES																										
		s_1	s_2	s_3	s_4	s_5	s_6	s_7	s_8	s_9	s_{10}	s_{11}	s_{12}	s_{13}	s_{14}	s_{15}	s_{16}	s_{17}	s_{18}	s_{19}	s_{20}	s_{21}	s_{22}	s_{23}	s_{24}	s_{25}	s_{26}	s_{27}
STATE OF NATURE	θ_1	0.0	0.2	0.5	0.6	0.8	1.1	1.5	1.7	2.0	1.2	1.4	1.7	1.8	2.0	2.3	2.7	2.9	3.2	3.0	3.2	3.5	3.6	3.8	4.1	4.5	4.7	5.0
	θ_2	4.0	3.5	3.0	3.7	3.2	2.7	3.4	2.9	2.4	3.8	3.3	2.8	3.5	3.0	2.5	3.2	2.7	2.2	3.6	3.1	2.6	3.3	2.8	2.3	3.0	2.5	2.0

Table 4 may be regarded as an extension of Table 1 with s_1, s_{14}, and s_{27} corresponding to a_1, a_2, and a_3. The increased size gives the appearance of greater complexity but implies that the archeologist has greater choice after performing the experiment than before. Since s_6 is obviously an improvement on s_{14}, having smaller expected losses under both θ_1 and θ_2, the experiment may be regarded as having succeeded in reducing the costly effects of ignorance.

The strategies and their expected losses are represented in Figure 1. Here each strategy is identified by a point whose coordinates are the expected losses under θ_1 and θ_2 respectively.

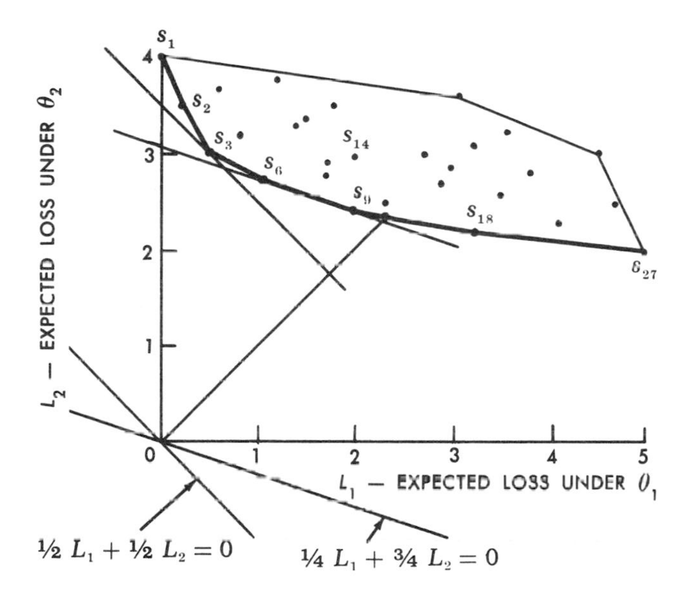

Figure 1 — Strategies and expected losses

The archeologist is not restricted to the 27 strategies mentioned above. He may decide to toss a fair coin and select s_1 if it falls heads and s_3 if it falls tails. Such a strategy is called a *randomized,* or *mixed,* strategy in contrast to the original *nonrandomized,* or *pure,* strategies. For the above mixed strategy the expected loss under θ_1 is 0 with probability $\frac{1}{2}$ and .5 with probability $\frac{1}{2}$ or .25, which is halfway between 0 and .5. In like manner, the expected loss under θ_2 is $\frac{1}{2}(4.0) + \frac{1}{2}(3.0) = 3.5$, which is halfway between 4.0 and 3.0. It follows that the point representing this randomized strategy is halfway between the points representing s_1 and s_3 in Figure 1. By replacing the coin with a more general randomizing device, it is possible to find a strategy corresponding to any point on the line segment connecting any two strategies. Thus the set of available strategies (randomized and nonrandomized) has the property that every point of the line segment connecting two points of the set is in the set. Such a set is called *convex.* The fact that the set of strategies is the smallest convex set containing the nonrandomized strategies plays a fundamental role in the mathematics of decision theory.

To minimize loss, the statistician seeks a strategy whose representative point in Figure 1 is as low and as far to the left as possible. A strategy is said to dominate another strategy when the first is never more costly than the second and sometimes less costly. In Figure 1, s dominates s^* if s is below and to the left of s^*, directly below s^*, or directly to the left of s^*. A strategy that is not dominated is called admissible. The admissible strategies lie on the heavy line $s_1s_2s_3s_6s_9s_{18}s_{27}$. The intuitively unreasonable strategy s_{22} is dominated by s_6; the strategy s_{14}, which ignores the result of the experiment, is also dominated by s_6.

Choice of good strategy. Clearly, the statistician can profitably confine his attention to the relatively few admissible strategies. How one should select from among these strategies is the subject of considerable controversy (Luce & Raiffa 1957, pp. 278–306). One criterion tentatively suggested by Wald was the *minimax criterion,* which may be described as follows: If nature is malevolent and the statistician picks s, nature will select the state θ that gives the greatest loss for s. Thus each strategy is evaluated by its maximum loss, and the statistician selects that strategy s for which the maximum loss is as small as possible. For the example, the minimax is found as the intersection of the line $L_1 = L_2$ with the segment s_9s_{18}, and consists in selecting s_9 with probability $\frac{5}{7}$ and s_{18} with probability $\frac{2}{7}$, yielding expected losses 2.34 under both θ_1 and θ_2. For illustration of minimax strategies that do not lie on the line $L_1 = L_2$, see Luce and Raiffa (1957, p. 404).

The minimax criterion is generally regarded as too pessimistic. In the theory of two-person, zero-sum games, where a player has a malevolent opponent who is anxious to ruin him, minimax has justification; it seems unreasonable, however, to regard nature as a malevolent opponent.

An alternative view is that if the archeologist is in complete ignorance he ought to act as though each state of nature were equally likely and select the strategy that minimizes the average of the two losses (Luce & Raiffa 1957, p. 284). This strategy is s_3 and may be derived graphically by moving the line $\frac{1}{2}L_1 + \frac{1}{2}L_2 = 0$ parallel to itself until it touches the convex set of strategies. This objective approach has been criticized on the grounds that rarely, if ever, does a scientist approach a problem in complete ignorance.

A third approach (Savage 1954) consists in proving, under relatively weak assumptions describing restrictions of consistent and rational behavior, that such behavior leads one to act as though he had certain prior subjective or psychological probabilities for the states of nature. Thus if the archeologist had prior probabilities $\frac{1}{4}$ and $\frac{3}{4}$ for θ_1 and θ_2, he would minimize $\frac{1}{4}L_1 + \frac{3}{4}L_2$. Graphically, this may be achieved by moving the line $\frac{1}{4}L_1 + \frac{3}{4}L_2 = 0$ parallel to itself until it touches the set of strategies. In fact this line first touches the set of strategies along the line segment connecting s_6 and s_9, which means that s_6, s_9, and all their mixtures are equally good for the archeologist.

Other statisticians refuse to be bound by such restrictions in matters as sensitive and complex as inductive inference.

The decision-theory controversy. The main objection to decision theory involves the losses. Some feel that in scientific problems where the only object is the satisfaction of scientific curiosity, it makes no sense to attach values or costs to the actions. Others, who are willing to accept a pragmatic approach, are overwhelmed at the difficulty in finding precise numbers to insert in the loss table. Still others feel that a statistician should not consider losses or actions but should confine himself to analyzing the inferential content of the data.

I feel that some conception of the value of consequences, be it based on personal gratification or on selfless consideration of the welfare of humanity, is necessary when faced with the problem of designing efficient experiments or when considering what data to gather and when to stop accumulating data. Since small deviations in the loss table have little effect on the relative merit of strategies when there is a substantial amount of data, precise values of losses are not vital; on the other hand, some rough conception of costs is required to understand the problem. By closing the gaps of previous formulations, decision theory has the advantage of focusing attention on all the major issues of a problem.

Another objection to decision theory is based on the requirement of a detailed model representing the preconception of the totality of possible states of nature. Common sense is necessary to avoid foolish results in the face of data indicating that the preconception was wrong. This issue is not peculiar to decision theory and lies at the heart of some of the current controversies over the comparative method in the social sciences.

Decision theory and classical statistics. Testing hypotheses, estimation, and confidence intervals are the three main classical theories of statistical inference. A hypothesis-testing problem corresponds to a two-action problem. The possible actions are to accept a hypothesis or to reject it. The considerable difficulty that troubled novices over the choice between one-tailed and two-tailed tests was largely resolved by Neyman–Pearson theory. Decision theory added clarification by indicating that some problems traditionally formulated as testing problems really involve three actions.

In estimation problems one attaches a loss to estimating θ by t. A popular special case is the loss $(t - \theta)^2$. Then a good procedure tends to minimize the mean squared error, $E(t - \theta)^2$, which is the traditional measure of the goodness of an estimate. (See Chernoff & Moses 1959, pp. 207–233.)

The theory of confidence intervals is mathematically elegant, appealing, and popular but does not fit naturally into the decision-theory framework. The introduction of costs in a reasonable fashion would undoubtedly lead to substantial modification of the theory of confidence intervals.

Decision theory is wider than these classical theories. For example, in the archeological example, decision theory permits the consideration of the compromise action a_2, which is not optimal for any state of nature. This possibility does not exist in hypothesis testing. Decision theory also accommodates formulations involving sequential design of experiments.

Complete classes and Bayes solutions. A complete class of strategies is one in which every strategy not in the class is dominated by some strategy in the class. A Bayes solution for certain prior probabilities is a strategy that minimizes the weighted average of the expected losses, with those prior probabilities as weights. In Figure 1, changing the prior probabilities changes the slope of the line that produces the Bayes solution; and by changing this slope, all admissible strategies are obtained. Thus, in the example, the class of all Bayes solutions is complete.

In many decision problems, in particular in all decision problems with a finite number of states of nature and a finite number of actions, the class of all Bayes solutions is complete. A consequence of this result is that one may restrict oneself to Bayes solutions whether one believes in psychological prior probabilities or not. The importance of this result is related to the fact that Bayes solutions are easily characterized using Bayes' theorem for posterior probabilities. If this theorem is used, the tedious task of listing all possible strategies before observations are obtained is no longer required to compute Bayes solutions [*see* PROBABILITY].

Compound decision problems. The Robbins theory of compound decision problems (Robbins 1951) can be partly illustrated by the following example: A subject is given a true–false question on an examination. An investigator wishes to classify the student as knowing the answer or not knowing it. The cost of misclassifying the subject is 1. It is assumed that a subject who knows the answer will answer correctly and that one who does not know the answer will guess at random. The traditional strategy of treating the subject as knowing the answer if he answers correctly seems reasonable.

Suppose now that several hundred students take the examination and about 50 per cent answer the question correctly. If a substantial proportion of the group knew the answer, one would expect considerably more than 50 per cent correct answers. Thus one may conclude that almost everyone should be treated as not knowing it.

This exotic approach illustrates that it is possible to use the data from many unrelated problems of the same type to improve over the results attained by analyzing each problem separately.

A variation of this approach yields Robbins' theory of empirical Bayes procedures (Robbins 1956), where past data on different examples of a problem are used to estimate empirical prior probabilities.

Further problems. Three other branches of decision theory in which interesting work has been done are comparison of experiments (Blackwell & Girshick 1954, chapter 12); invariance (Blackwell & Girshick 1954, pp. 223–236); the study of monotone procedures (Blackwell & Girshick 1954, pp. 179–195) and the extension of monotone procedures to problems involving Pólya type distributions (Karlin 1957–1958).

HERMAN CHERNOFF

[*Directly related are the entries* ESTIMATION; HYPOTHESIS TESTING; *and the biography of* WALD.]

BIBLIOGRAPHY

BLACKWELL, DAVID; and GIRSHICK, M. A. 1954 *Theory of Games and Statistical Decisions.* New York: Wiley.

BROSS, IRWIN D. J. 1953 *Design for Decision.* New York: Macmillan. → A paperback edition was published in 1965 by Free Press.

CHERNOFF, HERMAN; and MOSES, L. E. 1959 *Elementary Decision Theory.* New York: Wiley.

►GEORGESCU-ROEGEN, NICHOLAS 1968 Utility. Volume 16, pages 236–267 in *International Encyclopedia of the Social Sciences.* Edited by David L. Sills. New York: Macmillan and Free Press.

GIRSHICK, M. A. 1954 An Elementary Survey of Statistical Decision Theory. *Review of Educational Research* 24:448–466.

KARLIN, S. 1957–1958 Pólya Type Distributions. Parts 2–4. *Annals of Mathematical Statistics* 28:281–308, 839–860; 29:1–21.

LUCE, R. DUNCAN; and RAIFFA, HOWARD 1957 *Games and Decisions: Introduction and Critical Survey.* New York: Wiley.

ROBBINS, HERBERT E. 1951 Asymptotically Subminimax Solutions of Compound Statistical Decision Problems. Pages 131–148 in Berkeley Symposium on Mathematical Statistics and Probability, Second, University of California, 1950, *Proceedings.* Berkeley and Los Angeles: Univ. of California Press.

ROBBINS, HERBERT E. 1956 An Empirical Bayes Approach to Statistics. Volume 1, pages 157–163 in Berkeley Symposium on Mathematical Statistics and Probability, Third, University of California, 1954–1955, *Proceedings.* Berkeley and Los Angeles: Univ. of California Press.

SAVAGE, LEONARD J. 1954 *The Foundations of Statistics.* New York: Wiley.

VON NEUMANN, JOHN; and MORGENSTERN, OSKAR (1944) 1964 *Theory of Games and Economic Behavior.* 3d ed. New York: Wiley.

WALD, ABRAHAM 1939 Contributions to the Theory of Statistical Estimation and Testing Hypotheses. *Annals of Mathematical Statistics* 10:299–326.

WALD, ABRAHAM 1950 *Statistical Decision Functions.* New York: Wiley. → Reprinted in 1964.

Postscript

Lindley (1971), an elementary book, supports the Bayesian view that subjective probabilities must be used to avoid absurd or incoherent behavior.

HERMAN CHERNOFF

ADDITIONAL BIBLIOGRAPHY

LINDLEY, DENNIS V. 1971 *Making Decisions.* New York: Wiley.

DEGREES OF FREEDOM

See COUNTED DATA; LINEAR HYPOTHESES.

DEMOGRAPHY

► *The articles under this heading were first published in* IESS *as parts of the multiple-article entries "Population" and "Genetics."*

I
THE FIELD

This article was first published in IESS *as "Population: I. The Field of Demography" with six companion articles less relevant to statistics.*

Demography is the quantitative study of human populations. Its basic materials are censuses, vital statistics, and, increasingly, sample surveys. Its central concerns are the measurement and discovery of uniformities in the basic processes of human birth, death, population movement, and population growth; these phenomena are treated in both their socioeconomic and their biological contexts. The methods of demography are empirical and statistical and make as much use of advanced mathematics as those of any branch of the social sciences. Like anthropology and psychology, demography bridges the social and biological sciences.

The word "demography" was apparently first used by Achille Guillard in his *Éléments de statistique humaine, ou démographie comparée* in 1855. Its Greek origins are *demos* (people) and *graphein* (to draw, describe). Until recently, the word had not attained general usage among English-speaking peoples, and those using it were frequently the victims of typographers and casual audiences who insisted on confusing it with "democracy." The most widely used textbook on population problems in the United States during the period from 1930 to 1960 does not mention the word in the first three editions and refers to it only casually in the fourth and fifth (Thompson & Lewis 1930). However, the word is gaining both wider usage and wider meaning as describing the scientific aspect of population study.

Formal demography. A distinction is commonly made between "formal," or "pure," demography and broader population studies. Formal demography is a well-defined, technical subject with a highly developed mathematical methodology. It is concerned primarily with the measurement and analysis of the components of population change, especially births, deaths, and, to a smaller extent, migration. It is concerned with population structure—that is, the age, sex, and marital composition of the population—as it contributes to the understanding of population change.

Formal demography has provided some of the most fruitful uses of mathematical models in the social sciences. These include, the construction of life tables, which provide the terms within which life insurance and social security systems must operate; intrinsic rates of reproduction and other sophisticated measures of natality, which have contributed much to the understanding of the secular decline of birth rates in the West and their fluctuations after World War II; stable population analysis, which has provided methods of estimating birth rates and rates of population growth in the absence of reliable vital statistics and often with grossly inaccurate census data; and population projections, based on the analysis of trends in the components of population change. Such projections are in great demand for purposes of economic development, market research, city planning, educational planning, estimating future labor supply, and numerous other approaches to measuring people as producers and consumers.

For actuarial purposes, demography is usually defined in the narrow sense of formal demography (Cox 1950). This definition also prevails in certain European countries, notably Italy.

Broader usage. A broader and increasingly popular usage of the term "demography" includes studies of demographic variables in their social contexts as well as their biological contexts. In this approach, demographic changes are viewed as part of, and as both cause and effect of, their social environment. Demographic analysis has also come to mean "not only the statistical manipulation of population data, but more important, the study of such data as a method of solving empirical problems" (Spengler & Duncan [1956*a*] 1963, p. xiii).

In addition to the components of population change, demographic variables are rather arbitrarily assumed to include those mass measurements of population that are commonly enumerated in population censuses: the size and distribution of the population, its biological composition by age and sex, and certain of its more measurable socioeconomic characteristics. Among these variables, data on population size, geographical distribution (including rural–urban), age–sex composition, and marital status are generally considered to be more specifically the subject matter of demography. Other socioeconomic characteristics, such as race, language, religion, education, occupation, and income, are studied in their own right and often in relation to group differences in birth rates, death rates, and migration.

The broader definition of demography inevitably involves interdisciplinary aspects, since demographers and other social scientists may study human populations in relation to other variables. Thus, a study of rural–urban migration may be considered demographic if it is primarily concerned with measurement and quantitative uniformities, economic if it is concerned with economic causes and effects, and sociological if it deals with sociological causes and effects. There is no firm line of demarcation, nor should one be expected. Beyond the specific discipline of formal demography there is no distinguishing criterion of demography as such, other than its involvement in quantitative measures of human populations. Thus, it can be said that the subject matter of demography includes those population theories that are based on quantitative observation and generalization, but not those that are formulated a priori at a more abstract and polemical level.

The development of demography

Censuses in various forms were conducted early in human history. A Roman census was taken under Caesar Augustus the year Christ was born (thus forcing Joseph and Mary to journey to Bethlehem to be enrolled in the place of their ancestry); and a major census was conducted in China in the year 2 A.D., during the Han Dynasty.

While such scattered censuses are of great interest for historical demography, modern censuses usable for scientific pursuits are essentially of nineteenth-century origin. Regular decennial censuses were initiated in the United States in 1790 and in France and England in 1801. Important exceptions are the population registers in Scandinavia, which date back to 1686 in Sweden and 1769 in Denmark.

Origins in vital statistics. Empirical demography began in the context of vital statistics with the work of John Graunt. His *Natural and Political Observations Upon the Bills of Mortality*, published in 1662, is a study of current reports on burials and christenings for a population of about 500,000 persons in the vicinity of London. The very title of this early work illustrates the fact that demography has roots in both biological and social inquiries. In the 1670s, William Petty built on Graunt's work to write a treatise on what he described as "political arithmetick." Gradually demography was absorbed into the new and more general study of statistics to the point that, in the United States, for example, concern about the accuracy and implications of vital statistics in the city of Boston led to the establishment of the American Statistical Association in 1839.

The line of succession in the biostatistical tradition of demography is a notable one (see Lorimer 1959). Interest in biostatistics led to attempts to explain population change in terms of mathematical processes, particularly logistic curves. Population change was conceived in terms of mechanical models by nineteenth-century researchers such as Adolphe Quetelet and Pierre François Verhulst and in biological terms by Pearl and Reed (1920), Alfred J. Lotka (1925), and G. Udny Yule (1925). While mechanical and biologically deterministic models did not ultimately prove convincing, they did lead to major developments in the statistical analysis of the processes of birth and death. Outstanding among these were Lotka's contributions in computing reproduction rates and intrinsic, or true, rates of natural increase, holding constant the age structure of the population. The use of mathematical models, both stochastic and deterministic, has enjoyed a rebirth in demography, perhaps because of the vogue of model-building in social sciences generally, and certainly because of the capabilities of modern computer technology.

Malthusian theory. The economic approach to demography is commonly traced to the work of Thomas Malthus, although he readily admitted the contributions of his predecessors. His *Essay on the Principle of Population*, first published in 1798, is undoubtedly the most discussed work in the field of demography. While he modified his position in successive versions of the volume, Malthus held the general view that population tends to increase faster than the means of subsistence and thus to absorb all economic gains, except when checked by moral restraint, vice, and misery.

Malthus' influence was enormous, but his view was later challenged in three ways. First, during the nineteenth century economic growth far outdistanced population growth among European peoples, resulting in rising levels of living. Second, despite rises in the level of living, birth rates declined, first in France and Sweden, and by 1880 in western Europe generally. Finally, liberals and Marxists challenged the Malthusian view by asserting that poverty was the result of unjust social institutions rather than of population growth. Regardless of the validity of this argument, it caused population study to be associated with the conservative social philosophy of Malthus, an association which retarded the growth of demography as a scientific discipline. As one author says,

> The marriage of demography and economics while both were immature—"Parson" Malthus officiating—resulted in a stormy and unfruitful union. Both the dynamics of interactions among economic factors and the dynamics of vital trends in relation to population structure were long neglected in a hasty synthesis that placed undue emphasis on the relation of population to resources and the corollary theory of a hypothetical fixed "optimum." This led to the fallacious assumption that increase of population is necessarily advantageous in a country with low density of population and to an unwarranted pessimism about the possibilities of economic progress in densely populated countries. (Lorimer 1957, p. 21)

Dissatisfaction with the Malthusian approach led to the divorce of demography from economics and to a continuing suspicion among some economists that demography overemphasizes the force of population growth and that population control in underdeveloped areas is in some way a diversion from, and even a threat to, the central purpose of economic development. Therefore, the economics of population change in the contemporary world has not received the attention that its importance would seem to merit, although it has not been entirely neglected. Now largely separated from economics, demography in the English-speaking world is commonly taught as a branch of sociology.

The birth rate in Western society. In its sociological aspects, population study emerged as a science in the 1920s and 1930s. Up to that time, births, deaths, and natural increase had generally been regarded as biologically given, exogenous both to the economy and to the social system. But by the interwar period, the general reduction of the birth rate in Western society had become highly visible and was the principal focus of demographic inquiry. Demographic studies demonstrated the social, as opposed to the biological, correlates and determinants of the decline in the birth rate.

Thus, empirical studies showed that the decline was not the result of biological sterility but of voluntary restriction of births, predominantly by methods requiring a high level of motivation. It is perhaps not widely known even today that reduction of the birth rate through contraception was initially and predominantly the result of "male" methods, specifically the practice of coitus interruptus and, later, the use of the condom; the principal "female" method of birth control was abortion (see Tietze 1968). The feminist birth control movement was noisy, often courageous, and certainly psychologically important, but there is little empirical evidence that the methods it advocated or the services it provided ever played a large part in the reduction of the birth rate in the Western world. In much of the Western world the chief methods of birth control are still coitus interruptus and abortion, although these are being increasingly replaced by new methods that have become generally available only since 1960.

Another major concern of demographers was "differential fertility," that is, the general observation that birth control was first adopted by the more urbanized, the better educated, and the upper-income groups, with consequent rural–urban, educational, and class differences in birth rates. Demographers were reluctant to do more than measure these differences, but their observations reinforced concern about population "quality," since class differences in natality clearly seemed to favor the reproduction of the socially and economically unsuccessful, and possibly also the genetically inferior.

The obvious role of social environment and voluntary control in the reduction of the birth rate gave rise to a new realization that birth and death rates, like migration, are effects as well as causes of social changes—dependent as well as independent variables with reference to the society in which they occur. Some felt that the existing trends were disturbing and even alarming, threatening "race suicide," on the one hand, and deterioration of "quality," on the other.

The reaction to the declining birth rate was strongest in the totalitarian countries that for nationalistic reasons adopted "pronatalist" population policies. These policies included such measures as propaganda for larger families, family allowances, suppression of induced abortion, marriage loans with deductions for each child born within the period of repayment, "motherhood medals" for parents of large families, and various privileges in taxation and in the social insurance schemes. Prior to World War II such policies were adopted in fascist Italy, Nazi Germany, the Soviet Union, and Japan. Some of these measures, without the nationalistic overtones, were also adopted in France and Sweden (see Eldridge 1968).

In the English-speaking countries reaction to the declining birth rate was slower. In the United States the problem was given intensive study by the National Resources Committee (U.S. National Resources . . . 1938). In Great Britain and Canada family allowances were adopted after World War II; and in Great Britain a Royal Commission on Population was created and issued a report that is a landmark in British thinking on this point (Great Britain 1949–1954).

The effect of these pronatalist policies is controversial. The only clear cases of moderate success are rises in birth rates achieved in Nazi Germany and in the Soviet Union. In Germany the suppression of abortion and the promotion of earlier marriages and childbirth by marriage loans were contributing factors in the rise in the birth rate from 14.7 in 1933 to 20.3 in 1939 (Kirk 1942; *Demographic Yearbook* 1948, p. 262). But a substantial part of this increase was almost certainly due to the general improvement in economic conditions, especially the reduction of unemployment. In the Soviet Union the suppression of abortion was especially successful in raising birth rates in the cities, because abortion had formerly been legal and had even been provided as a health service of the state. In the mid-1950s abortion was again legalized and the birth rate in the Soviet Union is now quite low. In the period since World War II, French observers have expressed the view that family allowances have contributed to maintaining a higher birth rate in that country.

Theory of the vital revolution. In Western countries there were important regularities in the declines in death rates and birth rates, whether viewed in time or space. The spread of declines, first in death rates and later in birth rates, is a well-documented example of cultural diffusion. From these regularities were derived a number of general propositions concerning the sequence of demographic developments in the modern world. Taken together, these propositions have come to be called the "theory of the vital revolution," or the "theory of the demographic transition."

According to this theory, in premodern societies both birth and death rates are high; reproduction is inefficient because it is necessary to have high, relatively unrestricted natality in order to match the waste of a high death rate. The initial effect of modernization is a fall in the death rate, apparently owing to higher levels of living and to the introduction of epidemic controls and other elementary public health measures. Without a comparable decline in the birth rate, there is a growing margin of births over deaths and an accelerating rate of population growth. At a later stage of socioeconomic development, with the achievement of general literacy, urbanization, and industrialization (as in the Western countries and in Japan), the size of family is reduced by birth control and the birth rate falls, eventually reducing the rate of population growth.

The theory of the demographic transition gave a plausible interpretation of demographic events in Western countries up to World War II. Demographers sought to extrapolate it in time, by projecting the then existing trends, and in space, by assuming that in the process of modernization the rest of the world would follow the same sequence of events as in the West. Population "projections" based on extrapolation of natality and mortality trends gained wide currency and were commonly regarded as estimates rather than as the mechanical extrapolations that they really were.

It is ironic that demographers developed the techniques for projecting certain long-standing trends in the components of population growth, especially in natality, just at a time when these trends were about to dissolve. New attitudes favoring earlier marriage and more children appeared in the very societies where the great majority of families had been practicing birth control. The recovery of the birth rate in Western countries just before, during, and especially after World War II violated the projections of previous trends and those formulations of the demographic transition that considered Western countries to be approaching a stationary or declining population. While the prolonged "baby boom" after World War II has receded, all Western countries continue to have modest rates of population growth.

The demographic transition has not fully materialized in non-Western cultures, aside from Japan. The transition has largely run its course in countries inhabited by peoples of predominantly European race and heritage, including Europe, the Soviet Union, and the overseas countries of chiefly European background. As yet, very few non-European countries have clearly reached the stage of declining birth rates and slower rates of population growth. There has not been a continuum of countries moving through the stages of transition, as Western countries were doing at the time when the theory was developed. Instead, there is now a sharp division between countries that have passed through the "vital revolution" and now have modest birth rates and growth rates, and countries in which the death rates, but not the birth rates, have been going down, so that the rate of population growth has been accelerating. This division separates the developed from the underdeveloped world, and, aside from the important exception of Japan, it also separates the peoples of European race and tradition from the peoples of the rest of the world. But there are clear indications that other non-European countries such as Taiwan and Korea may now be entering the later phases of transition, and the development of government policies and especially of better contraceptives may well accelerate the transition. As yet, no country has become "modern"—in the sense of achieving mature socioeconomic development—without also undergoing a decline in its birth rate.

Later methodological developments. Two major developments have occurred in demography since World War II as a result of the problems and failures of previous population projections: the development of a complex methodology to analyze trends in natality and the use of field surveys to determine causal factors affecting the number and timing of births.

The attempts to obtain better forecasts of future population resulted in successive improvements in the measuring of reproduction. These included the refinement of the crude birth rate to a population standardized by age; Lotka's historic breakthrough in determining the true birth and death rates, given the continuation of existing age-specific natality and mortality rates; the analysis of births by duration of marriage, developed primarily in Europe (since such data were not available in the United States); the analysis of cohort fertility, that is, the number, spacing, and order of births per woman in her total reproductive experience (Whelpton et al. 1965); and stable population analysis, that is, the analysis of mathematical regularities in the interrelations of births, deaths, age structure, and population growth at fixed schedules of natality and mortality (Coale & Demeny 1966). Some of the methods in this area of growing demographic importance involve elaborate computations that are being facilitated by advances in computer technology.

Despite great technical virtuosity and much better understanding of the various ways in which natality may be measured, it is too early to tell whether these efforts have added much to the accuracy of population forecasts, although they have improved understanding of current natality trends. Cohort fertility analysis certainly provides meaningful interpretations of past trends and provides the basis for reasonable estimates of the total number of children a cohort of women will bear well before they have actually reached the end of their reproductive period. In the United States women born during the period 1905–1914 show the smallest average number of births and stand at the end of the long fertility decline that began in the early nineteenth century. It seems certain that later cohorts of women—that is, those born during the period 1915–1935—will show a larger average number of births (U.S. Department of Health . . . 1966, p. 6).

After World War II, couples in the United States married earlier, had their children earlier, and had more children than their parents. The result was the higher birth rate in the United States and in some other Western countries during the postwar period. Recent trends suggest that these forces are receding.

The official statistics used in the sophisticated methods of analysis referred to above have not given the reasons for the changes that have occurred. These have been sought in the second of the recent major developments in demographic methodology: field studies of attitudes, motivations, and expectations relating to family size. In the United States demographers have conducted a series of national sample surveys of attitudes and practices in relation to family planning. These surveys are documenting the changing aspirations of American families with regard to the size of family desired and the changing practices of marriage, spacing of children, and methods of family limitation (for example, see Whelpton et al. 1965; Westoff et al. 1961; 1963).

Demography and the population explosion

Since World War II, interest in population problems has re-emerged with great force. This is related to the rapid progress achieved in controlling or postponing deaths throughout the world, which has occurred without a compensating decline in the birth rate in the underdeveloped areas. The result has been the acceleration of population growth, dramatized in a vast and sometimes controversial literature about the "population explosion." The rate of world population growth, over 2 per cent per year, is accelerating (see Grauman 1968).

World population growth in the present period is historically unique, and there is a general realization that it leads to serious obstacles to the whole range of socioeconomic development, whether in food, health, education, housing, or the general level of living. Population growth is a major block to the satisfaction of rising aspirations in the underdeveloped countries (Coale & Hoover 1958).

In both developed and developing countries the rapid growth of metropolitan areas with their wide range of problems has also been classified as a "population problem" on the mistaken assumption that this is primarily due to the birth rate rather than to migration, which has been the principal factor. It is recognized that population growth and concentration have contributed to a wide range of problems involving the quality of education, the urban sprawl, the contamination of the atmosphere, and, according to some, the mediocrity of mass culture and the cacophony of much of modern life.

As logical results of these concerns, social scientists have become more interested in the population problem, whether defined as a social problem or as an area of scientific interest. Demography has, of course, provided analysis of the population trends and prospects that underlie this concern. Prior to World War II demographic studies were largely confined to countries with adequate censuses and vital statistics—that is, chiefly countries of the Western world. In recent years there has been growing interest in the demography of the underdeveloped world, where great ingenuity and methodological skill are often needed to make up for the absence of reliable data. This changing emphasis has been reflected, for example, in the much greater attention given the underdeveloped areas in the Second World Population Conference in Belgrade in 1965, as compared with the First World Population Conference in Rome in 1954 (World Population Conference 1966, p. 2).

The growth of applied demography. Demography has supplied much of the factual background for the population policies and family planning programs that have been adopted by several of the developing countries. Much of the demographic literature now relates to attitude studies and to action experiments designed to introduce family planning in these areas.

On the world scale, the field surveys of knowledge, attitudes, and practices (KAP studies) concerning family size and family planning are perhaps the broadest cross-cultural series of studies on any subject in the social sciences. Sample surveys of this type have been done in some twenty countries in Africa, Asia, and Latin America. These surveys have revealed the feasibility of obtaining useful information on such delicate subjects in all societies; the existence of a substantial "market" for contraceptive information and services almost everywhere; and actual contraceptive practice in close relation to the degree of modernization of the country or socioeconomic class concerned (see Tietze 1968).

Although demography has been described as "observational," as opposed to experimental, at least some demographers have embarked on controlled experiments in relation to efforts to reduce birth rates. In several instances, the KAP studies have included an action component, that is, experimentation in introducing family planning to the population concerned. These studies have met with varying success, depending on the characteristics of the populations and the methods of education and motivation employed. Such attempts have been successful where the population was already in the stream of modernization, had experienced increasing survival of infants, and had achieved some success in avoiding births through postponement of marriage and through contraception and abortion. Projects of this nature have been working in Taiwan, Korea, and Ceylon. However, where the population had not already made serious efforts to reduce family size, as in India and Pakistan, the results of such experiments have been limited (Conference . . . 1962; International Conference . . . 1966; *Studies in Family Planning*).

A number of countries have adopted national programs for introducing family planning: Communist China, South Korea, Taiwan, Malaysia, Singapore, Ceylon, Pakistan, Turkey, the United Arab Republic, Tunisia, Morocco, Barbados, Jamaica, Honduras, and Chile. The list grows each year. Most of these programs are as yet too new and too small-scale to have achieved measurable success in reducing the birth rate. However, their chances for success may change rapidly with the improvement of contraceptive technology, notably oral contraceptives and plastic intrauterine devices. All contraception involves motivation; the newer methods are effective because they help the less strongly motivated and those living in circumstances where earlier methods were impractical. It seems likely that these and further improvements will revolutionize the whole approach to family planning in developing and developed countries alike (International Conference . . . 1966).

There is controversy among demographers about the scientific relevance of applied studies in their field. There is less controversy about the role of getting better information on vital statistics and population growth, since these pursuits serve both academic and applied ends. In over half the world, including most of the countries that have adopted population policies, official vital statistics are inadequate for determining either the true levels or year-to-year changes in birth rates, death rates, and population growth. New approaches that are being undertaken to obtain better estimates include closely supervised registration in sample areas, repetitive surveys, and various combinations that have come to be called "population growth estimation studies." The real measure of the success of a national family planning program is reduction in the birth rate and in the rate of population growth, and demographers are therefore becoming increasingly involved in problems of measuring population growth, as well as in the evaluation of the results of family planning programs.

Demography in the social sciences

Demography is generally considered an interdisciplinary subject with strong roots in sociology and weaker, but still important, connections with economics, statistics, geography, human ecology, biology, medicine, and human genetics. It is rarely thought of as a completely separate discipline, but rather as an interstitial subject or as a subdivision of one of the major fields. In English-speaking countries persons who describe themselves professionally as "demographers" usually hold positions of broader definition as sociologists, economists, or statisticians, or are in offices and research organizations whose titles include the word "population" but only rarely "demography," a word that is more commonly used in other countries.

Interest in demography has led recently to the rapid multiplication of offices and institutes devoted specifically to population research. In the United States there are some ten such institutes associated with universities. There are important centers of demographic study in France, Italy, the United Kingdom, Australia, and Japan. The United Nations sponsors three regional centers of demographic training (in Santiago, Chile, for Latin America; in Bombay, India, for Asia; and in Cairo, United Arab Republic, for north Africa), and others are under consideration. Some of the most rapid progress in demographic work is occurring at newly established centers in eastern Europe and in the developing areas.

Demography is, in one sense, a general servant of other social sciences. It evaluates and initially digests the vast reservoirs of social data compiled in censuses and vital statistics. It provides substantial raw material for the study of social, political, and economic change. Moreover, in spite of past disappointments over their accuracy, projections and estimates are continually requested from demographers for a multitude of planning purposes.

Demography has not had a prominent place in social scientific theory except in that of economics, through the works of Malthus. It is quantitative, empirical, and methodologically rigorous. Its subject matter is statistical categories that only indirectly reflect real social groups. Its quantitative documentation of specific social change is not readily integrated into the structural–functional theories prevailing in sociology, which are commonly speculative and hard to subject to empirical tests.

Within its own domain demography has had its greatest successes in the analysis of mortality [*see* LIFE TABLES; Moriyama 1968]. Its greatest interest and virtuosity has been in the study of natality (see Freedman 1968). The stepchild of demography is migration, which up to now has defied the application of refined measurements comparable to those developed in the other two fields. Although international migration is no longer of great demographic importance, internal migration in the United States, for example, is the chief cause of variations in rates of population growth between states, cities, and regions. The importance of migration, as well as the advantages of the new computer technology, may bring greater talent and resources into this field (see Petersen 1968; Thomas 1968).

In recent years perhaps the most neglected area in demography has been on the important frontiers between demography and economics, in such areas as population growth in relation to capital formation; application of life table principles to labor force changes; the economics of health, morbidity, and mortality; and the economic implications of different levels of morbidity and mortality. There are also unexplored possibilities for more fruitful cross-disciplinary work between demography and the fields of anthropology, politics, geography, ecology, and genetics [*see* DEMOGRAPHY, *article on* DEMOGRAPHY AND POPULATION GENETICS]. There are beginnings, but as yet the joint efforts in these related fields are minimal.

DUDLEY KIRK

BIBLIOGRAPHY

BARCLAY, GEORGE W. 1958 *Techniques of Population Analysis.* New York: Wiley. → A nonmathematical presentation.

CARR-SAUNDERS, ALEXANDER (1936) 1965 *World Population: Past Growth and Future Trends.* New York: Barnes & Noble. → A classic work.

COALE, ANSLEY J.; and DEMENY, PAUL 1966 *Regional Model Life Tables and Stable Populations.* Princeton Univ. Press.

COALE, ANSLEY J.; and HOOVER, EDGAR M. 1958 *Population Growth and Economic Development in Low-income Countries: A Case Study of India's Prospects.* Princeton Univ. Press. → The most comprehensive work on this subject.

CONFERENCE ON RESEARCH IN FAMILY PLANNING, NEW YORK, *1960* 1962 *Research in Family Planning: Papers.* Edited by Clyde V. Kiser. Princeton Univ. Press.

COX, PETER R. (1950) 1959 *Demography.* 3d ed. Cambridge Univ. Press. → Emphasizes the needs of actuaries.

Demographic Yearbook. → Issued annually by the United Nations since 1948. The standard international compendium and source book for world demographic information. See especially the 1948 issue.

Demography. → Published irregularly since 1964 by the Population Association of America.

►ELDRIDGE, HOPE T. 1968 Population: VII. Population Policies. Volume 12, pages 381–388 in *International*

Encyclopedia of the Social Sciences. Edited by David L. Sills. New York: Macmillan and Free Press.

FREEDMAN, RONALD (editor) 1964 *Population: The Vital Revolution*. Garden City, N.Y.: Doubleday. → A symposium on current world demography.

►FREEDMAN, RONALD 1968 Fertility. Volume 5, pages 371–382 in *International Encyclopedia of the Social Sciences*. Edited by David L. Sills. New York: Macmillan and Free Press.

►GRAUMAN, JOHN V. 1968 Population: VI. Population Growth. Volume 12, pages 376–381 in *International Encyclopedia of the Social Sciences*. Edited by David L. Sills. New York: Macmillan and Free Press.

GREAT BRITAIN, ROYAL COMMISSION ON POPULATION 1949–1954 *Papers*. 6 vols. in 7. London: H.M. Stationery Office.

HAUSER, PHILIP M.; and DUNCAN, OTIS DUDLEY (editors) (1959) 1964 *The Study of Population: An Inventory and Appraisal*. Univ. of Chicago Press. → A major compendium of scholarly papers on the history, substance, and status of demography.

INTERNATIONAL CONFERENCE ON FAMILY PLANNING PROGRAMS, GENEVA, *1965* 1966 *Family Planning and Population Programs: A Review of World Developments*. Edited by Bernard Berelson et al. Univ. of Chicago Press.

KIRK, DUDLEY 1942 The Relation of Employment Levels to Births in Germany. *Milbank Memorial Fund Quarterly* 20:126–138.

LORIMER, FRANK 1957 General Survey. Part 1, pages 11–57 in David V. Glass (editor), *The University Teaching of Social Sciences: Demography*. Paris: UNESCO.

LORIMER, FRANK (1959) 1964 The Development of Demography. Pages 124–179 in Philip M. Hauser and Otis Dudley Duncan (editors), *The Study of Population: An Inventory and Appraisal*. Univ. of Chicago Press.

LOTKA, ALFRED J. (1925) 1957 *Elements of Mathematical Biology*. New York: Dover. → First published as *Elements of Physical Biology*.

►MORIYAMA, IWAO M. 1968 Mortality. Volume 10, pages 498–504 in *International Encyclopedia of the Social Sciences*. Edited by David L. Sills. New York: Macmillan and Free Press.

On Population: Three Essays, by Thomas R. Malthus, Julian Huxley, and Frederick Osborn. New York: Mentor, 1960. → A paperback publication. Contains "A Summary View of the Principle of Population" by Malthus, first published in 1830; "World Population" by Huxley, first published in 1956; and "Population: An International Dilemma" by Osborn, first published in 1958.

PEARL, RAYMOND; and REED, L. J. 1920 On the Rate of Growth of the Population of the United States Since 1790 and Its Mathematical Representation. National Academy of Sciences, *Proceedings* 6:275–288.

PETERSEN, WILLIAM (1961) 1964 *Population*. New York: Macmillan. → An introductory textbook.

►PETERSEN, WILLIAM 1968 Migration: I. Social Aspects. Volume 10, pages 286–292 in *International Encyclopedia of the Social Sciences*. Edited by David L. Sills. New York: Macmillan and Free Press.

Population. → Published since 1946 by the Presses Universitaires de France for the Institut National d'Études Démographiques.

Population Bulletin of the United Nations 1963 No. 7. → Discusses conditions and trends of fertility.

Population Index. → An annotated listing of current publications. Published since 1935 by Princeton University, Office of Population Research, and the Population Association of America, Inc.

Population Studies. → Published since 1947. Issued by the London School of Economics and Political Science, Population Investigation Committee.

SPENGLER, JOSEPH J.; and DUNCAN, OTIS DUDLEY (editors) (1956*a*) 1963 *Demographic Analysis: Selected Readings*. New York: Free Press.

SPENGLER, JOSEPH J.; and DUNCAN, OTIS DUDLEY (editors) 1956*b* *Population Theory and Policy: Selected Readings*. Glencoe, Ill.: Free Press.

Studies in Family Planning. → Published irregularly since 1963 by the Population Council.

►THOMAS, BRINLEY 1968 Migration: II. Economic Aspects. Volume 10, pages 292–300 in *International Encyclopedia of the Social Sciences*. Edited by David L. Sills. New York: Macmillan and Free Press.

THOMPSON, WARREN S.; and LEWIS, DAVID T. (1930) 1965 *Population Problems*. 5th ed. New York: McGraw-Hill.

►TIETZE, CHRISTOPHER 1968 Fertility Control. Volume 5, pages 382–388 in *International Encyclopedia of the Social Sciences*. Edited by David L. Sills. New York: Macmillan and Free Press.

UNITED NATIONS, DEPARTMENT OF SOCIAL AFFAIRS, POPULATION DIVISION 1953 *The Determinants and Consequences of Population Trends*. Population Studies, No. 17. New York: United Nations.

U.S. DEPARTMENT OF HEALTH, EDUCATION AND WELFARE, PUBLIC HEALTH SERVICE 1966 *Natality Statistics Analysis, United States, 1963*. National Center for Health Statistics, Series 21, No. 8. Washington: Government Printing Office.

U.S. NATIONAL RESOURCES COMMITTEE, SCIENCE COMMITTEE 1938 *The Problems of a Changing Population*. Report of the Committee on Population Problems. Washington: Government Printing Office.

WESTOFF, CHARLES F.; POTTER, ROBERT G.; and SAGI, PHILIP C. 1963 *The Third Child: A Study in the Prediction of Fertility*. Princeton Univ. Press.

WESTOFF, CHARLES F. et al. 1961 *Family Growth in Metropolitan America*. Princeton Univ. Press.

WHELPTON, PASCAL K. et al. 1965 *Fertility and Family Planning in the United States*. Princeton Univ. Press. → Report of a national sample survey conducted in 1960.

WORLD POPULATION CONFERENCE, SECOND, BELGRADE, *1965* 1966 *Proceedings*. Volume 1: Summary Report. New York: United Nations.

YULE, G. UDNY 1925 The Growth of Population and the Factors Which Control It. *Journal of the Royal Statistical Society* 88:1–58.

Postscript

Since the preparation of the main article, public interest in population problems has intensified, as reflected in numerous popular publications, a flood of alarmist pronouncements, and the initiation of population and family planning programs of greatly varying intensity in countries that include a majority of the world's population. These activities have been induced by recognition of the effects of continued population growth, presently averaging 2.4 per cent per annum among approximately 70 per cent of the world's population living in the traditional or underdeveloped societies. Fertility in most of the more developed countries has continued downward from postwar levels, and there have been efforts, notably in eastern European countries, to reverse the trend through suppression of abortion (Rumania) and differential family allowances for second and third children (Czechoslovakia and Hungary).

In birth rates, the rigid dichotomy of the postwar period—between the developed countries, all with relatively low birth rates, and the less developed countries, with almost universally high birth rates —no longer exists. It is being replaced by a continuum as birth rates fall as a result of socioeconomic progress and antinatalist policies in a number of the less developed countries, notably in east Asia and Latin America.

Reflecting popular interest, demographic scientific research and publication has expanded enormously with contributions too numerous to mention individually here. Journals specifically devoted to scientific demography and population study now appear in several languages. The most noteworthy in English are *Demography* and *Population Index* published by the Population Association of America, *Population Studies* (England), and *Demography* (India). Most valuable for current scientific reports oriented toward policy questions are the periodical publications of the Population Council (New York), including its *Studies in Family Planning, Reports on Population/Family Planning,* and *Country Profiles.* Comprehensive research reports by the U.S. Commission on Population Growth and the American Future (1972–1974) were prepared in seven volumes, including numerous studies of the implications of possible future population growth. These valuable studies have become less useful because of the unforeseen continued fertility decline in the United States, which has, for the time being at least, undermined the projections on which many of the studies were based.

Of several important international meetings, the United Nations World Population Conference held in Bucharest in August 1974 is worthy of special note. For the first time, official government representatives were brought together in a political, rather than scientific, meeting concerned with population problems. More typical of the periodic scientific meetings held to discuss population problems were the two sponsored by the United Nations in Bucharest in 1965 and in Rome in 1954. Proceedings are available of the International Population Conference (1973) sponsored in Liège by the International Union for the Scientific Study of Population and of the World Population Conference (1966–1967) sponsored by the United Nations in Belgrade in 1965.

The most ambitious international demographic research project ever undertaken is the World Fertility Survey of the 1970s. Sponsored by the International Statistical Institute (1973), it is designed to obtain scientific information on this subject by surveys in many countries and regions on a comparative basis.

DUDLEY KIRK

ADDITIONAL BIBLIOGRAPHY

INTERNATIONAL POPULATION CONFERENCE, LIÈGE, *1973* 1973 *Rapports/Proceedings.* 3 vols. Liège: Union Internationale Pour l'Étude Scientifique de la Population.

INTERNATIONAL STATISTICAL INSTITUTE 1973 World Fertility Survey. Occasional Papers, No. 1. → Mimeographed paper.

U.S. COMMISSION ON POPULATION GROWTH AND THE AMERICAN FUTURE 1972–1974 *Research Reports.* 7 vols. Washington: Government Printing Office.

WORLD POPULATION CONFERENCE, SECOND, BELGRADE, *1965* 1966–1967 *Proceedings.* 10 vols. New York: United Nations.

II

DEMOGRAPHY AND POPULATION GENETICS

This article was first published in IESS *as "Genetics: II. Demography and Population Genetics" with two companion articles less relevant to statistics.*

The best available definition of population genetics is doubtless that of Malécot: "It is the totality of mathematical models that can be constructed to represent the evolution of the structure of a population classified according to the distribution of its Mendelian genes" (1955, p. 240). This definition, by a probabilist mathematician, gives a correct idea of the "constructed" and abstract side of this branch of genetics; it also makes intelligible the rapid development of population genetics since the advent of Mendelism.

In its formal aspect this branch of genetics might even seem to be a science that is almost played out. Indeed, it is not unthinkable that mathematicians have exhausted all the structural possibilities for building models, both within the context of general genetics and within that of the hypotheses—more or less complex and abstract—that enable us to characterize the state of a population.

Two major categories of models can be distinguished: *determinist* models are those "in which variations in population composition over time are rigorously determined by (*a*) a known initial state of the population; (*b*) a known number of forces or 'pressures' operating, in the course of generations, in an unambiguously defined fashion" (Malécot 1955, p. 240). These pressures involve mutation, selection, and preferential marriages (by consanguinity, for instance). Determinist models, based on ratios that have been exactly ascertained from preceding phenomena, can be expressed only in terms of populations that are infinite in the mathematical sense. In fact, it is only in this type of population that statistical regularities can emerge (Malécot 1955). In these models the composition of each generation is perfectly defined by the composition of the preceding generation.

Stochastic models, in contrast to determinist ones, involve only finite populations, in which the gametes that, beginning with the first generation, are actually going to give birth to the new generation represent only a finite number among all possible gametes. The result is that among these active, or "useful," gametes (Malécot 1959), male or female, the actual frequency of a gene will differ from the probability that each gamete had of carrying it at the outset.

The effect of chance will play a prime role, and the frequencies of the genes will be able to drift from one generation to the other. The effects of random drift and of genetic drift become, under these conditions, the focal points for research.

The body of research completed on these assumptions does indeed form a coherent whole, but these results, in spite of their brilliance, are marked by a very noticeable formalism. In reality, the models, although of great importance at the conceptual level, are often too far removed from the facts. In the study of man, particularly, the problems posed are often too complex for the solutions taken directly from the models to describe concrete reality.

Not all these models, however, are the result of purely abstract speculation; construction of some of them has been facilitated by experimental data. To illustrate this definition of population genetics and the problems that it raises, this article will limit itself to explaining one determinist model, both because it is one of the oldest and simplest to understand and because it is one of those most often verified by observation.

A determinist model. Let us take the case of a particular human population: the inhabitants of an island cut off from outside contacts. It is obvious that great variability exists among the genes carried by the different inhabitants of this island. The genotypes differ materially from one another; in other words, there is a certain polymorphism in the population—polymorphism that we can define in genetic terms with the help of a simple example.

Let us take the case of autosome ("not connected with sex") gene a, transmitting itself in a monohybrid diallely. In relation to it individuals can be classified in three categories: *homozygotes* whose two alleles are a (a/a); heterozygotes, carriers of a and its allele a' (a/a'); and the homozygotes who are *noncarriers* of a (a'/a'). At any given moment or during any given generation, these three categories of individuals exist within the population in certain proportions relative to each other.

Now, according to Mendel's second law (the law of segregation), the population born out of a cross between an individual who is homozygote for a (a/a) and an individual who is homozygote for a' (a'/a') will include individuals a/a, a/a', and a'/a' in the following proportions: one-fourth a/a, one-half a/a', and one-fourth a'/a'. In this population the alleles a and a' have the same frequency, one-half, and each sex produces half a and half a'. If these individuals are mated randomly, a simple algebraic calculation quickly demonstrates that individuals of the generation following will be quantitatively distributed in the same fashion: one-fourth a/a, one-half a/a', and one-fourth a'/a'. It will be the same for succeeding generations.

It can therefore be stated that *the genetic structure of such a population does not vary from one generation to the other.* If we designate by p the initial proportion of a/a individuals and by q that of a'/a' individuals, we get $p + q = 1$, or the totality of the population. Applying this system of symbols to the preceding facts, it can be easily shown that the proportion of individuals of all three categories in the first generation born from a/a and a'/a' equals p^2, $2pq$, q^2. In the second and third generation the frequency of individuals will always be similar: p^2, $2pq$, q^2.

Until this point, we have remained at the individual level. If we proceed to that of the gametes carrying a or a' and to that of genes a and a', we observe that their frequencies intermingle. In the type of population discussed above, the formula p^2, $2pq$, q^2 still applies perfectly, therefore, to the gametes and genes. This model, which can be regarded as a formalization of the Hardy–Weinberg law, has other properties, but our study of it will stop here. (For a discussion of the study of isolated populations, see Sutter & Tabah 1951.)

Model construction and demographic reality. The Hardy–Weinberg law has been verified by numerous studies, involving both vegetable and animal species. The findings in the field of human blood groups have also been studied for a long time from a viewpoint derived implicitly from this law, especially in connection with their geographic distribution. Under the system of reproduction by sexes, a generation renews itself as a result of the encounter of the sexual cells (gametes) produced by individuals of both sexes belonging to the living generation. In the human species it can be said that this encounter takes place at random. One can imagine the advantage that formal population genetics can take of this circumstance, which can be compared to drawing marked balls by lot from two different urns. Model construction, already favored by these circumstances, is favored even further if the characteristics of the population utilized are artificially defined with the help of a certain number of hypotheses, of which the following is a summary description:

(1) Fertility is identical for all couples; there is no differential fertility.

(2) The population is closed; it cannot, therefore, be the locus of migrations (whether immigration or emigration).

(3) Marriages take place at random; there is no assortative mating.

(4) There are no systematic preferential marriages (for instance, because of consanguinity).

(5) Possible mutations are not taken into consideration.

(6) The size of the population is clearly defined.

On the basis of these working hypotheses, the whole of which constitutes panmixia, it was possible, not long after the rediscovery of Mendel's laws, to construct the first mathematical models. Thus, population genetics took its first steps forward, one of which was undoubtedly the Hardy–Weinberg law.

Mere inspection of the preceding hypotheses will enable the reader to judge how, taken one by one, they conflict with reality. In fact, no human population can be panmictic in the way the models are.

The following evidence can be cited in favor of this conclusion:

(1) Fertility is never the same with all couples. In fact, differential fertility is the rule in human populations. There is always a far from negligible sterility rate of about 18 per cent among the large populations of Western civilization. On the other hand, the part played by large families in keeping up the numbers of these populations is extremely important; we can therefore generalize by emphasizing that for one or another reason individuals carrying a certain assortment of genes reproduce themselves more or less than the average number of couples. That is what makes for the fact that in each population there is always a certain degree of *selection*. Hypothesis (1) above, essential to the construction of models, is therefore very far removed from reality.

(2) Closed populations are extremely rare. Even among the most primitive peoples there is always a minimum of emigration or immigration. The only cases where one could hope to see this condition fulfilled at the present time would be those of island populations that have remained extremely primitive.

(3) With assortative mating we touch on a point that is still obscure; but even if these phenomena remain poorly understood, it can nevertheless be said that they appear to be crucial in determining the genetic composition of populations. This choice can be *positive:* the carriers of a given characteristic marry among each other more often than chance would warrant. The fact was demonstrated in England by Pearson and Lee (1903): very tall individuals have a tendency to marry each other, and so do very short ones. Willoughby (1933) has reported on this question with respect to a great number of somatic characteristics other than height—for example, coloring of hair, eyes and skin, intelligence quotient, and so forth. Inversely, *negative* choice makes individuals with the same characteristics avoid marrying one another. This mechanism is much less well known than the above. The example of persons of violent nature (Dahlberg 1943) and of red-headed individuals has been cited many times, although it has not been possible to establish valid statistics to support it.

(4) The case of preferential marriages is not at all negligible. There are still numerous areas where marriages between relatives (consanguineous marriages) occur much more frequently than they would as the result of simple random encounters. In addition, recent studies on the structures of kinship have shown that numerous populations that do not do so today used to practice preferential marriage—most often in a matrilinear sense. These social phenomena have a wide repercussion on the genetic structure of populations and are capable of modifying them considerably from one generation to the other.

(5) Although we do not know exactly what the real rates of mutation are, it can be admitted that their frequency is not negligible. If one or several genes mutate at a given moment in one or several individuals, the nature of the gene or genes is in this way modified; its stability in the population undergoes a disturbance that can considerably transform the composition of that population.

(6) The size of the population and its limits have to be taken into account. We have seen that this is one of the essential characteristics im-

portant in differentiating two large categories of models.

Demography and population genetics

The above examination brings us into contact with the realities of population: fertility, fecundity, nuptiality, mortality, migration, and size are the elements that are the concern of demography and are studied not only by this science but also very often as part of administrative routine. Leaving aside the influence of size, which by definition is of prime importance in the technique of the models, there remain five factors to be examined from the demographic point of view. Mutation can be ruled out of consideration, because, although its importance is great, it is felt only after the passage of a certain number of generations. It can therefore be admitted that it is not of immediate interest.

We can also set aside choice of a mate, because the importance of this factor in practice is still unknown. Accordingly, there remain three factors of prime importance: fertility, migration, and preferential marriage. Over the last decade the progressive disappearance of consanguineous marriage has been noted everywhere but in Asia. In many civilized countries marriage between cousins has practically disappeared. It can be stated, therefore, that this factor has in recent years become considerably less important.

Migrations remain very important on the genetic level, but, unfortunately, precise demographic data about them are rare, and most of the data are of doubtful validity. For instance, it is hard to judge how their influence on a population of Western culture could be estimated.

The only remaining factor, fertility (which to geneticists seems essential), has fortunately been studied in satisfactory fashion by demographers. To show the importance of differential fertility in human populations, let us recall a well-known calculation made by Karl Pearson in connection with Denmark. In 1830, 50 per cent of the children in that country were born of 25 per cent of the parents. If that fertility had been maintained at the same rate, 73 per cent of the second-generation Danes and 97 per cent of the third generation would have been descended from the first 25 per cent. Similarly, before World War I, Charles B. Davenport calculated, on the basis of differential fertility, that 1,000 Harvard graduates would have only 50 descendants after two centuries, while 1,000 Rumanian emigrants living in Boston would have become 100,000.

Measurement of fertility. Human reproduction involves both *fecundity* (capacity for reproduction) and *fertility* (actual reproductive performance). These can be estimated for males, females, and married couples treated as a reproductive unit. Let us rapidly review the measurements that demography provides for geneticists in this domain.

Crude birth rate. The number of living births in a calendar year per thousand of the average population in the same year is known as the *crude birth rate*. The rate does not seem a very useful one for geneticists: there are too many different groups of childbearing age; marriage rates are too variable from one population to another; birth control is not uniformly diffused, and so forth.

General fertility rate. The ratio of the number of live births in one year to the average number of women capable of bearing children (usually defined as those women aged 15 to 49) is known as the *general fertility rate*. Its genetic usefulness is no greater than that of the preceding figure. Moreover, experience shows that this figure is not very different from the crude birth rate.

Age-specific fertility rates. Fertility rates according to the age reached by the mother during the year under consideration are known as *age-specific fertility rates*. Demographic experience shows that great differences are observed here, depending on whether or not the populations are Malthusian—in other words, whether they practice birth control or not. In the case of a population where the fertility is natural, knowledge of the mother's age is sufficient. In cases where the population is Malthusian, the figure becomes interesting when it is calculated both by age and by age group of the mothers at time of marriage, thus combining the mother's age at the birth of her child and her age at marriage. This is generally known as the *age-specific marital fertility rate*. If we are dealing with a Malthusian population, it is preferable, in choosing the sample to be studied, to take into consideration the age at marriage rather than the age at the child's birth. Thus, while the age at birth is sufficient for natural populations, these techniques cannot be applied indiscriminately to all populations.

Family histories. Fertility rates can also be calculated on the basis of family histories, which can be reconstructed from such sources as parish registries (Fleury & Henry 1965) or, in some countries, from systematic family registrations (for instance, the Japanese *koseki* or *honseki*). The method for computing the fertility rate for, say, the 25–29-year-old age group from this kind of data is first to determine the number of legitimate births in the group. It is then necessary to make a rigorous count of the number of years lived in wedlock between their 25th and 30th birthdays by all the women in the group; this quantity is known as the group's total "woman-years." The number of births is then divided by the number of "woman-years" to obtain the group's fertility rate. This method is very useful in the study of historical problems in genetics, since it is often the only one that can be applied to the available data.

Measurement of reproduction. Let us leave fertility rates in order to examine rates of *reproduction*. Here we return to more purely genetic considerations, since we are looking for the mechanism whereby one generation is replaced by the one that follows it. Starting with a series of fertility rates by age groups, a *gross reproduction rate* can be calculated that gives the average number of female progeny that would be born to an age cohort of women, all of whom live through their entire reproductive period and continue to give birth at the rates prevalent when they themselves were born. The gross reproduction rate obtaining for a population at any one time can be derived by combining the rates for the different age cohorts.

A gross reproduction rate for a real generation can also be determined by calculating the average number of live female children ever born to women of fifty or over. As explained above, this rate is higher for non-Malthusian than for Malthusian populations and can be refined by taking into consideration the length of marriage.

We have seen that in order to be correct, it is necessary for the description of fertility in Malthusian populations to be closely related to the date of marriage. Actually, when a family reaches the size that the parents prefer, fertility tends to approach zero. The preferred size is evidently related to length of marriage in such a manner that fertility is more closely linked with length of marriage than with age at marriage. In recent years great progress has been made in the demographic analysis of fertility, based on this kind of data. This should enable geneticists to be more circumspect in their choice of sections of the population to be studied.

Cohort analysis. Americans talk of *cohort analysis*, the French of analysis by *promotion* (a term meaning "year" or "class," as we might speak of the "class of 1955"). A cohort, or *promotion*, includes all women born within a 12-month period; to estimate fertility or mortality, it is supposed that these women are all born at the same moment on the first of January of that year. Thus, women born between January 1, 1900, and January 1, 1901, are considered to be exactly 15 years old on January 1, 1915; exactly 47 years old on January 1, 1947; and so forth.

The research done along these lines has issued in the construction of tables that are extremely useful in estimating fertility in a human population. As we have seen, it is more useful to draw up cohorts based on age at marriage than on age at birth. A fertility table set up in this way gives for each cohort the *cumulative* birth rate, by order of birth and single age of mother, for every woman surviving at each age, from 15 to 49. The progress that population genetics could make in knowing *real* genic frequencies can be imagined, if it could concentrate its research on any particular cohort and its descendants.

Demography of genic frequencies

This rapid examination of the facts that demography can now provide in connection with fertility clearly reveals the variables that population genetics can use to make its models coincide with reality. The models retain their validity for genetics because they are still derived from basic genetic concepts; their application to actual problems, however, should be based on the kind of data mentioned above. We have voluntarily limited ourselves to the problem of fertility, since it is the most important factor in genetics research.

The close relationship between demography and population genetics that now appears can be illustrated by the field of research into blood groups. Although researchers concede that blood groups are independent of both age and sex, they do not explore the full consequences of this, since their measures are applied to samples of the population that are "representative" only in a demographic sense. We must deplore the fact that this method has spread to the other branches of genetics, since it is open to criticism not only from the demographic but from the genetic point of view. By proceeding in this way, a most important factor is overlooked—that of *genic frequencies*.

Sample structure. Let us admit that the choice of a blood group to be studied is of little importance when the characteristic is widely distributed throughout the population—for instance, if each individual is the carrier of a gene taken into account in the system being studied (e.g., a system made up of groups A, B, and O). But this is no longer the case if the gene is carried only by a few individuals—in other words, if its frequency attains 0.1 per cent or less. In this case (and cases like this are common in human genetics) the structure of the sample examined begins to take on prime importance.

A brief example must serve to illustrate this cardinal point. We have seen that in the case of rare recessive genes the importance of consanguineous marriages is considerable. The scarcer that carriers of recessive genes become in the population as a whole, the greater the proportion of such carriers produced by consanguineous marriages. Thus if as many as 25 per cent of all individuals in a population are carriers of recessive genes, and if one per cent of all marriages in that population are marriages between first cousins, then this one per cent of consanguineous unions will produce 1.12 times as many carriers of recessive genes as will be produced by all the unions of persons not so related. But if recessive genes are carried by only one per cent of the total population, then the same proportion of marriages between first cousins will produce 2.13 times as many carriers as will be produced by all other marriages. This production ratio increases to 4.9 if the total frequency of carriers is .01 per cent, to 20.2 if it is .005 per cent, and to 226 if it is .0001 per cent. Under these conditions, one can see the importance of the sampling method used to estimate the frequency within a population, not only of the individuals who are carriers but of the gametes and genes themselves.

Genealogical method. It should be emphasized that genetic studies based on genealogies remain the least controversial. Studying a population where the degrees of relationship connecting individuals are known presents an obvious interest. Knowing one or several characteristics of certain parents, we can follow what becomes of these in the descendants. Their evolution can also be considered from the point of view of such properties of genes as dominance, recessiveness, expressivity, and penetrance. But above all, we can follow the evolution of these characteristics in the population over time and thus observe the effects of differential fertility. Until now the genealogical method was applicable only to a numerically sparse population, but progress in electronic methods of data processing permits us to anticipate its application to much larger populations (Sutter & Tabah 1956).

Dynamic studies. In very large modern populations it would appear that internal analysis of cohorts and their descendants will bring in the future a large measure of certainty to research in population genetics. In any case, it is a sure way to a dynamic genetics based on demographic reality. For instance, it has been recommended that blood groups should be studied according to age groups; but if we proceed to do so without regard for demographic factors, we cannot make our observations dynamic. Thus, a study that limits itself to, let us say, the fifty- to sixty-year-old age group will have to deal with a universe that includes certain genetically "dead" elements, such as unmarried and sterile persons, which have no meaning from the dynamic point of view. But if a study is made of this same fifty- to sixty-year-old age group and then of the twenty- to thirty-year-old age group, and if in the older group only those individuals are considered who have descendants in the younger group, the dynamic potential of the data is maximized. It is quite possible to subject demographic cohorts to this sort of interpretation, because in many countries demographic statistics supply series of individuals classified according to the mothers' age at their birth.

Other demographic factors. This discussion would not be complete if we did not stress another aspect of the genetic importance of certain demographic factors, revealed by modern techniques, which have truly created a demographic biology. Particularly worthy of note are the mother's age, order of birth, spacing between births, and size of family.

The mother's age is a great influence on fecundity. A certain number of couples become incapable of having a second child after the birth of the first child; a third child after the second; a fourth after the third; and so forth. This sterility increases with the length of a marriage and especially after the age of 35. It is very important to realize this when, for instance, natural selection and its effects are being studied.

The mother's age also strongly influences the frequency of twin births (monozygotic or dizygotic), spontaneous abortions, stillborn or abnormal births, and so on. Many examples can also be given of the influence of the order of birth, the interval between births, and the size of the family to illustrate their effect on such things as fertility, mortality, morbidity, and malformations.

It has been demonstrated above how seriously demographic factors must be taken into consideration when we wish to study the influence of the genetic structure of populations. We will leave aside the possible environmental influences, such as social class and marital status, since they have previously been codified by Osborn (1956/1957) and Larsson (1956–1957), among others. At the practical level, however, the continuing efforts to utilize vital statistics for genetic purposes should be pointed out. In this connection, the research of H. B. Newcombe and his colleagues (1965), who are attempting to organize Canadian national statistics for use in genetics, cannot be too highly praised. The United Nations itself posed the problem on the world level at a seminar organized in Geneva in 1960. The question of the relation between demography and genetics is therefore being posed in an acute form.

These problems also impinge in an important way on more general philosophical issues, as has been demonstrated by Haldane (1932), Fisher

(1930), and Wright (1951). It must be recognized, however, that their form of Neo-Darwinism, although it is based on Mendelian genetics, too often neglects demographic considerations. In the future these seminal developments should be renewed in full confrontation with demographic reality.

JEAN SUTTER

[*See* Freedman 1968; Ryder 1968; Tietze 1968; *see also* Harris 1968; Pressat 1968; Wynne-Edwards 1968.]

BIBLIOGRAPHY

BARCLAY, GEORGE W. 1958 *Techniques of Population Analysis*. New York: Wiley.

DAHLBERG, GUNNAR (1943) 1948 *Mathematical Methods for Population Genetics*. New York and London: Interscience. → First published in German.

DUNN, LESLIE C. (editor) 1951 *Genetics in the Twentieth Century: Essays on the Progress of Genetics During Its First Fifty Years*. New York: Macmillan.

FISHER, R. A. (1930) 1958 *The Genetical Theory of Natural Selection*. 2d ed., rev. New York: Dover.

FLEURY, M.; and HENRY, L. 1965 *Nouveau manuel de dépouillement et d'exploitation de l'état civil ancien*. Paris: Institut National d'Études Démographiques.

►FREEDMAN, RONALD 1968 Fertility. Volume 5, pages 371–382 in *International Encyclopedia of the Social Sciences*. Edited by David L. Sills. New York: Macmillan and Free Press.

GEPPERT, HARALD; and KOLLER, SIEGFRIED 1938 *Erbmathematik*. Leipzig: Quelle & Meyer.

HALDANE, J. B. S. 1932 *The Causes of Evolution*. London and New York: Harper.

►HARRIS, MARVIN 1968 Race. Volume 13, pages 263–269 in *International Encyclopedia of the Social Sciences*. Edited by David L. Sills. New York: Macmillan and Free Press.

HENRY, LOUIS 1953 *Fécondité des mariages: Nouvelle méthode de mesure*. Institut National d'Études Démographiques, Travaux et Documents, Cahier No. 16. Paris: The Institute.

LARSSON, TAGE 1956–1957 The Interaction of Population Changes and Heredity. *Acta genetica et statistica medica* 6:333–348.

L'HÉRITIER, PHILIPPE 1954 *Traité de génétique*. Volume 2: La génétique des populations. Paris: Presses Universitaires de France.

LI, CHING CHÜN (1948) 1955 *Population Genetics*. 2d ed. Univ. of Chicago Press.

MALÉCOT, GUSTAVE 1948 *Les mathématiques et l'hérédité*. Paris: Masson.

MALÉCOT, GUSTAVE 1955 La génétique de population: Principes et applications. *Population* 10:239–262.

MALÉCOT, GUSTAVE 1959 Les modèles stochastiques en génétique de population. Paris, Université, Institut de Statistique, *Publications* 8:173–210.

NEWCOMBE, H. B.; SMITH, M. E.; and SCHWARTZ, R. R. 1965 *Computer Methods for Extracting Kinship Data From Family Groupings of Records*. Chalk River (Ontario): Atomic Energy of Canada Limited.

OSBORN, F. 1956/1957 Changing Demographic Trends of Interest to Population Genetics. *Acta genetica et statistica medica* 6:354–362.

PEARSON, KARL; and LEE, ALICE 1903 On the Laws of Inheritance in Man. *Biometrika* 2:257–462.

PRESSAT, ROLAND 1961 *L'analyse démographique: Méthodes, résultats, applications*. Paris: Presses Universitaires de France.

►PRESSAT, ROLAND 1968 Nuptiality. Volume 11, pages 223–226 in *International Encyclopedia of the Social Sciences*. Edited by David L. Sills. New York: Macmillan and Free Press.

►RYDER, N. B. 1968 Cohort Analysis. Volume 2, pages 546–550 in *International Encyclopedia of the Social Sciences*. Edited by David L. Sills. New York: Macmillan and Free Press.

SUTTER, JEAN; and TABAH, LÉON 1951 Les notions d'isolat et de population minimum. *Population* 6:481–498.

SUTTER, JEAN; and TABAH, LÉON 1956 Méthode mécanographique pour établir la généalogie d'une population: Application à l'étude des esquimaux polaires. *Population* 11:507–530.

►TIETZE, CHRISTOPHER 1968 Fertility Control. Volume 5, pages 382–388 in *International Encyclopedia of the Social Sciences*. Edited by David L. Sills. New York: Macmillan and Free Press.

UNITED NATIONS 1961 *The Use of Vital and Health Statistics for Genetic and Radiation Studies*. New York: United Nations.

WHELPTON, PASCAL K. 1954 *Cohort Fertility: Native White Women in the United States*. Princeton Univ. Press.

WILLOUGHBY, RAYMOND R. 1933 Somatic Homogamy in Man. *Human Biology* 5:690–705.

WORLD POPULATION CONFERENCE 1965 *Proceedings*. New York: United Nations. → See especially "Recent Advances in the Theory of Population Genetics" by M. Kimura.

WRIGHT, SEWALL 1951 The Genetical Structure of Populations. *Annals of Human Genetics* 15:323–354.

►WYNNE-EDWARDS, V. C. 1968 Social Behavior, Animal: III. The Regulation of Animal Populations. Volume 14, pages 360–365 in *International Encyclopedia of the Social Sciences*. Edited by David L. Sills. New York: Macmillan and Free Press.

Postscript

The following publications help to update the above article by the late Jean Sutter.

ALBERT JACQUARD

ADDITIONAL BIBLIOGRAPHY

CAVALLI-SFORZA, LUIGI L.; and BODMER, W. F. 1971 *The Genetics of Human Populations*. San Francisco: Freeman.

CROW, JAMES F.; and KIMURA, MOTOO 1970 *An Introduction to Population Genetics Theory*. New York: Harper.

JACQUARD, ALBERT 1974*a* *The Genetic Structure of Populations*. Berlin and New York: Springer.

JACQUARD, ALBERT 1974*b* *Génétique des populations humaines*. Paris: Presses Universitaires de France.

MALÉCOT, GUSTAVE 1966 *Probabilités et hérédite*. Institut National d'Études Demographiques, Travaux et Documents, Cahier No. 47. Paris: Presses Universitaires de France.

STERN, CURT (1949) 1973 *Principles of Human Genetics*. 3d ed. San Francisco: Freeman.

DEPENDENCE

See ERRORS, *article on* EFFECTS OF ERRORS IN STATISTICAL ASSUMPTIONS; PROBABILITY; STATISTICS, DESCRIPTIVE, *article on* ASSOCIATION.

DESCRIPTIVE STATISTICS

See DATA ANALYSIS, EXPLORATORY; STATISTICS, DESCRIPTIVE.

DESIGN OF EXPERIMENTS

See under EXPERIMENTAL DESIGN.

DIMENSION REDUCTION

See CLUSTERING; FACTOR ANALYSIS AND PRINCIPAL COMPONENTS; LATENT STRUCTURE; SCALING, MULTIDIMENSIONAL.

DISCRETE DISTRIBUTIONS

See under DISTRIBUTIONS, STATISTICAL.

DISCRIMINANT FUNCTIONS

See MULTIVARIATE ANALYSIS, *article on* CLASSIFICATION AND DISCRIMINATION.

DISPERSION

See STATISTICS, DESCRIPTIVE, *article on* LOCATION AND DISPERSION; VARIANCES, STATISTICAL STUDY OF.

DISTRIBUTED LAGS

In theory, distributed lags arise when any economic cause, such as a price change or an income change, produces its effect (for example, on the quantity demanded) only after some lag in time, so that the effect is not felt all at once at a single point in time but is distributed over a period of time. Thus, when we say that the quantity of cigarettes demanded is a function of the price of cigarettes taken with a distributed lag, we mean essentially that the full effects of a change in the price of cigarettes is not felt immediately, and only after some passage of time does the quantity of cigarettes demanded show the full effect of the change in the price of cigarettes.

To consider the matter more concretely, let q_t be the quantity of cigarettes, say, demanded per unit time, and let p_t be the price per unit of cigarettes during time period t. Other things remaining constant, such as income, population, and the price of cough drops, we may express q_t as a function of current and past prices:

$$q_t = f(p_t, p_{t-1}, p_{t-2}, \cdots). \tag{1}$$

In particular, let us assume the demand function f to be linear with constant coefficients $a, b_0, b_1, \cdots,$

$$q_t = a + b_0 p_t + b_1 p_{t-1} + b_2 p_{t-2} + \cdots. \tag{2}$$

Suppose the price of cigarettes has been constant for a long time at a level p, so that $p = p_t = p_{t-1} = p_{t-2} = \cdots$. Then the quantity of cigarettes demanded per unit time will also be a constant:

$$q = a + b_0 p + b_1 p + b_2 p + \cdots = a + p \sum_{i=0}^{\infty} b_i. \tag{3}$$

Now let the price of cigarettes change from p to $p + \Delta p$ in period $t + 1$ and remain at this new level indefinitely. The effect of the change in period $t + 1$ will be to change the quantity demanded from q to $q + b_0 \Delta p$. But the effects of the price change do not stop here; in the next period, $t + 2$, the quantity demanded is further altered to $q + b_0 \Delta p + b_1 \Delta p$. In general, after θ periods, the change in q will be $\Delta p \sum_{i=0}^{\theta} b_i$. Thus, the effect of the change in price on demand is distributed over time: $b_0 \Delta p$ the first period, $b_0 \Delta p + b_1 \Delta p$ the second, and so on.

The example used to illustrate the concept of a distributed lag is taken from demand analysis, but the use of distributed lags is not restricted to analysis of problems of consumer demand. The wide application of distributed lags in econometrics may be indicated by a few examples: import demand (Tinbergen 1949); hyperinflation (Cagan 1956); investment (Koyck 1954; Jorgenson 1963; Eisner 1960); demand for chemical fertilizers in the United States (Griliches 1959); advertising of oranges by the two major U.S. orange grower cooperatives (Nerlove & Waugh 1961).

The causes of distributed lags in economic relationships are as varied as the variables entering such relationships. To illustrate, the cigarette example may be used again. When the price is increased, we expect, other things being equal and if cigarettes are not an inferior good, that the demand will fall off. *How* it falls off, that is, the time path that demand follows, is not discussed in the ordinary static theory of demand. At first, a price rise may induce many people to stop smoking, but the tenacity of the habit being what it is, some will return to it. Others may either temporarily or permanently reduce consumption. It is also conceivable that those who do not regard the price increase as permanent, and who do not react at first, may come eventually to believe in the permanence of the price change as it persists and be willing to go through the painful process of adjustment. Finally, some who are nonsmokers, but who might have begun smoking when prices were lower (for example, adolescents), may now never begin. Thus, we see that there are three general types of factors that cause the effects of the price change to be distributed over time: (1) habit persistence, (2) expectational rigidities, and (3) a semitechnological factor related to the age distribution of the population and the conventional time at which a person begins to smoke.

We could just as well deal with an income change or with the effects of a change in the price of a factor of production upon its employment in a certain branch of manufacturing. The exact causes and pattern of the distribution of lag would, of course, depend on the particular circumstances studied. The pattern (and causes) could be different for changes in one direction than for those in another. Indeed, as F. M. Fisher has pointed out (1962, p. 29), there is no need in theory even to assume, as we implicitly have, "that when a decision is made at given time, t, the events of a certain fixed period previous are given a certain weight, regardless of what those events were or of what happened before or since their occurrence." One can imagine the effect if a government report condemning smoking were issued just after the price of cigarettes declined. Nonetheless, for most practical purposes of econometrics, drastic simplifications must be made and the results can be regarded only as approximate.

History and forms. Irving Fisher (1925) was the first to use and discuss the concept of a distributed lag. In a later paper (1937, p. 323), he stated that the basic problem in applying the theory of distributed lags "is to find the 'best' distribution of lag, by which is meant the distribution such that . . . the total combined effect [of the lagged values of the variables taken with a distributed lag has] . . . the highest possible correlation with the actual statistical series . . . with which we wish to compare it." Thus, we wish to find the distribution of lag that maximizes the explanation of "effect" by "cause" in a statistical sense. There are unsophisticated, almost mechanical, approaches to this problem; and there are more sophisticated approaches involving considerable use of economic theory to develop underlying models of dynamic adjustment that in turn generate the distributed lags observed. There are also both more sophisticated and less sophisticated approaches to estimation.

No assumption may be made about the form of the distribution of lag, i.e., the relationships among the b's in an equation such as (2). This is the approach adopted by Tinbergen (1949) and Alt (1942). The procedure is then to estimate an equation such as (2), for example, by least squares. Since the number of observations is limited, only a finite number of lag terms may be included. Indeed, the coefficients of the lagged values may quickly become quite erratic because of the presence of strong serial correlation in most economic time series. Such difficulties led Alt, Irving Fisher (1925; 1937), and others to suggest that the parameters of a relationship such as (2) be constrained by specifying the form of the distribution of lag. Fisher (1937) suggests, for example, that all b's, after a certain point θ, are proportional to an arithmetical progression up to a certain maximal lag, θ'. Thus,

$$(4) \qquad b_i = b\left(1 - \frac{i-\theta}{\theta' - 1}\right), \qquad \theta \leqslant i \leqslant \theta'.$$

Fisher (1925) earlier had suggested that the weights of the distribution might plausibly be assumed to follow a logarithmic normal probability density function (with appropriate modification for the measurement of time in discrete units). Alt recommends a number of exponential forms. Koyck (1954) suggests weights that decline geometrically. Theil and Stern (1960) consider weights proportional to what are essentially approximations to the densities given by a gamma distribution with mean 2. Solow (1960) presents perhaps the ultimate flexibility and sophistication obtainable by a purely formal approach in suggesting the two-parameter family of unimodal lag distributions generated by the Pascal distribution. Thus, the weights are

$$(5) \qquad b_i = b\binom{r+i-1}{i}(1-\lambda)^r \lambda^i.$$

This family provides the discrete analogue of the exponential and gamma distributions. It represents a natural generalization of the geometrically declining weights of Koyck and may be thought of as r simple Koyck lags cascaded in series. The case $r = 1$ is the original Koyck lag. The distribution of lag is skewed; the larger the value of λ and the smaller the value of r, the greater is the degree of skewness. Both the center of gravity and the spread of the distribution increase with λ and r. This distribution offers a considerable gain in flexibility as compared with earlier suggestions.

A more complete review of the literature up to 1958 and some additional suggestions are contained in Nerlove (1958*a*, pp. 1–25).

Rather than directly specify the form of the lag, an alternative approach is to develop a dynamic model that leads to a distributed lag or lags in the observed relationships. This may be done at varying levels of generality, either by dealing with broad classes of causes in an attempt to derive models of wide applicability or by attempting to isolate dynamic features of a particular problem and show that these lead to certain forms of lag structure. Brown (1952) and Friedman (1957) develop different dynamic models designed to explain the behavior of total consumption expenditures. As Nerlove (1958*a*; 1958*c*) shows, both models lead essentially to a Koyck distribution of lag in the relation between income and consumption. The two models are examples of the two general classes of models leading to distributed lags extensively discussed by Nerlove (1958*a*; 1958*b*; 1958*c*): expectational models and dynamic adjustment models. Both classes in their simplest forms lead to Koyck distributions of lag but have quite different implications for estimation and are not at all the same in more complicated cases. Friedman's model is a member of the class of expectational models; and Brown's model is a member of the class of dynamic adjustment. The two models, however, were designed to be quite general and to apply to a variety of problems. Examples taken from the area of agricultural supply analysis are used here to illustrate each model (Nerlove 1958*b*, pp. 25–26).

An expectational model of supply response. Suppose that the quantity supplied in year t, x_t, is a linear function not of the current price p_t but of the price expected for year t in year $t-1$, p_t^* (for example, many agricultural crops are planted or planned far ahead):

$$x_t = a + bp_t^* + u_t, \tag{6}$$

where u_t is a stochastic residual. Suppose further that price forecasts are corrected each period by farmers in proportion to the error made:

$$p_t^* - p_{t-1}^* = \beta(p_{t-1} - p_{t-1}^*), \qquad 0 < \beta \leqslant 1, \tag{7}$$

where β is called the coefficient of expectations. A similar model is used explicitly by Cagan (1956) to generate expectations of changes in the general price level during hyperinflations. After some manipulation, it may be shown that

$$\begin{aligned} x_t = {} & a\beta + b\beta p_{t-1} \\ & + (1-\beta)x_{t-1} + u_t - (1-\beta)u_{t-1}. \end{aligned} \tag{8}$$

A dynamic adjustment model of supply response. In contrast, suppose that the desired or equilibrium quantity supplied, x_t^*, is linearly related to the price at the time of decision, p_{t-1}:

$$x_t^* = a + bp_{t-1} + u_t. \tag{9}$$

However, for a variety of technological and economic reasons (including plain habit), only a fraction, γ, of adjustment occurs each period:

$$x_t - x_{t-1} = \gamma(x_t^* - x_{t-1}), \qquad 0 < \gamma \leqslant 1, \tag{10}$$

hence

$$x_t = a\gamma + b\gamma p_{t-1} + (1-\gamma)x_{t-1} + u_t. \tag{11}$$

Apart from the residual terms, which differ in the two relations, both equations imply:

$$\begin{aligned} x_t = {} & a + b\alpha\,[p_{t-1} + (1-\alpha)p_{t-2} \\ & + (1-\alpha)^2 p_{t-3} + \cdots] + v_t, \end{aligned} \tag{12}$$

where α equals β for the expectational model, and α equals γ for the adjustment model. Equation (12) represents supply as a function of price taken with a Koyck distribution lag. In multiple-equation systems, however, the expectational and dynamic adjustment models do not lead to such similar results; in particular, in expectational models one can make use of the fact that the distribution of lag for the same variable in different equations should be the same (Nerlove 1958*a*, pp. 31–39).

Examples of further extension toward developing distributed lag relationships incidental to a more fundamental dynamic model of behavior are given by Jorgenson (1963) and Muth (1961). Jorgenson gives a model of investment behavior based on a theory of the demand for capital goods over time and on a theory of the relation between such demand and its translation into realized investment, which in turn rests on a distribution of times-to-completion of new investment projects. This model results in a case of a very general distribution of lag. Jorgenson calls the general case the "rational power series distribution"; the Pascal distribution discussed above is a member of this class. Jorgenson's distribution of lag, although general in form, results in fact from a highly particularized model of dynamic adjustment in the investment decision-and-realization process.

Muth elaborates on a model of expectation formation. In the highly simplified case of a single market with a nonstochastic demand function in which the quantity supplied is a linear function of expected price plus a residual generated by shocks of a permanent and transitory nature, the "rational expectations" that Muth develops can be shown to satisfy (7). Hence, the model leads to a distributed lag of the Koyck form. However, in more realistic models the rational expectations are no longer so simple.

Estimation problems. If one examines equations (8) or (11)—those that one might attempt to estimate by least-squares methods if one sought to determine both the distribution of lag and the long-run supply response—it can be seen that the presence of serial correlation in the residual terms will cause serious trouble. For example, least-squares estimates of autoregressions with serially correlated residual terms are known to be statistically inconsistent. Indeed, from (8) it appears that serially correlated residuals will be the rule. This is a subject that is too technical and complex for discussion here. The reader is referred to Koyck (1954, pp. 32–39); Griliches (1961); Klein (1958); Liviatan (1963); Malinvaud (1964); Phillips (1956); and especially Hannan (1964), who gives the most complete and fundamental discussion of this problem and its solution. While Hannan's results refer primarily to Koyck lags, they can be generalized.

MARC NERLOVE

BIBLIOGRAPHY

ALT, FRANZ L. 1942 Distributed Lags. *Econometrica* 10: 113–128.

BROWN, TILLMAN M. 1952 Habit Persistence and Lags in Consumer Behavior. *Econometrica* 20:355–371.

CAGAN, PHILLIP 1956 The Monetary Dynamics of Hyperinflation. Pages 23–117 in Milton Friedman (editor), *Studies in the Quantity Theory of Money.* Univ. of Chicago Press.

EISNER, ROBERT 1960 A Distributed Lag Investment Function. *Econometrica* 28:1–29.

FISHER, FRANKLIN M. 1962 Rigid Lags and the Estimation of "Long-run" Economic Reactions. Pages 21–47 in Franklin M. Fisher, *A Priori Information and Time Series Analysis: Essays in Economic Theory and Measurement.* Amsterdam: North-Holland Publishing.

FISHER, IRVING 1925 Our Unstable Dollar and the So-called Business Cycle. *Journal of the American Statistical Association* 20:179–202.

FISHER, IRVING 1937 Note on a Short-cut Method for Calculating Distributed Lags. International Statistical Institute, *Bulletin* 29, no. 3:323–328.

FRIEDMAN, MILTON 1957 *A Theory of the Consumption Function.* National Bureau of Economic Research, General Series, No. 63. Princeton Univ. Press.

GRILICHES, ZVI 1959 Distributed Lags, Disaggregation, and Regional Demand Functions for Fertilizer. *Journal of Farm Economics* 41:90–102.

GRILICHES, ZVI 1961 Note on Serial Correlation Bias in Estimates of Distributed Lags. *Econometrica* 29:65–73.

HANNAN, E. J. 1964 Estimation of Relationships Involving Distributed Lags. *Econometrica* 33:206–224.

JORGENSON, DALE W. 1963 Capital Theory and Investment Behavior. *American Economic Review* 53:247–259.

KLEIN, LAWRENCE R. 1958 The Estimation of Distributed Lags. *Econometrica* 26:553–565.

KOYCK, LEENDERT M. 1954 *Distributed Lags and Investment Analysis.* Amsterdam: North-Holland Publishing.

LIVIATAN, NISSAN 1963 Consistent Estimation of Distributed Lags. *International Economic Review* 4:44–52.

MALINVAUD, EDMOND 1964 Modèles à retards échelonnés. Pages 478–499 in Edmond Malinvaud, *Méthodes statistiques de l'économétrie.* Paris: Dunod.

MUTH, JOHN F. 1961 Rational Expectations and the Theory of Price Movements. *Econometrica* 29:315–335.

NERLOVE, MARC 1958*a* *Distributed Lags and Demand Analysis for Agricultural and Other Commodities.* U.S. Dept. of Agriculture, Handbook No. 141. Washington: Government Printing Office.

NERLOVE, MARC 1958*b* *The Dynamics of Supply: Estimation of Farmers' Response to Price.* Studies in Historical and Political Science, Series 76, No. 2. Baltimore: Johns Hopkins Press.

NERLOVE, MARC 1958*c* The Implications of Friedman's Permanent Income Hypothesis for Demand Analysis. *Agricultural Economics Research* 10:1–14.

NERLOVE, MARC; and WAUGH, FREDERICK V. 1961 Advertising Without Supply Control: Some Implications of a Study of the Advertising of Oranges. *Journal of Farm Economics* 43:813–837.

PHILLIPS, A. W. 1956 Some Notes on the Estimation of Time-forms of Reactions in Interdependent Dynamic Systems. *Economica* New Series 23:99–113.

SOLOW, ROBERT M. 1960 On a Family of Lag Distributions. *Econometrica* 28:393–406.

THEIL, H.; and STERN, ROBERT M. 1960 A Simple Unimodal Lag Distribution. *Metroeconomica* 12:111–119.

TINBERGEN, JAN 1949 Long-term Foreign Trade Elasticities. *Metroeconomica* 1:174–185.

Postscript

Although the main article was not published until 1968, it was written in 1963–1964, with references updated in 1964. Since that time there has been a vast amount of work on distributed lags, directed both to the theoretical problems of formulating models involving distributed lags and to the statistical problems of estimating the parameters of such models and making inferences about them. The existing literature is now too voluminous and detailed for a short postscript. Fortunately, three important surveys have been published.

Dhrymes (1971), despite his title, deals primarily with the statistical problems of estimation and inference in the context of distributed lag models of varying degrees of complexity and only peripherally with the theoretical bases of formulation of such models; he summarizes and expands upon much of the work completed on the subject prior to 1970, and presents a number of new results.

Nerlove (1972) argues strongly for a greater theoretical underpinning for distributed lag models and shows how much previous work has simply superimposed an ad hoc lag structure on what are basically static models, rather than developed true dynamic models of behavior that incorporate distributed lag relationships in an essential way. Emphasis is on the theoretical formulation of models, and an attempt is made to show how methods of statistical analysis flow naturally from those formulations, building on the mutually supportive roles of theory and measurement in economics.

Sims's masterful survey (1974) integrates the two approaches and brings the reader up to date with work completed by mid-1973. Sims covers (1) the relationship of distributed lag regressions to economic theory, extending Nerlove (1972) in several important ways; (2) the theory of estimation for regressions containing autoregressive terms, extending Dhrymes (1971); (3) frequency domain (or spectral) estimation methods; (4) specification errors with particular reference to seasonality and temporal aggregation; (5) collinearity problems and the use of Bayesian methods; (6) testing; and (7) systems of time series regressions involving feedback and mutual causality.

MARC NERLOVE

[*See also* TIME SERIES.]

ADDITIONAL BIBLIOGRAPHY

DHRYMES, PHOEBUS J. 1971 *Distributed Lags: Problems of Estimation and Formulation.* San Francisco: Holden-Day.

NERLOVE, MARC 1972 Lags in Economic Behavior. *Econometrica* 40:221–251.

SIMS, CHRISTOPHER A. 1974 Distributed Lags. Volume 2, chapter 5 in M. D. Intriligator and D. A. Kendrick (editors), *Frontiers of Quantitative Economics.* Amsterdam: North-Holland Publishing.

DISTRIBUTION PROCESSES

See SIZE DISTRIBUTIONS IN ECONOMICS. *Related material may be found in* RANK–SIZE RELATIONS.

DISTRIBUTION-FREE STATISTICS

See NONPARAMETRIC STATISTICS.

DISTRIBUTIONS, STATISTICAL

I SPECIAL DISCRETE DISTRIBUTIONS

The continuous distributions mentioned in this article, together with others, are described in detail in the companion article on SPECIAL CONTINUOUS DISTRIBUTIONS.

A discrete probability distribution gives the probability of every possible outcome of a discrete experiment, trial, or observation. By "discrete" it is meant that the possible outcomes are finite (or countably infinite) in number. As an illustration of the ideas involved, consider the act of throwing three coins onto a table. Before trying to assess the probability of various outcomes, it is necessary first to decide (as part of the definition of the experiment) what outcomes are possible. From the point of view of the dynamics of the falling coins, no two throws are exactly identical, and there are thus infinitely many possible outcomes; but if the experimenter is interested only in the heads and tails shown, he may wish to reduce the number of outcomes to the eight shown in Table 1.

Table 1 — Possible outcomes of the experiment of tossing three coins

Coin 1	Coin 2	Coin 3
Head	Head	Head
Head	Head	Tail
Head	Tail	Head
Head	Tail	Tail
Tail	Head	Head
Tail	Head	Tail
Tail	Tail	Head
Tail	Tail	Tail

If the coins are *fair*, that is, equally likely to show a head or tail, and if they behave independently, the eight possibilities are also all equally probable, and each would have assigned probability $\frac{1}{8}$. The list of eight $\frac{1}{8}$'s assigned to the eight outcomes is a discrete probability distribution that is exactly equivalent to the probabilistic content of the verbal description of the defining experiment.

The same physical activity (throwing three coins on a table) can be used to define another experiment. If the experimenter is interested only in the *number* of heads showing and not in which particular coins show the heads, he may define the possible outcomes to be only four in number: 0, 1, 2, 3. Since of the eight equally probable outcomes of the first experiment, one yields zero heads and one yields three heads while three yield one head and three yield two heads, the discrete probability distribution defined by the second experiment is

$$\begin{aligned} &\text{probability of no heads} = \tfrac{1}{8};\\ &\text{probability of one head} = \tfrac{3}{8};\\ &\text{probability of two heads} = \tfrac{3}{8};\\ &\text{probability of three heads} = \tfrac{1}{8}. \end{aligned}$$

It is frequently convenient to write such a list in

more compact form by using mathematical symbols. If p_n means the probability of n heads, the second experiment leads to the formula

$$(1) \qquad p_n = \frac{1}{8}\binom{3}{n}, \qquad n = 0, 1, 2, 3,$$

where $\binom{N}{n}$, the binomial coefficient, is the mathematical symbol for the number of combinations of N things taken n at a time,

$$\binom{N}{n} = \frac{N!}{n!(N-n)!},$$

$n!$ (which is read "n-factorial") being equal to $n(n-1)(n-2)\cdots 3\cdot 2\cdot 1$. Note that in (1) the form is given for the probabilities and then the domain of definition is stated.

A domain of definition of a probability distribution is merely a list of the possible outcomes of the experiment. In nearly every case, the experimenter finds it convenient to define the experiment so that the domain is numerical, as in the second experiment above. However, there is no reason why the domain should not be nonnumerical. In the first experiment above, the domain could be written (HHH), (HHT), (HTH), (HTT), (THH), (THT), (TTH), (TTT), where T means tails and H means heads.

An expression for p_n like the above needs only two properties to describe a valid discrete probability distribution. It must be nonnegative and the sum over all possible outcomes must be unity: $p_n \geqslant 0$, $\sum p_n = 1$. It is therefore possible to invent any number of discrete probability distributions simply by choosing some function p_n satisfying these simple conditions. In both theory and applications, it is more important to study certain probability distributions that correspond to basic probability experiments than to attempt a systematic classification of all functions satisfying these conditions. In the following list of important discrete distributions, the corresponding experiment is stated in very general terms, accompanied in many cases by concrete interpretations.

Discrete distributions

Binomial distribution. Suppose an experiment can have only two possible outcomes, conventionally called *success* and *failure*, and suppose the probability of success is p. If the experiment is repeated independently N times under exactly the same conditions, the number of successes, n, can be $0, 1, \cdots, N$. A composite experiment consisting of observing the number of successes in the N performances of the individual experiment can therefore have one of the following outcomes: $0, 1, 2, \cdots, N$. The probability of observing exactly n successes forms the binomial distribution:

$$(2) \qquad p_n = \binom{N}{n} p^n (1-p)^{N-n}, \qquad n = 0, 1, \cdots, N.$$

The second experiment mentioned in the introduction is of this form, where $N = 3$ and $p = \frac{1}{2}$.

The mean value of the binomial distribution is Np and its variance is $Np(1-p)$. The tails of the binomial distribution (and of the negative binomial discussed below) may be expressed as incomplete beta functions.

The following list of interpretations for the abstract words *success* and *failure* will indicate the breadth of application of the binomial distribution: male–female; infected–not infected; working–broken; pass–fail; alive–dead; head–tail; right–wrong. If a common test is known to give the right answer (using the last pair as an example) nine times out of ten on the average, and if it is performed ten times, what is the probability that it will give exactly nine right answers and one wrong one?

$$p_9 = \binom{10}{9}(.9)^9(.1)^1 = .38742.$$

In using the binomial distribution for this purpose, it is important to remember that the probability of a success must be constant. For example, if each of 30 students took an examination composed of 3 equally difficult parts and each separate part was graded pass or fail, the number of fails out of the 90 grades would not have a binomial distribution since the probability of failure varies from student to student.

If, for the binomial distribution, N is large and p is small, then the simpler Poisson distribution (see next section) often forms a good approximation. The sense of this approximation is a limiting one, with $N \to \infty$ and Np having a finite limit.

The normal distribution is often a good approximation to the binomial, especially when N is large and p not too close to 0 or 1. The limiting sense of the normal approximation is in terms of the *standardized* variable, n less its mean and divided by its standard deviation.

Poisson distribution. Imagine a machine turning out a continuous roll of wire or thread. Minor defects occur from time to time; their frequency and their statistical behavior are of interest. If the defects occur independently and if the risk of a defect is constant (that is, if the probability of a defect in a very small interval of the wire is a constant, λ, times the length of the interval), then the average number of defects per unit length is λ and the probability that a unit length contains exactly n defects is

$$(3) \qquad p_n = \frac{\lambda^n e^{-\lambda}}{n!}, \qquad n = 0, 1, 2, \cdots.$$

This distribution, named for the French mathematician Poisson, is an important example of a distribution defined over an infinite set of integers. Any value of n is possible, although large values (compared with λ) are unlikely. The mean and variance of the Poisson distribution are both λ. If the segment of unit length is replaced by a segment of length L, this is only a change in scale and one need only substitute $L\lambda$ for λ in (3) and in the moments. The tails of the Poisson distribution are incomplete gamma functions.

The Poisson distribution has a very wide range of application, not only because of its mathematical simplicity but because it represents one important concept of true randomness. In addition, the Poisson distribution, as has been noted, is useful as an approximation.

The most important interpretations of the Poisson distribution occur when the line is considered to be the axis of real time and the points to be times at which certain events take place. The following is a partial list of events that have been compared with the Poisson distribution: deaths by horse kick, radioactive particles entering a Geiger–Müller counter, arrival of patients into a doctor's waiting room, declarations of war, strikes of lightning, registration of vehicles over a road tape, demands for service at a telephone switchboard, demands for parking space, occurrence of accidents, demands for electric power, instants of nerve excitation, occurrence of suicide, etc. [*for a discussion of some of these applications, see* QUEUES].

On the other hand, the interpretation of the Poisson distribution as a criterion of perfect disorder remains valid in space of any number of dimensions, provided only that L is interpreted as length, area, or volume as the case may require. Space examples are perhaps not as numerous, but the following can be noted: one dimension, misprints per line, cars per mile; two dimensions, bomb bursts per square mile, weeds per square yard, stars per square unit of photographic plate; three dimensions, raisins per loaf of raisin bread, bacteria per cubic centimeter of fluid.

Morse distribution. The Poisson distribution is an example of a counting distribution. In a single dimension (whether time or space) it is characterized by the fact that the continuous density function (the gap distribution) of distance between consecutive points is negative exponential (with $\lambda = 1/\theta$). If the gap distribution is more general (for example, the Pearson Type III, with $\lambda = 1/\theta$), the counting distribution may be very complicated indeed. With Pearson Type III gaps, the counting probabilities are

$$p_n = (1+n)\,[\Gamma_{nr+r} - \Gamma_{nr}] + (1-n)\,[\Gamma_{nr} - \Gamma_{nr-r}] + (\lambda L/r)\,[-\Gamma_{nr+r-1} + 2\Gamma_{nr-1} - \Gamma_{nr-r-1}],$$

where

$$\Gamma_n = \frac{1}{(n-1)!}\int_{\lambda L}^{\infty} e^{-x} x^{n-1}\,dx.$$

This distribution takes its name from Philip Morse (1958), and reduces to the Poisson for $r = 1$. In data that fitted such a function, one would have a mixture of randomness and regularity, suggesting two factors at work.

Negative binomial distribution. There are two traditional important probability models leading to the negative binomial distribution, although others have been suggested. (*i*) If each member of a population experiences Poisson events (for example, accidents) but with the mean value λ varying statistically from member to member according to the Pearson Type III distribution, then the probability that n events occur in time T among all members of the population is

$$p_n = \binom{n+r-1}{r-1} p^r (1-p)^n, \qquad n = 0, 1, \cdots,$$

where $p = \theta/(T+\theta)$ and r and θ are the parameters of the Pearson Type III distribution. (*ii*) If experiments are performed in sequence with a fixed probability p of success (as for the binomial distribution), then the negative binomial distribution gives the probability that exactly n failures will precede the rth success. This is particularly useful for $r = 1$, since in many applications one calls a halt after the first "success," for example, in repeated dialings of a telephone that may be busy. The negative binomial distribution takes its name from its formal similarity to the binomial distribution.

The mean of the negative binomial distribution is $(1-p)r/p$ and its variance is $(1-p)r/p^2$. It is an example of a mixed distribution, obtained by assuming a parameter in one distribution to be itself subject to statistical fluctuation. [*Many other such distributions, including those called "contagious," are discussed in* DISTRIBUTIONS, STATISTICAL, *article on* MIXTURES OF DISTRIBUTIONS.]

Geometric distribution. In certain types of waiting lines, the probability of n persons in the queue is

$$p_n = \left(1 - \frac{\lambda}{\mu}\right)\left(\frac{\lambda}{\mu}\right)^n, \qquad n = 0, 1, 2, \cdots,$$

where λ is the mean arrival rate and μ is the mean service rate and $\lambda < \mu$. The mean value of the geometric distribution is $\lambda/(\mu - \lambda)$, and the variance is $\lambda\mu/(\mu - \lambda)^2$. The geometric distribution is a special case of the negative binomial distribution with $r = 1$ and $p = (\mu/\lambda) - 1$. [*These waiting line problems are discussed in* QUEUES.]

Hypergeometric distribution. Suppose a collection of N objects contains k of one kind and $N - k$ of another kind—for example, any of the dichotomies of the binomial distribution model. If exactly r objects are taken at random from the collection, the probability that n of them will be of the first kind is

$$(4) \quad p_n = \frac{\binom{k}{n}\binom{N-k}{r-n}}{\binom{N}{r}}, \quad n = 0, 1, \cdots, \min(r, k).$$

The mean of this distribution is kr/N and the variance is $kr(N - k)(N - r)/N^2$.

The binomial distribution and the hypergeometric distribution can be regarded as arising from analogous experiments, the former by sampling "with replacement" and the latter by sampling "without replacement." This can be seen from either of two facts: (*i*) if, in the probability experiment described for the hypergeometric distribution, each of the r objects is put back before the next one is chosen, the probability model leading to the binomial distribution would result; (*ii*) if $N \to \infty$ in equation (4), with $(N - k)/N$ replaced by p, the resulting limit is equation (2). Thus, if the total number of objects is very large and the proportions of the two types remain fixed, the hypergeometric model approaches the binomial model. If the number of objects of the first kind is very small in comparison with the total number of objects, the hypergeometric distribution will approach the Poisson. The hypergeometric distribution has important statistical application in Fisher's so-called exact test in a 2×2 table [*see* COUNTED DATA].

It is possible to obtain many other distributions, both discrete and continuous, as special or limiting cases of the hypergeometric. Karl Pearson used the probability model leading to the hypergeometric distribution as the basis for his analysis of density functions and obtained 12 separate types as special limiting cases. Of these, only the Pearson Type III is still frequently called by his designation, although many of the most common continuous densities (including the normal) fit into his classification system [*see* DISTRIBUTIONS, STATISTICAL, *article on* APPROXIMATIONS TO DISTRIBUTIONS].

Negative hypergeometric distribution. If the parameter p in a binomial distribution is itself beta distributed, then the unconditional distribution of the discrete variate is

$$p_n = \binom{N}{n}\frac{B(n + p, N + q - n)}{B(p, q)}, \quad n = 0, 1, \cdots, N,$$

where p and q are the parameters of the beta distribution and

$$B(p, q) = \int_0^1 x^{p-1}(1 - x)^{q-1}\,dx.$$

This model is the discrete time analog of the first interpretation of the negative binomial distribution; the events (such as accidents) would be counted in short time periods so that only a success or failure would be recorded.

Occupancy distribution. If each of k objects is thrown at random into one of N boxes, the probability that a given box contains n objects is

$$p_n = \binom{k}{n}\frac{(N - 1)^{k-n}}{N^k}, \qquad n = 0, 1, \cdots, k.$$

This distribution has mean k/N and variance $(k/N)(1 - 1/N)$. Problems of this type are called occupancy problems and are important in statistical mechanics.

Busy period distributions. A box contains r objects, and additional objects are thrown into the box at random instants, that is, in accordance with a Poisson distribution (mean λ). If one object is removed in every time interval of length $1/\mu$, then the probability that exactly n objects will pass through the box before it first becomes empty is

$$p_n = \frac{r}{(n - r)!}\, n^{n-r-1}\, e^{-\lambda n/\mu}\, (\lambda/\mu)^{n-r}, \qquad n = r, r + 1, r + 2, \cdots.$$

This is the Borel–Tanner distribution, with mean $r/(1 - \lambda/\mu)$ and variance $(\lambda r/\mu)/(1 - \lambda/\mu)^3$. If the objects are removed also in accordance with a Poisson distribution (mean μ), then the probability that n will pass through before the box first becomes empty is

$$p_n = \frac{r}{n}\binom{2n - r - 1}{n - 1}\frac{(\lambda/\mu)^{n-r}}{(1 + \lambda/\mu)^{2n-r}}, \qquad n = r, r + 1, \cdots,$$

which is called the Narayana distribution. This distribution has mean $r/(1 - \lambda/\mu)$ and variance $[(\lambda r/\mu)(1 - \lambda/\mu)]/(1 - \lambda/\mu)^3$.

Such distributions are important in the theory of queues, where the box represents the collection of people waiting. Then the number n, which has probability p_n, is the length of a busy period for the service mechanism, beginning with r in the system [*see* QUEUES].

Uniform distribution. An experiment with N equally likely outcomes corresponds to the discrete uniform (or discrete rectangular) distribution:

$$p_n = \frac{1}{N}, \qquad n = 1, 2, \cdots, N.$$

For a single throw of a die, $N = 6$. This distribution has mean $\frac{1}{2}(N+1)$ and variance $\frac{1}{12}(N^2 - 1)$.

Yule distribution. Some experimental data suggest that in a long list of names (for example, a telephone book) the probability that a randomly chosen name occurs n times is

$$p_n = \frac{k}{n}\binom{n+k}{k}^{-1}, \qquad n = 1, 2, \cdots.$$

Fisher distribution. In the distribution of frequency of plant species, the distribution

$$p_n = \frac{1}{n} a p^n, \qquad n = 1, 2, \cdots,$$

occurs, where $1/a = -\log_e(1-p)$. The Fisher distribution has mean $ap/(1-p)$ and variance $ap(1-ap)/(1-p)^2$. This distribution has also been called the logarithmic distribution because of its close relationship to the Taylor series for $-\log_e(1-p)$. Fisher's distribution can be obtained from the negative binomial by truncating the zero category and letting r approach zero.

Formation of distributions from others

The most important families of secondary distributions are the so-called mixed distributions, in which the parameter in a given distribution is itself subject to statistical fluctuation [*see* DISTRIBUTIONS, STATISTICAL, *article on* MIXTURES OF DISTRIBUTIONS]. Two examples above (the first case of the negative binomial and the negative hypergeometric) are also mixed. In addition, the following paragraphs illustrate other methods of obtaining secondary distributions.

Truncated distributions. It may happen that the domain of definition of a distribution does not exactly agree with some data, either for theoretical reasons or because part of the data is unobtainable, although the model is in other respects quite satisfactory. The most famous example of this is concerned with albinism. The Poisson distribution gives a satisfactory fit to the number of albino children born to parents genetically capable of producing albinos, except for the value of p_0, the probability of no albino children. This frequency cannot be observed, since such parents with no albino children are ordinarily indistinguishable from normal parents. Therefore, a new distribution is formed from the Poisson by removing the zero category, and dividing each probability by $1 - p_0$ so that the total sum remains one:

$$p_n = \frac{e^{-\lambda}\lambda^n}{n!\,(1-e^{-\lambda})}, \qquad n = 1, 2, 3, \cdots.$$

In certain applications, it has been useful to truncate similarly the binomial distribution. [*The practice, which has also been applied to continuous density functions, is discussed in* STATISTICAL ANALYSIS, SPECIAL PROBLEMS OF, *article on* TRUNCATION AND CENSORSHIP.]

Bissinger's system. In connection with an inventory problem, the distribution q_n, formed from a given distribution p_n by the transformation

$$q_n = \frac{1}{1-p_0}\sum_{j=n+1}^{\infty}\frac{p_j}{j}, \qquad n = 0, 1, \cdots,$$

has been useful. A more general transformation of this type is

$$q_n = \frac{\displaystyle\sum_{j=n+k}^{\infty}\frac{p_j}{j-k+1}}{\displaystyle\sum_{j=k}^{\infty}p_j}, \qquad n = 0, 1, 2, \cdots,$$

which reduces to the simpler form for $k = 1$. This transformation is discussed by Bissinger in Patil (1965).

Joint occupancy. Let p_n be the probability of an object of type I in a box, and let β be the probability that an object of type I brings an object of type II with it into the box. Then the probability q_n of a total of n objects in the box is

$$q_n = \sum_{i=0}^{[\frac{1}{2}n]}\binom{n-i}{i} p_{n-i}\,(1-\beta)^{n-2i}\beta^i, \qquad n = 0, 1, 2, \cdots,$$

where $[k]$ represents the integer part of k. This form of distribution has been applied to the number of persons in an automobile and to the number of persons in a group buying a railway ticket. The two types of objects might be male–female or adult–child.

Analogy with density functions. It is only necessary to adjust a constant in order to convert a continuous distribution into a discrete one. For example, the negative exponential density $\lambda e^{-\lambda x}$, usually defined over $0 < x < \infty$, can be applied to the domain $n = 0, 1, 2, \cdots$. However, since

$$\sum_{n=0}^{\infty} \lambda e^{-\lambda n} = \frac{\lambda}{1 - e^{-\lambda}},$$

the discrete probability distribution will be

$$p_n = (1 - e^{-\lambda}) e^{-\lambda n}, \qquad n = 0, 1, \cdots.$$

Several such discrete analogs, including the normal, have appeared in the literature.

Alternative descriptions of distributions

A probability distribution p_n can be transformed in many ways, and, provided the transformation is one-to-one, the result will characterize the distribution equally well. Some of the principal auxiliary functions are

$P_n = \sum_{i=0}^{n} p_i$, the cumulative distribution, or left tail;

$Q_n = \sum_{i=n+1}^{\infty} p_i = 1 - P_n$, the right tail of the distribution;

$\phi(s) = \sum_{i=0}^{\infty} p_i s^i$, the generating function;

$\alpha(s) = \sum_{i=0}^{\infty} p_i s^i / i!$, the exponential generating function;

$\beta(s) = \sum_{i=0}^{\infty} e^{is} p_i$, the moment generating function;

$\sigma(s) = \sum_{i=0}^{\infty} (1 + s)^i p_i$, the factorial moment generating function.

The last two functions are so named because in the corresponding power series expansion the coefficients involve, respectively, the central moments and the factorial moments.

For certain mathematical operations on discrete probability distributions, such as the calculation of moments, generating functions are extremely useful. In many cases the best way to obtain p_n from a defining experiment is to calculate $\phi(s)$ first. Many examples of this type of argument can be found in Riordan (1958).

Multivariate distributions

There have been a few discrete multivariate distributions proposed, but only one, the multinomial, has been very widely applied.

Multinomial distribution. If an experiment can have r possible outcomes, with probabilities $p_1, p_2, \cdots, p_r$ and if it is repeated under the same conditions N times, the probability that the jth outcome will occur n_j times, $j = 1, 2, \cdots, r$, is

$$\frac{N!}{n_1!\, n_2! \cdots n_r!} p_1^{n_1} p_2^{n_2} \cdots p_r^{n_r},$$

which reduces to the binomial for $r = 2$.

Negative multinomial distribution. If, in the scheme given above, trials stop after the rth outcome has been observed exactly k times, the joint probability that the jth outcome will be observed n_j times ($j = 1, \cdots, r-1$) is called the negative multinomial distribution.

Bivariate Poisson distribution. The joint probability

$$p_{n_1, n_2} = \exp[-(\lambda + \mu + \rho)] \cdot \lambda^{n_1} \mu^{n_2} \sum_{j=0}^{\min(n_1, n_2)} \frac{\rho^j}{(n_1 - j)!\,(n_2 - j)!\, j!}$$

has been proposed as a generalization of the Poisson; when $\rho = 0$, n_1 and n_2 are independently Poisson.

FRANK A. HAIGHT

[*Other relevant material may be found in* PROBABILITY *and in the biographies of* FISHER, R. A.; PEARSON; POISSON.]

BIBLIOGRAPHY

By far the best textbook on the theory and application of discrete distributions is Feller 1950. *Busy period distributions and other distributions arising from queueing theory will be found in* Takács 1962. Riordan 1958 *explains the various auxiliary functions and the relationships between them. An index to all distributions with complete references to the literature is given in* Haight 1961. *The most complete and useful volume of statistical tables is* Owen 1962. *Further references to tables may be found in* Greenwood & Hartley 1962. *Computer programs for generating tables are distributed on a cooperative basis by SHARE Distribution Agency, International Business Machines Corporation.* Patil 1965 *contains several research and expository papers discussing the probabilistic models, structural relations, statistical theory, and methods for many of the discrete distributions mentioned in the list above, together with a bibliography on the subject by the editor.*

FELLER, WILLIAM 1950–1966 *An Introduction to Probability Theory and Its Applications.* 2 vols. New York: Wiley. → A second edition of Volume I was published in 1957.

GREENWOOD, JOSEPH A.; and HARTLEY, H. O. 1962 *Guide to Tables in Mathematical Statistics.* Princeton Univ. Press.

HAIGHT, FRANK A. 1961 Index to the Distributions of Mathematical Statistics. U.S. National Bureau of Standards, *Journal of Research* Series B: Mathematics and Mathematical Physics 65B:23–60.

MORSE, PHILIP M. 1958 *Queues, Inventories and Maintenance: The Analysis of Operational Systems With Variable Demand and Supply.* New York: Wiley.

OWEN, DONALD B. 1962 *Handbook of Statistical Tables.* Reading, Mass.: Addison-Wesley. → A list of addenda and errata is available from the author.

PATIL, GANAPATI P. (editor) 1965 *Classical and Contagious Discrete Distributions.* Proceedings of the International Symposium held at McGill University, Montreal, Canada, August 15–August 20, 1963. Calcutta (India): Statistical Publishing Society; distributed by Pergamon Press. → See especially pages 15–17 on "A

Type-resisting Distribution Generated From Considerations of an Inventory Decision Model" by Bernard H. Bissinger.

RIORDAN, JOHN 1958 *An Introduction to Combinatorial Analysis.* New York: Wiley.

TAKÁCS, LAJOS 1962 *Introduction to the Theory of Queues.* New York: Oxford Univ. Press.

ADDITIONAL BIBLIOGRAPHY

HAIGHT, FRANK A. 1967 *Handbook of the Poisson Distribution.* New York: Wiley.

JOHNSON, NORMAN L.; and KOTZ, SAMUEL 1969 *Discrete Distributions.* Boston: Houghton Mifflin.

PATIL, GANAPATI P.; and JOSHI, S. W. 1968 *A Dictionary and Bibliography of Discrete Distributions.* Edinburgh: Oliver & Boyd.

II
SPECIAL CONTINUOUS DISTRIBUTIONS

This article describes, and gives the more important properties of, the major continuous distributions that arise in statistics. It is intended as both an overview for the reader generally interested in distributions and as a reference for a reader seeking the form of a particular distribution.

Technical terms, such as "density function," "cumulative distribution function," etc., are explained elsewhere [*see* PROBABILITY].

The present article is, with a few exceptions, restricted to univariate distributions. Further specific references to numerical tabulations of distributions are not generally given here; most of the distributions are tabulated in Owen (1962) or in Pearson and Hartley (1954), and full references to other tabulations are given in Greenwood and Hartley (1962). Tables of many functions discussed here and an extensive reference list to tables are given in Zelen and Severo (1964). An index to properties of distributions is given in Haight (1961).

Normal distributions. The most important family of continuous distributions is that of the normal (or Gaussian) probability distributions. [*See* PROBABILITY, *article on* FORMAL PROBABILITY, *for more of the many properties of the normal distributions.*]

The normal probability density function is

$$\frac{1}{\sigma\sqrt{2\pi}}\exp\left[-\tfrac{1}{2}\left(\frac{x-\mu}{\sigma}\right)^2\right], \qquad -\infty < x < +\infty,$$

where μ is the mean (and also here the median and the mode) and σ is the standard deviation. Figure 1 shows the shape of this density. Note that it is symmetric about μ, that is, $f(\mu + x) = f(\mu - x)$, and that the density is essentially zero for $x > \mu + 3\sigma$ and $x < \mu - 3\sigma$. The normal distribution is sometimes said to be "bell-shaped," but note that there are many nonnormal bell-shaped distributions. The normal distribution is "standardized" or "normalized" by the transformation $z = (x - \mu)/\sigma$, which gives the standard–normal density,

$$\frac{1}{\sqrt{2\pi}}\exp(-\tfrac{1}{2}z^2).$$

A standardized normal random variable is also referred to as a "unit" normal since the mean of the standardized form is zero and the variance is one. The cumulative distribution function for the normal distribution in standardized form is

$$\frac{1}{\sqrt{2\pi}}\int_{-\infty}^{z}\exp(-\tfrac{1}{2}y^2)dy.$$

Thus, to find the probability that a normal random variable with mean μ and standard deviation σ is less than x, first compute the number of standard deviations x is away from μ, that is, let $z = (x - \mu)/\sigma$. Probabilities associated with x then may be read from tables of the standardized distribution. Care must be exercised to determine what is tabulated in any particular table. Various tables give the following: the cumulative probability; the probability in the right-hand tail only; the sum of the probabilities in the two tails; the central probability, that is, the probability that the absolute value of the random variable is less than the argument; and others.

○ A great many probability distributions are derived from the normal distribution; see Eisenhart and Zelen (1958); Kendall and Stuart ([1958–1966] 1973–1977) Korn and Korn (1961); Zelen

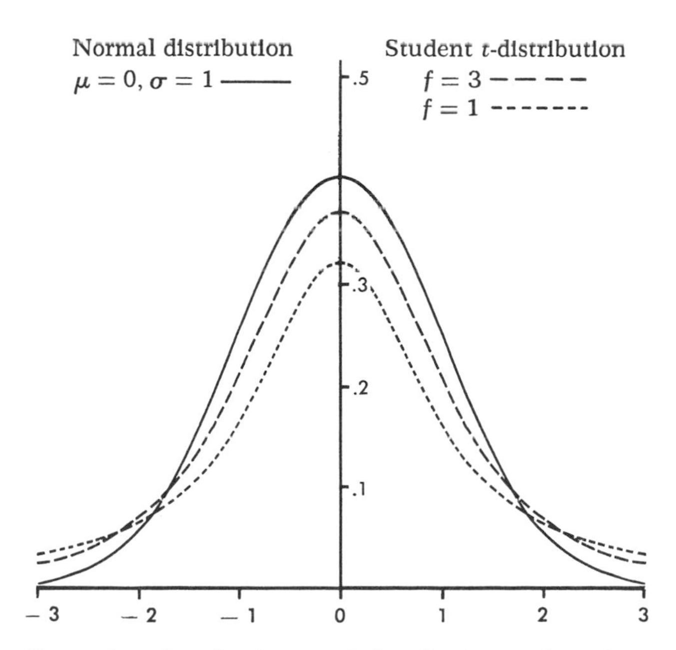

Figure 1 — Standard normal distribution and Student t-distribution showing density on the vertical scale corresponding to argument on the horizontal scale

and Severo (1964). These references also list additional continuous distributions not covered here.

Normal distributions are sometimes called Gaussian, sometimes Laplace–Gauss distributions, and sometimes distributions following Laplace's second law. This terminology reflects a tangled and often misstated history [*see* Walker 1929, chapter 2; *see also* GAUSS; LAPLACE].

Chi-square distributions. If X is unit normally distributed, X^2 has a chi-square distribution with one degree of freedom. If $X_1, X_2, \cdots, X_f$ are all unit normally distributed and independent, then the sum of squares $X_1^2 + X_2^2 + \cdots + X_f^2$ has a chi-square distribution with f degrees of freedom.

The probability density function for a chi-square random variable is

$$\frac{1}{2^{\frac{1}{2}f}\Gamma(\frac{1}{2}f)} x^{\frac{1}{2}(f-2)}e^{-\frac{1}{2}x}, \qquad 0 \leqslant x < \infty.$$

This is also a probability density function when f is positive, but not integral, and a chi-square distribution with fractional degrees of freedom is simply defined via the density function.

The mean of a chi-square distribution is f, and its variance is $2f$. For $f > 2$, the chi-square distributions have their mode at $f - 2$. For $0 < f \leqslant 2$, the densities are J-shaped and are maximum at the origin. Figure 2 shows the shape of the chi-square distributions.

If Y_1, Y_2 are independent and chi-square distributed with f_1, f_2 degrees of freedom, then $Y_1 + Y_2$ is chi-square with $f_1 + f_2$ degrees of freedom. This additivity property extends to any finite number of independent, chi-square summands.

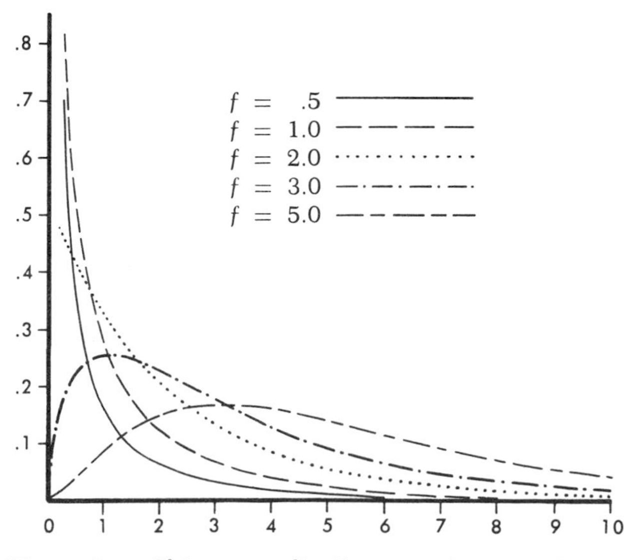

Figure 2 — Chi-square distributions showing density on the vertical scale corresponding to argument on the horizontal scale

If Y has a chi-square distribution with f degrees of freedom, then Y/f has a *mean-square distribution*, or a chi-square divided by degrees of freedom distribution.

The cumulative distribution function for the chi-square random variable with an even number of degrees of freedom (equal to $2a$) is related to the Poisson cumulative distribution function with mean λ as follows:

$$\sum_{j=0}^{a-1} \frac{e^{-\lambda}\lambda^j}{j!} = \frac{1}{2^{\frac{1}{2}f}\Gamma(\frac{1}{2}f)} \int_{2\lambda}^{\infty} x^{\frac{1}{2}(f-2)}e^{-\frac{1}{2}x}\,dx,$$

where $f = 2a$.

Gamma or Pearson Type III distributions. The gamma distribution, a generalization of the chi-square distribution, has probability density function

$$\frac{1}{\theta\Gamma(r)}\left(\frac{x}{\theta}\right)^{r-1} \exp\left(-\frac{x}{\theta}\right),$$

where $0 < \theta < \infty$, $0 < r < \infty$, and $0 < x < \infty$. Note that r does not have to be an integer. The mean of this distribution is $r\theta$, and the variance is $r\theta^2$. A simple modification permits shifting the left end-point from zero to any other value in this distribution and in several others discussed below. If $r = \frac{1}{2}f$ and $\theta = 2$, the gamma distribution reduces to a chi-square distribution. If X has a chi-square distribution with f degrees of freedom, then $Y = \frac{1}{2}\theta X$ has a gamma distribution with parameters θ, $\frac{1}{2}f$. If Y has a gamma distribution with parameters θ, r, then $2\theta^{-1}Y$ has a chi-square distribution with $2r$ degrees of freedom.

Negative exponential distributions. Negative exponential distributions are special cases of gamma distributions with $r = 1$. The probability density is $(1/\theta)\exp(-x/\theta)$ for $0 \leqslant x < \infty$, $0 < \theta < \infty$; the mean is θ and the variance is θ^2. If $\theta = 2$, the negative exponential distribution reduces to a chi-square distribution with 2 degrees of freedom.

The cumulative distribution is

$$1 - \exp(-x/\theta), \qquad 0 \leqslant x < \infty,\ 0 < \theta < \infty.$$

This distribution has been widely used to represent the distribution of lives of certain manufactured goods, for example, light bulbs, radio tubes, etc. [*see* QUALITY CONTROL, STATISTICAL *article on* RELIABILITY AND LIFE TESTING].

Suppose that the probability that an item will function over the time period t to $t + \Delta t$ is independent of t, given that the item is functioning at time t. In other words, suppose that the age of an item does not affect the probability that it continues to function over any specified length of future time provided the item is operating at present. In still other terms, if X is a random variable denoting length of life for an item, suppose that

$$Pr\{X \geqslant x \mid X \geqslant \xi\} = Pr\{X \geqslant x - \xi\}$$

for all ξ and $x > \xi$. This "constant risk property" obtains if, and only if, X has a negative exponential distribution. The underlying temporal process is called a stationary Poisson process; and, for a Poisson process, the number of failures (or deaths) in any given interval of time has a Poisson distribution [*see* QUEUES *and the biography of* POISSON].

Noncentral chi-square distributions. If X_1, X_2, $\cdots$, X_f are all normally distributed and independent, and if X_i has mean μ_i and variance one, then the sum $X_1^2 + X_2^2 + \cdots + X_f^2$ has a noncentral chi-square distribution with f degrees of freedom and noncentrality parameter $\lambda = \mu_1^2 + \mu_2^2 + \cdots + \mu_f^2$. This family of distributions can be extended to nonintegral values of f by noting that the density function (not given here) obtained for integral values of f is still a density function for the nonintegral values of f. The mean of the noncentral chi-square distribution is $f + \lambda$ and the variance is $2(f + 2\lambda)$.

Perhaps the main statistical use of the noncentral chi-square distribution is in connection with the power of standard tests for counted data [*see* COUNTED DATA].

The distribution also arises in bombing studies. For example, the proportion of a circular target destroyed by a bomb with a circular effects region may be obtained from the noncentral chi-square distribution with two degrees of freedom if the aiming errors follow a circular normal distribution. (A circular normal distribution is a bivariate normal distribution, discussed below, with $\rho = 0$ and $\sigma_X = \sigma_Y$.)

Noncentral chi-square distributions have an additivity property similar to that of (central) chi-square distributions.

Weibull distributions. A random variable X has a Weibull distribution if its probability density function is of the form

$$\frac{r}{\theta}\left(\frac{x}{\theta}\right)^{r-1} \exp\left(-\frac{x}{\theta}\right)^{r},$$

where $0 \leqslant x < \infty$, $0 < \theta < \infty$ and $r \geqslant 1$. This means that random variables with Weibull distributions can be obtained by starting with negative exponential random variables and raising them to powers $\geqslant 1$. If in particular $r = 1$, the Weibull distribution reduces to a negative exponential distribution. The Weibull distributions are widely used to represent the distribution of lives of various manufactured products. The mean of X is $\theta\Gamma[(r+1)/r]$, and $\theta^2\{\Gamma[(r+2)/r] - \Gamma^2[(r+1)/r]\}$ is the variance of X.

Student (or *t*-) distributions. If X has a unit normal distribution and if Y is distributed independently of X according to a chi-square distribution with f degrees of freedom, then $X/\sqrt{Y/f}$ has a Student (or t-) distribution with f degrees of freedom. Note that f need not be an integer since the degrees of freedom for chi-square need not be integral. The only restriction is $0 < f < \infty$. The density for a random variable having the Student distribution is

$$\frac{\Gamma(\frac{1}{2}f + \frac{1}{2})}{\sqrt{f\pi}\,\Gamma(\frac{1}{2}f)}\left(1 + \frac{x^2}{f}\right)^{-\frac{1}{2}(f+1)},$$

where $-\infty < x < +\infty$. A graph of the density functions for $f = 1$ and $f = 3$ is shown in Figure 1. Note that the density is symmetric about zero. As f approaches ∞, the Student density approaches the unit normal density.

The rth moment of the Student distribution exists if and only if $r < f$. Thus for $f \geqslant 2$, the mean is zero; and for $f \geqslant 3$, the variance is $f/(f-2)$. The median and the mode are zero for all f.

The Student distribution is named after W. S. Gosset, who wrote under the pseudonym Student. Gosset's development of the t-distribution, as it arises in dealing with normal means, is often considered to be the start of modern "small sample" mathematical statistics [*see* GOSSET].

The sample correlation coefficient, r, based on n pairs of observations from a bivariate normal population, may be reduced to a Student t-statistic with $n - 2$ degrees of freedom, when the population correlation coefficient is zero, by the transformation $t = (r\sqrt{n-2})/\sqrt{1-r^2}$ [*see* MULTIVARIATE ANALYSIS, *article on* CORRELATION METHODS].

Cauchy distributions. The Cauchy distribution is an example of a distribution for which no moments exist. The probability density function is

$$\frac{1}{\pi\beta} \cdot \frac{1}{1 + [(x-\lambda)/\beta]^2},$$

where $-\infty < x < +\infty$, $\beta > 0$, and $-\infty < \lambda < +\infty$. The cumulative probability distribution is

$$Pr\{X \leqslant x\} = \frac{1}{\pi}\arctan\left(\frac{x-\lambda}{\beta}\right) + \tfrac{1}{2}.$$

The median and the mode of this distribution are at $x = \lambda$. For $\beta = 1$ and $\lambda = 0$, the Cauchy distribution is also a Student t-distribution with $f = 1$ degree of freedom.

Noncentral *t*-distributions. If X is a unit normal random variable and if Y is distributed independently of X according to a chi-square distribution with f degrees of freedom, then $(X + \delta)/\sqrt{Y/f}$ has a noncentral t-distribution with f degrees of freedom and noncentrality parameter δ where $0 < f < \infty$ and $-\infty < \delta < \infty$. The mean for $f \geqslant 2$ of this distribution is $c_{11}\delta$ and the variance for $f \geqslant 3$ is $c_{22}\delta^2 + c_{20}$, where $c_{11} = \sqrt{f/2}\,\Gamma(\frac{1}{2}f - \frac{1}{2})/\Gamma(\frac{1}{2}f)$, $c_{22} = [f/(f-2)] - c_{11}^2$, and $c_{20} = f/(f-2)$. If $\delta = 0$, the noncentral t-distribution reduces to the Student t-distribution. Note that the moments do not exist for $f = 1$; despite this similarity, the noncentral t-distribution with $\delta \neq 0$ is not a Cauchy distribution. The noncentral t-distribution arises when considering one-sided tolerance limits on a normal distribution and in power computations for the Student t-test.

***F*-distributions.** If Y_1 has a chi-square distribution with f_1 degrees of freedom and Y_2 has a chi-square distribution with f_2 degrees of freedom, and Y_1 and Y_2 are independent, then $(Y_1/f_1)/(Y_2/f_2)$ has an F-distribution with f_1 degrees of freedom for the numerator and f_2 degrees of freedom for the denominator. The F-distributions are also known as Snedecor's F-distributions and variance ratio distributions. They arise as the distributions of the ratios of many of the mean squares in the analysis of variance [*see* LINEAR HYPOTHESES, *article on* ANALYSIS OF VARIANCE].

The density of F is

$$\frac{\Gamma(\frac{1}{2}f_1 + \frac{1}{2}f_2) f_1^{\frac{1}{2}f_1} f_2^{\frac{1}{2}f_2}}{\Gamma(\frac{1}{2}f_1)\Gamma(\frac{1}{2}f_2)} \cdot \frac{x^{(\frac{1}{2}f_1 - 1)}}{(f_2 + f_1 x)^{\frac{1}{2}(f_1+f_2)}},$$

where $0 < x < \infty$, $0 < f_1 < \infty$ and $0 < f_2 < \infty$. Figure 3 shows a plot of this density for four cases: $f_1 = f_2 = 1$, $f_1 = f_2 = 2$, $f_1 = 3$, $f_2 = 5$, and $f_1 = f_2 = 10$. Of these the cases of $f_1 = 3$, $f_2 = 5$, and $f_1 = f_2 = 10$ are most typical of F-distributions.

The mean of this distribution is $f_2/(f_2 - 2)$ for $f_2 > 2$; for $f_2 > 4$ the variance is given by $2f_2^2(f_1 + f_2 - 2)/[f_1(f_2-2)^2(f_2-4)]$; and the mode is $f_2(f_1 - 2)/[f_1(f_2+2)]$ for $f_1 > 2$. The F-distribution is J-shaped if $f_1 \leqslant 2$. When f_1 and f_2 are greater than 2, the F-distribution has its mode below $x = 1$ and its mean above $x = 1$ and, hence, is positively skew.

Let F_{f_1,f_2} represent a random variable having an F-distribution with f_1 degrees of freedom for the numerator and f_2 degrees of freedom for the denominator. Then $Pr\{F_{f_1,f_2} \leqslant c\} = 1 - Pr\{F_{f_2,f_1} \leqslant 1/c\}$. Hence the F-distribution is usually tabulated for one tail (usually the upper tail), as the other tail is easily obtained from the one tabulated.

The chi-square and t-distributions are related to the F-distributions as follows: $f_2/F_{\infty,f_2}$ has a chi-square distribution with f_2 degrees of freedom; $f_1F_{f_1,\infty}$ has a chi-square distribution with f_1 degrees

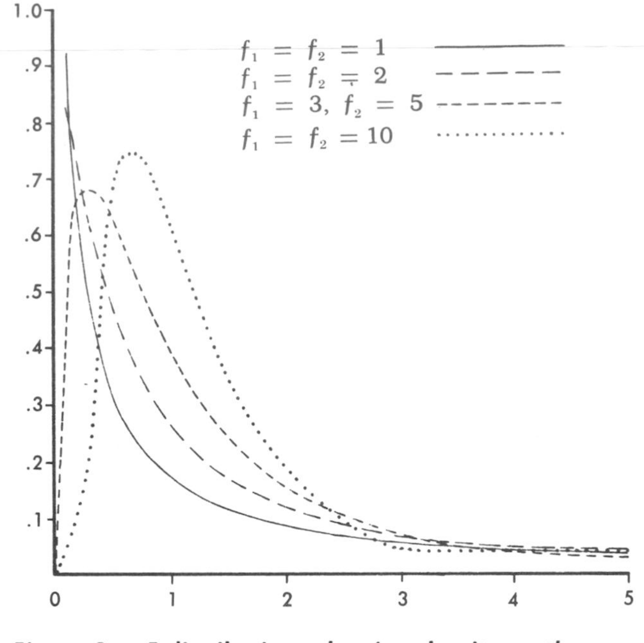

Figure 3 — F-distributions showing density on the vertical scale corresponding to argument on the horizontal scale

of freedom; F_{1,f_2} is distributed as the *square* of a Student-distributed random variable with f_2 degrees of freedom; $1/F_{f_1,\infty}$ is distributed as the square of a Student-distributed random variable with f_1 degrees of freedom.

Let $E(n,r,p) = \sum_{i=r}^{n} \binom{n}{i} p^i (1-p)^{n-i}$, that is, let $E(n,r,p)$ be the probability of r or more successes for a binomial probability distribution where p is the probability of success of a single trial [*see* DISTRIBUTIONS, STATISTICAL, *article on* SPECIAL DISCRETE DISTRIBUTIONS]. Let c_{v,f_1,f_2} be defined by $Pr\{F_{f_1,f_2} \leqslant c_{v,f_1,f_2}\} = v$. Then the following relationship exists between the binomial and F-distributions. If $E(n,r,p) = v$ then

$$p = \frac{r}{r + (n - r + 1)c_{1-v,2(n-r+1),2r}}.$$

A slight variation on the F-distribution obtained by the change of variable $Y = [f_1/f_2]X$ is often referred to as the inverted beta distribution, the beta distribution of the second kind, or the beta prime distribution. These names are also occasionally applied to the F-distribution itself. Another variation on the F-distribution is Fisher's z-distribution, which is that of $\frac{1}{2}\ln F$.

Beta distributions. A random variable, X, has a beta distribution with parameters p and q if its density is of the form

$$\frac{\Gamma(p+q)}{\Gamma(p)\Gamma(q)} x^{p-1}(1-x)^{q-1},$$

where $0 < x < 1$; $p,q > 0$. If the transformation

$X = f_1F_{f_1,f_2}/(f_2 + f_1F_{f_1,f_2})$ is made, then X has a beta distribution with $p = \frac{1}{2}f_1$ and $q = \frac{1}{2}f_2$.

The cumulative beta distribution is known as the incomplete beta function,

$$I_x(p,q) = \frac{\Gamma(p+q)}{\Gamma(p)\Gamma(q)} \int_0^x u^{p-1}(1-u)^{q-1}\,du.$$

The incomplete beta function has been tabulated by Karl Pearson (see Pearson 1934). The relationship $I_x(p,q) = 1 - I_{1-x}(q,p)$ is often useful. The mean of the beta distribution is $p/(p+q)$, and the variance is $pq/[(p+q)^2(p+q+1)]$.

The mode for $p \geqslant 1$ and $q \geqslant 1$, but p and q not both equal to 1, is $(p-1)/(p+q-2)$. For $p \leqslant 1$ and $q > 1$, the density is in the shape of a reversed J; for $p > 1$ and $q \leqslant 1$ the density is J-shaped; and for $p < 1$ and $q < 1$, the density is U-shaped. For both p and $q = 1$, the density takes the form of the density of the rectangular distributions with $a = 0$ and $b = 1$. The beta distribution is also known as the beta distribution of the first kind or the incomplete beta distribution.

Let $E(n,r,x) = \sum_{i=r}^{n} \binom{n}{i} x^i(1-x)^{n-i}$, that is, let $E(n,r,x)$ be the probability of r or more successes in n trials for a binomial probability distribution where x is the probability of success of a single trial. Then $I_x(p,q) = E(p+q-1,p,x)$. In other words, partial binomial sums are expressible directly in terms of the incomplete beta function. To solve $I_x(p,q) = \gamma$ for x, with γ, p, and q fixed, find x from $x = p/[p + qc_{1-\gamma,2q,2p}]$ where c_{γ,f_1,f_2} is defined by $Pr\{F_{f_1,f_2} \leqslant c_{\gamma,f_1,f_2}\} = \gamma$, that is, c_{γ,f_1,f_2} is a percentage point of the F-distribution.

If Y_1 and Y_2 are independent random variables having gamma distributions with equal values of θ, $r = n_1$ for Y_1, and $r = n_2$ for Y_2, then the variable $X = Y_1/(Y_1 + Y_2)$ has a beta distribution with $p = n_1$ and $q = n_2$.

Bivariate beta distributions. The random variables X and Y are said to have a joint bivariate beta distribution if the joint probability density function for X and Y is given by

$$\frac{\Gamma(f_1+f_2+f_3)}{\Gamma(f_1)\Gamma(f_2)\Gamma(f_3)}\, x^{f_1-1}y^{f_2-1}(1-x-y)^{f_3-1}$$

for $x,y > 0$ and $x + y < 1$, $0 < f_1 < \infty$, $0 < f_2 < \infty$, and $0 < f_3 < \infty$. This distribution is also known as the bivariate Dirichlet distribution. The mean of X is given by f_1/f_+, and the mean of Y by f_2/f_+, where f_+ represents $(f_1 + f_2 + f_3)$. The variance of X is $f_1(f_2+f_3)/f_+^2(f_+ + 1)$; the variance of Y is $f_2(f_1+f_3)/f_+^2(f_+ + 1)$; the correlation between X and Y is $-f_1f_2/\sqrt{f_1f_2(f_2+f_3)(f_1+f_3)}$. The conditional distribution of $X/(1-Y)$, given Y, is a beta distribution with $p = f_1$ and $q = f_3$. The sum $X + Y$ has a beta distribution with $p = f_1 + f_2$ and $q = f_3$.

Noncentral *F*-distributions. If Y_1 has a noncentral chi-square distribution with f_1 degrees of freedom and noncentrality parameter λ, and if Y_2 has a (central) chi-square distribution with f_2 degrees of freedom and is independent of Y_1, then $(Y_1/f_1)/(Y_2/f_2)$ has a noncentral F-distribution with f_1 degrees of freedom for the numerator and f_2 degrees of freedom for the denominator and noncentrality parameter λ. The cumulative distribution function of noncentral F may be closely approximated by a central F cumulative distribution function as follows:

$$Pr\{F^N_{f_1,f_2,\lambda} \leqslant c\} \cong Pr\left\{F_{f^*,f_2} \leqslant \frac{f_1c}{f_1+\lambda}\right\},$$

where $F^N_{f_1,f_2,\lambda}$ has a noncentral F-distribution and F_{f^*,f_2} has a central F-distribution and f^* denotes $(f_1+\lambda)^2/(f_1+2\lambda)$. The mean of the noncentral F-distribution is $f_2(f_1+\lambda)/[(f_2-2)f_1]$ for $f_2 > 2$, and the variance is

$$\frac{2f_2^2[(f_1+\lambda)^2 + (f_1+2\lambda)(f_2-2)]}{(f_2-2)^2f_1^2(f_2-4)}$$

for $f_2 > 4$. The means do not exist if $f_2 \leqslant 2$ and the variances do not exist if $f_2 \leqslant 4$.

The ratio of two noncentral chi-square random variables also arises occasionally. This distribution has not yet been given a specific name. The noncentral F-distribution is, of course, a special case of the distribution of this ratio. The noncentral F-distribution arises in considering the power of analysis of variance tests.

Bivariate normal distributions. To say that X and Y have a joint (nonsingular) normal distribution with means μ_X and μ_Y, variances σ_X^2 and σ_Y^2, and correlation ρ is to say that the joint probability density function for X and Y is

$$\frac{1}{2\pi\sigma_X\sigma_Y\sqrt{1-\rho^2}} \cdot \exp\left[\frac{\left(\frac{x-\mu_X}{\sigma_X}\right)^2 - 2\rho\left(\frac{x-\mu_X}{\sigma_X}\right)\left(\frac{y-\mu_Y}{\sigma_Y}\right) + \left(\frac{y-\mu_Y}{\sigma_Y}\right)^2}{-2(1-\rho^2)}\right].$$

The cumulative distribution function occurs in many problems; it is a special case (two-dimensional) of the multivariate normal distribution. A fundamental fact is that X and Y are jointly normal if and only if $aX + bY$ is normal for every a and b [*see* MULTIVARIATE ANALYSIS].

Distributions of the sum of normal variables. Let $X_1, X_2, \cdots, X_n$ be jointly normally distributed random variables so that X_i has mean μ_i and variance σ_i^2 and the correlation between X_i and X_j is ρ_{ij}. Then the distribution of the weighted sum, $a_1X_1 + a_2X_2 + \cdots + a_nX_n$, where $a_1, a_2, \cdots, a_n$ are any real constants (positive, negative, or zero), is normal with mean $a_1\mu_1 + a_2\mu_2 + \cdots + a_n\mu_n$ and variance

$$\sum_{i=1}^{n} a_i^2\sigma_i^2 + 2\sum_{i=1}^{n}\sum_{j=i+1}^{n} a_ia_j\sigma_i\sigma_j\rho_{ij}.$$

If the normality assumption is dropped, then the means and variances remain as stated, but the form of the distribution of the sum $a_1X_1 + a_2X_2 + \cdots + a_nX_n$ is often different from the distribution of the X's. If a linear function, $\sum_{i=1}^{n} a_iX_i$, $a_i \neq 0$, of a finite number of independent random variables is normally distributed, then each of the random variables $X_1, X_2, \cdots, X_n$ is also normally distributed. Note that in this instance independence of the random variables is required.

In particular, if X_1 and X_2 are jointly normally distributed, the sum $X_1 + X_2$ is normally distributed with mean $\mu_1 + \mu_2$ and variance $\sigma_1^2 + \sigma_2^2 + 2\rho_{12}\sigma_1\sigma_2$. The difference $X_1 - X_2$ is normally distributed with mean $\mu_1 - \mu_2$ and variance $\sigma_1^2 + \sigma_2^2 - 2\rho_{12}\sigma_1\sigma_2$. If $X_1, X_2, \cdots, X_n$ are jointly normally distributed with common mean μ and variance σ^2 and are independent, then the mean of the X's, that is, $\bar{X} = (1/n)\sum_{i=1}^{n} X_i$, is normally distributed with mean μ variance σ^2/n.

Rectangular distributions. The rectangular (or uniform) distribution has the following density: it is zero for $x < a$; it is $1/(b - a)$ for $a \leqslant x \leqslant b$; and it is zero for $x > b$, where a and b are real constants. In other words, it has a graph that is a rectangle with base of length $b - a$ and height of $1/(b - a)$. The cumulative distribution function is zero for $x < a$; it is $(x - a)/(b - a)$ for $a \leqslant x \leqslant b$; and it is one for $x > b$. The mean and the median of this distribution are both $\frac{1}{2}(a + b)$, and the variance is $(b - a)^2/12$.

One of the principal applications of the rectangular distribution occurs in conjunction with the probability integral transformation. This is the transformation $Y = F(X)$, where $F(X)$ is the cumulative distribution function for a continuous random variable X. Then Y is rectangularly distributed with $a = 0$ and $b = 1$. Many distribution-free tests of fit have been derived starting with this transformation [*see* GOODNESS OF FIT; NONPARAMETRIC STATISTICS].

If Y has the rectangular distribution with $a = 0$, $b = 1$, then $-2 \ln Y$ has the chi-square distribution with 2 degrees of freedom. It follows, from the additivity of chi-square distributions, that if $Y_1, Y_2, \cdots, Y_n$ are jointly and independently distributed according to rectangular distributions with $a = 0$ and $b = 1$, then the sum, $-2\sum_{i=1}^{n} \ln Y_i$, has a chi-square distribution with $2n$ degrees of freedom.

There is also a discrete form of the rectangular distribution.

Pareto distributions. The density functions for Pareto distributions take the form: zero for $x < b$; $(a/b)(b/x)^{a+1}$ for $b \leqslant x < \infty$, where a and b are positive real constants (not zero). The cumulative distribution function is zero for $x < b$ and is equal to $1 - (b/x)^a$ for $b \leqslant x < \infty$. For $a > 1$, the mean of the Pareto distribution is $ab/(a - 1)$, and for $a > 2$ the variance is $(ab^2)/[(a - 1)^2(a - 2)]$. The median for $a > 0$ is at $x = 2^{1/a}b$, and the mode is at $x = b$. The Pareto distribution is related to the negative exponential distribution by the transformation $Y = \theta a \ln(X/b)$ where Y has the negative exponential distribution and X has the Pareto distribution.

Pareto distributions have been employed in the study of income distribution (*see* Kravis 1968; Lebergott 1968).

Laplace distributions. A random variable X has the Laplace distribution if its probability density function takes the form

$$\frac{1}{2\theta}\exp\left(-\left|\frac{x-\lambda}{\theta}\right|\right),$$

where $-\infty < \lambda < +\infty$, $0 < \theta < \infty$ and $-\infty < x < +\infty$. The mean of this distribution is λ, and the variance is $2\theta^2$. This distribution is also known as the double exponential, since the graph has the shape of an exponential function for $x > \lambda$ and it is a reflection (about the line $x = \lambda$) of the same exponential function for $x < \lambda$. The Laplace distribution is sometimes called Laplace's first law of error, the second being the normal distribution.

Lognormal distributions. A random variable X is said to have a logarithmic normal distribution (or lognormal distribution) if the logarithm of the variate is normally distributed. Let $Y = \ln X$ be normally distributed with mean μ and variance σ^2. The mean of X is $\exp(\mu + \frac{1}{2}\sigma^2)$ and the variance of X is $(\exp\sigma^2 - 1)\exp(2\mu + \sigma^2)$. Note that $0 < X < \infty$, while $-\infty < Y < +\infty$. The base of the logarithm may be any number greater than one (or between zero and one), and the bases 2 and 10 are often used. If the base a is used, then the mean of X is $a^{\mu+\frac{1}{2}(\ln a)\sigma^2}$, and the variance of X is $(a^{\sigma^2 \ln a} - 1)a^{2\mu+(\ln a)\sigma^2}$.

Logistic distributions. The logistic curve $y = \lambda/[1 + \gamma e^{-\kappa x}]$ is used frequently to represent the growth of populations. It may also be used as a cumulative probability distribution function. A random variable, X, is said to have the logistic probability distribution if the density of X is given by

$$\frac{\pi}{4\sigma\sqrt{3}}\,\text{sech}^2\left[\left(\frac{x-\mu}{\sigma}\right)\frac{\pi}{2\sqrt{3}}\right],$$

where $-\infty < x < +\infty$; μ is the mean of X and σ^2 is the variance of X. As with the normal distribution, the transformation $z = (x-\mu)/\sigma$ gives a standardized form to the distribution. The cumulative distribution function for the standardized variable is

$$1/[1 + \exp(-\pi z/\sqrt{3})],$$

where $-\infty < z < +\infty$. The shape of this cumulative distribution so nearly resembles the normal distribution that samples from normal and logistic distributions are difficult to distinguish from one another.

The exponential family of distributions. The one-parameter, single-variate, *exponential density functions* are those of the form

$$c(\theta)\exp[\theta A(x) + B(x)],$$

where c, A, and B are functions usually taken to satisfy regularity conditions. If there are several parameters, $\theta_1, \cdots, \theta_r$, the exponential form is

$$c(\theta_1, \cdots, \theta_r)\exp\left[\textstyle\sum \theta_i A_i(x) + B(x)\right].$$

Analogous forms may be considered for the multivariate case and for discrete distributions. Most of the standard distributions (normal, binomial, and so on) are exponential, but reparameterization may be required to express them in the above form.

The exponential distributions are important in theoretical statistics, and they arise naturally in discussions of sufficiency. Under rather stringent regularity conditions, an interesting sufficient statistic exists if, and only if, sampling is from an exponential distribution; this relationship was first explored by Koopman (1936), Darmois (1935), and Pitman (1936), so that the exponential distributions are sometimes eponymously called the Koopman–Darmois or Koopman–Pitman distributions [*see* SUFFICIENCY].

A discussion of the exponential family of distributions, and its relation to hypothesis testing, is given by Lehmann (1959, especially pp. 50–54).

It is important to distinguish between the family of exponential distributions and that of negative exponential distributions. The latter is a very special, although important, subfamily of the former.

DONALD B. OWEN

BIBLIOGRAPHY

DARMOIS, GEORGES 1935 Sur les lois de probabilité à estimation exhaustive. Académie des Sciences, Paris, *Comptes rendus hebdomadaires* 200:1265–1266.

EISENHART, CHURCHILL; and ZELEN, MARVIN 1958 Elements of Probability. Pages 134–164 in E. U. Condon and Hugh Odishaw (editors), *Handbook of Physics.* New York: McGraw-Hill.

GREENWOOD, JOSEPH A.; and HARTLEY, H. O. 1962 *Guide to Tables in Mathematical Statistics.* Princeton Univ. Press.

HAIGHT, FRANK A. 1961 Index to the Distributions of Mathematical Statistics. U.S. National Bureau of Standards, *Journal of Research* Series B: Mathematics and Mathematical Physics 65B:23–60.

○KENDALL, MAURICE G.; and STUART, ALAN (1958–1966) 1973–1977 *The Advanced Theory of Statistics.* 3 vols. London: Griffin; New York: Macmillan. → Volume 1: *Distribution Theory,* 4th ed., 1977. Volume 2: *Inference and Relationship,* 3d ed., 1973. Volume 3: *Design and Analysis, and Time-series,* 3d ed., 1976. Early editions of Volumes 1 and 2, first published in 1943 and 1946, were written by Kendall alone. Stuart became a joint author on later, renumbered editions in the three-volume set.

KOOPMAN, B. O. 1936 On Distributions Admitting a Sufficient Statistic. American Mathematical Society, *Transactions* 39:399–409.

KORN, GRANIO A.; and KORN, THERESA M. 1961 *Mathematical Handbook for Scientists and Engineers: Definitions, Theorems, and Formulas for Reference and Review.* New York: McGraw-Hill. → See especially pages 521–586 on "Probability Theory and Random Processes" and pages 587–626 on "Mathematical Statistics."

►KRAVIS, IRVING B. 1968 Income Distribution: I. Functional Share. Volume 7, pages 132–145 in *International Encyclopedia of the Social Sciences.* Edited by David L. Sills. New York: Macmillan and Free Press.

►LEBERGOTT, STANLEY 1968 Income Distribution: II. Size. Volume 7, pages 145–154 in *International Encyclopedia of the Social Sciences.* Edited by David L. Sills. New York: Macmillan and Free Press.

LEHMANN, ERICH L. 1959 *Testing Statistical Hypotheses.* New York: Wiley.

OWEN, DONALD B. 1962 *Handbook of Statistical Tables.* Reading, Mass.: Addison Wesley. → A list of addenda and errata is available from the author.

PEARSON, EGON S.; and HARTLEY, H. O. (editors), (1954) 1958 *Biometrika Tables for Statisticians.* 2 vols., 2d ed. Cambridge Univ. Press. → See especially Volume 1.

PEARSON, KARL (editor) (1922) 1951 *Tables of the Incomplete Γ-function.* London: Office of Biometrika.

PEARSON, KARL (editor) 1934 *Tables of the Incomplete Beta-function.* London: Office of Biometrika.

PITMAN, E. J. G. 1936 Sufficient Statistics and Intrinsic Accuracy. Cambridge Philosophical Society, *Proceedings* 32:567–579.

U.S. NATIONAL BUREAU OF STANDARDS 1953 *Tables of Normal Probability Functions.* Applied Mathematics Series, No. 23. Washington: The Bureau.

U.S. NATIONAL BUREAU OF STANDARDS 1959 *Tables of the Bivariate Normal Distribution Function and Related Functions.* Applied Mathematics Series, No. 50. Washington: The Bureau.

WALKER, HELEN M. 1929 *Studies in the History of Statistical Method, With Special Reference to Certain Educational Problems.* Baltimore: Williams & Wilkins.

ZELEN, MARVIN; and SEVERO, NORMAN C. (1964) 1965 Probability Functions. Chapter 26 in Milton Abramowitz and I. A. Stegun (editors), *Handbook of Mathematical Functions: With Formulas, Graphs, and Mathematical Tables.* New York: Dover. → First published as National Bureau of Standards, Applied Mathematics Series, No. 55. A list of errata is available from the National Bureau of Standards.

Postscript

Further tabulations of distributions are given by Yamauti et al. (1972); references to such tabulations are extended by Gunst and Owen (1975); and a four-volume series on distributions in statistics is that by Johnson and Kotz (1969; 1970; 1972).

Notable works have been published on specific continuous distributions. Rao (1965) describes several situations that give rise to the normal distribution; an extensive treatise on the chi-square distribution is given by Lancaster (1969); and an extensive table of the noncentral chi-square distribution can be found in *Selected Tables in Mathematical Statistics* (1973). A power tabulation for the Student t-test is given in Owen (1965). See Tiku (1967; 1972) for tables of the noncentral F-distribution; see also Kastenbaum et al. (1970*a;* 1970*b*) for sample-size considerations in analysis of variance F-tests. More information on the lognormal distribution can be found in Aitchison and Brown (1957).

DONALD B. OWEN

ADDITIONAL BIBLIOGRAPHY

AITCHISON, JOHN; and BROWN, J. A. C. 1957 *The Lognormal Distribution With Special Reference to Its Uses in Economics.* Cambridge University, Department of Applied Economics, Monograph No. 5. New York: Cambridge Univ. Press.

GUNST, R. F.; and OWEN, DONALD B. 1975 The Availability of Tables Useful in Analyzing Linear Models. Pages 181–196 in International Symposium on Statistical Design and Linear Models, Colorado State University, 1973, *A Survey of Statistical Design and Linear Models: Proceedings.* Edited by Jagdish N. Srivastava. Amsterdam: North-Holland; New York: American Elsevier.

JOHNSON, NORMAN L.; and KOTZ, SAMUEL 1969 *Discrete Distributions.* Boston: Houghton Mifflin.

JOHNSON, NORMAN L.; and KOTZ, SAMUEL 1970 *Continuous Univariate Distributions.* 2 vols. Boston: Houghton Mifflin. → Reissued in 1972 by Wiley.

JOHNSON, NORMAN L.; and KOTZ, SAMUEL 1972 *Continuous Multivariate Distributions.* New York: Wiley.

KASTENBAUM, MARVIN A.; HOEL, DAVID G.; and BOWMAN, KIMIKO O. 1970*a* Sample Size Requirements: Oneway Analysis of Variance. *Biometrika* 57:421–430.

KASTENBAUM, MARVIN A.; HOEL, DAVID G.; and BOWMAN, KIMIKO O. 1970*b* Sample Size Requirements: Randomized Block Designs. *Biometrika* 57:573–577.

LANCASTER, HENRY O. 1969 *The Chi-squared Distribution.* New York: Wiley.

OWEN, DONALD B. 1965 The Power of Student's t-Test. *Journal of the American Statistical Association* 60: 320–333.

RAO, C. RADHAKRISHNA 1965 *Linear Statistical Inference and Its Applications.* New York: Wiley. → See especially pages 126–132.

Selected Tables in Mathematical Statistics. 1973 Edited by the Institute of Mathematical Statistics. Providence, R.I.: American Mathematical Society. → Jointly edited by H. Leon Harter and Donald B. Owen.

TIKU, M. L. 1967 Tables of the Power of the F-Test. *Journal of the American Statistical Association* 62: 525–539.

TIKU, M. L. 1972 More Tables of the Power of the F-Test. *Journal of the American Statistical Association* 67:709–710.

YAMAUTI, ZIRO et al. (editors) 1972 *Statistical Tables and Formulas With Computer Applications.* Tokyo: Japanese Standards Association.

III
APPROXIMATIONS TO DISTRIBUTIONS

The term *approximation* refers, in general, to the representation of "something" by "something else" that is expected to be a useful replacement for the "something." Approximations are sometimes needed because it is not possible to obtain an exact representation of the "something"; even when an exact representation is possible, approximations may simplify analytical treatment.

In scientific work, approximations are in constant use. For example, much scientific argument, and nearly all statistical analysis, is based on mathematical models that are essentially approximations. This article, however, is restricted to approximations to distributions of empirical data and to theoretical probability distributions.

When approximating empirically observed distributions—for example, a histogram of frequencies of different words in a sample of speech—the primary objectives are those of compact description and smoothing. These are also the primary objectives of much approximation in demographic and actuarial work [*see* LIFE TABLES].

On the other hand, approximation to a theoretical distribution is often needed when exact treatment is too complicated to be practicable. For example, an econometrician who has developed a new estimator of price elasticity may well find the exact distribution of his estimator quite intractable; he will probably resort to large-sample (asymptotic) methods to find an approximate distribution.

It may also happen that a distribution arising not from statistical considerations, but from another kind of mathematical model, requires approximation to improve understanding. For example, a psychologist may use a probabilistic model of the learning process, a model that leads to a theoretical distribution for the number of trials needed to reach a specified level of performance. This distribution may be so complicated that an approximate form will markedly increase appreciation of its meaning.

The final section of this article discusses a general requirement, the measurement of the goodness of any particular approximation; this is especially important in the comparison of different approximations.

Approximation to empirical distributions. This section deals with approximations to distributions of numerical data representing measurements on each of a group of individuals. (Usually the group is a sample of some kind.) Among important techniques not discussed here are those that are purely mathematical (such as numerical quadrature and iterative solutions of equations) and those associated with the analysis of time series (such as trend fitting and periodogram and correlogram analysis) [*see* TIME SERIES].

As a specific example, data on the distributions of diseases by frequency of diagnosis (for males) in the teaching hospitals of England and Wales for the year 1949 (based on Herdan 1957) are presented in Table 1. The figures mean, for example, that out of 718 different diseases of males reported during the year, 120 occurred between one and five times each, while two were reported between 4,000 and 5,000 times each.

The figures shown in the table have already been grouped but present a rather irregular appearance; it would be useful to summarize the data in a more readily comprehensible form. A quick way to do this is to group further and to form a histogram (as in Figure 1). [*Further information on this method is presented in* GRAPHIC PRESENTATION; STATISTICAL ANALYSIS, SPECIAL PROBLEMS OF, *article on* GROUPED OBSERVATIONS.]

Grouping is in itself a kind of approximation, since it does not reproduce *all* features of the original data. For concise description, however, representation by a formula can be more useful. This is effected by fitting a frequency curve. If the fitted curve is simple enough and the effectiveness of approximation ("goodness of fit") is adequate, considerable benefit can be derived by replacing a large accumulation of data, with its inevitable irregularities, with a simple formula that can be handled with some facility and is a conveniently brief way of summarizing the data.

The present-day decline in the importance of fitting observed frequency distributions may well

Table 1 — Diseases by frequency of diagnosis in the teaching hospitals of England and Wales for 1949

FREQUENCY OF DIAGNOSIS	NUMBER OF DISEASES Observed	Fitted
1–5	120	103
6–10	55	74
11–15	44	54
16–20	31	43
21–25	28	34
26–30	32	28
31–40	45	45
41–50	37	35
51–60	33	28
61–70	22	23
71–80	23	19
81–90	22	17
91–100	15	14
101–119	21	22
120–139	18	19
140–159	19	15
160–179	19	13
180–199	12	11
200–239	19	17
240–279	24	13
280–319	13	11
320–359	4	8.5
360–399	9	7.0
400–499	15	13.1
500–599	13	9.1
600–699	5	6.7
700–799	6	5.0
800–899	2	4.0
900–999	2	3.1
1,000–1,999	6	14.0
2,000–2,999	1	4.1
3,000–3,999	1	1.9
4,000–4,999	2	1.0
≥ 5,000	—	2.25
Total	718	717.75

Source: Herdan 1957.

be only a temporary phenomenon. In the years 1890 to 1915 (roughly) there was a need to demonstrate that statistical methods did apply to real physical situations. The χ^2 test developed by Karl Pearson (1900) demonstrated clearly that the normal distribution, which had previously been assumed to be of rather general application, was not applicable to much observed data. It was desirable, therefore, to show, if possible, that some reasonably simple mathematical formula could give an adequate fit to the data.

Subsequent development of the theory of mathematical statistics has, on the one hand, been very much concerned with clarification of the logical principles underlying statistical method (assuming that there is some fairly well-established mathematical representation of the distributions involved); and on the other hand it has produced "distribution free" procedures, particularly significance tests, that eliminate the need for considering

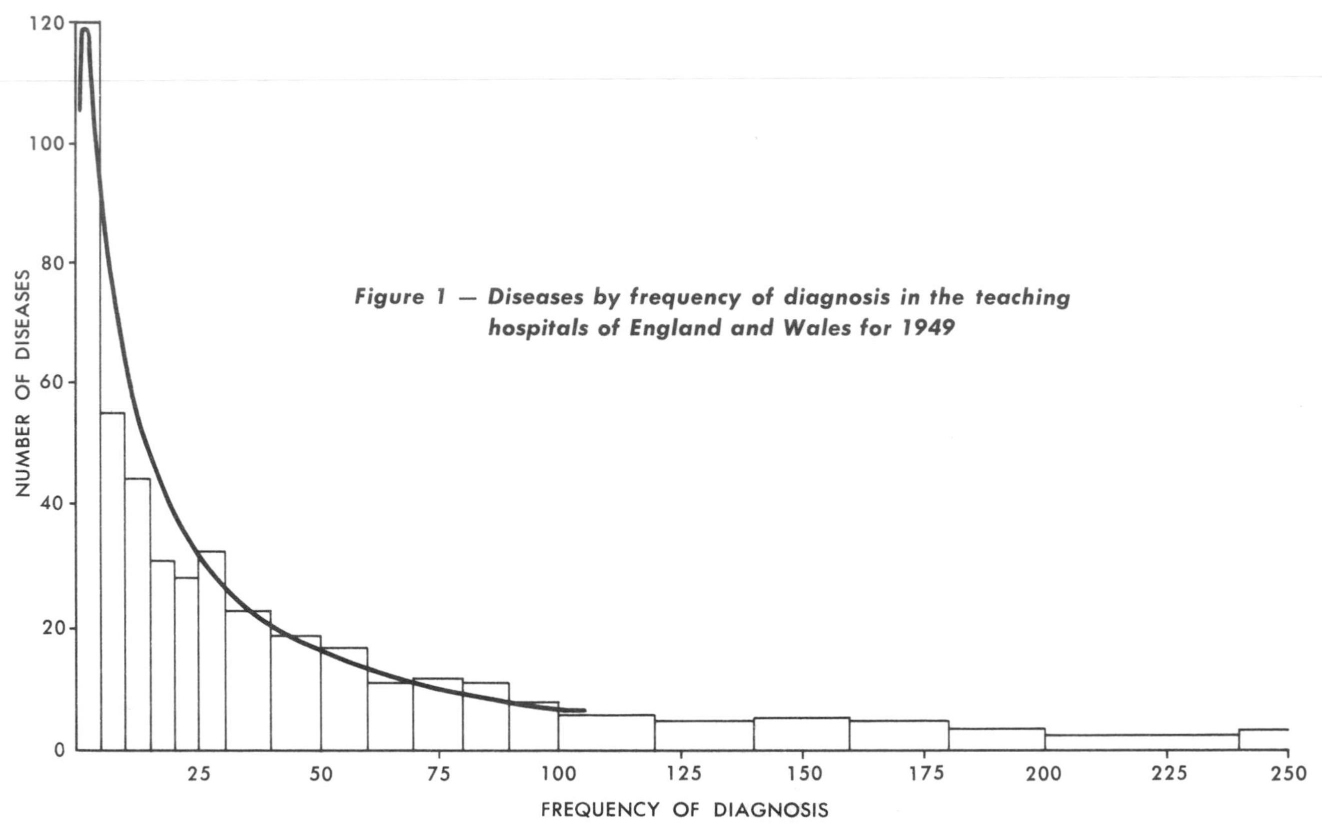

Figure 1 — Diseases by frequency of diagnosis in the teaching hospitals of England and Wales for 1949

Source: Herdan 1957.

the actual form of distribution (to any but the broadest detail) [*see* NONPARAMETRIC STATISTICS].

Both these lines of work tend to reduce interest in description of actual distributions in as precise detail as the data allow. However, from the more general viewpoint of scientific inquiry, the neglect of systematic study of distributional form in favor of application of formal techniques to arbitrarily hypothetical situations can represent a wasteful use of the data.

In the data of the example, the frequency of occurrences is naturally discrete—it takes only integer values—and, therefore, prime consideration should be given to formulas appropriate to discrete variables. In view of the wide range of variation of frequency, however, formulas appropriate to continuous variables may also give good approximations, and they are worth considering if they offer substantially simpler results.

A number of families of frequency curves have been found effective in approximating observed sets of data. Provided that a suitable form of curve has been chosen, fitting a maximum of *four* parameters gives a reasonably effective approximation. Fitting the curve is equivalent to estimating these parameters; this may be effected in various ways, among which are (*i*) the method of *maximum likelihood;* (*ii*) the method of *moments,* in which certain lower moments of the fitted curve are made to agree with the corresponding values calculated from the observed data; and (*iii*) the method of *percentile points*, in which the fitted curve is made to give cumulative proportions agreeing with those for the observed data at certain points (7%, 25%, 75%, and 93% are often recommended) [*see* ESTIMATION, *article on* POINT ESTIMATION].

In cases where a clear probabilistic model can be established, there is usually good reason to prefer the method of maximum likelihood. When it is not possible to establish such a model, another method is required.

Fitting is sometimes facilitated by using appropriately designed "probability paper," with the ordinate and abscissa scales such that if cumulative frequency (ordinate) is plotted against variable values (abscissa), a straight line relationship should be obtained [*see* GRAPHIC PRESENTATION].

Among the more commonly used families of frequency curves are those representing the normal, exponential and gamma distributions. These specify the proportion less than, or equal to, a fixed number x (the cumulative distribution function, often denoted by the symbol $F(x)$) as a function depending on two or three parameters (and also, of course, on x) [*see* DISTRIBUTIONS, STATISTICAL, *article on* SPECIAL CONTINUOUS DISTRIBUTIONS]. The best-known system of curves is the *Pearson* system (Elderton 1906), which is derived from solutions of the differential equation

$$\frac{d^2F}{dx^2} = -\frac{dF}{dx}\left[\frac{c(x+a_0)}{x^2+a_1x+a_2}\right].$$

Originally (Pearson 1895) this equation was arrived at by (i) considering limiting cases of sampling from a finite population and (ii) requiring the curve $y = dF/dx$ to satisfy certain natural conditions of unimodality, smoothness, etc. (Note that there are *four* parameters, c, a_0, a_1, a_2, in this equation.)

Methods have been worked out (Elderton 1906) for estimating the values of these parameters from the first four moments of the distributions. In certain cases, the procedures can be much facilitated by using Table 42 of the *Biometrika* tables (Pearson & Hartley 1954) or the considerably extended version of these tables in Johnson et al. (1963). Entering these tables with the values of the moment ratios ($\sqrt{\beta_1}$ and β_2) it is possible to read directly standardized percentage points (X'_P) for a number of values of P. The percentage points X_P (such that $\hat{F}(X_P) = P$) are then calculated as Mean $+ X'_P \times$ (Standard Deviation).

Other systems are based on "transformation to normality" with $z = a_1 + a_2 g[c(x+a_0)]$, where z is a normal variable with zero mean and unit standard deviation, and $g[\ ]$ is a fairly simple explicit function. Here, c, a_0, a_1, and a_2 are again parameters. If $g[y] = \log y$, one gets a *lognormal* curve. Such a curve has been fitted to the distribution of frequency of diagnosis in Figure 1 (for clarity, up to frequency 100 only). Since $z = a_1 + a_2 \log[c(x+a_0)]$ can be written as $z = a'_1 + a_2 \log(x+a_0)$, where now $a'_1 = a_1 + a_2 \log c$, there are in fact only *three* separate parameters in this case. In fitting the curve in Figure 1, a_0 has been taken equal to -0.5, leaving only a_1 and a_2 to be estimated. This has been done by the method of percentile points (making the fitted frequencies less than 11 and less than 200 agree with the observed frequencies—175 and 596 respectively). The fitted formula is (rounded)

$$\text{Number less than, or equal to, } x \text{ (an integer)} = 718\,\Phi(.55 \ln x - 1.95),$$

where Φ is the cumulative distribution function of the unit-normal distribution [*see* DISTRIBUTIONS, STATISTICAL, *article on* SPECIAL CONTINUOUS DISTRIBUTIONS]. Numerical values of the fitted frequencies are shown in the last column of Table 1.

While a reasonable fit is obtained in the center of the distribution, the fit is poor for the larger frequencies of diagnosis. As has been noted, although the data are discrete, a continuous curve has been fitted; however, this is not in itself reason for obtaining a poor fit. Indeed many standard approximations—such as the approximation to a binomial distribution by a normal distribution—are of this kind.

Another system of curves is obtained by taking the approximate value of $F(x)$ to be

$$\hat{F}(x) = \sum_{j=0}^{k} a_j\, \Phi^{(j)}\left(\frac{x-\xi}{\sigma}\right),$$

where $\Phi^{(j)}$ is the jth derivative of Φ, and where the parameters are $a_0, a_1, \cdots, a_k$, ξ, and $\sigma > 0$. The number of parameters depends on the value of k, which is usually not taken to be greater than 3 or 4. This system of curves is known as the Gram–Charlier system; a modified form is known as the Edgeworth system (Kendall & Stuart 1958–1966, chapter 6). If all a's after a_0 are zero (and a_0 is equal to one) then $F(x)$ is simply the normal cumulative distribution. The terms after the first can be regarded as "corrections" to the simple normal approximation. It should not automatically be assumed, however, that the more corrections added the better. If a high value of k is used, the curve may present a wavy appearance; even with a small value of k it is possible to obtain *negative* values of fitted frequencies (Barton & Dennis 1952).

Similar expansions can be constructed replacing $\Phi(x)$ by other standard cumulative distribution (see Cramér 1945, sec. 20.6). If the $\Phi^{(j)}$'s are replaced by cumulative distribution functions $\Phi_j(x)$, (with, of course, $a_0 + a_1 + a_2 + \cdots + a_k = 1$), then $F(x)$ is represented by a "mixture" of these distributions [*see* DISTRIBUTIONS, STATISTICAL, *article on* MIXTURES OF DISTRIBUTIONS]. When joint distributions of two or more variates are to be fitted, the variety of possible functional forms can be an embarrassment. A convenient mode of attack is to search for an effective normalizing transformation for each variate separately. If such a transformation can be found for each variate, then a joint multinormal distribution can be fitted to the set of transformed variates, using the means, variances, and correlations of the transformed variates. A discussion of some possibilities, in the case of two variates, will be found in Johnson (1949).

Approximation to one theoretical distribution by another. Approximation is also useful even when it is not necessary to deal with observed data. An important field of application is the replacement of complicated formulas for the theoretical distributions of statistics by simpler (or more thoroughly investigated) distributions. Just as when approximating to observed data, it is essential that the approximation be sufficiently effective to give useful results. In this case, however, the problem is more definitely expressible in purely mathematical terms; there are a number of results in the mathematical theory of probability that are often used in constructing approximations of this kind.

Among these results, the "Central Limit" group of theorems has the broadest range of applicability. The theorems in this group state that, under appropriate conditions, the limiting distribution of certain statistics, T_n, based on a number, n, of random variables, as n tends to infinity, is a normal distribution. [*See* PROBABILITY *for a discussion of the Central Limit theorems.*]

If the conditions of a Central Limit theorem are satisfied, then the distribution of the statistic $T'_n = [T_n - E(T_n)][\sigma(T_n)]^{-1}$, where $E(T_n)$ is the mean and $\sigma(T_n)$ the standard deviation of T_n, may be approximated by a "unit normal distribution." Then the probability $Pr[T_n \leqslant \tau]$ may be approximated by using tables of the normal integral with argument $[\tau - E(T_n)]/\sigma(T_n)$.

There are many different Central Limit theorems; the most generally useful of these relate to the special case when the X's are mutually independent and $T_n = X_1 + X_2 + \cdots + X_n$. The simplest set of conditions is that each X should have the *same* distribution (with finite expected value and standard deviation), but weaker sets of conditions can replace the requirement that the distributions be identical. These conditions, roughly speaking, ensure that none of the variables X_i have such large standard deviations that they dominate the distribution of T_n.

Central Limit results are used to approximate distributions of test criteria calculated from a "large" number of sample values. The meaning of "large" depends on the way in which the effectiveness of the approximation is measured and the accuracy of representation required. In turn, these factors depend on the use that is to be made of the results. Some of the problems arising in the measurement of effectiveness of approximation will be described in the next section.

Very often there is a choice of quantities to which Central Limit type results can be applied. For example, any power of a chi-square random variable tends to normality as the number of degrees of freedom increases. For many purposes, the one-third power (cube root) gives the most rapid convergence to normality (see Moore 1957), although it may not always be the best power to use. While many useful approximations can be suggested by skillful use of probability theory, it should always be remembered that sufficient numerical accuracy is an essential requirement. In some cases (for example, Mallows 1956; Gnedenko et al. 1961; Wallace 1959), theoretical considerations may provide satisfactory evidence of accuracy; in other cases, *ad hoc* investigations, possibly using Monte Carlo methods, may be desirable. Interesting examples of these kinds of investigation are described by Goodman and Kruskal (1963) and Pearson (1963). [*A discussion of Monte Carlo methods may be found in* RANDOM NUMBERS.]

Another type of result (Cramér 1945, sec. 20.6) useful in approximating theoretical distributions and, in particular, in extending the field of Central Limit results, may be stated formally: "if $\lim_{n\to\infty} Pr[X_n < \xi] = F(\xi)$ and Y_n tends to η in probability, then $\lim_{n\to\infty} Pr[X_n + Y_n < \xi] = F(\xi - \eta)$," or less formally, "the limiting distribution of $X_n + Y_n$ is the same as that of $X_n + \eta$." Similar statements hold for multiplication, and this kind of useful result may be considerably generalized. A fuller discussion is given in the Appendix of Goodman and Kruskal (1963).

A classic and important example of approximation of one theoretical distribution by another is that of approximating a binomial distribution by a normal one. If the random variable X has a binomial distribution with parameters p, n, then the simplest form of the approximation is to take $Pr\{X \leqslant c\}$ as about equal to

$$\Phi\left\{\frac{c - np}{[np(1-p)]^{\frac{1}{2}}}\right\},$$

where $c = 0, 1, 2, \cdots, n$. One partial justification for this approximation is that the mean and standard deviation of X are, respectively, np and $[np(1-p)]^{\frac{1}{2}}$; the asymptotic validity of the approximation follows from the simplest Central Limit theorem.

● The approximation may be generally improved by replacing $c - np$, in the numerator of the argument of Φ, with $c - np + \frac{1}{2}$. This modification is called a "continuity correction." Continuity corrections can often be used effectively to improve approximations at little extra computational cost. (See Molenaar 1970.)

The methods described in the section on approximating numerical data can also be used, with appropriate modification, in approximating theoretical distributions. Different considerations may arise, however, in deciding on the adequacy of an approximation in the two different situations.

Measuring the effectiveness of an approximation. The effectiveness of an approximation is its suitability for the purpose for which it is used. With sufficiently exhaustive knowledge of the properties of each of a number of approximations, it would be possible to choose the best one to use in any given situation, with little risk of making a bad choice. However, in the majority of cases, the attainment of such knowledge (even if possible) would entail such excessively time-consuming labor that a main purpose for the use of approximations —the saving of effort—would be defeated. Considerable insight into the properties of approximations can, however, be gained by the careful use of certain representative figures, or indexes, for their effectiveness. Such an index summarizes one aspect of the accuracy of an approximation.

Since a single form of approximation may be used in many different ways, it is likely that different indexes will be needed for different types of application. For example, the approximation obtained by representing the distribution of a statistic by a normal distribution may be used, *inter alia,* (*a*) for calculating approximate significance levels for the statistic, (*b*) for designing an experiment to have specified sensitivity, and (*c*) for combining the results of several experiments. In case (*a*), effectiveness will be related particularly to accuracy of representation of probabilities in the tail(s) of the distribution, but this will not be so clearly the case in (*b*) and (*c*). It is important to bear in mind the necessity for care in choosing an appropriate index for a particular application.

This article has been concerned with approximations to theoretical or empirical distribution functions. The primary purpose of such approximations is to obtain a useful representation, say $\hat{F}(x)$, of the actual cumulative distribution function $F(x)$. It is natural to base an index of accuracy on some function of the difference $\hat{F}(x) - F(x)$. Here are some examples of indexes that might be used:

$$(1)\quad \max_{x} |\hat{F}(x) - F(x)|$$

$$(2)\quad \max_{x_1,x_2} |\{\hat{F}(x_1) - \hat{F}(x_2)\} - \{F(x_1) - F(x_2)\}|$$

$$(3)\quad \max_{x} \frac{|\hat{F}(x) - F(x)|}{F(x)\{1 - F(x)\}}$$

$$(4)\quad \max_{x_1,x_2} \frac{|\{\hat{F}(x_1) - \hat{F}(x_2)\} - \{F(x_1) - F(x_2)\}|}{\{F(x_1) - F(x_2)\}[1 - \{F(x_1) - F(x_2)\}]}$$

The third and fourth of these indexes are modifications of the first and second indexes respectively. A further pair of definitions could be obtained by replacing the intervals (x_1, x_2) in (2) and (4) by sets, ω, of possible values, replacing $|\hat{F}(x_1) - \hat{F}(x_2)|$ and $|F(x_1) - F(x_2)|$ by the approximate and actual values of $Pr[X \text{ in } \omega]$, and replacing $\max_{x_1,x_2}$ by $\max_{\omega}$. Indexes of type (3) and (4) are based on the *proportional* error in the approximation, while indexes of type (1) and (2) are based on *absolute* error in the approximation. While the former are the more generally useful, it should be remembered that they may take infinite values. Very often, in such cases, the difficulty may be removed by choosing ω to exclude extreme values of x.

In particular instances ω may be quite severely restricted. For example, the 5% and 1% levels may be regarded as of paramount importance. In such a case, comparison of $F(x)$ and $\hat{F}(x)$ in the neighborhood of these values may be all that is needed. It has been suggested in such cases that actual values between 0.04 and 0.06, corresponding to a nominal 0.05, and between 0.007 and 0.013, corresponding to a nominal 0.01, can be regarded as satisfactory for practical purposes.

In measuring the accuracy of approximation to empirical distributions, it is quite common to calculate χ^2. This is, of course, a function of the differences $\hat{F}(x) - F(x)$. Even when circumstances are such that no probabilistic interpretation of the statistic is possible, indexes of relative accuracy of approximation have been based on the magnitude of χ^2. A similar index for measuring accuracy of approximations to theoretical distributions can be constructed. Such indexes are based on more or less elaborate forms of average (as opposed to maximum) size of error in approximation.

NORMAN L. JOHNSON

[*See also* GOODNESS OF FIT.]

BIBLIOGRAPHY

[1]BARTON, D. E.; and DENNIS, K. E. 1952 The Conditions Under Which Gram–Charlier and Edgeworth Curves Are Positive Definite and Unimodal. *Biometrika* 39: 425–427.

CRAMÉR, HARALD (1945) 1951 *Mathematical Methods of Statistics.* Princeton Mathematical Series, No. 9. Princeton Univ. Press.

[2]ELDERTON, W. PALIN (1906) 1953 *Frequency Curves and Correlation.* 4th ed. Washington: Harren.

GNEDENKO, B. V. et al. 1961 Asymptotic Expansions in Probability Theory. Volume 2, pages 153–170 in Berkeley Symposium on Mathematical Statistics and Probability, Fourth, University of California, 1960, *Proceedings.* Berkeley: Univ. of California Press.

GOODMAN, LEO A.; and KRUSKAL, WILLIAM H. 1963 Measures of Association for Cross-classifications: III. Approximate Sampling Theory. *Journal of the American Statistical Association* 58:310–364.

HERDAN, G. 1957 The Mathematical Relation Between the Number of Diseases and the Number of Patients in a Community. *Journal of the Royal Statistical Society* Series A 120:320–330.

JOHNSON, NORMAN L. 1949 Bivariate Distributions Based on Simple Translation Systems. *Biometrika* 36:297–304.

JOHNSON, NORMAN L.; NIXON, ERIC; AMOS, D. E.; and PEARSON, EGON S. 1963 Table of Percentage Points of Pearson Curves, for Given $\sqrt{\beta_1}$ and β_2, Expressed in Standard Measure. *Biometrika* 50:459–498.

○KENDALL, MAURICE G.; and STUART, ALAN (1958–1966) 1973–1977 *The Advanced Theory of Statistics.* 3 vols. London: Griffin; New York: Macmillan. → Volume 1: *Distribution Theory,* 4th ed., 1977. Volume 2: *Inference and Relationship,* 3d ed., 1973. Volume 3: *Design and Analysis, and Time-series,* 3d ed., 1976. Early editions of Volumes 1 and 2, first published in 1943 and 1946, were written by Kendall alone. Stuart became a joint author on later, renumbered editions in the three-volume set.

MALLOWS, C. L. 1956 Generalizations of Tchebycheff's Inequalities. *Journal of the Royal Statistical Society* Series B 18:139–168.

MOORE, P. G. 1957 Transformations to Normality Using Fractional Powers of the Variable. *Journal of the American Statistical Association* 52:237–246.

PEARSON, EGON S. 1963 Some Problems Arising in Approximating to Probability Distributions, Using Moments. *Biometrika* 50:95–112.

PEARSON, EGON S.; and HARTLEY, H. O. (editors) (1954) 1958 *Biometrika Tables for Statisticians.* 2d ed., 2 vols. Cambridge Univ. Press.

PEARSON, KARL 1895 Contributions to the Mathematical Theory of Evolution: II. Skew Variations in Homogeneous Material. Royal Society of London, *Philosophical Transactions* Series A 186:343–414.

PEARSON, KARL 1900 On the Criterion That a Given System of Deviations From the Probable in the Case of a Correlated System of Variables Is Such That It Can Be Reasonably Supposed to Have Arisen From Random Sampling. *Philosophical Magazine* 5th Series 50:157–175.

WALLACE, DAVID L. 1959 Bounds on Normal Approximations to Student's and the Chi-square Distributions. *Annals of Mathematical Statistics* 30:1121–1130.

ADDITIONAL BIBLIOGRAPHY

[1]DRAPER, NORMAN R.; and TIERNEY, DAVID E. 1972 Regions of Positive and Unimodal Series Expansion of the Edgeworth and Gram–Charlier Approximations. *Biometrika* 59:463–465.

[2]ELDERTON, W. PALIN; and JOHNSON, NORMAN L. 1969 *Systems of Frequency Curves.* New York: Cambridge Univ. Press. → A drastic revision of Elderton (1906), which is now out of print.

MOLENAAR, W. 1970 *Approximations to the Poisson, Binomial and Hypergeometric Distribution Functions.* Mathematical Centre Tracts No. 31. Amsterdam: Mathematisch Centrum.

IV
MIXTURES OF DISTRIBUTIONS

A mixture of distributions is a weighted average of probability distributions with positive weights that sum to one. The distributions thus mixed are called the *components* of the mixture. The weights themselves comprise a probability distribution called the *mixing distribution.* Because of this property of the weights, a mixture is, in particular, again a probability distribution.

As an example, suppose that the probability distribution of heights of 30-year-old men in New York is approximately a normal distribution, while that of 30-year-old women in New York is another approximately normal distribution. Then the probability distribution of heights of 30-year-old people in New York will be, to the same degree of approximation, a mixture of the two normal distributions. The two separate normal distributions are the components, and the mixing distribution is the simple one on the dichotomy male–female, with the weights given by the relative frequencies of men and women in New York.

Probability distributions of this type arise when observed phenomena can be the consequence of two or more related, but usually unobserved, phenomena, each of which leads to a different probability distribution. Mixtures and related structures often arise in the construction of probabilistic models; for example, models for factor analysis. Mixtures also arise in a number of statistical contexts. A general problem is that of "decomposing" a mixture on the basis of a sample, that is, of estimating the parameters of the mixing distribution and those of the components.

Mixtures occur most commonly when the parameter, θ, of a family of distributions, given, say, by the density or frequency functions $f(x;\theta)$, is itself subject to chance variation. The mixing distribution, say $g(\theta)$, is then a probability distribution on the parameter of the distributions $f(x;\theta)$.

The components of a mixture may be discrete, continuous, or some of each type. Mixtures are classified, in accordance with the number of their components, as *finite, countable,* or *noncountably infinite.*

The generic formula for the most common form of finite mixture is

$$\sum_{i=1}^{k} f(x;\theta_i)g(\theta_i); \tag{1}$$

the infinite analogue (in which g is a density function) is

$$\int f(x;\theta)g(\theta)\,d\theta. \tag{2}$$

As an illustration of the above ideas, consider the following simple example. Two machines produce pennies, which are fed into a common bin. The pennies are identical except that those produced by machine 1 have probability θ_1 of showing a head when tossed, while those produced by machine 2 have probability $\theta_2 \neq \theta_1$ of showing a head. Let α be the proportion of coins produced by machine 1, and $1-\alpha$ the proportion produced by machine 2.

A coin chosen at random from the bin is tossed n times. By the basic rules of probability theory, the probability of observing x heads is

$$\begin{aligned}(3)\quad & Pr\{x \text{ heads out of } n \text{ tosses}\} \\ &= Pr\{\text{penny from machine } 1\}\, Pr\{x \text{ heads given machine } 1\} \\ &\quad + Pr\{\text{penny from machine } 2\}\, Pr\{x \text{ heads given machine } 2\} \\ &= \alpha \binom{n}{x} \theta_1^x (1-\theta_1)^{n-x} \\ &\quad + (1-\alpha)\binom{n}{x} \theta_2^x (1-\theta_2)^{n-x}.\end{aligned}$$

This is a mixture of two binomial distributions (the components) with mixing distribution $g(\theta_1)=\alpha$, $g(\theta_2)=1-\alpha$. [*See* PROBABILITY, *article on* FORMAL PROBABILITY, *for a discussion of the rules giving rise to* (3).]

By contrast, if $\theta_1=\theta_2=\theta$, that is to say if the two machines produce exactly identical coins, then

$$Pr\{x \text{ heads out of } n \text{ tosses}\} = \binom{n}{x}\theta^x(1-\theta)^{n-x},$$

which is the simple binomial. Similarly if the experimenter selected a coin from a particular machine, say machine 2, rather than choosing a coin at random, the distribution of outcomes would be a binomial with parameter θ_2. A simple binomial also results if n distinct coins are chosen and each is tossed a single time. In this case, however, the binomial parameter is $\alpha\theta_1+(1-\alpha)\theta_2$.

A generalization of the above example is a mixture of two trinomial distributions. It is applicable to description of the sex distribution of twins (cf. Goodman & Kruskal 1959, p. 134; Strandskov & Edelen 1946). Twin pairs fall into three classes. *MM*, *MF*, and *FF*, where *M* denotes male and *F* female. This leads to the trinomial distribution. Since, in addition, twins may be dizygotic or monozygotic, a mixture of trinomials results. Because the sexes of individual dizygotic twins are independent, the corresponding trinomial has parameters p^2, $2p(1-p)$, and $(1-p)^2$, where p is the probability of a male. Monozygotic twins, however, are genetically identical, so that the corresponding trinomial has parameters p, 0, and q. The mixing distribution is determined by the relative frequencies of monozygotic and dizygotic twin births.

Some properties of mixtures. Mixtures are themselves probability distributions, and hence their density or frequency functions are nonnegative and sum or integrate to unity.

While the definitions (1) and (2) are given in terms of the density or frequency functions, precisely the same relationships hold with regard to the corresponding cumulative distribution functions. Similarly, the moments (about zero) of a mixture are weighted averages (that is, mixtures) of the moments of the component distributions. Thus, for a finite mixture, if $\mu_r'(\theta_i)$ is the rth moment about zero of the ith component, then the rth moment about zero of the mixture is

$$\sum_{i=1}^{k} g(\theta_i)\mu_r'(\theta_i).$$

The same is true of the factorial moments. A similar, although slightly more complicated, relationship holds for moments about the mean.

An important property of many common mixtures of discrete distributions is the relationship imposed upon the generating functions: if $\phi_1(s)$ and $\phi_2(s)$ are the generating functions of the mixing distribution and the component distributions, respectively, then $\phi_1(\phi_2(s))$ is the generating function of the mixture. This greatly simplifies certain calculations. [*See* DISTRIBUTIONS, STATISTICAL, *article on* SPECIAL DISCRETE DISTRIBUTIONS, *for a discussion of generating functions.*]

Mixtures of standard distributions may possess interesting geometric properties. They may be bimodal or multimodal and are ordinarily more dispersed than are the components. Properties such as these account, at least in part, for the fact that mixtures frequently fit data more satisfactorily than do standard distributions.

Conditionality. The concept of conditionality underlies the definitions (1) and (2), for the function $f(x;\theta)$ is the conditional probability distribution given the value of θ, while the product of $f(x;\theta)$ and $g(\theta)$ is the joint probability distribution of x and θ, and the sum (or integral) is the unconditional distribution of x.

The importance of conditionality in applications can be seen in the coin example: the machine on which the randomly chosen coin was produced may be thought of as an auxiliary chance variable (taking on the values 1 and 2 with respective probabilities α and $1-\alpha$); conditional on the value of the auxiliary chance variable, the distribution of heads is a simple binomial.

The auxiliary chance variable is usually unobservable and hence cannot play a direct role in data analysis. The merit of introducing the auxiliary chance variable in this context is, therefore, *not* that it results in simpler data analyses but that it yields a proper understanding of the underlying mathematical structure and simplifies the derivation of the probability distribution.

This situation is typical; the conditional distribution of the chance variable under investigation, given the value of a related unobservable chance variable, is either known exactly or is of a relatively simple form. Because of the inability to observe the related chance variable, however, the unconditional distribution is a mixture and is usually much more complex.

Some applications. Because of the immense complexity of living organisms, variables that the investigator can neither control nor observe frequently arise in investigations of natural phenomena. Some examples are attitude, emotional stability, skills, genotype, resistance to disease, etc. Since mixtures of distributions result when such variables are related to the variable under investigation, mixtures have many applications in the social and biological sciences.

Mixtures have played an important role in the construction of so-called contagious distributions. These distributions are deduced from models in which the occurrence of an event has the effect of changing the probability of additional such occurrences. An early example of a contagious distribution was given by Greenwood and Yule (1920) in an analysis of accident data. Greenwood and Yule derived the classical negative binomial distribution as a mixture of distributions. Other contagious distributions that are mixtures are given by Neyman (1939, the Neyman types A, B, and C) and Gurland (1958).

Mixtures of distributions also arise in dealing with unusual observations or outliers. One approach to the problem of outliers is predicated upon writing the underlying probability distribution as a mixture of distributions (see Dixon 1953). [*For further discussion, see* STATISTICAL ANALYSIS, SPECIAL PROBLEMS OF, *article on* OUTLIERS.]

Mixtures of distributions also arise in the Bayesian approach to statistical analysis. *Bayes' Procedures* are constructed under the assumption that the parameters of the underlying probability distribution are themselves chance variables. In this context, the mixing distribution is called the a priori distribution [*see* BAYESIAN INFERENCE].

The following are a few additional applications of mixtures:

(1) Life testing of equipment subject to sudden and delayed failures (Kao 1959).

(2) Acceptance testing, when the proportion of defectives varies from batch to batch (Vagholkar 1959).

(3) Latent stucture analysis based on mixtures of multivariate distributions with independence between the variates within each component (Lazarsfeld 1950; Green 1952). Factor analysis also comes under this general description. [*See* FACTOR ANALYSIS AND PRINCIPAL COMPONENTS; LATENT STRUCTURE.]

(4) Construction of a learning model as a weighted average of two simpler such models (Sternberg 1959).

(5) A model for spatial configurations in map analysis (Dacey 1964).

Special mixed distributions. Using the definitions (1) and (2), it is easy to generate large numbers of probability distributions. Furthermore, the process of mixing of distributions may be repeated; that is, new mixtures, in which the components are themselves mixtures of distributions, may be formed by repeated application of (1) and (2). Thus, it is possible to form an unlimited number of probability distributions by this relatively simple process. In addition, many classical probability distributions can be represented nontrivially as mixtures. Teicher (1960) gives two representations of the normal distribution as a mixture of distributions. [*See* DISTRIBUTIONS, STATISTICAL, *article on* SPECIAL DISCRETE DISTRIBUTIONS, *for both classical and mixture representations of the negative binomial distribution.*]

It is customary, when both component and mixing distribution are well known, to call the mixture by the names of the distributions involved.

The following are a few specific examples of mixtures:

Mixture of two normal distributions. The density function of a mixture of two normal distributions is

$$f(x) = \frac{1}{\sqrt{2\pi}}\left[\frac{\alpha_1}{\sigma_1}\exp\left\{-\frac{(x-\mu_1)^2}{2\sigma_1^2}\right\} + \frac{\alpha_2}{\sigma_2}\exp\left\{-\frac{(x-\mu_2)^2}{2\sigma_2^2}\right\}\right],$$

where $-\infty < x < \infty$, $0 < \alpha_1 < 1$, $0 < \alpha_2 < 1$, $\alpha_1 + \alpha_2 = 1$, and μ_1, μ_2 and σ_1^2, σ_2^2 are the respective means and variances of the component normal distributions. This example is due to Pearson (1894). The distribution has mean $\alpha_1\mu_1 + \alpha_2\mu_2$ and variance $\alpha_1\sigma_1^2 + \alpha_2\sigma_2^2 + \alpha_1\alpha_2(\mu_1 - \mu_2)^2$. An example of this mixture with $\mu_1 = 0$, $\mu_2 = 4$, $\sigma_1^2 = 1$, $\sigma_2^2 = 9$, and $\alpha_1 = 0.4$ is given in Figure 1.

Mixture of two Poisson distributions. The frequency function of a mixture of two Poisson distributions is

$$p(x) = \frac{1}{x!}\left[\alpha_1\lambda_1^x e^{-\lambda_1} + \alpha_2\lambda_2^x e^{-\lambda_2}\right], \qquad x = 0, 1, 2, \cdots,$$

where α_1 and α_2 are as before and $\lambda_1 \neq \lambda_2$ are the

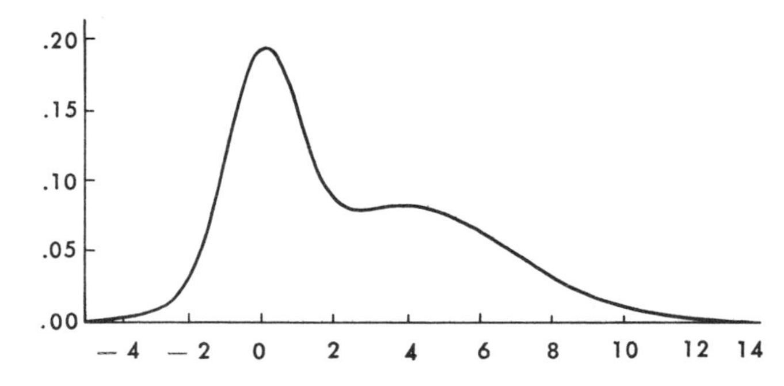

Figure 1 — A mixture of two normal distributions with $\mu_1 = 0$, $\mu_2 = 4$, $\sigma_1^2 = 1$, $\sigma_2^2 = 9$, and $\alpha_1 = 0.4$, showing f(x) on the vertical axis corresponding to x on the horizontal axis

parameters of the components. This distribution has mean $\alpha_1\lambda_1 + \alpha_2\lambda_2$ and variance $\alpha_1\lambda_1 + \alpha_2\lambda_2 + \alpha_1\alpha_2(\lambda_1 - \lambda_2)^2$. An example of this discrete mixed distribution with $\lambda_1 = 1$, $\lambda_2 = 8$, and $\alpha_1 = 0.5$ is given in Figure 2.

Poisson–binomial distribution. The frequency function of the Poisson–binomial distribution is

$$p(x) = e^{-\lambda} \sum_{i=0}^{\infty} \frac{\lambda^i}{i!} \binom{ni}{x} \theta^x (1-\theta)^{ni-x}, \qquad x = 0, 1, 2, \cdots,$$

where $0 < \theta < 1$, $0 < \lambda < \infty$, and n is an integer. The components are binomial distributions; the mixing distribution is Poisson. The probabilities can also be obtained by successive differentiation of the generating function

$$\phi(s) = \exp\{\lambda[(1 - \theta + \theta s)^n - 1]\},$$

or from the recursion formulas

$$p(0) = \exp\{\lambda[(1-\theta)^n - 1]\},$$

$$p(x+1) = \frac{\lambda n(1-\theta)^n}{x+1} \sum_{i=0}^{x} \binom{n-1}{x-i} \left(\frac{\theta}{1-\theta}\right)^{x+1-i} p(i).$$

The mean is $n\theta\lambda$; the variance is $n\theta\lambda[1 + (n-1)\theta]$.

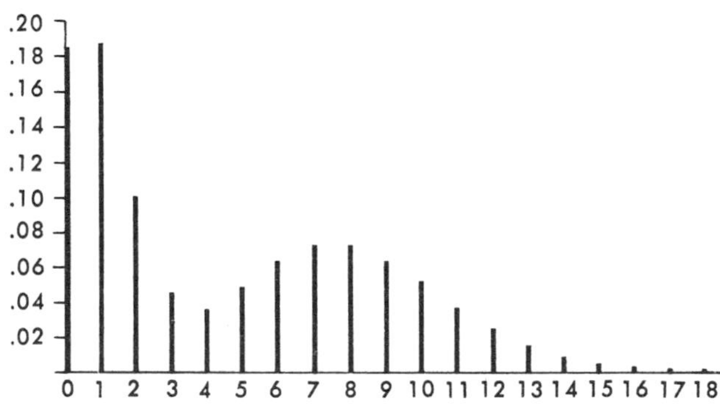

Figure 2 — A mixture of two Poisson distributions with $\lambda_1 = 1$, $\lambda_2 = 8$, and $\alpha_1 = 0.5$, showing p(x) on the vertical axis corresponding to x on the horizontal axis

Poisson–negative binomial distribution. The frequency function of the Poisson–negative binomial distribution is

$$p(x) = e^{-\lambda} \sum_{i=0}^{\infty} \frac{\lambda^i}{i!} \binom{ni + x - 1}{x - 1} \theta^{ni} (1-\theta)^x, \qquad x = 0, 1, 2, \cdots,$$

where $0 < \theta < 1$, $0 < \lambda < \infty$, and $0 < n < \infty$. This is a Poisson mixture of negative binomial distributions. Here

$$\phi(s) = \exp\{\lambda[\theta^n(1 - s + s\theta)^{-n} - 1]\};$$

the recursion formulas are

$$p(0) = \exp[-\lambda(1 - \theta^n)],$$

$$p(x+1) = \frac{\lambda}{x+1} \sum_{i=0}^{x} \left[\binom{n+x-1}{x} (x + 1 - i)\theta^n(1-\theta)^x - ip(i)\right].$$

The mean and variance are $n\lambda(1-\theta)/\theta$ and $n\lambda(1-\theta)(n - n\theta + 1)/\theta^2$, respectively.

Neyman Type A contagious distribution. The frequency function of the Neyman Type A distribution is

$$p(x) = e^{-\lambda_1} \frac{\lambda_2^x}{x!} \sum_{i=0}^{\infty} \frac{i^x}{i!} e^{-i\lambda_2} \lambda_1^i,$$

where $0 < \lambda_1 < \infty$ and $0 < \lambda_2 < \infty$. The Neyman Type A is a Poisson–Poisson. The generating function is

$$\phi(s) = \exp\{\lambda_1[e^{\lambda_2(s-1)} - 1]\};$$

the recursion formulas are

$$p(0) = \exp[\lambda_1(e^{-\lambda_2} - 1)],$$

$$p(x+1) = \frac{\lambda_1\lambda_2}{x+1} e^{-\lambda_2} \sum_{i=0}^{x} \frac{\lambda_2^i}{i!} p(x-i).$$

The mean is $\lambda_1\lambda_2$, and $\lambda_1\lambda_2(1 - \lambda_2)$ is the variance.

Point probability–negative exponential. An example of a mixture of a discrete and a continuous distribution is written in terms of its cumulative distribution function,

$$F(x) = \begin{cases} 0 & \text{if } x < 0 \\ \alpha_1 & \text{if } x = 0 \\ \alpha_1 + \alpha_2(1 - e^{-x/\theta}) & \text{if } x > 0, \end{cases}$$

where $\theta > 0$ and α_1 and α_2 are as in previous examples. An application in economics is given by Aitchison (1955). The distribution has mean $\alpha_2\theta$ and variance $\alpha_1(1 + \alpha_2)\theta^2$.

Identifiability. The subject of identifiability, that is, of unique characterization, is of concern on two levels. On both levels, identifiability, or lack thereof, has very important practical implications.

The first level is in model construction. Here the question revolves around the existence of a one-to-one correspondence between phenomena observed in nature and their corresponding mathematical models. That such one-to-one correspondences need not exist is apparent in the application of contagious distributions to accident data. Greenwood and Yule (1920) devised two models, *proneness* and *apparent proneness*, for accidents. Both lead to the negative binomial distribution. As noted by Feller (1943), even complete knowledge of the underlying negative binomial distribution therefore would not enable the experimenter to distinguish between proneness and apparent proneness as the cause of accidents.

The question of identifiability is an important consideration in mathematical modeling generally. It is particularly crucial in attempting to distinguish between competing theories. Such distinctions *can* be made through properly designed experiments. For example, proneness and apparent proneness can be distinguished in follow-up studies or by sampling in several time periods.

The question of identifiability also arises in a purely mathematical context. A mixture is called identifiable if there exists a one-to-one correspondence between the mixing distribution and the resulting mixture. Mixtures that are not identifiable cannot be expressed uniquely as functions of component and mixing distributions. This is true, for example, of a mixture of two binomials when $n = 2$. For example, both $\alpha_1 = .4$, $p_1 = .3$, $p_2 = .6$ and $\alpha_1 = .6$, $p_1 = .36$, $p_2 = .66$ result in identical mixed distributions.

The derivation of conditions under which mixtures are identifiable in this sense is difficult. Teicher (1963) derives some such conditions and discusses identifiability for many common mixtures. Identifiability in this sense has important statistical implications in that it is not possible to estimate or test hypotheses about the parameters of unidentifiable mixtures. [*For further discussion, see* STATISTICAL IDENTIFIABILITY.]

All of the mixtures listed in the section on "Special mixed distributions" are identifiable. The mixture of two binomial distributions given in equation (3) is identifiable if, and only if, $n \geqslant 3$.

Estimation. The construction of estimates for the parameters of mixtures of distributions is difficult. Procedures such as maximum likelihood and minimum χ^2, which are known to be optimal, at least for large sample sizes, require the solution of systems of equations that are typically intractable for mixtures. The problem has been somewhat alleviated with the advent of high-speed electronic computers, but it is by no means resolved. In any case the distributions of the estimators are li.:ely to be difficult to work with. [*For further discussion, see* ESTIMATION.]

Because of this complexity, it is not uncommon to choose estimation procedures almost solely on the basis of computational simplicity. Moment estimators are often used even though they may be inefficient in that they require larger sample sizes to attain a given degree of accuracy than do maximum likelihood estimators. Moment estimators are constructed by equating sample moments to population moments, the latter being written in terms of the parameters of the underlying distribution, then solving the resulting system of equations for the parameters. For example, for the Poisson–binomial distribution the moment estimators based on the sample mean, $\bar{x}$, and variance, s^2, are

$$\hat{\lambda} = \frac{(n-1)\bar{x}^2}{n(s^2-\bar{x})}, \qquad \hat{\theta} = \frac{s^2-\bar{x}}{(n-1)\bar{x}}.$$

Even moment estimators can become quite complex. Moment estimates for mixtures of binomial distributions can require formidable calculations (see Blischke 1964). Pearson's (1894) solution for a mixture of two normal distributions requires extraction of the roots of a ninth-degree equation. In the normal case, the problem is greatly simplified if the two normal components are assumed to have the same variance (see Rao 1952, section 8*b*.6).

Moment estimators and/or maximum likelihood estimators for the distributions mentioned above are given in the references. For a summarization of these results and some additional comments on the estimation problem, see Blischke (1963).

Another aspect of the estimation problem that can be troublesome in practice occurs if two components of a finite mixture are nearly identical. If this is the case, extremely large sample sizes are required to estimate the parameters of the mixture with any degree of accuracy. Such mixtures are, for all practical purposes, unidentifiable (cf. Chiang 1951).

Nomenclature. It is important to distinguish between the concepts of mixing of distributions and the distribution of a sum of random variables. The latter distribution is called a *convolution* and, except for special cases or particular notational conventions, is not a mixture of distributions.

There are, however, several additional terms

sometimes used as synonyms for "mixture of distributions." These include "compound distribution," "mixed distribution," "probability mixture," "superposition," "composite distribution," and "sum of distributions." The most common terms are "mixture," "mixed distribution," and "compound distribution." In addition, certain "generalized distributions" are mixtures of some specific structure (cf. Feller 1943).

The terms "dissection" and "decomposition" are sometimes used in connection with estimation for finite mixtures. These terms are descriptive since the estimates give information about the components from observations on the composite.

WALLACE R. BLISCHKE

BIBLIOGRAPHY

AITCHISON, JOHN 1955 On the Distribution of a Positive Random Variable Having a Discrete Probability Mass at the Origin. *Journal of the American Statistical Association* 50:901–908.

BLISCHKE, WALLACE R. (1963) 1965 Mixtures of Discrete Distributions. Pages 351–372 in Ganapati P. Patil (editor), *Classical and Contagious Discrete Distributions.* Proceedings of the International Symposium held at McGill University, Montreal, Canada, August 15–20, 1963. Calcutta: Statistical Publishing Society; Distributed by Pergamon Press.

BLISCHKE, WALLACE R. 1964 Estimating the Parameters of Mixtures of Binomial Distributions. *Journal of the American Statistical Association* 59:510–528.

CHIANG, CHIN LONG 1951 On the Design of Mass Medical Surveys. *Human Biology* 23:242–271.

DACEY, MICHAEL F. 1964 Modified Poisson Probability Law for a Point Pattern More Regular Than Random. Association of American Geographers, *Annals* 54:559–565.

DIXON, W. J. 1953 Processing Data for Outliers. *Biometrics* 9:74–89.

FELLER, W. 1943 On a General Class of "Contagious" Distributions. *Annals of Mathematical Statistics* 14: 389–400.

GOODMAN, LEO A.; and KRUSKAL, WILLIAM H. 1959 Measures of Association for Cross Classifications: II. Further Discussion and References. *Journal of the American Statistical Association* 54:123–163.

GREEN, BERT F. JR. 1952 Latent Structure Analysis and Its Relation to Factor Analysis. *Journal of the American Statistical Association* 47:71–76.

GREENWOOD, MAJOR; and YULE, G. UDNY 1920 An Inquiry Into the Nature of Frequency Distributions Representative of Multiple Happenings With Particular Reference to the Occurrence of Multiple Attacks of Disease or of Repeated Accidents. *Journal of the Royal Statistical Society* 83:255–279.

GURLAND, JOHN 1958 A Generalized Class of Contagious Distributions. *Biometrics* 14:229–249.

KAO, JOHN H. K. 1959 A Graphical Estimation of Mixed Weibull Parameters in Life-testing of Electron Tubes. *Technometrics* 1:389–407.

LAZARSFELD, PAUL F. 1950 The Logical and Mathematical Foundation of Latent Structure Analysis. Pages 362–412 in Samuel A. Stouffer et al., *Measurement and Prediction.* Studies in Social Psychology in World War II, Vol. 4. Princeton Univ. Press.

NEYMAN, J. 1939 On a New Class of "Contagious" Distributions, Applicable in Entomology and Bacteriology. *Annals of Mathematical Statistics* 10:35–57.

PEARSON, KARL 1894 Contributions to the Mathematical Theory of Evolution. Royal Society of London, *Philosophical Transactions* Series A 185:71–110.

RAO, C. RADHAKRISHNA 1952 *Advanced Statistical Methods in Biometric Research.* New York: Wiley.

STERNBERG, SAUL H. 1959 A Path-dependent Linear Model. Pages 308–339 in Robert R. Bush and William K. Estes (editors), *Studies in Mathematical Learning Theory.* Stanford Univ. Press.

STRANDSKOV, HERLUF H.; and EDELEN, EARL W. 1946 Monozygotic and Dizygotic Twin Birth Frequencies in the Total, the "White" and the "Colored" U.S. Populations. *Genetics* 31:438–446.

TEICHER, HENRY 1960 On the Mixture of Distributions. *Annals of Mathematical Statistics* 31:55–73.

TEICHER, HENRY 1963 Identifiability of Finite Mixtures. *Annals of Mathematical Statistics* 34:1265–1269.

VAGHOLKAR, M. K. 1959 The Process Curve and the Equivalent Mixed Binomial With Two Components. *Journal of the Royal Statistical Society* Series B 21: 63–66.

Postscript

As in many areas of statistical analysis, some of the most important advances with respect to the analysis of mixtures involve the computer. [*See* COMPUTATION.] This is particularly true with regard to the estimation problem, where most of the difficulties in the past have been computational rather than theoretical. In fact, Pearson's solution of a mixture of two normals (Pearson 1894) involves extraction of the roots of a ninth-degree equation! Some computerized techniques for estimating the parameters of finite mixtures of normals are discussed by Hasselblad (1966) and Gregor (1969). Mixtures of multivariate normal distributions and applications in cluster analysis are discussed by Day (1969), Marriott (1975), and Tarter and Silvers (1975). [*See* CLUSTERING.]

Other theoretical developments of interest include a method for classifying individuals in a sample from a mixture of two normal distributions (John 1970), tests for whether a population is indeed a mixture (Thomas 1969; Johnson 1973), and simplified techniques for estimating the parameters of many important discrete mixtures (Hinz & Gurland 1967).

WALLACE R. BLISCHKE

[*See also* ERRORS, *article on* EFFECTS OF ERRORS IN STATISTICAL ASSUMPTIONS.]

ADDITIONAL BIBLIOGRAPHY

DAY, N. E. 1969 Estimating the Components of a Mixture of Normal Distributions. *Biometrika* 56:463–474.

GREGOR, J. 1969 An Algorithm for the Decomposition of a Distribution Into Gaussian Components. *Biometrics* 25:79–93.

HASSELBLAD, VICTOR 1966 Estimation of Paramèters for a Mixture of Normal Distributions. *Technometrics* 8:431–444. → Discussion by A. Clifford Cohen on pages 445–446.

HINZ, PAUL; and GURLAND, JOHN 1967 Simplified Techniques for Estimating Parameters of Some Generalized Poisson Distributions. *Biometrika* 54:555–566.

JOHN, S. 1970 On Identifying the Population of Origin of Each Observation in a Mixture of Observations From Two Normal Populations. *Technometrics* 12: 553–563.

JOHNSON, NORMAN L. 1973 Some Simple Tests of Mixtures With Symmetrical Components. *Communications in Statistics* 1:17–26.

MARRIOTT, F. H. V. 1975 Separating Mixtures of Normal Distributions. *Biometrics* 31:767–770.

TARTER, MICHAEL; and SILVERS, ABRAHAM 1975 Implementation and Applications of Bivariate Gaussian Mixture Decomposition. *Journal of the American Statistical Association* 70:47–55.

THOMAS, E. A. C. 1969 Distribution Free Tests for Mixed Probability Distributions. *Biometrika* 56:475–484.

E

ECOLOGICAL ASSOCIATION

See FALLACIES, STATISTICAL; MULTIVARIATE ANALYSIS; STATISTICS, DESCRIPTIVE.

ECONOMETRIC MODELS, AGGREGATE

An *econometric model* is a set of equations designed to provide a quantitative explanation of the behavior of economic variables. This article discusses models that focus on the behavior of an economy in the *aggregate*, especially on the time paths of variables such as national income and product, consumption, investment, employment, the price level, the interest rate, etc. The pioneering model of this type was constructed by Jan Tinbergen (1939). The leading aggregate model builder for some years has been Lawrence R. Klein.

Aggregate econometric models grew out of a blend of several different streams of work. One is the mathematical stream springing from the work of Léon Walras, which represents the economy by a system of simultaneous equations. Another is the work of Ragnar Frisch and others in the theory of economic dynamics. A third is the work in statistical inference associated with Karl Pearson and his successors, showing how to estimate the value of unknown parameters with the aid of prior information and observed data. A fourth is the development by Willford King and Simon Kuznets and others of numerical estimates of national income and expenditure and their components (see Kendrick 1968). A fifth is the formulation of aggregative economic theories of income and employment, by R. F. Kahn, John Maynard Keynes, and others.

General features. The general characteristics of aggregate econometric models are described in the following sections. Then a very simple example is given, and contemporary models are discussed.

Definitional equations (identities). In any aggregate model some of the equations are *definitions* (usually called *identities*), of the type arising in national accounting; they are supposed to hold exactly and contain no unknown parameters. Examples are, "Consumption plus net investment plus government purchases plus exports minus imports equals net national product," and "Total money wage bill equals average money wage rate times quantity of labor input."

Stochastic equations. The remaining equations are stochastic. They are supposed to hold only approximately, and they contain disturbances that are assumed to be unobservable, small, and random, with expected values of zero. An example is, "Consumption during any period equals a constant proportion of that period's disposable income plus a constant proportion of the preceding period's consumption plus a third constant plus a random disturbance." In some models the disturbances take the form of random errors in the measurement of the variables. The assumption of *random* disturbances is very convenient for statistical estimation of the values of the unknown constants, called parameters. It is sometimes justifiable even if the disturbances have systematic components; for if those components are small and numerous and independent of each other, their total effect behaves approximately as if it were random. In formulating a model of this kind, one hopes to include explicitly in each equation all the important systematic influences that are present, so that the disturbances will be small and at least approximately random.

Structural equations. Some of the stochastic equations describe the *behavior* of a group in the economy, such as consumers (as in the foregoing example), investors in real capital goods, etc. Some describe institutional or technological *restraints*, such as the tax laws or the so-called production function, which indicates the maximum output that can be produced with any given quantities of inputs. Some describe *adjustment* processes that take place in particular markets (e.g., for labor or goods) when there is excess demand or supply. (A special case of an adjustment equation is an equilibrium condition asserting that demand equals supply.) These four types of equations (definitional, behavior, restraint, and adjustment) are called *structural* equations, for each is supposed to describe some more or less well-defined part of the structure of the economy.

Types of variables. In addition to constant parameters and unobservable random disturbances, the equations contain observable variables, usually more than there are equations in the model. Some of the variables are supposed to be determined by forces completely outside the model, and their values are assumed to be given; these are called *exogenous*. Variables often regarded as exogenous are government policy variables, population, foreign countries' actions, etc. The other variables, whose values are determined by the system when parameters, disturbances, and exogenous variables are given, are called *endogenous*. Typically, in a complete model there are just as many equations as endogenous variables. In many cases the equations for a given period will contain both current and lagged (i.e., past) values of the endogenous variables. The current endogenous variables are known as *jointly dependent* variables. The exogenous variables and lagged endogenous variables together are known as *predetermined* variables, for their values are determined as of any time period (either outside the system or by the past operation of the system) when the system goes to work to determine the jointly dependent variables for that time period.

The reduced form—forecasting. Suppose that the system of structural equations is solved for the jointly dependent variables, each being expressed as a function of structural-equation parameters, predetermined variables, and disturbances. The result is called the *reduced form* of the model. It could be used to forecast future values of the jointly dependent variables if its parameters and the future values of disturbances and predetermined variables were known in advance. In practice these are unknown, so that parameters must be estimated, future disturbances must be approximated by estimates of their expected values (zero is often used for these estimates, since the disturbances are assumed to have zero expected values), and future values of predetermined variables must be assumed. Thus, forecasts made from the reduced form are necessarily approximate.

When the exogenous and lagged endogenous variables are taken as given, reduced-form forecasts based on them are said to be *conditional* upon the values of the exogenous and lagged endogenous variables. For example, a model might forecast that *if* tax rates are cut 10 per cent at the end of this year and other predetermined variables are unchanged, then national income next year will be 7 per cent higher than this year; whereas *if* tax rates are not changed, other things being the same, then national income next year will be only 4 per cent higher than this year. When the unknown future values of predetermined variables are forecast in some way (for exogenous variables this involves using information from outside the model), reduced-form forecasts of jointly dependent variables are said to be *unconditional*. For example, one might forecast that tax rates *will* be cut by 10 per cent at the end of this year, and then use a model to forecast that next year's national income will be 7 per cent higher than this year's.

Dynamic features. If a model contains lagged endogenous variables, it has a dynamic character, for its jointly dependent variables are affected not only by parameters, disturbances, and exogenous variables but also by the past history of the system. Simple systems containing lagged values or year-to-year changes of endogenous variables can generate cycles and/or long-term growth or decline, even with no changes in parameters, disturbances, or exogenous variables. There are other devices for introducing dynamic effects, e.g., time-trend variables, derivatives with respect to time, and cumulative variables such as capital stock that is the sum of past net investments. [*See* TIME SERIES.]

Linearity versus nonlinearity. If the structural equations are linear in the jointly dependent variables and if the matrix of parameters of those variables is nonsingular, then the solution (the reduced form) is linear in the jointly dependent variables and is unique. If the structural equations contain nonlinearities in the jointly dependent variables, their solution will be nonlinear and may fail to be unique. In that case, one may use additional information to rule out the spurious solutions and find the one that represents the behavior of the economy (for example, any solution giving a negative national income would be spurious). Alternatively, a nonlinear model may be approximated by a linear one, with results that are acceptable as long as the range of variation of the variables being studied is small relative to their extreme values (this is likely to be so over short periods, but not over long periods). A model thus linearized has a linear reduced form. While models nonlinear in *variables* are fairly common, almost all models are built to be linear in unknown *parameters*, because that makes estimation of the parameters vastly simpler.

Model building and estimation. Numerical estimates of the unknown and supposedly constant parameters of reduced-form or structural equations are obtained by statistically fitting the equations to past data for the jointly dependent and predetermined variables, provided that the parameters have the property of *identifiability* [*see* SIMULTANEOUS EQUATION ESTIMATION; STATISTICAL IDENTIFIABILITY].

In this estimation process, one uses specifications indicating what variables appear in the model; which are endogenous and which are exogenous; what lags appear, if any; which variables appear in each of the structural equations; what the mathematical form of each structural equation is; and what properties the probability distribution of the random disturbances is assumed to have. In principle, these specifications are supposed to represent the model builder's prior knowledge of the economy—prior in the sense of arising from sources other than the observed data that are to be used in estimating the parameters. In practice, the model builder is often uncertain about some of the specifications of the model, and so he may try fitting several differently specified theoretically plausible models to the data and then choose the one that offers the best combination of (a) goodness of fit to the data and (b) consistency with any knowledge that may not have been incorporated into the formal specifications of the models. Such knowledge may come from economic theory, cross-section studies, results obtained for other countries or time periods, or other sources. Recently developed methods of Bayesian inference are suitable for incorporating probabilistic prior knowledge into the estimation process [*see* BAYESIAN INFERENCE].

Forecasting and model testing in practice. Conditional forecasts can be expected to be quite accurate if (a) the specification of the model is substantially correct for both the sample period and the forecast period; (b) the data sample and estimation technique used are such as to give approximately correct estimates of the reduced-form parameters under assumption (a); and (c) a highly accurate explanation of the sample-period jointly dependent variables is provided by the estimated reduced-form equations, in the sense that when the observed values of the predetermined variables are substituted into the reduced form to get calculated sample-period values of the jointly dependent variables, the calculated values are close to the observed values. Conditional forecasts can be expected to have substantial errors if the foregoing conditions are not met.

Condition (c), above, can readily be tested by substituting sample-period data into the estimated reduced form. Condition (b) can be tested in a probabilistic sense, with the aid of statistical inference techniques that reveal the degree of confidence one should have in the estimated parameters, on the assumption that the model is correctly specified. Condition (a), the correct specification of the model, is more difficult to test. Economic theory provides some information concerning the adequacy of a model's specifications, but the most powerful test is the indirect one based on the quality of forecasts that a model makes when conditions (b) and (c) are reasonably well satisfied. Individual *structural* equations can also be usefully tested by similar procedures, although the typical structural equation contains more than one jointly dependent variable and hence is not capable of making forecasts in the same sense as is the reduced form.

Unconditional forecasts, of course, can go wrong if the three foregoing conditions are not met, and also if the values of the predetermined variables used in the forecasts are not substantially correct.

A simple example. A very simple three-equation model will illustrate many of the foregoing points. Equation (1), below, is the accounting definition mentioned earlier: net national product (NNP) equals consumption plus net private domestic investment plus government purchases plus exports minus imports. Equation (2), below, is the consumer behavior equation mentioned earlier, specifying that consumption is a linear function of disposable income and lagged consumption plus a random disturbance. A third equation is needed to relate disposable income to NNP. Assume an economy in which (a) the whole of government revenue is raised by an income tax whose yield is a linear function of NNP; (b) there are no transfer payments; and (c) all business income is paid out to individuals, so that disposable income is also a linear function of NNP. Equation (3), below, expresses this. The three structural equations of this model are as follows (the notation and units are explained below):

(1) $y = c + i + g$,
(2) $c = \alpha + \beta d + \gamma c_{-1} + u \cong 5.7 + .69d + .25c_{-1}$,
(3) $d = y(1 - m) - h$.

This model specifies further that there are three endogenous variables (c = consumption, d = disposable income, and y = NNP); and that there are four exogenous variables (i = net private domestic investment, g = government purchases plus exports less imports, h = the fixed part of tax revenues independent of NNP, and m = the marginal tax rate on NNP). Lagged consumption is denoted by c_{t-1} or in brief by c_{-1}; α, β, and γ are three unknown parameters; and u is a random disturbance with zero mean. All these quantities are expressed in billions of real (i.e., deflated) dollars per year, except for m, β, and γ, which are pure numbers between 0 and 1.

The approximate numerical estimate of the consumption equation (2), above, was obtained by the two-stage least squares estimation method, from United States data (expressed in billions of 1954 dollars per year) for the years 1929–1941 and 1946–1959. The ratios of the three estimated parameters to their estimated standard errors are respectively 1.8, 12, and 3.5; and the standard error of estimate is 2.7 billion 1954 dollars per year (as compared with the maximum and minimum observed consumption values of 288.9 and 103.5 billion respectively).

The tax variable h may be negative if the income tax allows for a fixed total exemption. If equation (3) were to be applied to the United States economy, it would have to include a disturbance, for the United States tax structure is only *very roughly* described by a linear function of NNP.

The reduced form of this model is obtained by solving the three structural equations for y, d, and c, thus:

$$y = \frac{\alpha + \gamma c_{-1} + u + i + g - \beta h}{1 - \beta(1 - m)},$$

$$d = \frac{(1 - m)(\alpha + \gamma c_{-1} + u + i + g) - h}{1 - \beta(1 - m)},$$

$$c = \frac{\alpha + \gamma c_{-1} + u + \beta(1 - m)(i + g) - \beta h}{1 - \beta(1 - m)}.$$

Note that the model and its reduced form are linear in *endogenous* variables, but not in *all* variables because of the term containing ym in equation (3). If in the reduced form one substitutes estimated values for the three structural parameters, zero for the disturbance u, and numerical values for the five predetermined variables for a certain year, one obtains estimates of the expected values of the three jointly dependent variables y, d, and c for that year, conditional on the chosen values of the predetermined variables.

Medium-scale models. Most of the medium-scale aggregate econometric models published so far have from 14 to 48 equations and accordingly are much more detailed and complex than the simple example just given, although, of course, they still involve great simplifications of reality. These models differ from each other in significant ways, but the following features are typical of most of the medium-scale models listed in the bibliography.

There is an identity, substantially like equation (1), above, stating that national product equals consumption plus investment plus government purchases plus imports minus exports. Consumption, private investment, and imports (and in some cases exports) are endogenous variables, and there are behavior equations to explain them or their components. Consumption is sometimes divided into parts, with one equation for each—such as consumer durable goods, nondurable goods, and services—and is explained in terms of variables such as disposable income, past consumption, liquid asset holdings, consumer credit conditions, and income distribution. Investment is commonly separated into plant and equipment purchases and inventory investment, and in some models residential construction is treated in one or more separate equations. Plant and equipment investment is explained in terms of accelerator variables, such as output and capital stock, or profits or both. Inventory investment is usually specified to depend on lagged inventory holdings, sales or output, and other variables. Most import functions depend on income and exogenous import prices. If exports are not exogenous, they are a function of exogenous variables such as world income and world prices. Government purchases are regarded as exogenous. [*See* INVENTORY CONTROL THEORY; *see also* Eisner 1968; Stanback 1968; Tobin 1968.]

In the early models tax revenues and government transfer payments (such as social security benefits, unemployment compensation, and interest on the national debt) were specified as exogenous, but in more recent models the tax and transfer *schedules* are specified as exogenous and equations are provided to explain tax and transfer payments as endogenous variables depending upon national income, the number of retired persons, unemployment, the national debt, and so on, according to the exogenous schedules.

There is a production function, to explain the output of goods and services in terms of inputs of labor and capital. There is a demand-for-labor equation, in many cases expressing total real labor income in terms of total real output. There is an identity expressing property income as the difference between total income and labor income. In many cases there is an equation explaining the allocation of property income between retained income (which is not part of disposable income) and payments to individuals (including interest and dividends).

There is a wage-rate adjustment equation, commonly expressing the change in the money wage rate in terms of the unemployment rate and the rate of inflation. (But see Netherlands 1961 for a model in which the wage rate is exogenous, since it has been a policy variable in the Netherlands in recent years.) Unemployment of course is the difference between the labor force, typically exogenous, and employment.

The general price level is endogenous in most of these models; and in some cases other prices appear that, when endogenous, are usually expressed as functions of the general price level. Typically, real output and the price level can be thought of as determined by a pair of equations containing only these two variables plus predetermined variables, the equations being obtained by substituting into two important identities all the other equations of the model. One of these identities is that equating output to the sum of all expenditure components; the other expresses the total real wage bill as employment multiplied by the real wage rate.

Among the more commonly used exogenous policy variables are government purchases, government transfer payments, and tax revenues or tax rates. In some models there are few or no variables or equations describing interest rates and the supply and demand for money, and where they do appear, they are commonly rather loosely tied to the rest of the model. Thus, most models built so far are much better suited to an analysis of effects of fiscal policy (government purchases, taxes, and transfer payments) than the effects of monetary policy.

Among the more commonly used exogenous nonpolicy variables are population and its distribution (and sometimes labor force), import prices, exports (or world income and world prices), and time. In certain more recent models there are variables measuring attitudes or anticipations, obtained from surveys.

Behavior equations, in most cases, contain variables in real (deflated) terms rather than in money terms, to reflect the theoretical postulate that real economic behavior depends upon real tastes and real opportunities, unaffected by price changes that leave these things the same. Some consumer behavior equations are stated in per capita terms, to allow for the possibility that an increase in aggregate income may have different effects, depending upon how it is distributed between increases in population and in per capita income; but most behavior equations are stated in aggregate terms.

Nearly all models contain some nonlinearities, especially where identities of the form "value equals price times quantity" are involved; but nonlinearities in unknown parameters are rare in stochastic equations whose parameters have to be estimated. Tinbergen's models (1939, 1951) and the model of the Netherlands (1961) have had all their nonlinear equations linearized.

First differences (i.e., year-to-year changes) of the data are used occasionally (e.g., Suits 1962), but most models use ordinary data, without this transformation. Early models used annual data, but quarterly models are becoming more common as quarterly data become available. Time trends, lags, and cumulated variables are the main devices used for dynamic effects, although occasionally a ratchet-type variable is used for this purpose. An example of a ratchet-type variable is the value of disposable income at its previous peak, which is sometimes included in the consumption function.

Parameters of structural equations are estimated by a variety of methods. Least squares is common, in spite of its asymptotic bias in a simultaneous equations context. With the advent of electronic computers, consistent estimating methods have become cheap and are increasingly used, especially the limited-information and the two-stage least squares methods [*see* SIMULTANEOUS EQUATION ESTIMATION].

Large-scale models. Two large-scale models have appeared. One, dealing with the United States and having 219 equations, is sponsored by the Social Science Research Council (SSRC) and the Brookings Institution (Duesenberry et al. 1965). The other, dealing with Japan and having 164 equations, is the work of a group at Osaka University (Ichimura et al. 1964). In essential conception these models are similar to the medium-scale models discussed above; they innovate chiefly in providing a more detailed treatment of certain markets and sectors of the economy. The SSRC–Brookings model goes into detail particularly regarding consumption, housing, fixed and inventory investment, new orders, six nonagricultural production sectors, agriculture, government, and population and labor force. The Osaka model is particularly detailed regarding fixed and inventory investment, eight production sectors (including agriculture), foreign trade, and the monetary and financial sector. The SSRC–Brookings model contains a seven-by-seven input–output model corresponding to its seven production sectors, and plans are under way for a more detailed model to contain about 32 production sectors (see Leontief 1968).

Application and evaluation. Econometric model forecasts and their comparison with subsequent actual events have been all too uncommon but are becoming accepted as one important means of evaluating models. Some recent results have been quite good. The Klein–Goldberger model (1955) and its successors under Suits (1962) have been used each November since 1952 to make annual unconditional forecasts of real United States gross national product (GNP) for the following year. The percentage errors for the years 1953 to 1962 have been as follows (where, for example, −0.7 means that the model's forecast of GNP was too low by 0.7 per cent of the subsequently observed value, +0.5 means that it was 0.5 per cent too high, etc.): −0.7, +0.5, −6.4, +0.2, +0.5, +0.05, −4.0, −1.6, +0.7, −0.1.

Such models may also be used for simulation studies of the economy's stability and long-term behavior, as in Adelman and Adelman (1959), or of its reaction to policy changes. Simulation studies, as well as forecasts, acquire practical value to the extent that the models used can be shown to be accurate representations of the relevant aspects of economic behavior and not merely systems of equations that fit past data well. [*See* SIMULATION, *article on* ECONOMIC PROCESSES.]

CARL F. CHRIST

[*Directly related are the entries* BUSINESS CYCLES: MATHEMATICAL MODELS; PREDICTION AND FORECASTING, ECONOMIC. *See also* Dernburg 1968; Nataf 1968.]

BIBLIOGRAPHY

Following most of the entries is a notation giving the time period fitted, the number of equations, and the number of exogenous variables.

ADELMAN, IRMA; and ADELMAN, FRANK L. 1959 The Dynamic Properties of the Klein–Goldberger Model. *Econometrica* 27:596–625.

BROWN, T. M. 1964 A Forecast Determination of National Product, Employment, and Price Level in Canada, From an Econometric Model. Pages 59–86 in Conference on Research in Income and Wealth, *Models of Income Determination.* Studies in Income and Wealth, No. 28. Princeton Univ. Press. → 1926–1941 and 1946–1956 annually; 40 equations, 47 exogenous variables.

CHRIST, CARL F. 1951 A Test of an Econometric Model for the United States: 1921–1947. Pages 35–107 in *Conference on Business Cycles, New York, 1949.* New York: National Bureau of Economic Research. → 1921–1941 and 1946–1947 annually; 14 equations, 16 exogenous variables.

CHRIST, CARL F. 1956 Aggregate Econometric Models: A Review Article. *American Economic Review* 46, no. 3:385–408.

►DERNBURG, THOMAS 1968 Income and Employment Theory. Volume 7, pages 122–132 in *International Encyclopedia of the Social Sciences.* Edited by David L. Sills. New York: Macmillan and Free Press.

DUESENBERRY, JAMES S.; ECKSTEIN, OTTO; and FROMM, GARY 1960 A Simulation of the United States Economy in Recession. *Econometrica* 28:749–809.

DUESENBERRY, JAMES S., et al. (editors) 1965 *The Brookings Quarterly Econometric Model of the United States.* Chicago: Rand McNally; Amsterdam: North-Holland Publishing. → 1948–1962 quarterly; 219 independent equations (including about 150 estimated equations), over 100 exogenous variables.

►EISNER, ROBERT 1968 Investment: I. The Aggregate Investment Function. Volume 8, pages 185–194 in *International Encyclopedia of the Social Sciences.* Edited by David L. Sills. New York: Macmillan and Free Press.

ICHIMURA, SHINICHI et al. 1964 A Quarterly Econometric Model of Japan: 1952–1959. *Osaka Economic Papers* 12, no. 2:19–44. → 1952–1959 quarterly; 164 equations, about 130 exogenous variables. This paper presents only the equations and definitions of symbols; a book describing the model is scheduled to appear.

►KENDRICK, JOHN W. 1968 National Income and Product Accounts. Volume 11, pages 19–34 in *International Encyclopedia of the Social Sciences.* Edited by David L. Sills. New York: Macmillan and Free Press.

KLEIN, LAWRENCE R. 1950 *Economic Fluctuations in the United States: 1921–1941.* New York: Wiley. → 1921–1941 annually; 16 equations, 13 exogenous variables.

KLEIN, LAWRENCE R. 1961 A Model of Japanese Economic Growth: 1878–1937. *Econometrica* 29:277–292. → 1878–1937 quinquennially; 10 equations, 3 exogenous variables.

KLEIN, LAWRENCE R. 1964 A Postwar Quarterly Model: Description and Applications. Pages 11–36 in Conference on Research in Income and Wealth, *Models of Income Determination.* Studies in Income and Wealth, No. 28. Princeton Univ. Press. → A model of the United States. 1948–1958 quarterly; 34 equations, 19 exogenous variables.

KLEIN, LAWRENCE R.; and GOLDBERGER, A. S. 1955 *An Econometric Model of the United States: 1929–1952.* Amsterdam: North-Holland Publishing. → 1929–1941 and 1946–1952 annually; 20 or 25 equations, 20 exogenous variables.

KLEIN, LAWRENCE R.; and SHINKAI, Y. 1963 An Econometric Model of Japan: 1930–1959. *International Eco-*

nomic Review 4:1–28. → 1930–1936 and 1951–1958 annually; 22 equations, 15 exogenous variables.

KLEIN, LAWRENCE R. et al. 1961 *An Econometric Model of the United Kingdom.* Oxford: Blackwell. → 1948–1956 quarterly; 37 equations, 34 exogenous variables.

►LEONTIEF, WASSILY 1968 Input–Output Analysis. Volume 7, pages 345–354 in *International Encyclopedia of the Social Sciences.* Edited by David L. Sills. New York: Macmillan and Free Press.

LIEBENBERG, MAURICE; HIRSCH, ALBERT A.; and POPKIN, JOEL 1966 A Quarterly Econometric Model of the United States: A Progress Report. *Survey of Current Business* 46, no. 5:13–39. → 1953–1964 quarterly; 49 equations, 32 exogenous variables.

LIU, TA-CHUNG 1963 An Exploratory Quarterly Econometric Model of Effective Demand in the Postwar U.S. Economy. *Econometrica* 31:301–348. → 1947–1959 quarterly; 36 equations, 16 exogenous variables.

NARASIMHAN, NUTI V. A. 1956 *A Short-term Planning Model for India.* Amsterdam: North-Holland Publishing. → 1923–1948 annually; 18 equations, 13 exogenous variables.

►NATAF, ANDRÉ 1968 Aggregation. Volume 1, pages 162–168 in *International Encyclopedia of the Social Sciences.* Edited by David L. Sills. New York: Macmillan and Free Press.

NERLOVE, MARC 1962 A Quarterly Econometric Model for the United Kingdom: A Review Article. *American Economic Review* 52, no. 1:154–176.

NERLOVE, MARC 1966 A Tabular Survey of Macro-econometric Models. *International Economic Review* 7:127–175.

NETHERLANDS, CENTRAL PLANBUREAU 1961 *Central Economic Plan 1961.* The Hague: The Bureau. → 1923–1938 and 1949–1957 annually; 30 equations, 20 exogenous variables.

SMITH, PAUL E. 1963 An Econometric Growth Model of the United States. *American Economic Review* 53, no. 4:682–693. → 1910–1959 annually; 10 equations, 1 exogenous variable.

►STANBACK, THOMAS M. JR. 1968 Inventories: I. Inventory Behavior. Volume 8, pages 176–181 in *International Encyclopedia of the Social Sciences.* Edited by David L. Sills. New York: Macmillan and Free Press.

SUITS, DANIEL B. 1962 Forecasting and Analysis With an Econometric Model. *American Economic Review* 52, no. 1:104–132. → 1947–1960 annually; 32 equations, 21 exogenous variables.

TINBERGEN, JAN 1939 *Statistical Testing of Business-cycle Theories.* Volume 2: Business Cycles in the United States of America: 1919–1932. Geneva: Economic Intelligence Service, League of Nations. → 1919–1932 annually; 48 equations, 22 exogenous variables.

TINBERGEN, JAN 1951 *Business Cycles in the United Kingdom: 1870–1914.* Amsterdam: North-Holland Publishing. → 1870–1914 annually; 45 equations, 9 exogenous variables.

►TOBIN, JAMES 1968 Consumption Function. Volume 3, pages 358–369 in *International Encyclopedia of the Social Sciences.* Edited by David L. Sills. New York: Macmillan and Free Press.

UENO, HIROYA 1963 A Long-term Model of the Japanese Economy, 1920–1958. *International Economic Review* 4:171–193. → 1920–1936 and 1952–1958 annually; 38 equations, 35 exogenous variables.

VALAVANIS-VAIL, STEFAN 1955 An Econometric Model of Growth: USA 1869–1953. *American Economic Review* 45, no. 2:208–221. → 1869–1948 quinquennially; 20 equations, 7 exogenous variables.

Postscript

There are now many new aggregate econometric models for many countries. Project LINK, under the direction of Lawrence R. Klein, is designed to tie together models for several countries so that the international implications of economic policies and events in each participating country can be analyzed.

Models of the United States economy have been modified to give a larger role to monetary policy and the quantity of money, as more persuasive evidence has accumulated to support the view that these phenomena are important. The number of equations in the models has been increased, as has the degree of nonlinearity. Computer simulation techniques have been devised and applied, so that each model can now be used to generate a hypothetical future time-path for each of its endogenous variables under arbitrarily specified conditions.

CARL F. CHRIST

ADDITIONAL BIBLIOGRAPHY

BATTENBERG, DOUGLAS et al. 1975 MINNIE: A Small Version of the MIT–PENN–SSRC Econometric Model. *Federal Reserve Bulletin* Nov.:721–727. → An appendix entitled "Equation Listing" was issued in mimeographed form.

CHRIST, CARL F. 1975 Judging the Performance of Econometric Models of the U.S. Economy. *International Economic Review* 16:54–74.

CONFERENCE ON ECONOMETRIC MODELS OF CYCLICAL BEHAVIOR, HARVARD UNIVERSITY, 1969 1972 *Econometric Models of Cyclical Behavior.* 2 vols. Edited by Bert G. Hickman. Studies in Income and Wealth, No. 36. New York: Columbia Univ. Press, for the National Bureau of Economic Research.

DUESENBERRY, JAMES S. et al. (editors) 1969 *The Brookings Model: Some Further Results.* Chicago: Rand McNally; Amsterdam: North-Holland Publishing.

Econometric Model Performance: Comparative Simulation Studies of Models of the U.S. Economy. 1974–1975 *International Economic Review* 15:265–414, 541–654; 16:1–113. → A symposium of 17 papers dealing mainly with 11 leading U.S. models, namely, the Brookings, Bureau of Economic Analysis, Data Resources, Fair, Hickman–Coen, Liu–Hwa, Michigan, MIT–Penn, St. Louis Federal Reserve Bank, and two Wharton models.

FAIR, RAY C. 1971 *A Short-run Forecasting Model of the United States Economy.* Lexington, Mass.: Heath.

FAIR, RAY C. 1974 *A Model of Macroeconomic Activity.* Volume 1: *The Theoretical Model.* Cambridge, Mass.: Ballinger.

FROMM, GARY; and KLEIN, LAWRENCE R. 1973 A Comparison of Eleven Econometric Models of the United States. *American Economic Review* 63:385–393.

Fromm, Gary; and Klein, Lawrence R. 1976 The NBER/NSF Model Comparison Seminar: An Analysis of Results. *Annals of Economic and Social Measurement* 5:1–28.

Fromm, Gary; and Taubman, Paul 1968 *Policy Simulations With an Econometric Model.* Washington: Brookings.

McCarthy, Michael D. 1972 *The Wharton Quarterly Econometric Forecasting Model, Mark III.* Studies in Quantitative Economics, No. 6. Philadelphia: Wharton School of Finance and Commerce, Univ. of Pennsylvania.

McNees, Stephen K. 1975 An Evaluation of Economic Forecasts. Federal Reserve Bank of Boston, *New England Economic Review* Nov.–Dec.: 1–39.

ECONOMETRICS

Succinctly defined, econometrics is the study of economic theory in its relations to statistics and mathematics. The essential premise is that economic theory lends itself to mathematical formulation, usually as a system of relationships which may include random variables. Economic observations are generally regarded as a sample drawn from a universe described by the theory. Using these observations and the methods of statistical inference, the econometrician tries to estimate the relationships that constitute the theory. Next, these estimates may be assessed in terms of their statistical properties and their capacity to predict further observations. The quality of the estimates and the nature of the prediction errors may in turn feed back into a revision of the very theory by which the observations were organized and on the basis of which the numerical characteristics of the universe postulated were inferred. Thus, there is a reciprocating relationship between the formulation of theory and empirical estimation and testing. The salient feature is the explicit use of mathematics and statistical inference. Nonmathematical theorizing and purely descriptive statistics are not part of econometrics.

The union of economic theory, mathematics, and statistics has been more an aspiration of the econometrician than a daily achievement. Much of what is commonly known as econometrics is mathematical economic theory that stops short of empirical work; and some of what is known as econometrics is the statistical estimation of *ad hoc* relationships that have only a frail basis in economic theory. That achievement falls short of aspiration, however, ought not to be discouraging. It is part of the developmental process of science that theories may be advanced untested and that the search for empirical regularities may precede the systematic development of a theoretical framework. A consequence of this, however, is that although the word "econometrics" clearly implies measurement, much abstract mathematical theorizing that may or may not ultimately lend itself to empirical validation is often referred to as part of econometrics. The meaning of the word has frequently been stretched to apply to mathematical economics as well as statistical economics; and in common parlance the "econometrician" is the economist skilled and interested in the application of mathematics, be it mathematical statistics or not. In this article I shall accept this extended definition and consider both econometrics in its narrow sense and mathematical economic theory.

A brief history

The use of mathematics and statistics in economics is not of recent origin. In the latter part of the seventeenth century Sir William Petty wrote his essays on "political arithmetik" [*see the biography of* Petty]. This fledgling work, remarkable for its time, was econometric in its methodological framework, even from the modern point of view. Despite the fact that it was not referred to by Adam Smith, it had a discernible influence on later writers. In 1711 Giovanni Ceva, an Italian engineer, urged the adoption of the mathematical method in economic theory. Although many statistical studies appeared during the intervening years, the revolutionary impact of the mathematical method did not occur until the latter part of the nineteenth century. More than any other man, Léon Walras, professor at the University of Lausanne, is acknowledged to be the originator of general equilibrium economics, which is the basic framework of modern mathematical economics (see Jaffé 1968). His work, removed from any immediate statistical application, developed a comprehensive system of relationships between economic variables, including money, in order to explain the mutual determination of prices and quantities of commodities and capital goods produced and exchanged. Walras conceived of the economy as operating along the lines of classical mechanics, the state of the economy being determined by a balancing of forces between all market participants. His general equilibrium system was, however, essentially static because the values of the economic variables did not themselves determine their own time rates of change. For that reason the term "equilibrium" is something of a misnomer because, since Walras' general system was not explicitly dynamic, its solution cannot be described as an equilibrium state. Nevertheless, as is still true in much of economic theorizing, there were side discussions of the adjustment properties of the economy, and so, in a wider context, the solution can be regarded as the result of a balancing of dynamic forces of adjustment.

The significant combination of mathematical theory and statistical estimation first occurred in the work of Henry Luddell Moore, a professor at Columbia University during the early part of the twentieth century (see Stigler 1968). Moore did genuine econometric work on business cycles, on the determination of wage rates, and on the demand for certain commodities. His major publication, culminating some three decades of labor, was *Synthetic Economics*, which appeared in 1929. Incredibly, this work, of such seminal importance for the later development of a significant area of social science, sold only 873 copies (Stigler 1962).

Econometrics came to acquire its identity as a distinct approach to the study of economics during the 1920s. The number of persons dedicated to this infant field grew steadily, and, on December 29, 1930, they established an international association called the Econometric Society. This was achieved in large measure through the energy and persistence of Ragnar Frisch of the University of Oslo, with the assistance and support of the distinguished American economist, Irving Fisher, a professor at Yale University (see Allais 1968*a*). To call this small minority of economists a cult would impute to them too parochial and evangelical a view; nevertheless, they had a sense of mission "to promote studies that aim at a unification of the theoretical-quantitative and the empirical-quantitative approach to economic problems and that are penetrated by constructive and rigorous thinking similar to that which has come to dominate in the natural sciences" (Frisch 1933).

Their insights and ambitions were well founded. During the following years and through many a methodological controversy about the role of mathematics in economics (a topic now rather passé) their numbers grew and their influence within the wider profession of economics was steadily extended. Today all major university departments of economics in the Western world, including most recently those in the Soviet-bloc countries, offer work in econometrics, and many place considerable stress upon it. Specific courses in econometrics have been introduced even at the undergraduate level; textbooks have been written; the younger generation of economists entering graduate schools arrive with improved training in mathematics and statistical methods, gravitate in what appear to be increasing proportions toward specialization in econometrics, and soon excel their teachers in their command of econometric techniques. Membership in the Econometric Society increased from 163 in 1931 to over 2,500 in 1966. The society's journal, *Econometrica*, has virtually doubled in size over these years, and nearly all other scholarly journals in economics publish a regular fare of articles whose mathematical and statistical sophistication would have dazzled the movement's founders in the 1920s and 1930s.

Areas of application of econometrics within economics have been steadily widened. There is now scarcely a field of applied economics into which mathematical and statistical theory has not penetrated, including economic history. With the increasing interest and concentration in econometrics on the part of the economics profession, the very notion of specialization has become blurred. With its success as a major intellectual movement within economics, econometrics is losing its identity and is disappearing as a special branch of the discipline, becoming now nearly conterminous with the entire field of economics. These remarks must not be misunderstood, however. There remain many problems and much research in economics that is neither mathematical nor statistical, and although the modern economist's general level of training and interest in mathematics and statistics far exceeds that of his predecessors, a quite proper gradation of these skills and interests inevitably continues to exist. Moreover, to repeat, much of what is known as econometrics still falls short of the interrelating of the mathematical-theoretical and the statistical, which is the aspiration contained in the field's definition.

A survey of econometrics

Since econometrics is no longer a small enclave within economics, a survey of its subject matter must cover much of economics itself.

General equilibrium. Pursuing Walras' conception of a general economic equilibrium, mathematical economists have in recent years been engaged in a far more thorough analysis of the problem than Walras offered (see Arrow 1968). In the earlier work a general economic equilibrium was described by a system of equalities involving an indefinitely large number of economic variables, but a number equal to the number of independent equations. It was presumed that a system of simultaneous equations with the same number of unknowns as independent equations would have an "equilibrium" solution. This is loose mathematics, and in recent times economic theorists have been concerned to redevelop the earlier theory with greater rigor. Equality of equations and unknowns is neither a necessary nor a sufficient condition for either the existence or the uniqueness of a solution. Consequently, one cannot be sure that the early theory is adequate to explain the general equilibrium state to which the economy is postulated to converge. This might be because the theory does not impose conditions necessary to assure the existence of a general equilibrium state or because the theory might be indeterminate in that several different solutions are implied by it. The modern equilibrium theorist has therefore tried to nail down the necessary and sufficient conditions for the existence and uniqueness of the general economic equilibrium.

The concept of an *equilibrium* is that of a state in which no forces within the model, operating over time, tend to unbalance the system. Even if such a state can be demonstrated to exist within the framework of some general equilibrium model, there remains the question of whether it is stable or unstable, that is, whether, for any departure of the system from it, forces tend to restore the original equilibrium or to move the system further away. An analysis of these questions, which are rather more involved than suggested here, requires the explicit introduction of dynamical adjustment relationships.

Questions of the existence, uniqueness, and stability of an equilibrium are, in the present context, not questions about the actual economy, but questions regarding the properties of a theoretical model asserted to describe an actual economy. In this sense, their examination is oriented toward an improved understanding of the implications of alternate specifications of the theory itself rather than toward an improved empirical understanding of how our economy works.

Most of this work, moreover, has been restricted to an examination of the general equilibrium model of a *competitive* economy, which is a special case indeed. It is a case of particular interest, however, because, under idealized assumptions, welfare economists have imputed features to a competitive equilibrium that satisfy criteria which are regarded as interesting for a social evaluation of economic performance (see Mishan 1968). According to a concept of Pareto's, a state of the economy (not to be thought of as unique) is said to be optimal if there is no other state that is technologically feasible in which some individual would be in a position he prefers while no individual would be in a position that he finds worse (see Allais 1968*b*). The conditions under which a general economic equilibrium would be optimal in this sense have therefore been subject to rigid scrutiny. Thus, Pareto welfare economics is intimately involved in the modern examination of general equilibrium systems, but it is not well developed as an empirical study.

The positive economist, concerned with prediction, has also been concerned with general equilibrium systems in principle, but from a different point of view. His central question is, How does a change in an economic parameter (a coefficient or perhaps the value of some autonomous variable not itself determined by the system) induce a change in the equilibrium value of one or more other variables that are determined by the system? In short, How does the equilibrium solution depend upon the parameters? This is a problem in *comparative statics* which contrasts two different equilibria defined by a difference in the values of one or more parameters.

Comparative statics—partial equilibrium. It is in the problem of comparative statics—the comparison of alternative equilibrium states—that we can most ably distinguish between general equilibrium economics and *partial equilibrium economics*, a familiar contrast in the literature.

Suppose that, in the neighborhood of an equilibrium, a general system of simultaneous economic relationships is differentiated totally with respect to the change in a particular parameter, so that all direct and indirect effects of that change are accounted for. One might then hope to ascertain the direction of change of a particular economic variable with respect to that parameter. For example, if a certain tax rate is increased or if consumer preferences shift in favor of a particular commodity, will the quantity demanded of some other commodity increase, diminish, or stay the same? This question can sometimes be answered on the basis of the constellation of signs (plus, minus, zero) of many or all of the partial derivatives of the functions constituting the system (assuming here that they are continuously differentiable). Theoretical considerations or common sense may enable one to specify a priori the signs of these partial derivatives, for example, to assert that an elasticity of demand is negative or that a cross elasticity of demand is positive. In some cases, however, the theorist is not comfortable in making such assertions about a derivative, and hence some signs may be left unspecified. The question is whether the restrictions that the theorist is willing to impose a priori suffice to determine whether the total derivative of the economic variable of interest wtih respect to a given parameter is positive, negative, or zero. The formal consideration of the necessary and sufficient restrictions needed to resolve this question unambiguously constitutes the study of qualitative economics and presents a mathematical problem in its own right (Samuelson 1947; Lancaster 1965). In some situations it may be critical to know not only the signs of various partial derivatives but also their relative algebraic magnitudes. This points to the need for the statistical estimation of these derivatives, a task belonging to econometrics in its most narrow meaning.

At times it is also useful to know that certain derivatives are sufficiently close to zero that if they are assumed equal to zero the conclusion about the sign of the total derivative being investigated would not be affected. The trick or art of deciding when to regard certain partial derivatives as zero, that is, of deciding that certain economic variables do not enter in any significant way into certain relationships, is the essence of partial equilibrium analysis, so called because it tends to isolate a portion of the general system from other portions that have little interaction with it. Partial equilibrium analysis is, thus, a special case of general equilibrium analysis, in which more daring a priori restrictions have been introduced with the object of deducing more specific and meaningful results in comparative statics. Just as general equilibrium economics has been commonly associated with the name of Walras, so partial equilibrium economics has been associated with the work of Alfred Marshall (see Corry 1968).

In qualitative economics some light is shed upon the signs of the partial derivatives of the system by considering the dynamic stability of the model. With assumptions about the nature of the dynamic adjustment relationships, correspondences might be found between the conditions necessary for an equilibrium to be stable and the signs of the partial derivatives. Thus, just as stability depends upon assumptions about whether different variables enter a given relationship positively or negatively, so, also, the way in which those variables enter a given relationship may sometimes be inferred from the assumption that an equilibrium is stable. This is the famous correspondence principle, due to Samuelson. (See Baumol 1968.)

Spatial models. Most general equilibrium models have conceived of the economy as existing at a single point in space, thereby ignoring transportation costs, the regional specialization of resources, and locational preferences. Some studies, however, explicitly introduce the spatial dimension in which a general equilibrium occurs. This provides a framework for the study of interregional location, specialization, and interdependency in exchange. (See Moses 1968.) These models, because of their greater complexity, generally involve more-special assumptions, such as linearity of relationships and the absence of opportunities for substitution among factor services in production. They have also, however, lent themselves more directly to empirical work.

In the application of partial equilibrium analysis to problems of spatial economics, it is assumed, moreover, that the locations of certain economic activities are determined independently of the location decisions regarding other economic activities, and therefore the former can be regarded as fixed in the analysis of the latter. (For a discussion of this line of inquiry, see Hoover 1968.)

Aggregation and aggregative models. Since general equilibrium systems are conceived of as embracing millions of individual relationships, they obviously do not lend themselves to quantitative estimation. Much interest, therefore, inheres in reducing the dimensionality of the system, so that there is some possibility of econometric estimation. This means that relations of a common type, such as those describing the behavior of firms in a given industry or households of a certain character, need to be *aggregated* into a single relationship describing the behavior of a collectivity of comparable economic agents. The conditions needed to make such aggregation possible and the methods to be used are still in a rather preliminary stage of exploration. But a literature is developing on this subject. (See Nataf 1968.)

An older problem is simply that of aggregating into a single variable a multiplicity of similar variables. This is the familiar problem of "index numbers"—for example, how best to represent the prices of a great variety of different commodities by a single price index. The index number problem, therefore, has its theoretical aspects [*see* INDEX NUMBERS, *article on* THEORETICAL ASPECTS] as well as its statistical aspects [*see* INDEX NUMBERS, *articles on* PRACTICAL APPLICATIONS *and* SAMPLING]. The theory has been useful in guiding the interpretation of alternative statistical formulas.

The major efforts in the empirical study of general equilibrium systems that have to some limited degree been aggregated come under the heading *input–output analysis*. This approach, originated by Wassily Leontief in the late 1930s, consists essentially in considering the economy as a system of simultaneous *linear* relationships and regarding as constant the relative magnitudes of the inputs into a production process that are necessary to produce the process's output. These inputs may, of course, be the outputs of other processes. Thus, with fixity of coefficients relating the inputs and outputs of an integrated production structure, it is possible to determine what "bill of goods" can be produced, given an itemization of the quantities of various "primary" nonproduced inputs that are available. Alternatively, the quantities of primary inputs necessary to produce a given bill of goods can also be determined. The coefficients of such a system can be estimated by observing the ratios of inputs to outputs for various processes in a given year or by averaging these ratios over a sequence of years or by using engineering estimates. This may be done for an economy divided into a large number of different sectors (a hundred or more), or it may be done for portions of an economy, such as a metropolitan area. Moreover, the sectoring of the economy may be by regions as well as by industries, and the former makes the method applicable to the study of interregional or international trading relations. A great deal of empirical research has been done on input–output models, tables of coefficients having now been developed for over forty countries. The quantitative analysis of the workings of these models has, as one may readily surmise, required the availability of large-scale computers. (See Leontief 1968.)

Aggregative models in economics may be of either the partial or the general equilibrium type. Those of a partial equilibrium type deal with a single sector of the economy in isolation, under the assumption that the external economic variables that have an important impact on that sector are not in turn influenced by its behavior. Thus, for example, a market model of demand and supply for a particular commodity may regard the total income of consumers and its distribution as determined independently of the price and output of the particular commodity being studied. Yet, the market demand and supply functions are aggregates of the demand and supply functions of many individuals and firms. Aggregative models of the general equilibrium type may explain the mutual determination of many major economic variables that are aggregates of vast numbers of individual variables. Examples of the aggregate variables are total employment, total imports, total inventory investment, etc. These models are generally called *macroeconomic* models, in contrast to *microeconomic* models, which deal, in a partial equilibrium sense, with the individual household, firm, trade union, etc. Many macroeconomic models treat not only so-called *real* variables, which are physical stocks and flows of goods and productive services, but also *monetary* variables, such as price levels, the quantity of money, the value of total output, and the interest rate. Models of this sort have been especially common since 1936, having been stimulated by John Maynard Keynes's *General Theory of Employment, Interest and Money* and by the literature that devolved therefrom.

One type of aggregative, macroeconomic model is that which distinguishes a few important sectors of the economy or which relates macroeconomic variables of two or more economies interrelated in trade. Much of the theory of international trade deals with models of this sort (see McKenzie 1968). In fact, since this has been a natural mode for the analysis of international economic problems, international trade theory has historically been one of the liveliest areas for the development of economic theory, both mathematical and otherwise. More narrowly econometric studies in this area have focused on estimates of import demand elasticities.

Moreover, macroeconomic models have lent themselves to the study of economic change, and it is with these models that the most significant work in economic dynamics has occurred. Dynamical systems in economics are those in which the values of the economic variables at a given point in time determine either their own rates of change (continuous, differential equation models) or their values at a subsequent point in time (discrete, difference equation models). (For a general discussion of dynamic models see Baumol 1968.) Thus, dynamic models involve both variables and a measure of their changes over time. The former often occur as "stocks," and the latter as "flows." When both stocks and flows enter into a given model, there are complexities in reconciling the desired quantities of each. These problems become especially important when monetary variables are introduced, for example, when we consider the desire of individuals both to hold a certain value of monetary assets and to save (add to assets) at a certain rate. (Specific problems of stock-flow models are discussed in Clower 1968.)

Dynamic models arise both in the theory of long-run economic growth (see Morishima 1968), where both macroeconomic and completely disaggregated general equilibrium models have been employed, and in the theory of business fluctuations or business cycles [*see* BUSINESS CYCLES: MATHEMATICAL MODELS], where macroeconomic models are most common. Not all models intended to explain the level of business activity need be cyclical in character. The modern emphasis is more on macroeconomic models, cyclical or not, that explain the level of business activity and its change by a dynamical system that responds to external variables. These include variables of economic policy (government deficit, central bank policy, etc.) and other variables that, while having an important impact on the economy, have their explanation outside the bounds of the theory, for example, population growth and the rate of technological change. Thus, outside variables, known as exogenous or autonomous variables, play upon the dynamical economic system and generate fluctuations over time that need not be periodic. These models lend themselves to empirical investigation, and a great deal of work has been done in estimating them [*see* ECONOMETRIC MODELS, AGGREGATE]. The structure of these models has been refined and developed as a consequence of the empirical work.

The great advantage of aggregative models, of course, is that they substantially reduce the vast number of variables and equations that appear in general equilibrium systems and thereby make estimation possible. Even so, these models can be quite complex, either because they still contain a large number of variables and equations or because of nonlinearities in their functional forms. The modern computer makes it possible to estimate systems of this degree of complexity, however. But if one is interested in analyzing the dynamical behavior of these systems, the difficulties often transcend our capabilities in mathematical analysis. The computer once again comes to the rescue. It is possible, with the computer, to simulate complex systems of the type being considered, to drive them with exogenous variables, and to shock them with random disturbances drawn from defined probability distributions. In that way the performance of these systems under a variety of assumptions regarding the behavior of the exogenous variables and for a large sample of random variables can be surveyed. [*Simulation studies of this sort are discussed in* SIMULATION, *article on* ECONOMIC PROCESSES.]

Variables that commonly arise in macroeconomic models are aggregate consumer expenditure, inventory investment, and plant and equipment investment. Aggregate consumer expenditure, or consumption, reflects the behavior of households in deciding how much to spend on consumer goods, which in some studies may be further broken down into categories such as consumer durables, nondurables, and services. Using regression techniques, consumer expenditure is made to depend upon other variables, some of which are economic in character (consumer income, change in income, highest past income, the consumer price level and its rate of change, interest rates and terms of consumer credit, liquid assets, etc.) and some of which are demographic (race, family size, urban–rural residence, etc.). The empirical study of the dependency of consumer expenditure on variables of these kinds has been intensive during the past twenty years. (For a survey of this work, see Tobin 1968.)

The behavior of inventory investment has likewise been the object of intensive study, both in terms of how inventories have varied over time relative to the general level of business activity and in terms of how inventory investment has responded to such variables as the interest rate, sales changes, unfilled orders, etc. (This work is reviewed in Stanback 1968.) There are some subtle issues involved in formulating an inventory investment function. Sometimes inventories accumulate when firms intend they should, and other times they accumulate despite the desire of firms to reduce them, for example, when sales fall off rapidly relative to the capability of firms to alter their rates of output. Theoretical work concerned with the optimum behavior of firms in matters of inventory policy can therefore provide some underpinning to the selection and interpretation of the role of different variables in an inventory investment function. [*See* INVENTORY CONTROL THEORY.]

The dependency of plant and equipment investment upon such variables as business sales, sales changes, business profits, liquidity, etc., may also be studied by econometric techniques, and different theories have been advanced to support notions about the relative importance of these different variables. As with the consumption function and the determination of inventory investment, the plant and equipment investment function has also been the subject of intensive empirical research over the past couple of decades. (This work is reviewed in Eisner 1968.)

Decision making. Though it is methodologically proper for the economist to postulate *ad hoc* relationships between macroeconomic variables (Peston 1959), it is more gratifying, more unifying of economic theory, if the behavior of the macrovariables can be derived from elementary propositions regarding the behavior of the microvariables whose aggregates they are. This is the aggregation problem, referred to earlier. The aspiration is that an axiomatic theory of the behavior of the individual economic decision maker, most importantly the individual (or household) and the firm, can serve as a fundament to theories of the interaction of the aggregated macrovariables. Most of the behavioral theory of firms and households, however, is in the context of partial equilibrium analysis, because the individual economic agent does not bother to take account of the very slight influence that his own decisions exert on the market or on the economy as a whole. Thus, each household and each competitive (but not monopolistic) firm regards market prices as fixed and unaffected by its own choices. But in linking together such partial equilibrium models of the behavior of vast multitudes of individual households and firms, one cannot ignore the impact of their combined behavior on the very market variables they regard as constants. Thus, the partial equilibrium micromodels must be reorganized into more general models allowing for these individually unperceived but collectively important interactions.

Microeconomic theory is largely deductive, proceeding systematically from axioms regarding preference and choice to theorems regarding economic behavior. To proceed carefully through the logical intricacies of this deductive theory, formal mathematics is heavily invoked. Market decisions of economic agents are usually hypothesized to be prudent or rational decisions, by which is meant that they conform by and large to certain basic criteria of decision making that are thought to have wide intuitive appeal as precepts of prudent or rational choice. The situations in which a decision maker may be called upon to choose can be formulated in a variety of ways. There are "static" situations, where the decision is not assumed to have a temporal or sequential character. There are "dynamic" situations, in which a sequence of decisions must be made and in some consistent way. The decision problem may also be categorized according to the knowledge the decision maker believes he has about the consequences of his decisions. At one extreme is the case of complete certainty, where the consequences are thought to be completely known in advance. Other cases involve risk and arise when the decision maker is assumed to know only the probability distribution of the various outcomes that can result from the decision he makes. Finally, at the other extreme, the decision problem may be conceived of as involving almost complete uncertainty, in which case the decision maker knows what the possible outcomes are but has no a priori information about their probabilities. [*For a discussion of various criteria proposed for these different situations, see* DECISION MAKING, *article on* ECONOMIC ASPECTS.] Fundamental, however, is the notion that the decision maker has preferences and that he exercises these within the range of choice available to him. An index giving his preferences is commonly called *utility* and is conceived of as a function of the objects of his choice. In particular, where an individual with a fixed income is choosing among various "market baskets" of commodities, utility is commonly postulated to be a function of the components of the market basket. Axiomatic systems that are necessary and sufficient for the existence of such a function have been the object of intensive study by mathematical economists. (This central problem and many subtle aspects of it are considered in Georgescu-Roegen 1968.) Great effort, with perhaps little benefit for empirical economics, has gone into the refinement of the axiomatics of utility theory or of the theory of consumer choice; unfortunately, much less work has been done to strengthen the assumptions of the theory so as to increase its empirical content. From the theory of consumer behavior comes the concept of the demand function of the consumer for a particular commodity, depending characteristically on all prices and on income.

As for the theory of the firm, prudent, purposeful behavior is also assumed, and in the theory's most common formulation it is supposed that the firm wishes to maximize some measure of its preference among streams of future profits. This must be done subject to the prices that the firm must pay for factor services, the market opportunities it confronts when selling its products, and its internal technology of production. From this analysis comes the theory of production and of supply. (For the theory of production of the firm, see Smith 1968; for econometric studies of production relationships and of the cost of production, see Walters 1968; and for econometric studies of demand and supply, see Fox 1968.)

In the derivation of the theory of consumer behavior and the theory of the firm, purposeful and prudent behavior has characteristically been associated with the notion that the decision maker attempts to maximize some function subject to market and technological constraints. Thus, the mathematics of constrained maximization has served the economist as the most important tool of his trade. In an effort to develop models of maximizing behavior that would lend themselves better to quantitative formulation and solution, interest came to focus on problems where the function being maximized is linear and the constraints constitute a set of linear inequalities. Methods for solving such problems became known as linear programming. With further advances, nonlinearities and random elements were introduced, and the method came to be applied, as well, to problems of sequential decision making. The entire area is now known as mathematical programming [*see* PROGRAMMING]. Because of their practical usefulness, these methods lent themselves to the analysis of various specific planning and optimization problems, especially to problems internal to the operation of the firm. Stimulated by the availability of these techniques, as well as by advances in probability theory and some wartime experience in systems analysis, there has come to flourish a modern quantitative approach to the problems of production and business management. This is known as management science or operations research [*see* OPERATIONS RESEARCH]. This development is a case of fission, management science now being regarded as distinct from econometrics, although both fields have much in common and share many a professor and practitioner.

The most complex problems in the area of prudent decision making are those that involve strategical considerations. In its essence this means that the consequence of a decision or action taken by one participant depends upon the actions taken by others; but their actions in turn depend upon the actions of each of the other participants. Thus, the structure of the problem is not that of simple maximizing, even in the face of risk or uncertainty, but is that of the strategical game. [*See* GAME THEORY, *article on* THEORETICAL ASPECTS.] Based upon considerations of the prudent strategy of the individual participant and of the incentives for subsets of participants to form coalitions, the theory of games can be presented as a general equilibrium problem and has become intimately associated with the modern work in general equilibrium economics. In a more partial context, game theory has appeared applicable to the decision problems of firms in oligopolistic and bilateral monopoly situations. These are characterized by the fact that each firm, in choosing its best course of action, must take into account the effect of its action on the actions of other firms, which also perform in a prudent way. In general, the early enthusiasm for the application of game theory to these problems of industrial behavior has thus far been confirmed in only limited degree. [*For a review of the applications of game theory to business behavior, see* GAME THEORY, *article on* ECONOMIC APPLICATIONS.]

Distribution processes. A concern of long standing in economics has been the size distribution of economic variables. What determines the distribution of family incomes or the distribution of the assets or sales of firms in a given industry? In past years these problems have been dealt with descriptively by fitting frequency distributions to the data of different countries, different years, or different industries. A good fit to data from different sources could be declared an empirical "law"; thus, the Pareto law of income distribution. In more recent years the size distribution problem has been redefined. Econometricians now regard it as one of formulating a dynamic process of growth or decay with random elements. The task is to estimate the parameters of the process and to determine whether there is an equilibrium distribution of the size of units and what that distribution is. A good fit can thus have a theoretical mechanism behind it, and the parameters can be made to depend upon other economic variables that may change or may be controlled. [*In this connection, see* SIZE DISTRIBUTIONS IN ECONOMICS *and* MARKOV CHAINS.]

Statistical methods. In the natural sciences the investigator must make his own measurements. In economics, however, the economy itself generates data in vast quantities. Taxpayers, business firms, banks, etc., all record their operations, and in many cases these records are available to the economist. Unfortunately, these data are not always precisely the kind the economist wants, and they must frequently be adjusted for scientific purposes. In recent decades the government has been engaged increasingly in the accumulation and processing of economic data. This has been of tremendous help in the development of econometrics. Not only is this the case for the United States and western European governments but data are also accumulated in the planned economies, where they are of critical importance in the planning operation. (See Liu 1968; Ruggles 1968; Spulber 1968.) The absence of adequate data is felt most severely in the study of the underdeveloped economies, although, through the United Nations and other organizations, an increasing amount of data for those parts of the world is being gathered and collated.

A major form in which economic data occur is that of successive recordings of economic observations over time. Thus, there may be many years of price data for particular commodities, of employment data, etc. The econometrician, therefore, has traditionally been heavily concerned with time series analysis [*see* TIME SERIES] and especially with the use of regression methods, where the various observations are ordered in a temporal sequence. This has lent itself to the development of dynamical regression equations attempting to explain the observation of a particular date as a function not only of other variables but also of one or more past values of the same variable. Thus, the dynamic regression relationship is a difference equation incorporating a random term. When many past values of the variable are introduced into the difference equation, so that it is one of very high order, it becomes difficult to estimate the coefficients of these past variables without losing many degrees of freedom. As a result, the econometrician has tried to impose some pattern of relationship on these coefficients, so that they may all be estimated as functions of relatively few parameters. This is the technique of distributed lag regressions. [*See* DISTRIBUTED LAGS.]

The techniques just described have largely come to replace the older methods of time series decomposition, whereby a time series is split up into such components as trend, cycles of various lengths, a seasonal pattern of variation, and a random component. These methods implied the interaction of recurrent influences of regular periodicity and amplitude. With the move toward the difference equation and regression approach, exogenous variables have been introduced and random disturbances made cumulative in their effects. The temporal performance of a time series is thereby described less in terms of some inherent law of periodicity and more in terms of a succession of responses to random influences and to the temporal variation of other causal variables. Forecasting is not, therefore, the inexorable extrapolation of rhythms but is the revised projection, period by period, of an incremental relationship depending on present and past values, on exogenous variables, and on random elements. [*See* PREDICTION AND FORECASTING, ECONOMIC.]

Nevertheless, it has always been sensible to assume a rather strict periodicity for the seasonal component because of the recurrent nature of seasons, holidays, etc. As a result, when studying time series where the observations are daily, weekly, or monthly, it is customary first to estimate and remove the seasonal influence. [*Techniques for doing this are discussed in* TIME SERIES, *article on* SEASONAL ADJUSTMENT.]

The other kind of data that the economist uses is cross-sectional. For example, he may use a sample of observations, all made at approximately the same time, of assets, income, and expenditures of different households, firms, or industries. [*See* CROSS-SECTION ANALYSIS.] By observing differences in the behavior of the individuals in the sample and, again usually through regression analysis, ascribing these differences to differences in other variables beyond the control of these individuals, the econometrician attempts to infer how the behavior of similar economic units would change over time if the values of the independent variables were to alter. There are many pitfalls in this process of inferring change over time for a given firm or household on the basis of differences among firms and households at a given point of time. What becomes especially useful are data that are both cross sectional and time series in character, as, for example, when the budgets of a sample of households are observed, each over a number of successive years. To obtain usable information of a cross-section or of a cross-section and time-series sort commonly requires the design of a sample survey. (The application of survey methods in economics is discussed in Morgan 1968.)

A very common problem in econometrics arises when different variables are related in different ways. For example, aggregate investment depends on national income, but national income depends, in a different way, on aggregate investment. In demand and supply analysis, equilibrium quantity exchanged and market price must satisfy both a demand and a supply function simultaneously. This simultaneity of multiple relationships between the same variables presents special problems in the application of regression methods. These problems have been much studied over the past twenty years, and various devices for dealing with them are now available. These methods are often quite complex, but with advances in statistical theory and in the availability of data and with the use of the large-scale computer they have come into common use in estimating both partial equilibrium and macroeconomic models, sometimes of quite large dimension. Although touched upon only briefly here, this commanding problem in statistical methodology is perhaps the most central feature of econometric analysis and is the subject of a number of texts and treatises. It is also probably the largest block of material covered in most special courses in econometrics. [*See* SIMULTANEOUS EQUATION ESTIMATION.]

To those engaged in research at the frontiers of any science, progress seems always to be exceedingly slow; but a review of the accomplishments of econometricians both in the development of economic theory and in its quantitative estimation and testing over the past two or three decades gives one the feeling of great achievement. But as old problems are solved, new ones are invented. Thus, the advance of econometrics continues unabated.

ROBERT H. STROTZ

BIBLIOGRAPHY

○ *Works dealing with the nature and history of econometrics are* Divisia 1953; Frisch 1933; Tintner 1953, 1954. *Basic works in the field are* Allen 1956; Malinvaud 1964; Samuelson 1947; Theil 1971.

►ALLAIS, MAURICE 1968*a* Fisher, Irving. Volume 5, pages 475–485 in *International Encyclopedia of the Social Sciences.* Edited by David L. Sills. New York: Macmillan and Free Press.

►ALLAIS, MAURICE 1968*b* Pareto, Vilfredo: I. Contributions to Economics. Volume 11, pages 399–411 in *International Encyclopedia of the Social Sciences.*

Edited by David L. Sills. New York: Macmillan and Free Press.

ALLEN, R. G. D. (1956) 1963 *Mathematical Economics.* 2d ed. New York: St. Martin's; London: Macmillan.

►ARROW, KENNETH J. 1968 Economic Equilibrium. Volume 4, pages 376–389 in *International Encyclopedia of the Social Sciences.* Edited by David L. Sills. New York: Macmillan and Free Press.

►BAUMOL, WILLIAM J. 1968 Statics and Dynamics in Economics. Volume 15, pages 169–177 in *International Encyclopedia of the Social Sciences.* Edited by David L. Sills. New York: Macmillan and Free Press.

►CLOWER, ROBERT W. 1968 Stock-flow Analysis. Volume 15, pages 273–277 in *International Encyclopedia of the Social Sciences.* Edited by David L. Sills. New York: Macmillan and Free Press.

DIVISIA, FRANÇOIS 1953 La Société d'Économétrie a atteint sa majorité. *Econometrica* 21:1–30.

►EISNER, ROBERT 1968 Investment: I. The Aggregate Investment Function. Volume 8, pages 185–194 in *International Encyclopedia of the Social Sciences.* Edited by David L. Sills. New York: Macmillan and Free Press.

►FOX, KARL A. 1968 Demand and Supply: II. Econometric Studies. Volume 4, pages 104–111 in *International Encyclopedia of the Social Sciences.* Edited by David L. Sills. New York: Macmillan and Free Press.

[FRISCH, RAGNAR] 1933 Editorial. *Econometrica* 1:1–4.

►GEORGESCU-ROEGEN, NICHOLAS 1968 Utility. Volume 16, pages 236–267 in *International Encyclopedia of the Social Sciences.* Edited by David L. Sills. New York: Macmillan and Free Press.

►HOOVER, EDGAR M. 1968 Spatial Economics: I. The Partial Equilibrium Approach. Volume 15, pages 95–100 in *International Encyclopedia of the Social Sciences.* Edited by David L. Sills. New York: Macmillan and Free Press.

►JAFFÉ, WILLIAM 1968 Walras, Léon. Volume 16, pages 447–453 in *International Encyclopedia of the Social Sciences.* Edited by David L. Sills. New York: Macmillan and Free Press.

LANCASTER, K. J. 1965 The Theory of Qualitative Linear Systems. *Econometrica* 33:395–408.

►LEONTIEF, WASSILY 1968 Input–Output Analysis. Volume 7, pages 345–354 in *International Encyclopedia of the Social Sciences.* Edited by David L. Sills. New York: Macmillan and Free Press.

►LIU, TA-CHUNG 1968 Economic Data: III. Mainland China. Volume 4, pages 373–376 in *International Encyclopedia of the Social Sciences.* Edited by David L. Sills. New York: Macmillan and Free Press.

►MCKENZIE, LIONEL W. 1968 International Trade: II. Mathematical Theory. Volume 8, pages 96–104 in *International Encyclopedia of the Social Sciences.* Edited by David L. Sills. New York: Macmillan and Free Press.

MALINVAUD, EDMOND (1964) 1966 *Statistical Methods in Econometrics.* Chicago: Rand McNally. → First published in French.

►MISHAN, E. J. 1968 Welfare Economics. Volume 16, pages 504–512 in *International Encyclopedia of the Social Sciences.* Edited by David L. Sills. New York: Macmillan and Free Press.

►MORGAN, JAMES N. 1968 Survey Analysis: III. Applications in Economics. Volume 15, pages 429–436 in *International Encyclopedia of the Social Sciences.* Edited by David L. Sills. New York: Macmillan and Free Press.

►MORISHIMA, MICHIO 1968 Economic Growth: III. Mathematical Theory. Volume 4, pages 417–422 in *International Encyclopedia of the Social Sciences.* Edited by David L. Sills. New York: Macmillan and Free Press.

►MOSES, LEON N. 1968 Spatial Economics: II. The General Equilibrium Approach. Volume 15, pages 100–108 in *International Encyclopedia of the Social Sciences.* Edited by David L. Sills. New York: Macmillan and Free Press.

►NATAF, ANDRÉ 1968 Aggregation. Volume 1, pages 162–168 in *International Encyclopedia of the Social Sciences.* Edited by David L. Sills. New York: Macmillan and Free Press.

PESTON, M. H. 1959 A View of the Aggregation Problem. *Review of Economic Studies* 27, no. 1:58–64.

►RUGGLES, RICHARD 1968 Economic Data: I. General. Volume 4, pages 365–369 in *International Encyclopedia of the Social Sciences.* Edited by David L. Sills. New York: Macmillan and Free Press.

SAMUELSON, PAUL A. (1947) 1958 *Foundations of Economic Analysis.* Harvard Economic Studies, Vol. 80. Cambridge, Mass.: Harvard Univ. Press. → A paperback edition was published in 1965 by Atheneum.

►SMITH, VERNON L. 1968 Production. Volume 12, pages 511–519 in *International Encyclopedia of the Social Sciences.* Edited by David L. Sills. New York: Macmillan and Free Press.

►SPULBER, NICHOLAS 1968 Economic Data: II. The Soviet Union and Eastern Europe. Volume 4, pages 370–372 in *International Encyclopedia of the Social Sciences.* Edited by David L. Sills. New York: Macmillan and Free Press.

►STANBACK, THOMAS M. JR. 1968 Inventories: I. Inventory Behavior. Volume 8, pages 176–181 in *International Encyclopedia of the Social Sciences.* Edited by David L. Sills. New York: Macmillan and Free Press.

STIGLER, GEORGE J. 1962 Henry L. Moore and Statistical Economics. *Econometrica* 30:1–21.

►STIGLER, GEORGE J. 1968 Moore, Henry L. Volume 10, pages 479–481 in *International Encyclopedia of the Social Sciences.* Edited by David L. Sills. New York: Macmillan and Free Press.

►THEIL, HENRI 1971 *Principles of Econometrics.* New York: Wiley.

TINTNER, GERHARD 1953 The Definition of Econometrics. *Econometrica* 21:31–40.

TINTNER, GERHARD 1954 The Teaching of Econometrics. *Econometrica* 22:77–100.

►TOBIN, JAMES 1968 Consumption Function. Volume 3, pages 358–369 in *International Encyclopedia of the Social Sciences.* Edited by David L. Sills. New York: Macmillan and Free Press.

►WALTERS, A. A. 1968 Production and Cost Analysis. Volume 12, pages 519–523 in *International Encyclopedia of the Social Sciences.* Edited by David L. Sills. New York: Macmillan and Free Press.

ECONOMIC FORECASTING

See PREDICTION AND FORECASTING, ECONOMIC. *Related material may be found under* TIME SERIES.

ECONOMIC MAN

See DECISION MAKING, *article on* ECONOMIC ASPECTS.

EDGEWORTH, FRANCIS YSIDRO

Francis Ysidro Edgeworth (1845–1926) was raised on the family estate of Edgeworthstown, County Longford, Ireland. His father died when Edgeworth was two years old. Edgeworth studied under tutors and spent considerable time reading and memorizing the classics and English poetry. From an early age he also read widely in Spanish, French, German, and Italian. This interest in Continental literature may well have been strengthened by the fact that his mother was the daughter of a Spanish political refugee. One of Edgeworth's aunts was the famous novelist Maria Edgeworth, and the poet Thomas Lovell Beddoes was his cousin.

At the age of 17, Edgeworth entered Trinity College, Dublin, and then proceeded to Balliol College, Oxford, where he received first class honors in *litterae humaniores*. After several years of practicing law in London, he accepted a lectureship in logic at King's College and was later appointed Tooke professor of political economy.

In 1891 he became Drummond professor of political economy at Oxford and a fellow of All Souls College. Never marrying, he resided principally at All Souls for the remainder of his life, although he maintained rooms in London for over fifty years and inherited the family estate in 1911.

Although he was highly respected as a teacher and original thinker by both economists and statisticians of his time, Edgeworth does not seem to have profoundly influenced the then current thinking in either field. Present readers, however, find much that is stimulating and informative in Edgeworth and not uncommonly note with some surprise that ideas they believe to have originated recently are, sometimes clearly, sometimes vaguely, anticipated in Edgeworth's work.

When he became professor emeritus at Oxford in 1922, Edgeworth was serving his second term as president of the economic section of the British Association and was a vice-president of the Royal Economic Society. He had earlier served as president (1912) and council member of the Royal Statistical Society; he had been awarded its Guy medal in 1907. He was also a fellow of the British Academy.

Undoubtedly, Edgeworth's greatest professional contribution, in addition to teaching and writing, was his editorship of the *Economic Journal*. He became its first editor in 1891 and was later chairman of the editorial board. At the time of his death in 1926, Edgeworth was joint editor with Keynes.

He is reported (L. L. P. 1926) to have been an effective and stimulating teacher, attracting good students and encouraging many vigorous discussions. Frequently he would become one of the less active participants, intervening occasionally to expose an error or to bring a neglected but fruitful problem to the attention of his class.

Edgeworth's main written contributions to economics are contained in seven small books and numerous journal articles and reviews. In 1925 his principal articles and reviews were published in three volumes entitled *Papers Relating to Political Economy* (see 1891–1921) under his own editorship and under the sponsorship of the Royal Economic Society.

Mathematical Psychics, published in 1881, is probably Edgeworth's most important writing. His "contract curve" and its generalization, the set of "Edgeworth-allocations" (Debreu & Scarf 1963; Vind 1964) or "core," are still basic concepts for the theoretical study of exchange equilibrium and welfare economics; and his theory of barter has provided a valuable point of departure for recent research on dynamic economic adjustments (Uzawa 1962). Edgeworth discussed a number of interesting bargaining situations in light of the theory presented.

Edgeworth's arguments in the first section of *Psychics* against the common view that mathematics could be applied only to numerical phenomena were quite advanced and provocative then but will appear awkward and unnecessary to a current reader who has studied set theory and topology. Much of the remainder of the book is concerned with attempts to establish practical implications of utilitarian ethics. These views were, and remain, highly controversial.

Edgeworth's collected works include a group of related papers (1897*a*; 1897*b*; 1899; 1911; 1912) on taxation, price discrimination, and monopoly that involve original theoretical developments, which are still of considerable interest. In these papers are a number of famous paradoxes, such as: A tax on one of two monopolized commodities which are appropriately related in production, consumption, or both may cause prices of both commodities to fall; and the introduction of price discrimination into a competitive market might benefit both producers and consumers (where consumer benefit is taken to be aggregate consumer surplus). The 1912 paper contains a very good discussion of the difficulties of establishing a theoretical equilibrium for duopolists or monopolists who sell related commodities. A number of other papers deal with tax problems of Edgeworth's time and with utilitarian principles for determining equitable taxation. Quantitative problems encountered in contemplated applications of utilitarianism are also treated in his first book, *New and Old Methods of Ethics* (1877).

Edgeworth served several years as secretary of the Committee on Value of the Monetary Standard of the British Association for the Advancement of Science. This interest is reflected in ten articles on index numbers and the value of money. Two memoranda prepared for the British Association for the Advancement of Science (1887–1889) contain what can still be regarded as a fairly comprehensive discussion of theoretical problems of index number constructions.

Metretike draws a number of interesting parallels beween utility and its measurement and application, on the one hand, and probability and its measurement and application, on the other (1887*a*). Although there are some interesting observations, there is no fundamental interrelationship such as that later achieved through the axiom systems of Ramsey, Savage, and others.

The other four of Edgeworth's books are lectures on war finance delivered during the war and immediately afterward (1915*a*; 1915*b*; 1917; 1919). These works, like most of his writings on current policy issues, drew more criticism than praise from his colleagues: "[He] might descend eventually on one side or the other of the fence, but, . . . he kept himself so long poised evenly midway that the final movement when, and if, it happened, was apt to be unnoticed" (L. L. P. 1926). Keynes remarked:

> He feared a little the philistine comment on the strange but charming amalgam of poetry and pedantry, science and art, wit and learning, of which he had the secret; and he would endeavor, however unsuccessfully, to draw a veil of partial concealment over his native style, which only served, however, to enhance the obscurity and allusiveness and half-apologetic air with which he served up his intellectual dishes. (Keynes [1926] 1963, p. 227)

Some basis for these complaints must be admitted. Even in his theoretical and highly mathematical writings, Edgeworth's expository skill is notably less than his insight and logical ability. The overapologetic tone Keynes mentions is frequently apparent; it is illustrated by Edgeworth's description of his correction of a colleague's misinterpretation as: "points of detail on which the critical shoemaker corrected the masterpiece of the Grecian painter ([1891–1921] 1963, vol. 1, p. 143).

Another factor which frequently led to lengthy and sometimes complicated passages in Edgeworth and drew critical comments from his contemporaries was his desire to account explicitly for every logically possible combination of circumstances. Many current readers are more patient on this score. A more mathematically inclined generation is better aware of the advantages of carefully keeping track of the exact relations that exist between alternative conditions and consequences.

Among Edgeworth's other publications that should be noted are his article on the law of error in the 1902 *Encyclopædia Britannica*, his article on probability in the 1911 edition, and his biography of Mill and article on index numbers in *Palgrave's Dictionary of Political Economy*.

Edgeworth's best-known work on distribution theory is a series which gives an asymptotic approximation to a fairly general class of distribution functions and is still sometimes used (Cramer 1945, pp. 228–231). Under the heading "The law of error," Edgeworth used this series to derive several versions of the central limit theorem and to approximate a number of empirical frequency distributions (1883; 1898–1900; 1905; 1926; and with Bowley, 1902). He also extended Galton's work on correlation, and he developed a formula for the general multivariate normal density, and considered some of its properties (1893; 1905).

Following Bernoulli, Laplace, and Mill, Edgeworth was basically Bayesian in his approach to statistical inference. His "genuine inverse method" (1908–1909) consisted of finding a normal approximation to the posterior distribution of unknown parameters (*quaesita*, or frequency constants, in his terminology), assuming a diffuse prior distribution, and equating the means of the posterior distribution to their maximum likelihood estimates (called "most probable values"). He properly objected to forming a uniform prior on the basis of ignorance, since this would imply a specific nonuniform prior on any transformation of the original parameters. However, as Edgeworth also recognized, any reasonably smooth and unconcentrated prior will lead to results that differ little from those obtained from either a uniform distribution or a diffuse distribution when the number of observations is large. Thus Edgeworth's genuine inverse method was based on some of the same underlying ideas as the theory of stable estimation examined earlier by Fisher and Jeffreys and recently emphasized by Bayesian statisticians (Edwards et al. 1963).

Not all of Edgeworth's work on inference was developed from this approach. He believed the prior distribution would often be unknown and looked for other principles and devices. He discussed the difference of two means (1905; 1908–

1909) and the possible significance of an estimate of trend (1886) in language much like that of the Neyman–Pearson theory of hypothesis testing.

Several of Edgeworth's papers are concerned with the problem of the "best" mean (e.g., 1887*a*). This usually means an estimate of a population parameter, optimal according to some stated criterion; but the case of finding a suitable descriptive representative for a collection of empirical data is also considered. A criterion frequently employed is that of minimizing expected loss (called "least detriment" or "minimum disadvantage"), and a number of inquiries proceed under the assumption that all that is known of the loss function is that it is symmetric and an increasing function of the deviation of an estimate from the true value. This assumption led Edgeworth to criticize Laplace for advocating widespread use of least-squares. Edgeworth advocated use of the median in many circumstances. Noting that the median of a univariate sample minimizes the sum of absolute deviations, he developed a method for fitting a straight line to bivariate data to minimize this sum (1923).

Edgeworth never developed his many original contributions to either economics or statistics to make a comprehensive coordinated work. He dealt with questions only dimly understood in his time (many are still controversial and not fully developed), and although he made significant contributions, I suspect that he himself was not satisfied with the state in which he left most of his topics. He was still working with enthusiasm shortly before his death at the age of 81. His work, however, has retained much more interest than that of some who had greater impact on their contemporaries. Modern mathematical economists and statisticians find him a stimulating and reassuring intellectual forebear, surprisingly up-to-date in many respects and still instructive.

CLIFFORD HILDRETH

[*For discussion of the subsequent development of Edgeworth's ideas, see* BAYESIAN INFERENCE; *see also* Georgescu-Roegen 1968.]

WORKS BY EDGEWORTH

1877 *New and Old Methods of Ethics.* London: Parker.

(1881) 1953 *Mathematical Psychics: An Essay on the Application of Mathematics to the Moral Sciences.* New York: Kelley.

1883 The Law of Error. *London, Edinburgh and Dublin Philosophical Magazine and Journal of Science* Fifth Series 16:300–309.

1886 Progressive Means. *Journal of the Royal Statistical Society* [1886]:469–475.

1887*a* *Metretike: Or the Method of Measuring Probability and Utility.* London: Temple.

1887*b* Observations and Statistics: An Essay on the Theory of Errors of Observation and the First Principles of Statistics. Cambridge Philosophical Society, *Transactions* 14, part 2:138–169.

(1887–1889) 1963 Measurement of Change in the Value of Money. Volume 1, pages 195–297 in Francis Ysidro Edgeworth, *Papers Relating to Political Economy.* New York: Franklin. → Consists of two memoranda prepared for the British Association for the Advancement of Science.

(1891–1921) 1963 *Papers Relating to Political Economy.* 3 vols. New York: Franklin. → Contains and reviews articles which appeared in the *Economic Journal* . . . 1891–1921 inclusive.

1893 Exercises in the Calculation of Errors. *London, Edinburgh and Dublin Philosophical Magazine and Journal of Science* Fifth Series 36:98–111.

(1897*a*) 1963 The Pure Theory of Monopoly. Volume 1, pages 111–142 in Francis Ysidro Edgeworth, *Papers Relating to Political Economy.* New York: Franklin. → First published in Italian.

(1897*b*) 1963 The Pure Theory of Taxation. Volume 2, pages 63–125 in Francis Ysidro Edgeworth, *Papers Relating to Political Economy.* New York: Franklin.

1898–1900 On the Representation of Statistics by Mathematical Formulae. *Journal of the Royal Statistical Society* 61:671–700; 62:125–140, 373–385, 534–555; 63 (Supplement):72–81.

(1899) 1963 Professor Seligman on the Theory of Monopoly. Volume 1, pages 143–171 in Francis Ysidro Edgeworth, *Papers Relating to Political Economy.* New York: Franklin. → First published as "Professor Seligman on the Mathematical Method in Political Economy."

1902 EDGEWORTH, FRANCIS YSIDRO; and BOWLEY, A. L. Methods of Representing Statistics of Wages and Other Groups Not Fulfilling the Normal Law of Error. *Journal of the Royal Statistical Society* 65:325–354.

1905 The Law of Error. Cambridge Philosophical Society, *Transactions* 20:36–65; 113–141.

1908–1909 On the Probable Errors of Frequency Constants. *Journal of the Royal Statistical Society* 71: 381–397; 72:81–90.

(1911) 1963 Use of Differential Prices in a Regime of Competition. Volume 1, pages 100–107 in Francis Ysidro Edgeworth, *Papers Relating to Political Economy.* New York: Franklin. → First published as "Monopoly and Differential Prices."

(1912) 1963 Contributions to the Theory of Railway Rates. Part III. Volume 1, pages 172–191 in Francis Ysidro Edgeworth, *Papers Relating to Political Economy.* New York: Franklin.

1915*a* *The Cost of War and Ways of Reducing It Suggested by Economic Theory: A Lecture.* Oxford: Clarendon.

1915*b* *On the Relations of Political Economy to War.* Oxford Univ. Press; New York and London: Milford.

(1917) 1918 *Currency and Finance in Time of War: A Lecture.* Oxford: Clarendon; New York and London: Milford.

1919 *A Levy on Capital for the Discharge of Debt.* Oxford: Clarendon.

1923 On the Use of Medians for Reducing Observations Relating to Several Quantities. *London, Edinburgh and Dublin Philosophical Magazine and Journal of Science* Sixth Series 46:1074–1088.

1926 Mr. Rhodes' Curve and the Method of Adjustment.

Journal of the Royal Statistical Society New Series 89:129–143.

WORKS ABOUT EDGEWORTH

BOWLEY, ARTHUR L. 1928 *F. Y. Edgeworth's Contributions to Mathematical Statistics.* London: Royal Statistical Society.

KEYNES, JOHN MAYNARD (1926) 1963 Francis Ysidro Edgeworth: 1845–1926. Pages 218–238 in John Maynard Keynes, *Essays in Biography.* New York: Norton.

L. L. P. [Obituary of] F. Y. Edgeworth. 1926 *Journal of the Royal Statistical Society* 89:371–377.

[Obituary of] F. Y. Edgeworth. 1926 *London Times* February 16, p. 21, cols. 2–3.

UZAWA, HIROFUMI 1962 On the Stability of Edgeworth's Barter Process. *International Economic Review* 3:218–232.

VIND, KARL 1964 Edgeworth Allocations in an Exchange Economy With Many Traders. *International Economic Review* 5:165–177.

SUPPLEMENTARY BIBLIOGRAPHY

CRAMÉR, HARALD (1945) 1951 *Mathematical Methods of Statistics.* Princeton Mathematical Series, No. 9. Princeton Univ. Press.

DEBREU, GERARD; and SCARF, HERBERT 1963 A Limit Theorem on the Core of an Economy. *International Economic Review* 4:235–246.

EDWARDS, WARD; LINDMAN, HAROLD; and SAVAGE, LEONARD J. 1963 Bayesian Statistical Inference for Psychological Research. *Psychological Review* 70:193–242.

►GEORGESCU-ROEGEN, NICHOLAS 1968 Utility. Volume 16, pages 236–267 in *International Encyclopedia of the Social Sciences.* Edited by David L. Sills. New York: Macmillan and Free Press.

MARSHALL, ALFRED 1881 Review of *Mathematical Psychics. Academy* 19:457 only.

ROBERTSON, D. H. 1916 Review of *On the Relations of Political Economy to War. Economic Journal* 26:66–68.

Postscript

An excellent biographical article on Edgeworth (Kendall 1968) contains additional facts about his background, interesting excerpts from his letters to Karl Pearson, and brief reflections on his contributions to statistics. Kendall also notes that copies of Edgeworth's bibliography, compiled by Harry Johnson and believed to be complete, are in the libraries of the British Museum; the Royal Statistical Society; the London School of Economics; All Souls, Oxford; and the Bodleian Library and the University Library, Cambridge.

CLIFFORD HILDRETH

ADDITIONAL BIBLIOGRAPHY

KENDALL, MAURICE G. (1968) 1970 Francis Ysidro Edgeworth, 1845–1926. Volume 1, pages 257–263 in E. S. Pearson and Maurice G. Kendall (editors), *Studies in the History of Statistics and Probability.* London: Griffin; New York: Hafner. → First published in *Biometrika* 55:269–275.

EFFICIENCY

See ERRORS, *article on* EFFECTS OF ERRORS IN STATISTICAL ASSUMPTIONS; ESTIMATION.

ELABORATION FORMULA

See SURVEY ANALYSIS *and the biography of* LAZARSFELD.

EPIDEMIOLOGY

● *The article written for* IESS *by the late Edward A. Suchman was completely updated for this volume by Mervyn Susser.*

Epidemiology is the study of the distribution and causes of disease in populations and the techniques for establishing such knowledge. Epidemiology was, until recently, dominated by the study of acute epidemics and contagious diseases, and by the search for infectious agents. Today, the subject covers all types of disease—physical as well as mental, noninfectious as well as infectious—and epidemiologists study all types of population characteristics and agents—social and psychological as well as biological and physical—that may help to describe or explain the prevalence of disease.

Although epidemiology is the study of the distribution and determinants of health states in populations, that study is, of course, often for the purpose of prevention and control of health disorders. The discovery or description of these disorders is *descriptive* epidemiology, whereas analysis of the causal conditions producing the disorders is *explanatory*, or analytic, epidemiology. Originally used to describe the controlled introduction of epidemic conditions into populations of experimental animals in the laboratory (Greenwood 1932), the term "*experimental* epidemiology" now covers all planned interventions, including field experiments to test the efficacy of immunizing agents and of preventive measures. Experimental methods are used to establish causal relations, or the effectiveness of intervention, or both.

Epidemiology emphasizes relations between environmental factors and diseases, so it is a major branch of human ecology (LeRiche & Milner 1971). Three main sets of interacting factors are studied: (1) *the host*, or human individual, varying in genetic resistance, susceptibility, and degree of immunity to disease; (2) *the agent*, or initiator of the disease, including any adverse process (whether it be an excess, a deficiency, a microbe, a toxin, or a metabolic factor) that varies according to infectivity, virulence, dose, and pathogenesis; (3) *the environment*, or surrounding medium, social as well as biological and physical, which affects both the susceptibility of the host, the virulence of the agent, and the quantity and quality of contact between host and agent. These three sets of factors maintain a complex, ever changing balance. The occurrence of disease, especially mass disease, is the result of a multiplicity of causal factors, each of which contributes to, rather than accounts for, the appearance of the disease.

Epidemiological knowledge consists of facts and theories about the relationships between these three elements and disease entities or health states. *Social* epidemiology concentrates on the social, in contrast with the physical or biological, factors in the incidence and prevalence of disease. In the study of chronic diseases and mental and developmental disorders, all of which constitute prime subjects of modern epidemiology, distinctions among host, agent, and environmental factors, and among social, biological, and physical factors, can become blurred.

Historical background. Following the application and development of Pasteur's discoveries of micro-organisms and their effects by Koch and many others in the 1870s and later, the epidemiology that had flourished in the nineteenth century was forgotten, its practitioners fell into obscurity, and their contributions are only now being recognized and reevaluated (see Lilienfeld & Lilienfeld 1977).

Epidemiology as we know it was born in England in the late seventeenth century, when John Graunt in 1662 published the first tabular analysis of mortality. [*See the biography of* GRAUNT.] In the eighteenth century the experimental studies of James Lind established the nutritional basis of scurvy, and Henry Baker established the occupational basis of lead colic in brewers of cider, thereby demonstrating the applicability of epidemiological method to noninfectious diseases. It was not until the first half of the nineteenth century, however, that men like Jean-Pierre Louis and later William Farr in London used statistics systematically to study the distribution and determinants of disease. At the same time, Adolphe Quetelet, Jules Gaveret, and others produced the statistical methods for these developments. [*See the biography of* QUETELET.] William Farr, during his forty-year tenure (1839–1879) as the first statistical medical officer at the General Register Office in London, provided models for the contemporary epidemiological approach to diseases of all kinds (Farr 1885).

Ignaz Semelweiss, in his studies of puerperal fever, traced the source of a highly fatal disease to contagion and showed how to control it. The pioneering triumphs of epidemiology later in the nineteenth century related to communicable infectious diseases, or what has been called "the mass phenomena of infectious diseases" (see Frost 1941). The contributions of John Snow to the understanding of cholera, of Peter Panum to the understanding of measles, and of William Budd to the understanding of typhoid were notable among these. After Pasteur and Koch had demonstrated the existence and specificity of pathogenic micro-organisms, this emphasis was reinforced, and it persisted up to World War II. In the postwar period, with the dramatic conquest of the infectious diseases during the first half of the present century and the related growth in the importance of chronic and degenerative diseases, it became apparent that epidemiology could no longer be restricted to infectious agents.

The basis for the new trend had already been laid. Early in this century, Joseph Goldberger in the United States carried out his classic studies of pellagra (reprinted in Goldberger 1964), in which he eliminated the popular hypothesis of an infectious origin of the disorder and demonstrated its dietary cause. In Britain Sir Ronald Ross, discoverer of the mosquito as vector of malaria and an early Nobel prize winner, was applying his statistical "theory of happenings" in broad fashion (Ross 1910), and Major Greenwood studied the recurrence of accidents as an epidemiological phenomenon (Greenwood & Yule 1920). In the postwar period, the use of statistical associations based upon population surveys has become one of the foremost methods for studying the occurrence of all diseases (notably cancer, cardiovascular disease, and mental disorders) and for the difficult task of identifying causal agents.

Uses of epidemiology. As a standard tool of medical investigation, epidemiology has been brought to bear upon almost all aspects of the prevention and treatment of disease. Morris (1957) listed seven uses: the determination of individual risks (for example, the chances of a forty-year-old male getting cancer); the securing of data on subclinical and undetected cases; the identification of syndromes or recurring clusters of symptoms and signs; the determination of historical trends of disease; the diagnosis of community health needs and resources; program planning, operation, and evaluation; and the search for causes of disease. Epidemiology thus provides a large portion of the scientific base for public health practice as well as for the understanding of causality in medicine generally.

The major contribution of epidemiology to research is in the development and testing of hypotheses concerning specific factors that may influence the distribution of some particular disorder in a defined population. The epidemiologist identifies subgroups of the population with varying incidence rates of the disease under investigation. On the basis of existing knowledge, theory, or observation, hypotheses are generated about etiological factors that may be involved and that may differ among the subgroups being studied. By field surveys, the analysis of existing data, or more likely by a succession of studies of varied design, the direction and degree of association between the occurrence of the disease and the presence or absence of a group characteristic hypothesized as the etiological factor is tested. Health care is of concern to epidemiology as it influences the occurrence of disease, and the evaluation of the ultimate effects of health programs on health is a proper subject of epidemiological study.

Epidemiology and social science. Epidemiology has theoretical and methodological ties to the social sciences. Both the epidemiologist and the social scientist are concerned with demography and with ecology—the relationship of man to his environment. Health is but one dimension of the manifestations of social and economic structure. Any human environment includes sociocultural factors as possible causes of disease, either indirectly (as when poverty leads to malnutrition or unsanitary living conditions) or directly (as when emotional disturbance leads to mental disease or addiction to alcohol or drugs). All three basic components of epidemiology—host, agent, and environment—therefore take on important social dimensions. In the current era of the predominance of chronic disease, in which an individual's whole way of life may become more important than any single infectious agent in the disease process, social factors become a primary target for epidemiological investigation (see, for example, Gordon 1952; Cassel & Tyroler 1961; Susser & Watson 1962; King 1963).

Both the epidemiologist and the social scientist rely heavily upon population surveys and field experiments. Similar problems of research design confront both groups, while technical considerations such as sampling, questionnaire construction, interviewing, and multivariate analysis arouse common methodological interest (Wardwell & Bahnson 1964). When the sociologist selects for study the same dependent variable as the epidemiologist—namely, health states—he has entered the domain of social epidemiology. The epidemiologist no less than the social scientist is concerned with the family as a social unit and must develop the designs and analytic techniques to cope with its complexities (Fox 1974). In particular, both sciences have an interest in separating genetic from environmental contributions to manifestations of health states (Schull et al. 1970; Rosenthal 1974; Morton 1975).

Substantive research areas. All major diseases today are the subject of epidemiological research, and studies of almost all of these include, at the minimum, such social variables as sex, age, marital status and family composition, occupation, socioeconomic status, religion, and race. In addition, many studies are specifically aimed at the investigation of social factors, such as stressful life events, as possible etiological agents in the occurrence of disease.

Although substantive advances made by epidemiological approaches before World War II related primarily to infectious diseases (see Winslow 1943), in the postwar period, epidemiology has proved its continued usefulness in the study of infectious disease, for instance in the control of malaria (MacDonald 1957) and of poliomyelitis (Paul 1971), in the elucidation of the sources and transmission of infectious hepatitis, and most recently, in the historic achievement of the virtual elimination of smallpox from the world.

The major activities of contemporary epidemiologists, however, now relate to noninfectious diseases, particularly chronic disorders of the circulatory and central nervous systems, cancers, mental disorders, disorders of child development, and genetic problems. In the postwar period, epidemiology has changed the scientific paradigm that pertains to chronic disease. As these diseases became predominant after the conquest of infectious disease, the common view was that they arose from deteriorative processes intrinsic to aging. This view, indeed, was implied in the description of such diseases as degenerative. Epidemiological research changed the paradigm to a more hopeful concept of multiple causes, many of which were of environmental provenance and therefore subject to prevention, control, or amelioration.

Two disease categories, the cancers and the cardiovascular diseases, had come to dominate mortality, and among males, lung cancer and coronary heart disease took on epidemic proportions (Morris 1957). The shift in the scientific paradigm began with the demonstration in case–control studies (Wynder & Graham 1950; Doll & Hill 1952) of the relationship of tobacco smoking with lung cancer. The replication and reinforcement of this relationship in more than 35 subsequent case–control and cohort studies (U.S. Surgeon General's Advisory Committee 1964) made this hypothesis of an environmental cause of cancer virtually impregnable.

Further impetus was given to the shift in the paradigm by the consideration and study of coronary heart disease as an epidemic condition. Large-scale cohort studies were mounted, and several risk factors were successfully demonstrated (Morris 1957; Dawber & Kannel 1963; Kinch et al. 1964; Ostfeld et al. 1964). Among these, blood lipid levels, blood pressure, body weight, smoking, and exercise either had an environmental source or were amenable to manipulation and treatment.

Similar approaches were applied to other conditions: the whole range of cancers, neurological diseases, inherited disorders, developmental disorders, and many others. The view of mental disorder too was sharply altered by two major prevalence surveys that indicated environmental and social causes, one in Manhattan (Srole et al. 1962) and the other in Nova Scotia (Leighton et al. 1963).

The consequences of social stress have become the subject of intense inquiry. Although progress in this field has been slow, current research approaches have integrated sociobiological constructs (Hinkle & Wolff 1957; King & Cobb 1958; Cassel 1974). Instruments for measuring stressors have been developed (Holmes & Rahe 1967) and subjected to critical analysis and testing (Dohrenwend & Dohrenwend 1974; Goldberg & Comstock 1976). At least some measures of personality type have been adequate to the task of establishing with fair confidence that a particular spectrum of behavior is a substantial risk factor for coronary heart disease (Rosenman et al. 1975). At the same time, a better understanding of the sociology of sickness behavior (Mechanic 1968) has been absorbed into research design, analysis, and interpretation of epidemiological studies (Susser & Watson 1962). Epidemiological studies of the effects of social environment on health states have thus become broader and deeper (Cassel 1976; Marmot & Syme 1976).

In some chronic disorders, epidemiological studies indicate that the pathogenesis may turn once more on infectious agents. This seeming paradox underlines the common ground and necessary interchange of methods and ideas among epidemiologic subspecialties, as Lilienfeld (1973) has emphasized. Thus the viral theories of carcinogenesis have been given powerful support by studies of Burkitt's lymphoma, a childhood cancer shown to be concentrated in certain areas of central Africa in a distribution best explained by the coexistence of mosquito vectors and malaria. Likewise, as the social distribution of cervical cancer has come to be better understood, the likelihood that the agent is a sexually transmitted virus has grown strong, and viral studies can now be adduced in further support of the initial social studies. Further, we now know that developmental anomalies can be attributed to prenatal infection with rubella virus, cytomegalovirus, and toxoplasmosis.

A whole new perspective on viral pathogenesis opened with the demonstration of the etiology of kuru. The condition depends on the transmission, not of a pathological gene as was thought at first, but of a so-called slow virus. This chronic and fatal disorder of the central nervous system is found only among the women and children of certain New Guinea tribes. In these cannibalistic tribes, only women and children customarily eat the brain of the slain enemy. A close study of social behavior was necessary to eliminate the genetic hypothesis and substitute the viral hypothesis. Carleton Gajdusek, by patient experimental work in chimpanzees to complement his field studies of the disease in New Guinea, succeeded in transmitting the kuru virus after a wait of some three years before its effects were observed (Gajdusek & Gibbs 1975). Multiple sclerosis, another serious but relatively common central nervous system disease, also may be caused by a "slow virus" agent confined to temperate climes. Social evidence again plays its part. The disease occurs in individuals who migrate after a critical age from temperate climes to countries where the disease is rare; it is absent in those who migrate to temperate climes from nontemperate ones.

Some problems of research design. In research methods, epidemiology involves "the application of scientific principles to investigation of conditions affecting groups in the population" (Clark 1953, p. 65). In the main, this consists of the observation of the occurrence of disease and its associations under natural conditions in defined populations. By contrast, clinical or laboratory investigations observe disease in selected cases, or in a series of cases, without reference to the population from which they are drawn. Epidemiology, for the most part, uses the research techniques of the population survey to discover the relationship between the occurrence of disease and the presence of various biological, physical, and social factors. The proofs it seeks rest on statistical associations between the presumed causal factors and the occurrence of the health disorder.

The major problems of concept, method, and inference in epidemiology stem from its dependence upon the evidence of associations. The research design in epidemiology consists of the comparison of groups in terms of the frequency of disease and the relationships of those frequencies to hypothesized explanatory characteristics. Studies based on the secondary analysis of existing vital statistics comprise one form of this research. Other studies use data specially gathered for their purposes.

Descriptive studies that aim to establish frequency distributions or generate hypotheses will ordinarily be based on incidence or prevalence surveys. Both incidence and prevalence surveys use rates of cases identified in defined populations. *Incidence* is the frequency of cases—most usefully, new cases—that arise in a population over a given period of time. *Prevalence* is the frequency of cases existing in a population at a given point in time. Prevalence is a function of incidence and duration. Incidence surveys can better establish time order, and therefore causality, than prevalence studies; their disadvantage is that in studies of morbidity, if not mortality, case identification must in most instances depend on use of medical services. Such use is selective where illnesses are not severe, or where services are scarce or not equally accessible to all. Prevalence surveys better establish the amount of illness existing in a population. Given an agreed definition of the condition under study, a whole community can be reached. The disadvantage of such prevalence studies is that they will usually, although not invariably, lack the diagnostic depth attainable in incidence studies where the evaluation of a condition can be made in a service setting. Further, observing variables at the same point in time presents inherent barriers to establishing the time–order among them.

Exploratory or *explanatory* observational studies can be classified as either *case–control* or *cohort* studies (MacMahon & Pugh 1970). Case–control studies secure data from at least two groups, one a group of cases that manifest the disease and another a comparison group that does not. Cohort studies also involve at least two groups, one group exposed to the hypothesized causal factor and the other unexposed. Cohort studies are usually prospective and involve follow-up. Historical cohort studies, however, depend on the reconstruction of existing historical data without actual follow-up in the field.

In these study designs, the object is to determine statistical associations from which hypotheses and causal inferences can be drawn. The designs should be seen as complementary to each other; each has different strengths and weaknesses, and serves better for different purposes. Case–control studies, like prevalence surveys, suffer the disadvantage that they observe established cases post facto and, therefore, time order is difficult to establish securely. A central problem is that the investigator can never be certain that the control groups are comparable with the cases in all respects. Case–control studies are efficient in that they are economical and readily replicated, and they may afford the only possible means of studying rare diseases.

By contrast, cohort studies generally allow time order among variables to be established, and the universe from which the comparison groups are drawn is known from the outset. On the other hand, cohort studies are laborious and may stretch out over many years, and generally they do not provide a means for studying rare disease because of the huge starting populations required. The association between smoking and lung cancer provides an excellent example of the complementary and successful use of different types of study. The initial clinical observation was supported by demographic comparisons showing a much higher incidence of lung cancer among men than women. Case–control studies then revealed the correlation between smoking histories and the occurrence of lung cancer. Finally cohort studies showed on follow-up the greater frequency of lung cancer among smokers than among nonsmokers. Experimental studies have since been mounted to try to reduce smoking and its many hazards (Donovan 1975).

In observational studies, the epidemiologist cannot randomize the experimental and control groups or alter the characteristics of the experimental group. Unless the study is an experimental one, therefore, certain prerequisites must be satisfied if observational studies are to produce useful associations. Ideally, the generalizability of the samples and hence the study should be ascertainable. In practice, case-control studies proceed without meeting that desideratum. Epidemiologists have devised a variety of strategies and approximations to cope with less than ideal circumstances. [*See* EXPERIMENTAL DESIGN, *especially the article on* QUASI-EXPERIMENTAL DESIGN.] The definition of what is normal, or disease free, often presents a difficult problem in the study of chronic diseases, since manifestations may not be detectable until a late stage in the evolution of the pathological process. Diagnostic validity is especially difficult to establish in field studies, where the unfolding of a case over time under clinical observation is not available to remove error. Finally, the independent variable, that is, the hypothesized causal factor, must be similarly capable of objective definition and measurement.

These are difficult conditions to meet, and in establishing causal relations, epidemiologists must rely on rigorous design, careful analysis, and logical inference. The broad strategies available may be classified into five kinds (Susser 1973). The first kind, simplifying the conditions of observation, is essentially a matter of research design. Under this heading, devices that exclude or neutralize extraneous factors in the system under observation are considered. They include such procedures as sample selection, instrumentation, reliability and validity, randomization, and matching.

The second and third kinds of strategy contribute chiefly to analysis. The second, to screen a supposed causal relationship or extraneous variable, is to seek out two types of factors: those that can give rise to the illusion of a causal relationship and those that can suppress the appearance of a causal relationship. The third kind of strategy, to elaborate an observed association between variables, is to analyze its nature and validity; in this process, additional variables likely to be related to those found in association are introduced into the analysis. [*See* SURVEY ANALYSIS.] When these two kinds of strategy are applied to the study of associations, assumptions that weaken inferences are less likely to pass unrecognized.

The fourth kind of strategy is the application of notions of probability and statistical inference. [*See* STATISTICS, *article on* THE FIELD.] The fifth and final kind of strategy aids in interpreting the results of analysis. The determination of cause ultimately rests on subjective judgment, and epidemiologists have developed a number of criteria against which to test their judgments. These criteria include time order, specificity, strength of association, consistency on replication, and coherence [Sartwell 1959; Lilienfeld 1959; *see also* CAUSATION.]

The use of the epidemiological approach has expanded rapidly. Clinical and laboratory research has been complemented by epidemiology both as a source of hypotheses and as proof for them. No other means exist by which to establish environmental causes of health disorders, and our consciousness of the pervasiveness of these causes in the predominant chronic diseases has grown rapidly. It is commonplace now to think in terms of probabilities rather than certainties, of multiple causes and conditional relationships rather than specific causes and one-to-one relationships. The biological, social, and psychological forces that determine how man lives and how the life cycle unfolds are beginning to receive their due attention. Social epidemiologists have learned to operationalize both independent and dependent variables with precision for these purposes. They have now begun to test specific hypotheses about when, how, and why social factors are related to the origin and course of disease.

MERVYN SUSSER

BIBLIOGRAPHY

Findings of epidemiological surveys are to be found in the increasing numbers of journals chiefly or entirely devoted to the subject, for example, the American Journal of Epidemiology, British Journal of Preventive and Social Medicine, International Journal of Epidemiology, Journal of Chronic Diseases. *Epidemiological papers are to be found often also in general medical journals, for example,* Lancet, New England Journal of Medicine, British Medical Journal, Journal of the American Medical Association, *as well as in a large number of specialty journals, such as the* American Journal of Public Health, Public Health Reports, Public Health Reviews, Archives of Environmental Health, Journal of Infectious Diseases, Journal of Hygiene (*Cambridge*), Journal of Health and Human Behavior, Social Science and Medicine, Social Biology, Human Biology, Pediatrics, American Journal of Obstetrics and Gynecology, British Journal of Psychiatry, American Journal of Psychiatry, Social Psychiatry, Psychosomatic Research, Cancer, Circulation, British Heart Journal, Annals of Human Genetics, American Journal of Human Genetics, *and many others. A special epidemiological bibliography abstracted from the data collected for the* Index Medicus *is published monthly and cumulated annually in the* Current Bibliography of Epidemiology.

CASSEL, JOHN 1974 Psychosocial Processes and "Stress": Theoretical Formulation. *International Journal of Health Services* 4, no. 3:471–482.

CASSEL, JOHN 1976 The Contribution of the Social Environment to Host Resistance. *American Journal of Epidemiology* 104:107–123.

CASSEL, JOHN; and TYROLER, HERMAN A. 1961 Epidemiological Studies of Culture Change: I. Health Status and Recency of Industrialization. *Archives of Environmental Health* 3:25–33.

CLARK, E. GURNEY (1953) 1965 An Epidemiological Approach to Preventive Medicine. Chapter 3 in Hugh R. Leavell et al., *Preventive Medicine for the Doctor in His Community: An Epidemiologic Approach.* 3d ed. New York: McGraw-Hill.

CLAUSEN, JOHN A.; and KOHN, MELVIN L. 1954 The Ecological Approach in Social Psychiatry. *American Journal of Sociology* 60:140–151.

DAWBER, THOMAS R., and KANNEL, WILLIAM B. 1963 Coronary Heart Disease as an Epidemiological Entity. *American Journal of Public Health* 53:433–437.

DOHRENWEND, BARBARA SNELL; and DOHRENWEND, BRUCE P. (editors) 1974 *Stressful Life Events: Their Nature and Effects.* New York: Wiley.

DOLL, R.; and HILL, A. B. 1952 A Study of the Aetiology of Carcinoma of the Lung. *British Medical Journal* 2:1271–1286.

DONOVAN, J. W. et al. 1975 Routine Advice Against Smoking and Pregnancy. *Journal of the Royal College of General Practitioners* 25:264–268.

DORN, HAROLD F., and CUTLER, SIDNEY J. 1958 *Morbidity From Cancer in the United States.* U.S. Department of Health, Education, and Welfare, Public Health Service Publication No. 590. Washington: Government Printing Office.

DOUGLAS, J. W. B.; and BLOMFIELD, J. M. 1958 *Children Under Five.* London: Allen & Unwin.

FARR, WILLIAM 1885 *Vital Statistics: A Memorial Volume of Selections From the Reports and Writings of William Farr.* Edited for the Sanitary Institute of Great Britain by Noel A. Humphreys. London: The Institute. → Reprinted, with an introduction by M. Susser and A. Adelstein, in 1975 by Scarecrow. Also reprinted, in 1976, by Arno.

FLECK, ANDREW C.; and IANNI, FRANCIS A. J. 1958 Epidemiology and Anthropology: Some Suggested

Affinities in Theory and Method. *Human Organization* 16, no.4:38–40.

FOX, JOHN P. 1974 Family-based Epidemiologic Studies. *American Journal of Epidemiology* 99:165–179.

FROST, WADE HAMPTON 1941 *Papers of Wade Hampton Frost, M.D.: A Contribution to Epidemiological Method.* Edited by Kenneth F. Maxcy. New York: Commonwealth Fund; Oxford Univ. Press. → Essays dating from 1910 to 1939.

GAJDUSEK, CARLETON D.; and GIBBS, CLARENCE J. JR. 1975 Slow Virus Infection of the Nervous System and the Laboratories of Slow, Latent and Temperate Virus Infection. Volume 2, pages 113–135 in Donald B. Tower (editor), *The Nervous System.* Volume 2: *The Clinical Neurosciences.* New York: Raven.

GLOCK, CHARLES Y.; and LENNARD, HENRY L. 1956 Studies in Hypertension. *Journal of Chronic Diseases* 5:178–196.

GOLDBERG, E. L.; and COMSTOCK, G. W. 1976 Life Events and Subsequent Illness. *American Journal of Epidemiology* 104:146–158.

GOLDBERGER, JOSEPH 1964 *Goldberger on Pellagra.* Edited and introduced by Milton Terris. Baton Rouge: Louisiana State Univ. Press.

GORDIS, LEON 1973 *Epidemiology of Chronic Lung Diseases in Children.* Baltimore: Johns Hopkins Press.

GORDON, JOHN E. 1952 The Twentieth Century: Yesterday, Today and Tomorrow (1920—). Pages 114–167 in Franklin H. Top (editor), *The History of American Epidemiology.* St. Louis: Mosby.

GRAHAM, SAXON 1960 Social Factors in the Epidemiology of Cancer at Various Sites. New York Academy of Sciences, *Annals* 84:807–815.

GREENWOOD, MAJOR 1932 *Epidemiology, Historical and Experimental.* Baltimore: Johns Hopkins Press; Oxford Univ. Press.

GREENWOOD, MAJOR; and YULE, G. UDNY 1920 An Inquiry Into the Nature of Frequency Distributions Representative of Multiple Happenings With Particular Reference to the Occurrence of Multiple Attacks of Disease or of Repeated Accidents. *Journal of the Royal Statistical Society* 83:255–279.

HARBURG, E. et al. 1970 A Family Set Method for Estimating Heredity and Stress: I. Pilot Survey of Blood Pressure Among Negroes in High and Low Stress Areas, Detroit, 1966–1967. *International Journal of Chronic Diseases* 23:60–81.

HINKLE, LAWRENCE E. JR.; and WOLFF, HAROLD G. 1957 Health and Social Environment: Experimental Investigations. Pages 105–137 in Alexander H. Leighton, John A. Clausen, and Robert N. Wilson (editors), *Explorations in Social Psychiatry.* New York: Basic Books.

HOCH, PAUL H.; and ZUBIN, JOSEPH (editors) 1961 *Comparative Epidemiology of Mental Disorders.* New York: Grune & Stratton. → Proceedings of the 49th annual meeting of the American Psychopathological Association, Feb. 1959.

HOLMES, P. H.; and RAHE, R. H. 1967 The Social Readjustment Rating Scale. *Psychosomatic Medicine* 11:213–218.

JACO, E. GARTLY 1960 *The Social Epidemiology of Mental Disorders: A Psychiatric Survey of Texas.* New York: Russell Sage.

KINCH, S. H.; GITTELSOHN, A. M.; and DOYLE, J. T. 1964 Application of a Life Table Analysis in a Prospective Study of Degenerative Cardiovascular Disease. *Journal of Chronic Diseases* 17:503–514.

KING, STANLEY H. 1963 Social Psychological Factors in Illness. Pages 99–121 in Howard E. Freeman, Sol Levine, and Leo G. Reeder (editors), *Handbook of Medical Sociology.* Englewood Cliffs, N.J.: Prentice-Hall.

KING, STANLEY H.; and COBB, SIDNEY 1958 Psychosocial Factors in the Epidemiology of Rheumatoid Arthritis. *Journal of Chronic Diseases* 7:466–475.

KURLAND, LEONARD T.; KURTZKE, J. F.; and GOLDBERG, I. D. (editors) 1973 *Epidemiology and Neurologic and Sense Organ Disorders.* Vital and Health Statistics Monographs, APHA. Cambridge, Mass.: Harvard Univ. Press.

KURTZKE, J. F. et al. 1973 Multiple Sclerosis. Pages 64–107 in Leonard T. Kurland, J. F. Kurtzke, and I. D. Goldberg (editors), *Epidemiology of Neurologic and Sense Organ Disorders.* Vital and Health Statistics Monographs, APHA. Cambridge, Mass.: Harvard Univ. Press.

LEIGHTON, DOROTHEA CROSS et al. 1963 *The Character of Danger: Psychiatric Symptoms in Selected Communities.* The Stirling County Study of Psychiatric Disorder and Sociocultural Environment, Vol. 3. New York: Basic Books.

LERICHE, W. HARDING; and MILNER, JEAN 1971 *Epidemiology as Medical Ecology.* Baltimore: Williams & Wilkins; Edinburgh: Livingstone.

LILIENFELD, ABRAHAM M. 1959 On the Methodology of Investigations of Etiologic Factors in Chronic Diseases. *Journal of Chronic Diseases* 10:41–46.

LILIENFELD, ABRAHAM M. 1967 *Cancer Epidemiology: Methods of Study.* Baltimore: Johns Hopkins Press.

LILIENFELD, ABRAHAM M. 1973 Epidemiology of Infectious and Non-infectious Disease: Some Comparisons. *American Journal of Epidemiology* 97:135–147.

LILIENFELD, ABRAHAM M.; and LILIENFELD, DAVID E. 1977 What Else Is New? An Historical Excursion. *American Journal of Epidemiology* 105:169–179.

MACDONALD, GEORGE 1957 *The Epidemiology and Control of Malaria.* Oxford Univ. Press.

MACMAHON, BRIAN; and PUGH, THOMAS F. 1970 *Epidemiology: Principles and Methods.* Boston: Little, Brown.

MARMOT, M. G.; and SYME, S. L. 1976 Acculturation and Coronary Heart Disease in Japanese–Americans. *American Journal of Epidemiology* 104:225–247.

MECHANIC, DAVID (1968) 1978 *Medical Sociology.* 2d ed. New York: Free Press. → A comprehensive revision of the "selected view" in the first edition.

MORRIS, JEREMY N. 1975 *Uses of Epidemiology.* Baltimore: Williams & Wilkins; Edinburgh: Livingstone.

MORTON, NEWTON E. 1975 Analysis of Family Resemblance and Group Differences. *Social Biology* 22:111–116.

OSTFELD, ADRIAN M. et al. 1964 A Prospective Study of the Relationship Between Personality and Coronary Heart Disease. *Journal of Chronic Diseases* 17:265–276.

PAUL, JOHN R. 1971 *A History of Poliomyelitis.* Yale Studies in the History of Science and Medicine, No. 6. New Haven: Yale Univ. Press.

ROSENMAN, RAY H. et al. 1975 Coronary Heart Disease in Western Collaborative Group Study: Final Follow-up Experience of 8½ Years. *Journal of the American Medical Association* 233:872–877.

ROSENTHAL, D. A. 1974 Program of Research on Heredity in Schizophrenia. Pages 19–38 in Sarnoff A. Mednick et al. (editors), *Genetics, Environment and Psychopathology.* New York: American Elsevier; Amsterdam: North-Holland.

ROSS, RONALD 1910 *The Prevention of Malaria.* New York: Dutton.

SARTWELL, P. F. 1959 On the Methodology of Investigations of Etiologic Factors in Chronic Diseases. *Journal of Chronic Diseases* 11:61–63.

SCHULL, WILLIAM J. et al. 1970 A Family Set Method for Estimating Heredity and Stress: II. Preliminary Results of the Genetic Methodology in a Pilot Survey of Negro Blood Pressure, Detroit, 1966–1967. *Journal of Chronic Diseases* 23:83–92.

SROLE, LEO et al. (editors) 1962 *Mental Health in the Metropolis: The Midtown Manhattan Study.* New York: McGraw-Hill.

SUCHMAN, EDWARD A.; and SCHERZER, ALFRED L. 1960 Current Research in Childhood Accidents. Part 1 in Association for the Aid of Crippled Children, *Two Reviews of Accident Research.* New York: The Association.

SUSSER, MERVYN 1973 *Causal Thinking in the Health Sciences: Concepts and Strategies in Epidemiology.* New York: Oxford Univ. Press.

SUSSER, MERVYN; and WATSON, W. (1962) 1971 *Sociology in Medicine.* 2d ed. New York: Oxford Univ. Press.

TOP, FRANKLIN H. (editor) 1952 *The History of American Epidemiology.* St. Louis: Mosby.

U.S. SURGEON GENERAL'S ADVISORY COMMITTEE ON SMOKING AND HEALTH 1964 *Smoking and Health.* U.S. Department of Health, Education, and Welfare, Public Health Service Publication No. 1103. Washington: Government Printing Office.

WARDWELL, WALTER I.; and BAHNSON, CLAUS B. 1964 Problems Encountered in Behaviorial Science Research in Epidemiological Studies. *American Journal of Public Health* 54:972–981.

WINSLOW, C.-E. A. 1943 *The Conquest of Epidemic Disease: A Chapter in the History of Ideas.* Princeton Univ. Press. → A facsimile edition was published in 1967 by Hafner.

WYNDER, ERNEST L.; and GRAHAM, EVARTS A. 1950 Tobacco Smoking as a Possible Etiologic Factor in Bronchogenic Carcinoma: A Study of 684 Proved Cases. *Journal of the American Medical Association* 143: 329–336.

ERRORS

I
NONSAMPLING ERRORS

The view has sometimes been expressed that statisticians have laid such great emphasis on the study of sampling errors (the differences between the observed values of a variable and the long-run average of the observed values in repetitions of the measurement) that they have neglected or encouraged the neglect of other, frequently more important, kinds of error, called nonsampling errors.

Errors in conception, logic, statistics, and arithmetic, or failures in execution and reporting, can reduce a study's value below zero. The roster of possible troubles seems only to grow with increasing knowledge. By participating in the work of a specific field, one can, in a few years, work up considerable methodological expertise, much of which has not been and is not likely to be written down. To attempt to discuss every way a study can go wrong would be a hopeless venture. The selection of a kind of error for inclusion in this article was guided by its importance, by the extent of research available, by the ability to make positive recommendations, and by my own preferences.

Although the theory of sampling is generally well developed, both the theory and practice of the control of nonsampling errors are in a less satisfactory state, partly because each subject matter, indeed each study, is likely to face yet uncatalogued difficulties. Empirical results of methodological investigations intended to help research workers control nonsampling errors have accumulated slowly, not only because of myriad variables but also because the variables produce results that lack stability from one study to another.

This article deals mainly with techniques for reducing bias. The portions on variability are not exceptions, for they offer ways to avoid underestimating the amount of variability. The presentation deals, first, with the meaning of bias and with conceptual errors; second, with problems of nonsampling errors especially as they arise in the sample survey field through questionnaires, panel studies, nonresponse, and response errors; and, third, with errors occurring in the analysis of nearly any kind of quantitative investigation, errors arising from variability, from technical problems in analysis, in calculations, and in reporting. Some discussions of nonsampling errors restrict themselves to the field of sample surveys, where problems of bias and blunder have been especially studied, but this article also treats some nonsampling errors in experimental and observational studies.

Bias and conceptual errors

Bias and true values. What is bias? Most definitions of bias, or systematic error, assume that for each characteristic to be measured there exists a true value that the investigation ideally would produce. Imagine repeatedly carrying out the actual proposed, rather than the ideal, investigation, getting a value each time for the characteristic under study, and obtaining an average value from these many repetitions. The difference between that average value and the true value of the characteristic is called the bias. The difference between the outcome of *one* investigation and the true value is the sum of bias and sampling error. The point of averaging over many repetitions is to reduce the sampling error in the average value to a negligible amount. (It is assumed that for the process under study and for the type of average chosen, this reduction is possible.)

Is there a true value? The concept "true value" is most touchy, for it assumes that one can describe an ideal investigation for making the measurement. Ease in doing this depends upon the degree of generality of the question. For example, the measurement of "interventionist attitude" in the United States during World War II is discussed below. For such a broad notion, the concept of a true value seems vague, even admitting the possible use of several numbers in the description. It is easier to believe in a true value for the percentage of adults who would respond "Yes" to "Should we go to war now?" Even here the training of the interviewers, the rapidly changing fraction of the population holding given opinions, and the effect of the social class and opinions of the interviewer upon the responses of those interviewed must raise questions about the existence of a true value. At the very least, we wonder whether a true value could represent a time span and how its conditions of measurement could be specified. In designing an ideal sample survey, what kind of interviewer should be used?

Today some scientists believe that true values do not exist separately from the measuring process to be used, and in much of social science this view can be amply supported. The issue is not limited to social science; in physics, complications arise from the different methods of measuring microscopic and macroscopic quantities such as lengths. On the other hand, because it suggests ways of improving measurement methods, the concept of "true value" is useful; since some methods come much nearer to being ideal than others, the better ones can provide substitutes for true values. (See the discussion on describing response error in the section on "Response error," below.)

To illustrate further the difficulty of the notion of a true value, consider an example from one of the most quantitative social sciences. When the economist assesses the change in value of domestic product, different choices of weights and of base years yield different results. He has no natural or unique choice for these weights and years. He can only try to avoid extremes and unusual situations. While, as noted above, the belief in a true value independent of the measuring instrument must be especially weak in the area of opinion, similar weaknesses beset measures of unemployment, health, housing, or anything else related to the human condition. [*See* INDEX NUMBERS.]

Conceptual errors. Since the variety of sources of biases is practically unlimited, this article discusses only a few frequently encountered sources.

Target population–sampled population. Often an investigation is carried out on a sample drawn from a population—the sampled population—quite different from that to which the investigator wants to generalize—the target population. This mismatch makes the inference from sample to target population shaky. To match target and sampled population perfectly is usually impossible, but often the expenditure of time and money or the use of special skills or cooperation can patch what cannot be made whole.

Some examples of this process follow: (1) The psychologist wants to establish general laws of learning for all organisms, and especially for man, but he may choose to study only the college sophomore, usually in his own college and rarely outside his own country. His principal alternatives are the rat and the pigeon. Reallocation of time and money may extend the sampled population and bring him closer to the target he has in mind.

(2) The sociologist may want to study the actual organization of trade unions and yet be hard pressed to study in depth more than a single union. This limitation is impossible for an individual to overcome, but cooperative research may help. (For a remarkable cooperative anthropological study of child rearing, see *Six Cultures: Studies of Child Rearing* [Whiting 1963].)

(3) The historian or political scientist may want to exposit the whole climate of opinion within which an important decision is made, yet he must pick some facts and omit others, emphasize some and not others. The sampling of historical records offers a compromise between scanning everything, which may be impossible or unsatisfactorily superficial, and the case study of a single document or of a small collection.

(4) The man who generalizes on educational methods on the basis of his studies in one class, or one school subject, or one grade, or one school, or one school system, or even one country, needs to consider whether the bases of his investigations should be broadened.

(5) The investigator, especially in studies where he does not regard his investigation as based on a sample, but on a population or census, would be wise to consider what population he hopes his investigation applies to, whether the full breadth of it has had an appropriate chance to contribute

cases to his study and, if not, how he might get at the rest. He may be satisfied with describing the population under study, but often he is not.

(6) More narrowly, in sampling the membership of a professional society, the investigator may find his published membership list out of date by some years. For a fee the society may be willing to provide its current mailing list, which is probably as close as one can get to the target population. Obviously, the target population changes even while the study is being performed.

Incompatibility of meaning. While arguing for statistical thinking in the attempt to generalize one's results, one must not fall into the pit of statistical nonsense. Both anthropologists and historians call attention to mistakes that can come from regarding seemingly like objects, rituals, or behavior in different cultures as exchangeable commodities for statistical purposes. The notion of "father" without distinction between "pater" and "genitor" offers an example. In the Trobriands, a boy lives with his benign, biological father until he is nine or ten years old, then moves to his mother's brothers' village for training and discipline, and there he inherits property. In the United States the biological father theoretically plays the role of disciplinarian, and the uncles frequently play benign, indulgent roles.

Pilot studies. Toward the completion of a study, investigators usually feel that it would have been better done in some other way. But a study can be petted and patted so long that, before completion, its value is past. The huge, never-completed study usually damages the investigator's reputation, however wise the termination. Much can and must be learned by trying, and therefore nearly any investigation requires pilot work. Pilot work is little written about, perhaps because it is hard to summarize and perhaps because the results usually sound so obvious and often would be were they not hidden among thousands of other possible obvious results that did not occur. The whole spectrum from the tightest laboratory experiment to the loosest observational study requires careful pilot work. Pilot studies pinpoint the special difficulties of an investigation and, by encouraging initial action, overcome doctrines of omniscience that require a complete plan before starting. While it is true that the statistician can often give more valuable aid at the planning stage by preventing errors than by salvaging poor work through analysis, firm plans made in the absence of pilot studies are plans for disaster.

Hawthorne effects. Psychologists sadly say that even under the most carefully controlled conditions, laboratory animals do as they please. Humans do even worse. When Roethlisberger and Dickson (1939) carried out their experiments to find conditions that would maximize productivity of factory teams at the Hawthorne Works of Western Electric, they found that every change—increasing the lighting or reducing it, increasing the wage scale or reducing it—seemed to increase the group productivity. Paying attention to people, which occurs in placing them in an experiment, changes their behavior. This rather unpredictable change is called the Hawthorne effect. Instead of trying to eliminate this effect, it has been suggested that all educational efforts should be carried out as portions of experiments, so as to capitalize on the Hawthorne effect. No doubt boredom, even with experimentation, would eventually set in.

The existence of Hawthorne effects seriously restricts the researcher's ability to isolate variables that change performance in a consistent manner. Although experimenters, by adjusting conditions, may create substantial changes in behavior, what causes the changes may still be a mystery. Reliable repetition of results by different experimenters using different groups can establish results more firmly.

What treatment was applied? In experimental work with humans, it is especially difficult to know whether the treatment administered is the one that the experimenter had in mind. For example, in an unsuccessful learning experiment on the production of words by individuals, subjects in one group were instructed that every word in the class of words that they were seeking contained the same letters of the alphabet. When no differences in learning rates emerged between these subjects and those told nothing about the class of words being sought, further investigations were made. It turned out that few subjects listened to this particular instruction, and among those who did, several forgot it during the early part of the experiment. If a particular instruction is important, special efforts have to be made to ensure that the subject has received and appreciated it.

One approach to the problem of Hawthorne effects uses, in addition to experimental groups, two kinds of control groups: groups who are informed that they are part of an experiment and other groups who are not so informed. As always, the investigator has to be alert about the actual treatment of control and experimental groups. L. L. Thurstone told me about experimenting for the U.S. Army to measure the value of instruction during sleep for training in telegraphy. Thurstone had control squads who were not informed that they were in the study. The sergeants instructing these control squads felt that the "sleep learning" squads were getting favored treatment, and to keep their squads "even," they secretly instituted additional hours of wide-awake training for their own squads, thereby ruining the whole investigation.

Randomization. Generally speaking, randomization is a way to protect the study from bias in selecting subjects or in assigning treatments. It aids in getting a broad representation from the population into the sample. Randomization helps to communicate the objectivity of the study. It provides a basis for mathematical distribution theory that has uses in statistical appraisals and in simulations. [*See* EXPERIMENTAL DESIGN; RANDOM NUMBERS.]

Bad breaks in random sampling. Valuable as randomization is, chance can strike an investigator stunning blows. For example, suppose that a psychological learning experiment is intended to reinforce 5 randomly chosen responses in each burst of 20. If the randomization accidentally gives reinforcement to the first 5 responses in each burst of 20, the psychologist should notice this and realize that he has selected a special kind of periodic reinforcement. The objectivity of the random assignment cannot cure its qualitative failure.

Similarly, suppose that in preparing to study fantasy productions under two carefully controlled conditions the clinical psychologist observes that his randomizing device has put all his scientist subjects into one group and all his humanist subjects into another. In that case, he should reconsider the grouping.

In principle, one should write down, in advance, sets of assignments that one would not accept. Unfortunately, there are usually too many of these, and nobody is yet adept at characterizing them in enough detail to get a computer to list them, even if it could face the size of the task. One solution is to describe a restricted, but acceptable, set of assignments and to choose randomly from these. Omitting some acceptable assignments may help to make the description feasible while keeping the list satisfactorily broad.

If this solution is not possible, then one probably has either to trust oneself (admittedly risky) or else get a more impartial judge to decide whether a particular random assignment should be borne.

If there are many variables, an investigator cannot defend against all the bad assignments. By leaning upon subject matter knowledge and accepting the principle that the variables usually thought to be important are the ones to be especially concerned about, stratification, together with randomization, can still be of some assistance. For example, the stratification might enforce equal numbers of each sex, with individuals still randomly chosen from the potential pools of men and women. In studying bad breaks from randomization, the investigator can afford to consider rejecting only the assignments too closely related to proved first-order or main effects and not second-order effects or boomerang possibilities conceived in, but never observed from, armchairs.

Random permutations. Although arranging objects in a random order can easily be done by using an ordinary random number table, few people know how to do it. In any case, making these permutations is tedious, and it is worth noting the existence of tables of random permutations in some books on the design of experiments and in the book by Moses and Oakford (1963) that offers many permutations of sets numbering 9, 16, 20, 30, 50, 100, 200, 500, 1,000 elements. A set of any other size less than 1,000 can be ordered by using the permutations for the next larger size. With larger sets, some stratification is almost sure to be valuable.

Example. One (nonstratified) permutation for a set of 30 elements is shown in Table 1 (read left to right).

Table 1

11	5	29	26	3
1	19	14	4	20
24	25	27	6	9
30	15	13	18	12
28	22	7	8	17
2	21	16	10	23

To arrange at random the letters of the alphabet, we might assign the integers, starting with 1 for *a* and ending with 26 for *z*, to the positions of the letters in the alphabet. Then, according to the permutation of Table 1, 11 and 5 correspond to *k* and *e*, 29 is omitted, 26 corresponds to *z*. Continuing gives the permutation:

kezcasndtxyfiomrlvghqbupjw.

For a random sample of 5 letters from the alphabet, drawn without replacement, we could just take the first 5 listed.

Simulations for new statistical methods. Large-scale simulation of economic, political, and social processes is growing in popularity; social scientists who invent new statistics would often find it profitable to try these out on idealized populations, constructed with the aid of random number tables, to see how well they perform their intended functions under perfectly understood conditions. This sort of exploration should be encouraged as part of the pilot work. To illustrate the lack, many books and hundreds of papers have been written about factor analytic methods, yet in 1966 it is hard to point to more than a single published simulation (Lawley & Swanson 1954; Tucker 1964) of the methods proposed on artificially constructed populations with random error.

Nonsampling errors in sample surveys

Questionnaires. Questionnaires themselves present many sources of bias, of which the wording of questions and the options offered as answers are especially important. Some topics discussed below ("Panel studies," "Nonresponse," and "Response error") also treat questionnaire matters. [*See especially* SAMPLE SURVEYS; SURVEY ANALYSIS; *see also* INTERVIEWING IN SOCIAL RESEARCH.]

Wording. The wording and position of questions on questionnaires used in public opinion polls and other investigations illustrate the difficulties surrounding the notion of true value mentioned earlier. Rugg and Cantril's survey article (1944) analyzes and illustrates the effects of the manner of questioning on responses. For example, prior to the U.S. entry into World War II, variations on a question about American aid to Great Britain, asked of American citizens within a period of about six weeks, produced the following percentages in favor of the "interventionist" position: 76, 73, 58, 78, 74, 56. Here the interventionist position meant approval of "giving aid even at the risk of war." At much the same time, unqualified questions about "entering the war immediately" produced the following percentages in favor, 22, 17, 8, numbers substantially different from those in the previous set. Although one would be hard put to choose a number to represent degree of support for intervention, the interval 55 to 80 per cent gives a range; this range was clearly higher than that in support of entering the war immediately.

Pilot studies of the wordings of questions test their meaning and clarity for the intended population. Phillip Rulon recalls interviews with very bright second graders from a geography class to discuss a test item that they had "missed": "Wind-eroded rocks are most commonly found in the (*a*) deserts, (*b*) mountains, (*c*) valleys." They chose "valleys" because few people would *find* wind-eroded rocks in the mountains or the deserts, however many such rocks might be in those places. After a question previously found to be unsatisfactory is reworded, bitter experience advises the testing of the new version.

In single surveys, one needs to employ a variety of questions to get at the stability and meaning of the response. The use of "split ballots" (similar but modified questionnaires administered to equivalent samples of individuals) offers a way to experiment and to control for position and wording.

Changing opinions. To ignore the results of the polls because of the considerable variation in responses would be as big a mistake as to adopt their numbers without healthy skepticism. Since opinion in time of crisis may move rapidly, it is easy to misappraise the tenor of the times without a systematic measuring device. For example, between July 1940 and September 1941, the per cent of U.S. citizens saying that they were willing to risk war with Japan rather than to let it continue its aggression rose from 12 to 65 per cent. Again, although in June 1940 only 35 per cent thought it more important to help England than to keep out of war, by September 1940 the percentage had risen to the 50s (Cantril 1944, p. 222). In September 1940, President Roosevelt made a deal that gave Great Britain 50 destroyers in return for leases of bases (Leuchtenburg 1963, pp. 303–304); in the face of the fluctuations of public opinion a historian considering the destroyer deal might easily believe that Roosevelt acted against, rather than with, the majority. (As I recall from experience at the Office of Public Opinion Research, Roosevelt had his own personal polls taken regularly, with reports submitted directly to him, usually on a single question.)

Seemingly minor variations in questions may change the responses a good deal, and so to study changes over time, one needs to use one well-chosen question (or sequence) again and again. Naturally, such a question may come under attack as not getting at the "true value." If the question is to be changed, then, to get some parallel figures, it and the new question should be used simultaneously for a while.

Intercultural investigations. Considering the difficulty of getting at opinions and the dependence of responses upon the wording of the questions asked, even within a country, the problem of obtaining comparable cross-cultural or cross-national views looks horrendous. Scholars planning such studies will want to see three novel works. Kluckhohn and Strodtbeck's (1961) sociological and anthropological *Variations in Value Orientations* especially exploits ranking methods in the comparison of value orientations in Spanish-American, Mormon, Texas, Zuñi, and Navajo communities. Subjects describe the many values of their culture by ordering their preferences, for example, for ways of bringing up children: past (the old ways), present (today's ways), or future (how to find new ways to replace the old). Cantril's (1966) social-psychological and internationally oriented *Pattern of Human Concerns* uses rating methods and sample surveys to compare values and satisfactions in the populations of 15 nations. For example, the respondent's rating, on a scale of 0 to 10, expresses his view of how nearly he or his society has achieved the goal inquired about, and another rating evaluates how much either might expect to advance in five years. In international economics, measurements may be more easily compared, although the economist may be forced to settle for measuring the measurable as an index of what he would like to evaluate. Harbison and Myers' study, *Education, Manpower, and Economic Growth* (1964), illustrates this approach.

Panel studies. Although the single sample survey can be of great value, in some problems it is desirable to study the changes in the same people through time. The set of people chosen for repeated investigation is called a panel. One advantage of the panel study over the single survey is the deeper analysis available. For instance, when a net 5 per cent change takes place, does this mean that only 5 per cent of the people changed, or perhaps that 15 per cent changed one way and 10 per cent another? Second, additional measurement precision comes from matching responses from one interview to another. Third, panel studies offer flexibility that allows later inquiries to help explain earlier findings. [*See* PANEL STUDIES.]

Dropouts. Panel studies, even when they start out on an unbiased sample, have the bias that the less informed, the lower-income groups, and those not interested in the subject of the panel tend to drop out. Sobol (1959) suggested sampling these people more heavily to begin with, and she tried to follow movers. According to Seymour Sudman of the National Opinion Research Center, in national consumer panels and television rating panels where a fee is paid to the participant, the lower-income groups do not drop out.

Beginning effects. When new individuals or households first join a panel, their early responses may differ from their later ones. The "first-month" effect has unknown origins. For example, after the first month on the panel, the fraction of unemployed reported in private households decreases about 6 per cent. Over the course of several panel interviews, more houses become vacant and consumer buying decreases (Neter & Waksberg 1964*a*; Waksberg & Pearl 1964). Household repairs decreased by 9 per cent between the second and third interview. In consumer panels, the reports made during the first six or eight weeks of membership are usually not included in the analysis. The start-up differences are not clear-cut and emphatic but unsettling enough that the data are set aside, expensive as that is.

Sample surveys are not alone in these "first-time" effects. Doctors report that patients' blood pressures are higher when taken by a strange doctor. In the Peirce reaction-time data, presented in Table 4, the first day's average reaction time was about twice those of the other 23 days.

Long-run effects. A most encouraging finding in consumer panel studies has been the stability of the behavior of the panelists. By taking advantage of the process of enlarging two panels, Ehrenberg (1960) studied the effects of length of panel membership in Great Britain and in Holland. When he compared reports of newly recruited households (after their first few weeks) with those of "old" panel members, he found close agreement for purchasing rates, brand shares of market, and diary entries per week.

Panels do have to be adjusted to reflect changes in the universe, and panel families dissolve and multiply.

Nonresponse. The general problem of nonresponse arises because the properties of the nonrespondents usually differ to some degree from those of respondents. Unfortunately, nonresponse is not confined to studies of human populations. Physical objects can be inaccessible for various reasons. records may be lost, manholes may be paved over, a chosen area may be in dense jungle, or the object may be too small to be detected. One tries to reduce nonresponse, adjust estimates for it, and allow for it in measures of variability.

Mail questionnaires. The following advice, largely drawn from Levine and Gordon (1958–1959) and Scott (1961), is intended to increase response from mail questionnaires:

(1) Respondent should be convinced that the project is important.

(2) Preparatory letter should be on the letterhead of a well-known organization or, where appropriate, should be signed by a well-known person. In the United States and in Great Britain, governmental agencies are more likely to obtain responses than most organizations. Indeed, Scott (1961) reports 90 per cent response! Special populations respond to appeals from their organizations.

In pilot studies preparatory to using mailed census questionnaires in the initial stage of enumeration for the 1970 census (enumerator to follow up nonrespondents), the U.S. Bureau of the

Table 2

	PERCENTAGE OF RESPONSE	
	Long form	Short form
Cleveland, Ohio	78	80
Louisville, Kentucky	85	88

Census got the percentages of responses to the mailing shown in Table 2.

(3) Rewards may be used (gifts, trading stamps, sweepstakes). Do not offer a copy of the final report unless you are prepared to give it.

(4) Make questionnaire attractive (printing on good paper is preferred), easy to read, and easy to fill in, remembering that many people have trouble reading fine print. Longer questionnaires usually lower the response rate.

(5) Keep questions simple, clear, as short as possible, and where multiple-choice answers appear, make sure that they do not force respondent to choose answers that do not represent his position.

(6) Try to keep early questions interesting and easy; do not leave important questions to the end; keep related questions together, unless there are strong reasons to act otherwise.

(7) Use a high class of mail, first-class, airmail, and even special delivery, both for sending the questionnaire and on the return envelope. Do not expect respondent to provide postage. In Great Britain, Scott (1961) found that compared with a postcard a card to be returned in an envelope raised response.

(8) Follow hard-core resistance with repeat questionnaire (the sixth mailing may still be rewarding), telegram, long-distance phone call, or even personal interview, as discussed below. Small response from early mailings may be badly biased; for example, successful hunters respond more readily than unsuccessful ones to questions about their bag (Kish 1965, p. 547).

(9) Do not promise or imply anonymity and then retain the respondent's identity by subterfuge, however worthy the cause. Views on the effects of anonymity are mixed. If respondent's identity is needed, get it openly.

The principles set out above for mail questionnaires and those below for personal interviews may well be culture-bound for they are largely gathered from Western, English-speaking experience. For example, where paper is expensive, questionnaires on better paper may be less likely to be returned than those on poorer paper.

Sample surveys using personal interviews. In personal interviews, 80 to 90 per cent response has been attained even on intimate topics. In 1966, 85 per cent was regarded as rather good for predesignated respondents in household surveys. In addition to the relevant maxims given above for mail surveys, to reduce nonresponse in personal interview surveys Sharp and Feldt (1959), among others, suggest some of the following:

(1) Send preview letter; use press to announce survey. In three lengthy surveys on different topics, according to Reuben Cohen of the Opinion Research Corporation, a letter sent in advance led to an average gain of 9 per cent in reaching, after four calls, urban adult respondents randomly drawn from the household list. Cohen also suggests that follow-up letters, after unsuccessful interviewing attempts, can reduce urban nonresponse by about one-third. Some students of polling believe that the actual impact of the preview letter is largely on the interviewer, who thinks that obtaining cooperation will be easier because of it—and so it is.

(2) Use trained interviewers, that is, interviewers trained especially to handle opening remarks, to explain the need for full coverage, and to get information about profitable times to make later calls ("callbacks") to reach respondents who are initially not at home. Experienced interviewers have had 3 per cent to 13 per cent fewer nonrespondents than inexperienced ones.

(3) Be flexible about calling at convenience of respondent, even at his place of work or recreation, on evenings and on week ends.

(4) Allow interviewer to call back many times to locate assigned respondent.

(5) Employ interpreter when appropriate.

(6) In more esoteric situations, know the culture. Do not plan to interview farmers in the peak periods of farm activity. An anthropologist scheduled a survey of current sexual behavior among South Sea islanders during the season when women were taboo to fishermen—the natives, finding it a great joke, were slow to explain.

Extra effort. When a survey carried out in the usual way produces a surprisingly large nonresponse, an all-out effort may be mounted using many of the devices mentioned earlier. A rule of thumb is that the nonresponse can be reduced by about half.

Oversampling nonrespondents. Repeated callbacks are the traditional method for reducing nonresponse in personal interviews, and careful cost analysis has shown that their cost per completed interview is lower than was at first supposed when quota sampling was popular. Kish and Hess (1959) report a procedure for including in the current sample nonrespondents from similar previous surveys, so as to have in advance an oversupply of persons likely not to respond. Then the sample survey, although getting responses from these people at a lower rate than from others, more nearly fills out its quotas.

Subsampling nonrespondents. To reduce nonresponse in mail surveys, subsampling the nonrespondents and pursuing them with personal interviews has been used frequently (formulas for optimum design are given in Hansen & Hurwitz 1946). In methods thus far developed the assumption is made that nonrespondents can surely be interviewed. When this assumption is unjustified, the method is less valid.

Adjusting for respondents not at home. The next method adjusts for those not at home but does not handle refusals, which often come to about half the nonresponse. Bartholomew (1961) has got accurate results by assuming that most of the bias arises from the composition of the population available at the first call. By finding out when to call back, the interviewer reduces later biases from this source. The interviewer gets information either from others in the house or from neighbors. To illustrate, in empirical investigations of populations of known composition, Bartholomew studied the percentage of men in political wards of a city. In four wards, differences between first-call and second-call samples in percentage of men were 17 per cent, 29 per cent, 36 per cent, and 38 per cent, substantial differences. But the differences between the second-call percentage of men and the actual percentage of men not reached by the first call were only 6 per cent, 2 per cent, 2 per cent, and 2 per cent, supporting Bartholomew's point.

Suppose that proportion p of the population has the characteristic of interest. It is convenient to regard p as the weighted average $\rho p_1 + (1 - \rho)p_2$, where ρ is the proportion of first-call responders in the population, p_1 is the proportion of first-call responders having the characteristic, and p_2 is the proportion of others in the population having the characteristic. (It is assumed that p_2 is independent of response status after the first call.) Now if N, the total sample size, is expressed as $N = N_1 + N_2 + N_3$, where N_1 is the number of first-call responders in the sample, N_2 is the number of second-call (but not first-call) responders in the sample, and N_3 is the number of others, then ρ is naturally estimated by N_1/N (the proportion of first-call responders in the sample), and p_1 by n_1/N_1 (the proportion of first-call responders in the sample who have the characteristic) and p_2 by n_2/N_2 (the proportion of second-call responders in the sample having the characteristic). Putting these estimators in the weighted average gives, as estimator of p,

$$\frac{1}{N}\left[n_1 + \frac{n_2(N - N_1)}{N_2}\right].$$

For example, if the number of men in the first call is $n_1 = 40$ out of $N_1 = 200$ interviewed, the second-call data are $n_2 = 200$, $N_2 = 400$, and the original sample size is $N = 1{,}000$, then the estimate of the proportion of men is 0.44. Even if the theory were exactly true, some increase in variance would arise from using such weights instead of obtaining the whole sample (Kish 1965, secs. 11.7*B*, 11.7*C*).

Extrapolation. Hendricks (1956) suggests plotting the variable being measured against the percentage of the sample that has responded on successive waves and extrapolating to 100 per cent. This simple, sensible idea could profit from more research, empirical and theoretical.

Effect on confidence interval. In sample surveys, nonresponse increases the lengths of the confidence intervals for final estimates by unknown amounts. For dichotomous types of questions, the suggestion is often made that all the nonresponses be counted first as having the attribute, then as not having it. The effect on the 95 per cent confidence interval is shown in Table 3. When such extreme allowances are required, the result of, say, 20 per cent nonresponse is frequently disastrous. For large random samples, this treatment of nonresponse, as may be seen from Table 3, adds approximately the per cent of nonresponse to the total length of the confidence interval that would have been appropriate with 100 per cent response. For example, with a sample of 2,500 from a large population, a 95 per cent confidence interval from 58 per cent to 62 per cent would be lengthened by 20 per cent nonresponse to 48.5 per cent to 71.5 per cent. This

Table 3 — Allowance to be added to and subtracted from the observed percentage to give at least 95 per cent confidence of covering the true value*

PER CENT NONRESPONSE	SAMPLE SIZE 100	2,500	Infinite
0	9.8	2.0	0.0
5	10.7	4.1	2.5
10	12.9	6.6	5.0
15	15.1	9.0	7.5
20	17.4	11.5	10.0

* These numbers are approximately correct for percentages near 0.50; they are likely to be conservative otherwise.

Source: Cochran, Mosteller, and Tukey 1954, p. 280.

additional length gives motivation enough for wanting to keep nonresponse low.

No one believes that these "worst possible" limits represent the true state of affairs, nor should anyone believe the optimist who supposes that the nonrespondents are just like the respondents. In large samples, differences as large as 28 per cent in the fraction possessing a characteristic between the first 60 per cent interviewed and the next 25 per cent have been reported. To develop a more sensible set of limits in the spirit of Bayesian inference would be a useful research job for sample survey workers and theoretical statisticians. [*See* BAYESIAN INFERENCE.] This urgently needed work would require both empirical information (possibly newly gathered) and theoretical development.

The laboratory worker who studies human behavior rarely has a defined target population and he frequently works with a sample of volunteers. Under such circumstances we cannot even guess the extent of nonresponse. Again the hope is that the property being studied is independent of willingness or opportunity to serve as a subject—the position of the optimist mentioned above.

Since a few experimenters do sample defined populations, the argument that such sampling is impossible has lost some of its strength. The argument that such sampling is too expensive has to be appraised along with the value of inferences drawn from the behavior of undefined sampled populations.

Studying differences between groups offers more grounds for hope that bias from nonresponse works in the same direction and in nearly the same amount in both groups and that the difference may still be nearly right. This idea comes partly from physical measurements where sometimes knowledge can make such arguments about compensating errors rigorous. But, as Joseph Berkson warns, no general theorem states "Given any two wrong numbers, their difference is right."

Response error. When incorrect information about the respondent enters the data, a response error occurs. Among the many causes are misunderstandings, failures of memory, clerical errors, or deliberate falsehoods. The magnitudes of some of these errors and some ways to reduce them are discussed below.

Telescoping events. In reporting such things as amount of broken crockery or expenditures for household repairs, some respondents telescope the events of a considerable period into the shorter one under study. As a possible cure, Neter and Waksberg (1964*b*) have introduced a device called "bounded recall." In a study of household repairs, the respondent was interviewed twice, first under unbounded recall, during which the full story of the last month, including the telescoping from previous months, was recorded by the interviewer. Second, in an interview using bounded recall a month later, the respondent was deliberately aided by the record of repairs from the first interview. The magnitudes of the effects of telescoping are considerable, because the "unbounded" interview for household repairs gave 40 per cent more jobs and 55 per cent higher expenditure than did the "bounded" interview. Data from a "bounded" interview produce less bias.

Forgetting. Although telescoping occurs for some activities, chronic illness (Feldman 1960), which had already been clinically diagnosed, was reported only at a 25 per cent rate in household interviews. Others report rates in the 40 per cent to 50 per cent range. Feldman despairs of the household interview for this purpose; but if improvement in reports is to be attempted, he recommends more frequent interviews by competent, trained interviewers and, in a panel study, the use of a morbidity diary to improve self-reporting. One limitation, not attributable to forgetting, is set because physicians choose not to inform their patients of every illness they diagnose.

Sudman (1964) compared consumer panel reports based upon diary records with reports based upon unaided recall. First, he shows that for 72 grocery products (55 being food), the purchases recorded in the diary underestimate the amount shipped by the manufacturer (after adjustment for nonhousehold use) by a median of about 15 per cent. The underreporting was highly predictable, depending on both the properties of the product (frequency of purchase, where most often purchased) and its treatment in the diary (type size, page number, position on page). Second, when recall was compared with diary, the median ratio of purchases (purchases recalled divided by purchases in diary) was 1.05 for nonfood products, 1.83 for perishable food, 1.54 for staple foods. Leading nationally advertised brands have their market shares overstated under recall by 50 per cent compared to diary records, and chain brands are understated.

Use of experts. After respondents had valued their own homes, Kish and Lansing (1954) obtained expert appraisals for a sample of the homes. The comparison of the experts' appraisal with the homeowners' appraisal can be used to adjust the total valuation or to adjust valuations for groups of houses. For example, the homeowners may average a few per cent too high.

Editing records. Whenever a comparison of related records can be made, the accuracy of records can probably be improved. For example, Census Bureau editors, experienced in the lumber business, check annual sawmill production reports against those of the previous year, and large changes are rechecked with the sawmill.

Describing response errors. One common measure in the analysis of nonsampling errors puts bias and sampling variability into one index, called the mean square error. The larger the mean square error, the worse the estimate. The mean square error is the expected squared deviation of the observed value from the true value. This quantity can be separated into the sum of two parts, the variance of the observation around its own mean and the square of the bias. Although true values are not available, in the United States, the Bureau of the Census, for instance, tries to find standards more accurate than the census to get an estimate of the response bias to particular questions. For example, using the Current Population Survey as a standard, the Bureau of the Census not only finds out that the census underestimates the percentage in the labor force, but the bureau also gets data on the portions of the population not being satisfactorily measured, either because of variability or bias. Using such information, the bureau can profitably redesign its inquiries because it knows where and how to spend its resources.

Response uncertainty. In attitudinal studies (Katz 1946), the investigator must be especially wary of reports obtained by polling the public on a matter where opinion is not crystallized. The No Opinion category offers one symptom of trouble: for example, Katz reports that in 1945 only 4 per cent had No Opinion about universal military training in the United States, 13 per cent had No Opinion about giving the atom bomb secret to the United Nations, and 32 per cent had No Opinion about U.S. Senate approval of the United Nations charter. Even though the vote was 66 per cent to 3 per cent in favor of approving the charter, the 32 per cent No Opinion must suggest that the 69 per cent who offered an opinion contained a large subgroup who also did not hold a well-formed opinion.

Errors in analysis

Troubles with variability. In analyzing data, the presence of variability leads to many unsuspected difficulties and effects. In addition to treating some of the common traps, this section gives two ways to analyze variability in complicated problems where theoretical formulas for variance are either unavailable or should be distrusted.

Inflated sample size. The investigator must frequently decide what unit shall be regarded as independent of what other unit. For example, in analyzing a set of responses made by 10 individuals, each providing 100 responses, it is a common error to use as the sample size $10 \times 100 = 1{,}000$ responses and to make calculations, based perhaps on the binomial distribution, as if all these responses were independent. Unless investigation has shown that the situation is one in which independence does hold from response to response both within and between individuals, distrust this procedure. The analysis of variance offers some ways to appraise both the variability of an individual through time and the variation between individuals.

Use of matched individuals. Some investigators fail to take advantage of the matching in their data. Billewicz (1965, p. 623) reports that in 9 of 20 investigations that he examined for which the data were gathered from matched members in experimental and control groups, the analysis was done as if the data were from independent groups. Usually the investigator will have sacrificed considerable precision by not taking advantage in his analysis of the correlation in the data. Usually, but not always, the statistical significance of the results will be conservative. When matched data are analyzed as if independent, the investigator owes the reader an explanation for the decision.

Pooling significance tests. In the same vein, investigators with several small effects naturally wish that they could pool these effects to get more extreme levels of significance than those given by the single effects. Most methods of pooling significance tests depend upon independence between the several measures going into the pool. And that assumption implies, for example, that data from several items on the same sample survey cannot ordinarily be combined into a significance test by the usual pooling methods because independence cannot be assured. Correlation is almost certainly present because the same individuals respond to each item. Sometimes a remedy is to form a battery or scale that includes the several items of interest and to make a new test based upon the battery (Mosteller & Bush 1954, pp. 328–331). Naturally, the items would be chosen in advance for the purpose, not based *post hoc* upon their results. In the latter case, the investigator faces problems of multiplicity, discussed below.

Outlying observations. Frequently data contain suspicious observations that may be outliers, observations that cannot be rechecked, and yet that may considerably alter the interpretation of the data when taken at their face value. An outlier is an observation that deviates much more from the average of its mates or has a larger residual from a predicted value than seems reasonable on the basis of the pattern of the rest of the measurements. The classic example is given by the income distribution for members of a small college freshman class, exactly one member of which happens to be a multimillionaire. The arithmetic mean is not typical of the average member's income; but the median almost ignores an amount of income that exceeds the total for all the others in the class. Sometimes the outliers can be set aside for special study.

One current approach tailors the analysis to the type of outlier that is common in the particular kind of investigation by choosing statistics that are both appropriate and not especially sensitive to outliers. For example, as a measure of location, one might systematically use the median rather than the mean, or for more efficiency, the trimmed mean, which is the average of the measurements left after the largest and smallest 5 per cent (or 100α per cent) of the measurements have been removed (Mosteller & Tukey 1966, secs. *A5*, *B5*). In normal populations, the median has an efficiency of about 64 per cent, but the trimmed mean has most of the robustness of the median and an efficiency of about $1-\frac{2}{3}\alpha$, where α is the proportion trimmed off each end. For $\alpha=0.10$, the efficiency is 93 per cent. [*See* ERRORS, *article on the* EFFECTS OF ERRORS IN STATISTICAL ASSUMPTIONS; NONPARAMETRIC STATISTICS, *article on* ORDER STATISTICS; STATISTICAL ANALYSIS, SPECIAL PROBLEMS OF, *article on* OUTLIERS.]

Shifting regression coefficients. When one fits a regression equation to data, this regression equation may not forecast well for a new set of data. Among the reasons are the following:

(1) The fitted regression coefficients are not true values but estimates (sampling error).

(2) If one has selected the best from among many predictive variables, the selected ones may not be as good as they appeared to be on the basis of the sample (regression effect).

(3) Worse, perhaps none of the predictive variables were any good to start with (bad luck or poor planning).

(4) The procedure used to choose the form of the regression curve (linear, quadratic, exponential, . . .) has leaned too hard on the previously available data, and represents them too well as compared with the total population (wrong form).

(5) The new sample may be drawn from a population different from the old one (shifting population).

What are the effects of (2) and (5)? Consider the regression of height (Y) on weight (X) for a population of boys. Suppose that the true regression equation for this population is

$$E(Y)=a+b(X-\mu_X),$$

where a and b are unknown constants, $E(Y)$ is the expected value of Y for a given value of X, and μ_X is the mean of X. Suppose that an individual's height has a predictive error e that has mean 0, variance σ^2, and is unrelated to X, Y, and the true values of a and b.

Suppose that the experimenter chooses fixed values of X, x_i, such as 70, 80, 90, 100, 110, 120, 130, 140 pounds, obtaining boys having each of these weights and measuring their heights y_i. Then the data are paired observations (x_i, y_i), $i=1,2,\cdots,n$.

Estimating a and b from the sample by the usual least squares formulas one gets $\hat{a}$ and $\hat{b}$. Given a new sample with the same values of X from the same population, one can estimate the Y's for the new sample by

$$\hat{Y}_i=\hat{a}+\hat{b}(x_i-\bar{x}),$$

(where $\bar{x}$ is the average of the X_i) and then the expected mean square error of the estimates for the new sample is

$$\text{expected value of } \sum_i(Y'_i-\hat{Y}_i)^2/n=\sigma^2(1+2/n),$$

where Y'_i is the height for an individual in the new sample. Note that σ^2 is the expected mean square error that would obtain were a and b known exactly instead of having been estimated.

Suppose that in addition to this population, there is a new population with different values of a and b, say a' and b'. Both populations come from a group of several populations with a and b varying from population to population and having $\bar{a}$ and $\bar{b}$ as the mean values of a and b, respectively, and σ_a^2 and σ_b^2 as the variances of these sets of regression coefficients. For the example of the boys' weights, consider the distribution of values of a and b from one city to another.

The regression line fitted on the basis of the sample from one population and then used on another population yields expected mean square error

$$\left(1+\frac{2}{n}\right)\sigma^2+2(\sigma_a^2+\sigma_x^2\sigma_b^2),$$

where σ_x^2 is the variance of the chosen set of x's. The first term comes as before from ordinary sampling variation of the Y's around the fitted regression line (the 2 of $2/n$ being the dimension of the parameter space), but the $2(\sigma_a^2+\sigma_x^2\sigma_b^2)$ comes from drawing two sets of regression coefficients from the population of regression coefficients. This term may be substantial compared to $2\sigma^2/n$ or even σ^2.

We need extensive empirical results for such experiments to get a notion of the size of $2(\sigma_a^2 + \sigma_x^2\sigma_b^2)$ in various settings of interest to social and natural scientists. These investigations have not yet been carried out. The formulas for mean square error in this realistic situation must cause concern until more empirical studies are done. The existence of the added term should be recognized and an attempt made to assess its contribution numerically.

Uncontrolled sources of variation. Although the important formula $\sigma_{\bar{X}} = \sigma/\sqrt{n}$ for the standard deviation of a mean $\bar{X}$, a random variable, is correct when n uncorrelated measurements are drawn from a distribution with standard deviation σ, two difficulties arise. The measurements may not be uncorrelated, and the distribution may change from one set of measurements to another.

Peirce's data illustrate these difficulties. In an empirical study intended to test the appropriateness of the normal distribution, C. S. Peirce (1873) analyzed the time elapsed between a sharp tone stimulus and the response by an observer, who made about 500 responses each day for 24 days. Wilson and Hilferty (1929) reanalyzed Peirce's data. Table 4 shows sample means, $\bar{x}$, estimated standard deviations of the mean $s_{\bar{x}}$, and the ratio of the observed to the estimated interquartile range, $Q_3 - Q_1$. The observed interquartile range is based on percentage points of the observed distribution; the estimated interquartile range is based on the assumption of a normal distribution and has the value $2(0.6745s)$, where s is the sample standard deviation. In passing, note that the ratio is systematically much less than unity, defying the normality assumption. More salient for this discussion is the relation of day-to-day variation to the values of $s_{\bar{x}}$ based on within-day variation. The latter varies from 1.1 to 2.2 (after the first day's data, whose mean and standard deviation are obviously outliers, are set aside). These limits imply naive standard deviations of the difference between means for pairs of days ranging from 1.6 to 3.1. If these applied, most differences would have to be less than twice these, 3.2 to 6.2, and practically all less than 4.8 to 9.3. Table 4 shows that the actual differences −38, +2, −57, +27, ⋯, +11, −4, −8 impolitely pay little attention to such limitations.

Table 4 — Daily statistics from Wilson and Hilferty's analysis of C. S. Peirce's data

Day	$\bar{x} \pm s_{\bar{x}}$ (milliseconds)	$\frac{Q_3 - Q_1}{2(0.6745s)}$
1	475.6 ± 4.2	0.932
2	241.5 ± 2.1	0.842
3	203.1 ± 2.0	0.905
4	205.6 ± 1.8	0.730
5	148.5 ± 1.6	0.912
6	175.6 ± 1.8	0.744
7	186.9 ± 2.2	0.753
8	194.1 ± 1.4	0.840
9	195.8 ± 1.6	0.756
10	215.5 ± 1.3	0.850
11	216.6 ± 1.7	0.782
12	235.6 ± 1.7	0.759
13	244.5 ± 1.2	0.922
14	236.7 ± 1.8	0.529
15	236.0 ± 1.4	0.662
16	233.2 ± 1.7	0.612
17	265.5 ± 1.7	0.792
18	253.0 ± 1.1	0.959
19	258.7 ± 1.8	0.502
20	255.4 ± 2.0	0.521
21	245.0 ± 1.2	0.790
22	255.6 ± 1.4	0.688
23	251.4 ± 1.6	0.610
24	243.4 ± 1.1	0.730

Source: Wilson & Hilferty 1929.

In the language of analysis of variance, Peirce's data show considerable day-to-day variation. In the language of Walter Shewhart, such data are "out of control"—the within-day variation does not properly predict the between-days variation [*see* QUALITY CONTROL, STATISTICAL, *article on* PROCESS CONTROL]. Nor is it just a matter of the observer "settling down" in the beginning. Even after the twentieth day he still wobbles.

Need for a plurality of samples. The wavering in these data exemplifies the history of the "personal equation" problem of astronomy. The hope had been that each observer's systematic errors could be first stabilized and then adjusted for, thus improving accuracy. Unfortunately, attempts in this direction have failed repeatedly, as these data suggest they might. The observer's daily idiosyncrasies need to be recognized, at least by assigning additional day-to-day variation.

Wilson and Hilferty (1929, p. 125) emphasize that Peirce's data illustrate "the principle that we must have a plurality of samples if we wish to estimate the variability of some statistical quantity, and that reliance on such formula as $\sigma/\sqrt{n}$ is not scientifically satisfactory in practice, even for estimating unreliability of means" (see Table 4).

Direct assessment of variability. One way to get a more honest estimate of variability breaks the data into rational subgroups, usually of equal or nearly equal sizes. For each subgroup, compute the statistic (mean, median, correlation coefficient, spectral density, regression equation, or whatever), base the estimate for the whole group on the average of the statistic for the subgroups, and base the estimate of variability on Student's t with one degree of freedom less than the number of subgroups.

That is, treat the k group statistics like a sample of k independent measurements from a normal distribution. [*This method, sometimes called the method of interpenetrating samples, generalizes the method for calculating the sampling error for nonprobability samples described in* SAMPLE SURVEYS, *article on* NONPROBABILITY SAMPLING.]

At least five groups (preferably at least ten) are advisable in order to get past the worst part of the t-table. This suggestion encourages using more, not fewer, groups. For two-sided 5 per cent levels, see Table 5.

Table 5 — Two-sided 5 per cent levels for Student's t for selected degrees of freedom

Degrees of freedom	5 per cent critical point
1	12.7
2	4.3
3	3.2
5	2.6
10	2.23
20	2.09
60	2.00
500	1.96
∞	1.96

Two major difficulties with this direct assessment are (a) that it may not be feasible to calculate meaningful results for such small amounts of data as properly chosen groups would provide, or (b) even if the calculations yield sensible results, they may be so severely biased as to make their use unwise.

A method with wide application, intended to ameliorate these problems, is the *jackknife*, which offers ways to reduce bias in the estimate and to set realistic approximate confidence limits in complex situations.

Assessment by the jackknife. Again the data are divided into groups, but the statistic to be jackknifed is computed repeatedly on all the data except an omitted group. With ten groups, the statistic is computed each time for about 90 per cent of the data.

More generally, for the jackknife, the desired calculation is made for all the data, and then, after the data are divided into groups, the calculation is made for each of the slightly reduced bodies of data obtained by leaving out just one of the groups.

Let $y_{(j)}$ be the result of making the complex calculation on the portion of the sample that omits the jth subgroup, that is, on a pool of $k-1$ subgroups. Let y_{all} be the corresponding result for the entire sample, and define *pseudo values* by

$$(1)\qquad y_{*j} = ky_{\text{all}} - (k-1)y_{(j)}, \qquad j = 1, 2, \cdots, k.$$

These pseudo values now play the role played by the values of the subgroup statistics in the method of interpenetrating samples. For simple means, the jackknife reduces to that method.

As in the method of interpenetrating samples, in a wide variety of problems, the pseudo values can be used to set approximate confidence limits through Student's t, as if they were the results of applying some complex calculation to each of k independent pieces of data.

The jackknifed value y_*, which is the best single result, and an estimate, s_*^2, of its variance are given by

$$y_* = \frac{1}{k}(y_{*1} + \cdots + y_{*k}),$$

$$s^2 = \left[\sum y_{*i}^2 - ky_*^2\right] \Big/ (k-1),$$

$$s_*^2 = s^2/k.$$

If the statistic being computed has a bias that can be expressed as a series in the reciprocal of the sample size, N, the jackknife removes the leading term (that in $1/N$) in the bias. Specifically, suppose that $\hat{\mu}$, the biased estimate of μ, has expected value

$$E(\hat{\mu}) = \mu + \frac{a}{N} + \frac{b}{N^2} + \cdots,$$

where a, b, and so on are constants. If $\hat{\mu}_*$ is the jackknifed estimate, its expected value is

$$E(\hat{\mu}_*) = \mu + \frac{\alpha}{N^2} + \frac{\beta}{N^3} + \cdots,$$

where α, β, and so on are constants. To give a trivial example, $\sum(X_i - \bar{X})^2/N$ is a biased estimate of σ^2. Its expected value is $\sigma^2 - (\sigma^2/N)$, and so it has the sort of bias that would be removed by jackknifing.

To understand how the first-order bias terms are removed by jackknifing, one might compute the expected value of y_* for the special case where

$$E(\hat{\mu}) = \mu + a/N.$$

Then with the use of k groups of equal size, n, so that $kn = N$,

$$\begin{aligned} E(y_{(j)}) &= \mu + a/(k-1)n \\ E(y_{\text{all}}) &= \mu + a/kn \\ E(y_{*j}) &= E[ky_{\text{all}} - (k-1)y_{(j)}] \\ &= k\mu + a/n - [(k-1)\mu + a/n] \\ &= \mu. \end{aligned}$$

Finally,

$$E(y_*) = k\mu/k = \mu.$$

The leading term in the bias was removed in the construction of the y_{*j}'s. Even if the sample sizes are not equal, the leading term in the bias is likely to have its coefficient reduced considerably.

Example of the jackknife: ratio estimate. In expounding the use of ratio estimates, Cochran ([1953] 1963, p. 156) gives 1920 and 1930 sizes (number of inhabitants) for each city in a random sample of 49 drawn from a population of 196 large U.S. cities. He wishes to estimate the total 1930 population for these 196 cities on the basis of the results of the sample of 49, whose 1920 and 1930 populations are both known, and from the total 1920 population. The example randomly groups his 49 cities into 7 sets of 7 each. Table 6 shows their subtotals.

Table 6 — Subtotals in thousands for sets of 7 cities

	1920	1930
First 7	751	915
Second 7	977	1,122
Third 7	965	1,243
Fourth 7	385	553
Fifth 7	696	881
Sixth 7	830	937
Seventh 7	450	611
Total	5,054	6,262

Source: Cochran [1953] 1963, p. 156.

The formula for the ratio estimate of the 1930 population total is

$$\frac{(\text{1930 sample total})}{(\text{1920 sample total})} \times (\text{1920 population total}),$$

so that the logarithm of the estimated 1930 population total is given by log (1930 sample total) − log (1920 sample total) + log (1920 population total). Consequently the jackknife is applied to $z = \log(\text{1930 sample total}) - \log(\text{1920 sample total})$, since this choice minimizes the number of multiplications and divisions.

Further computation is shown in Table 7 where in the "all" column the numbers 5,054 and 6,262 come directly from the totals of the previous table, and in the "$i = 1$" column the numbers $4{,}303 = 5{,}054 - 751$ and $5{,}347 = 6{,}262 - 915$ are the results of omitting the first 7 cities, and so on for the other columns. Five-place logarithms have obviously given more than sufficient precision, so that the pseudo values of z are conveniently rounded to three decimals. From these are computed the mean z_* and the 95 per cent limits = mean ± allowance. Table 8 gives all the remaining details. The resulting point estimate is 28,300, about 100 lower than the unjackknifed estimate. (Since the correct 1930 total is 29,351, the automatic bias adjustment did not help in this instance. This is a reminder that bias is an "on the average" concept.) The limits on this estimate are ordinarily somewhat wider than would apply if each city had been used as a separate group, since the two-sided 95 per cent level for Student's t with 6 degrees of freedom is $|t_6|_{.95} = 2.447$, while with 47 degrees of freedom it is $|t_{47}|_{.95} = 2.012$. The standard error found here was .0125 in logarithmic units, which converts to about 840 in the final total ($4.360 + z_* + s_* = 4.464$; antilog $4.464 \cong 29{,}140$; $29{,}140 - 28{,}300 = 840$). The conversion from logarithmic units to original units for the confidence interval represent an ap-

Table 8 — Final computations for the ratio estimate*

	VALUE OF ESTIMATE	95 PER CENT CONFIDENCE INTERVALS
log ratio	$z_* = 0.092$	0.062 to 0.123
log total	$4.360 + z_* = 4.452$	4.422 to 4.483
total	antilog 4.452 = 28,300	26,000 to 30,400

* Base data: 1920 total = 22,919, log (1920 total) = 4.360
log total = log (1920 total) + log ratio

Table 7 — Details of jackknifing the ratio estimate

	all	$i=1$	$i=2$	$i=3$	$i=4$	$i=5$	$i=6$	$i=7$
$x_{(i)}$ (1920 sample)	5,054	4,303	4,077	4,089	4,669	4,358	4,224	4,604
$\log x_{(i)}$	3.70364	3.63377	3.61034	3.61162	3.66922	3.63929	3.62572	3.66314
$y_{(i)}$ (1930 sample)	6,262	5,347	5,140	5,019	5,709	5,381	5,325	5,651
$\log y_{(i)}$	3.79671	3.72811	3.71096	3.70062	3.75656	3.73086	3.72632	3.75213
$z_{(i)} = \log [y_{(i)}/x_{(i)}]$	.09307	.09434	.10062	.08900	.08734	.09157	.10060	.08899
$z_{*i} = 7z_{\text{all}} - 6z_{(i)}$	—	.08545	.04777	.11749	.12745	.10207	.04789	.11755
rounded z_{*i}	—	.085	.048	.117	.127	.102	.048	.118

Sum = .645; .645/7 ≅ .092 = mean = z_*

Sum Sq. = .065979; .065979 − $(.645)^2/7$ = .006547 = sum sq. deviations

$$\frac{.006547}{6 \times 7} \cong .00015588 = s_*^2$$

$$\sqrt{.00015588} \cong .0125 = s_*$$

$|t_6|_{.95} = 2.447$; (.0125) (2.447) = .0306 = allowance

proximation that may not always be appropriate [*see* STATISTICAL ANALYSIS, SPECIAL PROBLEMS OF, *article on* TRANSFORMATIONS OF DATA]. (Further material on the jackknife can be found in Mosteller and Tukey 1966, sec. *E*).

Analytical difficulties. In analyzing data or planning for its analysis, the choice of a base for rates is not always obvious; comparing many things leads to biases that need adjustment, selection reduces correlation, and selection for excellence leads to disappointments. This section treats these matters.

Bases for rates. The investigator should think about more than one possible base for a percentage or a rate and consider the value of reporting results using different bases. Examples from accident statistics may suffice. Are young women safer drivers than young men? Yes: in the United States in 1966 insurance rates for young women were ordinarily lower because they caused less expensive damage. On the other hand, these rates were based on total disbursements in a fixed period of time. Young women may well drive much less than young men, and if so, their accident rate per mile may be the higher.

Coppin, Ferdun, and Peck (1965) sent a questionnaire on driving in 1963 to a sample of 10,250 California drivers who were aged 16 to 19½ at the beginning of the period. Based on the information from the 65 per cent of questionnaires returned, where respondents estimated mileage driven per week, and on accidents reported in the respondents' Motor Vehicle Department files, the accident rates per 100,000 miles shown in Table 9 were found. On accidents, nonrespondents were very similar to respondents, but nonrespondents had considerably more violations. Since the mileage is estimated, the evidence is weak; but it seems to be the best available. Boys had more accidents per mile at 16, girls at 17, and after that their rates were nearly equal.

Table 9 — Accidents and violations per 100,000 miles

	ACCIDENTS		VIOLATIONS	
Age	Males	Females	Males	Females
16	2.9	2.1	7.1	4.9
17	1.9	2.3	6.1	4.2
18	1.4	1.4	5.6	3.7
19	1.4	1.5	5.4	3.5

Source: Coppin et al. 1965, pp. 27–28.

How should airplane safety (or danger) be assessed? Deaths per million passenger miles, deaths per trip, and casualties per hour flown suggest themselves, and each can be supported.

In general, different answers may be appropriate for different questions, as was the case in the insurance companies' view versus the accident-per-mile view of the safety of young drivers given above. Ease and economy may recommend giving several answers as well as the investigator's judgment about their merits. In some problems, no resolution may be possible, and then the investigator would do well to admit it.

Problems of multiplicity. When methods of appraisal designed for single comparisons are used to compare many things, the multiplicity may mislead. When means of two samples drawn from the same normal population are compared, they differ by more than twice the standard deviation of their difference in less than 5 per cent of the sample pairs. Among ten sample means from the same population, some pair is more likely than not to differ this much (Table 11). Although statistics has come a long way in providing honest methods of making comparisons when there are many to be made, it has largely done this in the framework of a closed system, where the particular items to be compared have already been specified. For example, many workers have offered suitable ways to measure the significance not only of all possible differences but also of all possible linear contrasts (weighted sums, the weights adding to zero) on the same data. [*See* LINEAR HYPOTHESES, *article on* MULTIPLE COMPARISONS.]

Statistics has not yet provided a way to test the significance of results obtained by peeking at large bodies of data and developing hypotheses as one goes along. The facility of the human brain for rationalizing almost any observed fact immediately after its realization is something that cannot yet be allowed for. This means that it is rarely possible to validate a hypothesis on the same body of data that suggested it and usually new studies are necessary to test hypotheses developed on completely different data (Mosteller & Tukey 1966, sec. *B*6).

Selection effects. Users of tests for purposes of selection (admission to college, personnel selection) often complain that the scores used to make the selection do not correlate well with the in-service performance of the individuals after selection. Possibly the chosen test does not give scores that correlate well with the performance being measured, but one must remember that when a population is truncated on one of its variables, the correlation of that dimension with the others is likely to be reduced toward zero. To illustrate, suppose that freshman calculus grades Y and pre-course examination grades X are bivariately normally distributed with correlation coefficient ρ.

Suppose that only individuals whose pretest scores exceed a certain value $X = x$ are admitted to the calculus course. This means that selection is based on the variable X with the criterion x. The new correlation ρ' between the grades of those taking the course and their pretest scores would be given (see Cochran 1951, p. 453) as

$$\rho' = \rho \sqrt{\frac{1 - A}{1 - A\rho^2}},$$

where

$$A = \frac{z}{p}\left(\frac{z}{p} - t\right),$$

p = proportion that the selected group is of the whole population,

t = standard normal deviate having proportion p to the right,

z = height of the standard univariate normal at the position t.

If the proportion selected $p = 0.05$, then $t = 1.645$, $z = 0.1031$. Values of ρ' for selected values of ρ and p are shown in Table 10. To return to the example, if pretest scores and calculus grades had originally been correlated $\rho = 0.8$, in the 5 per cent selected the correlation would drop to 0.44.

Table 10 shows that as the percentage truncated increases the correlation in the remaining population slowly decreases from its initial value. For initial correlations between .1 and .8 the reduction is between a half and a third of the original correlation when 75 per cent of the population has been removed. A very rough approximation for ρ' is $(.7p + .3)\rho$ for $p > .25$ and $0 \leqslant \rho \leqslant .7$. The new correlation decreases sharply for the higher initial correlation coefficients when more than 90 per cent of the population is deleted. Unfortunately, these results may be rather sensitive to the detailed shape of the bivariate population studied and so this bivariate normal example can only illustrate the possibilities.

Regression effect. Suppose that a fallible measure selects from many individuals a few that appear to be best. On a reassessment based on fresh performance data, the selected ones will ordinarily not do as well as they originally appeared to do on the selection test. The reason is that performance varies and on the occasion of the test some individuals accidentally perform much better than their average and are selected. Happily, individuals selected to be worst do not do as badly on reassessment. This phenomenon is known as regression toward the mean; instances are sometimes called regression effects or shrinkage effects. To illustrate, Mosteller and Wallace (1964, p. 209) selected words and obtained weights for their rates of use with intent to discriminate between the writings of Alexander Hamilton and James Madison. Writers differ in their rate of use of such words as *of*, *and*, *the*, *to*, and *upon*. On the basis of the writings used for the selection and weighting of the word counts, the two statesmen's writings were separated by 6.9 standard deviations. When the same words and weights were applied to fresh writings not used in selecting or weighting, the new writings were separated by 4.5 standard deviations—still good discrimination, but a loss of 2.4 standard deviations is substantial and illustrates well the effect. Losses are usually greatest among the poorer discriminants. Usages of the word *upon* originally separated the writings by 3.3 standard deviations, and did even better, 3.8, in the fresh validating materials; but a less effective set of words giving originally a separation of 1.3 standard deviations dropped to 0.3 on retesting. The lesson is that optimization methods (such as least squares and maximum likelihood) do especially well on just the data used to optimize. Plan for validation, and, where hopes are high for much gain from many small effects, prepare for disappointment.

Weights. If individuals are sampled to find out about their families, as in investigations carried out in schools, unless some account is taken of weights, a peculiar distribution may arise. For example, if a sample of girls is asked to report the

Table 10 — Values of ρ' for various values of $100(1-p)$, ρ pairs

100(1−p) PER CENT TRUNCATED	VALUES OF ρ							100p PER CENT SELECTED
	.100	.300	.500	.700	.800	.900	.950	
5	.090	.272	.461	.661	.768	.881	.939	95
10	.085	.256	.438	.637	.747	.867	.932	90
25	.073	.224	.389	.583	.698	.834	.912	75
50	.060	.186	.329	.509	.626	.780	.878	50
75	.049	.153	.273	.434	.548	.712	.831	25
90	.041	.128	.231	.374	.481	.647	.781	10
95	.037	.116	.210	.342	.444	.609	.749	5
99	.031	.098	.178	.293	.385	.542	.689	1

numbers of sons and of daughters (including themselves) in their families, it turns out that the average number of daughters observed in the sample is approximately one more than the average number of sons. (More precisely, mathematics not given here shows the difference to be: [*variance of number of daughters* minus *covariance of number of sons and daughters*] divided by [*average number of daughters*]. When the distribution of the number of daughters is approximately Poisson and the numbers of sons and of daughters are independent, the ratio is approximately unity.) Essentially, families of three girls report three times as often as families with one girl, and families with no girls do not report at all. If account is taken of the dependence of frequency of reporting upon the number of daughters, this matter can be adjusted, provided information about families with no daughters is available or is not needed.

Similarly, in studying the composition of special groups, unless the analysis is done separately for each family size, one needs to remember that more children are first-born than second, and so on.

Errors in calculation. A well-planned format for laying out calculations and careful checking aid in getting correct answers. To give a base line, a sample survey by the Internal Revenue Service (Farioletti 1952, pp. 65–78) found arithmetical errors in only 6 per cent of 160,000 personal income tax returns. Considering that the task is sometimes troublesome and often resented, this record appears good.

In scientific work, misreadings of numbers, misplaced decimals, errors in the application of formulas all take their toll. As a first step in the control of error, regard any unchecked calculation as probably wrong.

Overmechanization. Overmechanization of computing puts great pressure on the analyst to make one enormous run of the data and thereby economically get all the analyses he wished. Alas, one great sweep is never the way of good data analysis. Instead, we learn a little from each analysis and return again and again. To illustrate, in deciding whether to transform the data to square roots, logarithms, inverse sines, or reciprocals before launching on the major analysis, tests may be run for each function separately, leading to the choice of one or two transformations for use in the next stage. Otherwise the whole large calculation must be run too many times because there are many branch points in a large calculation with several choices available at each. Furthermore, data analysis requires extensive printout, little of which will be looked at; therefore the data analyst must resist the notion that the good computer user makes the machine do all the work internally and obtains very little printout. He must also resist the idea of having ever speedier programs at the cost of more and more time for programming and less and less for analysis. Fine programs are needed, but the cost of additional machine time from slow programs may be less than the cost of improvements in programming and of the waiting time before analysis can begin.

Possibly with the increase of time-sharing in high-speed computation and the handy packaging of general purpose programs for the analysis of data, the opportunities for making studied choices at each point in the analysis will become easier and less time-consuming.

Preserving data from erasure. After processing, data should usually be preserved in some form other than a single magnetic tape. Contrary to theory and rumor, magnetic tapes containing basic data are occasionally erased or made unusable in the high-speed computing process, and all the explanations in the world about how this could or could not have happened cannot restore a bit of information. One remedy is to have a spare tape with your data or program copied upon it. When disaster strikes, remember that few things seem more likely to recur than a rare event that has just happened, and so copy your spare tape before you submit it to the destroyer.

Hand copying. Since human copying is a major source of error, keep hand copying to a minimum and take advantage where possible of the high-speed computer's ability to produce tables in immediately publishable form and of mechanical reproduction processes. Editing can be done by cutting, pasting, and painting out. When copying is necessary, checks of both column totals and row totals are believed superior to direct visual comparisons of individual entries with the manuscript.

Checks. Checking the programming and calculations of a high-speed computer presents a major unsolved problem. One might suppose that once a machine began producing correct answers, it always would thereafter. Not at all. It may respond to stimuli not dreamed of by the uninitiated. To find, for example, that it throws a small error into the fifth entry in the fourth column of every panel is disconcerting and scary, partly because small errors are hard to find and partly because one wonders whether undetected errors may still be present. Thorough and systematic checking is advised. Some ways are through sample problems; through fuller printout of the details of a problem already worked by hand; by comparing corresponding parts of several problems, including special cases whose answers are known by outside means; and by solving the problem in more than one way.

In addition to the checks on the final calculations, check the input data. For input punched on cards, for example, some process of verifying the punching is required. Methods of checking will vary with the problem. Partly redundant checking may not be wasteful. Look for impossible codes in columns, look for interchanges of columns. Try to set up checks for inconsistency in cards. (In Western cultures, nursery school children are not married, wives don't have wives, and families with 42 children need verification.) Consider ways to handle blanks based on internal consistency.

In working with computers, be wary of the way symbols translate from keyboard to card or tape—dashes and minus signs or zeros, letter O's, and blanks are a few sources of confusion. In dealing with numbers using, say, a two-digit field, a number such as 6, unless written 06, may wind up as 60 or as a meaningless character. The possibilities here are endless, but in a given problem it is usually worth organizing systematic procedures to combat these difficulties.

Order-of-magnitude checking. When calculations are complete, order-of-magnitude checks are always valuable. Are there more people in the state of New York than in the United States? Does leisure plus work plus sleep take much more than 24 hours per day? Exercises in calculations of comparative orders of magnitude can be rewarding in themselves because new connections are sometimes made between the research and the rest of the subject matter.

Significant figures. Both hand and high-speed calculations require numbers to be carried to more places than seem meaningful and to more places than simple rules learned in childhood would suggest. These rules seem dedicated to rounding early so as not to exaggerate the accuracy of one's result. But they may erase the signal with the noise. About the only reassuring rule for complex calculations is that if the important digits are the same when the calculation is carried to twice as many places, enough accuracy has likely been carried.

The old rules for handling significant figures come from a simplified idea that a number can report both its value and its accuracy at the same time. Under such rules the numbers 3.26 and 0.0326 were thought of as correct to within half a unit in the last place. Sometimes in mathematical tables this approach is satisfactory. For data-based numbers, the uncertainty in a number has to be reported separately.

One-of-a-kind calculations. One-of-a-kind calculations, frequent in scientific reports, are especially error prone, both because the investigator may not set up a standard method of calculation, complete with checks, and because he does not have the aid of comparisons with other members of a long sequence. For example, some pollsters believed that their wrong forecast about a vote would have been close had proper weighting for household size been applied, a claim worth checking. Their ultimate error was in thinking that this claim was right. How did they make it? Pages of weightings carefully checked down to, but not including, the final estimate showed no error in their reanalysis. But their one-of-a-kind calculation leading to the final estimate was a ratio composed of an inappropriate numerator and an inappropriate denominator grabbed from the many column totals. By accident this meaningless quotient gave a number nearly identical with that produced by the voters. And who checks further an answer believed to be correct? Actually, the weighting for household size scarcely changed their original forecast. The moral is that the one-of-a-kind calculation offers grave danger.

Consequently, each new calculation can well be preceded by a few applications of the method to simple made-up examples until the user gets the feel of the calculation, of the magnitudes to be expected, and of a convenient way to lay out the procedure. Having someone else check the calculation independently requires that the investigator not teach the verifier the original mistakes. Yates has suggested that, in a large hand calculation, independence could nearly be preserved when different individuals calculate on separate machines in parallel in two different numerical units; for example, one computes in dollars, the other in pounds. At the end, the final answers are converted for comparison.

Gross errors in standard deviations. Since the sample range w (largest measurement minus smallest measurement) is easy to compute, it is often used to check the more complicated calculation of a sample standard deviation s. In the same sample, the ratio w/s must lie between the lower and upper bounds given in Table 11 or else the range, sample standard deviation, or quotient is in error. The table shows the 2.5 per cent and 97.5 per cent point of the distribution of w/s for a normal distribution. When calculations lead to ratios falling outside these limits but inside the bounds, they are not necessarily wrong; but further examination may pay.

Table 11 also shows the median of the distribution of the range of a sample of size n drawn from a standard normal distribution. It gives one an idea of the spread measured in standard deviations to be expected of the sample means of n equal-sized groups whose population means are identical. Note that through $n = 20$ a rough rule is that the median distance between the largest and smallest sample mean is $\sqrt{n}\sigma_{\bar{x}}$.

Table 11 — Bounds on the ratio: range/standard deviation[a]

n	Lower bound	w/s 2.5%	w/s 97.5%	Upper bound	Median of w/σ
2	1.41	1.41	1.41	1.41	.95[b]
3	1.73	1.74	2.00	2.00	1.59[b]
4	1.73	1.93	2.44	2.45	1.98
5	1.83	2.09	2.78	2.83	2.26
6	1.83	2.22	3.06	3.16	2.47
7	1.87	2.33	3.28	3.46	2.65
8	1.87	2.43	3.47	3.74	2.79
9	1.90	2.51	3.63	4.00	2.92
10	1.90	2.59	3.78	4.24	3.02
15	1.94	2.88	4.29	5.29	3.42
20	1.95	3.09	4.63	6.16	3.69
30	1.97	3.37	5.06	7.62	4.04
50	1.98	3.73	5.54	9.90	4.45
100	1.99	4.21	6.11	14.07	4.97
200	1.99	4.68	6.60	19.95	5.49[c]
500	2.00	5.25	7.15	31.59	6.07[c]
1,000	2.00	5.68	7.54	44.70	6.48[c]

a. Lower bound, 2.5% point, 97.5% point, and upper bound for the ratio: range/sample standard deviation (w/s); median of the distribution of the ratio: range/population standard deviation (w/σ); the sample size is n. The upper and lower bounds apply to any distribution and sampling method; the percentage points and the median are computed for random sampling from a normal distribution, but they should be useful for other distributions.

b. This is not an error. The median is expressed as the multiplier of the *population* standard deviation, whereas the bounds relate range and *sample* standard deviation.

c. The mean of the distribution is given as an approximation to the median because the latter is not available.

Sources: Pearson & Stephens 1964, p. 486, for lower and upper bounds and for 2.5% and 97.5% points; Harter 1963, pp. 162–164, for medians; Pearson & Hartley 1954, p. 174, for means.

Reporting. When writing the final report, remember that making clear the frame of reference of a study helps the reader understand the discussion.

Need for full reporting. In reporting on the investigation, be sure to give detailed information about the populations studied, the operational definitions used, and the exceptions to the general rules. Unless the details are carefully reported, they are quickly forgotten and are soon replaced by cloudy fancies. Discussions of accuracy, checks, and controls are needed in the final report.

Full and careful reporting can lead to ample prefaces, numerous appendixes, some jargon, and lengthy discussions. Shrink not from these paraphernalia, so amusing to the layman, for without them the study loses value; it is less interpretable, for it cannot be properly compared with other studies. Jargon may be the price of brevity.

The reader may object that editors will not allow such full reporting. Certainly the amount of detail required does vary with the sort of report to be made. Many studies that are published in short reports turn out to present a long sequence of short articles, and these, in one place or another, can give the relevant details.

Try to go beyond bare-bones reporting by giving readers your views of the sorts of populations, circumstances, or processes to which the findings of the study might apply. Warn the reader about generalizations that you are wary of but that he, on the basis of your findings, might reasonably expect to hold. While such discussions can be criticized as speculation, you owe it to the reader to do your best with them and to be as specific as you can be.

Beyond all this, where appropriate, do write as nontechnical a summary as you can for the interested public.

Suppression of data. In pursuit of a thesis, even the most careful may find it easy to argue themselves into the position that the exceptions to the desired proposition are based upon poorer data, somehow do not apply, would be too few to be worth reporting if one took the trouble to look them up, would mislead the simpleminded if reported, and therefore had best be omitted. Whether or not these views are correct, and some of them may well be, it is preferable to present the whole picture and then to present one's best appraisal of all the data. The more complete record puts readers in a much better position to consider both the judgments and the proposition.

FREDERICK MOSTELLER

[*Directly related are the entries* EXPERIMENTAL DESIGN; FALLACIES, STATISTICAL; SAMPLE SURVEYS.]

BIBLIOGRAPHY

BARTHOLOMEW, D. J. 1961 A Method of Allowing for "Not-at-home" Bias in Sample Surveys. *Applied Statistics* 10:52–59.

BILLEWICZ, W. Z. 1965 The Efficiency of Matched Samples: An Empirical Investigation. *Biometrics* 21:623–644.

CANTRIL, HADLEY (1944) 1947 The Use of Trends. Pages 220–230 in Hadley Cantril, *Gauging Public Opinion*. Princeton Univ. Press.

CANTRIL, HADLEY 1966 *The Pattern of Human Concerns*. New Brunswick, N.J.: Rutgers Univ. Press.

COCHRAN, WILLIAM G. 1951 Improvement by Means of Selection. Pages 449–470 in Berkeley Symposium on Mathematical Statistics and Probability, Second, *Pro-*

ceedings. Edited by Jerzy Neyman. Berkeley: Univ. of California Press.

COCHRAN, WILLIAM G. (1953) 1963 *Sampling Techniques*. 2d ed. New York: Wiley.

COCHRAN, WILLIAM G.; MOSTELLER, FREDERICK; and TUKEY, JOHN W. 1954 *Statistical Problems of the Kinsey Report on Sexual Behavior in the Human Male*. Washington: American Statistical Association.

COPPIN, R. S.; FERDUN, G. S.; and PECK, R. C. 1965 The Teen-aged Driver. California, Department of Motor Vehicles, Division of Administration, Research and Statistics Section, *Report* 21.

EHRENBERG, A. S. C. 1960 A Study of Some Potential Biases in the Operation of a Consumer Panel. *Applied Statistics* 9:20–27.

FARIOLETTI, MARIUS 1952 Some Results From the First Year's Audit Control Program of the Bureau of Internal Revenue. *National Tax Journal* 5, no. 1:65–78.

FELDMAN, JACOB J. 1960 The Household Interview Survey as a Technique for the Collection of Morbidity Data. *Journal of Chronic Diseases* 11:535–557.

HANSEN, MORRIS H.; and HURWITZ, WILLIAM N. 1946 The Problem of Non-response in Sample Surveys. *Journal of the American Statistical Association* 41:517–529.

HARBISON, FREDERICK; and MYERS, CHARLES A. 1964 *Education, Manpower, and Economic Growth: Strategies of Human Resource Development*. New York: McGraw-Hill.

HARTER, H. LEON 1963 The Use of Sample Ranges and Quasi-ranges in Setting Exact Confidence Bounds for the Population Standard Deviation. II. Quasi-ranges of Samples From a Normal Population—Probability Integral and Percentage Points; Exact Confidence Bounds for σ. → ARL 21, Part 2. Wright-Patterson Air Force Base, Ohio: U.S. Air Force, Office of Aerospace Research, Aeronautical Research Laboratories.

HENDRICKS, WALTER A. 1956 *The Mathematical Theory of Sampling*. New Brunswick, N.J.: Scarecrow Press.

KATZ, DANIEL 1946 The Interpretation of Survey Findings. *Journal of Social Issues* 2, no. 2:33–44.

KISH, LESLIE 1965 *Survey Sampling*. New York: Wiley.

KISH, LESLIE; and HESS, IRENE 1959 A "Replacement" Procedure for Reducing the Bias of Nonresponse. *American Statistician* 13, no. 4:17–19.

KISH, LESLIE; and LANSING, JOHN B. 1954 Response Errors in Estimating the Value of Homes. *Journal of the American Statistical Association* 49:520–538.

KLUCKHOHN, FLORENCE R.; and STRODTBECK, FRED L. 1961 *Variations in Value Orientations*. Evanston, Ill.: Row, Peterson.

LAWLEY, D. N.; and SWANSON, Z. 1954 Tests of Significance in a Factor Analysis of Artificial Data. *British Journal of Statistical Psychology* 7:75–79.

LEUCHTENBURG, WILLIAM E. 1963 *Franklin D. Roosevelt and the New Deal: 1932–1940*. New York: Harper. → A paperback edition was published in the same year.

LEVINE, SOL; and GORDON, GERALD 1958–1959 Maximizing Returns on Mail Questionnaires. *Public Opinion Quarterly* 22:568–575.

MOSES, LINCOLN E.; and OAKFORD, ROBERT V. 1963 *Tables of Random Permutations*. Stanford Univ. Press.

MOSTELLER, FREDERICK; and BUSH, ROBERT R. (1954) 1959 Selected Quantitative Techniques. Volume 1, pages 289–334 in Gardner Lindzey (editor), *Handbook of Social Psychology*. Cambridge, Mass.: Addison-Wesley.

MOSTELLER, FREDERICK; and TUKEY, JOHN W. 1966 Data Analysis, Including Statistics. Unpublished manuscript.

MOSTELLER, FREDERICK; and WALLACE, DAVID L. 1964 *Inference and Disputed Authorship: The Federalist*. Reading, Mass.: Addison-Wesley.

NETER, JOHN; and WAKSBERG, JOSEPH 1964*a* Conditioning Effects From Repeated Household Interviews. *Journal of Marketing* 28, no. 2:51–56.

NETER, JOHN; and WAKSBERG, JOSEPH 1964*b* A Study of Response Errors in Expenditures Data From Household Interviews. *Journal of the American Statistical Association* 59:18–55.

PEARSON, E. S.; and HARTLEY, H. O. (editors) (1954) 1958 *Biometrika Tables for Statisticians*. Volume 1, 2d ed. Cambridge Univ. Press.

PEARSON, E. S.; and STEPHENS, M. A. 1964 The Ratio of Range to Standard Deviation in the Same Normal Sample. *Biometrika* 51:484–487.

PEIRCE, CHARLES S. 1873 On the Theory of Errors of Observations. U.S. Coast and Geodetic Survey, *Report of the Superintendent* [1870]:200–224.

ROETHLISBERGER, FRITZ J.; and DICKSON, WILLIAM J. (1939) 1961 *Management and the Worker: An Account of a Research Program Conducted by the Western Electric Company, Hawthorne Works, Chicago*. Cambridge, Mass.: Harvard Univ. Press. → A paperback edition was published in 1964 by Wiley.

RUGG, DONALD; and CANTRIL, HADLEY (1944) 1947 The Wording of Questions. Pages 23–50 in Hadley Cantril, *Gauging Public Opinion*. Princeton Univ. Press.

SCOTT, CHRISTOPHER 1961 Research on Mail Surveys. *Journal of the Royal Statistical Society* Series A 124: 143–205.

SHARP, HARRY; and FELDT, ALLAN 1959 Some Factors in a Probability Sample Survey of a Metropolitan Community. *American Sociological Review* 24:650–661.

SOBOL, MARION G. 1959 Panel Mortality and Panel Bias. *Journal of the American Statistical Association* 54: 52–68.

SUDMAN, SEYMOUR 1964 On the Accuracy of Recording Consumer Panels: I and II. *Journal of Marketing Research* 1, no. 2:14–20; 1, no. 3:69–83.

TUCKER, LEDYARD R. 1964 Recovery of Factors From Simulated Data. Unpublished manuscript.

WAKSBERG, JOSEPH; and PEARL, ROBERT B. 1964 The Effects of Repeated Household Interviews in the Current Population Survey. Unpublished manuscript.

WHITING, BEATRICE B. (editor) 1963 *Six Cultures: Studies of Child Rearing*. New York: Wiley.

WILSON, EDWIN B.; and HILFERTY, MARGARET M. 1929 Note on C. S. Peirce's Experimental Discussion of the Law of Errors. National Academy of Sciences, Washington, D.C., *Proceedings* 15:120–125.

Postscript

Nonsampling errors in sample surveys. Several works have appeared on nonsampling errors. Zarkovich (1966) details all manner of such errors with suggestions for dealing with them, and includes examples from agricultural surveys. Szameitat and Deininger (1967) give a similar exposition in outline form. Marks, Seltzer, and Krotki (1974), who are concerned with determining vital rates in developing countries, describe a method of dual sampling that purportedly leads to improved estimates by adjusting for some nonsampling errors. The World Fertility Survey (1975) cites errors in sample implementation (undercoverage, duplication, incorrect or inadequate identification, and inadequate supplementary information in the sampling frame) and suggests how to prevent or ameliorate these errors. Naus (1975) presents methods especially suited to deal with gross errors in the handling of large amounts of data. Morgenstern ([1950] 1963) treats errors in economic data, with examples of problems with specific types of statistics, such as price and unemployment indexes. Bogue and Murphy (1964) study effects of errors in classification, particularly in the analysis of cross-classified data. Hollingsworth (1968) provides interesting examples of nonsampling errors, especially sources of demographic bias, in historical data. Bowley's 1897 paper is an early analysis of the effect of bias, missing or erroneous data, and wrong weights on the calculation of averages.

A related, more social, problem is what happens when times change. National Assessment of Educational Progress tries to sample the nation and give the same questions to children at two different times. Though the questions remain the same, the presentation (oral on tape as well as written) may change from year to year (as when a different announcer or different type size is used), and the context of the questions, or the connotations of the words used, may change. Also, background variables, such as "urban" and "rural" to describe neighborhoods, have different meanings now than they did forty years ago.

The U.S. Bureau of the Census has long had an interest in the analysis of survey errors. One particularly useful approach may be illustrated in terms of a question that elicits responses Y_{jt} from observed units $j = 1, 2, \cdots, n$ in a sample. The subscript "t" indicates that we think of the actual survey (at time t) as one of many possible in-principle-repeatable surveys under the same general conditions: questionnaire, training of interviewers, and so on. Thus Y_{jt} for unit j is a random quantity. Interest is centered in a true population quantity Z estimated by $\bar{Y}_t$, the average of the Y_{jt} over the first subscript. (If some other parameter were of major interest, or if a different sort of estimator were contemplated, the analysis might require change; this brief summary tries only to give the spirit of the approach.)

The deviation of $\bar{Y}_t$ from Z, $\bar{Y}_t - Z$, is decomposed linearly as follows:

$\bar{Y}_t - Z = (\bar{Y}_t - \bar{\mu}_s)$ where $\bar{\mu}_s = E\bar{Y}_t$ over hypothetical repetitions of the survey *with the same sample*, that is, $\bar{Y}_t - \bar{\mu}_s$ is random response error.

$+ (\bar{\mu}_s - \bar{\mu})$ where $\bar{\mu}$ is the expectation of $\bar{\mu}_s$ over hypothetical repeated samplings, that is, $\bar{\mu}_s - \bar{\mu}$ is sampling error.

$+ (\bar{\mu} - Z)$ bias: conceptual, measurement, and other.

The expected mean square error $E(\bar{Y}_t - Z)^2$ is the sum of three square terms (response variance, sampling variance, and square of the bias) and an interaction term between sampling and response error. The response variance can be written

$$E(\bar{Y}_t - \bar{\mu}_s)^2 = \frac{\sigma^2}{n}(1 + (n-1)\rho)$$

where σ^2 is the variance of the Y_{jt} over t, and ρ is the correlation of response errors within a sample. The correlation term may appear, for example, when the same interviewer observes several units, creating dependence among response errors. Even a modest correlation may affect substantially the mean square error.

Estimation of the components of mean square error is carried out in varying ways depending on circumstances. Such estimation can never be perfect, and the bias term, in particular, may present intractable difficulties.

Hansen et al. (1951) and Hansen et al. (1961) give mathematical details for the above approach. Bailar (1976) discusses components of error in estimating Z with examples of estimation by the U.S. Census and with a comprehensive bibliography. Bureau of the Census approaches for presenting sampling and nonsampling errors are described in Gonzalez et al. (1975), along with several examples.

Analysis. Three papers deal with assessment of variability using techniques similar to the jackknife. Finifter (1972) explains and encourages the use of replicated random subsamples from a sample to determine the confidence one may have in the values of the statistics used. McCarthy (1976) applies sampling techniques developed for complex sample survey designs to select such random subsamples.

Stone (1974) proposes using cross-validation in predicting situations, best explained by an example. Suppose there are n bivariate observations

(y, x), and the object is to find a regression equation to predict y given x. Assessing the variability of any procedure (for example, least squares) involves removing one observation from the data and using the equation based on the remaining $n - 1$ observations to predict the left out y from the left out x. This is done n times, leaving out each observation once. The average distance (for example, squared distance) between the observed and predicted left out values provides a measure of the variability of the procedure. Stone then finds the prediction function that minimizes the above assessment of variability, thus producing a new procedure. An application of cross-validation can be found in Mosteller and Tukey (1968, pp. 153–158, 160).

Cross-validation and replicated subsampling should be useful in dealing with lack of robustness to outliers, variance assessment in complex situations, and selection of a "best" subset of independent variables in multiple regression.

FREDERICK MOSTELLER

[*See also* DATA ANALYSIS, EXPLORATORY.]

ADDITIONAL BIBLIOGRAPHY

The International Statistical Institute plans to publish a bibliography on nonsampling errors by Tore Dalenius.

BAILAR, BARBARA A. 1976 Some Sources of Error and Their Effect on Census Statistics. *Demography* 13: 273–286.

BOGUE, DONALD J.; and MURPHY, EDMUND M. 1964 The Effect of Classification Errors Upon Statistical Inference: A Case Analysis With Census Data. *Demography* 1:42–55.

BOWLEY, A. L. 1897 Relations Between the Accuracy of an Average and That of Its Constituent Parts. *Journal of the Royal Statistical Association* 40:855–866.

FINIFTER, BERNARD M. 1972 The Generation of Confidence: Evaluating Research Findings by Random Subsample Replication. Pages 112–175 in Herbert L. Costner (editor), *Sociological Methodology 1972*. San Francisco: Jossey-Bass.

GONZALEZ, MARIA E. et al. 1975 Standards for Discussion and Presentation of Errors in Survey and Census Data. *Journal of the American Statistical Association* 70, part 2:5–23. → Revision of Bureau of the Census Technical Paper No. 32.

HANSEN, MORRIS H.; and WAKSBERG, JOSEPH 1970 Research on Non-sampling Errors in Censuses and Surveys. International Statistical Institute, *Review* 38: 317–332.

HANSEN, MORRIS H. et al. 1951 Response Errors in Surveys. *Journal of the American Statistical Association* 46:147–190.

HANSEN, MORRIS H. et al. 1961 Measurement Errors in Censuses and Surveys. International Statistical Institute, *Bulletin* 38, part 2:359–374.

HOLLINGSWORTH, T. H. 1968 The Importance of the Quality of the Data in Historical Demography. *Daedalus* 97:415–432.

MCCARTHY, PHILIP J. 1976 The Use of Balanced Half-sample Replication in Cross-validation Studies. *Journal of the American Statistical Association* 71:596–604.

MARKS, ELI S.; SELTZER, WILLIAM; and KROTKI, KAROL J. 1974 *Population Growth Estimation: A Handbook of Vital Statistics Measurement*. New York: Population Council.

MORGENSTERN, OSKAR (1950) 1963 *On the Accuracy of Economic Observations*. 2d ed. Princeton Univ. Press.

[1]MOSTELLER, FREDERICK; and TUKEY, JOHN W. 1968 Data Analysis Including Statistics. Volume 2, pages 80–203 in Gardner Lindzey and Elliot Aronson (editors), *Handbook of Social Psychology*. 2d ed. Volume 2: *Research Methods*. Reading, Mass.: Addison-Wesley. → The formal publication of Mosteller and Tukey (1966), cited in the main article.

NAUS, JOSEPH I. 1975 *Data Quality Control and Editing*. New York: Dekker.

STONE, M. 1974 Cross-validatory Choice and Assessment of Statistical Predictions. *Journal of the Royal Statistical Society* Series B 38:111–133. → Discussion on pages 133–147.

SZAMEITAT, KLAUS; and DEININGER, ROLF 1967 Some Remarks on the Problem of Errors in Statistical Results. International Statistical Institute, *Bulletin* 42, part 1:66–89.

WORLD FERTILITY SURVEY 1975 Errors in Implementing the Sample Design. Chapter 10, pages 61–66 in *Manual on Sample Design*. Basic Documentation No. 3. London: International Statistical Institute.

ZARKOVICH, SLOBODAN S. 1966 *Quality of Statistical Data*. Rome: Food and Agriculture Organization of the United Nations.

II

EFFECTS OF ERRORS IN STATISTICAL ASSUMPTIONS

● *This article was completely revised for this volume.*

This article discusses the effects of incorrect assumptions on the results of several statistical techniques, including the analysis of the matched pairs design, of the two-sample design and one-way analysis of variance, of the straight line regression design, and of simple multivariate designs. Alternative statistical techniques that are robust to departures from standard assumptions (that is, techniques for which the results are relatively insensitive to departures from these assumptions) are discussed, and some methods for detection of departures from assumptions are outlined.

The matched pairs design and the one-sample problem

In the matched pairs design, two treatments or conditions are compared by assigning the first treatment at random to one member of each pair of matched individuals and the second treatment to the remaining member of each pair. For example, a department of Slavic languages is interested in finding out whether one of two different teaching methods for a first-year language course is better than the other. A language aptitude and proficiency examination is given on the first day of class, and the scores on these tests are used to pair students who have approximately the same aptitude and proficiency. Then, from each pair of students, one student is randomly assigned to one teaching method and the other student is assigned to the other method. An examination is given at the end of the term to determine whether differences exist between the teaching methods.

Comparisons between the treatments are based on the difference between the responses to the treatment within each pair. Thus, the data consist of the n differences $X_1, X_2, \cdots, X_n$, where X_j denotes the difference between the response scores in the jth pair.

A scientist may want to make inferences about several features of the probability distribution of differences underlying the matched pairs experiment. How can one estimate the unknown mean or median difference? Which significance test can one use to test the unknown mean or median? How can one study the underlying variability among differences?

The matched pairs design is really a special instance of the so-called one-sample design. This arises when the investigator is interested in the frequency distribution of some numerical characteristic. For example, one might need to know how long it takes residents of San Francisco to reach the nearest medical clinic or hospital. The investigator might define a suitable target population and take a simple random sample from that population, so that the data consist of the n values $X_1, X_2, \cdots, X_n$, where X_j denotes the time it takes the jth individual in the sample to reach the nearest clinic or hospital. We call this design the one-sample design. As in the matched pairs design, the investigator may ask questions about, for example, the mean, median, or the spread of the distribution of times.

The following subsections discuss these two designs. First, we consider estimation of the "central value" of a distribution (for example, the population mean or median). We present the best estimator of the central value under standard assumptions, and then discuss likely departures from these standard assumptions and the effects of these departures. Next, we consider how to compare different estimators under a variety of assumptions. Some promising estimators are presented and compared, and recommendations are made. Following the discussion of point estimators, we turn to a similar discussion of significance tests and confidence intervals for the central value.

Robust estimation. An investigator is frequently faced with a dilemma in the choice of an estimator. Suppose that in the matched pairs design or in the one-sample design the aim is to estimate μ, the population mean of the X_j's. It is frequently assumed that $X_1, X_2, \cdots, X_n$ constitute a simple random sample from a population that has a normal distribution. If the investigator is willing to make these assumptions, then the sample mean $\bar{X} = (1/n)\sum X_j$ is in many cases the unique best estimator for μ [*see* ESTIMATION, *article on* POINT ESTIMATION]. (Since the normal distribution is symmetrical, the population mean, μ, and the population median are identical, so $\bar{X}$ is also the best estimator for the median.) Many extreme values might, however, create doubt about the normality assumption, even though the underlying distribution does appear to be symmetrical. [*See* STATISTICAL ANALYSIS, SPECIAL PROBLEMS OF, *article on* OUTLIERS.] (In the matched pairs design, symmetry is guaranteed by the randomization when there is no treatment effect.) For distributions of this "long-tailed" kind, the sample mean may have a very large variance relative to other estimators of μ. The sample median provides a reasonable estimator for long-tailed distributions but has a larger variance than the sample mean for nearly normal distributions. The problem then is how to achieve a reasonable compromise: an estimator that performs reasonably well under a variety of assumptions about the shape of the distribution.

Departures from the standard assumptions under which the sample mean is the best estimator are of two basic types: lack of normality as above, and departures from simple random sampling (that is, from independence and identity of distribution of the observations). We turn our attention at first to specifying in more detail alternatives to the assumption of normality and later discuss the effects of departures from simple random sampling.

Distributional models. The normal distributions form a particular family of continuous, symmetrical, unimodal, and bell-shaped curves that are unlimited in both directions. In specifying plausible alternatives to the assumption of normality, we restrict attention to continuous unimodal distributions. Within this framework, one major departure from normality is skewness (departure from symmetry); for example, income distributions will commonly be positively skewed with the bulk of the population having low to moderate incomes and a very few individuals having extremely high incomes. Theoretical distributions that are positively skewed include the chi square and the exponential. [*See* DISTRIBUTIONS, STATISTICAL, *article on* SPECIAL CONTINUOUS DISTRIBUTIONS.] Degree of skewness can be indexed by the parameter γ_1, the standardized third moment about the mean of the distribution, or by one of its variants. For symmetrical distributions $\gamma_1 = 0$, while in general $\gamma_1 > 0$ for distributions with a long right tail and $\gamma_1 < 0$ for distributions with a long left tail. The first column of Table 1 gives values of γ_1 for some standard distributions.

Alternatively, the assumption of symmetry might be plausible but the investigator might be concerned that the underlying distribution has "longer tails" than the normal distribution. Loosely speaking, a symmetrical distribution is "long tailed" with respect to the normal distribution if there are proportionately more extreme values—for example if more than 1 per cent of the distribution lies beyond the values $\mu \pm 2.58\sigma$, where σ is the standard deviation of the distribution. The kurtosis parameter, γ_2, has frequently been used in the literature as an index of tail length for symmetric distributions; it is the standardized fourth moment of the distribution. However, it must be remembered that γ_2 also depends on the peakedness or flatness of the center of the distribution, and there is no easy intuitive interpretation of the meaning of a γ_2 value of, say, 5.0 (see, for example, Hogg 1972; Ali 1974). No single parameter can summarize the varied meanings of tail length. For the normal distribution $\gamma_2 = 0$; values of $\gamma_2 > 0$ are said to denote a long-tailed distribution and $\gamma_2 < 0$, a short-tailed distribution. The second column of Table 1 shows values of γ_2 for several common long-tailed theoretical distributions such as Student's t and the compound normal (defined below).

Crow and Siddiqui (1967) suggest a measure of tail length, $C_{.95}$, which for a symmetrical distribution with mean zero is the ratio of the 95th percentile of the distribution to the 75th percentile; for a normal distribution $C_{.95}$ is $1.645/.674 = 2.4$. Table 1 shows $C_{.95}$ as well as $C_{.975}$ and $C_{.90}$ defined with the 97.5th percentile and the 90th percentile respectively instead of the 95th. Hogg (1974) suggests related measures of tail length, Q^* and Q_1^*, where for symmetrical distributions Q^* is the average of the upper 5 per cent of the distribution divided by the average of the upper 50 per cent, and Q_1^* is the average of the upper 20 per cent divided by the average of the upper 50 per cent.

One family of long-tailed symmetrical distributions widely used in the robustness literature is the compound normal distribution; it arises in the following way: (1) an observation is randomly drawn from one of two normal populations, (2) with probability $1 - \tau$ this observation is randomly drawn from a normal population with mean μ and variance σ^2, (3) with probability τ, this observation is randomly drawn from another normal population having mean μ and variance $K^2\sigma^2$. In short, the compound normal distributions considered here are *mixtures* of two normal distributions with a common mean but different variances.

Table 1 – Values of γ_1 and several measures of tail length for selected distributions

DISTRIBUTION	γ_1	γ_2	$C_{.975}$	$C_{.95}$	$C_{.90}$	Q^*	Q_1^*
Uniform	0	−1.2	1.9	1.8	1.6	1.9	1.6
Normal	0	0	2.9	2.4	1.9	2.58	1.75
Chi square (1df)	2.83	12					
Student's t[a]							
t_3	0		4.16	3.08	2.14	3.51	1.90
t_5	0	6	3.54	2.77	2.03	3.05	1.83
t_9	0	1.2	3.22	2.61	1.97	2.81	1.79
t_{10}	0	1	3.18	2.59	1.96	2.79	1.79
Exponential	2	6					
Double exponential	0	3	4.3	3.3	2.3	3.3	1.93
Compound normal							
$K=3$ $\tau=.05$	0	4.65	3.1	2.53	1.9	2.93	1.79
.10	0	5.33	3.5	2.64	2.0	3.19	1.83
.15	0	5.06	3.9	2.80	2.0	3.36	1.86
.20	0	4.54	4.4	2.98	2.0	3.47	1.88
.25	0	4.00	4.7	3.19	2.1	3.46	1.90

[a] "t_i" denotes the Student's t distribution with i degrees of freedom.

[*See* DISTRIBUTIONS, STATISTICAL, *article on* MIXTURES OF DISTRIBUTIONS.] Note that the ordering of the compound normal distributions as to tail length depends on which measure is used (see Table 1).

The emphasis in this article on long-tailed symmetrical distributions derives in part from a current lack of systematic work on short-tailed or asymmetrical distributions (especially in estimation), and in part from the complexity introduced by the fact that the mean and the median, in general, do not coincide in asymmetrical distributions.

Criteria for point estimation. A consideration in assessing the usefulness of an estimator is determining whether it is biased or unbiased. An estimator is said to be unbiased if its expected value is equal to the population parameter being estimated. A useful way to compare any two unbiased estimators of the same parameter is to compute their relative efficiency. The *efficiency* of estimator 1 relative to estimator 2 is defined as

$$e(1,2) = \frac{\text{variance of estimator 2}}{\text{variance of estimator 1}}.$$

A good estimator is one that compares favorably in terms of efficiency with its competitors over a range of plausible distributional assumptions, provided unbiasedness is not affected. An estimator has *robustness of efficiency* relative to another if the above ratio does not dip far below 1.0 for plausible alternatives. (This concept of efficiency can be generalized for comparisons of two biased estimators. Additional criteria for the comparison of estimators are discussed by Hampel 1971, for example.) We shall restrict attention to estimators with robustness of efficiency that are also relatively easy to compute.

Estimators of the mean. The sample mean always has relative efficiency greater than or equal to 1.0 with respect to other estimators of the mean in random samples from a normal distribution. Although the efficiency of the sample median with respect to the sample mean may be considerably greater than 1.0 for samples from long-tailed distributions, it is only about 0.64 in large samples from a normal distribution. One possible compromise between the sample mean and the sample median is the estimator

$$\frac{X_{(2)} + X_{(3)} + \cdots + X_{(n-1)}}{n-2},$$

where $X_{(j)}$ is the jth smallest observation. The procedure is to discard the largest and smallest observations and take the average of the remaining observations. [*See* NONPARAMETRIC STATISTICS, *article on* ORDER STATISTICS; STATISTICAL ANALYSIS, SPECIAL PROBLEMS OF, *article on* OUTLIERS.]

In general, an arbitrary (though less than $n/2$) number of the largest and smallest observations may be discarded before computing the mean. Define

$$(1) \quad \bar{X}_g = \frac{1}{(n-2g)} (X_{(g+1)} + \cdots + X_{(n-g)})$$

as the *trimmed mean* with the g largest and g smallest observations discarded; the proportion of observations discarded at each end is $P = g/n$. The definition of the trimmed mean can be extended to allow for the discarding of any proportion P of the largest and smallest observations (that is, P need not be an integral multiple of $1/n$). The general $100P$ per cent trimmed mean is defined as

$$(2) \quad \bar{X}_P = \frac{1}{n(1-2P)} (dX_{(f)} + X_{(f+1)} + \cdots + X_{(n-f)} + dX_{(n+1-f)}),$$

where $d = 1 + [Pn] - Pn$, $f = [Pn + 1]$, and $[Pn]$ denotes the greatest integer less than or equal to Pn. The 0 per cent trimmed mean, then, corresponds to the ordinary sample mean, $\bar{X}_0 = \bar{X}$, and the 50 per cent trimmed mean is defined as the sample median. In particular, $\bar{X}_g = \bar{X}_P$ with $P = g/n$.

The trimmed means thus constitute a family of estimators: for each fraction P there is a different trimmed mean. We confine discussion here to the trimmed means and inference procedures based on their use because the trimmed means as a family are easy to compute and interpret, include both the mean and the median, and inference procedures based on the trimmed means have received more thorough study than their competitors. The reader should note, however, that many robust alternatives to the trimmed mean are being investigated and the robustness literature is expanding rapidly (see Andrews et al. 1972).

Table 2, first row, gives large sample efficiencies for several trimmed means with respect to the sample mean, $\bar{X}$, when sampling is from a normal distribution. We note that the efficiency of the $100P$ per cent trimmed mean relative to the sample mean is roughly $1 - \frac{2}{3}P$ under normality. In short, even if the assumption of normality is warranted, little efficiency is lost by a modest amount of trimming.

Table 2 also shows the large sample relative efficiency of the trimmed means to the sample mean for sampling from selected long-tailed symmetric distributions. For these distributions, 10–25

Table 2 – Trimmed means: Large sample efficiencies with respect to the sample mean

DISTRIBUTION	$\bar{X}_{.125}$	$\bar{X}_{.25}$	$\bar{X}_{.50}$ (median)
Normal	.94	.84	.64
Compound normal			
$K=3$ $\tau=.01$	.98	.89	.66
.05	1.18	1.09	.74
Student's t			
t_3	1.91	1.97	1.62
t_5	1.24	1.20	.96
Double exponential	1.39	1.63	2.00

Source: Siddiqui & Raghunandanan 1967, p. 953.

per cent trimmed means may be markedly more efficient than $\bar{X}$. Work by Andrews et al. (1972) suggests that the large sample results shown in Table 2 are applicable for n as low as 20.

These computations (and many others) indicate that there is more to gain than to lose by discarding some extreme observations when long tails are possible. The study, rather than merely the discarding, of extreme observations, however, may often give important clues to the improvement of theory or of experimental and observational technique. [*See* STATISTICAL ANALYSIS, SPECIAL PROBLEMS OF, *article on* OUTLIERS.]

In practice, assessments of just how long-tailed the distribution appears to be may determine just how much trimming might be desirable. Such assessments might be based on considerable experience with the response measures involved, or, for large samples, on examination of histograms or normal probability plots. Dixon and Tukey (1968) and Hogg (1974) suggest formal procedures for choosing the trimming fraction when the sample size is small. If the distribution appears to be asymmetric, an appropriate transformation will often produce near symmetry and allow the use of techniques appropriate for symmetric distributions. [*See* STATISTICAL ANALYSIS, SPECIAL PROBLEMS OF, *article on* TRANSFORMATIONS OF DATA.]

Test and confidence interval criteria. A test is to be chosen to compare the null and alternative hypotheses

$$H_0\colon \mu = 0$$
$$H_1\colon \mu \neq 0 \quad (\text{or } H_1'\colon \mu > 0).$$

One may also wish to obtain a confidence interval for μ. (The null hypothesis $\mu = 0$ is natural for the matched pairs case; for the general one-sample case we may test $\mu = \mu_0$, where μ_0 is dictated by the problem at hand. For simplicity, we refer only to $\mu_0 = 0$ here.) The one-sided and two-sided Student's t tests are in many senses the best tests of H_0 against H_1' or H_1 if the X_j constitute a simple random sample from a normal distribution. [*See the biography of* GOSSET *and* SIGNIFICANCE, TESTS OF *for a definition of the* t *test.*] Our goal is to choose a good test under less stringent assumptions.

A good test should possess two important properties, robustness of validity and robustness of efficiency, over a range of plausible assumptions about the data (Box & Tiao 1964; Tukey 1962). These concepts are defined below.

A statistical test is said to have *validity* if the probability statements asserted are correct or nearly so for reasonable assumptions. Thus, from tables of the central t distribution we can see that the one-sided t test statistic with 9 degrees of freedom has probability .05 of exceeding 1.833 under the null hypothesis $\mu = 0$. This probability of .05 is referred to as the nominal significance level and symbolized α_N. The probability statement is valid when the assumptions of normality and simple random sampling hold. If, when the null hypothesis $\mu = 0$ is true, the probability that t exceeds 1.833, called the true significance level and symbolized α_T, is close to .05 for plausible underlying distributions and sampling schemes, and the same situation holds for other degrees of freedom and other significance levels, then the one-sided t test is said to have robustness of validity of significance level with respect to the set of plausible assumptions. We note later (see Table 3) that if the set of plausible assumptions includes the possibility that the data are a simple random sample from some highly skewed distribution, then the t test will not have robustness of validity.

In addition to validity of significance level, validity of the power function is important. For example, for a two-sided 5 per cent level t test, a sample size of 54 is needed to assure a power of 95 per cent against the alternative hypothesis that the mean is one half of a standard deviation unit from the null hypothesis when the underlying distribution is normal (see Dixon & Massey 1969, table A-12c). If a sample size of 54 still results in a power near 95 per cent against this alternative over a range of plausible distributions, and if actual power is close to nominal power for other sample sizes, other alternative hypotheses, and other power levels, we would say that the t test has robustness of validity of its power function.

In addition to having robustness of validity, a test should be a good discriminator between the hypotheses; that is, the probability of rejecting H_0 given that H_1 (or H_1') is true should be high. [*See* HYPOTHESIS TESTING.] Both the one-sided and the two-sided t tests have the strong property that their power is higher than the power of any other reasonable test if the assumptions of simple random sampling and normality obtain; this need not be true when the underlying distribution is nonnormal or alternative sampling procedures are used.

Thus we seek tests with good validity properties and plan to choose from among these a test with good power relative to its competitors. It is natural to compare the power of two tests in terms of their relative efficiency, $e(1,2)$, that is, the relative efficiency of test 1 with respect to test 2; if $e(1,2)$ is greater than 1.0, test 1 is more powerful than test 2. Efficiency of tests and estimators are related concepts. [*See* NONPARAMETRIC STATISTICS *for a discussion of efficiency*.] A test is said to have *robustness of efficiency* if its efficiency relative to its competitors is not appreciably below 1.0 for the range of plausible alternative distributions.

The concepts of robustness of validity and robustness of efficiency extend naturally to a study of confidence intervals. A confidence interval procedure with nominal confidence level $100(1-\alpha_N)$ per cent would be said to have robustness of validity if the true confidence level were close to the nominal level under a range of assumptions. Comparisons of efficiency may be based on a confidence interval analogue of power, or on expected lengths of confidence intervals. For procedures based on the standard t test there is a direct correspondence. For nonparametric procedures or procedures based on robust estimators the extension to confidence intervals may not be so direct. Comparisons of test procedures typically indicate results for comparisons of the corresponding confidence interval procedures.

Tests and confidence intervals. A wide variety of tests and confidence interval procedures have been developed for the one-sample or matched pairs case. We discuss and compare the validity and efficiency properties of the t test, the Wilcoxon signed-ranks test [*see* NONPARAMETRIC STATISTICS *for a definition of this test*], and the trimmed t test when used on data obtained from a simple random sample or from a matched pairs design.

The g level trimmed t statistic is

$$(3)\qquad t_g = (\bar{X}_g-\mu)/\sqrt{SSD_g/h(h-1)},$$

where $\bar{X}_g$ is the g level trimmed mean (see equation (1)), and $\sqrt{SSD_g/h(h-1)}$ is an estimate of the standard error of X_g. Here

$$(4)\qquad SSD_g = (g+1)(X_{(g+1)}-\bar{X}_{wg})^2 + (X_{(g+2)}-\bar{X}_{wg})^2 + \cdots + (X_{(n-g-1)} - \bar{X}_{wg})^2 + (X_{(n-g)} - \bar{X}_{wg})^2(g+1),$$

$$(5)\qquad \bar{X}_{wg} = \frac{1}{n}[(g+1)X_{(g+1)} + X_{(g+2)} + \cdots + X_{(n-g-1)} + (g+1)X_{(n-g)}],$$

$$(6)\qquad h = n - 2g.$$

[*SSD_g and $\bar{X}_{wg}$ are called Winsorized statistics and are discussed in* NONPARAMETRIC STATISTICS, *article on* ORDER STATISTICS.] For random sampling from a normal distribution t_g has approximately a t distribution with $h-1$ degrees of freedom (see Shorack 1974; Yuen & Dixon 1973).

Validity. The two-sided Student's t test is valid in large samples under a variety of distributional assumptions, although this validity does not extend to very small significance levels such as .001 or .0001 or to a very high power level such as .999 (see Hotelling 1961).

For small or moderate sample sizes, however, the true and nominal significance levels of Student's t test may be rather different when the underlying distribution is not normal. Validity studies for the t test have concentrated on characterizing its behavior for varying values of γ_1, γ_2 in the context of certain types of underlying distributions. Although its behavior may not be identical for differing distributions with the same γ_1 or γ_2 and although other aspects of the underlying distribution may also be relevant, these results provide a useful picture of the effects of nonnormality on the validity of the t test. When sampling is from a symmetric distribution, the true significance level of a 5 per cent one- or two-sided t test will be greater than .05 for short-tailed distributions ($\gamma_2 < 0$) and less than .05 for long-tailed distributions ($\gamma_2 > 0$). For a fixed value of γ_2 the true significance level will approach the nominal level as the sample size increases. Empirical results for the uniform, normal, double exponential, and exponential distributions appear in Table 3 and for t distributions with 3, 5, or 9 degrees of freedom in Table 4. These figures have been obtained from Monte Carlo simulation rather than by exact

Table 3 – One-sample t test: Empirical significance levels for uniform, normal, double exponential, and exponential distributions*

DISTRIBUTION	γ_2	$n = 5$	$n = 20$
Uniform	−1.2	.065	.052
Normal	0	.050	.049
Double exponential	3	.037	.048
Exponential	6	.112	.082

* Two-sided nominal significance level, $\alpha_N = .05$.

Source: Chase & Bulgren 1971, p. 501.

Table 4 – One-sample t test: Empirical significance levels for t distributions with 3, 5, and 9 degrees of freedom*

DISTRIBUTION		SAMPLE SIZE 5	10	15	20
	t_3	.037	.039	.042	.043
Student's t	t_5	.042	.044	.046	.047
	t_9	.046	.047	.048	.049

* Two-sided nominal significance level, $\alpha_N = .05$.

Source: Yuen & Murthy 1974, p. 496.

theoretical calculation (which accounts for the value of .049 for the normal distribution).

When sampling is from a skewed distribution, we must distinguish between the effects on the one-sided and two-sided t test. Skewness in the underlying population causes t to be skewed in the opposite direction, so for a 5 per cent nominal level test of H_0: $\mu \leqslant 0$ versus H_1: $\mu > 0$ for $n = 10$ in which we reject if $t \geqslant 1.833$, the true significance level will be greater than .05 for negative skewness and less than .05 for positive skewness. Table 5 shows true significance levels for sampling from an Edgeworth population with parameters γ_1, γ_2 for $n = 10$, $\alpha_N = .05$. This pattern will be reversed for a one-sided test of H_0: $\mu \geqslant 0$ versus H_1: $\mu < 0$. For a two-sided t the true significance level increases slightly as the value of $|\gamma_1|$ increases. For $|\gamma_1| = .2$, $\alpha_T = .051$; for $|\gamma_1| = .6$, $\alpha_T = .054$.

Table 5 – One-sample t test: True significance levels for Edgeworth distributions*

		γ_1: −0.6	−0.2	0	0.2	0.6
	−0.1	.072	.057	.051	.045	.036
	0.0	.071	.056	.050	.044	.035
γ_2	1.0	.070	.055	.049	.043	.034
	2.0	.069	.054	.048	.043	.033

* Probability that $t \geqslant 1.833$, $n = 10$, one-sided nominal significance level, $\alpha_N = .05$.

Source: Srivastava 1958, p. 427.

For the joint effects of skewness and kurtosis for the two-sided t, we first note that the parameters are related according to $\gamma_2 \geqslant \gamma_1^2 - 2$. Illustrative tables from Gayen (1949) and Srivastava (1958) for samples from Edgeworth populations with parameters γ_1, γ_2 show a general pattern in which for fixed skewness, γ_1^2, the true significance level decreases as γ_2 increases, and for fixed tail length, γ_2, the true significance level increases as the amount of skewness, γ_1^2, increases. Thus, for the exponential distribution ($\gamma_1 = 2$, $\gamma_2 = 6$) the empirical significance level is about 11 per cent for $n = 5$ and the nominal 5 per cent level, see Table 3. For certain combinations of γ_1 and γ_2 we see that the compensating effects of the two parameters will produce a true significance level quite close to the nominal one.

Monte Carlo investigations for small samples ($n = 6, 9, 16$) from compound normal distributions with $K = 3$ and $\tau = 0.2, 0.4$ yielded values between 4 per cent and 6 per cent for the empirical probability of the t being greater than the one-sided 5 per cent point (see Afifi, Elashoff, & Langley 1968).

Significance levels of the trimmed t are only approximate under normality, but we conjecture that they should be less affected by the length of the tails of the population distribution than are the significance levels for Student's t (see, for example, Table 10).

The Wilcoxon signed-rank test has perfect robustness of validity with respect to significance level when based on a simple random sample from a symmetrical distribution. More generally, we need only assume that the observations X_j are independent and that each X_j was obtained from a symmetrical distribution with the same mean or median μ.

Efficiency. Large sample relative efficiencies for the trimmed t with respect to Student's t test are the same as those shown in Table 2 for the trimmed means. Large sample efficiency of the Wilcoxon test with respect to Student's t test and two trimmed t's are shown in Table 6. The large sample efficiency of the Wilcoxon with respect to Student's t test is .96 for samples from a normal distribution,

Table 6 – One-sample Wilcoxon test: Large sample efficiencies with respect to Student's t test and trimmed t tests

DISTRIBUTION	STUDENT'S t TEST	TRIMMED t P = .125	TRIMMED t P = .25
Normal	.96	1.01	1.14
Compound normal			
K = 3 τ = .01	1.01	1.03	1.13
.05	1.20	1.01	1.10
.10	1.37		
.20	1.57		
Student's t			
t_3	1.90	.99	.96
t_5	1.24	1.00	1.03
Double exponential	1.50	1.08	.92

Source: Siddiqui & Raghunandanan 1967.

is greater than .86 for sampling from any continuous symmetric population, and is greater than 1.0 for samples from long-tailed symmetric distributions like those shown. Note that for the distributions shown, the 12.5 per cent trimmed t and the Wilcoxon show quite similar behavior. Thompson, Govindarajulu, and Doksum (1967) report small sample relative power calculations for the Wilcoxon and Student's t test for $n = 20$ of from .93 to .99 for samples from the normal and from 1.01 to 1.15 for samples from the double exponential ($\alpha_N = .025$, one-sided).

Conclusions. The preceding validity and efficiency studies suggest that if the underlying symmetrical distribution might have longer tails than the normal, the Wilcoxon test or trimmed t test would be preferable to Student's t test. If validity of significance level or ease of hand calculation is very important, use the Wilcoxon signed-rank test. If a point or interval estimate of the mean is of interest use the 10 to 25 per cent trimmed t procedure.

Departures from simple random sampling in the one-sample case. The preceding discussion of robustness of one-sample procedures dealt only with alternatives to the normality assumption. Departures from simple random sampling will now be discussed; they take many forms, in particular (1) the observations may constitute a biased or selected sample, (2) the observations may not be independent, and (3) the observations may not have a common distribution.

The observations constitute a biased sample if, for example, in estimating the average length of time it takes residents to get to the nearest medical clinic in San Francisco, observations come only from those households where someone is at home during the day. This could result in an overrepresentation in our sample of retired persons or families with small children who might rely more heavily on slow public transportation than residents of the city in general. In this type of biased sampling the sample mean (or indeed any of the trimmed means) might be a systematic overestimate of the population mean, μ. No estimator or test based on manipulation of the observed data without bringing in supplementary information would perform well in such a situation.

A detailed example of the effects of lack of independence is given in the section on regression. In the example of a matched pairs design for assessing the difference between teaching methods, cheating between subjects of different pairs could produce positive correlations among the outcome scores. The effects of lack of independence among the observations can be quite serious; positive correlations can result in a serious inflation of the true significance level of any procedure that is based on the assumption of independence.

In the matched pairs design, each difference, X_j, between the two treatments might well have a different population mean or variance, in which case the X_j's would not constitute a simple random sample from a common distribution. In our example, the mean and variance of the differences between the two teaching methods might depend on the aptitude level of the pair being tested. (Note that the matched pairs design is a special case of the randomized block design and that different mean differences for each pair would constitute a block $\times$ treatment interaction [*see* EXPERIMENTAL DESIGN].)

In this same example of comparing two methods of teaching, a simple model in which the observations do not have a common mean is one in which the size of treatment differences for a pair j depends on the average aptitude level of the pair, z_j. Thus, we might assume the structure

$$X_j = \mu + \beta(z_j - \bar{z}) + e_j, \quad (7)$$

where X_j is the difference in achievement scores between the two teaching methods, z_j is the aptitude level of the jth pair, $\bar{z}$ is the mean aptitude level taken over the sample pairs, μ is the population mean difference between the teaching methods, β is the unknown population slope or regression coefficient, and e_j represents random error. For simplicity in computation we assume that aptitude levels can be scored so that $z_j = j/n$, $j = 1, \cdots, n$. We suppose that the investigator is unaware of, or decides to ignore, the dependence of the mean difference on aptitude level and wants to use Student's t test to test H_0: $\mu = 0$ against H_1: $\mu \neq 0$ using the ordinary sample mean and sample variance of the X_j's. Table 7 gives the true probability that $|t| \geqslant 1.96$ when $\mu = 0$ for various values of $|\beta|$ and σ_e^2 for nominal significance level $\alpha_N = .05$ and n large. In computing Table 7 it was assumed that the e_j's are independent normal random variables with zero mean and common variance σ_e^2.

Table 7 – One-sample t test: True significance levels with varying slope*

		$\lvert\beta\rvert$			
		0	.25	.50	1.0
σ_e	.25	.05	.0414	.0223	.0028
	.50	.05	.0477	.0414	.0238
	1.00	.05	.0488	.0477	.0414
	2.00	.05	.0500	.0488	.0477

* When slope differs from zero, observations do not have a common mean; probability that $|t| \geqslant 1.96$, two-sided nominal significance level $\alpha_N = .05$.

Two important effects of incorrectly assuming that $\beta = 0$ are apparent from Table 7: the behavior of the t statistic depends on the unknown variance of the observations as well as on β, and the true significance level is always less than .05. Large sample work by Walsh (1951) shows that for the one-sided test of level α_N the true significance level, α_T, is given by $Z_{\alpha_T} = Z_{\alpha_N}\sqrt{1+\theta}$ where $\theta = \beta^2/12\sigma_e^2$ in this example and Z_α is the 100αth percentile of the normal distribution.

Another situation in which the X_j's do not have a common distribution is one in which their variances differ from pair to pair. That is, each X_j might come from a normal distribution with common mean μ but variance $\sigma_j^2 = w_j^2\sigma^2$. For example, half the pairs might be male and half female, with the variance of the difference larger for boys than for girls, say $w_j = 1.0$ for girls and $w_j = 3.0$ for boys. (This model for inequality of variances resembles the compound normal model described in our discussion of long-tailed distributions.) A trimmed t or a Wilcoxon procedure would be preferable in these circumstances to Student's t test for testing H_0: $\mu = 0$. (See Sen 1968.)

In short, departures from simple random sampling may cause serious difficulties in interpreting the results of standard statistical methods: the exact nature of the effects often depends heavily on the assumed model underlying the observations.

Two-sample problems

Two-sample problems arise when, without matching as before, each individual is assigned at random and independently to one of two treatments or conditions or when they come from two populations we are comparing. For example, in the first section, we described a matched pairs experiment to compare two teaching methods. This study could also have been done without matching by assigning students at random to each of the two teaching methods.

In a two-sample design, the data consist of the N response scores X_{ij} ($i = 1,2$; $j = 1, \cdots, n_i$) where the first subscript, i, denotes the sample and the second indexes observations within the samples. Sample sizes are n_1 and n_2 with $n_1+n_2 = N$. The investigator might wish to estimate the means, μ_i, and the variances, σ_i^2, of scores in the two conditions, to establish a confidence interval on the difference, $\mu_1 - \mu_2$, or to test hypotheses about the μ_i and σ_i^2. (Procedures for comparison of variances will be discussed later.)

The individual means, μ_i, could be estimated by their respective sample means, medians, or trimmed means as discussed in the preceding section. A natural way to estimate the difference $\mu_1 - \mu_2$ is by the difference between the estimators of μ_1 and μ_2: $\bar{X}_1 - \bar{X}_2$ or $\bar{X}_{1,.10} - \bar{X}_{2,.10}$, and so on. These estimators of $\mu_1 - \mu_2$ will have the same large sample relative efficiencies as do their one-sample counterparts for random sampling from two independent populations.

Tests and confidence intervals for differences between means. To discriminate between the null and alternative hypotheses

$$H_0: \mu_1 = \mu_2, \tag{8}$$

$$H_1: \mu_1 \neq \mu_2, \tag{9}$$

or, to construct a confidence interval for $\mu_1 - \mu_2$, the standard procedure is to use Student's t statistic:

$$t = \frac{\bar{X}_1 - \bar{X}_2 - (\mu_1 - \mu_2)}{\sqrt{[\sum\sum(X_{ij} - \bar{X}_i)^2/(N-2)]}\sqrt{(1/n_1 + 1/n_2)}} \tag{10}$$

[*See* SIGNIFICANCE, TESTS OF.]

Use of the t statistic provides a good test of H_0 against H_1 or a good confidence interval for $\mu_1 - \mu_2$ if

(1) the observations are independent random samples from their population distributions,
(2) the population distributions for sample 1 and 2 differ only in their means,
(3) the population distributions are normal.

Under these assumptions, t has the Student's t distribution with $n_1 + n_2 - 2$ degrees of freedom.

In the following sections we discuss the effects of violating assumptions (2) and (3) on the validity of procedures based on the two-sample t statistic; we also discuss alternative procedures. Effects of violation of the assumption of random sampling are similar to those in the one-sample case.

Competitors to Student's t statistic include the Wilcoxon rank–sum test, the Kolmogorov–Smirnov test, the median test and other nonparametric procedures, together with the two-sample trimmed t and other procedures based on robust estimators of the mean. Discussion here will focus on the behavior of Student's t, the Wilcoxon test [*see* NONPARAMETRIC STATISTICS, *article on* RANKING METHODS] and the trimmed t; this contrasts the standard normal theory procedure, an efficient and frequently used nonparametric competitor, and a procedure based on a simple robust estimator.

The two-sample trimmed t studied by Yuen and Dixon (1973) is the direct generalization of the one-sample trimmed t. That is, for $n_1 = n_2 = n$ and g observations trimmed from each end in both samples, the trimmed t statistic is defined as

$$(11) \qquad t_g = \frac{\bar{X}_{1g} - \bar{X}_{2g} - (\mu_1 - \mu_2)}{\sqrt{(SSD_{1g} + SSD_{2g})/h(h-1)}},$$

where $h = n - 2g$ and $\bar{X}_{ig}$ and SSD_{ig} are defined as in equations (1) and (4). Under assumptions (1), (2), and (3), the trimmed t statistic, t_g, is distributed approximately as t with $2(h-1)$ degrees of freedom for $n \geqslant 7$; for $n < 7$, the user should refer to empirical percentage points given in Yuen and Dixon (1973).

For $n_1 \neq n_2$, Yuen (1974) suggests using

$$(12) \qquad t_g = \frac{(\bar{X}_{1g} - \bar{X}_{2g} - (\mu_1 - \mu_2))\sqrt{h_1 h_2/(h_1 + h_2)}}{\sqrt{(SSD_{1g} + SSD_{2g})/(h_1 + h_2 - 2)}},$$

where $h_i = n_i - 2g_i$ and the g_i are chosen such that g_1/n_1 and g_2/n_2 are approximately equal. Preliminary evaluation of this statistic for $n_1 = 10$, $n_2 = 20$ and $n_1 = 5$, $n_2 = 20$ with $g_1/n_1 = g_2/n_2$ indicates that the null distribution of this statistic is approximately a t distribution with $h_1 + h_2 - 2$ degrees of freedom.

We must begin our comparisons of these test procedures by defining precisely the null hypothesis tested by each procedure. Student's t is a test of the null hypothesis of means (8) for any two distributions; the trimmed t test is also a test of equality of means if both distributions are identical in shape and variance or if both distributions are symmetrical. The Wilcoxon test is a test of H_0: $P(X_1 > X_2) = \frac{1}{2}$ against H_1: $P(X_1 > X_2) \neq \frac{1}{2}$, which is equivalent to (8) if both populations have the same shape and equal variances or if both populations are symmetrical.

The two populations are normal. We consider first the situation in which all three of the assumptions necessary for the validity of the t test are satisfied. Under these conditions, the Student's t and Wilcoxon tests have true significance levels and actual power exactly equal to nominal and tabled values. Although significance levels for trimmed t are only approximate, they are quite close to the nominal values even in small samples (see Table 8).

The efficiency of the two-sample Wilcoxon test relative to Student's t test under normality is .96 in large samples just as in the one-sample case. The relative efficiencies of the 100P per cent two-sample trimmed t's to Student's t test under normality are also the same as for the corresponding one-sample tests and estimators (see Table 2). The small sample power efficiencies are quite close to the large sample values for samples of twenty or more (see Table 8 for small sample power efficiency of special trimmed t's with respect to Student's t test).

Table 8 – Two-sample trimmed t test: Empirical significance levels and power efficiencies (P.E.) relative to Student's t test for normal distributions*

n	g	g/n	α_T	P.E. for $\frac{\lvert\mu_1-\mu_2\rvert}{\sigma\sqrt{2/n}} = 3$
5	1	.20	.050	.71
7	1	.14	.052	.86
10	1	.10	.050	.92
	2	.20	.052	.82
20	1	.05	.050	.97
	2	.10	.050	.94
	3	.15	.050	.87
	4	.20	.051	.86
	5	.25	.051	.80

* Two-sided nominal significance level, $\alpha_N = .05$; $n_1 = n_2 = n$, $g_1 = g_2 = g$.

Source: Yuen & Dixon 1973, p. 373.

The two populations have the same shape and equal variances but are nonnormal. First we discuss validity and then turn to efficiency comparisons. In large samples the two-sample Student's t statistic (10) provides a valid test of the hypothesis $\mu_1 = \mu_2$ except at stringent significance levels (for example, $p = .001$). Robustness of validity to nonnormality of the two-sample t test in small and moderate samples shows the same basic pattern as for the one-sample t (although the degree to which $(1/n_1) + (1/n_2)$ differs from $4/(n_1 + n_2)$ also influences the true significance level in the two-sample case). When $n_1 = n_2$, the true significance level, α_T, decreases as γ_2 increases and increases as γ_1^2 increases, although when skewness is the same in both populations, the two-sided two-sample t is much less sensitive to skewness than the two-sided one-sample t. In fact, findings by Gayen (1950) indicate that one would not expect the true significance level to be greater than .055 for a nominal .05 level test with equal n's for any values of γ_1, γ_2.

Three Monte Carlo studies provide illustrative values of α_T, the true significance level of Student's t test, when the two-sided nominal significance level is .05. Hsu and Feldt (1969) found that $.0403 \leqslant \alpha_T \leqslant .0518$ for $n_1 = n_2 = 11$ for Monte Carlo sampling from discrete distributions with 2 to 5 scale points and negative kurtosis. When sampling is from an exponential or a lognormal distribution (both distributions are markedly skewed with long tails) Table 9 shows $.031 \leqslant \alpha_T \leqslant .049$ for equal samples of sizes 4 to 32. Results for the exponential illustrate the lessened effect of skewness in the two-sample case: true significance levels in the one-sample case were roughly twice the values obtained here (see Table 3). The rows with $g = 0$ in Table 10 show α_T for the two-sample Student's t test for samples from compound normal distributions in which 80 per cent of the observations are from a normal distribution with variance σ^2 and 20 per cent from a normal distribution with variance $9\sigma^2$. Note that they are consistently less than .05.

Table 9 – Two-sample t test: Empirical significance levels for exponential and lognormal distributions*

n	Exponential $\gamma_1 = 2, \gamma_2 = 6$	Lognormal $\gamma_1 = 2, \gamma_2 = 37$
4	.042	.031
8	.044	.035
16	.046	.038
32	.049	.046

* Two-sided nominal significance level, $\alpha_N = .05$; $n_1 = n_2 = n$.

Source: Donaldson 1968, p. 665.

Table 10 – Two-sample t test and several trimmed t's: Empirical significance levels and power efficiencies (P.E.) relative to t test for mixture of normal distributions*

n	g	g/n	α_T	P.E. for $\frac{\mid\mu_1 - \mu_2\mid}{\sigma\sqrt{2/n}} = 3$
5	0		.036	
	1	.20	.047	.85
10	0		.042	
	1	.10	.045	1.39
	2	.20	.052	1.38
20	0		.046	
	1	.05	.044	1.33
	3	.15	.048	1.55
	5	.25	.052	1.47

* Two-sided nominal significance level, $\alpha_N = .05$; $n_1 = n_2 = n$; $.8n$ observations are from $N(0, \sigma^2)$ and $.2n$ are from $N(0, 9\sigma^2)$.

Source: Yuen & Dixon 1973, p. 373.

The Wilcoxon test has perfect robustness of validity with respect to significance levels and power for sampling from any continuous distribution.

The validity of the trimmed t for samples from a nonnormal distribution has been studied only for long-tailed symmetric distributions. The remaining rows of Table 10 show true significance levels for the trimmed t. For this example, the significance levels of the trimmed t's are generally somewhat closer to .05 than those of Student's t test.

The large sample relative efficiences of the two-sample Wilcoxon test and the two-sample Student's t test and trimmed t test are the same as those shown in Tables 2 and 6 for the one-sample case. The Wilcoxon and 10 to 25 per cent trimmed t tests are more efficient than the Student's t test for sampling from long-tailed symmetrical distributions with the same shape and equal variances. Some small sample power efficiencies of the trimmed t to Student's t test are shown in Table 10. Afifi and Kim (1972) report that for equal samples of sizes 10 to 100 from various compound normal distributions ($\tau = .1, .3$; $K = 1, 2, 3$; and shifts of $a = 0, .5, 1.0, 2.6$ in the normal distribution with the larger variance) the Wilcoxon test was more powerful than Student's t test for cases in which $\gamma_2 > 1.5$ or $|\gamma_1| > .63$.

The two populations have equal variances, but may have different skewness or kurtosis. Little information is available in the literature about the effects of differences in the shapes of the two populations. If the two populations differ in degree of skewness, the trimmed t's and the Wilcoxon are no longer tests of the null hypothesis of equal means and the three procedures are no longer comparable. If the two populations are symmetrical with equal variances but just differ in degree of kurtosis, Student's t test and the trimmed t's will probably do well in large samples; but the true significance level for the Wilcoxon test will not be equal to the nominal level even in large samples. For example, if one sample is drawn from a normal distribution and the other from a double exponential with the same variance, the true significance level for a nominal 5 per cent Wilcoxon test is .089 in large samples. For $n_1 = n_2$ the effect of differing kurtosis will be to inflate α_T for the Wilcoxon test from .05 to as much as .11; for unequal n's, α_T may be either increased or decreased.

Inequality of variances. Next, we consider the validity of the preceding tests and some further tests when the assumption of equal variances is dropped. The normality assumption is retained unless otherwise indicated since the possible effects of inequality of variances are far stronger than those of moderate nonnormality; in addition the effects of nonnormality appear to be additive with those due to inequality of variance (see Glass et al. 1972).

It is necessary at this point to examine the rationale for carrying out a test. Suppose a random sample of N mental patients is drawn; n_1 are assigned to treatment 1, and n_2 $(= N - n_1)$ are assigned to treatment 2. After a period of treatment, each patient is tested and given a score that is assumed to be normally distributed in each population. The null hypothesis tested is that $\mu_1 = \mu_2$ (μ_i is the mean for treatment i); suppose that the conclusion is that $\mu_1 < \mu_2$. If high scores are indicative of improvement, a decision is made to use the second treatment. Why?

When variances are equal, the treatment with the higher mean is more likely to give rise to scores greater than or equal to any given score w. Thus the significance test may provide useful results, especially if a score of w or above means release from the psychiatric hospital.

Suppose, however, that the treatments have different variances and that $\mu_1 < \mu_2$. Then it is not uniformly true that the treatment with the higher mean is more likely to give rise to scores greater than or equal to a particular release score of w. For example, suppose that the scores for treatments 1 and 2 follow normal distributions with $\mu_1 = 0, \sigma_1^2 = 9$ and $\mu_2 = 1, \sigma_2^2 = 1$, respectively; see Figure 1. In this case Table 11 gives the probability that a randomly chosen score on treatment i exceeds $w_j, i = 1, 2; j = 1, \cdots, 4$.

Table 11 – Probability that a randomly chosen score on treatment i exceeds w_j (normal model with means 0 and 1, variances 9 and 1)

			TREATMENT 1	TREATMENT 2
RELEASE SCORE	w_1	1.0	.3707	.5000
	w_2	1.5	.3085	.3085
	w_3	2.0	.2514	.1587
	w_4	2.5	.2033	.0668

Nonetheless, it is often appropriate to test equality of means even when the variances may be different, and the remainder of this section deals with that case. (For example, we may ask whether a drug does or does not produce a higher average response than the control.) This problem is often called the Behrens–Fisher problem. Consider, first, the large sample validity of the two-tailed, two-sample Student's t test given by equation (10). Unless either $\sigma_1^2 = \sigma_2^2$ or $n_1 = n_2$, the denominator of t does not approach the true standard error of $\bar{X}_1 - \bar{X}_2$ as the sample size increases. The resulting nonrobustness of t to inequality of variances in the two populations is illustrated in Table 12, which gives the large sample probability that $|t| \geqslant 1.96$ for different values of $\theta = \sigma_1^2/\sigma_2^2$ and $R = n_1/n_2$

Table 12 – Two-sample t test: True significance levels for normal distributions when means are equal but variances differ*

		$\theta = \sigma_1^2/\sigma_2^2$ .20	.50	1.0	2.0	5.0
$R = n_1/n_2$	1	0.050	0.050	0.050	0.050	0.050
	2	0.120	0.080	0.050	0.029	0.014
	5	0.220	0.120	0.050	0.014	0.002

* Large sample probability $|t| \geqslant 1.96$, two-sided nominal significance level, $\alpha_N = .05$.

Source: Scheffé 1959, p. 340.

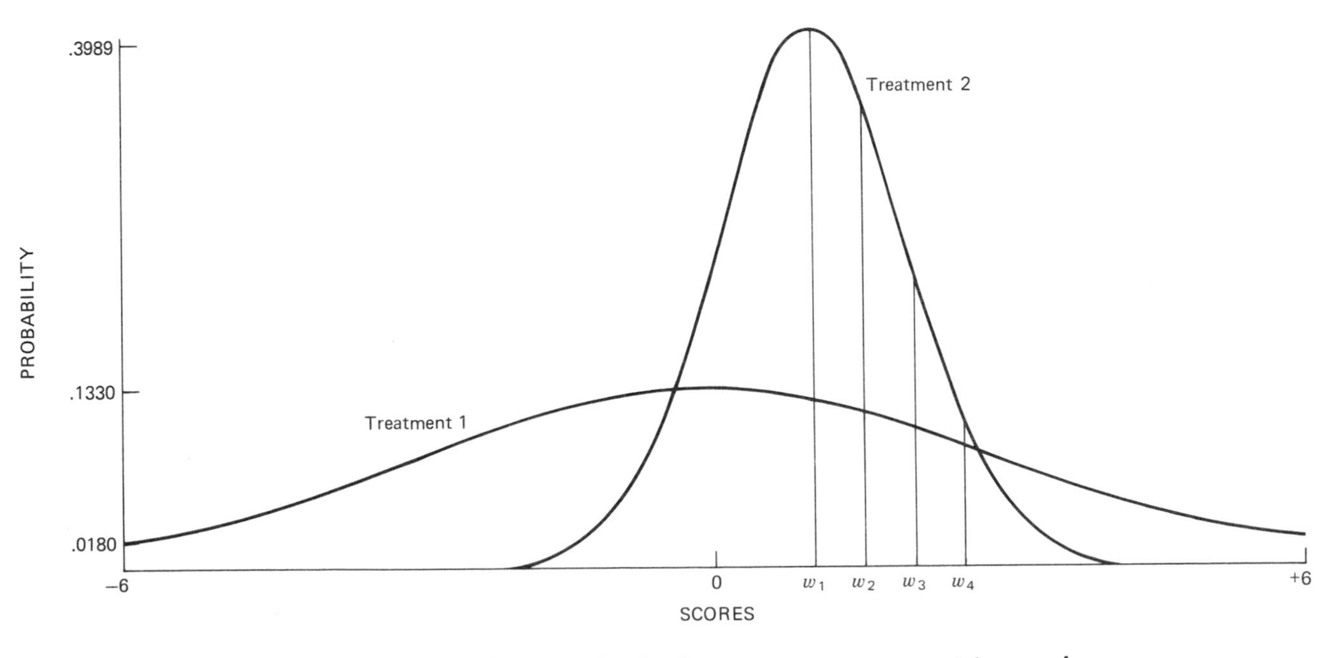

Figure 1 – Probability that a randomly chosen score on treatment i exceeds w_j (normal model with means 0 and 1, variances 9 and 1)*

* Treatment 1 has mean 0, variance 9; treatment 2 has mean 1, variance 1. The area under each curve to the right of w_j represents required probability.

when the null hypothesis of equal means holds. Table 12 indicates the importance of equal sample sizes in controlling the effects of unequal variances: the significance level remains at .05 irrespective of the value of θ if $R = 1$. Moreover, if $\theta < 1$ and $R > 1$ so that the more variable population has the smaller sample, the true significance level is always larger than .05 and may be seriously so. On the other hand, if $\theta > 1$ and $R > 1$, the true significance level is always less than .05. These results are essentially independent of γ_1 and γ_2 because of the large sample sizes. This lack of robustness of significance-level validity extends to power. The small sample validity of t follows along the lines of the large sample theory. Even for small samples the true significance level of Student's t test is not far from the nominal level for widely varying θ when $n_1 = n_2$ (see Bhattacharjee 1968).

Inequality of variances also affects the significance level of the Wilcoxon rank–sum test. Van der Vaart (1961) and Pratt (1964), assuming normal distributions and large samples, show that inequality of variances has an effect on W similar to that on Student's t, although W is sensitive to inequality of variances even in the case of equal sample sizes (see Table 13).

Table 13 – Two-sample Wilcoxon test: True significance levels when variances and sample sizes differ*

		$\theta = \sigma_1^2/\sigma_2^2$				
		.2	.5	1.0	2.0	5.0
$R = n_1/n_2$	1	.062	.053	.050	.053	.062
	2	.093	.067	.050	.039	.033
	5	.124	.082	.050	.026	.010

* Two-sided nominal significance level, $\alpha_N = .05$, n large.

Source: Pratt 1964.

The effect of inequality of variances on the trimmed t can be expected to be much the same as on Student's t test (see Yuen 1974).

Since equal sample sizes, while clearly desirable, are sometimes difficult to obtain, even approximately, considerable research has been focused upon alternative ways to test $\mu_1 = \mu_2$ versus $\mu_1 \neq \mu_2$. Transformation of the response variable may achieve equality of variance for the transformed variable, so that t may be used. But then a hypothesis on the means of the *transformed* variates is being tested. [*See* STATISTICAL ANALYSIS, SPECIAL PROBLEMS OF, *article on* TRANSFORMATIONS OF DATA.]

Welch (1938; 1947) investigates the natural alternative test statistic using s_1^2 and s_2^2 as separate estimators of the two population variances, $s_i^2 = \sum_j (X_{ij} - \bar{X}_i)^2/(n_i - 1)$. He proposes

$$v = \frac{\bar{X}_1 - \bar{X}_2}{\sqrt{s^2/n_1 + s^2/n_2}},$$

which is the same as t if $n_1 = n_2$. He shows that the approximate significance level of v may be obtained from tables of the t distribution with f degrees of freedom, where

$$f = \frac{[(s_1^2/n_1) + (s_2^2/n_2)]^2}{s_1^4/[n_1^2(n_1-1)] + s_2^4/[n_2^2(n_2-1)]}.$$

The v test thus defined is robust for both significance level and power in large samples and is much less sensitive to differences in variances than the usual t in small samples. For example, if $\theta = 1$ and $n_1 = 5$, $n_2 = 15$, the exact probability that $|v| \geqslant 5.2$ is .05. The probability that $|v| \geqslant 5.2$ for any other θ value is always between .035 and .085. In addition, even if $\theta = 1$, so that the t test is valid, the v test is nearly as efficient as t. When both n_1 and n_2 are small, s_1^2 and s_2^2 will have low precision. In these situations, compute f as

$$f = \frac{[(1/n_1) + (1/n_2)]^2}{1/[n_1^2(n_1-1)] + 1/[n_2^2(n_2-1)]}.$$

(For other two-sample tests, see Scheffé 1970; Scott & Smith 1971; Mehta & Srinivasan 1970; Pfanzagl 1974.)

A conservative version of the Wilcoxon statistic for use when variances are unequal has been studied by Potthoff (1963), Leaverton and Birch (1969), and Hettmansperger (1973). A trimmed t analogue to Welch's statistic in which s_i^2 is replaced by s_{ig}^2 and n_i by h_i in the definition of the statistic and in the formula for the approximate degrees of freedom was evaluated by Yuen (1974). The relative performance of Welch's v and its trimmed t analogue follows the same pattern as for Student's t test and the regular trimmed t. Under nonnormality the actual significance level of Welch's t decreases as the length of the tails increases for a fixed nominal level; the effect on the actual significance level for the trimmed t analogue is similar but less marked. The trimmed t version of Welch's statistic is less powerful under normality but more powerful for long-tailed distributions.

Conclusions. What procedure should an investigator use to make inferences about $\mu_1 - \mu_2$ in the two-sample case? The recommendations made here apply only if the observations are independent random samples from their population distributions. Since Student's t test, the trimmed t's, and the Wilcoxon test all exhibit better robustness properties for equal n's, sample sizes should be made equal, or nearly so, whenever possible. If it can be assumed that the two population distributions have identical shapes and equal variances, the Wilcoxon or the 10 to 25 per cent trimmed t have better validity of significance level and higher efficiency than Student's t test when the underlying distribution has longer tails than the normal. If the distributions may differ in kurtosis or variance the Wilcoxon does not have robustness of validity in large samples even for equal n's. If sample sizes are unequal, use of Welch's v or its trimmed t analogues will provide protection against the possibility of unequal variances; the choice among them will depend on the tail length. If the populations differ in degree of skewness, the various inference procedures described here are no longer comparable. The reader should remember that if shapes (skewness, kurtosis, and so on) and/or variances differ between two populations, the rationale for carrying out a test on means may be questionable.

How can an investigator decide what to do in a particular problem? It is no easy task to make recommendations; statisticians are only beginning the detailed studies necessary to support specific recommendations; much remains to be done. If the investigator has large samples or has seen a lot of data of this same type, these may supply some information on tail lengths and possible differences between shapes; if such information is not available the investigator may decide to use a conservative procedure "just in case," or to use a standard procedure because he cannot afford the insurance premium or because he is optimistic.

The one-way analysis of variance

The one-way analysis of variance may be used when, without matching, N individuals are independently and randomly assigned to k treatments. The data consist of the n_i response scores X_{ij} ($i = 1, \cdots, k; j = 1, \cdots, n_i$) with $\sum_{i=1}^{k} n_i = N$. The observations are assumed to be independent, and the probability distributions of the response variable are assumed identical for individuals receiving the same treatment.

This one-way layout is frequently the basis for estimation of the means, μ_i, and the variances, σ_i^2, of the treatments, for confidence intervals on the differences, $\mu_i - \mu_j$, and for testing hypotheses about the μ_i and σ_i^2. The point estimation problems present no essentially new questions. The discussion in this section is limited to the so-called *fixed-effects* model for a one-way analysis of variance. Random effect models will be briefly discussed later.

Tests and confidence intervals for the mean. To discriminate betwen the null and the alternative hypotheses

$$H_0: \mu_1 = \mu_2 = \cdots = \mu_k, \tag{13}$$

$$H_1: \mu_i \neq \mu_j \text{ for some } i, j, \tag{14}$$

one ordinarily employs the F test or the Kruskal–Wallis H test [*see* NONPARAMETRIC STATISTICS, *article on* RANKING METHODS; LINEAR HYPOTHESES, *article on* ANALYSIS OF VARIANCE].

No k-sample analogue of the trimmed t has yet appeared in the literature although such an extension would appear to be straightforward.

Generally speaking, the robustness of the F test and comparisons of it with the Kruskal–Wallis H test can be expected to follow closely those laid out in detail for their two-sample versions, the two-sided Student's t test and the Wilcoxon rank–sum test. See Glass et al. (1972) for an extensive review of the literature on the robustness of the fixed-effects analysis of variance.

Tests of the null hypothesis (13) that are robust to inequality of variances have been developed along the lines of Welch's v test for general k by Welch (1947) and James (1951); Scott and Smith (1971) evaluate some additional proposals. Welch's approach seems to perform well for sample sizes as small as five per group (see Brown & Forsythe 1974*a*; 1974*b*).

For the simplest confidence intervals, the prior one- and two-sample discussions hold. For confidence regions corresponding to interesting multiple-comparison regions, there appears to have been little discussion of robustness. One investigation of the robustness of the Studentized range method of multiple comparisons to inequality of variances and to nonnormality is by Brown (1974). [*See* LINEAR HYPOTHESES, *article on* MULTIPLE COMPARISONS.]

Inference procedures for variances

Of the several measures of spread reported in the statistical literature, the most common are the population variance, σ^2 (and its square root the standard deviation), the mean absolute deviation, and the interquartile range. The discussion here is only about estimators and confidence intervals for the population variance, σ^2.

When the observations constitute a random sample from a normal distribution, the sample variance, s^2, is the best estimator of σ^2 and a confidence interval for σ^2 is easily obtainable from the chi-square distribution [*see* VARIANCES, STATISTICAL STUDY OF]. When the population is symmetrical but long-tailed, s^2 is unfortunately no longer a good estimator of σ^2, and the standard confidence interval for σ^2 based on χ^2 is no longer valid; see Table 14. An approximate confidence interval for σ^2 that has satisfactory validity behavior can be obtained by using the jackknife method (see Miller 1974). Development of estimators and confidence intervals for σ^2 that have robustness of efficiency remains an open problem.

Table 14 – 95 per cent confidence interval for σ^2; true probability under nonnormality that interval does not cover σ^{2*}

	γ_2						
	−1.0	−.5	0	.5	1	2	4
Probability	.006	.024	.05	.08	.11	.17	.26

* *n* large, $\gamma_1 = 0$.

Source: Scheffé 1959, p. 337.

Tests for equality of variances. Another problem of interest in a two-sample design or a one-way layout with k treatments is comparison of variability; thus an investigator may wish to test

$$H_0: \sigma_1^2 = \sigma_2^2 = \cdots = \sigma_k^2$$

against

$$H_1: \sigma_i^2 \neq \sigma_j^2 \text{ for some } i, j,$$

or find confidence intervals for all ratios σ_j^2/σ_i^2, $i \neq j$ [*see* VARIANCES, STATISTICAL STUDY OF].

The standard normal theory procedures for this problem are Bartlett's test for homogeneity of variances and confidence intervals derived from Bartlett's test (see Dixon & Massey 1969, p. 308, for details). Bartlett's M test requires normality and has almost no robustness of validity, as may be seen from Table 15, where the nonnormality is characterized by the γ_2 parameter ($\gamma_1 = 0$). Indeed, as Scheffé (1959) remarks, "for some populations with $\gamma_2 > 0$ the test has a sensitivity to non-normality comparable to that of standard tests for non-normality."

Table 15 – M, statistic for Bartlett's test: True significance levels under nonnormality*

		γ_2			
		−1	0	1	2
Number of populations (k)	2	.0056	.05	.11	.166
	3	.0025	.05	.136	.224
	5	.0008	.05	.176	.315
	10	.001	.05	.257	.489

* Nominal significance level, $\alpha_N = .05$, $\gamma_1 = 0$.

Source: Box 1953, p. 320.

The disastrous behavior of M for long-tailed symmetrical distributions may be explained by the following suggestive argument by Box (1953). Let T denote a statistic; for example, $\bar{X}$, s_1^2/s_2^2, and M are statistics. Then, in large samples, T divided by its estimated standard error is usually approximately normally distributed, by virtue of the central limit theorem and associated mathematical arguments. Thus, even though sampling may not be from a normal distribution, $\bar{X}/(s/\sqrt{n})$ is approximately normally distributed. This result explains the robustness to nonnormality of the t test in the matched-pairs design and the F test in the one-way analysis of variance when the populations have the same shape. But Bartlett's M test does not have this structure of T divided by its standard error; hence it does not find protection via the central limit theorem.

The lack of robustness of Bartlett's test motivates an alternative test for general use. Cochran's test and Hartley's test can be expected to show the same lack of robustness to nonnormality as Bartlett's test (see Gartside 1972).

Miller (1968) reviewed a variety of test procedures for comparison of variances and investigated asymptotic performances. Table 16 shows Monte Carlo results for sample sizes of 10 and 25 for several promising alternatives to the standard normal theory F test. The procedures shown are the Box–Andersen test, also called the *APF* test in Shorack (1969), which uses an estimate of the

Table 16 – Two-sample tests on variances: Empirical power functions for uniform, normal, and double exponential distributions*

	$\theta = \sigma_1^2/\sigma_2^2$	$n_1 = n_2 = 10$			$n_1 = n_2 = 25$		
DISTRIBUTION	TEST	1[a]	2	6	1[a]	2	6
Uniform	F	.017	.18	.91	.007	.50	
	Box–Andersen	.077	.36	.90	.053	.79	
	Jackknife	.036	.28	.91	.029	.79	
	Levene	.052	.36	.89	.041	.71	
Normal	F	.063	.26	.83	.060	.50	
	Box–Andersen	.075	.26	.77	.047	.48	
	Jackknife	.062	.22	.75	.050	.47	
	Levene	.050	.23	.69	.041	.45	
Double exponential	F	.125	.32	.75	.127	.49	.95
	Box–Andersen	.077	.20	.60	.053	.32	.88
	Jackknife	.074	.21	.56	.069	.31	.85
	Levene	.045	.17	.45	.043	.28	.82

* Nominal significance level, $\alpha_N = .05$.
[a] These columns show empirical significance levels.

Source: Miller 1968, p. 576, 578.

kurtosis parameter, γ_2, to adjust the degrees of freedom of the F test; the jackknife procedure based on combining the n_1 values $\ln s_{1j}^2$ and the n_2 values $\ln s_{2j}^2$ where s_{ij}^2 is the sample variance for sample i computed with observation j excluded from the sample [*see* ERRORS, *article on* NONSAMPLING ERRORS], and Levene's s test, an analysis of variance of the scores $(X_{ij} - \bar{X}_i)^2$. All three procedures require considerable computational effort. Table 16 and additional results from Shorack (1969) indicate that these procedures have true probability levels reasonably close to the nominal .05 level for samples of size 25 and appear to have comparable efficiencies. (The Levene test appears to be slightly less powerful in general but its true significance level is also generally somewhat smaller than those for the other two tests.)

For the k-sample case, Layard (1973) investigated the performance of a chi-square test that uses an estimate of the kurtosis parameter in defining the test statistic, and the jackknife (see Table 17). These procedures perform reasonably well for samples of 25 or more per group.

Model II analysis of variance. The extreme sensitivity of the standard F test and of Bartlett's M test for equality of variances to the value of γ_2 suggests that one may expect trouble with the normal theory analysis of the random effects model for analysis of variance, sometimes called Model II [*see* LINEAR HYPOTHESES, *article on* ANALYSIS OF VARIANCE; *see also* Dixon & Massey 1969, p. 302]. Real difficulties do exist with such analyses, even in large samples. A jackknife procedure for variance components in the one-way random effects analysis of variance is described by Arvesen (1969) and Arvesen and Schmitz (1970).

Comparison of correlated variances. Comparison of variances may also be of interest in the matched-pairs design or its k-sample analogue. For the case of two variances, the standard normal theory test of H_0: $\sigma_1^2 = \sigma_2^2$ is based on noting that the Pearson product–moment correlation between $X_{1j} + X_{2j}$ and $X_{1j} - X_{2j}$ is $\rho = \sigma_1^2 - \sigma_2^2$ and thus a test of $\sigma_1^2 = \sigma_2^2$ is equivalent to a test of $\rho = 0$ [*see* VARIANCES, STATISTICAL STUDY OF]. Han (1968) discussed four normal theory test procedures for the k-sample case. Layard (1969) showed that the normal theory tests are not asymptotically robust to departures from normality and examined the large sample properties of several large sample robust procedures applicable to the problem.

Conclusions. Standard normal theory inference procedures for population variances are not robust to nonnormality. The true significance level may be considerably larger than the nominal level when the underlying population is even moderately long tailed. Robust procedures based on the jackknife or on estimation of the kurtosis parameter appear to be promising for moderate to large sample sizes.

Multivariate problems

Multivariate analogues of the univariate one-, two-, and k-sample problems discussed in earlier sections arise when several outcome variables are measured on each individual. For example, in an experiment to compare two methods of teaching arithmetic, the investigator might administer a speeded computation test, a word-problem test, and an abstract reasoning test to each student and wish to make comparisons of the two teaching methods simultaneously on all three outcome mea-

Table 17 – k-sample tests on variances: Empirical power functions for uniform, normal, and double exponential distributions*

		$n_1 = n_2 = n_3 = n_4 = 10$		$n_1 = n_2 = n_3 = n_4 = 25$	
	Variance ratio	1:1:1:1[a]	1:1:4:4	1:1:1:1[a]	1:1:4:4
DISTRIBUTION	TEST				
Uniform	M	.004	.61	.002	1.0
	Chi square	.078	.74	.050	1.0
	Jackknife	.026	.74	.030	1.0
Normal	M	.054	.58	.048	.98
	Chi square	.090	.60	.062	.98
	Jackknife	.050	.52	.040	.98
Double exponential	M	.260	.70	.336	.96
	Chi square	.088	.42	.046	.73
	Jackknife	.084	.35	.066	.73

* Nominal significance level, $\alpha_N = .05$.
[a] These columns show empirical significance levels.

Source: Layard 1973, pp. 197, 198.

sures. The number of such outcome variables considered will be called p.

Broadly speaking, the robustness of multivariate procedures follows the same general lines as the robustness of their univariate counterparts, although the behavior of multivariate procedures tends to deteriorate as the number of variables or the number of groups increases. Little development or evaluation of the small sample performance of nonparametric or robust alternatives to normal-theory multivariate methods has been made.

One-sample problems. The multivariate analogue of the one-sample t test is Hotelling's T^2 [*see* MULTIVARIATE ANALYSIS, *overview article*]. Table 18 shows empirical significance levels for T^2 for samples of sizes 5 and 20 from various bivariate distributions ($p = 2$). Comparison of these results with those shown in Table 3 for the univariate Student's t test shows essentially the same pattern of decreasing α_T with increasing tail length, increasing α_T with increasing skewness, and of α_T approaching α_N as n increases. The major difference is that the effects of nonnormality are somewhat more pronounced in the bivariate case than in the univariate case with comparable sample size and that the size of ρ may have an effect on α_T in the bivariate case.

Table 18 – Hotelling's T^2: Empirical significance levels for various bivariate distributions with correlation ρ*

DISTRIBUTION	ρ	n	
		5	20
Bivariate normal	0	.051	.046
	.25	.051	.050
	.50	.051	.051
	.75	.049	.046
Bivariate uniform	0	.068	.053
	.25	.067	.051
Bivariate exponential	0	.135	.100
	.25	.133	.103
	.50	.117	.104
	.75	.072	.092
Bivariate double exponential	0	.034	.044

* Two-sided nominal significance level, $\alpha_N = .05$.

Source: Chase & Bulgren 1971, p. 501.

Two-sample problems. Similarly, the multivariate analogue of the two-sample Student's t is the two-sample Hotelling's T^2.

Monte Carlo work for the bivariate two-sample T^2 by Hopkins and Clay (1963) for samples from the compound normal when the two variables are independent yields values of α_T very similar to those obtained for the univariate Student's t.

Hopkins and Clay (1963), Ito (1969), and Holloway and Dunn (1967) have investigated the robustness of the p-variate T^2 to the inequality of covariance matrices $\boldsymbol{\Sigma}_1$ and $\boldsymbol{\Sigma}_2$ for samples from the multivariate normal. Investigations have concentrated on the case in which all eigenvalues of $\boldsymbol{\Sigma}_2\boldsymbol{\Sigma}_1^{-1}$ are equal, say, to d. (For $p = 1$, the single eigenvalue is σ_2^2/σ_1^2.)

Although in the univariate case, for $\alpha_N = .05$, equality of sample sizes ($n_1 = n_2 = n$) maintains α_T very close to .05 for n as small as 5 across a variety of assumptions, in the multivariate case as the number of variables increases the sample size necessary to achieve α_T close to .05 increases very rapidly. When $d = 3$, $p = 7$, for example, the empirical significance level is greater than .08 until n is greater than 12.

When $n_1 \neq n_2$ but total sample size and d are fixed, the discrepancy between nominal and empirical significance levels increases as the number of variables increases. Reading the charts in Holloway and Dunn (1967), we find that for $n_1 + n_2 = 50$, $d = 3$ and ratios of n_1 to n_2 from .5 to 2, empirical significance levels for $\alpha_N = .05$ are approximately .017 and .14, respectively, for two variables and .01 and .24, respectively, for seven variables.

Even when $n_1 = n_2$, so that the significance level is not much affected, the power of the test may be considerably reduced; for example, when $n_1 = n_2 = 25$, $p = 2$, and d is as small as 1.5, the empirical power is .70 when the nominal power is .80. As d increases, the distortion of power increases markedly.

Multivariate k-sample tests on means. A variety of test criteria have been proposed for testing equality of means in the k-sample case. The relative power of the various test criteria depends on the nature of the alternative hypothesis. Their relative robustness to nonnormality or inequality of covariance matrices, although generally similar to that seen for Hotelling's T^2 in the two-sample case, does depend on the nature of the departure from assumptions. Olson (1974) made a Monte Carlo study of six criteria, including Roy's largest root, Wilks's likelihood ratio (the k-sample extension of Hotelling's T^2), T_0^2 and the Pillai–Bartlett trace criteria for k groups with equal n's, under various departures from normality or equality of covariance matrices. He concluded that the Pillai–Bartlett trace performs reasonably well for a variety of departures from assumptions.

Multivariate procedures have been developed for the case where both sample sizes and covariance matrices may differ among the groups—the multivariate Behrens–Fisher problem. Subrahmaniam and Subrahmaniam (1973) evaluated several proposals: empirical results for significance level and power in the two-sample case indicate that the Yao (1965) procedure, which reduces to Welch's v test for $p = 1$, provides a reasonable choice.

Comparison of variances. Standard normal theory tests for equality of covariance matrices are not asymptotically robust to nonnormality (see, for example, Hopkins & Clay 1963 for empirical significance levels of the multivariate M test for bivariate samples from the compound normal). Asymptotically robust tests based on estimating the fourth moments of the distribution or on the jackknife have been investigated by Layard (1972).

Linear regression

Suppose that a psychologist observes a response, Y_t, at times $t = 1, 2, \cdots, n$. The psychologist assumes that the response delay has a linear regression over time, that is

$$Y_t = \alpha + \beta\left(\frac{t}{n} - \frac{n+1}{2n}\right) + e_t. \tag{15}$$

(The use of $(t/n) - (n+1)/(2n)$ instead of just t represents merely a convenient coding of the t values. In particular $(n+1)/(2n)$ is just the mean of the (t/n)'s: $(n+1)/(2n) = (1/n)\sum_{t=1}^{n}(t/n)$.) The standard statistical model for linear regression is based on the following assumptions (in addition to the linearity of equation (15)): (1) α, β are unknown constants and no previous information is available about their values; (2) the time at which Y_t is observed is exactly time t, that is, the independent variable t is measured without error; (3) the e_t, which may represent either errors of measurement of Y or errors in the assumption of linear regression, are random quantities; (4) the error terms, e_t, are jointly independent; (5) the random quantities e_t have the same unknown variance, σ^2; (6) the e_t follow a normal distribution with zero expectation.

The goal is to estimate α and the slope β, and to obtain confidence intervals (or tests) for α and β. Before proceeding to discuss the estimation and testing problems engendered by violations of the above assumptions, we make some general remarks. Descriptive and probability plots of the data are a necessary first step in assessing whether the assumption of a linear relation between Y_t and t is reasonable (see Daniel & Wood 1971 for examples of probability plots and their interpretation). In addition, possible sources of measurement error in both Y and t should be reviewed to check on the tenability of assumption (2). [*See* LINEAR HYPOTHESES, *article on* REGRESSION, *for comments on the effects of errors in both variables.*] If the same study has been run on other individuals under the same conditions (that is, if assumption (1) is untrue), the results of the preceding studies might be combined in some way with those of the present study (see, for example, Swamy 1971).

We now consider in more detail the effects of departures from the assumptions of independence (4), homoscedasticity (5), and normality (6).

Dependence among the observations. We suppose that equation (15) and assumptions (1)–(3), (5), and (6) hold, but that assumption (4) does not. It should be pointed out here that correlations among the errors can occur even when the independent variable does not represent time. A time-series problem merely represents the most obvious context in which the problem is likely to occur.

Suppose, at first, the psychologist believes that there is no dependence among the observations. Then reasonable estimators for α and β are found by the method of unweighted least squares, which gives the standard results

$$\hat{\alpha} = \sum_{t=1}^{n} Y_t/n,$$

$$\hat{\beta} = \sum_{t=1}^{n} (Y_t - \bar{Y})\frac{t}{n} \Big/ \sum_{t=1}^{n}\left(\frac{t}{n} - \frac{n+1}{2n}\right)^2.$$

If no dependence exists, then these estimators have minimum variance among all linear unbiased estimators. The variances of $\hat{\alpha}$ and $\hat{\beta}$ are

$$\text{var}\,\hat{\alpha} = \sigma^2/n$$
$$\text{var}\,\hat{\beta} = \sigma^2 \Big/ \sum_{t=1}^{n}\left(\frac{t}{n} - \frac{n+1}{2n}\right)^2 \tag{16}$$

where σ^2 is the variance of e_t.

Now, suppose that dependence exists among the observations and assume that the correlation between Y_t and Y_s, denoted by ρ_{st}, is given by

$$\rho_{st} = \rho^{|s-t|}, \tag{17}$$

where $|s - t|$ denotes the absolute value of the difference $s-t$; this is a simple mathematical model for the common situation in which the correlation decreases as the distance between observations increases. What are the effects of nonzero values of ρ on the estimators $\hat{\alpha}$ and $\hat{\beta}$? First, while these estimators are still unbiased, they are in this case no longer the minimum variance linear unbiased estimators (see Johnston 1963 for the way to construct the latter estimators). The estimator $\hat{\beta}$ is less efficient than the minimum variance linear unbiased estimator, β^*, computed under the assumption that ρ is known, in small samples.

A second effect of nonzero ρ is that the variances of $\hat{\alpha}$ and $\hat{\beta}$ given by (16) are incorrect. The correct large sample variances when (17) holds are

$$\operatorname{var} \hat{\alpha} = \frac{\sigma^2}{n}\left(1 + \frac{2\rho}{1-\rho}\right)$$

$$\text{(18)} \quad \operatorname{var} \hat{\beta} = \frac{\sigma^2}{\sum_t \left(\frac{t}{n} - \frac{n+1}{2n}\right)^2}\left(1 + \frac{2\rho}{1-\rho}\right).$$

These variances may depart radically from (16). Table 19 shows the ratio of the standard error from formula (16) to that from formula (18), the quantity $\sqrt{(1-\rho)/(1+\rho)}$, as a function of ρ. Since the standard errors from (16) may be in serious error, it is clear that the standard error of prediction, that is, the standard error of the quantity $\hat{\alpha} + \hat{\beta}[(t/n)-(n+1)/(2n)]$, may also be very wrong.

Table 19 – Ratios of incorrect to correct standard error of β when ρ is not zero

	ρ						
	−.50	−.20	−.10	0	.10	.20	.50
$\frac{\text{s.e. } \hat{\beta} \text{ from eq. (16)}}{\text{s.e. } \hat{\beta} \text{ from eq. (18)}}$	1.73	1.22	1.11	1	.90	.82	.58

The third effect of correlation between observations concerns s^2, the conventional estimator of the underlying variance, σ^2,

$$s^2 = \sum_{t=1}^{n}\left[Y_t - \hat{\alpha} - \hat{\beta}\left(\frac{t}{n} - \frac{n+1}{2n}\right)\right]^2 \Big/ (n-2).$$

If $\rho \neq 0$, then s^2 is a biased estimator of σ^2. Some sampling experiments by Cochrane and Orcutt (1949) suggest that the expected value of s^2 is less than σ^2, although for large n, the bias is negligible.

In testing hypotheses about α and β, a primary concern is with the behavior of the t statistic

$$t = \hat{\beta}\sqrt{\sum \left(\frac{t}{n} - \frac{n+1}{2n}\right)^2} \Big/ \sqrt{s^2}$$

to examine H_0: $\beta = 0$ against the alternative H_1: $\beta \neq 0$ in the presence of the correlation model (17). Table 20 gives the probability that $|t| \geqslant 1.96$ if H_0: $\beta = 0$ is true when the sample size is large (if $\rho = 0$ this probability is $\alpha_N = .05$). Table 20 vividly demonstrates the sensitivity of the standard t test to nonzero correlations between the observations when (15) and (17) hold. The nonrobustness of t comes primarily from the use of an incorrect standard error of $\hat{\beta}$ in the denominator of t. The probability computations in Table 20 also hold when H_0': $\alpha = 0$ is tested against H_1': $\alpha \neq 0$ using the statistic $t = \sqrt{n}\hat{\alpha}/\sqrt{s^2}$. In many circumstances $\rho > 0$; the null hypothesis would be incorrectly rejected more often than the nominal 5 per cent level in such situations.

Table 20 – t tests for regression coefficients: True significance levels when observations are correlated*

	ρ				
	−.40	−.20	0	.20	.40
Probability	.00026	.0164	.0500	.1096	.1994

* $\rho_{st} = \rho^{|s-t|}$; when $\rho = 0$, large sample probability that $|t| \geqslant 1.96$, nominal significance level, $\alpha_N = .05$.

It is important to remember that the magnitude of effects on a statistical technique from dependence among the observations is a function of the technique, the model of dependence, and the values of correlational parameters. The econometric literature (for example, Malinvaud 1964) is rich in discussions of this question.

Tests robust against the correlation structure (17) are discussed by Hannan (1955).

Heteroscedasticity (unequal variances). We now investigate the effects of unequal variances on estimation and hypothesis testing for α and β. At this point, we consider a slightly more general regression model

$$\text{(19)} \quad Y_{ij} = \alpha + \beta X_i + e_{ij}, \quad \text{where } j = 1, \cdots, K, \; i = 1, \cdots, n.$$

Note that the α and β in equation (19) are not the same α and β as in equation (15), and X_i is any predictor variable. Further, we assume that (1) α and β are unknown constants, (2) an independent group of K subjects is observed for each value of X, (3) the e_{ij} are jointly independent random quantities, (4) the random quantities e_{ij} have variances σ_i^2 that may differ for different values of

X, (5) the e_{ij} are normally distributed. It is (4) that provides the major difference from the last section.

Whenever the variance of e_{ij} is σ^2 for all i and j, the ordinary least squares estimators of α and β (designated OLS) are the minimum variance unbiased linear estimators of α and β, and the tests of hypotheses based upon the OLS estimators have optimal properties.

When $\sigma_i^2 \neq \sigma^2$ for some i, then the OLS estimators of α and β remain unbiased, but their variances and the tests and confidence intervals computed on the assumption of equality of variances may be markedly incorrect; also the OLS estimators may not possess efficiency robustness. Jacquez, Mather, and Crawford (1968) examined several models for σ_i^2, various spacings of X on a scale of 1 to 10, and from 6 to 10 distinct values of X. The variance of the OLS estimator of α computed assuming $\sigma_i^2 = \sigma^2$ may be from two to three times the true variance depending upon the number of points (n), the number of replications (K), the model for σ_i^2, and the spacing of points X. On the other hand, the variance of the OLS estimator of β computed assuming $\sigma_i^2 = \sigma^2$ is smaller than the true variance of the estimator. The covariance between the OLS estimators is between 1.5 and 3 times larger (in absolute value) than the true covariance. These results suggest that the usual t tests will not have their stated Type 1 and 2 errors.

We now turn to efficiency considerations. Rao (1970) introduced "minque" estimation to obtain improved estimators for α and β compared to OLS and weighted least squares (WLS) estimators (see Neter & Wasserman 1974). Simulation studies by Rao and Subrahmanian (1971) and Jacquez and Norusis (1973) comparing OLS, minque, and weighted least squares estimators show that the relative efficiencies of the three types of estimators depend on the model for σ_i^2, the pattern of points X, and the relationship between K and n. When $\sigma_i^2 = X_i$, $K > 1$, X is equally spaced on the interval 1 to 10, and n runs from 6 to 10, the OLS estimators are more efficient for $K/n < \frac{1}{2}$, and the minque estimators are more efficient for $1 \geqslant K/n \geqslant \frac{1}{2}$; no results are available for $K/n > 1$. Tests of hypotheses for α and β using minque estimators can be obtained by dividing by the appropriate variances (see Rao & Subrahmaniam 1971). For the case of only one observation at each point X ($K = 1$), Rutemiller and Bowers (1968) derive estimators and tests based on a general model for σ_i^2.

Nonnormality. Suppose that Y_{ij} and X_i are related by equation (19) and that conditions (1)–(4) are true, and that $\sigma_i^2 = \sigma^2$ for all i, but that the assumption (5) that the e_{ij} are normally distributed is no longer plausible. Instead, we suppose that the e_{ij} are sampled from a long-tailed symmetric distribution. Although the OLS estimators of α and β remain unbiased with variances and covariances given by the standard formulas, the estimators need no longer be efficient.

Further, the standard t test of H_0: $\beta = 0$ versus H_1: $\beta \neq 0$ may no longer be efficient or have robustness of validity in small and medium-sized samples. Box and Watson (1962) show that the true significance level of the test may be either much greater or much less than the nominal level, depending on the kurtosis of the distribution of the e_{ij}'s and on the distribution of the independent variable X. Interestingly, if a histogram for the values of the independent variable X is constructed and the histogram can be well approximated by a normal curve (recall that the variable X is not random), then the test will be valid for e_{ij}'s from a wide variety of long-tailed symmetric distributions. (See Box & Draper 1974 for further remarks on the importance of the distribution of the independent variable in controlling robustness in regression problems.)

A wide variety of approaches to obtaining tests and confidence intervals for α and β with reasonable robustness of validity and efficiency for nonnormal e_{ij}'s have appeared in the literature. [*See, for example,* NONPARAMETRIC STATISTICS; *further details are given by* Hollander & Wolfe 1973.] Miller (1974) has extended the jackknife technique to simple and multiple regression problems and obtained tests and confidence intervals.

A promising general approach to obtaining robust estimators of α and β and the regression line as well as robust tests and confidence intervals is the method of M-estimation. Andrews et al. (1972) studied several M-estimators for the population mean. Huber (1973), Hill and Holland (1974), Andrews (1974), and Harter (1975) report on recent work in linear and multiple regression problems. The essential idea underlying M-estimation is the following: while the ordinary least squares estimators for α and β are chosen to minimize $\sum(Y_{ij} - \alpha - \beta X_i)^2$, the M-estimators involve iterative minimization of functions of the residuals, $Y_{ij} - \alpha - \beta X_i$, that place less weight on extreme residuals. Results from studies referred to above indicate that any one of several M-estimators is nearly as efficient as the OLS estimators when the e_{ij} are normally distributed and can be considerably more efficient than the OLS estimators when the e_{ij} come from a long-tailed distribution. Tests and confidence intervals can be based on the M-estimators.

JANET D. ELASHOFF AND ROBERT M. ELASHOFF

BIBLIOGRAPHY

AFIFI, A. A.; ELASHOFF, ROBERT M.; and LANGLEY, P. G. 1968 An Investigation Into the Small Sample Properties of a Two Sample Test of Lehmann's. *Journal of the American Statistical Association* 63:345–363.

AFIFI, A. A., and KIM, P. J. 1972 Comparison of Some Two-sample Location Tests for Non-normal Alternatives. *Journal of the Royal Statistical Society* Series B 34:448–455.

ALI, MUKHTAR M. 1974 Stochastic Ordering and Kurtosis Measure. *Journal of the American Statistical Association* 69:543–545.

ANDREWS, DAVID F. 1974 A Robust Method for Multiple Linear Regression. *Technometrics* 16:523–532.

ANDREWS, DAVID F. et al. 1972 *Robust Estimates of Location: Survey and Advances.* Princeton Univ. Press.

ARVESEN, JAMES N. 1969 Jackknifing U-statistics. *Annals of Mathematical Statistics* 40:2076–2100.

ARVESEN, JAMES N.; and SCHMITZ, THOMAS H. 1970 Robust Procedures for Variance Component Problems Using the Jackknife. *Biometrics* 26:677–686.

BHATTACHARJEE, G. P. 1968 Non-normality and Heterogeneity in Two-sample t-test. *Annals of the Institute of Statistical Mathematics* 20:239–254.

BOX, GEORGE E. P. 1953 Non-normality and Tests on Variance. *Biometrika* 40:318–335.

BOX, GEORGE E. P.; and DRAPER, NORMAN R. 1974 *Robust Designs.* U.S. Army, Mathematics Research Center, Technical Summary Report No. 1420. Madison: Univ. of Wisconsin.

BOX, GEORGE E. P.; and TIAO, GEORGE C. 1964 A Note on Criterion vs. Inference Robustness. *Biometrika* 51: 168–173.

BOX, GEORGE E. P.; and WATSON, G. S. 1962 Robustness to Non-normality of Regression Tests. *Biometrika* 49: 93–106.

BROWN, M. B.; and FORSYTHE, A. B. 1974*a* The Small Sample Behavior of Some Statistics Which Test the Equality of Several Means. *Technometrics* 16:129–132.

BROWN, M. B.; and FORSYTHE, A. B. 1974*b* The ANOVA and Multiple Comparisons for Data With Heterogenous Variances. *Biometrics* 30:719–724.

BROWN, R. A. 1974 Robustness of the Studentized Range Statistic. *Biometrika* 61:171–175.

CHASE, G. R.; and BULGREN, W. G. 1971 A Monte Carlo Investigation of the Robustness of T^2. *Journal of the American Statistical Association* 66:499–502.

COCHRANE, DONALD; and ORCUTT, G. H. 1949 Application of Least Squares Regression to Relationships Concerning Auto-correlated Error Terms. *Journal of the American Statistical Association* 44:32–61.

CROW, EDWIN L.; and SIDDIQUI, M. M. 1967 Robust Estimation of Location. *Journal of the American Statistical Association* 62:353–389.

DANIEL, CUTHBERT; and WOOD, FRED S. 1971 *Fitting Equations to Data: Computer Analysis of Multifactor Data for Scientists and Engineers.* New York: Wiley.

DIXON, WILFRID J.; and MASSEY, FRANK J. JR. (1951) 1969 *Introduction to Statistical Analysis.* 3d ed. New York: McGraw-Hill.

DIXON, WILFRID J.; and TUKEY, JOHN W. 1968 Approximate Behavior of the Distribution of Winsorized t (Trimming/Winsorization 2). *Technometrics* 10:83–98.

DONALDSON, THEODORE S. 1968 Robustness of the F-test to Errors of Both Kinds and the Correlation Between the Numerator and Denominator of the F-ratio. *Journal of the American Statistical Association* 63:660–676.

GARTSIDE, P. S. 1972 A Study of Methods for Comparing Several Variances. *Journal of the American Statistical Association* 67:342–346.

GAYEN, A. K. 1949 The Distribution of "Student's" t in Random Samples of Any Size Drawn From Non-normal Universes. *Biometrika* 36:353–369.

GAYEN, A. K. 1950 The Distribution of the Variance Ratio in Random Samples of Any Size Drawn From Non-normal Universes. *Biometrika* 37:236–255.

GLASS, GENE V.; PECKHAM, P. D.; and SANDERS, J. R. 1972 Consequences of Failure to Meet Assumptions Underlying the Analysis of Variance and Covariance. *Review of Educational Research* 42:237–288.

HAMPEL, FRANK R. 1971 A General Qualitative Definition of Robustness. *Annals of Mathematical Statistics* 42:1887–1896.

HAN, CHIEN-PAI 1968 Testing the Homogeneity of a Set of Correlated Variances. *Biometrika* 55:317–326.

HANNAN, EDWARD J. 1955 An Exact Test for Correlation Between Time Series. *Biometrika* 42:316–326.

HARTER, H. LEON 1975 The Method of Least Squares and Some Alternatives. Part 3. *International Statistical Revue* 43:1–44.

HETTMANSPERGER, THOMAS P. 1973 A Large Sample Conservative Test for Location With Unknown Scale Parameter. *Journal of the American Statistical Association* 68:466–468.

HILL, R. W.; and HOLLAND, P. W. 1974 A Monte Carlo Study of Two Robust Alternatives to Least Squares Regression Estimation. National Bureau of Economic Research Working Paper No. 58. New York: The Bureau.

HOGG, ROBERT V. 1972 More Light on the Kurtosis and Related Statistics. *Journal of the American Statistical Association* 67:422–424.

HOGG, ROBERT V. 1974 Adaptive Robust Procedures: A Partial Review and Some Suggestions for Future Applications and Theory. *Journal of the American Statistical Association* 69:909–923.

HOLLANDER, MYLES; and WOLFE, DOUGLAS A. 1973 *Nonparametric Statistical Methods.* New York: Wiley.

HOLLOWAY, L. N.; and DUNN, O. J. 1967 The Robustness of Hotelling's T^2. *Journal of the American Statistical Association* 62:124–136.

HOPKINS, J. W.; and CLAY, P. P. F. 1963 Some Empirical Distributions of Bivariate T^2 and Homoscedasticity Criterion M Under Unequal Variance and Leptokurtosis. *Journal of the American Statistical Association* 58:1048–1053.

HOTELLING, HAROLD 1961 The Behavior of Some Standard Statistical Tests Under Nonstandard Conditions. Volume 1, pages 319–359 in Berkeley Symposium on Mathematical Statistics and Probability, Fourth, University of California, 1960, *Proceedings.* Berkeley: Univ. of California Press.

HSU, T. C.; and FELDT, L. S. 1969 The Effect of Limitation on the Number of Criterion Score Values on the Significance Level of the F-test. *American Educational Research Journal* 6:515–527.

HUBER, PETER J. 1973 Robust Regression: Asymptotics, Conjectures and Monte Carlo. *Annals of Statistics* 1:799–821. → The 1972 Wald lecture.

ITO, KIYOSHI 1969 On the Effect of Heteroscedasticity

and Non-normality Upon Some Multivariate Test Procedures. Pages 87–120 in International Symposium on Multivariate Analysis, Second, Wright State University, 1968, *Multivariate Analysis: Proceedings*. Edited by Paruchuri R. Krishnaiah. New York: Academic Press.

JACQUEZ, JOHN A.; MATHER, F. J.; and CRAWFORD, C. R. 1968 Linear Regression With Non-constant, Unknown Error Variances: Sampling Experiments With Least Squares, Weighted Least Squares and Maximum Likelihood Estimators. *Biometrics* 24:607–626.

JACQUEZ, JOHN A.; and NORUSIS, MARIJA 1973 Sampling Experiments on the Estimation of Parameters in Heteroscedastic Linear Regression. *Biometrics* 29: 771–780.

JAMES, G. S. 1951 The Comparison of Several Groups of Observations When the Ratios of the Population Variances Are Unknown. *Biometrika* 38:324–329.

JOHNSTON, JOHN 1963 *Econometric Methods*. New York: McGraw-Hill.

LAYARD, M. W. J. 1969 Asymptotically Robust Tests About Covariance Matrices. Ph.D. dissertation, Stanford Univ.

LAYARD, M. W. J. 1972 Large Sample Tests for the Equality of Two Covariance Matrices. *Annals of Mathematical Statistics* 43:123–141.

LAYARD, M. W. J. 1973 Robust Large-sample Tests for Homogeneity of Variances. *Journal of the American Statistical Association* 68:195–198.

LEAVERTON, PAUL; and BIRCH, JOHN J. 1969 Small Sample Power Curves for the Two-sample Location Problem. *Technometrics* 11:299–307.

MALINVAUD, EDMOND (1964) 1970 *Statistical Methods of Econometrics*. 2d ed. New York: American Elsevier. → First published in French.

MEHTA, J. S.; and SRINIVASAN, R. 1970 On the Behrens–Fisher Problem. *Biometrika* 57:649–655.

MILLER, RUPERT G. 1968 Jackknifing Variances. *Annals of Mathematical Statistics* 39:567–582.

MILLER, RUPERT G. 1974 The Jackknife: A Review. *Biometrika* 61:1–16.

NETER, JOHN; and WASSERMAN, WILLIAM 1974 *Applied Linear Statistical Models: Regression, Analysis of Variance, and Experimental Designs*. Homewood, Ill.: Irwin.

OLSON, C. L. 1974 Comparative Robustness of Six Tests in Multivariate Analysis of Variance. *Journal of the American Statistical Association* 69:894–908.

PFANZAGL, J. 1974 On the Behrens–Fisher Problem. *Biometrika* 61:39–47.

POTTHOFF, R. F. 1963 Use of the Wilcoxon Statistic for a Generalized Behrens–Fisher Problem. *Annals of Mathematical Statistics* 34:1596–1599.

PRATT, JOHN W. 1964 Robustness of Some Procedures for the Two-sample Location Problem. *Journal of the American Statistical Association* 59:665–680.

RAO, C. RADHAKRISHNA 1970 Estimation of Heteroscedastic Variances in Linear Models. *Journal of the American Statistical Association* 65:161–172.

RAO, J. N. K.; and SUBRAHMANIAM, KATHLEEN 1971 Combining Independent Estimators and Estimation in Linear Regression With Unequal Variances. *Biometrics* 27:971–990.

RUTEMILLER, HERBERT C.; and BOWERS, DAVID A. 1968 Estimation in a Heteroscedastic Regression Model. *Journal of the American Statistical Association* 63: 552–557.

SCHEFFÉ, HENRY 1959 *The Analysis of Variance*. New York: Wiley.

SCHEFFÉ, HENRY 1970 Practical Solutions of the Behrens–Fisher Problem. *Journal of the American Statistical Association* 65:1501–1508.

SCOTT, A. J.; and SMITH, T. M. F. 1971 Interval Estimates for Linear Combinations of Means. *Applied Statistics* 20:276–285.

SEN, PRANAB KUMAR 1968 On a Further Robustness Property of the Test and Estimator Based on Wilcoxon's Signed Rank Statistic. *Annals of Mathematical Statistics* 39:282–285.

SHORACK, GALEN R. 1969 Testing and Estimating Ratios of Scale Parameters. *Journal of the American Statistical Association* 64:999–1013.

SHORACK, GALEN R. 1974 Random Means. *Annals of Statistics* 2:661–675.

SIDDIQUI, M. M.; and RAGHUNANDANAN, K. 1967 Asymptotically Robust Estimators of Location. *Journal of the American Statistical Association* 62:950–953.

SRIVASTAVA, A. B. L. 1958 Effect of Non-normality on the Power Function of *t*-test. *Biometrika* 45:421–429.

SUBRAHMANIAM, KOCHERLAKOTA; and SUBRAHMANIAM, KATHLEEN 1973 On the Multivariate Behrens–Fisher Problem. *Biometrika* 60:107–111.

SWAMY, P. A. V. B. 1971 *Statistical Inference in Random Coefficient Regression Models*. Lecture Notes in Operations Research and Mathematical Economics No. 55. Berlin and New York: Springer.

THOMPSON, R.; GOVINDARAJULU, Z.; and DOKSUM, K. A. 1967 Distribution and Power of the Absolute Normal Scores Test. *Journal of the American Statistical Association* 62:966–975.

TUKEY, JOHN W. 1962 The Future of Data Analysis. *Annals of Mathematical Statistics* 33:1–67, 812.

U.S. NATIONAL CENTER FOR HEALTH STATISTICS 1972 *Annotated Bibliography on Robustness Studies of Statistical Procedures*. Vital and Health Statistics, Series 2: Data Evaluation and Methods Research, No. 51. Department of Health, Education, and Welfare Publication No. (HSM) 72–1051. Washington: Government Printing Office.

VAN DER VAART, H. ROBERT 1961 On the Robustness of Wilcoxon's Two Sample Test. Pages 140–158 in Symposium on Quantitative Methods in Pharmacology, University of Leiden, 1960, *Quantitative Methods in Pharmacology: Proceedings*. Amsterdam: North-Holland.

WALSH, JOHN E. 1951 A Large Sample *t*-statistic Which Is Insensitive to Non-randomness. *Journal of the American Statistical Association* 46:79–88.

WELCH, B. L. 1938 The Significance of the Difference Between Two Means When the Population Variances Are Unequal. *Biometrika* 29:350–362.

WELCH, B. L. 1947 The Generalization of "Student's" Problem When Several Different Population Variances Are Involved. *Biometrika* 34:28–35.

WETHERILL, G. B. 1960 The Wilcoxon Test and Non-null Hypotheses. *Journal of the Royal Statistical Society* Series B 22:402–418.

YAO, YING 1965 An Approximate Degrees of Freedom Solution to the Multivariate Behrens–Fisher Problem. *Biometrika* 52:139–147.

YUEN, KAREN K. 1974 The Two-sample Trimmed *t* for Unequal Population Variances. *Biometrika* 61:165–170.

YUEN, KAREN K.; and DIXON, WILFRID J. 1973 The Approximate Behaviour and Performance of the Two-sample Trimmed *t*. *Biometrika* 60:369–374.

YUEN, KAREN K.; and MURTHY, V. K. 1974 Percentage Points of the Distribution of the *t* Statistic When the Parent Is Student's *t*. *Technometrics* 16:495–497.

ESTIMATION

I. POINT ESTIMATION — D. L. Burkholder
II. CONFIDENCE INTERVALS AND REGIONS — J. Pfanzagl

I POINT ESTIMATION

How many fish are in this lake? What proportion of the voting population favors candidate A? How much paint is needed for this particular room? What fuel capacity should this airplane have if it is to carry passengers safely between New York and Paris? How many items in this shipment have the desired quality? What is the specific gravity of this metal? Questions like these represent problems of point estimation. In present-day statistical methodology, such problems are usually cast in the following form: A mathematical model describing a particular phenomenon is completely specified except for some unknown quantity or quantities. These quantities must be estimated. Galileo's model for freely falling bodies and many models in learning theory, small group theory, and the like provide examples.

Exact answers are often impossible, difficult, expensive, or merely inconvenient to obtain. However, approximate answers that are quite likely to be close to the exact answer may be fairly easily obtainable. The theory of point estimation provides a guide for obtaining such answers; above all, it makes precise, or provides enough framework so that one could make precise, such phrases as "quite likely to be close" and others such as "this estimator is better than that one."

As an introduction to some of the problems involved, consider estimating the number N of fish in a given lake. Suppose that M fish are taken from the lake, marked, and returned to the lake unharmed. A little later, a random sample of size n of fish from the lake is observed to contain x marked fish. A little thought suggests that probably the ratio x/n is near M/N or that the unknown N and the ratio Mn/x (defined only if $x > 0$) are not too far apart. For example, if $M = 1{,}000$, $n = 1{,}000$, and $x = 20$, it might be reasonable to believe that N is close to 50,000. [*A similar example, concerning moving populations of workers, is discussed in* SAMPLE SURVEYS.]

Clearly, this procedure *may* lead one badly astray. For example, it is possible, althoughly highly unlikely, that the same value $x = 20$ could be obtained, and hence, using the above procedure, N be estimated as 50,000, even if N is actually as small as 1,980 or as large as 10,000,000. Clearly, considerations of probability are basic here. If $L(N)$ denotes the probability of obtaining 20 marked fish when N fish are in the lake, it can be shown that $0 = L(1{,}979) < L(1{,}980) < \cdots < L(49{,}999) = L(50{,}000)$ and $L(50{,}000) > L(50{,}001) > \cdots$; that is, $N = 50{,}000$ maximizes the *likelihood* of obtaining 20 marked fish.

Design of experiments. What values of M and n are most satisfactory in the above experiment? Clearly, the bigger n is, the better it is for estimation purposes, but the more expensive the experiment [*see* EXPERIMENTAL DESIGN]. A balance has to be reached between the conflicting goals of minimizing error and minimizing expense. Also, perhaps another experimental design might give better results. In the above problem, let $M = 1{,}000$, but instead of pulling a fixed number of fish out of the lake, pull out fish until exactly x marked fish have been obtained, where x is fixed in advance. Then n, the sample size, is the observation of interest [*see* SEQUENTIAL ANALYSIS]. Which design, of all the possible designs, should be used? This kind of question is basic to any estimation problem.

Testing hypotheses. An altogether different problem would arise if one did not really want the value of N for its own sake but only as a means of deciding whether or not the lake should be restocked with small fish. For example, it might be desirable to restock the lake if N is small, say less than 100,000, and undesirable otherwise. In this case, the problem of whether or not the lake should be restocked is equivalent to testing the hypothesis that N is less than 100,000 [*see* HYPOTHESIS TESTING]. In general, a good estimator does not necessarily lead to a good test.

Confidence intervals. The value of an estimator, that is, a point estimate, of N for a particular sample is a number, hopefully one close to N; the value of a confidence interval, that is, an interval estimate, of N for a particular sample is an interval, hopefully one that is not only small but that also contains N [*see* ESTIMATION, *article on* CONFIDENCE INTERVALS AND REGIONS]. The problem of finding a good interval estimate is more closely related to hypothesis testing than it is to point estimation.

Note that certain problems are clearly point estimation problems rather than problems of interval estimation: when deciding what the fuel capacity of an airplane should be, the designers must settle on one particular number.

Steps in solving an estimation problem

The first step in the solution of an estimation problem, as suggested above, is to design an experiment (or method of taking observations) such that the outcome of the experiment—call it x—is affected by the unknown quantity to be estimated, which in the above discussion was N. Typically, x is related to N probabilistically rather than deterministically. This probability relation must be specified. For example, the probability of obtaining x marked fish in a sample of size n is given by the hypergeometric distribution,

$$\binom{M}{x}\binom{N-M}{n-x}\Big/\binom{N}{n},$$

provided the sample has been drawn randomly without replacement [*see* DISTRIBUTIONS, STATISTICAL, *article on* SPECIAL DISCRETE DISTRIBUTIONS]. (The denominator is the number of combinations of N things taken n at a time, and so forth.) If the randomness assumption is not quite satisfied, then the specified probability relation will be only approximately true. Such specification problems and their implications will be discussed later. Next, after the experiment has been designed and the probability model specified, one must choose a function f defined for each possible x such that if x is observed, then $f(x)$, the value of the function f at x, is to be used as a numerical estimate of N. Such a function f is called an *estimator* of N. The problem of the choice of f will be discussed later. Finally, after a particular estimator f has been tentatively settled on, one might want to calculate additional performance characteristics of f, giving further indications of how well f will perform on the average. If the results of these calculations show that f will not be satisfactory, then changes in the design of the experiment, for example, an increase in sample size, might be contemplated. Clearly, there is a good deal of interplay among all the steps in the solution of an estimation problem outlined here.

Terminological note. Some authors distinguish terminologically between the *estimator*, the function f, and its numerical value for a particular sample, the *estimate*. Another distinction is that between a random variable and a generic value of the random variable. (Some authors use X for the former and x for the latter.) Such distinctions are sometimes important, but they are not generally made in this article, although special comments appear in a few places. Otherwise it should be clear from context whether reference is made to a function or its value, or whether reference is made to a random variable or its value.

Choice of estimator

As a means of illustrating the various considerations influencing the choice of an estimator, a few typical examples will be discussed.

Example 1. Let x be the number of successes in n independent trials, the probability of a success on an individual trial being p. (For example, x might be the number of respondents out of n questioned in a political poll who say they are Democrats, and p is the probability that a randomly chosen individual in the population will say he is a Democrat.) Here p is unknown and may be any number between 0 and 1 inclusive. An estimator f of p ideally should be such that $f(x)$ is close to p no matter what the unknown p is and no matter what the observation x is. That is, the error $f(x)-p$ committed by using $f(x)$ as an approximation to p should always be small. This is too much to expect since x can, by chance, be quite misleading about p. However, it is not too much to expect that the error be small in some average sense. For example, the mean squared error,

$$E_p(f-p)^2=\sum_{x=0}^{n}[f(x)-p]^2\binom{n}{x}p^x(1-p)^{n-x},$$

should be small no matter what the unknown p is, or the mean absolute error $E_p|f-p|$ should be small no matter what p is, or the like. For the time being, estimators will be compared only on the basis of their mean squared errors. A more general approach, the underlying ideas of which are well illustrated in this special case, will be mentioned later. The first question that arises is, Can one find an estimator f such that, for every p satisfying $0\leqslant p\leqslant 1$, the mean squared error of f at p is smaller than (or at least not greater than) the mean squared error at p of any other estimator? Obviously, such an estimator would be best in this mean squared error sense. Unfortunately, and this is what makes the problem of choosing an estimator a nontrivial problem, a best estimator does not exist. To see this, consider the estimates f_1 and f_2 defined by $f_1(x)=x/n$ and $f_2(x)=\frac{1}{2}$. It is not hard to show that $E_p(f_1-p)^2=p(1-p)/n$, and clearly, $E_p(f_2-p)^2=(\frac{1}{2}-p)^2$. If a best estimator f existed it would have to satisfy $E_p(f-p)^2\leqslant E_p(f_2-p)^2$. But the latter quantity is zero for $p=\frac{1}{2}$, implying that $f=f_2$. However, f_2 is not best since $E_p(f_1-p)^2$ is smaller than $E_p(f_2-p)^2$ for p near 0 or 1.

Although no best estimator exists, many good estimators exist. For example, there are many estimators f satisfying $E_p(f-p)^2\leqslant 1/(4n)$ for $0\leqslant p\leqslant 1$. The estimator f_1, defined above, is such an estimator. The estimator f_3 defined by $f_3(x)=[\sqrt{n}(x/n)+\frac{1}{2}]/(\sqrt{n}+1)$ with mean squared error $E_p(f_3-p)^2=1/[4(1+\sqrt{n})^2]$ is another. If n is large, the mean squared error of any such estimator is small for each possible value of p. In this problem, as is typical, any one of many good available estimators would no doubt be reasonable to use in practice. Only by adding further assumptions, for example, assumptions giving some information about the unknown p, can the class of reasonable estimators be narrowed. Note that estimators are still being compared on the basis of their mean squared errors only.

The estimator f_3 is *minimax* in the sense that f_3 minimizes $\max_{0 \leqslant p \leqslant 1} E_p(f-p)^2$ with respect to f. The minimax approach focuses attention on the worst that can happen using f and chooses f accordingly [*see* DECISION THEORY]. Note that the estimator f_1 $(f_1(x) = x/n)$ does have slightly larger mean squared error than does f_3 for values of p near $\frac{1}{2}$; for values of p near 0 or 1 the advantage lies wholly with f_1. Other properties of these estimators will be discussed later.

Example 2. Suppose that $x_1, x_2, \cdots, x_n$ are observations on n independent random variables, each having the Poisson distribution with parameter λ, where λ is unknown and may be any nonnegative number [*see* DISTRIBUTIONS, STATISTICAL, *article on* SPECIAL DISCRETE DISTRIBUTIONS]. For example, x_k could be the number of occurrences during the kth time interval of unit length of any phenomenon occurring "randomly" over time, possibly telephone calls coming into an exchange, customers coming into a store, and so forth [*see* QUEUES]. Knowing that λ is both the mean and variance of the Poisson distribution, it might not be unreasonable to suppose that both the sample mean,

$$m(x) = m(x_1, \cdots, x_n) = \sum_{k=1}^{n} x_k/n = \bar{x},$$

and the sample variance,

$$s^2(x) = \sum_{k=1}^{n} (x_k - \bar{x})^2/(n-1)$$

(here, one must assume that $n > 1$), provide good estimators of the unknown λ. It is not hard to show that m is *better* than s^2, that is, $E_\lambda(m-\lambda)^2 \leqslant E_\lambda(s^2-\lambda)^2$ for all $\lambda \geqslant 0$, with strict inequality for some $\lambda \geqslant 0$.

An estimator is *inadmissible* (with respect to a given criterion like mean squared error) if a better one exists; accordingly, the estimator s^2 is here inadmissible. An estimator is admissible if it is not inadmissible. Although it is not obvious, the estimator m is admissible. In fact the class of admissible estimates is very large here, as is typically the case. In example 1, all three estimators discussed, f_1, f_2, and f_3, are admissible.

Example 3. Let $x_1, x_2, \cdots, x_n$ be observations on n independent random variables each having the normal distribution with mean μ and variance σ^2, where both μ and σ^2 are unknown; μ may be any real number and σ^2 may be any positive number. One might be interested in estimating only μ, only σ^2, the pair (μ, σ^2), or perhaps some combination such as μ/σ.

Example 4. Let $x_1, x_2, \cdots, x_n$ be observations on n independent random variables each having the uniform distribution over the set of integers $\{1, 2, \cdots, N\}$, where N may be any positive integer. For example, in a state where automobile license plates are numbered from 1 to N, each x_i would be the number of a randomly chosen license plate. What is a good estimator of N?

Sufficient statistics. A simple and effective way to narrow the class of estimators that one ought to consider when choosing a good estimator is to identify a sufficient statistic for the problem and to consider only those estimators that depend on the sufficient statistic [*see* SUFFICIENCY]. Roughly speaking, if t is a sufficient statistic, knowing $t(x)$ is as useful as knowing x. The following result is important. If t is a sufficient statistic and f is an estimator with finite mean squared error and f does not depend on t (that is, f is not essentially expressible as $f = h(t)$ for some function h), then there is another estimator f_0 that does depend on t and such that f_0 is better than f (in the technical sense defined above). One f_0 that works is the conditional expectation of f relative to t.

In example 2, m is a sufficient statistic, hence only estimators depending on m need be considered. In particular s^2, which does not depend on m in the sense defined above, need not be considered. In example 3, the ordered pair (m, s^2), where m and s^2 are defined as in example 2, is a sufficient statistic. In example 4, the estimator $2m$ might seem at first to be a plausible estimator of N. However, it does not depend on the sufficient statistic u defined by $u(x) =$ the largest of the x_k. Much better estimators than $2m$ exist. For example, the rather complicated

$$f_4 = \frac{u^{n+1} - (u-1)^{n+1}}{u^n - (u-1)^n}$$

is such an estimator. (Note that f_4 is approximately equal to $(n+1)u/n$.)

Further criteria for choice of estimator. So far estimators have been compared on the basis of their mean squared errors only. Since no best estimator exists, a unique solution to the problem of choosing an estimator is generally not obtainable by this approach. This is not really too regrettable since many good estimators usually exist. Even demanding that an estimator be minimax, not necessarily always a reasonable demand, does not always lead to a unique estimator. In example 2, every estimator of λ has unbounded mean squared error; and in example 3 every estimator of μ has unbounded mean squared error. Hence, in these two examples, all estimators of λ and μ, respectively, are minimax, but the concept loses all interest. In example 1, demanding minimaxity does lead to the unique minimax estimator f_3. A unique minimax estimator is clearly admissible.

The strong intellectual and psychological tendency of human beings to be satisfied only with unique answers has often led to further demands being placed on estimators in addition to the one that their mean squared errors be small.

Unbiasedness. An estimator is *unbiased* if the mean value of the estimator is equal to the quantity being estimated. In example 1, f_1 is unbiased since $E_p f_1 = p$, $0 \leqslant p \leqslant 1$. Both m and s^2 are unbiased estimators of λ in example 2. In example 3, m is an unbiased estimator of μ and s^2 is an unbiased estimator of σ^2. In example 4, both $2m$ and f_4 are unbiased estimators of N. The search for a best unbiased estimator often leads to a unique answer. In example 2, an estimator f would be best unbiased (or minimum variance unbiased) if it is unbiased and satisfies

$$E_\lambda(f-\lambda)^2 \leqslant E_\lambda(f^*-\lambda)^2$$

for every $\lambda \geqslant 0$ and every unbiased estimator f^*. The estimator m is such an estimator for this problem and is the only such estimator. The estimator f_1 is the best unbiased estimator of p in example 1; the estimator m is the best unbiased estimator of μ in example 3; and the estimator f_4 is the best unbiased estimator of N in example 4. The *relative efficiency* of two unbiased estimators is the ratio of their reciprocal variances. Relative efficiency may well depend on the parameter value.

Unbiased estimators fail to exist in some important problems. Using the first design mentioned in the problem of estimating the number of fish in a lake, N, no unbiased estimator of N exists. Although in example 3, s^2 is a best unbiased estimator of σ^2, another estimator of σ^2, $(n-1)s^2/(n+1)$, despite being biased, is actually better than s^2 in the sense of mean squared error. This shows that placing extra demands on estimators can actually come into conflict with the small mean squared error demand. Of course, the relative importance of the various properties an estimator may have will no doubt be judged slightly differently by different reasonable individuals.

Invariance. Notions of invariance can sometimes be invoked so that a best invariant estimator exists. For example, if $x = (x_1, \cdots, x_n)$, b is a real number, and $y = (x_1 + b, \cdots, x_n + b)$, the estimator m is invariant in the sense that it satisfies $m(y) = m(x) + b$. It turns out that among all the estimators of μ in example 3 with this property of scale invariance, the estimator m is best in the usual mean squared error sense. The argument for invariance may be stated rather loosely as follows. Irrelevancies in the data (for example, whether time is measured from 12 noon New York time or from 12 noon Greenwich time) should not make a fundamental difference in the results obtained from the analysis of the data.

A different kind of invariance problem can be troublesome in some circumstances. Suppose in example 1 that interest centers not on p but on some function of p, say $1/p$. If f is a satisfactory estimator of p, it need not follow that $1/f$ is a satisfactory estimator of $1/p$, for properties like unbiasedness, mean squared error functions, etc., can change drastically under nonlinear transformations. Fortunately, in many problems the parameter itself, or a single function of it, is of central interest, so that this kind of noninvariance is not serious.

Specification problems. So far, estimators have been chosen relative to given probability models. If an estimator seems satisfactory for a given probability model, it may be relevant to ask if this estimator is also good for probability models closely related to the given one. For example, it is too much to expect that a model postulating normal distributions describes exactly the practical situation of interest. Fortunately, in many common problems slight changes in the probability model will not materially affect the goodness of an estimator reasonable for the original model [*see* ERRORS, *article on the* EFFECTS OF ERRORS IN STATISTICAL ASSUMPTIONS]. For example, the estimator m of μ in example 3 is actually a fairly reasonable estimator of the population mean μ in a large variety of cases, particularly if the population variance σ^2 is finite and the sample size n is not too small, as can be seen from the formula for its mean squared error, σ^2/n. Circumstances arise, however, in which alternative estimators, for example, the sample median, not so much affected by slight changes in the tails of the distribution, may need to be considered. [*A process for arriving at other such estimators, called Winsorization, is discussed in* NONPARAMETRIC STATISTICS, *article on* ORDER STATISTICS; *the closely related concept of trimming is discussed in* ERRORS, *article on the*

EFFECTS OF ERRORS IN STATISTICAL ASSUMPTIONS.]

More than one parameter. Most of the material of this article deals with estimation of a single parameter. The multiparameter case is, of course, also important and multiparameter analogues of all the topics in this article exist. They are treated in the references, for example by Kendall and Stuart (1961), Cramér (1945), and Wilks (1962).

Constructive estimation methods

Maximum likelihood estimators. In example 1, the estimator f_1 is the maximum likelihood estimator of p: For each x, $f_1(x)$ is that value of p maximizing $\binom{n}{x} p^x(1-p)^{n-x}$, the probability of obtaining x. In example 2, m is the maximum likelihood estimator of λ. In example 3, $(m,[n-1]s^2/n)$ is the maximum likelihood estimator of (μ,σ^2). In example 4, u is the maximum likelihood estimator of N. In the problem of estimating the number of fish in a lake, N, using the first design, no maximum likelihood estimator exists since no such estimate can be defined for $x = 0$, although for $x > 0$ no trouble occurs. In some examples, there is no unique maximum likelihood estimator.

Maximum likelihood estimators are often easy to obtain. A maximum likelihood estimator does not necessarily have small mean squared error nor is it always admissible. So the maximum likelihood principle can sometimes conflict with the small mean squared error principle. Nevertheless, maximum likelihood estimators are often quite good and worth looking at. If the sample size is large, they tend to behave nearly as nicely as the estimator m of μ in example 3.

Maximum likelihood estimation is often constructive, that is, the method provides machinery that often gives a unique estimating function. There are other constructive methods, three of which are described here: the method of moments, least squares, and Bayes estimation. One or another of these constructive methods may provide a simpler or a better behaved estimator in any particular case.

The method of moments. The approach of the method of moments (or of expected values) is to set one or more sample moments equal to the corresponding population moments and to "solve," if possible, for the parameters, thus obtaining estimators of these parameters. The method is particularly appropriate for simple random sampling. In example 1, if the sample is regarded as made up of n observations, the kth being a 1 (success) or 0 (failure), the sample mean is x/n and the population mean is p, so the resulting method of moments estimator is x/n. In example 4, the method of moments, as it would ordinarily be applied, leads to a poor estimator. The method can, nonetheless, be very useful, especially in more complex cases with several parameters.

Least squares. The least squares approach is especially useful when the observations are not obtained by simple random sampling. One considers the formal sum of squares $\sum_k(x_k - EX_k)^2$, where x_k is an observation on the random variable X_k with expectation EX_k (depending on the parameters to be estimated). Then one attempts to minimize the sum of squares over possible values of the parameters. If a unique minimum exists, the minimizing values of the parameters are the values of their *least square estimators*.

The method is particularly appropriate when the X_k are independent and identically distributed except for translational shifts that are given functions of the parameters. If the X_k all have the same expectation, as in examples 1–4, the least squares estimator of that expectation is the sample mean. Least squares estimation, without modification or extension, does not provide estimators of parameters (like σ^2 in example 3) that do not enter into expectations of observations. [*A fuller treatment of this topic appears in* LINEAR HYPOTHESES, *article on* REGRESSION.]

Bayes estimation. Consider example 1 again, this time supposing that the unknown p is itself the outcome of some experiment and that the probability distribution underlying this experiment is known. For example, x could be the number of heads obtained in n tosses of a particular coin, the probability of a head for the particular coin being p where p is unknown, but where the coin has been picked randomly from a population of coins with a known distribution of p values. Then it would be reasonable to choose an estimator f that minimizes the mean value of the squared error $[f(x) - p]^2$ where the averaging is done with respect to the *known* joint distribution of x and p. Such a minimizing estimator is called a *Bayes estimator*; of course it depends on the distribution assigned to p [*see* BAYESIAN INFERENCE]. A distribution may be assigned to p merely as a technical device for obtaining an estimator and completely apart from the question of whether p actually is the outcome of an experiment. This is the spirit in which Bayes estimators are often introduced, as a way of obtaining an estimator that may or may not have good properties. On the other hand, one may assign a distribution to p in such a way that those values of p that seem more likely to obtain are given greater weight. Of course, different individuals might assign different distributions, for this is a matter of judgment. However, this approach does provide one possible method for using any previously obtained information about p that may be available. It would be rather rare that the *only* information available about p before the experiment is that $0 \leqslant p \leqslant 1$.

Examples of Bayes estimators include the estimator f_3 of example 1, obtained by assigning a certain beta distribution to p, and the estimator f_5 of p, defined by $f_5(x) = (x+1)/(n+2)$, obtained by assigning to p the uniform distribution on the interval between 0 and 1. [*See* DISTRIBUTIONS, STATISTICAL, *article on* SPECIAL CONTINUOUS DISTRIBUTIONS, *for discussions of these specific distributions.*] Even f_2 is a Bayes estimator. However, f_1 is not a Bayes estimator but is rather the limit of a sequence of Bayes estimators.

Restricting attention to estimators that are Bayes or the limits (in a certain sense) of sequences of Bayes estimators usually assures one of not overlooking any admissible estimator. Bayes methods frequently prove useful as technical devices in solving for minimax estimators and in many other situations.

Asymptotic estimation theory

Because it is often difficult to compare estimators for small sample sizes, much research on point estimation is in terms of large sample sizes, working with limits as the sample size goes to infinity. In this context, an estimator itself is not considered, but rather a *sequence* of estimators, each member of which corresponds to a single sample size. For example, consider the sequence of sample means $m_1, m_2, \cdots$, where $m_n(x_1, \cdots, x_n) = \sum_{k=1}^{n} x_k/n$. If a sequence of estimators has desirable properties in a limiting large sample sense, it is often presumed that particular members of the sequence will to some extent partake of these desirable properties.

Consistency. An asymptotic condition that is often regarded as essential is that of *consistency*, in the sense that the sequence of estimators is close to the true value of the parameter, with high probability, for large sample sizes. More precisely if $\{t_n\}$ is the sequence of estimators, and if θ is the parameter being estimated, the sequence $\{t_n\}$ is said to estimate θ consistently if, for every interval I containing θ in its interior, the probability that the value of t_n belongs to I approaches 1 as n approaches infinity, no matter what the value of θ is. (There is also a nonasymptotic concept of consistency, closely related to the above. Both ideas, and their applications, originated with R. A. Fisher.)

Comparison of estimators. For simplicity, consider now independent identically distributed random variables with common distribution depending on a single parameter, θ. Let ϕ_θ be the density function (or frequency function) corresponding to that common distribution for the parameter value θ. A large number of regularity conditions are traditionally, and often tacitly, imposed on ϕ_θ; for example, distributions like those of example 4 do not come under the standard theory here. In this brief summary, the regularity conditions will not be discussed. With almost no modifications, the discussion applies to qualitative, as well as numerically valued, random quantities.

Two sequences of estimators, competing as estimators of θ, are often compared by considering the ratios of their asymptotic variances, that is, the variances of limit distributions as n approaches infinity. In particular, one or both sequences may have the lowest possible asymptotic variance. In discussing such matters, the following constructs, invented and named by R. A. Fisher, are important.

Score, Fisher information, and efficiency. The *score* of the single observation x_k is a function of both x_k and θ, defined by

$$s_\theta(x_k) = \frac{\partial}{\partial\theta} \ln \phi_\theta(x_k),$$

and it provides the relative change in ϕ (for each possible value of x_k) when θ is slightly changed. Two basic facts about the score are $E_\theta s_\theta = 0$, $\operatorname{var}_\theta s_\theta = -E_\theta(\partial s_\theta/\partial\theta)$. The quantity, $-E(\partial s_\theta/\partial\theta)$, is often called the *Fisher information* contained in a single observation and is denoted by $I(\theta)$.

For the entire sample, $x = \{x_1, x_2, \cdots, x_n\}$, the *sample score* is just the sum of the single observation scores,

$$s_{\theta n}(x) = \sum_{k=1}^{n} s_\theta(x_k).$$

The Fisher information $I_n(\theta)$ contained in the entire sample is defined as above with $s_{\theta n}$ replacing s_θ; it is just the sum of the Fisher information values for the n single observations. Under the assumptions, each observation contributes the same amount to total information—that is, $I(\theta)$ is the same for each observation—so that $I_n(\theta) = n\,I(\theta)$.

Except for sign, $I_n(\theta)$ is the curvature of the likelihood function near the true value of θ. Roughly speaking, sharp curvature of the likelihood function corresponds to sharper estimation, or lower variance of estimation. The *information inequality* says that, for sequences of estimators $\{t_n\}$ such that $\sqrt{n}(t_n - \theta)$ converges in distribution to a distribution with mean zero and variance σ^2,

$$\sigma^2 \geqslant \frac{1}{I(\theta)}.$$

Nonasymptotic variants of this inequality have been explored by Darmois, Dugué, Cramér, Rao, and others. The basic variant, for an unbiased estimator t_n, based on a sample of size n, is

$$\text{var}_\theta t_n \geqslant \frac{1}{I_n(\theta)}.$$

(This is usually called the Cramér–Rao inequality.) Under the tacit regularity conditions, this inequality becomes an equality just when

$$t_n = \theta + \frac{s_{\theta n}}{I_n(\theta)}.$$

This can happen only if the right side is not a function of θ, and this in turn occurs (under regularity) when and only when the distributions given by ϕ_θ form an exponential family [*see* DISTRIBUTIONS, STATISTICAL, *article on* SPECIAL CONTINUOUS DISTRIBUTIONS].

The maximum likelihood estimator of θ based on $x = (x_1, \cdots, x_n)$, say $\hat{\theta}_n$, is (under regularity) the solution of the *likelihood equation* $s_{\theta n}(x) = 0$. Under these circumstances, $s_{\theta n}/\sqrt{I_n(\theta)}$ and $\sqrt{I_n(\theta)}\,[\hat{\theta}_n - \theta]$ are both asymptotically normal with zero mean and variance unity. Further, the difference between these two quantities converges to zero in probability as n increases.

Thus the maximum likelihood estimator is *asymptotically efficient*, in the sense that its asymptotic variance is as low as possible, for it satisfies the asymptotic information inequality. In general there exist other (sequences of) estimators also satisfying the information inequality; these are called *regular best asymptotically normal* (RBAN) estimators. The RBAN estimators are those that are indistinguishable from the maximum likelihood estimator in terms of asymptotic distribution, as it is traditionally construed. Often some RBAN estimator distinct from the maximum likelihood estimator is easier to compute and work with.

The word "regular," used above, refers in part to regularity conditions on the estimators themselves, considered as functions of the sample. Without that restriction, somewhat strange *superefficient* estimators can be constructed.

The concept of asymptotic treatment has been extended recently in other directions than those summarized above, in particular by the work of R. R. Bahadur and C. R. Rao.

A more general approach to estimation

So far the discussion has been based largely on comparing estimators through their mean squared errors. The mean absolute error could, of course, have been used. More generally, suppose that $W(\theta,d)$ is the *loss* incurred when the numerical estimate d is used as if it were the value $g(\theta)$. Here θ is the unknown parameter of the probability distribution underlying the outcome x of an experiment, and $g(\theta)$ is to be estimated. If f is an estimator, x has been observed, and $f(x)$ is used as if it were the value of $g(\theta)$, then the loss incurred is $W[\theta,f(x)]$. The mean loss, $E_\theta W(\theta,f)$, denoted by $r(\theta,f)$, a function of both θ and f, is of interest. The function r is called the *risk function*. Now such terms as *better, admissible, minimax, Bayes*, and so forth could be defined using the risk function, r, rather than mean squared error. For example, f is better than f^* (relative to the loss W) if $r(\theta,f) \leqslant r(\theta,f^*)$ for all θ with strict inequality for some θ [*see* DECISION THEORY].

In the earlier discussion, W was taken to be $W(\theta,d) = [d - g(\theta)]^2$ and r was therefore mean squared error.

In the more general multiparameter context mentioned earlier, $\boldsymbol{\theta}$ is a vector of more than one ordinary (scalar) parameter, and so may be $\boldsymbol{g}(\boldsymbol{\theta})$, the quantity to be estimated. For example, in example 3, $\boldsymbol{\theta} = (\mu,\sigma^2)$, $\boldsymbol{g}(\boldsymbol{\theta})$ could be $\boldsymbol{\theta}$, and $W(\boldsymbol{\theta},\boldsymbol{d})$ could be $(d_1 - \mu)^2 + (d_2 - \sigma^2)^2$, where $\boldsymbol{d} = (d_1,d_2)$ is an ordered pair of real numbers. Or consider the following example in which an infinite number of quantities are simultaneously estimated.

Example 5. Let $x_1, x_2, \cdots, x_n$ be observations on n independent random variables each having the same distribution function F, where F may be any distribution function on the real line. The problem is to estimate the whole function F, that is, to estimate $F(a)$ for each real number a. Here $\boldsymbol{\theta} = F$, $\boldsymbol{g}(\boldsymbol{\theta}) = F$, $\boldsymbol{d}$ may be any distribution function, and $W(\boldsymbol{\theta},\boldsymbol{d})$ may be given, for example, by $\sup |d(a) - F(a)|$, where the supremum (least upper bound) is taken over all real a. A quite satisfactory estimator, the sample distribution function, exists here. For large n, its risk function is near 0. For $x = (x_1, x_2, \cdots, x_n)$, the value of the sample distribution function is that distribution that places probability $1/n$ on each of $x_1, x_2, \cdots, x_n$ if these values are distinct, with the obvious differential weighting otherwise.

One difficulty with the more general approach to estimation outlined here is that the loss function W is often hard to define realistically, that is, in such a way that $W(\boldsymbol{\theta},\boldsymbol{d})$ approximates the actual loss incurred when $\boldsymbol{d}$ is used as if it were the value of $\boldsymbol{g}(\boldsymbol{\theta})$. Fortunately, an estimator that is good relative to one loss function, say squared error, is often good relative to a wide class of loss functions.

Perhaps the key concept in estimation theory is *better*. Once it has been decided what "this estimator is better than that one" should mean, a large part of the theory follows naturally. Many definitions of *better* are possible. Several others besides the one mentioned here appear in the literature, but none has been so deeply investigated.

History

The theory of point estimation has a long history and a huge literature. The Bernoullis, Moivre, Bayes, Laplace, and Gauss contributed many important ideas and techniques to the subject during the eighteenth century and the early part of the nineteenth century. Karl Pearson stressed the method of moments and the importance of computing approximate variances of estimators. During the early twentieth century, no one pursued the subject with more vigor than R. A. Fisher. His contributions include the development of the maximum likelihood principle and the introduction of the important notion of sufficiency. Neyman's systematic study of interval estimation appeared in 1937. Although the possibility of a loss function approach to statistical problems had been mentioned by Neyman and E. S. Pearson in 1933, its extensive development was not initiated until the work of Abraham Wald in 1939 [*see the biographies of* BAYES; BERNOULLI FAMILY; FISHER, R. A.; GAUSS; LAPLACE; MOIVRE; PEARSON; WALD].

New and nonstandard estimation problems requiring new and nonstandard techniques of solution will no doubt continue to arise. Remarkable solutions to two such problems have recently been proposed under the general name of *stochastic approximation* [*see* SEQUENTIAL ANALYSIS].

Ideally, scientific constructs should possess not only great explanatory power but simplicity as well. The search for both will, no doubt, encourage more and more mathematical model building in the social sciences. Moreover, it is quite likely that these models will have to become more and more probabilistic if they are to achieve these aims. As a consequence, the statistical problems involved, checking the goodness of fit of the model, estimating the unknown parameters, and so forth, will have to be handled with ever-increasing care and knowledge.

D. L. BURKHOLDER

[*See also* STATISTICS, DESCRIPTIVE.]

BIBLIOGRAPHY

Many elementary textbooks on statistical theory discuss the rudiments of point estimation, for example, Hodges & Lehmann 1964. *Fuller treatments will be found in* Cramér 1945, Wilks 1962, *and* Kendall & Stuart 1961. *Large sample theory is treated at length in* LeCam 1953. *Further discussion of estimation from the loss function point of view will be found in Chapter 5 of* Wald 1950. Lehmann 1959 *treats sufficiency and invariance in some detail. Chapter 15 of* Savage 1954 *contains many illuminating comments on the problem of choosing a good estimator.*

CRAMÉR, HARALD (1945) 1951 *Mathematical Methods of Statistics.* Princeton Mathematical Series, No. 9. Princeton Univ. Press.

FISHER, R. A. (1922) 1950 On the Mathematical Foundations of Theoretical Statistics. Pages 10.308*a*–10.368 in R. A. Fisher, *Contributions to Mathematical Statistics.* New York: Wiley. → First published in Volume 222 of the *Philosophical Transactions,* Series A, of the Royal Society of London.

FISHER, R. A. (1925) 1950 Theory of Statistical Estimation. Pages 11.699*a*–11.725 in R. A. Fisher, *Contributions to Mathematical Statistics.* New York: Wiley. → First published in Volume 22 of the *Proceedings* of the Cambridge Philosophical Society.

HODGES, JOSEPH L. JR.; and LEHMANN, E. L. 1964 *Basic Concepts of Probability and Statistics.* San Francisco: Holden-Day.

○KENDALL, MAURICE G.; and STUART, ALAN (1961) 1973 *The Advanced Theory of Statistics.* Volume 2: *Inference and Relationship.* 3d ed. London: Griffin; New York: Hafner. → Early editions of Volume 2, first published in 1946, were written by Kendall alone. Stuart became a joint author on later, renumbered editions in the three-volume set.

KIEFER, J.; and WOLFOWITZ, J. 1952 Stochastic Estimation of the Maximum of Regression Function. *Annals of Mathematical Statistics* 23:462–466.

LECAM, LUCIEN 1953 On Some Asymptotic Properties of Maximum Likelihood Estimates and Related Bayes' Estimates. California, University of, *Publications in Statistics* 1:277–329.

LEHMANN, ERICH L. 1959 *Testing Statistical Hypotheses.* New York: Wiley.

NEYMAN, JERZY 1937 Outline of a Theory of Statistical Estimation Based on the Classical Theory of Probability. Royal Society of London, *Philosophical Transactions* Series A 236:333–380.

PITMAN, E. J. G. 1939 The Estimation of the Location and Scale of Parameters of a Continuous Population of Any Given Form. *Biometrika* 30:391–421.

ROBBINS, HERBERT; and MONRO, SUTTON 1951 A Stochastic Approximation Method. *Annals of Mathematical Statistics* 22:400–407.

SAVAGE, LEONARD J. 1954 *The Foundations of Statistics.* New York: Wiley.

WALD, ABRAHAM 1939 Contributions to the Theory of Statistical Estimation and Testing Hypotheses. *Annals of Mathematical Statistics* 10:299–326.

WALD, ABRAHAM (1950) 1964 *Statistical Decision Functions.* New York: Wiley.

WILKS, SAMUEL S. 1962 *Mathematical Statistics.* New York: Wiley.

Postscript

There have been many new developments in the theory and practice of point estimation since the main article was prepared. One line of development, the search for robust estimates, is in response to the increasing demand for statistical tools that work reasonably well even if the probability mechanism generating the data is not quite normal or otherwise as expected. The remarkable book by Andrews et al. (1972) contains a discussion of the problem of robust estimation from several different points of view and an enormous amount of information about many different families of estimates. Also, the review papers by Huber (1972) and Hampel (1973) should be consulted. All three of these works contain many other references to the literature. [*For further discussion of robust estimation, see* ERRORS, *article on* EFFECTS OF ERRORS IN STATISTICAL ASSUMPTIONS.]

An important new step in the asymptotic theory of estimation was taken by Weiss and Wolfowitz with their introduction of maximum probability estimators. The stringent regularity conditions required by the older maximum-likelihood theory are not needed in the new theory, and there are many other advantages as well. A key reference is the paper by Wolfowitz (1971), where earlier references can be found.

Some progress in the theory of sequential estimation has been made by Bickel, Yahav, and others. See the paper by Linnik (1971) for a discussion and further references. [*See also* SEQUENTIAL ANALYSIS.]

Progress has continued on questions of admissibility and other classical topics, some related to the work of Charles Stein on the estimation of the mean of a multivariate normal random vector. The work of Farrell, Brown, Strawderman, and others mentioned in Brown (1971) should be consulted as should the paper by Efron and Morris (1975).

D. L. BURKHOLDER

ADDITIONAL BIBLIOGRAPHY

ANDREWS, D. F. et al. (editors) 1972 *Robust Estimates of Location: Survey and Advances*. Princeton Univ. Press.

BROWN, L. D. 1971 Admissible Estimators, Recurrent Diffusions, and Insoluble Boundary Value Problems. *Annals of Mathematical Statistics* 42:855–903.

EFRON, BRADLEY; and MORRIS, CARL 1975 Data Analysis Using Stein's Estimator and Its Generalizations. *Journal of the American Statistical Association* 70:311–319.

HAMPEL, FRANK R. 1973 Robust Estimation: A Condensed Partial Survey. *Zeitschrift für Wahrscheinlichkeitstheorie* 27:87–104.

HUBER, PETER J. 1972 Robust Statistics: A Review. *Annals of Mathematical Statistics* 43:1041–1067. → The 1972 Wald lecture.

LINNIK, YU. V. 1971 Some Recent Developments in the Sequential Estimation Theory. Volume 3, pages 255–258 in International Congress of Mathematicians, Nice, 1970, *Proceedings*. Paris: Gauthier-Villars.

WOLFOWITZ, J. 1971 Asymptotically Efficient Tests and Estimators. Volume 3, pages 259–263 in International Congress of Mathematicians, Nice, 1970, *Proceedings*. Paris: Gauthier-Villars.

II
CONFIDENCE INTERVALS AND REGIONS

Confidence interval procedures—more generally, *confidence region procedures*—form an important class of statistical methods. In these methods, the outcome of the statistical analysis is a subset of the set of possible values of unknown parameters. Confidence procedures are related to other kinds of standard statistical methods, in particular to point estimation and to hypothesis testing. In this article such relationships will be described and contrasts will be drawn between confidence methods and superficially similar methods of other kinds, for example, Bayesian estimation intervals [*see* BAYESIAN INFERENCE; ESTIMATION, *article on* POINT ESTIMATION; HYPOTHESIS TESTING].

As an example of this sort of procedure, suppose the proportion of voters favoring a candidate is to be estimated on the basis of a sample. The simplest possible answer is to give a single figure, say 47 per cent; this is the type of procedure called *point estimation*. Since this estimate of the proportion is derived from a sample, it will usually be different from the true proportion. How far off the true value is this estimate likely to be? This question can be answered by supplementing the estimate with error bounds, say $\pm .5$ per cent. Thus, one might say that the true proportion lies between 46.5 per cent and 47.5 per cent. This statement might be false. One task of the statistician is to develop a procedure for the computation of such intervals, a procedure that guarantees that the statements are true in, say, 99 per cent of all applications of this procedure. Such procedures are called confidence procedures.

Estimation by confidence intervals. It is perhaps easiest to begin with a simple example from normal sampling theory.

Example 1. Let $X_1, \cdots, X_n$ be a random sample of size n from a normal distribution with unknown mean μ and known variance σ^2. Then the sample mean $\bar{X} = \sum_i X_i/n$ is a reasonable point estimator of μ. Hence $\pm 2.58\sigma/\sqrt{n}$ are reasonable error bounds in the following sense: the estimator $\bar{X}$ lies between $\mu - 2.58\sigma/\sqrt{n}$ and $\mu + 2.58\sigma/\sqrt{n}$ with probability .99. In other words, the interval $(\mu - 2.58\sigma/\sqrt{n},\ \mu + 2.58\sigma/\sqrt{n})$ contains the estimator $\bar{X}$ with probability .99; that is, whatever the value of μ really is,

$$P_\mu\{\mu - 2.58\sigma/\sqrt{n} < \bar{X} < \mu + 2.58\sigma/\sqrt{n}\} = .99.$$

This probability statement follows directly from the facts that $(\bar{X}-\mu)/(\sigma/\sqrt{n})$ has a unit normal distribution and that a unit normal random variable lies in the interval $(-2.58, +2.58)$ with probability .99.

This statement can be given a slightly different but equivalent form: the interval $(\bar{X}-2.58\sigma/\sqrt{n}, \bar{X}+2.58\sigma/\sqrt{n})$ covers μ with probability .99, or whatever μ really is,

$$P_\mu\{\bar{X}-2.58\sigma/\sqrt{n}<\mu<\bar{X}+2.58\sigma/\sqrt{n}\}=.99.$$

The interval $(\bar{X}-2.58\sigma/\sqrt{n}, \bar{X}+2.58\sigma/\sqrt{n})$ is called a confidence interval for μ with confidence coefficient (or confidence level) .99. The confidence interval is a random interval containing the true value with probability .99. Note that it would be incorrect to say, after computing the confidence interval for a particular sample, that μ will fall in this interval with probability .99; for μ is an unknown constant rather than a random variable. It is the confidence interval itself that is subject to random variations.

Generally speaking, there is an unknown parameter, say θ, to be estimated and an estimator $f(X)$ depending on the sample $X=(X_1, \cdots, X_n)$. In example 1, θ is called μ and $f(X)$ is $\bar{X}$. As this estimator f is based on a random sample, it is itself subject to random variations. If f is a good estimator, its probability distribution will be concentrated closely around the true value, θ. From this probability distribution of f, one can often derive an interval, with lower bound $\underline{c}(\theta)$ and upper bound $\bar{c}(\theta)$, containing the estimator $f(X)$ with high probability β (for example, $\beta=.99$). That is, whatever the actual value of θ,

$$(1) \qquad P_\theta\{\underline{c}(\theta)<f(X)<\bar{c}(\theta)\}=\beta.$$

Often these inequalities can be inverted, that is, two functions $\underline{\theta}(X)$ and $\bar{\theta}(X)$ can be specified such that $\underline{\theta}(X)<\theta<\bar{\theta}(X)$ if and only if $\underline{c}(\theta)<f(X)<\bar{c}(\theta)$. Then, whatever θ really is,

$$(2) \qquad P_\theta\{\underline{\theta}(X)<\theta<\bar{\theta}(X)\}=\beta.$$

This means that the interval $(\underline{\theta}(X), \bar{\theta}(X))$ contains the true value θ with probability β. Quantities like $\underline{\theta}(X)$ and $\bar{\theta}(X)$ are often called confidence limits. In example 1, the bounds $\underline{c}(\theta), \bar{c}(\theta)$ and $\underline{\theta}(X), \bar{\theta}(X)$ are given by $\mu \pm 2.58\sigma/\sqrt{n}$ and $\bar{X}\pm 2.58\sigma/\sqrt{n}$, respectively.

It is also possible to develop the concept of a confidence region procedure in general, without reference to point estimation. Denote by P_θ the assumed probability distribution depending on a parameter θ (which may actually be a vector of several univariate, that is, real valued, parameters). Let $\boldsymbol{\Theta}$ be the set of all possible parameter values θ. By a confidence procedure is meant a rule for assigning to each sample X a subset of the parameter space, say $\Theta(X)$. If $\Theta(X)$ contains the true value θ with probability β, regardless of the true value of θ (that is, if for all $\theta \in \boldsymbol{\Theta}$, $P_\theta\{\theta \in \Theta(X)\}=\beta$), then $\Theta(X)$ is called a confidence region for θ. The probability β that the true parameter value is covered by $\Theta(X)$ is called the confidence coefficient.

In example 1, the interval $(\bar{X}-2.58\sigma/\sqrt{n}, \bar{X}+2.58\sigma/\sqrt{n})$ is the confidence region for the sample $X=(X_1, \cdots, X_n)$ with confidence coefficient .99.

The probability specified by the confidence coefficient has the following frequency interpretation: If a large number of confidence regions are computed on different, independent occasions, each with a confidence coefficient β, then, in the long run, a proportion β of these confidence regions will contain the true parameter value. There is some danger of misinterpretation. This occurs if θ itself is erroneously considered as a random variable and the confidence statement is given the following form: the probability is β that θ falls into the computed confidence set $\Theta(X)$. It should be clear that $\Theta(X)$ is the random quantity and not θ.

In the simplest applications, θ is a real parameter and the confidence region $\Theta(X)$ is either a proper interval $(\underline{\theta}(X), \bar{\theta}(X))$ or a semi-infinite interval: $(-\infty, \bar{\theta}(X))$ or $(\underline{\theta}(X), +\infty)$. If for all θ, $P_\theta(\theta<\bar{\theta}(X))=\beta$, then $\bar{\theta}(X)$ is called an upper confidence bound for θ with confidence coefficient β. Similarly, $\underline{\theta}(X)$ is a lower confidence bound.

Let $\underline{\theta}(X)$ and $\bar{\theta}(X)$ be lower and upper confidence bounds with confidence coefficients β_1 and β_2, and suppose that $\underline{\theta}(X)<\bar{\theta}(X)$ for all samples X. Then the interval $(\underline{\theta}(X), \bar{\theta}(X))$ is a confidence interval with confidence coefficient $\beta_1+\beta_2-1$. If $\beta_1=\beta_2$, that is, if $P_\theta\{\bar{\theta}(X)<\theta\}=P_\theta\{\theta<\underline{\theta}(X)\}$, the confidence interval $(\underline{\theta}(X), \bar{\theta}(X))$ is called central.

Example 2. As in example 1, let $X=(X_1, \cdots, X_n)$ be a sample of n independent normally distributed random variables with unknown mean μ and known variance σ^2. Then $\bar{\theta}(X)=\bar{X}+2.33\sigma/\sqrt{n}$ is an upper confidence bound for μ at confidence level .99. Thus $(-\infty, \bar{X}+2.33\sigma/\sqrt{n})$ is a semi-infinite confidence interval for μ with confidence coefficient .99, as is $(\bar{X}-2.33\sigma/\sqrt{n}, +\infty)$. Hence $(\bar{X}-2.33\sigma/\sqrt{n}, \bar{X}+2.33\sigma/\sqrt{n})$ is a central confidence interval for μ with confidence coefficient $.98=.99+.99-1$. This central confidence interval differs from that in example 1 in that the latter has confidence coefficient .99 and is correspondingly wider.

Example 3. Let $X = (X_1, \cdots, X_n)$ be a random sample from a normal distribution with known mean $\mu = 0$ and unknown variance σ^2. In this case $S_1^2 = \sum_i X_i^2/n$ is a reasonable estimator of σ^2. (A subscript is used in "S_1^2" because "S^2" will later denote a more common, related, but different quantity.) Suppose $n = 10$. Then the central confidence interval for σ^2 with confidence coefficient .98 is given by $(10S_1^2/23.21,\ 10S_1^2/2.56)$. The constants 23.21 and 2.56 are readily obtained from a table of quantiles for the chi-square distribution, for nS_1^2/σ^2 has a chi-square distribution with 10 degrees of freedom. This example shows that the endpoints of a confidence interval are generally not symmetric around the usual point estimator.

Relation to point estimation. The computation of confidence intervals is often referred to as *interval estimation,* in contrast to *point estimation.* As outlined above, in many practical cases, interval estimation renders information about the accuracy of point estimates. The general definition of confidence intervals is, however, independent of the problem of point estimation.

In many cases, a particular point estimator is related to the set of central confidence intervals. One forms the estimator for a given sample by thinking of the progressively narrowing intervals as the confidence level decreases toward zero. Except in pathological cases, the interval will squeeze down to a point, whose numerical value furnishes the estimator. Such an estimator is, for continuous distributions, median unbiased; that is, it is equally likely to be above and below the parameter under estimation.

Relation to hypothesis testing. The theory of confidence intervals is closely related in a formal way to the theory of hypothesis testing [*see* HYPOTHESIS TESTING].

Example 4. In example 1, the confidence interval for μ with confidence coefficient .99 was given by $\bar{X} - 2.58\sigma/\sqrt{n} < \mu < \bar{X} + 2.58\sigma/\sqrt{n}$. To test the hypothesis $\mu = \mu_0$ against the alternative $\mu \neq \mu_0$ at significance level .01, accept the hypothesis if

$$(3) \quad \mu_0 - 2.58\sigma/\sqrt{n} < \bar{X} < \mu_0 + 2.58\sigma/\sqrt{n};$$

reject it otherwise. This is the customary two-sided test.

Observe that, given $\bar{X}$, the confidence interval consists of all those values μ_0 for which the hypothesis $\mu = \mu_0$ would be accepted. In other words, the confidence interval consists of all μ_0 whose acceptance region contains the given $\bar{X}$.

On the other hand, given the confidence interval with confidence coefficient .99, it is easy to perform a test of a hypothesis $\mu = \mu_0$: Accept the hypothesis if the hypothetical value μ_0 belongs to the confidence interval; otherwise reject the hypothesis. Proceeding in this way, the pattern is precisely that of testing the hypothesis $\mu = \mu_0$, since μ_0 belongs to the confidence interval if and only if (3) is fulfilled, that is, if the hypothesis $\mu = \mu_0$ would be accepted according to the test procedure.

This duality is illustrated generally in Figure 1. The figure is directly meaningful when there is a single (real) parameter θ and when the sample can be reduced to a single (real) random variable. The latter reduction can frequently be accomplished via a sufficient statistic [*see* SUFFICIENCY]. When the problem is more complex, the figure is still of schematic use.

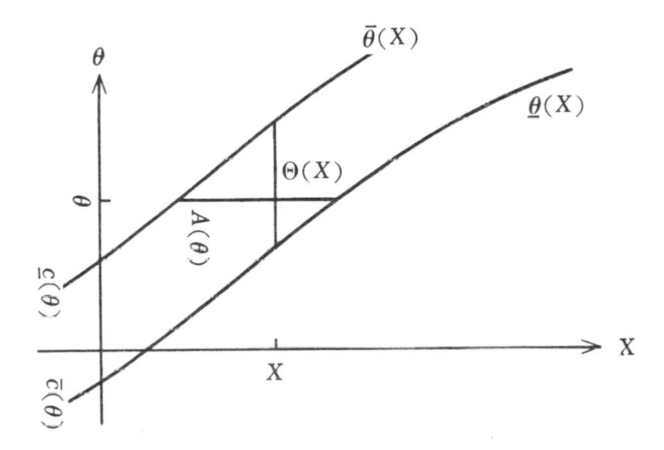

Figure 1 — A confidence region for the parameter θ

The figure shows that for each value of θ there is an acceptance region, $A(\theta)$, illustrated as an interval. The two curves determine the lower and upper bounds of this interval respectively. The set of all those θ for which $A(\theta)$ contains a given X, $\Theta(X)$, is the interval on the vertical through X between the two curves.

If the graphic representation is considered in a horizontal way (in terms of the X axis), the lower curve represents the lower confidence bound $\underline{\theta}(X)$ as a function of X, and similarly the upper curve represents the upper confidence bound $\bar{\theta}(X)$. If it is considered from the left (in terms of the θ axis), the functions $\underline{\theta}(X)$ and $\bar{\theta}(X)$ depending on X are inverted into the functions $\bar{c}(\theta)$ and $\underline{c}(\theta)$ respectively, depending on θ. (For this reason the letters are turned.)

The general duality between the testing of simple hypotheses and confidence procedures may be described as follows: Let $\boldsymbol{\Theta}$ be the set of unknown parameter values and assume that to each sample X a confidence set $\Theta(X)$ is assigned, such that $P_\theta\{\theta \in \Theta(X)\} = \beta$ for all $\theta \in \boldsymbol{\Theta}$. On the basis of such a confidence procedure, a test for any hypothesis $\theta = \theta_0$ can easily be defined as follows: Let $A(\theta)$ be the set of all X, such that $\theta \in \Theta(X)$. Then the events $X \in A(\theta)$ and $\theta \in \Theta(X)$ are equivalent, whence $P_\theta\{X \in A(\theta)\} = P_\theta\{\theta \in \Theta(X)\} = \beta$. Therefore, if $A(\theta_0)$ is taken as the acceptance region for testing the hypothesis $\theta = \theta_0$, a test with acceptance probability β (or significance level $\alpha = 1 - \beta$) is obtained. On the other hand, given a family of acceptance regions (that is, for each hypothesis $\theta \in \boldsymbol{\Theta}$ an acceptance region $A(\theta)$ contains the sample X with probability β when θ is the case), it is possible to define a confidence procedure by assigning to the sample X the set $\Theta(X)$ of all θ for which $A(\theta)$ contains X (that is, the set of all parameter values θ for which the hypothesis θ would be accepted on the evidence X). Then, again $\theta \in \Theta(X)$ if and only if $X \in A(\theta)$, whence $P_\theta\{\theta \in \Theta(X)\} = P_\theta\{X \in A(\theta)\} = \beta$. These remarks refer only to the case of simple hypotheses. In practice the more important case of composite hypotheses arises if several real parameters are present and the hypothesis consists in specifying the value of one of these. (This case is dealt with in "Nuisance parameters," below.)

Under exceptional circumstances the confidence set $\Theta(X)$ may show an unpleasant property: For some X, $\Theta(X)$ might be empty, or it might be identical with the whole parameter space, $\boldsymbol{\Theta}$. Those cases are usually of little practical relevance.

Thus a confidence statement contains much more information than the conclusion of a hypothesis test: The latter tells only whether a specified hypothesis is compatible with the evidence or not, whereas the confidence statement gives compatibility information about *all* relevant hypotheses.

[1]**Optimality.** The duality between confidence procedures and families of tests implies a natural correspondence between the optimum properties of confidence procedures and optimum properties of tests.

A confidence procedure with confidence region $\Theta'(X)$ is called *most accurate* if $\Theta'(X)$ covers any value different from the true value with lower probability than any other confidence region $\Theta(X)$ with the same confidence coefficient:

$$P_{\theta_0}\{\theta \in \Theta'(X)\} \leqslant P_{\theta_0}\{\theta \in \Theta(X)\} \qquad \text{for any } \theta \neq \theta_0.$$

Another expression occasionally used instead of "most accurate" is "most selective." The term "shortest," originally introduced by Neyman, is now unusual because of the danger of confusing shortest confidence intervals and confidence intervals of minimum length.

The family of tests corresponding to most accurate confidence procedures consists of uniformly most powerful tests: Let $A'(\theta)$ and $A(\theta)$ be the acceptance regions corresponding to the confidence regions $\Theta'(X)$ and $\Theta(X)$ respectively; then

$$\begin{aligned} P_{\theta_0}\{X \in A'(\theta)\} &= P_{\theta_0}\{\theta \in \Theta'(X)\} \\ &\leqslant P_{\theta_0}\{\theta \in \Theta(X)\} = P_{\theta_0}\{X \in A(\theta)\} \\ &\qquad \text{for any } \theta \neq \theta_0. \end{aligned}$$

Therefore, by using the acceptance region A' the false hypothesis θ is accepted with lower probability than by using A.

Uniformly most powerful tests exist only in exceptional cases. Therefore, the same holds true for most accurate confidence procedures. If, however, the class of tests is restricted (to unbiased tests or invariant tests, for example), the restricted class often contains a uniformly most powerful test within that class. Similarly, tests most powerful against a restricted class of alternatives can often be obtained. In the case of a real parameter a test for the hypothesis θ_0 that is most powerful against all $\theta > \theta_0$ may typically be found. All these restricted optimum properties of tests lead to corresponding restricted optimum properties of confidence procedures.

A confidence procedure is called unbiased if the confidence region covers no parameter value different from the true value with probability higher than its probability of covering the true value. The corresponding property of tests is also called unbiasedness. Therefore, families of uniformly most powerful unbiased tests lead to most accurate unbiased confidence procedures, that is, confidence procedures that are most accurate among the unbiased confidence procedures: No other unbiased confidence procedures exist leading to confidence regions that contain any value different from the true value with lower probability. The confidence interval given in example 1 is unbiased and most accurate among all unbiased confidence procedures with confidence coefficient .99. On the other hand, the confidence interval given in example 3 is not unbiased.

The optimum properties discussed above are related to concepts of optimality derived from the duality to the testing of hypotheses. A completely different concept is that of minimum length. For instance, the confidence interval given in example 1 is of minimum length. In general the length of the confidence interval is itself a random variable, as in example 3. It is therefore natural to consider a confidence procedure as optimal if the expected length of the confidence intervals is minimal. This concept is appropriate for two-sided confidence intervals. For one-sided confidence intervals the concept is not applicable immediately, as in this case the length is infinite. However, the expected value of the boundary value of the one-sided confidence interval can be substituted for expected length.

In general, confidence intervals with minimum expected length are different from, for example, most accurate unbiased confidence intervals (where such intervals exist). Under special circumstances, however (including the assumption that the distributions of the family have the same shape and differ only in location), invariant confidence procedures are of minimum expected length. The confidence procedure given in example 3 is not of minimum expected length.

Two objections that may be raised against the use of expected length as a criterion are (1) when a confidence interval fails to cover the true parameter value, a short interval is undesirable in that it pretends great accuracy when there is none, and (2) expected length depends strongly on the mode of parameterization, for example, there is no sharp relation between the expected length of a confidence interval for θ and that of the induced interval for θ^3.

Discrete distributions. In the general consideration above it was assumed that there exists a confidence procedure with confidence coefficient β in the sense that, for all θ in Θ, the probability of covering the parameter θ is exactly β when θ is the true parameter. This means that for each θ there exists an acceptance region $A(\theta)$ such that $P_\theta\{A(\theta)\} = \beta$. This is, however, in general true only for distributions of the continuous type, not for discrete distributions such as the binomial and Poisson distributions [*see* DISTRIBUTIONS, STATISTICAL]. Thus acceptance regions $A(\theta)$ of probability approximately β must be chosen, with the degree of approximation depending on θ. In practice the acceptance region is selected such that $P_\theta\{A(\theta)\}$ approximates β as closely as possible, either with or without the restriction $P_\theta\{A(\theta)\} \geqslant \beta$. These acceptance regions $A(\theta)$ define the confidence regions $\Theta(X)$ with (approximate) confidence coefficient β. When the restriction $P_\theta\{A(\theta)\} \geqslant \beta$ is made, the term "bounded confidence region" is often used, and the region is said to have bounded confidence level β.

Example 5. Let X be the number of successes in n independent dichotomous trials with constant probability p of success. Then X is binomially distributed, that is, $P_p\{X=k\} = \binom{n}{k}p^k(1-p)^{n-k}$. Choose the confidence coefficient $\beta = .99$. Choose for each p, $0 \leqslant p \leqslant 1$, the smallest integer $c(p)$ such that

$$\sum_{k=0}^{c(p)} \binom{n}{k} p^k(1-p)^{n-k} \geqslant 0.99.$$

Inverting the bound $c(p)$ one obtains one-sided confidence intervals of confidence coefficient .99 for p.

As an illustration, let $n = 20$ and $p = .3$. Since $P\{X \leqslant 11\} = .995$ and $P\{X \leqslant 10\} = .983$, the smallest integer such that $P\{X \leqslant c(p)\} \geqslant 0.99$ is $c(p) = 11$. Troublesome computations of $c(p)$ can be avoided by use of one of the tables or figures provided for this purpose. For references see Kendall and Stuart (1961, p. 118).

[2]Nuisance parameters. In many practical problems, more than one parameter is involved. Often the interest is concentrated on one of these parameters, say θ, while the others are regarded as nuisance parameters. The aim is to make a confidence statement about θ that is true with high probability regardless of the values of the nuisance parameters. The corresponding test problem is that of testing a composite hypothesis that specifies the value of θ without making any assertion about the nuisance parameters. The test is required to have significance level less than or equal to a prescribed α regardless of the nuisance parameters. The corresponding confidence procedure will yield confidence intervals that cover the true value at least with probability $1-\alpha$ regardless of the nuisance parameters, that is, confidence intervals with bounded confidence level $1-\alpha$. A special role is played by the so-called similar tests, having exactly significance level α for all values of the nuisance parameters. They lead to confidence intervals covering the true value with probability exactly $1-\alpha$ regardless of the nuisance parameters.

Example 6. Let $X_1, \cdots, X_n$ be a random sample from a normal distribution with unknown mean μ and unknown variance σ^2. The variance σ^2 is to be considered a nuisance parameter. Let $\bar{X} = \sum_i X_i/n$ and $S^2 = \sum_i(X_i - \bar{X})^2/(n-1)$. For $n = 10$ a (similar) confidence interval for μ with confidence coefficient .99 is given by $\bar{X} - 3.17\ S/\sqrt{10} <$

$\mu < \bar{X} + 3.17\, S/\sqrt{10}$. For general n, the confidence interval with confidence coefficient .99 is given by $\bar{X} - t_{.005,n-1}\, S/\sqrt{n} < \mu < \bar{X} + t_{.005,n-1}\, S/\sqrt{n}$, where $t_{.005,n-1}$ is the upper .005 point of the tabled t-distribution with $n-1$ degrees of freedom, for example $t_{.005,9} = 3.17$. Hence the above confidence procedure corresponds to the usual t-test. As for large n, because $t_{.005,n-1}$ is close to 2.58, the confidence interval given here corresponds for large n to the confidence interval given in example 1. The confidence procedure given here is most accurate among unbiased confidence procedures.

Example 7. Consider μ in example 6 as the nuisance parameter. Define S^2 as in example 6 and again take $n = 10$. Then a one-sided confidence interval for σ^2 of confidence coefficient .99 is given by $\sigma^2 \leqslant 9\, S^2/2.09$. In general, the one-sided confidence interval for σ^2 with confidence coefficient .99 is given by $\sigma^2 \leqslant (n-1)S^2/\chi^2_{.01,n-1}$, where $\chi^2_{.01,n-1}$ is the lower .01 point of the chi-square distribution with $n-1$ degrees of freedom. Observe that here the number of degrees of freedom is $n-1$ while in example 3 it is n.

Confidence coefficient. The expected length of the confidence interval depends, of course, on the confidence coefficient. If a higher confidence coefficient is chosen, that is, if a statement that is true with higher probability is desired, this statement has to be less precise; the confidence interval has to be wider.

It is difficult to give general rules for the selection of confidence coefficients. Traditional values are .90, .95, and .99 (corresponding to significance levels of .10, .05, and .01, respectively). The considerations to be made in this connection are the same as the considerations for choosing the size of a test [*see* HYPOTHESIS TESTING].

Nested confidence procedures. One would expect the wider confidence interval (belonging to the higher confidence level) to enclose the narrower confidence interval (belonging to the lower confidence level). A confidence procedure with this property is called "nested." All the usual confidence procedures are nested, but this is not a fully general property of confidence procedures.

Sample size. Given the confidence coefficient, the expected length of the confidence interval depends, of course, on the sample size. Larger samples contain more information and therefore lead to more precise statements, that is, to narrower confidence intervals.

Given a specific problem, the accuracy that it is reasonable to require can be determined. In order to estimate the number of housewives knowing of the existence of the superactive detergent X, a confidence interval of ±5 per cent will probably be sufficiently accurate. If, on the other hand, the aim is to forecast the outcome of elections and the percentage of voters favoring a specific party was 48 per cent in the last elections, an accuracy of ±5 per cent would be quite insufficient. In this case, a confidence interval of length less than ±1 per cent would probably be required.

Given the accuracy necessary for the problem at hand, the sample size that is necessary to achieve this accuracy can be determined. In general, however, the confidence interval (and therefore the necessary sample size as well) depends on nuisance parameters. Assume that a confidence interval for the unknown mean μ of a normal distribution with unknown variance σ^2 is needed. Although in example 6 a confidence interval is given for which no information about σ^2 is needed, such information is needed to compute the expected length of the confidence interval: The length of the confidence interval is $2t_{.005,n-1}\, S/\sqrt{n}$, the expected value for large n is therefore nearly equal to $2t_{.005,n-1}\, \sigma/\sqrt{n}$. Therefore, in order to determine the necessary sample size n, some information about σ^2 is needed. Often everyday experience or information obtained from related studies will be sufficient for this purpose. If no information whatsoever is at hand, a relatively small pilot study will yield a sufficiently accurate estimate for σ^2. This idea is treated rigorously in papers on sequential procedures for obtaining confidence intervals of given length (Stein 1945). In the case of the binomial distribution, no prior information at all is needed, for $\sigma^2 = p(1-p) \leqslant \frac{1}{4}$, whatever p might be. Using $\frac{1}{4}$ instead of σ^2 can, however, lead to wastefully large samples if p is near 0 or 1.

Robustness—nonparametric procedures. Any statistical procedure starts from a basic model on the underlying family of distributions. In example 1, for instance, the basic model is that of a number of independent normally distributed random variables. Since it is never certain how closely these basic assumptions are fulfilled in practice, desirable statistical procedures are those that are only slightly influenced if the assumptions are violated. Statistical procedures with this property are called *robust* [*see* HYPOTHESIS TESTING]. Another approach is to abandon, as far as possible, assumptions about the type of distribution leading to nonparametric procedures.

As the duality between families of tests and confidence procedures holds true in general, robust or nonparametric tests lead to robust or nonparametric confidence procedures, respectively. [*Examples showing the construction of confidence inter-*

vals for the median of a distribution from the sign test and from Wilcoxon's signed rank test are given in NONPARAMETRIC STATISTICS.]

Relationship to Bayesian inference. If the parameter is not considered as an unknown constant but as the realization of a random variable with given prior distribution, Bayesian inference can be used to obtain estimating intervals containing the true parameter with prescribed probability [*see* BAYESIAN INFERENCE].

Confidence statements can be made, however, without assuming the existence of a prior distribution, and hence confidence statements are preferred by statisticians who do not like to use "subjective" prior distributions for Bayesian inference. A somewhat different, and perhaps less controversial, application of subjective prior distributions is their use to define so-called subjective accuracy. Subjectively most accurate confidence procedures are defined in analogy to the most accurate ones by averaging the probability of covering the fixed parameter with respect to the subjective prior distribution. It can be shown that a most accurate confidence procedure is subjectively most accurate under any prior distribution with a positive density function (Borges 1962).

Relation to fiducial inference. Fiducial inference was introduced by R. A. Fisher (1930). This paper and succeeding publications of Fisher contain a rule for determining the fiducial distribution of the parameter on the basis of the sample X [*see* FIDUCIAL INFERENCE].

As in Bayesian inference, this distribution can be used to compute "fiducial intervals," giving information about the parameter θ. The fiducial interval is connected with a probability statement, which admits, however, no frequency interpretation (although some advocates of fiducial methods might disagree).

For many elementary problems, fiducial intervals and confidence intervals are identical. But this is not true in general. One of the attractive properties of fiducial inference is that it leads to solutions even in cases where the classical approach failed until now, as in the case of the Behrens–Fisher problem.

Many scholars, however, find it difficult to see a convincing justification for Fisher's rule of computing fiducial distributions and to find an intuitive interpretation of probability statements connected with fiducial intervals.

A reasonable interpretation of fiducial distributions would be as some sort of posterior distributions for the unknown parameter. It can be shown, however, that fiducial distributions cannot be used as posterior distributions in general; a Bayesian inference, starting from two independent samples and using the fiducial distribution of the first sample as prior distribution to compute a posterior distribution from the second sample, would in general lead to a result different from the fiducial distribution obtained from both samples taken together. For the comparison of fiducial and Bayesian method, see Richter (1954) and Lindley (1958).

Prediction intervals, tolerance intervals. Whereas confidence intervals give information about an unknown parameter, prediction intervals give information about future independent observations. Hence prediction intervals are subsets of the sample space whereas confidence intervals are subsets of the parameter space.

Example 8. If $X_1, \cdots, X_n$ is a random sample from a normal distribution with unknown mean μ and unknown variance σ^2, the interval given by $(\bar{X} - t_{\alpha,n-1}S\sqrt{(n+1)/n}, \bar{X} + t_{\alpha,n-1}S\sqrt{(n+1)/n})$ is a prediction interval containing a future independent observation X_{n+1} with probability $1-2\alpha$, if $t_{\alpha,n-1}$ is the upper α point of the t-distribution with $n-1$ degrees of freedom. Note that the probability of the event

$$\bar{X} - t_{\alpha,n-1}S\sqrt{(n+1)/n} < X_{n+1} < \bar{X} + t_{\alpha,n-1}S\sqrt{(n+1)/n}$$

is $1-2\alpha$ before the random variables $X_1, \cdots, X_n$ are observed. For further discussion of this example see Proschan (1953); for discussion of a similar example, see Mood and Graybill (1950, pp. 220–244, 297–299).

The prediction interval, computed in example 8 above, must not be interpreted in the sense that it covers a proportion α of the population. In a special instance, the interval computed according to this formula might cover more or less than the proportion α. Only on the average will the proportion be α.

In many cases, there is a need for intervals covering a proportion γ with high probability, say β. This is, however, not possible. In general, it is possible only to give rules for computing intervals covering *at least* a proportion γ with high probability β. Intervals with this property are called γ-proportion tolerance regions with confidence coefficient β. In the normal case, one might, for example, seek a constant c, for given γ and β, such that, whatever the values of μ and σ,

$$P\left\{\int_{\bar{X}-cS}^{\bar{X}+cS} f(u;\mu,\sigma)\,du \geqslant \gamma\right\} = \beta,$$

where $f(u; \mu, \sigma)$ is the normal density with mean μ and variance σ^2.

The constants c, leading to a γ-proportion tolerance interval $(\bar{X} - cS, \bar{X} + cS)$ with confidence coefficient β, cannot be expressed by one of the standard distributions (as was the case in the example of the prediction interval dealt with above). Tables of c can be found in Owen (1962, p. 127 ff.). For further discussion see Proschan (1953), and for nonparametric tolerance intervals see Wilks (1942). [*See also* NONPARAMETRIC STATISTICS.]

Confidence regions. In multivariate problems, confidence procedures yielding intervals are generalized to those yielding confidence regions.

Example 9. Let X and Y be two normally distributed random variables with unknown means μ and ν, known variances 2 and 1, and covariance -1. A confidence region for (μ, ν) with confidence coefficient .99 is given by $(X-\mu)^2 + 2(X-\mu)(Y-\nu) + 2(Y-\nu)^2 \leqslant 9.21$. The figure 9.21 is obtained from a chi-square table, since the quadratic form on the left is distributed as chi-square with two degrees of freedom. The confidence region is an ellipse with center (X, Y). When such a region is described in terms, say, of pairs of parallel tangent lines, the result may usefully be considered in the framework of multiple comparisons. [*See* LINEAR HYPOTHESES, *article on* MULTIPLE COMPARISONS.]

J. PFANZAGL

BIBLIOGRAPHY

The theory of confidence intervals is systematically developed in Neyman 1937; 1938b. *Prior to Neyman, this concept had been used occasionally in a rather vague manner by a number of authors, for example, by* Laplace 1812, *section 16, although in a few cases the now current meaning was clearly stated, perhaps first by* Cournot 1843, *pp. 185–186. A precise formulation without systematic theory is given in* Hotelling 1931. *A more detailed account of the history is given in* Neyman 1938a.

BORGES, RUDOLPH 1962 Subjektivtrennscharfe Konfidenzbereiche. *Zeitschrift für Wahrscheinlichkeitstheorie* 1:47–69.

COURNOT, ANTOINE AUGUSTIN 1843 *Exposition de la théorie des chances et des probabilités.* Paris: Hachette.

FISHER, R. A. (1930) 1950 Inverse Probability. Pages 22.527*a*–22.535 in R. A. Fisher, *Contributions to Mathematical Statistics.* New York: Wiley. → First published in Volume 26 of the *Proceedings* of the Cambridge Philosophical Society.

FISHER, R. A. 1933 The Concepts of Inverse Probability and Fiducial Probability Referring to Unknown Parameters. Royal Society of London, *Proceedings* Series A 139:343–348.

HOTELLING, HAROLD 1931 The Generalization of Student's Ratio. *Annals of Mathematical Statistics* 2:360–378.

○KENDALL, MAURICE G.; and STUART, ALAN (1961) 1973 *The Advanced Theory of Statistics.* Volume 2: *Inference and Relationship.* 3d ed. London: Griffin; New York: Hafner. → See especially pages 103–140 on "Interval Estimation: Confidence Levels" and pages 537–540 on "Distribution-free Tolerance Intervals." Early editions of Volume 2, first published in 1946, were written by Kendall alone. Stuart became a joint author on later, renumbered editions in the three-volume set.

LAPLACE, PIERRE SIMON DE (1812) 1820 *Théorie analytique des probabilités.* 3d ed., rev. Paris: Courcier. → Laplace's mention of confidence intervals first appeared in the 2d (1814) edition.

LEHMANN, ERICH L. 1959 *Testing Statistical Hypotheses.* New York: Wiley. → See especially pages 78–83, 173–180, and 243–245.

LINDLEY, D. V. 1958 Fiducial Distributions and Bayes' Theorem. *Journal of the Royal Statistical Society* Series B 20:102–107.

MOOD, ALEXANDER M.; and GRAYBILL, FRANKLIN A. (1950) 1963 *Introduction to the Theory of Statistics.* 2d ed. New York: McGraw-Hill. → See especially pages 220–244 on "Interval Estimation." (Mood was the sole author of the 1950 edition.)

NEYMAN, JERZY 1937 Outline of a Theory of Statistical Estimation Based on the Classical Theory of Probability. Royal Society of London, *Philosophical Transactions* Series A 236:333–380.

NEYMAN, JERZY (1938*a*) 1952 *Lectures and Conferences on Mathematical Statistics and Probability.* 2d ed. Washington: U.S. Dept. of Agriculture. → See especially Chapter 4, "Statistical Estimation."

NEYMAN, JERZY 1938*b* L'estimation statistique traitée comme un problème classique de probabilité. *Actualités scientifiques et industrielles* 739:26–57.

OWEN, DONALD B. 1962 *Handbook of Statistical Tables.* Reading, Mass.: Addison-Wesley. → A list of addenda and errata is available from the author.

PROSCHAN, FRANK 1953 Confidence and Tolerance Intervals for the Normal Distribution. *Journal of the American Statistical Association* 48:550–564.

RICHTER, HANS 1954 Zur Grundlegung der Wahrscheinlichkeitstheorie. *Mathematische Annalen* 128:305–339. → See especially pages 336–339 on "Konfidenzschluss and Fiduzialschluss."

SCHMETTERER, LEOPOLD 1956 *Einführung in die Mathematische Statistik.* Berlin: Springer. → See especially Chapter 3 on "Konfidenzbereiche."

STEIN, CHARLES 1945 A Two Sample Test for a Linear Hypothesis Whose Power Is Independent of the Variance. *Annals of Mathematical Statistics* 16:243–258.

WILKS, S. S. 1942 Statistical Prediction With Special Reference to the Problem of Tolerance Limits. *Annals of Mathematical Statistics* 13:400–409.

Postscript

[1]The relationship between different concepts of optimality is discussed in Zacks (1971, secs. 10.1 and 10.3).

[2]The general form of a confidence set in the presence of nuisance parameters may be found in Lehmann (1959, chapter 5. secs. 4 and 5) and in Zacks (1971, sec. 10.2). A satisfactory procedure for obtaining approximately optimal confidence intervals in the presence of nuisance parameters is

available only in the asymptotic theory (that is, for large sample sizes). By using asymptotic expansions, such procedures become useful also for moderate sizes. For a survey of such developments, see Pfanzagl (1974).

A problem may arise with confidence regions—and indeed with many modes of statistical inference—of side information from the sample that affects *conditionally* the probabilistic interpretation of the confidence interval. Suppose, to take one of the simplest examples, that one is dealing with a random sample from a uniform distribution of range one but of unknown position. If the range of the sample is near one, then, conditionally on that, we know that the true midrange is very close to the sample midrange, with a corresponding statement for the confidence interval. On the other hand, a sample with small range will, conditionally, provide a confidence interval of lowered coverage probability. Robinson (1975) presents a number of such examples and references to the literature; other important references in this area include Cox (1958, pp. 360–361) and Barndorff-Nielsen (1974).

J. PFANZAGL

ADDITIONAL BIBLIOGRAPHY

BARNDORFF-NIELSEN, OLE 1973 Exponential Families and Conditioning: II. On Conditioning in Statistical Inference. Ph.D. dissertation, Copenhagen Univ.

COX, D. R. 1958 Some Problems Connected With Statistical Inference. *Annals of Mathematical Statistics* 29:357–372.

PFANZAGL, J. 1974 Asymptotically Optimum Estimation and Test Procedures. Volume 1, pages 201–272 in Symposium on Asymptotic Statistics, Prague, 1973, *Proceedings*. Edited by Jaroslav Hájek. Prague: Charles Univ.

PRATT, JOHN W. 1961 Length of Confidence Intervals. *Journal of the American Statistical Association* 56: 549–567.

ROBINSON, G. K. 1975 Some Counterexamples to the Theory of Confidence Intervals. *Biometrika* 62:155–161.

ZACKS, SHELEMYAN 1971 *The Theory of Statistical Inference.* New York: Wiley.

ETHICAL ISSUES IN THE SOCIAL SCIENCES

This article was first published in IESS *as "Ethics: II. Ethical Issues in the Social Sciences" with a companion article less relevant to statistics.*

Ethics is concerned with standards of conduct among people in social groups; for this reason, research in social science is inextricably bound up in ethical problems. The initial choice of a problem for investigation by the social scientist is often value-laden. The process of inquiry in the social sciences, engaging as it frequently does the lives of people, must meet moral as well as scientific standards. And the product of inquiry constantly adds new data and new theories requiring the revision of established ethical systems. Ethics and social science thus move in contrapuntal relationship, each adding to the character of the other (Shils 1959).

Old issues and new. There are a number of principles of ethics in social science research that are so widely recognized and honored that they do not need detailed discussion. Among these are maintaining highest standards of work, reporting procedures and results faithfully, protecting information given in confidence, giving appropriate credit to co-workers, making appropriate acknowledgment of other writers' materials, representing accurately one's own qualifications, and acknowledging, when appropriate, sources of financial support. The central issue in all of these is integrity, as indeed it is in every step of a true research endeavor. For this reason some social scientists have objected to proposals to define ethical standards for research, arguing that the canons of science are an exacting and sufficient guide to conduct. However, new problems arise as scientists move into new areas under new auspices; old problems appear in new contexts and require new solutions. Ethical standards must be redefined continually to keep them relevant to contemporary situations. Below are several issues that are subjects of concern and of lively debate as this article is written. If these issues are soon dated and no longer lively, it is probably a healthy sign that consensus is being reached on them and that new issues are capturing concern.

Deception in social science research. In many experiments or inquiries in the social sciences, it is necessary, or has been widely considered necessary, to disguise the nature of the task assigned to the subject. The procedure arises usually from the need to control the "set" or "expectancy" with which the subject approaches the task, since set is known to be an important determinant of responses. While in most instances the consequences are trivial, in some instances they may not be trivial at all. In all instances the issue is raised, Is deception ever justified?

Clearly, scientists think that deception is sometimes required to achieve a good that would not otherwise be achievable. For example, it is common practice in medical research to administer a placebo to a control group in order to assess the effects of a drug. No harm is done; the control subjects might still be given the drug if it proves effective. But the outcome of deception is not always benign. In one of the classical experiments on deceit, the investigators tempted children to steal and deceived them into believing that their action could not be detected. Some children did indeed steal. The investigators concluded that honesty is often influenced by the situation, a point demonstrated as much in their own behavior as in that of the children (Hartshorne & May 1928). In a second well-known investigation, social psychologists infiltrated a religious group, posing as converts (Festinger et al. 1956); their conduct has been questioned (Smith 1957). In an experiment on the effects of group pressure on judgment, five coworkers of the experimenter were represented as uninstructed subjects, just like the person whose resistance to social pressure was to be tested (Asch 1948). Both the deception and the stress generated thereby may be questioned, from an ethical viewpoint. Russian psychologists investigating the same problem have avoided the need for deception by using all naive subjects and analyzing the data for trends that occur naturally, accepting the loss in experimental efficiency.

A reasonable ethical standard for such a situation would be that the investigator has an obligation to inform his prospective subject of any aspect of the experiment that might be considered an important factor in the subject's decision to serve. While such an ethical policy obviously has much to commend it, the losses would be great; many experiments concerned with the dynamics of human behavior would be made impossible. Ethics aside, there are pragmatic arguments in favor of a policy of full disclosure of intent. With growing sophistication, the public may come to regard all social science experiments as situations in which deception is to be expected. At this point even truth is suspect. The problem is not simple, nor is it unimportant. Perhaps a minimum obligation of the social scientist is to make the public aware of the problem.

Stress in social science research. While many experimenters have subjected participants in research to stress, one investigator has been taken to task for his seeming insensitivity to the excruciating ordeal his subjects were going through and for his failure to see the larger implications of his methodology. The critic (Baumrind 1964) very reasonably questioned the ethics of subjecting people to extreme stress and pointed out the moral parallels to historical situations in which innocent people have been tortured in the interest of science. The experimenter's rejoinder (Milgram 1964) provides further instruction in the complexity of the problem and demonstrates the value of a continuing debate of ethical issues in research.

The customary routine is to talk with the subject after an experiment involving stress, to explain the procedure, and to try to relieve any residual discomfort. This procedure may suffice in many investigations, but there are others reported in the literature in which the stress is so severe that one could not realistically hope to repair the damage by such a postsession conference. A suitable topic for cross-disciplinary research would be an investigation of possible lingering or delayed effects of experiments involving stress or deception.

It has been proposed that there is already enough stress in life arising from natural causes and that social scientists should not add to it. An alternative is to study stress reactions in natural settings. Many of these are unpredictable and are amenable only to observational study after the event, but some excellent research has been done following disasters, such as tornadoes and earthquakes, by sociologists who were prepared to take advantage of an unpredictable event. There are also predictable and necessary stressful situations that are a normal part of living and could be used in research. A first-grade classroom on the first day of school and the father's waiting room at a maternity hospital are settings where stress can be studied without the investigator's causing it. Webb and his coworkers (1966) have provided an imaginative and useful examination of methodological options in "nonreactive research in the social sciences," including attention to ethical problems.

Protection of research data. The right of the clinician to keep data confidential is widely (though not universally) recognized by custom and in some states and countries by law. But the scientific investigator does not as clearly enjoy such protection. For example, the social scientist engaged in survey research may encounter a serious ethical problem, and lack clear guidelines for conduct, when his evidence is introduced in a court as legal testimony. The court or either contending party may have a legitimate interest in the reliability of the survey and may demand that respondents be identified in order to call them as witnesses. But survey data are generally obtained with assurances of anonymity; a violation of this pledge would not only involve a betrayal of confidence but would also impair the survey method as a research technique by diminishing public confidence in agencies that use the procedure. In at least one ruling, a court has sustained the right of a survey agency to keep confidential the names of persons interviewed, but other judges may rule differently. Obviously, the social scientist engaged in survey research has a minimum obligation to inform himself on the issues involved so that he can behave responsibly toward people who supply him with information (King & Spector 1963). He might also be expected to anticipate such problems in the planning stages of a study and to take protective measures against a number of contingencies. The issue of proper protection of data, here discussed with reference to surveys, may be equally relevant in other kinds of research. The problem is complicated by the investigator's obligation to keep his work open for scrutiny by competent scientists.

The invasion of privacy. Privacy is a most cherished right of the individual in a free society, and it may well be an important condition for the integration of experience and the achievement of autonomous selfhood. Social scientists are engaged in a number of enterprises that can lead to a reduction of individual privacy. The ethical issue that seems most frequently involved is that information about a person or his family may be collected, and perhaps used officially, without the individual's being aware of what is happening. The use of personality tests for appraising prospective employees, screening school children, and so on has recently attracted public attention. In some instances, restrictive regulations have been imposed to prevent what is seen as an undue invasion of privacy.

Privacy is not always an individual matter but may involve social institutions which depend for their effectiveness on assurances against intrusion; such is true of the jury system in the United States. In 1955, some sociological investigators, with the permission of the trial judge and the contending lawyers, concealed microphones in a jury room and recorded the jury's deliberation. Although the information obtained was treated with scrupulous care by the investigators, the incident created a national furor. The jurors had clearly been deceived and were appropriately indignant. An issue of broader concern involved in this instance was the appropriateness of scientific inquiry into an established social institution; the social scientist who undertakes such studies must be uncommonly concerned with ethical issues, since damage may be done both to social science and to the institutions studied by social scientists.

As computers become increasingly available and efficient in both storage and processing capacities, we face the prospect of an invasion of privacy of quite a different sort. With various agencies collecting diverse data about an individual over a sufficient period of time, and with the data centrally stored and processed, the possibility is imminent that extensive and reliable inferences can be made about an individual that far exceed his intentions of disclosure. The protection of privacy that has come from fragmentation of information or from the sheer tedium and expense of analysis may indeed be lost.

One example may suffice to indicate the further significance of technological developments: it is now possible to obtain, by mail order, a detailed analysis of an individual's responses to the Minnesota Multiphasic Personality Inventory; the evaluation that once required the services of a highly skilled clinician can be provided now, in much shorter time, by a computer. The ethical implications of advances in computer technology are yet to be explored.

The invasion of privacy issue arises at the point of intersection of two highly valued social goods: the need for knowledge about problems, opinions, motivations, and expectancies of people and the need for preservation of personal rights. While the conflict of social values involved is an ancient one (the rack and screw were information-obtaining devices), the problem is of notable contemporary importance because of the steady increase in amount of, and reliance on, social science research, on the one hand, and the advances in the technology of inquiry, including electronic listening devices, recorders, cameras, computers, personal inventories, projective techniques, and planted informers or confederates, on the other hand.

Among the issues that must be considered in achieving a proper balance of conflicting social and individual interest are the importance of the investigation, the informed consent of subjects, the preservation of confidentiality, and the judicious use of records of research. The individual scientist's decisions about these moral issues must be harmonious with the opinion of his peers or with a community consensus. As the social scientist comes to have more of value to offer the community, he can expect more community understanding and support of the unavoidable violation of privacy attendant upon much social science research. (For an informed and sophisticated analysis of the problem of privacy, see Ruebhausen & Brim 1965.)

The issue of informed consent. In medical research it has generally been the practice to obtain the informed consent of a patient as a condition for his participation in an investigation; however, loose definitions of what is meant by *informed* have permitted great latitude in practice. In a decision that will have implications for all research involving human subjects, the Board of Regents of the University of the State of New York in 1966 stringently defined expectations for medical investigators:

> No consent is valid unless it is made by a person with legal and mental capacity to make it, and is based on a disclosure of all material facts. Any fact which might influence the giving or withholding of consent is material. A patient has the right to know he is being asked to volunteer and to refuse to participate in an experiment for any reason, intelligent or otherwise, well-informed or prejudiced. A physician has no right to withhold from a prospective volunteer any fact which he knows may influence the decision. It is the volunteer's decision to make, and the physician may not take it away from him by the manner in which he asks the question or explains or fails to explain the circumstances. (Langer 1966, p. 664)

In this statement the words *social scientist* might be substituted for *physician* and *subject* for *patient* to arrive at an important guideline for research in the social sciences.

But again the issue is not simple. Is a patient in a control group in a medical experiment to be told that the treatment he will receive is known to have no physiological effect but will be administered to control for psychological effects? If such candor were required, much medical research would be impossible. And so it is with social science research, where possible gains in socially valuable knowledge must be weighed against possible losses of individual prerogatives. For a clear joining of the issue, in regard to psychological research, see the correspondence of Miller and Rokeach (1966). Rokeach wrote, to define the complexity of the problem: "What is typically involved in making a decision about moral values, whether in or out of science, is not a choice between good and evil but a choice between two or more positive values, or a choice between greater and lesser evils" (1966, p. 15). All-or-none solutions are seldom satisfactory.

Cross-cultural studies. The many ethical issues involved in cross-cultural and transnational investigations, long a concern of the professional anthropologist (see, for example, Redfield 1953 and also the "Statement on Ethics of the Society for Applied Anthropology" 1963–1964), were thrust into public prominence, in 1965, by the debacle of Project Camelot, an inquiry sponsored by the U.S. Department of Defense into "the causes of revolutions and insurgency in underdeveloped areas of the world." Exposure of the project in a South American country led to protests from the U.S. ambassador, a Congressional investigation, the cancellation of the project, and a policy requiring that all government-sponsored, foreign-area research be approved by the U.S. Department of State. The fact that Camelot became a national and international *cause célèbre* involving ambassadors, senators, cabinet members, newspapermen, university officials, social scientists, and the president himself, and that it was interpreted as a cloak-and-dagger operation in spite of the sincerity and good will of the participating scientists, has served to obscure the ethical issues involved, issues that demand serious and sophisticated consideration by the social scientist, whether involved in cross-cultural studies or not.

Among the ethical issues are these: Should the intentions of a sponsoring agency be the concern of a social scientist even when he is personally allowed full freedom of inquiry? Should the social scientist be concerned with the uses to which the results of his studies will be put? What is the responsibility of the social scientist for ensuring that the very process of inquiry does not have a deleterious effect on the people being studied? Does the social scientist have an obligation to preserve access to people for subsequent investigators? Is there a point at which inadequacies of design or procedure, or lack of scientific merit in a study, become intrinsically ethical issues by virtue of their imposition on others? These and similar questions may appear to have easy answers, but a sympathetic study of Project Camelot will show their complexity and emphasize the need for social scientists to consider them anew in the context of every proposed investigation (Horowitz 1965).

Social science and social issues. Social science may often have relevance to crucial matters of public policy. With increasing frequency advocates of diverse political and social policies turn to the social scientist for support of their position. Or the social scientist himself, exercising the prerogative of a citizen to make public statements on social and political issues, may find his statements given credence beyond what could be supported by data, by virtue of his being recognized as a scientist, regardless of his competence on the particular topic. Drawn into such an unaccustomed arena, the social scientist must be especially mindful of how he presents his qualifications and of the ethical implications of his statements. Issues related to racial characteristics, for example, have so conjoined science and public policy that they have been made the subject of study by the Committee on Science in the Promotion of Human Welfare of the American Association for the Advancement of Science ("Science and the Race Problem" 1963).

Care of animals in research. The psychologist has relied heavily on animals—rats, dogs, birds, primates—as subjects in research. To protect laboratory animals from neglect or abuse, formal regulations governing the management of animal laboratories have been developed. These require the provision of adequate food, water, and medical care, the maintenance of sanitary living quarters, the use of anesthetics to prevent pain in operations and other procedures, the provision of postoperative care, and the destruction of animals by humane means. Committees on care of laboratory animals review problems periodically. The U.S. Public Health Service publishes a booklet entitled "Guide for Laboratory Animal Facilities and Care" (Animal Care Panel 1963) and requires recipients of grant support to observe the requirements to assure proper and humane treatment of research animals. The American Psychological Association requires posting in "all rooms where animals are housed and where animal experimentation is conducted" of regulations titled "Guiding Principles for the Humane Care and Use of Animals." In spite of these efforts to assure highest ethical standards in the care of laboratory animals, there is a perennial demand for federal legislation to control practices, especially with respect to dogs and cats. In 1964 there were eight bills introduced in the 88th Congress of the United States, two of which would have been severely restrictive. Although there are occasional cases of negligence or of needless infliction of pain, animals are generally well cared for, and the Congress has shied away from enacting legislation on the matter (Brayfield 1963).

Communication in social science research. Marin Mersenne promoted science in seventeenth-century France by copious letter writing; the problem of communication in science has since become exceedingly complex, with many attendant ethical issues. Ethical problems have involved such issues as plagiarism, misrepresentation of data, the betrayal of confidence, claiming undue credit, and other clearly unacceptable behavior. With the development of what has been called "big science" with extensive government support, problems of a new and more subtle character have emerged. For example, the assignment of credit for research accomplished by a large organization seems to be solved neither by crediting the director alone, as has been done and protested, nor by crediting 30 contributors, as was done in a recent listing of authors. Although promotions may depend on publications, there is a growing need to limit publication to significant findings likely to be of value to others. The sheer volume of reports threatens to overwhelm our most efficient systems for coding, storing, and finding information. Thus, for an investigator to impose the same findings twice on about the same audience constitutes an offense to the development and dissemination of knowledge. The following statement has been proposed to control the volume of publication: ". . . *scientific publication* [*should*] *be considered a privilege consequent upon the finding of something which people may need to read, rather than as a duty consequent upon the spending of time and money.* . . . Furthermore . . . no paper [should] be committed more than once to the published literature without very special pleading" (Price 1964).

Research on moral development. Thus far certain theoretical and practical problems relating to ethics and social science research have been considered. It should be noted now that social science research itself is a potential major source of understanding of ethical conduct, of the origins and development of moral standards. Pioneer work was done by Hartshorne and May (1928). Piaget (1932) provided a theoretical matrix for illuminating stages in the moral development of the child. Anthropologists and social psychologists (Whiting 1963) have studied the influence of the family on character formation in different cultures. Russian pedagogical specialists are working explicitly to provide educational experiences to instill communist values in children (Bronfenbrenner 1962). In the United States, the establishment of the National Institute of Child Health and Human Development, to promote research on normal development, can be expected to encourage basic research on the problem.

Social control of scientific inquiry. Various professional, trade, labor, and fraternal groups exert a major influence on the behavior of individuals in contemporary society. Perhaps because of their very diversity they escape attention as instruments of social control, yet it has been contended that they speak with more authority today than do organized religious groups and, further, that they influence day-by-day conduct even more than do local, state, and national governments.

Many of these associations have formal codes of ethics. For the most part these codes have been found to have little effect on the behavior of members of the group (American Academy of Political and Social Science 1955). They are one of the appurtenances of associations and are designed with an eye to building public confidence. However, the traditions, mores, and expectancies that are generated in professional groups do affect behavior, often holding members to extraordinarily high standards of conduct. When codes of ethics are in harmony with long-established tradition (as in *The Principles of Medical Ethics*) or when they are backed up by effective machinery for enforcement, they can be powerful instruments of social control.

The American Psychological Association has applied social science theory and methodology to the task of developing a code of ethics (Hobbs 1948). The critical incidents technique was used to obtain the basic data for the construction of the code. Members of the association were asked to supply descriptions of situations in which a psychologist took some action that either upheld or violated ethical standards. From over a thousand such incidents a committee extracted the principles that appeared to be involved in the behavior reported. The result is two documents: a succinct code (American Psychological Association 1963) and a book-length statement (American Psychological Association 1953) of ethical standards that includes principles, discussions of issues, and illustrations drawn from the collection of critical incidents. Now underway is a new inquiry directed specifically at ethical issues in psychological research; the critical incident technique is again being used to develop basic data from which ethical principles will be derived.

The psychologists' statement of ethical standards is being augmented by a collection of case studies drawn from the files of ethics committees responsible for the enforcement of the code. The assumption is made that the definition of ethical standards is an ongoing, never-finished process and that participation in the process by members of the association may be more important than the written code itself in nurturing high ethical standards in the profession. The Committee on Cooperation Among Scientists of the American Association for the Advancement of Science is collecting similar descriptions, not necessarily to prepare a code of ethics but to illuminate the ethical problems encountered by scientists in all fields.

When scientists fail to regulate their own behavior to the satisfaction of informed members of the community, one can confidently predict that controls will be imposed by legislation or by administrative regulations. In 1965–1966, two major federal agencies adopted procedures governing ethical issues in research supported by their grants. One agency requires that tests, questionnaires, and other data-gathering devices be approved in Washington by a special review group composed of staff members, with the assistance of consultants. The other agency has established a requirement that grant requests involving possible ethical issues must be reviewed by a recognized local committee of peers of the investigator. The second solution appears to offer protection to research subjects on the basis of competent review without the danger of overcentralized control of scientific inquiry. However, there are responsible investigators who contend that a prescribed review by local peers is an invidious requirement implying incompetence and guilt when competence and rectitude should be assumed, with intervention indicated only when there is some evidence to the contrary. Here again a social process to define appropriate procedures is underway, with the proper resolution still unclear.

It can be expected that society will develop, in time, a productive balance between its need for knowledge and the individual's need for protection against intrusion, inconvenience, or discomfort. A dialectic tension involving values fundamental to a democracy must be resolved, both in terms of general principles and in terms of particular instances. For example, freedom of inquiry must be balanced against rights of privacy, both cherished values in our society. While the issues are complex, resolution is possible. The accommodation, both in substance and in process, will probably be comparable in character to rules governing the right of eminent domain and the right of the individual to own property.

The individual investigator is not without common-sense guidelines. While the answers may not always be clear, some of the questions are: Is the knowledge to be gained worth the imposition involved in obtaining it? Would another design be equally productive but less intrusive? Has fullest advantage been taken of the subject's informed willingness to cooperate? Has the proposed inquiry been designed to minimize effects on the subject population so that subsequent investigators will not be handicapped? To what extent are the proposed procedures consonant with emerging standards, or a calculated departure from them?

Nor is the investigator without criteria to assess and perhaps discover the adequacy of his answers to such questions: first, his own standards as an investigator, concerned quite as much with ethical as with statistical elegance of design; then the approbation of other competent scientists; and, finally, the appreciation of the larger community, or of significant sections of it, whose support is essential to the continued development of the social sciences.

It is of greatest importance to keep ethical problems under continuing scrutiny and debate, in journals, in training programs, in public forums, with social scientists themselves taking the initiative in the process, in order to provide increasingly instructive principles for clarifying ethical issues in social science research.

NICHOLAS HOBBS

[*See also* SCIENCE; Barber 1968; Hagstrom 1968; Kaplan & Storer 1968; Kohlberg 1968; Price 1968; Simmel 1968.]

BIBLIOGRAPHY

AMERICAN ACADEMY OF POLITICAL AND SOCIAL SCIENCE 1922 *The Ethics of the Professions and of Business.* Edited by Clyde L. King. Annals, Vol. 101, no. 190. Philadelphia: The Academy.

AMERICAN ACADEMY OF POLITICAL AND SOCIAL SCIENCE 1955 *Ethical Standards and Professional Conduct.* Edited by Benson Y. Landis. Annals, Vol. 297. Philadelphia: The Academy.

AMERICAN PSYCHOLOGICAL ASSOCIATION 1953 *Ethical Standards of Psychologists.* Washington: The Association.

AMERICAN PSYCHOLOGICAL ASSOCIATION 1963 Ethical Standards of Psychologists. *American Psychologist* 18:56–60.

AMERICAN STATISTICAL ASSOCIATION, BOSTON, DECEMBER, *1951* 1952 Standards of Statistical Conduct in Business and Government. *American Statistician* 6, no. 1:6–20.

ANIMAL CARE PANEL, ANIMAL FACILITIES STANDARDS COMMITTEE 1963 *Guide for Laboratory Animal Facilities and Care.* U.S. Public Health Service, Publication No. 1024. Washington: Government Printing Office.

ASCH, SOLOMON E. 1948 The Doctrine of Suggestion, Prestige and Imitation in Social Psychology. *Psychological Review* 55:250–276.

►BARBER, BERNARD 1968 Science: III. The Sociology of Science. Volume 14, pages 92–100 in *International Encyclopedia of the Social Sciences.* Edited by David L. Sills. New York: Macmillan and Free Press.

BARNES, JAMES A. 1963 Some Ethical Problems in Modern Fieldwork. *British Journal of Sociology* 14:118–134.

BAUMRIND, DIANA 1964 Some Thoughts on Ethics of Research: After Reading Milgram's *Behavioral Study of Obedience. American Psychologist* 19:421–423.

BRAYFIELD, ARTHUR H. 1963 Humane Treatment of Laboratory Animals. *American Psychologist* 18:113–114.

BRONFENBRENNER, URIE 1962 Soviet Methods of Character Education: Some Implications for Research. *American Psychologist* 17:550–564.

BURGESS, ROBERT W. 1947 Do We Need a "Bureau of Standards" for Statistics? *Journal of Marketing* 11: 281–282.

COMMISSION DE DÉONTOLOGIE DE LA SOCIÉTÉ FRANÇAISE DE PSYCHOLOGIE 1960 Projet de code déontologie à l'usage des psychologues. *Psychologie française* 5:1–27.

FESTINGER, LEON; RIECKEN, H. W.; and SCHACHTER, STANLEY 1956 *When Prophecy Fails.* Minneapolis: Univ. of Minnesota Press.

FREEMAN, WILLIAM W. K. 1963 Training of Statisticians in Diplomacy to Maintain Their Integrity. *American Statistician* 17, no. 5:16–20.

►HAGSTROM, WARREN O. 1968 Science: V. Scientists. Volume 14, pages 107–111 in *International Encyclopedia of the Social Sciences.* Edited by David L. Sills. New York: Macmillan and Free Press.

HARTSHORNE, HUGH; and MAY, MARK A. 1928 *Studies in Deceit.* 2 parts. New York: Macmillan.

HOBBS, NICHOLAS 1948 The Development of a Code of Ethical Standards for Psychology. *American Psychologist* 3:80–84.

HOBBS, NICHOLAS 1959 Science and Ethical Behavior. *American Psychologist* 14:217–225.

HOROWITZ, I. L. 1965 The Life and Death of Project Camelot. *Trans-action,* 3, no. 1:3–7, 44–47.

►KAPLAN, NORMAN; and STORER, NORMAN W. 1968 Science: VI. Scientific Communication. Volume 14, pages 112–117 in *International Encyclopedia of the Social Sciences.* Edited by David L. Sills. New York: Macmillan and Free Press.

KELMAN, HERBERT C. 1965 Manipulation of Human Behavior: An Ethical Dilemma for the Social Scientist. *Journal of Social Issues* 21:31–46.

KIMBALL, A. W. 1957 Errors of the Third Kind in Statistical Consulting. *Journal of the American Statistical Association* 52:133–142.

KING, ARNOLD J.; and SPECTOR, AARON J. 1963 Ethical and Legal Aspects of Survey Research. *American Psychologist* 18:204–208.

►KOHLBERG, LAWRENCE 1968 Moral Development. Volume 10, pages 483–494 in *International Encyclopedia of the Social Sciences.* Edited by David L. Sills. New York: Macmillan and Free Press.

LANGER, ELINOR 1966 Human Experimentation: New York Verdict Affirms Patient's Rights. *Science* 151:663–666.

MILGRAM, STANLEY 1964 Issues in the Study of Obedience: A Reply to Baumrind. *American Psychologist* 19:848–852.

MILLER, SAMUEL E.; and ROKEACH, MILTON 1966 [Letters] Psychology Experiments Without Subjects' Consent. *Science* 152:15 only.

PIAGET, JEAN (1932) 1948 *The Moral Judgment of the Child.* Glencoe, Ill.: Free Press. → First published in French.

PRICE, DEREK J. DE SOLLA 1964 Ethics of Scientific Publication. *Science* 144:655–657.

►PRICE, DON K. 1968 Science: IV. Science–Government Relations. Volume 14, pages 100–107 in *International Encyclopedia of the Social Sciences.* Edited by David L. Sills. New York: Macmillan and Free Press.

REDFIELD, ROBERT 1953 *The Primitive World and Its Transformations.* Ithaca, N.Y.: Cornell Univ. Press.

RUEBHAUSEN, OSCAR M.; and BRIM, ORVILLE G. JR. 1965 Privacy and Behavioral Research. *Columbia Law Review* 65:1184–1211.

SAUVY, ALFRED 1961 La responsibilité du statisticien devant l'opinion et les pouvoirs publics. International Statistical Institute, *Bulletin* 38, no. 2:573–578.

Science and the Race Problem: Report of the AAAS Committee on Science in the Promotion of Human Welfare. 1963 *Science* 142:558–561.

SHAKOW, DAVID 1965 Ethics for a Scientific Age: Some

Moral Aspects of Psychoanalysis. *Psychoanalytic Review* 52:335–348.

SHILS, EDWARD 1959 Social Inquiry and the Autonomy of the Individual. Pages 114–157 in Daniel Lerner (editor), *The Human Meaning of the Social Sciences*. New York: Meridian.

SILVERT, K. H. 1965 American Academic Ethics and Social Research Abroad: The Lesson of Project Camelot. American Universities Field Staff [*Reports From Foreign Countries*]: *West Coast South America Series* 12, no. 3.

►SIMMEL, ARNOLD 1968 Privacy. Volume 12, pages 480–487 in *International Encyclopedia of the Social Sciences*. Edited by David L. Sills. New York: Macmillan and Free Press.

SMITH, M. BREWSTER 1957 Of Prophecy and Privacy: [A Book Review of] *When Prophecy Fails*, by L. Festinger, H. W. Riecken, and S. Schachter. *Contemporary Psychology* 2, no. 4:89–92.

SNOW, CHARLES P. 1961 The Moral Un-neutrality of Science. *Science* 133:255–262. → With comments by Warren Weaver, Theodore M. Hesburgh, and William O. Baker.

Statement on Ethics of the Society for Applied Anthropology. 1963–1964 *Human Organization* 22:237 only.

U.S. OFFICE OF SCIENCE AND TECHNOLOGY 1967 *Privacy and Behavioral Research*. Washington: Government Printing Office.

WEBB, EUGENE et al. 1966 *Unobtrusive Measures: Nonreactive Research in the Social Sciences*. Chicago: Rand McNally.

WHITING, BEATRICE B. (editor) 1963 *Six Cultures: Studies of Child Rearing*. New York: Wiley.

ZIRKLE, CONWAY 1954 Citation of Fraudulent Data. *Science* 120:189–190.

Postscript

Formerly a concern mainly of professional people and specialists, the topic of ethics in research has now captured the attention of the general public, engaging the lively interest of policy makers, editorial writers, and even television commentators. In matters of professional ethics and conduct, widespread and informed debate is crucial. Ethics codes and guidelines are important in shaping scientific and professional conduct, but there is nothing so effective as heightened awareness and a quickened conscience. As a result of lively discussions, there has been progress in sharpening ethical issues about such crucial matters as privacy, informed consent, confidentiality of records, right to treatment and to due process when rights are intruded upon, and dealing with stress precipitated by experimentation. Perhaps the greatest advances have been made in defining appropriate ground rules for the conduct of research.

The most dramatic development has been government intervention to safeguard the rights of participants in experiments. The conscience of the community has been made evident through legislation, administrative guidelines for sponsored research programs, and court orders. In reaction to the injection of cancer cells in aged patients (see Beecher 1966), and to the withholding of treatment of patients with syphilis for experimental purposes (see Curran 1973), federal granting agencies now require systematic review of procedures for all research grants involving human subjects to ensure compliance with ethical standards and to protect participants from harm. The requirements embrace social science research as well as biological research. Research proposals must now be reviewed by a special committee at the investigator's institution, by the staff of the funding agency, and by the appropriate national advisory council. In similar spirit, requirements have been promulgated to ensure due process, confidentiality of records, and informed consent both for participants in research and for patients, prisoners, school children, and others whose lives have been intruded upon by the state for any reason. As early as 1954, the seemingly simple but brilliant ruling by Judge David L. Bazelon, chief judge of the U.S. Court of Appeals for the D.C. Circuit, stated that, in order to hold a patient involuntarily, mental hospital officials must report what they are doing for him. Subsequently there has been a series of rulings assuring right to treatment, right to education, freedom from institutional peonage, and a number of related similar protections. Even the use of intelligence tests in ways that perpetuate discrimination in schools has been proscribed. Government action is responsive to public concerns, of course, but in this instance, especially, government action has been remarkably successful in raising the general level of awareness of individual rights in settings where social scientists have been involved for many years, notably in schools, hospitals, and prisons.

Is government intrusion into the ethical practices of professions unwarranted and, as has been suggested, "counterproductive"? Probably not. In a democracy, government is the instrument through which the people assert their moral and ethical commitment as well as their pragmatic concerns. One might wish that government correctives had not been made necessary by the moral insensitivity of some researchers, but, in the long run, an informed and critical public is the ethical scientist's best friend.

A second development of major importance is the work of the Ad Hoc Committee on Ethical Standards in Psychological Research of the American Psychological Association (1973). Following procedures described in the main article, the committee collected several thousand critical incidents involving practices in research judged by psychologists to involve ethical issues. Building on these reports of actual cases, the committee formulated a number of ethical principles covering such topics as the locus of responsibility for ethical practices, the imperative of consultation with professional colleagues in doubtful situations, requirements of disclosure, limits on use of deception, freedom of the subject not to participate in an experiment, confidentiality of research protocols, and other matters. The report aroused complaints of unnecessary and unwarranted restraint on investigators. But an assumption underlying the work of the committee was that widespread involvement in developing ethical standards is essential, and that the process of developing a code of ethics is quite as important as the final product. The committee's work was successful in meeting this goal as well as in providing the most thoughtful guide available for helping social scientists design and conduct studies with informed sensitivity to the rights of participants. A revised statement on ethical standards extending beyond research concerns has been published in an editorial (1977) by the American Psychological Association.

A third development of consequence was the flowering of behavior modification and its application to almost every conceivable situation in an effort to replace art and intuition by precise procedures for making people behave properly (an interesting ethical issue itself: What is "proper" behavior?). This development is important to discussion on ethics and morality precisely because it requires serious address to the idea of *man as machine*. The metaphor, advanced by Descartes and made explicit by LaMetrie, has been around for a long time, but behavior modification has stripped the concept of its metaphorical character and made it an operational reality (under certain conditions). Control of behavior by other means, by surgery or drugs, for example, does not seem to join the issue of man as machine as explicitly as does behavior modification, which replaces individual "will" by individual "reinforcement history."

Behavior modification is based on the assumption that the behavior of both mice and men is controlled by environmental contingencies, and further, that these contingencies can often be externally controlled with precision (Skinner 1969). Literally thousands of experiments support this mechanistic view, in principle. Knowledge of the experiments had already created anxiety in philosophical and theological camps when B. F. Skinner's *Beyond Freedom and Dignity* (1971) appeared. Since then the debate has raged furiously, with moral and ethical considerations foremost. The arguments have to do with the assumptions about the nature of man, with various assessments of the efficacy of behavior controlled technology, with free will and determinism, and with the responsibilities of controllers.

Fourth in this inventory of developments that have advanced understanding of the ethics and morality of designed intrusion into the lives of individuals should be put the important worldwide debate on the issue of termination of life, including abortion, experimentation on the fetus, and the extended prolongation of life by extraordinary measures. Although such issues appear at first glance to be far removed from social science research, their implications are direct. The debate on morality and ethics has moved out of the academy into living rooms, market places, and legislative forums. Social science investigators will be fortunate indeed if their work evokes such widespread concern in the years ahead. Comparable public concern would be for the social scientist the surest protection against egregious error, against the hubris of men and women obedient to the technologies they presume to command.

In the design of research, the social scientist must frequently face the question of how much instrusion, how much inconvenience, or even how much harm a participant in research can be expected to accede to in the interest of contributing to the gaining of new knowledge. Relevant to such decisions are the extraordinarily restraining guidelines for research promulgated in 1976 by the U.S. National Institutes of Health, setting boundaries on research on recombinant DNA. Heretofore, popes and princes and lesser mortals have attempted to impose limits on man's inquiry into nature. This restriction on a segment of biological research appears to be the first time in history that scientists themselves have stated publicly, after careful deliberation, that research must be limited because a problem may be too dangerous to explore. Should research in social science ever lead to the kind of precision and control now possible in biology, will comparable restraints on social science research be required?

In this inventory of important recent events bearing upon ethics and social science research, attention should be called to several symposia, institutes, journals, and professorships originated since 1968 and dedicated to the exploration of ethical and moral issues associated with advances in science. Notable among institutes are the Institute of Society, Ethics, and Life Sciences in New

York and The Joseph P. Kennedy Institute of Bioethics at Georgetown University. Journals include *Soundings*, *Philosophy and Public Affairs*, *Journal of Medicine and Philosophy*, *Bioethics Digest*, *Journal of Moral Education*, *Journal of Medical Ethics*, and the *Hastings Center Report*. The Hastings Center (New York City) also publishes the *Bibliography of Society, Ethics, and the Life Sciences*. The 1973 Loyola Symposium (Kennedy 1975) also addressed human rights and psychological research. At the 1976 meeting of the American Psychological Association, a number of papers on human freedom were presented by Kenneth E. Clark, D. O. Hebb, David McClelland, Carl Rogers, B. F. Skinner, and others.

In the late 1960s and early 1970s there occurred in most of the social science professional groups a strong challenge to the conventional modes of work, to the values associated with the disciplines, and to the relationship of social science to society. The movement, which opponents viewed as an effort to radicalize the disciplines and which advocates viewed as an effort to make disciplines aware of their social responsibilities, has passed its peak in intensity. Its influence, while clearly substantial, has not been systematically assessed. Associated with the movement are the names of Ivan Illich, in education (1971); Thomas Szasz, in psychiatry (1970); Alvin Gouldner, in sociology (1970); Samuel Bowles and Herbert Gintis, in economics and education (1976); and Charles McCoy and John Playford, in political science (1968).

It is necessary to record, sadly, the occurrence of two serious and widely publicized incidents involving falsification of research data. A young investigator in a cancer research institute "painted black patches on the skin of white mice to make it appear that he had successfully transplanted skin between genetically incompatible animals" (Culliton 1974, p. 644). An eminent psychologist (see Wade 1976) was found to have invented data and even names of joint investigators in a largely fabricated report of a research program on human intelligence. It has been suggested that these incidents and others reflect the moral temper of the times, an interpretation I reject. More probably, they appear to be manifestations of ever present human frailty. The incidents also attest to the eventual self-correction of consequential errors in the scientific enterprise; science carries an intrinsic ethical imperative.

Ethical standards for research in social science should ultimately be founded in some theoretical or philosophical system; they should represent more than a description of current practice. Several contributions have been made to the development of such a foundation in recent years. An important event was the publication of John Rawls's *Theory of Justice* (1976). Working in the rational–intuitive tradition of Locke and Kant, Rawls proposes a social contract based on the concept of "justice as fairness" in which parties of assumed equal power negotiate the ground rules of their association on a rational basis. Closer to the traditions of empirical social science is the work of John F. Scott (1971) and Lawrence Kohlberg (1969). Scott holds that moral commitment is learned and that operant conditioning is the basis of learning. He elaborates this formula into an interesting social psychology of ethics. Kohlberg, in the tradition of Piaget, uses children's verbal responses to moral problems as a basis for constructing a developmental scale of moral maturity. Drawing on economics, sociology, political science, and psychology, Duncan MacRae, Jr., in an important synthesis of social science and ethics, argues that "scientific propositions and ethical assertions, while clearly distinguishable, may be fruitfully combined in academic disciplines concerned with the study of man and society" (1976, p. 5).

If one may speculate about the future as well as record the past, two shifts in the locus of concern on ethical standards seem likely.

First, a commitment to others, to the larger community, seems called for. In the past decade, discussions of ethics have strongly emphasized the rights of individuals. Indeed, in the society as a whole we may be in a period of overemphasis on individual rights. Self-gratification is ascendent; altruism is in decline (Marin 1975; Bell 1976). Survival of the species (an ultimate criterion of ethical systems) may depend on the nurturance of altrusitic impulse (Wilson 1976). Some movement back toward the center that would give appropriate weight to the legitimate needs of the social order may likely occur. Ethical thought would be the better for it.

Second, deliberations about ethics in the past decade have emphasized due process in the protection of individual rights, and important achievements have resulted therefrom. But emphasis on process is not sufficient; it is simply easier to deal with than more fundamental issues. The fundamental question of ethics in research involves the human costs and the human benefits of acquiring new knowledge. Due process, a basic precondition of ethics research, should be assumed in the calculus. How much intrusion, deprivation, or pain should the individual be expected to contribute in order that his fellow man may benefit from new knowledge that may be acquired through research? How beneficial must the research be and to whom? Always one must go beyond process and ask the central issue in ethics: Which is the lesser and which the better good?

Ethical issues in statistics. Ethical issues arise in statistical practice and influence statistical theory. For example, statistical problems of confidentiality and privacy have been much discussed; a consideration of statistical, as contrasted with administrative, data is given in *Records, Computers, and the Rights of Citizens* (U.S. Department of Health, Education, and Welfare 1973). On the more technical side, papers such as Fellegi (1972) treat problems and procedures of maintaining confidentiality when both the producer and consumer of statistical compilations have access to modern computational equipment. Procedures to permit linkage or matching of data, so that various pieces of information about the same individual can be brought together, have been much discussed, along with devices to maintain confidentiality without losing the scientific value of linkage (see Boruch 1972).

Integrity of government and other statisticians has, from time to time, been both questioned and defended. Like any other professional, a statistician may face conflict between professional or scientific codes and the mission—or perceived mission—of an employer. Some pertinent discussions are Hauser (1973), American Statistical Association–Federal Statistics Users' Conference (1973), and Chambers (1965). Helpful treatments of the integrity problem are also given by Deming (1965) and Freeman (1963).

Statisticians also have special concern with suppressed or fraudulent data. Some references are given in the main article (for example, Zirkle 1954). Another is Culliton (1974). Analyses of the fraudulent data from psychological research referred to above are also germane. (See, for example, Wade 1976 and Kamin 1977.)

The term "informed consent" refers to an explanation to prospective experimental subjects or to prospective survey respondents of the purposes, procedures, analyses, uses, and possible effects of the experiment or survey. Much writing in this area urges the ethical importance of informed consent in varying degrees and stringency. There is a qualifying counterargument that the more strongly one requires informed consent, the more special will be the effective populations of volunteer experiment subjects and of survey respondents. Not very much is known empirically about the effects of informed consent, but there is active, on-going work. Some proponents of informed consent go so far as to urge it even for what would otherwise be unobtrusive measurement (Webb et al. 1966). The argument is that resulting analyses and publication might affect or reflect on social groups to which the observed persons belong. Opponents say that such overscrupulousness would so hinder scientific advance that all social groups would suffer.

Some ethical issues arise in connection with properly controlled and randomized experiments. [*See* EXPERIMENTAL DESIGN; PUBLIC POLICY AND STATISTICS.] Is it fair, for example, to deny some patients an experimental medical treatment? There is another way of asking that question: Is it fair to subject some patients to experimental treatment that may do them harm? Does it mitigate or exacerbate these questions if assignment to treatment or control is made at random? If assignment is not made with random input, or if there is not a control group, then conclusions may be wrong or misleading, and thus future patients may be at unnecessary peril. Discussions of these important problems are given by Chalmers et al. (1972), Gilbert, Light, and Mosteller (1975), and Meier (1975). For discussion in the context of social psychological experiments, see Aronson (1972, chapter 9).

NICHOLAS HOBBS

ADDITIONAL BIBLIOGRAPHY

Reviews of ethical issues focusing on social research include Freund 1970 *and* Rivlin & Timpane 1975.

AMERICAN PSYCHOLOGICAL ASSOCIATION, AD HOC COMMITTEE ON ETHICAL STANDARDS IN PSYCHOLOGICAL RESEARCH 1973 *Ethical Principles in the Conduct of Research With Human Participants.* Washington: The Association.

AMERICAN STATISTICAL ASSOCIATION–FEDERAL STATISTICS USERS' CONFERENCE, COMMITTEE ON THE INTEGRITY OF FEDERAL STATISTICS 1973 Maintaining the Professional Integrity of Federal Statistics. *American Statistician* 27, no. 2:58–67.

ARONSON, ELLIOT (1972) 1976 *The Social Animal.* 2d ed. San Francisco: Freeman. → See especially Chapter 9.

BEECHER, HENRY K. 1966 Ethics and Clinical Research. *New England Journal of Medicine* 274, no. 24:1354–1360.

BELL, DANIEL 1976 *The Cultural Contradictions of Capitalism.* New York: Basic Books.

BORUCH, ROBERT F. 1972 Strategies for Eliciting and Merging Confidential Social Research Data. *Policy Sciences* 3:275–297.

BOWLES, SAMUEL; and GINTIS, HERBERT 1976 *Schooling in Capitalist America: Educational Reform and the Contradictions of Economic Life.* New York: Basic Books.

CHALMERS, THOMAS C.; BLOCK, JEROME B.; and LEE, STEPHANIE 1972 Controlled Studies in Clinical Cancer Research. *New England Journal of Medicine* 287, no. 2:75–78.

Chambers, S. Paul 1965 Statistics and Intellectual Integrity. *Journal of the Royal Statistical Society* Series A 128:1–16.

Culliton, Barbara J. 1974 The Sloan–Kettering Affair. Parts 1 and 2. *Science* 184:644–650; 1154–1157. → Part 1: A Story Without a Hero. Part 2: An Uneasy Resolution.

Curran, William J. 1973 The Tuskegee Syphilis Study. *New England Journal of Medicine* 289, no. 14:730–731.

Deming, W. Edwards 1965 Principles of Professional Statistical Practice. *Annals of Mathematical Statistics* 36:1883–1900.

Editorial 1977 Revised Ethical Standards of Psychologists. *APA Monitor* 8:22–23.

Fairley, William B.; and Mosteller, Frederick (editors) 1977 *Statistics and Public Policy*. Reading, Mass.: Addison-Wesley.

Fellegi, I. P. 1972 On the Question of Statistical Confidentiality. *Journal of the American Statistical Association* 67:7–18.

Freeman, William W. K. 1963 Training of Statisticians in Diplomacy to Maintain Their Integrity. *American Statistician* 17, no. 5:16–20.

Freund, Paul A. (editor) 1970 *Experimentation With Human Subjects*. New York: Braziller.

Gilbert, John P.; Light, Richard J.; and Mosteller, Frederick 1975 Assessing Social Innovations: An Empirical Base for Policy. Pages 39–193 in Carl A. Bennett and Arthur A. Lumsdaine (editors), *Evaluation and Experiment: Some Critical Issues in Assessing Social Programs*. New York: Academic Press. → A shortened version is reprinted in Fairley and Mosteller (1977, pp. 185–241).

Gouldner, Alvin W. 1970 *The Coming Crisis of Western Sociology*. New York: Basic Books.

Hauser, Philip M. 1973 Statistics and Politics. *American Statistician* 27, no. 2:68–71.

Hobbs, Nicholas 1948 The Development of a Code of Ethical Standards for Psychology. *American Psychologist* 3:80–84.

Illich, Ivan D. 1971 *Deschooling Society*. New York: Harper & Row.

Kamin, Leon J. 1977 Burt's IQ Data. *Science* 195:246–248. → Letter to the editor.

Kennedy, Eugene C. (editor) 1975 *Human Rights and Psychological Research: A Debate on Psychology and Ethics*. New York: Crowell. → Based on the Loyola Symposium on Psychology and Ethics, May 2, 1973.

Kohlberg, Lawrence L. 1969 Stage and Sequence: The Cognitive Developmental Approach to Socialization. Pages 347–480 in David A. Goslin (editor), *Handbook of Socialization Theory and Research*. Chicago: Rand McNally.

Kurtz, Paul et al. 1977 *The Ethics of Teaching and Scientific Research*. Buffalo, N.Y.: Prometheus.

McCoy, Charles A.; and Playford, John (editors) 1968 *Apolitical Politics: A Critique of Behavioralism*. New York: Crowell.

MacRae, Duncan Jr. 1976 *The Social Function of Social Science*. New Haven: Yale Univ. Press.

Marin, Peter 1975 The New Narcissism. *Harper's Magazine* 251, Oct.:45–56.

Medical Research: Statistics and Ethics. 1977 *Science* 198:677–705. → Comprises articles adapted from lectures and discussions presented at the Birnbaum Memorial Symposium held at the Memorial Sloan–Kettering Cancer Center, New York, May 27: Valerie Miké and Robert A. Good, "Old Problems, New Challenges"; John W. Tukey, "Some Thoughts on Clinical Trials, Especially Problems of Multiplicity"; John P. Gilbert, Bucknam McPeek, and Frederick Mosteller, "Statistics and Ethics in Surgery and Anesthesia"; Victor Herbert, "Acquiring New Information While Retaining Old Ethics"; Jerome Cornfield, "Carcinogenic Risk Assessment"; and André Cournand, "The Code of the Scientist and Its Relationship to Ethics."

Meier, Paul 1975 Statistics and Medical Experimentation. *Biometrics* 31:511–529.

Rawls, John 1971 *A Theory of Justice*. Cambridge, Mass.: Belknap Press.

Rivlin, Alice M.; and Timpane, Paul M. (editors) 1975 *Ethical and Legal Issues of Social Experimentation*. Washington: Brookings.

Scott, John F. 1971 *Internalization of Norms: A Sociological Theory of Moral Commitment*. Englewood Cliffs, N.J.: Prentice-Hall.

Skinner, B. F. 1969 *Contingencies of Reinforcement: A Theoretical Analysis*. New York: Appleton-Century-Crofts. → A paperback edition was published in 1972 by Prentice-Hall.

Skinner, B. F. 1971 *Beyond Freedom and Dignity*. New York: Knopf. → A paperback edition was published in 1972 by Bantam.

Szasz, Thomas S. 1970 *The Manufacture of Madness: A Comparative Study of the Inquisition and the Mental Health Movement*. New York: Harper & Row.

U.S. Department of Health, Education, and Welfare, Secretary's Advisory Committee on Automated Personnel Data Systems 1973 *Records, Computers, and the Rights of Citizens: Report*. DHEW Publication No. (OS) 73-94. Washington: Government Printing Office.

Wade, Nicholas 1976 IQ and Heredity: Suspicion of Fraud Beclouds Classic Experiment. *Science* 194:916–919.

Wilson, Edward O. 1976 *Sociobiology, the New Synthesis*. Cambridge, Mass.: Belknap Press.

EVALUATION RESEARCH

Ours is an age of social-action programs, where large organization and huge expenditures go into the attempted solution of every conceivable social problem. Such programs include both private and public ventures and small-scale and large-scale projects, ranging in scope from local to national and international efforts at social change. Whenever men spend time, money, and effort to help solve social problems, someone usually questions the effectiveness of their actions. Sponsors, critics, the public, even the actors themselves, seek signs that their program is successful. Much of the assessment of action programs is irregular and, often by necessity, based upon personal judgments of supporters or critics, impressions, anecdotes, testimonials, and miscellaneous information available for the evaluation. In recent years, however, there has been a striking change in attitudes toward evaluation activities and the type and quality of evidence that is acceptable for determining the relative success or failure of social-action programs.

Two trends stand out in the modern attitude toward evaluation. First, evaluation has come to be expected as a regular accompaniment to rational social-action programs. Second, there has been a movement toward demanding more systematic, rigorous, and objective evidence of success. The application of social science techniques to the appraisal of social-action programs has come to be called evaluation research.

Examples of the applications of evaluation research are available from a wide variety of fields. One of the earliest attempts at building evaluation research into an action program was in the field of community action to prevent juvenile delinquency. The 1937 Cambridge–Somerville Youth Study provided for an experimental and a control group of boys, with the former to receive special attention and advice from counselors and other community agencies. The plan called for a ten-year period of work with the experimental group followed by an evaluation that would compare the record of their delinquent conduct during that decade with the record of the control group. The results of the evaluation (see Powers & Witmer 1951) showed no significant differences in conduct favorable to the program. A subsequent long-term evaluation of the same program failed to find new evidence of less criminal activity by persons in the experimental group but added a variety of new theoretical analyses to the evaluation (McCord et al. 1959).

Several evaluations of programs in citizenship training for young persons have built upon one another, thus providing continuity in the field. Riecken (1952) conducted an evaluation of summer work camps sponsored by the American Friends Service Committee to determine their impact on the values, attitudes, and opinions of the participants. His work was useful in specifying those areas in which the program was successful or unsuccessful as well as pointing up the importance of measuring unsought by-products of action programs. Subsequently, Hyman, Wright, and Hopkins carried out a series of evaluations of another youth program, the Encampment for Citizenship (1962). Their research design was complex, including a comparison of campers' values, attitudes, opinions, and behavior before and after a six-week program of training; follow-up surveys six weeks and four years after the group left the program; three independent replications of the original study on new groups of campers in later years; and a sample survey of alumni of the program. These various studies demonstrated the effectiveness of the program in influencing campers' social attitudes and conduct; they also examined the dynamics of attitudinal change.

Evaluations have been made in such varied fields as intergroup relations, induced technological change, mass communications, adult education, international exchange of persons for training or good will, mental health, and public health. Additional examples of applications of evaluation research, along with discussions of evaluation techniques, are presented by Klineberg and others in a special issue of the *International Social Science Bulletin* (1955) and in Hyman and Wright (1966).

Defining characteristics

A scientific approach to the assessment of a program's achievements is the hallmark of modern evaluation research. In this respect evaluation research resembles other kinds of social research in its concern for objectivity, reliability, and validity in the collection, analysis, and interpretation of data. But it can be distinguished as a special form of social research by its purpose and the conditions under which the research must be conducted. Both of these factors affect such components of the research process as study design and its translation into practice, allocation of research time and other resources, and the value or worth to be put upon the empirical findings.

The primary purpose of evaluation research is "to provide objective, systematic, and comprehensive evidence on the degree to which the program achieves its intended objectives plus the degree to which it produces other unanticipated consequences, which when recognized would also be regarded as relevant to the agency" (Hyman et al. 1962, pp. 5–6). Evaluation research thus differs in its emphasis from such other major types of social research as exploratory studies, which seek to formulate new problems and hypotheses, or explanatory research, which places emphasis on the testing of theoretically significant hypotheses, or descriptive social research, which documents the existence of certain social conditions at a given moment or over time (Selltiz et al. 1959). Since the burden is on the evaluator to provide firm evidence on the effects of the program under study, he favors a study design that will tend toward maximizing such evidence and his confidence in conclusions drawn from it. Although good evaluation research often seeks explanations of a program's success or failure, the first concern is to obtain basic evidence on effectiveness, and therefore most research resources are allocated to this goal.

The conditions under which evaluation research is conducted also give it a character distinct from other forms of social research. Evaluation research is applied social research, and it differs from other modes of scholarly research in bringing together an outside investigator to guarantee objectivity and a client in need of his services. From the initial formulation of the problem to the final interpretation of findings, the evaluator is duty-bound to keep in mind the very practical problem of assessing the program under study. As a consequence he often has less freedom to select or reject certain independent, dependent, and intervening variables than he would have in studies designed to answer his own theoretically formulated questions, such as might be posed in basic social research. The concepts employed and their translation into measurable variables must be selected imaginatively but within the general framework set by the nature of the program being evaluated and its objectives (a point which will be discussed later). Another feature of evaluation research is that the investigator seldom has freedom to manipulate the program and its components, i.e., the independent variable, as he might in laboratory or field experiments. Usually he wants to evaluate an ongoing or proposed program of social action in its natural setting and is not at liberty, because of practical and theoretical considerations, to change it for research purposes. The nature of the program being evaluated and the time at which his services are called upon also set conditions that affect, among other things, the feasibility of using an experimental design involving before-and-after measurements, the possibility of obtaining control groups, the kinds of research instruments that can be used, and the need to provide for measures of long-term as well as immediate effects.

The recent tendency to call upon social science for the evaluation of action programs that are local, national, and international in scope (a trend which probably will increase in future years) and the fact that the application of scientific research procedures to problems of evaluation is complicated by the purposes and conditions of evaluation research have stimulated an interest in methodological aspects of evaluation among a variety of social scientists, especially sociologists and psychologists. Methodological and technical problems in evaluation research are discussed, to mention but a few examples, in the writings of Riecken (1952), Klineberg (1955), Hyman et al. (1962), and Hayes (1959).

While it is apparent that the specific translation of social-science techniques into forms suitable for a particular evaluation study involves research decisions based upon the special nature of the program under examination, there are nonetheless certain broad methodological questions common to most evaluation research. Furthermore, certain principles of evaluation research can be extracted from the rapidly growing experience of social scientists in applying their perspectives and methods to the evaluation of social-action programs. Such principles have obvious importance in highlighting and clarifying the methodological features of evaluation research and in providing practical, if limited, guidelines for conducting or appraising such research. The balance of this article will discuss certain, but by no means all, of these compelling methodological problems.

Methodological steps and principles

The process of evaluation has been codified into five major phases, each involving particular methodological problems and guiding principles (see Hyman et al. 1962). They are (1) the conceptualization and measurement of the objectives of the program and other unanticipated relevant outcomes; (2) formulation of a research design and the criteria for proof of effectiveness of the program, including consideration of control groups or alternatives to them; (3) the development and application of research procedures, including provisions for the estimation or reduction of errors in measurement; (4) problems of index construction and the proper evaluation of effectiveness; and (5) procedures for understanding and explaining the findings on effectiveness or ineffectiveness. Such a division of the process of evaluation is artificial, of course, in the sense that in practice the phases overlap and it is necessary for the researcher to give more or less constant consideration to all five steps. Nevertheless it provides a useful framework for examining and understanding the essential components of evaluation research.

Conceptualization. Each social-action program must be evaluated in terms of its particular goals. Therefore, evaluation research must begin with their identification and move toward their specification in terms of concepts that, in turn, can be translated into measurable indicators. All this may sound simple, perhaps routine, compared with the less structured situation facing social researchers engaged in formulating research problems for theoretical, explanatory, descriptive, or other kinds of basic research. But the apparent simplicity is deceptive, and in practice this phase of evaluation research repeatedly has proven to be both critical and difficult for social researchers working in such varied areas as mental health (U.S. Dept. of Health, Education & Welfare 1955), juvenile delinquency (Witmer & Tufts 1954), adult education (Evaluation Techniques 1955), and youth programs for citizenship training (Riecken 1952; Hyman et al. 1962), among others. As an example, Witmer and Tufts raise such questions about the meaning of the concept "delinquency prevention" as: What is to be prevented? Who is to be deterred? Are we talking only about "official" delinquency? Does prevention mean stopping misbehavior before it occurs? Does it mean reducing the frequency of misbehavior? Or does it mean reducing its severity?

Basic concepts and goals are often elusive, vague, unequal in importance to the program, and sometimes difficult to translate into operational terms. What is meant, for example, by such a goal as preparing young persons for "responsible citizenship"? In addition, the evaluator needs to consider possible effects of the program which were unanticipated by the action agency, finding clues from the records of past reactions to the program if it has been in operation prior to the evaluation, studies of similar programs, the social-science literature, and other sources. As an example, Carlson (1952) found that a mass-information campaign against venereal disease failed to increase public knowledge about these diseases; nevertheless, the campaign had the unanticipated effect of improving the morale of public health workers in the area, who in turn did a more effective job of combating the diseases. The anticipation of both planned and unplanned effects requires considerable time, effort, and imagination by the researcher prior to collecting evidence for the evaluation itself.

Research design. The formulation of a research design for evaluation usually involves an attempt to approximate the ideal conditions of a controlled experiment, which measures the changes produced by a program by making comparisons of the dependent variables before and after the program and evaluating them against similar measurements on a control group that is not involved in the program. If the control group is initially similar to the group exposed to the social-action program, a condition achieved through judicious selection, matching, and randomization, then the researcher can use the changes in the control group as a criterion against which to estimate the degree to which changes in the experimental group were probably caused by the program under study. To illustrate, suppose that two equivalent groups of adults are selected for a study on the effects of a training film intended to impart certain information to the audience. The level of relevant information is measured in each group prior to the showing of the film; then one group sees the film while the other does not; finally, after some interval, information is again measured. Changes in the amount of information held by the experimental group cannot simply be attributed to the film; they may also reflect the influence of such factors in the situation as exposure to other sources of information in the interim period, unreliability of the measuring instruments, maturation, and other factors extraneous to the program itself. But the control group presumably also experienced such nonprogrammatic factors, and therefore the researcher can subtract the amount of change in information demonstrated by it from the changes shown by the experimental group, thereby determining how much of the gross change in the latter group is due to the exclusive influence of the program.

So it is in the ideal case, such as might be achieved under laboratory conditions. In practice, however, evaluation research seldom permits such ideal conditions. A variety of practical problems requires alterations in the ideal design. As examples, suitable control groups cannot always be found, especially for social-action programs involving efforts at large-scale social change but also for smaller programs designed to influence volunteer participants; also ethical, administrative, or other considerations usually prevent the random assignment of certain persons to a control group that will be denied the treatment offered by the action programs.

In the face of such obstacles, certain methodologists have taken the position that a slavish insistence on the ideal control-group experimental research design is unwise and dysfunctional in evaluation research. Rather, they advocate the ingenious use of practical and reasonable alternatives to the classic design (see Hyman et al. 1962; and Campbell & Stanley 1963). Under certain conditions, for example, it is possible to estimate the amount of change that could have been caused by extraneous events, instability of measurements, and natural growth of participants in a program by examining the amount of change that occurred among participants in programs similar to the one being evaluated. Using such comparative studies as "quasi-control" groups permits an estimate of the relative effectiveness of the program under study, i.e., how much effect it has had over and above that achieved by another program and assorted extraneous factors, even though it is impossible to isolate the specific amount of change caused by the extraneous factors. Another procedure for estimating the influence of nonprogrammatic factors is to study the amount of change which occurs among a sample of the population under study during a period of time prior to the introduction of the action program, using certain of the ultimate participants as a kind of control upon themselves, so to speak. Replications of the evaluation study, when possible, also provide safeguards against attributing too much or too little effect to the program under study. Admittedly, all such practical alternatives to the controlled experimental design have serious limitations and must be used with judgment; the classic experimental design remains preferable whenever possible and serves as an ideal even when impractical. Nevertheless, such expedients have proven useful to evaluators and have permitted relatively rigorous evaluations to be conducted under conditions less perfect than those found in the laboratory.

Error control. Evaluation studies, like all social research, involve difficult problems in the selection of specific research procedures and the provision for estimating and reducing various sources of error, such as sampling bias, bias due to nonresponse, measurement errors arising in the questions asked or in recording of answers, deliberate deception, and interviewer bias. The practices employed to control such errors in evaluation research are similar to those used in other forms of social research, and no major innovations have been introduced.

Estimating effectiveness. To consider the fourth stage in evaluation, a distinction needs to be made between demonstrating the effects of an action program and estimating its effectiveness. Effectiveness refers to the extent to which the program achieves its goals, but the question of just how much effectiveness constitutes success and justifies the efforts of the program is unanswerable by scientific research. It remains a matter for judgment on the part of the program's sponsors, administrators, critics, or others, and the benefits, of course, must somehow be balanced against the costs involved. The problem is complicated further by the fact that most action programs have multiple goals, each of which may be achieved with varying degrees of success over time and among different subgroups of participants in the program. To date there is no general calculus for appraising the over-all net worth of a program.

Even if the evaluation limits itself to determining the success of a program in terms of each specific goal, however, it is necessary to introduce some indexes of effectiveness which add together the discrete effects within each of the program's goal areas. Technical problems of index and scale construction have been given considerable attention by methodologists concerned with various types of social research (see Lazarsfeld & Rosenberg 1955). But as yet there is no theory of index construction specifically appropriate to evaluation research. Steps have been taken in this direction, however, and the utility of several types of indexes has been tentatively explored (see Hyman et al. 1962). One type of difficulty, for example, arises from the fact that the amount of change that an action program produces may vary from subgroup to subgroup and from topic to topic, depending upon how close to perfection each group was before the program began. Thus, an information program can influence relatively fewer persons among a subgroup in which, say, 60 per cent of the people are already informed about the topic than among another target group in which only 30 per cent are initially informed. An "effectiveness index" has been successfully employed to help solve the problem of weighting effectiveness in the light of such restricted ceilings for change (see Hovland et al. 1949; and Hyman et al. 1962). This index, which expresses actual change as a proportion of the maximum change that is possible given the initial position of a group on the variable under study, has proven to be especially useful in evaluating the *relative* effectiveness of different programs and the relative effectiveness of any particular program for different subgroups or on different variables.

Understanding effectiveness. In its final stage, evaluation research goes beyond the demonstration of a program's effects to seek information that will help to account for its successes and failures. The reasons for such additional inquiry may be either practical or theoretical.

Sponsors of successful programs may want to duplicate their action program at another time or under other circumstances, or the successful program may be considered as a model for action by others. Such emulation can be misguided and even dangerous without information about which aspects of the program were most important in bringing about the results, for which participants in the program, and under what conditions. Often it is neither possible nor necessary, however, to detect and measure the impact of each component of a social-action program. In this respect, as in others noted above, evaluation research differs from explanatory survey research, where specific stimuli are isolated; and from experimental designs, where isolated stimuli are introduced into the situation being studied. In evaluation research the independent variable, i.e., the program under study, is usually a complex set of activities no one of which can be separated from the others without changing the nature of the program itself. Hence, explanations of effectiveness are often given in terms of the contributions made by certain gross features of the program, for example, the total impact of didactic components versus social participation in a successful educational institution.

Gross as such comparisons must be, they nevertheless provide opportunities for testing specific hypotheses about social and individual change, thereby contributing to the refinement and growth of social science theories. It is important to remember, however, that such gains are of secondary concern to evaluation research, which has as its primary goal the objective measurement of the effectiveness of the program.

Certain forms of research design promise to yield valuable results both for the primary task of evaluation and its complementary goal of enlarging social knowledge. Among the most promising designs are those that allow for *comparative* evaluations of different social-action programs, *replication* of evaluations of the same program, and *longitudinal* studies of the long-range impact of programs. Comparative studies not only demonstrate the differential effectiveness of various forms of programs having similar aims but also provide a continuity in research which permits testing theories of change under a variety of circumstances. Replicative evaluations add to the confidence in the findings from the initial study and give further opportunity for exploring possible causes of change. Longitudinal evaluations permit the detection of effects that require a relatively long time to occur and allow an examination of the stability or loss of certain programmatic effects over time and under various natural conditions outside of the program's immediate control.

Viewed in this larger perspective, then, evaluation research deserves full recognition as a social science activity which will continue to expand. It provides excellent and ready-made opportunities to examine individuals, groups, and societies in the grip of major and minor forces for change. Its applications contribute not only to a science of social planning and a more rationally planned society but also to the perfection of social and psychological theories of change.

CHARLES R. WRIGHT

[*See also* EXPERIMENTAL DESIGN; SURVEY ANALYSIS.]

BIBLIOGRAPHY

CAMPBELL, DONALD T.; and STANLEY, JULIAN C. 1963 Experimental and Quasi-experimental Designs for Research on Teaching. Pages 171–246 in Nathaniel L. Gage (editor), *Handbook of Research on Teaching.* Chicago: Rand McNally.

CARLSON, ROBERT O. 1952 The Influence of the Community and the Primary Group on the Reactions of Southern Negroes to Syphilis. Ph.D. dissertation, Columbia Univ.

Evaluation Techniques. 1955 *International Social Science Bulletin* 7:343–458.

HAYES, SAMUEL P. 1959 *Measuring the Results of Development Projects: A Manual for the Use of Field Workers.* Paris: UNESCO.

HOVLAND, CARL I.; LUMSDAINE, ARTHUR A.; and SHEFFIELD, FREDERICK D. 1949 *Experiments on Mass Communication.* Studies in Social Psychology in World War II, Vol. 3. Princeton Univ. Press.

[1]HYMAN, HERBERT H.; and WRIGHT, CHARLES R. 1966 Evaluating Social Action Programs. Unpublished manuscript.

HYMAN, HERBERT H.; WRIGHT, CHARLES R.; and HOPKINS, TERENCE K. 1962 *Applications of Methods of Evaluation: Four Studies of the Encampment for Citizenship.* Berkeley: Univ. of California Press.

KLINEBERG, OTTO 1955 Introduction: The Problem of Evaluation. *International Social Science Bulletin* 7: 346–352.

[2]LAZARSFELD, PAUL F.; and ROSENBERG, MORRIS (editors) 1955 *The Language of Social Research: A Reader in the Methodology of Social Research.* Glencoe, Ill.: Free Press.

MCCORD, WILLIAM; MCCORD, JOAN; and ZOLA, IRVING K. 1959 *Origins of Crime: A New Evaluation of the Cambridge–Somerville Youth Study.* New York: Columbia Univ. Press.

POWERS, EDWIN; and WITMER, HELEN L. 1951 *An Experiment in the Prevention of Delinquency.* New York: Columbia Univ. Press; Oxford Univ. Press.

RIECKEN, HENRY W. 1952 *The Volunteer Work Camp: A Psychological Evaluation.* Reading, Mass.: Addison-Wesley.

SELLTIZ, CLAIRE et al. (1959) 1962 *Research Methods in Social Relations.* New York: Holt.

U.S. DEPT. OF HEALTH, EDUCATION & WELFARE, NATIONAL INSTITUTES OF HEALTH 1955 *Evaluation in Mental Health: Review of Problem of Evaluating Mental Health Activities.* Washington: Government Printing Office.

WITMER, HELEN L.; and TUFTS, EDITH 1954 *The Effectiveness of Delinquency Prevention Programs.* Washington: Government Printing Office.

Postscript

A virtual explosion in evaluation activities, as forecast in the main article, has resulted in a large number of new research projects, conferences, and books about evaluation research. None of these recent developments, however, has changed the basic conceptual and methodological principles originally set forth.

Much of the impetus for evaluation research has come from the federal government's efforts to evaluate its programs in a variety of fields such as education, housing, employment, equal opportunities, and health (see, for examples, Wholey et al. 1970). Nonfederal organizations also have expanded their evaluation activities. This expansion has enlisted scholars from many disciplines besides sociology and psychology, including social welfare, education, public health, and communications, among others. Concern over the large sums of money spent on federal action programs, for example, has spurred interest in benefit–cost analysis, a specialty of economists, which attempts to assess the relative gains achieved for money and other resources spent. An initial burst of enthusiasm for benefit–cost analysis has been followed by a more qualified skepticism about its usefulness in evaluation research. Many of the valued outcomes of social action programs are not solely economic or easily translated into dollar values, for example, good health and quality of education (see Rivlin 1971). Regardless of disciplinary background, evaluators face the persistent problems of conceptualizing a program's output and of devising relevant and satisfactory indicators of achievement.

An increased recognition of the value of continuing social statistics—as social indicators of the state of the society and of its welfare—has sparked movements by statisticians and other social scientists for the regular collection of such data and for the systematic storage and retrieval of available social statistics, to aid in the evaluation of the long-range impact of action programs and other social forces. Secondary analysis of previously collected data from social surveys and from other sources of statistical data provides one new and promising method for the evaluation of long-range or enduring effects of changes in larger social institutions, such as education and health delivery systems (for an example, see Hyman et al. 1975).

As the demand for evaluation research has grown, concerned social scientists have pressed for a more active role in the planning of action programs. They desire, at the least, to help design social action programs in ways that permit scientific evaluation, through field experiments or other research designs, rather than to accept the role of technician called in to evaluate existing programs that were established without concern for the feasibility of testing their results (for examples, see Campbell 1969; Zurcher & Bonjean 1970). A committee of the Social Science Research Council has recently advocated social experimentation as a method for planning and evaluating social intervention (see Riecken & Boruch 1974; Social Science Research Council Conference on Social Experiments 1976). The committee views social experimentation "as a cycle that begins with program analysis; proceeds through the planning of an intervention, its development, experimental trial, and evaluation; and ends in either program implementation or in replanning the intervention" (1974, pp. 13–14). The book discusses some of the strengths and costs of social experimentation, its methodology, and various practical, social, and cultural factors affecting such research and the utilization of its findings. Whether or not social scientists in the future will be called upon to play a significant role in the planning and conduct of social action programs, for the purpose of scientific evaluation research by experimental or other methods, remains to be seen.

The development of the field was also signaled by the founding of a new journal, *Evaluation Quarterly.*

CHARLES R. WRIGHT

[*See also* PUBLIC POLICY AND STATISTICS.]

ADDITIONAL BIBLIOGRAPHY

CAMPBELL, DONALD T. 1969 Reforms as Experiments. *American Psychologist* 24:409–429.

CARO, FRANCIS G. (editor) 1971 *Readings in Evaluation Research.* New York: Russell Sage.

[1]HYMAN, HERBERT H.; and WRIGHT, CHARLES R. 1967 Evaluating Social Action Programs. Pages 741–782 in Paul F. Lazarsfeld, William H. Sewall, and Harold Wilensky (editors), *The Uses of Sociology.* New York: Basic Books. → A formal publication of Hyman and Wright (1966).

HYMAN, HERBERT H.; WRIGHT, CHARLES R.; and REED, JOHN SHELTON 1975 *The Enduring Effects of Education.* Univ. of Chicago Press.

[2]LAZARSFELD, PAUL F.; PASANELLA, ANN K.; and ROSENBERG, MORRIS (editors) 1972 *Continuities in the Language of Social Research.* New York: Free Press. → Forty-eight essays, continuing the theme of those in Lazarsfeld and Rosenberg (1955).

MULLEN, EDWARD J. et al. (editors) 1972 *Evaluation of Social Intervention.* San Francisco: Jossey-Bass. → Proceedings of a symposium held on the Lincoln Center Campus of Fordham University, Jan. 14–15, 1971.

RIECKEN, HENRY W.; and BORUCH, ROBERT F. 1974 *Social Experimentation: A Method for Planning and Evaluating Social Intervention.* New York: Academic Press.

RIVLIN, ALICE M. 1971 *Systematic Thinking for Social Action.* Washington: Brookings. → The H. Rowan Gaither Lectures in Systems Science.

ROSSI, PETER H.; and WILLIAMS, WALTER (editors) 1972 *Evaluating Social Programs: Theory, Practice, and Politics.* New York: Seminar Press.

SOCIAL SCIENCE RESEARCH COUNCIL CONFERENCE ON SOCIAL EXPERIMENTS, BOULDER, *1974* 1976 *Experimental Testing of Public Policy: Proceedings.* Edited by Robert F. Boruch and Henry W. Riecken. Boulder, Colo.: Westview.

WEISS, CAROL H. 1972 *Evaluation Research: Methods of Assessing Program Effectiveness.* Englewood Cliffs, N.J.: Prentice-Hall.

WHOLEY, JOSEPH S. et al. 1970 *Federal Evaluation Policy: Analyzing the Effects of Public Programs.* Washington: Urban Institute.

ZURCHER, LOUIS A. JR.; and BONJEAN, CHARLES M. (editors) 1970 *Planned Social Intervention: An Interdisciplinary Anthology.* Scranton, Pa.: Chandler.

EVIDENCE, LEGAL

See STATISTICS AS LEGAL EVIDENCE.

EXPERIMENTAL DESIGN

I
THE DESIGN OF EXPERIMENTS

In scientific research, the word "experiment" often denotes the type of study in which the investigator deliberately introduces certain changes into a process and makes observations or measurements in order to evaluate and compare the effects of different changes. These changes are called the *treatments*. Common examples of treatments are different kinds of stimuli presented to human subjects or animals or different kinds of situations with which the investigator faces them, in order to see how they respond. In exploratory work, the objective may be simply to discover whether the stimuli produce any measurable responses, while at a later stage in research the purpose may be to verify or disprove certain hypotheses that have been put forward about the directions and sizes of the responses to treatments. In applied work, measurement of the size of the response is often important, since this may determine whether a new treatment is practically useful.

A distinction is often made between a controlled experiment and an uncontrolled observational study. In the latter, the investigator does not interfere in the process, except in deciding which phenomena to observe or measure. Suppose that it is desired to assess the effectiveness of a new teaching machine that has been much discussed. An observational study might consist in comparing the achievement of students in those schools that have adopted the new technique with the achievement of students in schools that have not. If the schools that adopt the new technique show higher achievement, the objection may be raised that this increase is not necessarily caused by the machine, as the schools that have tried a new method are likely to be more enterprising and successful and may have students who are more competent and better prepared. Examination of previous records of the schools may support these criticisms. In a proper experiment on the same question, the investigator decides which students are to be taught by the new machine and which by the standard technique. It is his responsibility to ensure that the two techniques are compared on students of equal ability and degree of preparation, so that these criticisms no longer have validity.

The advantage of the proper experiment over the observational study lies in this increased ability to elucidate cause-and-effect relationships. Both types of study can establish associations between a stimulus and a response; but when the investigator is limited to observations, it is hard to find a situation in which there is only one explanation of the association. If the investigator can show by repeated experiments that the same stimulus is always followed by the same response and if he has designed the experiments so that other factors that might produce this response are absent, he is in a much stronger position to claim that the stimulus causes the response. (However, there are many social science fields where true experimentation is not possible and careful observational investigations are the only source of information.) [*See, for example,* EXPERIMENTAL DESIGN, *article on* QUASI-EXPERIMENTAL DESIGN; SURVEY ANALYSIS. *See also* Becker 1968; Wax 1968.]

Briefly, the principal steps in the planning of a controlled experiment are as follows. The treatments must be selected and defined and must be relevant to the questions originally posed. The *experimental units* to which the treatments are to be applied must be chosen. In the social sciences, the experimental unit is frequently a single animal or human subject. The unit may, however, be a group of subjects, for instance, a class in comparisons of teaching methods. An important point is that the choice of subjects and of the environmental conditions of the experiment determine the range of validity of the results.

○ The next step is to determine the size of the sample—the number of subjects or of classes. In general, the precision of the experiment increases as the sample size increases, but usually a balance must be struck between the precision desired and the costs involved. The method for allocating treatments to subjects must be specified, as must the detailed conduct of the experiment. Other factors that might influence the outcome must be controlled (by *blocking* or *randomization*, as discussed later) so that they favor each treatment equally. The responses or criteria by which the treatments will be rated must be defined. These may be simple classifications or measurements on a discrete or continuous scale. Like the treatments, the responses must be relevant to the questions originally posed.

► When data from the experiment become available, the statistical analysis of these data and the preparation of a report on the results of the experiment are final steps. This report faces a number of questions. Have differences in the effects of different treatments been clearly shown? What can be said about the sizes of these differences? To what types of experimental unit, for example, of human subjects, can the reader safely apply the reported results?

History. The early history of ideas on the planning of experiments appears to have been but little studied (Boring 1954). Modern concepts of experimental design are due primarily to R. A. Fisher, who developed them from 1919 to 1930 in the planning of agricultural field experiments at the Rothamsted Experimental Station in England. The main features of Fisher's approach are as follows (*randomization*, *blocking*, and *factorial experimentation* will be discussed later):

(1) The requirement that an experiment itself furnish a meaningful estimate of the underlying variability to which the measurements of the responses to treatments are subject.

(2) The use of randomization to provide these estimates of variability.

(3) The use of blocking in order to balance out known extraneous sources of variation.

(4) The principle that the statistical analysis of the results is determined by the way in which the experiment was conducted.

(5) The concept of factorial experimentation, which stresses the advantages of investigating the effects of different factors or variables in a single complex experiment, instead of devoting a separate experiment to each factor.

These ideas were stated very concisely by Fisher in 1925 and 1926 but more completely in 1935.

Experimental error

Some sources of experimental error. A major problem in experimentation is that the responses of the experimental units are influenced by many sources of variation other than the treatments. For example, subjects differ in their ability to perform a task under standard conditions: a treatment that is allotted to an unusually capable group of subjects will appear to do well; the instruments by which the responses are measured may be liable to errors of measurement; both the applied treatment and the environment may lack uniformity from one occasion to another.

In some experiments, the effects of subject-to-subject variation are avoided by giving every treatment to each subject in succession, so that comparisons are made within subjects. Even then, however, learning, fatigue, or delayed consequences of previously applied treatments may influence the response actually measured after a particular treatment.

The primary consequence of extraneous sources of variation, called *experimental errors*, is a masking of the effects of the treatments. The observed difference between the effects of two treatments is the sum of the true difference and a contribution due to these errors. If the errors are large, the experimenter obtains a poor estimate of the true difference; then the experiment is said to be of low precision.

Bias. It is useful to distinguish between random error and error due to bias. A bias, or systematic error, affects alike all subjects who receive a specific treatment. Random error varies from subject to subject. In a child growth study in which children were weighed in their clothes, a bias would arise if the final weights of all children receiving one treatment were taken on a cold day, on which heavy clothing was worn, while the children receiving a second treatment were weighed on a mild day, on which lighter clothing was worn. In general, bias cannot be detected in the analysis of the results, so that the conclusions drawn by statistical methods about the true effects of the treatments are misleading.

It follows that constant vigilance against bias is one of the requisites of good experimentation. The devices of randomization and blocking, if used intelligently, do much to guard against bias. Additional precautions are necessary in certain types of experiments. If the measurements are subjective evaluations or clinical judgments, the expectations and prejudices of the judges and subjects may influence the results if it is known which treatment any of the subjects received. Consequently, it is important to ensure, whenever it is feasible, that neither the subject nor the person taking the measurement knows which treatment the subject is receiving; this is called a "double blind" experiment. For example, in experiments that compare different drugs taken as pills all the pills should look alike and be administered in the same way. If there is a no-drug treatment, it is common practice to administer an inert pill, called a *placebo*, in order to achieve this concealment.

►Even in well-planned experiments, bias may enter in more subtle forms. In education, a new method of teaching may be highly successful in early experiments as compared with standard methods, partly because it is a welcome change from a standard routine. It may be much less successful when it has been widely adopted and is no longer a novelty. Many experiments on human behavior and resistance to stress and pain are carried out on subjects who are volunteers in one sense or another; because they may behave or react differently from nonvolunteers, conclusions from the experiments have limited applicability. In their book on the volunteer subject, Rosenthal and Rosnow (1975) quote McNemar's earlier statement (1946) "The existing science of human behavior is largely the science of the behavior of sophomores."

Methods for reducing experimental error. Several devices are used to remove or decrease bias and random errors due to extraneous sources of variation that are thought to be substantial. One group of devices may be called refinements of technique. If the response is the skill of the subject in performing an unfamiliar task, a major source of error may be that subjects learn this task at different rates. An obvious precaution is to give each subject enough practice to reach his plateau of skill before starting the experiment. The explanation of the task to the subjects must be clear; otherwise, some subjects may be uncertain what they are supposed to do. Removal from an environment that is noisy and subject to distractions may produce more uniform performance. The tasks assigned to the subjects may be too easy or too hard so that all perform well or poorly under any treatment, making discrimination between the treatments impossible. The reduction of errors in measurement of the response often requires prolonged research. In psychometrics, much of the work on scaling is directed toward finding superior instruments of measurement [see SCALING].

Blocking. In many experiments involving comparisons between subjects, the investigator knows that the response will vary widely from subject to subject, even under the same treatment. Often it is possible to obtain beforehand a measurement that is a good predictor of the response of the subject. A child's average score on previous tests in arithmetic may predict well how he will perform on an arithmetic test given at the end of a teaching experiment. Such initial data can be used to increase the precision of the experiment by forming blocks consisting of children of approximately equal ability. If there are three teaching methods, the first block contains the three children with the best initial scores. Each child in this block is assigned to a different teaching method. The second block contains the three next best children, and so on. The purpose of the blocking is to guarantee that each teaching method is tried on an equal number of good, moderate, and poor performers in arithmetic. The resulting gain in precision may be striking.

The term "block" comes from agricultural experimentation in which the block is a compact piece of land. With human subjects, an arrangement of this kind is sometimes called a *matched pairs* design (with two treatments) or a *matched groups* design (with more than two treatments).

A single blocking can help to balance out the effects of several different sources of variation. In a two-treatment experiment on rats, a block comprising littermates of the same sex equalizes the two treatments for age and sex and to some extent for genetic inheritance and weight also. If the conditions of the experiment are subject to uncontrolled time trends, the two rats in a block can be tested at approximately the same time.

Adjustments in the statistical analysis. Given an initial predictor, x, of the final response, y, an alternative to blocking is to make adjustments in the statistical analysis in the hope of removing the influence of variations in x. If x and y represent initial and final scores in a test of some type of skill, the simplest adjustment is to replace y by $y - x$, the improvement in score, as the measure of response. This change does not always increase precision. The error variance of $y - x$ for a subject may be written $\sigma_y^2 + \sigma_x^2 - 2\rho\sigma_y\sigma_x$, where ρ is the correlation between y and x. This is less than σ_y^2 only if ρ exceeds $\sigma_x/2\sigma_y$.

A more accurate method of adjustment is given by the analysis of covariance. In this approach, the measure of response is $y - bx$. The quantity b, computed from the results of the experiment, is an estimate of the average change in y per unit increase in x. The adjustment accords with common sense. If the average x value is three units higher for treatment A than for treatment B, and if b is found to be $\frac{2}{3}$, the adjustment reduces the difference between the average y values by two units.

If the relation between y and x is linear, the use of a predictor, x, to form blocks gives about the same increase in precision as its use in a covariance analysis. For a more detailed comparison in small experiments, see Cox (1957). Blocking by means of x may be superior if the relation between y and x is not linear. Thus, a covariance adjustment on x is helpful mainly when blocking has been used to balance out some other variable or when blocking by means of x is, for some reason, not feasible. One disadvantage of the covariance adjustment is that it requires considerable extra computation. A simpler adjustment such as $y - x$ is sometimes preferred even at some loss of precision.

Randomization. Randomization requires the use of a table of random numbers, or an equivalent device to decide some step in the experiment, most frequently the allotment of treatments to subjects [see RANDOM NUMBERS].

Suppose that three treatments—A, B, C—are to be assigned to 90 subjects without blocking. The subjects are numbered from 1 to 90. In a two-digit column of random numbers, the numbers 01 to 09 represent subjects 1 to 9, respectively; the numbers 10 to 19 represent subjects 10 to 19, respectively, and so on. The numbers from 91 to 99 and the number 00 are ignored. The 30 subjects whose numbers are drawn first from the table are assigned to treatment A, the next 30 to B, and the remaining 30 to C.

In the simplest kind of blocking, the subjects or experimental units are arranged in 30 blocks of three subjects each. One in each block is to receive A, one B, and one C. This decision is made by randomization, numbering the subjects in any block from 1 to 3 and using a single column of random digits for the draw.

Unlike blocking, which attempts to eliminate the effects of an extraneous source of variation, randomization merely ensures that each treatment has an equal chance of being favored or handicapped by the extraneous source. In the blocked experiment above, randomization might assign the best subject in every block to treatment A. The probability that this happens is, however, only 1 in 3^{30}. Whenever possible, blocking should be used for all major sources of variation, randomization being confined to the minor sources. The use of randomization is not limited to the allotment of treatments to subjects. For example, if time trends are suspected at some stage in the experiment, the order in which the subjects within a block are processed may be randomized. Of course, if time trends are likely to be large, blocking should be used for them as well as randomization, as illustrated later in this article by the crossover design.

In his *Design of Experiments*, Fisher illustrated how the act of randomization often allows the investigator to carry out valid tests for the treatment means without assuming the form of the frequency distribution of the data (1935). The calculations, although tedious in large experiments, enable the experimenter to free himself from the assumptions required in the standard analysis of variance. Indeed, one method of justifying the standard methods for the statistical analysis of experimental results is to show that these methods usually give serviceable approximations to the results of randomization theory [*see* NONPARAMETRIC STATISTICS; *see also* Kempthorne 1952].

Size of experiment. An important practical decision is that affecting the number of subjects or experimental units to be included in an experiment. For comparing a pair of treatments there are two common approaches to this problem. One approach is to specify that the observed difference between the treatment means be correct to within some amount $\pm d$ chosen by the investigator. The other approach is to specify the power of the test of significance of this difference.

Consider first the case in which the response is measured on a continuous scale. If σ is the standard deviation per unit of the experimental errors and if each treatment is allotted to n units, the standard error of the observed difference between two treatment means is $\sqrt{2}\,\sigma/\sqrt{n}$ for the simpler types of experimental design. Assuming that this difference is approximately normally distributed, the probability that the difference is in error by more than $d = 1.96\,\sqrt{2}\,\sigma/\sqrt{n}$ is about 0.05 (from the normal tables). The probability becomes 0.01 if d is increased to $2.58\,\sqrt{2}\,\sigma/\sqrt{n}$. Thus, although there is no finite n such that the error is certain to be less than d, nevertheless, from the normal tables, a value of n can be computed to reduce the probability that the error exceeds d to some small quantity α such as 0.05. Taking $\alpha = 0.05$ gives $n = 7.7\sigma^2/d^2 \cong 8\sigma^2/d^2$. The value of σ is usually estimated from previous experiments or preliminary work on this experiment.

If the criterion is the proportion of units that fall into some class (for instance, the proportion of subjects who complete a task successfully), the corresponding formula for n, with $\alpha = 0.05$, is

$$n \cong 4[p_1(1-p_1) + p_2(1-p_2)]/d^2,$$

where p_1, p_2 are the true proportions of success for the two treatments and d is the maximum tolerable error in the observed difference in proportions. Use of this formula requires advance estimates of p_1 and p_2. Fortunately, if these lie between 0.3 and 0.7 the quantity $p(1-p)$ varies only between 0.21 and 0.25.

The choice of the value of d should, of course, depend on the use to be made of the results, but an element of judgment often enters into the decision.

The second approach (specifying the power) is appropriate, for instance, when a new treatment is being compared with a standard treatment and when the investigator intends to discard the new treatment unless the test of significance shows that it is superior to the standard. He does not mind discarding the new treatment if its true superiority is slight. But if the true difference (new − standard) exceeds some amount, Δ, he wants the probability of finding a significant difference to have some high value, β (perhaps 0.95, 0.9, or 0.8).

With continuous data, the required value of n

is approximately

$$n \cong 2\sigma^2(\xi_\alpha + \xi_{1-\beta})^2/\Delta^2,$$

where

ξ_α = normal deviate corresponding to the significance level, α, used in the test of significance,

and

$\xi_{1-\beta}$ = normal deviate for a *one-tailed* probability $1 - \beta$.

For instance, if the test of significance is a one-tailed test at the 5% level and β is 0.9, so that $\xi_\alpha = 1.64$ and $\xi_{1-\beta} = 1.28$, then $n \cong 17\sigma^2/\Delta^2$. The values of Δ, α, and β are chosen by the investigator.

With proportions, an approximate formula is

$$n \cong 2(\xi_\alpha + \xi_{1-\beta})^2\bar{p}\bar{q}/(p_2 - p_1)^2,$$

where $\bar{p} = (p_1 + p_2)/2$ and $\bar{q} = 1 - \bar{p}$ and $p_2 - p_1$ is the size of difference to be detected. One lesson that this formula teaches is that large samples are needed to detect small or moderate differences between two proportions. For instance, with $p_1 = 0.3$, $p_2 = 0.4$, $\alpha = 0.05$ (two-tailed), and $\beta = 0.8$, the formula gives $n = 357$ in each sample, or a total of 714 subjects.

More accurate tables for n, with proportions and continuous data, are given in Cochran and Cox (1950) and a fuller discussion of the sample size problem in Cox (1958).

If the investigator is uncertain about the best values to choose for Δ, it is instructive to compute the value of Δ that will be detected, say with probability 80% or 90%, for an experiment of the size that is feasible. Some experiments, especially with proportions, are almost doomed to failure, in the sense that they have little chance of detecting a true difference of the size that a new treatment is likely to produce. It is well to know this before doing the experiment.

Controls. Some experiments require a *control*, or comparison, treatment. For a discussion of the different meanings of the word "control" and an account of the history of this device, see Boring (1954). In a group of families having a prepaid medical care plan, it is proposed to examine the effects of providing, over a period of time, additional free psychiatric consultation. An intensive initial study is made of the mental health and social adjustment of the families who are to receive this extra service, followed by a similar inventory at the end. In order to appraise whether the differences (final − initial) can be attributed to the psychiatric guidance, it is necessary to include a control group of families, measured at the beginning and at the end, who do not receive this service. An argument might also be made for a second control group that does not receive the service and is measured only at the end. The reason is that the initial psychiatric appraisal may cause some families in the first control group to seek psychiatric guidance on their own, thus diluting the treatment effect that is to be studied. Whether such disturbances are important enough to warrant a second control is usually a matter of judgment.

The families in the control groups, like those in the treated group, must be selected by randomization from the total set of families available for the experiment. This type of evaluatory study presents other problems. It is difficult to conceal the treatment group to which a family belongs from the research workers who make the final measurements, so that any preconceptions of these workers may vitiate the results. Second, the exact nature of the extra psychiatric guidance can only be discovered as the experiment proceeds. It is important to keep detailed records of the services rendered and of the persons to whom they were given.

Factorial experimentation

In many programs of research, the investigator intends to examine the effects of several different types of variables on some response (for example, in an experiment on the accuracy of tracking, the effect of speed of the object, the type of motion of the object, and the type of handle used by the human tracker). In factorial designs, these variables are investigated simultaneously in the same experiment. The advantages of this approach are that it makes economical use of resources and provides convenient data for studying the interrelationships of the effects of different variables.

These points may be illustrated by an experiment with three factors or variables, A, B, and C, each at two levels (that is, two speeds of the object, etc.). Denote the two levels of A by a_1 and a_2, and similarly for B and C. The treatments consist of all possible combinations of the levels of the factors. There are eight combinations:

(1) $a_1b_1c_1$	(3) $a_1b_2c_1$	(5) $a_1b_1c_2$	(7) $a_1b_2c_2$
(2) $a_2b_1c_1$	(4) $a_2b_2c_1$	(6) $a_2b_1c_2$	(8) $a_2b_2c_2$

Suppose that one observation is taken on each of the eight combinations. What information do these give on factor A? The comparison (2) − (1), that is, the difference between the observations for combinations (2) and (1), is clearly an estimate of the difference in response, $a_2 - a_1$, since the factors B and C are held fixed at their lower levels. Similarly, (4) − (3) gives an estimate of $a_2 - a_1$, with B held at its higher level and C at its lower level. The differences (6) − (5) and (8) − (7) supply two further estimates of $a_2 - a_1$. The average of these four differences provides a comparison of a_2 with a_1 based on two samples of size four and is called the *main effect* of A.

Turning to B, it may be verified that (3) − (1), (4) − (2), (7) − (5), and (8) − (6) are four comparisons of b_2 with b_1. Their average is the main effect of B. Similarly, (5) − (1), (6) − (2), (7) − (3), and (8) − (4) provide four comparisons of c_2 with c_1.

Thus the testing of eight treatment combinations in the factorial experiment gives estimates of the effects of each of the factors A, B, and C based on samples of size four. If a separate experiment were devoted to each factor, as in the "one variable at a time" approach, 24 combinations would have to be tested (eight in each experiment) in order to furnish estimates based on samples of size four. The economy in the factorial approach is achieved because every observation contributes information on all factors.

In many areas of research, it is important to study the relations between the effects of different factors. Consider the following question: Is the difference in response between a_2 and a_1 affected by the level of B? The comparison

$$(a_2b_2 - a_1b_2) - (a_2b_1 - a_1b_1),$$

where each quantity has been averaged over the two levels of C, measures the difference between the response to A when B is at its higher level and the response to A when B is at its lower level. This quantity might be called the effect of B on the response to A. The same expression rearranged as follows,

$$(a_2b_2 - a_2b_1) - (a_1b_2 - a_1b_1),$$

also measures the effect of A on the response to B. It is called the *AB two-factor interaction*. (Some writers introduce a multiplier, $\frac{1}{2}$, for conventional reasons.) The AC and BC interactions are computed similarly.

The analysis can be carried further. The AB interaction can be estimated separately for the two levels of C. The difference between these quantities is the effect of C on the AB interaction. The same expression is found to measure the effect of A on the BC interaction and the effect of B on the AC interaction. It is called the ABC three-factor interaction.

The extent to which different factors exhibit interactions depends mostly on the way in which nature behaves. Absence of interaction implies that the effects of the different factors are mutually additive. In some fields of application, main effects are usually large relative to two-factor interactions, and two-factor interactions are large relative to three-factor interactions, which are often negligible. Sometimes a transformation of the scale in which the data are analyzed removes most of the interactions [*see* STATISTICAL ANALYSIS, SPECIAL PROBLEMS OF, *article on* TRANSFORMATIONS OF DATA]. There are, however, many experiments in which the nature and the sizes of the interactions are of primary interest.

The factorial experiment is a powerful weapon for investigating responses affected by many stimuli. The number of levels of a factor is not restricted to two and is often three or four. The chief limitation is that the experiment may become too large and unwieldy to be conducted successfully. Fortunately, the supply of rats and university students is large enough so that factorial experiments are widely used in research on learning, motivation, personality, and human engineering (see, for example, Cattell 1968).

Several developments mitigate this problem of expanding size. If most interactions may safely be assumed to be negligible, good estimates of the main effects and of the interactions considered likely to be important can be obtained from an experiment in which a wisely chosen fraction (say $\frac{1}{2}$ or $\frac{1}{3}$) of the totality of treatment combinations is tested. The device of *confounding* (see Cochran & Cox 1950, chapter 6, esp. pp. 183–186; Cox 1958, sec. 12.3) enables the investigator to use a relatively small sized block in order to increase precision, at the expense of a sacrifice of information on certain interactions that are expected to be negligible. If all the factors represent continuous variables $(x_1, x_2, \cdots)$ and the objective is to map the *response surface* that expresses the response, y, as a function of $x_1, x_2, \cdots$, then one of the designs specially adapted for this purpose may be used. [*For discussion of these topics, see* EXPERIMENTAL DESIGN, *article on* RESPONSE SURFACES; *see also* Cox 1958; Davies 1954.]

In the remainder of this article, some of the commonest types of experimental design are outlined.

Randomized groups. The randomized group arrangement, also called the one-way layout, the simple randomized design, and the completely randomized design, is the simplest type of plan. Treatments are allotted to experimental units at random, as described in the discussion of "Randomization," above. No blocking is used at any stage of the experiment; and, since any number of treatments and any number of units per treatment may be employed, the design has great flexibility. If mishaps cause certain of the responses to be missing, the statistical analysis is only slightly complicated. Since, however, the design takes no advantage of blocking, it is used primarily when no criteria for blocking are available, when criteria previously used for blocking have proved ineffective, or when the response is not highly variable from unit to unit.

Randomized blocks. If there are v treatments and the units can be grouped into blocks of size v, such that units in the same block are expected to give about the same final response under uniform treatment, then a randomized blocks design is appropriate. Each treatment is allotted at random to one of the units in any block. This design is, in general, more precise than randomized groups and is very extensively used.

Sometimes the blocks are formed by assessing or scoring the subjects on an initial variable related to the final response. It may be of interest to examine whether the comparative effects of the treatments are the same for subjects with high scores as for those with low scores. This can be done by an extension of the analysis of variance appropriate to the randomized blocks design. For example, with four treatments, sixty subjects, and fifteen blocks, the blocks might be classified into three levels, *high*, *medium*, or *low*, there being five blocks in each class. A useful partition of the degrees of freedom (df) in the analysis of variance of this "treatments × levels" design is as follows:

	df
Between levels	2
Between blocks at the same level	12
Treatments	3
Treatments × levels interactions	6
Treatments × blocks within levels	36
Total	59

The mean square for interaction is tested, against the mean square for treatments × blocks within levels, by the usual F-test. Methods for constructing the levels and the problem of testing the over-all effects of treatments in different experimental situations are discussed in Lindquist (1953). [*See* LINEAR HYPOTHESES, *article on* ANALYSIS OF VARIANCE.]

The crossover design. The crossover design is suitable for within-subject comparisons in which each subject receives all the treatments in succession. With three treatments, for example, a plan in which every subject receives the treatments in the order *ABC* is liable to bias if there happen to be systematic differences between the first, second, and third positions, due to time trends, learning, or fatigue. One design that mitigates this difficulty is the following: a third of the subjects, selected at random, get the treatments in the order *ABC*, a third get *BCA*, and the remaining third get *CAB*. The analysis of variance resembles that for randomized blocks except that the sum of squares representing the differences between the over-all means for the three positions is subtracted from the error sum of squares.

The Latin square. A square array of letters (treatments) such that each letter appears once in every row and column is called a Latin square. The following are two 4 × 4 squares.

(1)

C	*A*	*B*	*D*
A	*B*	*D*	*C*
B	*D*	*C*	*A*
D	*C*	*A*	*B*

(2)

A	*B*	*C*	*D*
B	*C*	*D*	*A*
D	*A*	*B*	*C*
C	*D*	*A*	*B*

This layout permits simultaneous blocking in two directions. The rows and columns often represent extraneous sources of variation to be balanced out. In an experiment that compared the effects of five types of music programs on the output of factory workers doing a monotonous job, a 5 × 5 Latin square was used. The columns denoted days of the week and the rows denoted weeks. When there are numerous subjects, the design used is frequently a group of Latin squares.

For within-subject comparisons, the possibility of a residual or carry-over effect from one period to the next may be suspected. If such effects are present (and if one conventionally lets columns in the above squares correspond to subjects and rows correspond to order of treatment) then square (1) is bad, since each treatment is always preceded by the same treatment (*A* by *C*, etc.). By the use of square (2), in which every treatment is preceded once by each of the other treatments, the residual effects can be estimated and unbiased estimates obtained of the direct effects (see Cochran & Cox 1950, sec. 4.6*a*; Edwards 1950, pp. 274–275). If there is strong interest in the residual effects, a more suitable design is the *extra-period Latin square*. This is a design like square (2), in which the treatments *C*, *D*, *A*, *B* in the fourth period are given again in a fifth period.

Balanced incomplete blocks. When the number of treatments, v, exceeds the size of block, k, that appears suitable, a balanced incomplete blocks design is often appropriate. In examining the taste preferences of adults for seven flavors of ice cream in a within-subject test, it is likely that a subject can make an accurate comparison among only three flavors before his discrimination becomes insensitive. Thus $v = 7$, $k = 3$. In a comparison of three methods of teaching high school students, the class may be the experimental unit and the school a suitable block. In a school district, it may be possible to find twelve high schools each having two classes at the appropriate level. Thus $v = 3$, $k = 2$.

Balanced incomplete blocks (BIB) are an extension of randomized blocks that enable differences among blocks to be eliminated from the experimental errors by simple adjustments performed in the statistical analysis. Examples for $v = 7$, $k = 3$ and for $v = 3$, $k = 2$ are as follows (columns are blocks):

$v = 7, k = 3$

A	B	C	D	E	F	G
B	C	D	E	F	G	A
D	E	F	G	A	B	C

$v = 3, k = 2$

A	B	C
B	C	A

The basic property of the design is that each pair of treatments occurs together (in the same block) equally often.

In both plans shown, it happens that each row contains every treatment. This is not generally true of BIB designs, but this extra property can sometimes be used to advantage. With $v = 7$, for instance, if the row specifies the order in which the types of ice cream are tasted, the experiment is also balanced against any consistent order effect. This extension of the BIB is known as an *incomplete Latin square* or a *Youden square*. In the high schools experiment, the plan for $v = 3$ would be repeated four times, since there are twelve schools.

Comparisons between and within subjects. Certain factorial experiments are conducted so that some comparisons are made within subjects and others are made between subjects. Suppose that the criterion is the performance of the subjects on an easy task, T_1, and a difficult task, T_2, each subject attempting both tasks. This part of the experiment is a standard crossover design. Suppose further that these tasks are explained to half the subjects in a discouraging manner, S_1, and to the other half in a supportive manner, S_2. It is of interest to discover whether these preliminary suggestions, S, have an effect on performance and whether this effect differs for easy and hard tasks. The basic plan, requiring four subjects, is shown in the first three lines of Table 1, where O denotes the order in which the tasks are performed.

Table 1

Subject	1		2		3		4	
Order	O_1	O_2	O_1	O_2	O_1	O_2	O_1	O_2
Treatment	T_1S_1	T_2S_1	T_2S_1	T_1S_1	T_1S_2	T_2S_2	T_2S_2	T_1S_2
$(T_2 - T_1)$	−	+	+	−	−	+	+	−
$(S_2 - S_1)$	−	−	−	−	+	+	+	+
TS	+	−	−	+	−	+	+	−
TO	+	+	−	−	+	+	−	−

The comparison $T_2 - T_1$, which gives the main effect of T, is shown under the treatments line. This is clearly a within-subject comparison since each subject carries a + and a −. The main effect of suggestion, $S_2 - S_1$, is a between-subject comparison: subjects 3 and 4 carry + signs while subjects 1 and 2 carry − signs. The TS interaction, measured by $T_2S_2 - T_1S_2 - T_2S_1 + T_1S_1$, is seen to be a within-subject comparison.

Since within-subject comparisons are usually more precise than between-subject comparisons, an important property of this design is that it gives relatively high precision on the T and TS effects at the expense of lower precision on S. The design is particularly effective for studying interactions. Sometimes the between-subject factors involve a classification of the subjects. For instance, the subjects might be classified into three levels of anxiety, A, by a preliminary rating, with equal numbers of males and females of each degree included. In this situation, the factorial effects A, S (for sex), and AS are between-subject comparisons. Their interactions with T are within-subject comparisons.

The example may present another complication. Subjects who tackle the hard task after doing the easy task may perform better than those who tackle the hard task first. This effect is measured by a TO interaction, shown in the last line in Table 1. Note that the TO interaction turns out to be a between-subject comparison. The same is true of the TSO three-factor interaction.

In designs of this type, known in agriculture as *split-plot* designs, separate estimates of error are calculated for between-subject and within-subject

Table 2

Source	df
Between subjects	
S	1
TO	1
TSO	1
Error b	$4(n - 1)$
Within subjects	
O	1
T	1
TS	1
SO	1
Error w	$4(n - 1)$

comparisons. With $4n$ subjects, the partition of degrees of freedom in the example is shown in Table 2 (if it is also desired to examine the *TO* and *TSO* interactions).

Plans and computing instructions for all the common types of design are given in Cochran and Cox (1950); and Lindquist (1953), Edwards (1950), and Winer (1962) are good texts on experimentation in psychology and education.

WILLIAM G. COCHRAN

[*Directly related are the articles under* LINEAR HYPOTHESES.]

BIBLIOGRAPHY

►BECKER, HOWARD S. 1968 Observation: I. Social Observation and Social Case Studies. Volume 11, pages 232–238 in *International Encyclopedia of the Social Sciences.* Edited by David L. Sills. New York: Macmillan and Free Press.

BORING, EDWIN G. 1954 The Nature and History of Experimental Control. *American Journal of Psychology* 67:573–589.

CAMPBELL, DONALD T.; and STANLEY, JULIAN C. 1963 Experimental and Quasi-experimental Designs for Research on Teaching. Pages 171–246 in Nathaniel L. Gage (editor), *Handbook of Research on Teaching.* Chicago: Rand McNally.

►CATTELL, RAYMOND B. 1968 Traits. Volume 16, pages 123–128 in *International Encyclopedia of the Social Sciences.* Edited by David L. Sills. New York: Macmillan and Free Press.

COCHRAN, WILLIAM G.; and COX, GERTRUDE M. (1950) 1957 *Experimental Designs.* 2d ed. New York: Wiley.

COX, D. R. 1957 The Use of a Concomitant Variable in Selecting an Experimental Design. *Biometrika* 44: 150–158.

COX, D. R. 1958 *Planning of Experiments.* New York: Wiley.

DAVIES, OWEN L. (editor) (1954) 1956 *The Design and Analysis of Industrial Experiments.* 2d ed., rev. Edinburgh: Oliver & Boyd; New York: Hafner.

EDWARDS, ALLEN (1950) 1960 *Experimental Design in Psychological Research.* Rev. ed. New York: Holt.

○FISHER, R. A. (1925) 1970 *Statistical Methods for Research Workers.* 14th ed., rev. & enl. New York: Hafner; Edinburgh: Oliver & Boyd.

FISHER, R. A. (1926) 1950 The Arrangement of Field Experiments. Pages 17.502a–17.513 in R. A. Fisher, *Contributions to Mathematical Statistics.* New York: Wiley. → First published in Volume 33 of the *Journal of the Ministry of Agriculture.*

○FISHER, R. A. (1935) 1971 *The Design of Experiments.* 9th ed. New York: Hafner; Edinburgh: Oliver & Boyd.

KEMPTHORNE, OSCAR 1952 *The Design and Analysis of Experiments.* New York: Wiley.

LINDQUIST, EVERET F. 1953 *Design and Analysis of Experiments in Psychology and Education.* Boston: Houghton Mifflin.

►MCNEMAR, QUINN 1946 Opinion Attitude Methodology. *Psychological Bulletin* 43:289–374.

►ROSENTHAL, ROBERT; and ROSNOW, R. L. 1975 *The Volunteer Subject.* New York: Wiley.

►WAX, ROSALIE HANKEY 1968 Observation: II. Participant Observation. Volume 11, pages 238–241 in *International Encyclopedia of the Social Sciences.* Edited by David L. Sills. New York: Macmillan and Free Press.

WINER, B. J. 1962 *Statistical Principles in Experimental Design.* New York: McGraw-Hill.

Postscript

Social experimentation. When a social program that is intended to solve or alleviate a social, economic, or health problem has been in progress for some time, an effort to measure the effects of the program as a guide to a policy decision about its future is a natural step. These efforts are important because many programs are costly, and experience has suggested that only a minority of new programs are successful (Light et al. 1971). The traditional method of evaluating the effects of the programs has been a post hoc observational study. But owing to such factors as poor retrospective measurement of the state of the problem before the program started, the absence of any comparable control group of people who do not receive the program, and vested interests in the continuation of the program, the evaluations were often badly biased.

Since the mid-1960s a group of social scientists have been stressing two points. Wherever feasible, an evaluation of the effects of a new social program should be conducted as a study in itself before a policy decision is made whether to put the program into effect on a large scale. Second, much more strenuous efforts should be made to use randomized experimentation with a control group in this evaluation. Experimentation on the effects of social programs involves many difficulties—a few are mentioned in the experiment on the effects of family psychiatric care described in the main article. Problems of planning, measurement, record keeping, managerial resources, cost, and interpretation of results are encountered, as is a delay—usually of some years—of the decision whether to institute the program itself on a large scale. In 1971 a committee of the Social Science Research Council decided to prepare a book, *Social Experimentation* (Riecken & Boruch 1974), on the available knowledge about the use of randomized experiments in planning and evaluating social programs. This book also contains examples of well-planned social experiments that have been completed—for example, on fertility control programs, on types of training for police recruits, on mental health rehabilitation programs, on teaching self-care to nursing home patients, on negative taxation, and on the absence of a bond as a requisite for bail. A companion volume, *Experimental Testing of Public Policy* (Social Science Research Council Conference on Social Experiments 1976), has also appeared. The future of this development will be important and interesting to observe.

WILLIAM G. COCHRAN

[*See also* EVALUATION RESEARCH; PUBLIC POLICY AND STATISTICS; *and the biography of* YOUDEN.]

ADDITIONAL BIBLIOGRAPHY

LIGHT, RICHARD J.; MOSTELLER, FREDERICK; and WINOKUR, H. S. JR. 1971 Using Controlled Field Studies to Improve Public Policy. Volume 2, chapter 6 in U.S. President's Commission on Federal Statistics, *Federal Statistics: Report.* Washington: Government Printing Office.

RIECKEN, HENRY W.; and BORUCH, ROBERT F. 1974 *Social Experimentation: A Method for Planning and Evaluating Social Intervention.* New York: Academic Press.

SOCIAL SCIENCE RESEARCH COUNCIL CONFERENCE ON SOCIAL EXPERIMENTS, BOULDER, *1974* 1976 *Experimental Testing of Public Policy: Proceedings.* Edited by Robert F. Boruch and Henry W. Riecken. Boulder, Colo.: Westview.

II RESPONSE SURFACES

Response surface methodology is a statistical technique for the design and analysis of experiments; it seeks to relate an average response to the values of quantitative variables that affect response. For example, response in a chemical investigation might be yield of sulfuric acid, and the quantitative variables affecting yield might be pressure and temperature of the reaction.

In a psychological experiment, an investigator might want to find out how a test *score* achieved by certain subjects depended upon *duration* of the period during which they studied the relevant material and the *delay* between study and test. In mathematical language, the psychologist is interested in the presumed *functional relationship* $\eta = f(\xi_1, \xi_2)$ that expresses the *response score*, η, as a function of the two *variables* duration, ξ_1, and delay, ξ_2. If repeated experiments were made at any fixed set of experimental conditions, the measured response would nevertheless vary because of measurement errors, observational errors, and variability in the experimental material. We regard η therefore as the *mean response* at particular conditions; y, the response actually observed in a particular experiment, differs from η because of an (all-inclusive) error e. Thus $y = \eta + e$, and a mathematical model relating the observed response to the levels of k variables can be written in the form

$$y = f(\xi_1, \cdots, \xi_k) + e. \tag{1}$$

The appropriate investigational strategy depends heavily on the state of ignorance concerning the functional form, f. At one extreme the investigator may not know even which variables, ξ, to include and must make a preliminary screening investigation. At the other extreme the true functional form may actually be known or can be deduced from a mechanistic theory.

Response surface methods are appropriate in the intermediate situation; the important variables are known, but the true functional form is neither known nor easily deducible. The general procedure is to approximate f locally by a suitable function, such as a polynomial, which acts as a "mathematical French curve."

Geometric representation of response relationships. The three curves of Figure 1A, showing a hypothetical relationship associating test score with study period for three different periods of delay, are shown in Figure 1B as sections of a *response surface*. This surface is represented by its response *contours* in Figure 1C. Figure 1D shows how a third variable may be accommodated by the use of three-dimensional *contour surfaces*.

Local graduation. It is usually most convenient to work with coded variables like $x_1 = (\xi_1 - \xi_1^0)/S_1$, $x_2 = (\xi_2 - \xi_2^0)/S_2$ in which ξ_1^0, ξ_2^0 are the coordinates of the center of a region of current interest and S_1 and S_2 are convenient scale factors.

Let $\hat{y}$ represent the calculated value of the response obtained by fitting an approximating function by the method of least squares [*see* LINEAR HYPOTHESES, *article on* REGRESSION]. In a region like R_1 in Figure 1C an adequate approximation can be obtained by fitting the first-degree polynomial

$$\hat{y} = b_0 + b_1x_1 + b_2x_2\,. \tag{2}$$

The response contours of such a fitted plane are, of course, equally spaced parallel straight lines. In a region like R_2 a fair approximation might be achieved by fitting a second-degree polynomial

$$\hat{y} = b_0 + b_1x_1 + b_2x_2 + b_{11}x_1^2 + b_{22}x_2^2 + b_{12}x_1x_2. \tag{3}$$

Flexibility of functions like those in (2) and (3) is greatly increased if the possibility is allowed that y, x_1, and x_2 are suitable transformed values of the response and of the variable. For example, it might be appropriate to analyze log score rather than score itself. [*Ways of choosing suitable transformations are described in* STATISTICAL ANALYSIS,

SPECIAL PROBLEMS OF, *article on* TRANSFORMATIONS OF DATA; *and in* Box & Cox 1964 *and* Box & Tidwell 1962.]

Uses of response surface methodology

A special pattern of points at which observations are to be made is called an experimental design. In Figure 1C are shown a first-order design in R_1, suitable for fitting and checking a first-degree polynomial, and a second-order design in R_2, suitable for fitting and checking a second-degree polynomial. Response surface methodology has been applied (*a*) to provide a description of how the response is affected by a number of variables over some already chosen region of interest and (*b*) to study and exploit multiple response relationships and constrained extrema. In drug therapy, for example, the true situation might be as depicted in Figure 2. First-degree approximating functions fitted to *each* of the three responses—η_1, therapeutic effect, η_2, nausea, and η_3, toxicity—could approximately locate the point P where maximum therapeutic effect is obtained with nausea and toxicity maintained at the acceptable limits $\eta_2 = 5$, $\eta_3 = 30$. Response surface methodology has also been applied (*c*) to locate and explore the neighborhood of maximal or minimal response. Because problems in (*c*) often subsume those in (*a*) and (*b*), only this application will be considered in more detail.

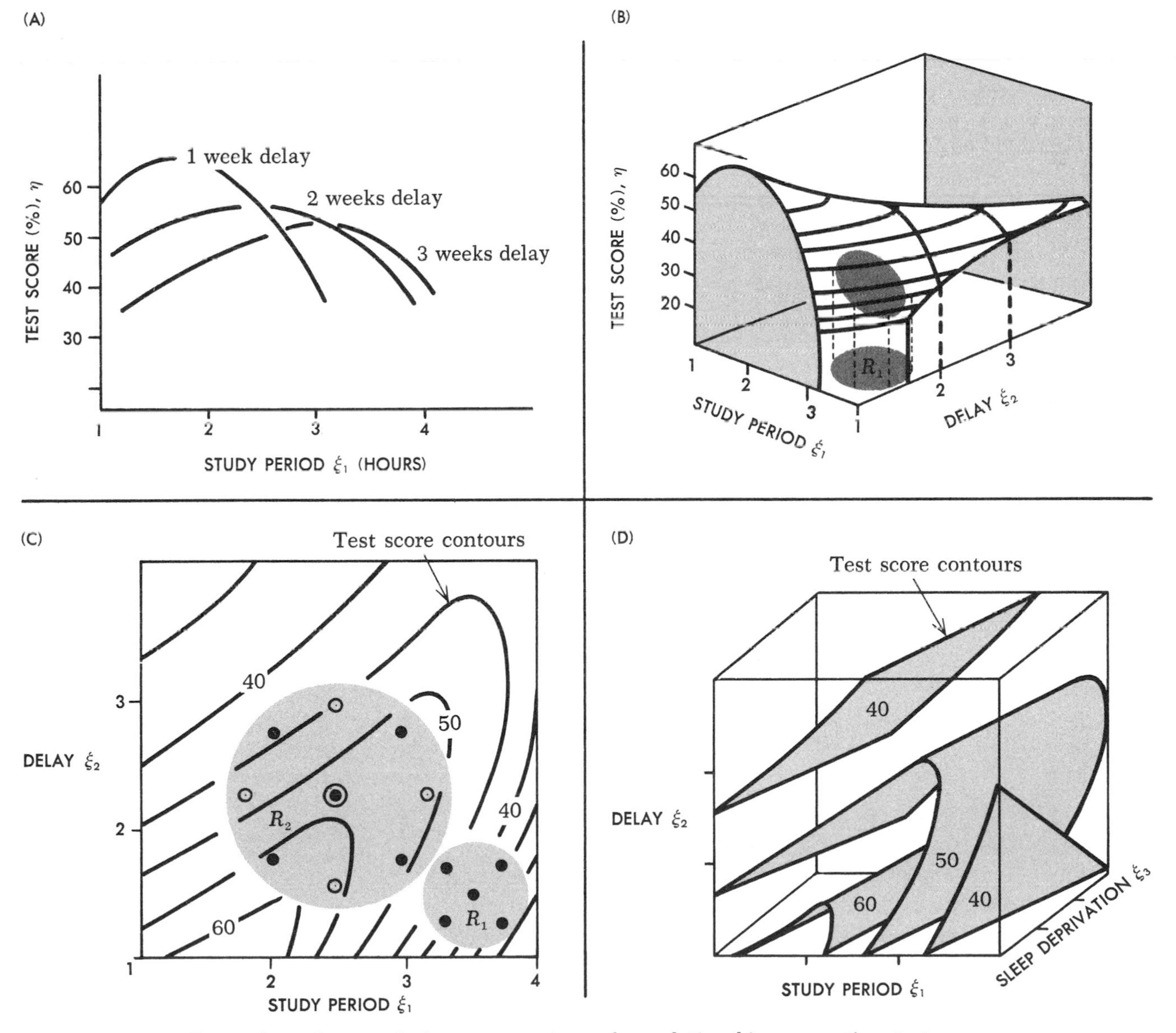

Figure 1 — Geometrical representations of a relationship connecting test scores with study period, delay, and sleep deprivation

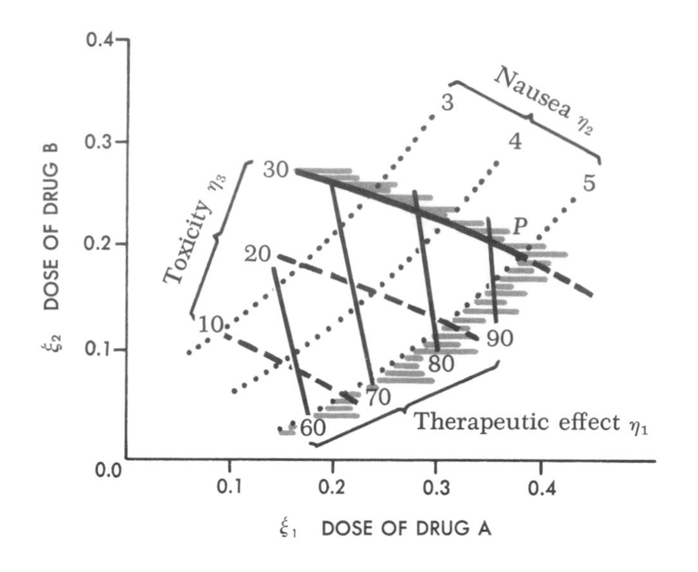

Figure 2 — Dependence of three responses on combined dosages of drugs

Location and exploration of a maximal region. Various tactics have been proposed to deal with the problem of finding where the response surface has its maximum or minimum and of describing its shape nearby. Because the appropriateness of a particular tactic usually depends upon factors that are initially unknown, an adaptive strategy of multiple iteration must be employed, that is, the investigator must put himself in a position to learn more about each of a number of uncertainties as he proceeds and to modify tactics accordingly. It is doubtful whether an adaptive strategy could be found that is appropriate to every conceivable response function. One such procedure, which has worked well in chemical applications and which ought to be applicable in some other areas, is as follows: When the initially known experimental conditions are remote from the maximum (a parallel strategy applies in the location of a minimum) rapid progress is often possible by locally fitting a sloping plane and moving in the indicated direction of greatest slope to a region of higher response. This tactic may be repeated until, when the experimental sequence has moved to conditions near the maximizing ones, additional observations are taken and a quadratic (second-order) fit or analysis is made to indicate the approximate shape of the response surface in the region of the maximum.

An example. In this example iteration occurs in (A) the amount of replication (to achieve sufficient accuracy), (B) the location of the region of interest, (C) the scaling of the variables, (D) the transformation in which the variables are considered, and (E) the necessary degree of complexity of approximating functions and of the corresponding design. The letters A, B, C, etc., are used parenthetically to indicate the particular type of iteration that is being furthered at any stage. Suppose that, unknown to the experimenter, the true dependence of percentage yield on temperature and concentration is as shown in Figure 3A and the experimental error standard deviation is 1.2 per cent.

A first-degree approximation. Suppose that five initial duplicate runs made in random order at points labeled 1, 2, 3, 4, and 5 in Figure 3B yield the results $y_1 = 24$, $y_1' = 27$, $y_2 = 38$, $y_2' = 40$, $y_3 = 42$, $y_3' = 42$, $y_4 = 42$, $y_4' = 41$, $y_5 = 50$, $y_5' = 53$. The average yields at the five points are then $\bar{y}_1 = 25.5$, $\bar{y}_2 = 39$, $\bar{y}_3 = 42$, $\bar{y}_4 = 41.5$, $\bar{y}_5 = 51.5$. At this stage it is convenient to work with the coded variables $x_1 = (\text{temp.} - 70)/10$ and $x_2 = (\text{conc.} - 42.5)/2.5$. Using standard least squares theory the coefficients b_0, b_1, b_2 of equation (2) are then easily estimated (for example, $b_1 = \frac{1}{4}\{-\bar{y}_1 + \bar{y}_2 - \bar{y}_4 + \bar{y}_5\} = 5.9$) and the locally best-fitting plane is

$$(4) \qquad \hat{y} = 39.9 + 5.9\,x_1 + 7.1\,x_2 .$$

The differences in the duplicate runs provide an estimate $s = 1.5$, with five degrees of freedom, of σ, the underlying standard deviation. The standard errors of b_0, b_1, and b_2 are then estimated as 0.5, and no further replication (A) appears necessary to obtain adequate estimation of y.

Checking the fit. To check the appropriateness of the first-degree equation it would be sensible to look at the size of second-order effects. For reason of experimental economy a first-order design usually contains points at too few distinct levels to allow separate estimation of all second-order terms. The design may be chosen, however, so as to allow estimates of "specimen" second-order coefficients or combinations thereof. In the present case estimates can be made of $b_{12} = \frac{1}{4}(\bar{y}_1 - \bar{y}_2 - \bar{y}_4 + \bar{y}_5) = -0.9 \pm 0.5$ and $(b_{11} + b_{22}) = \frac{1}{4}(\bar{y}_1 + \bar{y}_2 + \bar{y}_4 + \bar{y}_5) - \bar{y}_3 = -2.6 \pm 1.2$. Some inadequacy of the first-degree equation is indicated, therefore, but this is tentatively ignored because of the dominant magnitude of b_1 and b_2.

Steepest ascent. It is now logical to explore (B) higher temperatures and concentrations. The points 6, 7, and 8 are along a steepest ascent path obtained by changes proportional to $b_1 \times S_1 = 5.9 \times 10° = 59°$ in temperature and $b_2 \times S_2 = 7.1 \times 2.5\% = 17.75\%$ in concentration. Suppose that $y_6 = 59$, $y_7 = 63$, and $y_8 = 50$. Graphical interpolation indicates that the highest yield on this path is between runs 6 and 7, and this is chosen (B) as the center of the new region to be explored.

The path calculated as above is at right angles

to contours of the fitted plane when 10-degree units of temperature and 2.5 per cent units of concentration are represented by the same distances. That the experimenter currently regards these units as appropriate is implied by his choice of levels in the design.

Scaling correction. To correct unsuitable scaling (C) the investigator can adopt the rule that if a variable produces an effect that is small compared with that produced by the other variables, the center level for that variable is moved away from the calculated path and a larger change is made for this variable in the next set of runs. No change of relative scaling is indicated here, but progress up the surface would normally be accompanied by reduction in the sizes of b_1 and b_2. Also, the checks have already indicated that second-order effects can scarcely be estimated with adequate accuracy in the present scaling. Thus, wider ranges in both variables should be employed in a second design.

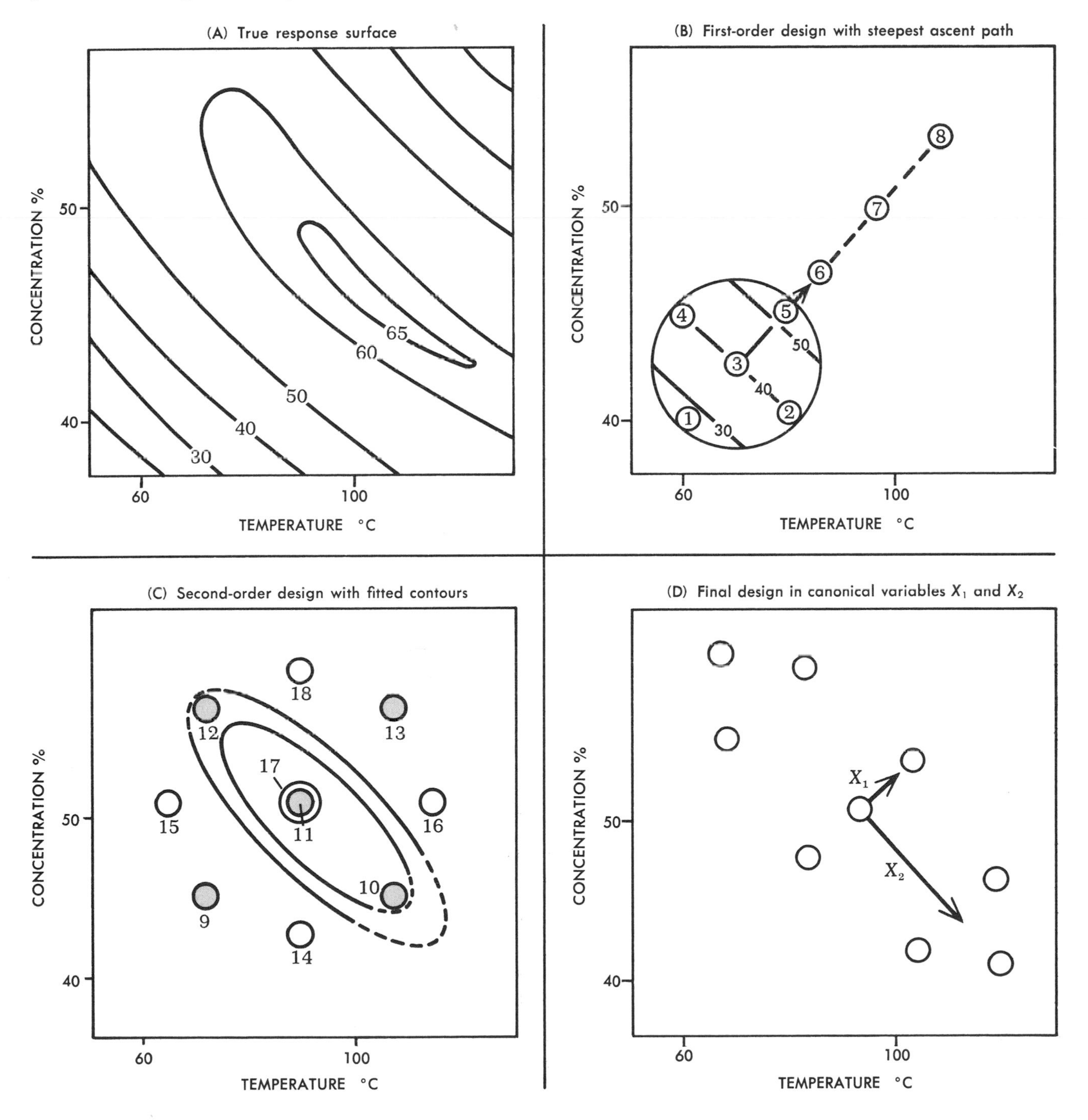

Figure 3 — A true response surface and three successive design patterns

A second-degree approximation. A widened first-order design at the new location might give $y_9 = 50$, $y_{10} = 66$, $y_{11} = 66$, $y_{12} = 63$, and $y_{13} = 52$, as in Figure 3C.

Then $\hat{y} = 59.4 + 1.3\,x_1 - 0.3\,x_2$ is the best-fitting plane, in which x_1 is given by (temp. − 90)/15, x_2 is given by (conc. − 18.75)/3.75, and the estimated standard error of the coefficients is about 0.8. In the new scaling the check quantities are now $b_{12} = -6.75 \pm 0.8$ and $b_{11} + b_{22} = 8.25 \pm 1.7$. It is clear, without this time duplicating the design, that first-order terms no longer dominate, and no worthwhile further progress can be made by ascent methods. To make possible the fitting and checking of a second-degree polynomial (E), five additional observations might be taken, say $y_{14} = 54$, $y_{15} = 54$, $y_{16} = 57$, $y_{17} = 65$, $y_{18} = 55$. The last ten observations now form a second-order design. A second-degree equation fitted to these observations gives

$$(5) \qquad \hat{y} = 65.50 + 1.16\,x_1 + 0.05\,x_2 - 4.31\,x_1^2 - 4.81\,x_2^2 - 6.75\,x_1x_2 .$$

The design allows a check on the adequacy (E) of the second-degree equation by providing estimates of certain "specimen" combinations of third-order terms

$$b_{111} - b_{122} = \tfrac{1}{4}\{y_9 - y_{10} + y_{12} - y_{13} - \sqrt{2}y_{15} + \sqrt{2}y_{16}\} = -0.19 \pm 1.03$$

and

$$b_{222} - b_{112} = \tfrac{1}{4}\{y_9 + y_{10} - y_{12} - y_{13} - \sqrt{2}y_{14} + \sqrt{2}y_{18}\} = -0.60 \pm 1.03.$$

The estimated standard errors of the linear coefficients b_1 and b_2, of the quadratic coefficients b_{11} and b_{22}, and of the interaction coefficient b_{12} are, respectively, 0.52, 0.62, and 0.73.

Before an attempt is made to interpret equation (5) there must be some assurance (A) that the change in response it predicts is large compared with the standard error of that prediction. For a design requiring N observations and an approximating equation containing p constants, the average variance of the N calculated responses $\hat{y}$ is $(p/N)s^2 = (6/12) \times 2.1 = 1.1$ for this example. The square root (1.0 for this example) gives an "average" standard error for $\hat{y}$. This may be compared with the range of the predicted $\hat{y}$'s, which is 17.08, the highest predicted value being $\hat{y}_{11} = \hat{y}_{17} = 65.50$ and the lowest $\hat{y}_9 = 48.42$.

A more precise indication of adequacy may be obtained by an application of the analysis of variance, but a discussion of this is outside the scope of the present account [*see* LINEAR HYPOTHESES, *article on* ANALYSIS OF VARIANCE]. It is to be noted, however, that bare statistical significance of the regression would *not* ensure that the response surface was *estimated* with sufficient accuracy for the interpretation discussed below.

Interpretation. Once adequate fit and precision have been obtained, a contour plot of the equation over the region of the experimental design is helpful in interpretation. Especially where there are more than two variables, interpretation is further facilitated by writing the second-degree equation in canonical form (D). In most cases, this means that the center of the quadratic system is chosen as a new origin, and a rotation of axes is performed to eliminate cross-product terms.

In a final group of experiments the new canonical axes and scales could be used to position the design. In Figure 3D the design points are chosen so that they roughly follow a contour and make a rather precise final fitting possible.

It might be asked, Why not simply use the twenty or so experimental points to cover the region shown in Figure 3A with some suitable grid in the first place? The answer is that it is not known initially that the region of interest will be in the area covered by that diagram. The "content" of the space to be explored goes up rapidly as the number of dimensions is increased.

Suitable designs. From the foregoing discussion it will be clear that the arrangements of experimental points suitable for response surface study should satisfy a number of requirements. Ideally, a response surface design should (1) allow $\hat{y}(x)$ to be estimated throughout the region of interest, R; (2) ensure that $\hat{y}(x)$ is as "close" as possible to $\eta(x)$; (3) give good detectability of lack of fit; (4) allow transformations to be fitted; (5) allow experiments to be performed in blocks; (6) allow designs of increasing order to be built up sequentially; (7) provide an internal estimate of error; (8) be insensitive to wild observations; (9) require a minimum number of experimental points; (10) provide patterning of data allowing ready visual appreciation; (11) ensure simplicity of calculation; and (12) behave well when errors occur in settings of the x's.

A variety of designs have been developed, many of which have remarkably good over-all behavior with respect to these requirements. When maximum economy in experimentation is essential, designs that fail to meet certain of these criteria may have to be used at some increased risk of being misled.

G. E. P. BOX

BIBLIOGRAPHY

ANDERSEN, S. L. 1959 Statistics in the Strategy of Chemical Experimentation. *Chemical Engineering Progress* 55:61–67.

BOX, GEORGE E. P. 1954 The Exploration and Exploitation of Response Surfaces: Some General Considerations and Examples. *Biometrics* 10:16–60.

Box, George E. P. 1957 Integration of Techniques in Process Development. Pages 687–702 in American Society for Quality Control, National Convention, Eleventh, *Transactions*. Detroit, Mich.: The Society.

Box, George E. P. 1959 Fitting Empirical Data. New York Academy of Sciences, *Annals* 86:792–816.

Box, George E. P.; and Cox, D. R. 1964 An Analysis of Transformations. *Journal of the Royal Statistical Society* Series B 26:211–252. → Contains eight pages of discussion.

Box, George E. P.; and Tidwell, Paul W. 1962 Transformations of the Independent Variables. *Technometrics* 4:531–550.

Box, George E. P.; and Wilson, K. B. 1951 On the Experimental Attainment of Optimum Conditions. *Journal of the Royal Statistical Society* Series B 13:1–45. → Contains seven pages of discussion.

Davies, Owen L. (editor) (1954) 1956 *The Design and Analysis of Industrial Experiments*. 2d ed., rev. New York: Hafner; London: Oliver & Boyd.

Hill, William G.; and Hunter, William G. 1966 A Review of Response Surface Methodology: A Literature Survey. *Technometrics* 8:571–590.

Hotelling, Harold 1941 Experimental Determination of the Maximum of a Function. *Annals of Mathematical Statistics* 12:20–45.

III
QUASI-EXPERIMENTAL DESIGN

The phrase "quasi-experimental design" refers to the application of an experimental mode of analysis and interpretation to bodies of data not meeting the full requirements of experimental control. The circumstances in which it is appropriate are those of experimentation in social settings—including planned interventions such as specific communications, persuasive efforts, changes in conditions and policies, efforts at social remediation, etc.—where complete experimental control may not be possible. Unplanned conditions and events may also be analyzed in this way where an exogenous variable has such discreteness and abruptness as to make appropriate its consideration as an experimental treatment applied at a specific point in time to a specific population. When properly done, when attention is given to the specific implications of the specific weaknesses of the design in question, quasi-experimental analysis can provide a valuable extension of the experimental method.

History of quasi-experimental design. While efforts to interpret field data as if they were actually experiments go back much further, the first prominent methodology of this kind in the social sciences was Chapin's ex post facto experiment (Chapin & Queen 1937; Chapin 1947; Greenwood 1945), although it should be noted that because of the failure to control regression artifacts, this mode of analysis is no longer regarded as acceptable. *The American Soldier* volumes (Stouffer et al. 1949) provide prominent analyses of the effects of specific military experiences, where it is implausible that differences in selection explain the results. Thorndike's efforts to demonstrate the effects of specific coursework upon other intellectual achievements provide an excellent early model (for example, Thorndike & Woodworth 1901; Thorndike & Ruger 1923). Extensive analysis and review of this literature are provided elsewhere (Campbell 1957; 1963; Campbell & Stanley 1963) and serve as the basis for the present abbreviated presentation.

True experimentation. The core requirement of a true experiment lies in the experimenter's ability to apply experimental treatments in complete independence of the prior states of the materials (persons, etc.) under study. This independence makes resulting differences interpretable as effects of the differences in treatment. In the social sciences the independence of experimental treatment from prior status is assured by randomization in assignments to treatments. Experiments meeting these requirements, and thus representing true experiments, are much more possible in the social sciences than is generally realized. Wherever, for example, the treatments can be applied to individuals or small units, such as precincts or classrooms, without the respondents being aware of experimentation or that other units are getting different treatments, very elegant experimental control can be achieved. An increased acceptance by administrators of randomization as the democratic method of allocating scarce resources (be these new housing, therapy, or fellowships) will make possible field experimentation in many settings. Where innovations are to be introduced throughout a social system and where the introduction cannot, in any event, be simultaneous, a use of randomization in the staging can provide an experimental comparison of the new and the old, using the groups receiving the delayed introduction as controls.

Validity of quasi-experimental analyses. Nothing in this article should be interpreted as minimizing the importance of increasing the use of true experimentation. However, where true experimental design with random assignment of persons to treatments is not possible, because of ethical considerations or lack of power, or infeasibility, application of quasi-experimental analysis has much to offer.

The social sciences must do the best they can with the possibilities open to them. Inferences must frequently be made from data obtained under circumstances that do not permit complete control. Too often a scientist trained in experimental method rejects any research in which complete control is lacking. Yet in practice no experiment is perfectly executed, and the practicing scientist overlooks those imperfections that seem to him to offer no plausible rival explanation of the results. In the light of modern philosophies of science, no experiment ever *proves* a theory, it merely *probes* it. Seeming proof results from that condition in which there is no available plausible rival hypothesis to explain the data. The general program of quasi-experimental analysis is to specify and examine those plausible rival explanations of the results that are provided by the uncontrolled variables. A failure to control that does not in fact lend plausibility to a rival interpretation is not regarded as invalidating.

It is well to remember that we do make assured causal inferences in many settings not involving randomization: the earthquake caused the brick building to crumble; the automobile crashing into the telephone pole caused it to break; the language patterns of the older models and mentors caused this child to speak English rather than Kwakiutl; and so forth. While these are all potentially erroneous inferences, they are of the same type as experimental inferences. We are confident that were we to intrude experimentally, we could confirm the causal laws involved. Yet they have been made assuredly by a nonexperimenting observer. This assurance is due to the effective absence of other plausible causes. Consider the inference about the crashing auto and the telephone pole: we rule out combinations of termites and wind because the other implications of these theories do not occur (there are no termite tunnels and debris in the wood, and nearby weather stations have no records of heavy wind). Spontaneous splintering of the pole by happenstance coincident with the auto's onset does not impress us as a rival, nor would it explain the damage to the car, etc. Analogously in quasi-experimental analysis, tentative causal interpretation of data may be made where the interpretation in question is consistent with the data and where other rival interpretations have been rendered implausible.

Dimensions of experimental validity. A set of twelve dimensions, representing frequent threats to validity, have been developed for the evaluation of data as quasi-experiments. These may be regarded as the important classes of frequently plausible rival hypotheses that good research design seeks to rule out. Each will be presented briefly even though not all are employed in the evaluation of the designs used illustratively here.

Fundamental to this listing is a distinction between *internal validity* and *external validity*. Internal validity is the basic minimum without which any experiment is uninterpretable: Did in fact the experimental treatments make a difference in this specific experimental instance? External validity asks the question of *generalizability:* To what populations, settings, treatment variables, and measurement variables can this effect be generalized? Both types of criteria are obviously important, even though they are frequently at odds in that features increasing one may jeopardize the other. While internal validity is the *sine qua non*, and while the question of external validity, like the question of inductive inference, is never completely answerable, the selection of designs strong in both types of validity is obviously our ideal.

Threats to internal validity. Relevant to internal validity are eight different classes of extraneous variables that if not controlled in the experimental design might produce effects mistaken for the effect of the experimental treatment. These are the following. (1) *History:* other specific events in addition to the experimental variable occurring between a first and second measurement. (2) *Maturation:* processes within the respondents that are a function of the passage of time per se (not specific to the particular events), including growing older, growing hungrier, growing tireder, and the like. (3) *Testing:* the effects of taking a test a first time upon subjects' scores in subsequent testing. (4) *Instrumentation:* the effects of changes in the calibration of a measuring instrument or changes in the observers or scorers upon changes in the obtained measurements. (5) *Statistical regression:* operating where groups of subjects have been selected on the basis of their extreme scores. (6) *Selection:* biases resulting in differential recruitment of respondents for the comparison groups. (7) *Experimental mortality:* the differential loss of respondents from the comparison groups. (8) *Selection–maturation interaction:* in certain of the multiple-group quasi-experimental designs, such as the nonequivalent control group design, an interaction of maturation and differential selection is confounded with, that is, might be mistaken for, the effect of the experimental variable [*see* LINEAR HYPOTHESES, *article on* REGRESSION; SAMPLE SURVEYS].

Threats to external validity. Factors jeopardizing external validity or *representativeness* are: (1) The *reactive or interaction effect of testing,* in which a pretest might increase or decrease the respondent's sensitivity or responsiveness to the experimental variable and thus make the results obtained for a pretested population unrepresentative of the effects of the experimental variable for the unpretested universe from which the experimental respondents were selected. (2) *Interaction* effects between *selection* bias and the *experimental variable*. (3) *Reactive effects of experimental arrangements*, which would preclude generalization about the effect of the experimental variable for persons being exposed to it in nonexperimental settings. (4) *Multiple-treatment interference*, a problem wherever multiple treatments are applied to the same respondents, and a particular problem for one-group designs involving equivalent time samples or equivalent materials samples.

Types of quasi-experimental design. Some common types of quasi-experimental design will be outlined here.

One-group pretest–posttest design. Perhaps the simplest quasi-experimental design is the one-group pretest–posttest design, $O_1\ X\ O_2$ (O represents measurement or observation, X the experimental treatment). This common design patently leaves uncontrolled the threats to internal validity of history, maturation, testing, instrumentation, and, if subjects were selected on the basis of extreme scores on O_1, regression. There may be situations in which the investigator could decide that none of these represented plausible rival hypotheses in his setting: A log of other possible change-agents might provide no plausible ones; the measurement in question might be nonreactive (Campbell 1957), the time span too short for maturation, too spaced for fatigue, etc. However, the sources of invalidity are so numerous that a more powerful quasi-experimental design would be preferred. Several of these can be constructed by adding features to this simple one.

[2] *Interrupted time-series design*. The interrupted time-series experiment utilizes a series of measurements providing multiple pretests and posttests, for example:

$$O_1\ O_2\ O_3\ O_4\ X\ O_5\ O_6\ O_7\ O_8\,.$$

If in this series, $O_4{-}O_5$ shows a rise greater than found elsewhere, then maturation, testing, and regression are no longer plausible, in that they would predict equal or greater rises for $O_1{-}O_2$, etc. Instrumentation may well be controlled too, although in institutional settings a change of administration policy is often accompanied by a change in record-keeping standards. Observers and participants may be focused on the occurrence of X and may speciously change rating standards, etc. History remains the major threat, although in many settings it would not offer a plausible rival interpretation.

Multiple time-series design. If one had available a parallel time series from a group not receiving the experimental treatment, but exposed to the same extraneous sources of influence, and if this control time series failed to show the exceptional jump from O_4 to O_5, then the plausibility of history as a rival interpretation would be greatly reduced. We may call this the multiple time-series design.

[3] *Nonequivalent control group*. Another way of improving the one-group pretest–posttest design is to add a "nonequivalent control group." (Were the control group to be randomly assigned from the same population as the experimental group, we would, of course, have a true experimental design not a quasi-experimental design.) Depending on the similarities of setting and attributes, if the nonequivalent control group fails to show the gain manifest in the experimental group, then history, maturation, testing, and instrumentation are controlled. In this popular design, the frequent effort to "correct" for the lack of perfect equivalence by matching on pretest scores is *absolutely wrong* (e.g., Thorndike 1942; Hovland et al. 1949; Campbell & Clayton 1961), because it introduces a regression artifact. Instead, one should accept any initial pretest differences, using analysis of covariance, gain scores, or graphic presentation. (This, of course, is not to reject blocking on pretest scores in true experiments where groups have been assigned to treatments at random.) Remaining uncontrolled is the selection–maturation interaction, that is, the possibility that the experimental group differed from the control group not only in initial level but also in its autonomous maturation rate. In experiments on psychotherapy and on the effects of specific coursework this is a very serious rival. Note that it can be rendered implausible by use of a time series of pretest for both groups thus moving again to the multiple time-series design.

[4] *Other quasi-experimental designs*. There is not space here to present adequately even these four quasi-experimental designs, but perhaps the strategy of adding specific observations and analyses to check on specific threats to validity has been illustrated. This is carried to an extreme in the recurrent institutional cycle design (Campbell & McCormack 1957; Campbell & Stanley 1963), in which longitudinal and cross-sectional measurements are combined with still other analyses to assess the impact of indoctrination procedures, etc. through exploiting the fact that essentially similar treatments are being given to new entrants year after year or cycle after cycle. Other quasi-experimental designs are covered in Campbell and Stanley (1963), Campbell and Clayton (1961), Campbell (1963), and Pelz and Andrews (1964).

[5] *Correlational analyses*. Related to the program of quasi-experimental analysis are those efforts to achieve causal inference from correlational data. Note that while correlation does not prove causation, most causal hypotheses imply specific cor-

relations, and examination of these thus probes, tests, or edits the causal hypothesis. Furthermore, as Blalock (1964) and Simon (1947–1956) have emphasized, certain causal models specify uneven patterns of correlation. Thus the $A \rightarrow B \rightarrow C$ model implies that r_{AC} be smaller than r_{AB} or r_{BC}. However, their use of partial correlations or the use of Wright's path analysis (1920) are rejected as tests of the model because of the requirement that the "cause" be totally represented in the "effect." In the social sciences it will never be plausible that the cause has been measured without unique error and that it also totally lacks unique systematic variance not shared with the effect. More appropriate would be Lawley's (1940) test of the hypothesis of *single factoredness.* Only if single factoredness can be rejected would the causal model, as represented by its predicted uneven correlation pattern, be the preferred interpretation [*see* MULTIVARIATE ANALYSIS, *article on* CORRELATION METHODS.]

Tests of significance. A word needs to be said about tests of significance for quasi-experimental designs. It has been argued by several competent social scientists that since randomization has not been used tests of significance assuming randomization are not relevant. On the whole, the writer disagrees. However, some aspects of the protest are endorsed: Good experimental design is needed for any comparison inferring change, whether or not tests of significance are used, even if only photographs, graphs, or essays are being compared. In this sense, experimental design is independent of tests of significance. More importantly, tests of significance have mistakenly come to be taken as thoroughgoing *proof.* In vulgar social science usage, finding a "significant difference" is apt to be taken as *proving* the author's basis for predicting the difference, forgetting the many other plausible rival hypotheses explaining a significant difference that quasi-experimental designs leave uncontrolled. Certainly the valuation of tests of significance in some quarters needs demoting. Further, the use of tests of significance designed for the evaluation of a single comparison becomes much too lenient when dozens, hundreds, or thousands of comparisons have been sifted. And in a similar manner, an experimenter's decision as to which of his studies is publishable and the editor's decision as to which of the manuscripts are acceptable further bias the sampling basis. In all of these ways, reform is needed.

However, when a quasi-experimenter has, for example, compared the results from two intact classrooms employed in a sampling of convenience, a chance difference is certainly *one*, even if only one, of the many plausible rival hypotheses that must be considered. If each class had but 5 students, one would interpret the fact that 20 per cent more in the experimental class showed increases with less interest than if each class had 100 students. In this case there is available an elaborate formal theory for the plausible rival hypothesis of chance fluctuation. This theory involves the assumption of randomness, which is quite appropriate when the null model of random association is rejected in favor of a hypothesis of systematic difference between the two groups. If a "significant difference" is found, the test of significance will not, of course, reveal whether the two classes differed because one saw the experimental movie or for some selection reason associated with class topic, time of day, etc., that might have interacted with rate of autonomous change, pretest instigated changes, reactions to commonly experienced events, etc. But such a test of significance will help rule out what can be considered as a ninth threat to internal validity; that is, that there is no difference here at all that could not be accounted for as a vagary of sampling in terms of a model of purely chance assignment. Note that the statement of probability level is in this light a statement of the plausibility of this one rival hypothesis, which always has some plausibility, however faint.

DONALD T. CAMPBELL

[*Other relevant material may be found in* HYPOTHESIS TESTING; PSYCHOMETRICS; SURVEY ANALYSIS. *See also* Johnson 1968; Santostefano 1968.]

BIBLIOGRAPHY

BLALOCK, HUBERT M. JR. 1964 *Causal Inferences in Nonexperimental Research.* Chapel Hill: Univ. of North Carolina Press.

CAMPBELL, DONALD T. 1957 Factors Relevant to the Validity of Experiments in Social Settings. *Psychological Bulletin* 54:297–312.

CAMPBELL, DONALD T. 1963 From Description to Experimentation: Interpreting Trends as Quasi-experiments. Pages 212–242 in Chester W. Harris (editor), *Problems in Measuring Change.* Madison: Univ. of Wisconsin Press.

CAMPBELL, DONALD T.; and CLAYTON, K. N. 1961 Avoiding Regression Effects in Panel Studies of Communication Impact. *Studies in Public Communication* 3: 99–118.

CAMPBELL, DONALD T.; and MCCORMACK, THELMA H. 1957 Military Experience and Attitudes Toward Authority. *American Journal of Sociology* 62:482–490.

CAMPBELL, DONALD T.; and STANLEY, JULIAN C. 1963 Experimental and Quasi-experimental Designs for Research on Teaching. Pages 171–246 in Nathaniel L. Gage (editor), *Handbook of Research on Teaching.* Chicago: Rand McNally.

CHAPIN, FRANCIS S. (1947) 1955 *Experimental Designs in Sociological Research.* Rev. ed. New York: Harper.

CHAPIN, FRANCIS S.; and QUEEN, S. A. 1937 *Research Memorandum on Social Work in the Depression.* New York: Social Science Research Council.

GREENWOOD, ERNEST 1945 *Experimental Sociology: A Study in Method.* New York: Columbia Univ. Press.

HOVLAND, CARL I.; LUMSDAINE, ARTHUR A.; and SHEFFIELD, FREDERICK D. 1949 *Experiments on Mass Communication.* Studies in Social Psychology in World War II, Vol. 3. Princeton Univ. Press.

►JOHNSON, DONALD M. 1968 Reasoning and Logic. Volume 13, pages 344–350 in *International Encyclopedia of the Social Sciences.* Edited by David L. Sills. New York: Macmillan and Free Press.

LAWLEY, D. N. 1940 The Estimation of Factor Loadings by the Method of Maximum Likelihood. Royal Society of Edinburgh, *Proceedings* 60:64–82.

PELZ, DONALD C.; and ANDREWS, F. M. 1964 Detecting Causal Priorities in Panel Study Data. *American Sociological Review* 29:838–848.

►SANTOSTEFANO, SEBASTIANO 1968 Personality Measurement: IV. Situational Tests. Volume 12, pages 48–55 in *International Encyclopedia of the Social Sciences.* Edited by David L. Sills. New York: Macmillan and Free Press.

SIMON, HERBERT A. (1947–1956) 1957 *Models of Man: Social and Rational; Mathematical Essays on Rational Human Behavior in a Social Setting.* New York: Wiley.

STOUFFER, SAMUEL A. et al. 1949 *The American Soldier.* Studies in Social Psychology in World War II, Vols. 1 and 2. Princeton Univ. Press. → Volume 1: *Adjustment During Army Life.* Volume 2: *Combat and Its Aftermath.*

THORNDIKE, EDWARD L.; and RUGER, G. J. 1923 The Effect of First-year Latin Upon Knowledge of English Words of Latin Derivation. *School and Society* 18: 260–270.

THORNDIKE, EDWARD L.; and WOODWORTH, R. S. 1901 The Influence of Improvement in One Mental Function Upon the Efficiency of Other Functions. *Psychological Review* 8:247–261, 384–395, 553–564.

THORNDIKE, R. L. 1942 Regression Fallacies in the Matched Groups Experiment. *Psychometrika* 7:85–102.

WRIGHT, S. 1920 Correlation and Causation. *Journal of Agricultural Research* 20:557–585.

Postscript

[1]A list of threats to validity can never be complete. The 12 listed in the main article are still the ones most frequently cited in discussions of quasi-experimental design. Bracht and Glass (1968), Snow (1974), and Campbell (1969) have suggested others. Cook, as expressed in Cook and Campbell (1976), provides the most complete list to date. To the eight threats to *internal validity* listed in the main article, he adds six others shared by both randomized and quasi-experimental designs. In a new category of *statistical conclusion validity* he lists six threats. His *construct validity*, with ten threats, covers those aspects of what are called in the main article *external validity*, which have to do with the theoretical interpretation of the treatment and the measurements (that is, generalization to other treatments and measures) plus six or eight new threats. His more narrowly defined *external validity* has five threats, bringing the total to 35. Although this proliferation of threats has its disadvantages, each one of Cook's 35 categories is a true danger. Many of the new additions are particularly salient in settings of program evaluation.

[2]The interrupted time-series design, with or without a comparison series, has proven to be one of the most useful of designs for social policy purposes (Campbell 1969; 1975; Riecken et al. 1974, chapter 4; Ross 1973; Baldus 1973). Tests of significance have been developed for it (Box & Tiao 1965; 1975; Glass et al. 1975). (The tests of significance suggested in Campbell 1963 are wrong.) Because of the large number of time points needed, its use in social analyses is largely limited to administrative statistics.

[3]The discussion of the nonequivalent control group design in the main article is now wrong in recommending the use of covariance analysis to adjust for pretest differences between the experimental group and control group. Covariance adjustments and regression adjustments produce a regression artifact of the same magnitude as does matching, more pernicious because of the many more degrees of freedom (Lord 1960; Porter 1967; Campbell & Erlebacher 1970; Campbell & Boruch 1975). What I now recommend is Porter's (1967) reliability corrected covariance adjustment, using the pretest–posttest correlation as the reliability (if the internal consistency reliabilities of the pretest and posttest are known to be or assumed to be the same). Equally acceptable are the procedures suggested by Kenny (1975).

[4]One quasi-experimental design not mentioned in the main article now seems worthy of special attention for program evaluation purposes. The *regression–discontinuity design* (Thistlethwaite & Campbell 1960; Campbell & Stanley 1963; Campbell 1969; Sween 1971; Riecken & Boruch 1974; Cook & Campbell 1976), is usable where presumably ameliorative programs have a surplus of eligible applicants, and where random assignment to treatment and control conditions from this pool of eligibles is not accepted. The design requires that degrees of eligibility be quantified, if only by ratings or rankings, and that those most eligible on this quantified score be admitted to the program. (See also Goldberger 1972 for a discussion of assignment to treatments on the basis of a fallible observed score.)

[5]Causal inference from *correlational analysis* is increasingly dominant in the sociology literature (for example, Blalock 1971, and any issue of jour-

nals initiated in the 1970s such as *Social Science Research* and *Sociological Methods and Research*), and is being integrated with econometric methods (Goldberger & Duncan 1973). My reservations still hold (Cook & Campbell 1976) except for those structural models including error and unique determinants for each observed variable (Jöreskog 1973) —models not yet typical of many applications. Where the same variables have been measured on the same persons on two or more occasions, I recommend cross-lagged panel correlations (Kenny 1973; Crano et al. 1972; Cook & Campbell 1976).

Other major developments in quasi-experimental design have dealt with estimation of the impact of specific, presumably ameliorative, programs initiated by government. Suchman (1967), in founding the field, designated experimental or quasi-experimental modes of inference as optimal for this area. Overview treatments from this point of view include Campbell (1969), Riecken and Boruch (1974), Campbell (1975); Cook and Campbell (1976). Because of the great potential utility of government records for program evaluation when combined with proper experimental and quasi-experimental designs, procedures for linking record files that assure confidentiality and preclude file merger become very important (Campbell et al. 1975).

DONALD T. CAMPBELL

ADDITIONAL BIBLIOGRAPHY

BALDUS, D. C. 1973 Welfare as a Loan: An Empirical Study of the Recovery of Public Assistance Payments in the United States. *Stanford Law Review* 25, no. 2:123–250.

BLALOCK, HUBERT M. JR. (editor) 1971 *Causal Models in the Social Sciences*. Chicago: Aldine-Atherton.

BOX, GEORGE E. P.; and TIAO, GEORGE C. 1965 A Change in Level of a Non-stationary Time Series. *Biometrika* 52:181–192.

BOX, GEORGE E. P.; and TIAO, GEORGE C. 1975 Intervention Analysis With Applications to Economic and Environmental Problems. *Journal of the American Statistical Association* 70:70–79.

BRACHT, GLENN H.; and GLASS, GENE V. 1968 The External Validity of Experiments. *American Educational Research Journal* 5:437–474.

CAMPBELL, DONALD T. 1969 Reforms as Experiments. *American Psychologist* 24, no. 4:409–429.

CAMPBELL, DONALD T. 1975 Assessing the Impact of Planned Social Change. Pages 3–45 in Gene M. Lyons (editor), *Social Research and Public Policies: The Dartmouth/OECD Conference*. Hanover, N.H.: Public Affairs Center, Dartmouth College.

CAMPBELL, DONALD T.; and BORUCH, ROBERT F. 1975 Making the Case for Randomized Assignment to Treatments by Considering the Alternatives: Six Ways in Which Quasi-experimental Evaluations in Compensatory Education Tend to Underestimate Effects. Pages 195–216 in Carl A. Bennett and Arthur A. Lumsdaine (editors), *Evaluation and Experiment: Some Critical Issues in Assessing Social Programs*. New York: Academic Press.

CAMPBELL, DONALD T.; and ERLEBACHER, ALBERT 1970 How Regression Artifacts in Quasi-experimental Evaluations Can Mistakenly Make Compensatory Education Look Harmful. Volume 3, pages 185–210 in Jerome Hellmuth (editor), *The Disadvantaged Child*. Volume 3: *Compensatory Education: A National Debate*. New York: Brunner/Mazel.

CAMPBELL, DONALD T. et al. 1975 Confidentiality-preserving Modes of Access to Files and to Interfile Exchange for Useful Statistical Analysis. Appendix A in National Research Council, Committee on Federal Agency Evaluation Research, *Protecting Individual Privacy in Evaluation Research*. Washington: National Academy of Sciences.

COOK, THOMAS D.; and CAMPBELL, DONALD T. 1976 The Design and Conduct of Quasi-experiments and True Experiments in Field Settings. Pages 223–326 in Marvin D. Dunnette (editor), *Handbook of Industrial and Organizational Psychology*. Chicago: Rand McNally. → An expanded version of this article was published in a paperback edition by Rand McNally in 1978: Donald T. Campbell and Thomas D. Cook, *The Design and Analysis of Quasi-experiments for Field Settings*.

CRANO, WILLIAM D.; KENNY, D. A.; and CAMPBELL, DONALD T. 1972 Does Intelligence Cause Achievement? A Cross-lagged Panel Analysis. *Journal of Educational Psychology* 63:258–275.

GLASS, GENE V.; WILSON, V. L.; and GOTTMAN, J. M. 1975 *Design and Analysis of Time-series Experiments*. Boulder: Colorado Associated Univ. Press.

GOLDBERGER, ARTHUR S. 1972 Selection Bias in Evaluating Treatment Effects: Some Formal Illustrations. Discussion Paper No. 123-72. Madison: Institute for Research on Poverty, Univ. of Wisconsin.

GOLDBERGER, ARTHUR S.; and DUNCAN, OTIS DUDLEY (editors) 1973 *Structural Equation Models in the Social Sciences*. New York: Academic Press.

JÖRESKOG, KARL G. 1973 A General Method for Estimating a Linear Structural Equation System. Pages 85–112 in Arthur S. Goldberger and Otis Dudley Duncan (editors), *Structural Equation Models in the Social Sciences*. New York: Academic Press.

KENNY, D. A. 1973 Cross-lagged and Synchronous Common Factors in Panel Data. Pages 153–156 in Arthur S. Goldberger and Otis Dudley Duncan (editors), *Structural Equation Models in the Social Sciences*. New York: Academic Press.

KENNY, D. A. 1975 A Quasi-experimental Approach to Assessing Treatment Effects in the Nonequivalent Control Group Design. *Psychological Bulletin* 82:345–362.

LORD, FREDERIC M. 1960 Large-sample Covariance Analysis When the Control Variable Is Fallible. *Journal of the American Statistical Association* 55:307–321.

PORTER, ANDREW C. 1967 The Effects of Using Fallible Variables in the Analysis of Covariance. Ph.D. dissertation, Univ. of Wisconsin.

RIECKEN, HENRY W.; and BORUCH, ROBERT F. 1974 *Social Experimentation: A Method for Planning and Evaluating Social Intervention*. New York: Academic Press.

ROSS, H. LAURENCE 1973 Law, Science, and Accidents: The British Road Safety Act of 1967. *Journal of Legal Studies* 2:1–78. → Also published as a pamphlet by the American Bar Foundation.

SNOW, RICHARD E. 1974 Representative and Quasi-representative Designs for Research in Teaching. *Review of Educationl Research* 44:265–291.

SUCHMAN, EDWARD A. 1967 *Evaluative Research: Principles and Practice in Public Service and Social Action Programs.* New York: Russell Sage.

SWEEN, JOYCE A. 1971 The Experimental Regression Design: An Inquiry Into the Feasibility of Nonrandom Treatment Allocation. Ph.D. dissertation, Northwestern Univ.

THISTLETHWAITE, D. L.; and CAMPBELL, DONALD T. 1960 Regression–Discontinuity Analysis: An Alternative to the Ex Post Facto Experiment. *Journal of Educational Psychology* 51, no. 6:309–317.

EXPLANATION

See CAUSATION; SCIENTIFIC EXPLANATION.

EXPLORATION OF DATA

See DATA ANALYSIS, EXPLORATORY; STATISTICS; SURVEY ANALYSIS, *article on* METHODS OF SURVEY ANALYSIS.

EXPLORATORY DATA ANALYSIS

See DATA ANALYSIS, EXPLORATORY.

EXTREME VALUES

See NONPARAMETRIC STATISTICS, *article on* ORDER STATISTICS; STATISTICAL ANALYSIS, SPECIAL PROBLEMS OF, *article on* TRUNCATION AND CENSORSHIP.

F

FACTOR ANALYSIS AND PRINCIPAL COMPONENTS

I BILINEAR METHODS

► *This article was specially written for this volume.*

I believe this article to be the first fully integrated account of factor analysis and principal components analysis. The mathematical level is carefully limited, yet the fundamental conceptual issues are addressed directly. Several vigorous controversies are explored. What I consider a fruitless controversy over the true meaning of the phrase "factor analysis" is avoided, however, by not defining this phrase as such, but only defining specific varieties of factor analysis. Another view of these methods is presented in this volume [*see* FACTOR ANALYSIS AND PRINCIPAL COMPONENTS, *article on* FACTORING CORRELATION MATRICES.]

Table 5 lists two varieties of principal components analysis and many varieties of factor analysis, and shows their distinguishing characteristics. (Experts may wish to look at this table immediately, perhaps together with section 5, which tersely describes it, and section 4, which explains the distinguishing characteristics of the various methods.) All these methods rest on a bilinear model, although they stem from linear conceptions, as I shall explain at length. Since this common foundation seems more important to me than the distinctions among the methods, I refer to them all under the nonstandard phrase "bilinear methods."

Principal components analysis and factor analysis enjoy the unusual status of being the only general-purpose nonlinear models in widespread statistical use (at least, until the last few years). This may partly explain the common feeling that these methods are somehow different from other general-purpose methods, such as analysis of variance and time-series analysis.

Bilinear methods deal primarily with *two-way data.* (I shall not cover the few exceptions, such as Gabriel 1971.) This means that the data consist of a two-way array or matrix $\mathbf{X}$ of values x_{ij} with I rows and J columns. Here $i = 1$ to I labels the row and $j = 1$ to J labels the column. Such data never exist in a vacuum: the rows are associated with a set of I objects, units, variables, or other entities that we call U_i, and the columns are associated with J variables, objects, or other entities that we call V_j. For example, the U_i might be countries and the V_j might be variables such as median income, average educational level, percentage of population engaged in farming, and so on. The data often result from observation of several variables V_j on several objects U_i, and that is the viewpoint from now on.

Consider some data from Boynton and Gordon (1965) shown in Table 1. Here the U_i are 23 pure spectral colors and the V_j are 4 color names (red, yellow, blue, green). Roughly speaking, the datum x_{ij} is the frequency with which three highly experienced subjects chose color name V_j as appropriate when the color U_i was presented. The entries in each row of $\mathbf{X}$ sum to 225. (Actually, each color was presented 25 times to each subject, who was required to pick a first-choice name and a second-choice

Table 1 – Boynton–Gordon data on names for spectral colors*

		COLOR NAME			
		Blue	*Green*	*Yellow*	*Red*
WAVE LENGTH (in mm) OF SPECTRAL COLOR	*440*	159	—	—	66
	450	165	—	—	60
	460	196	3	2	24
	470	204	13	5	3
	480	177	41	7	—
	490	130	94	1	—
	500	80	143	2	—
	510	24	172	29	—
	520	2	173	50	—
	530	2	166	57	—
	540	1	153	71	—
	550	—	139	86	—
	560	—	117	107	1
	570	—	76	144	5
	580	—	41	161	23
	590	—	8	163	54
	600	—	1	130	94
	610	—	—	106	119
	620	—	—	73	152
	630	7	—	48	170
	640	7	—	23	195
	650	10	—	6	209
	660	7	—	4	214

* The values in this table were derived by summing the values from the three graphs in Figure 2 of Boynton and Gordon (1965), which pertain to illumination of 100 trolands. The values thus differ slightly from the original experimental results.

Source: Data from Boynton & Gordon 1965, p. 82.

name. The first choice was given a value of 2, and the second choice a value of 1. The datum x_{ij} is the sum over the 75 presentations of color i of these values for name j. For further explanation and for the reason that pooling the three subjects seemed desirable, see Boynton & Gordon 1965.) I return to these data later.

Just because data have the form of a two-way array does not mean that a bilinear method is necessarily an appropriate way to analyze them. There are many methods of analysis available, and the choice among them is one of the most difficult decisions facing the data analyst and one for which no easy rules can be given. A wise choice often rests on understanding the ideas motivating the various methods. Other methods for analyzing two-way data include multivariate analysis [*see* MULTIVARIATE ANALYSIS] and the analysis of contingency tables [*see* COUNTED DATA]. Furthermore, multivariate analysis and the bilinear methods rest on the same linear conception, with one key difference, as I shall explain below.

Examination of the literature suggests that three different uses are made of the bilinear methods:

(1) To reveal relationships among the objects and among the variables by *dimension reduction*—reducing the number of dimensions

(2) To discover *linear relationships among the variables* and *linear relationships among the objects*

(3) To suggest what the *basic underlying variables* might be.

Dimension reduction is the most important of these uses, and the only one I shall cover fully in a separate section. All three, however, are discussed briefly in the following paragraphs. How dimension reduction can help reveal relationships is illustrated at the end of this section.

Dimension reduction. In the given data matrix $\mathbf{X}$, each object U_i is represented by J coordinates, $x_{i1}, \cdots, x_{iJ}$. Dimension reduction on the objects means representing the objects by a smaller number of coordinates, say, $a_{i1}, \cdots, a_{iR}$ with R smaller than J, which in some sense retain approximately the same information as the original coordinates. (The meaning of these phrases will be clarified later.) Similarly in the given data matrix $\mathbf{X}$, each variable V_j is represented by I coordinates $x_{1j}, \cdots, x_{Ij}$. Dimension reduction on the variables means representing the variables by a smaller number of coordinates, say $b_{1j}, \cdots, b_{Rj}$ with R smaller than I, which in some sense retain approximately the same information as the original coordinates. Factor analysis uses the so-called factor pattern matrix for the b_{rj} and the so-called factor score matrix for the a_{ir}. Principal components analysis uses the first (or largest) few so-called principal components themselves for the a_{ir}, and coefficients with respect to the principal components for the b_{rj}.

As an example of dimension reduction, consider data of Adelman and Morris (1967) on 74 countries U_i covering 17 economic variables, 12 sociocultural variables, and 12 political variables, for a total of 41 variables V_j. Thus these data may be regarded as 74 rows or points, each having 41 coordinates, that is, 74 points in 41-dimensional space. It is difficult to study such a high-dimensional array of data directly. Yet the configuration of 74 points may perhaps be well approximated by a configuration in far fewer dimensions, say, four. (Using a set of 25 variables selected from the 41 above, Adelman and Morris found four factors to give an adequate approximation to the original configuration.) The reduction to four dimensions greatly simplifies the task of studying the relationships by scatter plots or otherwise, although it still requires substantial effort. Furthermore, as Gabriel (1971) has emphasized in his biconfiguration and biplot method, dimension reduction also helps study the relationship *between* the objects and the variables.

Linear relationships. Principal components analysis uses each of the last (or smallest) few principal components to provide a possible linear relationship among the objects. (Each principal component is itself a linear combination of the variables.) An analogous procedure is used to discover possible linear relationships among the variables. Factor analysis has apparently not been adapted to discover linear relationships.

Dimension reduction (on the objects) and the discovery of linear relationships (among the variables) are intimately connected, and in a mathematical sense they are much the same. In particular, dimension reduction can be thought of as discovering and discarding (mathematically, projecting to 0) many linear combinations of the variables that do not vary very much. But the distinction between *discovering* linear relationships and finding them merely to *discard* them leads to an important distinction in practice. Many more linear relationships are implicitly discarded in dimension reduction than deserve mention as linear relationships worth thinking about explicitly.

Underlying variables. Factor analysis uses the so-called method of simple structure as a way of discovering what linear combinations of the observed variables might conceivably correspond to basic underlying variables. The first (or largest) few principal components are also sometimes used for the same purpose. Simple structure is a geometric concept about the reduced factor diagram (this term is explained later), which leads to the idea that the factor pattern matrix (or the factor loading matrix) should contain many coefficients that are close to 0. For further discussion of simple structure, see, for example, Mulaik (1972, chapters 9–11). The possibility of finding basic underlying variables, and the method of simple structure for this purpose, have been the subject of considerable controversy. I believe that the methods just mentioned have frequently been misused, and that any candidate linear combination suggested by one of these methods requires substantial further verification before being accepted as basic or intrinsic. This requirement is often dispensed with, perhaps because verification is hard to come by. These methods are nonetheless sometimes useful, but caution is required. Incidentally, a more frequently successful method for discovering basic underlying variables is the INDSCAL method [*see* SCALING, MULTIDIMENSIONAL], but this method requires three-way rather than two-way data. [*See also* LATENT STRUCTURE.]

An example of the uses. To give concrete meaning to the vague idea of studying relationships among the objects, consider the Boynton–Gordon data of Table 1. Because this example is small enough to display easily, it does not well illustrate dimension reduction; instead it indicates benefits attainable after the dimensionality has been reduced. The 23 pure spectral colors constitute the objects U_i and the four color names correspond to the variables V_j. It turns out in this case that three dimensions (or factors, or components) are appropriate from a scientific viewpoint (see discussion by Shepard & Carroll 1966), but only two are used here for illustration. A particular bilinear method, "modern principal components," was used to produce two dimensions. The resulting coordinates for the colors and for the color names are shown in Table 2. A graphical presentation of the same coordinates, called the *reduced diagram*, is shown in Figure 1, in which the colors are represented by points and the color names by vectors (that is, points to which lines from the origin have been drawn). (For reasons discussed later, it is conventional to indicate the points corresponding to variables by vectors.) We see that the 23 color points lie along a triangle with a curved bottom edge and a gap on the left edge. The order of the points on this broken triangle is in accord with their order in the spectrum. The three corners of the triangle appear to correspond to three of the vectors (blue, red, and green), and the curve along

Table 2 – Coordinates for reduced diagram in two dimensions of Boynton–Gordon data

Color	Coordinates	
1	−.049	.316
2	−.036	.329
3	.043	.391
4	.097	.399
5	.129	.324
6	.182	.200
7	.228	.066
8	.247	−.096
9	.241	−.164
10	.232	−.166
11	.214	−.175
12	.196	−.184
13	.165	−.192
14	.105	−.205
15	.027	−.202
16	−.071	−.185
17	−.152	−.148
18	−.198	−.124
19	−.257	−.091
20	−.287	−.052
21	−.332	−.027
22	−.356	−.004
23	−.366	−.008

Color name	Coordinates	
Blue	75.9	347.2
Green	266.6	−124.8
Yellow	7.4	−198.4
Red	−350.0	−24.0

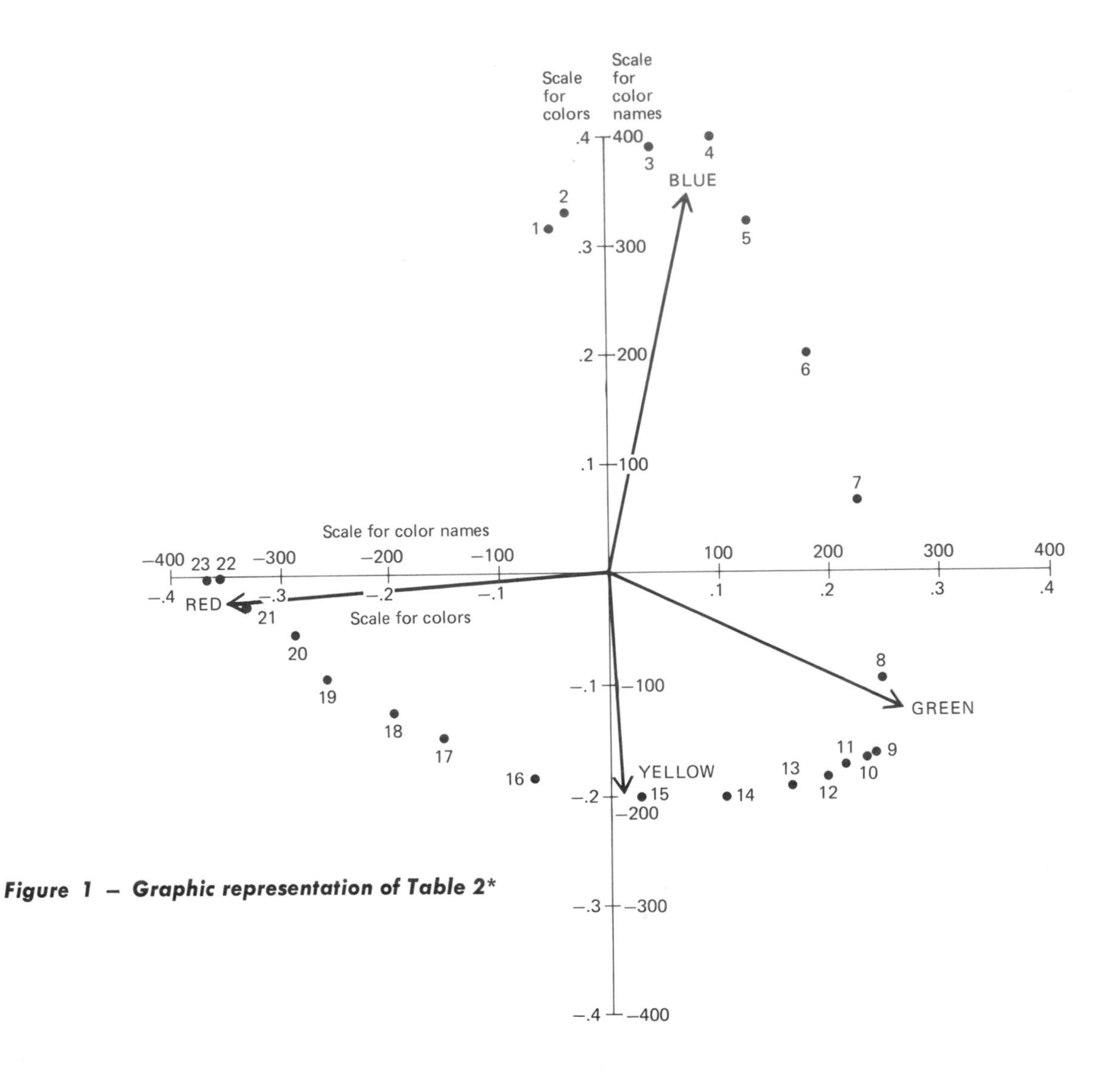

Figure 1 – Graphic representation of Table 2*

* This is a two-dimensional "reduced diagram" for the data of Table 1.

the bottom may be possibly considered as a "weak corner" corresponding to the vector for yellow. (The three-dimensional bilinear analysis shows a clear corner for yellow (Shepard & Carroll 1966), as does a two-dimensional analysis by multidimensional scaling) [*see* SCALING, MULTIDIMENSIONAL]. Figure 1 reveals, although in distorted form, the well-known color circle. It is much easier to see this using two dimensions than in the original four dimensions required for the raw data. Since the corners are associated with the color name vectors, the distortion is already hinted at, even if we had no prior knowledge of the color circle.

1. Conceptual foundations

The phrase "bilinear methods" is used here for methods with three characteristics. In practice, this definition essentially restricts the phrase to factor analysis and principal components analysis.

(1) Bilinear methods deal with two-way data, generally resulting from the observation of several variables V_j on several objects U_i.

(2) Bilinear methods use a linear conception (explained below) in approaching the data.

(3) The linear conception leads to a bilinear model (also explained below) for describing the data.

Of the linear conceptions underlying the bilinear methods, three different ones will be described, plus some variations. A similar linear conception underlies multivariate analysis [*see* MULTIVARIATE ANALYSIS], but with one key difference, as will be explained. Methods for analyzing two-way contingency tables and related types of data [*see* COUNTED DATA] sometimes use a bilinear model. These methods, however, do not use a linear conception and are not covered here.

Factor analysis is often described as a method for analyzing correlation matrices. Subsequent sections will make it clear how this viewpoint arises, but I consider it rather narrow, and feel it is better to consider factor analysis as being used on the data from which the correlations were calculated.

From now on we suppose that x_{ij} is obtained by measuring variable V_j for object U_i,

$$x_{ij} = V_j(U_i) + \text{disturbance},$$

where the disturbance includes errors of observation, measurement, and so forth, which would vary on repeated measurement of the same quantity (if repeated measurement is possible). The disturbance may also contain a systematic component that does not vary, as illustrated below. If repeated measurements are available, additional methods of analysis are possible, but they are outside the scope of this article.

Arrow notation. In order to deal with the rows and columns of $\mathbf{X}$ (the matrix of values x_{ij}) and of other matrices, I introduce a nonstandard notation illustrated in Table 4 to indicate them: $x_{i\rightarrow}$ (pronounced "x sub i row" or "x i row") indicates the ith row of $\mathbf{X}$ and $x_{\downarrow j}$ (pronounced "x sub column j" or "x column j") indicates the jth column of $\mathbf{X}$. The horizontal arrow in $x_{i\rightarrow}$ reminds us that a row runs horizontally from left to right, and the vertical arrow in $x_{\downarrow j}$ reminds us that a column runs vertically from top to bottom. In each case the arrow is written in place of the missing subscript.

Temperature example. To introduce the linear conceptions, I offer a simple example: suppose that we have measured (without error) the temperature at five places in both Fahrenheit (F) and Celsius (C) degrees. Places form the objects U_i, and temperature scales the variables V_j, so that x_{ij} is the temperature at place U_i in scale V_j. Here i takes the values 1,2,3,4,5 and j the values 1,2. The measurements appear in Table 3. Note that F and C have a linear relationship, which may be described in various ways, for example,

$$F = 32 + (9/5)\ C,$$

or

$$C = -(160/9) + (5/9)\ F.$$

As a result, note that the five rows $x_{i\rightarrow}$, regarded as points in the plane, fall on a straight line, and that the above equations are also ways of describing this line. These equations are the natural mode of expression if we think of x_{i1} as a linear function of x_{i2}, or vice versa.

A central motivating idea of the bilinear methods is that linear relationship(s) like this one exist in two-way data in real cases, but unlike

Table 3 – Hypothetical temperature measurements on two scales at five places

		SCALE	
		$V_1 = F$	$V_2 = C$
PLACE	U_1	−4	−20
	U_2	14	−10
	U_3	32	0
	U_4	50	10
	U_5	59	15

this artificial example, we don't know how many relationships there are, or what they are, in advance. The bilinear methods attempt to make use of or discover such relationships. Let us consider how such relationships can be manipulated and handled systematically.

It is helpful to describe the linear relationship in a more nearly symmetrical way, which does not place either scale in a special role. This can be done, for example, by writing

$$5\,F - 9\,C = 160.$$

The same relationship has thus been expressed in three different ways, and there are many more, for example,

$$10\ F - 18\ C = 320,$$

or

$$-15\ F + 27\ C = -484.$$

The alternative ways to express the same information become much more varied when there are not just two variables but many, and not just one relationship but several that hold simultaneously. The existence of such alternatives is called indeterminacy. Indeterminacy is a serious problem: it can become difficult even to decide whether two sets of relationships do or do not express the same information.

Another symmetrical description is by so-called parametric equations, in which F and C are separately expressed in terms of a new variable or parameter (more generally, in terms of several new variables or parameters). For example, we might use absolute temperature, in degrees Kelvin (K), as the new variable, and write

$$C = -273.18 + K,$$
$$F = [32 + (9/5)(-273.18)] + (9/5)\ K.$$

Of course, many other parametric equations for F or C are possible. For example, we can use C itself (or alternatively F itself) as the "new" variable, which would yield the parametric equations

$$C = C,$$
$$F = 32 + (9/5)\ C,$$

so that the parametric equations approach encompasses the first descriptions of the linear relationship. We can also use an arbitrary combination such as $(C + F)/2$ or $7.2\,C - 1.3\,F$ as the new variable, which would yield other linear parametric equations. (We can even find parametric equations that are trivially nonlinear, for example, if we replace K above by S^2 where S is the square root of K.) Thus the variety of different ways of expressing essentially the same information is even more varied here than in the preceding paragraph. When there are many variables, not just two, and several relationships that hold simultaneously, the indeterminacy becomes even more complicated. Whether or not two different sets of parametric equations express the same information may be far from obvious.

The branch of mathematics including matrix theory and linear algebra contains tools that are quite adequate to deal with the indeterminacy problem, so it is easier for mathematicians and others familiar with these topics to learn about the bilinear methods. I believe that it is possible, however, to present the essential concepts comprehensibly while assuming only elementary familiarity with vectors and matrices, and this article is an attempt to do so. Nevertheless it is very clear historically that the difficulty of *understanding* indeterminacy has substantially hindered the use and development of the bilinear methods.

The parametric equations approach to describing linear relationships turns out to have such significant advantages of mathematical and computational convenience (which it is outside my scope to describe), that it is heavily used in all detailed descriptions of all bilinear methods (although the phrase "parametric equations" is seldom used). Other approaches, however, are combined with it to varying degrees. It will be the focus of attention below.

The parametric equations approach also has some conceptual values. Suppose that we did not know in advance the reason for the relationship between F and C. We might guess that there is some variable V, not yet measured or understood, possibly of particular conceptual significance, which intervenes through parametric equations or mappings of the form

$$F = c_1 + d_1V,$$
$$C = c_2 + d_2V.$$

(We do not rule out the possibility that V will turn out to be F or C itself, though we would not usually expect this to happen.) The advantage of this approach is that it may lead us to important new concepts as we try to understand V, and later perhaps to measure it independently.

The difficulty in finding V has been foreshadowed, namely, the indeterminacy problem. Even when there is only a single new variable V (and in general there would be several new variables), many different alternative versions of the new variable are possible. For example, V suffers from an indeterminacy of scaling that is only a minor annoyance here, but becomes much more troublesome in general. In our simple case, one sees immediately that the same linear relationship is described whether we use the above parametric equations or an alternative such as

$$F = c_1 + (5d_1)(V/5),$$
$$C = c_2 + (5d_2)(V/5),$$

so that $V/5$ (and in fact any nonzero constant times V) does just as well as V itself. Thus from the observed data one cannot distinguish between V and $V/5$ as underlying variables. (Nor does it make much difference, if any, which is used.) It is traditional and helpful to use conventions to eliminate or reduce the indeterminacy. For example, one might specify that the standard deviation of V is required to be unity (which would still leave $+V$ and $-V$ as alternative variables), or instead one might specify that d_1 (that is, the coefficient of V in the equation for F) is unity.

There is also indeterminacy of location. For example, the first parametric equations in V might have been written equally well as

$$F = (c_1 + 6d_1) + d_1(V - 6),$$
$$C = (c_2 + 6d_2) + d_2(V - 6),$$

so that $V - 6$ (and in fact V plus any constant) does just as well as V itself. A common convention to eliminate this ambiguity is to require the expected value of V to be 0.

Fortunately, it turns out that the two types of indeterminacy just described completely cover the indeterminacy of the parametric equations approach, in a certain sense. In the case of linear parametric equations using a single new variable V, it can be proved that for *any* other linear parametric equations in a single new variable W that describe the same linear relationship, W is related to V by a combination of change of scale and change of location, that is, by $W = \text{constant}_1 + \text{constant}_2\, V$. Similar but more complicated results exist in general.

In real examples, measurement and sampling error must be considered, and points will rarely if ever fall exactly on a straight line. Note that some kinds of error will simply shift a point along the

Table 4 – Artificial data on five wooden balls (I = 5) and two variables (J = 2)

OBJECT (diam. in cm)	VARIABLES Wt. of ball (g)	Wt. of paint (mg)		DATA MATRIX X v_1	v_2	COLUMNS OF X $x_{\downarrow 1}$	$x_{\downarrow 2}$	ROWS OF X	
"Tennis ball" (6.5)	275	676	u_1	275	676	275	676	$x_{1\rightarrow}$	(275, 676)
"Ping-Pong ball" (3.7)	51	219	u_2	51	219	51	219	$x_{2\rightarrow}$	(51, 219)
"Baseball" (7.3)	389	853	u_3	389	853	389	853	$x_{3\rightarrow}$	(389, 853)
"Orange" (8)	512	1024	u_4	512	1024	512	1024	$x_{4\rightarrow}$	(512, 1024)
"Handball" (4.8)	111	369	u_5	111	369	111	369	$x_{5\rightarrow}$	(111, 369)

line, and consequently will not obscure the existence of the line, while other kinds of error act quite differently. In the present example, warmth from the observer's hand might affect both F and C commensurately, and simply shift $x_{i\rightarrow}$ along the line. If, however, observers read the temperatures on different thermometers at slightly different times, the points will in practice surely deviate somewhat from a straight line. If the readings are sufficiently crude—merrymaking meteorologists on New Year's Eve with tiny, cheap thermometers—then the points $x_{i\rightarrow}$ may be so scattered that the underlying line is far from obvious.

Wooden balls example. In another artificial yet helpful example, the underlying relationship deviates somewhat from linearity. Suppose we have several balls, all made from the same wood, with one the size of a tennis ball, one the size of a Ping-Pong ball, and so forth. They are listed in Table 4, together with their diameters in centimeters. The weight of each ball (in grams) is measured, and gives the first variable. The weight of paint (in milligrams) required to cover each ball is measured, and gives the second variable. (This measurement could be made experimentally independent of the first one by dipping each ball in a can of paint, weighing the can before and after dipping, and subtracting.) The 5×2 table is the data matrix $\mathbf{X}$. For simplicity, no random or systematic error has been incorporated in these values, though scientific data generally contain such error, sometimes of substantial magnitude.

The five rows $x_{i\rightarrow}$ are plotted as points in Figure 2. If d is the diameter of the ball, then ball weight is proportional to the volume and hence to d^3, while the paint weight is proportional to the surface area and hence d^2. (In fact, these pseudo-data were generated by using d^3 for the first column and $16d^2$ for the second.) Thus all possible balls of the same wood correspond to points on the curve. In fact, we can describe this curve by a mapping (*not* the mapping displayed by the curved arrows) from one-dimensional points (that is, numbers) d into two-dimensional points given by

$$d \rightarrow (d^3, 16d^2).$$

Note that this mapping is not linear. We can also describe this curve by the equation $y = 16x^{2/3}$ where x and y are the usual horizontal and vertical coordinates. In real situations, we would not know the formulas, the curves, the diameters of the spheres, the relationship between the variables, and so forth. Thus suppose that all we know is the identity of the five objects, the identity of the two variables, and the table of values $\mathbf{X}$. (Incidentally, if we transformed the data by taking their loga-

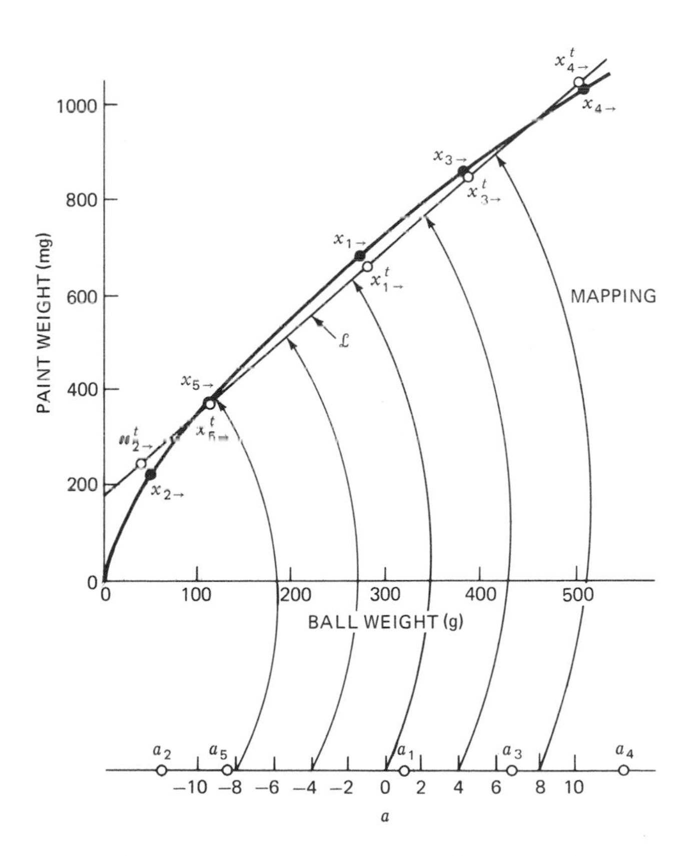

Figure 2 – Weight of wooden balls by weight of paint*

* The solid points $x_{i\rightarrow}$ show the data. The curve through these points shows the true theoretical relationship. The other items show the hypothetical structure postulated by the bilinear methods.

rithms, the relationship between the two variables would become precisely linear. Transforming the data before subsequent analysis is often helpful) [*for further information, see* STATISTICAL ANALYSIS, SPECIAL PROBLEMS OF, *article on* TRANSFORMATIONS OF DATA].

Conception A. The first linear conception is that the points $x_{i\rightarrow}$ lie approximately on a straight line, such as line $\mathfrak{L}$ in Figure 2. This geometrical conception, due to Karl Pearson (1901), is closely linked to the idea of dimension reduction. If these data were not exact formula values but contained random disturbances of substantial size, then the smooth curve connecting the points would not be evident, and the straight-line conception would seem plausible. This conception is quite similar to that of linear regression. In contrast to regression, however, there is no distinction between dependent and independent variables: the two variables are treated alike. [*See* LINEAR HYPOTHESES, *article on* REGRESSION.]

Where there are many variables instead of two, the straight line above might be replaced by its two-dimensional relative, a plane, or by some still higher-dimensional relative, referred to in general as a flat subspace, or simply a *flat.* (A one-dimensional flat is a line contained in some larger space; a two-dimensional flat is a plane contained in some larger space; and so on.) Using a one-dimensional flat, as above, corresponds to using one principal component or extracting one factor. In general, the dimensionality R of the flat corresponds to the number of principal components or the number of factors.

Conception A′. A variation is that the observed points, $x_{i\rightarrow}$, arise from true underlying points, $x^t_{i\rightarrow}$ (here "t" stands for "true"),

$$x_{i\rightarrow} = x^t_{i\rightarrow} + \text{disturbance},$$

and that the true underlying points, $x^t_{i\rightarrow}$, lie exactly on a straight line or in general, a flat. The disturbance is meant to include random errors of measurement (which are not present in the example) as well as systematic deviations, like those in the example, where a straight line is not really correct although it may be an adequate approximation over a limited range of ball sizes. Whether the true underlying points should be estimated at all, and if so how, is a matter on which bilinear methods differ.

Conception B. This conception is clearly the motivating idea of Hotelling (1933), which introduced principal components analysis. Conceptions B and B′ have generally been presented as the rationale for factor analysis at least since the 1930s. (The original conception of factor analysis in terms of correlations, due to Spearman (1904*b*), is generally derived from conception B or B′, but even today is seen by many factor analysts as somehow basic in itself, a view I do not share.) This conception has four parts. The first part is a new space (like the one shown at the bottom of Figure 2 whose points are indicated here by a), outside the space containing the data points. The second part is a linear mapping L (like the one shown by the curved arrows in Figure 2) that maps the new space into the space containing the data points. The third part is the image of L (like the line $\mathfrak{L}$). The fourth part is that the image of L lies near all the points $x_{i\rightarrow}$. Thus conception B contains the line of conception A, but obtains it from a linear mapping on a new space. More generally, the new space would be R-dimensional, and its image would be an R-dimensional flat. Historically it is interesting that Spearman and his successors first believed that a single factor ($R = 1$) was sufficient to explain intelligence. The subsequent extension to more factors was accepted only with reluctance, so the desire to keep R small has an old history.

Conception B is closely related to parametric equations, which serve as an algebraic description of the mapping. The new variable(s) used in the parametric equations are the coordinate(s) of the new space. For the parametric equations to be linear means exactly the same thing as for the mapping to be linear.

Conception B′. Conception B also has a variation based on the idea of true underlying points, $x^t_{i\rightarrow}$. This conception is that there is a linear mapping $a \rightarrow L(a)$ like that above, and a value a_i for each wooden ball, such that

$$x_i = L(a_i) + \text{disturbance}.$$

Obviously $L(a_i)$ has the same conceptual meaning here that $x^t_{i\rightarrow}$ does above. (More generally the $a_{i\rightarrow}$ would be elements of an R-dimensional space.)

Every straight line can be described as the image of a linear mapping, that is, by linear parametric equations. For example, to find parametric equations for the line $\mathfrak{L}$ of Figure 2, first pick an arbitrary fixed point on $\mathfrak{L}$, say (268, 628). Then pick a second point, say (306, 692), and subtract the first point to yield (38, 64). Using a as a new variable, $\mathfrak{L}$ is described by these equations:

$$x = 268 + 38\,a,$$
$$y = 628 + 64\,a.$$

To say the same thing in different language, $\mathfrak{L}$ is the image of the mapping L that maps

$$a \to (268+38a, 628+64a) = (268, 628) + a(38, 64).$$

When a is 0, the image of a is the first arbitrary point; when a is 1, its image is the second; and as a varies its image sweeps along the line. (More generally, every R-dimensional flat can be described as the image of a linear mapping on an R-dimensional space, and a similar construction can be used to show this.)

Indeterminacy of *L*. While every line can be described as the image of a linear mapping, it is also an elementary mathematical fact that every (nontrivial) linear mapping (of a one-dimensional space) has a straight line as an image. Thus conception B is very closely related to conception A, but has extra machinery (whose value is discussed below). However, each linear mapping yields only *one* straight line, although each straight line is the image of *many different linear mappings* (since the two arbitrary points can be chosen in many ways). The characteristic that many linear mappings yield the same line carries over to flats in general and is another description of the troublesome indeterminacy mentioned above.

Note that a straight line (or a flat) can also be obtained as the image of a great many *non*linear mappings, so that straightness of the observed points does not imply that there is a linear underlying mechanism, although a linear underlying mechanism does imply straightness of the points. For example, the straight line in the illustration is also the image of the mapping $a \to (268, 628) + a_3 (38, 64)$. Fortunately, it is not necessary to work with these nonlinear mappings. Noticing that they exist, however, reminds us that the linear mapping we reconstruct from the data may have little to do with the true underlying mechanism that generates the data, as is illustrated by the wooden balls example.

Use of *L*. For uses (1) and (2) of the three main uses of bilinear methods, dimension reduction and finding linear relationships, the straight line (or flat) used in conception A is all that we actually get from the data, and the linear mapping L of conception B serves no other role than as a way to describe it. Use of this linear mapping leads to the many difficulties caused by indeterminacy. Nevertheless, the machinery of conception B is widely, almost universally, used. What is the reason for this curious state of affairs? The answer is that by far the easiest mathematical and computational way to describe and work with a straight line (or a flat) is by a linear mapping like the one in conception B. For this reason, every bilinear method and every computational approach yields a linear mapping as prime output. (What is actually printed out is usually a matrix, but the meaning of the matrix is that it describes the linear mapping.) Different linear mappings that describe the same straight line (or flat) provide the same information for uses (1) and (2). The choice among different linear mappings that describe the same straight line (or flat) is based on secondary considerations. Among such considerations are conventions, like those mentioned before.

For use (3), finding basic underlying variables (a controversial idea), the linear mapping of conception B plays a central role. In fact, the bilinear methods now available for use (3) proceed by first finding a line (or flat) in the same manner as for use (1), and then as a separate, extra step go on carefully to select one particular linear mapping from those yielding that line (or flat). Factor analysts refer to this step as "rotating in the factor space," since it can be described in terms of a rotation from one set of coordinates to another in the new space.

Conception C. In terms of the wooden balls example, a third conception is that there is an approximate linear relationship between the two variables, in this case

$$64\, x_{i1} - 38\, x_{i2} + 6712 \cong 0,$$

for all i, or in more informal terms,

$$64 \times \frac{\text{ball wt.}}{\text{(in g)}} - 38 \times \frac{\text{paint wt.}}{\text{(in mg)}} \cong -6712.$$

This conception is closely linked with use (2), discovering linear relationships. Since a linear equation between two variables defines a straight line and since a straight line can always be described by a linear equation, the first and third conceptions are very closely related, but with a difference in emphasis. I shall not consider the third conception any further.

Generalization of the conclusions. Users frequently desire to generalize conclusions from the given set of objects to other similar objects, or from the given set of variables to other similar variables. The illustration provides useful guidance about generalizing to other similar objects. Since we know the underlying mechanism in this case, we can see that balls of the same general size, made of the same kind of wood, should yield points near the same line. If we consider balls made of denser or less dense material than the original wood, or if we consider balls whose surface holds paint differently, or if we consider wooden cubes or wooden pyramids, then we cannot expect these objects to behave like the ones analyzed. Furthermore, if we consider balls whose size is considerably larger or considerably smaller than those in the sample, we must also expect different behavior, because our analysis assumed a linear relationship, which is a valid approximation only over a limited region. This illustration typifies many realistic situations with respect to generalizability.

Summary of the conceptions. Returning to conceptions A and B, I now restate them for the general case involving an R-dimensional flat instead of a line (which is a one-dimensional flat). Consider the I rows $x_{i\rightarrow}$ of $\mathbf{X}$ as points in J-dimensional space. Readers may visualize this in terms of Figure 2, which corresponds to $R = 1$ and $J = 2$. To visualize a useful picture for $R = 2$ it is necessary to use $J \geqslant 3$; for $R = 2$ and $J = 3$ readers may visualize a set of I points in three-dimensional space near some plane. In the following discussion, keep in mind that R (which is the dimensionality of the flat, the number of principal components being used, and/or the number of factors being extracted) is generally quite a bit smaller than J, and that always $R \leqslant J$. A one-dimensional flat is a line in a larger space, a two-dimensional flat is a plane in a larger space, a three-dimensional flat is a three-dimensional subspace in a larger space, and an R-dimensional flat is something hard to visualize for $R > 3$.

Conception A is that the I points, $x_{i\rightarrow}$, lie approximately on some R-dimensional flat. Conception A′ is that the observed points, $x_{i\rightarrow}$, arise from some true underlying points, $x^t_{i\rightarrow}$, which lie exactly on an R-dimensional flat. Conception B is that there is a new R-dimensional space, and a linear mapping $a_{\rightarrow} \rightarrow L(a_{\rightarrow})$ that takes this space into J-dimensional space so that the image (an R-dimensional flat) lies near all the points $x_{i\rightarrow}$. Conception B′ is that there is a point $a_{i\rightarrow} = (a_{i1}, \cdots, a_{iR})$ for each object U_i such that

$$x_{i\rightarrow} = L(a_{i\rightarrow}) + \text{disturbance},$$

where the disturbance terms are not large. Of course, $L(a_{i\rightarrow})$ has the same significance as $x^t_{i\rightarrow}$. Any particular R-dimensional flat is the image of many different linear mappings (what they all are I shall clarify later), while each linear mapping yields a single R-dimensional flat. Any linear mapping $a_{\rightarrow} \rightarrow L(a_{\rightarrow})$ can be described using an arbitrary point in its image, say $b_{0\rightarrow} = (b_{01}, \cdots, b_{0J})$, together with a properly chosen $R \times J$ matrix; call it $\boldsymbol{B}$. Using $b_{0\rightarrow}$ and $\boldsymbol{B}$, $L(a_{\rightarrow})$ has the form

$$L(a_{\rightarrow}) = b_{0\rightarrow} + a_{\rightarrow}\boldsymbol{B},$$

where $a_{\rightarrow}\boldsymbol{B}$ means the matrix product. Of course,

$$L(a_{i\rightarrow}) = b_{0\rightarrow} + a_{i\rightarrow}\boldsymbol{B}.$$

The important part of this description, on which interest is focused, is the matrix $\boldsymbol{B}$. This matrix is the same regardless of which arbitrary point $b_{0\rightarrow}$ was chosen.

It turns out that linear conceptions A, B, and C all lead to the same mathematical model, which is bilinear. To demonstrate this would take too long, but we use conception B to derive the model. Recalling that $b_{\downarrow j}$ is the jth column of $\boldsymbol{B}$, and that the jth coordinate of $a_{i\rightarrow}\boldsymbol{B}$ is $a_{i\rightarrow}b_{\downarrow j}$, we see that

$$\begin{aligned} x_{ij} &= (L(a_{i\rightarrow}))_j + \text{disturbance}, \\ &= b_{0j} + a_{i\rightarrow}b_{\downarrow j} + \text{disturbance}, \\ &= b_{0j} + a_{i1}b_{1j} + a_{i2}b_{2j} + \cdots + a_{iR}b_{Rj} \\ &\quad + \text{disturbance}, \\ &= b_{0j} + \sum_{r=1}^{R} a_{ir}b_{rj} + \text{disturbance}. \end{aligned}$$

The summation is called a *pure bilinear expression* because the variables can be divided into two groups (the a's and the b's) so that the expression is purely linear in each group of variables separately. ("Purely linear" means linear homogeneous, that is, linear with no constant term.) The summation plus the term b_{0j} is a *bilinear expression* because it is the sum of a pure bilinear expression plus an expression that happens to be linear.

Multivariate analysis is based on the same kind of data and on similar conceptions to those above. To understand the differences between multivariate analysis and bilinear methods, consider one of the conceptions underlying the former: there is a new space, called the parameter space, and a linear mapping whose image (a flat) lies near the data points. This is similar to conception B. The difference is that in multivariate analysis the linear mapping is fixed and known in advance, not derived from the data but from a priori substantive considerations. The matrix (corresponding to $\boldsymbol{B}$ above) that specifies this mapping is called the design matrix. Because the design matrix is known, the data values are linear functions of the unknown parameters (corresponding to our a_{ir} and our b_{0r}), which is what is meant by calling the multivariate analysis model a linear model. Because our corresponding matrix $\boldsymbol{B}$ is not known but must be estimated, the data values in our case are bilinear functions of the unknown parameters a_{ir} and b_{jr} (including the b_{0r}); this is why the methods of this article can be called bilinear.

2. Dimension reduction through the object and factor spaces

"Dimension reduction" means portraying the objects and the variables or both in a space of relatively low dimensionality, R. Historically, factor analysis has given much greater attention to dimension reduction on the variables, and principal components analysis has tended to emphasize the objects. In any given application there may be good reason to emphasize one aspect or the other, but there is no intrinsic reason to continue the historical imbalance of emphasis. Rather, one should realize the symmetry and utility of both aspects, and look at one or the other or both according to whether one wants to study the relationship among the objects, among the variables, or both.

In the absence of an adequate terminology for the unified view presented here, I have created the necessary terms, which I shall connect with existing terminology as I go along.

Object space. Let us first consider the problem of dimension reduction for the objects. Here it is necessary to keep in mind three spaces: the *row space*, the *object space*, and the *reduced object space*. In the wooden balls example, the row space is the space shown in the upper part of Figure 2; the object space is the line $\mathfrak{L}$ contained in the row space; and the reduced object space is the line at the bottom of Figure 2. (Those already familiar with principal components analysis may note that the space of the first—largest—few principal components is used somewhat ambiguously there for both the object space and the reduced object space.) In general for an $I \times J$ data matrix $\mathbf{X}$ the *row space*, denoted by $\mathfrak{U}$, is J-dimensional space, and each row $x_{i\rightarrow}$ of $\mathbf{X}$ denotes a point in $\mathfrak{U}$. The name $\mathfrak{U}$ reminds us that these points correspond to the objects U_i.

Recall conception A, which assumes that for some relatively small R the object points $x_{i\rightarrow}$ lie near some R-dimensional flat (that is, flat subspace) of the row space $\mathfrak{U}$. Were it not for the inevitable disturbances in real data, the object points would, according to this assumption, lie exactly on some flat called the *true object space*. In practice, this flat is not known, and it is necessary to work with some estimated flat $\mathfrak{L}$, which is called the (*fitted*) *object space*. As in conception B, the flat $\mathfrak{L}$ can be described by the use of a new R-dimensional space, here called the *reduced object space*, together with a linear mapping L from it into $\mathfrak{U}$ whose image is $\mathfrak{L}$. Figure 3a, which is similar to Figure 2, shows these concepts without reference to any specific example, using $J = 2$ and $R = 1$ for pictorial purposes. For real data J and R are usually larger, so the situation becomes difficult to visualize and mathematical methods are needed.

The estimation process for many bilinear methods can be thought of in the following way. A range of reasonable values of R is selected by judgment. For each value of R, $\mathfrak{L}_R$ is selected to be the R-dimensional flat in $\mathfrak{U}$ that is nearest to all the object points simultaneously in some suitable sense. (One major distinction among different bilinear methods lies in how they assess nearness for this purpose.) Goodness of fit of $\mathfrak{L}_R$ necessarily improves as R increases, but how fast it increases is considered and one of the $\mathfrak{L}_R$ is

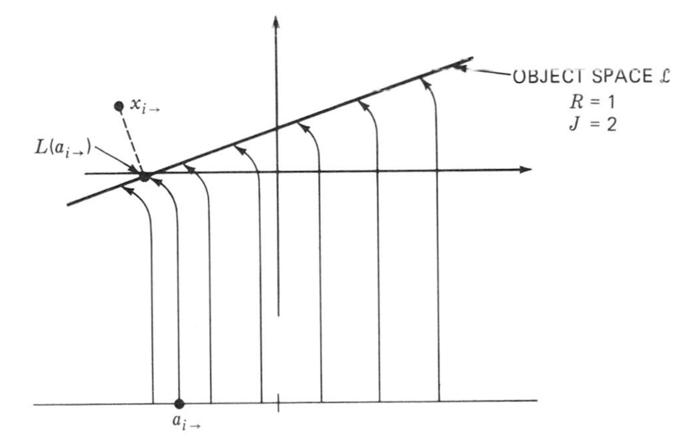

Figure 3a – A two-dimensional row space, a one-dimensional object space $\mathfrak{L}$, and a one-dimensional reduced object space*

* The curved arrows indicate the mapping L.

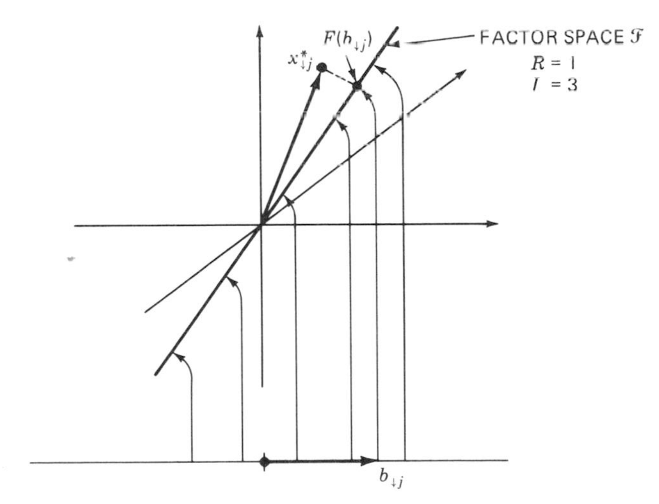

Figure 3b – A three-dimensional column space, a one-dimensional factor space $\mathfrak{F}$, and a one-dimensional reduced factor space*

* The curved arrows indicate the mapping F.

selected, generally with the aid of subjective considerations. This dimensionality selection process is referred to as "deciding how many factors to retain (or to extract)" in factor analysis and as "deciding how many principal components to use" in principal components analysis. Among the tools commonly used to aid selection is "the proportion of variance accounted for," to be discussed below, and the closely associated "eigenvalues" or "roots" for the factors. Although a number of papers touch on dimensionality selection, such as Bartlett (1950), Guttman (1954), Linn (1968), and Wachter (1975), no comprehensive treatment has yet been published.

In addition to fitting an R-dimensional flat $\mathcal{L}$ jointly to the observed points $x_{i\to}$, many bilinear methods fit points $x^f_{i\to}$ ("f" for fitted) to the $x_{i\to}$. Of course, these points lie exactly on $\mathcal{L}$, and in some bilinear methods, $x^f_{i\to}$ is the point on $\mathcal{L}$ that is nearest to $x_{i\to}$ in some suitable sense. The fitted points $x^f_{i\to}$ can be used as estimates of the true underlying points, $x^t_{i\to}$, discussed in conception A′.

Recall from the previous section that any linear mapping $a_\to \to L(a_\to)$ from the reduced object space into $\mathcal{U}$ has the form

$$L(a_\to) = b_{0\to} + a_\to \boldsymbol{B},$$

where $b_{0\to} = (b_{01}, \cdots, b_{0J})$ is a properly chosen point in $\mathcal{U}$ and $\boldsymbol{B} = \|b_{rj}\|$ is a properly chosen $R \times J$ matrix. As $(a_\to)$ varies over all points in the reduced object space, $L(a_\to)$ varies over all points of the (fitted) object space $\mathcal{L}$, and L provides a one-to-one linear mapping between these two spaces. Hence there are points denoted by $a_{i\to}$ such that $L(a_{i\to}) = x^f_{i\to}$ exactly.

Introducing the Greek letters $\alpha_{i\to}$ and Λ for the true underlying values used in conception B′ and previously denoted by $a_{i\to}$ and L, we obtain the following parallel equations:

$$x_{i\to} = x^t_{i\to} + \text{disturbance}, \qquad \text{where } x^t_{i\to} = \Lambda(\alpha_{i\to}),$$

$$x_{i\to} = x^f_{i\to} + \text{residual}, \qquad \text{where } x^f_{i\to} = L(a_{i\to}).$$

It is natural to take the $a_{i\to}$ and L as estimates of the $\alpha_{i\to}$ and Λ. Further introducing the Greek letters $\beta_{0\to}$ and $\mathbf{B}$ as the true underlying values of $b_{0\to}$ and $\boldsymbol{B}$, we have

$$x^t_{i\to} = \Lambda(\alpha_{i\to}) = \beta_{0\to} + \alpha_{i\to}\mathbf{B},$$

$$x^f_{i\to} = L(a_{i\to}) = b_{0\to} + a_{i\to}\boldsymbol{B}.$$

Because each object point $x_{i\to}$ is near the corresponding fitted point $x^f_{i\to}$, the object points are arranged approximately the same way in $\mathcal{U}$ that the fitted points are in $\mathcal{L}$. Because the linear mapping L links them together, the fitted points are arranged the same way in $\mathcal{L}$ that the points $a_{i\to}$ are in the reduced object space. This is why it is sensible to study the relationship among the objects by using the points $a_{i\to}$. The *reduced object diagram* is defined as a diagram showing the points $a_{i\to}$ in the reduced object space. (Figure 1 shows a reduced object diagram with some added vectors.)

The vectors $b_{r\to}$ (which may be considered the rows of $\boldsymbol{B}$, but see below) together with $b_{0\to}$ yield an effective description of the object space $\mathcal{L}$, since this space consists of all points of form

$$b_{0\to} + \text{a linear combination of } \{b_{1\to}, \cdots, b_{R\to}\}.$$

However, b_0 provides only positional information about the space, which is usually less important, while the $b_{r\to}$ provide the vital information about its angular orientation. The rows of $\boldsymbol{B}$ (this excludes $b_{0\to}$) form a *coordinate system* for $\mathcal{L}$. The entries of $a_{i\to}$ are the coordinates of $x^f_{i\to}$ in $\mathcal{L}$ because $x^f_{i\to} = b_{0\to} + \sum_r a_{ir}b_{r\to}$. Thus $a_{i\to}$ provides coordinates associated with U_i both in $\mathcal{L}$ and in the reduced object space.

A similarity to linear regression is apparent in connection with Figure 3a, since linear regression (in its simplest form) also is based on the idea that the points lie near a straight line, and fits such a line to the points. For bilinear methods, however, no distinction is made between dependent and independent variables: all variables are treated exactly alike. A more fundamental difference concerns the dimensionality of the object space; it appears when J is larger than 2. In linear regression involving a total of J variables ($J-1$ independent and 1 dependent), the space analogous to the object space is the graph of the function we are fitting and *must* be $(J-1)$-dimensional, while in the bilinear methods the object space is R-dimensional, and R is seldom as large as $J-1$.

Factor space. Next let us consider the problem of dimension reduction for the variables. Here it will be necessary to keep in mind three spaces: the *column space,* the *factor space,* and the *reduced factor space.* These are analogous to the spaces used in dimension reduction for the objects, and they are illustrated abstractly in Figure 3b. Furthermore the reduced factor space and the reduced object space will turn out to be closely related. For an $I \times J$ data matrix $\mathbf{X}$ the column space, denoted by $\mathcal{V}$, is I-dimensional space, and each column $x_{\downarrow j}$ of $\mathbf{X}$ corresponds to a point $x^*_{\downarrow j}$ in $\mathcal{V}$. (The meaning of the asterisk is explained below.) It is helpful to draw a vector or arrow from the origin to each of these points, and to talk of these points as vectors. The name $\mathcal{V}$ reminds us that these vectors correspond to the variables V_j.

Unfortunately it is not possible to portray the I-dimensional space of Figure 3b as helpfully as the J-dimensional space of Figure 3a, because illustrating the geometric ideas adequately seems to require $I \geqslant 5$ or at least $I \geqslant 4$, but only $J \geqslant 3$ or $J \geqslant 2$. Figure 3b makes the uneasy compromise of using $I = 3$, which is hard to visualize and not too useful after one succeeds.

A mathematical fact is needed. If $\mathbf{X}$ is any matrix, define $\mathbf{X}^*$ to be $\mathbf{X}$ after it is column-centered, that is,

$$x^*_{ij} = x_{ij} - \bar{x}_j , \qquad \text{where } \bar{x}_j = \sum_i x_{ij}/I.$$

The assumption of conception A′ is that the points $x^t_{i\rightarrow}$ lie exactly on an R-dimensional flat in $\mathcal{U}$. It can be proved in a paragraph that this holds if and only if the vectors $x^{t*}_{\downarrow j}$ lie exactly on some R-dimensional flat in $\mathcal{V}$ that contains the origin. Note two asymmetries with the object situation: the columns used are those of $\mathbf{X}^*$ rather than those of $\mathbf{X}$; and the flat in $\mathcal{V}$ must contain the origin where no such requirement was imposed in $\mathcal{U}$. These asymmetries are closely related to the fact that the equations above contain a term $b_{0\rightarrow}$ but no corresponding term $a_{0\rightarrow}$. They also reflect the fact that the bilinear methods as usually presented, and as described in this article, are not sensitive to an additive constant change in a variable (that is, adding a constant to a column), but are sensitive to an additive change in an object (that is, adding a constant to a row). Variations of the bilinear methods with other properties can in some cases be appropriate, but the subject is complex and outside the scope of this article.

Now conception A can be described in terms of the column space. It is equivalent to assuming that for some relatively small R the vectors $x^*_{\downarrow j}$ lie near some R-dimensional flat in $\mathcal{V}$ that contains the origin. If it were not for the inevitable disturbances in real data, the vectors $x^*_{\downarrow j}$ would, according to the assumption, lie exactly on some flat through the origin called the *true factor space*. In practice this flat is not known, and it is necessary to work with some estimated flat $\mathfrak{F}$, which is called the *(fitted) factor space*. (In factor analysis this is often called the "common factor space," but as presented here there is no reason for the word "common." By analogy with the object space it would be natural to call it the "variable space," but this term can be too easily misunderstood.) The flat $\mathfrak{F}$ can be described by the use of a new R-dimensional space, here called the *reduced factor space*, together with a linear mapping F from it into $\mathcal{V}$ whose image is $\mathfrak{F}$.

The estimation process for many bilinear methods can be thought of just as well in terms of the factor space as in terms of the object space. Furthermore, use of the factor space is much closer to the spirit in which factor analysis is usually presented than use of the object space. A description of the estimation process in terms of the factor space would be entirely parallel to the description above in terms of the object space, and is thus omitted.

In addition to fitting an R-dimensional flat $\mathfrak{F}$, many bilinear methods fit points $x^{*f}_{\downarrow j}$ ("f" for fitted) to the $x^*_{\downarrow j}$. Of course these lie exactly on $\mathfrak{F}$, and in some bilinear methods $x^{*f}_{\downarrow j}$ is the vector in $\mathfrak{F}$ that is nearest to $x^*_{\downarrow j}$ in some suitable sense. The fitted points can be used as estimates of their true underlying counterparts, which have not been previously discussed explicitly but have the natural significance.

Consider the linear mapping F from the reduced space into $\mathcal{V}$ whose image is $\mathfrak{F}$. We denote the elements of the reduced factor space by R-tuples written as columns, and refer to these elements as vectors instead of points. As $b_\downarrow$ varies over all vectors in the reduced factor space, $F(b_\downarrow)$ varies over all vectors in $\mathfrak{F}$, and F provides a one-to-one linear mapping between the two spaces. It can be shown that $F(b_\downarrow)$ has the form

$$F(b_\downarrow) = \mathbf{A}b_\downarrow$$

for some suitably chosen matrix $\mathbf{A}$. The reason that there is no term here analogous to $b_{0\rightarrow}$ in the equation for L is that the image $\mathfrak{F}$ of this mapping is required to pass through the origin. For each vector $x^{*f}_{\downarrow j}$ there is a vector $b_{\downarrow j}$ such that $F(b_{\downarrow j}) = x^{*f}_{\downarrow j}$ exactly. Introducing the Greek letters $\beta_{\downarrow j}$ and Φ for the true underlying values of $b_{\downarrow j}$ and F, we obtain the following parallel equations:

$$x^*_{\downarrow j} = x^{*t}_{\downarrow j} + \text{disturbance}, \qquad \text{where } x^{*t}_{\downarrow j} = \Phi(\beta_{\downarrow j}),$$

$$x^*_{\downarrow j} = x^{*f}_{\downarrow j} + \text{residual}, \qquad \text{where } x^{*f}_{\downarrow j} = F(b_{\downarrow j}).$$

It is natural to take the $b_{\downarrow j}$ and F as estimates of the $\beta_{\downarrow j}$ and Φ. Further introducing $\mathbf{A}$ as the true underlying value of $\mathbf{A}$, we have

$$x^{*t}_{\downarrow j} = \Phi(\beta_{\downarrow j}) = \mathbf{A}\beta_{\downarrow j}$$

$$x^{*f}_{\downarrow j} = F(b_{\downarrow j}) = \mathbf{A}b_{\downarrow j}.$$

A similar chain of reasoning to that used for the objects shows that the data vectors $x^*_{\downarrow j}$ are arranged approximately in $\mathcal{V}$ as the vectors $b_{\downarrow j}$ are arranged in the reduced factor space: the $x^*_{\downarrow j}$ are linked to the $x^{*f}_{\downarrow j}$ by closeness, and the latter to the $b_{\downarrow j}$ by the linear mapping F. Thus it is sensible to study the relationship among the variables by using the vectors $b_{\downarrow j}$. The *reduced factor diagram* is defined as a diagram showing the vectors $b_{\downarrow j}$ in the reduced factor space, while the *reduced diagram* is a combination of the reduced object diagram and the reduced factor diagram, as illustrated by Figure 1.

Notation. Several letters such as $\boldsymbol{A}$ and $\boldsymbol{B}$ have implicitly been given two separate meanings, one dealing with the object space and one dealing with the factor space. For example, $\boldsymbol{B}$ helps describe the linear mapping L for the object space, and the vectors $b_{\downarrow j}$, which appear to be the columns of $\boldsymbol{B}$, have been defined separately in connection with the factor space. Similarly, $\boldsymbol{A}$ helps describe the linear mapping F for the factor space while the points $a_{r\rightarrow}$, which appear to be the rows of $\boldsymbol{A}$, have been defined separately in connection with the object space. Similar situations exist for $\mathbf{B}$ and $\mathbf{A}$. This notation is not accidental. Some bilinear methods estimate $\boldsymbol{A}$ and $\boldsymbol{B}$ simultaneously in such a way that the two different meanings coalesce, an approach I favor and recommend. Other methods, including the most widely used forms of factor analysis, follow a more complex route of estimation.

3. Illustrations of use

Although the most famous and the most numerous applications of the bilinear methods come from traditional factor analysis, I do not illustrate the bilinear methods from this tradition. The best-known applications are probably to the study of mental abilities (for example, see Guilford 1967) and in the semantic differential technique (for example, see Snider & Osgood 1969), which uses factor analysis as a central tool to study the affective dimensions of words and concepts. Hundreds of other studies in this tradition from journals in psychology and closely related fields have been individually summarized by Bolton, Hinman, and Tuft (1973).

I do not wish to use an illustration of this kind for several related reasons. First, the factor analysis is done for use (3), finding basic underlying variables, and this use of the bilinear methods is not well accepted in many fields. Second, there is a strong tendency in this tradition to a one-paradigm approach, with little attempt to support the factor analytic results by other experimental tasks and other methods of analysis. Third, the scientific significance of many of these applications is questioned in many quarters. Fourth, while I do not hold with the severer critics, neither am I enthusiastic about these applications.

My first illustration is a psychoacoustical application by McDermott (1969). She processed speech through $I = 22$ different circuits, each of which introduced a different type or degree of distortion. She obtained quality ratings x_{ij} from $J = 31$ listeners, and applied principal components analysis. She used the first three principal components, and rotated them to an acoustically meaningful position subjectively. She also obtained similarity ratings for each *pair* of circuits from 30 other listeners, analyzed them by multidimensional scaling [*see* SCALING, MULTIDIMENSIONAL], and obtained a three-dimensional solution very similar to the reduced objects diagram from the principal components analysis. The concordance between different methods of analysis based on different tasks supports the validity of both. (Actually, the quality ratings x_{ij} were derived from preference judgments on each *pair* of circuits by counting how often circuit i was preferred to another circuit by listener j. Furthermore, in her published paper, McDermott does not cite the principal components analysis described but an analysis she did later by a specialized method applied directly to the pairwise preference judgments. However, the results were essentially the same.)

Although she provided tentative interpretation for the three factors she found, these factors are only a minor aspect of her paper, since they are so heavily dependent on the specific circuits used. The most significant aspect of her article was the unequivocal evidence it provided that listeners differ strongly on the evaluation of types of distortion, even though (as indicated by the multidimensional scaling) they perceive the distortion in much the same way. This raised serious questions about the previously conventional method of averaging preferences across listeners, and ignoring interindividual differences.

My second illustration comes from gas chromatography, a major chemical method for separating complex mixtures by diffusing the vaporized mixture (together with a carrier gas) through a liquid absorbent. Substances to be separated (called solutes) having different retention times in the liquid absorbent (called liquid phase) move through the column at different speeds. Wold and Andersson (1973) analyzed a matrix $\mathbf{X}$ of "retention indices" (transformed retention times) published by McReynolds. This 226×10 matrix covered $I = 226$ liquid phases and $J = 10$ solutes, which were intended to be typical of the liquid phases and solutes in practical use. Wold and Andersson deliberately restricted their principal components analysis to 3 components, despite the fact that previous factor analyses of similar matrices by other workers had required from 5 to 8 factors. Instead of using many components, they used several preliminary principal components analyses to remove the liquid phases that had unduly large residuals (that is, they removed liquid phases whose object points were particularly far from the three-dimensional object flat.) After removing these 16 rows from the data matrix and reanalyzing, they obtained adequately good fit. (The large residuals were themselves of interest, and some of them could be explained.) At this point the third principal component was chemically interpretable, and rotating in the space of the first two components gave one coordinate (but not the other) a chemical interpretation.

The chief result of this paper is the 210×3 table showing the three (rotated) principal components (that is, the matrix $\boldsymbol{A}$). Like the original table with 10 columns, this table has the property that liquid phases that have similar values in the table perform similarly in gas chromatography. It is, however, much easier to work with triples of numbers (for example, by plotting) than with 10-tuples.

4. Distinctions among bilinear methods

In this section I often refer to individual bilinear methods by name before I describe them, simply to illustrate the distinctions. The reader may find it helpful to refer to Table 5 when individual methods are cited.

Bilinear methods assume that the data values x_{ij} arise from the *basic conceptual equation*,

$$x_{ij} = \beta_{0j} + \alpha_{i1}\beta_{1j} + \alpha_{i2}\beta_{2j} + \cdots + \alpha_{iR}\beta_{Rj} + \text{disturbance}$$
$$= \beta_{0j} + \sum_{r=1}^{R} \alpha_{ir}\beta_{rj} + \text{disturbance}.$$

(Greek letters continue to be used for true values, and Latin letters for estimates of them.) In factor analysis the columns $\beta_{\downarrow j}$ are referred to as the *factors* or *common factors*, although these terms are also used to refer to estimates.

Further, a_{ir} and b_{rj} are used to indicate estimates of the α_{ir} and the β_{rj}; $\boldsymbol{A}$ to indicate the matrix of values a_{ir}, which has I rows and R columns; and $\boldsymbol{B}$ to indicate the matrix of values b_{rj}, which has R rows and J columns. In factor analysis, $\boldsymbol{A}$ is called the *factor score matrix* and $\boldsymbol{B}$ the *factor pattern.* ($\boldsymbol{B}$ is also commonly called the *factor loading matrix,* but this term is unfortunately ambiguous since it is also often used for a conceptually distinct matrix to which $\boldsymbol{B}$ is equal in many but not all situations.) Principal components analysis has names for these concepts only when the indeterminacy in $\boldsymbol{A}$ and $\boldsymbol{B}$ (see below) has been used to make them satisfy the conventions that $\boldsymbol{BB}'$ equals the identity matrix and $\boldsymbol{A}'\boldsymbol{A}$ equals a diagonal matrix whose diagonal entries are arranged in order of descending magnitude. If these conventions hold, then the rth column $a_{\downarrow r}$ of $\boldsymbol{A}$ is the rth *principal component of the data matrix,* and the rth row $b_{r\rightarrow}$ contains the *coefficients with respect to the* rth *principal component.* I shall be talking about $\boldsymbol{A}$ and $\boldsymbol{B}$ and their rows ($a_{i\rightarrow}$ and $b_{r\rightarrow}$) and columns ($a_{\downarrow r}$ and $b_{\downarrow j}$) frequently. Dimension reduction works by using the R coordinates of $a_{i\rightarrow}$ rather than the J coordinates of $x_{i\rightarrow}$ to describe the object U_i; similarly, it uses the R coordinates of $b_{\downarrow j}$ rather than the I coordinates of $x_{\downarrow j}$ to describe the variable V_j. Since R is typically much less than I and J, discussion and insight are improved.

1. One issue that distinguishes among bilinear methods is whether the basic conceptual equation is fit *directly* or *indirectly*. In the direct approach, used for example by Eckart–Young factor analysis, the estimates a_{ir} and b_{rj} are found by a fitting process based directly on this equation. The indirect approach rests on the "covariance conceptual equation" derived below; several prominent methods of factor analysis use the indirect approach, including, for example, "MinRes" (minimum residuals) and principal factor analysis. Since the new equation involves only the β_{rj} and not the α_{ir}, the indirect approach does not immediately lead to values for the a_{ir}. In some methods a subsequent step is used to find estimates a_{ir}, either by fitting them with the aid of the basic conceptual equation and taking the b_{rj} as already known, or by some other method.

2. A second issue that distinguishes bilinear methods is whether or not the fitting is conceptually based on optimizing some objective function (mostly it is), and if so whether the objective function is a sum of squared residuals, a likelihood function, or, rarely, some other alternative. For example, both types of principal components analysis and Eckart–Young factor analysis are based on minimizing a sum of squared residuals, while maximum likelihood factor analysis does what its name implies. Both major types of objective function can be used with both the direct and indirect approaches to fitting, although historically likelihood functions have received little discussion in connection with the direct approach. [*See* ESTIMATION.]

An important objective function for the direct approach to fitting is based on the residuals from fitting $\boldsymbol{X}^*$ (the column-centered data matrix) by $\boldsymbol{AB}$, that is, on the entries in $\boldsymbol{X}^* - \boldsymbol{AB}$. The sum of squares of these residuals is the objective function that is explicitly minimized in Eckart–Young factor analysis. This sum is often described as the "residual variance" or the "variance *unac-*

counted for," while the sum of squares of the entries in $\mathbf{X}^*$ is often called the "total variance." Their difference is referred to as the "variance accounted for," or by other similar phrases such as "variance explained." The "proportion (or fraction) of variance accounted for" (variance accounted for divided by total variance) is an important and frequently cited quantity, whose value is one useful index of how well the data have been fitted. Happily it turns out that a surprising variety of natural objective functions, including the proportion of variance accounted for and the residual variance, are equivalent in the sense that optimizing any one of them yields the same result as optimizing any other. (One such objective function is the sum of the squared distances from the data points $x_{i\rightarrow}$ to the object space $\mathfrak{L}$, which is used in Karl Pearson 1901.) The variance accounted for is frequently decomposed (in a natural way) into portions associated with or "accounted for" by the various principal components (or the various factors); each of these is also frequently expressed as a proportion of total variance. The first principal component always accounts for the largest amount of variance, and each succeeding principal component accounts for less than (or as much as) the preceding one.

An important objective function for the indirect approach is based on the residuals from fitting *not* the data matrix but its sample correlation matrix (or sample covariance matrix), as explained below. Modern principal components analysis is frequently based on this objective function, while several varieties of factor analysis and Hotelling principal components analysis are based on modified versions of it. It is a curious fact that this objective function (when not modified) is one of the objective functions equivalent to proportion of variance accounted for. Because of this equivalence, modern principal components analysis can be derived either from the direct or from the indirect approach to fitting.

3. A third issue that distinguishes among bilinear methods is whether the method allows for so-called *unique factors*. These are used only in some types of factor analysis. (Indeed, some authors use unique factors as the central distinguishing characteristic between factor analysis and principal components analysis, which they often call simply *component(s) analysis*.) When unique factors are used, there is one unique factor for each observed variable, in addition to the ordinary factors, which are referred to as *common* factors. The unique factors are assumed to be statistically independent of each other, of the common factors, and of the error terms in the model (except in image analysis, which treats the unique factors in a special way). These factors, which are not seen as being of interest in themselves, appear to me to be largely a theoretical construct to account for some "missing" variance. Specifically, suppose (contrary to the situation elsewhere in this article) that repeated experimentally independent measurements are available for the variables on the objects U_i. The total variance for a variable can be analyzed into parts. One part is due to the common factors and one part is due to the within-object variability (measurement–remeasurement variability), which is ascribed to error. These two parts are commonly insufficient to account for the entire variance, and the unique factors are invoked to explain the difference. Where repeated measurements are not available, however, and this is the situation factor analysis has mostly dealt with, the unique factor can be swallowed up by the error term without loss of generality. Where repeated measurements *are* available, the unique factors can be seen as simply generalizing the most common error model of statistics slightly. This particular generalization seems to be supported largely by tradition. It is easy to extend the error model in other, equally appealing, ways that would explain the missing variance, but I am not aware that such generalizations have been studied and compared.

4. A fourth issue distinguishes only among bilinear methods explicitly derived from precise probabilistic assumptions. (In the literature this means in practice methods based on the maximum likelihood approach, although some of the least-squares methods could be so derived.) This issue is whether we treat the objects U_i as fixed or as a random sample from a population having a known form of distribution (usually normal). Under the random objects assumption we treat the α_{ir} as random variables; we may nevertheless want to estimate the values that the α_{ir} have taken on in our data (although they are not parameters of the model in the classical sense of mathematical statistics, and some purists consider such estimation inappropriate or invalid, or prefer to describe it by some word other than "estimation"). Historically, the random objects model has generally been analyzed by the indirect fitting approach, though there is no logical or mathematical necessity for this connection. For the bilinear methods, the distribution of the α_{ir} is always taken in practice to be a normal distribution. (An apparent exception occurs in the analysis of two-way contingency tables. As explained above, however, the present definition

of bilinear methods excludes such uses of the bilinear model.)

The variables are almost always treated as being fixed in applications of the bilinear methods, and I shall make this assumption. There have been a few attempts to develop an approach in which the variables are treated as randomly drawn from a population of possible variables, including one in Hotelling (1933), but little use of this approach has occurred.

5. A fifth issue that distinguishes bilinear methods is how they approach the problem of "scaling the variables." To scale (or rescale) a variable means to divide the values of that variable (the numbers in that column) by a constant; for example, conversion from grams to kilograms, to pounds, to ounces, or to arbitrary units is scaling. (This use of the word has no connection other than historical with multidimensional scaling.) When several variables are scaled, different constants may be used for different variables. (Scaling of several variables is sometimes used in a more general sense, which permits a multivariate linear transformation, but I exclude that meaning here.) If we scale a data matrix, then apply a fitting procedure, it is simple to adjust the results to be relevant to the original data matrix: it is merely necessary to multiply the columns of the factor pattern by the same constants used as divisors for scaling.

Scaling a variable is usually part of "standardizing" it, which means first subtracting a central value (generally the mean), and then rescaling it. If we standardize the variables in a data matrix and then apply a fitting procedure, it is simple to adjust the results to be relevant to the original data matrix. First the columns are multipled, as above, and then the same constants previously subtracted from the columns are added to them.

Consider the compound procedure consisting of scaling plus some fitting procedure plus adjustment of the results (as just described). Conceptually, this seems as defensible as the use of the fitting procedure alone, since there is seldom reason to believe that the original units have any special validity. For some but not all fitting procedures the results are affected by the constants used in the scaling step. How to choose the constants is the problem of "scaling the variables."

This issue (of how to scale the variables) exists for both direct and indirect fitting procedures. In connection with indirect fitting it should be noted that scaling the variables changes the covariance matrix in an easily described way. Thus scaling the variables can be and often is described in terms of changes to the covariance matrix. In particular, changing from the covariance matrix to the correlation matrix corresponds to scaling each variable by its sample standard deviation.

Bilinear methods, both direct and indirect, may be classified into three categories.

(1) Some methods include scaling and adjustment as an integral part of the method itself, and obtain results that are intrinsically invariant to any rescaling that might be applied prior to use of the method. A prominent representative of this approach is so-called maximum likelihood factor analysis.

(2) Some methods include scaling and adjustment as an integral part of the method itself, and obtain results that vary only slightly under rescaling applied prior to use of the method. Such methods are almost invariably extended to incorporate a separate preliminary step of standardizing the variables, so that the extended method is invariant to prior rescaling of the variables, but not for the same conceptually satisfying reason as in (1). Prominent representatives of this approach include principal factor analysis and MinRes factor analysis.

(3) Some methods are based on an explicit or implicit assumption that the disturbance terms in all the variables have the same magnitude, and incorporate no rescaling at all. For such methods, rescaling prior to using the method is very important, and is frequently based on standardizing the variables as a step external to the method. Prominent representatives of this approach include modern principal components analysis and Eckart–Young factor analysis. Unfortunately, expositions of such methods frequently fail to give sufficient emphasis to the practical importance of scaling the variables appropriately before applying the method.

Direct approach to fitting. As we consider the alternative approaches to finding the estimates a_{ir} and b_{rj}, it is important always to remember the *indeterminacy phenomenon*, which I have touched on before. This phenomenon is that the values of x^t_{ij} and the equations

$$x^t_{ij} = \beta_{0j} + \sum_{r=1}^{R} \alpha_{ir}\beta_{rj}$$

for all i and j do not determine the values of the parameters α_{ir} and the β_{rj}. Different values of the parameters can give the same values for these expressions. Except in connection with the use (3) of the bilinear methods (finding basic underlying variables), different values of the parameters that yield the same x^t_{ij} are equivalent for most practical purposes, as previously described.

In the direct approach to fitting the basic conceptual equation, we seek estimates a_{ir} and b_{rj} of the α_{ir} and the β_{rj} so as to satisfy the basic conceptual equation as well as possible in some sense. This leads to the *first direct fitting equation,*

$$x_{ij} = b_{0j} + \sum_r a_{ir}b_{rj} + \text{residual}.$$

It would be natural to fit this equation all at once using least squares, maximum likelihood, or some other objective function, and indeed many commonly used methods are mathematically equivalent to such an approach. They are not, however, described in this way, and I am not aware that the mathematical equivalence just referred to has been published. Almost always, the b_{0j} are determined by a preliminary procedure (whose scientific rationale is not made clear), and then an explicit fitting procedure is used to determine the remaining parameters. In the great majority of cases, b_{0j} is taken to be the mean (or average) of the jth column of $\mathbf{X}$,

$$b_{0j} = \text{mean of } x_{\downarrow j} = \sum_i x_{ij}/I.$$

This is often stated as if it were an assumption or an arbitrary definition, although it is possible to prove for the least-squares approach and several maximum likelihood approaches that the optimum solution to the first direct fitting equation is the same, up to the indeterminacy phenomenon, as a solution where the b_{0j} have the mean values.

After the values b_{0j} have been chosen, they are "taken out" by defining modified data $\mathbf{X}^*$ by

$$x^*_{ij} = x_{ij} - b_{0j}.$$

This leads to the *second direct fitting equation,*

$$x^*_{ij} = \sum_r a_{ir}b_{rj} + \text{residual},$$

or, in more compact form using matrix notation,

$$\mathbf{X}^* = \mathbf{AB} + \text{residual matrix}.$$

Methods for direct fitting. The method of factor analysis due to Eckart and Young (1936) is to fit this equation by the procedure of least squares. "Method II," which is the focus of Lawley (1941) but which has received relatively little attention since then, fits this equation by maximum likelihood. (Note that this method is distinct from so-called maximum likelihood factor analysis, which was introduced in a prior paper and is referred to in the cited paper as Method I.) Actually Lawley starts with the first direct fitting equation and simply asserts in effect, quite correctly but without explicit proof, that the second equation can be used without loss of generality.

Indirect approach to fitting. The indirect fitting approach seeks not to match the data matrix $\mathbf{X}$ (or $\mathbf{X}^*$), but instead to match its *sample covariance matrix*, which is defined by

$$\mathbf{C} = \mathbf{X}^{*\prime}\mathbf{X}^*/I.$$

(Recall that I equals the number of objects U_i.) In practice, it is far more common to match the sample correlation matrix. Because that matrix is simply the sample covariance matrix of the variables after they have been suitably rescaled, however, the distinction between covariance matrix and correlation matrix comes down to part of the scaling problem. To those unfamiliar with bilinear methods, the introduction and use of the covariance matrix may seem strange and roundabout, while to veteran users this step seems so natural that they may no longer question its rationale.

The reason for this step can be seen partly in historical terms. The invention of factor analysis can reasonably be attributed to Spearman (1904*a*; 1904*b*), in which the whole object was to explain correlations, then a fascinating new tool. By the 1930s it had become usual to give conception B or B′ as the rationale for factor analysis, but the correlation matrix still played a special role, and fitting it remains a primary goal, although one not necessarily stated in print. (It is fascinating to see how Thurstone 1947, which otherwise provides extremely lucid exposition, slips into using the correlation matrix without ever explaining why it is introduced.) Another historical reason for using the correlation matrix is the great computational advantage it provided, in the days before automatic computers, when (as usual) there were many objects (or subjects) and few variables.

The question of what approach is best at this time is subject to strong and conflicting opinions, built on whole lifetimes of intellectual investment. Personally, I am not aware that any clear advantage in performance has been established either way. However the direct and indirect approaches each offer certain conveniences. The indirect approach lends itself to certain generalizations that Jöreskog (1969; 1970) has effectively exploited. The direct approach avoids several difficulties concerning the meaning and estimation of the a_{ir} (the factor scores), which offer both theoretical and practical difficulties to the indirect approach. Personally, I prefer the simplicity of the direct approach, and I hope that values such as those developed by Jöreskog can somehow be made available in that framework.

The covariance conceptual equation, derived below, is based on the expected value of the sample covariance matrix. In order to derive this equation in its usual form, it is necessary to discuss the *fundamental indeterminacy*, the most puzzling aspect of the indeterminacy phenomenon, although space permits only a brief formal description. The heart of the second direct fitting equation is the matrix product $\boldsymbol{AB}$. If $\boldsymbol{T}$ is any $R \times R$ matrix that is nonsingular (has an inverse), then

$$\boldsymbol{AB} = (\boldsymbol{AT}^{-1})(\boldsymbol{TB}).$$

Thus if we define new matrices by

$$\boldsymbol{A}_1 = \boldsymbol{AT}^{-1} \quad \text{and} \quad \boldsymbol{B}_1 = \boldsymbol{TB},$$

then $\boldsymbol{A}_1\boldsymbol{B}_1$ has precisely the same entries as $\boldsymbol{A}$ and $\boldsymbol{B}$, and hence fits $\boldsymbol{X}^*$ just as well by many reasonable methods of assessing fit. It can be proved that as $\boldsymbol{T}$ runs through all nonsingular matrices, $\boldsymbol{A}_1$ and $\boldsymbol{B}_1$ run through all pairs of matrices (of their size and shape) whose product equals $\boldsymbol{AB}$. (This assumes that $R \leqslant \text{rank}(\boldsymbol{X}^*)$, which generally holds.)

The fundamental indeterminacy thus creates what is often called the *rotational problem:* once we have a solution to either direct fitting equation, which alternative pair of matrices shall we use to describe what is essentially the same solution? For some purposes, it does not make much difference which pair we use, but for other purposes it does.

Now we derive the covariance conceptual equation. (In this paragraph I must assume greater mathematical background.) First we rewrite the basic conceptual equation in the form

$$x_{ij} = \beta_{0j} + \alpha_{i\rightarrow}\beta_{\downarrow j} + v_{ij},$$

where v_{ij} is the disturbance. Then we make these further assumptions (where E refers to expected value):

$\alpha_{i\rightarrow}$ are random, identically distributed points from R-space ($i = 1$ to I), (this is the random objects assumption);

$\alpha_{1\rightarrow}, \cdots, \alpha_{I\rightarrow}, \quad v_{11}, \cdots, v_{IJ}$ are all independent;

$E(\alpha_{i\rightarrow}) = 0_{\rightarrow}$ (the population of random objects is centered at the origin of R-dimensional space);

$E(v_{ij}) = 0$ and $E(v_{ij}^2) = \sigma_j^2$ (the random disturbances are unbiased and their variance depends only on which variable V_j is involved).

Then using $\mathbf{A}$ for the matrix of α_{ir} and $\mathbf{B}$ for the matrix of β_{rj}, it can be proved that the covariance matrix $\boldsymbol{C}$ has expectation

$$E(\boldsymbol{C}) = \mathbf{B}'\left(\frac{I-1}{I}\mathbf{A}'\mathbf{A}\right)\mathbf{B} + \begin{bmatrix} \sigma_1^2 & \cdot & \cdot & \cdot & 0 \\ \cdot & \cdot & & & \cdot \\ \cdot & & & & \cdot \\ \cdot & & & \cdot & \cdot \\ 0 & \cdot & \cdot & \cdot & \sigma_J^2 \end{bmatrix}.$$

Now we "use up" part of the fundamental indeterminacy to simplify this equation. If we let $\mathbf{A}_1 = \mathbf{A}\boldsymbol{T}^{-1}$ and $\mathbf{B}_1 = \boldsymbol{T}\mathbf{B}$ so that $\mathbf{AB} = \mathbf{A}_1\mathbf{B}_1$, then it is not hard to show that

$$\mathbf{B}'\left(\frac{I-1}{I}\mathbf{A}'\mathbf{A}\right)\mathbf{B} = \mathbf{B}_1'\left(\frac{I-1}{I}\mathbf{A}_1'\mathbf{A}_1\right)\mathbf{B}_1.$$

By a well-known theorem it is possible to pick $\boldsymbol{T}$ in such a way that $((I-1)/I)\mathbf{A}_1'\mathbf{A}_1$ is the identity matrix. Using this in the equation above, using $\mathbf{\Sigma}$ for the diagonal matrix with values σ_j and dropping the subscript from $\mathbf{B}$, we find that

$$E(\boldsymbol{C}) = \mathbf{B}'\mathbf{B} + \mathbf{\Sigma}^2.$$

From this we get the *covariance conceptual equation,*

$$\boldsymbol{C} = \mathbf{B}'\mathbf{B} + \mathbf{\Sigma}^2 + \text{disturbance matrix},$$

where the disturbance matrix has expected value $\mathbf{0}$. If we assume in addition that the $\alpha_{i\rightarrow}$ and v_{ij} are normally distributed, then $\boldsymbol{C}$ is distributed according to the much studied Wishart distribution.

From the covariance conceptual equation we get the *indirect fitting equation*

$$\boldsymbol{C} = \boldsymbol{B}'\boldsymbol{B} + \boldsymbol{S}^2 + \text{residual matrix},$$

where $\boldsymbol{C}$ is the $J \times J$ sample covariance matrix, $\boldsymbol{B}$ is an estimate of $\mathbf{B}$, and $\boldsymbol{S}^2$ is a diagonal matrix whose diagonal entries s_j^2 are estimates of the σ_j^2. Alternatively, if we rescale the variables by dividing by the sample standard deviations of the $x_{\downarrow j}$, then $\boldsymbol{C}$ is the sample correlation matrix, and the entries of $\boldsymbol{S}^2$ are generally called the "uniquenesses." (For explanation see, for example, Harman 1967, sec. 2.4.)

Recall that when $\boldsymbol{B}$ is fitted by the indirect approach, $\boldsymbol{A}$ is fitted (if at all) by a separate subsequent step, which is sometimes based on the direct fitting equation but taking $\boldsymbol{B}$ as fixed and given there. Many people agree that the least-squares fitting is the most natural method to use during this step, although other methods with certain advantages are also used. See Harman (1967) and Mulaik (1972) for accounts of the various methods used for this step.

It is important to keep in mind that we have "used up" some of the fundamental indeterminacy during the derivation of the covariance conceptual equation, by requiring that $((I-1)/I)\mathbf{A}_1'\mathbf{A}_1$ equal the identity matrix. (This may be interpreted geometrically in terms of the "ellipsoid of concentration" of the distribution of $\alpha_{i\rightarrow}$: it requires this ellipsoid to be a sphere.) As a result, factor patterns $\boldsymbol{B}$ based on the indirect fitting equation have less indeterminacy than factor patterns based on fitting the direct fitting equation. If $\boldsymbol{T}$ is any *orthogonal* matrix (that is, $\boldsymbol{T}'\boldsymbol{T}$ = identity matrix = $\boldsymbol{T}\boldsymbol{T}'$), and we define $\boldsymbol{B}_1 = \boldsymbol{T}\boldsymbol{B}$, then

$$\boldsymbol{B}_1'\boldsymbol{B}_1 = \boldsymbol{B}'\boldsymbol{B}.$$

Furthermore, it can be proved that as $\boldsymbol{T}$ runs through all orthogonal matrices, $\boldsymbol{B}_1$ runs through all matrices (of the same size and shape as $\boldsymbol{B}$) such that this equation holds. Thus the fundamental indeterminacy is modified here.

It is important, however, to keep in mind the *reason* for this modified indeterminacy: it is based on a convenient but *arbitrary* restriction about the "shape" of the distribution of $\alpha_{i\rightarrow}$. If we abandon this restriction, as is sometimes necessary, then $\boldsymbol{B}$ is subject to the full indeterminacy. For example, suppose we compare the factor patterns $\boldsymbol{B}_1$ and $\boldsymbol{B}_2$ from the same variables measured on samples drawn from two different populations, say mental tests on college students and on prison inmates. In making the comparison, it would be very unrealistic to suppose that the two samples have the same distribution on the underlying factors: either *one* of these samples can be arbitrarily set as spherical, but not both at once. Lack of a clear understanding of this situation long delayed the Thurstone school of factor analysis from a valid approach to making such comparisons, which requires that at least one of the matrices $\boldsymbol{B}_1$ and $\boldsymbol{B}_2$ be permitted the full range of indeterminacy.

Methods for indirect fitting. As we can see from Table 5, many of the bilinear methods can be defined as either least-squares or maximum likelihood fitting of the indirect fitting equation with various modifications and variations. This includes the most prominent modern forms of factor analysis. For the methods where rescaling of the covariance matrix makes a difference, it is generally assumed (as part of the named method) to have been rescaled as the sample correlation matrix. Modern descriptions of principal components analysis often describe it in terms of least-squares fitting (to the indirect fitting equation) with $\boldsymbol{C}$ as the correlation matrix, and $\boldsymbol{S}^2$ *eliminated* from the equation. (See below for discussion of such an apparently unreasonable step as eliminating this term.) In the original introduction of the principal components method (Hotelling 1933), several alternative methods are mentioned, but it is fair to represent them by least-squares fitting where $\boldsymbol{C}$ is taken to be the correlation matrix and the jth diagonal entry of $\boldsymbol{S}^2$ is

$$1 - \text{reliability}$$

of the jth variable. (The reliability of a variable is the measurement–remeasurement correlation.) Principal factor analysis is least-squares fitting where the jth diagonal entry of $\boldsymbol{S}^2$ is set to be the uniqueness of the jth variable, that is, the fraction of the total variance of the jth variable that is ascribed to the unique factor for the jth variable, as estimated by some preliminary procedure. MinRes (minimum residuals) factor analysis is least-squares fitting, using the correlation matrix for $\boldsymbol{C}$, which cleverly dodges the troublesome diagonal terms by minimizing the sum of the squared *off-diagonal* terms. In addition, there are other less prominent methods, similar to some of those above, that minimize a *weighted* sum of squares, divided by some suitable scale factor, where the weights are generally estimated from the data and are intended to reflect the relative values of the σ_j. Mostly, these methods give essentially the same result whether the correlation or the covariance matrix is used, since they implicitly rescale the variables on the basis of internal evidence.

So-called maximum likelihood factor analysis fits the covariance equation by maximizing the likelihood. For this purpose an assumption is needed beyond those used to derive the equation, namely, that the $\alpha_{i\rightarrow}$ and the disturbance terms v_{ij} are normally distributed. This prominent method was introduced by Lawley (1940) and is discussed in many places, including this volume [*see* FACTOR ANALYSIS AND PRINCIPAL COMPONENTS, *article on* FACTORING CORRELATION MATRICES; *see also* Anderson & Rubin 1956; Lawley & Maxwell 1963]. This method served as the starting place for work by Jöreskog, who developed computer programs that make this method practical, and who has substantially extended it. Another maximum likelihood method, which is described in Lawley (1941) and attributed to Thompson, could be described as the maximum likelihood solution under the same assumptions but with the added assumption that the σ_j are all equal.

Although the modern principal components method may seem unreasonable in setting the intrinsically nonnegative matrix $\boldsymbol{S}^2 = \mathbf{0}$, this method is mathematically equivalent to a direct fitting approach by least squares, which seems entirely reasonable. Thus it would appear that in some sense it is the step of going from direct fitting to indirect fitting that introduces the term $\boldsymbol{S}^2$ into the covariance equation, and the question of reasonability is shifted to whether direct or indirect fitting is more

Table 5 – Bilinear methods

Abbrev.	Name or description	Fit	Objective function	Unique factors	Objects	Scaling	Modification of diagonal of **S**
KP	*Karl Pearson's* method (1901) Math. equ. to EY & MPC	Dir	SSQ	N	——	Ext	——
EY	*Eckart—Young* factor analysis Math. equ. to KP & MPC	Dir	SSQ	N	——	Ext	——
	Implicit in Lawley (1941) description of PC. Math. equ. to MPC, etc.	Dir	ML	N	Fix	Ext	——
	Method II of Lawley (1941)	Dir	ML	N	Fix	Ind	——
	Unpublished. Math. equ. to TM	Dir	ML	N	Ran	Ext	——
	Unpublished. Math. equ. to MLFA	Dir	ML	N	Ran	Ind	——
MPC	*Modern principal components* Math. equ. to EY & KP	Ind	SSQ	N	Fix	Ext	None
HPC	*Hotelling's principal components* Includes several alternatives	Ind	SSQ	N	Fix?	Ext	External reliabilities
PFA	*Principal factor analysis*	Ind	SSQ	Y	Ran?	Dep	Preliminary communalities
MRFA	*MinRes* (Harman)	Ind	SSQ	Y?	Ran?	Ind	Ignore diagonal
TM	"Thompson's method," implicit in Lawley (1941)	Ind	ML	N	Ran	Ext	None
MLFA	*Maximum likelihood factor analysis* Method I of Lawley (1940; 1941)	Ind	ML	N	Ran	Ind	None
IA	*Image analysis* (Guttman)	Ind	——	Y	——	Ind	(see text)
CFA	*Canonical factor analysis* (Rao) Math. equ. to special form of MLFA	Ind	Cor	N	Ran	Ind	None
CA	*Correspondence analysis* Math. equ. to scaling + SVD	Dir	SSQ?	N	Fix?	Ind?	——
	ASSOCIATED NUMERICAL PROCEDURES						
SVD	*Singular value decomposition* Use first R singular vectors of **X**; Math. equ. to EY, MPC, & CA; also to EV applied to **S**	Dir					
EV	*Eigenvectors* Use first R eigenvectors; If applied to **S** itself, math. equ. to EY, MPC, and SVD; If applied to suitably modified **S**, math. equ. to HPC, PFA	Ind					

reasonable. The main considerations, discussed above, do not really seem to settle this question.

5. Summary of bilinear methods

Table 5 lists many bilinear methods, together with a number of significant characteristics, including all the issues mentioned in the previous section. After explaining the columns, I shall identify the methods and discuss choice among them.

The abbreviations of names of methods are used primarily for cross-referencing within the table, though some are commonly used. The names given are those commonly used, where possible. Some methods without names are included for theoretical completeness.

Characteristics. "Fit" indicates whether direct ("Dir") or indirect ("Ind") fitting is used, that is, whether the data matrix **X** or its sample covariance matrix **C** is the object to be fitted. "Objective function" indicates what sort of objective function is used to define the desired solution: a sum of squared residuals ("SSQ"), a likelihood function ("ML" for maximum likelihood), a function based on certain correlations ("Cor"). In this and other columns, a long dash (——) is used to indicate that the characteristic is not applicable. In this column it indicates that the method is not defined by use of an objective function.

"Unique factors" indicates whether or not the concept of unique factors (as distinct from random error variables) plays an essential role in the main fitting step of the method. (For methods using indirect fitting, nothing is specified as to whether unique factors play a role in any subsequent fitting to find factor scores.) "Y" indicates that it does, "N" that it does not.

"Objects" indicates whether the method treats the objects as fixed ("Fix") or randomly selected ("Ran") from an underlying population. In some cases the stochastic assumptions are not fully specified, so that this characteristic may not be entirely clear. In this case the prevailing attitude is indicated, with doubt indicated by a question mark.

"Scaling" indicates how the scaling of variables is handled. The three categories are described in the previous section: intrinsically independent of scaling ("Ind"), scaling is integrated into the method but is not intrinsically independent of the original scaling ("Dep"), and scaling must be handled as a step external to the method ("Ext"). The distinction between the categories is not always so clear. "Modification of diagonal of **S**" indicates, for indirect fitting methods only, what information is used to modify the diagonal of the covariance matrix before fitting.

Methods. Karl Pearson's method (KP), introduced in 1901, was based on a geometric viewpoint. Eckart–Young factor analysis (EY), introduced in Eckart and Young (1936), is closely related, though the viewpoint is algebraic. The next four methods indicated by a bracket are a natural quartet formed by using fixed or random objects with independent or external scaling of the variables. The first of these is essentially principal components analysis as described by Lawley (1941). The last two of these, which use random objects, are mathematically equivalent to other methods (I am not aware this fact has been published): one to MLFA and the other to TM. In this section, methods are considered mathematically equivalent if they always yield the same factor pattern matrix, b_{rj}, up to the approximate indeterminacy, even though they yield different factor score matrices, a_{ir}, or one of them yields such a matrix and the other does not.

Modern principal components (MPC) is described in many sources, such as Anderson (1958). It is not generally recognized that principal components analysis as introduced by Hotelling in 1933 (HPC) is not quite the same as MPC, and I have not located evidence that Hotelling ever changed his mind about the definition. It is a surprising and by no means trivial fact that EY and MPC are mathematically equivalent. Note that one involves least-squares fit to the *data* matrix and the other to its *covariance* matrix.

Principal factor analysis (PFA) is one of the commonest methods of factor analysis, but it is not generally credited to any one individual; see Harman (1967) for a description. MinRes ("minimum residuals factor analysis," MRFA), was first rendered practical by Harman and Jones (1966) and is described in Harman (1967) and Mulaik (1972). "Thompson's method" so far as I know has never received more than a brief mention in Lawley (1941). Maximum likelihood factor analysis, so called, was introduced by Lawley (1940) and has achieved practical importance since Jöreskog (1967) rendered it practical. Image analysis was introduced by Guttman (1953) and is well described in Mulaik (1972). This approach, which treats the unique factors quite differently from other methods, does not fit perfectly into my framework but can be seen as a way of first estimating the unique factors and then using these estimates to modify the covariance matrix prior to the main fitting step. Canonical factor analysis (CFA) was introduced by Rao (1955). Both IA and CFA have had greater theoretical than practical importance.

Correspondence analysis (CA) does not properly meet my definition of a bilinear method, since it is not based on a linear conception, but it is so closely related that it seems worth mentioning. For a good discussion, see Hill (1974), who gives references to a surprisingly large number of independent discoveries of this method, to which I add Hayashi (1952). This method is sometimes based on a sum of squares objective function, and sometimes developed without any objective function at all. Finally I include two numerical methods (closely related to each other and closely associated with the bilinear methods) that are very important in this field.

Choices among the bilinear methods. Which of the bilinear methods should one use? Does it make any difference? Since there are heated disagreements on this topic, I venture only a few limited opinions and restrict my attention to the true bilinear methods. One obviously important consideration is the practical availability of a computer program, and the amenities it provides. Incidentally, caution is in order when choosing a program; it is not unusual to discover that the content differs significantly from the description.

The next consideration is the purpose of the analysis. Of the three uses mentioned in the introductory section, I restrict attention to use (1), dimension reduction. First consider dimension reduction on the variables, which replaces the data columns $x^{\bullet}_{\downarrow i}$ by the columns $b_{\downarrow j}$ plotted as vectors in the reduced factor diagram. For most sets of data, I believe that the reduced factor diagram is very insensitive to choice of method. However, it should be clearly understood that this is true only if such diagrams differing by the *full* fundamental indeterminacy, that is, multiplication of the $b_{\downarrow j}$ by an *arbitrary* nonsingular linear transformation, are considered the same. (Recall that modern principal components analysis builds in the assumption that $\boldsymbol{BB'}$ equals the identity matrix, whereas several methods of factor analysis build in the assumption that $\boldsymbol{A'A}$ equals the identity matrix. These assumptions, which are merely conventions of convenience, will in general conflict with each other. Either one by itself leads to a modified form of the fundamental indeterminacy—see section 4—but in comparing reduced diagrams from different methods it is obviously necessary to drop conflicting conventions.)

Now consider dimension reduction on the objects, which replaces the data rows $x_{i\rightarrow}$ by the rows $a_{i\rightarrow}$, plotted as points in the reduced object diagram. Here the situation is more complicated. Indirect bilinear methods only accomplish this step, if at all, by a supplementary stage of fitting, and I have not discussed the methods used in this supplementary stage. My opinion is that *direct* fitting methods (which accomplish dimension reduction on the objects without a further step) are superior *for this purpose* to existing combinations of indirect fitting and supplementary fitting. Each of the methods used for the supplementary stage has some disadvantage. For most sets of data, I believe that the reduced object diagram is very insensitive to choice of method within the class of direct methods. Of course, it is understood that the full fundamental indeterminacy, that is, multiplication of the $a_{i\rightarrow}$ by an *arbitrary* nonsingular matrix, is permitted, for the same reason mentioned above. For many sets of data, dimension reduction through a combination of indirect method and supplementary fitting would not be much inferior in practice to direct methods.

There is one more important comparison among methods, namely, how helpful they are or might be in choosing a good dimensionality. Although methods may perhaps differ greatly in this respect, no one yet knows which methods are superior.

JOSEPH B. KRUSKAL

BIBLIOGRAPHY

ADELMAN, IRMA; and MORRIS, CYNTHIA TAFT 1967 *Society, Politics, and Economic Development: A Quantitative Approach.* Baltimore: Johns Hopkins Press.

ANDERSON, T. W. 1958 *An Introduction to Multivariate Statistical Analysis.* New York: Wiley.

ANDERSON, T. W.; and RUBIN, HERMAN 1956 Statistical Inference in Factor Analysis. Volume 5, pages 111–150 in Berkeley Symposium on Mathematical Statistics and Probability, Third, *Proceedings.* Edited by Jerzy Neyman. Berkeley: Univ. of California Press.

BARTLETT, M. S. 1950 Tests of Significance in Factor Analysis. *British Journal of Psychology* Statistical Section 3:77–85.

BOLTON, BRIAN; HINMAN, SUKI; and TUFT, SHIRLEY 1973 *Annotated Bibliography: Factor Analytic Studies 1941–1970.* 4 vols. Fayetteville: Univ. of Arkansas, Arkansas Rehabilitation Research and Training Center. → Tuft did not collaborate on Volumes 3 and 4.

BOYNTON, ROBERT M.; and GORDON, JAMES 1965 Bezold–Brücke Hue Shift Measured by Color-naming Technique. *Journal of the Optical Society of America* 55:78–96.

ECKART, CARL; and YOUNG, GALE 1936 The Approximation of One Matrix by Another of Lower Rank. *Psychometrika* 1:211–218.

GABRIEL, K. RUBEN 1971 The Biplot Graphic Display of Matrices With Application to Principal Component Analysis. *Biometrika* 58:453–467.

GUILFORD, JOY P. 1967 *The Nature of Human Intelligence.* New York: McGraw-Hill.

GUTTMAN, LOUIS 1953 Image Theory for the Structure of Quantitative Variates. *Psychometrika* 18:277–296.

GUTTMAN, LOUIS 1954 Some Necessary Conditions for Common Factor Analysis. *Psychometrika* 19:149–161.

HARMAN, HARRY H. (1960) 1976 *Modern Factor Analysis.* 3d ed., rev. Univ. of Chicago Press.

HARMAN, HARRY H.; and JONES, WAYNE H. 1966 Factor Analysis by Minimizing Residuals (Minres). *Psychometrika* 31:351–368.

HAYASHI, CHIKIO 1952 On the Prediction of Phenomena From Qualitative Data and the Quantification of Qualitative Data From the Mathematico-statistical Point of View. Institute of Statistical Mathematics, *Annals* 3, no. 2:69–98.

HILL, M. O. 1974 Correspondence Analysis: A Neglected Multivariate Method. *Applied Statistics* 23:340–354.

HOTELLING, HAROLD 1933 Analysis of a Complex of Statistical Variables Into Principal Components. *Journal of Educational Psychology* 24:417–441, 498–520.

JÖRESKOG, K. G. 1967 Some Contributions to Maximum Likelihood Factor Analysis. *Psychometrika* 32:443–482.

JÖRESKOG, K. G. 1969 A General Approach to Confirmatory Maximum Likelihood Factor Analysis. *Psychometrika* 34:183–202.

JÖRESKOG, K. G. 1970 A General Method for Analysis of Covariance Structures. *Biometrika* 57:239–251.

LAWLEY, D. N. 1940 The Estimation of Factor Loadings by the Method of Maximum Likelihood. Royal Society of Edinburgh, Section A, *Proceedings* 60:64–82.

LAWLEY, D. N. 1941 Further Investigations in Factor Estimation. Royal Society of Edinburgh, Section A, *Proceedings* 61:176–185.

LAWLEY, D. N.; and MAXWELL, A. E. (1963) 1971 *Factor Analysis as a Statistical Method.* 2d ed. London: Butterworth.

LINN, ROBERT L. 1968 A Monte Carlo Approach to the

Number of Factors Problem. *Psychometrika* 33:37–71.

McDermott, Barbara J. 1969 Multidimensional Analyses of Circuit Quality Judgments. *Journal of the Acoustical Society of America* 45:774–781.

Mulaik, Stanley A. 1972 *The Foundations of Factor Analysis.* New York: McGraw-Hill.

Pearson, Karl 1901 On Lines and Planes of Closest Fit. *Philosophical Magazine* 6:559–572.

Rao, C. Radhakrishna 1955 Estimation and Tests of Significance in Factor Analysis. *Psychometrika* 20: 93–111.

Shepard, Roger N.; and Carroll, J. Douglas 1966 Parametric Representation of Nonlinear Data Structures. Pages 561–592 in International Symposium on Multivariate Analysis, Dayton, Ohio, 1965, *Multivariate Analysis: Proceedings.* Edited by Paruchuri R. Krishnaiah. New York: Academic Press.

Snider, James G.; and Osgood, Charles E. (editors) 1969 *Semantic Differential Techniques: A Source Book.* Chicago: Aldine.

Spearman, C. E. 1904*a* The Proof and Measurement of Association Between Two Things. *American Journal of Psychology* 15:72–101.

Spearman, C. E. 1904*b* "General Intelligence" Objectively Determined and Measured. *American Journal of Psychology* 15:201–293.

Thurstone, Louis L. 1947 *Multiple-factor Analysis: A Development and Expansion of the Vectors of Mind.* Univ. of Chicago Press.

Wachter, Kenneth W. 1975 User's Guide and Tables to Probability Plotting for Large Scale Principal Components Analysis. Harvard University, Statistics Department, Research Report W-75-1.

Wold, Svante; and Andersson, Kurt 1973 Major Components Influencing Retention Indices in Gas Chromatography. *Journal of Chromatography* 80:43–59.

II

FACTORING CORRELATION MATRICES

This article was first published in IESS *as "Factor Analysis: I. Statistical Aspects" with a companion article less relevant to statistics.*

In many fields of research—for example, agriculture (Banks 1954), psychology (Burt 1947), economics (Geary 1948), medicine (Hammond 1944; 1955), and the study of accidents (Herdan 1943), but notably in psychology and the other social sciences—an experimenter frequently has scores for each member of a sample of individuals, animals, or other experimental units on each of a number of variates, such as cognitive tests, personality inventories, sociometric and socioeconomic ratings, and physical or physiological measures. If the number of variates is large, or even moderately so, the experimenter may wish to seek some reduction or simplification of his data. One approach to this problem is to search for some hypothetical variates that are weighted sums of the observed variates and that, although fewer in number than the latter, can be used to replace them. The statistical techniques by which such a reduction of data is achieved are known collectively as *factor analysis*, although it is well to note here that the principal component method of analysis discussed below (see also Kendall & Lawley 1956) has certain special features. The derived variates are generally viewed merely as convenient descriptive summarizations of the observed data. But occasionally their composition is such that they appear to represent some general basic aspects of everyday life, performance or achievement, and in such cases they are often suitably labeled and are referred to as *factors*. Typical examples from psychology are such factors as "numerical ability," "originality," "neuroticism," and "toughmindedness." This article describes the statistical procedures in general use for arriving at these hypothetical variates or factors.

Preliminary concepts. Suppose that for a random sample of size N from some population, scores exist on each of p jointly normally distributed variates x_i $(i = 1, 2, \cdots, p)$. If the scores on each variate are expressed as deviations from the sample mean of that variate, then an unbiased estimator of the variance of x_i is given by the expression

$$a_{ii} = (N-1)^{-1} \sum x_i^2,$$

summation being over the sample of size N. Similarly, an unbiased estimator of the covariance between variates x_i and x_j is given by

$$a_{ij} = (N-1)^{-1} \sum x_i x_j .$$

Note that this is conventional condensed notation. A fuller, but clumsier, notation would use $x_{i\nu}$ for the deviation $(\nu = 1, \cdots, N)$ so that $\sum x_i^2$ really means $\sum_{\nu=1}^{N} x_{i\nu}^2$.

In practice, factor analysis is often used even in cases in which its usual assumptions are known to be appreciably in error. Such uses make the tacit presumption that the effect of the erroneous assumptions will be small or negligible. Unfortunately, nearly nothing is known about the circumstances under which this robustness, or nonsensitivity to errors in assumptions, is justified. Of course, the formal manipulations may always be carried out; the assumptions enter crucially into distribution theory and optimality of the estimators.

The estimated variances and covariances between the p variates can conveniently be written in square matrix form as follows:

$$\mathbf{A} = \begin{bmatrix} a_{11} & a_{12} & \cdots & a_{1p} \\ a_{21} & a_{22} & \cdots & a_{2p} \\ \vdots & \vdots & & \vdots \\ a_{p1} & a_{p2} & \cdots & a_{pp} \end{bmatrix}$$

Since $a_{ij} = a_{ji}$, the matrix $\mathbf{A}$ is symmetric about its main diagonal.

From the terms of $\mathbf{A}$, the sample correlations, r_{ij}, between the pairs of variates may be obtained from

$$r_{ij} = \frac{a_{ij}}{(a_{ii}a_{jj})^{\frac{1}{2}}},$$

with $r_{ii} = 1$. The corresponding matrix is the *correlation matrix.*

The partial correlation concept is helpful here. If, to take the simplest case, estimates of the correlations between three variates are available, then the estimated correlation between any two, say x_i and x_j, for a given constant value of the third, x_k, can be found from the expression

$$\frac{r_{ij} - r_{ik}r_{jk}}{(1 - r_{ik}^2)^{\frac{1}{2}}(1 - r_{jk}^2)^{\frac{1}{2}}}$$

and is denoted by $r_{ij \cdot k}$.

In terms of a correlation matrix, the aim of factor analysis can be simply stated in terms of partial correlations (see Howe 1955). The first question asked is whether a hypothetical random variate f_1 exists such that the partial correlations $r_{ij \cdot f_1}$, for all i and j, are zero, within the limits of sampling error, after the effect of f_1 has been removed. (If this is so, it is customary to say that the correlation matrix, apart from its diagonal cells, is of rank *one*, but details will not be given here.) If the partial correlations are not zero, then the question is asked whether *two* hypothetical random variates, f_1 and f_2, exist such that the partial correlations between the variates are zero after the effects of both f_1 and f_2 have been removed from the original matrix, and so on. (If f_1 and f_2 reduce the partial correlations to zero, then the matrix, apart from its diagonal cells, is said to be of rank *two*, and so on.) The aim of the procedure is to replace the observed variates with a set of derived variates that, although fewer in number than the former, are still adequate to account for the correlations between them. In other words, the derived variates, or factors, account for the variance common to the observed variates.

Historical note. Factor analysis is generally taken to date from 1904, when C. E. Spearman published an article entitled " 'General Intelligence' Objectively Determined and Measured." Spearman postulated that a single hypothetical variate would in general account for the intercorrelations of a set of cognitive tests, and this variate was his famous factor "g." For the sets of tests that Spearman was considering, this hypothesis seemed reasonable. As further matrices of correlations became available, however, it soon became obvious that Spearman's hypothesis was an oversimplification of the facts, and multiple factor concepts were developed. L. L. Thurstone, in America, and C. Burt and G. H. Thomson, in Britain, were the most active pioneers in this movement. Details of their contributions and references to early journal articles can be found in their textbooks (Thurstone 1935; 1947; Burt 1940; Thomson 1939). These writers were psychologists, and the statistical methods they developed for estimating factors were more or less approximate in nature. The first rigorous attempt by a mathematical statistician to treat the problem of factor estimation (as distinct from principal components) came with the publication in 1940 of a paper by D. N. Lawley entitled "The Estimation of Factor Loadings by the Method of Maximum Likelihood." Since 1940, Lawley has published other articles dealing with various factor problems, and further contributions have been made by Howe (1955), by Anderson and Rubin (1956), and by Rao (1955), to mention just a few. Modern textbooks on factor analysis are those of Harman (1960) and Lawley and Maxwell (1963).

While methods of factor analysis, based on the above model, were being developed, Hotelling in 1933 published his *principal components* model, which, although it bears certain formal resemblances to the factor model proper, has rather different aims. It is widely used today and is described below.

The basic factor equations. The factor model described in general correlational terms above can be expressed more explicitly by the equations

$$(1) \qquad x_i = \sum_{s=1}^{k} l_{is}f_s + e_i, \qquad k < p.$$

In these equations k (the number of factors) is specified; f_s stands for the factors (generally referred to as *common* factors, since they usually enter into the composition of more than one variate). The factors are taken to be normally distributed and, without loss of generality, to have zero means and unit variances; to begin with, they will be assumed to be independent. The term e_i refers to a residual random variate affecting only the variate x_i. There are p of these e_i, and they are assumed to be normally distributed with zero means and to be independent of each other and of the f_s. Their variances will be denoted by v_i; the diagonal matrix of the v_i is called $\mathbf{V}$. The l-values are called *loadings* (weights), l_{is} being the loading of the ith variate on the sth factor. The quantities l_{is} and v_i are taken to be unknown parameters that have to be estimated. If a subscript for individual were introduced, it would be added to x_i and f_s, but not to l_{is} or v_i.

If the population variance–covariance matrix corresponding to the sample matrix $\mathbf{A}$ is denoted by $\mathbf{C}$, with elements c_{ij}, then it follows from the model that

$$c_{ii} = \sum_{s=1}^{k} l_{is}^2 + v_i , \tag{2}$$

and

$$c_{ij} = \sum_{s=1}^{k} l_{is} l_{js} , \qquad (i \neq j). \tag{3}$$

If the loadings for p variates on k factors are denoted by the $p \times k$ matrix $\mathbf{L}$, with transpose $\mathbf{L}'$, eqs. (2) and (3) can be combined in the single matrix equation

$$\mathbf{C} = \mathbf{LL}' + \mathbf{V}. \tag{4}$$

Estimating the parameters in the model. Since the introduction of multiple factor analysis, various approximate methods for estimating the parameters l_{is} and v_i have been proposed. Of these, the best known is the *centroid*, or *simple summation*, method. It is well described in the textbooks mentioned above, but since the arithmetic details are unwieldy, they will not be given here. The method works fairly well in practice, but there is an arbitrariness in its procedure that makes statistical treatment of it almost impossible (see Lawley & Maxwell 1963, chapter 3). For a rigorous approach to the estimation of the factor parameters, I turn to the method of maximum likelihood, although this decision requires some justification. The maximum likelihood method of factor estimation has not been widely used in the past for two reasons. First, it involves very onerous calculations which were well-nigh prohibitive before the development of electronic computers. Second, the arithmetic procedures available, which were iterative, frequently did not lead to convergent estimates of the loadings. But recently, largely because of the work of the Swedish statistician K. G. Jöreskog, quick and efficient estimation procedures have been found. These methods are still being perfected, but a preliminary account of them is contained in a recent paper (Jöreskog 1966). When they become better known, it is likely that the maximum likelihood method of factor analysis will become the accepted method. An earlier monograph by Jöreskog (1963) is also of interest. In it he links up work by Guttman (1953) on *image theory* with classical factor analytic concepts (see also Kaiser, in Harris 1963). (The image of a variate is defined as that part of its variance which can be estimated from the other variates in a matrix.)

The first point to note about eqs. (1) is that since the p observed variates x_i are expressed in terms of $p + k$ other variates, namely, the k common factors and the p residual variates, which are not observable, these equations are not capable of direct verification. But eq. (4) implies a hypothesis, H_0, regarding the covariance matrix $\mathbf{C}$, which can be tested, that it can be expressed as the sum of a diagonal matrix with positive diagonal elements and a symmetric positive semidefinite matrix with at most k latent roots: these matrices are respectively $\mathbf{V}$ and $\mathbf{LL}'$. The value postulated for k must not be too large; otherwise, the hypothesis would be trivially true. If the v_i were known, it would only be necessary to require $k < p$, but in the more usual case, where they are unknown, the condition can be shown to be $(p + k) < (p - k)^2$. Since the x_i are assumed to be distributed in a multivariate normal way, the log-likelihood function, omitting a function of the observations, is given by

$$L = -\tfrac{1}{2} n \ln |\mathbf{C}| - \tfrac{1}{2} n \sum_{i,j} a_{ij} c^{ij}, \tag{5}$$

where $n = N - 1$, $|\mathbf{C}|$ is the determinant of the matrix $\mathbf{C}$, and c^{ij} is the element in the ith row and jth column of its inverse, $\mathbf{C}^{-1}$. To find maximum likelihood estimators of l_{is} and v_i, (5) is differentiated with respect to them and the results are equated to zero. A difficulty arises, however, when $k > 1$, for there are then too many parameters in the model for them to be specified uniquely. This can be seen by an examination of eq. (4), for if $\mathbf{L}$ is postmultiplied by an orthogonal matrix $\mathbf{M}$, the value of $\mathbf{LL}'$, which is now given by $\mathbf{LMM}'\mathbf{L}'$, is unaltered since $\mathbf{MM}' = \mathbf{I}$, the identity matrix. This means that the maximum likelihood method, although it provides a unique set of estimates of the c_{ij}, leads to equations for estimating the l_{is} which are satisfied by an infinity of solutions, all equally good from a statistical point of view.

In this situation all the statistician can do is to select a particular solution, one that is convenient to find, and leave the experimenter to apply whatever rotation he thinks desirable. Thus the custom is to choose $\mathbf{L}$ in such a way that the $k \times k$ matrix $\mathbf{J} = \mathbf{L}'\mathbf{V}^{-1}\mathbf{L}$ is diagonal. It can be shown that the successive elements of $\mathbf{J}$ are the latent roots, in order of magnitude, of the matrix $\mathbf{V}^{-\frac{1}{2}} (\mathbf{A} - \mathbf{V}) \mathbf{V}^{-\frac{1}{2}}$, so that for a given value of $\mathbf{V}$, the determination of the factors in the factor model resembles the determination of the principal components in the component model.

The maximization of eq. (5) with the above diagonalization side condition leads to the equations

$$\hat{c}_{ii} = a_{ii}$$

or

$$\hat{v}_i = a_{ii} - \sum_{s=1}^{k} \hat{l}_{is}^2 , \tag{6}$$

and

$$\hat{\mathbf{L}}' = \hat{\mathbf{J}}^{-1}\hat{\mathbf{L}}'\hat{\mathbf{V}}^{-1}(\mathbf{A} - \hat{\mathbf{V}}), \tag{7}$$

where circumflex accents denote estimates of the parameters in question. Eq. (7) can usually be solved by iterative methods and details of those in current use can be found in Lawley and Maxwell (1963), Howe (1955), and Jöreskog (1963; 1966). The calculations involved are onerous, and when p is fairly large, say 12 or more, an electronic computer is essential.

A satisfactory property of the above method of estimation, which does not hold for the centroid and principal component methods, is that it can be shown to be independent of the metric used. A change of scale of any variate x_i merely introduces proportional changes in its loadings.

Testing hypotheses on number of factors. In the factor analysis of a set of data the value of k is seldom known in advance and has to be estimated. To begin with, some value of it is assumed and a matrix of loadings $\mathbf{L}$ for this value is estimated. The effects of the factors concerned are now eliminated from the observed covariance (or correlation) matrix, and the residual matrix, $\mathbf{A} - \mathbf{LL}'$, is tested for significance. If it is found to be statistically significant, the value of k is increased by one and the estimation process is repeated. The test employed is of the large sample chi-square type, based on the likelihood ratio method of Neyman and Pearson, and is given by

$$X^2 = n \ln(|\hat{\mathbf{C}}|/|\mathbf{A}|), \tag{8}$$

with $\frac{1}{2}\{(p-k)^2 - (p+k)\}$ degrees of freedom. A good approximation to expression (8), and one easier to calculate, is

$$X^2 = n \sum_{i<j} (a_{ij} - \hat{c}_{ij})^2/(\hat{v}_i\hat{v}_j). \tag{9}$$

There is also some evidence to suggest that the test can be improved by replacing n by $n' = n - \frac{1}{6}(2p+5) - \frac{2}{3}k$.

Factor interpretation. As already mentioned, the matrix of loadings, $\mathbf{L}$, given by a factor analysis is not unique and can be replaced by an equivalent set $\mathbf{LM}$ where $\mathbf{M}$ is an orthogonal matrix. This fact is frequently used by experimenters when interpreting their results, a matrix $\mathbf{M}$ being chosen that will in some way simplify the pattern of loadings or make it more intuitively meaningful. For example, $\mathbf{M}$ may be chosen so as to reduce to zero, or nearly zero, as many loadings as possible in order to reduce the number of parameters necessary for describing the data. Again, $\mathbf{M}$ may be chosen so as to concentrate the loadings of variates of similar content, say verbal tests, on a single factor so that this factor may be labeled appropriately. Occasionally, too, the factors are allowed to become correlated if this seems to lead to more meaningful results.

It is now clear that given a matrix of loadings from some analysis, different experimenters might choose different rotation matrices in their interpretation of the data. This subjective element in factor analysis has led to a great deal of controversy. To avoid subjectivity, various empirical methods of rotation have been proposed which, while tending to simplify the pattern of loadings, also lead to unique solutions. The best known of these are the *varimax* and the *promax* methods (for details see Kaiser 1958; Hendrickson & White 1964). But another approach to the problem, proposed independently by Howe (1955), Anderson and Rubin (1956), and Lawley (1958), seems promising. From prior knowledge the experimenter is asked to postulate in advance (*a*) how many factors he expects from his data and (*b*) which variates will have zero loadings on the several factors. In other words, he is asked to formulate a specific hypothesis about the factor composition of his variates. The statistician then estimates the nonzero loadings and makes a test of the "goodness of fit" of the factors structure. In this approach the factors may be correlated or uncorrelated, and in the former case estimates of the correlations between them are obtained. The equations of estimation and illustrative examples of their application can be found in Howe (1955) and in Lawley and Maxwell (1963; 1964); the latter gives a quick method of finding approximate estimates of the nonzero loadings.

Estimating factor scores. As the statistical theory of factor analysis now stands, estimation is a twofold process. First, the factor structure, as described above, of a set of data is determined. In practice, however, it is often desirable to find, in addition, equations for estimating the scores of individuals on the factors themselves. One method of doing this, developed by Thomson, is known as the "regression method." In it the l_{is} are taken to be the covariances between the f_s and the x_i, and then for uncorrelated factors the estimation equation is

$$\hat{\mathbf{f}} = \mathbf{L}'\mathbf{C}^{-1}\mathbf{x}, \tag{10}$$

or, more simply from the computational viewpoint,

$$\hat{\mathbf{f}} = (\mathbf{I} + \mathbf{J})^{-1}\mathbf{L}'\mathbf{V}^{-1}\mathbf{x}, \tag{11}$$

where $\hat{\mathbf{f}} = \{\hat{f}_1, \hat{f}_2, \cdots, \hat{f}_k\}'$, $\mathbf{x} = \{x_1, x_2, \cdots, x_p\}'$, and, as before, $\mathbf{J} = \mathbf{L}'\mathbf{V}^{-1}\mathbf{L}$, and $\mathbf{I}$ is the identity matrix. If sampling errors in $\mathbf{L}$ and $\mathbf{V}$ are neglected, the covariance matrix for the errors of estimates of the factor scores is given by $(\mathbf{I} + \mathbf{J})^{-1}$.

If the factors are correlated and their estimated

correlation matrix is denoted by $\boldsymbol{P}$, then eqs. (10) and (11) become, respectively,

(12) $$\hat{\boldsymbol{f}} = \boldsymbol{PL'C}^{-1}\boldsymbol{x},$$

and

(13) $$\hat{\boldsymbol{f}} = (\boldsymbol{P}^{-1} + \boldsymbol{J})^{-1}\boldsymbol{L'V}^{-1}\boldsymbol{x},$$

while the errors of estimates are given by $(\boldsymbol{P}^{-1} + \boldsymbol{J})^{-1}$.

An alternative method of estimating factor scores is that of Bartlett (1938). Here, the principle adopted is the minimization, for a given set of observations, of $\sum_i e_i^2/v_i$, which is the sum of squares of standardized residuals. The estimation equation now is

(14) $$\hat{\boldsymbol{f}}^* = \boldsymbol{J}^{-1}\boldsymbol{L'V}^{-1}\boldsymbol{x}.$$

It is of interest to note that although the sets of estimates gotten by the two methods have been reached by entirely different approaches, a comparison shows that they are simply related. For uncorrelated factors the relationship is

$$\hat{\boldsymbol{f}}^* = (\boldsymbol{I} + \boldsymbol{J}^{-1})\hat{\boldsymbol{f}};$$

for correlated factors it is

$$\hat{\boldsymbol{f}}^* = (\boldsymbol{I} + \boldsymbol{J}^{-1}\boldsymbol{P}^{-1})\hat{\boldsymbol{f}}.$$

Comparing factors across populations. If factors can be viewed as representing "permanent" aspects of behavior or performance, ways of identifying them from one population to another are required. In the past, identification has generally been based on the comparison of matrices of loadings. In the case of two matrices, a common approach, developed by Ahmavaara (1954) and Cattell and Hurley (1962), is to rotate one into maximum conformity in the least square sense with the other. For example, the matrix required for rotating $\boldsymbol{L}_1$ into maximum conformity with $\boldsymbol{L}_2$, when they both involve the same variates, is obtained by calculating the expression $(\boldsymbol{L}_1'\boldsymbol{L}_1)^{-1}\boldsymbol{L}_1'\boldsymbol{L}_2$ and normalizing it by columns. The factors represented by $\boldsymbol{L}_1$ in its transformed state are likely to be more or less correlated, but estimates of the correlations between them are given by $(\boldsymbol{M}_0'\boldsymbol{M}_0)^{-1}$, standardized so that its diagonal cells are unity, where $\boldsymbol{M}_0 = (\boldsymbol{L}_1'\boldsymbol{L}_1)^{-1}\boldsymbol{L}_1'\boldsymbol{L}_2$ and $\boldsymbol{M}_0'$ is its transpose. This procedure is fairly satisfactory when the sample *covariance* matrices involved do not differ significantly. When they do, the problem of identifying factors is more complicated.

A possible approach to it has been suggested by Lawley and Maxwell (1963, chapter 8), who make the assumption that although two covariance matrices, $\boldsymbol{C}_1$ and $\boldsymbol{C}_2$, involving the same variates may be different, they may still have the same $\boldsymbol{L}$-matrix. This could occur if the two $k \times k$ covariance matrices $\boldsymbol{\Gamma}_1$ and $\boldsymbol{\Gamma}_2$ between the factors themselves were different. To keep the model fairly simple, they assume that the residual variances in the populations are in each case $\boldsymbol{V}$ and then set up the equations

(15) $$\boldsymbol{C}_1 = \boldsymbol{L}\boldsymbol{\Gamma}_1\boldsymbol{L'} + \boldsymbol{V},$$
$$\boldsymbol{C}_2 = \boldsymbol{L}\boldsymbol{\Gamma}_2\boldsymbol{L'} + \boldsymbol{V}.$$

For this model Lawley and Maxwell show how estimates of $\boldsymbol{L}$, $\boldsymbol{V}$, $\boldsymbol{\Gamma}_1$, and $\boldsymbol{\Gamma}_2$ may be obtained from two sample covariance matrices $\boldsymbol{A}_1$ and $\boldsymbol{A}_2$. They also supply a test for assessing the significance of the difference between the estimates of $\boldsymbol{\Gamma}_1$ and $\boldsymbol{\Gamma}_2$, and also for testing the "goodness of fit" of the model.

The method of principal components

The principal component method of analyzing a matrix of covariances or correlations is also widely used in the social sciences. The components correspond to the latent roots of the matrix, and the weights defining them are proportional to the corresponding latent vectors.

The model can also be stated in terms of the observed variates and the derived components. An orthogonal transformation is applied to the x_i $(i = 1, 2, \cdots, p)$ to produce a new set of uncorrelated variates $y_1, y_2, \cdots, y_p$. These are chosen such that y_1 has maximum variance, y_2 has maximum variance subject to being uncorrelated with y_1, and so on. This is equivalent to a rotation of the coordinate system so that the new coordinate axes lie along the principal axes of an ellipsoid closely related to the covariance structure of the x_i. The transformed variates are then standardized to give a new set, which will be denoted z_s. When this method is used, no hypothesis need be made about the nature or distribution of the x_i. The model is by definition linear and additive, and the basic equations are

(16) $$x_i = \sum_{s=1}^{p} w_{is} z_s, \qquad i, s = 1, 2, \cdots, p,$$

where z_s stands for the sth component, and w_{is} is the weight of the sth component in the ith variate. In matrix notation eqs. (16) become

$$\boldsymbol{x} = \boldsymbol{Wz},$$

where $\boldsymbol{x} = \{x_1, x_2, \cdots, x_p\}'$, $\boldsymbol{z} = \{z_1, z_2, \cdots, z_p\}'$, and $\boldsymbol{W}$ is a square matrix of order p with elements w_{is}.

Comparison of eqs. (16) with eqs. (1) shows that in the principal component model residual variates do not appear, and that if all p components are obtained, the sample covariances can be reproduced exactly, that is, $\boldsymbol{A} = \boldsymbol{W'W}$. Indeed, there is a simple reciprocal relationship between the observed variates and the derived components.

A straightforward iterative method for obtaining the weights w_{is} is given by Hotelling in his original papers; the details are also given in most textbooks on factor analysis. In practice, all p components are seldom found, for a small number generally accounts for a large percentage of the variance of the variates and can be used to summarize the data. There is also a criterion, developed by Bartlett (1950; 1954), for testing the equality of the remaining latent roots of a matrix after the first k have been extracted; this is sometimes used to help in deciding when to stop the analysis.

The principal component method is most useful when the variates x_i are all measured in the same units. Otherwise, it is more difficult to justify. A change in the scales of measurement of some or all of the variates results in the covariance matrix being multiplied on both sides by a diagonal matrix. The effect of this on the latent roots and vectors is very complicated, and unfortunately the components are not invariant under such changes of scale. Because of this, the principal component approach is at a disadvantage in comparison with the proper factor analysis approach.

A. E. MAXWELL

[*See also* CLUSTERING; DISTRIBUTIONS, STATISTICAL, *article on* MIXTURES OF DISTRIBUTIONS; LATENT STRUCTURE; STATISTICAL IDENTIFIABILITY.]

BIBLIOGRAPHY

AHMAVAARA, Y. 1954 Transformational Analysis of Factorial Data. Suomalainen Tiedeakatemia, Helsinki, *Toimituksia: Annales* Series B 88, no. 2.

ANDERSON, T. W.; and RUBIN, HERMAN 1956 Statistical Inference in Factor Analysis. Volume 5, pages 111–150 in Berkeley Symposium on Mathematical Statistics and Probability, Third, *Proceedings*. Edited by Jerzy Neyman. Berkeley: Univ. of California Press.

BANKS, CHARLOTTE 1954 The Factorial Analysis of Crop Productivity: A Re-examination of Professor Kendall's Data. *Journal of the Royal Statistical Society* Series B 16:100–111.

BARTLETT, M. S. 1938 Methods of Estimating Mental Factors. *Nature* 141:609–610.

BARTLETT, M. S. 1950 Tests of Significance in Factor Analysis. *British Journal of Psychology* (Statistical Section) 3:77–85.

BARTLETT, M. S. 1954 A Note on the Multiplying Factor for Various χ^2 Approximations. *Journal of the Royal Statistical Society* Series B 16:296–298.

BURT, CYRIL 1940 *The Factors of the Mind: An Introduction to Factor-analysis in Psychology*. Univ. of London Press.

BURT, CYRIL 1947 Factor Analysis and Physical Types. *Psychometrika* 12:171–188.

CATTELL, RAYMOND B.; and HURLEY, JOHN R. 1962 The Procrustes Program: Producing Direct Rotation to Test a Hypothesized Factor Structure. *Behavioral Science* 7:258–262.

GEARY, R. C. 1948 Studies in Relationships Between Economic Time Series. *Journal of the Royal Statistical Society* Series B 10:140–158.

GIBSON, W. A. 1960 Nonlinear Factors in Two Dimensions. *Psychometrika* 25:381–392.

HAMMOND, W. H. 1944 Factor Analysis as an Aid to Nutritional Assessment. *Journal of Hygiene* 43:395–399.

HAMMOND, W. H. 1955 Measurement and Interpretation of Subcutaneous Fats, With Norms for Children and Young Adult Males. *British Journal of Preventive and Social Medicine* 9:201–211.

HARMAN, HARRY H. 1960 *Modern Factor Analysis*. Univ. of Chicago Press. → A new edition was scheduled for publication in 1967.

HARRIS, CHESTER W. (editor) 1963 *Problems in Measuring Change: Proceedings of a Conference*. Madison: Univ. of Wisconsin Press. → See especially "Image Analysis" by Henry F. Kaiser.

HENDRICKSON, ALAN E.; and WHITE, PAUL O. 1964 Promax: A Quick Method for Rotation to Oblique Simple Structure. *British Journal of Statistical Psychology* 17:65–70.

HERDAN, G. 1943 The Logical and Analytical Relationship Between the Theory of Accidents and Factor Analysis. *Journal of the Royal Statistical Society* Series A 106:125–142.

HORST, PAUL 1965 *Factor Analysis of Data Matrices*. New York: Holt.

HOTELLING, HAROLD 1933 Analysis of a Complex of Statistical Variables Into Principal Components. *Journal of Educational Psychology* 24:417–441, 498–520.

HOWE, W. G. 1955 Some Contributions to Factor Analysis. Report No. ORNL-1919, U.S. National Laboratory, Oak Ridge, Tenn. Unpublished manuscript.

JÖRESKOG, K. G. 1963 *Statistical Estimation in Factor Analysis: A New Technique and Its Foundation*. Stockholm: Almqvist & Wiksell.

JÖRESKOG, K. G. 1966 Testing a Simple Hypothesis in Factor Analysis. *Psychometrika* 31:165–178.

KAISER, HENRY F. 1958 The Varimax Criterion for Analytic Rotation in Factor Analysis. *Psychometrika* 23:187–200.

KENDALL, M. G.; and LAWLEY, D. N. 1956 The Principles of Factor Analysis. *Journal of the Royal Statistical Society* Series A 119:83–84.

LAWLEY, D. N. 1940 The Estimation of Factor Loadings by the Method of Maximum Likelihood. Royal Society of Edinburgh, Section A, *Proceedings* 60:64–82.

LAWLEY, D. N. 1953 A Modified Method of Estimation in Factor Analysis and Some Large Sample Results. Pages 35–42 in *Uppsala Symposium on Psychological Factor Analysis, March 17–19, 1953*. Nordisk Psykologi, Monograph Series, No. 3. Uppsala (Sweden): Almqvist & Wiksell.

LAWLEY, D. N. 1958 Estimation in Factor Analysis Under Various Initial Assumptions. *British Journal of Statistical Psychology* 11:1–12.

[1]LAWLEY, D. N.; and MAXWELL, A. E. 1963 *Factor Analysis as a Statistical Method*. London: Butterworth.

LAWLEY, D. N.; and MAXWELL, A. E. 1964 Factor Transformation Methods. *British Journal of Statistical Psychology* 17:97–103.

MAXWELL, A. E. 1964 Calculating Maximum-likelihood

Factor Loadings. *Journal of the Royal Statistical Society* Series A 127:238–241.

RAO, C. RADHAKRISHNA 1955 Estimation and Tests of Significance in Factor Analysis. *Psychometrika* 20: 93–111.

SPEARMAN, C. E. 1904 "General Intelligence" Objectively Determined and Measured. *American Journal of Psychology* 15:201–293.

THOMSON, GODFREY H. (1939) 1951 *The Factorial Analysis of Human Ability.* 5th ed. Boston: Houghton Mifflin.

THURSTONE, LOUIS L. 1935 *The Vectors of Mind: Multiple-factor Analysis for the Isolation of Primary Traits.* Univ. of Chicago Press.

THURSTONE, LOUIS L. 1947 *Multiple-factor Analysis.* Univ. of Chicago Press. → A development and expansion of Thurstone's *The Vectors of Mind,* 1935.

Postscript

A long-standing difficulty of applying the maximum likelihood method of estimation in factor analysis was the lack of satisfactory methods of obtaining numerical solutions. This difficulty has now been almost entirely removed. The first breakthrough came with a paper by Jöreskog (1966), followed by other papers (for example Jöreskog 1967; Jöreskog & Lawley 1968; Clarke 1970) in which methods of steepest descent, other optimization methods given by Fletcher and Powell (1963), and the Newton–Raphson method were employed for the minimization of a function of many variables. These methods are fully described and illustrated in the second edition of the book by Lawley and Maxwell (1971). Jöreskog and Goldberger (1972) have also provided a method for estimating the parameters in the factor model by generalized least squares, using the inverse of the observed covariance matrix as a weight matrix, and have shown that the estimates are scale free and asymptotically efficient.

On the more theoretic side the main contribution has been the provision of formulas for the standard errors of factor loading and residual variance estimates found by the maximum likelihood method. The basic paper is by Lawley (1967), and the theory is fully described and illustrated in Lawley and Maxwell (1971). Lawley's work has since been extended by Archer and Jennrich (1973), Jennrich (1973), and Jennrich and Thayer (1973) to obtain standard errors for analytically rotated loadings, and in the course of this a slight error in the original formulas was detected and corrected. Other theoretic work of special interest is that by McDonald (1967) on nonlinear factor analysis, some new results on factor indeterminacy by Schönemann and Wang (1972), and a general method for the analysis of covariance structures (Jöreskog 1970), in which first- and second-order factor analyses appear as special cases. A paper of tangential interest is one by Lawley and Maxwell (1973) on regression and factor analysis. In this paper they show that, in assessing the effects of errors of measurement in the independent variates on the estimates of the parameters in the multiple linear regression model, it is frequently advantageous to express the latter in factor analytic terms. That is to say, assume that the independent variates are linearly related to a set of "instrumental variates" which are unobservable and which we will call factor scores. Imposing the usual conditions (see Lawley & Maxwell 1971, pp. 1–13) for identifiability of the factor analysis model on the independent variates and factor scores, we can estimate the factor scores consistently. Consequently, either using standard instrumental variable estimators or the relation between the factor loadings and the parameters of the multiple linear regression, the regression parameters can be estimated consistently. For a special case they also show how a "best" subset of any size of the indpendent variates can readily be determined for predicting the dependent variate. [*The relationship between factor analysis and other multidimensional procedures is discussed in* FACTOR ANALYSIS AND PRINCIPAL COMPONENTS, *article on* BILINEAR METHODS.]

Preeminent among recent papers of an applied nature is McGaw and Jöreskog (1971). It is concerned with the thorny problem of comparing factors derived from different populations. A model, in which the same factor pattern was assumed to hold in each of four populations, was fitted to data on 12 aptitude and achievement tests for an over-all sample of 11,743 subjects. Goodness-of-fit indices suggested that the model fitted satisfactorily. Differences across the four populations in the factor dispersion matrices and in the mean factor scores are also examined and discussed there.

A. E. MAXWELL

[*See also* SCALING, MULTIDIMENSIONAL.]

ADDITIONAL BIBLIOGRAPHY

ARCHER, CLAUDE O.; and JENNRICH, ROBERT I. 1973 Standard Errors for Rotated Factor Loadings. *Psychometrika* 38:581–592.

CLARKE, M. R. B. 1970 A Rapidly Convergent Method for Maximum-likelihood Factor Analysis. *British Journal of Mathematical and Statistical Psychology* 23: 43–52.

FLETCHER, R.; and POWELL, M. J. D. 1963 A Rapidly Convergent Descent Method for Minimization. *Computer Journal* 6:163–168.

JENNRICH, ROBERT I. 1973 Standard Errors for Obliquely Rotated Factor Loadings. *Psychometrika* 38:593–604.

JENNRICH, ROBERT I.; and THAYER, DOROTHY T. 1973 A Note on Lawley's Formulas for Standard Errors in Maximum Likelihood Factor Analysis. *Psychometrika* 38:571–580.

JÖRESKOG, K. G. 1966 Testing a Simple Structure Hypothesis in Factor Analysis. *Psychometrika* 31:165–178.

JÖRESKOG, K. G. 1967 Some Contributions to Maximum Likelihood Factor Analysis. *Psychometrika* 32:443–482.

JÖRESKOG, K. G. 1970 A General Method for Analysis of Covariance Structures. *Biometrika* 57:239–251.

JÖRESKOG, K. G.; and GOLDBERGER, ARTHUR S. 1972 Factor Analysis by Generalized Least Squares. *Psychometrika* 37:243–260.

JÖRESKOG, K. G.; and LAWLEY, D. N. 1968 New Methods in Maximum Likelihood Factor Analysis. *British Journal of Mathematical and Statistical Psychology* 21:85–96.

LAWLEY, D. N. 1967 Some New Results in Maximum Likelihood Factor Analysis. Royal Society of Edinburgh, Section A, *Proceedings* 67:256–264.

LAWLEY, D. N.; and MAXWELL, A. E. (1963) 1971 *Factor Analysis as a Statistical Method.* 2d ed. London: Butterworth.

LAWLEY, D. N.; and MAXWELL, A. E. 1973 Regression and Factor Analysis. *Biometrika* 60:331–338.

MCDONALD, RODERICK P. 1967 Factor Interaction in Nonlinear Factor Analysis. *British Journal of Mathematical and Statistical Psychology* 20:205–215.

MCGAW, BARRY; and JÖRESKOG, K. G. 1971 Factorial Invariance of Ability Measures in Groups Differing in Intelligence and Socioeconomic Status. *British Journal of Mathematical and Statistical Psychology* 24:154–168.

SCHÖNEMANN, PETER H.; and WANG, MING-MEI 1972 Some New Results on Factor Indeterminacy. *Psychometrika* 37:61–91.

FACTORIAL EXPERIMENT

See EXPERIMENTAL DESIGN.

FALLACIES, STATISTICAL

This article will be mainly concerned with statistical fallacies, but it should be noted that most other fallacious types of reasoning can be carried over into statistics.

Most fallacies seem foolish when pinpointed, but they are not the prerogative of fools and statisticians. Great men make mistakes, and when they admit them remorsefully, they reveal a facet of their greatness. The reason for mentioning the mistakes of eminent people in this article is to make it more fun to read.

Many fallacies, statistical or otherwise, have their origin in wishful thinking, laziness, and busyness. These conditions lead to oversimplification, the desire to win an argument at all costs (even at the cost of over*complication*), failure to listen to the opposition, too-ready acceptance of authority, too-ready rejection of it, too-ready acceptance of the printed word (even in newspapers), too-great reliance on a machine or formal system or formula (*deus ex machina*), and too-ready rejection of them (*diabolus ex machina*). These emotionally determined weaknesses are not themselves fallacies, but they provoke them. For example, they provoke special pleading, the use of language in more than one sense without notice of the ambiguity (if the argument leads to a desirable conclusion), the insistence that a method used successfully in one field of research is the only appropriate one in another, the distortion of judgment, and the forgetting of the need for judgment.

A logical or syntactical fallacy. We begin with an example of a fallacious argument in which the conclusion is correct:

> "No cat has no tail. One cat has one more tail than no cat. Therefore one cat has one tail."

A good technique for exposing fallacious reasoning is to use the same form of argument in order to deduce an obviously false result:

> "No cat has eight tails. One cat has one more tail than no cat. Therefore one cat has nine tails."

The fallacy can be explained by careful attention to syntax, specifically by noting that the following two propositions have been confused: (1) It is false that any cat has eight tails, and (2) the object named "no cat" has eight tails. P. M. S. Blackett once said, exaggerating somewhat, that a physicist is satisfied with an argument if it leads to a result that he believes to be true.

Arguments from authority. The book *Popular Fallacies* by Alfred S. E. Ackermann is more concerned with fallacies of fact than of reasoning, and here many fallacies depend on the acceptance of authority. It is interesting to see that the author was himself misled by authority on at least two occasions.

First, he argues that it is a fallacy "that cigarette smoking is especially pernicious," appealing to the opinions of several authorities: for example, "Of the various forms of smoking, cigarette smoking is the most wholesome, preferably without a holder," according to Sir Robert Armstrong-Jones, F.R.C.P., in the *Daily Mail*, January 1, 1927 (Ackermann [1907] 1950, pp. 174–175). (The current medical opinion is that, of cigarettes, cigars, and pipes, cigarettes are the least wholesome, at any rate in regard to lung cancer. Of course, Armstrong-Jones *might* be right after all.)

Then Ackermann refers to the thesis "that there

is a prospect of atomic energy being of practical use." Lord Rutherford is quoted, from the *Evening News*, September 11, 1933, as saying that "anyone who expects a source of power from the transformation of these atoms is talking moonshine" (*ibid.*, pp. 708–709).

It would be unfair to blame Ackermann for relying on these authorities, but it is useful to hold in mind that the highest authorities can be wrong, even when they are emphatic in their opinions. Their desire not to seem too academic should sometimes be allowed for, especially when they hold an administrative appointment.

What should the question really be? When Gertrude Stein was on her deathbed, one of her friends asked her, "What is the answer?" After a few seconds she whispered back, "What is the question?"

It is important for the statistician to satisfy himself that a *right* question is being asked, by inquiring into the purposes behind the question. Chambers (1965) states that when a member of Parliament asked for some inland revenue figures that were not available from the published statistics, his invariable rule was to find out the purpose for which the information was needed. More often than not he found that the figures sought were irrelevant, that other figures already published were more helpful, or that the M.P. was misguided over the whole business.

It is often reasonable to make exploratory investigations without a clear purpose in mind. The fallacy we have just pointed out is the assumption that the questioner necessarily asks for information that is very relevant to his purposes, whether those purposes are clear or vague. The fallacy of giving the "right answer to the wrong question" is further discussed by A. W. Kimball (1957).

Ignoring the "exposure base." Consider, for example, the reports on traffic deaths that are issued after public holidays. Many readers conclude from the increased number of deaths that it is more dangerous to drive on public holidays than on ordinary days. The conclusion may or may not be correct, but the reasoning is fallacious, since it ignores the fact that many more people drive automobiles on public holidays and thus more people are exposed to the possibility of an accident. If holidays and ordinary days are compared on the basis of deaths per passenger mile, it might turn out that the holiday death rate is lower, since the reduction of the average speed caused by the volume of traffic also may reduce the seriousness, if not the number, of accidents.

"Deus ex machina"—the precision fallacy. When we know a machine or formal system that can produce an exact answer to a question, we are tempted to provide an answer and inquire no further. But exact methods often produce exact answers to wrong questions.

One of the main aims of statistical technique is to fight the danger of wishful thinking and achieve a measure of objectivity in probability statements. But absolute objectivity and precision are seldom if ever attainable: there is always, or nearly always, a need for judgment in any application of statistical methods. (This point is especially emphasized in Good 1950.) Inexperienced statisticians often overestimate the degree of precision and objectivity that can be attained. An elementary form of the precision fallacy, which is less often committed by statisticians than by others, is the use of an average without reference to "spread." A related trap is to gauge closeness by some measure of spread but to ignore systematic errors (bias) [*see* ERRORS, *article on* NONSAMPLING ERRORS].

The use of an average without a measure of spread can be especially misleading, or even comic, when the sample is small and the spread is therefore large. But even if the mean and standard deviation of the *population* are given, they can be misleading if the population is very skew. For skew populations it is better to give some of the quantiles [*see* STATISTICS, DESCRIPTIVE, *article on* LOCATION AND DISPERSION].

Randomization. An example of the precision fallacy occurs in connection with the important technique of randomization. Let us consider the famous tea-tasting experiment (see Fisher 1935, chapter 1; Good 1956). A lady claims to be able to tell by tasting, with better than random chance, whether the milk is put into her tea first or last. We decide to test her by giving her twenty cups of tea to taste, in ten of which the milk is poured first, and in ten last. If the lady gets many more than ten of her assertions right and if we have not randomized the order of the twenty trials, we might suspect that whatever sequence we selected for some psychological reason, the lady might have tended to guess for similar psychological reasons. So we randomize the order and can then apparently make use of the hypergeometric tail-area probability as a precise, objective, and effectively complete summary of the statistical significance of the experiment [*see* EXPERIMENTAL DESIGN *and* RANDOM NUMBERS].

But suppose we now examine the random sequence and spot some pattern in it—for example, that cups 3, 6, 9, 12, 15, and 18 had the milk in first. This would at once undermine the precise validity of our result in relation to its relevance to the hypothesis under investigation. We can partly remove this difficulty by means of "restricted randomization," but we cannot completely remove it, because every finite sequence exhibits some special features, however recondite. Only by judgment, necessarily subjective, can we decide that any given feature is unimportant. The only way we can preserve the precision is to make use of a "statistician's stooge" to perform the experiment for us, including the randomization (Good 1960–1961). He must report the number of successes to us but must on no account tell us the randomized order. Precision in a randomized experiment can be obtained only at the price of suppressing some of the information. The point made here is not universally accepted by statisticians.

If the number of cups of tea is very large, then the importance of the above criticism of randomization will usually be negligible. But long experiments are expensive, and in the statistical design of experiments the expense can never be ignored.

Randomization in itself is by no means a fallacious technique; what is fallacious is the notion that without the suppression of information, it can lead to a precise tail-area probability relevant only to a null hypothesis.

[2]The suppression of information. In its crudest forms the suppression of information is often at least as wicked as an outright lie. We shall later refer to some of the cruder forms. But we have just seen that randomization loses its precision unless some information is suppressed, and we shall now argue more forcibly that it is a fallacy to suppose that the suppression of information is always culpable.

One way of seeing this is in terms of digital communication. When an electrical "pulse" of a given shape is liable to have been attenuated and distorted by noise, some circuitry is often incorporated for the purpose of re-forming ("regenerating") the pulse. This circuitry, as it were, accepts the hypothesis that the pulse is supposed to be present rather than absent. Since noise in the electronic system makes the probability less than one that the supposed pulse is present, the regeneration loses some information. But allowing for the nature of the *subsequent* communication channel, it can be proved that the loss is often more than compensated. (This will not be proved here, but it is not surprising to common sense.) In pedagogy the corresponding principle is that simplification is necessary when teaching beginners. In statistics the corresponding device is known as the *reduction of the data*, that is, the reduction of a mass of data to a more easily assimilable form. If the statistics are "sufficient," then there is no loss of information, but we often have to be satisfied with "insufficient" statistics in order to make an effective reduction of the data. Thus it is fallacious to say that the suppression of information is always a statistical crime. (Note that apparently sufficient statistics might not really be so if the model is wrong. People sometimes publish, say, only a mean and variance of a sample, and this prevents readers from checking the validity of the model for which these statistics would be sufficient.) [*See* SUFFICIENCY.]

Terminological ambiguities. Important examples of terminological ambiguities occur both in the philosophy of statistics and in its practical applications. Often they are as obvious as the "no cat" ambiguity, once they are pointed out. But before they are pointed out, they lead to a great deal of argument at cross purposes. Thus, many of the problems in the philosophy of probability clear themselves up as soon as we distinguish between various kinds of probability (see Good 1959*a*). (We shall not discuss here whether they can all be reduced to a single kind or whether they all "exist." But they are all talked about.) [*See* PROBABILITY, *article on* INTERPRETATIONS.]

There is tautological, or mathematical, probability, which occurs in mathematical theories and requires no operational definition. It occurs also in the definition of a "simple statistical hypothesis," which is a hypothesis for which some probabilities of the form $P(E|H)$ are assigned *by definition*. (Here E represents an event or a proposition asserting that an event obtains, and the vertical stroke stands for "given" or "assuming.") There are *physical*, or material, probabilities, or *chances*. These relate to tautological probabilities by means of the *linguistic axiom* that to say that H is true is to say that the physical probability of E is $P(E|H)$, for some class of events, E. There are *logical* probabilities, or credibilities. There are *subjective*, or personal, probabilities, which are the intensities of conviction that a man will use for betting purposes, after mature consideration. There are *multisubjective* probabilities, belonging to groups of people. And there are *psychological* probabilities, which are the probabilities that people behave as if they accept, even before applying any criterion to test their consistency. By confusing pairs of these six kinds of probability, *fifteen different kinds of fallacy can be generated*. For example, it is often said that there is no sense in talking about the probability that a population parameter has a certain value, for "it either has the value or it does not, and therefore the probability is either 0 or 1." It need hardly be mentioned, to those who are not choked with emotion, that the probability need not be 0 or 1 when it is interpreted not as a physical probability but as a logical or as a subjective probability.

Even physical probabilities can be confused with each other, since they can be mistakenly referred to the same event. For example, apparent variations in the incidence of some crime or disease from one place or time to another are very often found to be due to variations in the methods of classification. Adultery would appear to increase enormously if Christ's definition were suddenly to be accepted in the law—"Whosoever looketh on a woman to lust after her hath committed adultery with her already in his heart" (Matthew 5.28). Since partners to adultery do not often turn in official reports, perhaps a better example is that of crime records. An example with documentation, quoted by Wallis and Roberts (1956), is that of felonies in New York. It was alleged that there had been an increase of 34.8 per cent from 1949 to 1950, but later it appeared that this was at least largely due to a revised method of classification.

This class of practical statistical fallacies is extremely common in the social sciences, and one should be very much on guard against it. As a further example, two standard definitions of the number of unemployed in the United States differ by a factor of over 3, namely, the "average monthly rate" and the "total annual rate." Putting it roughly, one measure for any given year is the average monthly number of people unemployed; the other, larger measure is the number of people who were unemployed at any time during the year. (See Gordon 1968.)

[3]An example of a terminological fallacy is the confusion of "some" and "all." It is perpetrated by John Hughlings Jackson in the following excerpt: "To coin the word, verbalising, to include all ways in which words serve, I would assert that both halves of the brain are alike in that each serves in *verbalising*. That the left half does is evident, because damage of it makes a man speechless. That the right does is inferable, because the speechless man understands all I say to him in ordinary matters" (quoted in Penfield & Roberts 1959, p. 62). Some damage to the left hemisphere seems here to have been confused with destruction of all of it.

Another example of the "some and all" fallacy is to assume that since some poems are better than others in the opinion of any reasonable judge, then, given any set of poems, one of them must be the best. More generally, the possibility of *partial* ordering is easily overlooked. But sometimes the assumption of partial ordering, although truer than complete ordering, is too complicated for a given application. In a beauty competition, for example, each girl might be the best of her kind, but it might be essential to award the prize to only one of them.

In some social surveys respondents are asked to rank several objects in order of merit. An alternate design, which will often be less watered down by the need to reach decisions in doubtful cases, is to ask for comparisons of pairs of objects but to permit "no comparison" as a response for any given pair.

[4]**Ignoring a relevant concomitant variable.** "The death rate in the American Army in peacetime is lower than that in New York City. Therefore leave New York and join the army." The fallacy is that the methods of selection for the army are biased toward longevity, both by age and by health, and these clearly relevant variables have been ignored. The fallacy can also be categorized as failure to control for exposure, since many inhabitants of New York City are subject to the possibility of death from infant diseases, chronic diseases, and old age, whereas very few men in the army are so exposed. The example can also be regarded as one of "biased sampling," a category of fallacy to be considered later.

[5]*Ignoring half of a contingency table.* It is commonly believed that government scientists in the United Kingdom earn more on the average than university teachers. But, as Rowe pointed out (1962), the average age of university teachers is less than that of government scientists, because a large proportion of lecturers leave the universities before their mid-thirties. Rowe showed that the median earning of university teachers above the age of 35 is greater than that of government scientists. But he did not estimate what these men *would* have earned in government service. Thus, although he refuted the original argument, he did not ask the really relevant question. This question is very difficult to answer. A possible approach, which would shed some light, would be to find out the distribution of salaries as a function of job, age, and intelligence quotient.

The perennial problem here is which covariates to choose and where to stop choosing them, since the list of possibilities is typically impracticably large. A related issue is that the greater the number of conditioning classificatory variables (dimensionality of the contingency table), the fewer the cases in the relevant cross-classification cell. This is one of the unsolved problems of actuarial science, where the problem of estimating probabilities in multidimensional contingency tables is philosophically basic (for some discussion of this problem, with references, see Good 1965).

It sometimes happens that a fact is almost universally ignored, although in retrospect it is clearly highly relevant. In the late 1950s people in the

United States were arguing that the standard of college teaching staffs was deteriorating, since the proportion of newly employed teachers who held a doctorate was decreasing. What was overlooked was that the proportion of teachers who took their doctorates after becoming teachers was increasing. Cartter (1965) states that Bernard Berelson was almost alone in his correct interpretation of the situation.

Biased sample. At one time, most known quasi-stellar radio sources lay approximately in a plane, and this seemed to one writer to have deep cosmological significance. But these radio sources could not be definitely identified with optical sources unless they were located with great accuracy, and for this purpose they had to be occluded by the moon. Also, as it happened, most of the observations had been made from the same observatory. Hence there was a very strong bias in the sampling of the sources (this was mentioned by D. W. Dewhurst in a lecture in Oxford on January 28, 1965). As Sir Arthur Eddington once pointed out, if you catch fish with a net having a 6-inch mesh, you are liable to formulate the hypothesis that all fish are more than 6 inches in length. [*See* ERRORS, *article on* NONSAMPLING ERRORS.]

It is sometimes overlooked that atrocity stories usually form a biased sample. Newspapers tend to report the atrocities of political opponents more than those of friends. An exception was the Nazi atrocities, which were so great that the evidence for them had to be overwhelming before they could be believed. (For example, there appears to be no reference to them in the 1951 edition of the *Encyclopaedia Britannica.*)

Sometimes inferences from a sample are biased because of seasonal variations. According to Starnes (1962), Democratic Secretary of Labor Willard Wirtz stated just before an election that over "four and a half million more Americans have jobs than when this Administration took office in January of 1961." Wirtz later admitted that the figure should have been 1.224 million, and he said, "It isn't proper to compare January figures with October figures without a seasonal adjustment." Similarly, the Republican governor of New York, Nelson D. Rockefeller, once referred to a "net increase of 450,000 jobs" since he had taken office. The figure is worthless because it again ignores the adjustment for seasonal variations.

Bias is difficult to avoid in social surveys, for example, in the use of questionnaires, where poor wording is frequent and where one sometimes (especially in political and commercial surveys) finds tendentious wording.

Even with an unbiased sample, it is possible to get a biased conclusion by computing the significance level of various tests of the null hypothesis and selecting the one most favorable to one's wishes. Although these tests are based on the same sample and are therefore statistically interdependent, there will be a reasonable probability that one out of twenty such tests will reach a 5 per cent significance level. A suggestion of how to combine such "parallel" tests is given by Good (1958*a*).

[6]**The suppression of the uninteresting.** Suppose we have done an experiment, and it reaches a significance level of 5 per cent. Should we reject the null hypothesis? Perhaps the experiment has been performed by others without significant results. If these other experiments were taken into account, the total significance of all the experiments combined might be negligible. Moreover, the other results might have been unpublished because they were nonsignificant and therefore uninteresting. This explains why some apparent medical advances do not fulfill their early promise. The published statistics are biased in favor of what is interesting. As one physician said, "Hasten to use the remedy before it is too late" (Good 1958*b*, p. 283; Sterling 1959).

Sample too small. One of the most frequent and elementary statistical fallacies is the reliance on too small a sample. In 1933 Meduna, believing that schizophrenia and epilepsy were incompatible because of the rarity of their joint occurrence, started to induce convulsions in mental patients by chemical means. Consequently, the beneficial effect of convulsions on depressives was eventually accidentally discovered. Meduna's sample was too small, and in fact it has now been found that schizophrenia and epilepsy are *positively* correlated (Slater & Beard 1963). One moral of this story is that experiments can be worth trying without theoretical reason to believe that they might be successful.

Misleading use of graphs and pictures. Graphs and pictures are often used in newspapers in the hope of misleading readers who are not experienced in interpreting them. Sometimes graphs are inadequately labeled; sometimes the scale is chosen so as to make a small slope appear large; sometimes the graph is drawn on a board and the board is pictured in perspective so as to accentuate the most recent slope; sometimes too little of a time series is shown, and the graph is started at a trough (a device that is useful for salesmen of stocks, when they wish the public to invest in a particular equity).

A useful method for misleading with pictures is

to depict, say, salaries by means of objects such as cash boxes whose *linear* dimensions are proportional to the salaries. In this way an increase is made to appear much larger than it really is. Another useful method for misleading the public is attributed by Huff (1954) to the First National Bank of Boston. The bank represented governmental expenditure by means of a map of the United States in which states of low population densities were shaded to indicate that total government spending was equal to the combined income of the people of those states. The hope was that the reader would get the impression that federal spending, as a fraction of the total income of the United States, was equal to the total area of the shaded states divided by the whole area of the country. [*See* GRAPHIC PRESENTATION.]

"Smaller" versus "smaller than necessary." The confusion of "smaller" with "smaller than necessary" will be illustrated in a hereditary context, and an oversimplification of the theory of natural selection will be pointed out. Let us suppose that it is true that intelligent people tend to have fewer children than less intelligent people and that the level of intelligence is hereditary. (We are not here concerned with whether and where this supposition is true, nor with the precise interpretation of "intelligent.") It then appears to follow that the average level of intelligence will necessarily decline. This fallacy will be perpetrated on most readers of Chapter 5 of the book by the eminent zoologist Peter B. Medawar (1960, p. 86), in spite of the words italicized by us in the following quotation: "If innately unintelligent people tend to have larger families, then, *with some qualifications*, we can infer that the average level of intelligence will decline." In order to show that the argument without the qualification is invalid it is sufficient to use a mathematical model that, for other purposes, would be much oversimplified (see Behrens 1963). Imagine a population in which 10 per cent of men are intelligent and 90 per cent are unintelligent and that, on the average, 100 intelligent fathers have 46 sons, of whom 28 are intelligent and 18 unintelligent, whereas 100 unintelligent fathers have 106 sons, of whom 98 are unintelligent and 8 are intelligent. It will be seen from Table 1 that the proportion of intelligent males would remain steady in expectation.

But now it must be determined whether the right question is being asked. Suppose we were convinced that the general level of intelligence was decreasing, and we made suggestions accordingly for encouraging the more intelligent to have more children. Should we not put these suggestions forward even if the general level of intelligence were *increasing*? Would we not like to see the rate of increase also increase? Yes, of course. Looking from this point of view, we might not fully agree with Medawar's arguments, but we might well agree with some of his recommendations.

Table 1 — Hypothetical proportions of intelligent and unintelligent sons

		SONS	
		Intelligent	*Unintelligent*
FATHERS	*100 intelligent*	28	18
	900 unintelligent	72	882
	Total	100	900

"Regression fallacy." If we select a short or tall person at random, the chances are that his relatives will be closer in average height than he is to the mean height of the population. Francis Galton described this phenomenon as "regression." If now we consider the heights of the sons of tall men and of short men, we might infer that the variability of heights is decreasing with time. This would be an example of the regression fallacy. One way of seeing that the argument must be fallacious is by considering the heights of the parents of short and tall people: we would then infer that the variability of heights is *increasing* with time!

Wallis and Roberts (1956) mention several other examples of the regression fallacy. One is the widespread belief that the second year in the major leagues is an unlucky one for new baseball players who have successfully finished their first year.

Invalid use of formulas or theorems. The use of formulas or theorems in situations where they are not valid is a special case of the *deus ex machina* class of fallacies and is very frequent. The following are a few examples.

Implicit assumption of independence. In an experiment consisting of n trials, each successful with probability p, is the variance (the square of the standard deviation) of the number of successes equal to $np(1-p)$, as it would be if independence held? (An example would be the quality inspection of items on an assembly line.) The formula is so familiar that it is tempting to assume that it is always a good approximation. But familiarity breeds mistakes. For a Markov chain the variance can be quite different (see, for example, Good 1963), as it can also be when sampling features of children in families or fruit on trees.

Another example of a fallacious assumption of independence relates to the variability of physiological traits. Why, even if there were only eight traits, each trichotomized into equal thirds, only one person out of $3^8 = 6561$ would be in the middle (normal) group for all eight traits!

Assuming form determines distribution. Let n_{ij} be the frequency of the "dinome," that is, pair of adjacent digits (i,j), in a sequence of N random sampling digits $(i,j = 0,1,2, \cdots, 9)$. Clearly $\sum n_{ij} = N - 1$. Let

$$\psi^2 = \frac{10}{N-1} \sum_{i,j} \left(n_{ij} - \frac{N-1}{100} \right)^2.$$

It has been erroneously assumed at least four times in the statistical literature that ψ^2 has asymptotically (for large N) a tabular chi-squared distribution. In one case this led to the unfair rejection of a method of producing pseudo random numbers. Presumably the erroneous distribution arose from the typographical identity of the expression for ψ^2 with the familiar statistic of the chi-square test. (For references to three of these papers and to a paper that gives a correct method of using ψ^2, see Good 1963.) The misapplication of the above so-called serial test is particularly disastrous when working with binary digits (0 and 1), that is, with base 2.

Assuming the winner leads half the time. There is a fallacy in assuming that in a long sequence of statistically independent fair games of chance between two players the ultimate winner will be in the lead about half the time. This is a misapplication of the law of large numbers. That it *is* a fallacy depends on one of the most surprising theorems in the theory of probability, the so-called arc sine law. In fact, the probability that a specified player will be in the lead for less than a fraction x of the time is approximately $(2/\pi)$ arc sin $x^{\frac{1}{2}}$ (see, for example, Feller 1950–1966, vol. 1, p. 251). This implies that however long the game, it is much more likely that a specified player will be ahead most of the time or behind most of the time than that he will be about even; for example, the probability that a specific player will be ahead 90 per cent or more of the time, or behind 90 per cent or more of the time, is about .40, while the probability that the player will be ahead between 40 and 60 per cent of the time is only about .13. As Feller says, the arc sine law "should serve as a warning to those who easily discover 'obvious' secular trends" in economic and social phenomena.

The "maturity of the chances." An elementary misapplication of the law of large numbers, or "law of averages," is known as the maturity of the chances. In World War I many soldiers took shelter in bomb craters on the grounds that two bombs seldom hit the same spot. For the same reason P. S. Milner-Barry, the British chess master, decided to retain his London flat after it was bombed in World War II. As a matter of fact it *was* bombed again. At roulette tables, it is said, the chips pile up on the color that has not occurred much in recent spins. Of course, in practice, if a coin came down heads fifty times running, it would be *more* likely than not, in logical probability, to come down heads on the next spin, not *less* likely. In fact, it would probably be double-headed. There are circumstances, of course, when an event is less likely to occur soon after it has just occurred: this would be true for some kinds of accidents and in many situations where one is sampling without replacement. Usually the question is basically empirical, but the expression "maturity of the chances," or "Monte Carlo fallacy," usually refers to sequences of events that are statistically independent, at least to a good approximation.

Law of large numbers misapplied to pairs. A mnemonic for the fallacy of misapplying the law of large numbers when considering pairs of objects selected from a set is the well-known "birthday problem." If 24 people are selected at random, then it is more likely than not that at least one pair of them will have the same birthday (that is, month and day). This is simple to prove, but a good intuitive "reason" for it is that the number of pairs of people in a group of 24 people is 276, and $\exp(-276/365) < \frac{1}{2}$. (The crude argument here is based on a Poisson approximation to the probability of no "successes" in 276 roughly independent trials with common success probability 1/365.) The result is true a fortiori if births are not distributed uniformly over days of the year.

C. R. Hewitt, in a "Science Survey" program of the British Broadcasting Corporation in February, 1951, stated that the probability is less than 1/64,000,000,000 that two fingerprints of different people will be indistinguishable. From this he inferred that no two people have indistinguishable fingerprints and thus committed the birthday fallacy. The argument is fallacious even if we ignore resemblances of fingerprints among relatives, since the number of *pairs* of people in the world exceeds 4,000,000,000,000,000,000. The conclusion might be correct.

A similar fallacy arises in connection with precognition. Suppose, entirely unrealistically, that there is just one remarkable and well-documented case of somebody in the world having an apparently precognitive dream. How small must the apparent probability be in order that the report, if true, should by itself convince us of precognition? Presumably its reciprocal should be at least of the order of the population of the world times the number of dream experiences of a man times the number of his waking experiences. This triple product might be as large as 1,000,000,000,000,000,000,000,000,000 (or 10^{27}). This informal application of statistics should discourage a too ready assumption that the evidence from apparently precognitive dreams is overwhelming. A formal application of statistical methods to this problem is very difficult. This discussion is not intended to undermine a belief in the possibility of precognition, but it is a plea for a better evaluation of the evidence.

Failure to use precise notation. An example of the fallacy of failure to use sufficiently explicit notation is given by the "fiducial argument." The purpose of R. A. Fisher's fiducial argument (1956, pp. 52–54) was to produce a final (posterior) distribution for a parameter without assuming an initial (prior) distribution for it. This was ambitious, to say the least, since *de nihilo nihilum.*

The argument starts off from a parametric distribution for a random variable, X. Fisher selected an example of which the following is a special case. For each positive number x_0, suppose that

$$P(X > x_0 \mid \theta) = \exp(-x_0\theta),$$

where θ is a positive parameter in whose value and final distribution we are interested. Writing $x_0 = u/\theta$, we get $P(X\theta > u \mid \theta) = \exp(-u)$. From this it can be proved, using the usual axioms of probability theory (although Fisher omitted the proof), that $P(X\theta > u) = \exp(-u)$ for any positive number, u, provided that an initial distribution for θ is assumed to exist. (It is not necessary to assume that this distribution is in any sense known.) Hence $P(\theta > \theta_0) = \exp(-x\theta_0)$, where $\theta_0 = u/x$. Fisher infers from this that

$$P(\theta > \theta_0 \mid x) = \exp(-x\theta_0),$$

where θ_0 is any real positive number. But this last equation does not follow from the axioms of probability unless the initial probability density of θ is proportional to $1/\theta$. The fallacy in the fiducial argument was due to Fisher's failure to indicate what is "given" in his probability notation. So great was Fisher's authority that there are still many statisticians who make use of the fiducial argument; thus the analysis given here is currently considered controversial [*see* FIDUCIAL INFERENCE].

[7]*Assuming order of operations reversible.* An example of the fallacy of assuming that the order of two mathematical operations can be interchanged is the assumption that the expectation of a square is equal to the square of the expectation. This occurs in M. J. Moroney (1951, p. 250), where he says that evidently the expected value of chi-square for a multinomial distribution is zero.

Correlation and causation. *Positive correlation does not imply causation, either way round.* There is a positive correlation between the number of maiden aunts one has and the proportion of calcium in one's bones. But you cannot acquire more maiden aunts by eating calcium tablets. (Younger people tend to have more maiden aunts and more bone calcium.) In New Hebrides people in good health are lousier than people with fever. The advice to acquire lice cannot be rationally given, since lice avoid hot bodies [*see* Huff 1954, p. 99; *see also* CAUSATION].

Zero correlation does not imply statistical independence, although it does so for a bivariate normal distribution and for some other special families of distributions.

If there is a positive correlation between A and B and also between B and C, this does not imply that the correlation between A and C is positive, even for a trivariate normal distribution. But the implication *does* follow if the sum of the squares of the first two correlation coefficients exceeds unity.

If the time order is wrong, then causation is unlikely, to say the least. In one survey vaccination was found to be positively correlated with various infectious diseases, when one looked at different districts in India. This was used by antivaccinationists for propaganda. If they had not been emotionally involved, they would probably have noticed that in several districts increased vaccination had *followed* an increase in the incidence of disease (Chambers 1965).

Post hoc, ergo propter hoc ("after this, therefore because of this"). D. O. Moberg, in a lecture in Oxford on February 2, 1965, stated that premarital intercourse seemed to be positively correlated with divorce and inferred that the propensity to divorce was increased by premarital intercourse. The inference *might* be correct, but an equally good explanation is that premarital intercourse and divorce are both largely consequences of the same attitude toward the institution of matrimony. It is also possible that untruthful responses are associated with a propensity to divorce or with a propensity to avoid divorce.

Ecological correlation. Suppose we find that in American *cities* the illiteracy rate and the percentage of foreign-born are associated. This does not imply the same association for *individuals* (see Goodman 1959). It would even be possible that every foreign-born person was highly literate. Cities might attract foreign-born people and also attract or produce illiteracy.

Wrong criteria for suboptimization. Granted that in most decision problems it is not so much a matter of optimization as of "suboptimization," that is, of approximate optimization, there is still an acute problem in choosing *what* to suboptimize. Various fallacies arise through choosing a wrong criterion or through not using a criterion at all (see Koopman 1956; Good 1962). Often a criterion is selected from too narrow a point of view, ignoring questions of consistency with higher-level criteria. For example, when coeducation at New College, Oxford, was being discussed at another Oxford college, the question of the relative requirements for education of men and women was ignored, but the effect on the atmosphere of the senior common room was mentioned. Another fallacy is to ignore the "spillover," or side effects, of some project. Sometimes, when an urgent decision is required, the cost in delay of detailed theory is unjustifiably ignored. At other times the cost of the theory is said to be too heavy, and the fact is overlooked that the results of this theory might be valuable in similar circumstances in the future and that the training of the theoretician is important. Sometimes the criterion of profitability is given too little weight, sometimes too much (see also McKean 1958; Hitch & McKean 1954).

[8] **Statistics of statistical fallacies.** There is some unpublished work by Christopher Scott on the statistics of statistical fallacies and errors for the specialized field of sample surveys conducted by mail. Scott read the 117 articles and research reports that had been written in English on this topic up to the end of 1960. He excluded 22 of the reports either because they were duplicates of others or because they gave almost no details of method. Of the remaining 95 articles, he found one or more definite errors in 54 and definite shortcomings in another 13. Among the definite errors there were 14 cases in which the experimental variable was not successfully isolated, that is, a change in technique was reported as causing a change in the result, whereas the latter change could reasonably be ascribed to variation in some concomitant variable. There were 9 cases in which obviously relevant data, such as sample size or response rate, were not reported, and 7 cases in which a necessary significance test was not given. There is not space here for further details, and hopefully they will be published elsewhere.

For misuses of the chi-square test, see Lewis and Burke (1949).

Good fallacies. It is a fallacy to suppose that all fallacies are bad. A clearly self-contradictory epigram can be a neat way of conveying truth or advice, to everybody except quibblers. For example:

"Only a half-truth can be expressed in a nutshell."

"Everything in moderation."

"It would be a *non sequitur* if it were not a tautology."

"Races in which people were immortal became extinct by natural selection."

"There's nothing wrong with chess players that not being people wouldn't put right."

In this article it has been necessary to omit reference to many kinds of fallacies. A more complete listing is given in the categorization of logical and statistical fallacies by Good (1959*b*).

IRVING JOHN GOOD

[*See also* ERRORS, *article on* NONSAMPLING ERRORS.]

BIBLIOGRAPHY

Further literature on fallacies is mentioned in Good 1959b. *In particular,* Thouless 1932 *for fallacies in ordinary reasoning and chapter 3 of* Wallis & Roberts 1956 *for fallacies in statistics are both very useful.* Wagemann 1935 *also gives an interesting general treatment.*

ACKERMANN, ALFRED S. E. (1907) 1950 *Popular Fallacies: A Book of Common Errors, Explained and Corrected With Copious References to Authorities.* 4th ed. London: Old Westminster Press.

BEHRENS, D. J. 1963 High IQ, Low Fertility? Statistical "Non Sequitur." *Mensa Correspondence* (London) no. 50:6 only.

CARTTER, ALLAN M. 1965 A New Look at the Supply of College Teachers. *Educational Record* 46:267–277.

CHAMBERS, S. PAUL 1965 Statistics and Intellectual Integrity. *Journal of the Royal Statistical Society* Series A 128:1–15.

○FELLER, WILLIAM (1950–1960) 1968–1971 *An Introduction to Probability Theory and Its Applications.* 2 vols. New York: Wiley. → The third edition of Volume 1 was published in 1968, the second edition of Volume 2 in 1971.

○FISHER, R. A. (1935) 1971 *The Design of Experiments.* 9th ed. New York: Hafner; Edinburgh: Oliver & Boyd.

○FISHER, R. A. (1956) 1973 *Statistical Methods and Scientific Inference.* 3d ed., rev. & enl. New York: Hafner; Edinburgh: Oliver & Boyd.

GOOD, I. J. 1950 *Probability and the Weighing of Evidence.* London: Griffin.

GOOD, I. J. 1956 Which Comes First, Probability or Statistics? *Journal of the Institute of Actuaries* 82:249–255.

GOOD, I. J. 1958*a* Significance Tests in Parallel and in Series. *Journal of the American Statistical Association* 53:799–813.

GOOD, I. J. 1958*b* How Much Science Can You Have at Your Fingertips? *IBM Journal of Research and Development* 2:282–288.

GOOD, I. J. 1959*a* Kinds of Probability. *Science* New Series 129:443–447.

GOOD, I. J. (1959*b*) 1962 A Classification of Fallacious Arguments and Interpretations. *Technometrics* 4:125–132. → First published in Volume 11 of *Methodos.*

GOOD, I. J. 1960–1961 The Paradox of Confirmation. *British Journal for the Philosophy of Science* 11:145–149; 12:63–64.

GOOD, I. J. (1962) 1965 How Rational Should a Manager Be? Pages 88–98 in *Executive Readings in Management Science*. Edited by Martin K. Starr. New York: Macmillan. → First published in Volume 8 of *Management Science*.

GOOD, I. J. 1963 Quadratics in Markov-chain Frequencies, and the Binary Chain of Order 2. *Journal of the Royal Statistical Society* Series B 25:383–391.

GOOD, I. J. 1965 *The Estimation of Probabilities: An Essay in Modern Bayesian Methods*. Cambridge, Mass.: M.I.T. Press.

GOODMAN, LEO A. 1959 Some Alternatives to Ecological Correlation. *American Journal of Sociology* 64:610–625.

►GORDON, R. A. 1968 Employment and Unemployment. Volume 5, pages 49–60 in *International Encyclopedia of the Social Sciences*. Edited by David L. Sills. New York: Macmillan and Free Press.

HITCH, CHARLES; and MCKEAN, RONALD 1954 Suboptimization in Operations Problems. Volume 1, pages 168–186 in *Operations Research for Management*. Edited by Joseph F. McCloskey and Florence N. Trefethen. Baltimore: Johns Hopkins Press.

HUFF, DARREL 1954 *How to Lie With Statistics*. New York: Norton. → Also published in paperback edition.

JACKSON, JOHN H. (1931) 1958 *Selected Writings of John Hughlings Jackson*. Vol. 2. Edited by James Taylor. New York: Basic Books.

KIMBALL, A. W. 1957 Errors of the Third Kind in Statistical Consulting. *Journal of the American Statistical Association* 52:133–142.

KOOPMAN, B. O. 1956 Fallacies in Operations Research. *Journal of the Operations Research Society of America* 4:422–426.

LEWIS, D.; and BURKE, C. J. 1949 The Use and Misuse of the Chi-square Test. *Psychological Bulletin* 46:433–489. → Discussions of the article may be found in subsequent issues of this bulletin: 47:331–337, 338–340, 341–346, 347–355; 48:81–82.

MCKEAN, RONALD N. 1958 The Criterion Problem. Pages 25–49 in Ronald N. McKean, *Efficiency in Government Through Systems Analysis*. New York: Wiley.

MEDAWAR, PETER B. 1960 *The Future of Man*. New York: Basic Books; London: Methuen.

MORONEY, M. J. (1951) 1958 *Facts From Figures*. 3d ed., rev. Harmondsworth (England): Penguin.

PENFIELD, WILDER; and ROBERTS, LAMAR 1959 *Speech and Brain-mechanisms*. Princeton Univ. Press.

ROWE, P. 1962 What the Dons Earn. *The Sunday Times* (London) October 21.

SLATER, ELIOT; and BEARD, A. W. 1963 The Schizophrenia-like Psychoses of Epilepsy: Psychiatric Aspects. *British Journal of Psychiatry* 109:95–112.

STARNES, RICHARD 1962 Age of Falsehood. *Trenton Evening Times* December 19.

STERLING, THEODORE D. 1959 Publication Decisions and Their Possible Effects on Inferences Drawn From Tests of Significance—Or Vice Versa. *Journal of the American Statistical Association* 54:30–34.

THOULESS, ROBERT H. (1932) 1947 *How to Think Straight*. New York: Simon & Schuster. → First published as *Straight and Crooked Thinking*.

WAGEMANN, ERNST F. (1935) 1950 *Narrenspiegel der Statistik; Die Umrisse eines statistischen Weltbildes*. 3d ed. Salzburg (Austria): Verlag "Das Bergland-Buch."

WALLIS, W. ALLEN; and ROBERTS, HARRY V. 1956 *Statistics: A New Approach*. Glencoe, Ill.: Free Press. → An abridged paperback edition was published in 1965 by The Free Press.

Postscript

[1]**Ignoring the "exposure base."** Compare the headline of a 1966 Chicago newspaper article, "Motorcycle fatalities up sixty per cent in three years," with the comment in its last paragraph: "Deaths per 100,000 vehicles declined from 118 to 115 between 1961 and 1964. The number of registered motor driven cycles increased. . . ." In another Chicago newspaper story, Sydney J. Harris (1969) also commits the fallacy of ignoring the exposure base when he says, "You are far safer in a lonely park late at night than in your own home. . . . Lust, jealousy, rage, injured feelings, resentments and frustrations—these account for more corpses in a month than all the killings for simple gain in a year." I wonder if any of his readers, or even Harris himself, decided to sleep on park benches during the summer.

[2]**The suppression of (relevant) information.** In 1969 an advertisement appeared for a drug called Preludin stating that 93 obese patients were given this drug or a placebo, alternating between these every four weeks. "Sixty-one per cent of the patients [57 patients] lost more weight on Preludin than on placebo. In fact they lost an average 1.9 pounds per week—almost four times as much as when they were on placebo!" Yes, but what about the other 36 patients when they took Preludin? Did they lose weight, gain weight, stay the same, suffer side effects, or what?

[3]**Terminological ambiguities.** Rensberger (1975) mentions some research by Evan Zaidel and Roger Sperry that is relevant to the discussion in the main article concerning speech and the brain. It appears that the right hemisphere of the adult's brain, "once thought to lack much language ability, has the vocabulary of a 14-year old and the syntactical ability of a 5-year old." So Hughlings Jackson may have been more correct in his conclusion than I had supposed, although his reasons were weak.

The mixing of two distinct populations when estimating a percentage: the Cohen–Nagel paradox. Suppose that a treatment of some disease is harmful for both women and men in the sense of increasing the probability of death, as compared with no treatment. Suppose further that two samples, S_1 and S_2, are taken and that treatment is applied only in S_2. Then it can happen that the fraction of fatalities is lower in S_2, although the samples are so large that the effects of random variation on the percentages can be ignored. This "paradox" comes about if the probability of death among women is lower than for men and if S_2 contains a higher percentage of women than does S_1. The paradox might be due to Cohen and Nagel (1934, p. 449). It would not arise if S_1 and S_2 were selected by the same random sampling procedure. An exposition of this paradox, and variations, is given by Gardner (1976) and by Blyth (1972*a*; 1972*b*); a discussion in the context of alleged sex discrimination by universities is presented by Bickel et al. (1975).

[4]**Ignoring a relevant concomitant variable.** Uhr (1959) obtained evidence that women were less good drivers than men, but he did not consider whether women have less practice.

[5]*Ignoring half of a contingency table.* Bishop, Fienberg, and Holland (1975) is an important reference for multidimensional contingency tables.

[6]**The suppression of the uninteresting and the interpretation of significance tests.** Tullock (1959) argues that "a man who has devoted a good deal of time to testing a hypothesis with a given experimental design and has been forced to conclude that the hypothesis is incorrect will normally simply file his results. If, however, he sees an article reporting a similar experiment with significant results, he will at least write to the editor of the journal in which it appears. Since this type of comment on articles is rare, I judge that the duplication of experimental designs is rare."

Quite often a tail-area of 0.05 corresponds in a Bayesian analysis to multiplying the odds of the non-null hypothesis by some 4 or 5. Let x be the proportion of experiments that are performed for which the null hypothesis is (approximately) true. If x exceeds 0.8, a published tail-area probability of exactly 0.05 would often make the non-null hypothesis only about as probable as the null hypothesis. On any given occasion there would be more information on which to base a Bayesian analysis, including some formulation of the non-null hypotheses, but it would be interesting to know more about the numerical value of x. Since experiments that fail to reject a null hypothesis are not usually published, a survey for estimating x cannot readily be based on published literature.

Berkson (1942) argues that it is a fallacy to say that a low P-value is evidence against the null hypothesis unless an alternative can be suggested that would make this P-value more probable. He argues further that middle values of P, such as 0.6, can support the null hypothesis. His views agree with the conclusions typically reached in a Bayesian analysis.

[7]**Assuming order of operations reversible.** Mackie (1967) points out that many fallacies can be classified under this heading, for example, the following fallacious proof of determinism: "Necessarily either you will go or you will stay; so either you will go necessarily or you will stay necessarily."

[8]**Statistics of statistical fallacies.** Lana and Lubin (1962) say that there is no change in the proportion of articles that use the χ^2 test in the *Journal of Experimental Psychology*, since the article by Lewis and Burke (1949), but that there is an apparent increase in the proportion of correct usages from 21 to 64 per cent.

The fallacy of false opposition. Ingle (1972) says, "It is an error to assume that because a given theory expresses an important truth every other theory is false. . . . During the early 1930's there was a hypothesis that the hormone of the adrenal cortex affected the metabolism of carbohydrates and the rival hypothesis that the hormone affected electrolyte metabolism. . . . There was an important truth in each hypothesis. . . ."

For complex systems there is the danger in the too ready application of "Ockham's lobotomy." For example, people have been tempted to assume that the operation of the brain is either serial or parallel, but there are reasons for supposing it is both (Good 1971; Noton & Stark 1971).

Incorrect inductive reasoning. Scientific induction is still a controversial philosophical area and provides examples of subtle fallacious reasoning.

Vanishing prior probabilities. Popper (1959, p. 363) argued that "in an infinite universe . . . the probability of any (nontautological) universal law will be zero." The argument is refuted by Good (1975, p. 49).

The paradox of confirmation. Let H be a hypothesis of the form that all x's belong to a class C, for example, that all crows are black. It is usually assumed that seeing an instance of H (a black crow) necessarily supports H. That this is false is shown by Good (1967; 1968). Another useful reference is Schlesinger (1975).

Misunderstanding of conditional probability. It is said there is a man who carries a bomb when he travels by air, on the grounds that it is extremely unlikely that there would be two independent bombs on the same plane.

Prospective versus retrospective studies. When studying whether circumstances Q tend to cause others U it is often *easier* to sample cases of U and work backwards (retrospective studies) than the other way round (prospective studies), but the latter are often much more informative per sampling unit. The mere choice of the symbols "Q" and "U" is mnemonic, for in English "U" nearly always follows "Q", but it does not follow that "Q" nearly always precedes "U". The tendency of Q to "cause" U is large only if the probability that U follows Q is close to 1 while the probability that U follows not-Q is much less close to 1. One is often interested in the values of these two conditional probabilities, $P(U|Q)$ and $P(U|\text{not-}Q)$, without wishing to assign a quantitative measure to the degree of causation. These probabilities cannot be inferred solely from the values of $P(Q|U)$ and $P(Q|\text{not-}U)$.

Bortkiewicz (1911) makes the point in the following manner when attacking some work by Gerring (on the breakup of estates): "We might sample a group of idiots and find that their grandparents were mostly normal. It would certainly be absurd to infer from this that most of the grandchildren of normal people are idiotic."

But retrospective studies do not need to be used as idiotically as that. A retrospective study can involve so much less effort than a prospective one that its value per unit effort can be greater than a prospective study, especially of course when a prospective study is impracticable. In formal terms, $P(U|Q)/P(U) = P(Q|U)/P(Q)$, so if we discover by a retrospective study that $P(Q|U) > P(Q)$ then we can deduce that $P(U|Q) > P(U)$, and therefore that Q is a probabilistic cause of U, but not necessarily a strong one.

A probabilistic cause need not be a strong one in order to be important. Joshua Lederberg has mentioned to me the striking example of work by Herbst et al. (1971; 1975), who found that many mothers of daughters with vaginal cancer (event U) had been treated with DES during pregnancy (event Q). The word "many" here must be taken to imply that $P(Q|U)/P(Q)$ is much larger than 1, so that $P(U|Q)/P(U)$ is much larger than 1. Although this does not show that $P(U|Q)$ is close to 1, it does show (if the observed facts are clear and without measurement artifact) that the administration of DES during pregnancy can cause much damage.

Another special circumstance occurs when Q and U are the same feature of relatives, such as round-headedness of both grandparent and grandchild. For some features, $P(Q) \cong P(U)$, which implies that $P(Q|U) \cong P(U|Q)$; if this probability is close to 1, then prospective and retrospective studies are roughly equally informative. Of course, $P(Q)$ and $P(U)$ are not necessarily equal for all features, because some features are strongly associated with fertility, positively or negatively, so that their relative frequency in the population changes over generations.

Remarks on references. De Morgan (1847) gives examples of fallacious reasoning in ordinary language and especially condemns inaccurate quotations, which he thinks ought to be clarified by saying, for example, "Cicero (cited by Bacon) says, '. . .'." I think inaccurate quotations are acceptable if it is made clear that they are inaccurate, perhaps by using a new kind of quotation mark. De Morgan (1872) provides a rambling, entertaining collection of articles, including book reviews, from the weekly periodical the *Athenaeum*, on literary, social, scientific, and mathematical questions, including many examples of fallacious reasoning. He is somewhat obsessed by circle squarers and by the Royal Society, to which he was unjustly never elected.

Fischer (1970) gives an account of fallacies in historical writings, including a few statistical fallacies. This book is very interesting and well written but is unfortunately often sarcastic and dogmatic; one wonders whether any historical writing could escape its strictures.

Ingle (1972) lists about a hundred kinds of fallacy perpetrated in biological and medical science, including a few statistical ones, together with relevant hilarious quotations from Lewis Carroll. He gives only a few precise examples, but there is no reason to doubt that most of his classes of fallacy actually occur.

Mackie (1967) provides a highly succinct general account of fallacies in reasoning, with special emphasis on formal logic and philosophy, and mentions some further useful references in his bibliographical summary.

Moran (1973) gives a useful list of fifteen statistical fallacies, some of which are less well known than many of those in the present article.

IRVING JOHN GOOD

ADDITIONAL BIBLIOGRAPHY

BENTHAM, JEREMY (1824) 1971 *Bentham's Handbook of Political Fallacies*. Edited by Harold A. Larrabee. New York: Apollo. → Also published by Peter Smith.

BERKSON, JOSEPH 1942 Tests of Significance Considered as Evidence. *Journal of the American Statistical Association* 37:325–335.

BICKEL, P. J.; HAMMEL, E. A.; and O'CONNELL, J. W. 1975 Sex Bias in Graduate Admissions: Data from Berkeley. *Science* 187:398–404.

BISHOP, YVONNE M. M.; FIENBERG, STEPHEN E.; and HOLLAND, PAUL W. 1975 *Discrete Multivariate Analysis*. Cambridge, Mass.: M.I.T. Press.

BLYTH, COLIN R. 1972*a* On Simpson's Paradox and the Sure-thing Principle. *Journal of the American Statistical Association* 67:364–366.

BLYTH, COLIN R. 1972b Some Probability Paradoxes in Choice From Among Random Alternatives. *Journal of the American Statistical Association* 67:366–373. → Comments and a rejoinder by Blyth appear on pages 373–381.

BORTKIEWICZ, LADISLAUS VON 1911 Discussion in a Conference. *Schriften des Verein(s) für Sozialpolitik* 138: 175–176.

COHEN, MORRIS R.; and NAGEL, ERNEST 1934 *An Introduction to Logic and Scientific Method.* New York: Harcourt.

DE MORGAN, AUGUSTUS (1847) 1926 *Formal Logic.* Edited by A. E. Taylor. London: Open Court.

DE MORGAN, AUGUSTUS (1872) 1915 *A Budget of Paradoxes.* 2 vols. 2d ed. Edited by David Eugene Smith. Chicago and London: Open Court. → A facsimile edition is published by Books for Libraries.

FISCHER, DAVID HACKETT 1970 *Historians' Fallacies: Toward a Logic of Historical Thought.* New York: Harper & Row.

GARDNER, MARTIN 1976 Mathematical Games: On the Fabric of Inductive Logic, and Some Probability Paradoxes. *Scientific American* 234, March:119–124.

GOOD, I. J. 1961–1962 A Causal Calculus. *British Journal for the Philosophy of Science* 11:305–318, 12:43–51; 13:88 only.

GOOD, I. J. 1967 The White Shoe Is a Red Herring. *British Journal for the Philosophy of Science* 17:322 only.

GOOD, I. J. 1968 The White Shoe *qua* Herring Is Pink. *British Journal for the Philosophy of Science* 19:156–157.

GOOD, I. J. 1971 [Letter.] *Scientific American* 225, Oct.:8 only.

GOOD, I. J. 1975 Explicativity, Corroboration, and the Relative Odds of Hypotheses. *Synthese* 30:39–73.

HAMBLIN, CHARLES L. 1970 *Fallacies.* London: Methuen.

HARRIS, SYDNEY J. 1969 You're Safer in a Lonely Park. *Chicago Daily News,* March 31.

HERBST, ARTHUR L. et al. 1971 Adenocarcinoma of the Vagina: Association of Maternal Stilbestrol Therapy With Tumor Appearance in Young Women. *New England Journal of Medicine* 284, no. 16:878–881.

HERBST, ARTHUR L. et al. 1975 Prenatal Exposure to Stilbestrol: A Prospective Comparison of Exposed Female Offspring With Unexposed Controls. *New England Journal of Medicine* 292, no. 7:334–339.

HILL, AUSTIN BRADFORD (1937) 1971 *Principles of Medical Statistics.* 9th ed., rev. & enl. London: The Lancet. → See especially pages 274–308.

INGLE, DWIGHT J. 1972 Fallacies and Errors in the Wonderlands of Biology, Medicine, and Lewis Carroll. *Perspectives in Biology and Medicine* 15:254–281.

LANA, ROBERT E.; and LUBIN, ARDIE 1962 Chi Square Revisited. *American Psychologist* 17:793 only.

MACKIE, J. L. 1967 Fallacies. Volume 3, pages 169–179 in *The Encyclopedia of Philosophy.* Edited by Paul Edwards. New York: Macmillan and Free Press.

MILL, JOHN STUART (1843) 1974 On Fallacies. Volume 8, pages 733–830 in *The Collected Works of John Stuart Mill.* Edited by J. M. Robson. Univ. of Toronto Press; London: Routledge & Kegan Paul.

MORAN, P. A. P. 1973 Problems and Mistakes in Statistical Analysis. *Communications in Statistics* 2:245–257.

Motorcycle Fatalities Up Sixty Per Cent in Three Years. 1966 *Chicago Tribune,* Aug. 1, sec. 3, p. 7, col. 4.

NOTON, DAVID; and STARK, LAWRENCE 1971 [Letter.] *Scientific American* 225, Oct.:8 only.

POPPER, KARL R. (1935) 1959 *The Logic of Scientific Discovery.* New York: Basic Books; London: Hutchinson. → First published as *Logik der Forschung.*

RENSBERGER, BOYCE 1975 Language Ability Found in Right Side of the Brain. *New York Times,* July 1, sec. 1, p. 14, cols. 1–2.

SCHLESINGER, GEORGE 1975 *Confirmation and Confirmability.* New York: Oxford Univ. Press.

TULLOCK, GORDON 1959 Publication Decisions and Tests of Significance: A Comment. *Journal of the American Statistical Association* 54:593 only.

UHR, LEONARD 1959 Sex as a Determinant of Driving Skills: Women Drivers! *Journal of Applied Psychology* 43:35 only.

FAST FOURIER TRANSFORM

See TIME SERIES.

FEEDBACK

See MODELS, MATHEMATICAL; SIMULATION; *and the biography of* WIENER.

FIDUCIAL INFERENCE

R. A. Fisher (1930) proposed a statistical method for obtaining from observed data a probability distribution concerning a parameter value; he called the distribution a *fiducial probability distribution.* The theory of confidence intervals, as developed by J. Neyman, was initially presented in the literature as a clarification and development of fiducial probability [*see* ESTIMATION, *article on* CONFIDENCE INTERVALS AND REGIONS]. Fisher denied the equivalence and in his subsequent theoretical papers developed and extended fiducial probability.

As an example, consider a sample of independent measurements, $X_1, \cdots, X_n$, on a physical characteristic μ, and suppose that the measurement error is normally distributed with mean 0 and known variance σ_0^2. Fisher requires that fiducial inference be based on the simplest statistic containing all the information about the parameter, in this case, on the sample mean $\bar{X}$. [*The sample mean here is a minimal sufficient statistic; see* SUFFICIENCY.] The expression $W = \bar{X} - \mu$, involving the variable $\bar{X}$ and the characteristic μ, has a known distribution: normal with mean 0 and variance σ_0^2/n. In an application of the method, the value of $\bar{X}$ is obtained and substituted in the expression $W = \bar{X} - \mu$; the expression is solved for μ in terms of W, giving $\mu = \bar{X} - W$; the fiducial distribution of μ then derives from the known distribution of W:

μ is normal with mean $\bar{X}$ and variance σ_0^2/n. A 95 per cent fiducial interval is $\bar{X} \pm 1.96\sigma_0/\sqrt{n}$.

Fisher claimed that a fiducial probability statement has the same meaning as an ordinary probability statement. In the example, suppose that the 95 per cent fiducial interval as calculated in a specific application is 163.9 ± 0.8. The fiducial statement is that there is 95 per cent probability that the unknown value of μ lies in this interval. The interval 163.9 ± 0.8 is also a 95 per cent confidence interval, but as a confidence interval its interpretation is different [*see* ESTIMATION, *article on* CONFIDENCE INTERVALS AND REGIONS]. Confidence methods and fiducial methods do not, however, always lead to the same numerical results.

The proponents of the confidence method claim that in this context probability statements concerning μ cannot be made; the value of μ is something that exists: either it *is* in the interval or it *is not*, and we don't know which.

The proponents of the fiducial method reply that probabilities concerning realized values are commonplace: in the play of card games, for example, a player may observe his own hand and perhaps other cards (say, those already played) and make a probability statement concerning the distribution of cards in the concealed hands.

The rejoinder is that μ did not arise from a random process such as the card shuffling and dealing. The relevance of this rejoinder is perhaps the key element to criticisms of the fiducial method.

In more complex problems the fiducial method may give a result different than the confidence method. In one prominent problem mentioned below, the Behrens–Fisher problem, the fiducial method gives an answer where confidence methods have not yet produced an entirely satisfactory result.

In his original paper on the fiducial method, Fisher (1930) considered a statistic, T, obtained by the maximum likelihood method for estimating a parameter, θ. Let $F(T, \theta)$ be the cumulative distribution function for the statistic T. The probability density function for T is obtained by differentiating with respect to T: $f(T, \theta) = \partial F(T, \theta)/\partial T$. Correspondingly, the fiducial density function for θ, given an observed value for T, is obtained by differentiating with respect to θ: $g(\theta, T) = \partial F(T, \theta)/\partial \theta$. Fisher illustrated the method with the correlation coefficient r of a sample from a bivariate normal distribution with population correlation coefficient ρ.

For more complex problems, Fisher proposed the use of a *pivotal quantity*, $W = h(T, \theta)$, a function of the statistic T and the parameter θ that has a fixed known distribution regardless of the value of θ. For the first example, $W = \bar{X} - \mu$ is a pivotal quantity. In an application, the observed value of the statistic T is substituted in the expression $W = h(T, \theta)$; the parameter θ is expressed in terms of W; and the fiducial distribution of θ is obtained from the known distribution of W.

Fisher's original method for obtaining a fiducial distribution is a special case of the pivotal method. As a function of a continuous statistic, T, the cumulative distribution function $W = F(T, \theta)$ has a uniform distribution on the interval $(0, 1)$; this relationship is called that of the *probability integral transformation*. The fiducial density of θ for fixed T is obtained from the uniform distribution of W, in the same way as the density of T for fixed θ is obtained by differentiation.

As a second example, consider a random sample, $X_1, \cdots, X_n$, from a normal distribution with mean μ and variance σ^2, both unknown, and suppose that interest centers on the parameter μ. The quantity $t = \sqrt{n}(\bar{X} - \mu)/s_X$, using the sample mean $\bar{X}$ and sample standard deviation s_X, has a known distribution, the *t*-distribution on $n - 1$ degrees of freedom. In an application, the values of $\bar{X}$ and s_X are substituted and the parameter is solved for $\mu = \bar{X} - ts_X/\sqrt{n}$. This equation gives a fiducial distribution for μ that is of *t*-distribution form ($n - 1$ degrees of freedom), located at $\bar{X}$ and scaled by the factor $s_X/\sqrt{n}$.

The Behrens–Fisher problem is an extension of this example. Consider a first random sample, $X_1, \cdots, X_n$, from a normal distribution with mean μ_X and variance σ_X^2, and a second independent random sample, $Y_1, \cdots, Y_m$, from a normal distribution with mean μ_Y and variance σ_Y^2. The Behrens–Fisher problem concerns inference about the parameter difference, $\mu_X - \mu_Y$. The fiducial method gives a distribution described by $\bar{X} - t_1 s_X/\sqrt{n}$ for μ_X and a distribution described by $\bar{Y} - t_2 s_Y/\sqrt{m}$ for μ_Y. (Here t_1 and t_2 are independent t variables with $n - 1$, $m - 1$ degrees of freedom.) The fiducial distribution for $\mu_X - \mu_Y$ is the difference, that of $\bar{X} - \bar{Y} - (t_1 s_X/\sqrt{n} - t_2 s_Y/\sqrt{m})$; some percentage points are given in Fisher and Yates ([1938] 1949, p. 44).

For many problems involving normal and chi-square distributions, the fiducial distribution has the form of a Bayesian posterior distribution as based on a prior distribution with uniformity characteristics [*see* BAYESIAN INFERENCE].

Fisher (1956) considers a wide range of statistical problems and derives the corresponding fiducial distributions.

A central criticism of the fiducial method has been concerned with whether fiducial probabilities are in fact probabilities in an acceptable sense. Some recent analysis, mentioned later, has clarified this question.

Other criticism seems to fall under three headings. First, fiducial probabilities in some examples may not add or integrate to a total of 1. James (1954) and Stein (1959) produce examples that can yield fiducial distributions that do not integrate to 1. Second, in some examples more than one reasonable pivotal quantity may be present; these can lead to several inconsistent fiducial distributions (see Creasy 1954; Fieller 1954; Mauldon 1955). In other examples no reasonable pivotal quantity may be present. Third, if a fiducial distribution from a collection of data is used as a prior distribution for a Bayesian analysis on a second collection of data, the resulting distribution may be different from the fiducial distribution based on the combined collection of data [*see* Lindley 1958; *see also* BAYESIAN INFERENCE].

Fraser (1961) uses transformations to investigate fiducial probability. The transformation approach applies to a large proportion of Fisher's examples, and it introduces an additional range of problems for which fiducial distributions can be obtained.

In a later paper (Fraser 1966) the emphasis in the transformation approach is focused on error variables. Consider the example involving a sample of measurements, $X_1, \cdots, X_n$, on a physical quantity, μ. Let e be a variable describing the error introduced by the measuring instrument: in the example, e is normally distributed with mean 0 and variance σ_0^2. A measurement, X_i, can then be expressed in the form $X_i = \mu + e_i$. Correspondingly, the sample mean takes the form $\bar{X} = \mu + \bar{e}$, where $\bar{e}$ is normally distributed with mean 0 and variance σ_0^2/n. Now consider an application and suppose there is *no information* concerning μ. With no information concerning μ, there is no information concerning $\bar{e}$ other than that describing its distribution. Probability statements can then be made concerning the unknown $\bar{e}$ just as the card player makes statements concerning the realized but unrevealed cards in his opponents' hands.

Suppose the normal distribution of $\bar{e}$ with variance σ_0^2/n gives a 95 per cent probability for $\bar{e}$ lying in the interval 0.0 ± 0.8. Then, *with an observed* $\bar{x} = 163.9$, the probability statement concerning $\bar{e}$ is equivalent to the statement that μ is in the interval 163.9 ± 0.8 with probability 95 per cent.

This analysis involving error variables applies to many of Fisher's examples, and it extends to other problems. The name structural probability has been introduced (Fraser 1966) to distinguish it in cases where the method conflicts with the fiducial method. None of the criticisms mentioned concerning fiducial probability apply to structural probability.

D. A. Sprott (1964) uses a more general class of transformations to analyze a wider range of fiducial distributions.

Some alternative methods have been proposed for obtaining probability distributions concerning parameter values: Dempster (1963; 1966) proposes *direct probabilities*, and Verhagen (1966) proposes *induced probabilities*.

D. A. S. FRASER

[*See also* ESTIMATION, *article on* CONFIDENCE INTERVALS AND REGIONS.]

BIBLIOGRAPHY

A survey of fiducial methods and criticisms may be found in Fraser 1964.

CREASY, MONICA A. 1954 Limits for the Ratio of Means. *Journal of the Royal Statistical Society* Series B 16:186–194.

DEMPSTER, A. P. 1963 On Direct Probabilities. *Journal of the Royal Statistical Society* Series B 25:100–110.

DEMPSTER, A. P. 1966 New Methods for Reasoning Towards Posterior Distributions Based on Sample Data. *Annals of Mathematical Statistics* 37:355–374.

FIELLER, E. C. 1954 Some Problems in Interval Estimation. *Journal of the Royal Statistical Society* Series B 16:175–185.

FISHER, R. A. (1930) 1950 Inverse Probability. Pages 22.527a–22.535 in R. A. Fisher, *Contributions to Mathematical Statistics*. New York: Wiley. → First published in Volume 26 of the *Proceedings* of the Cambridge Philosophical Society.

○FISHER, R. A. (1956) 1973 *Statistical Methods and Scientific Inference*. 3d ed., rev. & enl. New York: Hafner; Edinburgh: Oliver & Boyd.

○FISHER, R. A.; and YATES, FRANK (1936) 1963 *Statistical Tables for Biological, Agricultural and Medical Research*. 6th ed., rev. & enl. New York: Hafner; Edinburgh: Oliver & Boyd.

FRASER, D. A. S. 1961 The Fiducial Method and Invariance. *Biometrika* 48:261–280.

FRASER, D. A. S. 1964 On the Definition of Fiducial Probability. International Statistical Institute, *Bulletin* 40, part 2:842–856.

FRASER, D. A. S. 1966 Structural Probability and a Generalization. *Biometrika* 53:1–9.

JAMES, G. S. 1954 Discussion on the Symposium on Interval Estimation. *Journal of the Royal Statistical Society* Series B 16:214–218.

LINDLEY, D. V. 1958 Fiducial Distributions and Bayes' Theorem. *Journal of the Royal Statistical Society* Series B 20:102–107.

MAULDON, J. G. 1955 Pivotal Quantities for Wishart's and Related Distributions, and a Paradox in Fiducial

Theory. *Journal of the Royal Statistical Society* Series B 17:79–85.

Sprott, D. A. 1961 Similarities Between Likelihoods and Associated Distributions a Posteriori. *Journal of the Royal Statistical Society* Series B 23:460–468.

Sprott, D. A. 1964 A Transformation Model for the Investigation of Fiducial Distributions. International Statistical Institute, *Bulletin* 40, part 2:856–869.

Stein, Charles 1959 An Example of Wide Discrepancy Between Fiducial and Confidence Intervals. *Annals of Mathematical Statistics* 30:877–880.

Verhagen, A. M. W. 1966 The Notion of Induced Probability in Statistical Inference. Division of Mathematical Statistics, Technical Paper No. 21. Unpublished manuscript, Commonwealth Scientific and Industrial Research Organisation, Melbourne, Australia.

FISHER, R. A.

Ronald Aylmer Fisher (1890–1962) achieved world-wide recognition during his lifetime as a statistician and geneticist. He continued the work begun in England by Karl Pearson at the beginning of the twentieth century, but he developed it in new directions. Others also contributed to the tremendous surge in the development of statistical techniques and their application in biology; but these two men, by their energetic research and example, in turn held the distinction of dominating the statistical scene for a generation.

Fisher was born in East Finchley, near London. Apart from a twin brother who did not live long, he was the youngest of seven children. His father, George Fisher, was an auctioneer; no particular scientific ability is evident in the achievements of his relatives, except perhaps those of an uncle who, like Ronald Fisher, was a wrangler in the mathematical tripos at Cambridge. Fisher attended school at Stanmore Park and then went to Harrow, where he was encouraged in his mathematical interests and won a scholarship to Gonville and Caius College, Cambridge. His leanings in mathematics followed the English tradition in natural philosophy, and his university student years, from 1909 to 1913, culminated in his receiving first a distinction in optics for his degree papers in 1912 and then a studentship in physics during his postgraduate year. He had, however, already noticed from studying Karl Pearson's *Mathematical Contributions to the Theory of Evolution* that natural philosophy need not stop with the physical sciences.

After leaving Cambridge, Fisher spent a short time with the Mercantile and General Investment Company. When World War I broke out in 1914, his very bad myopia prevented him from joining the army, and he taught mathematics and physics for four years at various English public schools. In 1917 he married Ruth Eileen Guinness, who was to bear him eight children.

Fisher did not really begin his full-time statistical and biological career until 1919, when he became statistician at Rothamsted Experimental Station, an agricultural research institute in Harpenden, Hertfordshire. His earlier years had, however, been a valuable gestation period. In a short paper published in 1912 while he was still at Cambridge, he had already proposed the method of maximum likelihood for fitting frequency curves. Two more solid papers established his permanent reputation for research. The first was his remarkable paper on the sampling distribution of the correlation coefficient, published in Karl Pearson's journal *Biometrika* in 1915, in which his geometrical powers of reasoning were first fully displayed. The second, published in 1918, examined the correlation between relatives on the basis of Mendelian inheritance and exhibited his ability to resolve crucial problems of statistical genetics. Fisher received from Karl Pearson an offer of a post at University College at the same time that he received the Rothamsted offer, but he wisely chose Rothamsted, largely because of the much greater scope and independence of this new statistical post but also, perhaps, because his contacts with Pearson had not been particularly promising. Pearson was apt to bulldoze his way into research problems without worrying unduly about territorial rights. Having been previously stymied by the correlation distribution problem, he took over Fisher's solution with enthusiasm but without the further close consultation that professional etiquette would seem to require. Fisher was aggrieved by this treatment, and it may well have been the start of the long and bitter feud that developed over the years. Fisher had reason to criticize much of Karl Pearson's work, but the personal animosity that developed between them was something more than a substantive disagreement. As late as 1950, when a selection of Fisher's best statistical papers was published, the omission of the 1915 *Biometrika* paper was a silent reminder of Fisher's feelings.

The period at Rothamsted, from 1919 to 1933, was the most brilliant and productive of Fisher's career. The institute, with its teams of biologists and congenial research atmosphere, was precisely the environment Fisher needed. His own wide range of biological interests enabled him to understand his colleagues' problems and to discuss their statistical aspects constructively with them. His statistical activities were represented by the publication of his best-known book, *Statistical Methods for Research Workers* (1925*a*), which has been published in 13 English editions and also translated into several foreign languages. Fisher's varied accomplishments included doing much of his own computing and initiating many of his own genetical experiments on poultry, snails, and mice, although it is for his creative theoretical ability that he will be remembered. By 1929 he had been elected a fellow of the Royal Society of London. He published his classic on population genetics, *The Genetical Theory of Natural Selection* (1930*a*), in which he did much to reconcile Darwin's theory of evolution by natural selection with Mendel's genetical principles, which were unknown to Darwin. In the later chapters of this work he discussed the theory, first suggested by Francis Galton, of the evolution of a genetic association between infertility and ability. This could result from marriages between those successful because of high innate ability and those successful because of social advantages due to relatively infertile parents' having concentrated their material resources on one or two children. His energetic views included proposals for family allowances proportional both to size of family and to size of income, which would offset the penalty imposed on children by parental fertility.

Fisher's genetic and eugenic interests were soon to be reflected in his move to London in 1933, when he was appointed Galton professor at University College as successor to Karl Pearson. This move, however, no doubt fanned the flames of their feud. When Pearson retired, the college isolated the teaching of statistics in a new department of statistics, under his son Egon; the Galton professor was left with eugenics and biometry. In spite of Egon Pearson's greater tolerance and appreciation of Fisher's new statistical techniques, which emphasized precise methods of analysis in small samples, Fisher felt frustrated, as he indicated at the time in a letter to W. S. Gosset. Moreover, Jerzy Neyman, who held a post in the statistics department from 1935 to 1938, incurred Fisher's wrath by publishing work that Fisher regarded as unnecessary or misguided; their proximity in the same building at University College exacerbated Fisher's sense of injury. The recurrence of feuds of this kind was by now beginning to be as much a manifestation of Fisher's own temperament as of his antagonists'. His wide interests and strong personality made him a charming and lively companion when he chose to be and a generous colleague to those who were in sympathy with his work, as many have testified. But his emotions as well as his intellect were too bound up in his work for him to tolerate criticism, to which he replied in vigorous and sometimes quite unfair terms. Apart from such lapses from objectivity, Fisher proceeded to consolidate his scientific reputation both by the development of the study of genetics (especially human genetics) in his department and by the continued publication of statistical works, such as *The Design of Experiments* (1935*a*), *Statistical Tables for Biological, Agricultural, and Medical Research* (with F. Yates, in 1938), and further original papers in his departmental journal (the *Annals of Eugenics*) and elsewhere.

The third main phase of Fisher's scientific career was his appointment to the Arthur Balfour chair of genetics at Cambridge, from 1943 until his retirement in 1957. During this time he wrote two more books—*The Theory of Inbreeding*, in 1949, and *Statistical Methods and Scientific Inference*, in 1956—and also edited the collection of his papers published in 1950 (see 1920–1945); but most of his important work was already under way. Honors continued to accumulate, including three medals from the Royal Society (a royal medal in 1938, a Darwin medal in 1948, and a Copley medal in 1955), a knighthood in 1952. Shortly before he retired, he became master of his Cambridge college. He was an honorary member of the American Academy of Arts and Sciences, a foreign associate of the National Academy of Sciences, a member of the Pontifical Academy of Sciences, and a foreign member of the Royal Danish Academy of Sciences and Letters and of the Royal Swedish Academy of Sciences.

After retirement he visited the division of mathematical statistics of the Commonwealth Scientific and Industrial Research Organisation in Adelaide, Australia, where he was a research fellow at the time of his death.

Contributions to statistics and genetics

Statistics. To turn in somewhat more detail to Fisher's original work, a formal listing of his contributions to statistics item by item might result in an underemphasis of the strength of their impact, which was due to their simultaneous variety and depth. Moreover, a formal listing is unsatisfactory since the intimate relation of Fisher's contributions to the practical problems arising from his professional environment sometimes meant that their academic presentation was incomplete or late, or both. Work of great value, such as the technique of analysis of variance, received inadequate discussion in *Statistical Methods for Research Workers* because it had hardly reached any degree of finality when this book was published in 1925, and it was still rather cluttered with ideas of intraclass correlation; apparently, Fisher never bothered to redraft the discussion for the later editions.

Excluding Fisher's mathematical work in genetics, it is nevertheless convenient to try to list his chief contributions to statistical theory under two main headings: (1) fundamental work in statistical inference, and (2) statistical methodology and technique. The first group would include his important work on statistical estimation, mainly represented by two papers, one published in the *Philosophical Transactions* of the Royal Society (1922), and the other published in the *Proceedings* of the Cambridge Philosophical Society (1925*b*). Before writing these papers, Fisher had already been much concerned with precise inference in small samples for familiar quantities, such as the correlation coefficient, chi-square, etc., and had produced a steady flow of papers on their sampling distributions, of which his 1915 paper on the correlation coefficient is the best known. He was very careful to distinguish between an unknown population parameter and its sample estimate. When the sampling distribution of the estimate was available in numerical form, a test of the significance of any hypothetical value of the parameter became possible. These precise tests of significance—for example, of an apparent correlation r in the sample on the "null hypothesis" of no real correlation ($\rho = 0$)—were particularly valuable at the time because of the tendency among biologists and other research workers not to bother with them. Fisher's own emphasis on them was, however, rather inconsistent with his subsequent attack on the Neyman–Pearson theory of testing hypotheses, especially since an unthinking use of these significance tests by some workers, as in the failure to recognize that a nonsignificant result does not imply the truth of the hypothesis tested (e.g., $\rho = 0$), caused some reaction against their use later on.

It is evident that Fisher had begun to think about his general theory of estimation before 1922; apart from his advocacy of maximum likelihood in 1912, the notion of sufficiency had also arisen in the special case of the root mean square deviation as an estimate of the true standard deviation σ in the case of a normal, or Gaussian, sample. Nevertheless, the general theory was first systematically developed in the two papers cited, and it included a discussion of the concept of consistency and a heuristic derivation of the asymptotic properties of maximum likelihood estimates in large samples [*see* ESTIMATION, *article on* POINT ESTIMATION]. It also included a crystallization of the concept of information on a parameter θ in the formula

$$I = E\left\{\left(\frac{\partial L}{\partial \theta}\right)^2\right\},$$

where $L = \log p(S|\theta)$ is the logarithm of the probability of the sample S when the parameter has true value θ. The importance of this work lay in (1) examining the actual sampling properties of maximum-likelihood estimates, particularly in large samples (the *method* of maximum likelihood goes back quite a long way, at least as it is analogous to maximum a posteriori probability by Bayes' inverse-probability theorem on the assumption of a uniform a priori distribution); and (2) emphasizing that a sample provides, in some appropriate statistical sense, a definite amount of information on a parameter. Fisher's concept of information, preceding Shannon's, which was introduced in quite a different context (see Shannon 1948), was especially appropriate for large samples because of the possible ordering of normally distributed estimates in terms of their variances (squared standard deviations). (See Pollack 1968.) Fisher justified his concept more generally in the use of small samples by thinking in terms of many such samples; it is curious that he missed the exact inequality relating the information function and the variance of an unbiased estimate known as the Cramér–Rao inequality. In any case, however, the arbitrariness of the variance and unbiasedness remains; and in small samples Fisher introduced the general notion of a sufficient statistic, which, by rendering conditional distributions of any other sample quantities independent of the unknown parameter, exhausted the "information" in the sample [*see* SUFFICIENCY].

The next remarkable contribution in this general area came with Fisher's brief paper entitled "Inverse Probability" (1930*b*). Fisher had always been derisory of the estimates and inferences resulting from the Bayes inverse-probability approach. He felt that a unique and more objective system of inferences should be possible in fields where statistical probabilities operate and noted that an exact sampling distribution involving a sample quantity or statistic T and an unknown parameter θ leads (under appropriate regularity and monotonicity properties) to the feasibility of assigning what he termed a "fiducial interval," with a known fiducial probability that the parameter θ is contained in the interval. The interpretation accepted at the time, and implied by Fisher's own wording in this paper, was that this interval, which is necessarily a function of the statistic T, is in consequence a random interval and that fiducial probability is a statistical probability with the usual frequency connotation. Referring to the case of a true correlation coefficient ρ, he said, "We know that if we take a number of samples of 4, from the same or different populations, and for each calculate the fiducial 5 per cent value for ρ, then in 5 per cent of cases, the true value of ρ will be less than the value we found" ([1930*b*] 1950, pp. 22, 535). With some restrictions (for example, to sufficient statistics) it was thus apparently identical with the theory of confidence intervals developed about the same time by Neyman (1937). [*See* ESTIMATION, *article on* CONFIDENCE INTERVALS AND REGIONS; FIDUCIAL INFERENCE.]

On inductive inference questions, Fisher often did not make it clear what he was claiming; but it should be stressed that regardless of what his interpretation was or of its relevance to the problem at hand, it still was formulated in terms of an assumed statistical framework. Nevertheless, its formal bypassing of Bayes' theorem was a masterly stroke which received attention outside statistical circles (cf. Eddington's remark, "We can never be sure of particular inferences; therefore we should aim at a system of inference that will give conclusions of which in the long run not more than a stated proportion, say $1/q$, will be wrong" [1935*a*] 1960, p. 126).

Later, Fisher attempted to extend fiducial theory to more than one parameter. His first paper (1935*b*) discussing this extension took as one example the problem of inferring the difference in population means of two samples coming from normal populations with different variances. The difficulty here (which Fisher may not have realized at the time, since he himself never examined in detail the logical relations of sufficient statistics in the case of more than one parameter) is that the effect of the unknown variance ratio cannot be segregated in the absence of a "sufficient" quantity for it that does not involve unwanted parameters, such as the individual population means. In rejecting a criticism along these lines by the present author (Bartlett 1936; for further details see, for example, Bartlett 1965), Fisher explicitly gave up the orthodox frequency interpretation for fiducial probability which he appeared to have assumed earlier. He and others attempted to formulate a theory for several parameters that would be both unique and self-consistent, but this has yet to be achieved in any generality, and to many this search is misguided in that it does not eliminate from fiducial theory the arbitrariness that Fisher had so strongly criticized in the Bayes approach.

A rather different and somewhat more technical estimation problem that Fisher solved in 1928 is the derivation of sample statistics that are unbiased estimates of the corresponding population quantities and of the sampling moments of these statistics. The population quantities are the cumulants or semi-invariants first introduced by Thiele (1903), and Fisher's combinatorial rules for obtaining the appropriate sample statistics and their own cumulants constituted a striking example of Fisher's intuitive mathematical powers. Another paper published in 1934 is worth noting as an original and independent contribution to the theory of games developed about the same time by von Neumann and Morgenstern. [*See* GAME THEORY, *article on* THEORETICAL ASPECTS.]

Fisher's work on the design of experiments is so important logically as well as practically that it may be regarded as one of his most fundamental contributions to the science of statistical inference. It is, however, convenient to consider it in the second general area of statistical methodology and technique, in conjunction with analysis of variance. Fisher perceived the simultaneous simplicity and efficiency of balanced and orthogonal experimental designs in agriculture. Replication of the same treatment in different plots is essential if any statistical assessment of error is to be made, and formally equal numbers of plots per treatment are desirable. However, simplification in the statistical analysis is illusory if the analysis is not valid. When observations are collected haphazardly, the most sensible assumptions about statistical variability have to be made. In controlled experiments there is the opportunity for deliberately introducing randomness into the design so that systematic variation can be separated from purely random error. This is the first vital point Fisher made, the second naturally accompanied it. With the analysis geared to the design, all variation not attributable to the treatments does not have to inflate the error. With equal numbers of plots per treatment, each complete replication can be contained in a separate block, and only variability among plots in the same block is a source of error; variability between different blocks can be automatically removed in the analysis as irrelevant. The third point arose from treatment combinations, such as different fertilizer ingredients. For example, if nitrogenous fertilizer (N) and phosphate (P) are to be tested, the recommended set of treatment combinations is

Control (no fertilizer), N, P, NP,

where NP denotes the treatment consisting of both the ingredients N and P (each in the same amount as when given alone). This design maintains simplicity and may improve efficiency, for if phosphate has no effect, or even if its effect is purely additive, the plots are balanced for nitrogen and doubled in number, and similarly for phosphate. Moreover, if both ingredients do not act additively, an interaction term can be defined that measures the difference in effect of N (or P) in the presence and absence of the other ingredient. Such a definition, although to some extent arbitrary, completes the specification of the treatment effects; and the whole technique of *factorial experimentation* typified by the above example is of the utmost importance both in principle and in practice. As Fisher put it:

> The modifications possible to any complicated apparatus, machine, or industrial process must always be considered as potentially interacting with one another, and must be judged by the probable effects of such interactions. If they have to be treated one at a time this is not because to do so is an ideal scientific procedure, but because to test them simultaneously would sometimes be too troublesome, or too costly. In many instances . . . this impression is greatly exaggerated. (1935*a*, p. 97)

A further device that naturally arose in factorial designs was that of *confounding*, by which some of the higher-order interaction effects in designs with three or more factors are assumed to be unimportant and are deliberately arranged to coincide in the analysis with particular block contrasts. This enables the number of plots per block to be smaller and the accuracy of the remaining treatment effects thereby to be increased.

To a large extent the practical value of these experimental methods was not dependent on the statistical analysis, but the simplicity and clarity of the analysis greatly contributed to the worldwide popularity of these designs. This analysis was in principle classical least-squares theory, but the orthogonality of the design rendered the estimation problem trivial, and the concomitant assessment of error was systematized by the technique of "analysis of variance." Basically, this technique is a breakdown of the total sum of squares of the observations into relevant additive parts containing any systematic terms ascribable to treatments, blocks, and so on. Once the technique was established and the appropriate tests of significance were available (on the assumption of normality and of homogeneity of error variance) from Fisher's derivation and tabulation of the "variance-ratio" distribution, it could handle more complicated least-squares problems, such as more complex and even nonorthogonal experimental designs or linear and curvilinear regression problems. One useful extension was the adjustment of observed experimental quantities, such as final agricultural yield, by some observed quantity measured prior to the application of treatments. This technique was referred to as analysis of covariance, although this last term seems more appropriate for the simultaneous analysis of two or more variables—that is, the technique of multivariate analysis.

Fisher was active in the development of multivariate analysis. Earlier workers had of course encountered multivariate problems in various contexts, and Fisher had followed up his geometrical distribution of the correlation coefficient with a derivation of the distribution of the multiple correlation coefficient (1928*b*), again brilliantly using his geometrical approach. The problem exercising Harold Hotelling in the United States and P. C. Mahalanobis in India, as well as Fisher, was the efficient use of several correlated variables for discriminatory and regression problems. Fisher's name is particularly associated with the concept of the discriminant function, some function of the variables that will efficiently distinguish from the measurements of these variables for a single individual whether he came from one or another of two different populations.

Contributions to genetics. Fisher's work in genetics was comparable in importance to his purely statistical contributions and equally reflected his originality and independence of outlook. In the first decade of the twentieth century, Mendelian genetics was still a new subject, and its quantitative consequences were not yet properly appreciated. It was in dispute whether they were consistent with Darwin's theory of evolution by natural selection or even with the observed inheritance of metrical characters. Fisher took the second and lesser problem first, and in his 1918 paper he gave a penetrating theoretical analysis of correlation, breaking it down into nongenetic effects, additive gene action, and further complications, such as genic interaction and dominance. He was thus able to demonstrate the consistency of Mendelian principles with the observed correlations between sibs or between parents and offspring. Then, in his book on the genetic theory of natural selection, he tackled the larger problem. He pointed out that the atomistic character of gene segregation (in contrast to Darwin's hypothetical "blending" theory of heredity) is essential for maintaining variability, which in turn is the basis of the process of natural selection. He clarified the theoretical role of mutations, showing that mutations provide a reservoir from which eventually only favorable ones can survive. He emphasized the possibility of the selection of modifier genes, for example, in rendering the action of many mutant genes recessive by modifying the heterozygote phenotypically toward the natural wild type. The relative importance of this theory of the evolution of dominance was queried, for example, by Sewall Wright, among others, but this did not prevent it from being a relevant thesis that stimulated further research. The effect of modifier genes was also shown to be important in the phenomenon of mimicry.

At the Galton laboratory, Fisher's work in human genetics included linkage studies and the initiation of serology research on the human blood groups. An exciting moment in the work on serology came when G. L. Taylor and R. R. Race studied the Rhesus blood groups. Fisher was able to predict, from the experimental results to date, the effective triple structure of the gene, and hence two more anti-sera and one more allele, which were soon successfully traced. Such predictions may be compared with Fisher's earlier theoretical predictions in the theory of evolution, for example, on the evolution of dominance. They made possible the maintenance of a healthy link with experimental and observational work, which was often initiated or encouraged by Fisher himself. It was in this spirit that he collaborated for several years with E. B. Ford in sampling studies of natural populations.

In retrospect, Fisher's wholehearted immersion in his own research problems, fundamental and broad as these were, did cause him to ignore some important theoretical trends. His neglect of purely mathematical probability, which had been rigorously formulated by A. N. Kolmogorov (1933), seemed to extend to developments in the theory of random or stochastic processes, although these were very relevant to some of his own problems in evolutionary genetics. In England, A. G. McKendrick had published some brilliant papers on stochastic processes in medicine, and G. U. Yule on the analysis of time series, but Fisher never appeared to appreciate this work; in particular, his own papers on the statistical analysis of data recorded in time sometimes showed a rather over-rigid adherence to classical and unduly narrow assumptions. In appraising Fisher's work, one must consider, in addition to these general boundaries that demarcate it, his occasional specific errors and, more importantly, his temperamental bias in controversy. Fisher's scientific achievements are, however, so varied and so penetrating that such lapses cannot dim their luster or reduce his ranking as one of the great scientists of this century.

M. S. BARTLETT

[*For the historical context of Fisher's work, see* STATISTICS, *article on* THE HISTORY OF STATISTICAL METHOD; *and the biographies of* GALTON; GOSSET; PEARSON; YULE. *For discussion of the subsequent development of Fisher's ideas, see* ESTIMATION; EXPERIMENTAL DESIGN; FIDUCIAL INFERENCE; HYPOTHESIS TESTING; LINEAR HYPOTHESES, *article on* ANALYSIS OF VARIANCE; MULTIVARIATE ANALYSIS.]

WORKS BY FISHER

1912 On an Absolute Criterion for Fitting Frequency Curves. *Messenger of Mathematics* New Series 41: 155–160.

1915 Frequency Distribution of the Values of the Correlation Coefficient in Samples From an Indefinitely Large Population. *Biometrika* 10:507–521.

1918 The Correlation Between Relatives on the Supposition of Mendelian Inheritance. Royal Society of Edinburgh, *Transactions* 52:399–433.

(1920–1945) 1950 *Contributions to Mathematical Statistics.* New York: Wiley.

(1922) 1950 On the Mathematical Foundations of Theoretical Statistics. Pages 10.308a–10.368 in R. A. Fisher, *Contributions to Mathematical Statistics.* New York: Wiley. → First published in Volume 222 of the *Philosophical Transactions,* Series A, of the Royal Society of London.

[1](1925*a*) 1958 *Statistical Methods for Research Workers.* 13th ed., rev. New York: Hafner. → Previous editions were also published by Oliver & Boyd.

(1925*b*) 1950 Theory of Statistical Estimation. Pages 11.699a–11.725 in R. A. Fisher, *Contributions to Mathematical Statistics.* New York: Wiley. → First published in Volume 22 of the *Proceedings* of the Cambridge Philosophical Society.

(1928*a*) 1950 Moments and Product Moments of Sampling Distributions. Pages 20.198a–20.237 in R. A. Fisher, *Contributions to Mathematical Statistics.* New York: Wiley. → First published in Volume 30 of the *Proceedings* of the London Mathematical Society.

(1928*b*) 1950 The General Sampling Distribution of the Multiple Correlation Coefficient. Pages 14.653a 14.763 in R. A. Fisher, *Contributions to Mathematical Statistics.* New York: Wiley. → First published in Volume 121 of the *Proceedings* of the Royal Society of London.

(1930*a*) 1958 *The Genetical Theory of Natural Selection.* 2d ed., rev. New York: Dover.

(1930*b*) 1950 *Inverse Probability.* Pages 22.527a–22.535 in R. A. Fisher, *Contributions to Mathematical Statistics.* New York: Wiley. → First published in Volume 26 of the *Proceedings* of the Cambridge Philosophical Society.

1934 Randomization and an Old Enigma of Card Play. *Mathematical Gazette* 18:294–297.

[2](1935*a*) 1960 *The Design of Experiments.* 7th ed. New York: Hafner. → Previous editions were also published by Oliver & Boyd.

(1935*b*) 1950 The Fiducial Argument in Statistical Inference. Pages 25.390a–25.398 in R. A. Fisher, *Contributions to Mathematical Statistics.* New York: Wiley. → First published in Volume 6, part 4 of the *Annals of Eugenics.*

(1938) 1963 FISHER, R. A.; and YATES, FRANK *Statistical Tables for Biological, Agricultural, and Medical Research.* 6th ed., rev. & enl. New York: Hafner.

[3]1949 *The Theory of Inbreeding.* Edinburgh: Oliver & Boyd; New York: Hafner.

[4](1956) 1959 *Statistical Methods and Scientific Inference.* 2d ed., rev. New York: Hafner. → Previous editions were also published by Oliver & Boyd.

SUPPLEMENTARY BIBLIOGRAPHY

BARTLETT, M. S. 1936 The Information Available in Small Samples. Cambridge Philosophical Society, *Proceedings* 32:560–566.

BARTLETT, M. S. 1965 R. A. Fisher and the Last Fifty Years of Statistical Methodology. *Journal of the American Statistical Association* 60:395–409.

EDDINGTON, ARTHUR STANLEY 1935 *New Pathways in Science.* New York: Macmillan.

KOLMOGOROV, ANDREI N. (1933) 1956 *Foundations of the Theory of Probability.* New York: Chelsea. → First published in German.

NEYMAN, JERZY 1937 Outline of a Theory of Statistical Estimation Based on the Classical Theory of Probability. Royal Society of London, *Philosophical Transactions* Series A 236:333–380.

NEYMAN, JERZY 1967 R. A. Fisher (1890–1962): An Appreciation. *Science* 156:1456–1462. → Includes a two-page "Footnote" by William G. Cochran.

►POLLACK, IRWIN 1968 Information Theory. Volume 7, pages 331–337 in *International Encyclopedia of the Social Sciences.* Edited by David L. Sills. New York: Macmillan and Free Press.

SHANNON, C. E. 1948 Mathematical Theory of Communication. *Bell System Technical Journal* 27:379–423, 623–656.

THIELE, THORWALD N. 1903 *Theory of Observations.* London: Layton.

YATES, F.; and MATHER, K. 1963 Ronald Aylmer Fisher. Volume 9, pages 91–129 in Royal Society of London, *Biographical Memoirs of the Fellows of the Royal Society.* London: The Society. → Contains a bibliography on pages 120–129.

Postscript

Since Fisher's death, the increasing output of papers in statistics and population genetics show both indebtedness to his ideas and some emancipation from them. Whether or not Fisher would have approved of some developments in statistics, such as a renewed interest in some quarters in Bayesian methods, or developments in genetics of theories of random drift at the molecular level more related to earlier views of Sewall Wright than to his own, such continual divergences of view are not unexpected or unhealthy in these lively scientific areas, nor do they diminish the importance of his own work. This may now be consulted in an authoritative edition of Fisher's collected papers (1971–1974).

M. S. BARTLETT

OTHER WORKS BY FISHER

Posthumous editions of several works have been published, including new material and clarifications introduced from notes that Fisher made for this purpose before his death.

[1](1925) 1970 *Statistical Methods for Research Workers.* 14th ed., rev. & enl. New York: Hafner.

[2](1935) 1971 *The Design of Experiments.* 9th ed. New York: Hafner.

[3](1949) 1965 *The Theory of Inbreeding.* 2d ed. New York: Academic Press.

[4](1956) 1973 *Statistical Methods and Scientific Inference.* 3d ed., rev. & enl. New York: Hafner.

Collected Papers of R. A. Fisher. 5 vols. Edited by J. H. Bennett. Univ. of Adelaide, 1971–1974.

ADDITIONAL BIBLIOGRAPHY

SAVAGE, LEONARD J. 1976 On Rereading R. A. Fisher. *Annals of Statistics* 4:441–483. → Discussion on pages 483–500.

FORECASTING

See DEMOGRAPHY; PREDICTION; PREDICTION AND FORECASTING, ECONOMIC. *Related material may be found under* TIME SERIES.

FRACTIONAL REPLICATION

See EXPERIMENTAL DESIGN.

FREQUENCY CURVES

See DISTRIBUTIONS, STATISTICAL; STATISTICS, DESCRIPTIVE.

G

GALTON, FRANCIS

Francis Galton was born in 1822 and died in 1911. He was educated successively at home, at a dame school, at Boulogne, and at Kenilworth. In 1835, at the age of 13, he entered King Edward's School at Birmingham, where he stayed for two years. He spent two years as a medical student, the first at the General Hospital, Birmingham and the second at King's College, London. In 1840 he entered Trinity College, Cambridge, as a mathematics student, but was content to take a poll degree in 1843, when his health broke down.

Galton's father, Samuel Tertius Galton, was a banker. His mother, Violetta Darwin, was the daughter of Erasmus Darwin by his second wife. One of Erasmus Darwin's grandsons through his first wife was Charles Robert Darwin; there was a certain physical resemblance between the two cousins. His mental development is interesting: he is credibly reported to have read a simple book before he reached the age of three, and his restless ingenuity with regard to machinery dates from his early youth. He did not enjoy school, however, nor did he find the profession of medicine, which was chosen for him, congenial. In spite of his interest in mechanics and mathematics, he was not successful in his Cambridge studies.

When his father died in 1844, Galton immediately forsook any idea of continuing his medical career. He found himself the possessor of a more than adequate income and proceeded to spend his time and energy "hunting with a set chiefly noteworthy for their extravagance and recklessness . . . the strange thing [being] that it [shooting] seemed to absorb his whole nature, and to be done not for the sake of the experience, but in the pure pursuit of occupation" (Pearson 1914–1930, vol. 1, pp. 208–209).

It was after these fallow years, as Pearson called them, that Galton carried out the explorations for which he was later awarded the gold medal of the Royal Geographical Society in 1853. Even before going to Cambridge, Galton had taken an extended trip down the Danube and on to Smyrna, which had perhaps awakened the young man to the delights of foreign scenes and strange peoples. After his father's death he set off again for Egypt, Khartoum, and Syria, but he "was still touring for the boyish fun of movement and of new scenes. He had not yet thoughts of the language, habits, or archaeology of the people he mingled with" (*ibid.*, p. 205). It was not until after four years of idleness in England that he set out on a trip to tropical Africa, the results of which showed that he had come to terms with life and with himself.

In 1850 Galton set off for the Cape and spent two years upcountry exploring from Walvis Bay to Lake Ngami, territory of which little was known. He composed 15 brief laws for the Hottentot chiefs who governed the Damaras of the plain and compiled a rudimentary dictionary for the English who wished to use the local tongue. He returned to England early in 1852 and read a paper to the Royal Geographical Society, which awarded him its gold medal the following year. This award was followed in 1854 by a silver medal from the French Geographical Society. Early in 1853 he met Louisa Butler and married her in August. After an extended honeymoon tour of Europe, punctuated by visits to England, the Galtons finally settled in London, and in 1855 Galton really began to work.

Early publications. As might be anticipated, Galton's first publication was on exploration, and in 1855 *The Art of Travel* was published. There were signs that his scientific curiosity was developing in new directions, since in *Vacation Tourists and Notes of Travel* (1861–1864), which he meant to be an annual magazine, there is a description of the eclipse of the sun in 1860, with a drawing of the curved rays of the corona that he had observed. Galton's first piece of fruitful research was on the weather. He started to plot wind and pressure maps and noted, from very scanty data, that centers of high pressure are associated with clockwise directions of winds around the calm center. He coined the name "anticyclone" for such systems in 1863. Several other papers followed, in which he was clearly feeling his way toward the concepts of correlation and regression.

He tried to determine a linear prediction formula for the velocity of the wind, given the pressure, temperature, and humidity. He did not succeed, possibly because of his failure to realize that the prediction formula for pressure from velocity was not the same as the prediction formula for velocity from pressure. The realization that there are two regression lines was still in the future, as was the concept of correlation. In 1870 he read a paper at the British Association entitled "Barometric Predictions of Weather," in which he was fumbling toward a multiple regression, trying to predict the wind from pressure, temperature, and humidity. He failed in his objective at the time, but he posed the problem for others who were to succeed.

Intellectual influences. In assessing the intellectual influences on Galton, continuing uncertainty exists as to the extent of Quetelet's influence. Pearson tended to minimize the significance of Quetelet for Galton; he wrote, "I am very doubtful how far [Galton] owed much to a close reading of the great Belgian statistician" (Pearson 1914–1930, vol. 2, p. 12), and he placed perhaps undue weight on the fact that Galton possessed no copy of Quetelet's *Letters . . . on the Theory of Probabilities* (1846). Pearson further remarked that Galton "was never a great student of other men's writings: he was never an accumulator like his cousin Charles Darwin" (Pearson 1914–1930, vol. 1, p. 209). Now Pearson was closer to Galton's time and actually knew him, so that some weight must be given to his opinions. Nevertheless, Pearson would appear to have underestimated the influence of Quetelet; he himself pointed out that Galton's work seemed to flow naturally out of that of Quetelet. Further, Galton's obsession with the normal curve of error which, to a certain extent, has unduly influenced the development of statistical method, can only have stemmed from Quetelet. One of Quetelet's great achievements was to consider all human experience as ultimately capable of being described numerically, which was fundamentally Galton's attitude also.

The other great influence on Galton during the period in which he was establishing himself as a research worker affected the whole of the scientific world in the second half of the nineteenth century —the publication in 1859 of Charles Darwin's *The Origin of Species*. The effect of this work on Galton was not immediately apparent in his writings, but there can be no doubt that the book was responsible for transforming him from a geographer into an anthropologist and eugenist. He began with the article "Hereditary Talent and Character" in 1865 and proceeded through *Hereditary Genius* (1869); *English Men of Science: Their Nature and Nurture* (1874); *Inquiries Into Human Faculty* (1883); and *Natural Inheritance* (1889), by which time he was 67 years of age. As Pearson said, "We see that his researches in heredity, in anthropometry, in psychometry and statistics, were not independent studies; they were all auxiliary to his main object, the improvement in the race of man."

Application of statistics. In *Hereditary Genius*, Galton claimed that his discussion of heredity was the "first to treat the subject in a statistical manner" ([1869] 1952, p. vi). He clearly owed much to Quetelet and paralleled Quetelet's use of the normal curve for anthropometric measurements by using it to grade intellectual ability. He was quite explicit about this: "The law is an exceedingly general one. M. Quetelet, the Astronomer-Royal of Belgium, and the greatest authority on vital and social statistics, has largely used it in his inquiries. He has also constructed numerical tables, by which the necessary calculations can be easily made, whenever it is desired to have recourse to the law" (*ibid.*, p. 23).

Galton supplemented Quetelet's tables by a short table of the abscissas of the unit normal curve corresponding to percentiles of area (1889). He examined the abilities of the kin of persons who had achieved eminence of some kind—judges, generals, scientists, statesmen, painters, poets, and clerics. He was concerned with distinguishing between general ability and special ability and regarded each individual personality as a combination of natural ability and the advantages accruing from early environment, i.e., nature and nurture.

This idea of nature and nurture recurs in his

writings. Thus we find in *English Men of Science* (1874), "It is, I believe, owing to the favourable conditions of their early training that an unusually large proportion of the sons of the most gifted men of science become distinguished in the same career. They have been nurtured in an atmosphere of free enquiry. . . ." The thesis is that heredity tends to produce eminence in *some* area and that environment tends to be the deciding factor in specifying *what* this area shall be. Galton tried to go beyond this in *Inquiries Into Human Faculty and Its Development* (1883), the book that possibly holds most interest for students of the history of psychology, in which he discussed preliminary results that he had obtained in the psychometric field.

In 1876, at the exhibition of scientific instruments at South Kensington, Galton exhibited his "Whistles for Determining the Upper Limits of Audible Sounds in Different Persons." Both before and after this time he was active in proposing tests for the measurement of muscular sensitivity by weight discrimination, for the perception of differences of tint, for reaction time, for acuteness of hearing, for keenness of vision and judgment of length by the eye, and for the senses of smell and touch. In an attempt to describe the skewed distributions that often resulted from the application of his tests, Galton hypothesized that in some frequency distributions, such as, for example, judgment of length, the geometric mean, rather than the arithmetic mean, is the best "medium" for the distribution, and he wrote a paper on "The Geometric Mean in Vital and Social Statistics" (1879). As usual the mathematical conceptualization was beyond him, and he took the problem to Sir Donald Macalister, who derived what is now known as the log-normal distribution.

At this stage of his work, he was associated with the American psychologist James McKeen Cattell, who on his return to the University of Pennsylvania (and later at Columbia University), began to teach statistical psychology, giving his first course in 1887. Through Cattell, Galton's ideas and experiments exerted possibly the greatest single influence upon American psychology during the last years of the nineteenth century.

From the statistical point of view, *Natural Inheritance* is probably the most important of Galton's writings. As can be seen from his earlier works, the ideas in it had been fermenting in his mind for some time, but it was their expression in *Natural Inheritance* that excited the interest of those whom today we might call the practitioners of applied mathematics. Again he was influenced by the fact that Quetelet was using the normal curve to describe anthropometric data and by the interest in the problems of inheritance aroused in him by *The Origin of Species*.

He began the book with a summary of those properties of the normal curve that appealed to him. He had previously suggested representing a frequency distribution by using grades or percentiles, and he elaborated on this suggestion here, pointing out that the normal distribution is completely determined from a knowledge of the median and one other quantile. Galton had observed that many measured characteristics can be closely described by a normal curve. He used the "quincunx," first shown in print with the publication of his lecture "Typical Laws of Heredity," delivered at the Royal Institution (1877), to illustrate the build-up of the normal curve: He had noticed that a normal curve is reproduced by lead shot falling vertically through a harrow of pins and he tried to explain the stability of measured characteristics by this mechanical device. In this paper he had almost reached the concepts of both regression and correlation but must have felt the need for further thought, since it was at this time that he began to collect data bearing on inheritance in man. Galton published nothing further on heredity for eight years. The foundation of his ideas on regression and correlation did not perhaps become clear to him until a short time before the publication of *Natural Inheritance*.

The regression line arose naturally out of measurements of the sizes of the seeds of mother and daughter sweet pea plants. The sizes of the seeds of daughter plants appeared to "revert" to the mean (the word "revert" was soon replaced by "regress"). This inspired him to look at a bivariate frequency table of the heights of fathers and sons, in which he found a regression to "mediocrity." The arguments he used became familiar ones with the analysis of variance put forward by R. A. Fisher some forty to fifty years later. Suppose, Galton said, that we want to predict the height of brother A, given the height of brother B. We take, therefore, all the individuals who have heights the same as B and form a collection of the heights of all of their brothers. These brothers as a group Galton called a cofraternity, and he proceeded to discuss the variation in height of all individuals about the grand mean, the variation of the cofraternity means about the grand mean, and the variation of the individuals of the cofraternities about their respective cofraternity means. This splitting up of variation had been done previously by Lexis in Germany and Dormoy in France, but Galton was possibly the first to carry out this type of analysis with the idea of assigning the variation.

While studying the bivariate frequency table of heights of fathers and sons, Galton was struck by the observation that the contours of equal frequency in the table were similar and similarly situated ellipses. He also found the lines that fitted the medians of the arrays (possibly drawing them by eye) and the slopes of these lines eventually became his regression coefficients. This early work, as is inevitable with a pioneering effort, is confused and difficult to evaluate, not least because Galton himself was not explicit. When, however, he had determined that he had what would now be termed linearity of regression and homoscedasticity in the arrays of the table, his mathematical powers were not sufficient to enable him to form a mathematical model for his surface, and he took the problem to Hamilton Dickson, a Cambridge mathematician. Dickson's mathematical formulation was published in an appendix to Galton's paper "Family Likeness in Stature," presented to the Royal Society in 1886 Galton was troubled by the fact that the slope of the regression line depends on the variability of the margins, and this concern led to his search for a unit-free measure of association.

Some time earlier, in 1882, Alphonse Bertillon had put forward a scheme for classifying criminals according to 12 physical measurements that was adopted by the prefecture of police in Paris. Galton became interested in this scheme and pondered for some time over which measurements would be the most descriptive—that is, which would discriminate one man most effectively from his fellows. It was from these considerations that he was led to the realization that some measurements might be so highly correlated with other measurements as to be useless for the prescribed purposes and finally to the necessity for describing how any two measurements are related. The slope of the regression line is not adequate for this, since it depends on both the scales of measurement and the choice of dependent variables. However, the regression line fitted between the variables that Galton used (1888) after dividing the heights (reduced by their median) by a measure of their variability (their semi-interquartile range) and similarly dealing with forearm length provides a unit-free measure of association. Given the problem and 65 years of subsequent statistical development, the correlation coefficient may now appear to have been inevitable. There can be no question, however, that at the time at which Galton wrote, 1888–1889, the production of a measure of association that was independent of location and scale was an immense contribution to statistical methodology.

The Bertillon system of measurement also started Galton wondering about the whole procedure of personal identification. In the paper for the Royal Institution in which he discussed bertillonage, he also drew incidental attention to fingerprints. In his book *Finger Prints* (1892), he referred to the work of Jan Purkinje, Kollman, William Herschel, and Henry Faulds, who had preceded him in this study, but it is clear that at the time he wrote little was known. As he himself said:

> It became gradually clear that three facts had to be established before it would be possible to advocate the use of finger prints for criminal or other investigations. First it must be proved, not assumed, that the pattern of a finger print is constant throughout life. Secondly that the variety of patterns is really very great. Thirdly, that they admit of being so classified, or lexiconised, that when a set of them is submitted to an expert, it would be possible for him to tell, by reference to a suitable dictionary or its equivalent, whether a similar set had already been registered. These things I did, but they required much labor.

As a result of Galton's book and his evidence to a committee set up by the Home Office in 1893, a fingerprint department was established, the forerunner of many such throughout the world. Galton himself, as might be expected from his previous work and interest, turned to studying the inheritance of fingerprints, a study which was carried on for many years in the laboratory that he founded and that was named after him.

Eugenics. The term "eugenics" was introduced by Galton in his book *Inquiries Into Human Faculty* (1883) and soon won general acceptance. The study of human inheritance and the possibility of improving human stock were undoubtedly linked in his mind, as his public lectures and papers witness. He did more than lecture, however. In 1904 he founded a research fellowship in national eugenics at the University of London which was to develop in a few years into the Galton Laboratory of National Eugenics, with Karl Pearson as its first director. Pearson was succeeded by R. A. Fisher, and the now vast complex of statistical theory and method developed there thus owes its origin to Galton.

It was inevitable that Galton's work should attract the interest of young men able in the mathematical and in the biological fields, and the late 1880s saw Karl Pearson, and W. F. R. Weldon—the one a professor of applied mathematics and the other a professor of zoology and both at University College, London—working in the field of "biometry," i.e., the application of mathematics to problems of biological inheritance. Galton himself said, "The primary object of Biometry is to afford material that shall be exact enough for the discovery of incipient changes in evolution which are too small to be otherwise apparent." Pearson and Weldon met difficulties in their attempts to publish papers relating to biometry in existing journals and determined to start their own. A guarantee was required. Galton, on being asked to help, not only guaranteed the whole amount but followed it up with an additional gift that enabled his admirers to go their way in freedom; the journal *Biometrika*, the first to be devoted to both the theory and practice of statistics, was established on a firm footing. In the last decade of his life, Galton played the part of counselor and adviser to the younger men, but he still worked away at his own problems, as his continued output of letters and papers indicates.

During his last years many honors came his way. He had been elected a fellow of the Royal Society in 1856, receiving a gold medal in 1886, the Darwin medal in 1902, and the much-prized Copley medal in 1910, the year before his death. He was awarded the Huxley medal by the Anthropological Institute in 1901 and the Darwin–Wallace medal by the Linnean Society in 1908. He received honorary degrees from both Oxford and Cambridge universities and became an honorary fellow of Trinity College, Cambridge, his old college, in 1902. The citation for the Darwin medal said, in part, "It may safely be declared that no one living has contributed more definitely to the progress of evolutionary study, whether by actual discovery or by the fruitful direction of thought, than Mr. Galton." Mr. Galton's private comment was, typically, "Well, I am very pleased except that I stand in the way of younger men" (quoted in Pearson 1914–1930, vol. 3A, p. 237).

F. N. DAVID

[*For the historical context of Galton's work, see the biography of* QUETELET; *see also* Gillispie 1968. *For discussion of the subsequent development of Galton's ideas, see* LINEAR HYPOTHESES, *article on* REGRESSION; MULTIVARIATE ANALYSIS, *article on* CORRELATION METHODS; *and the biographies of* FISHER; PEARSON; *see also* Allen 1968; Gates 1968.]

WORKS BY GALTON

(1855) 1856 *The Art of Travel: Or, Shifts and Contrivances Available in Wild Countries.* 2d ed., rev. & enl. London: Murray.

1861–1864 GALTON, FRANCIS (editor) *Vacation Tourists and Notes of Travel in 1860 [1861, 1862–1863].* London: Macmillan.

1863 *Meteorographica: Or, Methods of Mapping the Weather.* London: Macmillan.

1865 Hereditary Talent and Character. *Macmillan's Magazine* 12:157–166, 318–327.

(1869) 1952 *Hereditary Genius: An Inquiry Into Its Laws and Consequences.* New York: Horizon Press. → A paperback edition was published in 1962 by World.

1870 Barometric Predictions of Weather. British Association for the Advancement of Science, *Report* 40 [2]:31–33.

1874 *English Men of Science: Their Nature and Nurture.* London: Macmillan.

1876 Whistles for Determining the Upper Limits of Audible Sounds in Different Persons. Page 61 in South Kensington Museum, London, *Conferences Held in Connection With the Special Loan Collection of Scientific Apparatus, 1876.* Volume 2: Physics and Mechanics. London: Chapman.

(1877) 1879 Typical Laws of Heredity. Royal Institution of Great Britain, *Proceedings* 8:282–301. → First published in Volume 15 of *Nature.*

1879 The Geometric Mean in Vital and Social Statistics. Royal Society of London, *Proceedings* 29:365–367.

(1883) 1952 *Inquiries Into Human Faculty and Its Development.* London: Cassell.

1886 Family Likeness in Stature. Royal Society of London, *Proceedings* 40:42–63. → Supplemented with an appendix by J. D. Hamilton Dickson on pages 63–72.

1888 Co-relations and Their Measurement, Chiefly From Anthropomorphic Data. Royal Society of London, *Proceedings* 45:135–145.

1889 *Natural Inheritance.* London and New York: Macmillan.

1892 *Finger Prints.* London and New York: Macmillan.

1908 *Memories of My Life.* London: Methuen.

SUPPLEMENTARY BIBLIOGRAPHY

►ALLEN, GORDON 1968 Eugenics. Volume 5, pages 193–197 in *International Encyclopedia of the Social Sciences.* Edited by David L. Sills. New York: Macmillan and Free Press.

BURT, CYRIL 1962 Francis Galton and His Contributions to Psychology. *British Journal of Statistical Psychology* 15:1–49.

DARWIN, GEORGE H. (1912) 1939 Sir Francis Galton. Volume 2, pages 70–73 in *Dictionary of National Biography: Second Supplement.* Oxford Univ. Press.

►GATES, ARTHUR I. 1968 Cattell, James McKeen. Volume 2, pages 344–346 in *International Encyclopedia of the Social Sciences.* Edited by David L. Sills. New York: Macmillan and Free Press.

►GILLISPIE, CHARLES C. 1968 Darwin, Charles. Volume 4, pages 7–14 in *International Encyclopedia of the Social Sciences.* Edited by David L. Sills. New York: Macmillan and Free Press.

NEWMAN, JAMES R. 1956 Commentary on Sir Francis Galton. Volume 2, pages 1167–1172 in James R. Newman (editor), *The World of Mathematics: A Small Library of the Literature of Mathematics From A'h-mosé the Scribe to Albert Einstein.* New York: Simon & Schuster.

PEARSON, KARL 1914–1930 *The Life, Letters and Labours of Francis Galton.* 3 vols. Cambridge Univ.

Press. → Includes a comprehensive bibliography of Galton's works.

Quetelet, Adolphe (1846) 1849 *Letters Addressed to H. R. H. the Grand Duke of Saxe-Coburg and Gotha, on the Theory of Probabilities, as Applied to the Moral and Political Sciences.* London: Layton. → First published in French.

ADDITIONAL BIBLIOGRAPHY

Cowan, Ruth Schwartz 1972 Francis Galton's Statistical Ideas: The Influence of Eugenics. *Isis* 63:509–528.

Cowan, Ruth Schwartz 1977 Nature and Nurture: The Interplay of Biology and Politics in the Work of Francis Galton. Volume 1, pages 133–208 in William Coleman and Camille Limoges (editors), *Studies in History of Biology.* Baltimore and London: Johns Hopkins Press.

Forrest, Derek W. 1974 *Francis Galton: The Life and Work of a Victorian Genius.* New York: Taplinger. → Includes a bibliography of Galton's published works.

GAME THEORY

I
THEORETICAL ASPECTS

The theory of games is a mathematical discipline designed to treat rigorously the question of optimal behavior of participants in games of strategy and to determine the resulting equilibria. In such games each participant is striving for his greatest advantage in situations where the outcome depends not only on his actions alone, nor solely on those of nature, but also on those of other participants whose interests are sometimes opposed, sometimes parallel, to his own. Thus, in games of strategy there is conflict of interest as well as possible cooperation among the participants. There may be uncertainty for each participant because the actions of others may not be known with certainty. Such situations, often of extreme complexity, are found not only in games but also in business, politics, war, and other social activities. Therefore, the theory serves to interpret both games themselves and social phenomena with which certain games are strictly identical. The theory is normative in that it aims at giving advice to each player about his optimal behavior; it is descriptive when viewed as a model for analyzing empirically given occurrences. In analyzing games the theory does not assume rational behavior; rather, it attempts to determine what "rational" can mean when an individual is confronted with the problem of optimal behavior in games and equivalent situations.

The results of the interlocking individual actions are expressed by numbers, such as money or a numerically defined utility for each player transferable among all. Games of strategy include games of chance as a subcase; in games of chance the problem for the player is merely to determine and evaluate the probability of each possible outcome. In games of strategy the outcome for a player cannot be determined by mere probability calculations. Specifically, no player can make mere statistical assumptions about the behavior of the other players in order to decide on his own optimal strategy.

But nature, when interfering in a game through chance events, is assumed to be indifferent with regard to the player or players affected by chance events. Since the study of games of chance has given rise to the theory of probability, without which modern natural science could not exist, the expectation is that the understanding of the far more complicated games of strategy may gradually produce similar consequences for the social sciences.

History. In 1710 the German mathematician–philosopher Leibniz foresaw the need and possibility of a theory of games of strategy, and the notion of a minimax strategy (see section on "Two-person, zero-sum games," below) was first formulated two years later by James Waldegrave. (See the letter from Waldegrave in the 1713 edition of Montmort 1708; see also Baumol & Goldfeld 1967.) The similarity between games of strategy and economic processes was occasionally mentioned, for example, by Edgeworth in his *Mathematical Psychics* (1881). Specialized theorems, such as Ernst Zermelo's on chess, were stated for some games; and Émile Borel developed a limited minimax strategy, but he denied the possibility of a general theorem. It was not until John von Neumann (1928) proved the fundamental theorem that a true theory of games emerged (see section on "Two-person, zero-sum games," below). In their *Theory of Games and Economic Behavior*, von Neumann and Morgenstern (1944) extended the theory, especially to games involving more than two players, and gave applications of the theory in economics. Since then, throughout the world a vast literature has arisen in which the main tenets of the theory have been widened and deepened and many new concepts and ideas introduced. The four-volume *Contributions to the Theory of Games* (Kuhn & Tucker 1950–1959) and *Advances in Game Theory* (Dresher, Shapley, & Tucker 1964) give evidence of this continuing movement. These works contain extensive bibliographies, but see especially Volume 4 of *Contributions to the Theory of Games*.

Game theory concepts

Games are described by specifying possible behavior within the rules of the game. The rules are in each case unambiguous; for example, certain moves are allowed for specific pieces in chess but are forbidden for others. The rules are also inviolate. When a social situation is viewed as a game, the rules are given by the physical and legal environment within which an individual's actions may take place. (For example, in a market individuals are permitted to bargain, to threaten with boycotts, etc., but they are not permitted to use physical force to acquire an article or to attempt to change its price.) The concrete occasion of a game is called a play, which is described by specifying, out of all possible, allowable moves, the sequence of choices actually made by the players or participants. After the final move, the umpire determines the payments to each player. The players may act singly, or, if the rules of the game permit it and if it is advantageous, they may form coalitions. When a coalition forms, the distribution of the payments to the coalition among its members has to be established. All payments are stated in terms of money or a numerically defined utility that is transferable from one player to another. The payment function is generally assumed to be known to the players, although modifications of this assumption have been introduced, as have other modifications for example, about the character of the utilities and even about the transferability of payments.

The "extensive" form of a game, given in terms of successive moves and countermoves, can be represented mathematically by a game tree, which describes the unfolding of the moves, the state of information of the players at the moment of each choice, and the alternatives for choices available to each player at each occasion. This description can, in a strict mathematical sense, be given equivalently in a "normalized" form: each player, uninformed about the choices made by any other player, chooses a single number that identifies a "strategy" from his given finite or infinite set of strategies. When all personal choices and a possible random choice are made (simultaneously), the umpire determines the payments. Each strategy is a complete plan of playing, allowing for all contingencies as represented by the choices and moves of all other players and of nature. The payoff for each player is then represented by his mathematical expectation of the outcome for himself. The final description of the game therefore involves only the players' strategies and no further chance elements.

The theory explicitly assumes that each player, besides being completely informed about the alternative payoffs due to all moves made or strategies chosen, can perform all necessary computations needed to determine his optimal behavior. (This assumption of complete information is also commonplace in current economic theory, although seldom stated explicitly.)

The payments made by all players may add up to zero, as in games played for entertainment. In this case the gains of some are exactly balanced by the losses of others. Such games are called zero-sum games. In other instances the sum of all payments may be a constant (different from zero) or may be a variable; in these cases all players may gain or lose. Applications of game theory to economic or political problems require the study of these games, since in a purchase, for example, both sides gain. An economy is normally productive so that the gains outweigh any losses, whereas in a war both sides may lose.

If a player chooses a particular strategy as identified by its number, he selects a *pure* strategy; if he allows a chance mechanism, specified by himself, to make this selection for him, he chooses a *mixed* or *statistical* strategy. The number of pure strategies for a player normally is finite, partly because the rules of games bring the play to an end after a finite number of moves, partly because the player is confronted with only a finite number of alternatives. However, it is possible to treat cases with infinitely many strategies as well as to consider even the borderline case of games with infinitely many players. These serve essentially to study pathological examples or to explore certain mathematical characteristics.

Game theory uses essentially combinatorial and set-theoretical concepts and tools, since no specific calculus has as yet evolved—as happened when differential and integral calculus were invented simultaneously with the establishment of classical mechanics. Differential calculus is designed to determine maxima and minima, but in games, as well as in politics, these are not defined, because the outcome of a player's actions does not depend on his actions alone (plus nature). This applies to all players simultaneously. A maximum (or minimum) of a function can be achieved only when all variables on which the maximum (minimum) depends are under the complete control of the would-be maximizer. *This is never the case in games of strategy.* Therefore, in the equivalent business, political, or military operations there obtains no maximum (minimum) problem, whether with or without side conditions, as assumed in the classical literature of these fields; rather one is confronted there with an entirely different conceptual structure, which the theory of games analyzes.

Two-person, zero-sum games

The simplest game of strategy is a two-person, zero-sum game, in which players A and B each have a finite number of strategies and make their choices unknown to each other. Let P be the payoff to the first player, and let $-P$ be the payoff to the second player. Then P is greater than, equal to, or less than 0, depending on whether A wins, draws, or loses. Let $A_1, A_2, \cdots, A_n$ be the strategies available to player A and $B_1, B_2, \cdots, B_m$ be the strategies available to player B. In the resulting $n \times m$ array of numbers, each row represents a pure strategy of A, each column a pure strategy of B. The intersections of the rows and columns show the payoffs to player A from player B. The first player wishes to maximize this payoff, while the second wishes to minimize it. This array of numbers is called the payoff matrix, an example of which is presented in Table 1, where payments go from B to A. Player A's most desirable payoff is 8; B's is -10. Should player A pick strategy A_1, either of these two events may happen depending on B's action. But if A picks A_1, B in his own interest would want to pick B_3, which would mean that A would have to pay 10 units to B instead of receiving 8. The row minima represent the worst that could happen to A for each of his strategies, and it is natural that he would want to make as great as possible the least gain he can expect from each; that is, he seeks the maximum of the row minima, or the *maximin*, which in Table 1 is -1 (strategy A_3). Conversely, B will wish to minimize the column maxima—that is, seek the *minimax*—which is also -1 (strategy B_2). We would say that each player is using a minimax strategy—that is, each player selects the strategy that minimizes his maximum loss. Any deviation from the optimal strategies A_3 and B_2 is fraught with danger for the deviating player, so that each will choose the strategy that contains the so-called *saddle point of the payoff function.* The saddle point is defined as the point at which the maximin equals the minimax. At this point the least that A can secure for himself is equal to the most that B may have to part with. (In the above example A has to pay one unit to B.) If there is more than one saddle point in the payoff matrix, then they are all equal to each other. Games possessing saddle points in pure strategies are called *specially strictly determined.* In these games it is immaterial whether the choice of the pure strategy by either player is made openly before the other makes his choice. Games of *perfect* information—that is, games in which each player at each move is always informed about the entire previous history of the play, so that what is preliminary to his choice is also anterior to it—are always specially strictly determined. Chess belongs in this class; bridge does not, since each of the two players (one "player" being the north–south team, the other the east–west team) is not even completely informed about himself—for example, north does not know precisely what cards south holds.

Table 1 — Payoff matrix for a two-person, zero-sum game

B's strategy / A's strategy	B_1	B_2	B_3	Row minima
A_1	8	−3	−10	−10
A_2	0	−2	6	−2
A_3	4	−1	5	−1
Column maxima	8	−1	6	

Most games will have no saddle points in pure strategies; they are then said to be not strictly determined. The simplest case is matching pennies. The payoff matrix for this game is presented in Table 2. Here, if one player has to choose openly before the other does, he is sure to lose. Each player will therefore strive to prevent information about his choice from flowing to the other. This is accomplished by the player's choice of a chance mechanism, which selects from among the available pure strategies with probabilities determined by the player. In matching pennies, the chance mechanism should select "heads" with probability $\frac{1}{2}$ and "tails" with probability $\frac{1}{2}$. This randomization may be achieved by tossing the coin before showing it. If there is a premium, say on matching heads over matching tails, the payoff matrix would reflect this, and the probabilities with which the two sides of the coin have to be played in order to prevent disclosure of a pattern of playing to the benefit of the opponent would no longer be $\frac{1}{2}$ for heads and $\frac{1}{2}$ for

Table 2 — Payoff matrix for matching pennies

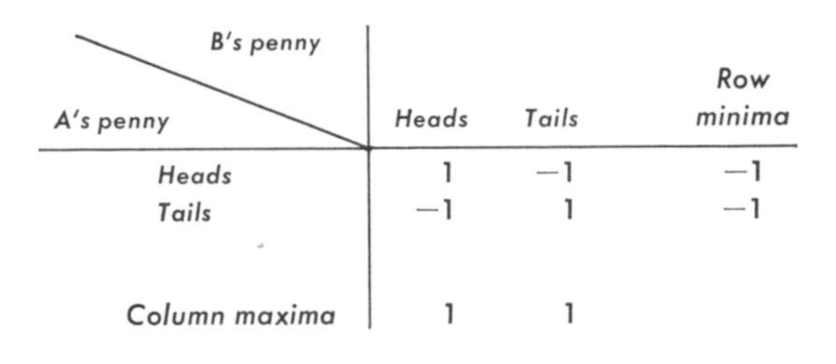

B's penny / A's penny	Heads	Tails	Row minima
Heads	1	−1	−1
Tails	−1	1	−1
Column maxima	1	1	

tails. Thus, when there is no saddle point in pure strategies a randomization by a chance mechanism is called for. The players are then said to be using mixed, or statistical, strategies. This does *not* transform a game of strategy into a game of chance: the strategic decision is the specification of the randomization device and the assignment of the proper probabilities to each available pure strategy. Whether pure or mixed strategies are needed to assure a saddle point, the theory at no point requires that the players make assumptions about each other's intelligence, guesses, and the like. The choice of the optimal strategy is independent of all such considerations. Strategies selected in this way are perfect from the defensive point of view. A theory of true offensive strategies requires new ideas and has not yet been developed.

Von Neumann proved that each matrix game can be made strictly determined by introducing mixed strategies. This is the *fundamental theorem* of game theory. It shows that each zero-sum, two-person game has a saddle point in mixed strategies and that optimal mixed strategies exist for each of the two players. The original proof of this theorem made use of rather complex properties of set theory, functional calculus, and combinatorics. Since the original proof was given, a number of alternative, simplified versions have been given by various authors. The numerical solution of a matrix game with m columns and n rows demands the solution of a system of linear inequalities of $m + n + 1$ unknowns, the $m + n$ probabilities for the strategies of players A and B and the minimax value. There exist many techniques for solving such systems; notably, an equivalence with solving dual linear programs has proved to be of great importance [*see* PROGRAMMING]. High-speed computers are needed to cope with the rapid rise of the required arithmetical operations. A more modest view of mixed strategies is the notion of behavioral strategies, which are the probability distributions over each player's information sets in the extensive form of the game. For games such as chess, even the optimal pure strategy cannot be computed, although the existence of a saddle point in pure strategies can be proved and either white or black has a winning pure strategy no matter what the other does (or both have pure strategies that enforce a draw). The problems of finding further computational techniques are actively being investigated.

n-Person, zero-sum games

When the number of players increases to $n \geqslant 3$, new phenomena arise even when the zero-sum restriction remains. It is now possible that cooperation will benefit the players. If this is not the case, the game is called inessential. In an essential game the players will try to form *coalitions* and act through these in order to secure their advantage. Different coalitions may have different strength. A winning coalition will have to divide its proceeds among its members, and each member must be satisfied with the division in order that a stable solution obtains [*see* COALITIONS, THE STUDY OF; *see also* Gamson 1968].

Any possible division of payments among all players is called an *imputation*, but only some of all possible imputations will be contained in a *solution*. An inessential game has precisely one imputation that is better than any other, that is, one that *dominates* all others. This unique imputation forms the solution, but this uniqueness is trivial and applies only to inessential games. There is no cooperation in inessential games.

A solution of an essential game is characteristically a nonempty set of several imputations with the following properties: (1) No imputation in the set is dominated by another imputation in the set. (2) All imputations not in the set are dominated by an imputation contained in the set. There may be an infinite number of imputations in a solution set, and there may be several solution sets, each of which has the above properties. Furthermore, it should be noted that every imputation in a solution set is dominated by some imputation not in that set, but property (2) assures that such a dominating imputation is, in turn, dominated by an imputation in the solution set.

To be considered as a member of a coalition, a player may have to offer *compensations* or side payments to other prospective members. A compensation or side payment may even take the form of giving up privileges that the rules of the game may attribute to a player. A player may be admitted to a coalition under terms less favorable than those obtained by the players who form the initial core of a coalition (this happens first when $n = 4$). Also, coalitions of different strength can be distinguished. *Discrimination* may occur; for example, some players may consider others "taboo"—that is, unworthy as coalition partners. This leads to the types of discriminatory solutions that already occur when $n = 3$. Yet discrimination is not necessarily as bad for the affected player as defeat is for a nondiscriminated player, because cooperation against the discriminated player may not be perfect. A player who by joining a coalition does not contribute more to it than what he can get by playing for himself merely has the role of a dummy.

The fundamental fact of cooperation is that the

players in a coalition can each obtain more than they could obtain by playing alone. This expresses the nonadditivity—specifically, the superadditivity—of value, the explanation of which has long been recognized as a basic problem in economics and sociology. In spite of many efforts, no solution was found, but it is now adequately described by the characteristic function $v(S)$, a numerical set function that states for any cooperative n-person game the proceeds of the coalition S, and an imputation that describes the distribution of all payments among all players (von Neumann & Morgenstern 1944, chapter 6).

Since there may be many solutions to a cooperative (essential) n-person game, the question arises as to which of them will in fact prevail. Each solution may correspond to a specific mode of behavior of the players or a specific form of social organization. This expresses the fact that in the same physical setting different types of social organization can be established, each one consistent in itself but in contradiction with other organizations. For example, we observe that the same technology allows the maintenance of varying economic systems, income distributions, and so on. If a *stable standard of behavior* exists (a mode of behavior accepted by society), then it can be argued that the only relevant solution is the one corresponding to this standard.

The choice of an imputation *not* in the solution set, while advantageous to each of those in the particular coalition that is able to enforce this imputation, cannot be maintained because another coalition can enforce another imputation, belonging to the solution set, that dominates the first one. Hence, a standard is set and proposals for imputations that are not in the solution will be rejected. The theory cannot state which imputation of all those belonging to the standard of behavior actually will be chosen—that is, which coalition will form. Work has been done to introduce new assumptions under which this may become feasible. No imputation contained in the solution set guarantees stability by itself, since each is necessarily dominated from the outside. But in turn each imputation is always protected against threats by another one *within* the solution set that dominates the imputation *not* in the solution set.

Since an imputation is a division of proceeds among the players, these conditions define a certain fairness, such that the classical problems of fair division (for example, cutting a cake) become amenable to game-theoretic analysis.

This conceptual structure is more complicated than the conventional view that society could be organized according to some simple principle of maximization. The conventional view would be valid only if there were inessentiality—that is, if there were no advantage in cooperation, or if cooperation were forbidden, or, finally, if a supreme authority were to do away with the entire imputation problem by simply assigning shares of income to the members of the society. Inessentiality would be the case for a strictly communistic society, which is formally equivalent to a Robinson Crusoe economy. This, in turn, is the only formal setup under which the classical notion of marginal utility is logically valid. Whether cooperation through formation of coalitions is advantageous to participants in a society, whether such cooperation, although advantageous, is forbidden, or whether compensations or side payments are ruled out by some authority although coalitions may be entered—these are clearly empirical questions. The theory should take care of all eventualities, and current investigations explore the different avenues. In economic life, mergers, labor unions, trade associations, cartels, etc., express the powerful tendencies toward cooperation. The cooperative case with side payments is the most comprehensive, and the theory was originally designed to deal with this case. Important results have been obtained for cooperative games without side payments (Aumann & Peleg 1961), and the fruitful idea of "bargaining sets" has been introduced (Aumann & Maschler 1964).

All indications point overwhelmingly to the benefits of cooperation of various forms and hence to the empirical irrelevance of those noncooperative, inessential games with uniquely determined solutions consisting only of one single imputation dominating all others (as described in the Lausanne school's general economic equilibrium).

Cooperation may depend on a particular flow of information among the players. Since the required level may not in fact be attainable, noncooperative solutions become important. Economic markets in which players act independently and have no incentive to deviate from a given state have been studied (Nash 1950). *Equilibrium points* can be determined as those points for which unilateral changes in strategy are unprofitable to everyone. As Nash has shown, every finite game, or the domain of mixed strategies, has at least one equilibrium point. If there is more than one equilibrium point, an intermixture of strategy choices need not give another equilibrium point, nor is the payoff to players the same if the points differ from each other.

There is no proof, as yet, that every cooperative n-person, zero-sum game for any $n > 4$ has a solution of the specified kind. However, every individual game investigated, even with arbitrarily large n, has been found to possess a solution. The indications are that the proof for the general case will eventually be given. Other definitions of solutions—still differing from that of the Lausanne–Robinson Crusoe convention—are possible and somewhat narrow the field of choices. They are inevitably based on further assumptions about the behavior of the participants in the game, which have to be justified from case to case.

Simple games

In certain n-person games the sole purpose is to form a *majority* coalition. These games are the "simple" games in which voting takes place. Ties in voting may occur, and weights may differ from one player to another; for example, the chairman of a committee may have more than one vote. A player's presence may therefore mean the difference between victory or defeat. Games of this nature can be identified with classical cases of production, where the players represent factors of production. It has been proven that even in relatively simple cases, although complete substitutability among players may exist, substitution rates may be undetermined and values are attributed to the players (factors) only by virtue of their *relation* to each other and not by virtue of their individual contribution. Thus, contrary to current economic doctrine, substitutability does not necessarily guarantee equality as far as value is concerned.

Simple games are suited for interpretation of many political situations in that they allow the determination of the weights, or power, of participants in decision processes. A particular power index has been proposed by Shapley. It is based on the notion of the average contribution a player can make to the coalitions to which he may belong, even considering, where necessary, the order in which he joins them. The weight of a senator, a congressman, and the president in the legislative process has been calculated for the United States. The procedure is applicable to other political systems—for example, the Security Council of the United Nations (Shapley 1953).

Composition of games

Every increase in the number of players brings new phenomena: with the increase from two to three players, coalitions become possible, from three to four, ties may occur among coalitions, etc. There is no guarantee that for very large n an asymptotic convergence of solutions will occur, since coalition formation always reduces large numbers of individual players to small numbers of coalitions acting upon each other. Thus, the increase in the number of players does not necessarily lead to a simplification, as in the case of an enlargement of the numbers of bodies in a physical system, which then allows the introduction of classical methods of statistical averages as a simplification. (When the game is inessential, the number of participants is irrelevant in any case.)

An effective extension of the theory by the enlargement of numbers can be achieved by viewing games played separately as one composite game and by introducing contributions to, or withdrawals from, the proceeds of a given game by a group of players outside the game under consideration. These more complicated notions involve constant-sum games and demonstrate, among other things, how the coalition formation, the degree of cooperation among players, and consequently the distribution of the proceeds among them are affected by the availability of amounts in excess of those due to their own strategies alone. Strategy is clearly greatly influenced by the availability of greater payments than those that can be made by only the other players. Thus, coalitions—namely, social structures—cannot be maintained if outside contributions become larger than specified amounts, such that as a consequence no coalition can exhaust the amounts offered. It can also be shown that the outside source, making contributions or withdrawals, can never be less than a group of three players.

These concepts and results are obviously of a rather complicated nature; they are not always directly accessible to intuition, as corresponds to a truly mathematical theory. When that level is reached, confidence in the mathematical results must override intuition, as the experience in the natural sciences shows. The fact that solutions of n-person games are not single numbers or single sets of numbers—but that the above-mentioned, more complicated structures emerge—is not an imperfection of the theory: it is a fundamental property of social organization that can be described only by game-theoretic methods.

Nonzero-sum games

Nonzero-sum games can be reduced to zero-sum games—which makes that entire theory applicable—by the introduction of a fictitious player, so that an n-person, nonzero-sum game becomes equivalent to an $(n + 1)$-person, zero-sum game. The fictitious player is either winning or losing, but since he is fictitious he can never become a member of a coalition. Yet he can be construed as proposing alternative imputations, thereby influencing the players' strategies and thus the course of the play. He will lose according to the degree of cooperation among the players. If the players cooperate perfectly, the maximum social benefit will be attained. In these games there is an increased role of threats, and their costs to the threatening player, although threats already occur in the zero-sum case.

The discriminatory solutions, first encountered for the three-person, zero-sum game, serve as instruments to approach these problems. Most applications to economics involve gains by the community—an economy being productive and there being no voluntary exchange unless both sides profit—while many other social phenomena fall under the domain of zero-sum games. The non-zero-sum theory is so far the part of game theory least developed in detail, although its foundations seem to be firmly established by the above procedure.

Applications

Game theory is applicable to the study of those social phenomena in which there are agents striving for their own advantage but not in control of all the variables on which the outcome depends. The wide range of situations of which this is true is obvious: they are economic, political, military, and strictly social in nature. Applications have been made in varying degree to all areas; some have led to experiments that have yielded important new insights into the theory itself and into special processes such as bargaining. Finally, the possibility of viewing the basic problem of statistics as a game against nature has given rise to modern statistical decision theory (Wald 1950). The influence of game theory is also evident in philosophy, information theory, cybernetics, and even biology.

OSKAR MORGENSTERN

[*See also the biography of* VON NEUMANN.]

BIBLIOGRAPHY

AUMANN, R. J.; and PELEG, B. 1961 Von Neumann–Morgenstern Solutions to Cooperative Games Without Side Payments. American Mathematical Society, *Bulletin* 66:173–179.

AUMANN, R. J.; and MASCHLER, M. 1964 The Bargaining Set for Cooperative Games. Pages 443–476 in M. Dresher, L. S. Shapley, and A. W. Tucker (editors), *Advances in Game Theory.* Princeton Univ. Press.

BAUMOL, WILLIAM J.; and GOLDFELD, STEPHEN M. (editors) 1967 Precursors in Mathematical Economics. Unpublished manuscript. → To be published in 1967 or 1968 by the London School of Economics and Political Science. Contains the letter from Waldegrave to Rémond de Montmort, first published in the second (1713) edition of Montmort (1708), describing his formulation, and a discussion by Harold W. Kuhn of the identity of Waldegrave.

BERGE, CLAUDE 1957 *Théorie générale des jeux à n personnes.* Paris: Gauthier-Villars.

BLACKWELL, DAVID; and GIRSHICK, M. A. 1954 *Theory of Games and Statistical Decisions.* New York: Wiley.

BRAITHWAITE, RICHARD B. 1955 *Theory of Games as a Tool for the Moral Philosopher.* Cambridge Univ. Press.

BURGER, EWALD (1959) 1963 *Introduction to the Theory of Games.* Englewood Cliffs, N.J.: Prentice-Hall. → First published in German.

DRESHER, MELVIN 1961 *Games of Strategy: Theory and Applications.* Englewood Cliffs, N.J.: Prentice-Hall.

DRESHER, MELVIN; SHAPLEY, L. S.; and TUCKER, A. W. (editors) 1964 *Advances in Game Theory.* Annals of Mathematic Studies, Vol. 32. Princeton Univ. Press.

EDGEWORTH, FRANCIS Y. (1881) 1953 *Mathematical Psychics: An Essay on the Application of Mathematics to the Moral Sciences.* New York: Kelley.

FRÉCHET, MAURICE; and VON NEUMANN, JOHN 1953 Commentary on the Three Notes of Émile Borel. *Econometrica* 21, no. 1:118–127.

►GAMSON, WILLIAM A. 1968 Coalitions: II. Coalition Formation. Volume 2, pages 529–534 in *International Encyclopedia of the Social Sciences.* Edited by David L. Sills. New York: Macmillan and Free Press.

KARLIN, SAMUEL 1959 *Mathematical Methods and Theory in Games, Programming and Economics.* 2 vols. Reading, Mass.: Addison-Wesley.

KUHN, HAROLD W.; and TUCKER, A. W. (editors) 1950–1959 *Contributions to the Theory of Games.* 4 vols. Princeton Univ. Press.

LUCE, R. DUNCAN; and RAIFFA, HOWARD 1957 *Games and Decisions: Introduction and Critical Survey.* A Study of the Behavioral Models Project, Bureau of Applied Social Research, Columbia University. New York. → First published in 1954 as *A Survey of the Theory of Games*, Columbia University, Bureau of Applied Social Research, Technical Report No. 5.

MCKINSEY, JOHN C. C. 1952 *Introduction to the Theory of Games.* New York: McGraw-Hill.

[MONTMORT, PIERRE RÉMOND DE] (1708) 1713 *Essay d'analyse sur les jeux de hazard.* 2d ed. Paris: Quillau. → Published anonymously.

MORGENSTERN, OSKAR 1963 *Spieltheorie und Wirtschaftswissenschaft.* Vienna: Oldenbourg.

NASH, JOHN F. JR. 1950 Equilibrium in n-Person Games. National Academy of Sciences, *Proceedings* 36:48–49.

PRINCETON UNIVERSITY CONFERENCE 1962 *Recent Advances in Game Theory.* Princeton, N.J.: The Conference.

SHAPLEY, L. S. 1953 A Value for n-Person Games. Volume 2, pages 307–317 in Harold W. Kuhn and A. W. Tucker (editors), *Contributions to the Theory of Games.* Princeton Univ. Press.

SHAPLEY, L. S.; and SHUBIK, MARTIN 1954 A Method for Evaluating the Distribution of Power in a Committee System. *American Political Science Review* 48: 787–792.

SHUBIK, MARTIN (editor) 1964 *Game Theory and Related Approaches to Social Behavior: Selections.* New York: Wiley.

SUZUKI, MITSUO 1959 *Gemu no riron.* Tokyo: Keisho Shobo.

VILLE, JEAN 1938 Sur la théorie générale des jeux ou intervient l'habilité des joueurs. Pages 105–113 in

Émile Borel (editor), *Traité du calcul des probabilités et de ses applications*. Volume 4: Applications diverses et conclusion. Paris: Gauthier-Villars.

VOGELSANG, RUDOLF 1963 *Die mathematische Theorie der Spiele*. Bonn: Dümmler.

VON NEUMANN, JOHN (1928) 1959 On the Theory of Games of Strategy. Volume 4, pages 13–42 in Harold W. Kuhn and A. W. Tucker (editors), *Contributions to the Theory of Games*. Princeton Univ. Press. → First published in German in Volume 100 of the *Mathematische Annalen*.

VON NEUMANN, JOHN; and MORGENSTERN, OSKAR (1944) 1953 *Theory of Games and Economic Behavior*. 3d ed., rev. Princeton Univ. Press. → A paperback edition was published in 1964 by Wiley.

VOROB'EV, N. N. (editor) 1961 *Matrichnye igry*. Moscow: Gosudarstvennoe Izdatel'stvo Fiziko–Matematicheskoi Literatury. → A collection of translations into Russian from foreign-language publications.

WALD, ABRAHAM (1950) 1964 *Statistical Decision Functions*. New York: Wiley.

WILLIAMS, JOHN D. 1954 *The Compleat Strategyst: Being a Primer in the Theory of Games and Strategy*. New York: McGraw-Hill.

Postscript

Many new concepts in game theory (apart from mere terminological changes such as replacing the word "solution" with "stable set") have been developed, especially in the theory of *n*-person games. The new concepts are too numerous to be discussed briefly.

Problems of cooperation and competition naturally dominate, and efforts are being made to arrive at solution concepts that are useful in practical applications to economics, sociology, and politics. Notably the idea of a "core" (a subset of the solution set) has found wide entry into economic literature (for a survey, see Schotter 1973). All variations in approaches that have been considered are based on the original von Neumann–Morgenstern theory. One result is of particular importance: W. F. Lucas has found a ten-person game that can be interpreted as a market game and that has no solution. This discovery has given rise to the search for new general principles that may govern social organizations, and it is proof of the enormous complexity of the social world. At the same time, a convergence toward more familiar notions of equilibrium has been observed when the number of players goes toward infinity—an important mathematical result, but hardly of great empirical relevance.

Mathematical interest in game theory has begun to spread over even wider areas. This interest creates new possibilities for social sciences, leading ever farther away from neoclassical economics, which is constrained by very narrow underlying assumptions. A new line of thought is to study further differences in the "power" of individual players and to allow for the possibility that the same player may be simultaneously involved in different power games, so that issues arise of optimal allocation of his resources among the games.

In the area of utility theory, closely related to game theory, important work has also been done (see Fishburn 1970).

OSKAR MORGENSTERN

ADDITIONAL BIBLIOGRAPHY

A collection of articles edited by Maurice Allais and Ole Hagen is planned for publication by Reidel as Rational Decisions Under Uncertainty. *Of special interest is "Some Reflections on Utility" by Oskar Morgenstern.*

ARNASZUS, HELMUT 1974 *Spieltheorie und Nutzenbegriff aus Marxistischer Sicht: Ein Kritik aktueller ökonomischer Theorien*. Frankfurt am Main: Suhrkamp.

AUMANN, ROBERT J.; and SHAPLEY, LLOYD S. 1974 *Values of Non-atomic Games*. Princeton Univ. Press.

BLAQUIÈRE, AUSTIN (editor) 1973 *Topics in Differential Games*. Amsterdam: North-Holland.

BOUZITAT, JEAN (1965) 1970 *Présentation synthétique de la théorie des jeux*. 2d ed. Cahiers du Bureau Universitaire de Recherche Opérationnelle, No. 7. Institut de Statistique, Université de Paris.

BUCHLER, IRA R.; and NUTINI, HUGO G. (editors) 1969 *Game Theory in the Behavioral Sciences*. Univ. of Pittsburgh Press.

DAVIS, MORTON D. 1972 *Game Theory: A Nontechnical Introduction*. New York: Basic Books.

FISHBURN, PETER C. 1970 *Utility Theory for Decision Making*. New York: Wiley.

FISHBURN, PETER C. 1973 *The Theory of Social Choice*. Princeton Univ. Press.

International Journal of Game Theory. → Published since 1971 by the Institute of Advanced Studies, Vienna.

KLAUS, GEORG 1968 *Spieltheorie in philosophischer Sicht*. Berlin: Deutscher Verlag der Wissenschaften.

KUHN, HAROLD W.; and SZEGÖ, G. P. (editors) 1971 *Differential Games and Related Topics*. Amsterdam: North-Holland; New York: American Elsevier.

LUCAS, WILLIAM F. 1971 Some Recent Developments in *n*-Person Game Theory. *Society for Industrial and Applied Mathematics Review* 13:491–523.

LUCAS, WILLIAM F. 1972 An Overview of the Mathematical Theory of Games. *Management Science* 18: 3–19. → Includes a bibliography.

MAY, FRANCIS B. 1970 *Introduction to Games of Strategy*. Boston: Allyn & Bacon.

MORGENSTERN, OSKAR 1972 Strategic Allocation and the Integral Games. Working Paper No. 3. Center of Applied Economics, New York University.

MUNIER, BERTRAND 1973 *Jeux et marchés*. Paris: Presses Universitaires.

PARTHASARATHY, T.; and RAGHAVAN, T. E. S. 1971 *Some Topics in Two-person Games*. New York: American Elsevier.

RAPOPORT, ANATOL 1966 *Two-person Game Theory: The Essential Ideas*. Ann Arbor: Univ. of Michigan Press.

RAPOPORT, ANATOL 1970 *n-Person Game Theory: Concepts and Applications*. Ann Arbor: Univ. of Michigan Press.

ROSENMÜLLER, JOACHIM 1971 *Kooperative Spiele und Märkte.* Berlin: Springer.

SCHOTTER, ANDREW 1973 Core Allocations and Competitive Equilibrium: A Survey. *Zeitschrift für Nationalökonomie* 33:281–313.

SCHWODIAUER, G. 1971 Glossary of Game Theoretical Terms. Working Paper No. 1. Center for Applied Economics, New York University.

SHUBIK, MARTIN 1975 *The Uses and Methods of Gaming.* New York: American Elsevier.

SUZUKI, MITSUO (editor) 1970 *Kyōsō shakai no gēmu no riron* (The Development of Game Theory in Competitive Society). Tokyo: Keisō Shobō.

II
ECONOMIC APPLICATIONS

The major economic applications of game theory have been in oligopoly theory, bargaining theory, and general equilibrium theory. Several distinct branches of game theory exist and need to be identified before our attention is limited to economic behavior. John von Neumann and Oskar Morgenstern, who first explored in depth the role of game theory in economic analysis (1944), presented three aspects of game theory which are so fundamentally independent of one another that with a small amount of editing their opus could have been published as three independent books.

The first topic was the description of a game, or interdependent decision process, in extensive form. This provided a phraseology ("choice," "decision tree," "move," "information," "strategy," and "payoff") for the precise definition of terms, which has served as a basis for studying artificial intelligence, for developing the behavioral theory of the firm (Cyert & March 1963), and for considering statistical decision making [*see* DECISION THEORY]. The definition of "payoff" has been closely associated with developments in utility theory (see Georgescu-Roegen 1968).

The second topic was the description of the two-person, zero-sum game and the development of the mathematical theory based upon the concept of the minimax solution. This theory has formal mathematical connections with linear programming and has been applied successfully to the analysis of problems of pure conflict; however, its application to the social sciences has been limited because pure conflict of interests is the exception rather than the rule in social situations [*see* PROGRAMMING].

The third subject to which von Neumann and Morgenstern directed their attention was the development of a static theory for the *n*-person ($n \geqslant 3$), constant-sum game. They suggested a set of stability and domination conditions which should hold for a cooperative solution to an *n*-person game. It must be noted that the implications of this solution concept were developed on the assumption of the existence of a transferable, interpersonally comparable linear utility which provides a mechanism for side payments. Since the original work of von Neumann and Morgenstern, twenty to thirty alternative solution concepts for the *n*-person, nonconstant-sum game have been suggested. Some have been of purely mathematical interest, but most have been based on considerations of bargaining, fair division, social stability, and other aspects of human affairs. Many of the solution concepts do not use the assumption of transferable utility.

Oligopoly and bargaining

Markets in which there are only a few sellers (oligopoly), two sellers (duopoly, a special case of oligopoly), one seller and one buyer (bilateral monopoly), and so on, lend themselves to game-theoretic analyses because the fate of each participant depends on the actions taken by the other participant or participants. The theory of games has provided a unifying basis for the mathematical and semimathematical works dealing with such situations and has also provided some new results. The methodology of game theory requires explicit and detailed definition of the strategies available to the players and of the payoffs associated with the strategies. This methodology has helped to clarify the different aspects of intent, behavior, and market structure in oligopolistic markets (Shubik 1957). So-called conjectural variations and lengthy statements regarding an oligopolist's (or duopolist's or bargainer's) moves and countermoves can be investigated in a unified way when expressed in terms of strategies.

Oligopoly. Perhaps the most pervasive concept underlying the writings on oligopoly is that of a noncooperative equilibrium. A group of individuals is in a state of noncooperative equilibrium if, in the individual pursuit of his own self-interest, no one in the group is motivated to change his strategy. This concept is basic in the works of Cournot, Bertrand, Edgeworth, Chamberlin, von Stackelberg, and many others. Nash (1951) has presented a general theory of noncooperative games, based on the equilibrium-point solution. This theory is directly related to Chamberlin's theory of monopolistic competition, among others.

The outcome given by a solution is called Pareto optimal if no participant can be made better off without some other participant's being made worse off. Noncooperative solutions, whose outcomes need not be Pareto optimal, have been distinguished from cooperative solutions, whose outcomes must be Pareto optimal. Also, equilibrium points are distinguished on the basis of whether the oligopoly model studied is static or dynamic. In much of the literature on oligopoly, quasi-cooperative solutions have been advanced and quasi-dynamic models have been suggested. Thus, while the Chamberlin large-group equilibrium can be interpreted as the outcome of a static noncooperative game, the small-group equilibrium and the market resolution suggested by Fellner (1949) are cast in a quasi-dynamic, quasi-cooperative framework. A limited amount of development of games of survival (Milnor & Shapley 1957) and games of economic survival (Shubik & Thompson 1959) has provided a basis for the study of multiperiod situations and for an extension of the noncooperative equilibrium concept to include quasi-cooperative outcomes.

New results. The recasting of oligopoly situations into a game-theory context has produced some new results in oligopoly theory (see, for example, Mayberry, Nash, & Shubik 1953; Shubik 1959*a*). Nash (1953) and Shubik (1959*a*) have developed the definition of "optimum threat" in economic warfare. The kinky oligopoly demand curve and the more general problem of oligopolistic demand have been re-examined and interpreted. Other results concern stability and the Edgeworth cycle in price-variation oligopoly; duopoly with both price and quantity as independent variables; and the development of diverse concepts applicable to cartel behavior, such as blocking coalitions (Scarf 1965), discriminatory solutions, and decomposable games.

Selten (1965) has been concerned with the problem of calculating the noncooperative equilibria for various classes of oligopolistic markets. His work has focused on both the explicit calculation and the uniqueness of equilibrium points. Vickrey (1961), Griesmer and Shubik (1963), and others have studied a class of game models applicable to bidding and auction markets. Working from the viewpoint of marketing and operations research, Mills (1961) and others have constructed several noncooperative game-theoretic models of competition through advertising. Jacot (1963) has considered problems involving location and spatial competition.

Behavioristic findings. Game theory can be given both a normative and a behavioristic interpretation. The meaning of "rational behavior" in situations involving elements of conflict and cooperation is not well defined. No single set of normative criteria has been generally accepted, and no universal behavior has been validated. Closely related to and partially inspired by the developments in game theory, there has been a growth in experimental gaming, some of which has been in the context of economic bargaining (Siegel & Fouraker 1960) or in the simulated environment of an oligopolistic market (Hoggatt 1959). Where there is no verbal or face-to-face communication, there appears, under the appropriate circumstances, to be some evidence in favor of the noncooperative equilibrium.

Bargaining. The theory of bargaining has been of special interest to economists in the context of bilateral monopoly, which can involve two firms, a labor union and a firm, or two individuals engaged in barter in the market place or trying to settle a joint estate. Any two-person, nonconstant-sum situation, be it haggling in the market or international negotiations, can be formally described in the same game-theoretic framework. However, there are several substantive problems which limit application of this framework and which have resulted in the development of different approaches. In nonconstant-sum games communication between the players is of considerable importance, yet its role is exceedingly hard to define. In games such as chess and even in many oligopolistic markets, a move is a well-defined physical act—moving a pawn in a definite manner or changing a price or deciding upon a production rate; in bargaining it may be necessary to interpret a statement as a move. The problem of interpreting words as moves in negotiation is critical to the description and understanding of bargaining and negotiation processes. This "coding" problem has to be considered from the viewpoint of many other disciplines, as well as that of game theory.

A desirable property of a theoretical solution to a bargaining problem is that it predicts a unique outcome. In the context of economics this would be a unique distribution of resources (and unique prices, if prices exist at all). Unfortunately, there are few concepts of solution pertaining to economic affairs which have this property. The price system and distribution resulting from a competitive market may in general not be unique; Edgeworth's solution to the bargaining problem was the contract curve, which merely predicts that the outcome will be some point among an infinite set of possibilities.

The contract curve has the property that any point on it is jointly optimal (both bargainers cannot improve their position simultaneously from a point on this curve) and individually rational (no point gives an individual less than he could obtain without trading). The Pareto-optimal surface is larger than the contract curve, for it is restricted only by the joint optimality condition. If it is assumed that a transferable comparable utility exists, then the Pareto-optimal surface (described in the space of the traders' utilities) is flat; if not, it will generally be curved. Any point on the Pareto-optimal surface that is individually rational is called an imputation. In the two-person bargain the Edgeworth contract curve coincides with two game-theoretic solutions, the *core* and the *stable set*. The core consists of all undominated imputations (it may be empty). A stable set is a set of imputations which do not dominate each other but which together dominate all other imputations. An imputation, α, is said to *dominate* another imputation, β, if (1) there exists a coalition of players who, acting jointly but independently of the others, could guarantee for themselves at least the amounts they would receive if they accepted α, and (2) each player obtains more in α than in β. The core and stable-set solutions can be defined with or without the assumption of transferable utilities. Neither of these solution concepts predicts a unique outcome.

One approach to bilateral monopoly has been to regard it as a "fair-division" problem, and several solution concepts, each one embodying a formalization of concepts of symmetry, justice, and equity, have been suggested (Nash 1953; Shapley 1953; Harsanyi 1956). These are generally known as *value* solutions, since they specify the amount that each participant should obtain. For the two-person case, some of the fair-division or arbitration schemes do predict unique outcomes. The Nash fair-division scheme assumes that utilities of the players are measurable, but it does not need assumptions of either comparability or transferability of utilities (Shubik 1966). Shapley's scheme does utilize the last two assumptions. Other schemes have been suggested by Raiffa (1953), Braithwaite (1955), Kuhn (in Shubik 1967), and others.

Another approach to bargaining is to treat it in the extensive form, describing each move explicitly and showing the time path taken to the settlement point. This involves attempting to parametrize qualities such as "toughness," "flexibility," etc. Most of the attempts to apply game theory in this manner belong to studies in social psychology, political science, and experimental gaming. However, it has been shown (Harsanyi 1956) that the dynamic process suggested by Zeuthen (1930) is equivalent to the Nash fair-division scheme.

General equilibrium

Game theory methods have provided several new insights in general equilibrium economics. Under the appropriate conditions on preferences and production, it has been proved that a price system that clears the market will exist, provided that each individual acts as an independent maximizer. This result holds true independently of the number of participants in the market; hence, it cannot be interpreted as a limiting phenomenon as the number of participants increases. Yet, in verbal discussions contrasting the competitive market with bilateral monopoly, the difference generally stressed is that between the market with many participants, each with little if any control over price, and the market with few participants, where the interactions of each with all the others are of maximum importance.

The competitive equilibrium best reflects the spirit of "the invisible hand" and of decentralization. The use of the word "competitive" is counter to both game-theoretic and common-language implications. It refers to the case in which, if each individual considers himself an isolated maximizer operating in an environment over which he has no control, the results will be jointly optimal.

Game-theoretic solutions. The power and appeal of the concept of competitive equilibrium appears to be far greater than that of mere decentralization. This is reflected in the finding that under the appropriate conditions the competitive equilibrium may be regarded as the limit solution for several conceptually extremely different game-theoretic solutions.

○ *Convergence of the core.* It has been noted that for bilateral monopoly the Edgeworth contract curve is the core. Edgeworth had suggested and presented an argument to show that if the number of traders is increased on both sides of the market, the contract curve would shrink (interpreted appropriately, given the change in dimensions). Shubik (1959*b*) observed the connection between the work of Edgeworth and the core; he proved the convergence of the core to the competitive equilibrium in the special case of the two-sided market with transferable utility and conjectured that the result would be generally true for any number of markets without transferable utility. This result was proved by Scarf (the proof, although achieved earlier, is described in Scarf 1965); Debreu and Scarf improved upon it (1963). Using the concept of a continuum of players (rather than considering a limit by replicating the finite number of players in each category, as was done by Shubik, Scarf, and Debreu), Aumann (1966) proved the coincidence of the core and competitive equilibrium. When transferable utility is assumed, the core converges to a single point and the competitive equilibrium is unique. Otherwise it may split and converge to the set of competitive equilibria.

The convergence of the core establishes the existence of a price system as a result of a theory which makes no mention of prices. The theory's prime concern is with the power of coalitions. It may be looked upon as a formalization of countervailing power, inasmuch as it rules out imputations which can be dominated by any group in the society.

Shapley and Shubik (1966) have shown the convergence of the value in the two-sided market with transferable utility. In unpublished work Shapley has proved a more general result for any number of markets, and Shapley and Aumann have worked on the convergence of a nontransferable utility value recently defined by Shapley. Harsanyi (1959) was able to define a value that generalized the Nash two-person fair-division scheme to situations involving many individuals whose utilities are not transferable. This preceded and is related to the new value of Shapley, and its convergence has not been proved.

There are several other value concepts (Selten 1964), all of which make use of symmetry axioms and are based upon some type of averaging of the contributions of an individual to all coalitions.

If one is willing to accept the value as reflecting certain concepts of symmetry and fairness, then in an economy with many individuals in all walks of life, and with the conditions which are required for the existence of a competitive equilibrium satisfied, the competitive equilibria will also satisfy these symmetry and fairness criteria.

Noncooperative equilibrium. One of the important open problems has been the reconciliation of the various noncooperative theories of oligopolistic competition with general equilibrium theory. The major difficulty is that the oligopoly models are open in the sense that the customers are usually not considered as players with strategic freedom, while the general equilibrium model considers every individual in the same manner, regardless of his position in the economy. Since the firms are players in the oligopoly models, it is necessary to specify the domain of the strategies they control and their payoffs under all circumstances. In a general equilibrium model no individual is considered a player; all are regarded as individual maximizers. Walras' law is assumed to hold, and supply is assumed to equal demand.

When an attempt is made to consider a closed economic model as a noncooperative game, considerable difficulties are encountered in describing the strategies of the players. This can be seen immediately by considering the bilateral monopoly problem; each individual does not really know what he is in a position to buy until he finds out what he can sell. In order to model this type of situation as a game, it may be necessary to consider strategies which do not clear the market and which may cause a player to become bankrupt—i.e., unable to meet his commitments. Shapley and Shubik (in Shubik 1967) have successfully modeled the closed two-sided two-commodity market without side payments and have shown that the noncooperative equilibrium point converges from below the Pareto-optimal surface to the competitive equilibrium point. They also have considered more goods and markets on the assumption of the existence of a transferable (but not necessarily comparable) utility.

When there are more than two commodities and one market, the existence of a unique competitive equilibrium point appears to be indispensable in defining the strategies and payoffs of players in a noncooperative game. No one has succeeded in constructing a satisfactory general market model as a noncooperative game without using a side-payment mechanism. The important role played by the side-payment commodity is that of a strategy decoupler. It means that a player with a supply of this type of "money" can decide what to buy even though he does not know what he will sell.

In summary, it appears that, in the limit, at least three considerably different game-theoretic solutions are coincidental with the competitive equilibrium solution. This means that by considering different solutions we may interpret the competitive market in terms of decentralization, fair division, the power of groups, and the attenuation of power of the individual.

The stable-set solution of von Neumann and Morgenstern, the bargaining set of Aumann and Maschler (1964), the "self-policing" properties of certain imputation sets of Vickrey (1959), and several other related cooperative solutions appear to be more applicable to sociology, and possibly anthropology, than to economics. There has been no indication of a limiting behavior for these solutions as numbers grow; on the contrary, it is conjectured that in general the solutions proliferate. When, however, numbers are few, as in cartel arrangements and in international trade, these other solutions provide insights, as Nyblén has shown in his work dealing with stable sets (1951).

Nonexistence of competitive equilibrium. When conditions other than those needed for the existence of a competitive equilibrium hold, such as external economies or diseconomies, joint ownership, increasing returns to scale, and interlinked tastes, then the different solutions in general do not converge. There may be no competitive equilibrium; the core may be empty; and the definition of a noncooperative game when joint property is at stake will call for a statement of the laws concerning damages and threats. (Similarly, even though the conditions for the existence of a competitive equilibrium are satisfied, the various solutions will be different if there are few participants.) When the competitive equilibrium does not exist, we must seek another criterion to solve the problem of distribution or, if possible, change the laws to reintroduce the competitive equilibrium. The other solutions provide different criteria. However, if a society desires, for example, to have its distribution system satisfy conditions of decentralization and fair division, or of fair division and limits on power of groups, it may be logically impossible to do so.

Davis and Whinston (1962), Scarf (1964), and Shapley and Shubik (1964) have investigated applications of game theory to external economies, to increasing returns to scale, and to joint ownership. In the case of joint ownership the relation between economics and politics as mechanisms for the distribution of the proceeds from jointly owned resources is evident.

It must be noted that the "many solutions" approach to distribution is in contrast to the type of welfare economics that considers a community welfare function or social preferences, which are not necessarily constructed from individual preferences.

Other applications

Leaving aside questions of transferable utility, there is a considerable difference between an economy in which there is only barter or a passive shadow price system and one in which the government, and possibly others, have important monetary strategies. Faxen (1957) has considered financial policy from a game-theoretic viewpoint.

There have been some diverse applications of game theory to budgeting and to management science, as can be seen in the articles by Bennion (1956) and Shubik (1955).

Nyblén (1951) has attempted to apply the von Neumann and Morgenstern concept of stable set to problems of macroeconomics. He notes that the Walrasian system bypasses the problem of individual power by assuming it away. He observes that in game theory certain simple aggregation procedures do not hold; thus, the solutions to a four-person game obtained by aggregating two players in a five-person game may have little in common with the solutions to the original five-person game. He outlines an institutional theory of the rate of interest based upon a standard of behavior and (primarily at a descriptive level) links the concepts of discriminatory solution and excess to inflation and international trade.

MARTIN SHUBIK

[*The reader who is not familiar with oligopoly theory and general equilibrium theory should consult* Arrow 1968; Markham 1968; Mishan 1968.]

BIBLIOGRAPHY

►ARROW, KENNETH J. 1968 Economic Equilibrium. Volume 4, pages 376–389 in *International Encyclopedia of the Social Sciences*. Edited by David L. Sills. New York: Macmillan and Free Press.

AUMANN, ROBERT J. 1966 Existence of Competitive Equilibria in Markets With a Continuum of Traders. *Econometrica* 34:1–17.

AUMANN, ROBERT J.; and MASCHLER, MICHAEL 1964 The Bargaining Set for Cooperative Games. Pages 443–476 in Melvin Dresher, Lloyd S. Shapley, and A. W. Tucker (editors), *Advances in Game Theory*. Princeton Univ. Press.

BENNION, E. G. 1956 Capital Budgeting and Game Theory. *Harvard Business Review* 34:115–123.

BRAITHWAITE, RICHARD B. 1955 *Theory of Games as a Tool for the Moral Philosopher*. Cambridge Univ. Press.

CYERT, RICHARD M.; and MARCH, JAMES G. 1963 *A Behavioral Theory of the Firm*. Englewood Cliffs, N.J.: Prentice-Hall.

DAVIS, OTTO A.; and WHINSTON, A. 1962 Externalities, Welfare, and the Theory of Games. *Journal of Political Economy* 70:241–262.

DEBREU, GERARD; and SCARF, HERBERT 1963 A Limit Theorem on the Core of an Economy. *International Economic Review* 4:235–246.

FAXEN, KARL O. 1957 *Monetary and Fiscal Policy Under Uncertainty*. Stockholm: Almqvist & Wiksell.

FELLNER, WILLIAM J. 1949 *Competition Among the Few: Oligopoly and Similar Market Structures*. New York: Knopf.

►GEORGESCU-ROEGEN, NICHOLAS 1968 Utility. Volume 16, pages 236–267 in *International Encyclopedia of the Social Sciences*. Edited by David L. Sills. New York: Macmillan and Free Press.

GRIESMER, JAMES H.; and SHUBIK, MARTIN 1963 Towards a Study of Bidding Processes. *Naval Research Logistics Quarterly* 10:11–21, 151–173, 199–217.

HARSANYI, JOHN C. 1956 Approaches to the Bargaining Problem Before and After the Theory of Games. *Econometrica* 24:144–157.

HARSANYI, JOHN C. 1959 A Bargaining Model for the Cooperative *n*-Person Game. Volume 4, pages 325–356 in Harold W. Kuhn and A. W. Tucker (editors), *Contributions to the Theory of Games*. Princeton Univ. Press. → Volume 4 was edited by A. W. Tucker and R. Duncan Luce.

HOGGATT, A. C. 1959 An Experimental Business Game. *Behavioral Science* 4:192–203.

JACOT, SIMON-PIERRE 1963 *Stratégie et concurrence de l'application de la théorie des jeux à l'analyse de la concurrence spatiale*. Paris: SEDES.

►MARKHAM, JESSE W. 1968 Oligopoly. Volume 11, pages 283–290 in *International Encyclopedia of the Social Sciences*. Edited by David L. Sills. New York: Macmillan and Free Press.

MAYBERRY, J. P.; NASH, J. F.; and SHUBIK, MARTIN 1953 A Comparison of Treatments of a Duopoly Situation. *Econometrica* 21:141–154.

MILLS, H. D. 1961 A Study in Promotional Competition. Pages 245–301 in Frank M. Bass et al. (editors), *Mathematical Models and Methods in Marketing.* Homewood, Ill.: Irwin.

MILNOR, JOHN W.; and SHAPLEY, LLOYD S. 1957 On Games of Survival. Volume 3, pages 15–45 in Harold W. Kuhn and A. W. Tucker (editors), *Contributions to the Theory of Games.* Princeton Univ. Press. → Volume 3 was edited by M. Dresher, A. W. Tucker, and P. Wolfe.

►MISHAN, E. J. 1968 Welfare Economics. Volume 16, pages 504–512 in *International Encyclopedia of the Social Sciences.* Edited by David L. Sills. New York: Macmillan and Free Press.

NASH, JOHN F. JR. 1951 Non-cooperative Games. *Annals of Mathematics* 54:286–295.

NASH, JOHN F. JR. 1953 Two-person Cooperative Games. *Econometrica* 21:128–140.

NYBLÉN, GÖREN 1951 *The Problem of Summation in Economic Sciences.* Lund (Sweden): Gleerup.

RAIFFA, HOWARD 1953 Arbitration Schemes for Generalized Two-person Games. Volume 2, pages 361–387 in Harold W. Kuhn and A. W. Tucker (editors), *Contributions to the Theory of Games.* Princeton Univ. Press.

SCARF, HERBERT E. 1964 Notes on the Core of a Productive Economy. Unpublished manuscript, Yale Univ., Cowles Foundation for Research in Economics.

[2]SCARF, HERBERT E. 1965 The Core of an n-Person Game. Unpublished manuscript, Yale Univ., Cowles Foundation for Research in Economics.

SELTEN, REINHARD 1964 Valuation of n-Person Games. Pages 577–626 in M. Dresher, Lloyd S. Shapley, and A. W. Tucker (editors), *Advances in Game Theory.* Princeton Univ. Press.

SELTEN, REINHARD 1965 Value of the n-Person Game. → Paper presented at the First International Game Theory Workshop, Hebrew University of Jerusalem.

SHAPLEY, LLOYD S. 1953 A Value for n-Person Games. Volume 2, pages 307–317 in Harold W. Kuhn and A. W. Tucker (editors), *Contributions to the Theory of Games.* Princeton Univ. Press.

SHAPLEY, LLOYD S.; and SHUBIK, MARTIN 1964 *Ownership and the Production Function.* RAND Corporation Research Memorandum, RM-4053-PR. Santa Monica, Calif.: The Corporation.

SHAPLEY, LLOYD S.; and SHUBIK, MARTIN 1966 *Pure Competition, Coalition Power and Fair Division.* RAND Corporation Research Memorandum, RM-4917. Santa Monica, Calif.: The Corporation.

SHUBIK, MARTIN 1955 The Uses of Game Theory in Management Science. *Management Science* 2:40–54.

SHUBIK, MARTIN 1957 Market Form, Intent of the Firm and Market Behavior. *Zeitschrift für Nationalökonomie* 17:186–196.

SHUBIK, MARTIN 1959*a* *Strategy and Market Structure: Competition, Oligopoly, and the Theory of Games.* New York: Wiley.

SHUBIK, MARTIN 1959*b* Edgeworth Market Games. Volume 4, pages 267–278 in Harold W. Kuhn and A. W. Tucker (editors), *Contributions to the Theory of Games.* Princeton Univ. Press. → Volume 4 was edited by A. W. Tucker and R. Duncan Luce.

SHUBIK, MARTIN 1966 Measureable, Transferable, Comparable Utility and Money. Unpublished manuscript, Yale Univ., Cowles Foundation for Research in Economics.

SHUBIK, MARTIN (editor) 1967 *Essays in Mathematical Economics in Honor of Oskar Morgenstern.* Princeton Univ. Press. → See especially Harold W. Kuhn, "On Games of Fair Division"; and Lloyd S. Shapley and Martin Shubik, "Concept and Theories of Pure Competition."

SHUBIK, MARTIN; and THOMPSON, GERALD L. 1959 Games of Economic Survival. *Naval Research Logistics Quarterly* 6:111–123.

SIEGEL, S.; and FOURAKER, L. E. 1960 *Bargaining and Group Decision Making: Experiments in Bilateral Monopoly.* New York: McGraw-Hill.

VICKREY, WILLIAM 1959 Self-policing Properties of Certain Imputation Sets. Volume 4, pages 213–246 in Harold W. Kuhn and A. W. Tucker (editors), *Contributions to the Theory of Games.* Princeton Univ. Press. → Volume 4 was edited by A. W. Tucker and R. Duncan Luce.

VICKREY, WILLIAM 1961 Counterspeculation, Auctions and Competitive Sealed Tenders. *Journal of Finance* 16:8–37.

○VON NEUMANN, JOHN; and MORGENSTERN, OSKAR (1944) 1953 *Theory of Games and Economic Behavior.* 3d ed., rev. Princeton Univ. Press. → A paperback edition was published in 1964 by Wiley.

ZEUTHEN, F. 1930 *Problems of Monopoly and Economic Warfare.* London: Routledge.

Postscript

[1]Since I noted that "no one has succeeded in constructing a satisfactory general market model as a noncooperative game," Shubik (1972*a*; 1972*b*), Shubik and Whitt (1973), and Shapley (1976) have produced several no-side-payment models of a closed economy with trade in either a fiat or a commodity money. This development appears to be a strong link between the theory of games and a microeconomic theory of money. It provides a basis for a mathematical institutional economics in the sense that the specification of all the rules of the game to be played in strategic form is tantamount to the definition of the workings of a monetary, credit, and banking system.

Considerable development in the study of games with a continuum of players is evidenced by the work of Aumann and Shapley (1974), Drèze et al. (1972), Shitovitz (1973), and others.

MARTIN SHUBIK

ADDITIONAL BIBLIOGRAPHY

AUMANN, ROBERT J.; and SHAPLEY, LLOYD S. 1974 *Values of Non-atomic Games.* Princeton Univ. Press.

DRÈZE, JACQUES et al. 1972 Cores and Prices in an Exchange Economy With an Atomless Sector. *Econometrica* 40:1091–1108.

FOLEY, DUNCAN K. 1970 Lindahl's Solution and the

Core of an Economy With Public Goods. *Econometrica* 38:66–72.
ROTHKOPF, MICHAEL A. 1969 A Model of Rational Competitive Bidding. *Management Science* 15:362–373.
[2]SCARF, HERBERT E. 1967 The Core of an *n*-Person Game. *Econometrica* 35:50–69. → A formal publication of Scarf (1965).
SELTEN, REINHARD 1973 A Simple Model of Imperfect Competition Where Four Are Few and Six Are Many. *International Journal of Game Theory* 2, no. 3:141–201.
SHAPLEY, LLOYD S. 1976 Noncooperative General Exchange. Pages 155–175 in Steven A. Y. Lin (editor), *Theory and Measurement of Economic Externalities.* New York: Academic Press.
SHAPLEY, LLOYD S.; and SHUBIK, MARTIN 1969 On Market Games. *Journal of Economic Theory* 1:9–25.
SHITOVITZ, BENYAMIN 1973 Oligopoly in Markets With a Continuum of Traders. *Econometrica* 41:467–501.
SHUBIK, MARTIN 1972*a* A Theory of Money and Financial Institutions: Fiat Money and Noncooperative Equilibrium in a Closed Economy. *International Journal of Game Theory* 1:243–268.
SHUBIK, MARTIN 1972*b* Commodity Money, Oligopoly Credit and Bankruptcy in a General Equilibrium Model. *Western Economic Journal* 10:24–38.
SHUBIK, MARTIN; and WHITT, WARD 1973 Fiat Money in an Economy With One Nondurable Good and No Credit (A Noncooperative Sequential Game). Pages 401–448 in Austin Blaquière (editor), *Topics in Differential Games.* Amsterdam: North-Holland; New York: American Elsevier.
SMITH, VERNON L. 1967 Experimental Studies of Discrimination Versus Competition in Sealed-bid Auction Markets. University of Chicago, Graduate School of Business, *Journal of Business* 40:56–84.
TELSER, LESTER G. 1972 *Competition, Collusion, and Game Theory.* London: Macmillan; Chicago: Aldine.

GAMING

See SIMULATION.

GAUSS, CARL FRIEDRICH

Carl Friedrich Gauss (1777–1855), greatest of German mathematicians, was born in Brunswick on April 30, 1777. (He was baptized Johann Friedrich Carl, but he later dropped his first name and reversed the second and third.) Ranked with Archimedes and Newton as one of the three greatest mathematicians of all time, he combined in a most unusual way a pure mathematician's interest in abstract ideas and logical rigor, a theoretical physicist's interest in the creation of mathematical models of the physical world, an astronomer's talent for keen observation, and an experimentalist's skill in the application and invention of methods of measurement. He was blessed with a stupendous faculty for mental calculation, which enabled him to explore numerical relationships experimentally and to carry out extensive or involved routine computations quickly and accurately; he also had a gift for learning ancient and modern languages, which became his hobby. He contributed mightily to every branch of pure and applied mathematics that existed in his day, some of which he had founded, and he made major contributions in astronomy, geodesy, physics, and metrology. Each of his many interests had its principal, but not exclusive, season: before 1800, philology and number theory; 1800–1820, astronomy; 1820–1830, geodesy, differential geometry, and conformal mapping; 1830–1840, geomagnetism, electromagnetism, and general theory of inverse-square forces; 1840–1855, topology and the geometry of functions of a complex variable. He died in Göttingen on February 23, 1855. Of his many contributions, those used most widely in the physical, biological, and social sciences today relate to the method of least squares, his first formulation of which dates from 1795 to 1798 and his second from 1821 to 1823.

Gauss was the only child of Gebhard Dietrich Gauss, a bricklayer and gardener, by his second wife, née Dorothea Benze, daughter of a stonemason. Gebhard Gauss, being skilled in writing and calculating, kept accounts for a local insurance company; he was esteemed by the townspeople, but at home he was harsh and uncouth, which repelled his brilliant son. Gauss's mother had no special schooling, could not write, and could barely read, but she was cheerful, intelligent, and of strong character. She had a genial and extraordinarily intelligent younger brother, Johann Friedrich, a skilled weaver of artistic damasks, who quickly spotted his nephew's unusual talents and capacities and enjoyed sharpening his wits on those of his sister's young genius. Gauss, as a small boy, thought highly of his uncle and later, lamenting his untimely death in 1809, often declared that a born genius had been lost in him.

[1]Gauss's precocity is unequaled in the history of mathematics. Before he was three, while watching his father's payroll calculations he detected an error and announced the correct result. He was admitted at age ten to the beginners' class in arithmetic at St. Katherine's Volksschule in Brunswick, where the speed and accuracy of his mental calculations so astonished the schoolmaster that he purchased for the boy the best obtainable textbook on arithmetic, which Gauss quickly mastered, convincing the schoolmaster that Gauss had gone beyond him. Luckily the schoolmaster had an assistant who was interested in mathematics. He and Gauss studied algebra and the rudiments of calculus together in the evenings, helping each

other over difficulties and amplifying the textbook proofs. Thus in his eleventh year Gauss became acquainted with the binomial theorem and, finding the textbook "proof" unsatisfactory when the exponent n is not a positive integer, devised his own proof of the convergence of the infinite series involved, which established him as one of the first of the "rigorists" and served as an inspiration for some of his greatest later work.

Early in 1791 Gauss's amazing powers came to the attention of Carl Wilhelm Ferdinand, duke of Brunswick, who became Gauss's patron, paid the expenses of his education at Caroline College from 1792 to 1795 and at the University of Göttingen from 1795 to 1798, and until his death in 1806 gave Gauss considerable additional financial support. At Caroline College, Gauss devoted himself with equal success to classical literature, philosophy, and advanced mathematics. He studied carefully the original works of Newton, Euler, and Lagrange. In March 1795 (Dunnington 1955, p. 391) he rediscovered that invaluable principle of number theory, the law of quadratic reciprocity, which Legendre had published in 1785, and of which Gauss was to publish the first rigorous proof in 1801. Yet, on entering the University of Göttingen in October 1795, he was still undecided whether to make mathematics or philology his career.

[2]Mathematics won on March 30, 1796, the day Gauss discovered that a regular polygon of 17 sides is amenable to straightedge-and-compass construction; such a construction had eluded mathematicians for two thousand years. By June 1, 1796, he had discovered much more: a regular polygon with an odd number of sides is amenable to such construction if and only if the number of its sides is a prime Fermat number, $F_n = 2^{2^n} + 1$, or a product of different prime Fermat numbers (five of which are known today). In 1796 he also developed the first rigorous proof of the fundamental theorem of algebra (i.e., that every nonconstant polynomial has a root), which became the subject of the doctoral dissertation for which he was awarded a PH.D. *in absentia* by the University of Helmstedt in 1799. In the autumn of 1798 Gauss polished the final draft of his greatest masterpiece, *Disquisitiones arithmeticae*, the printing of which began in 1799 but was not completed until September 1801, owing to the sale of the original print shop. Gauss brought together in this work his own original contributions to the theory of integral numbers and rational fractions and all of the principal related, but somewhat disconnected, results of his predecessors, so enriching the latter by rigorous reformulation and blending into a unified whole that this great book is regarded as marking the beginning of the theory of numbers as a separate, systematic branch of mathematics.

Among Gauss's early achievements was his reduction of the ecclesiastical calendar's extremely complicated computational procedure for finding the date of Easter (see *Encyclopaedia Britannica*, 11th ed., vol. 4, pp. 991–999) to a set of simple formulas from which the answer for any given year can be found in a few minutes. His motivation stemmed from his mother's inability to recall the exact date of his birth—only that it was a Wednesday, eight days before Ascension Day. As first published (1800), his procedure gave an incorrect date for Easter in 1734 and would have been incorrect again in 1886. Gauss supplied the necessary correction in 1807 and later provided an additional correction needed from 4200 on. (A semipopular exposition of Gauss's procedure, with a worked example, has been provided by H. Herbert Howe [1954].)

Gauss's first formulation of the method of least squares dates from his student days. In the autumn of 1794 he read (Galle 1924, p. 5) Lambert's discussion (1765, pp. 428–488) of the determination of the coefficients of a linear relationship $y = \alpha + \beta x$ from a set of n (>2) observational points (Y_i, x_i) by the method of averages. In 1795 Gauss conceived the simpler and more objective procedure of taking for α and β the values a and b that minimize the sum of squared residuals, $\sum_i (Y_i - a - bx_i)^2$, and worked out the computational details, except for the weighting of observations of unequal precision (Galle 1924, p. 7). In 1797 he attempted to justify his minimum-sum-of-squared-residuals (mssr) technique by means of the theory of probability but "found out soon that determination of the most probable value of an unknown quantity is impossible unless the probability distribution of errors of observation is known explicitly" ([1821*a*] 1880, p. 98). Therefore, "it seemed to him most natural to proceed the other way around" and to seek the probability distribution of errors that "in the simplest case would result in the rule, generally accepted as good, that the arithmetic mean of several values for the same unknown quantity obtained from equally reliable observations shall be considered the most probable value" ([1821*a*] 1880, p. 98).

The concept of a probability distribution, or "law," of errors originated with Thomas Simpson (1755): he studied the sampling distribution of the arithmetic mean of n independent and identically distributed errors subject to a discrete rectangular or a discrete triangular law of error, and concluded that "the more observations or experiments are made, the less will the conclusion be subject to err, provided they admit of being made under the same circumstances" (p. 93). In 1757 he extended his analysis to samples of n from a continuous triangular distribution. Studies of other continuous laws of error followed: rectangular, by Lagrange, published in 1774; double-exponential ($Ce^{-m|x|}, -\infty < x < +\infty$), by Laplace in 1774; semicircular ($C\sqrt{r^2 - x^2}, -r \leqslant x \leqslant +r$), by Daniel Bernoulli in 1778; and double-logarithmic ($C \log a/|x|, -a \leqslant x \leqslant +a$), by Laplace in 1781 (for further details, see Eisenhart 1964; Todhunter 1865; Merriman 1877).

Using inverse probability, Gauss found $f(v) = (h/\sqrt{\pi}) \exp(-h^2v^2)$ to be the required probability distribution of errors (1809, art. 175–177). He noted that h ($= 1/(\sigma\sqrt{2})$ in modern notation, where σ is the *root-mean-square error*, or *standard deviation*, of the distribution) "can be considered as the measure of precision of the observations" (art. 178). Although De Moivre in 1733 had adduced the *function* e^{-t^2} to approximate sums of successive terms of the binomial expansion of $(a+b)^n$ when n is large (for what he actually wrote, see Smith [1929] 1959, pp. 566–575), and Laplace, in a series of papers published during the period 1774–1786, had explored in great detail the use of this function and its derivatives to approximate various probability distributions arising in games of chance, notably the binomial and hypergeometric distributions and the corresponding incomplete beta-function forms obtained through application of Bayes' theorem, and had suggested tabulation of its integral for use in such problems (see Todhunter [1865] 1949, art. 890–911) neither De Moivre nor Laplace seems to have considered $C \exp(-h^2x^2)$, or an equivalent expression, as a *law of error* or as a *probability distribution* in its own right.

[3]Gauss then went on to deduce his mssr technique from $f(v)$, by what we would today call the method of maximum likelihood, finding that when the respective observations, $Y_i, i = 1, 2, \cdots, n$, are of unequal precision, each residual should be multiplied by the corresponding measure of precision, $h_i, i = 1, 2, \cdots, n$, before squaring (art. 179); derived the formula for the precision, $h(\overline{\overline{Y}}_0)$, of any weighted arithmetic mean, $\overline{\overline{Y}}$, of n independent observations (art. 181), finding that the precision, $h(\bar{Y})$, of the unweighted arithmetic mean, $\bar{Y}$, of n equally precise independent observations is proportional to $\sqrt{n}$ (art. 173); derived for the case of observations on linear functions of several unknown quantities $\alpha, \beta, \cdots$ the rule of formation of (what we call) the *normal equations* that jointly determine optimal estimators $A, B, \cdots$ (art. 180); outlined (art. 182) his famous method of elimination for solving the normal equations—most widely known today through a modification published by M. H. Doolittle (1878)—that provides as a byproduct an easily evaluated expression (art. 182, sec. 5) for the corresponding minimum sum of squared residuals in terms of quantities found in the course of their solution; and gave similar expressions (art. 183) for evaluating the precisions $h_A, h_B, \cdots$ of $A, B, \cdots$ as determinations of $\alpha, \beta, \cdots$.

An entry in Gauss's diary (see *Werke*, vol. 10, p. 533) indicates that he completed the foregoing development of the method of least squares on June 17, 1798. A fragment of his first application of these procedures in the spring of 1799 has been found (see *Werke*, vol. 12, pp. 64–68), and a short note ([1799] 1900, p. 136) dated August 24, 1799, on another application was published in October 1799.

In 1801 Gauss became deeply involved in the development and application of new astronomical methods which, in conjunction with his least squares techniques, resulted in amazingly accurate predictions, despite the scanty data available, of the orbits of some newly discovered planets; the success of his new methods brought about Gauss's immediate recognition as a first-rank astronomer (Bell 1937; Dunnington 1955, pp. 49–57). In 1805 Gauss began preparation (in German) of his second masterpiece, *Theoria motus corporum coelestium* . . . (1809), in which he was to give a complete system of formulas and procedures for computing the motion of a body whose orbit is a conic section, and a general method for determining the orbit of a comet or a planet from only three observed positions. The third section (art. 172–189) he devoted to a detailed exposition of the theory and application of his least squares methods. The German text was completed in 1806 (Galle 1924, p. 10). When the work was offered for publication in 1807, the publisher accepted it only on condition that Gauss translate it into Latin, because of the very unsettled political situation in Germany following the disastrous defeat of the Prussian army (under the leadership of Gauss's patron, the duke of Brunswick) by the Napoleonic forces at the battle of Auerstedt in October 1806. Consequently, full details of Gauss's first formulation of the method of least squares were not published until the Latin translation appeared in 1809. In the meantime, Legendre's independent formulation of his methods of least squares had appeared in print (1805); this went little beyond what

Gauss had developed in 1795—no probability considerations are involved, and there is no discussion of precision or weighting of observations. A controversy followed, bitter on Legendre's part, in which many persons became involved, and which, as Bell explains, "was most unfortunate for the future development of mathematics" ([1937] 1956, p. 331).

[3]In his 1809 presentation, Gauss regarded the measures of precision $h_1, h_2, \cdots$ of the respective observations involved as known quantities and said nothing about how to determine their values in practice. F. W. Bessel took the first step in this direction: in 1815 he introduced the *probable error* as a measure of imprecision (1815, p. 234), which he later defined (1816, p. 142) as the magnitude, r, that an error has an equal chance of exceeding (or being less than) in absolute value; and from 48 determinations of the right ascension of the polestar he obtained a value for the probable error of such determinations (1815, p. 234) by means of the formula $R = 0.8453\sum_i|Y_i - \bar{Y}|/n$, where $\bar{Y} = \sum_i Y_i/n$. (The numerical result given conforms to this formula [cf. Gauss 1816, art. 8], which is not stated.) The following year, Bessel (1816, pp. 141–142) showed that the probable error for Gauss's error distribution, $f(v)$, is given by $r = 0.476936h^{-1} = 0.8453\epsilon = 0.6745\sigma$. In this formulation $\epsilon = 1/(h\sqrt{\pi})$ represents the mean absolute error of the distribution and $\sigma = 1/(h\sqrt{2})$ is the root-mean-square error of the distribution, respectively. Gauss immediately pointed out (1816, art. 3) that given m independent errors V_1, V_2, $\cdots$, V_m distributed according to his error function, $f(v)$, then for "m large or small" the "most probable values" (or, in modern terminology, the *maximum likelihood estimators*) of h and r are $\hat{H} = \sqrt{m/2\sum_i V_i^2}$ and $\hat{R} = 0.6744897\sqrt{\sum_i V_i^2/m}$, respectively. He showed (art. 4) that for large m, $\hat{H} - h$ and $\hat{R} - r$ are distributed approximately as $f(v)$, with $h_H = \sqrt{m}/h$ and $h_R = \sqrt{m}/r$, and gave (*loc. cit.*) explicit expressions, in terms of $\hat{H}$ and $\hat{R}$, for the "probable limits" of (in modern terminology, *50 per cent confidence limits* for) "the true values of h and r." Next he considered (art. 5) the estimation of r by $R_p = C_p\sqrt[p]{\sum_i|V_i|^p/m}$, with $C_p = C_p(m,h)$ chosen so that r is the mean of the large-sample distribution of R_p, and found R_2 to be the most precise, noting (art. 6) that R_2 for $m = 100$ is as precise a measure of r as is R_1 for $m = 114$, R_3 for $m = 109$, R_4 for $m = 133$, R_5 for $m = 178$, and R_6 for $m = 251$. Finally, he considered (art. 7) the estimation of r by the "middlemost" (i.e., the *median*) of the absolute errors $|V_i|$ when m is odd and found that this procedure requires $m = 249$ in order to achieve precision comparable to R_2 for $m = 100$. Gauss then evaluated (art. 8) R_1, R_2, R_3, and R_4 for the 48 determinations considered by Bessel (1815), taking the residuals, $Y_i - \bar{Y}$, as measures of the corresponding errors, V_i, and concluded that Bessel had used the formula R_1.

On February 15, 1821, Gauss presented to the Royal Academy of Sciences in Göttingen a completely new formulation of the method of least squares that was entirely free from dependence on any particular probability distribution of errors. Entitled "Theoria combinationis observationum erroribus minimis obnoxiae. Pars prior" (1821*a*), it stemmed from the earlier work of Laplace: Three years before Gauss's birth, Laplace ([1774] 1891, pp. 41–48) had said that to estimate a parameter θ one ought to use that function $T = T(Y_1, Y_2, \cdots)$ of the observations, $Y_1, Y_2, \cdots$, for which the mean (or expected) absolute error of estimation, $E\{|T - \theta|\}$, is a *minimum* for the given probability distribution of errors, and, for the case of independent identically distributed observations, Laplace gave an explicit procedure for finding such a function for estimating the location parameter of their common distribution, $f(y - \tau)$, when this is completely specified except for the value of τ. Finally, Laplace showed that in the case of independent observations of equal (or unequal) precision, Gauss's technique of minimizing the sum of squared residuals leads to the same estimators as Laplace's own procedure for minimizing the mean absolute error of estimation when and only when (Laplace [1812] 1820, book 2, chapter 4, sec. 23) the errors $X_i = Y_i - \tau$ of the respective observations are distributed in accordance with Gauss's law of error,

$$\frac{h_i}{\sqrt{\pi}} \exp\left(-h_i^2 x_i^2\right), \qquad i = 1, 2, \cdots.$$

Gauss ([1821*a*] 1880, art. 6–7) proposed using instead the mean *square* error of an observation, $E\{(Y - \eta)^2\}$, or of an estimator $E\{(T - \theta)^2\}$, as a better measure of "uncertainty" ("incertitudo") and derived (art. 9) his remarkable inequalities (see Savage 1961, eq. 9) for the probability that a random variable Z with a continuous unimodal probability distribution will differ (positively or negatively) from its modal value z_0 by more than λ times its root-mean-square error measured from z_0. Then he showed (art. 18–23) that when the mean values $E\{Y_i\} = \eta_i$ of a set of n independent observations $Y_1, Y_2, \cdots, Y_n$ are linear functions of k ($< n$) unknown parameters θ_j ($j = 1, 2, \cdots, k$) and (in modern terminology) the *variances*

$E\{(Y_i - \eta_i)^2\} = \sigma_i^2$ of the Y_i are all finite, then the mssr technique yields estimators T_j of the θ_j having minimum mean-square error $(j = 1, 2, \cdots, k)$, whatever the distribution(s) of errors of the observations. In a second memoir (1823), he extended the preceding result to estimators L_r of linear functions $\Lambda_r(\theta_1, \theta_2, \cdots, \theta_k)$ of the θ's $(r = 1, 2, \cdots, m; m \leqslant k)$ and showed (art. 37–40) that the resultant minimum sum of squared residuals is strictly equivalent to the sum of $n - m$ independent errors (i.e., has $n - m$ degrees of freedom). These results, at one time attributed erroneously to A. A. Markov (for explanation, see Neyman ([1938] 1952, p. 228), are derived and discussed by various recent authors (e.g., Graybill 1961; Scheffé 1959; Zelen 1962) as the "Gauss–Markov theorem," and the method of least squares is thereby endorsed as a procedure for obtaining *minimum variance linear unbiased estimators*.

Gauss definitely preferred his 1821 formulation of the method of least squares above all others (for his statement to this effect on February 26, 1821, see [1821*b*] 1880, p. 99). In a letter to F. W. Bessel (Gauss & Bessel 1880) dated February 28, 1839 (an excerpt from which is given in his *Werke*, vol. 8, pp. 146–147), he remarked that he had never made a public statement of his reasons for abandoning the metaphysical approach of his first formulation but that a decisive reason was his belief that maximizing the probability of a zero error is less important than minimizing the probability of committing large errors.

In his later years, Gauss took special pride in his contributions to the development of the method of least squares and, despite a lifelong aversion to teaching, gave a course on this subject at the University of Göttingen each year from 1835 until his death. During his lifetime, the method of least squares became a basic tool in astronomy and geodesy throughout the world and has remained one to this day. And when Karl Pearson and G. Udny Yule began to develop the mathematical theory of correlation in the 1890s, they found that much of the mathematical machinery that Gauss devised for finding "best values" for the parameters of empirical formulas by the method of least squares was immediately applicable in correlation analysis, in spite of the fact that the aims of correlation analysis are the very antithesis of those of the theory of errors. As Galton remarked in his *Memories of My Life* (1908, p. 305), "The primary objects of the Gaussian Law of Error were exactly opposed, in one sense, to those to which I applied them. They were to get rid of, or to provide a just allowance for errors. But these errors or deviations were the very things I wanted to preserve and to know about." In consequence, Gauss's contributions to the method of least squares embody mathematics essential to statistical theory and its applications in almost every field of science today.

[4]Gauss was visited in 1829 by the Belgian astronomer–physicist–statistician Adolphe Quetelet, who was engaged in making geomagnetic measurements in Holland, Germany, Italy, and Switzerland. Gauss had been interested in geomagnetism from around the turn of the century and had remarked, in a letter to Heinrich Olbers dated March 1, 1803, "I believe that this offers a greater field for the application of mathematics than has yet been supposed" (Olbers & Gauss, *Briefwechsel* . . .); his intense activity in other fields had thus far prevented his undertaking research in geomagnetism, despite the strenuous efforts of Alexander von Humboldt, in 1804 and again in 1828, to persuade him to do so. When Quetelet arrived, he found Gauss studying Russian for relaxation: "I have been very fatigued," he said, "from occupying myself with astronomy, geodesy, and other subjects that I know fairly well; I wanted to turn my attention to a language that I did not know at all, and now I am reading Russian" (Quetelet 1866, p. 646). Measurement of the intensity of the earth's magnetism was new to Gauss, and he was eager to know how such measurements were made and the precision that could be achieved (p. 645). Quetelet set up his apparatus in Gauss's yard, and together they conducted a series of experiments, taking observations simultaneously but in slightly different manners. The agreement of the results obtained astonished Gauss, who exclaimed: "But these observations conform to the precision of astronomical observations" (p. 646).

From January 1831 on, geomagnetic measurements were made regularly at Göttingen. By February 1832, Gauss was deeply involved in research on geomagnetism and had found that he could express the intensity of geomagnetism in what he called "absolute units," that is, in terms of units of the three fundamental physical quantities: length, mass, and time. On December 15, 1832, he presented his findings to the Royal Academy of Sciences in Göttingen in a paper entitled "Intensitas vis magneticae terrestris ad mensuram absolutam revocata" (1832), which was promptly recognized as one of the most important papers of the century.

Gauss made electromagnetic measurements for the first time in October 1832. By Easter 1833, he and his young physicist colleague Wilhelm Weber had put into operation an electromagnetic telegraph between Gauss's observatory and Weber's physics laboratory (Dunnington 1955, pp. 147–148, 395). They sent only individual words at first, and later complete sentences. Plans were drawn up in 1835–1836 for its use on the Leipzig–Dresden railroad but were dropped when the railroad authorities declared that the wires would have to be put underground. A monument showing Gauss and Weber discussing their telegraph was erected on the campus of the University of Göttingen.

In 1832 Gauss had begun preparation of his "Allgemeine Theorie des Erdmagnetismus" ("General Theory of Terrestrial Magnetism"); completion was delayed by lack of experimental material. At Gauss's suggestion a magnetic observatory (of nonmagnetic construction) was erected at the University of Göttingen in 1833. By 1836 Göttingen had become the principal European center for research on geomagnetism; and an association of magnetic observatories, known as the Göttingen Magnetic Union, was formed to coordinate simultaneous measurement of geomagnetic phenomena throughout Europe. In 1837 Gauss and Weber collaborated in the invention of a galvanometerlike device, the bifilar magnetometer, for measuring magnetic field intensities; and with it, Gauss verified the inverse-square law of magnetic attraction to which he had already been led by theory. His "Allgemeine Theorie . . ." appeared in 1839 and was followed in 1840 by his (with Weber) "Atlas des Erdmagnetismus" (1840) and his great treatise "Allgemeine Lehrsätze in Beziehung auf die im verkehrten Verhältnisse . . ." (1840)—i.e., what today we call "potential theory"—which marked the peak of his work in physics and the close of his work on magnetism. Near the end of his life, Gauss, like Laplace, Fourier, and Poisson, turned his attention to the social sciences and to the help that they might derive from the physical sciences. In particular, he took an active interest in application of the theory of probability to social laws. Thus, in 1847, he corresponded with the Danish astronomer Heinrich Christian Schumacher on the laws of mortality and on the construction of mortality tables (Quetelet 1866, pp. 653–655).

Called "the prince of mathematicians" even in his lifetime, Gauss received a steady stream of honors (listed chronologically in Dunnington 1955, appendix B), beginning with his election in 1802 as a corresponding member of the Imperial Academy of Arts and Sciences in St. Petersburg. He was elected a fellow of the Royal Society of London in 1804 and received the Copley Medal in 1838; he became a full member of the Royal Academy of Sciences in Berlin in 1810, a fellow of the American Academy of Arts and Sciences in Boston, Massachusetts, in 1822, and a member of the American Philosophical Society in Philadelphia in 1853. In 1842 the highest order conferred by the kingdom of Prussia, *Pour le mérite*, was awarded to him. On July 16, 1849, exactly 50 years after receipt of his doctorate, a Gauss jubilee was held in Göttingen at which honors were showered upon him, including honorary citizenship of Brunswick and Göttingen, which he prized above all the rest.

CHURCHILL EISENHART

[*For the historical context of Gauss's work, see the biographies of* LAPLACE; MOIVRE; *for discussion of the subsequent development of his ideas, see* ESTIMATION; LINEAR HYPOTHESES, *article on* REGRESSION; MULTIVARIATE ANALYSIS, *article on* CORRELATION METHODS; *and the biographies of* PEARSON; QUETELET; YULE.]

A BIBLIOGRAPHY OF GAUSS

WORKS BY GAUSS

(1799) 1900 Zur Geschichte der Entdeckung der Methode der kleinsten Quadrate. Volume 8, pages 136–141 in *Carl Friedrich Gauss Werke.* Göttingen: Dieterichsche Universitäts-Druckerei.

(1800) 1874 Berechnung des Osterfestes. Volume 6, pages 73–79 in *Carl Friedrich Gauss Werke.* Göttingen: Dieterichsche Universitäts-Druckerei.

(1801) 1966 *Disquisitiones arithmeticae.* English translation by Arthur A. Clarke. New Haven: Yale Univ. Press.

(1807) 1874 Noch etwas über die Bestimmung des Osterfestes. Volume 6, pages 82–86 in *Carl Friedrich Gauss Werke.* Göttingen: Dieterichsche Universitäts-Druckerei.

(1809) 1963 *Theory of Motion of the Heavenly Bodies Moving About the Sun in Conic Sections.* New York: Dover. → First published as *Theoria motus corporum coelestium. . . .*

(1816) 1880 Bestimmung der Genauigkeit der Beobachtungen. Volume 4, pages 109–117 in *Carl Friedrich Gauss Werke.* Göttingen: Dieterichsche Universitäts-Druckerei. → An English translation appears in *Gauss's Work (1803–1826).* See also section 3 of Whittaker and Robinson 1924.

(1821*a*) 1880 Theoria combinationis observationum erroribus minimis obnoxiae. Pars prior. Volume 4, pages 1–26 in *Carl Friedrich Gauss Werke.* Göttingen: Dieterichsche Universitäts-Druckerei. → An English translation appears in *Gauss's Work (1803–1826).*

(1821*b*) 1880 Anzeigen: Theoria combinationis observationum erroribus minimis obnoxiae. Pars prior. Volume 4, pages 95–100 in *Carl Friedrich Gauss Werke.* Göttingen: Dieterichsche Universitäts-Druckerei.

(1823) 1880 Theoria combinationis observationum erroribus minimis obnoxiae. Pars posterior. Volume 4, pages 27–53 in *Carl Friedrich Gauss Werke.* Göttingen: Dieterichsche Universitäts-Druckerei. → An English translation is in *Gauss's Work (1803–1826).*

(1825–1827) 1965 *General Investigations of Curved Surfaces.* Hewlett, N.Y.: Raven Press. → Translations of Gauss's 1827 paper "Disquisitiones generales circa superficies curvas," his abstract of it, and his 1825 fragment "Neue allgemeine Untersuchungen über die krummen Flächen."

(1832) 1877 Intensitas vis magneticae terrestris ad mensuram absolutam revocata. Volume 5, pages 79–118 in *Carl Friedrich Gauss Werke*. Göttingen: Dieterichsche Universitäts-Druckerei. → Magie 1935 contains English excerpts.

(1839) 1966 General Theory of Terrestrial Magnetism. Volume 2, pages 184–251 in Richard Taylor (editor), *Scientific Memoirs, Selected From the Transactions of Foreign Academies of Science, and Learned Societies, and From Foreign Journals*. New York: Johnson. → First published in German.

(1840) 1966 General Propositions Relating to Attractive and Repulsive Forces Acting in the Inverse Ratio of the Square of the Distance. Volume 3, pages 153–196 in Richard Taylor (editor), *Scientific Memoirs, Selected From the Transactions of Foreign Academies of Science, and Learned Societies, and From Foreign Journals*. New York: Johnson. → First published in German.

(1840) 1929 GAUSS, CARL FRIEDRICH; and WEBER, WILHELM Atlas des Erdmagnetismus nach den Elementen der Theorie entworfen: Supplement zu den Resultaten aus den Beobachtungen des magnetischen Vereins unter Mitwirkung von C. W. B. Goldschmidt. Volume 12, pages 335–408 in *Carl Friedrich Gauss Werke*. Göttingen: Dieterichsche Universitäts-Druckerei. → Charts following page 408.

COLLECTIONS OF GAUSS'S WORKS

1855 *Méthode des moindres carrés: Mémoires sur la combinaison des observations*. Translated by J. Bertrand, and published with the authorization of the author. Paris: Mallet-Bachelier.

Abhandlungen zur Methode der kleinsten Quadrate. Thesaurus mathematicae, Vol. 5. Würzburg: Physica-Verlag, 1964.

Carl Friedrich Gauss Werke. 12 vols. Göttingen: Dieterichsche Universitäts-Druckerei, 1870–1933.

Gauss's Work (1803–1826) on the Theory of Least Squares. Translated by Hale F. Trotter. Statistical Techniques Research Group, Technical Report, No. 5. Princeton, N.J. Princeton Univ., 1957. → Prepared from Gauss 1855.

WORKS CONTAINING EXTRACTS BY GAUSS

MAGIE, WILLIAM F. (1935) 1963 *A Source Book in Physics*. Cambridge, Mass.: Harvard Univ. Press. → See especially extracts from Gauss 1832.

MIDONICK, HENRIETTA O. (editor) 1965 *The Treasury of Mathematics: A Collection of Source Material in Mathematics*. New York: Philosophical Library.

SMITH, DAVID EUGENE (1929) 1959 *A Source Book in Mathematics*. 2 vols. New York: Dover.

TAYLOR, RICHARD (editor) 1966 *Scientific Memoirs, Selected From the Transactions of Foreign Academies of Science, and Learned Societies, and From Foreign Journals*. 7 vols. New York: Johnson.

CORRESPONDENCE

Briefe von C. F. Gauss an B. Nicolai. Karlsruhe: Braun, 1877.

GAUSS, CARL FRIEDRICH; and BESSEL, FRIEDRICH W. *Briefwechsel zwischen Gauss und Bessel*. Leipzig: Engelmann, 1880.

GAUSS, CARL FRIEDRICH; and BOLYAI, WOLFGANG *Briefwechsel zwischen Carl Friedrich Gauss und Wolfgang Bolyai*. Leipzig: Teubner, 1899.

GAUSS, CARL FRIEDRICH; and GERLING, CHRISTIAN L. *Briefwechsel zwischen Carl Friedrich Gauss und Christian Ludwig Gerling*. Berlin: Elsner, 1927.

GAUSS, CARL FRIEDRICH; and SCHUMACHER, HEINRICH C. *Briefwechsel zwischen C. F. Gauss und H. C. Schumacher*. 6 vols. Altona: Esch, 1860–1865.

HUMBOLDT, ALEXANDER VON; and GAUSS, CARL F. *Briefe zwischen A. v. Humboldt und Gauss*. Leipzig: Engelmann, 1877.

OLBERS, WILHELM; and GAUSS, CARL F. *Briefwechsel zwischen Olbers und Gauss*. 2 parts. Berlin: Springer, 1900–1909. → Published as Volume 2 of Wilhelm Olbers, *Sein Leben und seine Werke*.

SUPPLEMENTARY BIBLIOGRAPHY

BELL, ERIC T. (1937) 1956 Gauss: The Prince of Mathematicians. Volume 1, pages 295–299 in James R. Newman (editor), *The World of Mathematics: A Small Library of the Literature of Mathematics From A'h-mosé the Scribe to Albert Einstein*. New York: Simon & Schuster. → A paperback edition was published in 1960.

BESSEL, FRIEDRICH W. 1815 Ueber den Ort des Polarsterns. *Astronomisches Jahrbuch* [1818]:233–241.

BESSEL, FRIEDRICH W. 1816 Untersuchungen über die Bahn des Olbersschen Kometen. Akademie der Wissenschaften, Berlin, Mathematische Klasse, *Abhandlungen* [1812–1813]:119–160.

DOOLITTLE, M. H. 1878 [Method Employed in This Office in the Solution of Normal Equations and in the Adjustment of a Triangulation.] U.S. Coast and Geodetic Survey, *Report of the Superintendent* [1878], Appendix 8, Paper 3:115–120.

DUNNINGTON, G. WALDO 1955 *Carl Friedrich Gauss, Titan of Science: A Study of His Life and Work*. New York: Hafner. → Contains a comprehensive bibliography of works by and about Gauss.

EISENHART, CHURCHILL 1964 The Meaning of "Least" in Least Squares. *Journal of the Washington Academy of Sciences* 54:24–33.

Festschrift zur Feier der Enthüllung des Gauss–Weber-Denkmals in Göttingen. 2 vols. 1899 Leipzig: Teubner.

GALLE, A. 1924 Über die geodätischen Arbeiten von Gauss. Volume 11, part 2, pages 1–165 in *Carl Friedrich Gauss Werke*. Göttingen: Dieterichsche Universitäts-Druckerei.

GALTON, FRANCIS 1908 *Memories of My Life*. London: Methuen.

GRAYBILL, FRANKLIN A. 1961 *An Introduction to Linear Statistical Models*. Vol. 1. New York: McGraw-Hill.

HÄNSELMANN, LUDWIG 1878 *Karl Friedrich Gauss: Zwölf Kapitel aus seinem Leben*. Leipzig: Duncker & Humblot.

HOWE, H. HERBERT 1954 How to Find the Date of Easter. *Sky and Telescope* 13:196.

KLEIN, FELIX et al. 1911–1920 *Materialien für eine wissenschaftliche Biographie von Gauss*. 8 vols. Leipzig: Teubner.

LAMBERT, JOHANN HEINRICH 1765 *Beyträge zum Gebrauche der Mathematik und deren Anwendung*. Vol. 1. Berlin: Buchladen der Realschule.

LAPLACE, PIERRE SIMON DE (1774) 1891 Mémoire sur la probabilité des causes par les événements. Volume 8, pages 27–65 in *Oeuvres complètes de Laplace*. Paris: Gauthier-Villars. → See especially "Problème III: Déterminer le milieu que l'on doit prendre entre trois observations données d'un même phénomène" on pages 41–48.

LAPLACE, PIERRE SIMON DE (1812) 1820 *Théorie analytique des probabilités*. 3d ed., rev. Paris: Courcier. → Smith 1929 contains English extracts from Book 2.

LAPLACE, PIERRE SIMON DE *Oeuvres complètes de Laplace.* 14 vols. Paris: Gauthier-Villars, 1878–1912.

LEGENDRE, ADRIEN M. (1805) 1959 On the Method of Least Squares. Volume 2, pages 576–579 in David Eugene Smith, *A Source Book in Mathematics.* New York: Dover. → First published as "Sur la méthode des moindres quarrés" in Legendre's *Nouvelles méthodes pour la détermination des orbites des comètes.*

MACK, HEINRICH (editor) 1927 *Carl Friedrich Gauss und die Seinen: Festschrift zu seinem 150. Geburtstage.* Brunswick: Appelhans.

MERRIMAN, MANSFIELD 1877 A List of Writings Relating to the Method of Least Squares, With Historical and Critical Notes. Connecticut Academy of Arts and Sciences, *Transactions* 4:151–232.

MOIVRE, ABRAHAM DE (1733) 1959 A Method of Approximating the Sum of the Terms of the Binomial $\overline{a+b}\backslash^n$ Expanded Into a Series, From Whence Are Deduced Some Practical Rules to Estimate the Degree of Assent Which Is to Be Given to Experiments. Volume 2, pages 566–575 in David Eugene Smith, *A Source Book in Mathematics.* New York: Dover. → First published as "Approximatio ad summam terminorum binomii $\overline{a+b}\backslash^n$ in seriem expansi."

NEYMAN, JERZY (1938) 1952 *Lectures and Conferences on Mathematical Statistics and Probability.* 2d ed. Washington: U.S. Department of Agriculture. → See especially Chapter 4, "Statistical Estimation," in the second edition.

QUETELET, ADOLPHE 1866 Charles-Frédéric Gauss. Pages 643–655 in Adolphe Quetelet, *Sciences mathématiques et physiques chez les Belges au commencement du XIXe siècle.* Brussels: Van Buggenhoudt.

SARTORIUS VON WALTERSHAUSEN, WOLFGANG 1856 *Gauss zum Gedächtniss.* Leipzig: Hirzel.

SAVAGE, I. RICHARD 1961 Probability Inequalities of the Tchebycheff Type. U.S. National Bureau of Standards, *Journal of Research, B. Mathematics and Mathematical Physics* 65 B:211–222.

SCHEFFÉ, HENRY 1959 *The Analysis of Variance.* New York: Wiley.

SCHERING, ERNST (1877) 1909 Carl Friedrich Gauss' Geburtstag nach hundertjähriger Wiederkehr. Pages 176–213 in Ernst Schering, *Gesammelte mathematische Werke.* Berlin: Mayer & Müller.

SIMPSON, THOMAS 1755 A Letter to the Right Honourable George Earl of Macclesfield, President of the Royal Society, on the Advantage of Taking the Mean of a Number of Observations, in Practical Astronomy. Royal Society of London, *Philosophical Transactions* 49, part 1:82–93.

SIMPSON, THOMAS 1757 An Attempt to Show the Advantage Arising by Taking the Mean of a Number of Observations in Practical Astronomy. Pages 64–75 in Thomas Simpson, *Miscellaneous Tracts on Some Curious and Very Interesting Subjects in Mechanics, Physical-astronomy, and Speculative Mathematics.* London: Nourse.

TODHUNTER, ISAAC (1865) 1949 *A History of the Mathematical Theory of Probability From the Time of Pascal to That of Laplace.* New York: Chelsea.

WHITTAKER, E. T.; and ROBINSON, G. (1924) 1944 *The Calculus of Observations: A Treatise on Numerical Mathematics.* 4th ed. Princeton, N.J.: Van Nostrand. → Section 103 provides a digest of Gauss 1816.

WINNECKE, F. A. T. 1877 *Gauss: Ein Umriss seines Lebens und Wirkens.* Brunswick: Vieweg.

WITTSTEIN, THEODOR 1877 *Gedächtnissrede auf Carl Friedrich Gauss zur Feier des 30. April 1877.* Hanover: Hahn.

ZELEN, MARVIN 1962 Linear Estimation and Related Topics. Pages 558–584 in John Todd (editor), *Survey of Numerical Analysis.* New York: McGraw-Hill.

Postscript

[1]The statement crediting Gauss with a schoolboy proof of convergence of the infinite series form of the binomial theorem for "n . . . not a positive integer" may possibly be an overstatement traceable to the account of E. T. Bell ([1937] 1956, p. 299), which in turn, no doubt, derives from that of Wolfgang Sartorius von Walterschausen, professor of geology at the University of Göttingen and close friend of Gauss in his later years: "Bartels [the schoolmaster] was able to procure some useful books on mathematics which the two young people studied together. Gauss thus became fully acquainted with the Binomial Theorem in complete generality and soon thereafter with the Theory of Infinite Numbers [*sic: unendlichen Reihen* = infinite series] which opened the way for him into Higher Analysis" ([1856] 1966, p. 5).

[2]To avoid ambiguity the formula for a Fermat number should be written

$$F_n = 2^{(2^n)} + 1.$$

[3]The identical mathematical form of the expressions derived by the method of *maximum likelihood* and by the method of *inverse probability* when a uniform prior distribution is adopted has been a source of continuing confusion. Thus, there is no doubt that Gauss did derive (in *Theoria motus corporum coelestium* . . . , art. 181) the probability density function

$$\frac{h\sqrt{n}}{\sqrt{\pi}} \exp\left(-nh^2 (\bar{x} - \mu)^2\right),$$

which, with the differential element $d\mu$ appended thereto, he interpreted to be the a posteriori probability distribution of μ *given* the observed value $\bar{x}$ of the sample mean and (assumed) validity of the uniform *prior* distribution of μ that he had adopted at the outset; and $\bar{x}$ was evidently the "most probable" value of μ under these circumstances. Today, by going through very much the same steps, we derive this function, append thereto the differential element $d\bar{x}$, and refer to the result as "the *sampling distribution* of the sample mean $\bar{X}$ *in repeated random samples* from a normal population with *fixed* mean, μ, and (fixed) standard deviation given by $\sigma = 1/(h\sqrt{2})$," and call $\bar{X}$ the *maximum likelihood estimator* of μ. Thus, speaking loosely, one is inclined today to attribute to Gauss the original (or "first") derivation of this particular "sampling distribution." On the other hand, I am quite sure that in 1809 Gauss had no conception of a "sampling distribution" of $\bar{X}$ "in repeated samples" with μ *fixed*. In the same manner, the "standard errors" given by Gauss in his 1816 paper were undeniably derived via the method of *inverse probability* and strictly speaking are the standard deviations of the a posteriori probability distributions of the parameters concerned, given the observed values of the particular functions of sample values considered. On the other hand, by virtue of the above-mentioned equivalence of form, Gauss's 1816 formulas can be recognized as giving the "standard errors," that is, the standard deviations of the sampling distributions, of the functions of sample values involved for *fixed* values of the corresponding population parameters. Thus, again speaking loosely, one is inclined today to attribute to Gauss the original ("first") derivation of these "standard error" formulas, even though he may have had (in 1816) no conception of the "sampling distribution," for *fixed* values of a population parameter, of a sample function used to estimate the value of this parameter. In contrast, in his 1821 paper he unequivocally studies the sampling distribution of s, the standard deviation of a sample of independent values, as an estimator of the population standard deviation, σ, and not as the a posteriori distribution of σ given s.

[4]Kurt-R. Biermann of Berlin has pointed out to me that Quetelet makes no mention of Gauss's reading Russian in the notes on his visit to Göttingen (Quetelet 1830, pp. 175–178) that he published soon after his "scientific voyage" through Germany in the summer of 1829—for good reason—it was not until a decade later that Gauss began his study of Russian. In a letter of August 17, 1839, to H. C. Schumacher, Gauss said: "Considering the acquisition of a new skill as a kind of rejuvenation [*Verjüngung*], I began at the start of last spring to occupy myself with the Russian language . . . and found considerable interest in it" (Gauss & Schumacher, *Briefwechsel*, vol. 3, 1861, pp. 242–243; quoted in Biermann 1964, p. 44). Schumacher responded by sending Gauss a Russian astronomical calendar on August 22 and, astonished that Gauss would choose to learn a new language at the age of 62 as a means of rejuvenation, inquired whether Gauss, seeking diversion, would not rather take refuge in the game of chess. Gauss thanked him for the calendar on September 8 and replied that the game of chess was too similiar to his other occupations to afford him any relaxation [*Erhohlung*] (Gauss & Schumacher, *Briefwechsel*, vol. 3, 1861, pp. 247–248, 269; Biermann 1964, p. 44). Quetelet mentions the six volumes of Gauss and Schumacher's *Briefwechsel*, and says they are "very interesting," on page 639 of his biography of Schumacher (Quetelet 1866, pp. 620–642), and quotes from Volume 5 on page 653 of his biography of Gauss. Consequently, as Biermann indicated to me, Quetelet undoubtedly learned of Gauss's study of Russian from his reading of the particular letters cited above, then apparently forgot when and how he had acquired this information, and, at the age of 70, when writing his biography of Gauss, actually believed that Gauss had told him personally about it on the occasion of his visit to Göttingen.

CHURCHILL EISENHART

ADDITIONAL BIBLIOGRAPHY

BIERMANN, KURT-R. 1964 Einige Episoden aus den russichen Sprachstudien des Mathematikers C. F. Gauss. *Forschungen und Fortschritte* 38:44–46.

COURANT, RICHARD (1955) 1969 Gauss and the Present Situation of the Exact Sciences. Pages 141–155 in T. L. Saaty and J. F. Weyl (editors), *The Spirit and Uses of Mathematics*. New York: McGraw-Hill. → English translation of an address given at the Gauss Centennial of the Academy and University of Göttingen, Feb. 19, 1955.

MAY, KENNETH O. 1972 Carl Friedrich Gauss. Volume 5, pages 208–215 in *Dictionary of Scientific Biography*. Edited by Charles C. Gillispie. New York: Scribner's. → Contains extensive annotated bibliographies of Gauss's works, correspondence, reprints, and translations and of works about Gauss arranged by topic.

QUETELET, ADOLPHE 1830 Notes extraites d'un voyage scientifique, fait en Allemagne pendant l'été de 1829. Part 2. Volume 6, pages 161–178 in *Correspondence mathématique et physique*. Brussels: Hayez.

SARTORIUS VON WALTERSHAUSEN, WOLFGANG (1856) 1966 *Gauss, a Memorial*. Colorado Springs, Colo.: Privately published. → English translation by Gauss's great-granddaughter, Helen Worthington Gauss.

GAUSSIAN DISTRIBUTION (NORMAL DISTRIBUTION)

See DISTRIBUTIONS, STATISTICAL, *article on* SPECIAL CONTINUOUS DISTRIBUTIONS, *and the biographies of* GAUSS; LAPLACE; MOIVRE.

GENETICS

See DEMOGRAPHY, *article on* DEMOGRAPHY AND POPULATION GENETICS.

GEOGRAPHY, STATISTICAL

This article was first published in IESS *as "Geography: VI. Statistical Geography" with five companion articles less relevant to statistics.*

Statistical geography is to geography what econometrics is to economics, sociometrics to sociology, psychometrics to psychology, or even jurimetrics to jurisprudence—an approach to the field rather than a subdivision of that field. Like these other approaches, it is of recent development—the manifestation within geography of the trend to a more quantitative approach that has characterized all the social sciences since the end of World War II. As is also the case with these other approaches, many of the pioneering contributions to statistical geography have come from workers in other fields. The parallel term "geometrics" is not used to describe the work of the statistical geographer, however, not so much because it provides an inappropriate picture of the statistical geographer's attempts to identify and measure regularities observable in spatial distributions as because the branch of mathematics that originated in Greek attempts to measure the earth has a two-thousand-year priority in its right to the name. In addition, mathematical geography, concerned as it is with map projections, is the branch of geography that today relies most heavily and directly upon geometry.

The syndrome that characterizes statistical geography today includes a more formal theoretical orientation than was true of geography in the past, a reliance upon statistical inference and numerical analysis in empirical research, the use of mathematical programming and simulation procedures in applied research, a basic concern with model construction, and an involvement with high-speed computers, mass-data banks, and automated mapping devices.

Yet this assemblage of interests is not monolithic. There are differences in emphasis, corresponding to each of the four main traditions of geographic research: spatial, area studies, man–land, and earth science (Pattison 1964; National Research Council 1965). Statistical geography originated within the spatial tradition, with its emphasis upon analysis of spatial distributions and associations, and initially relied heavily upon exercises in distribution fitting and upon regression and correlation analysis. It spread to area studies when the spatial tradition began using multivariate analysis, then to the man–land tradition via behavioral studies of environmental perception and individual decision making, and to the earth-science tradition as spatial studies and related systematic sciences began using systems analysis. To understand these several facets of statistical geography, with their differences in use of theory, choices, and timing of applications of statistical methods, and their contacts with the rest of science, therefore requires some understanding of the four research traditions. These, in turn, can best be viewed within a formal overview of approaches to regional analysis (Berry 1964*a*).

Approaches to regional analysis

Virtually any sort of regional analysis may be considered as starting from information about one or more *places*. For each place, one or more *properties*, or characteristics, may be measured, and the measurements may be made at one or more points in *time*. It is often convenient to think in terms of a three-dimensional array $\mathbf{X}$ of order $v \times p \times t$. Cell x_{ijk} of this array (or matrix) records the value of variable i at place j in time k. Rows of the array—vectors of array values for fixed i, k—record the distribution of variables over places at some point in time. Column vectors inventory the properties of places at time t. Time vectors report on the t states of variable i at place j.

Operational specification of variables, places, and times varies with the interests of the particular analyst, but a matrix such as $\mathbf{X}$ is (albeit usually implicitly) the object of all forms of geographic study. A row vector of $\mathbf{X}$ is a spatial distribution that can be mapped. A column vector is a locational inventory. Such row and column vectors are the bases of systematic (topical) and regional geography, respectively. Time vectors report changes in spatial distributions and at locations and are basic to historical geography.

The data. Cell x_{ijk} may contain a variety of different records, depending upon specification of i, j, and k. Columns, for example, may be defined as places with area, such as countries, states, counties, census tracts, or quarter-mile-square cells of a half-mile grid. Alternatively, they may be dimensionless points; three such examples are triangulation points used for measuring altitude, weather stations, and soil-sample-core locations (Kao 1963). Rows may be defined as properties of the places (scalar quantities, such as population residing in each of a set of census tracts or altitude at each of a set of triangulation points), or they may refer to connections between places (vector quantities, such as flow of coal from southern Illinois to each of the Midwestern counties). Rows also might be airline traffic flow between cities; in this case columns would represent *pairs* of cities. In addition, any level of measurement may appear, whether nominal, ordinal, interval, or ratio [*see* STATISTICS, DESCRIPTIVE, *article on* LOCATION AND DISPERSION, *for a discussion of levels of measurement*].

Spatial tradition. Geography's spatial tradition is founded upon cartographic portrayal and subsequent study of spatial distributions, thus having as its base row-wise analysis. A few examples of the kinds of spatial distributions studied follow:

(1) A land-use map—scalar, nominal, of areas, an areal distribution.

(2) A map showing the location of major cities by dots—scalar, nominal, at points, a point distribution.

(3) An airline-route map—vector, nominal, joining points.

(4) A map of soil quality—scalar, ordinal, of areas.

(5) A map showing average annual temperature—scalar, interval. If the map shows averages by states or districts, it is also discontinuous and areal (a choropleth map, if different shadings are applied to the units), but if it shows the temperature varying over the country as a surface, it is a continuous generalization. The generalization is of point observations (usually isopleth, since the surface will be depicted by contours) if, for example, weather-station data are used, or of areal data if interpolations were made, for example, with respect to the district averages treated as points central to each of the districts.

(6) A highway map, with routes classified by quality—vector, ordinal, joining points.

(7) A map of city-to-city air-passenger movements—vector, ratio, joining points.

Two dominant themes emerge in spatial analysis of such maps: evaluation of the *pattern* of scalar distributions and of similarities in pattern over a number of such distributions and evaluation of the *connectivity* evidenced by vector distributions and of similarities in the connectivity of several such distributions. Apparently, the fundamental properties of pattern include absolute location (position), relative location (geometry), and scale, with a family of interesting derived properties, including density and density gradients, spacing, directional orientation, and the like. Similarly, accessibility is central to the study of connectivity, and from it are derived such properties as centrality itself, relative dominance, degree of interdependence, etc. The two themes merge in *spatial-systems* analysis, where pattern and connectivity are examined in their association. For example, urban land values decline with increasing distance from the city center, and type of farming varies with distance from market. Such examples are readily generalized to the dynamic, that is, time-dependent, case.

Area-studies (chorographic) tradition. Just as row-wise analysis is the basis of the spatial tradition, so columnwise analysis provides the base for geography's area-studies tradition. The essential problems of this tradition are those of regional intelligence: the characterization of place in terms of the associations between characteristics localized in that place. This approach is often restricted to those features of place that are directly observable as landscape but, especially among the French school of human geographers, is also extended to an evaluation of both tangible and intangible aspects of "regional character" and of the differentiation of places (the study of "areal differentiation") (see Mikesell 1968).

An appetite for information, a penchant for the peculiar, emphasis upon field work, attachment to the people and language of a particular part of the world, a strong literary bent, a companionship with history and great reliance upon historical modes of explanation (in contrast to the functional, deterministic, and probabilistic modes of the other geographic traditions)—all these serve to identify the work of the student in the area-studies tradition, whatever the areas examined: countries or continents, regions or culture areas.

Man–land tradition. In the tradition of medieval philosophy, classical geographers distinguished between two major sets of variables: the physical (inorganic plus biotic) and the cultural. The associated methodological argument was that these provided the bases of the two major segments of the field: physical geography and cultural geography. Finer groupings of variables within these categories led to the variety of systematic branches of the subject, such as the geographies of landforms, of plants, of industry, of cities, or of language (the many topical fields and subfields might thus be identified as nested subsets of rows of the matrix $\mathbf{X}$).

The classical modes of thought, however, also led to a particular tradition in geography, the man–land tradition, in which the relationships between physical variables and human characteristics and activities were examined. In combination with the social Darwinism of the late nineteenth century, simple one-way studies were made of the effects of environment on man. These were later complemented by studies of the effects of man on environment, from which emerged much of the original thinking in the field of conservation. More recently the ancient dichotomy has been relaxed with, for example, studies of the effects of environmental perception on resource evaluation and decision making in resources management (Kates 1962) and with the adoption of a systems-analysis frame (Ackerman 1963).

Earth-science tradition. During the eighteenth century, geography was an integral and substantial part of natural philosophy. At that time geographical study embraced all aspects of the earth, air, and waters. Since then, however, most aspects of these studies have branched off as separate systematic sciences, and geography's earth-science tradition has been left with such concerns as the study of landforms and their evolution (geomorphology), descriptive climatology, and certain aspects of the geographies of soils, plants, and animals, together with the attempt to achieve some spatial synthesis of these in order to identify "natural environmental complexes." It is to this latter end that systems-analysis procedures have recently been employed in this research tradition (Chorley 1962).

Antecedents and stimuli

Prior to World War II few papers of a statistical nature had been published by geographers. Perhaps the only contributions worthy of note were Matui's (1932) fitting of the Poisson distribution to quadrat counts of settlements in a portion of Japan and Wright's (1937) discussion of Lorenz measures of concentration in the spatial case. However, two general antecedents can be distinguished, in addition to the pioneering contributions of workers in other fields: centrography and social physics. From the former came the idea of developing a special family of descriptive statistics for spatial distributions, and from the latter the recognition of certain classes of regularities in such distributions.

Centrography. During the early part of the twentieth century there was a lively debate among statisticians concerning such measures as the center of population. Part of the debate stemmed from publication by the U.S. Bureau of the Census of a piece entitled *Center of Population and Median Lines* . . . (U.S. Bureau of the Census 1923). Many articles were published, notably in the *Journal of the American Statistical Association* and in *Metron*, concerning the relative advantages of alternative centers, the center of gravity, the spatial median, and the center of minimum aggregate travel. (The U.S. Census Bureau's geography branch still reports on the center of gravity of the United States' population after each census.) This debate gradually subsided in the United States, but in the Soviet Union centrography flourished during the 1920s and 1930s. A centrographic laboratory was founded at Leningrad, under the auspices of the Russian Geographic Society, in 1925, and its director, E. E. Sviatlovsky, pursued studies of the "actual" and "proper" centers and the distributions of all manner of phenomena. However, the set of "proper" centers of economic activities prepared for the Gosplan of 1929 was at odds with the second Five-Year Plan, and as a result, the laboratory was finally disbanded. Porter (1963) provides a fairly complete bibliography of the relevant literature on centrography. Recently, the Israeli statistician Bachi (1963) has attempted to revive centrography, with the development of a variety of measures of dispersion and association of spatial distributions. Current interest within geography is slight, however, except as embraced by the social physicists (Stewart & Warntz 1958).

Social physics. The attempt to describe human phenomena in terms of physical laws has a long history in every social science [*see* RANK–SIZE RELATIONS].

In geographical studies this has been expressed in two major ways: (1) by use of "gravity models" to describe spatial interaction; and (2) by use of "potential models" as general summaries of interdependency between all places in large areas. Such models are said by their advocates to summarize a wide variety of social and economic distributions in economically advanced societies. Gravity models were first used in a relatively formal way by E. G. Ravenstein, in his seminal study "The Laws of Migration" (1885; 1889). Thereafter, these models found wide application, for example, in marketing geography—Reilly's "law of retail gravitation" (1931)—and in urban transportation studies describing interzonal travel. Carrothers (1956) reviews this work and the basic postulate that interactions or movements between places are proportional to the product of the masses and inversely proportional to some exponent of distance: that is, $I_{ij} \propto M_i \cdot M_j / d_{ij}^x$. Such gravity analogs were generalized by the astronomer J. Q. Stewart (1947) to the case of the potential surface, which simultaneously describes the interactions of each place and every other. Thus, the potential at any point i is given by

$$P_i = \sum_{j=1}^{j=n} (P_j / d_{ij}^x).$$

The surface is interpolated from such measures for a sample set of points. There is still considerable interest within geography in social physics (Stewart & Warntz 1958), and new applications are continually being developed. Mackay (1959), for example, used gravity models to translate the depressing effects upon telephone communications of the French–English language boundary in Canada and of the United States–Canadian political boundary into their physical-distance equivalents, thus showing how social space can be transformed into the metric of physical space. Similar applications are to be found in all branches of cultural geography today.

Pioneering contributions from elsewhere. Workers in other fields provided several significant examples of the application of statistical methods to geographic problems, identified the major statistical problems of regional analysis, and prepared the first text on statistical geography, thereby doing much to set the pace and tone of statistical geography today. A statistician, M. G. Kendall (1939), for example, showed how principal-components analysis could be used to develop a multivariate index that would portray the geographical distribution of crop productivity in England. M. D. Hagood (1943), an agricultural statistician, used multiple-factor analysis to define multivariable uniform regions. An economist, C. Clark (1951), showed that the negative exponential distribution fitted population density patterns within cities. G. K. Zipf (1949), a philologist, developed the rank–size distribution of cities. A mathematical social scientist, H. A. Simon (1955), showed the bases of this distribution in simple stochastic processes, and sociologists found that a repetitive three-factor structure characterized the social geography of cities (Berry 1964*b*; 1965). G. U. Yule and M. G. Kendall ([1911] 1950) identified the problem of *modifiable units*: if data are of areas rather than at points, results of any analysis will be in part dependent upon the nature of the areal units of observation utilized. W. S. Robinson (1950) provided the relevant relationship between individual and ecological (areal, set-type) correlations, subsequently extended by Goodman (1959) in the context of ecological regression. A second problem, that of *contiguity*, or spatial autocorrelation, has been examined by Moran (1948) in the nominal case and by Geary (1954) more generally. Geary also applied his measures of contiguity to evaluating lack of independence of residuals from regression in studies of spatial association. These studies are reviewed in *Statistical Geography* (Duncan et al. 1961), the first general book in the field, written by sociologists. Further examination of autocorrelation in spatial series is to be found in "Spatial Variation" (Matérn 1960), a major contribution to areal sampling by a mathematician. Spatial analysis remains basic to quantitative plant ecology and to epidemiology, and from these fields have come many of the ideas used today in the study of pattern in point distributions and of spatial diffusion processes.

Statistical geography—1950 to 1965

Centrography, social physics, and external stimuli, facilitated by developments in computer technology that for the first time enabled the mass data of geographic problems to be handled conveniently, combined to stimulate workers in geography's spatial tradition to work quantitatively. The older forms of cartographic analysis provided firm bases for this development, and many of the early studies were simply quantitative extensions of analyses cartographically conceived and executed. Arthur H. Robinson (1962), a cartographer, for example, utilized correlation and regression analysis to improve the ways by which he could map spatial associations. McCarty and his associates (1956) used similar procedures to replace older cartographic means of comparison. Thomas (1960) showed the various ways in which residuals from regression could be treated cartographically so as to draw upon traditional geographical means of map analysis in model reformulation and refinement. King (1962) applied the "nearest-neighbor" methods of the quantitative plant ecologist to the study of pattern in point distributions, with, like the 1932 Matui study, expectations derived from the Poisson distribution. These represent but a few examples of the spatial studies concerned with distribution fitting as a means of studying spatial pattern or with uses of correlation and regression in studies of spatial association. Many other examples are to be found in uses of regression to fit gravity models and obtain the distance exponents for different phenomena (Carrothers 1956) or to fit negative exponential distributions to urban population densities and the like (Berry 1965).

These kinds of studies represent the beginnings, from which statistical geography has grown rapidly. Dacey and Tung (1962) have made major advances in point-pattern analysis, for example, by transforming the distribution-fitting exercise into an explicit hypothesis-testing frame, with relevant expectations derived from settlement theory. Curry (1964) views many urban phenomena as the outcome of known-probability mechanisms. The Swedish geographer Hägerstrand (1953) was the first to show that many spatial patterns might be considered as the outcome of diffusion processes that could be simulated, using Monte Carlo methods, and his work led to a burst of similar simulation studies in the United States (Morrill 1963).

New approaches to spatial analysis have also been developed. Most of the examples outlined above use scalar data. Garrison (1960) showed that the mathematical theory of graphs provided an excellent base from which to examine vector distributions, and Nystuen and Dacey (1961) extended his argument to the case of organizational regions, using graph-theoretic measures of accessibility of places to communications networks to define relatively independent subsets of relatively interdependent places. Tobler (1963) showed how a generalization of map projections, traditionally studied by the mathematical geographer, could be used as the basis for mapping social, economic, cultural, or political space into physical space, as a further means for merging geographical applications of various statistical and mathematical methods with the more traditional means of geographical analysis. Finally, in addition to developments of the descriptive kind, statistical geography has extended its work to embrace investigations of a prescriptive nature. Garrison and Morrill (1960), for example, applied the techniques of spatial price-equilibrium analysis to determining what should be the patterns of interregional trade in wheat and flour in the United States. Other research workers are now much concerned with the procedures of spatial programming. Haggett (1965) has provided an excellent review of the substance of the first decade of quantitative work in the spatial tradition.

With the use of multivariate analysis, statistical geography has spread from the spatial tradition to that of area studies. A traditional geographic problem in this latter tradition is that of regionalization—the attempt to derive areas relatively uniform in terms of a complex of associated characteristics and also relatively different from other areas in terms of that complex. Such problems, involving mass-data analysis, were traditionally handled by overlaying maps. This earlier procedure has been replaced, however, by the use of the modern computer, applying such multivariate procedures as factor analysis to reduce many variables to a few factors representing "complexes" of associated characteristics, and the application of numerical taxonomy to get optimal classification (minimizing within-group variance) of observations into regions on the basis of the distances between observations in the factor space (Berry & Ray 1966). Output from the entire procedure of data analysis and reduction includes the complexes of characteristics that define "regional character," measures of the similarity of the observations, and the regions [*see* CLUSTERING].

Statistical work also characterizes the man–land tradition, largely by virtue of either simple correlation and regression studies that include physical variables, on one side, and cultural variables, on the other (for example, correlations of annual precipitation and population densities in the high plains), or through uses of probability theory. It is the latter, indeterministic type of study that represents new departures. Curry (1962*a*), for example, shows how livestock management in the intensive grassland-farming areas of New Zealand is related to probabilities of fodder availability, which in turn are derived from probabilities of requisite climatic conditions. Much of the basic research goes into establishment of the relevant probabilities, in this case, of the probabilities of repetitive events that play a central role in farm management. Kates (1962), on the other hand, examined relations of management of flood-plain property to flood hazards, rare events. He found management practices to be conditioned, not by reasonably precise evaluations of the situation, as in the case of the New Zealand farmers, but by a widely varying set of preconceptions, at variance with the actual probability mechanisms.

Work in the earth-science tradition of geography has, also, become statistical and ranges from the attempt to reformulate the geography of landforms generated by fluvial mechanisms in the framework of general-systems theory (Chorley 1962) through studies of climatic change as a random series (Curry 1962*b*) to the analysis of precipitation climatology using harmonic methods (Sabbagh & Bryson 1962) or to the development of linear models predictive of some characteristic through prior multivariate analysis, so as to satisfy the assumptions of the model ultimately to be produced (Wong 1963). There is today perhaps more work of a statistical kind in the earth-science tradition than in either the area-studies or man–land tradition.

Statistical geography—analysis of both the statistical and the mathematical kind—is to be found in all branches of geography today. However, in the methods utilized, certain differences between geography's four main research traditions are to be noted. In the spatial tradition, distribution fitting, correlation and regression analysis, uses of such methods as the mathematical theory of graphs, and prescriptive uses of spatial programming dominate, along with uses of probability mechanisms to study diffusion processes. The area-studies tradition relies upon multivariate analysis, particularly factor analysis, and upon numerical taxonomy, to facilitate mass-data analysis. In the man–land tradition, a neat contrast is to be noted between those of traditional deterministic outlook, who use regression methods, and those concerned with decision making in resources management, who focus upon probabilities of the a priori and a posteriori kinds.

Finally, in the earth-science tradition those procedures that facilitate systems analysis have been those most rapidly adopted and used. At the end of World War II geography was nonquantitative. Statistical geography has played an integral, even critical, part in the transformation of geography into a modern social science in the postwar years.

BRIAN J. L. BERRY

[*Other relevant material may be found in* CLUSTERING; FACTOR ANALYSIS AND PRINCIPAL COMPONENTS; MULTIVARIATE ANALYSIS. *See also* Berry & Harris 1968; Isard & Reiner 1968; A. H. Robinson 1968.]

BIBLIOGRAPHY

ACKERMAN, EDWARD A. 1963 Where Is a Research Frontier? Association of American Geographers, *Annals* 53:429–440.

BACHI, ROBERTO 1963 Standard Distance Measures and Related Methods for Spatial Analysis. Regional Science Association, *Papers* 10:83–132.

BERRY, BRIAN J. L. 1964*a* Approaches to Regional Analysis: A Synthesis. Association of American Geographers, *Annals* 54:2–11.

BERRY, BRIAN J. L. 1964*b* Cities as Systems Within Systems of Cities. Pages 116–137 in John Friedmann and William Alonso (editors), *Regional Development and Planning: A Reader.* Cambridge, Mass.: M.I.T. Press.

BERRY, BRIAN J. L. 1965 Research Frontiers in Urban Geography. Pages 403–430 in Philip M. Hauser and L. F. Schnore (editors), *The Study of Urbanization.* New York: Wiley.

►BERRY, BRIAN J. L.; and HARRIS, CHAUNCEY D. 1968 Central Place. Volume 2, pages 365–370 in *International Encyclopedia of the Social Sciences.* Edited by David L. Sills. New York: Macmillan and Free Press.

BERRY, BRIAN J. L.; and RAY, MICHAEL 1966 Multivariate Socio-economic Regionalization: A Pilot Study in Central Canada. Unpublished manuscript.

CARROTHERS, GERALD A. P. 1956 An Historical Review of the Gravity and Potential Concepts of Human Interaction. *Journal of the American Institute of Planners* 22:94–102.

CHORLEY, RICHARD J. 1962 Geomorphology and General Systems Theory. U.S. Geological Survey, *Professional Paper* 500B.

CHORLEY, RICHARD J. 1963 Geography and Analogue Theory. Association of American Geographers, *Annals* 54:127–137.

CLARK, COLIN 1951 Urban Population Densities. *Journal of the Royal Statistical Society* Series A 114:490–496.

CURRY, LESLIE 1962*a* The Climatic Resources of Intensive Grassland Farming: The Waikato, New Zealand. *Geographical Review* 52:174–194.

CURRY, LESLIE 1962*b* Climatic Change as a Random Series. Association of American Geographers, *Annals* 52:21–31.

CURRY, LESLIE 1964 The Random Spatial Economy: An Exploration in Settlement Theory. Association of American Geographers, *Annals* 54:138–146.

DACEY, MICHAEL F.; and TUNG, TSE-HSIUNG 1962 The Identification of Randomness in Point Patterns. *Journal of Regional Science* 4:83–96.

DUNCAN, OTIS DUDLEY; CUZZORT, RAY P.; and DUNCAN, BEVERLY 1961 *Statistical Geography: Problems in Analyzing Areal Data.* New York: Free Press.

GARRISON, WILLIAM L. 1960 Connectivity of the Interstate Highway System. Regional Science Association, *Papers* 6:121–137.

GARRISON, WILLIAM L.; and MORRILL, RICHARD L. 1960 Projections of Interregional Patterns of Trade in Wheat and Flour. *Economic Geography* 36:116–126.

GEARY, R. C. 1954 The Contiguity Ratio and Statistical Mapping. *Incorporated Statistician* 5:115–145. → Includes four pages of discussion.

GOODMAN, LEO A. 1959 Some Alternatives to Ecological Correlation. *American Journal of Sociology* 64:610–625.

HÄGERSTRAND, TORSTEN 1953 *Innovationsförloppet ur korologisk synpunkt.* Lund (Sweden): Gleerupska Universitetsbokhandeln.

HAGGETT, PETER (1965) 1966 *Locational Analysis in Human Geography.* New York: St. Martins.

HAGOOD, MARGARET D. 1943 Statistical Methods for Delineation of Regions Applied to Data on Agriculture and Population. *Social Forces* 21:287–297.

►ISARD, WALTER; and REINER, THOMAS A. 1968 Regional Science. Volume 13, pages 382–390 in *International Encyclopedia of the Social Sciences.* Edited by David L. Sills. New York: Macmillan and Free Press.

KAO, RICHARD C. 1963 The Use of Computers in the Processing and Analysis of Geographic Information. *Geographical Review* 53:530–547.

KATES, ROBERT W. 1962 *Hazard and Choice Perception in Flood Plain Management.* Department of Geography Research Paper No. 78. Univ. of Chicago Press.

KENDALL, M. G. 1939 The Geographical Distribution of Crop Productivity in England. *Journal of the Royal Statistical Society* Series A 102:21–62.

KING, LESLIE J. 1962 A Quantitative Expression of Patterns of Urban Settlements in Selected Areas of the United States. *Tijdschrift voor economische en sociale geografie* 53:1–7.

MCCARTY, HAROLD H.; HOOK, J. C.; and KNOS, D. S. 1956 *The Measurement of Association in Industrial Geography.* Iowa City: State Univ. of Iowa, Department of Geography.

MACKAY, J. ROSS 1959 The Interactance Hypothesis and Political Boundaries in Canada: A Preliminary Study. *Canadian Geographer* 11:1–8.

MATÉRN, BERTIL 1960 Spatial Variation: Stochastic Models and Their Applications to Some Problems in Forest Surveys and Other Sampling Investigations. Sweden, Statens Skogsforskningsinstitut, *Meddelanden* 49, no. 5.

MATUI, ISAMU 1932 Statistical Study of the Distribution of Scattered Villages in Two Regions of the Tonami Plain, Toyama Prefecture. *Japanese Journal of Geology and Geography* 9:251–256.

►MIKESELL, MARVIN W. 1968 Landscape. Volume 8, pages 575–580 in *International Encyclopedia of the Social Sciences.* Edited by David L. Sills. New York: Macmillan and Free Press.

MORAN, P. A. P. 1948 The Interpretation of Statistical Maps. *Journal of the Royal Statistical Society* Series B 10:245–251.

MORRILL, RICHARD L. 1963 The Development of Spatial Distributions of Towns in Sweden: A Historical–Predictive Approach. Association of American Geographers, *Annals* 53:1–14.

MOSER, CLAUS A.; and SCOTT, WOLF 1961 *British Towns: A Statistical Study of Their Economic and Social Differences*. Edinburgh: Oliver.

NATIONAL RESEARCH COUNCIL, AD HOC COMMITTEE ON GEOGRAPHY 1965 *The Science of Geography: Report*. National Research Council Publication No. 1277. Washington: National Academy of Sciences–National Research Council.

NYSTUEN, JOHN D.; and DACEY, MICHAEL F. 1961 A Graph Theory Interpretation of Nodal Regions. Regional Science Association, *Papers* 7:29–42.

PATTISON, WILLIAM D. 1964 The Four Traditions of Geography. *Journal of Geography* 63:211–216.

PORTER, P. W. 1963 What Is the Point of Minimum Aggregate Travel? Association of American Geographers, *Annals* 53:224–232.

RAVENSTEIN, E. G. 1885 The Laws of Migration. *Journal of the Royal Statistical Society* Series A 48:167–235. → Includes seven pages of discussion.

RAVENSTEIN, E. G. 1889 The Laws of Migration: Second Paper. *Journal of the Royal Statistical Society* Series A 52:241–305. → Includes three pages of discussion.

REILLY, WILLIAM J. 1931 *The Law of Retail Gravitation*. New York: Putnam.

ROBINSON, ARTHUR H. 1962 Mapping the Correspondence of Isarithmic Maps. Association of American Geographers, *Annals* 52:414–429.

►ROBINSON, ARTHUR H. 1968 Cartography. Volume 2, pages 325–329 in *International Encyclopedia of the Social Sciences*. Edited by David L. Sills. New York: Macmillan and Free Press.

ROBINSON, W. S. 1950 Ecological Correlations and the Behavior of Individuals. *American Sociological Review* 15:351–357.

SABBAGH, MICHAEL A.; and BRYSON, REID A. 1962 Aspects of the Precipitation Climatology of Canada Investigated by the Method of Harmonic Analysis. Association of American Geographers, *Annals* 52:426–440.

SIMON, HERBERT A. 1955 On a Class of Skew Distribution Functions. *Biometrika* 42:425–440.

STEWART, JOHN Q. 1947 Empirical Mathematical Rules Concerning the Distribution and Equilibrium of Population. *Geographical Review* 37:461–485.

STEWART, JOHN Q.; and WARNTZ, WILLIAM 1958 Physics of Population Distribution. *Journal of Regional Science* 1:99–123.

THOMAS, EDWIN N. 1960 *Maps of Residuals From Regression: Their Characteristics and Uses in Geographic Research*. Iowa City: State Univ. of Iowa, Department of Geography.

TOBLER, WALDO R. 1963 Geographical Area and Map Projections. *Geographical Review* 53:59–78.

U.S. BUREAU OF THE CENSUS 1923 *Fourteenth Census of the United States; 1920: Center of Population and Median Lines and Centers of Area, Agriculture, Manufactures, and Cotton*. Washington: Government Printing Office.

WONG, SHUE TUCK 1963 A Multivariate Statistical Model for Predicting Mean Annual Flood in New England. Association of American Geographers, *Annals* 53:298–311.

WRIGHT, JOHN K. 1937 Some Measures of Distributions. Association of American Geographers, *Annals* 27: 177–211.

YULE, G. UDNY; and KENDALL, MAURICE G. 1950 *An Introduction to the Theory of Statistics*. 14th ed., rev. & enl. London: Griffin; New York: Hafner. → See especially pages 310–325, "Correlation and Regression: Some Practical Problems." Yule was the sole author of the first edition (1911). Kendall, who became a joint author on the eleventh edition (1937), revised the current edition and added new material to the 1965 printing.

ZIPF, GEORGE K. 1949 *Human Behavior and the Principle of Least Effort: An Introduction to Human Ecology*. Reading, Mass.: Addison-Wesley.

Postscript

Since 1965 the revolutionary impacts of statistical geography upon the discipline have been codified into routine. Statistical training is becoming a more common part of a geographer's education. New journals have appeared devoted to theoretical–quantitative studies, notably *Geographical Analysis* in the United States and *Environment and Planning* in Britain, and traditional journals have become more receptive to statistical studies.

Professional statisticians and mathematicians have been showing serious interest in geographic questions (for example, Bartlett 1975; Matheron 1975), and there has been increasing dialogue between such professionals and geographers, as evidenced in three issues (1974; 1975) of the *Statistician* devoted mainly to geography and statistics.

The International Geographical Union established a Commission on Quantitative Methods in 1964 to stimulate the world-wide communication of more formal research methods, and annual meetings of the commission have resulted both in the rapid spread of the "quantitative revolution" and in a flow of influential proceedings volumes, usually focused on special research methods (for example, *Comparative Factorial Ecology* 1971). By 1975, the revolution could justifiably be said to be over, with statistical geographers embarking instead on the difficult tasks of moving from borrowed methods to the development of techniques that address the uniquely geographical problems that afflict quantitative research in the field (Berry 1971). Among these problems are those of spatial autocorrelation (see Cliff & Ord 1973), inference in the investigation of spatial point processes (see Rogers 1974), and the complexities of analyzing space–time series (Chisholm et al. 1971). With growing numbers of able investigators being attracted to the field, prospects seem good for major advances.

BRIAN J. L. BERRY

ADDITIONAL BIBLIOGRAPHY

Three issues of the Statistician *are devoted mainly to geography and statistics: 23 [1974], no. 3 (Sept.), no. 4 (Dec.); 24 [1975], no. 3 (Sept.).*

BARTLETT, MAURICE S. 1975 *The Statistical Analysis of Spatial Pattern.* London: Chapman & Hall; New York: Wiley.

BERRY, BRIAN J. L. 1971 Problems of Data Organization and Analytical Methods in Geography. *Journal of the American Statistical Association* 66:510–523.

CHISHOLM, MICHAEL; FREY, ALLAN E.; and HAGGETT, PETER (editors) 1971 *Regional Forecasting.* Colston Papers No. 22. London: Butterworth; Hamden, Conn.: Archon. → Proceedings of the 22d symposium of the Colston Research Society held at the University of Bristol, April 6–10, 1970.

CLIFF, ANDREW D.; and ORD, J. K. 1973 *Spatial Autocorrelation.* London: Pion; New York: Academic Press.

Comparative Factorial Ecology. 1971 *Economic Geography* 47, no. 2 (supplement; June). → Special issue.

MATHERON, GEORGES 1975 *Random Sets and Integral Geometry.* New York: Wiley.

ROGERS, ANDREI 1974 *Statistical Analysis of Spatial Dispersion: The Quadrat Method.* London: Pion; New York: Academic Press.

GINI, CORRADO

The work of Corrado Gini (1884–1965) has had a profound impact on many fields within the social sciences. The son of a manufacturer, Gini was born in Motta di Livenza, Treviso Province. He studied law at the University of Bologna, where he also took courses in mathematics. The broad scope of his scholarly contributions is reflected in the variety of subjects he taught, successively, at the universities of Cagliari, Padua, and Rome: statistics, political economy, demography, biometrics, constitutional law, and sociology.

Gini also furthered the development of social science in other than academic capacities. From 1926 to 1932 he was chairman of the Central Institute of Statistics. He directed several scientific expeditions to study the demographic, anthropometric, and medical characteristics of particular ethnic groups in Fezzan, Palestine, Mexico, Poland and Lithuania, Calabria, and Sardinia, as well as the processes of assimilation of immigrant groups generally. He was the founder and editor of the international journal of statistics *Metron*, in 1920, and of the journal *Genus*, the organ of the Italian Committee for the Study of Population Problems.

In recognition of his achievements he was awarded honorary doctorates, one in economic science—from the Catholic University in Milan, in 1932—and another in sociology—from the University of Geneva, in 1934; and he was awarded an honorary doctor of science degree by Harvard at the 1936 tercentenary celebration of that university.

The main currents of thought in Italian culture at the time that Gini began his scientific activity were positivism and idealism. Gini's commitment to positivism was limited. His works manifest a remarkable eclecticism, arising from his systematic study of a variety of disciplines: law, economics, statistics, mathematics, and biology. He always placed the phenomena he was studying in a more general context, bringing essential social, economic, and demographic elements to bear on the topic of his immediate concern.

Gini thoroughly understood the importance of descriptive statistics, as is shown by his particular admiration for Luigi Bodio among Italian statisticians, but at the same time he deeply felt the need to study the procedures of inferential statistics and to find suitable criteria for evaluating those procedures. To this end he studied the works of Bernoulli, Lexis, and Czuber, as well as those of the Italians Angelo Messedaglia and Rodolfo Benini.

Contributions to statistical method. Gini's first scientific works deal with the statistical regularity of rare events. To the same period—1908 to 1911—belongs his initial work on the concept and measurement of probability, with special attention and application to the human sex ratio at birth. His findings were contained in a book (1908) in which he presented analyses of extensive empirical materials on sex ratios from different countries. In the same work, he developed a theory of dispersion, that is, of the quantitative structure of the scatter or variability among measured quantities of the same kind. He later returned to the topic of dispersion (1940; 1941).

Gini made a lengthy critique of the principles of statistics in order to reinforce the logical basis of statistics and to rid it of naïve empiricism. He wished to raise statistics to the level of an independent science, one of whose tasks would be the systematic investigation of the advantages and limitations of various indexes, or descriptive statistics, under differing structures of error.

In an address entitled "I pericoli della statistica" (1939; "The Dangers of Statistics"), inaugurating the first meeting of the Società Italiana di Statistica, Gini warned statisticians about the faulty logical foundations of some statistical methods and stressed the importance of estimating meaningful parameters. He published further papers on this general subject (1943; 1947). In the latter paper, "Statistical Relations and Their Inversions," he pointed out that poor statistical notation may lead to misunderstanding and error. For example, if one writes $y = 3 + 7x$, as a mathematical relationship between the real variables x and y, one may just as well write $x = -3/7 + 1/7y$; this is called invertibility. On the other hand, in a statistical context of regression, one may loosely write $y = 3 + 7x$, meaning that the expectation (or average) of a random variable Y is thus linearly related to possible values of another random variable X. For clarity, one should use a more complete notation like the now conventional

$$E(Y|X = x) = 3 + 7x,$$

and, in general, $E(X|Y = y) \neq -3/7 + 1/7 y$. On the other hand, taking complete averages, $EY = 3 + 7EX$; this situation is called one of subinvertibility.

In a paper written in collaboration with Luigi Galvani, "Di talune estensioni dei concetti di media ai caratteri qualitativi" (1929; "Extension of Mean Value Theory to Qualitative Characteristics"), Gini made an original contribution in another area. The problem here was to find a meaningful analogue of the usual arithmetic mean for a purely qualitative or nominal distribution. By defining a measure of deviation between any two classes of the qualitative classification and by a minimization argument, Gini showed how to choose one of the classes of the classification as a kind of mean.

The theory of generalized means was of continuing interest to Gini, and he was particularly concerned with its very general expression and with a wide variety of mean values (1938). In another article, again with Galvani, he discussed extensively the topic of location parameters for bivariate distributions, noting especially geographical problems (see Gini et al. 1933).

Next to generalized means, measures of variability are of great importance, and Gini discussed these at length, giving special attention to descriptive statistics not intrinsically tied to the normal distribution (1912). He introduced a new measure of variability, the *mean difference*, which is essentially the expected absolute value of $X_1 - X_2$, where X_1 and X_2 are independent random variables with the same distribution of interest. (Gini thought primarily in terms of the sample version of the mean difference.)

Gini also worked on measures of variability for qualitative or nominal distributions and on measures of variability that are tied to the concentration curve and are considered important in the economic analysis of income and wealth (1921).

Nontypicality of typical census areas. Gini set himself the task of selecting a sample of typical census areas in Italy. To this end, he showed that it is not adequate simply to select the sample in such a way that the averages of various characteristics calculated for the sample are equal to those calculated for the entire territory of Italy. Even when the examination is confined to only the characteristics used in selecting the sample, the sample proves to be unrepresentative with respect to the variability and concentration of those characteristics and to the relations existing between them. Accordingly, Gini included in the criteria of representativeness these last two aspects of the distributions of characteristics, and he examined the conditions to which each of the proposed criteria subjects the other criteria.

Theory of price indexes. To the theory of price indexes, which is of great importance in economic statistics, Gini contributed a classic paper, "Quelques considérations au sujet des nombres indices des prix et des questions analogues" (1924), in which he solved some logical and technical problems connected with elimination methods.

Distinguishing between simple and complex indices, Gini showed how, in general, the methods for obtaining complex indexes can be considered as particular cases of the method of elimination (which consists in separating the different circumstances affecting a phenomenon in order to make comparisons *ceteris paribus* with similar phenomena at different times or places). The problem was to keep the phenomenon of "price variation" distinct from the phenomenon of "quantity variation."

Operating with the general method of the "typical population" organized into four special methods, Gini obtained various formulas for index numbers, at the same time providing criteria for choosing that one of the formulas best suited to the aims of particular problems. (Another of his papers on the same topic is "Methods of Eliminating the Influence of Several Groups of Factors," 1937).

Theory of distributions. Gini made substantial contributions to the study of relationships between two probability distributions and between two random variables with a joint distribution. Particularly important are the following:

(*a*) *Transvariation.* Suppose that one considers two probability distributions, represented for convenience by (independent) random variables X, Y. Suppose further that $EX < EY$; it may still be quite probable that $X > Y$, that is, the difference between the observations X and Y may show a different sign than the corresponding difference between expected values. Gini proposed a specific measure for this phenomenon, which he called transvariation (1953; 1959*a*; see also Kruskal 1957).

(*b*) *Distance between two distributions.* Gini was one of the first to consider the problem of measuring meaningfully how much two probability distributions differ in gross (1914*a*). He proposed a measure of distance, based on the two cumulative distribution functions, that was in fact a metric in the technical mathematical sense (see Fréchet 1947).

(c) *Association between two random variables with a joint distribution.* Gini made the basic distinction between *connection* (any kind of statistical dependence) and *concordance* (dependence in which it makes sense to speak of one variable tending to increase—or decrease—along with the other). The correlation ratio, for example, is a measure of connection, while the coefficient of correlation is a measure of concordance. Gini analyzed previously suggested measures of association and proposed new ones both for the case of numerical variates and for that of qualitative variates (see Goodman & Kruskal 1959, which contains relevant references to Gini, Weida, and Pietra).

Demography and biometrics. Gini pioneered in studies that relate demographic phenomena to social and biological phenomena. Thus, he studied the sources of differential fertility (1949), and he related the phenomenon of migration to broader social and demographic considerations (1946*b*).

Gini also proposed a unified research program for analyzing the eugenic and dysgenic effects of war, with special emphasis on the measurement of the effects of war on mortality (1915–1920). Also related to demographic phenomena are Gini's studies of various populations in phases of expansion, regression, and extinction (1934).

Economic and sociological research. From the earliest years of his career, Gini tackled the difficult problem of estimating national wealth. He made a constructive criticism of the method of the interval of devolution and the first calculation by a direct method of the private wealth of Italy (1914*b*). He followed a methodological organization of the data with a comparative study of the qualitative composition of wealth, clarifying the conditions that make nations wealthy (1959*b*).

Gini originated the idea of including the value of human capital in the calculation of wealth (see Schultz 1968). This makes it possible to indicate the preponderant cause for the incommensurability in time and territory of the wealth and economic well-being of various collectivities (1956*a*). The value-as-property of the man who labors becomes a factor in estimating wealth, just as the share derived from human labor is included in income (1914*a*). This conception clearly links economic considerations with sociological and demographic ones.

Related to Gini's conception of human capital is his cyclical theory of population. Observing the differential rates of reproduction of social classes, Gini formulated a theory of social metabolism that is based on an analogy to organic metabolism: the upper classes, having low rates of reproduction, will tend to extinction unless they get new members from the lower classes, which have higher rates of reproduction (1927). Gini intended this theory to replace Pareto's concept of the circulation of elites.

Gini also developed a theory of phases of population growth, these being related to social and economic phenomena (1923). According to his theory, a particular population grows rapidly in the first phase of its development, then has a diminishing rate of growth until it is approximately stationary, and finally begins a decline that may even lead to total extinction. In the first phase, capital is very scarce and there is little differentiation between social classes; later, as capital is accumulated, classes become increasingly differentiated, and there is a concomitant differentiation of reproductive behavior. As birth control spreads, a deterioration in the quality of the population can be avoided only by the influence of external factors, such as the immigration of young people from populations in the phase of demographic expansion.

Gini's neo-organicist approach makes it possible to distinguish normal economic processes from pathological ones—those in which the state of a society is unbalanced—and to initiate economic processes of restoration. In general, he believed that states of imbalance could be remedied by the regulatory activities of political agencies and of economic organizations.

In Gini's view, an important aspect of the development of society is the transition from forced to free labor and then to spontaneous labor (1956*b*). This last represents the final stage in a psychological process that raises labor from a primordial means of obtaining a living to a free and autonomous activity that enriches the human personality.

Gini's pre-eminent place in the development of Italian statistics is based on a dual contribution: his work in scientific statistics, which included important teaching and editorial activities; and his efforts toward the development of official statistics, notably, the consolidation in a single institute of all the various agencies for the collection of data and the extension of the number of items about which data are collected.

TOMMASO SALVEMINI

[*Directly related are the entries* SAMPLE SURVEYS; STATISTICS, DESCRIPTIVE; VARIANCES, STATISTICAL STUDY OF. *Other relevant material may be found in* CENSUS; DEMOGRAPHY, *article on* THE FIELD; INDEX NUMBERS, *article on* THEORETICAL ASPECTS;

and in the biographies of BENINI; BERNOULLI FAMILY; LEXIS. *See also* Allais 1968; Broadhurst 1968; Grauman 1968; Lebergott 1968; Mayer 1968; Petersen 1968; Sauvy 1968; Spengler 1968; Thomas 1968; Whitney 1968.]

WORKS BY GINI

1908 *Il sesso dal punto di vista statistico: Le leggi della produzione dei sessi.* Milan: Sandron.

(1912) 1955 *Memorie di metodologia statistica.* Volume 1: Variabilità e concentrazione. 2d ed. Rome: Veschi.

1914*a* *Di una misura della dissomiglianza fra due gruppi di quantità e delle sue applicazioni allo studio delle relazioni statistiche.* Venice: Ferrari.

(1914*b*) 1962 *L'ammontare e la composizione della ricchezza della nazioni.* 2d ed. Turin: Bocca.

(1915–1920) 1921 *Problemi sociologici della guerra.* Bologna: Zanichelli. → A collection of previously published articles.

1921 Measurement of Inequality of Incomes. *Economic Journal* 31:124–126.

(1923) 1952 *Patologia economica.* 5th ed., rev. & enl. Turin: Unione Tipografico–Editrice Torinese.

1924 Quelques considérations au sujet de la construction des nombres indices des prix et des questions analogues. *Metron* 4:3–162.

1927 *Il neo-organicismo: Prolusione al corso di sociologia.* Catania: Studio Editoriale Moderno.

1929 GINI, CORRADO; and GALVANI, LUIGI Di talune estensioni dei concetti di media ai caratteri qualitativi. *Metron* 8, no. 1/2:3–209.

1933 GINI, CORRADO; BERARDINIS, L. DE; and GALVANI, L. Sulla selettività delle cause di morte durante l'infanzia. *Metron* 11, no. 1:163–183.

1934 Ricerche sulla popolazione. *Scientia* 55:357–373.

1937 Methods of Eliminating the Influence of Several Groups of Factors. *Econometrica* 5:56–73.

1938 Di una formula comprensiva delle medie. *Metron* 13:3–22.

1939 I pericoli della statistica. *Rivista di politica economica* 29:901–924.

1940 Sur la théorie de la dispersion et sur la vérification et l'utilisation des schémas théoriques. *Metron* 14:3–29.

1941 Alle basi del metodo statistico: Il principio della compensazione degli errori accidentali e la legge dei grandi numeri. *Metron* 14:173–240.

1943 I testi di significatività. Società Italiana di Statistica, *Atti* [1943]:241–279.

1946*a* Actualidades demográficas. *Revista internacional de sociologia* 4:147–169.

1946*b* Los efectos demográficos de las migraciones internacionales. *Revista internacional de sociologia* 4:351–388.

1947 Statistical Relations and Their Inversions. International Statistical Institute, *Revue de l'Institut International de Statistique* 15:24–42.

1948 Evoluzione della psicologia del lavoro e della accumulazione. Banca Nazionale del Lavoro, *Moneta e credito* Whole No. 2.

1949 Vecchie e nuove osservazioni sulle cause della natalità differenziale e sulla misura della fecondità naturale delle coniugate. *Metron* 15:207–358.

1950 Metodologia statistica: La misura dei fenomeni collettivi. Volume 3, part 2, pages 245–321 in *Enciclopedia delle matematiche elementari.* Milan: Hoepli.

1951 Caractère des plus récents développements de la méthodologie statistique. *Statistica* 11:3–11.

1952 On Some Symbols That May Be Usefully Employed in Statistics. International Statistical Institute, *Bulletin* 33, no. 2:249–282.

1953 The Measurement of the Differences Between Two Quantity Groups and in Particular Between the Characteristics of Two Populations. *Acta genetica et statistica medica* 4:175–191.

1955 Sur quelques questions fondamentales de statistique. Paris, Université, Institut Henri Poincaré, *Annales* 14:245–364.

1956*a* Valutazione del lavoro e del capitale nell' *Economia lavorista. Rivista bancaria* New Series 12:522–530.

1956*b* *Economia lavorista: Problemi del lavoro.* Turin: Tipografia Sociale Torinese.

(1956*c*) 1966 *Statistical Methods.* Rome: Biblioteca del Metron. → Lectures delivered at the International Center for Training in Agricultural Economics and Statistics, Rome.

1958 Logic in Statistics. *Metron* 19, no. 1/2:1–77.

1959*a* *Transvariazione.* Rome: Libreria Goliardica.

1959*b* *Ricchezza e reddito.* Turin: Unione Tipografico–Editrice Torinese.

1959*c* Mathematics in Statistics. *Metron* 19, no. 3/4:1–9.

SUPPLEMENTARY BIBLIOGRAPHY

►ALLAIS, MAURICE 1968 Pareto, Vilfredo: I. Contributions to Economics. Volume 11, pages 399–411 in *International Encyclopedia of the Social Sciences.* Edited by David L. Sills. New York: Macmillan and Free Press.

►BROADHURST, P. L. 1968 Genetics: I. Genetics and Behavior. Volume 6, pages 96–107 in *International Encyclopedia of the Social Sciences.* Edited by David L. Sills. New York: Macmillan and Free Press.

CASTELLANO, VITTORIO 1965 Corrado Gini: A Memoir With the Complete Bibliography of His Works. *Metron* 24; no. 1–4.

FRÉCHET, MAURICE 1947 Anciens et nouveaux indices de corrélation: Leur application au calcul des retards économiques. *Econometrica* 15:1–30, 374–375.

FRÉCHET, MAURICE 1947–1948 Le coefficient de connexion statistique de Gini–Salvemini. *Mathematica* 23:46–51.

FRÉCHET, MAURICE 1957 Sur la distance de deux lois de probabilité. Paris, Université, Institut de Statistique, *Publications* 6:183–198.

GALVANI, LUIGI 1947 À propos de la communication de M. Thionet: "L'école moderne de statisticiens italiens." Société de Statistique de Paris, *Journal* 88:196–203. → A bitter attack on the 1945–1946 article by Pierre Thionet. See pages 203–208 for Thionet's reply to Galvani.

GOODMAN, LEO A.; and KRUSKAL, WILLIAM H. 1959 Measures of Association for Cross-classifications: II. Further Discussion and References. *Journal of the American Statistical Association* 54:123–163.

►GRAUMAN, JOHN V. 1968 Population: VI. Population Growth. Volume 12, pages 376–381 in *International Encyclopedia of the Social Sciences.* Edited by David L. Sills. New York: Macmillan and Free Press.

KRUSKAL, WILLIAM H. 1957 Historical Notes on the Wilcoxon Unpaired Two-sample Test. *Journal of the American Statistical Association* 52:356–360.

►LEBERGOTT, STANLEY 1968 Income Distribution: II.

Size. Volume 7, pages 145–154 in *International Encyclopedia of the Social Sciences*. Edited by David L. Sills. New York: Macmillan and Free Press.

►MAYER, KURT B. 1968 Population: IV. Population Composition. Volume 12, pages 362–370 in *International Encyclopedia of the Social Sciences*. Edited by David L. Sills. New York: Macmillan and Free Press.

NEYMAN, JERZY (1938) 1952 *Lectures and Conferences on Mathematical Statistics and Probability*. 2d ed., rev. & enl. Washington: U.S. Department of Agriculture, Graduate School.

►PETERSEN, WILLIAM 1968 Migration: I. Social Aspects. Volume 10, pages 286–292 in *International Encyclopedia of the Social Sciences*. Edited by David L. Sills. New York: Macmillan and Free Press.

SALVEMINI, TOMMASO 1943 La revisione critica di Gini ai fondamenti della metodologia statistica. *Statistica* 3:46–59.

►SAUVY, ALFRED 1968 Population: II. Population Theories. Volume 12, pages 349–395 in *International Encyclopedia of the Social Sciences*. Edited by David L. Sills. New York: Macmillan and Free Press.

►SCHULTZ, THEODORE W. 1968 Capital, Human. Volume 2, pages 278–287 in *International Encyclopedia of the Social Sciences*. Edited by David L. Sills. New York: Macmillan and Free Press.

►SPENGLER, JOSEPH J. 1968 Population: III. Optimum Population Theory. Volume 12, pages 358–362 in *International Encyclopedia of the Social Sciences*. Edited by David L. Sills. New York: Macmillan and Free Press.

THIONET, PIERRE 1945–1946 L'école moderne de statisticiens italiens. Société de Statistique de Paris, *Journal* 86:245–255; 87:16–34.

►THOMAS, BRINLEY 1968 Migration: II. Economic Aspects. Volume 10, pages 292–300 in *International Encyclopedia of the Social Sciences*. Edited by David L. Sills. New York: Macmillan and Free Press.

►WHITNEY, VINCENT H. 1968 Population: V. Population Distribution. Volume 12, pages 370–376 in *International Encyclopedia of the Social Sciences*. Edited by David L. Sills. New York: Macmillan and Free Press.

GIRSHICK, MEYER A.

Meyer Abraham Girshick (1908–1955), American statistician, was born in Russia and immigrated to the United States in 1922. After graduating from Columbia College in 1932, he did graduate work in statistics under Harold Hotelling at Columbia University from 1934 until 1937. (Several years later he received a doctorate from Columbia.) In 1937–1939 he was a statistician with the Bureau of Home Economics, U.S. Department of Agriculture, where he participated in a pioneer study of body measurements of 147,000 American children that helped manufacturers to develop improved sizing of garments. At the same time, he gave evening courses in statistics at the U.S. Department of Agriculture Graduate School, courses that had a profound influence on the dissemination of sound statistical methods in government work and encouraged many new research workers. In 1939–1944 and again in 1945–1946 he was principal statistician for the Bureau of Agricultural Economics. For one year during World War II he participated in the work of the Statistical Research Group at Columbia University, a panel composed of statisticians and designed to develop statistical methods appropriate to wartime problems. Working with others on the panel, particularly Abraham Wald, had a decisive influence on Girshick's subsequent career. After a brief stay at the Bureau of the Census, he became a research statistician and mathematician at the RAND Corporation in Santa Monica, California (a research organization primarily devoted to work for the U.S. Air Force). From 1948 until his untimely death in 1955 he was professor of statistics at Stanford University. He was elected president of the Institute of Mathematical Statistics (one of the principal learned societies for theoretical statistics) in 1951.

His early papers, written between 1936 and 1944, were primarily concerned with problems of multivariate statistical analysis. His major achievement was to find the distribution of the roots and characteristic vectors associated with certain determinantal equations that are used in testing the null hypothesis that two sets of variates are independent (1939). [*See* MULTIVARIATE ANALYSIS.]

Subsequently he also found the distribution of these roots when the null hypothesis is not true. From this the power function of certain tests can in principle be obtained (1941).

His interest in multivariate analysis and his substantive work at the Bureau of Agricultural Economics combined to turn his attention to newly developing methods of estimation of simultaneous equations in economics. These methods were originated by Trygve Haavelmo and continued by Tjalling Koopmans, T. W. Anderson, and Herman Rubin. In collaboration with Haavelmo, Girshick conducted one of the first major empirical studies on an econometric model for the agricultural sector of the United States economy (1947), and through unpublished work he contributed importantly to the development of the limited-information approach to simultaneous equation estimation. [*See* SIMULTANEOUS EQUATION ESTIMATION.]

During the period that Girshick was associated with the Statistical Research Group, Wald was originating two major concepts of statistics: sequential analysis and statistical decision theory [*see* DECISION THEORY; SEQUENTIAL ANALYSIS; *and the biography of* WALD]. Girshick's subsequent work was almost exclusively confined to these two areas (see Blackwell & Girshick 1954).

His earliest work in sequential analysis concerned the testing of composite hypotheses in which the mean of one process is less than, or equal to, the mean of another, the alternative being that it is greater. He showed that the power functions of such tests were constant on certain curves in the parameter space, and he was able to use this information to derive approximate sequential tests for the parameters of the exponential family of distributions (1946*a*). He also found exact relations for tests when the variables take on only a finite set of values. As a by-product of his interest in sequential analysis he found, in collaboration with Frederick Mosteller and L. J. Savage, a method of getting unbiased estimates of a parameter from data that have been generated by a sequential, or similar, sampling scheme (Girshick et al. 1946*b*). He and David Blackwell found a lower bound for the variance of such estimates (Blackwell & Girshick 1947).

His work in decision theory was developed in a series of papers, most of them coauthored with Blackwell, Rubin, Savage, or Arrow. The major results of these studies were systematically presented in *Theory of Games and Statistical Decisions* (1954), written jointly with Blackwell. This book today represents a major study of statistical method using the concepts of the theory of games from the decision-theory point of view. The theory of games, which is closely allied to decision theory, is also given prominence in the book. [*See* GAME THEORY.] Among its most noteworthy accomplishments are a study of the interrelations of various criteria for complete classes of solutions and related concepts and a systematic treatment of Bayesian and related procedures in statistical contexts. Other special features include rigorous analysis of the concepts of sufficiency and of invariance, a clear exposition and characterization of sequential probability-ratio tests and their optimal properties, a study of Bayesian estimation procedures with special loss functions, particularly the quadratic and the absolute value, and the theory of comparison of experiments.

Girshick's influence on the development of statistical theory occurred as much through his direct personal relations, his enthusiasm and intelligent guidance, as through his published work, important as the latter is.

KENNETH J. ARROW

WORKS BY GIRSHICK

1939 On the Sampling Theory of Determinantal Equations. *Annals of Mathematical Statistics* 10:203–224.

1941 The Distribution of the Ellipticity Statistic L_e When the Hypothesis Is False. *Terrestrial Magnetism and Atmospheric Electricity* 46:455–457.

1946*a* Contributions to the Theory of Sequential Analysis. *Annals of Mathematical Statistics* 17:123–143, 282–298.

1946*b* GIRSHICK, MEYER A.; MOSTELLER, FREDERICK; and SAVAGE, L. J. Unbiased Estimates for Certain Binomial Sampling Problems With Applications. *Annals of Mathematical Statistics* 17:13–23.

1947 GIRSHICK, MEYER A.; and HAAVELMO, TRYGVE Statistical Analysis of the Demand for Food: Examples of Simultaneous Estimation of Structural Equations. *Econometrica* 15:79–110.

1947 BLACKWELL, DAVID; and GIRSHICK, MEYER A. A Lower Bound for the Variance of Some Unbiased Sequential Estimates. *Annals of Mathematical Statistics* 18:277–280.

1954 BLACKWELL, DAVID; and GIRSHICK, MEYER A. *Theory of Games and Statistical Decisions.* New York: Wiley.

SUPPLEMENTARY BIBLIOGRAPHY

BLACKWELL, DAVID; and BOWKER, ALBERT H. 1955 Meyer Abraham Girshick: 1908–1955. *Annals of Mathematical Statistics* 26:365–367.

GOODNESS OF FIT

A goodness of fit procedure is a statistical test of a hypothesis that the sampled population is distributed in a specific way, for example, normally with mean 100 and standard deviation 15. Corresponding confidence procedures for a population distribution also fall under this topic. Related tests are for broader hypotheses, for example, that the sampled population is normal (without further specification). Others test hypotheses that two or more population distributions are the same.

Populations arise because of variability, of which various sources (sometimes acting together) can be distinguished. First, there is inherent variability among experimental units, for example, the heights, IQ's, or ages of the students in a class each vary among themselves. Then there is measurement error, a more abstract or conceptual notion. The age of a student may have negligible measurement error, but his IQ does not; it depends on a host of accidental factors: how the student slept, the particular questions chosen for the test, and so on. There are also other conceptual populations, not properly thought of in terms of measurement error—the population of subject responses, for example, in the learning experiment below.

The distribution of a numerical population trait is often portrayed by a histogram, a density function, or some other device that shows the proportion of cases for which a particular value of the numerical trait is achieved (or the proportion within a small interval around a particular value). The shape of the histogram or density function is important; it may or may not be symmetrical. If it is not, it is said to be *skew*. If it is symmetrical, it may have a special kind of shape called *normal*. For example, populations of scores on intelligence tests are often assumed normally distributed by psychologists. Indeed, the construction of the test may aim at normality, at least for some group of individuals. Again, lifetimes of machines may be assumed to have negative exponential distributions, meaning that expected remaining life does not vary with age. [*See* DISTRIBUTIONS, STATISTICAL, *article on* SPECIAL CONTINUOUS DISTRIBUTIONS; PROBABILITY; STATISTICS, DESCRIPTIVE.]

It is technically often convenient, especially in connection with goodness of fit tests, to deal with the *cumulative distribution function* (c.d.f.) rather than with the density function. The c.d.f. evaluated at x is the proportion of cases with numerical values less than or equal to x; thus, if $f(x)$ is a density function, the corresponding c.d.f. is

$$F(x) = \int_{-\infty}^{x} f(t)\,dt.$$

For explicitness, a subscript will be added to F, indicating the population, distribution, or random variable to which it applies. It is a matter of convention that cumulation is from the left and that it is based on "less than or equal to" rather than just "less than."

The *sample c.d.f.* is the steplike function whose value at x is the proportion of observations less than or equal to x. Many goodness of fit procedures are based on geometrically suggested measures of discrepancy between sample and hypothetical population c.d.f.'s. Some informal procedures use "probability" graph paper, especially normal paper (on which a normal c.d.f. becomes a straight line)

For nominal populations (for example, proportions of people expressing allegiance to different religions or to none) there is no concept corresponding to the c.d.f. The main emphasis of this article is on numerical populations.

Although goodness of fit procedures address themselves principally to the shape of population c.d.f.'s, the term "goodness of fit" is sometimes applied more generally than in this article. In particular, some authors write of goodness of fit of observed regressions to hypothetical forms, for example, to a straight line. [*This topic is dealt with in* LINEAR HYPOTHESES, *article on* REGRESSION.]

Hypotheses—simple, composite, approximate. A test of goodness of fit, based on a sample from a population, assesses the plausibility that the population distribution has specified form; in brief, *tests* the hypothesis that F_X has shape F_0. The specification may be complete, that is, the population distribution may be specified completely, in which case the hypothesis is called *simple*. Alternatively, the form may be specified only up to certain unknown parameters, which often are the parameters of location and scale. In this case the hypothesis is called *composite*. Still another type of hypothesis is an *approximate* one, which is composite in a certain sense. Here one specifies first what one would consider a material departure from a hypothesized shape (Hodges & Lehmann 1954). For example, in the case of a *simple approximate* hypothesis, one might agree that F_X departs materially from F_0 if the maximum vertical deviation between the actual and hypothesized cumulative distribution functions exceeds .07. The approximate hypothesis then states that the actual and hypothesized distributions do not differ materially in this sense.

Approximate hypotheses specialize to the others, so that a complete theory of testing for the former would be desirable. This is especially true since, as has been pointed out by Karl Pearson (1900) and Joseph Berkson (1938), tests of "exact" hypotheses, being as a rule *consistent*, have problematical logical status: unless the exact hypothesis is exactly correct and all of the sampling assumptions are exactly met, rejection of the hypothesis is assured (for fixed significance level) when sample size is large. Unfortunately, such a complete theory does not now exist, but the strong early interest in "exact" hypotheses was not misspent: The testing and "acceptance" of "exact" hypotheses concerning F_X seems to have much the same status as the provisional adoption of physical or other "laws." If the latter has helped the advancement of science, so has no doubt the former; this is true notwithstanding that old hypotheses or theories will almost surely be discarded as additional data become available. This point has been made by Cochran (1952) and Chapman (1958). Cochran also suggests that the tests of "exact" hypotheses are "invertible" into confidence sets, in the usual manner, thus providing statistical procedures somewhat similar in intent to tests of approximate hypotheses [*see* ESTIMATION, *article on* CONFIDENCE INTERVALS AND REGIONS].

Conducting a test of goodness of fit. Many tests of goodness of fit have been developed; as with statistical tests generally, a test of goodness of fit is conveniently conducted by computing from the sample a *statistic* and its *sample significance level* [*see* HYPOTHESIS TESTING]. In the case of a test of goodness of fit, the statistic will measure the discrepancy between what the sample in fact *is* and what a sample from a population of hypothesized form *ought to be*. The sample significance level of an observed measure of discrepancy, d_0, is, at least for all the standard goodness of fit procedures, the probability, $Pr\{d \geqslant d_0\}$, that d exceeds d_0 under random sampling from a population of hypothesized form. In other words, it is the proportion of like discrepancy measures, d, exceeding d_0, computed on the basis of many successive hypothetical random samples of the same size from a population of hypothesized form. For many tests of goodness of fit, there exist tables (for extensive bibliography see Greenwood & Hartley 1962) that give those values of d_0 corresponding to given significance level and sample size (n). Many of these standard tests are *nonparametric*, which means that $Pr\{d \geqslant d_0\}$ is the same for a very large class of hypotheses F_0, so that only one such tabulation is required [*see* NONPARAMETRIC STATISTICS].

If, as is usual, the relevant alternative population distributions (more generally, alternative probabilistic models for the generation of the sample at hand) tend to encourage large values of d_0, the hypothesized population distribution will be judged implausible if the sample significance level is small (conventionally .05 or less). If the sample significance level is not small, it means that the statistic has a value unsurprising under the null hypothesis, so that the test gives no reason to reject the null hypothesis. If, however, the sample significance level is very large, say .95 or more, one may construe this as a warning of possible trouble, say, that an overzealous proponent of the hypothesis has slanted the data or that the sampling was not random. Note here an awkward usage prevalent in statistics generally: an observed measure of discrepancy d_0 with *low probability* $Pr\{d \geqslant d_0\}$ usually is described as *highly significant*.

Choosing a test of goodness of fit. Choosing a test of goodness of fit amounts to deciding in what sense the discrepancy between the hypothesized population distribution and the sample is to be measured: The sample c.d.f. may be compared directly with the hypothesized population c.d.f., as is done in the case of tests of the Kolmogorov–Smirnov type. For example, the original Kolmogorov–Smirnov test itself, as described below, summarizes the discrepancy by the maximum absolute deviation between the hypothesized population c.d.f., F_0, and the sample c.d.f. Alternatively, one may compare uncumulated frequencies, as for the χ^2 test. Again, a standard shape parameter, such as skewness, may be computed for the sample and for the hypothesized population and the two compared.

Any reasonable measure of discrepancy will of course tend to be small if the population yielding the sample conforms to the null hypothesis. A good measure of discrepancy will, in addition, tend to be large under the likely alternative forms of the population distribution, a property designated technically by the term *power*. For example, the sample skewness coefficient might have good power if the hypothesized population distribution were normal (zero population skewness coefficient) and the relevant alternative distributional forms were appreciably skew.

Two general considerations. Two general considerations should be kept in mind. First it is important that the particular goodness of fit test used be selected without consideration of the sample at hand, at least if the calculated significance level is to be meaningful. This is because a measure of discrepancy chosen in the light of an observed sample anomaly will tend to be inordinately large. Receiving license plate 437918 hardly warrants the inference that, this year, the first and second digits add to the third, and the fifth and sixth to the fourth. It may of course be true, in special instances, that some adjustment of the test procedure in the light of the data does not affect the significance computations appreciably—as, for example, when choosing category intervals, based on the sample mean and variance, for the χ^2 test (Watson 1957).

Second, a goodness of fit test, like any other statistical test, leads to an inference from a *sample* to the *population sampled*. Indeed, the usual hypothesis under test is that the sample is in fact a *random sample* from an *infinite population* of hypothesized form, and the tabulated probabilities, $Pr\{d \geqslant d_0\}$, almost always presuppose this. (In principle, one could obtain goodness of fit tests for more complex kinds of probability samples than random ones, but little seems to be known about such possibilities.) It is therefore essential that the sample to which a standard test is applied can be thought of as a random sample. If it cannot, then one must be prepared either to do one's own nonstandard significance probability computations or to defend the adequacy of the approximation involved in using the standard tabulations. Consider, for example, starting with a random sample involving considerable repetition, say the sample of response words obtained from a panel of subjects taking a psychological word association test or the sample of nationalities obtained from a survey of the United Nations. Suppose now that one tallies the number of items in the sample (response words, nationalities) appearing exactly once, exactly twice, etc. There results a new set of data, consisting of a certain number of one's, a certain number of two's, etc. This collection of integers has the outward appearance of a random sample, and the literature contains instances of the application of the standard tests of goodness of fit to such observed frequencies. Yet the probability mechanism that generates these integers has no resemblance whatever to random sampling, and the standard probability tabulations cannot be assumed to apply. Other examples arise when the data are generated by time series; for some of these the requisite nonstandard probability computations have been done (Patankar 1954), while, in other cases, special devices have made the standard computations apply. For example, in the case of the learning experiment by Suppes and his associates (1964), the sample consists of the time series of a subject's responses to successive stimuli. Certain theories of learning predict a particular bimodal long-run response population distribution; but the goodness of fit test of this hypothesized shape, on the basis of a series of subject responses, is hampered by the statistical dependence of neighboring responses. However, theory suggests, and a test of randomness confirms, that the subsample consisting of every fifth response is effectively random, enabling a standard χ^2 test of goodness of fit to be carried out on the basis of this subsample. Whether four-fifths of the sample is a reasonable price to pay for validly carrying out a standard procedure is of course a matter of debate.

Tests of simple hypotheses

The χ^2 test. The χ^2 test was first proposed in 1900 by Karl Pearson. To apply the test, one first divides the possible range of numbers (number pairs in the bivariate case) into k regions. For example, if only nonnegative numbers are possible, one might use the categories 0 to .2, .2 to .5, .5 to .7, and .7 and beyond. Next, one computes the probabilities, p_i, associated with each of these regions (intervals in the example just given) under the hypothesized F_0. This is often done by subtracting values of F_0 from each other: for example, when F_0 is the exponential cumulative distribution function $1 - e^{-x}$,

$$\begin{aligned} p_1 &= F_0(.2) - F_0(0) = F_0(.2) = 1 - e^{-.2} &&= .18\\ p_2 &= F_0(.5) - F_0(.2) \qquad = e^{-.2} - e^{-.5} &&= .21\\ p_3 &= F_0(.7) - F_0(.5) \qquad = e^{-.5} - e^{-.7} &&= .11\\ p_4 &= F_0(\infty) - F_0(.7) \qquad = e^{-.7} &&= .50 \end{aligned}$$

The expected numbers E_i of observations in each category are (under the null hypothesis) $E_i = np_i$, where n is the size of the random sample.

After the sample has been collected, there also will be observed numbers, O_i, of sample members in each category. The chi-square measure of discrepancy d_{χ^2} is then computed by summing squared differences of class frequencies, weighted in such a way as to bring to bear standard distribution theory,

$$d_{\chi^2,0} = \sum_{i=1}^{k} (O_i - E_i)^2/E_i ,$$

where the subscript 0 indicates the specific sample value of d_{χ^2}. (Often "X^2" or "χ^2" is used to denote this statistic.)

As is shown, for example, by Cochran (1952), the probability distribution of d_{χ^2}, when $F_X = F_0$, can be approximated by the chi-square distribution with $k-1$ degrees of freedom, χ^2_{k-1}. This fact, to which the test owes its name, was first demonstrated by Karl Pearson. The larger the expectations E_i, the better is the approximation; this has been pointed out, for example, by Mann and Wald (1942). Hence, the significance, $Pr\{d_{\chi^2} \geqslant d_{\chi^2,0}\}$, is evaluated to a good approximation by consulting a tabulation of the χ^2_{k-1} distribution. For example, if k, as above, equals 4, and $d_{\chi^2,0}$ had happened to be 4.6, then $Pr\{d_{\chi^2} \geqslant d_{\chi^2,0}\} \cong .20$. With a sample significance level of .20, most statisticians would not question the plausibility of F_0. However, were $d_{\chi^2,0}$ larger, and the corresponding significance equal to .05 or less, the consensus would be reversed.

At what point is the distributional approximation endangered by small E_i? An early study of this problem, performed by Cochran in 1942 (referred to in Cochran 1952), shows that a few E_i near 1 among several large ones do not materially affect the approximation. Recent studies, by Kempthorne (1966) and by Slakter (1965), show that this is true as well when *all* E_i are near 1.

These and other studies indicate that, although some care must be taken to avoid very small E_i, much latitude remains for choosing categories. How is this to be done? To begin with, in keeping with the spirit of remarks by Birnbaum (1953), if the relevant alternatives F^* to F_0 are such that

$$d_{\chi^2}(F^*, F_0) = \sum_{i=1}^{k} (E_i^* - E_i)^2/E_i$$

is large for a certain choice of k categories, it is these categories that should be selected. Among various sets of k categories, those yielding large $d_{\chi^2}(F^*, F_0)$ are preferred.

In the absence of detailed knowledge of the alternatives, the usual recommendation, at least in the one-dimensional case, is to use intervals of equal

E_i. There remains the question of how many such intervals there should be. The typical statistical criterion for this is power, that is, the likelihood that the value of d_{χ^2} will be large enough to warrant rejection of the hypothesis F_0 when the population is in fact a relevant alternative one. If large power is desired for *all* alternative population c.d.f.'s departing from F_0 at some x by at least a given fixed amount, Mann and Wald (1942) recommend a number of categories of the order of $4n^{2/5}$. Williams (1950) has shown that this figure can easily be halved.

The χ^2 test is versatile; it is readily adapted to problems involving nominal rather than numerical populations [*see* COUNTED DATA]. It can also be adapted to bivariate and multivariate problems, as, for example, by Keats and Lord (1962), where the joint distribution of two types of mental test scores is considered. As opposed to many of its competitors, the χ^2 test is not biased, in the sense that there are no alternatives F^* to F_0 under which acceptance of F_0 is more likely than it is under F_0 itself. It is readily adapted to composite and approximate testing problems. Also, it seems to be true that the χ^2 test is in the best position among its competitors with regard to the practical computation of power. As is pointed out by Cochran (1952), such computations are performed by means of the *noncentral* chi-square distribution with $k-1$ degrees of freedom.

Modifications of the χ^2 test. Important modifications of the χ^2 test, intended to increase its power against specific families $\mathcal{G}$ of distributions alternative to F_0, are given by Neyman (1949) and by Fix, Hodges, and Lehmann (1954). Here $\mathcal{G}$ is assumed to include F_0 and to allow differentiable parametric representation of the category expectations E_i. Note that the inclusion of F_0 in $\mathcal{G}$ differs from the point of view adopted, for example, by Mann and Wald (1942). These modifications are essentially *likelihood ratio* tests of F_0 versus $\mathcal{G}$ and are similar to procedures used to test composite and approximate hypotheses.

Another modification, capable of orientation against specific "smooth" alternatives, Neyman's ψ^2 test, was introduced in 1937. Other important modifications are described in detail in Cochran (1954).

Other procedures. When $(X_1, \cdots, X_n)$ is a random sample from a population distributed according to a *continuous* c.d.f. F_0, then $(U_1, \cdots, U_n) = (F_0(X_1), \cdots, F_0(X_n))$ has all the probabilistic properties of a random sample from a population distributed uniformly over the numbers between zero and one. (If the population has a density function, the c.d.f. is continuous.) No matter what the hypothesized F_0, the initial application of this *probability integral* transformation thus reduces all probability computations to the case of this *uniform* population distribution and gives a *nonparametric* character to any procedure based on the transformed sample $(U_1, \cdots, U_n)$. Most goodness of fit tests of simple hypotheses are nonparametric in this sense, including the χ^2 test itself, when categories are chosen so as to assign specified values, for example, the constant value $1/k$, to the category probabilities p_i.

Another common test making use of the transformation $U = F_0(X)$ is the Kolmogorov–Smirnov test, first suggested by Kolmogorov (1933) and explained in detail by Goodman (1954) and Massey (1951). The test bears Smirnov's name, as well as Kolmogorov's, presumably because Smirnov (as Doob and Donsker did later) gave an alternate derivation of its asymptotic null distribution, tabulated this distribution, and also extended the test to the two-sample case discussed below (1939*a*). Denote by $F_n(x)$ the *sample c.d.f.*, that is, $F_n(x)$ is the proportion of sample values less than or equal to x. The test is based on the maximum absolute vertical deviation between $F_n(x)$ and $F_0(x)$,

$$d_K = d_K(F_n, F_0) = \sqrt{n}\,(\max_{-\infty<x<+\infty} |F_n(x) - F_0(x)|),$$

the dependence of d_K on the quantities $U_i = F_0(X_i)$ being best brought out by the alternate formula

$$d_K = \sqrt{n}\max\left[\max_i\left(\frac{i}{n} - u_i\right), \max_i\left(u_i - \frac{i-1}{n}\right)\right],$$

where u_1 is the smallest U_i, u_2 is the next to smallest, etc.; the equivalence of the two formulas is made clear by a sketch. As Kolmogorov noted in his original paper, the probabilities tabulated for d_K are *conservative* when F_0 is not continuous, in the sense that, for discontinuous F_0, actual probabilities of $d_K \geqslant d_{K,0}$ will tend to be less than those obtained from the tabulations, leading to occasional unwarranted acceptance of F_0.

Computations (Shapiro & Wilk 1965) suggest that this test has low power against alternatives with mean and variance equal to those of the hypothesized distribution. It has, however, been argued, for example, by Birnbaum (1953) and Kac, Kiefer, and Wolfowitz (1955), that the test yields good minimum power over classes of alternatives F^* satisfying $d_K(F^*, F_0) \geqslant \delta$; these, as the reader will note, are precisely the classes of alternatives envisaged by Mann and Wald (1942) in optimizing the number of categories used in the

χ^2 test. A detrimental feature of the Kolmogorov–Smirnov test is its bias, pointed out in Massey (1951).

An important feature of the test is that it can be "inverted" in keeping with the usual method to provide a confidence band for $F_0(x)$ centered on $F_n(x)$, which, except for the narrowing caused by the restriction $0 \leqslant F_0(x) \leqslant 1$, has constant width [*see* ESTIMATION, *article on* CONFIDENCE INTERVALS AND REGIONS]. The construction of such a band has been suggested by Wald and Wolfowitz and is described by Goodman (1954). Attaching a significance probability to an observed $d_{K,0}$ amounts to ascertaining the band width required in order just to include wholly the hypothesized F_0 in the confidence band.

The Kolmogorov–Smirnov test has been modified in several ways; the first of these converts the test into a "one-sided" procedure based on the discrepancy

$$d_{K^+} = \sqrt{n}\,[\max_{-\infty<x<+\infty}(F_n(x) - F_0(x))]$$
$$= \sqrt{n}\left[\max_i \left(\frac{i}{n} - u_i\right)\right].$$

A useful feature of this modification is the simplicity of the large sample computation of significance probabilities associated with observed discrepancies $d_{K^+,0}$; abbreviating the latter to d, one has $Pr\{d_{K^+} \geqslant d\} \cong e^{-2d^2}$. It is verified by Chapman (1958) that d_{K^+} yields good minimum power over those classes of alternatives F^* that satisfy $d_{K^+}(F^*, F_0) = \sqrt{n}\max_x(F^*(x) - F_0(x)) \geqslant \delta$.

Other, more complex modifications provide greater power against special alternatives, as in the weight function modifications (Darling 1957), which provide greater power against discrepancies from F_0 in the tails. Another sort of modification, introduced and tabulated by Kuiper in 1960, calls for a measure of discrepancy d_V that is especially suited to testing goodness of fit to hypothesized *circular* distributions, being invariant under arbitrary choices of the angular origin. This property could be important, for example, in psychological studies involving responses to the color wheel, or in the learning experiment mentioned above. The measure d_V also has been singled out by E. S. Pearson (1963) as the generally most attractive in competition with d_K and the discrepancy measures d_{ω^2} and d_U mentioned below.

A second general class of procedures also making use of the transformation $U = F_0(x)$ springs from the discrepancy measure

$$d_{\omega^2} = n\int_{-\infty}^{+\infty}[F_n(x) - F_0(x)]^2\,dF_0(x)$$
$$= \frac{1}{12n} + \sum_{i=1}^{n}\left(\frac{2i-1}{2n} - u_i\right)^2,$$

first proposed by Cramér in 1928 and also by von Mises in 1931 (see Darling 1957). Marshall (1958) has verified a startling agreement between the asymptotic and small sample distributions of d_{ω^2} for sample sizes n as low as 3. Power considerations for d_{ω^2} are similar to those expressed for d_K, and are discussed also in the sources cited by Marshall; the test based on d_{ω^2} can be expected to have good minimum power over classes of alternatives F^* satisfying the conditions $d_{\omega^2}(F^*, F_0) = n\int_{-\infty}^{+\infty}[F^*(x) - F_0(x)]^2\,dF_0(x) \geqslant \delta$. However, the test is biased (as is that based on d_K).

As in the case of d_K, d_{ω^2} has weight function modifications for greater power selectivity, and also a modification d_U, analogous to the modification d_V of d_K and introduced by Watson (1961), which does not depend on the choice of angular origin and is thus also suited for testing the goodness of fit to hypothesized circular distributions.

Other procedures include those based on the Fisher–Pearson measures $d_{FP}^{(1)} = -2\sum_{i=1}^{n}\ln u_i$ and $d_{FP}^{(2)} = -2\sum_{i=1}^{n}\ln(1 - u_i)$, apparently first suggested in connection with goodness of fit in 1938 by E. S. Pearson. As pointed out by Chapman (1958), the tests based on $d^{(1)}$ and $d^{(2)}$ are uniformly most powerful against polynomial alternatives to $F_U(x) = x$ of form x^k and $(1 - x)^k$, $k > 1$, and hence are "smooth" in the sense of Neyman's ψ^2 test. Computations by Chapman suggest that, dually to d_K, $d^{(2)}$ has good maximum power over classes of alternatives F^* satisfying $d_K(F^*, F_0) \leqslant \delta$.

Another set of procedures, discussed and defended by Pyke (1965) and extensively studied by Weiss (1958), are based on functions of the spacings, $u_{i+1} - u_i$ or $u_i - (i + 1)^{-1}$, of the u's, from each other or from their expected locations under F_0. Still another criterion (Smirnov 1939*b*) examines the number of crossings of $F_n(x)$ and $F_0(x)$.

An important modification, applicable to all of the procedures in this section, is suggested in Durbin (1961). This modification is intended to increase the power of any procedure based on the transforms U_i, against a certain class of alternatives described in that paper.

Since there are multivariate probability integral transformations, applying an initial "uniformizing" transformation is possible in the multivariate case as well. However, one of several possible transformations must now be chosen, and, related to this nonuniqueness, the direct analogues of the univariate discrepancy measures are no longer functions of uniformly distributed transforms and do not lead to nonparametric tests (Rosenblatt 1952).

Tests of composite hypotheses

The χ^2 test. In the composite case, the null hypothesis specifies only that $F_X(x)$ is a member of a certain parametric class $\{F_\theta(x)\}$. Typically, but not necessarily, θ is the pair (μ, σ), μ a parameter of location, and σ a parameter of scale, in which case $F_\theta(x)$ may be written $F_0[(x-\mu)/\sigma]$. In any event, there arises the question of modifying the measure d_{χ^2} of discrepancy between the sample and a particular cumulative distribution function into a measure D_{χ^2} of discrepancy between the sample and the *class* $\{F_\theta(x)\}$. A natural approach is to set

$$D_{\chi^2} = \min_\theta d_{\chi^2}.$$

If θ is composed of m parameters, it can be shown that, under quite general conditions, D_{χ^2} is approximately distributed according to the χ^2_{k-1-m} distribution when $F_X(x)$ equals any one of the $F_\theta(x)$. Hence significance probability computations can once again be referred to tabulations of the χ^2_ν distribution. The requisite minimization with respect to θ can be cumbersome, and several modifications have been proposed, for example, the following by Neyman (1949):

Suppose that one defines $d_{\chi^2}(\theta)$ as the discrepancy d_{χ^2} between the observed sample and the particular distribution $F_\theta(x)$. Then D is defined also by

$$D_{\chi^2} = d_{\chi^2}(\tilde{\theta}),$$

with the estimator $\tilde{\theta}$ computed from

$$d_{\chi^2}(\tilde{\theta}) = \min_\theta d_{\chi^2}(\theta),$$

that is, with $\tilde{\theta}$ the *minimum chi-square estimator* of θ. The suggested modifications involve using estimators of θ alternate to $\tilde{\theta}$ in this last definition of D_{χ^2}, that is, estimators that "essentially" minimize $d_{\chi^2}(\theta)$; among these are the so-called grouped-data or partial-information maximum likelihood estimators.

Frequently used but *not equivalent* estimators are the ordinary "full-information" maximum likelihood estimators $\hat{\theta}$ of θ, for example, $(\bar{x}, s)$ for (μ, σ) in the normal case. These do *not* "essentially" minimize d_{χ^2} and consequently tend to inflate D_{χ^2} beyond values predicted by the χ^2_{k-1-m} distribution, leading to some unwarranted rejections of the composite hypothesis. However, it is indicated by Chernoff and Lehmann (1954), and also by Watson (1957), that no serious distortion will result if the number of categories is ten or more.

Composite analogues of other tests. Adaptation of the tests based on the probability integral transformation to the composite case proceeds much as in the case of χ^2. With definitions of $d_{\omega^2}(\theta)$ and $d_K(\theta)$ analogous to that of $d_{\chi^2}(\theta)$, Darling (1955) has investigated the large sample probability distribution of $D_{\omega^2} = d_{\omega^2}(\hat{\theta})$ and of $D_K = d_K(\hat{\theta})$ for efficient estimators $\hat{\theta}$ of θ analogous to the estimators $\tilde{\theta}$ for χ^2. Note that in the absence of any χ^2-like categories, the ordinary full-information maximum likelihood estimators now *do* qualify as estimators θ.

A major problem now is, however, that the modified procedures are no longer nonparametric. Thus a special investigation is required for every composite hypothesis. This is done by Kac, Kiefer, and Wolfowitz (1955) for the normal scale-location family, and the resulting large sample distribution is partly tabulated.

Tests based on special characteristics. The alternatives of concern sometimes differ from a composite null hypothesis in a manner easily described by a standard shape parameter. Special tests have been proposed for such cases. For example, the sample skewness coefficient has been suggested (Geary 1947) for testing normality against skew alternatives. Again, for testing Poissonness against alternatives with variance unequal to mean, R. A. Fisher has recommended the variance-to-mean ratio $\sum_{i=1}^n (X_i - \bar{X})^2/\bar{X}$ (see Cochran 1954). This measure is approximately distributed as χ^2_{n-1} when Poissonness in fact obtains, for $\lambda > 1$ and $n > 15$ (Sukhatme 1938), which follows from the fact that the denominator is then a high-precision estimate of λ, and the numerator is approximately distributed as $\lambda\chi^2_{n-1}$. Analogous recommendations apply to testing binomiality. Essentially the same point of view underlies tests of normality based on the ratio of mean deviation, or of the range, to the standard deviation.

Transforming into simple hypotheses. Another interesting approach to the composite problem, advocated by Hogg and also by Sarkadi (1960), is to transform certain composite hypotheses into equivalent simple ones.

Specifically, there are location-scale parametric families $\{F_0[(x-\mu)/\sigma]\}$ with the following property: A random sample from any particular $F_0[(x-\mu)/\sigma]$ is reducible by a transformation T to a new set of random variables, $Y = T(X)$, constituting in effect a random sample from a distribution $G(y)$ involving no unknown parameters at all. Moreover, *only* random samples from distributions $F_0[(x-\mu)/\sigma]$ lead to $G(y)$ when operated on by T.

It then follows that testing the composite hypothesis H that $(X_1, \cdots, X_n)$ is a random sample from a distribution $F_0[(x-\mu)/\sigma]$ with some μ and some σ is equivalent to testing the hypothesis H' that $(Y_1, \cdots, Y_m)$ is a random sample from the distribution $G(y)$. Any of the tests for simple hypotheses is then available for testing H'. An example is provided by a negative exponential F_0 and uniform G, in which case the ordered exponential random sample $(X_{(1)}, \cdots, X_{(n)})$ is transformed into an ordered uniform random sample $(Y_{(1)}, \cdots, Y_{(n-2)})$ by the transformation

$$Y_{(i)} = \frac{(X_{(n)} - X_{(n-1)}) + \cdots + i\,(X_{(n-i+1)} - X_{(n-i)})}{(X_{(n)} - X_{(n-1)}) + \cdots + (n-1)\,(X_{(2)} - X_{(1)})}.$$

Conditioning. Another way of neatly doing away with the unknown parameter is to consider the conditional distribution of the sample, given a sufficient estimate of it. This method is advocated, at least for testing Poissonness, in Fisher (1950).

[2]Tests related to probability plots. S. S. Shapiro and M. B. Wilk have quantified in various ways the departure from linearity of the sorts of probability plots mentioned above, in particular of the plot of the ordered sample values against the expected values of the standardized order statistics [*see* NONPARAMETRIC STATISTICS, *article on* ORDER STATISTICS]. This new approach bears some similarity to one given in Darling (1955), which is based on the measure $d_\omega{}^2$ modified for the composite case. Both approaches, in a sense, compare adjusted observed order statistics with standardized order statistic expectations. But the approach of Shapiro and Wilk is tailored more explicitly to particular scale-location families, by using their particular order statistic variances and covariances. It is no wonder that preliminary evaluations of this sort of approach (for example, by Shapiro & Wilk 1965) have shown exceptional promise. As an added bonus, the procedure is *similar* over the entire scale-location family; that is, its probability distribution is independent of location and scale.

Approximate hypotheses

The first, and seemingly most practically developed, attempt to provide the requisite tests of approximate hypotheses is found in Hodges and Lehmann (1954). Hodges and Lehmann assume the k typical categories of the χ^2 test and formulate the approximate *simple* hypothesis in terms of the discrepancy $d(p, p_0)$ between the category probabilities p_i under F_X and the category probabilities $p_{0,i}$ under a *simple* hypothesis F_0. A very tractable discrepancy measure of this type is ordinary distance, for which the approximate hypothesis takes the form

$$d(p, p_0) = \sum_{i=1}^{k} (p_i - p_{0,i})^2 \leqslant \delta.$$

Denoting O_i/n by o_i, the suggested test reduces, essentially, to the *one-sided* test of the hypothesis $d(p, p_0) = \delta$ based on the approximately normal statistic $[d(o, p_0) - \delta]/\hat{\sigma}$, where $\hat{\sigma}$ is the standard deviation, estimated from the sample o_i, of $d(o, p_0)$. For example, when F_0 specifies k categories with $p_{0,i} = 1/k$, one treats as unit normal (under the null hypothesis) the statistic

$$S = \frac{\sqrt{n}\,[d(o, p_0) - \delta]}{2\sqrt{\sum o_i^3 - (\sum o_i^2)^2}}$$

and uses an upper-tail test. Thus a value of S of 1.645 leads to a sample significance level of .05. This approach lends itself easily to the computation of power and is extended as well by Hodges and Lehmann to the testing of approximate *composite* hypotheses.

Extension of other tests for simple hypotheses to the testing of approximate hypotheses has been considered by J. Rosenblatt (1962) and by Kac, Kiefer, and Wolfowitz (1955).

Further topics

That the sample is random may itself be in doubt, and tests have been designed to have power against specific sorts of departure from randomness. For example, tests of the hypothesis of randomness against the alternative hypothesis that the data are subject to a Markov structure are given by Billingsley (1961) and Goodman (1959); the latter work also covers testing that the data have Markov structure of a given order against the alternative that the data have Markov structure of higher order, and the testing of hypothesized values of transition probabilities when a Markov structure of given order is assumed [*see* MARKOV CHAINS].

Many of the tests described in this article can be extended to *several-sample* procedures for testing the hypothesis that several populations are in fact distributed identically; thus, as first suggested in Smirnov (1939*a*), if $G_m(x)$ denotes the proportion of values less than or equal to x, in an independent random sample $(Y_1, \cdots, Y_m)$ from a second population, $d_K(F_n, G_m)$ provides a natural test of the hypothesis that the two continuous population distribution functions F_X and G_Y coincide. Many of these extensions are functions only of the *relative ranks* of the two samples and, as such, are nonparametric, that is, their null probability distributions do not depend on the common functional form of F_X and G_Y. [*Several-sample nonparametric procedures are discussed in* NONPARAMETRIC STATISTICS.]

Another topic is that of tests of goodness of fit as preliminary tests of significance, in a sense discussed, for example, by Bancroft (1964). That tests of goodness of fit are typically applied in this sense is recognized by Chapman (1958), and the probabilistic properties of certain "nested" sequences of tests beginning with a test of goodness of fit have been considered by Hogg (1965). The Bayes and information theory approaches to χ^2 tests of goodness of fit are also important (see Lindley 1965; Kullback 1959).

H. T. DAVID

[*Directly related are the entries* HYPOTHESIS TESTING; SIGNIFICANCE, TESTS OF. *Other relevant material may be found in* COUNTED DATA; ESTIMATION; NONPARAMETRIC STATISTICS.]

BIBLIOGRAPHY

BANCROFT, T. A. 1964 Analysis and Inference for Incompletely Specified Models Involving the Use of Preliminary Test(s) of Significance. *Biometrics* 20:427–442.

BERKSON, JOSEPH 1938 Some Difficulties of Interpretation Encountered in the Chi-square Test. *Journal of the American Statistical Association* 33:526–536.

BILLINGSLEY, PATRICK 1961 Statistical Methods in Markov Chains. *Annals of Mathematical Statistics* 32:12–40.

BIRNBAUM, Z. W. 1953 Distribution-free Tests of Fit for Continuous Distribution Functions. *Annals of Mathematical Statistics* 24:1–8.

CHAPMAN, DOUGLAS G. 1958 A Comparative Study of Several One-sided Goodness-of-fit Tests. *Annals of Mathematical Statistics* 29:655–674.

CHERNOFF, HERMAN; and LEHMANN, E. L. 1954 The Use of Maximum Likelihood Estimates in χ^2 Tests for Goodness of Fit. *Annals of Mathematical Statistics* 25:579–586.

COCHRAN, WILLIAM G. 1952 The χ^2 Test of Goodness of Fit. *Annals of Mathematical Statistics* 23:315–345.

COCHRAN, WILLIAM G. 1954 Some Methods for Strengthening the Common χ^2 Tests. *Biometrics* 10:417–451.

DARLING, D. A. 1955 The Cramér–Smirnov Test in the Parametric Case. *Annals of Mathematical Statistics* 26:1–20.

DARLING, D. A. 1957 The Kolmogorov–Smirnov, Cramér–von Mises Tests. *Annals of Mathematical Statistics* 28:823–838.

DURBIN, J. 1961 Some Methods of Constructing Exact Tests. *Biometrika* 48:41–55.

FISHER, R. A. 1924 The Conditions Under Which χ^2 Measures the Discrepancy Between Observation and Hypothesis. *Journal of the Royal Statistical Society* 87:442–450.

FISHER, R. A. 1950 The Significance of Deviations From Expectation in a Poisson Series. *Biometrics* 6:17–24.

FIX, EVELYN; HODGES, J. L. JR.; and LEHMANN, E. L. 1954 The Restricted Chi-square Test. Pages 92–107 in Ulf Grenander (editor), *Probability and Statistics.* New York: Wiley.

GEARY, R. C. 1947 Testing for Normality. *Biometrika* 34:209–242.

GOODMAN, LEO A. 1954 Kolmogorov–Smirnov Tests for Psychological Research. *Psychological Bulletin* 51:160–168.

GOODMAN, LEO A. 1959 On Some Statistical Tests for *m*th Order Markov Chains. *Annals of Mathematical Statistics* 30:154–164.

GREENWOOD, JOSEPH A.; and HARTLEY, H. O. 1962 *Guide to Tables in Mathematical Statistics.* Princeton Univ. Press. → A sequel to the guides to mathematical tables produced by and for the Committee on Mathematical Tables and Aids to Computation of the National Academy of Sciences National Research Council of the United States.

HODGES, J. L. JR.; and LEHMANN, E. L. 1954 Testing the Approximate Validity of Statistical Hypotheses. *Journal of the Royal Statistical Society* Series B 16:261–268.

HOGG, ROBERT V. 1965 On Models and Hypotheses With Restricted Alternatives. *Journal of the American Statistical Association* 60:1153–1162.

KAC, M.; KIEFER, J.; and WOLFOWITZ, J. 1955 On Tests of Normality and Other Tests of Goodness of Fit Based on Distance Methods. *Annals of Mathematical Statistics* 26:189–211.

KEATS, J. A.; and LORD, FREDERIC M. 1962 A Theoretical Distribution for Mental Test Scores. *Psychometrika* 27:59–72.

[3]KEMPTHORNE, OSCAR 1966 The Classical Problem of Inference: Goodness of Fit. Unpublished manuscript. → Paper presented at the Berkeley Symposium on Mathematical Statistics and Probability, Fifth.

KOLMOGOROV, A. N. 1933 Sulla determinazione empirica di une legge di distribuzione. Istituto Italiano degli Attuari, *Giornale* 4:83–99.

KUIPER, NICOLAAS H. 1960 Tests Concerning Random Points on a Circle. Akademie van Wetenschappen, Amsterdam, *Proceedings* Series A 63:38–47.

KULLBACK, S. 1959 *Information Theory and Statistics.* New York: Wiley.

LINDLEY, DENNIS V. 1965 *Introduction to Probability and Statistics From a Bayesian Viewpoint.* Volume 2: *Inference.* Cambridge Univ. Press.

MANN, H. B.; and WALD, A. 1942 On the Choice of the Number of Class Intervals in the Application of the Chi Square Test. *Annals of Mathematical Statistics* 13:306–317.

MARSHALL, A. W. 1958 The Small Sample Distribution of $n\omega_n^2$. *Annals of Mathematical Statistics* 29:307–309.

MASSEY, FRANK J. JR. 1951 The Kolmogorov–Smirnov Test for Goodness of Fit. *Journal of the American Statistical Association* 46:68–78.

NEYMAN, JERZY 1937 "Smooth Test" for Goodness of Fit. *Skandinavisk aktuarietidskrift* 20:149–199.

NEYMAN, JERZY 1949 Contribution to the Theory of the χ^2 Test. Pages 239–273 in Berkeley Symposium on Mathematical Statistics and Probability, First, *Proceedings*. Berkeley: Univ. of California Press.

PATANKAR, V. N. 1954 The Goodness of Fit of Frequency Distributions Obtained From Stochastic Processes. *Biometrika* 41:450–462.

PEARSON, E. S. 1938 The Probability Integral Transformation for Testing Goodness of Fit and Combining Independent Tests of Significance. *Biometrika* 30: 134–148.

PEARSON, E. S. 1963 Comparison of Tests for Randomness of Points on a Line. *Biometrika* 50:315–325.

PEARSON, KARL 1900 On the Criterion That a Given System of Deviations From the Probable in the Case of a Correlated System of Variables Is Such That It Can Be Reasonably Supposed to Have Arisen From Random Sampling. *Philosophical Magazine* 5th Series 50:157–175.

PYKE, RONALD 1965 Spacings. *Journal of the Royal Statistical Society* Series B 27:395–449.

ROSENBLATT, JUDAH 1962 Testing Approximate Hypotheses in the Composite Case. *Annals of Mathematical Statistics* 33:1356–1364.

ROSENBLATT, MURRAY 1952 Remarks on a Multivariate Transformation. *Annals of Mathematical Statistics* 23: 470–472.

SARKADI, KÁROLY 1960 On Testing for Normality. Magyar Tudományos Akadémia, Matematikai Kutató Intézet, *Közlemények* Series A 5:269–274.

SHAPIRO, S. S.; and WILK, M. B. 1965 An Analysis of Variance Test for Normality (Complete Samples). *Biometrika* 52:591–611.

SLAKTER, MALCOLM J. 1965 A Comparison of the Pearson Chi-square and Kolmogorov Goodness-of-fit Tests With Respect to Validity. *Journal of the American Statistical Association* 60:854–858.

SMIRNOV, N. V. 1939*a* On the Estimation of the Discrepancy Between Empirical Curves of Distribution for Two Independent Samples. Moscow, Universitet, *Bulletin mathématique* Série Internationale 2, no. 2: 3–26.

SMIRNOV, N. V. 1939*b* Ob ukloneniiakh empiricheskoi krivoi raspredeleniia (On the Deviations of the Empirical Distribution Curve). *Matematicheskii sbornik* New Series 6, no. 1:1–26. → Includes a French résumé.

SUKHATME, P. V. 1938 On the Distribution of χ^2 in Samples of the Poisson Series. *Journal of the Royal Statistical Society* 5 (Supplement):75–79.

SUPPES, PATRICK et al. 1964 Empirical Comparison of Models for a Continuum of Responses With Noncontingent Bimodal Reinforcement. Pages 358–379 in R. C. Atkinson (editor), *Studies in Mathematical Psychology*. Stanford Univ. Press.

WATSON, G. S. 1957 The χ^2 Goodness-of-fit Test for Normal Distributions. *Biometrika* 336–348.

WATSON, G. S. 1961 Goodness-of-fit Tests on a Circle. *Biometrika* 48:109–114.

WEISS, LIONEL 1958 Limiting Distributions of Homogeneous Functions of Sample Spacings. *Annals of Mathematical Statistics* 29:310–312.

WILLIAMS, C. ARTHUR JR. 1950 On the Choice of the Number and Width of Classes for the Chi-square Test of Goodness of Fit. *Journal of the American Statistical Association* 45:77–86.

Postscript

Principal developments have been in the area of the Kolmogorov–Smirnov and Cramér–von Mises tests and in the area of the Shapiro–Wilk test. As indicated in the main article, these two areas are concerned, respectively, with tests based on the sample c.d.f. and with tests based on the linearized sample c.d.f.

[1]Work in the first area has addressed the testing of composite hypotheses; for example, that the sample has been drawn from some member of the normal scale-location family (that is, testing *normality*). Lilliefors (1967; 1969) has considered testing normality and testing exponentiality, and Sukhatme (1972) has considered the general composite hypothesis. Other work has dealt with the power of tests based on the sample c.d.f. Steck (1969) has studied the power of one-sided two-sample Kolmogorov–Smirnov tests under Lehmann alternatives (tests of H_0: $G = F$ versus H_1: $G = F^k$, using the two-sample version of d_{K^+} based on the two sample c.d.f.'s G_n and F_n). Stephens (1974) has tabulated the powers of several of the sample c.d.f. tests against a variety of alternatives. Tables, not of power but of significance levels, also are provided by Kiefer (1959) for two-sample Kolmogorov–Smirnov and Cramér–von Mises and by Lauschbach et al. (1967) for the two-sample Kolmogorov–Smirnov tests. Additional work in the area of tests based on the sample c.d.f. includes Durbin's extension to the bivariate case (Durbin 1970) and Vincze's work on the case when the population c.d.f. is discontinuous (Vincze 1970). An excellent overview of the area is Durbin (1973).

[2]Developments in the second area, concerning tests based on the linearized sample c.d.f., deal with simplifications of the numerator of the original Shapiro–Wilk statistic; D'Agostino (1971), Shapiro and Francia (1972), and Filliben (1975) are notable. In addition, the Shapiro–Wilk idea has been extended to the multivariate case by Malkovich and Afifi (1973), Andrews et al. (1973), and others. Most of these papers contain power comparisons and related comments on choosing the appropriate test; such comments also appear in Sneyers (1974).

H. T. DAVID

[*See also* DATA ANALYSIS, EXPLORATORY, *for a discussion of less formal methods of examining goodness of fit.*]

ADDITIONAL BIBLIOGRAPHY

ANDREWS, DAVID F.; GNANADESIKAN, RAMANATHAN; and WARNER, J. L. 1973 Methods for Assessing Multivariate Normality. Pages 95–116 in International Symposium on Multivariate Analysis, Third, Wright State University, 1972, *Multivariate Analysis: Proceedings*. Edited by Paruchuri R. Krishnaiah. New York: Academic Press.

D'AGOSTINO, RALPH B. 1971 An Omnibus Test of Normality for Moderate and Large Size Samples. *Biometrika* 58:341–348.

DURBIN, J. 1970 Asymptotic Distributions of Some Statistics Based on the Bivariate Sample Distribution Function. Pages 435–449 in International Symposium on Nonparametric Techniques in Statistical Inference, Indiana University, 1969, *Nonparametric Techniques in Statistical Inference*. Edited by Madan Lal Puri. Cambridge Univ. Press. → Discussion by Oscar Kempthorne on pages 450–451.

DURBIN, J. 1973 *Distribution Theory for Tests Based on the Sample Distribution Function*. Regional Conference Series in Applied Mathematics No. 9. Philadelphia: Society for Industrial and Applied Mathematics.

FILLIBEN, JAMES J. 1975 The Probability Plot Correlation Coefficient Test for Normality. *Technometrics* 17: 111–117.

[3]KEMPTHORNE, OSCAR 1967 The Classical Problem of Inference: Goodness of Fit. Volume 1, pages 235–249 in Berkeley Symposium on Mathematical Statistics and Probability, Fifth, *Proceedings*. Berkeley: Univ. of California Press. → A formal publication of Kempthorne (1966).

KIEFER, J. 1959 k-Sample Analogues of the Kolmogorov–Smirnov and Cramér–v. Mises Tests. *Annals of Mathematical Statistics* 30:420–447.

LAUSCHBACH, HANS et al. 1967 *Tabellen der Verteilungsfunktion zum Zwei-Stichproben–Smirnoff–Kolmogoroff Test*. Berichte aus dem Institut für Statistik und Wirtschaftsmathematik und aus dem Institut für angewandte Statistik der Freien Universität Berlin No. 3. Würzburg (West Germany): Physica-Verlag.

LILLIEFORS, HUBERT W. 1967 On the Kolmogorov–Smirnov Test for Normality With Mean and Variance Unknown. *Journal of the American Statistical Association* 62:399–402.

LILLIEFORS, HUBERT W. 1969 On the Kolmogorov–Smirnov Test for the Exponential Distribution With Mean Unknown. *Journal of the American Statistical Association* 64:387–389.

MALKOVICH, J. F.; and AFIFI, A. A. 1973 On Tests for Multivariate Normality. *Journal of the American Statistical Association* 68:176–179.

SHAPIRO, S. S.; and FRANCIA, R. S. 1972 An Approximate Analysis of Variance Test for Normality. *Journal of the American Statistical Association* 67:215–216.

SNEYERS, R. 1974 Sur les tests de normalité. *Revue de statistique appliquée* 22:29–36.

STECK, G. P. 1969 The Smirnov Two Sample Tests as Rank Tests. *Annals of Mathematical Statistics* 40: 1449–1466.

STEPHENS, M. A. 1974 EDF Statistics for Goodness of Fit and Some Comparisons. *Journal of the American Statistical Association* 69:730–737.

SUKHATME, SHASHIKALA 1972 Fredholm Determinant of a Positive Definite Kernel of a Special Type and Its Application. *Annals of Mathematical Statistics* 43: 1914–1926.

VINCZE, ISTVÁN 1970 On Kolmogorov–Smirnov Type Distribution Theorems. Pages 385–401 in International Symposium on Nonparametric Techniques in Statistical Inference, Indiana University, 1969, *Nonparametric Techniques in Statistical Inference*. Edited by Madan Lal Puri. Cambridge Univ. Press.

GOSSET, WILLIAM SEALY

The impact of W. S. Gosset (1876–1937) on the social sciences was entirely indirect. He was, however, one of the pioneers in the development of modern statistical method and its application to the design and analysis of experiments. He is far better known to the scientific world under the pseudonym of "Student" than under his own name. Indeed all his papers except one appeared under the pseudonym.

He was the son of Colonel Frederic Gosset of the Royal Engineers, the descendant of an old Huguenot family that left France after the revocation of the Edict of Nantes. Gosset was a scholar of Winchester—that is, a boy who was awarded a prize on the basis of a competitive examination to pay for part or all of his education—which shows that his exceptional mental powers had developed early. From Winchester he went, again as a scholar, to New College, Oxford, where he obtained first class degrees in mathematics and natural science.

On leaving Oxford in the autumn of 1899 he joined the famous brewing firm of Guinness in Dublin. He remained with Guinness all his life, ultimately becoming, in 1935, chief brewer at Park Royal, the firm's newly established brewery in London.

At that time scientific methods and laboratory determinations were beginning to be seriously applied to brewing, and this naturally led Gosset to study error functions and to see the need for adequate methods to deal with small samples in examining the relations between the quality of the raw materials of beer, such as barley and hops, the conditions of production, and the finished article. The importance of controlling the quality of barley ultimately led him to study the design of agricultural field trials.

In 1904 he drew up for the directors the first report on "The Application of the Law of Error." This emphasized the importance of the theory of probability in setting "an exact value on the results of our experiments; many of which lead to results which are probable but not certain." He used only the classical theory of errors, such as is found in G. B. Airy's *On the Algebraical, and Numerical*

Theory of Errors of Observation (1861) and M. Merriman's *A Text-book on the Method of Least Squares* (1884). But he observed that if X and Y are both measured from their mean, there are often considerable differences between $\sum(X+Y)^2$ and $\sum(X-Y)^2$; in other words, he was feeling his way toward the notion of correlation, although he had not yet heard of the correlation coefficient.

His first meeting with Karl Pearson took place in 1905, and in 1906/1907 he was sent for a year's specialized study in London, where he worked at, or in close contact with, the biometric laboratory at University College.

Mathematical statistics. Gosset was once described by Sir Ronald Fisher as the "Faraday of statistics." The comparison is apt, for he was not a profound mathematician but had a superb intuitive faculty that enabled him to grasp general principles and see their relevance to practical ends.

His first mathematical paper was "On the Error of Counting With a Haemacytometer" ([1907] 1943, pp. 1–10); here he derived afresh the Poisson distribution as a limiting form of the binomial and fitted it to four series of counts of yeast cells. The derivation presented no particular difficulty (it had in fact been obtained before by several investigators), but it was characteristic of him to see immediately the correct method of dealing with a practical problem. One of these series has become world famous owing to its inclusion as an example in Fisher's *Statistical Methods for Research Workers* (1925).

His next paper, "The Probable Error of a Mean" ([1908] 1943, pp. 11–34), brought him more fame, in the course of time, than any other work that he did, for it provided the basis of Student's *t*-test.

In his work at the brewery he had been struck by the importance of knowing the accuracy of the mean of a small sample. The usual procedure at the time was to compute the sample average and standard deviation, $\bar{x}$ and s, and to proceed as if $\bar{x}$ were normally distributed with the same mean as that of the population and with standard deviation $s/\sqrt{n}$, where n is sample size. The difficulty here is that s is a fallible estimate of the true population standard deviation. Gosset's intuition told him that the usual procedure, based on large sample considerations, would, for small samples, give a spuriously high impression of how accurately the population mean is estimated.

By a combination of exceptional clearheadedness and simple algebra, he obtained the first four moments of the distribution of s^2. He then proceeded to fit the Pearson curve that has these moments. His results showed that the curve has to be of Type III (essentially the gamma or χ^2 distribution), and he found the distribution of s^2 to be $C(s^2/\sigma^2)^{(n-3)/2}\exp[(-ns^2/2\sigma^2)]d(s^2/\sigma^2)$. He then showed that the correlation coefficient between $\bar{x}^2$ and s^2 was zero, and assuming absolute independence (which does not necessarily follow but was true in this case), he deduced the probability distribution of $z=(\bar{x}-\mu)/s$, where μ is the true mean. With a mere change of notation this is the *t*-distribution. Here s is defined to be $(1/n)\sum(x-\bar{x})^2$ so that

$$t=(\bar{x}-\mu)\Big/\sqrt{\frac{\sum(x-\bar{x})^2}{n(n-1)}}=\frac{\bar{x}-\mu}{s}\sqrt{(n-1)}.$$

He then checked the adequacy of this distribution by drawing 750 samples of 4 from W. R. Macdonell's data on the height and middle-finger length of 3,000 criminals and by working out the standard deviations of both variates in each sample (see Macdonell 1902). This he did by shuffling 3,000 pieces of cardboard on which the results had been written, possibly the earliest work in statistical research that led to the development of the Monte Carlo method.

Later in his paper on "Probable Error of a Correlation Coefficient" ([1908] 1943, pp. 35–42), Gosset used the 750 correlation coefficients of the two variables. Here his remarkable intuition again led him to a correct answer. By correlating the height measurements of one sample with the middle-finger lengths of the next, he was able to obtain 750 values of r, the sample correlation coefficient, for which ρ, the true population correlation coefficient, was presumably zero. He noticed that the observed distribution of r was approximately rectangular. If it were a Pearson curve it would have to be Type II, that is, $C(1-r^2)^{\lambda}$, and his result from the 750 samples suggested $\lambda=k(n-4)$. He guessed that $k=\frac{1}{2}$ and confirmed the result by taking 750 samples of 8 to which $C(1-r^2)^2$ gave an excellent fit. Six years later Fisher proved that all these brilliant conjectures for the distribution of s^2, t, and r when $\rho=0$ were indeed correct.

The correlation coefficient between the two measurements in the 3,000 criminals was 0.66. Gosset also examined two sets of 750 samples of sizes 4 and 8 and one set of 100 samples of 30, for which the true value must have been close to 0.66. He could see from his results that the standard deviation given by $(1-\rho^2)/\sqrt{n}$ was too small and that the distribution could not be of Pearson type except when $\rho=0$. He succeeded in obtaining the exact repartition of r for any ρ in samples of 2, but the general solution for $\rho\neq 0$ had to await the publication of Fisher's famous paper in 1915.

Gosset's fourth paper ([1909] 1943, pp. 43–48) dealt with the distribution of the means of samples not drawn at random. His brewing experience had repeatedly drawn his attention to the fact that successive observations were not uncorrelated. Here he supposed a sample of n values to be drawn in such a way that the correlation between every pair of observations is the same (say ρ), so that ρ is effectively an intraclass correlation. He used the algebraical methods of his second paper to determine the first four moments of the mean, in this case employing as an illustration some data published by Greenwood and White (1909) in which 2,000 phagocytic counts had been grouped in samples of 25. Both the original counts and the distribution of means could be fitted by Pearson Type I (Beta) curves. However, the observed values of β_1 and $(\beta_2 - 3)$ for the distribution of means were bigger than would have been anticipated if the usual theory for independent observations had been valid. The modified theory produced much better agreement.

Gosset published five more mathematical papers between 1909 and 1921 ([1913; 1914; 1917; 1919; 1921] 1943, pp. 53–89). With the possible exception of the first of these, they are still of interest. The 1921 paper gave for the first time the correction for ties in calculating Spearman's rank correlation coefficient.

Agricultural and other biometric studies. It was natural, owing to the high importance of barley quality in brewing, that Gosset should have become interested in agricultural problems. His active interest seems to have started in 1905 when he was first asked for advice by E. S. Beaven, a maltster who had started experimental work in the 1890s. From then onward there was a constant interchange of correspondence and ideas between them, in which the mathematical insight of the younger man supplemented the experimental experience of the older.

Gosset's first meeting with Fisher was at Rothamsted in August 1922; each had the greatest admiration for the other's work and doubtless each had considerable influence on the development of the other's ideas on experimental design. Toward the end of Gosset's life they had a difference of opinion about the relative methods of random and systematic arrangements, but this did not affect the high regard that they always had for one another.

In 1911 Gosset examined the results of some uniformity trials carried out by Mercer and Hall at Rothamsted ([1911] 1943, pp. 49–52). In the most important of these, an acre of wheat had been harvested in 1/500-acre plots. Gosset showed from the results how advantage could be taken of the correlation between the yields of adjacent plots to increase the accuracy of varietal comparisons, and he showed that for a given acreage greater accuracy could be obtained with smaller plots rather than with larger plots.

As early as 1912 and 1913 Beaven had invented the "chessboard" design, and experiments had been laid down, each with eight varieties of barley on yard-square plots, in three centers. These were essentially "block designs," with each variety occurring once in each block; but within the block, the arrangement was balanced rather than random. At this time Gosset discovered the correct estimate of error per plot for the varietal comparisons, precisely the same result as would be obtained from an analysis of variance. He compared every possible pair of varieties and calculated for each pair $\sum(d - \bar{d})^2$, d being the difference in one block. He added these results together for all n varieties and divided by $\frac{1}{2}n(n-1)(m-1)$, where m is the number of blocks. These experiments were discontinued during World War I, but in 1923 Gosset and Fisher discovered, independently, the analysis of variance method of obtaining the result. In a letter to Gosset, Fisher proved the algebraical equivalence of Gosset's original method and the new one.

These chessboard designs were small-scale work. For field trials, Gosset and Beaven favored the "half-drill strip method," in which two varieties were compared on an area of about an acre. In this method, the two varieties are sown in long strips—*CAACCAAC*, etc.—there being an integral number of "sandwiches" (such as *CAAC*).

The error of the varietal comparison was obtained from the variances of the differences $(C - A)$ either in individual strips or in sandwiches. In one such experiment, described by Gosset, on some thing more than an acre, the standard error of a varietal mean was found to be about 0.6 per cent. Gosset was later criticized by Fisher for preferring this method to randomized strips or randomized sandwiches. Gosset welcomed the advances in the science of agricultural experimentation that came from Fisher and his school. His own attitude was a very practical one, based on his extensive experience in Ireland experimenting with barley.

A good account of much of this kind of work is given in Gosset's most important paper on agricultural experimentation, "On Testing Varieties of Cereals" ([1923] 1943, pp. 90–114). The paper also describes some large-scale work carried out by the department of agriculture in Ireland during 1901–1906 to find the best variety of barley to grow in that country. Here two varieties, Archer and Goldthorpe, were carried right through the whole period and each tested on two-acre plots in a large number of centers. With 50 pairs of plots of this size, the standard error of the comparison was still about 10 per cent to 15 per cent. However, the result was based on wide experience. In the half-drill strip experiment the corresponding standard error was only 1 per cent, but the result applied only to an acre, in one place, under very particular conditions of soil and season.

While it was important to plan yield trials in such a way as to reduce experimental error and to obtain an accurate estimate of it, it was only by comparison and analysis of the results from a number of soils, seasons, and climates that one could judge the relative value of different varieties or different treatments. Further, products must also be subjected to tests of quality. *Conclusions drawn in one center could in any case be applicable only to the particular conditions under which the trials were carried out.* While he insisted that "experiments must be capable of being considered to be a *random* sample of the population to which the conclusions are to be applied," in an individual center he often preferred balanced (that is, systematic) arrangements to randomized ones. He liked the Latin square, because of its combination of balance (to eliminate soil heterogeneity) with a random element, thus conforming to all the principles of allowed witchcraft ([1926] 1943, pp. 199–215). He was less happy about randomized blocks because he felt that a balanced arrangement within the blocks often gave a greater accuracy than did a random one. Further, he was unwilling to accept the result of the toss of a coin, or its equivalent, if the arrangement so obtained was biased in relation to already available knowledge of the fertility gradients of the experimental area. In his last paper, "Comparison Between Balanced and Random Arrangements of Field Plots" ([1938] 1943, pp. 193–215), he wrote:

It is of course perfectly true that *in the long run*, taking all possible arrangements, exactly as many misleading conclusions will be drawn as are allowed for in the tables, and anyone prepared to spend a blameless life in repeating an experiment would doubtless confirm this; nevertheless it would be pedantic to continue with an arrangement of plots known beforehand to be likely to lead to a misleading conclusion. (p. 202)

He thought that an experimenter with a knowledge of his job could arrange the treatments within a block so that *real error*, that is, the variance of the different treatment means that would be obtained with dummy treatments in a uniformity trial, would be less than if the treatments had been randomized. This statement was no doubt often true in the domain in which he worked, but its general validity has often been questioned. He distinguished between the *real error* as here defined and the *calculated error*, that is, the error variance of the treatment mean, that would be obtained from usual analysis of variance procedures. He maintained, perfectly correctly, that if the real error were reduced by balancing, the calculated error would be too high. In his last paper, he showed, in addition, that in this situation experiments that have a real error less than the calculated one fail to give as many "significant" results as those that have a greater error, if the real treatment differences are small. When, however, the real treatment differences are large, the reverse is the case. Therefore, if balanced arrangements have a small real error, they will less often miss large real differences and more often miss small ones. He regarded this as a positive advantage; where real differences in a particular center were small, he was satisfied to have an upper limit to his error because he thought that only by collating results from different centers could he arrive at the truth. Where real differences were small, even if statistically significant, the results at different centers were likely to be conflicting.

This last paper was written in reply to one by Barbacki and Fisher (1936), which purported to show that the half-drill strip method is less accurate than the corresponding randomized arrangement. Gosset was right in maintaining that these authors were in error, for they had not compared like with like in the actual data they had examined —a uniformity trial carried out by Wiebe (1935). However the data were not very good for deciding the question, for as subsequently shown by Yates (1939), owing to defective drilling they contained a periodic fluctuation, two drill-widths wide. Gosset would almost certainly have welcomed the combination of balance and randomization achieved by some of the designs invented since his day, which are likely to give a gain in accuracy similar to that obtained by his systematic designs over randomized blocks and at the same time are free from difficulties in error estimation.

In an article on Gosset, Sir Ronald Fisher praised "Student's" work on genetical evolutionary theory (see Gosset [1907–1938] 1943, pp. 181–191). He concluded: "In spite of his many activities it is the 'Student' of *'Student's' test of significance* who has won, and deserved to win, a unique place in the history of scientific method" (Fisher 1939, p. 8).

J. O. IRWIN

[*For the historical context of Gosset's work, see* DISTRIBUTIONS, STATISTICAL; STATISTICS, *article on* THE HISTORY OF STATISTICAL METHOD; *and the biographies of* FISHER, R. A.; and PEARSON. *For discussion of the subsequent development of his ideas, see* ESTIMATION; EXPERIMENTAL DESIGN; HYPOTHESIS TESTING.]

WORKS BY GOSSET

(1907–1938) 1943 *"Student's" Collected Papers.* Edited by E. S. Pearson and John Wishart. London: University College, Biometrika Office. → William S. Gosset wrote under the pseudonym "Student." The 1943 edition contains all the articles cited in the text.

SUPPLEMENTARY BIBLIOGRAPHY

AIRY, GEORGE B. (1861) 1879 *On the Algebraical and Numerical Theory of Errors of Observations and the Combination of Observations.* 3d ed. London: Macmillan.

BARBACKI, S.; and FISHER, R. A. 1936 A Test of the Supposed Precision of Systematic Arrangements. *Annals of Eugenics* 7:189–193.

FISHER, R. A. 1915 Frequency Distribution of the Values of the Correlation Coefficient in Samples From an Indefinitely Large Population. *Biometrika* 10:507–521.

○FISHER, R. A. (1925) 1970 *Statistical Methods for Research Workers.* 14th ed., rev. & enl. New York: Hafner; Edinburgh: Oliver & Boyd.

FISHER, R. A. 1939 "Student." *Annals of Eugenics* 9: 1–9.

GREENWOOD, M. JR.; and WHITE, J. D. C. 1909 On the Frequency Distribution of Phagocytic Counts. *Biometrika* 6:376–401.

MACDONELL, W. R. 1902 On Criminal Anthropometry and the Identification of Criminals. *Biometrika* 1: 177–227.

MERRIMAN, MANSFIELD (1884) 1911 *A Text-book on the Method of Least Squares.* 8th ed. New York: Wiley.

WIEBE, G. A. 1935 Variation and Correlation in Grain Yield Among 1,500 Wheat Nursery Plots. *Journal of Agricultural Research* 50:331–357.

YATES, F. 1939 The Comparative Advantages of Systematic and Randomized Arrangements in the Design of Agricultural and Biological Experiments. *Biometrika* 30:440–466.

Postscript

A collection of nearly two hundred letters from Gosset to R. A. Fisher was issued, for limited circulation, in 1967. Covering the period 1915–1936, the letters are of great interest in relation to the history of statistical theory and practice. Unfortunately, few of the corresponding letters from Fisher to Gosset survived to be included in the collection. Material from a number of those missing letters, however, appears in Gosset ([1907–1938] 1943, pp. 181–191). The correspondence is particularly interesting and stimulating, as well as exemplifying Gosset's well-developed, though delicate, sense of humor.

J. O. IRWIN

OTHER WORK BY GOSSET

1967 *Letters from W. S. Gosset to R. A. Fisher: 1915–1936.* Dublin: Guinness. → With a foreword by L. McMullen. Published by Arthur Guinness Son & Co., Ltd.

GOVERNMENT STATISTICS

Statistical data generated by government sources are referred to and cited in many articles in this encyclopedia, including CENSUS; FALLACIES, STATISTICAL; PUBLIC POLICY AND STATISTICS; SAMPLE SURVEYS; VITAL STATISTICS; *and the biographies of* QUETELET *and* WILLCOX. *This article is confined to over-all aspects: the relation of statistics to the establishment of national states; the range of substantive matters on which statistics are collected; the ways in which that collection is organized in different parts of the world; the business and professional environment within which government statisticians work; the problems and difficulties they face in both developed and underdeveloped countries.*

Government statistics in the modern state are an essential part of a wider information system. The first compilers of statistics did not make a sharp distinction between numerical and other facts, and it is still true that numerical data complement other kinds of information in the process by which decisions are made in private and public undertakings. These data are as indispensable in a centralized as in a pluralistic society; as important in the deliberations of government itself as they are in the firms, commissariats, or other groupings concerned with getting out a product. The effectiveness of the market mechanism of a free society depends on the quality of the information on which its entities base their decisions. Western European economic planning consists in large measure in the provision of supplementary information to private enterprise. A tightly controlled, centralized system needs, more than anything else, a feedback—of which the statistics it collects are a pivotal element —if it is not to be the unwitting victim of its own concentration of power.

Statistics and the state

The history, the present condition, and the problems yet to be solved in government statistics reflect the circumstances of modern national states. A governmental administrative apparatus is a principal user of official statistics, and its existence and efficiency are a main condition for securing them. While sporadic attempts at counting people or goods were known in classical times and in the ancient empires of the Middle East, there is virtually no continuity between these and modern official compilations.

Modern history. Like so much else that pertains to the modern state, its statistical system emerged about the time of the French Revolution. One of the early acts of the Constituent Assembly was to see to the publication of a statistical account of the resources of France, prepared by Lavoisier. In 1800 the Bureau de Statistique was created in Paris. The assembly required a census (Articles I and II of the Law of July 22, 1791, according to Faure 1918, p. 277), which was duly taken in 1791. François de Neufchâteau, in a circular dated the 15th of Fructidor of the year VI (presumably September 1, 1799), saw the census—with characteristic revolutionary exaggeration—as "the measure of the strength, the source of the wealth, the political thermometer of the power of states" (Faure 1918, p. 284).

Just as British nationality erupted less violently than did that of France, so its statistical system had a more gradual inception. Early landmarks were the *Domesday Book* of William the Conqueror in 1086; the record of customs dues collected in the Port of London in the time of Edward III; the Tudor counts of men and resources in the face of the danger of war; and the registration of deaths, initiated by Henry VIII in 1532, when there was widespread fear of the plague.

The dependence of statehood on a statistical system has become more and more clear with the passing of time. The mercantilist writers of the epoch of the absolute monarchs were concerned with the power of the state in peace and war; the monarch and his advisers had to have measures of the stock of men and other resources (see Viner 1968). The British and French colonial administrations in North America, and later the British in India, took censuses, that of the Canadian province of Quebec in 1666 being the first of modern censuses (Linder 1959, p. 330). With democracy there came to be other reasons for statistical compilations. In the first of the major federal constitutions, that of the United States, a means of determining the political weight of the contracting entities was required; their relative populations seemed to provide this, and the first of the regular series of decennial censuses of the United States was taken in 1790, although a printed schedule was not employed until 1830. Confederation in Canada in 1867 was based on a similar provision for representation by population and was followed by a census in 1871; the Australian colonies were united in 1900, and the formation of a statistical office followed in 1905.

Since World War II dynastic, colonial, tribal, and other political forms have been displaced by national organizations throughout the world. Along with attempts to provide themselves with constitutions, elections, and the beginnings of modern industry, new countries have set up statistical systems, both as a precondition for development and as a symbol of nationhood. The new statistical style is far more ambitious than the colonial model that preceded, in proportion as over-all national aims are more extended than colonial aims. An important role in promoting the extension and quality of statistical work was played by the League of Nations, through its Committee of Statistical Experts, and subsequently by the United Nations, through its statistical and population commissions. Under UN auspices a world population census was attempted about 1950 and again in 1960, with a high degree of national cooperation. In the decade of the 1860s censuses were taken in which 17 per cent of the estimated population of the planet was counted; over ten times as many people were counted in censuses around 1960, and these constituted 67 per cent of the population of the globe (*Demographic Yearbook 1962*, p. 1). Results are not uniformly satisfactory; a study of the quality of statistical organization and statistical output of the various countries would undoubtedly show a close relation to the quality of governmental administration in general. Chile is better organized statistically than Burma; Burma than Cambodia.

Government statistics classified. A wide range of statistical data has come to be regarded as appropriate for government compilation and publication.

Classification by source. One way to classify government statistics is according to source: households for population censuses, family-budget surveys, employment and unemployment counts; business establishments for production and employment data; incorporated and unincorporated firms for profit figures; national-government revenue departments for foreign trade and income tax accounts. The categories are not exclusive; households and commercial establishments often provide complementary data bearing on the same matter.

Where economic statistics are essentially a summary record of transactions, they are in principle available through questions addressed to either of the parties to the transaction; retail prices may be ascertained through surveys of retail stores with respect to goods sold or through household surveys with respect to goods bought. If the business establishment is more often used, this is a matter of the greater availability of records and the larger number of transactions implicitly covered by one report; in underdeveloped countries, where business is less organized, questions addressed to the consumer are relatively more favored. For data on employment and unemployment, the household seems on the whole as satisfactory a source as the business establishment. In North America the contest between rival figures of unemployment has been an important stimulus to improvement not only of the sampling and questioning techniques used in household surveys but also of the administration of unemployment insurance itself, in respects that go far beyond statistics.

Classification by time of publication. A grosser form of classification is by the indication of temporal trends that series give. Some are issued promptly and tell the latest news: weekly railway carloadings, department-store sales, stocks of wheat in central elevators, stocks of the several metals. At the other extreme are full censuses of industry and population, taken only at long intervals; their results are released over a period of time, beginning within a few months after enumeration and continuing for some years after the data are collected, and they are valuable for the cross-sectional relations that they reveal. Because of this configuration of prompt summary data and delayed details, the national economic picture of a country with respect to any moment of time is only gradually filled in, over the five or ten years subsequent to that time.

Government statistical activities include analysis and interpretation as well as data collection. Although knowledge of analysis helps in gathering data, and vice versa, the two specializations are different. An example of the division of labor on this basis is that existing between the United States Bureau of the Census and the Department of Labor on statistics of employment and unemployment; the former agency has the responsibility for gathering the material, the latter for the official analysis.

Expansion in modernization. A statistical system seems to start with population censuses and foreign trade as the main items; other kinds of data—for example, counts of starts and completions of residential construction—are added as time goes on. As a country develops, the need for statistics mounts with the variety and difficulty of decisions required by an increased division of labor in the economy. A demand for factual justification, in terms of which people can explain their decisions to themselves and to others, seems to have characterized the North American continent from an early date. General Francis A. Walker, speaking before the International Statistical Institute, pointed this out in 1893: "A strong passion for statistics early developed itself in the life of our people, and such statesmen and publicists as Hamilton . . . became working statisticians. . . . No government in the world has ever lavished money and labor . . . more cheerfully and patiently in this respect" (quoted in Cummings 1918, p. 573). But this culture of facts is no longer confined to the United States, to those of English speech, or to Europeans; it is becoming world-wide.

The proposition that the statistics collected by any government are a function of its interests and responsibilities can be exemplified in the process of decolonization. When much of the world was under the hegemony of the states of western Europe, the statistical system centered on foreign trade. What was important for British administrators was the amount of rice exported from Burma, of jute exported from Bengal; the Netherlands wanted to know the amounts of coffee, sugar, and rubber exported from Java in the periods when each of these commodities was at its height. Another phase of colonial development was the land tax; the land tax made it both necessary and possible to have figures on acreages and production of the main crops. A further stage was some rudimentary concern with people, expressed by population censuses. The building of railways in India and Java, for instance, was followed by the collection of transport statistics—data that no one would have collected when oxcarts, proas, and the backs of men were the principal means of shipping commodities. And the evolution that occurred in the colonies in Asia was paralleled in Latin America, where the several governments, although independent, had interests nearly as restricted as those of the colonial powers.

The advent of independence in Asia and of the welfare state everywhere has increased the range of government statistics. Every country concerned with development is trying to expand its educational system, and statistics of schools, teachers, and pupils are nearly universal. In England and Wales educational statistics of a kind date back to the 1820s and 1830s, when public money was first given to the schools (Baines 1918, p. 377). Vital statistics are being gradually improved, a process that requires the inculcation of the habit of recording births and deaths not only on the part of the hierarchy of the civil registration system but also on the part of the medical profession and the public at large; such institutionalization will take at least a generation or two in the new countries. Meanwhile, sample surveys are being introduced, to observe the birth rate and its changes, as well as many other population and economic phenomena. Family-budget surveys, which had been foreshadowed by colonial governments in their moments of welfare-mindedness (a coolie-budget survey was made in the Netherlands Indies in the 1930s), are becoming more frequent; the government of India, through the Indian Statistical Institute in Calcutta, has been particularly active in this and other types of household survey.

Organization of statistical services

Every government has sooner or later to consider the organizational framework within which its statistics will be collected. When left to themselves, individual departments of the central government tend to collect whatever data their administration generates, and they hardly separate the collection of such information from administrative control of their operations. When the act regulating the collection of duties in the United States was passed in 1789 (Cummings 1918, p. 579), it required the collector of customs to record ships' manifests and other information connected with trade; the series of foreign trade statistics for the United States accordingly dates from 1789. The British Post Office collects and publishes the returns of the postal service; the British Department of Inland Revenues, established in 1849, collects the estate and stamp duties, land taxes, and income taxes. But with increase of scale, such departments come to separate their statistical from their administrative work; the British Board of Trade formed within itself a statistical branch as early as 1833 (Baines 1918, p. 374). Subsequently, labor, local departments of agriculture, and many other departments or boards came to have their own statistical units, not only in the United Kingdom but in the United States, the British dominions, and the countries of western Europe. The Interstate Commerce Commission (ICC) of the United States was from the start insistent on securing data from the railways it regulated, and ICC publications constitute an important series.

Sometimes one of these agencies would be assigned the taking of a decennial census, a periodic task required of the registrar-general in England and Wales, whose continuing work was the civil register and vital statistics. In Canada the decennial census provided for in the British North America Act was the charge of an office established in the Department of Agriculture in 1905. But in all countries, increasing attention to statistics led to proposals for centralization in a specialized agency.

Degree of centralization. With the establishment of even small statistical offices within government departments, the work of statistics began to benefit from some degree of separation from the day-to-day exigencies of administration.

Centralization in national government. The collection into a single office of all the statistical activities of the national government was considered, in order to secure even more of the gains of specialization. To the argument of efficiency (concentration of specialized personnel and equipment to deal with the large scale of centralized statistical work) was added that of avoidance of duplication, a perennial hazard of government work and one which appeared in statistical work from the earliest days. Separating statistical work from the other work of government lessens the pressure on the statistician to distort his results to protect political or administrative interests—reference is made below to the judicial aspect of the statistician's function. The better coordination of a central system, moreover, ought to permit the recognition of gaps in data. It should also make easier the adoption of uniform classifications; it is highly desirable that production and exports, for instance, be recorded on the same commodity classification, so that comparisons may be possible. But anyone who knows the tendency to autonomy of the divisions of a governmental organization will realize that having the several compilations under a single roof is no automatic guarantee of uniformity; the several sections can pursue different courses. One special danger of a centralized organization is that it will become so self-contained as to be immune to the needs of the users of its data. Centralization offers high returns when the central organization is well coordinated internally and alert to the changing requirements of outside users.

The administrative argument for centralization was early carried into effect, in one degree or another, in the Netherlands, Canada, Australia, Germany, and Italy. About the middle of the 19th century the Netherlands took important steps in this direction; in 1848 a statistical bureau was established in the Department of Home Affairs whose responsibilities were far wider than the tabulation of data generated in the department. Canada and Australia committed themselves to a statistical system that was centralized in the double sense that statistics were at least as much the affair of the central government as of the states or provinces or of any lower level of government and that among the agencies of the central government there was one with pre-eminence in the collection and tabulation of statistics in a number of different fields. The Canadian system has been in principle entirely centralized since the founding of the Dominion Bureau of Statistics in 1918. A degree of centralization was arranged in Germany with the establishment of the Imperial Statistical Office in 1872, the year after the establishment of the German Empire; in Italy centralization was begun in 1861, the year of the constitution of the monarchy, with a directorate of statistics that was the ancestor of the present Istituto Centrale di Statistica.

In the United States and the United Kingdom practice has evolved closer to the decentralized pole of the centralized–decentralized continuum. The means of coordination in the United Kingdom is an office in the Cabinet Secretariat that operates through a series of understandings with the heads of statistical divisions in the Board of Trade, the Ministry of Food, the Ministry of Agriculture, etc. All the statistical divisions in departments are directed by statistician members of the Professional, Scientific, and Technical Class, one of the five classes of the British civil service; appointments anywhere are open to members of the statistician class irrespective of where they are serving. The Treasury approves the expenditures of statistical departments, and on new expenditures they are advised by the Central Statistical Office.

The American arrangement is more formal; the Office of Statistical Standards of the Bureau of the Budget not only reviews all requests for funds to carry out statistical work within the federal government; its approval is required before any federal government questionnaire can be sent to ten or more respondents. This requirement dates back to the Federal Reports Act of 1942 and was intended to reduce the burden on respondents and to eliminate duplication. Where duplication exists, it not only is wasteful of government funds but also arouses the fiercest resentment of respondents who are required to answer the same questions more than once.

The best coordination does not consist exclusively of the negative injunctions of an enforcement agency. An example of a more positive kind is the collaboration of the U.S. Social Security Administration, essentially an operating agency, and the Bureau of the Census in the creation of a publication showing county business patterns based on Social Security Administration records. The U.S. Bureau of the Census has used tax records to eliminate business-census questionnaires for a million retail establishments without employees. By using Social Security records as a source of lists, the Bureau of the Census is able to take the economic censuses at reduced cost.

Coordination requires fine judgment at a thousand points. Should both the U.S. Bureau of Mines and the Bureau of the Census continue to secure mineral statistics? The Bureau of the Census collects benchmark data on an establishment basis every five years from all establishments within the scope of the census, assigning to the mining industry the output of all those that are classifiable as mining. The Bureau of Mines collects information more frequently, on the output of minerals as well as on engineering and technical matters. These data are thought to be independently valuable and different enough not to seem to constitute duplication.

Division of responsibility. Along quite a different dimension is the division of responsibility between the national government and the local governments of states or provinces and cities. In this dimension the United Kingdom is highly centralized, while Germany has problems because of the division of statistical activities among the *Länder*, and even in the Netherlands there is some devolution of statistical activities to organizations in the cities.

Every national statistical system must depend on local sources for its information. At the one extreme is complete central control, as found in the national census of countries such as the United States and Canada, in which the local officers are appointed and instructed by the center through a training ladder that inculcates definitions and procedures that are, at least in principle, fully determined centrally, and which is backed by a central budget from which all participants are paid. More decentralized is the census in countries such as Argentina, where the plans are made by a national statistical office but where the budget does not include the funds for paying enumerators, who are government employees on the payroll of other departments of the center or the provinces (for example, schoolteachers), co-opted for one or several days' work on the enumeration. Coordination of such an operation requires feats of diplomacy. Further down on the scale of decentralization is the system of vital registration, used in the United States and other countries, in which local registry officials report to a state authority and whatever national uniformity exists is the result of negotiations in which the national office can take leadership more by virtue of its persuasiveness than because of any legal or budgetary power.

Comparability and change. The government statistical agency must have both continuity and adaptation of its output to changing needs, a combination of virtues easier to prescribe than to follow. In some instances an agency is very rigid; the disposition to continue collecting the same data in the same fashion year after year is both the strength and the weakness of government agencies. But at other times it changes arbitrarily; A. L. Bowley charged that the United Kingdom census of 1921 was deliberately made noncomparable with anything that went before, a statement challenged by Greenwood et al. (1932, p. 279). Again, judgment is required; the elimination of old series and the initiation of new ones and the modification of definitions that make a given series more useful for the future but lessen its comparability with the past must be discussed on a more specific level than this article can attain.

Practices that are good in one place and time may not be so in a different economy and society. Attempts to transfer categories of statistical compilation from a country in which money economy dominates to one in which households produce for their own use rather than for exchange may not be satisfactory. Some of the greatest difficulties in collecting statistics within the developed countries are in the sector of subsistence agriculture. The relation of statistics to industrial practice is brought home especially strikingly by the present situation in the United States, where the National Bureau of Economic Research and the Bureau of the Census, among other agencies, are trying to find how individual series can be improved or replaced. With the shift of traffic to trucks and airlines, carloadings have become much less useful. Department-store sales mean less when many of the lines of merchandise are also sold by discount houses and drugstores. The inventive spirit that discovered department-store sales and carloadings as key figures in the economy is needed now to go beyond these.

Interactions with environment. The government statistical agency is by no means a free agent but depends on the administrative practices of other governmental agencies and of business concerns. Statistics of imports depend on valuations made for customs purposes; comparing the exports of newsprint or metals from Canada to the United States, for instance, with the imports of nominally the same commodities into the United States from Canada will show how substantial is the effect of varying definitions of commodities as well as of sources and destinations. Since definitions are often part of procedures for customs valuation, and these are embedded in laws around which substantial material interests have developed, discrepancies between countries are likely to persist despite many conventions and attempts at accommodation, from that of Brussels in 1910 to those of the United Nations Statistical Commission.

Effects of accounting practices. One important network in which the statistical agency is enmeshed is the accounting practices of industry, themselves in part the consequence of the regulations of the government income tax department. The art of statistical collection includes finding modes of definition of the entities about which inquiry is made that are most in conformity with the accounting practices of business and provide the most useful information to the public. The industrial establishment, as against the firm, is generally taken as the primary element for collection of industrial statistics and is defined in practice as the smallest production unit that maintains more or less complete records. The statistical agency does not always play a purely passive role; it is often in a position to exercise influence and initiative. In Canada the federal agency concerned with financial statistics of the 5,000 municipalities of the country was able to persuade them to adopt a standard accounting procedure, which benefited them as well as national statistics.

National professional associations. The environment within which a statistics-producing office of government has to work includes, besides the operating agencies of the same government and the productive units of the country at large, a host of professional associations, as well as individuals interested in statistics. The American Statistical Association in 1844 petitioned the Congress of the United States to recompile the census of 1840; its policy of protesting poor work and supporting good in the federal statistical field has operated to the present day, although the need for its intervention is not as great as when the government service was staffed almost exclusively by amateurs (Bowman 1964).

International professional associations. To the national associations have now been added important international ones. First among these was the International Statistical Institute, from the nineteenth century on a persistent and wholesome influence in favor of high standards and comparable classifications. The Economic and Social Council of the United Nations, through its statistical and population commissions, has had an effect in stimulating countries to take censuses; to adopt a minimum set of questions in these, for which a serious attempt would be made to attain international comparability; to use modern techniques, in the interest of accuracy and promptness of release of data. Sensitive to the desire of nations to be independent of outside interference in statistics as in other fields, the council has made its chief weapon the argument of international comparability in statistics of production, as well as of population. Also of influence have been the regional agencies of the United Nations for Asia, Europe, Latin America, and now Africa; the Inter-American Statistical Association; the specialized agencies of the United Nations, including the United Nations Educational, Scientific, and Cultural Organization (UNESCO), the Food and Agriculture Organization (FAO), the International Labour Organisation (ILO), etc., in their respective fields. The resolutions and the conventions that these have drawn up often suffer from being stated in highly abstract terms, but they are supplemented by technical assistance services to less developed members, including the sending out of advisers and the providing of training facilities. The publications of international bodies have not only been directly useful to scholars and others but also have been an incentive to members to secure data. FAO conferences of groups of neighboring countries have constituted an effective form of pressure for improvement of agricultural statistics.

Other influences. Series of government statistics have often been initiated from the outside. The Metropolitan Life Insurance Company began collecting statistics of labor turnover in the United States in 1926, and these compilations were later taken over by the United States Department of Labor (Hauser & Leonard 1946, p. 363). In Canada statistics of wage rates collected by the Bell Telephone Company for its own purposes have influenced federal government collections. Statistics of stocks on hand have often been started by associations of manufacturers and then continued by a public statistical agency. In the Netherlands one of the influences that led to the centralized system was the private Union for Statistics, founded in 1856 (Stuart 1918, p. 435), which published many volumes of data, although it did no primary collecting. Later this private union was subsidized by the government, and ultimately all of its activities were taken over. Especially well known is the pressure that Quetelet, the Belgian statistician, exerted through his researches as a private citizen between about 1825 and 1841 (when he became a public official). His avid interest in the regularity of certain social phenomena, e.g., the "budget of the scaffold," finds its monument in the extensive series of criminal statistics produced in Belgium and many other countries (Julin 1918, p. 128). In England the first edition of the essay by Malthus that attracted so much attention to population was followed within three years by the census of 1801 [*see the biography of* QUETELET; *see also* Blaug 1968].

There are many other examples of the influence of men and ideas—as well as outside organizations—on government statistics. The collection of economic data during the present generation has been altered by the recognition of national accounts as a general framework. The concept was developed by scholars in universities in the United States and Great Britain, following suggestions of J. M. Keynes. The notion of national income makes it possible to arrange a great variety of existing series in the pigeonholes of a national-accounts framework; the output of the statistical systems ceases to appear arbitrary, for each portion measures a contribution to the gross national product; certain gaps become visible—for instance, personal services—and there is pressure to fill them. Furthermore, the examination in this fashion of economic statistics as a whole permits the grand total of the country's economic activities to be calculated in at least three ways—by value added, by income received, and by expenditure—which have important elements of independence as to source of data. The confrontation of calculations of a total derived from more or less independent statistical sources has been an incentive to improve the accuracy of all components.

The current prestige of economic planning has had important effects on government statistical work. Planning may be good or bad, effective or ineffective, but in all instances it requires data. One cannot even go through the motions of planning without statistics. Government planning agencies requiring data are in an especially strong position to see that the necessary resources are allocated to their collection.

The statistical profession and government. Underlying recent developments is a rapid professionalization of statistics. Times have changed greatly since Francis A. Walker lamented, in an address before the American Statistical Association in 1896, "I do not know of a single man now holding, or who has ever held, a position in this country as the head of a statistical bureau, or as chief of a statistical service, or as a statistician, who had any elementary training for his work" (quoted in Cummings 1918, p. 574). A recent Bureau of the Census survey of its staff showed 576 employees in professional statistical positions, of whom all but 22 had an academic degree, including 36 PH.D.s and 102 M.A.s (Bowman 1964, p. 14). The change has come about through the extension of the field. Statistical method is applied to acceptance sampling and quality control in industry, to experimental design in agricultural trials, to bio-assay, operations research, and sample surveys of national populations. The interchange among these applications and between each application and a rapidly expanding mathematical statistical theory has had a decisive effect. Backed by a theory based on probability and relatively well-established methods of application of the theory; with a rapidly growing body of literature in books and professional journals; with departments of statistics in a number of universities, and courses in nearly all; there is less and less need for the government statistical agency to depend on the gifted amateur like Quetelet in Belgium, Knibbs in Australia, or Coats in Canada. And yet, present-day expansion and progress can be matched in quality by some of the early work. Florence Nightingale had much effect on the British War Office through her statistics showing deaths from battle and from disease separately. Her data on the army hospitals in India led to important reforms. It is not that the modern professional is better than the old-fashioned gifted amateur, but the former is more certain to be on hand when needed and commands the tools that have now been created.

Sampling in the new professionalization. Among the large-scale sample surveys now conducted are those on a monthly basis in the United States and Canada, quarterly in the Federal Republic of Germany and other European countries, and virtually continuously in India. The topics of survey are almost as wide as the topics of statistics itself [*see* SAMPLE SURVEYS].

Like any important change in technique, sampling does not merely attain the earlier objectives at lower cost; it brings a radically new viewpoint on the whole process of data collection. Instead of thinking of himself as charged with passively compiling given data, the statistician orients himself to the purpose of the compilation and the degree of accuracy required if the purpose is to be served; thus, he is led to take account of the nonsampling as well as the sampling error to which his work is subject. He sees himself at his professional best in measuring error, controlling quality, and evaluating results. He comes to measure information, not by the mass of data he can turn out, but rather by the accuracy attained in a single figure. The percentage of the labor force unemployed, for example, may be the occasion for a major decision in regard to the economy. If there is an allotment to improve the figure, he must decide what amount should be spent to improve the accuracy of response and what to increase the sample. In other words, he must decide if there will be more return from applying a given effort to the sampling or to the nonsampling error. The changed attitude toward error—the view that it is both inevitable and to be constantly combated—is seen in the United States Bureau of the Census, where the 1950 and subsequent population censuses have been accompanied by an official estimate of the degree of underenumeration, an estimate in part made by having superior enumerators survey a sample of areas. This practice has been applied in Canada since 1951. It is the mark of the professional statistician that he does not assert that his results are exact; even when he has done the survey by the best means that are available, he is satisfied to consider his one survey as an arbitrary selection from all the surveys that might have been carried out at about the same time and with equally acceptable definitions [*see* ERRORS, *article on* NONSAMPLING ERRORS].

Enduring problems of official statistics. A survey of the statistical agencies of any government, even those of a very advanced country, would show the continuance of some out-of-date practices, the publication of some figures that are not usable because they are based on purely administrative definitions of the entities counted or because the error is not stated by the agency but may be presumed to be large. Besides the valuable statistics they contain, library shelves are weighted with irrelevant, useless, and inaccurate statistics, a situation to be deplored in proportion as good techniques are available.

Government statistics' judicial function. Especially pernicious are those inaccuracies intended to serve some political purpose. The strongest argument for statistical centralization relates to what may be called the judicial function of the official statistician. The statistics of foreign trade may show that the government's trade policy is going badly; the statistics of prices may show that its monetary policy is leading to inflation. Protection of the public requires that the agency responsible for the policy not be the one with exclusive control over the statistics; to give it such control is to ask of it superhuman strength to withstand temptation. In a hundred minor decisions of statistical compilation, an operating agency may be swayed by noting how the result will come out. Presentation also requires impartiality. On issues such as whether a price index ought to be released now, when it has been scheduled to appear, or next week, after a national election has taken place, hangs the virtue of the statistician. Without this virtue and the resulting public confidence in statistics, much of the cost of collecting data is wasted; statistics that are not believed are of no use. In a statistician, as in a judge, ignorance is less dangerous than corruption. The statistician will be most useful to the community when his statistical honor is not in conflict with some nonstatistical responsibility. At a higher administrative level, the political head of the department that includes the statistical office must accept the self-discipline of permitting decisions regarding statistics to be made on statistical considerations, not because such decisions are always correct but because, when they are wrong, they are wrong in a disinterested way. One method that lessens the need for self-discipline is to have a commission entirely outside of politics control the statistical office, as is done in the Netherlands.

Much less subtle than the issues of the preceding paragraph are the distortions as a matter of policy that have been seen in some totalitarian countries. During the Stalin period in the Soviet Union, statistics were not considered to pertain to science but to revolution and to mobilization of the people for national construction. If the statistics do not show progress, the masses will become discouraged and their effort will flag. The same attitude was found in China in the 1950s (Li 1962). The competition between districts to be in the van of socialist production was extended to competition with respect to the figures of production, and this local zeal for declaring high figures reached its peak in 1960, when total production of cereals was announced as 275 million tons, apparently an exaggeration of over 100 per cent. It had been thought democratic to bring the masses into the work of compiling statistics; the consequent socialist rivalry seems to have genuinely deceived the regime, as well as its subjects, with subsequent disastrous consequences. There are clear signs that in the Soviet Union, as in older industrial societies, trustworthy statistics are now recognized as something more than a bourgeois luxury; whether the dependency of good decisions on good statistics is now understood in China is not clear.

Technical contributions

Extensive contributions have been made to technique in nearly every division of statistics by agencies attempting to implement efficiently their legislative assignment to collect data. One may mention the methods of calculating life tables; the deseasonalization of economic time series; the theory and practice of index-number calculation, as it has been developed in the United States Bureau of Labor Statistics and in the Dominion Bureau of Statistics in Canada; the classification of occupations, industries, and commodities (Coats 1925). Particularly important for the growth of demography were the technical contributions during the nineteenth century by William Farr and other statisticians of the United Kingdom General Register Office [*see* INDEX NUMBERS; LIFE TABLES; DEMOGRAPHY; TIME SERIES; *see also* Grauman 1968; Mayer 1968; Whitney 1968].

The most striking single example of these contributions is the intimate relation of the U.S. Bureau of the Census to the development of tabulating and computing equipment, from the primitive punch-card devices of the 1890s, built in the Census Bureau itself by Herman Hollerith, to the Univac, built privately with bureau encouragement and financing and technical help from the Bureau of Standards. The Canadian census office also pioneered in this, with some highly original electric and compressed-air tabulators, the result of the ingenuity of Fernand Belisle, a lifelong employee of the Canadian Bureau of Statistics. Both Canadian and United States census offices have in recent years built or stimulated the building of input devices to avoid the keypunching of data, the most tedious part of the processing of large-scale surveys.

Besides having had a part in the development of computers, government statistical agencies have shown great initiative in using them. Such use, whether in an insurance office, an oil refinery, or a government statistical office, requires extensive rethinking of processes that have developed in the course of decades. Several of the more advanced of the world's statistical offices have drastically reorganized their work in order to exploit electronic computation, with consequences of more extensive cross tabulation, better control of error in processing, and economy. In the United States the demand for machine-readable results has increased as users of statistics have acquired their own computers. The Bureau of the Census has an extensive catalogue of results available on punch cards or electronic tape; this mode of publication is sure to become widespread.

Publication of results

Prompt and full publication is the highest of virtues in government statistics—provided that the data published are accurate as well as relevant.

This follows from the place of statistics in a decision-making process: the appropriateness of the decision—for example, to build a factory—will depend on the situation that exists at some future date; and if the statistician is not in a position to count the population of five or 25 years hence, he can at least describe what the situation was up to as recent a moment as possible. On the content of what is published—what cross classifications of distributions, what percentages, what time comparisons, what charts—the judgment of the government statistician becomes better the greater his contact with users of his results. With the multiplicity of governmental and private sources, the need increases for good indexing within any publication, as well as for cross referencing of comparable data in other publications.

Government statistical offices have entered the field of social science in their efforts to interpret their data. This runs from comment on statistical tables that enables the reader to know how the entities counted have been defined, along with measures of sampling and nonsampling error, through calculations distributing not-stated ages and deseasonalizing economic series, to full-scale monographs that have become a regular part of the census, in the United States and Canada going back at least a full generation. India is preparing a set of monographs based on the 1961 census, and Pakistan is planning to do likewise. If one thinks of the continuum, from the collection of raw statistical data to its final use as an ingredient either in scientific investigation or in the making of decisions, then there is room for differences of opinion as to where the role of the government statistical office ends and that of the user begins. The vital minimum is that the government statistician describe all aspects of his collection procedures that can possibly affect the interpretation of the results. It is to be hoped that in some future time no survey will be issued without realistic estimates of the accuracy of its figures, but that time still seems to be far in the future.

Notwithstanding all this, a certain abstemiousness is forced on the official statistician by virtue of his position; he cannot afford to take the sorts of risks in the interpretation of data that newspapermen can take so freely. Public confidence in the accuracy of his figures, and hence their usefulness, would be jeopardized by palpable arbitrariness in judgments expressed in his text. This consideration often drives him to the writing of text that is no more than a gloss on what is obviously revealed in the tables and is perhaps less clear than the figures themselves. On the other hand, the government statistician can argue with justice that interpretation, lively or dull, made without his intimate knowledge of the basis of the figures is likely to be misleading. The answer is that, as a minimum, he give enough detail on his procedures to put the reader in a position to interpret correctly.

Secrecy. The matter of secrecy enters into statistical work in two ways, to which exactly opposite principles apply. In the report made by the individual to a government agency, it is generally considered that the most honest reporting will be secured if a guarantee of confidentiality is provided, giving the supplier assurance that his contribution will not be identified outside the statistical office and that only the statistical aggregates of which it is a part will be made public. The government statistician is endowed with the power to enforce reporting on the part of the persons or corporate bodies to which he addresses his questionnaires; the obverse of this power is the obligation to keep individual returns secret (Dobrovits 1947). In the large number of government surveys for which reporting is not mandatory, the degree of voluntary response may be proportional to the public confidence in provisions for maintaining secrecy.

On the other hand, the aggregate results of every survey ought to be given the widest possible publicity. One of the difficulties of government publication is that the government does not have complete access to the distribution channels of the private book trade; sometimes a solution is possible through private publication, as was done for the monographs on the 1950 census of the United States and as is done in the Netherlands and Germany. The principle of equal access of the public to statistics is important; this is safeguarded by preannounced release times, at which statistics are simultaneously made available to all who are interested. In many countries there is much diffusion of government statistical results through nongovernmental intermediate sources, including trade yearbooks and scientific journals; it is customary to allow republishing without specific permission.

A serious problem arises from the desire of government to keep certain statistical results secret. This was a regrettable necessity in the United States and Canada during World War II in respect, for instance, to detailed foreign trade figures, because the information would have been used by the Axis powers in their submarine campaign against Allied shipping. Aside from such exceptional cases, it is fair to say that there are no instances in which the public as a whole benefits from concealment of statistical reports that provide accurate data, although governments and particular departments within them may indeed benefit at the expense of the public. The discussion of the publication of government statistics merges at this point with the wider question of the free flow of information in the society as a whole.

NATHAN KEYFITZ

[*See also* Kemp 1968; Liu 1968; Michaely 1968; Rokkan 1968; Ruggles 1968; Spulber 1968.]

BIBLIOGRAPHY

BAINES, ATHELSTANE 1918 The History and Development of Statistics in Great Britain and Ireland. Pages 365–389 in John Koren (editor), *The History of Statistics*. New York: Macmillan.

►BLAUG, MARK 1968 Malthus, Thomas Robert. Volume 9, pages 549–552 in *International Encyclopedia of the Social Sciences*. Edited by David L. Sills. New York: Macmillan and Free Press.

BOWMAN, RAYMOND T. 1964 The American Statistical Association and Federal Statistics. *Journal of the American Statistical Association* 59:1–17.

COATS, ROBERT H. 1925 The Classification Problem in Statistics. *International Labor Review* 11:509–525.

CUMMINGS, JOHN 1918 Statistical Work of the Federal Government of the United States. Pages 573–689 in John Koren (editor), *The History of Statistics*. New York: Macmillan.

Demographic Yearbook. → Issued by the United Nations since 1948.

DOBROVITS, ALEXANDRE 1947 Sur le secret en statistique. Volume 3, pages 769–778 in International Statistical Conference, Washington, D.C., *Proceedings*. Calcutta: Eka.

FAURE, FERNAND 1918 The Development and Progress of Statistics in France. Pages 217–329 in John Koren (editor), *The History of Statistics*. New York: Macmillan.

GODFREY, ERNEST H. 1918 The History and Development of Statistics in Canada. Pages 179–198 in John Koren (editor), *The History of Statistics*. New York: Macmillan.

►GRAUMAN, JOHN V. 1968 Population: VI. Population Growth. Volume 12, pages 376–381 in *International Encyclopedia of the Social Sciences*. Edited by David L. Sills. New York: Macmillan and Free Press.

GREENWOOD, MAJOR et al. 1932 Discussion on the Quantity and Quality of Official Statistical Publications. *Journal of the Royal Statistical Society* 95:279–302.

HAUSER, PHILIP M. 1963 Statistics and Society. *Journal of the American Statistical Association* 58:1–12.

HAUSER, PHILIP M.; and LEONARD, WILLIAM R. (editors) (1946) 1956 *Government Statistics for Business Use*. New York: Wiley.

HOLMES, OLIVER W. 1960 "Public Records"—Who Knows What They Are? *American Archivist* 23:3–26.

JULIN, ARMAND 1918 The History and Development of Statistics in Belgium. Pages 125–175 in John Koren (editor), *The History of Statistics*. New York: Macmillan.

►KEMP, M. C. 1968 International Trade: III. Terms of Trade. Volume 8, pages 105–108 in *International Encyclopedia of the Social Sciences*. Edited by David L. Sills. New York: Macmillan and Free Press.

KNIBBS, GEORGE H. 1918 The History and Development of the Statistical System of Australia. Pages 55–81 in John Koren (editor), *The History of Statistics*. New York: Macmillan.

KOREN, JOHN (editor) 1918 *The History of Statistics*. New York: Macmillan.

LEONARD, WILLIAM R. 1958 An Outlook Report. *Journal of the American Statistical Association* 53:1–10.

LI, CHOH-MING 1962 *The Statistical System of Communist China*. Berkeley: Univ. of California Press.

LINDER, FORREST E. 1959 World Demographic Data. Pages 321–360 in Philip M. Hauser and Otis Dudley Duncan (editors), *The Study of Population: An Inventory and Appraisal*. Univ. of Chicago Press.

►LIU, TA-CHUNG 1968 Economic Data: III. Mainland China. Volume 4, pages 373–376 in *International Encyclopedia of the Social Sciences*. Edited by David L. Sills. New York: Macmillan and Free Press.

►MAYER, KURT B. 1968 Population: IV. Population Composition. Volume 12, pages 362–370 in *International Encyclopedia of the Social Sciences*. Edited by David L. Sills. New York: Macmillan and Free Press.

MEITZEN, AUGUST (1886) 1891 *History, Theory, and Technique of Statistics*. Philadelphia: American Academy of Political and Social Science. → First published in German.

►MICHAELY, MICHAEL 1968 International Trade: IV. Patterns of Trade. Volume 8, pages 108–113 in *International Encyclopedia of the Social Sciences*. Edited by David L. Sills. New York: Macmillan and Free Press.

MILLS, FREDERICK C.; and LONG, CLARENCE D. 1949 *The Statistical Agencies of the Federal Government*. Washington: Government Printing Office.

MORGENSTERN, OSKAR (1950) 1963 *On the Accuracy of Economic Observations*. 2d ed. Princeton Univ. Press.

NIXON, JAMES W. 1960 *A History of the International Statistical Institute: 1885–1960*. The Hague: International Statistical Institute.

NORTH, S. N. D. 1918 Seventy-five Years of Progress in Statistics: The Outlook for the Future. Pages 15–49 in John Koren (editor), *The History of Statistics*. New York: Macmillan.

►ROKKAN, STEIN 1968 Elections: II. Electoral Systems. Volume 5, pages 6–21 in *International Encyclopedia of the Social Sciences*. Edited by David L. Sills. New York: Macmillan and Free Press.

►RUGGLES, RICHARD 1968 Economic Data: I. General. Volume 4, pages 365–369 in *International Encyclopedia of the Social Sciences*. Edited by David L. Sills. New York: Macmillan and Free Press.

►SPULBER, NICOLAS 1968 Economic Data: II. The Soviet Union and Eastern Europe. Volume 4, pages 370–372 in *International Encyclopedia of the Social Sciences*. Edited by David L. Sills. New York: Macmillan and Free Press.

STERN, JOHANNA 1958 [Review of] Th. L. Galland's *Statistik der Beschaeftigten und Arbeitslosen*. *Journal of the American Statistical Association* 53:1040–1043.

STUART, C. A. V. 1918 The History and Development of Statistics in the Netherlands. Pages 429–444 in John Koren (editor), *The History of Statistics*. New York: Macmillan.

UNITED NATIONS, STATISTICAL OFFICE 1954 *Handbook of Statistical Organization*. Studies in Methods, Series F, No. 6. New York: United Nations.

U.S. BUREAU OF THE BUDGET, OFFICE OF STATISTICAL STANDARDS (1947) 1963 *Statistical Services of the United States Government*. Rev. ed. Washington: Government Printing Office.

►VINER, JACOB 1968 Economic Thought: II. Mercantilist Thought. Volume 4, pages 435–443 in *International Encyclopedia of the Social Sciences*. Edited by David L. Sills. New York: Macmillan and Free Press.

WESTERGAARD, HARALD L. 1932 *Contributions to the History of Statistics*. London: King.

►WHITNEY, VINCENT H. 1968 Population: V. Population Distribution. Volume 12, pages 370–376 in *International Encyclopedia of the Social Sciences*. Edited by David L. Sills. New York: Macmillan and Free Press.

WURZBURGER, EUGENE 1918 The History and Development of Official Statistics in the German Empire. Pages 333–362 in John Koren (editor), *The History of Statistics*. New York: Macmillan.

Postscript

In the later stages of industrial development the population becomes numerate as well as literate, and digests an increasing volume of statistical data. Each year the demand for new series becomes more insistent. Where once numbers of births registered and tons of steel imported typified social and economic statistics, and the questions asked of official agencies were much the same from decade to decade, now a shift in statistical needs accompanies every shift of public attention, and these are many.

Government agencies have responded sensitively. When crime became a conspicuous issue, much effort went into making more complete and comparable the tally of those crimes reported to the police. With a ceiling set on this line of improvement by the large number of crimes not known to the police, a new primary source was tried: the victims of crimes, whose direct questioning has been initiated by the National Opinion Research Center (under contract with the U.S. government) with the instrumentalities of sample surveys, and then continued on a larger scale by the Department of Justice. Almost as soon as pollution entered public debate, the data on the mass of contaminants in the air of the main American cities were greatly expanded and began to appear in the *Statistical Abstract of the United States*. When fuel shortages were first projected, statistics concerning miles per vehicle per gallon obtained on American roads were presented, along with number of passengers per vehicle. Such challenges to statistical ingenuity, and one hopes equally effective responses, will continue to arise as long as new economic and social problems come to face government and society, which is to say, for a long time. Every such expansion of the materials of statistics tends to expand its clientele.

Aside from new primary data sought and provided, pressure for more highly processed calculations is exerted by the wider audience and the new relevance to policy. This demand occurred in respect to economic data during the 1940s when the national accounts came into existence, especially the Gross National Product, a series that has taken its place alongside the population census as a mark of nationhood for old and new countries alike. The national accounts are aggregated in a degree far beyond such previous compilations as the index of industrial production. National income gained acceptance as the measure of social progress during the switch from household to factory production; it measured this process along with the increase in total output, and, since both trends were looked on with favor, no one then objected much to the confounding. The official bias toward market production and toward economic growth reflected that of the public, and few perceived any departure from objectivity. Yet the decisions on what to include and what to omit from GNP affect economic perspectives.

The shift to market production is now almost complete in advanced countries (less than 5 per cent of their labor force is in agriculture, and even farm families produce little for their own needs), and gross differences of opinion have come to exist on the direction of further progress. Those wanting more goods are opposed by those wanting to preserve the environment. Stress is laid on the quality of life, a concept largely unknown to official statistics prior to 1970. Both public and private effort is now spent on devising social indicators.

Differences of viewpoint among the social groups served are nothing new to the official statistician, whose aim has always been to provide basic materials to contending parties. Impartiality, that no group be favored by the definition of a series or even by its date of release, has always been a condition of the official statistician's success in free societies, and departures have usually drawn effective protest. Statistical work is of maximum usefulness when it commands the confidence of both sides in many disputes: between those who favor conservation and those who want more production; between employers and trade unions when inflation makes price indexes important to both; between nationalists and internationalists in a time of pressure for defenses against foreign competition; between government and opposition always. Debate should revolve around the interpretation of the numbers rather than their impartiality.

The expansion of the government statistician's clientele has generated a new demand for interpretation of the numbers. Masses of users do not have the patience or the knowledge to profit from the traditional austerity and massiveness of government publications. The uncertainty that arose in the 1970s as to the direction of economic and social progress, makes objective interpretation especially difficult. The official statistician has always been badgered on the one side by those who want him to tell what his figures mean, and on the other by those who call him biased when the meaning he presents is not the one that they perceive. Some part of the meaning is determined in the process of collection, and he best knows it. For example, he knows the errors that arise in his primary data, and this part of the interpretation is his clear obligation. But how much beyond this? His professional discretion must decide how far to go in the prevailing circumstances.

Much depends on the official statistician continuing to perform effectively amid many cross-pressures, and on government, however hard pressed it may be, continuing to see the advantage of supporting his neutral professionalism.

NATHAN KEYFITZ

[*See also* PUBLIC POLICY AND STATISTICS *and the biography of* MAHALANOBIS.]

ADDITIONAL BIBLIOGRAPHY

CASSEDY, JAMES H. 1969 *Demography in Early America: Beginnings of the Statistical Mind, 1600–1800.* Cambridge, Mass.: Harvard Univ. Press.

ECKLER, A. ROSS 1972 *The Bureau of the Census.* New York: Praeger.

KAHN, ELY J. JR. 1974 *The American People: The Findings of the 1970 Census.* New York: Weybright & Talley.

KRUSKAL, WILLIAM H. 1973 The Committee on National Statistics. *Science* 180:1256–1258.

U.S. OFFICE OF MANAGEMENT AND BUDGET, STATISTICAL POLICY DIVISION 1976 *Federal Statistics: Coordination, Standards, Guidelines.* Washington: Government Printing Office.

U.S. PRESIDENT'S COMMISSION ON FEDERAL STATISTICS 1971 *Federal Statistics: Report.* 2 vols. Washington: Government Printing Office.

GRAPHIC PRESENTATION

Graphic presentation represents a highly developed body of techniques for elucidating, interpreting, and analyzing numerical facts by means of points, lines, areas, and other geometric forms and symbols. Graphic techniques are especially valuable in presenting quantitative data in a simple, clear, and effective manner, as well as facilitating comparisons of values, trends, and relationships. They have the additional advantages of succinctness and popular appeal; the comprehensive pictures they provide can bring out hidden facts and relationships and contribute to a more balanced understanding of a problem.

The choice of a particular graphic technique to present a given set of data is a difficult one, and no hard and fast rules can be made to cover all circumstances. There are, however, certain general goals that should always be kept in mind. These include completeness, clarity, and honesty; but there is often conflict between the goals. For instance, completeness demands that all data points be included in a chart, but often this can be done only at some sacrifice of clarity. Such problems can be mitigated by the practice (highly desirable on other grounds as well) of indicating the source of the data from which the chart was constructed so that the reader himself can investigate further. Another problem occurs when it is necessary to break an axis in order to fit all the data in a reasonable space; clarity is then served, but honesty demands that attention be strongly called to the break.

A choice among graphic techniques also depends upon the proposed use to which the chart will be put. As Schmid (1954) has pointed out, graphs that are satisfactory in memoranda for private circulation may be inappropriate for a published book or paper.

In classifying charts and graphs, criteria of purpose, of circumstances of use, of type of comparison, and of form have been used. On the basis of form, charts and graphs may be classified as (1) rectilinear coordinate graphs; (2) semilogarithmic charts; (3) bar and column charts; (4) frequency graphs and related charts; (5) maps; (6) miscellaneous charts, including pie diagrams, scattergrams, fan charts, ranking charts, etc.; (7) pictorial charts; and (8) three-dimensional projection charts.

In graphic presentation, three different basic geometrical forms can be utilized for purposes of comparing magnitudes of coordinate items: (1) linear or one-dimensional, (2) areal or two-dimensional, and (3) cubic or three-dimensional. The simplest and most exact comparisons are those made on a linear basis; comparison of relative sizes of areas is more difficult and of volumes most difficult. Accordingly, where possible, the use of areal and cubic forms should be avoided in graphic presentation.

Rectilinear coordinate graphs. Perhaps the best known form of graphic presentation, and certainly one of the most frequently used, is the simple arithmetic line chart, one of several types of the *rectilinear* (or *Cartesian*) coordinate graph.

The basic form of this type of graph is derived by plotting one or more series of figures on a coordinate surface in which the successive plotting points are joined together in the form of a continuous line, customarily referred to as a "curve." A curve on a graph of this kind is not necessarily smooth and regular but instead may be straight and angular. Figure 1 indicates the basic structural characteristics of a rectilinear coordinate system.

Figure 2 portrays the characteristic features and basic standards of the rectilinear coordinate graph. Many of the essential elements and specifications of the rectilinear coordinate graph are also applicable to other graphic forms.

Semilogarithmic charts. The semilogarithmic chart is often superior to the arithmetic chart (as used in Figure 1), for the former can show relative changes clearly; hence it is sometimes referred to as a ratio chart. Sometimes the semilogarithmic chart has the merit of representing as nearly straight lines functions that otherwise would be appreciably curved.

The essential feature of the semilogarithmic chart is that one scale is logarithmic and the other arithmetic, so that the chart effectively is a convenient device for plotting log y against x, or log x against y. There is no zero line on the logarithmic scale, since the logarithm of zero is minus infinity. The logarithmic scale consists of one or more sets of rulings calibrated in terms of logarithmic values; each complete set of rulings is referred to as a "deck," "cycle," "bank," or "tier." The rulings for each deck are the same, but the scale values change from one to the other. For example, if one deck runs from 1 to 10, the adjacent deck above will vary from 10 to 100, the third from 100 to 1,000, and so forth. On the other hand, the adjacent deck below the one from 1 to 10 would vary from 0.1 to 1.0. The logarithmic scale can thus be extended upward or downward indefinitely.

In a semilogarithmic chart, relative rate-of-change comparisons can be made readily between different parts of a single series or between two or more series. The relative rate of change of y with respect to x is the slope of y as a function of x, divided by x; equivalently it is the slope of log y except for a constant depending on the logarithmic base. The slope of the logarithmic scale variable is the relative rate of change of the variable. If the slope is steep, the relative rate of change is great. It makes no difference on what part of a semilogarithmic chart a curve is located; the same slope means the same relative rate of change. This is particularly convenient if the two series being compared have very different ranges of values.

This situation is illustrated in Figure 3, where the same data have been plotted on both arithmetic and semilogarithmic grids, placed in juxtaposition. On the arithmetic grid, the relative rate of growth of the city in comparison to that of the entire state cannot be easily determined; by contrast, the curves on the semilogarithmic chart portray relative rate of growth clearly and correctly. Note, however, that this clarity is achieved only at the expense of some distortion; careful labeling is imperative to prevent the reader from mistakenly concluding that the absolute growth of Dallas is greater than that of the state as a whole.

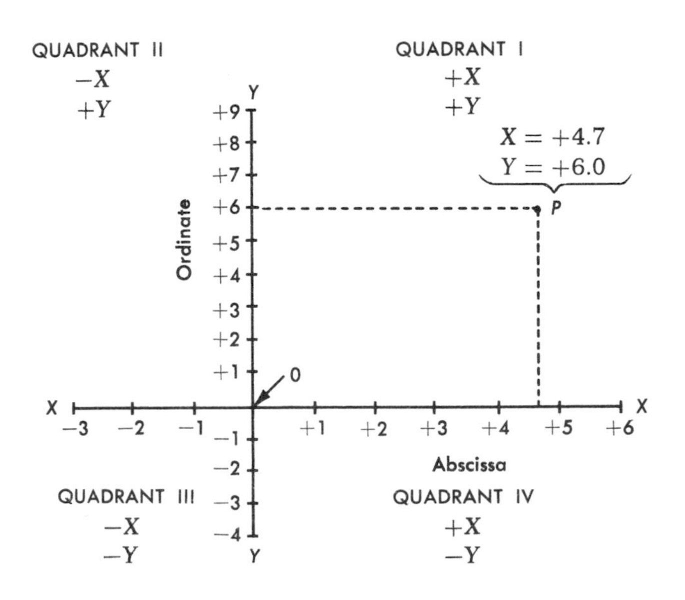

Figure 1 — Basic structural features of a rectangular coordinate graph

Other special graph papers exist, such as double logarithmic, normal, and hyperbolic.

Bar and column charts. Bar and column charts are simple, flexible, and effective techniques for comparing the size of coordinate values or of parts of a total. The basis of comparison is linear or one-dimensional; the length of each bar or column is proportional to the value portrayed. Bar and column charts are very much alike; they differ mainly in that the bars are arranged horizontally in a bar chart and the columns are arranged vertically in a column chart. In addition, the bar chart is seldom used for depicting time series, whereas the column chart is often used for that purpose.

There are several different kinds of bar and column charts. Four of them are illustrated in Figure 4.

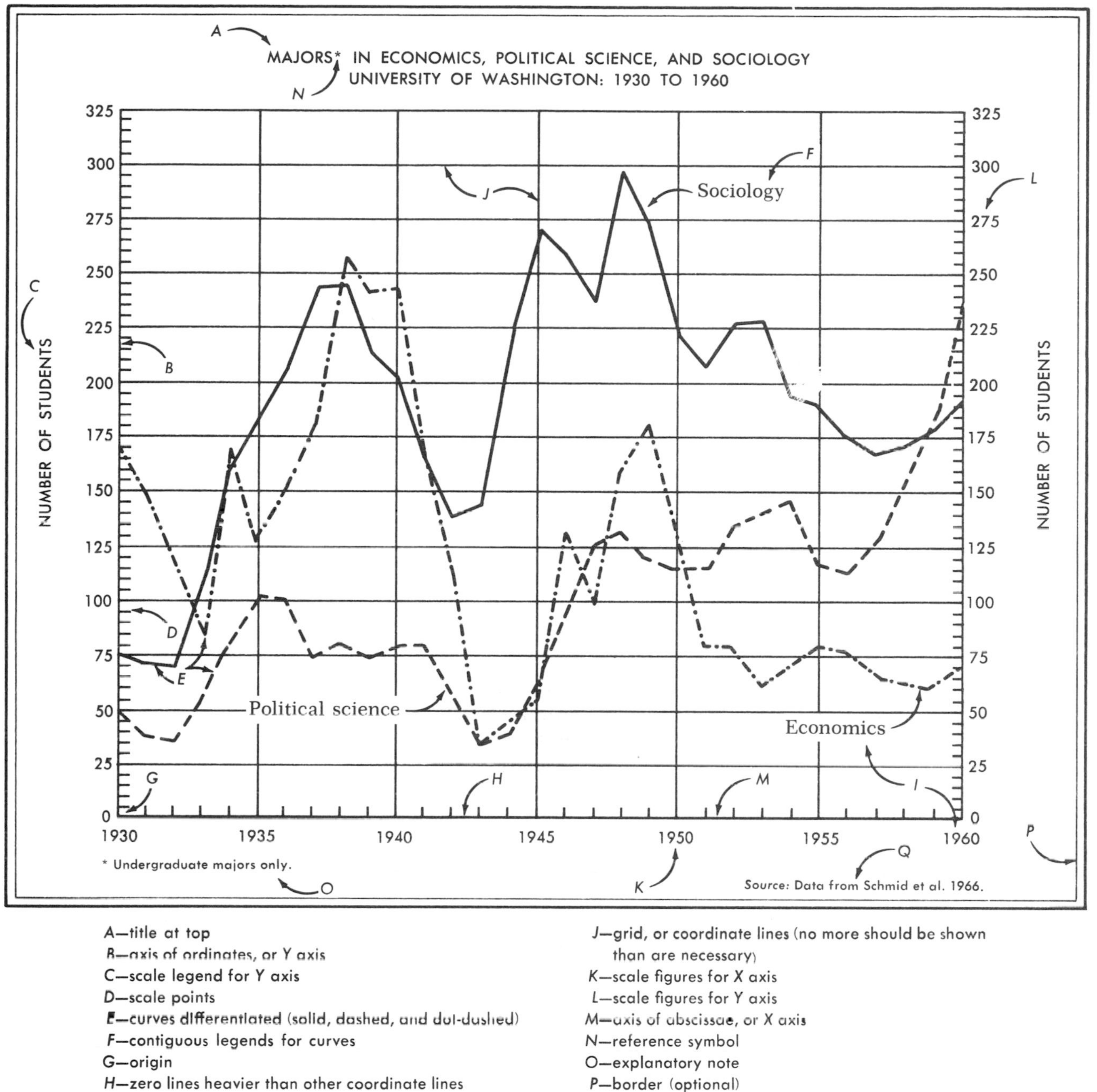

Figure 2 — Essential characteristics and specifications of a rectilinear coordinate graph

Frequency graphs and related charts. The most common graphic forms for portraying simple frequency distributions are the frequency polygon, the histogram, and the smoothed frequency curve. Frequency graphs are usually drawn on rectilinear coordinates, with the *Y* axis representing frequencies and the *X* axis representing the class intervals. The *Y* axis always begins with zero, and under no circumstances is it broken. The horizontal scale does not have to begin with zero unless, of course, the lower limit of the first class interval is zero. In laying out a frequency polygon the appropriate frequency of each class customarily is located at the midpoint of the interval, and the plotting points are then connected by straight lines. The typical histogram is constructed by erecting vertical lines at the limits of the class intervals and forming a series of contiguous rectangles or columns. The area of each rectangle represents the respective class frequencies. (If the class intervals are unequal, special care must be taken.) Smoothed frequency curves may be constructed by either mathematical or graphic techniques. The main purpose of smoothing a frequency graph is to remove accidental irregularities resulting mainly from sampling errors; the data from which the smoothed curve was obtained should always be presented as should some indication of the method of smoothing employed.

For some purposes, the cumulative-frequency curve, or ogive, is more useful than the simple frequency graph. In a cumulative-frequency distribution, the frequencies of the successive class intervals are accumulated, beginning at one end of the distribution. If the cumulation process is from the lesser to the greater, it is referred to as a "less than" type of distribution; if from the greater to the lesser, it is known as a "more than" type of distribution. In constructing an ogive, the cumulative frequencies are represented by the vertical axis and the class intervals by the horizontal axis. The cumulated frequencies are plotted either at the lower or upper end of the respective class intervals, depending on whether the cumulation is of the "less than" or "more than" type. A common configuration of the cumulative-frequency curve is that of an elongated S.

Concentration curves. A special kind of graph, related to cumulative frequency graphs, is known as a concentration curve or a Lorenz curve. Such a graph is used to portray the nature of nonuniformity in the distributions of inherently positive quantities like wealth, income, amount of retail sales, etc. The graph shows the proportions of total wealth (to be definite) held by various proportions of the relevant population; one reads from the graph that the least wealthy 10 per cent of the population holds, say, 1 per cent of total wealth, that the most wealthy 5 per cent of the population holds, say, 30 per cent of total wealth, etc. Thus the graph is a curve running through points given parametrically by cumulative relative frequencies and cumulative relative totals; it is usually presented in square form, with both axes taking values from 0 to 1 (for proportions) or 0 to 100 (for per cents). If wealth, or whatever, is uniformly (equally) distributed in the population, the concentration curve is a straight line, the diagonal of

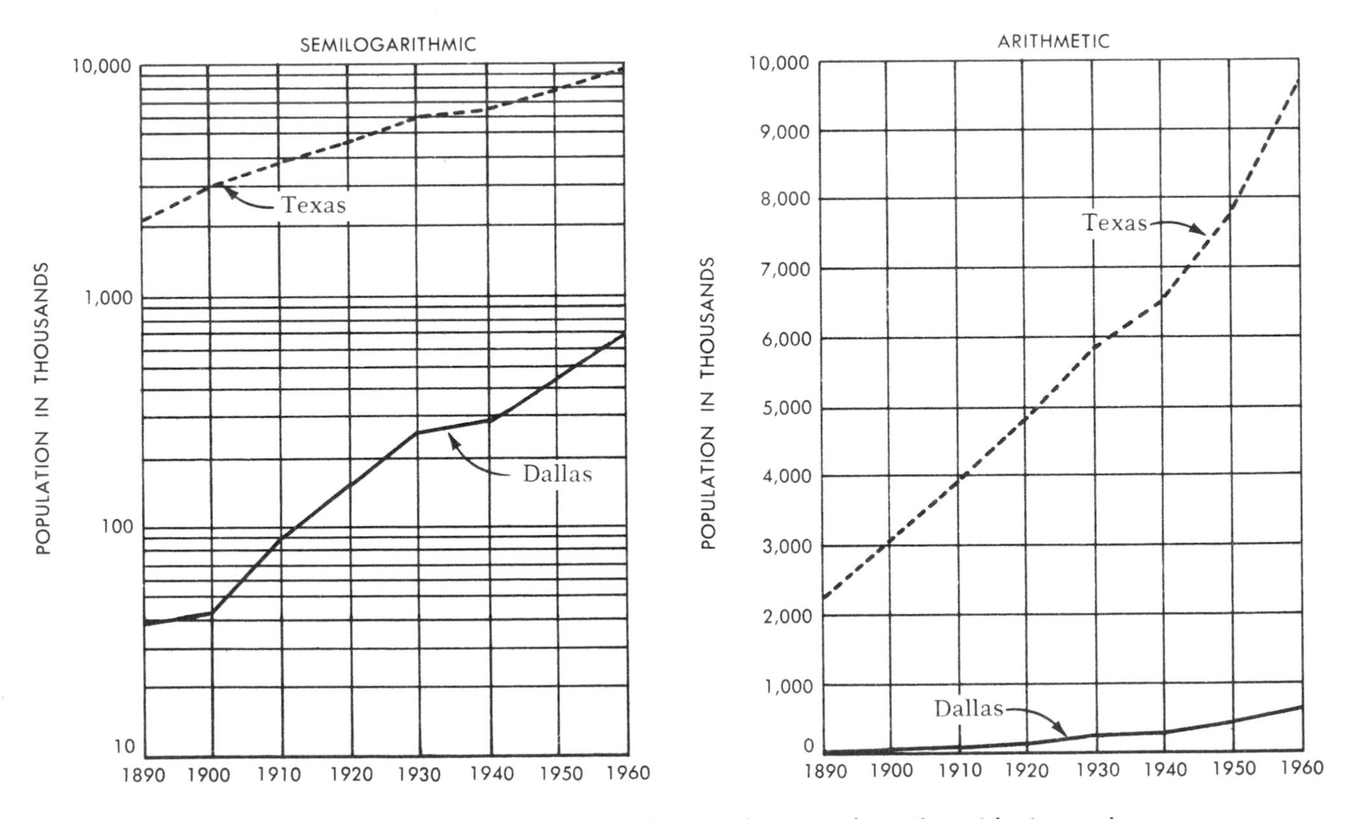

Figure 3 — Comparison* of rectangular coordinate and semilogarithmic graphs

* Comparison clearly demonstrates the superiority of the semilogarithmic graph in portraying relative rates of change.

Source: Data from U.S. Bureau of the Census 1963, pp. 45-19, 45-22.

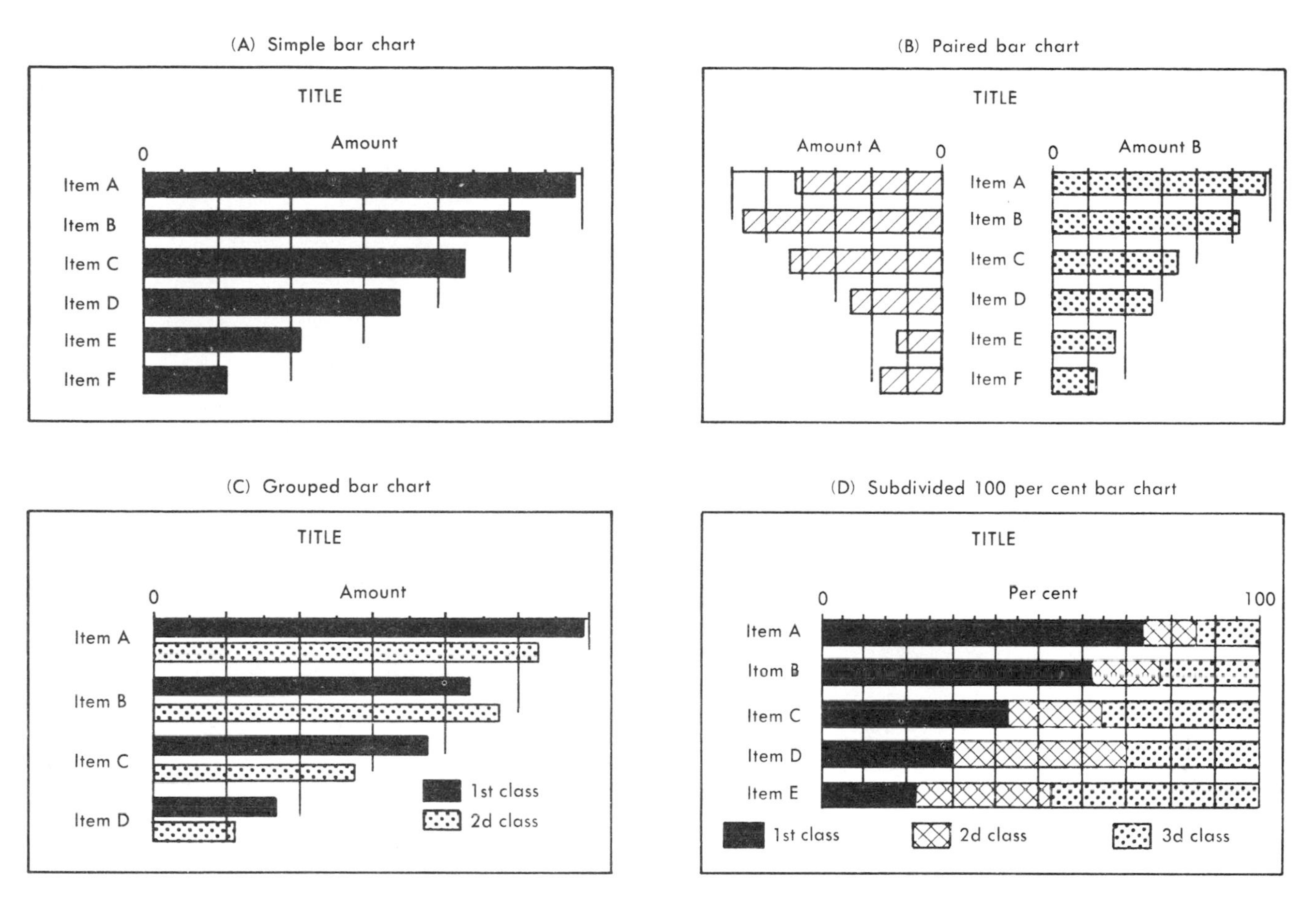

Figure 4 — Four types of bar charts

the square from (0,0) to (1,1). The more the concentration curve deviates from the straight line the greater is inequality.

Statistical maps. There are many varieties of maps used in portraying statistical data. They can be grouped under the following basic types: (1) crosshatched or shaded maps; (2) spot or point-symbol maps; (3) isoline maps; (4) maps with one or more types of graphs superimposed, such as the bar, column, line, flow, or pictorial forms (see Figure 7); and (5) a combination of two or more of the preceding types. (For an overview of map making, see Robinson 1968.)

Crosshatched maps. The crosshatched or shaded map, characteristically, is used to portray rates and ratios that are based on clearly delineated areal units such as regions, nations, states, counties, or census tracts. Value ranges of rates and percentages are represented by a graded series of crosshatchings. Figure 5 is an illustration of a crosshatched map. Since this figure is for illustration only, the medians were computed by the simplest formula, with no attempt made to correct for the fact that most school dropouts occur at the end of a school year.

Spot maps. In spot or point-symbol maps emphasis is placed on frequencies or absolute amounts rather than on rates or proportions as in the crosshatched map. Although there may be some overlapping, spot maps may be differentiated into five types on the basis of the symbols used. Symbols may stress (1) size, (2) number, (3) density, (4) shading, or (5) form. In the first type of map the size of each symbol is proportional to the frequency or magnitude of the phenomena represented. Symbols may be either two-dimensional or three-dimensional and are normally in the form of circles or spheres rather than rectangles, cubes, or irregular forms. In the second type of spot map the basic criterion is not size but number or frequency of spots or point symbols. The spots are uniform in size, each representing a specific value. The spots are designed and arranged to make them as readily countable as possible. The third type of map is also a multiple-spot variety; but instead of emphasizing countable frequencies, comparative density and distribution are emphasized. Figure 6 is an example of the third type of spot map, in which over-all density patterns are stressed. In the fourth type of spot map the criterion is shading. The size

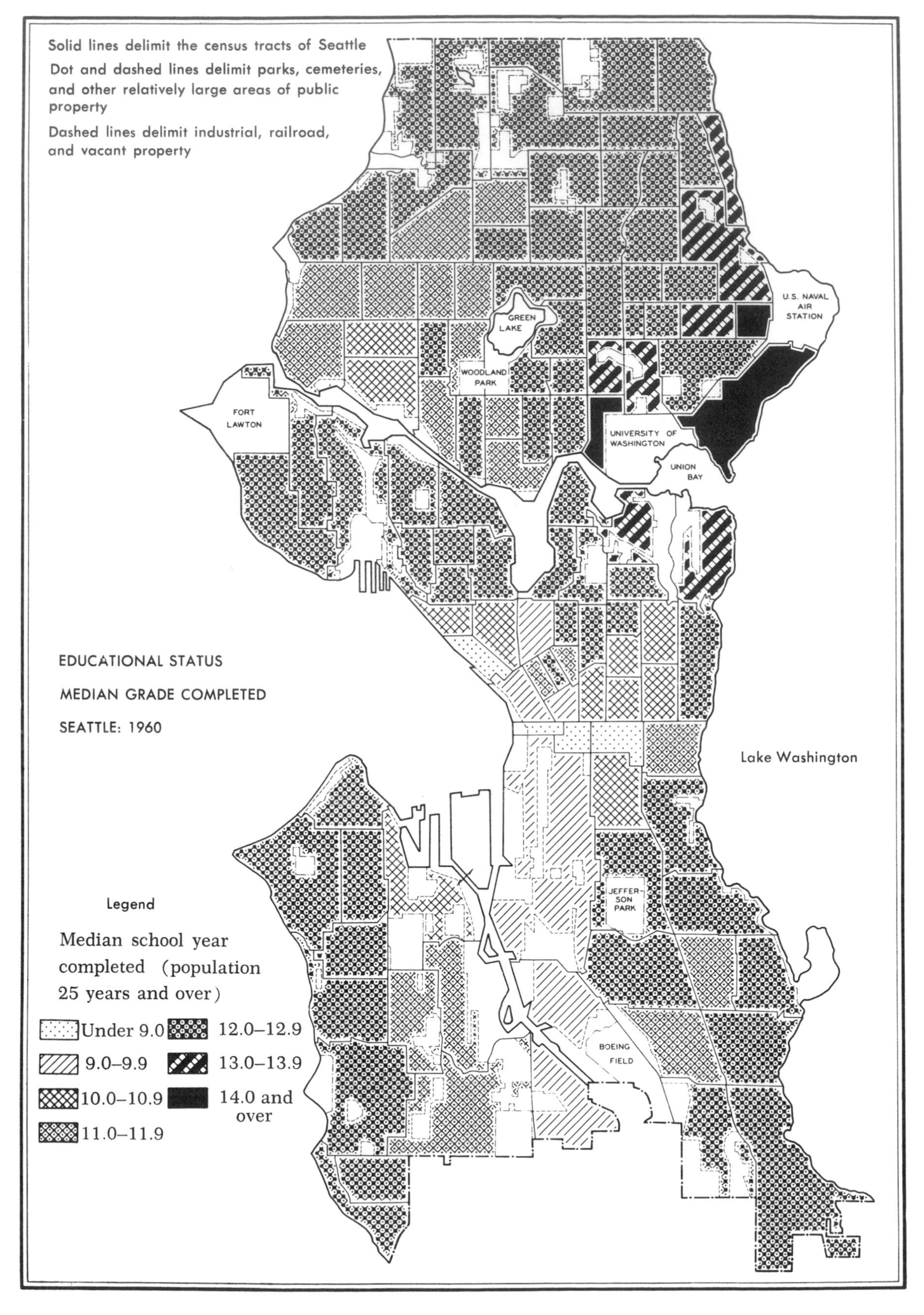

Figure 5 — Crosshatched map*

* Note hatching gradation from light to dark. Each type of crosshatching represents a value interval in a series of averages.

Source: Data from U.S. Bureau of the Census 1962, pp. 15–24.

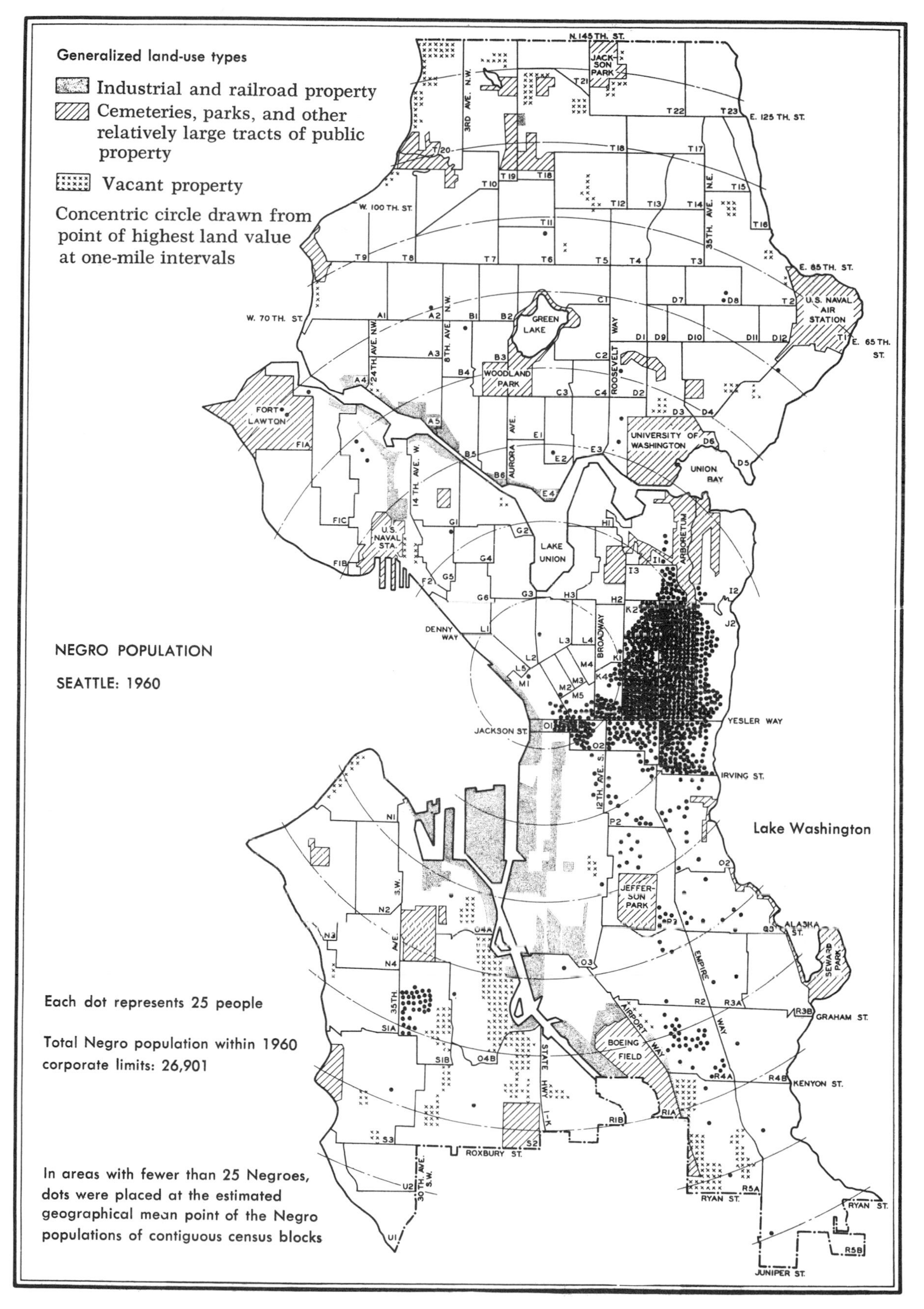

Figure 6 — Spot map*

* This type of spot map is designed to emphasize comparative density and basic distributional patterns. Note also census tracts, concentric circles, and generalized land-use features on base map. *Source:* Schmid & McVey 1964, p. 9.

of the symbols is uniform, but the amount of shading is indicative of the magnitude or value represented. The form of the symbol in the fifth type of map represents certain qualities or attributes rather than quantities, as in the previous types. For example, if the dichotomy male–female is to be portrayed on a map, one type of symbol would represent male and another symbol would represent female.

Isoline maps. There are two fundamental types of isoline (from the Greek *isos*, meaning equal) maps: the isometric map, in which the lines are drawn through points of equal value or intensity, and the isopleth map, in which the lines connect equal rates or ratios for specific areas. The isopleth map is particularly valuable in the social sciences.

Miscellaneous graphic forms. Although not as basic or as widely used as the graphs and charts discussed in the preceding sections, there are certain other graphic forms that possess advantages for certain problems.

Pie charts. The pie or sector chart is widely used to portray proportions of an aggregate or total. Given the several proportions of the total, a pie chart is made by dividing a circle into pieces, one for each separate proportion, by boundary radii. Each proportion corresponds to a single slice, or sector, thus formed; and the central angles (equivalently, the circumference arcs, or again the sector areas) are proportional to the magnitudes of the proportions. Frequently, shading or coloring is used to help distinguish the sectors, and of course proper labeling is essential. Although a pie chart can be very effective in simple situations, it becomes difficult to use if there are more than four or five sectors or if one wishes to compare corresponding proportions for several pie charts.

Trilinear charts. The trilinear chart (or barymetric coordinate system) is used to portray simultaneously three nonnegative variables with a fixed sum. Usually they are percentages and the sum is 100. It is drawn in the form of an equilateral triangle, each side of which is calibrated in equal percentage divisions ranging from 0 to 100.

Scatter diagrams. The scatter diagram (scattergram) and other types of correlation charts portray in graphic form the degree and type of relationship

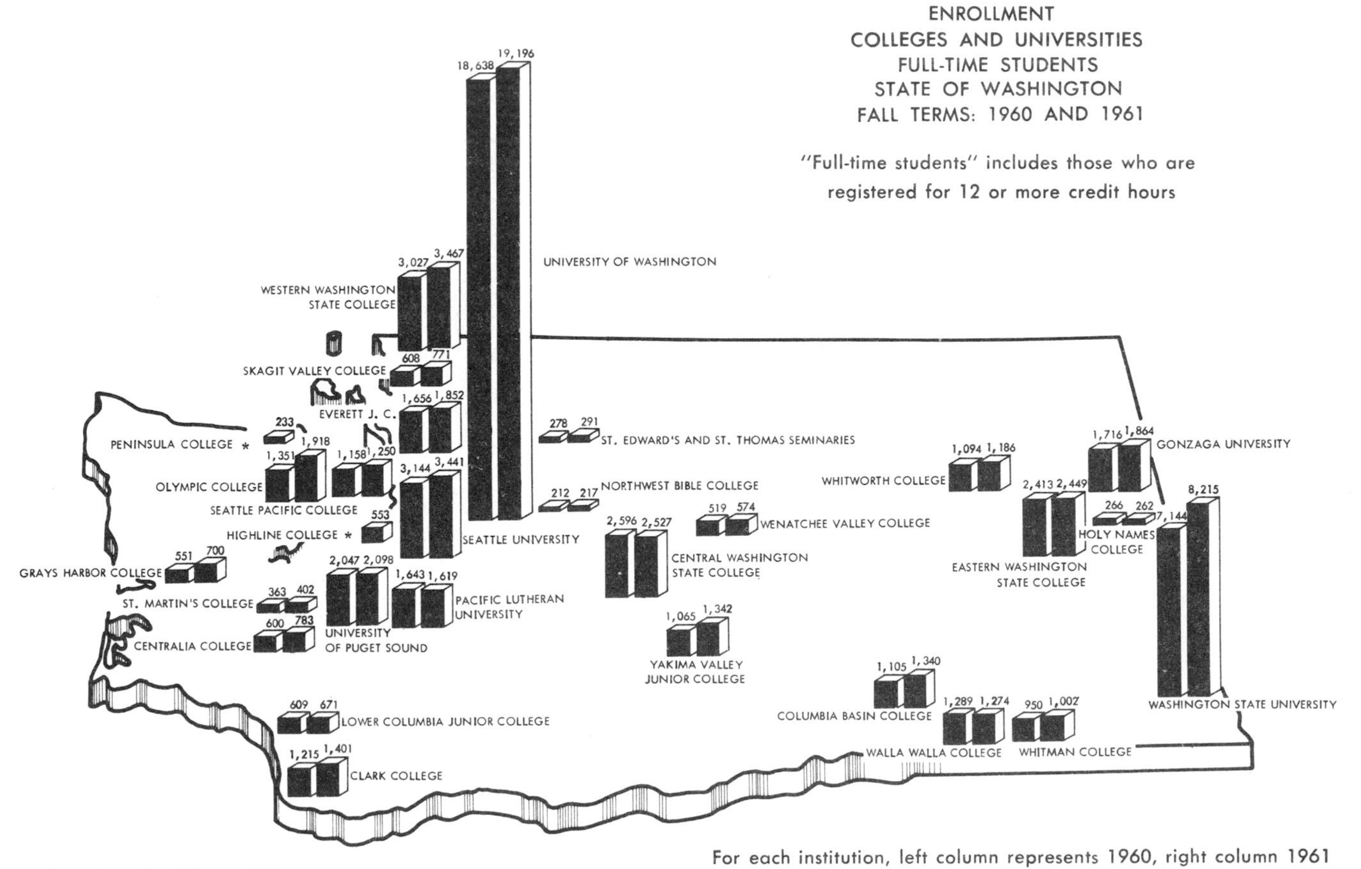

Figure 7 — Chart in three-dimensional projection [a]

a. Note especially that columns are drawn in oblique projection.

Source: Schmid et al. 1962, p. 9.

or covariation between two series of data. The scatter diagram shows a two-way or bivariate frequency distribution. Customarily, arithmetic scales are used in the construction of scatter diagrams, although semilogarithmic or double-logarithmic scales sometimes may be more appropriate.

Pictorial charts. Pictorial graphs are used mainly because of their popular appeal, although they rarely convey more information than do more conventional graphic forms. In general, there are four basic types of charts in which pictorial symbols are used, distinguished by criteria of purpose and emphasis.

The four types of pictorial charts are (1) charts in which the size of the pictorial symbol is proportional to the values portrayed; (2) pictorial unit graphs, in which each symbol represents a definite and uniform value; (3) cartoon and sketch charts, in which the basic graphic form, such as a curve or bar, is portrayed as a picture; and (4) charts with pictorial embellishments ranging from a single pictorial filler to elaborate and detailed pictorial backgrounds.

► Through the initiative, skill, and dedication of Otto Neurath (1973) and his co-workers, the pictorial unit chart has had a significant impact, not only on statistical graphics, but also on graphic communication generally. The pictorial unit chart was first used extensively in connection with social and economic exhibits organized by Neurath, who was director of the Gesellschafts-und-Wirtschafts Museum in Vienna from 1924 to 1934. During the 1930s, the pictorial unit chart enjoyed widespread popularity, both in Europe and in America, but since World War II, interest in this type of chart has waned considerably.

► After leaving Vienna in 1934, Neurath became involved increasingly with visual education, specifically with the development of an international picture language. The picture language that he envisaged was an auxiliary one, not intended to supplant verbal language. The international significance of Neurath's pictorial language is reflected by the acronym *isotype* (International System Of TYpographic Picture Education), which he used to describe his system. At present, many ideas derived from Neurath's philosophy and techniques are used throughout the world, perhaps often without knowledge of their source, for road signs, in journalism, and in air terminals and other public places. Edwards et al. (1975) deals with some uses of isotype.

Three-dimensional projection charts. In recent years it has become common practice to portray various kinds of graphs and charts in axonometric, oblique, and perspective projection. Charts in three-dimensional form, with depth and other picturelike qualities, unquestionably possess definite popular appeal.

The design of charts and graphs in three-dimensional form should be based on technically acceptable principles of axonometric, oblique, and perspective projection. Axonometric and oblique projections are the most satisfactory for three-dimensional graphs. Although perspective projection is perhaps the most realistic of the three, it possesses serious limitations as a technique in graphic presentation. Charts constructed in perspective projection are generally distorted; they do not portray exact distance, shape, or size. Figure 7 illustrates a map in three-dimensional form.

CALVIN F. SCHMID

[*See also* TABULAR PRESENTATION.]

BIBLIOGRAPHY

REFERENCES ON GRAPHIC PRESENTATION

BRINTON, WILLARD C. 1939 *Graphic Presentation.* New York: Brinton.

►EDWARDS, J. A. et al. 1975 *Graphic Communication Through Isotype.* Reading (England): Univ. of Reading.

FUNKHOUSER, H. G. 1937 Historical Development of the Graphical Representation of Statistical Data. *Osiris* 3:269–404.

HUFF, DARREL; and GEIS, IRVING 1954 *How to Lie With Statistics.* New York: Norton. → Contains a discussion of graphical fallacies. Also published in paperback.

JENKS, G. F.; and BROWN, D. A. 1966 Three-dimensional Map Construction. *Science* 154:857–864.

MODLEY, RUDOLF; and LOWENSTEIN, DYNO 1952 *Pictographs and Graphs: How to Make and Use Them.* New York: Harper.

►NEURATH, OTTO 1973 *Empiricism and Sociology.* Edited by Marie Neurath and Robert S. Cohen. Dordrecht (Netherlands) and Boston: Reidel. → A selection of essays and biographical and autobiographical sketches originally written in German between 1919 and 1945.

PÈPE, PAUL 1959 *Présentation des statistiques.* Paris: Dunod.

►ROBINSON, ARTHUR H. 1968 Cartography. Volume 2, pages 325–329 in *International Encyclopedia of the Social Sciences.* Edited by David L. Sills. New York: Macmillan and Free Press.

ROYSTON, ERICA 1956 Studies in the History of Probability and Statistics. III. A Note on the History of the Graphical Representation of Data. *Biometrika* 43:241–247.

SCHMID, CALVIN F. 1954 *Handbook of Graphic Presentation.* New York: Ronald.

SCHMID, CALVIN F. 1956 What Price Pictorial Charts? *Estadistica: Journal of the Inter-American Statistical Institute* 15:12–25.

SCHMID, CALVIN F.; and MACCANNELL, EARLE H. 1955 Basic Problems, Techniques, and Theory of Isopleth Mapping. *Journal of the American Statistical Association* 50:220–239.

SOURCES OF FIGURES AND DATA

SCHMID, CALVIN F.; and MCVEY, WAYNE W. JR. 1964 *Growth and Distribution of Minority Races in Seattle, Washington.* Seattle Public Schools.

SCHMID, CALVIN F. et al. 1962 *Enrollment Statistics, Colleges and Universities: State of Washington, Fall Term, 1961.* Seattle: Washington State Census Board.

SCHMID, CALVIN F. et al. 1966 *Studies in Enrollment Trends and Patterns.* Part 1: Regular Academic Year: 1930 to 1964. Seattle: Univ. of Washington.

U.S. BUREAU OF THE CENSUS 1962 *U.S. Censuses of Population and Housing: 1960.* Census Tracts, Final Report PHC (1) 142: Seattle, Wash. Washington: Government Printing Office. → See especially pages 15–34, Table P-1, "General Characteristics of the Population, by Census Tracts: 1960."

U.S. BUREAU OF THE CENSUS 1963 *U.S. Census of Population: 1960.* Vol. 1, Part 45: Texas. Washington: Government Printing Office.

Postscript

Computer graphics. An unprecedented development has been the use of electronic computers and auxiliary equipment in the preparation of statistical graphics. This development has occurred since about 1955, with the period since 1970 encompassing spectacular changes. In fact, "during the early 1970's computer graphic technology progressed far more rapidly than many of its most ardent proponents had hoped was possible" (U.S. Bureau of the Census 1973, p. 1). Further application and development of computer technology in graphic presentation will surely come.

Computerized techniques occupy an essential place in graphic presentation, but they have constraints and limitations. Complex technology cannot fully replace personal insight and expertise in designing statistical charts. Shortage of appropriate equipment and of skilled personnel, along with such economic factors as time and cost, place substantial constraints on the general use of computer graphics.

Computer techniques are advantageous for elaborate and mathematically derived maps and charts, when extensive calculations and data processing are required. Electronic computers are also especially valuable when many charts of the same type are required. In such instances the computer represents advantages in speed, economy, accuracy, and flexibility.

The most significant and productive application of computer techniques in graphic presentation has been to statistical mapping. There has been a proliferation of computer mapping systems and technologies throughout the world.

Most computer techniques for graphic presentation represent a highly skilled and complicated process. For example, in developing a computer mapping capability several prerequisites are essential.

(1) A geographic base file, including street address information and coordinates, is required unless only small volumes of data are involved and predefined areal units such as census tracts are used. Geographic coding and manipulation of data are by far the most time-consuming aspects of the computer mapping process.

(2) Data to be used for computer mapping must be encoded with computer-readable addresses that can be matched with the base file for the area from which the data have been drawn.

(3) Access to mapping programs and computer equipment is essential. Fortunately, programs for the preparation of various types of statistical map are available from a substantial number of computer centers. Particular programs are adapted to certain types of electronic equipment.

(4) Adequate staff expertise is indispensable.

At this stage of development, most charts produced by computers generally lack the artistry, clarity, and appeal of well-designed, well-constructed, handmade charts. The characteristic coarseness, woodenness, and crudity of computer-produced charts are inherent in the hardware that is used. The printing of the first charts of this kind were produced on computer-driven typewriters, and at the present time such well-known programs as Symap, Linmap, and Grids are adapted to the line printer where type symbols or combinations of symbols are used. The computer-driven drum or flatbed plotter equipped with a pen is also used, but the product from a qualitative point of view is still inferior. An entirely new approach to computer graphics was developed in 1974, producing charts of a higher quality, at a lower cost, and with less clerical work than the older processes. The key to the new approach is the use of micrographics on a COM (Computer Output on Microfilm) unit. These devices plot a picture or write alpha-numerics on a high precision CRT (cathode ray tube) at a very high rate of speed. This image is then photographed on microfilm. Conventional processing procedures are used for enlargement and other necessary steps so that the completed chart is of very high quality and ready for printing. (See Schweitzer 1975; Broome 1975.)

An important contribution to computer-generated graphics has been the innovative operational techniques developed by the U.S. Bureau of the Census in connection with the construction of hundreds of charts for *Status*, a monthly periodical produced as a collaborative effort by various agencies of the federal government. After extensive experimentation, including the publication of prototype material, the first issue of *Status* was released in July 1976. As a consequence of insufficient funds, however, its publication was suspended after the October 1976 issue. *Status* was projected as a monthly chart book of social and economic trends, both for the general public and for decision makers in business, government, and education. From a technical point of view, significant features of this new computer-generated system of graphics are threefold: (1) simplified general-purpose plotting programs; (2) speed of production; and (3) relatively high quality of charts. (See U.S. Bureau of the Census 1976; Barabba 1976.)

Kinostatistics. A development for improved graphic social reporting has been proposed by Albert D. Biderman and his associates (Biderman 1971; Feinberg, 1976). Although the development is now formative and tentative, the basic hope is for radical improvement via techniques that transcend the limits of the printed page. The use of television and motion pictures to bring in a time line is fundamental to kinostatistics.

CALVIN F. SCHMID

ADDITIONAL BIBLIOGRAPHY

BACHI, ROBERTO 1968 *Graphical Rational Patterns: A New Approach to Graphical Presentation of Statistics.* Jerusalem: Israel Universities Press.

BARABBA, VINCENT P. 1976 Automating Statistical Graphics: A Tool for Communication. Part 1, pages 82–88 in American Statistical Association, Social Statistics Section, *Proceedings.* Washington: The Association.

BERTIN, JACQUES 1968 *Sémiologie graphique: Les diagrammes, les réseaux, les cartes.* Paris: Gauthier-Villars.

BIDERMAN, ALBERT D. 1971 Kinostatistics for Social Indicators. *Educational Broadcasting Review* 5, no. 5: 13–19.

BROOME, R. 1975 Micrographics: A New Approach to Cartography at the Census Bureau. Pages 28–39 in U.S. Bureau of the Census, *Geographic Base (DIME) System: A Local Program.* Series GE 60, No. 6. Washington: Government Printing Office.

DICKINSON, G. C. (1963) 1973 *Statistical Mapping and the Presentation of Statistics.* 2d ed. London: Arnold; New York: Crane, Russak.

FEINBERG, BARRY M. 1976 *Kinostatistics: Communicating a Social Report to the Nation.* Washington: Bureau of Social Science Research.

FEINBERG, BARRY M. et al. (editors) 1975 *Social Graphics Bibliography.* Washington: Bureau of Social Science Research.

INTERNATIONAL SYMPOSIUM ON COMPUTER-ASSISTED CARTOGRAPHY, WASHINGTON, *1975* 1977 *Proceedings.* Washington: Government Printing Office. → Symposium conducted by the U.S. Bureau of the Census, Sept. 21–25.

KRUSKAL, WILLIAM H. 1977 Visions of Maps and Graphs. Pages 27–36 in International Symposium on Computer-assisted Cartography, Washington, 1975, *Proceedings.* Washington: Government Printing Office.

LOCKWOOD, ARTHUR 1969 *Diagrams: A Visual Survey of Graphs, Maps, Charts, and Diagrams for the Graphic Designer.* London: Studio Vista; New York: Watson-Guptill.

MONKHOUSE, F. J.; and WILKINSON, H. R. (1952) 1971 *Maps and Diagrams: Their Compilation and Construction.* 3d ed. London: Methuen.

MUEHRCKE, PHILIP 1972 *Thematic Cartography.* Association of American Geographers, Commission on College Geography, Resource Paper No. 19. Washington: The Association.

NEWMAN, WILLIAM M.; and SPROULL, ROBERT F. 1973 *Principles of Interactive Computer Graphics.* New York: McGraw-Hill.

PEDRONI, FERNANDO 1968 *Rappresentazioni statistiche.* Milan (Italy): Hoepli.

PEUCKER, THOMAS K. 1972 *Computer Cartography.* Association of American Geographers, Commission on College Geography, Resource Paper No. 17. Washington: The Association.

ROBINSON, ARTHUR H.; and SALE, RANDALL D. (1953) 1969 *Elements of Cartography.* 3d ed. New York: Wiley.

SCHMID, CALVIN F. 1976 The Role of Standards in Graphic Presentation. Part 1, pages 74–81 in American Statistical Association, Social Statistics Section, *Proceedings.* Washington: The Association.

SCHWEITZER, RICHARD H. JR. 1975 Micrographics: A New Approach to Automated Cartography and Graphics. Pages 305–316 in International Communication Conference, 22d, 1975, *Proceedings.* New York: Society for Technical Communication.

U.S. BUREAU OF THE CENSUS 1969 *Census Use Study: Computer Mapping.* Report No. 2. Washington: Government Printing Office.

U.S. BUREAU OF THE CENSUS 1972 *Grids: A Computer Mapping System.* By Mathew A. Jaro. Washington: Government Printing Office.

U.S. BUREAU OF THE CENSUS 1973 *Mapping Urban America With Automated Cartography.* By Richard H. Schweitzer, Jr. Washington: Government Printing Office.

U.S. BUREAU OF THE CENSUS 1975 *Geographic Base (DIME) System: A Local Program.* Series GE 60, No. 6. Washington: Government Printing Office. → Papers of a conference held in Washington, Nov. 18 19, 1974.

U.S. BUREAU OF THE CENSUS 1976 *Status: A Monthly Chartbook of Social and Economic Trends.* Washington: Government Printing Office. → Discontinued.

GRAUNT, JOHN

John Graunt (1620–1674) is generally regarded as having laid the foundations of demography as a science with the publication of his *Natural and Political Observations Made Upon the Bills of Mortality* (1662*a*). Graunt began, as many scientists have done, by exercising idle curiosity—"Now having (I know not by what accident) engaged my thoughts upon the *Bills of Mortality* . . ." ([1662*a*] 1939, p. 3)—and proceeding to "observations, which I happened to make (for I designed them not) . . ." (*ibid.*, p. 5). Characteristically, he did not believe that curiosity should remain idle: "finding some *Truths*, and not commonly-believed Opinions . . . I proceeded further, to consider what benefit the knowledge of the same would bring the World" (*ibid.*, p. 18).

Graunt's application of the statistical method to raw material had two essential elements: (1) *classification* of like with like, so as to break the data down into homogeneous groups, and (2) the *comparison* of those groups in order to recognize significant differentials. Another attribute of Graunt as a statistician was his ability to apply logic to his arithmetic. He never accepted a calculation if it offended his common sense; nor did he accept it without a test, if he had the means of testing it.

As to classification, we may read in Chapter 2 of his *Observations* his very sound remarks about the identification of causes of death, then dependent on the observation of the "ancient matron" searchers (*ibid.*, pp. 27–32). Deaths at advanced ages, he argued, can hardly be safely attributed to one specific cause, and not much worthwhile information can be expected about sudden deaths beyond their suddenness. As to comparison, we find in the same chapter his observation about the stability of proportionate mortality rates for certain causes, and in Chapter 4 (*ibid.*, pp. 45–48) his observation that in plague years, deaths from causes other than plague were inflated to such an extent as to lead to his conclusion that of deaths from plague fully one-fifth were attributed to some other cause (in order to avoid the closure of the plague-infested dwelling and the virtual incarceration of the surviving members of the household). Again in Chapter 3 (*ibid.*, pp. 33–44), we may see how comparison of mortality rates for rickets with mortality rates referring to possible similar names for the same disease led Graunt, after examining the figures, to conclude that rickets was a new disease and not a new name for an old one. (This was probably a wrong conclusion, but Graunt was not a physician and would not have known that medical attention was at that time being drawn to the disease.)

The method of comparison was again employed when Graunt estimated the population of London in Chapter 11 (*ibid.*, pp. 67–70). He used three methods and reconciled them. First, he estimated that a fertile woman had a baby every other year, that fertile women numbered half the married women, and that there were seven other members of a family (husband, 3 children, 3 servants); therefore, the population was 32 times the number of annual births. This gives 384,000. Next he estimated that in certain parishes there were annually three deaths from every 11 families; therefore, the number of deaths (13,000) multiplied by 11/3 yielded 48,000 families, or 384,000 persons. Finally, he observed 54 families per 100 square yards within the walls (12,000 families) and guessed that there were three times as many families outside the walls.

Graunt's most important contribution to demography was his rudimentary life table. Its importance lies not in the table itself, which is indeed defective, but in the novelty of presenting mortality in terms of survivorship. Graunt began with only two observations—the proportion of births surviving to age 6 (.64) and the proportion surviving to age 76 (.01). He then assumed that a constant multiplier is involved in proceeding from the first proportion to the last by decennial intervals of age. He did not make clear how he got his multiplier (or "mean proportion," as he called it), but it was obviously about .6. Michel Ptouka (1938) has put forward the hypothesis that the multiplier was $(64-1)/100 = .63$. D. V. Glass has suggested (1950) that Graunt made a second difference interpolation from $l_0 = 100$, $l_6 = 64$, and $l_{76} = 1$. The method and any error in it is not important. (Halley was not to make the correct calculation for another sixty years or so.) So much has been developed from this simple but immensely powerful thought that actuarial criticism would be quite out of place.

Graunt is rarely considered apart from another English scientist of the time, Sir William Petty. Both were of Hampshire stock; Graunt was born in 1620 and Petty in 1623. They became acquainted in or before 1650. It appears that their relationship was initially that of client and patron (Graunt being Petty's patron; see Greenwood 1941–1943), but the roles were reversed after the fire of London in 1666.

Graunt was the son of a city tradesman and became a haberdasher and a man of substance. Petty, who had sampled the merchant navy, studied mathematics in France, spent a short time in the Royal Navy, returned to the Continent to study anatomy, spent some time in business in London, and went to Oxford in 1649, becoming a doctor of medicine by dispensation. Later he rose to be professor of anatomy and vice-principal of Brasenose. He became a candidate for a Gresham professorship in London and made contact with Graunt. We do not know what interests they shared at that time or whether Graunt (or his father) had any influence in Petty's subsequent appointment as Gresham professor of music. Soon after, Petty (who seems to have had no professional duties) went to Ireland and made a fortune. For many years Graunt remained a prosperous tradesman, but the fire of 1666 destroyed his business. A little later he became a Roman Catholic convert and was apparently not interested in rebuilding his business, for he was soon bankrupt. It was now Petty's turn to be the patron, and Graunt, the client—and apparently not an easy one. Despite Petty's attempts to help him, Graunt's financial troubles were with him until he died in 1674.

Petty's vital statistical work was on a different level from that of Graunt. Petty had inspiration and brilliance, many ideas and a breadth of vision, but he did not pursue his ideas as persistently as Graunt did. It was Petty who proposed a central government statistical office and a system of census taking. He anticipated William Farr in estimating the economic loss due to mortality. Many of his calculations, however, do not stand up to the tests of consistency that Graunt would have applied. It seems to be generally agreed that Graunt made an immeasurably greater contribution to demography.

Essentially, Graunt was a man with an inquiring mind whose ideas laid the foundation of the political arithmetic that we now accept as part of good social organization and government. The Royal Society of London, of which Graunt was elected a fellow, commemorated the tercentenary of the publication of his *Observations* by holding a special series of scientific meetings to which leading demographers in England and other countries contributed papers (Glass 1963). The Institute of Actuaries of England paid its own tribute to Graunt by republishing, in a more modern format but without abridgment or alteration, his original *Observations*.

B. BENJAMIN

[*For the historical context of Graunt's work, see* VITAL STATISTICS *and the biography of* PETTY; *see also* Rosen 1968. *For discussion of the subsequent development of his ideas, see* LIFE TABLES *and the biography of* KŐRÖSY, *see also* Moriyama 1968; Spengler 1968.]

WORKS BY GRAUNT

(1662*a*) 1939 *Natural and Political Observations Made Upon the Bills of Mortality.* Edited with an introduction by Walter F. Willcox. Baltimore: Johns Hopkins Press. → See especially Chapter 2, "General Observations Upon the Casualties"; Chapter 3, "Of Particular Casualties"; Chapter 4, "On the Plague"; Chapter 11, "Of the Number of Inhabitants."

(1662*b*) 1964 Natural and Political Observations Mentioned in a Following Index, and Made Upon the Bills of Mortality: With Reference to the Government, Religion, Trade, Growth, Ayre, Diseases, and Several Changes of the Said City [London]. *Journal of the Institute of Actuaries* (London) 90:1–61. → Includes a three-page introduction by B. Benjamin.

SUPPLEMENTARY BIBLIOGRAPHY

GLASS, D. V. 1950 Graunt's Life Table. *Journal of the Institute of Actuaries* (London) 76:60–64.

GLASS, D. V. 1963 John Graunt and His *Natural and Political Observations.* Royal Society of London, *Proceedings* Series B 159:1–38. → Includes six pages of discussion.

GREENWOOD, MAJOR (1941–1943) 1948 *Medical Statistics From Graunt to Farr.* Cambridge Univ. Press. → First published in *Biometrika.*

►MORIYAMA, IWAO M. 1968 Mortality. Volume 10, pages 498–504 in *International Encyclopedia of the Social Sciences.* Edited by David L. Sills. New York: Macmillan and Free Press.

NEWMAN, JAMES R. 1956 Commentary on an Ingenious Army Captain and on a Generous and Many-sided Man. Volume 3, pages 1416–1419 in James R. Newman (editor), *The World of Mathematics: A Small Library of the Literature of Mathematics From A'h-mosé the Scribe to Albert Einstein.* New York: Simon & Schuster.

PTOUKA, MICHEL 1938 John Graunt, fondateur de la démographie: 1620–1674. Volume 2, pages 61–74 in International Congress for Studies on Population, Paris, 1937, *Congrès international de la population.* Paris: Hermann.

►ROSEN, GEORGE 1968 Public Health. Volume 13, pages 164–170 in *International Encyclopedia of the Social Sciences.* Edited by David L. Sills. New York: Macmillan and Free Press.

►SPENGLER, JOSEPH J. 1968 Lotka, Alfred J. Volume 9, pages 475–476 in *International Encyclopedia of the Social Sciences.* Edited by David L. Sills. New York: Macmillan and Free Press.

SUTHERLAND, IAN 1963 John Graunt: A Tercentenary Tribute. *Journal of the Royal Statistical Society* Series A 126:537–556.

GROUPED OBSERVATIONS

See under STATISTICAL ANALYSIS, SPECIAL PROBLEMS OF.

H

HISTORY OF SCIENCE

See under SCIENCE.

HISTORY OF SOCIAL RESEARCH

See SOCIAL RESEARCH, THE EARLY HISTORY OF.

HISTORY OF STATISTICAL METHOD

See under STATISTICS.

HOTELLING, HAROLD

► *This article was specially written for this volume.*

Harold Hotelling, a leader in the field of multivariate statistical analysis, played a prominent part in the spectacular growth of mathematical statistics in the United States that began in the 1930s and helped bring about the revival of mathematical economics in the late 1920s.

Hotelling was born in Fulda, Minnesota, in 1895. Although his B.A. from the University of Washington was in journalism, his M.A., also from the University of Washington, was in mathematics. During his undergraduate years Eric Temple Bell had detected his mathematical talent and encouraged him to turn to mathematics. He continued his mathematical studies at Princeton, where he earned his PH.D. in 1924.

The next seven years he spent at Stanford University, first as a research associate in the Food Research Institute and then as associate professor of mathematics. A leave of absence, from June to December 1929, which he spent with R. A. Fisher at the Rothamsted Experimental Station at Harpenden in England, helped to establish Hotelling's lifelong interest in mathematical statistics. [*See the biography of* FISHER.]

In 1931, Hotelling moved to Columbia University as professor of economics. It was during his fifteen years at Columbia that his most important work in mathematical statistics and mathematical economics appeared. He was able to attract to Columbia a group of outstanding workers in these two disciplines, as well as many students who later became leading scholars. During World War II, to help the war effort, Hotelling organized the Statistical Research Group at Columbia University. (Abraham Wald developed his theory of sequential analysis while he was consultant to that group.) [*See* SEQUENTIAL ANALYSIS *and the biography of* WALD.]

In 1946 Hotelling left Columbia to organize a department of mathematical statistics at the University of North Carolina in Chapel Hill, which then became an important center of statistical research and teaching. He retired in 1966 and died in 1973.

Unlike Fisher, Jerzy Neyman, and Wald, Hotelling was not actively concerned with laying and developing the foundations of mathematical statistics. In his many contributions to statistics, he took the existing foundations, mainly those laid by Fisher, for granted. His broad range of interests, encompassing mathematical economics as well as diverse branches of statistics, and his sustained effort in the development of multivariate analysis are the hallmarks that characterize the scope of his scientific activity. To these must be added the zeal with which he spread the gospel of mathematical statistics.

It is noteworthy that Hotelling's first three scholarly publications were in the areas of topology

(1925*a*), mathematical economics (1925*b*), and mathematical statistics (1925*c*). Although he soon concentrated his attention on the latter two disciplines, his training in pure mathematics with a geometrical emphasis was reflected in his striving, in all of his scientific work, for mathematical rigor, often combined with skillful use of geometric arguments. Geometric reasoning also permeated the statistical work of Fisher, and undoubtedly Hotelling was influenced by Fisher in this as in other respects (see Levene 1974). But where Fisher relied on geometrical intuition, Hotelling used geometry as a mathematical discipline in a rigorous form.

Hotelling never lost touch with applications in his statistical work. Most of his theoretical contributions were motivated by problems that arose in practice (see Levene 1974). Not content with giving analytical solutions, he proceeded to methods of computation, followed by illustrations using real data. Apart from statistical theory, he published papers dealing with substantive problems such as birth-rate fluctuations, duration of pregnancy, and measurements of pelves of American Indians in the Southwest.

Hotelling's first important contribution to statistics (1931*b*) generalized to the multivariate case Student's test for the mean of a univariate normal distribution. [*See* MULTIVARIATE ANALYSIS.] In deriving what is now known as Hotelling's generalized T^2 test, he was guided by ideas of invariance, which were later recognized as having wide applicability in statistics. To get at the problems of factor analysis, as they arise, for example, in educational testing, Hotelling proposed the method of principal components (1933). [*See* FACTOR ANALYSIS AND PRINCIPAL COMPONENTS.] Suggested by considerations of n-dimensional geometry, the principal components are linear functions of multivariate observations that have greatest variability in a stepwise sense. A similar mathematical idea underlies Hotelling's theory of canonical correlations (1936*b*) for analysis of the interrelations between *two sets* of variates.

Among Hotelling's other contributions to statistics, the following are the most historically important. His paper on differential equations subject to error (1927) was one of the first publications dealing with statistical problems of stochastic processes. His paper (written jointly with Holbrook Working) on the interpretation of trends (Working & Hotelling 1929) contains one of the first examples of a confidence region and the idea of multiple comparisons. [*See* ESTIMATION, *article on* CONFIDENCE INTERVALS AND REGIONS; LINEAR HYPOTHESES, *article on* MULTIPLE CONPARISONS.] His attempt at a rigorous proof of the consistency and asymptotic normality of Fisher's maximum likelihood estimates (1930), although not entirely successful, was one of the first steps forward in the history of this thorny problem, which was not finally solved until much later. His derivation (done with Margaret Richards Pabst) of the large-sample distribution of Spearman's rank correlation coefficient (Hotelling & Pabst 1936) initiated a sequence of increasingly more general results by other authors on the asymptotic normality of linear rank statistics. His work on the experimental determination of the maximum of a function (1941) was an early attack on a problem to which, starting a decade later, much research activity has been devoted. (For a more detailed survey of Hotelling's work in statistics, see Anderson 1960.)

Hotelling was one of the leaders in the revival of mathematical economics in the late 1920s. As early as his first economic paper (1925*b*), dealing with depreciation, he stressed the importance of maximizing principles to economics. His ideas on location theory ("Stability in Competition" 1929) exerted a great influence on later writers. His article on the economics of exhaustible resources (1931*c*) has also attracted increasing attention in more recent years. His detailed analysis of interrelated demand and supply functions of profit maximizers (1932) formed the basis for much subsequent work. His paper on welfare economics (1938), with its rigorous proof of the marginal cost pricing theorem, was his last major contribution to mathematical economics, and possibly the most important one. (For an analysis of Hotelling's work in mathematical economics, see Samuelson 1960.)

Hotelling's remarkable gift of identifying genuine talent and of attracting it to the sphere of his interests played an important part in the raising of standards in statistical research work and in the transformation of mathematical statistics into a respected academic discipline. He was an indefatigable and vigorous advocate of competence in the teaching of statistics (see, for example, Hotelling 1940). This activity had a great impact on the academic community and contributed to the establishment of some of the first departments of statistics in American universities. (On the educational and organizational activity of Hotelling, see Neyman 1960.)

Hotelling's students have paid warm tribute to his excellence as a teacher (see Levene 1974; Madow 1960*b*). Their comments on the lucidity and polished style of his lectures apply equally to his publications.

Hotelling was a member of the National Academy of Sciences and of the Accademia Nazionale dei Lincei, an honorary fellow of the Royal Statistical Society, a distinguished fellow of the American Economic Association, and a fellow of the Econometric Society, the Institute of Mathematical Statistics, and the Royal Economic Society. He served as president of the Econometric Society in 1936–1937 and of the Institute of Mathematical Statistics in 1941. He was awarded an honorary L.L.D. by the University of Chicago in 1955 and an honorary D.SC. by the University of Rochester in 1963.

WASSILY HOEFFDING

WORKS BY HOTELLING

For a bibliography of Hotelling's work until 1959, see pages 25–32 in Olkin et al. 1960. *A biography of Hotelling by Walter L. Smith, including an updated bibliography, is planned for publication in the* Annals of Statistics.

1925*a* Three-dimensional Manifolds of States of Motion. American Mathematical Society, *Transactions* 27: 329–344.

1925*b* A General Mathematical Theory of Depreciation. *Journal of the American Statistical Association* 20: 340–353.

1925*c* The Distribution of Correlation Ratios Calculated From Random Data. National Academy of Science, *Proceedings* 11:657–662.

1927 Differential Equations Subject to Error, and Population Estimates. *Journal of the American Statistical Association* 22:283–314.

1929 Stability in Competition. *Economic Journal* 39:41–57.

1929 WORKING, HOLBROOK; and HOTELLING, HAROLD. Applications of the Theory of Error to the Interpretation of Trends. *Journal of the American Statistical Association* 24 (March supp.): 73–85.

1930 The Consistency and Ultimate Distribution of Optimum Statistics. American Mathematical Society, *Transactions* 32:847–859.

1931*a* The Economics of Exhaustible Resources. *Journal of Political Economy* 39:137–175.

1931*b* The Generalization of Student's Ratio. *Annals of Mathematical Statistics* 2:360–378.

1932 Edgeworth's Taxation Paradox and the Nature of Demand and Supply Functions. *Journal of Political Economy* 40:577–616.

1933 Analysis of a Complex of Statistical Variables Into Principal Components. *Journal of Educational Psychology* 24:417–441, 498–520.

1935 Demand Functions With Limited Budgets. *Econometrica* 3:66–78.

1936 Relations Between Two Sets of Variates. *Biometrika* 28:321–377.

1936 HOTELLING, HAROLD; and PABST, MARGARET RICHARDS Rank Correlation and Tests of Significance Involving No Assumption of Normality. *Annals of Mathematical Statistics* 7:29–43.

1938 The General Welfare in Relation to Problems of Taxation and of Railway and Utility Rates. *Econometrica* 6:242–269. → Presidential address to the Econometric Society at the meeting in Atlantic City, N.J., Dec. 28, 1937.

1940 The Teaching of Statistics. *Annals of Mathematical Statistics* 11:457–470.

1941 Experimental Determination of the Maximum of a Function. *Annals of Mathematical Statistics* 12:20–45.

SUPPLEMENTARY BIBLIOGRAPHY

ANDERSON, T. W. 1960 Harold Hotelling's Research in Statistics. *American Statistician* 14, no. 3:17–21.

LEVENE, HOWARD 1974 In Memoriam: Harold Hotelling, 1895–1973. *American Statistician* 28, no. 2:71–73.

MADOW, WILLIAM G. 1960*a* Harold Hotelling. Pages 3–5 in Ingram Olkin et al. (editors), *Contributions to Probability and Statistics: Essays in Honor of Harold Hotelling*. Stanford Univ. Press.

MADOW, WILLIAM G. 1960*b* Harold Hotelling as a Teacher. *American Statistician* 14, no. 3:15–17.

NEYMAN, JERZY 1960 Harold Hotelling: A Leader in Mathematical Statistics. Pages 6–10 in Ingram Olkin et al. (editors), *Contributions to Probability and Statistics: Essays in Honor of Harold Hotelling*. Stanford Univ. Press.

OLKIN, INGRAM et al. (editors) 1960 *Contributions to Probability and Statistics: Essays in Honor of Harold Hotelling*. Stanford Univ. Press.

SAMUELSON, PAUL A. 1960 Harold Hotelling as Mathematical Economist. *American Statistician* 14, no. 3: 21–25.

HYPOTHESIS TESTING

The formulation of hypotheses and their testing through observation are essential steps in the scientific process. A detailed discussion of their role in the development of scientific theories is given by Popper (1935). On the basis of observational evidence, a hypothesis is either accepted for the time being (until further evidence suggests modification) or rejected as untenable. (In the latter case, it is frequently desirable to indicate also the direction and size of the departure from the hypothesis.)

It is sometimes possible to obtain unequivocal evidence regarding the validity of a hypothesis. More typically, the observations are subject to chance variations, such as measurement or sampling errors, and the same observations could have occurred whether the hypothesis is true or not, although they are more likely in one case than in the other. It then becomes necessary to assess the strength of the evidence and, in particular, to decide whether the deviations of the observations from what ideally would be expected under the hypothesis are too large to be attributed to chance. This article deals with methods for making such decisions: the *testing of statistical hypotheses*.

Probability models. A quantitative evaluation of the observational material is possible only on the basis of quantitative assumptions regarding the errors and other uncertainties to which the observations are subject. Such assumptions are conveniently formulated in terms of a probability model for the observations. In such a model, the observations appear as the values of random variables, and the hypothesis becomes a statement concerning the distribution of these variables [*see* PROBABILITY, *article on* FORMAL PROBABILITY].

The following are examples of some simple basic classes of probability models. Some of the most important applications of these models are to samples drawn at random from large populations and possibly subject to measurement errors.

Example 1—binomial model. If X is the number of successes in n independent dichotomous trials with constant probability p of "success," then X has a binomial distribution. This model is applicable to large (nominally infinite) populations whose members are of two types (voters favoring one of two candidates, inmates of mental institutions who are or are not released within one year) of which one is conventionally called "success" and the other "failure." The trials are the drawings of the n members of the population to be included in the sample. This model is realistic only if the population is large enough so that the n drawings are essentially independent.

Example 2—binomial two-sample model. To compare two proportions referring to two different (large) populations (voters favoring candidate A in two different districts, mental patients in two different institutions), a sample is drawn from each population. If the sample sizes are m and n, the observed proportions in the samples are X/m and Y/n, and the proportions in the populations are p_1 and p_2, then the model may assume that X and Y have independent binomial distributions. The same model may also be applicable when two samples are drawn from the same large population and subjected to different treatments.

Example 3—multinomial model. If a sample of size n is drawn from a (large) population whose members are classified into k types and the number of members in the sample belonging to each type is $X_1, \cdots, X_k$, respectively, then an appropriate model may assign to $(X_1, \cdots, X_k)$ a multinomial distribution.

Example 4—normal model. If $Z_1, \cdots, Z_n$ are measurements of the same characteristic taken on the n members of a sample (for example, test scores on a psychological test for n subjects or skull width for n skulls), an appropriate model may assume that $Z_1, \cdots, Z_n$ are independently and normally distributed with common mean μ and variance σ^2.

Example 5—normal two-sample model. To study the effect of a treatment (for example, the effect of training or of a drug on a test score) two independent samples may be obtained, of which the first serves as control (is not treated) and the second receives the treatment. If the measurements of the untreated subjects are $X_1, \cdots, X_m$ and those of the treated subjects are $Y_1, \cdots, Y_n$, it may be reasonable to assume that $X_1, \cdots, X_m; Y_1, \cdots, Y_n$ are all independently normally distributed—the X's with mean μ_X and variance σ_X^2, the Y's with mean μ_Y and variance σ_Y^2. Frequently it may be realistic to make the additional assumption that $\sigma_Y^2 = \sigma_X^2$, that the variance of the measurements is not affected by the treatment.

Example 6—nonparametric one-sample model. If the normality assumption in example 4 cannot be justified, it may instead be assumed only that $Z_1, \cdots, Z_n$ are independently distributed, each according to the same continuous distribution, F, about which no other assumption is made.

Example 7—nonparametric two-sample model. If the normality assumption in example 5 cannot be justified, it may instead be assumed only that $X_1, \cdots, X_m; Y_1, \cdots, Y_n$ are independently distributed, the X's according to a continuous distribution F, the Y's according to G. It may be realistic to suppose that the treatment has no effect on the shape or spread of the distribution but only on its location.

In a testing problem the model is never completely specified, for if there were no unknown element in the model, it would be known whether the hypothesis is true or false. One is thus dealing not with a single model but rather with a class of models, say Ω. For example, in a problem for which the models of example 1 are appropriate, Ω may consist of all binomial models corresponding to n trials and with p having any value between 0 and 1. If the model is specified except for certain parameters (the probabilities $p_1, \cdots, p_k$ in example 3, the mean, μ, and the variance, σ^2, in example 4), the class Ω is called *parametric;* otherwise, as in examples 6 and 7, it is *nonparametric* [*see* NONPARAMETRIC STATISTICS].

Statistical hypotheses. A hypothesis, when expressed in terms of a class of probability models, becomes a statement imposing additional restrictions on the class of models or on the distributions specified by the models.

Example 8. The hypothesis that the probability p_2 of a cure with a new treatment is no higher than the probability p_1 of a cure with the standard treatment, in the model of example 2, states that the parameters p_1, p_2 satisfy H: $p_2 \leqslant p_1$.

Example 9. Consider the hypothesis that the rate at which a rat learns to run a maze is unaffected by its having previously learned to run a different maze. If $X_1, \cdots, X_m$ denote the learning times required by m control rats who have not previously run a maze and $Y_1, \cdots, Y_n$ denote the learning times of n rats with previous experience on another maze, and if the model of example 7 is assumed, then the hypothesis of no effect states that the distributions F and G satisfy $H: G = F$. (Since, as in the present example, hypotheses frequently state the absence of an effect, a hypothesis under test is sometimes referred to as the *null hypothesis*.)

Hypotheses about a single parameter. Hypotheses in parametric classes of models frequently concern only a single parameter, such as μ in example 4 or $\mu_Y - \mu_X$ in example 5, the remaining parameters being "nuisance parameters." The most common hypotheses concerning a single parameter θ either (a) completely specify the value of the parameter, for example, state that $p = \frac{1}{2}$ in example 1, that $\mu = 0$ in example 4, or that $\mu_Y - \mu_X = 0$ in example 5—in general, such a hypothesis states that $\theta = \theta_0$ where θ_0 is the specified value; or (b) state that the parameter does not exceed (or does not fall short of) a specified value, for example, the hypothesis $p \leqslant \frac{1}{2}$ in example 1 or $\mu_X - \mu_Y \leqslant 0$ in example 5—the general form of such a hypothesis is $H_1: \theta \leqslant \theta_0$ (or $H_2: \theta \geqslant \theta_0$).

Two other important, although not quite so common, hypotheses state (c) that the parameter θ does not differ from a specified value θ_0 more than a given amount Δ: $|\theta - \theta_0| \leqslant \Delta$ or, equivalently, that θ lies in some specified interval $a \leqslant \theta \leqslant b$; or ($d$) that the parameter θ lies outside some specified interval.

Hypotheses about several parameters. In a parametric model involving several parameters, the hypothesis may of course concern more than one parameter. Thus, in example 3, one may wish to test the hypothesis that all the probabilities $p_1, \cdots, p_k$ have specified values. In example 5, the hypothesis might state that the point (μ_X, μ_Y) lies in a rectangle $H: a_1 \leqslant \mu_X \leqslant a_2, b_1 \leqslant \mu_Y \leqslant b_2$, or that it lies in a circle $H: (\mu_X - \mu_X^0)^2 + (\mu_Y - \mu_Y^0)^2 \leqslant c$, etc.

Hypotheses in nonparametric models. The variety of hypotheses that may arise in nonparametric models is illustrated by the following hypotheses, which have often been considered in connection with examples 6 and 7. In example 6, (1) F is the normal distribution with zero mean and unit variance; (2) F is a normal distribution (mean and variance unspecified); and (3) F is symmetric about the origin. In example 7, (1) $G = F$; (2) $G(x) \leqslant F(x)$ for all x; and (3) for no x do $G(x)$ and $F(x)$ differ by more than a specified value Δ.

Simple and composite hypotheses. A hypothesis, by imposing restrictions on the original class Ω of models, defines the subclass Ω_H of those models of Ω that satisfy the restrictions. If the hypothesis H completely specifies the model, so that Ω_H contains only a single model, then H is called *simple;* otherwise it is *composite*. Examples of simple hypotheses are the hypothesis $p = \frac{1}{2}$ in example 1 and the hypothesis that F is the normal distribution with zero mean and unit variance in example 6. Examples of composite hypotheses are the hypothesis $p_2 \leqslant p_1$ in example 2, the hypothesis $\mu = 0$ in example 4 when σ^2 is unknown, and the hypothesis that F is a normal distribution (mean and variance unspecified) in example 6.

Tests of hypotheses. A test of a hypothesis H is a rule that specifies for each possible set of values of the observations whether to accept or reject H, should these particular values be observed. It is therefore a division of all possible sets of values (the so-called sample space) into two groups: those for which the (null) hypothesis will be accepted (the *acceptance region*) and those for which it will be rejected (the *rejection region* or *critical region*).

Tests are typically defined in terms of a test statistic T, extreme values of which are highly unlikely to occur if H is true but are not surprising if H is false. To be specific, suppose that large values of T (and no others) are surprising if H is true but are not surprising if it is false. It is then natural to reject H when T is sufficiently large, say when

$$(1) \qquad T \geqslant c,$$

where c is a suitable constant, called the *critical value*.

The above argument shows that the choice of an appropriate test does not depend only on the hypothesis. The choice also depends on the ways in which the hypothesis can be false, that is, on the models of Ω not satisfying H (not belonging to Ω_H); these are called the *alternatives* (or *alternative hypotheses*) to H. Thus, in example 8, the alternatives consist of the models of example 2 satisfying $p_2 > p_1$; in example 9, they consist of the models of example 7 satisfying $G \neq F$.

The following two examples illustrate how the choice of the values of T for which H is rejected and the choice of T itself depend on the class of alternatives.

Example 10. Consider in example 1 the hypothesis $H: p = \frac{1}{2}$ and the three different sets of alternatives: p is less than $\frac{1}{2}$, p is greater than $\frac{1}{2}$, or p is different from (either less or greater than) $\frac{1}{2}$. Since one expects the proportion X/n of successes to be close to p, it is natural to reject H against the alternative $p < \frac{1}{2}$ if X/n is too small. (Very small values of X/n would be surprising under H but not under the alternatives.) Similarly, one would reject H against the alternative $p > \frac{1}{2}$ if X/n is too large. Finally, H would be rejected against the alternative $p \neq \frac{1}{2}$ if X/n is either too large or too small, for example, if $|(X/n) - \frac{1}{2}| > c$.

Alternatives of the first two types of this example and the associated tests are called *one-sided;* those corresponding to the third type are called *two-sided.*

Example 11. In example 7, consider the hypothesis $H: G = F$ that the Y's and X's have the same distribution against the alternatives that the Y's tend to be larger than the X's. A standard test for this problem is based on the Wilcoxon statistic, W, which counts the number among the mn pairs (X_i, Y_j) for which Y_j exceeds X_i. The hypothesis is rejected if W is too large [*see* NONPARAMETRIC STATISTICS].

Suppose instead that the alternatives to H state that the Y's are more spread out than the X's, or only that G and F are unequal without specifying how they differ. Then W is no longer an appropriate test statistic, since very large (or small) values of W are not necessarily more likely to occur under such alternatives than under the hypothesis.

Significance. To specify the test (1) completely, it is still necessary to select a critical value. This selection is customarily made on the basis of the following consideration. The values $T \geqslant c$, for which the hypothesis will be rejected, could occur even if the hypothesis H were true; they would then, however, be very unlikely and hence very surprising. A measure of how surprising such values are under H is the probability of observing them when H is true. This probability, $P_H(T \geqslant c)$, is called the *significance level* (or *size*) of the test. The traditional specification of a critical value is in terms of significance level. A value α (typically a small value such as .01 or .05) of this level is prescribed, and the critical value c is determined by the equation

$$P_H(T \geqslant c) = \alpha. \tag{2}$$

Values of T that are greater than or equal to c, and for which the hypothesis is therefore rejected, are said to be (statistically) *significant* at level α. This expresses the fact that although such extreme values could have occurred under H, this event is too unlikely (its probability being only α) to be reasonably explained by random fluctuations under H.

Tests and hypotheses suggested by data. In stating that the test determined by (1) and (2) rejects the hypothesis H with probability α when H is true, it is assumed that H and the rejection region (1) were determined before the observations were taken. If, instead, either the hypothesis or the test was suggested by the data, the actual significance level of the test will be greater than α, since then other sets of observations would also have led to rejection. In such cases, the prescribed significance levels can be obtained by carrying out the test on a fresh set of data. There also exist certain multiple-decision procedures that permit the testing, at a prescribed level, of hypotheses suggested by the data [*see* LINEAR HYPOTHESES, *article on* MULTIPLE COMPARISONS].

Determination of critical value. The actual determination of c from equation (2) for a given value of α is simple if there exists a table of the distribution of T under H. In cases where a complete table is not available, selected percentage points of the distribution, that is, the values of c corresponding to selected values of α, may have been published. If, instead, c has to be computed, it is frequently convenient to proceed as follows.

Let t be the observed value of the test statistic T. Then the probability $\hat{\alpha}$ of obtaining a value at least as extreme as that observed is called the *significance probability* (also *P-value, sample significance level,* and *descriptive level of significance*) of the observed value t and is given by

$$P_H(T \geqslant t) = \hat{\alpha}. \tag{3}$$

For the observed value t, the hypothesis is rejected if $t \geqslant c$ (and hence if $\hat{\alpha} \leqslant \alpha$) and is otherwise accepted. By computing $\hat{\alpha}$, one can therefore tell whether H should be rejected or accepted at any given significance level α from the rule

$$\text{reject } H \text{ if } \hat{\alpha} \leqslant \alpha; \text{ accept } H \text{ if } \hat{\alpha} > \alpha. \tag{4}$$

This rule, which is equivalent to the test defined by (1) and (2), requires only the computation of the probability (3); this is sometimes more convenient than determining c from (2).

When publishing the result of a statistical test, it is good practice to state not only whether the hypothesis was accepted or rejected at the chosen significance level (particularly since the choice of level is typically rather arbitrary) but also to publish the significance probability. This enables others to perform the test at the level of their choice by applying (4). It also provides a basis for combining the results of the test with those of other independent tests that may be performed at other times. (Various methods for combining a number of independent significance probabilities are discussed by Birnbaum 1954.) If no tables are available for the distribution of T but the critical values c of (2) are tabled for a number of different levels α, it is desirable at least to give the largest tabled value α at which the observations are nonsignificant and the smallest level at which they are significant, for example, "significant at 5 per cent, nonsignificant at 1 per cent." Actually, whenever possible, some of the basic data should be published so as to permit others to carry the statistical analysis further (for example, to estimate the size of an effect, to check the adequacy of the model, etc.).

In addition to the above uses, the significance probability—by measuring the degree of surprise at getting a value of T as extreme as or more extreme than the observed value t—gives some indication of the strength of the evidence against H. The smaller $\hat{\alpha}$ is, the more surprising it is to get this extreme a value under H and, therefore, the stronger the evidence against H.

The use of equation (2) for determining c from α involves two possible difficulties:

(*a*) If H is composite, the left-hand side of (2) may have different values for different distributions of Ω_H. In this case, equation (2) is replaced by

$$(5) \qquad \max_H P_H(T \geqslant c) = \alpha;$$

that is, the significance level or size is defined as the maximum probability of rejection under H. As an illustration, let T be Student's t-statistic for testing H: $\mu_Y \leqslant \mu_X$ in example 5 [*see* LINEAR HYPOTHESES, *article on* ANALYSIS OF VARIANCE]. Here the maximum probability of rejection under H occurs when $\mu_Y = \mu_X$, so that c is determined by the condition $P_{\mu_Y=\mu_X}(T \geqslant c) = \alpha$. This example illustrates the fact that $P_H(T \geqslant c)$ typically takes on its maximum on the boundary between the hypothesis and the alternatives.

(*b*) If the distribution of T is discrete, there may not exist a value c for which (2) or (5) holds. In practice, it is then usual to replace the originally intended significance level by the closest smaller (or larger) value that is attainable. In theoretical comparisons of tests, it is sometimes preferable instead to get the exact value α through randomization—namely, to reject H if $T > c$, to accept H if $T < c$, and if $T = c$ to reject or accept with probability ρ and $1 - \rho$ respectively, where ρ is determined by the equation $P_H(T > c) + \rho P_H(T = c) = \alpha$.

Power and choice of level. Suppose that a drug is being tested for possible beneficial effect on schizophrenic patients, with the hypothesis H stating that the drug has no such effect. Then a small significance level, by controlling the probability of falsely rejecting H when it is true, gives good protection against the possibility of falsely concluding that the drug is beneficial when in fact it is not. The test may, however, be quite unsatisfactory in its ability to detect a beneficial effect when one exists. This ability is measured by the probability of rejecting H when it is false, that is, by the probability P_A(rejecting H), where A indicates an alternative to H (in the example, an average effect of a given size). This probability, which for tests of the form (1) is equal to $P_A(T \geqslant c)$, is called the *power* of the test against the alternative A. The probability of rejecting H and the complementary probability of accepting H, as functions of the model or of the parameters specifying the model, are called respectively the *power function* and the *operating characteristic* of the test. Either of these functions describes the performance of the test against the totality of models of Ω.

Unfortunately, the requirements of high power and small level are in conflict, since the smaller the significance level, the larger is c and the smaller, therefore, is the power of the test. When choosing a significance level (and hence c), it is necessary to take into account the power of the resulting test against the alternatives of interest. If this power is too low to be satisfactory, it may be necessary to permit a somewhat larger significance level in order to provide for an increase in power. To increase power without increasing significance level, it is necessary to find a better test statistic or to improve the basic structure of the procedure, for example, by increasing sample size.

The problem of achieving a balance between significance level and power may usefully be considered from a slightly different point of view. A test, by deciding to reject or to accept H, may come to an erroneous decision in two different ways: by rejecting when H is true (error of the first kind) or by accepting when H is false (error of the second kind). The probabilities of these two kinds of errors are

$$(6) \qquad P(\text{error of first kind}) = P_H(\text{false rejection}) = \alpha$$

and

$$(7) \qquad P(\text{error of second kind}) = P_A(\text{false acceptance}) = 1 - \text{power}.$$

It is desirable to keep both the first probability and the second probability low, and these two requirements are usually in conflict. Any rule for balancing them must involve, at least implicitly, a weighing of the seriousness of the two kinds of error.

Choice of test. The constant c determines the size of the test (1) not only in the technical sense of (2) or (5) but also in the ordinary sense of the word, since the larger c is, the smaller is the rejection region. In the same sense, the test statistic T determines the shape of the test, that is, of the rejection region. The problem of selecting T is one of the main concerns of the theory of hypothesis testing.

A basis for making this selection is provided by the fact that of two tests of the same size, the one that has the higher power is typically more desirable, for with the same control of the probability of an error of the first kind, it gives better protection against errors of the second kind. The most satisfactory level α test against a particular alternative A is therefore the test that, subject to (5), maximizes the power against A: the *most powerful* level α test against A.

The fundamental result underlying all derivations of optimum tests is the Neyman–Pearson lemma, which states that for testing a simple hypothesis against a particular alternative A, the statistic T of the most powerful test is given for each possible set x of the observations by

$$T = \frac{P_A(x)}{P_H(x)} \tag{8}$$

(or by any monotone function of T), where P_A and P_H denote the probabilities (or probability densities) of x under A and H respectively.

In most problems there are many alternatives to H. For example, if the hypothesis specifies that a treatment has no effect and the alternatives specify that it has a beneficial effect, a different alternative will correspond to each possible size of this effect. If it happens that the same test is simultaneously most powerful against all possible alternatives, this test is said to be *uniformly most powerful* (UMP). However, except for one-tailed tests (and for tests of the hypothesis that the parameter lies outside a specified interval) in the simplest one-parameter models, a UMP test typically does not exist; instead, different tests are most powerful against different alternatives.

If a UMP test does not exist, tests may be sought with somewhat weaker optimum properties. One may try, for example, to find a test that is UMP among all tests possessing certain desirable symmetry properties or among all *unbiased* tests, a test being unbiased if its power never falls below the level of significance. Many standard tests have one or the other of these properties.

A general method of test construction that frequently leads to satisfactory results is the *likelihood ratio* method, which in analogy to (8) defines T by

$$T = \frac{\max_{\Omega} P(x)}{\max_{\Omega_H} P(x)}. \tag{9}$$

Here the denominator is the maximum probability of x when H is true, while the numerator is the over-all maximum of this probability. If the numerator is sufficiently larger than the denominator, this indicates that x has a much higher probability under one of the alternatives than under H, and it then seems reasonable to reject H when x is observed. The distribution of the test statistic (9) under H has a simple approximation when the sample sizes are large, and in this case the likelihood ratio test also has certain approximate optimum properties (Kendall & Stuart 1961, vol. 2, p. 230; Lehmann 1959, p. 310).

The specification problem. For many standard testing problems, tests with various optimum properties have been worked out and can be found in the textbooks. As a result, the principal difficulty, in practice, is typically not the choice of α or T but the problem of *specification*, that is, of selecting a class Ω of models that will adequately represent the process generating the observations. Suppose, for example, that one is contemplating the model of example 5. Then the following questions, among others, must be considered: (a) May the experimental subjects reasonably be viewed as randomly chosen from the population of interest? (b) Are the populations large enough so that the X's and Y's may be assumed to be independent? (c) Does the normal shape provide an adequate approximation for the distribution of the observations or of some function of the observations? (d) Is it realistic to suppose that $\sigma_X = \sigma_Y$? The answers to such questions, and hence the choice of model, require considerable experience both with statistics and with the subject matter in which the problem arises. (Protection against some of the possible failures of the model of example 5 in particular may be obtained through the method of randomization discussed below.)

Problems of robustness. The difficulty of the specification problem naturally raises the question of how strongly the performance of a test depends on the particular class of models from which it is derived. There are two aspects to this question, namely robustness (insensitivity to departures from assumption) of the size of the test and robustness of its power [*see* ERRORS, *article on* EFFECTS OF ERRORS IN STATISTICAL ASSUMPTIONS].

'Two typical results concerning robustness of size of a test are the following: (*a*) In example 5, assuming $\sigma_Y = \sigma_X$, the size of Student's *t*-test for testing $H: \mu_Y = \mu_X$ is fairly robust against non-normality except for very small sample sizes. (*b*) In example 5, the *F*-test for testing $\sigma_Y = \sigma_X$ is very nonrobust against nonnormality.

The second aspect, the influence of the model on power, may again be illustrated by the case of Student's *t*-test for testing $H: \mu_Y = \mu_X$ in example 5. It unfortunately turns out—and this result is typical—that the power of the test is not robust against distributions with heavy tails (distributions that assign relatively high probability to very large positive and negative values), for example, if normality is disturbed by the presence of gross errors. This difficulty can be avoided by the use of nonparametric tests such as the Wilcoxon test defined in example 11, which give only limited weight to extreme observations. The Wilcoxon test, at the expense of a very slight efficiency loss (about 5 per cent) in the case of normality, gives much better protection against gross errors. In addition, its size is completely independent of the assumption of normality, so that it is in fact a test of the hypothesis $G = F$ in the nonparametric model of example 7.

Design. Analysis of data, through performance of an appropriate test, is not the only aspect of a testing problem to which statistical considerations are relevant. At least of equal importance is the question of what observations should be taken or what type of experiment should be performed. The following are illustrations of some of the statistical problems relating to the proper *design* of an investigation [*see* EXPERIMENTAL DESIGN].

Sample size. Once it has been decided in principle what kind of observations to take, for example, what type of sampling to use, it is necessary to determine the number of observations required. This can be done by fixing, in addition to the significance level α, the minimum power β that one wishes to achieve against the alternatives of interest. When the sample size is not fixed in advance, a compromise in the values of α and β is no longer necessary, since both can now be controlled. Instead, the problem may arise of balancing the desired error control against the cost of sampling. If the sample size n required to achieve a desired α and β is too large, a compromise between n and the values of α and β becomes necessary.

Sequential designs. Instead of using a fixed sample size, it may be more advantageous to take the observations one at a time, or in batches, and to let the decision of when to stop depend on the observations. (The stopping rule, which states for each possible sequence of observations when to stop, must of course be specified before any observations are taken.) With such a *sequential* design, one would stop early when the particular values observed happen to give a strong indication of the correct decision and continue longer when this is not the case. In this way, it is usually possible to achieve the same control of significance level and power with a (random) number of observations, which on the average is smaller than that required by the corresponding test of fixed sample size. When using a sequential test, account must be taken of the stopping rule to avoid distortion of the significance level [*see* SEQUENTIAL ANALYSIS].

Grouping for homogeneity. The power of tests of the hypothesis $\mu_Y = \mu_X$ in example 5 depends not only on the size of the difference $\mu_Y - \mu_X$ but also on the inherent variability of the subjects, as measured by σ_X and σ_Y. Frequently, the power can be increased by subdividing the subjects into more homogeneous groups and restricting the comparison between treatment and control to subjects within the same group [*see* EXPERIMENTAL DESIGN, *article on* THE DESIGN OF EXPERIMENTS].

An illustration of such a design is the method of *paired comparisons*, where each group consists of only two subjects (for example, twins) chosen for their likeness, one of which receives the treatment and the other serves as control. If a sample of n such pairs is drawn and the difference of the measurements in the treated and control subjects of the ith pair is $Z_i = Y_i - X_i$, then $Z_1, \cdots, Z_n$ are distributed as in example 4. The appropriate test is Student's one-sample *t*-test, which is now, however, based on fewer degrees of freedom than for the design of example 5. To determine in any specific case whether the design of example 5, the paired-comparison design, or some intermediate design with group size larger than two is best, it is necessary to balance the reduction of variability due to grouping against the loss in degrees of freedom in the resulting *t*-test.

Randomization. When testing the effect of a treatment by comparing the results on n treated subjects with those on m controls, it may not be possible to obtain the subjects as a random sample from the population of interest. A probabilistic basis for inference and, at the same time, protection against various biases can be achieved by assigning the subjects to treatment and control not on a systematic or haphazard basis but *at random*, that is, in such a way that all possible such assignments are equally likely. Randomization is possible also in a paired-comparisons situation where within each pair one of the two possible assignments of treatment and control to the two subjects is chosen, for example, by tossing a coin. In the model resulting from such randomization, it is possible to carry out a test of the hypothesis of no treatment effect without any further assumptions.

Relation to other statistical procedures. The formulation of a problem as one of hypothesis testing may present a serious oversimplification if, in case of rejection, it is important to know which alternative or type of alternative is indicated. The following two situations provide typical and important examples.

(*a*) Suppose a two-sided test of the hypothesis $H: \theta = \theta_0$ rejects H when a test statistic T is either too small or too large. In case of rejection, it is usually not enough to conclude that θ differs from θ_0; one would wish to know in addition whether θ is less than or greater than θ_0. Here a three-decision procedure of the following form is called for:

conclude that $\theta < \theta_0$	if $T \leqslant c_1$
conclude that $\theta > \theta_0$	if $T \geqslant c_2$,
accept H	if $c_1 < T < c_2$.

The constants c_1, c_2 can be determined by specifying the two error probabilities $\alpha_1 = P_{\theta_0}(T \leqslant c_1)$ and $\alpha_2 = P_{\theta_0}(T \geqslant c_2)$, whose sum is equal to the error probability, α, of the two-sided test that rejects for $T \leqslant c_1$ and $T \geqslant c_2$. How the total error probability, α, is divided between α_1 and α_2 would depend on the relative seriousness of the two kinds of error involved and on the relative importance of detecting when θ is in fact less than or greater than θ_0. If concern is about equal between values of $\theta > \theta_0$ and values of $\theta < \theta_0$, the procedure with $\alpha_1 = \alpha_2 = \frac{1}{2}\alpha$ may be reasonable. It is interesting to note that this three-decision procedure may be interpreted as the simultaneous application of two tests: namely $T \leqslant c_1$ as a test of the hypothesis $H_1: \theta \geqslant \theta_0$ at level α_1 and $T \geqslant c_2$ as a test of $H_2: \theta \leqslant \theta_0$ at level α_2.

(*b*) If $\theta_1, \cdots, \theta_c$ denote the (average) effects of c treatments, one may wish to test the hypothesis $H: \theta_1 = \cdots = \theta_c$ to see if there are any appreciable differences. In case of rejection, one might wish to determine which of the θ's is largest, or to single out those θ's that are substantially larger than the rest, or to obtain a complete ranking of the θ's, or to divide the θ's into roughly comparable groups, etc.

Under suitable normality assumptions, the last of these objectives can be achieved by applying a t-test to each of the hypotheses $H_{ij}: \theta_i = \theta_j$, or rather by applying to each the three-decision procedure based on t-tests of the type discussed in (*a*). Combining the conclusions ($\theta_i < \theta_j$, $\theta_i = \theta_j$, or $\theta_i > \theta_j$) obtained in this way leads to a grouping of the kind desired. In determining the significance level, say α', at which the individual t-test should be performed, one must of course relate it to the significance level α that one wishes to achieve for the original hypothesis H. (For further details and references regarding this procedure, see Lehmann 1959, p. 275; Mosteller & Bush 1954, p. 304.)

These two examples illustrate how a procedure involving several choices may sometimes be built up by the simultaneous consideration of a number of situations involving only two choices, that is, a number of testing problems. A similar approach also leads to the method of estimation by confidence sets [*see* ESTIMATION, *article on* CONFIDENCE INTERVALS AND REGIONS].

The simultaneous consideration of a number of different tests also arises in other contexts. Frequently, investigators wish to explore a number of different aspects of the same data and for this purpose carry out multiple tests. This raises serious difficulties, since the stated significance levels relate to a single test without relation to others. Essentially the same difficulties arise when testing hypotheses *suggested by the data*, since this may be viewed as testing a (possibly large) number of potential hypotheses but reporting only the most significant outcome. This, of course, again invalidates the stated significance level. [*Methods for dealing with such problems are discussed in* LINEAR HYPOTHESES, *article on* MULTIPLE COMPARISONS.]

History. Isolated examples of tests, as statements of (statistical) significance or nonsignificance of a set of observations, occurred throughout the eighteenth and nineteenth centuries. A systematic use of hypothesis testing, but without explicit mention of alternatives to the hypothesis being tested, began with the work of Karl Pearson (1900) and owes much of its further development to R. A. Fisher (1925; 1935). That the choice of an appropriate test must depend on the alternatives as well as on the hypothesis was first explicitly recognized by Neyman and Pearson (1928; 1933), who introduced the concept of power and made it the cornerstone of the theory of hypothesis testing described here. Two other approaches to the subject, based on concepts of probability other than that of long-run frequency, which has been implicitly assumed here, are the Bayesian approach and that of Jeffreys [1939; *see also* BAYESIAN INFERENCE].

E. L. LEHMANN

[*See also* ESTIMATION; LINEAR HYPOTHESES; SIGNIFICANCE, TESTS OF.]

BIBLIOGRAPHY

BIRNBAUM, ALLAN 1954 Combining Independent Tests of Significance. *Journal of the American Statistical Association* 49:559–574.

○FISHER, R. A. (1925) 1970 *Statistical Methods for Research Workers*. 14th ed., rev. & enl. New York: Hafner; Edinburgh: Oliver & Boyd.

○FISHER, R. A. (1935) 1971 *The Design of Experiments*. 9th ed. New York: Hafner; Edinburgh: Oliver & Boyd.

○HODGES, JOSEPH L. JR.; and LEHMANN, E. L. (1964) 1970 *Basic Concepts of Probability and Statistics*. 2d ed. San Francisco: Holden-Day. → Includes a more leisurely exposition of the basic concepts of hypothesis testing.

JEFFREYS, HAROLD (1939) 1961 *Theory of Probability*. 3d ed. Oxford: Clarendon.

○KENDALL, MAURICE G.; and STUART, ALAN (1961) 1973 *The Advanced Theory of Statistics*. Volume 2: *Inference and Relationship*. 3d ed. London: Griffin; New York: Hafner. → Early editions of Volume 2, first published in 1946, were written by Kendall alone. Stuart became a joint author on later, renumbered editions in the three-volume set.

LEHMANN, E. L. 1959 *Testing Statistical Hypotheses*. New York: Wiley.

MOSTELLER, FREDERICK; and BUSH, ROBERT R. (1954) 1959 Selected Quantitative Techniques. Volume 1, pages 289–334 in Gardner Lindzey (editor), *Handbook of Social Psychology*. Cambridge, Mass.: Addison-Wesley.

NEYMAN, JERZY; and PEARSON, E. S. 1928 On the Use and Interpretation of Certain Test Criteria for Purposes of Statistical Inference. *Biometrika* 20A:175–240, 263–294.

NEYMAN, JERZY; and PEARSON, E. S. 1933 On the Problem of the Most Efficient Tests of Statistical Hypotheses. Royal Society of London, *Philosophical Transactions* Series A 231:289–337.

PEARSON, KARL 1900 On the Criterion That a Given System of Deviations From the Probable in the Case of a Correlated System Can Be Reasonably Supposed to Have Arisen From Random Sampling. *Philosophical Magazine* Fifth Series 50:157–175.

POPPER, KARL R. (1935) 1959 *The Logic of Scientific Discovery*. New York: Basic Books; London: Hutchinson. → First published as *Logik der Forschung*. A paperback edition was published by Harper in 1965.

Postscript

[1]The discussion of robustness of power in the main article is misleading. In example 5, the power of the *t*-test is fairly robust against nonnormality (a not necessarily desirable property), but *relative* power or required sample size is highly nonrobust against heavy-tailed distributions. It is this latter property that is much better for the Wilcoxon test.

E. L. LEHMANN

I

IDENTIFIABILITY, STATISTICAL

See STATISTICAL IDENTIFIABILITY.

IMAGE ANALYSIS

See FACTOR ANALYSIS AND PRINCIPAL COMPONENTS.

INDEPENDENCE

See ERRORS, *article on* EFFECTS OF ERRORS IN STATISTICAL ASSUMPTIONS; PROBABILITY; STATISTICS, DESCRIPTIVE, *article on* ASSOCIATION.

INDEX NUMBERS

I. THEORETICAL ASPECTS	*Erik Ruist*
II. PRACTICAL APPLICATIONS	*Ethel D. Hoover*
III. SAMPLING	*Philip J. McCarthy*

I
THEORETICAL ASPECTS

An index number measures the magnitude of a variable relative to a specified value of the variable. For example, suppose that a certain type of apple sold at an average price of 10 cents last year but sells at an average price of 11 cents this year. If last year's price is chosen as a base and is arbitrarily set equal to 100 (to be thought of as a pure number or as 100 per cent), then the index number identified with this year's price would be 110 (that is, $[11/10] \times 100$), which indicates that this year's price is 10 per cent higher than last year's price. Rather than comparing the price at two different dates, we might wish to compare the current price of this type of apple in New York City with that in Chicago, in which case we would choose the price in one of those two cities as a base and express the price in the other city relative to the base. Thus, index numbers can be used to make comparisons over both time and space.

If index numbers were used only to compare such variables as the price of a *single* commodity at different dates or places, there would be little need for a special theory of index numbers. However, we might wish to compare, for example, the general price levels of commodities imported by the United States in two different years. The prices of some commodities will have risen, and the prices of others will have fallen. The problem that arises is how to combine the relative changes in the prices of the various commodities into a single number that can meaningfully be interpreted as a measure of the relative change in the general price level of imported commodities. This example illustrates perhaps the major problem dealt with in index number theory, and this article discusses primarily the various solutions that have been proposed.

Two approaches. The possibility of using an index number as an aggregate measure of the price change of several commodities seems to have been recognized in the eighteenth century, but deliberate theoretical discussions did not begin until the middle of the nineteenth century. Among the formulas suggested then were those of the German economists Étienne Laspeyres and Hermann Paasche, which are still used extensively. The choice of formula was made according to what was considered to be "fair." A major step forward in the development of criteria by which to judge the various formulas was the set of tests suggested by Irving Fisher (1922). From the 1920s on, however, greater care was taken to place the index in an economic context. Thinking mainly of cost-of-living comparisons, investigators defined the price index as the relative change in income necessary to maintain an unchanged standard of living.

The two lines of thought, the mainly statistical one and the economic one, still exist side by side. The economic approach starts with some economic considerations but does not arrive at a definite result. In contrast to this, the statistical approach starts with some subjectively chosen rules but arrives at formulas that may be used directly in practical work. Fortunately, the index formulas derived in these two ways are usually very similar. There is, however, still need of a more unified theory of index numbers.

The statistical theory

The pure statistical theory of index numbers is general in the sense that it could be applied to any index without regard to what the index represents. In order to avoid confusion, this discussion will refer mainly to price indexes and will revert to quantity indexes only when necessary for the development of the price index theory.

Consider the relations p_{i1}/p_{i0} $(i = 1, \cdots, n)$ between the prices of n commodities at two points of time, t_0 and t_1. These relations, called price ratios or price relatives, could be considered as elements having a certain distribution, the central measure of which is sought. Thus, it is natural to construct a weighted arithmetic, harmonic, or geometric average of the price ratios.

By choosing the weights in different ways it is possible to arrive at many types of indexes. Unweighted arithmetic averages were long used. However, by the first half of the nineteenth century, Arthur Young, Joseph Lowe, and G. Poulett Scrope, all of England, used weighted arithmetic averages of the price ratios. The weight w_i given to the price ratio p_{i1}/p_{i0} was

$$w_i = \frac{p_{i0}q_i}{\sum_{j=1}^{n} p_{j0}q_j},$$

where q_i is a quantity of commodity i showing its general importance in the list of commodities. If q_i is specified to be the quantity of commodity i traded during a period around t_0, so that $q_i = q_{i0}$, the resulting formula is the one suggested in 1864 by Laspeyres, namely,

$$P_{01}^{L} = \sum_{i=1}^{n} \frac{p_{i0}q_{i0}}{\sum_{j=1}^{n} p_{j0}q_{j0}} \cdot \frac{p_{i1}}{p_{i0}} = \frac{\sum_{i=1}^{n} p_{i1}q_{i0}}{\sum_{i=1}^{n} p_{i0}q_{i0}}.$$

The subscripts on P^L simply indicate that the index measures the price level at date t_1 relative to that at the base date t_0.

There is a complication here that does not seem to have attracted much attention. In practice, prices are observed at *points* of time, whereas quantities generally have to be taken as referring to *periods* of time. However, the denominator in Laspeyres's formula is usually interpreted as the actual value of transactions during the base *period* t_0. This implies either that prices have been constant within the period or that the p_i are to be interpreted as average prices. To avoid ambiguity it will be assumed in the following discussion that the periods are very short, so that the prices can be regarded as constant during each period.

A formula equivalent to Laspeyres's but using q_{i1} instead of q_{i0} in the weights naturally suggested itself and seemed equally justifiable. It was advocated in 1874 by Paasche.

Alfred Marshall suggested that instead of using quantities referring to one of the two points of time compared by the index, an average of the corresponding quantities should be used—that is,

$$P_{01}^{E} = \frac{\sum_{i=1}^{n} p_{i1}(q_{i0} + q_{i1})}{\sum_{i=1}^{n} p_{i0}(q_{i0} + q_{i1})}.$$

Because this index formula was strongly advocated by F. Y. Edgeworth, it is often called Edgeworth's formula. The use of a geometric average of the quantities associated with the two dates has also been suggested.

Fisher's tests. The various formulas gave results that were sometimes widely different, and it was evident that criteria for judging the quality of the formulas were needed. Fisher suggested a set of criteria that have remained the most extensive and widely used. Before describing Fisher's criteria, let us define P_{rs} as the index number given by formula P expressing the change in the general price level between dates r and s relative to the price level at date r; that is, date r is the base.

The *time reversal test* states that in comparing the prices at two dates, a formula should give the same result regardless of which of the two dates is chosen as the base. For example, if a formula indicates that the price level in 1965 was double that in 1964 when 1964 is taken as the base, then it should indicate that the price level in 1964 was one half that in 1965 when 1965 is taken as the base. Symbolically, the test requires that

$$P_{rs}P_{sr} = 1, \qquad \text{for all } r \text{ and } s.$$

The *circular test* requires that

$$P_{rs}P_{st} = P_{rt}, \qquad \text{for all } r, s, \text{ and } t;\ r \neq t.$$

For example, if a formula indicates that the price level doubled between 1963 and 1964 and then doubled again between 1964 and 1965, it should indicate that the price level in 1965 was four times that in 1963 when 1965 and 1963 are compared directly. Although this test has great intuitive appeal, Fisher argued that it is not an essential test and even suggested that a formula that satisfies it exactly should generally be rejected.

The *factor reversal test* presupposes that the weights used in the price index formula are functions of quantities. It states that if the price index P_{01} is multiplied by a corresponding quantity index Q_{01}, derived by interchanging p and q in the index formula, the result should equal the value ratio—that is,

$$P_{01}Q_{01} = \frac{\sum_{i=1}^{n} p_{i1}q_{i1}}{\sum_{i=1}^{n} p_{i0}q_{i0}}.$$

Fisher's tests have been found to be inconsistent with each other. However, Fisher classified all the formulas he tested according to which tests they fulfilled. He found only four formulas that deserved a "superlative" rating, all of which used quantities from both t_0 and t_1. They included Edgeworth's formula, noted above, and the corresponding formula with geometric instead of arithmetic means of quantities. Also "superlative" were the arithmetic and geometric means of Laspeyres's and Paasche's formulas. Fisher called the geometric mean of these formulas the "ideal" index.

Chain indexes. The discussion so far has dealt with a comparison of prices between only two points of time. In practice, however, indexes are calculated for many dates at regular intervals—for example, monthly or annually. Thus, the question that arises is how a series of index numbers should be calculated. For convenience, it is customary to use a fixed base, say t_0, and calculate successively $P_{01}, P_{02}, \cdots, P_{0k}$. This means that if Laspeyres's formula is used, a comparison of prices at t_{k-1} and t_k is in fact made using quantities that do not refer to either of these dates but refer instead to t_0.

Since very often the comparison of the prices of one date with those of the immediately preceding date is more important than the comparison with the base date, it may be useful to compute for every t_r an index $P_{r-1,r}$ and then define

$$P_{0k} = P_{01} \cdot P_{12} \cdot \cdots \cdot P_{k-1,k}.$$

This *chain index*, originally suggested by Marshall in 1887, satisfies the circular test for any $r < s < t$. In spite of Fisher's doubts, satisfaction of the circular test is a very attractive property, and chain indexes are often used. Any index formula could be applied for the index $P_{r-1,r}$ within the links.

The development of the chain index may be said to have originated in a desire to find an index with the properties expressed by Fisher's circular test. It is interesting to note that a somewhat similar result may be obtained by reasoning from the factor reversal test. Following Divisia (1925) and Törnqvist (1937), we start from the criterion

$$P_{0t}Q_{0t} = \frac{\sum_{i=1}^{n} p_{it}q_{it}}{\sum_{i=1}^{n} p_{i0}q_{i0}}.$$

Taking logarithms of both sides and differentiating with respect to t, we have

$$\frac{1}{P_{0t}} \cdot \frac{dP_{0t}}{dt} + \frac{1}{Q_{0t}} \cdot \frac{dQ_{0t}}{dt} = \frac{1}{\sum_{i=1}^{n} p_{it}q_{it}} \left(\sum_{i=1}^{n} q_{it}\frac{dp_{it}}{dt} + \sum_{i=1}^{n} p_{it}\frac{dq_{it}}{dt} \right).$$

To obtain symmetry between the price index and the quantity index, the terms may be equated pairwise—that is, the first term on the left-hand side of the equality sign may be set equal to the first term on the right-hand side of the equality sign, and similarly for the second terms.

Integrating the equations, we obtain for P_{01}

$$\text{(1)} \qquad \log P_{01} = \sum_{i=1}^{n} \int_{t_0}^{t_1} c_i(t)\, d\log p_{it},$$

where $c_i(t) = p_{it}q_{it} / \sum_{i=1}^{n} p_{it}q_{it}$. Thus, the value of P_{01} depends on the development of the $c_i(t)$ between t_0 and t_1, which are the proportions of the different commodities in the total budget. If an assumption is made about this development, the integral can be solved explicitly. Thus, if the $c_i(t)$ are assumed to be constant over time—that is, if all commodities are assumed to have a price elasticity of demand equal to one, we obtain the geometric index formula

$$P_{01} = \prod_{i=1}^{n} \left(\frac{p_{i1}}{p_{i0}} \right)^{c_i}.$$

If no assumptions are made about the c_i, the mean value theorem of the integral calculus may be applied to (1) to obtain

$$\text{(2)} \qquad \log P_{01} = \sum_{i=1}^{n} \bar{c}_i (\log p_{i1} - \log p_{i0}),$$

where $\bar{c}_i$ is a weighted mean of the $c_i(t)$ over the period from t_0 to t_1. If this period is not too long, $\bar{c}_i$ may be approximated by the share of commodity i in the total expenditure during the period. The definite integral in (1) can be split up into a sum of several integrals, each covering a period short enough to make this approximation satisfactory. The resulting series of price indexes corresponds to a chain index composed of weighted geometric means of price changes.

Best linear indexes. A further development of the idea that an index does not compare only two points of time has been suggested by Theil (1960). He argues that in the calculation of an index the situation during all observed points of time should influence the result symmetrically.

In terms of the notation used above, Theil starts from an arithmetic mean index with fixed weights —that is,

$$P_{0t} = \sum_{i=1}^{n} w_i \frac{p_{it}}{p_{i0}}.$$

If $\alpha_i p_{i0} / \sum_{i=1}^{n} \alpha_i p_{i0}$ is substituted for w_i, this index may be written

$$P_{0t} = \frac{P_t}{P_0} = \frac{\sum_{i=1}^{n} \alpha_i p_{it}}{\sum_{i=1}^{n} \alpha_i p_{i0}}.$$

Here P_t and P_0 may be called absolute indexes to contrast with the relative index P_{0t}.

Using the absolute indexes, each individual price could be represented as

$$P_{it} = a_i P_t + v_{it},$$

where v_{it} is a disturbance or error term. If all prices moved proportionately, the v_{it} could be made zero. In general, however, this is not the case. But if the v_{it} cannot be made zero, the parameters a_i and α_i can be determined so that the v_{it} are minimized in some sense. If the parameters are determined in this way, the resulting index is called a best linear index.

If a quantity index is defined in a similar way, symmetric best linear price and quantity indexes with some interesting optimal properties can be derived. By a suitable choice of minimization procedure, the indexes could be made to minimize the sum of squares of "cross-value discrepancies," that is, they could be made to minimize

$$\sum_{r,s} (P_r Q_s - \sum_{i=1}^{n} p_{ir} q_{is})^2,$$

where r and s take on all observed values of t. This is a kind of generalization of the factor reversal test. The factor reversal test could be said to specify that

$$(P_0 Q_0 - \sum_{i=1}^{n} p_{i0} q_{i0})^2 + (P_1 Q_1 - \sum_{i=1}^{n} p_{i1} q_{i1})^2 = 0.$$

It has been found that the best linear index tends to be biased in the sense that the differences $(P_r Q_r - \sum_{i=1}^{n} p_{ir} q_{ir})$ are systematically positive. To construct a best linear unbiased index, Kloek and de Wit (1961) introduced the condition

$$\sum_{r} (P_r Q_r - \sum_{i=1}^{n} p_{ir} q_{ir}) = 0.$$

Fulfillment of this condition means that the factor reversal test is satisfied "on the average." As in the case of the best linear index, the weights are obtained by finding the largest latent root of a certain matrix.

The economic theory

The economic theory of index numbers is most often discussed in terms of a consumer price index. The object is to measure the changes in the cost of living of a person or of a group of persons who have identical tastes for goods. A utility, U (not necessarily a cardinal number), is associated with every combination of quantities of goods that is under consideration. The person is assumed to be well adapted to the prevailing price situation, so that given his income he chooses the set of quantities that gives him the highest level of utility. For each point of time t (defining a set of prices) there exists for each level of utility U an expenditure $\mu(t,U)$ which is the lowest possible expenditure to attain U.

The constant utility index. At the point of time t_0 the person's choice of quantities and his expenditure $\mu(t_0,U_0)$ are observed. The equivalent observations are made at t_1. If it can be stated that for the person observed $U_0 = U_1$, the price index for this level of utility is

$$P_{01}(U_0) = \frac{\mu(t_1,U_0)}{\mu(t_0,U_0)}.$$

This index, sometimes referred to as the "indifference defined index" or "the constant utility index," was first discussed by Konüs ([1924] 1939). It is to be noted that the price index is associated with a certain value of U—in this case U_0. For other values of U the price index may be different.

It is very seldom known whether U_0 is greater than, equal to, or less than U_1; that is, the value of $\mu(t_1, U_0)$ is generally not known. Thus, to estimate $\mu(t_1, U_0)$ it is necessary to make certain assumptions or approximations. Alternatively, it is possible to find an upper and/or a lower limit to the value of the constant utility index.

Upper and lower limits. It can be shown that a Laspeyres index gives an upper bound to P_{01}. The Laspeyres index shows the change in expenditure necessary to buy the same quantities of all goods at t_1 as those actually bought at t_0. However, this corresponding expenditure at t_1 may under the new price structure be disposed of differently so as to give a level of utility higher than U_0. The utility U_0 would then be attainable with a lower expenditure, and the indifference defined index for U_0 would be lower than the Laspeyres index.

In a similar way it is possible to show that a Paasche index gives a lower limit to a constant utility index for the utility level U_1. Thus,

$$(3a) \qquad P_{01}^{L} \geqslant P_{01}(U_0)$$

and

$$(3b) \qquad P_{01}^{P} \leqslant P_{01}(U_1).$$

However, since without further assumptions nothing is known about the relation between $P_{01}(U_0)$ and $P_{01}(U_1)$, these rules do not, contrary to what has sometimes been believed, give upper and lower limits for the same index.

Several attempts have been made to arrive at simultaneous upper and lower limits. Thus, Staehle (1935) tried to find a utility level U' such that the set of quantities corresponding to $\mu(t_1, U')$ would cost as much at t_0 prices as $\mu(t_0, U_0)$. Since clearly $U' \leqslant U_0$ (the budget was available at t_0 but was not chosen), $\mu(t_1, U') \leqslant \mu(t_1, U_0)$ and

$$\frac{\mu(t_1, U')}{\mu(t_0, U_0)} \leqslant \frac{\mu(t_1, U_0)}{\mu(t_0, U_0)} = P_{01}(U_0).$$

Thus, this index constitutes a lower limit to $P_{01}(U_0)$, and since P_{01}^{L} gives an upper limit, the desired result is obtained. There remains, however, the problem of determining $\mu(t_1, U')$. Under certain circumstances it can be determined from family budget data.

Ulmer (1949) took a quite different approach, which can be described as follows. Let D_u be the difference between $P_{01}(U_0)$ and $P_{01}(U_1)$. It could be positive or negative. The difference between the Laspeyres and Paasche indexes can then be written

$$\begin{aligned} &P^L - P^P \\ &= [P^L - P(U_0)] + [P(U_1) - P^P] + [P(U_0) - P(U_1)] \\ &= \qquad D_L \qquad + \qquad D_P \qquad + \qquad D_u, \end{aligned}$$

where the subscript "01" on the P's has been dropped for convenience. Using this identity together with the inequalities (3a) and (3b), it can be shown that

$$P^L - D_L - D_P \leqslant P(U_0) \leqslant P^L$$

and that

$$P^P \leqslant P(U_1) \leqslant P^P + D_L + D_P.$$

Hence, both constant utility indexes are given upper and lower bounds.

Although $(D_L + D_P)$ is not directly observable, $(D_L + D_P + D_u) = (P^L - P^P)$ is, and Ulmer argued that it is usually reasonable to suppose that

$$\max_t (D_L + D_P) \leqslant \max_t (D_L + D_P + D_u)$$

and therefore that $\max_t (P^L - P^P)$ would be a conservative estimate of the difference between the upper and lower bounds on $P(U_0)$ or $P(U_1)$.

Point estimates. Several attempts have been made to arrive at a point estimate of a constant utility index by using family budget data. If such data were available for each period for which the index was to be calculated, criteria could be developed to find in each period a family or group of families with a given level of utility. By comparing their expenditures in different periods, an index could be calculated. These methods have not been used in practice.

An approximation to a constant utility index has been found by Theil (1965). Following Theil, it may be shown that

$$\frac{\partial \log \mu(t, U_0)}{\partial \log P_{it}} = c_i(t),$$

where $c_i(t)$ is, as before, the value share of commodity i in the total budget at time t. Using this relation and applying the Taylor expansion to $\log \mu(t, U_0)$ as a function of $\log p_{1t}$, $\log p_{2t}$, $\cdots$, $\log p_{nt}$, we obtain (keeping terms up to the second degree) the relation

$$\begin{aligned} &\log \mu(t, U_0) \\ &= \log \mu(t_0, U_0) + \frac{1}{2} \sum_{i=1}^{n} [c_i(t_0) + c_i(t)] \log \frac{p_{it}}{p_{i0}}, \end{aligned}$$

which gives, for example,

$$P_{01} = \frac{\mu(t_1, U_0)}{\mu(t_0, U_0)} = \prod_{i=1}^{n} \left(\frac{p_{i1}}{p_{i0}}\right)^{\frac{1}{2}[c_i(t_0) + c_i(t_1)]}$$

This is very similar to formula (2), which was obtained by purely statistical reasoning.

Thus, the economic and statistical lines of thought point to similar formulas. There is, however, still very little interaction between the two approaches.

ERIK RUIST

[*See also* Georgescu-Roegen 1968; Nataf 1968.]

BIBLIOGRAPHY

DIVISIA, FRANÇOIS 1925 L'indice monétaire et la théorie de la monnaie. *Revue d'économie politique* 39:842–861.

FISHER, IRVING (1922) 1927 *The Making of Index Numbers: A Study of Their Varieties, Tests, and Reliability.* 3d ed., rev. Boston: Houghton Mifflin.

FRISCH, RAGNAR 1936 Annual Survey of General Economic Theory: The Problem of Index Numbers. *Econometrica* 4:1–38.

►GEORGESCU-ROEGEN, NICHOLAS 1968 Utility. Volume 16, pages 236–267 in *International Encyclopedia of the Social Sciences.* Edited by David L. Sills. New York: Macmillan and Free Press.

INTERNATIONAL STATISTICAL INSTITUTE 1956 *Bibliographie sur les nombres indices; Bibliography on Index Numbers.* The Hague: Mouton.

KLOEK, T.; and DE WIT, G. M. 1961 Best Linear and Best Linear Unbiased Index Numbers. *Econometrica* 29:602–616.

KLOEK, T.; and THEIL, H. 1965 International Comparisons of Prices and Quantities Consumed. *Econometrica* 33:535–556.

KONÜS, A. A. (1924) 1939 The Problem of the True Index of the Cost of Living. *Econometrica* 7:10–29. → First published in Russian.

►NATAF, ANDRÉ 1968 Aggregation. Volume 1, pages 162–168 in *International Encyclopedia of the Social Sciences.* Edited by David L. Sills. New York: Macmillan and Free Press.

STAEHLE, H. 1935 A Development of the Economic Theory of Price Index Numbers. *Review of Economic Studies* 2:163–188.

THEIL, H. 1960 Best Linear Index Numbers of Prices and Quantities. *Econometrica* 28:464–480.

THEIL, H. 1965 The Information Approach to Demand Analysis. *Econometrica* 33:67–87.

TÖRNQVIST, LEO 1937 Finlands Banks konsumtionsprisindex. *Nordisk tidskrift for teknisk økonomi* 8:79–83.

ULMER, MELVILLE J. (1949) 1950 *The Economic Theory of Cost of Living Index Numbers.* New York: Columbia Univ. Press.

Postscript

Discussion on index number theory has tended to narrow the gap between the statistical and the economic approach. The Divisia index, equation (1) in the main article, or its discrete time approximation, equation (2) in the main article, has been shown to have some desirable properties as a quantity index in connection with the measurement of technological change (see Richter 1966). A Divisia index has, however, one drawback: its change from one point of time to another depends ordinarily not only on the endpoint prices and quantities, but on the entire path these have described between these points. In particular, the index does not necessarily come back to its original value even if all prices and quantities do.

Hulten (1973) showed, however, that the index may be path independent if certain conditions are fulfilled, and that these conditions have an economic interpretation. Thus, a necessary condition is that there is an economic aggregate associated with each point in the relevant quantity space. For example, in order to construct a path independent Divisia index of capital stock, it is necessary that the corresponding capital aggregate exists.

ERIK RUIST

ADDITIONAL BIBLIOGRAPHY

HULTEN, CHARLES R. 1973 Divisia Index Numbers. *Econometrica* 41:1017–1025.

RICHTER, MARCEL K. 1966 Invariance Axioms and Economic Indexes. *Econometrica* 34:739–755.

II
PRACTICAL APPLICATIONS

The search for a measure of the effect on the purchasing power of money of the influx of precious metals into Europe after the discovery of America resulted in the first index number of price changes, as far as we know today. In 1764 an Italian nobleman, Giovanni Rinaldo Carli, calculated the ratios of prices for three commodities—grain, wine, and oil—for dates close to 1500 and 1750. A simple average of these three ratios constituted his measure of the price change that had occurred over the 250-year period. This idea of isolating the effect of price changes in the measures of value changes in economic life has been a dominant and continuing theme in the development and use of index numbers.

Measures of changes in prices and changes in quantities have become a familiar and useful part of current economic life. In most countries of the world official agencies now issue regular reports on one or more of the following kinds of index numbers: wholesale prices, retail prices (often called cost-of-living), prices of goods in foreign trade, quantities of goods produced, and quantities of goods in foreign trade. These indexes are frequently supplemented with indexes for domestic trade and for specialized types of goods, such as agricultural commodities and raw materials.

In most cases these official indexes are designed as general-purpose indexes, and their availability leads to their use for many and varied purposes. "General-purpose" indexes are at variance with the first principle advocated by many serious students of the making of index numbers, stated by Wesley C. Mitchell as "defining the purpose for which the final results are to be used" ([1915] 1938, p. 23). Irving Fisher, who examined various methods of computing index numbers, disagreed with this principle and thought that ". . . from a practical standpoint, it is quite unnecessary to discuss the fanciful arguments for using 'one formula for one purpose and another for another,' in view of the great practical fact that all methods (if free of freakishness and bias) *agree!*" ([1922] 1927, p. 231). Melville Ulmer and others put forth the point of view, now generally accepted in principle by most economists, that the making of index numbers should be tied to economic concepts and that these concepts should be expressed in operational terms (Ulmer 1949, pp. 23–24). But despite the massive amount of discussion and the long history of index number practice, the empirical difficulties of closing the gap between theory and practice have not been overcome in many cases.

This article is concerned with some of the practical aspects of index number making—the general characteristics of indexes, the kinds of data employed, and the problems and difficulties encountered because we do not live in a static economy. It is intended as a nontechnical guide to those index number practices that an index user should review to help determine whether a specific index is the appropriate one for his purpose and is likely to be an adequate approximation for the use he intends (or more realistically, what limitations are likely to result from the use of the only index available). Although the discussion is oriented largely to price indexes, the major problems and procedures also apply to quantity indexes.

Index designations. Most of the indexes of prices labeled "wholesale" refer to the primary market level—the prices charged by manufacturers or producers to wholesalers and other buyers. In a few countries these indexes relate to the wholesale level of distribution.

Indexes of changes in retail prices are popularly referred to as cost-of-living indexes. However, the indexes in all countries are basically measures of price change (with minor variations), and official titles attempt to indicate this. In France the index is called the general retail price index; in the Federal Republic of Germany, the price index of living; in the United Kingdom, the index of retail prices. Although the name "cost-of-living index" is still retained by some countries, the most common name is "consumer price index." This name grew out of the controversy over the United States index during World War II, when it was an important factor in wage stabilization. "Consumer price index" was adopted to help clarify interpretation of what the index measured.

The names for other indexes may vary to a minor degree, but in general they are self-explanatory: e.g., index numbers of industrial production; index numbers of the volume of wholesale (or retail) trade; quantity and unit value of commodities in external trade.

Calculation formulas. In most countries, indexes of wholesale and retail prices are measures of price changes with "fixed" quantity weights as of some earlier period; that is, the calculation framework is that of a Laspeyres formula, (1) below, or some modification of it [*see* INDEX NUMBERS, *article on* THEORETICAL ASPECTS]. The use of this form for approximating changes in living costs is particularly controversial, although the practical difficulties of translating a "welfare" or "utility" concept into practice are generally recognized. The use of a consumer price index with fixed weights for the escalation of wages or for determining wage policy is an attempt to maintain the real purchasing power of the workers' dollar and might be interpreted as providing the income required to maintain the base-period utility level. However, the base-period utility level can be attained without keeping the kinds and quantities of goods fixed. When there are differential changes in prices among items, consumers are likely to buy a different collection of goods. If they increase their purchases of products whose relative prices have declined, escalation according to a fixed weight index will allow an increase in utility.

Whatever the limitations of a consumer price index on a conceptual basis, the present systems of index numbers do provide a measure of the average changes in prices for the quantities of the earlier year. The corresponding quantity indexes provide a measure of average changes in quantities at fixed prices, those of the base year. The relationships are

$$\text{price ratio for the current period} = \frac{\text{current-period price} \times \text{base-period quantity}}{\text{base-period price} \times \text{base-period quantity}} = \frac{p_i \times q_0}{p_0 \times q_0}$$

and

$$\text{quantity ratio for the current period} = \frac{\text{base-period price} \times \text{current-period quantity}}{\text{base-period price} \times \text{base-period quantity}} = \frac{p_0 \times q_i}{p_0 \times q_0}.$$

These simplified forms refer only to an individual commodity. When they are expressed for aggregations of commodities, the Greek capital sigma (Σ) is used to indicate the sums, and the formulas for the Laspeyres indexes become

$$\text{(1)} \qquad \text{Price index} \quad I_{pi} = \frac{\Sigma p_i q_0}{\Sigma p_0 q_0}, \qquad \text{Quantity index} \quad I_{qi} = \frac{\Sigma p_0 q_i}{\Sigma p_0 q_0}.$$

The denominator of both expressions is the total value in the base period of the aggregation of goods in the index. The numerator of the price index is the value of the base-period goods at current prices. The numerator of the quantity index is the value of current quantities at base-period prices.

The corresponding forms for Paasche indexes, which use current weights, are

$$\text{(2)} \qquad \text{Price index} \quad I_{pi} = \frac{\Sigma p_i q_i}{\Sigma p_0 q_i}, \qquad \text{Quantity index} \quad I_{qi} = \frac{\Sigma p_i q_i}{\Sigma p_i q_0}.$$

The numerator of these expressions is the total expenditure in the current period. Since it is unusual to have separate quantity data available to use in these formulas for every item in the index, an algebraic equivalent (which is in effect a weighted average of relatives, or price ratios) is generally employed, to permit the use of value data: e.g., the Laspeyres price index becomes

$$I_{pi} = \frac{\Sigma (p_i/p_0)(p_0 q_0)}{\Sigma (p_0 q_0)}.$$

Since it is also unusual for the values of consumption, production, or other variables to relate exactly to the period selected as a reference base for the index series, many indexes use the following form for a price index with fixed quantity weights:

$$I_{pi} = \frac{\Sigma (p_i/p_0)(p_a q_a)}{\Sigma (p_a q_a)},$$

where the subscript a refers to the date for which the values are available.

These formal calculation formulas are further modified in practice to accommodate changes in samples of items, sources of reports, incomplete data, and similar unpredictable problems of noncomparability. The usual procedure, called linking, is to calculate an index for each period, with the preceding period as a base, and multiply successive indexes together to obtain an index on a fixed base.

The reference-base period. Index numbers have two kinds of base dates: the date to which the consumption or production weights refer (date a in the preceding formula) and the date on which the price-change comparisons are based (date 0 in this formula). The former may be called the weight base, and the latter the reference base. It is customary to set the index equal to 100 at the reference-base date.

The initiation of index numbers and the deep interest in their fluctuations indicate that the needs for such economic measures were engendered by unusual periods of economic activity, such as inflations, depressions, and wars. This is probably the reason why there historically has been so much emphasis on the choice of a "normal" period for a reference base. Such a choice is extremely difficult because few, if any, periods can be said to be normal for all segments of the economy. The emphasis is perhaps justified to the extent that an index series is more meaningful when the weight base and the reference base are identical or not widely separated in time. In practice compromises must be made, and the two base dates seldom correspond exactly.

For the most part indexes are more reliable for short-period comparisons and, theoretically, more reliable for periods close to the weight base. But because of the extensive resources required to keep the weights continuously up to date and because in practice moderate weight changes have less influence on an index than have price changes, a common practice is to change the reference-base period more frequently than the weighting structure.

An index with a fixed reference base makes direct comparisons between the base selected and any one of the succeeding dates. Comparisons with any period other than the reference base require conversion to a new base. Such conversions, essentially, are made every time a percentage change is obtained, as shown in Table 1.

If the conversions are done correctly, percentage changes from one date to another will aways be the same for the original series and for the converted series (except for minor rounding differences, as shown in two instances in Table 1).

In about half the countries that maintain retail price indexes the practice has been to select a reference-base period reasonably close to the weight base and to change the reference base when the indexes are revised. This has the advantage of alerting the user to the fact that revisions have been made and of getting him to consider whether such revisions

Table 1 — Example of the conversion of an index to a new reference base

	ORIGINAL BASE 1950 = 100		CONVERTED BASE 1955 = 100		CONVERTED BASE 1960 = 100	
Year	Index	Per cent change from preceding date	Index	Per cent change from preceding date	Index	Per cent change from preceding date
1950	100.0		90.9		87.0	
1955	110.0	10.0	100.0	10.1	95.7	10.0
1960	115.0	4.5	104.5	4.5	100.0	4.5
1965	120.0	4.3	109.1	4.4	104.3	4.3

are so important that the indexes before and after the revision cannot be considered comparable. In Hungary the index is computed with the preceding year as base, and in the United Kingdom with January of each year as base. In the United States the present policy is to update the reference base about every ten years and to convert most economic index series to a uniform base to facilitate their use in economic analysis.

Uniform base periods for various economic series and for various components of the same index must be used with some discrimination, in order to present facts in proper perspective. Since there is seldom one period that can be considered "normal" for all segments of an economy, conversions of indexes to other periods for supplementary exposition is often required. For example, the change in the base period for United States indexes after World War II gave rise to complaints that the consumer price index gave a distorted picture. A postwar reference base (1947–1949 average) was substituted for a prewar base (1935–1939 average) at a time when the price rise for the commodity sector was leveling off but the rise for services, prices for which had remained fairly stable during the war, was beginning. The postwar base period thus highlighted a major price rise for medical care and other services throughout the 1950s. A fairer picture of the changes for commodities relative to services was afforded by comparisons with prewar prices. The rise from 1939 for services did not equal that for commodities until 1962.

Weights. The system of weights for price indexes relates to the level of distribution for which measures are desired.

For wholesale price indexes, censuses and similar surveys are utilized to derive the total value of sales, exclusive of taxes, for all commodities produced or processed by the private sector of the economy and sold in primary markets. In some cases both imports and exports are included. In some countries weights are limited to sales of goods for domestic consumption, thus including imports but not exports. Generally the principal exclusions from the "universe" covered by wholesale price indexes are business services, construction and real estate, sales by government, military production, securities, and goods produced and consumed within the same plant.

For consumer price indexes, weights representing the importance of individual goods and services are usually derived from special surveys of expenditures by the groups in the population for which price changes are measured, e.g., urban wage earners, farmers, low-income families, families of two or more, single consumers, etc. Generally the weights include all taxes directly associated with the purchase or ownership of specific goods and services, such as sales and excise taxes, property taxes, car registration fees, and the like. The principal exclusions are direct taxes, such as income taxes; expenditures for investments; contributions to churches and other organizations; and goods and services received without direct cash outlay, such as gifts received, home-produced foods, fringe benefits paid for by employers, and services supplied by government agencies without payment of a special tax or fee.

Indexes of industrial production generally use weights of the "census value-added" type; that is, the weights are usually proportional to value added at factor cost in different industries, as given by census data, and are derived by reducing values of gross output by costs of raw materials, fuels, containers, industrial services, etc. In a few countries weights represent gross value of production.

Price data. The sample of commodities and services priced regularly for the computation of price indexes is usually selected with considerable care, so that the items are well distributed among the major classification groups. The number of individual items priced varies considerably from country to country but in most cases is large enough to provide fairly reliable indexes, for groups and subgroups, of commodities and services.

Prices for the selected items in the wholesale price indexes are obtained from a sample of manufacturers or other producers and refer to the form in which the item enters commercial markets. Thus, as raw materials are processed into semifinished or finished goods, prices at each successive stage of processing may be included if the product is sold in primary markets in that form. Prices are usually reported for a precise specification, a specific class of buyer, and a specific level of distribution, quantity of purchase, and set of delivery terms. Prices are usually net of discounts, allowances, and excise taxes. Practices vary with regard to transportation costs; in the United Kingdom "delivered" prices are used for imported goods purchased by industry and "ex-works" prices for domestically produced goods; in the United States prices exclude delivery costs unless it is the normal custom to quote on a delivered basis.

Price data for consumer price indexes are usually obtained from a sample of stores and service establishments in a sample of communities that provides a good geographical coverage of the country. In a few cases the index is confined to one city or a few large cities. The prices quoted are generally cash prices for goods as offered for sale to the consumer. Usually all sales and other taxes applicable to the purchase of the specific item are added to the quoted price. In the United States and a few other countries, concessions and discounts are deducted.

The care with which data are gathered determines in large measure whether a price index for a particular period is good, bad, or indifferent. The essence of the collection process is to obtain "comparable" prices for successive periods, so that changes in the index refer to price changes only, not to a mixture of price, quality, and marketing changes. Comparability of outlets or producers is obtained by using a matched sample for each two successive time periods.

The description or specification of the individual quality of the item for which data are requested plays a key role. The precision of the descriptions used by various countries differs in degree, but the principle of comparability is generally adhered to. For Ireland's consumer price index, for example, the requirement is that the item priced conform to a general commodity description and that it be in substantial demand in the area for which prices are being reported. Comparability is achieved by obtaining prices of the identical item in the same store for two successive periods. In most countries, however, detailed descriptions or specifications are developed to identify the quality of the item to be priced. These detailed descriptions define quality in terms of physical features, such as the kind and grade of materials, parts, construction and workmanship, size or capacity, strength, packaging, and similar factors or identification characteristics. The item is described as it enters into transactions in the market, and the assumption is made that the physical makeup determines performance characteristics.

The specifications adopted generally allow some latitude for minor variations in quality, in recognition of the many small differences from firm to firm or store to store. Within limits explicitly stated, all articles are considered comparable, and the price of the one specific item sold in largest volume each period is usually reported. Substitutions from time to time of items that fall outside the stated quality limits require special comparison procedures (as discussed for quality change below).

Major problems and limitations. There are many problems in the making of index numbers that materially affect the precision with which they can be applied. The unpredictable difficulties that occur as a regular part of the collection-and-comparison process were referred to briefly above. The major problems and limitations pervading most indexes are discussed in the following paragraphs.

Definition of "universe." One limitation that has relevance to the uses made of the indexes is the definition of the "universe" to which the indexes relate, that is, which segment of the population, which categories of business, etc. Retail or consumer price indexes for city families may be inappropriate for estimating the change in prices paid by farm families, not only because farm families may purchase their goods in different places but also because food and housing are a less important part of their expenditures than of the expenditures of city families. An index of wholesale prices of commodities, regardless of how good a measure it is, tells nothing about changes in other costs of doing business, such as wages and salaries, costs of printing, advertising, and other business services. Nor can it be assumed that wholesale prices in commercial markets are a good indicator of changes in prices paid by governments, because of special contract arrangements typical of government purchases. In such circumstances, when lack of an appropriate series forces the use of an index that is available, the user must evaluate its limitations for the specific purpose.

Sampling error. Sampling of some kind is a requirement for practically all indexes. Sampling, as opposed to complete coverage, introduces the familiar sampling error—a difficult factor to measure. In the absence of measures of sampling error, we can only judge precision intuitively, by knowing the composition of the sample. Although measurement of sampling error was being attempted in the United States for its consumer price index [*see* INDEX NUMBERS, *article on* SAMPLING], actual measures were still lacking. It is probable that understatements and overstatements of price change are quite small for the comprehensive "all-items" indexes in most countries but are larger at the group and subgroup levels.

Quality changes. The problems that occupy the greater part of the time and effort of those who compile indexes are the identification and measurement of the effect of quality changes in the items in the market, the introduction of new items and variations and disappearance of old, changes in the importance of various products and services from one time to another, changes in the types of establishments through which goods flow during the marketing process, and many of the other facets of change in an economy that is not standing still. The measurement instrument must seek to disentangle the changes in prices from the effects on price of all the other changes that take place.

The problems of eliminating the effect of quality changes are particularly difficult, and the extent to which adjustments are made for them varies considerably from country to country. Where they are taken into account, market price valuations for quality changes are generally obtained in one of three ways: (1) by assuming that the price difference is all due to quality change; (2) by estimating the value of the quality difference associated with changes in physical characteristics; or (3) by estimating the value of quality changes through operating characteristics.

When two varieties of an item are selling in volume simultaneously, the assumption that a difference in price between them is entirely due to a quality difference is realistic and reasonable. When two or more varieties do not sell simultaneously, it is uncertain whether a price difference results entirely from price change, entirely from quality change, or from a mixture of both. Thus, for automobiles and other highly complex products, the prices of new models are seldom compared directly with those of the old, since it is a common market practice to introduce price changes simultaneously with model changes. In these cases producers and sellers aid in identifying the changes made in the physical characteristics and provide production costs or estimated market prices, to permit the development of quality adjustment factors. In a few countries the adjustments for quality changes for some of the complex products are applied to one or more operating or use characteristics. For example, for turbines, in the Soviet Union the price or cost per unit of potential power is obtained. In Sweden the estimate of the relative worth of different automobile models is based on results of engineering, road, and other tests.

Considerable ingenuity and experimental work has been devoted to the quality problem in index numbers. But the very elusive nature of the quality concept, combined with the difficulties of detecting quality improvement and deterioration and of deriving objective values for these changes, means that considerable judgment and discrimination must be exercised. In some cases it is likely that the mechanics of the index make too large an adjustment for quality changes, and since most of the changes are labeled "improvement," some downward bias may be introduced. On the other hand, it is also probable that insufficient allowance has been made for quality improvement of other items. In any one monthly, quarterly, or annual interval, the total index will be made up largely of commodities and services that are unchanged in quality, and the effect of incomplete measurement for individual changes up and down is likely to be unimportant. For longer periods of time the influence may be greater, particularly for specific goods, where small quality changes cumulate from year to year and may not be detected. An evaluation of how much the factor of quality change may have influenced the movement of an index must be based on fairly detailed knowledge of both the timing and degree of quality change during the period under consideration and of the way in which they were accounted for in the index calculation.

New products. Closely related to the quality problem is the problem of timing the introduction of new products. Modifications or new varieties of older products are generally put into an index, when they have been in the market long enough to sell in substantial volume, by adjusting for quality change in the manner indicated above. But truly new products—those, such as television, that have no earlier counterpart—are generally introduced into an index only at major revisions, when new weights are available to reflect their impact throughout the weighting system. New products frequently enter the market in small volume at relatively high prices. As production and sales volume increase, price reductions are generally made. Some upward bias in an index can result if new items are introduced after major price reductions have occurred. However, the specific timing for introducing new items and the method of handling volume changes in the weights are still matters of some disagreement between index technicians.

Long-term comparisons and revision. Here the phrase "long-term comparisons" is used to mean comparisons over a period that encompasses a revision of an index. These present special problems. Theoretically, revisions in the conceptual structure, coverage, system of weights and/or operational aspects of index construction result in an index different from the previous one. If, however, an agency always presented a revised index as a new measure, individual users would have to provide some kind of a bridge from one index to the other to obtain a long-term perspective. Consequently, the issuing agencies usually "chain" the different indexes together to form a seemingly continuous series. This practice is common in most countries and provides many advantages. But users must also recognize, through a study of the changes in the makeup of the two or more separate indexes that have been chained together, the limitations involved in such comparisons.

The retail price index of the United Kingdom and the consumer price index of Sweden are examples of indexes with annual changes in weights. Technically speaking, the index for each year is different from the index for the preceding year because the weights represent a different level of living. Studies of the effect of changes in weights on the index indicate that year-to-year comparisons are so nearly the same with the old and the new weights that for all practical purposes the effect of the weight differences can be ignored for comparisons over two or three years. Over a period of five or ten years the differences may be more significant, since the continuous series includes the net effects of changes in the distribution of living expenditures over the longer period. In the United Kingdom it is felt that short-term comparisons are the most important to index users. In the United States, the importance of short-term comparisons is recognized but it is felt that the available resources should be devoted to maintaining the adequacy of current price data. The net effects of changes in living habits in the United States over approximately ten years are introduced at one time, during a major revision, rather than in smaller increments. Hence, the problems to be considered by users over periods of more than ten years are the same in both countries.

A more serious question would be raised if the conceptual structure of an index were changed. If a consumer price index were revised to measure the constant level of utility defined in economic theory or if weights at wholesale were changed from total value of shipments at each stage of processing to "value added" weights, such changes would have major influences on the index, and "long-term" comparisons would be practically meaningless.

A practical guide to users on the effect of changes made during revisions is usually provided by issuing agencies, in the form of concurrent indexes on both the old and new basis, either through continuation of the old after the new has been issued or (less frequently) through the recalculation retroactively of the preceding index.

ETHEL D. HOOVER

BIBLIOGRAPHY

CARLI, GIOVANNI (1760) 1785 Del valore e della proporzione de' metalli monetati con i generi in Italia prima delle scoperte dell' Indie, col confronto del valore e della proporzione de' tempi nostri. Volume 7, pages 1–190 in Giovanni Carli, *Delle opere*. Milan: Nell' Imperial Monistero di S. Ambrogio Maggiore.

Definitions and Explanatory Notes. 1964 United Nations, Statistical Office, *Monthly Bulletin of Statistics* [1963], no. 5 (Supplement). → This supplement is a good general reference for descriptions of the statistical series for the various countries of the world in the *Monthly Labor Bulletin*. The descriptions are necessarily brief but include the main features, as well as a reference to the various national publications, where greater detail may be found.

FISHER, IRVING (1922) 1927 *The Making of Index Numbers: A Study of Their Varieties, Tests, and Reliability*. 3d ed., rev. Boston: Houghton Mifflin.

GILBERT, MILTON 1961*a* Quality Changes and Index Numbers. *Economic Development and Cultural Change* 9:287–294.

GILBERT, MILTON 1961*b* The Problem of Quality Changes and Index Numbers. U.S. Bureau of Labor Statistics, *Monthly Labor Review* 84:992–997.

GILBERT, MILTON 1962 Quality Change and Index Numbers: The Reply. U.S. Bureau of Labor Statistics, *Monthly Labor Review* 85:544–545.

GRILICHES, ZVI 1962 Quality Change and Index Numbers: A Critique. U.S. Bureau of Labor Statistics, *Monthly Labor Review* 85:542–544.

HOFSTEN, ERLAND VON 1952 *Price Indexes and Quality Changes*. Stockholm: Forum.

HOOVER, ETHEL D. 1961 The CPI and Problems of Quality Change. U.S. Bureau of Labor Statistics, *Monthly Labor Review* 84:1175–1185.

INTERNATIONAL LABOR OFFICE 1962 *Computation of Consumer Price Indices: Special Problems*. Report No. 4. Geneva: The Office.

MITCHELL, WESLEY C. (1915) 1938 *The Making and Using of Index Numbers*. 3d ed. U.S. Bureau of Labor Statistics, Bulletin No. 656. Washington: Government Printing Office. → The preface contains a discussion of the 1915 and 1921 editions.

MUDGET, BRUCE D. 1951 *Index Numbers*. New York: Wiley.

ORGANIZATION FOR EUROPEAN ECONOMIC CO-OPERATION 1956 *Quantity and Price Indexes in National Accounts*, by Richard Stone. Paris: The Organization.

ULMER, MELVILLE J. 1949 *The Economic Theory of Cost of Living Index Numbers*. New York: Columbia Univ. Press.

UNITED NATIONS, ECONOMIC AND SOCIAL COUNCIL 1965 The Gathering and Compilation of Statistics of Prices. E/CN.3/328. Mimeographed.

U.S. BUREAU OF THE CENSUS 1960 *Historical Statistics of the United States, Colonial Times to 1957: A Statis-*

tical Abstract *Supplement.* Washington: Government Printing Office.

U.S. Bureau of the Census 1965 *Historical Statistics of the United States, Colonial Times to 1957: Continuation to 1962 and Revisions.* Washington: Government Printing Office.

U.S. Bureau of Labor Statistics 1964 *Computation of Cost-of-living Indexes in Developing Countries.* Bureau of Labor Statistics, Report No. 283. Washington: The Bureau.

U.S. Congress, Joint Economic Committee 1961 *Government Price Statistics.* 2 parts. Hearings before the Sub-committee on Economic Statistics. Washington: Government Printing Office. → Part 1 contains the report prepared by the Price Statistics Review Committee of the National Bureau of Economic Research, *The Price Statistics of the Federal Government: Review, Appraisal, and Recommendations.* Part 2 contains the comments of witnesses before the Joint Economic Committee on the contents of the report.

III
SAMPLING

No matter how one resolves the conceptual and practical problems described in the two accompanying articles, the actual construction of an index number will almost always be based on sampling. The quality of an index will, therefore, depend upon the nature of the sampling process. Although this dependence has long been recognized (see King 1930, for discussion and references to earlier work), the sampling aspects of index number construction have been relatively neglected.

This article emphasizes the following points: (1) most economic index numbers have a complex sampling structure; (2) the sampling precision of an index number can be defined, even though conceptual and practical problems are not fully solved; (3) estimates of sampling error are required both for the analytic use of index numbers and for reasonable allocation of resources in designing the data-gathering procedure on which the index number is based; and (4) most components of sampling error can be estimated only by use of replication.

In order to make the discussion specific, it is framed in terms of Laspeyres indexes of consumer prices, with special reference to the consumer price index of the United States Bureau of Labor Statistics. Nearly all the discussion, however, is readily applicable to other kinds of index numbers.

Sampling aspects of a price index. The various points at which sampling must be employed in order to provide data for constructing a consumer price index are easily identifiable.

A Laspeyres price index for an individual consumer can be viewed as a weighted average of price ratios for the commodities and services purchased by the individual, where the weights are proportions of total expenditures in the base period for the different items of the index. The weights may be called *base year value weights.* An index for a *group* of consumers, for example, those living in a particular city or geographic area, then involves the following major sampling problems: (1) average base year value weights must be estimated from a sample of consumers; (2) since it is impossible to price all of the goods and services purchased by all of the consuming units in the population, this list of goods and services must be sampled; and (3) an average price for any good or service, in either the base year or at a current point in time, must be estimated from a sample of outlets.

The foregoing sampling problems relate to the production of an index for a particular city or geographic region. If one wishes to construct an index that relates to a country, then it becomes necessary to select a sample of cities or regions and combine their individual results into an over-all index. Finally, prices must be collected at repeated points in time, and thus temporal sampling is involved.

Some general views on sampling

Although price indexes are based on highly complex sampling structures, measures of sampling precision are not available for any of the currently prepared indexes, although I understand that the Bureau of Labor Statistics is planning to provide such information about the consumer price index starting in early 1967 (see Wilkerson 1964). Three related arguments have been set forth to justify the absence of such reporting.

The Laspeyres index follows the prices of a sample of goods and services through time. Because the universe of goods and services available to the consumer is continually changing (some items change in quality, others disappear, and new items enter the universe), it is necessary to make a variety of adjustments in the sample items and in observed prices. Since there exists no "best" procedure for making these adjustments, the index is subject to a *procedural* error. It is then argued that the sampling error is probably small in relation to the procedural error and that it is therefore neither necessary nor desirable to attempt to estimate its magnitude.

Because of the complexity of the adjustment procedures, it is frequently stated that it is impossible to define and estimate that portion of the sampling variability of an index that arises from the sampling of commodities. Hence it is impossible to define or estimate the sampling precision of the index itself.

A third argument admits that it might be possible to employ probability sampling for all components of a price index. But the great complexity of the design and data-gathering operations are then stressed and the conclusion is reached that the attainment of this goal would require the use of more or less unlimited resources. These views have been expressed by Hofsten (1952, p. 42; 1959, p. 403) and Jaffe (1961), among others; and direct quotations from these articles are provided by McCarthy (1961, pp. 205–209).

Definition of sampling precision. The argument that it is impossible to discuss sampling precision because of the changing nature of the universe of commodities is clearly basic to a consideration of the other two arguments. Adelman (1958) seems to accept this view and, as a consequence, sets forth a method of index number construction that is more directly in line with modern sampling theory as described by Hansen, Hurwitz, and Madow (1953). She suggests periodic stratified resamplings of the changing commodity universe, together with the use of a chain index in place of the Laspeyres index. The meaningfulness of the Adelman approach to index number construction will not be argued here. Rather, it will be argued that it is quite reasonable to talk about the sampling precision of a Laspeyres index, provided (1) that a very general view of sampling precision, similar to that described by Stephan and McCarthy (1958, pp. 226–229), is adopted and (2) that one does not always expect to measure this precision by the application of standard formulas from the theory of sampling.

Replication to estimate sampling error. Assume the existence of a set of adjustment procedures that are used to follow a sample of goods and services through time, so that sampling variability arises only from the fact that a sample of items is selected at time zero. If one now thinks of drawing an indefinitely large number of independent samples in accordance with the same sampling procedure and of independently following each of these through to time t in accordance with the defined adjustment procedures, the resulting values of the index will define the sampling distribution of the index with respect to the sampling of items. The variance of this distribution is an acceptable measure of sampling precision for the index, and it includes a component for any inherent variability of the adjustment procedure. Furthermore, an estimate of this variance can easily be obtained by actually drawing two or more independent samples of items and independently following them through time, that is, through the use of replicated samples. It should be observed that the use of two independent samples, for example, does not mean that each sample must be as large as the desired over-all sample of commodities. Each sample may be only half as large as the over-all sample, and the published index would be the average of the two resulting indexes. Of course the reliability of the estimate of variance would improve as the number of independent samples increases. It should also be noted that in practice the independence of the samples would be difficult to preserve as time goes on.

Bias. The measure of sampling precision just defined is obviously taken about the mean of the sampling distribution of the index. If the population value of this index at time t is denoted by $R_p^{(t)}$, where $R_p^{(t)}$ would be obtained by applying the adjustment procedures to *all* commodities, then the difference between the expected value of the index and $R_p^{(t)}$ is the bias of the estimate arising from the sampling and estimation procedures. If the selection were based on expert judgment, then such bias might arise because all the experts might, consciously or unconsciously, not consider for selection items having a different form of price behavior from those items considered for selection. [*See* SAMPLE SURVEYS, *article on* NONPROBABILITY SAMPLING.]

Imperfection of adjustment procedures. In addition, one usually questions the adjustment procedures and therefore views $R_p^{(t)}$ as only an approximation to the index that would be obtained through the use of a "perfect" adjustment procedure.

Three components of total error. As a result of the foregoing, the total error in a single estimate can be viewed as the sum of three components. The first component represents the error of variability that arises from the use of sampling (plus possible contributions from variability in applying adjustment procedures); the second component represents the bias arising from the sampling and estimation procedures; and the third component represents the bias that arises through inherent imperfection in the adjustment procedures. Other errors may of course arise from interviewing, in clerical work, or from computations; but these will not be treated in this article. [*See* ERRORS, *article on* NONSAMPLING ERRORS.]

It would appear that at least some of the differences in opinion on sampling for index numbers can be traced to a failure to distinguish carefully among these three components of error, particularly between the first and third components. All writers agree that it is unlikely that anyone will ever be able to devise a "perfect" set of rules for treating quality changes and for introducing new items into the index, but this does not mean that it is impossible or unnecessary to estimate the values of all three components.

Importance of sampling error. Next consider the argument that this precision is dominated by the procedural error and can therefore be ignored. Some investigations reported by McCarthy (1961) suggest that the procedural error of current consumer price indexes may indeed dominate the sampling error, although empirical investigations of the over-all effect of procedural error are almost as lacking as those of sampling error. This does not necessarily mean that sampling error can be ignored. It remains important for several reasons, in particular:

(1) If the goal is to estimate the *level* of the "true" index at various points in time and if resources are fixed, then the most efficient way of improving the accuracy of these estimates would be to divert resources from the maintenance of a relatively large sample of commodities and to use these resources in basic research aimed at reducing the magnitude of the procedural error. It is clear that good estimates of sampling precision and of bounds on the procedural error are required in order to make judgments of this kind.

(2) If the goal is to estimate short-term *changes* in the level of the "true" index, then it appears likely that sampling error will be more important than procedural error and hence an estimate of sampling error becomes essential.

(3) The construction of a price index involves not only a set of adjustment procedures and the sampling of commodities but also the sampling of localities and the sampling of price reporters within these localities. There must be a balance between these errors and the sampling errors arising from the other parts of the design. Again it is impossible to discuss such a balancing operation unless some attempt is made to measure these components of error.

Reporting estimates of error. Estimates of error for the various components of a price index should also be available in published form to assist those who wish to use the indexes in a critical fashion. When one considers that small monthly changes in important indexes may lead to major policy decisions, and that these indexes are basic tools in much economic analysis, the necessity for having measures of error becomes apparent. Kruskal and Telser (1960) have emphasized the latter point.

Probability sampling for index numbers

In order to guard against nonmeasurable biases from sampling and estimation, it seems reasonable that some appropriate form of probability sampling should be utilized in the selection of each sample that enters an index design. The selection of a sample of consumers, from which to estimate the base year expenditure weights, and the selection of a sample of cities or regions, from which to obtain current price data, should cause no more difficulty than is encountered in the ordinary large-scale sample survey. The sampling of goods and services does, however, pose an especially difficult problem. Nevertheless, the following are some of the convincing reasons for attempting to use probability sampling methods in the original selection of items: (1) The replicated sample approach can provide an estimate of sampling precision for almost any type of sampling procedure, but it cannot even indicate the existence of bias. The only way to ensure that biases due to sampling and estimation are small or nonexistent is to use appropriate probability sampling methods. (2) A probability model will make clear the manner in which one can obtain two or more independent samples of goods and services. (3) Even the mere attempt to make the sampling of goods and services conform to some appropriate probability model will force one to make definite decisions about problems of definition and estimation that exist no matter how such a sample is chosen but that can too easily be ignored with judgment procedures.

Probability sampling of goods and services. Although probability sampling of goods and services has not been the practice in the past, the general format that possible procedures would probably follow can be indicated. Items of expenditure would be divided into major groups, then into subgroups, sub-subgroups, and so on. Ultimately this subdivision process leads to what may be termed specific items, for example, one item might be mattresses for single beds. These specific items could be grouped into strata, using any available information about substitutability, similarity of price movements, and other related variables. The first sampling operation would then consist of selecting one or more specific items out of each stratum.

Drawing a specific item into the sample usually draws an entire cluster of specified-in-detail items into the sample. One or more specified-in-detail items must be chosen from the cluster defined by each of the selected specific items, and this is the second sampling operation to be considered. For example, with the single-bed mattress item, one might specify number of coils; gauge of wire; type of cover; padding material; and so on. The chosen specified-in-detail items are the ones on which price quotations are to be obtained. At this second level of sampling, the problems become much more difficult than at the first level. Complete lists of specified-in-detail items will be difficult, if not impossible, to obtain; some specified-in-detail items may not be purchased by the consumer group to which the index is supposed to refer; and expenditure weights may not be available for many of the items. Possibly anything that one can do at this level (for example, using a restricted list of specified-in-detail items instead of a complete list or assuming equal base year expenditure weights when actual weights are unequal) is going to be only an approximation to what one would like to do; but at least this type of approach can be described accurately, and it should be possible to investigate the effects of some of the approximations that are used. The U.S. Bureau of Labor Statistics has experimented with this approach in connection with a recently completed revision of the U.S. consumer price index, and their experiences will be available as a guide to others in the future. Banerjee (1960) has written on this aspect of the sampling problem.

Probability sampling of outlets. It might also be observed that the probability sampling of outlets, from which to obtain current price reports, is a much more troublesome problem than might appear at first sight. Lists of outlets are difficult to obtain; many different commodities will ordinarily be priced in the same store and this introduces correlation among the price quotations; in addition, the maintenance of a panel of price reporters is complicated by the birth and death of firms.

The production of an index number obviously involves a highly complex network of samples. Even though probability sampling could be used for all components, it would be extremely difficult, or even impossible, to apply on a routine basis ordinary variance estimating procedures.

The difficulties involved in the determination of sampling variability for estimates derived from complex sample surveys are not unique to index number problems. The necessity for obtaining "simple" procedures for the routine estimation of sampling error has long been recognized and has been discussed by many authors under such titles as "interpenetrating samples," "replicated samples," "ultimate clusters," and "random groups" (Deming 1960; Hansen, Hurwitz, & Madow 1953, vol. 1, p. 440; Mahalanobis 1946; Stephan & McCarthy 1958, pp. 226–229). This matter was discussed briefly here in connection with the sampling of goods and services, but a more detailed treatment of the application of the principles of replication to sampling for index numbers has been given by McCarthy (1961).

PHILIP J. MCCARTHY

[*See also* SAMPLE SURVEYS.]

BIBLIOGRAPHY

ADELMAN, IRMA 1958 A New Approach to the Construction of Index Numbers. *Review of Economics and Statistics* 40:240–249.

BANERJEE, K. S. 1960 Calculation of Sampling Errors for Index Numbers. *Sankhyā* 22:119–130.

DEMING, W. EDWARDS 1960 *Sample Design in Business Research.* New York: Wiley.

HANSEN, MORRIS H.; HURWITZ, WILLIAM N.; and MADOW, WILLIAM G. (1953) 1956 *Sample Survey Methods and Theory.* Vol. 1. New York: Wiley.

HOFSTEN, ERLAND VON 1952 *Price Indexes and Quality Changes.* Stockholm: Bokförlaget Forum.

HOFSTEN, ERLAND VON 1959 Price Indexes and Sampling. *Sankhyā* 21:401–403.

JAFFE, SIDNEY A. 1961 The Consumer Price Index: Technical Questions and Practical Answers. Part 2, pages 603–611 in U.S. Congress, Joint Economic Committee, *Hearings: Government Price Statistics.* 87th Congress, 1st Session. Washington: Government Printing Office.

KING, WILLFORD I. 1930 *Index Numbers Elucidated.* New York: Longmans.

KRUSKAL, WILLIAM H.; and TELSER, LESTER G. 1960 Food Prices and the Bureau of Labor Statistics. *Journal of Business* 33:258–279.

MCCARTHY, PHILIP J. 1961 Sampling Considerations in the Construction of Price Indexes With Particular Reference to the United States Consumer Price Index. Part 1, pages 197–232 in U.S. Congress, Joint Economic Committee, *Hearings: Government Price Statistics.* 87th Congress, 1st Session. Washington: Government Printing Office.

MAHALANOBIS, P. C. 1946 Recent Experiments in Statistical Sampling in the Indian Statistical Institute. *Journal of the Royal Statistical Society* Series A 109: 326–378. → Contains eight pages of discussion.

STEPHAN, FREDERICK F.; and MCCARTHY, PHILIP J. (1958) 1963 *Sampling Opinions: An Analysis of Survey Procedure.* New York: Wiley.

WILKERSON, MARVIN 1964 Measurement of Sampling Error in the Consumer Price Index: First Results. American Statistical Association, Business and Economic Statistics Section, *Proceedings* [1964]:220–233.

Postscript

Wilkerson (1967) explains how the replicated sampling notions discussed in the main article are being used to compute values for the reliability of per cent changes in the CPI. Such values are published as part of the Bureau of Labor Statistics *Monthly Report on the Consumer Price Index.*

Revisions of the CPI have been undertaken, using sampling from a different (and perhaps broader) population of consumers to obtain expenditure weights. Both the old and the new index are published (Shiskin 1974).

PHILIP J. MCCARTHY

ADDITIONAL BIBLIOGRAPHY

SHISKIN, JULIUS 1974 Updating the Consumer Price Index: An Overview. *Monthly Labor Review* 97, no. 7:3–20.

[1]WILKERSON, MARVIN 1967 Sampling Error in the Consumer Price Index. *Journal of the American Statistical Association* 62:899–914. → A more formal publication of Wilkerson (1964).

INDIVIDUAL DIFFERENCE SCALING

See SCALING, MULTIDIMENSIONAL.

INFERENCE, BAYESIAN

See BAYESIAN INFERENCE.

INFERENCE, FIDUCIAL

See FIDUCIAL INFERENCE.

INFERENCE, STATISTICAL

See STATISTICS.

INSURANCE

See LIFE TABLES.

INTERACTION, STATISTICAL

See EXPERIMENTAL DESIGN; LINEAR HYPOTHESES, *article on* ANALYSIS OF VARIANCE.

INTERPENETRATING SAMPLES

See SAMPLE SURVEYS *and the biography of* MAHALANOBIS.

INTERVIEWING IN SOCIAL RESEARCH

This article was first published in IESS *as "Interviewing: I. Social Research" with two companion articles less relevant to statistics.*

The interview has been defined as a conversation with a purpose, and the purposes for which interviews are conducted are many and various. They include the purpose of therapeutic change, as in the psychiatric interview; the purpose of instruction and appraisal, as in the interviews initiated by a supervisor with a subordinate; and the purposes of selection and assessment, as in the interviews conducted with applicants for jobs or with students applying for admission to universities. In all these situations, there is the transaction of giving and getting information, and the understanding of this transaction as the immediate task of interviewer and respondent. This immediate task, however, is embedded in a larger cycle of purposive activities that define the roles of interviewer and respondent more exactly, reflect the motives of both for undertaking the interview, and stipulate the consequences of the interview for other aspects of their lives.

To label an interview "psychiatric" or "therapeutic," for example, implies that it probably has been initiated by the respondent (or patient) and that his motivation in doing so is to obtain relief from certain symptoms or strains of a mental or emotional sort. Moreover, the interviewer is seen not only as an information getter but also as a direct and powerful source of help; the interview is seen not only as an informational transaction but also as part of the therapeutic experience.

By contrast, the research interview, to which this article is addressed, may be defined as a two-person conversation that is initiated by the interviewer for the specific purpose of obtaining information that is relevant to research. Such an interview is focused on content specified by the usual research objectives of systematic description, prediction, or explanation. Other characteristics of the research interview are more variable. Typically, however, the differentiation of roles between interviewer and respondent is pronounced. The interviewer has not only initiated the conversation; he presents each topic by means of specific questions, and he decides when the conversation on a topic has satisfied the research objectives (or the specific criteria which represent them) and when another topic shall be introduced. In the research interview the respondent is led to restrict his discussion to the questions posed.

The consequences of the research interview for an individual respondent are often minimal and almost always removed in time, space, and person from the interview experience itself. The respondent is asked to provide information about himself, his experiences, or his attitudes to an interviewer who has no direct power or intention to provide therapy, instruction, a job, or any other major tangible reward. If the research interview does contribute to such a reward, it does so through a sequence of events that involves the aggregation of responses from numerous interviews, some process of data reduction and inference, and some additions to the description or explanation of social facts. From this enlarged base of knowledge may come applications or decisions of policy that have great importance for the respondent, but the sequence is complex and often uncertain. Nevertheless, the prospect of ultimate benefit, public or personal, from the accumulation of knowledge is one major basis for respondent agreement to participate in a research interview.

Perhaps the prototypical example of research interviews is provided by the national census. Most countries of the world conduct some kind of population count, and in many countries the census has been expanded to provide with regularity an inventory of social resources and problems. Census interviews usually make only modest demands on interviewer and respondent. They are brief; they ask for demographic data well within the respondent's knowledge and not of a kind that he is likely to regard as confidential. Moreover, the information is requested under circumstances familiar to, or expected by, most respondents, and the request is backed by the legitimate power of the national government.

Similar to the census in most of these respects is a whole class of brief, officially sponsored, information-getting interview surveys. In the United States alone, hundreds of thousands of such interviews are conducted by agencies of government in randomly selected homes each year to provide continuing data on family income, employment and unemployment, health and illness, and other aspects of economic and social welfare.

Almost as widely known as census taking and other government-sponsored research that involves interviewing are the activities of those private agencies which conduct recurrent interview studies of public opinion on national and international affairs, family life, and other subjects of public interest. The Gallup Poll is typical of private organizations that conduct such surveys and, in the United States and much of Europe, its name has become a general term for describing them. The interviews conducted by such polls resemble those of the census in brevity, simplicity, and the avoidance of very private material. However, public opinion interviewing differs from most government-sponsored surveys in dealing with matters of attitude rather than of fact and in depending on interviewer persuasiveness rather than on legal authority and prestige to obtain respondent cooperation. Market research and studies of readership are usually of like simplicity and brevity, although more elaborate and indirect techniques of interviewing have often been used in such studies.

It is likely that the most ambitious and demanding use of the interview as a research technique has been made by social scientists in the course of psychological, sociological, political, and economic investigations. Such studies often involve interviews of an hour or more, on subjects that may raise difficult problems of recall, potential embarrassment, and self-awareness. Consider as examples the recurring studies of consumer behavior and family income (Katona 1960), the studies of fertility and family planning (Freedman et al. 1959), the studies of sexual behavior (Kinsey et al. 1948), the studies of mental health and illness (Gurin et al. 1960), of political behavior (Michigan . . . 1960), and the many studies of supervisor–subordinate relations and worker attitudes (Argyris 1964; Herzberg et al. 1959; Kahn et al. 1964; Likert 1961).

These examples suggest a conclusion that can hardly be questioned: much of the data of social science is generated by means of the interview. Sociologists, psychologists, anthropologists, political scientists, and economists depend on interviews to obtain data for describing the phenomena of interest to them and for testing their theories and hypotheses about those phenomena.

Moreover, the use of the interview in these disciplines is not limited to surveys and other studies done in the field; interviewing is a necessary element in laboratory research as well. Laboratory experiments in psychology and other social sciences typically involve a situation contrived by the experimenter in order to introduce some factor into the experience of the people who are his experimental subjects, and to do so under conditions in which their reactions can be closely studied. That study often requires interviewing as well as observation, physiological measures, and the like. The experimenter depends on the subject to report anxiety or elation, increased confidence or reduced self-esteem, and feelings of acceptance or rejection (see, for example, Asch 1952; Milgram 1965). The interview helps the experimenter to learn whether the intended manipulation of a variable really "took," and if so, whether it had the predicted effects.

In short, the social scientist, from the first nineteenth-century British surveys of poverty (Booth et al. 1889–1891) and the early psychophysical experiments in the laboratory (Boring 1929), has been, willy-nilly, an interviewer. Whether he has been too much an interviewer (Webb et al. 1966) is a question of prime importance for the strategy of social science but not for the present discussion.

Like other scientists, the social researcher has attempted to measure rather than merely to describe in qualitative terms, and for him the interview has been the most useful instrument in the measurement process.

The interview as measurement

We have defined the interview as a conversation with a purpose and have further specified that the purpose with which we are concerned is information getting. The research interview, however, is not after mere information; it has to do with that particular quantitative form of information getting called *measurement*. The interview is one part, and a crucial one, in the measurement process as it is conducted in much of social research. Thus, the use of the interview is subject to the laws of measurement; it can be properly judged by the standards of measurement, and it suffers from the limitations of all measurement processes in degrees peculiar to itself.

The key concept for thinking about the adequacy of measurement is *validity,* defined as the extent to which an instrument and the rules for its use in fact measure what they purport to measure (Kaplan 1964; Selltiz et al. 1959). Inferences about validity, however, particularly about interview validity, are too often made on the basis of *face validity*, that is, whether the questions asked look as if they are measuring what they purport to measure. A preferable way of thinking about validity, and a basis for developing tests of validity, is the question of what a given measurement will do. Does the measure do the things that theory and experiment have convinced us it should do (Campbell 1957; Cronbach 1946; Coombs 1964)? For example, does a test that purports to measure intelligence enable us to predict scholastic achievement to some significant degree?

A similar approach to the validation of interview measures involves the comparison of the interview measure with some other measure that has already met the test of validity. This kind of comparison has been called *convergent validity*, in part to distinguish it from other approaches to the validation of constructs. If the two measures in fact agree, there is a presumption of validity for the measure being tested at least as great as that of the measure taken as standard. Thus hospitalization data obtained from interviews might be validated against hospital admission records. When the standard measure has already been validated, the method of convergent validity is powerful. For example, when the results of political surveys correspond to election statistics, or when the results of consumer surveys agree with the volume of actual purchases, we have confidence in the survey measures. The problem of validation becomes more difficult, however, in the case of interview measures of attitudes, for which no independent and objective measures exist in quite the same terms.

Questions of validity and invalidity are only part of the problem of measurement adequacy. A measure is invalid to the extent that it measures something more than or less than it purports to measure. Put another way, the mark of invalidity is bias, which is a systematic or persistent tendency to make errors in the same direction, that is, to overstate or understate the "true value" of an attribute. Scarcely less important than validity is *reliability*, which has to do with the stability and equivalence of a measure (Cronbach 1949). The reliability of an interview measure is defined by such questions as these: If the measure is used repeatedly in the same circumstances, will it yield the same results? If it is used by different interviewers to measure the same attribute, will it produce the same results? Methods for determining the reliability of a measure are to arrange for repetitions of it in identical circumstances ("test–retest reliability") or, if the measure involves numerous items, to compare the results obtained on the basis of one half of the items, randomly selected, with the results obtained by using the other half of the items. The latter method is called the "split-half reliability" test.

The relationship between validity and reliability is complex. For the interview method, however, it is important to remind ourselves that a measure may be valid without being reliable; that is, it may measure what it purports to, but do so badly. On the average, such a measure obtains the "true value," but its variance is large. Repeated use of the measure in the same circumstances produces values which are random about the mean but vary from it by large amounts.

The question of how to insure or achieve measurement adequacy by means of interview procedures can be answered in several ways. One solution, of course, is to restrict oneself to measures already developed, for which pedigrees of reliability and validity have been established. Many paper and pencil tests and some interview scales have been developed with enough attention to methodological considerations so that such data are available. Examples include numerous personality scales, tests of intelligence and reading ability, the census procedures for ascertaining labor-force status and occupation, measures of political party identification, and others. Unfortunately, in the social sciences standard, well-validated measures are not available for most concepts. As a result the investigator is commonly faced with the need to develop his own measures.

For the investigator who must create his own interview measures, there exists a considerable accumulation of general principles and specific procedures to guide him in the preparation of questions, scales or sets of questions, and questionnaires (Payne 1951; Kahn & Cannell 1957; Richardson et al. 1965). All of these help to achieve validity in measurement. It is also possible to improve validity by including in the data collection measures of potential sources of bias, so that there can be an after-the-fact assessment of the extent of their intrusion and a correction in the raw data. The "lie scale" of the Minnesota Multiphasic Personality Inventory, the Edwards Social Acceptability Scale, and the Mandler–Sarason Test Anxiety Scale are examples of measures that are used for such ex post facto statistical corrections with some success. The achievement of high reliability depends on the same basic principles but is particularly enhanced by specifying the exact wording of the questions to be asked in the interview, as well as the forms and range of behavior that may be used in the interview to evoke response, and by using multiple questions rather than single questions for the measurement of each concept or variable.

In short, adequacy of measurement by means of interviews requires knowledge of the conditions for a successful interview, and the skill to meet those conditions both in the construction of questionnaires or interview schedules and in the conduct of the interview itself. These issues are the subjects of the following sections of this article.

Conditions for successful interviewing

Interviews are not uniformly successful. Respondents differ in ability and motivation; interviewers differ in skill; and interview content differs in feasibility. While many approaches have been taken to these problems, three broad concepts seem to comprise much of the available research and advice. These are *accessibility* of the required data to the respondent, *cognition* or understanding by the respondent of his role and the informational transaction required of him, and *motivation* of the respondent to take the role and fulfill its requirements. These are not independent factors; they can be thought of as a set of interrelated conditions for attaining an adequate interview, and most of the specific techniques of interviewing (to be discussed in the following section) can be thought of as means of meeting these conditions.

Accessibility. The simplest condition for interviewing occurs when the datum which the interviewer requires is completely accessible to the respondent, that is, when he has the information in conscious form, clearly conceptualized in the terms used by the interviewer. This condition is typically met for simple demographic data—age, family size, and the like. To the extent that the required data are accessible to the respondent, the interviewer can turn his attention to problems of cognition and motivation, making sure that the respondent understands what is asked of him and that he is willing to provide the information which he possesses. To the extent that the data are inaccessible to the respondent, this inaccessibility constitutes the first problem in interviewing.

Three major reasons for inaccessibility can be distinguished, each with its own implications for the formulation of questions and the conduct of the interview. First, the material may simply have been forgotten (Bartlett 1932). The respondent may once have been in conscious possession of the required information, but it has receded from conscious recollection. A second kind of inaccessibility has to do with repression; an event is important or recent enough to be remembered, but it involves sufficient emotional stress to have been obliterated from conscious memory. The third category of inaccessibility has to do not with the intrinsic content of the material sought but with the terms or categories in which the interview requires recollection and communication. Problems of language, vocabulary, and understanding are involved, as well as differences in social class, subculture, and region. This aspect of inaccessibility is related to the cognitive conditions for successful interviewing. For example, an interviewer may be so ill-advised as to ask workers to recount recent experiences on the job which were "ego-enhancing" and "ego-threatening." A respondent may be quite insightful and observant, quite able to describe in some detail his experience on the job, and yet be unable to respond meaningfully to this question. Moreover, his difficulties are likely to persist even after the words "ego-enhancing" and "ego-threatening" have been defined for him, because he has not thought of his experiences in terms of such categories.

Cognitive conditions. A second requirement for successful interviewing has to do with respondent cognition or understanding. The respondent role is by definition an active, self-conscious one, and the respondent can meet its expectations best when he understands them fully. Specifically, he needs to know what constitutes successful completion of the role requirements and to know the concepts or terms of reference by means of which he is being asked to provide data. Without this understanding, data accessible to the respondent are nevertheless likely to remain unreported because interviewer and respondent lack a common frame of reference, a common conceptual language, or common standards of response adequacy and excellence.

How much understanding of the research enterprise by the respondent is appropriate will vary with the demands being made in the interview, and how much effort the interviewer must devote to the development of such understanding will depend on the extent of the interview demand and the sophistication of the respondent. Typically the respondent need not understand the nature of the measurement being attempted, the construction of scales, or the plans for computer analysis. He should understand the requirements of his own role in the interview—the demands to be made on him, and the criteria for relevance and completeness. To understand these things and to be motivated to accept the role may, in turn, require acquaintance with the over-all aims of the research enterprise, information about its compatibility with the respondent's own goals, and reassurance about its risks. These issues bring us to the third condition for successful interviewing.

Respondent motivation. There is general agreement among students and practitioners of interviewing that respondent motivation or willingness to report is a prime condition for successful data collection; it could hardly be otherwise. There is little agreement, however, about the theory or model of motivation that is most appropriate to the interview, and about the major sources of respondent motivation. Kahn and Cannell (1957) propose a dual emphasis: intrinsic motivation, because the experience and relationship with the interviewer are valued by the respondent; and instrumental motivation, because the respondent sees that the enterprise of which the interview is a part is congruent with his own goals and values. Kinsey and his colleagues (1948) stress altruism as the initial source of respondent motivation, although many of their interview descriptions seem to rely more on the assumption of legitimate authority and the use of medical–scientific prestige. Richardson, Dohrenwend, and Klein (1965) explain respondent motivation in terms of altruism, emotional satisfaction, and intellectual satisfaction. Such partial agreements and discontinuities are hardly surprising; they reflect the more general diversity of motivational theories. The interview is one form of complex molar behavior; attempts to understand it will inevitably share the contemporary strengths and weaknesses of motivational theory as a whole.

Despite the lack of agreement on any one motivational model, the research evidence on the interviewing process (Hyman et al. 1954; Riesman 1958; Kahn & Cannell 1957; Richardson et al. 1965) strongly urges that respondent motivation be conceptualized in terms that take account of the social situation of interviewer and respondent, the nature of the transaction between them, their perceptions of each other and of their joint task, and the effects of such perceptions. In short, the evidence argues in favor of a motivational model that treats the interview as a social process and regards the interview product as a social outcome. One such model is presented in Figure 1.

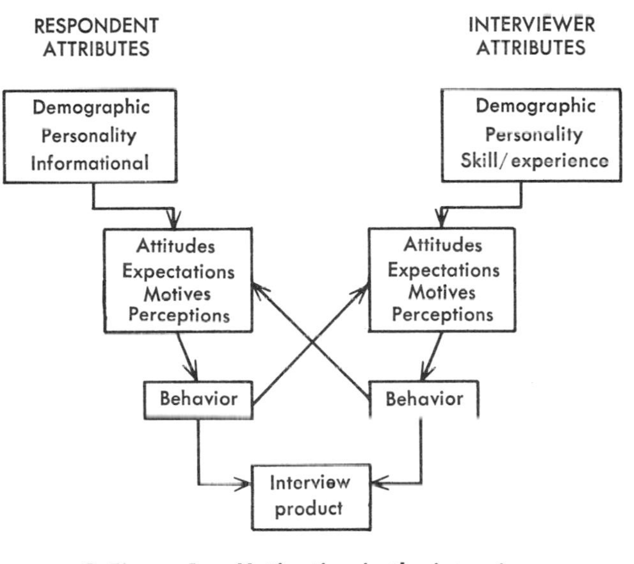

○ ***Figure 1 — Motivation in the interview: a process model***

This model is compatible with the role-oriented view of the interview. It stipulates that the interview product or outcome is the immediate and joint result of interviewer and respondent behavior; that the behavior of both interviewer and respondent stems from their attitudes, motives, expectations, and perceptions; and that these, in turn, can be understood as reflecting more enduring attributes of demography and personality. The model also emphasizes the interaction of respondent and interviewer. The behavior of the interviewer is perceived by the respondent, and it generates or modifies his attitudes and his motivation to continue the interaction. The respondent is reacting not only to the interviewer's behavior as such, how-

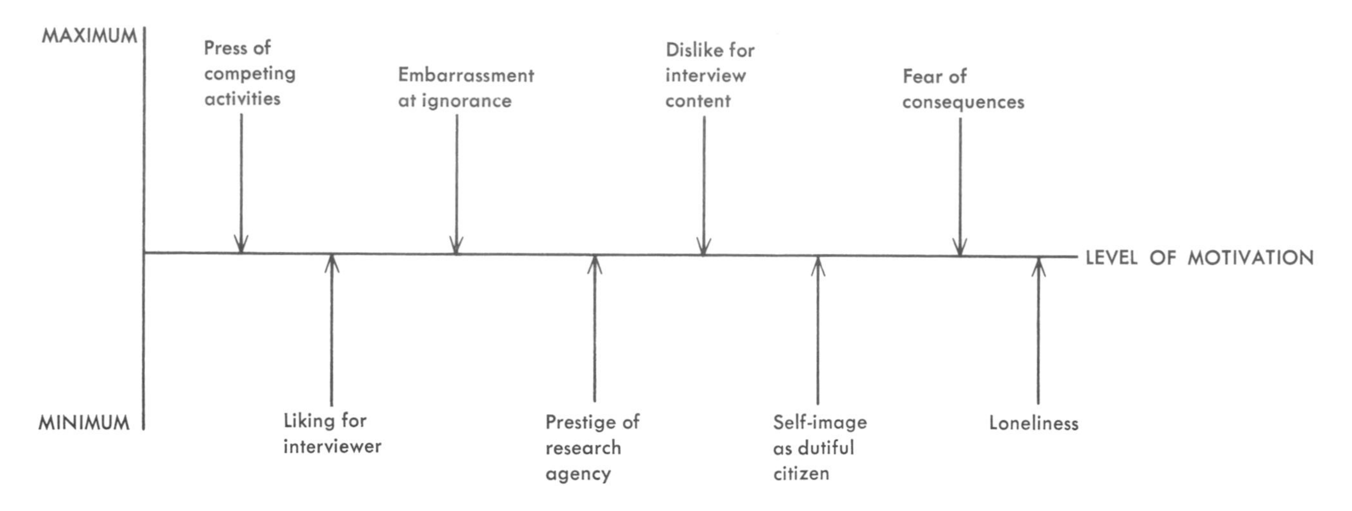

Figure 2 — Opposing motivational forces in the interview

ever; he is reacting to it as a cue that evokes role behavior already familiar in other contexts, as well as attitudes already formed. Thus, the respondent may behave toward the interviewer as a polite stranger, a hospitable host, a dutiful citizen, a fellow research worker and scientist, or even an obedient servant or intimidated inferior. The interviewer's expectations and behavior in turn are a mixed product of his own personality and experience, combined with his immediate reaction to what the respondent is doing.

Such a general model of the interview requires additional specification. It requires, among other things, some means for representing the moment-by-moment state of the respondent's motivation to provide complete and accurate data. The various forces tending, at any point in the interview, to increase or decrease respondent motivation to work in this sense can be well represented by the Lewinian model of the "quasi-stationary equilibrium" (Lewin 1947), in which each factor urging compliance or resistance is depicted as an arrow (force), and the level of motivation is depicted as a horizontal line that is the resultant of the opposing forces. In Figure 2 the factors identified as opposing forces have been taken from current research on the interview process (Fowler 1965; U.S. Department . . . 1965*a*).

Techniques of data collection

It is clear that the conceptualizations of the interview and of the conditions for success have implications for the specific techniques and procedures to be advocated. We have emphasized the conditions of accessibility, cognition, and motivation, and the role relationship of the interviewer and respondent. It follows, for example, that if the dominant problem in a particular case is accessibility, one thinks in terms of ways in which the demand for data might be limited, of records or documents which might evoke associated recollections, and the like.

The technique of data collection by interview can be usefully separated into two main and related aspects—question formulation (developing a measurement instrument) and interviewing itself (using the instrument). The close relationship of these two aspects of technique is most apparent when the same person is performing both functions in rapid succession, as anthropologists do habitually and other social research workers do at least in the early phases of investigation. Even when question formulation and data collection are conspicuously separated in time and space (as in survey research and opinion polling), the connection remains. The instrument both limits and assists the interviewer, and the interviewer necessarily modifies the instrument and the accompanying rules in the very act of using them.

To relate each bit of detailed practical advice on question formulation and interviewing technique to the above theoretical material would be desirable, but lengthy and difficult. The following sections are based, however, on the same model of the interviewing situation, with special emphasis on respondent–interviewer interaction, role taking, and the attainment of accessibility, understanding, and motivation.

The formulation of questions. Of the many ways of describing decisions that the interviewer or research worker must make about the form of questions, two seem particularly useful and important: questions may be open or closed; and they may be direct or indirect in their relationship to objectives. Openness has to do with the form of the response required by a question; an open question invites the respondent to reply in his own words, and a closed question asks him to select, from a series of alternatives, the answer that best approximates his views. The closed question thus controls the form, length, and content of the possible response. The classic example is the trial lawyer's instruction to the witness on the stand that he "answer 'Yes' or 'No.'"

Directness and indirectness have to do with the relationship between the question and the objective, that is, the interviewer's purpose in asking the question. For example, when a respondent is shown an ambiguous picture and asked to tell a story about its meaning for him, so that the story can subsequently be used to infer the intensity of the respondent's need for achievement (Atkinson 1958), we consider the question to be indirect. A direct (but ill-advised) question to meet the same objective would ask the respondent how achievement-oriented he considers himself to be.

Open versus closed questions. The intolerant advocacy of one or another of these question forms has been largely abandoned, as Lazarsfeld (1944) long ago proposed it should be. But there remains the problem of how and when to choose one type of question rather than another. At least five considerations are relevant to the choice between open and closed questions: interview objectives, the respondent's information level, the strength of the respondent's opinions on the topic, the respondent's motivation to communicate on the topic, and the interviewer's initial knowledge of these characteristics of the respondent.

With respect to the objectives of the interview, the general principle is that closed questions tend to be more appropriate for straightforward categorizing of respondents according to their agreement or disagreement with some stated point of view. If the interviewer's objectives go beyond such description and include explanatory aims, such as discovering the respondent's particular frame of reference or the process by which he came to his present views, an open question will almost certainly be superior to a single closed one, and perhaps to a combination of several closed questions.

The choice between open and closed questions should also be guided by the probable degree of structuring of the respondent's opinion on or experience with the topic. To the extent that the respondent has done his cognitive work in advance, so to speak, and has formulated his ideas in terms close to those of the question, the closed form is appropriate, as well as economical in terms of the interviewer's time and the respondent's effort. On the other hand, to the extent that the respondent's thoughts are less structured on the topic in question, the interviewer must assist the respondent to recall, order, and perhaps evaluate his experience.

Definitive research on the motivational advantages of open versus closed questions is yet to be done. The closed question demands less effort from the respondent, as well as less self-revelation, and may therefore be less threatening. To the extent that the closed question incorporates extreme alternative responses, it may also make these extremes more admissible (Kahn 1952; Metzner & Mann 1951). On the other hand, the closed question may be restrictive and may also invite an easy, invalid response instead of the more difficult "don't know." Finally, the choice between the closed form and the open form should take into account the interviewer's (or research worker's) advance knowledge of the situation. If he knows relatively little about the range or terms of response he will encounter, he is obviously in a poor position to formulate closed questions, which are meaningful and successful only when their limited alternatives match the respondent's experience, vocabulary, and frame of reference. If these conditions are not met, the open question is likely to be preferable.

Direct and indirect questions. A direct question, as we have said, simply asks for information stipulated by the objectives of the interview; there is a congruence between objective and question that is obvious on inspection. An indirect question is not congruent with its objective in the same sense; it is related by some inference or theory. One of the major reasons for using indirect questions is to obtain information about the respondent that he is incapable of providing directly because it is beyond the limits of his conscious insight. Indirect questions are also used to get around unwillingness or inability to report certain kinds of material directly.

The forms of indirection include the use of the third person, since people are sometimes willing to impute to others feelings and opinions that they will not admit as their own. They also include questions that have the appearance of directness but are interpreted as measuring attributes which the respondent is not aware of as being measured (see, for example, the *F*-scale for the measurement of authoritarianism, as described by Adorno et al. 1950). Finally, indirection may take the form of a purposefully ambiguous stimulus—an ink blot, a picture of uncertain meaning, or the beginning of a sentence which the respondent is asked to talk about spontaneously, to interpret, or to complete.

The indirect approach clearly makes accessible data that would otherwise be inaccessible. The disadvantages of indirection have mainly to do with the problem of validation. Face validity is always risky, but with indirect measures it quickly becomes meaningless.

Apart from the decisions regarding the open or closed form of questions and the direct or indirect means of questioning, there are a number of considerations in question wording and question sequence which bear on the attainment of cognitive understanding, accessibility, and respondent motivation. These are considered briefly below.

Language. The primary criterion in the choice of language is complete and accurate communication of ideas to the respondent. The language of the question must therefore conform to vocabulary that is available to him. This does not mean that the respondent's colloquialisms or regionalisms should be imitated; such efforts are more often ludicrous than convincing. However, it does mean that the basis for communication between interviewer and respondent consists of their shared vocabulary and that questions should be formulated in these terms.

Frame of reference. Inevitably each respondent interprets and replies to a question in terms of his own experience and of his own present concerns and interests. Even the casual greeting "How are things?" leads one person to think of health, another of financial matters, a third of family affairs. A common but undesirable practice in question formulation is to assume the respondent's frame of reference is the same as the interviewer's. More desirable practice is to ascertain the respondent's frame of reference by additional questioning ("Why is that?" "What did you have in mind?") whenever his reply to a major question has not made clear his frame of reference. An alternative procedure is to stipulate the desired frame of reference as part of the question itself. Thus the unqualified "How are things?" might become "How are things going—financially, I mean?" The respondent's frame of reference may be so powerful, however, that it becomes necessary to allow him to answer the question in his own terms before attempting to impose a different frame of reference on him (Bancroft & Welch 1946).

Avoiding ambiguity. A common fault in question formulation is the inclusion of two or more propositions in such a way that the meaning of the response is ambiguous. For example, consider the survey question "Do you favor or oppose raising real estate taxes for new schools or highways?" A direct response of "Favor" or "Oppose" is ambiguous with respect to the referent; it may be tax increases in general, schools, highways, or all of these. Avoidance of such ambiguity is simple: the researcher need only keep in mind that each question should have a single and unambiguous referent and should test, with respect to that referent, a single proposition or point of view.

Recognition versus recall. Where recall is a problem for respondents, questions can usefully be formulated in terms that require only recognition. Experiments in the psychological laboratory have shown that in studies of memory the process of recognition is an easier task for the subject than the process of recall. Moreover, there is some evidence (Kahn 1952) that the presentation of alternatives covering a wide range also increases the likelihood of respondents' choosing extreme statements and admitting to socially unacceptable opinions or behaviors. Unfortunately, little research has been done to explore the disadvantages of such questions or to document unequivocally their relative reliability and validity under varying conditions of data accessibility.

Sanctioning. The problem of defensiveness can be dealt with in many ways, primarily in the interpersonal technique of interviewing rather than in the formulation of questions. Nevertheless, the wording of questions can facilitate or inhibit the respondent's admission to facts or opinions that are in some fashion ego-threatening to him. One means of making such material admissible is to include it in the hypothetical range of alternatives presented to the respondent, as suggested above. Another is to build into the question some phrase of reassurance, some reminder of the purpose of the inquiry, or some factual indication that the "unacceptable" is common and, in this context, acceptable.

Leading questions. Avoiding "biased" or "leading" questions is a standard and oversimplified piece of advice in the formulating of questions. By a "leading question" we mean one so formulated that respondents find it easier or more acceptable to answer in one way than another, or to choose one alternative over another. When such questions are formulated inadvertently and the responses are interpreted without regard to the asymmetrical tendency of the questions themselves, the results are biased. The most common example in survey research is the yes–no type of questions, in which respondents who answer "Yes" are then subjected to a long series of additional questions about time, place, reason, and reaction, while respondents who answer "No" are asked no more on the topic. People learn quickly that it is easy and brief to say "No." Still more crude is the use of question wording that assumes a particular answer and thus forces the respondent to contradict the interviewer in order to formulate a response of his own. The generally accepted principle regarding such questions is that they should be asked only for the purpose of imposing some additional stress on the respondent, for example, in order to identify people who feel strongly enough to assert that the interviewer has assumed wrongly (Smith et al. 1956; Litwak 1956; Kahn & Cannell 1957; Kinsey et al. 1948). The rule of thumb justly favors balanced wording, questions designed to equalize the amount of work and the degree of social acceptability regardless of which alternative a respondent chooses, and avoidance of emotionally loaded words and phrases.

Organization and sequence of questions. Issues of organization and sequence of questions, like so much else in interviewing, cannot be settled by generalization but must be resolved in relation to the interview objectives and the characteristics of the population from whom information is sought. Generally, a battery of questions is preferable to a single question, for reasons of both reliability and validity. The more complex the issue and the less tested the approach, the more important it becomes to use multiple questions. When multiple closed questions are used, and the respondent is asked to indicate agreement or disagreement with a stated proposition, it is important to randomize the form of the statements. The purpose of doing so is to randomize the tendency of some respondents to be chronic "yea-sayers" or "nay-sayers" (Couch & Keniston 1960). Similarly, if lists of items are being presented to respondents, the order should be varied, since some respondents show a tendency to select the alternative first presented. (See Messick 1968.)

The sequence of the topics themselves should be planned to make the total interview experience as meaningful as possible by giving it a beginning, a middle, and an end. More specifically, the early questions should serve to engage the respondent's interest—without threatening or taxing him before he is really committed to the transaction—and to teach him the kind of task the interview represents. The most demanding of questions might well be placed later in the interview, at a point when the respondent's commitment can be presumed to have peaked but fatigue has not yet set in.

Interviewing technique. Two major qualifications must be made of any specific set of recommendations respecting interviewing technique. First, the technique of the interviewer is not really separable from the formulation of questions. In almost all situations, the interviewer is in some degree a formulator of questions. In much case study and anthropological work the interviewer formulates and asks questions almost simultaneously; he interacts spontaneously with the respondent. At the other extreme is the injunction to census enumerators to ask each question precisely as worded, to do no improvising of additional questions, and to respond to requests for explanation by repeating the question. As methodological research has shown, there is considerable variation in interviewer technique even under such constrained circumstances (Hansen et al. 1951). The second qualification to recommendations about technique is that the most appropriate advice must differ with the situation—as defined by respondent characteristics, interviewer characteristics, and above all by the task requirements of the interview itself. For example, to develop a close and trusting relationship as a preliminary to a few questions for a school census is ludicrous, whereas to neglect the development of such a relationship in psychiatric diagnosis is equally bad.

It follows that recommendations about technique must be stated in relation to such situational factors. The following recommendations accordingly are based on the situational factors most common in social research (including but not limited to survey research). In such research the respondent is free to give information or refuse it. Moreover, the power of an interviewer over a respondent is limited. He can neither impose formal penalties for nonresponse nor offer a prize for response. The demands on respondent time and effort are significant but not overwhelming, perhaps as little as one-half hour or as much as two hours. The interviewer approaches the respondent as a stranger; the interviewer is identified with some sponsoring agency (university, research institute, or the like) that has at least modest prestige value without possessing legitimate authority or coercive power to demand information. The interviewer must generate and maintain sufficient respondent motivation to meet the interview objectives, and he must direct and control the communication process in the service of those objectives. In general, he does these things by describing the purpose of the interview, treating the respondent with some reasonable show of warmth and interest, indicating directly and approvingly those responses which are relevant and complete, and letting the respondent know also when he is being irrelevant or fragmentary in his answers. These things the interviewer does by building on the specific questions that have been prepared in advance of the interview. He adds supplementary or "probe" questions; he comments on the completeness or inadequacy of a response; he nods, murmurs, and in other ways exerts control over the communications process. Let us consider these behaviors sequentially and specifically.

○ The introduction of the interview to the respondent is basically the negotiation of informed consent. An appropriate introduction would include a statement of the purpose of the interview and identification of the sponsoring agency. In some situations this may be introduction enough. If the data requirements and time demands are more than trivial, however, or if the respondent shows curiosity or reluctance, the introduction should also include reassurances to the respondent with respect to the manner in which he came to be selected for interview, the protection and confidentiality which will be accorded his statements, and the specific ways in which his statements will be used. There is a related misunderstanding that the interviewer must be alert for during the early moments of an interview. Many respondents lack experience and knowledge of interviewing for research purposes, but they have no lack of experience at being interrogated for other purposes. To them the appearance of a stranger who wishes to ask questions suggests truant officers, bill collectors, policemen, unwanted salesmen, and a variety of other sources of threat or annoyance. The interviewer will need to look for signs of such misidentification and be prepared to explain and, on occasion, to document his true identity and function.

○ A remaining problem that arises in the introductory moments of an interview has to do with the ethics of persuasion, particularly in describing the purpose of the interview and the uses to which it will be put. To counsel absolute and complete truthfulness is easy and irreproachable, but there are circumstances in which the effect of such completeness would negate the purpose of the interview. It seems consistent with the ethics of the social sciences in general to resolve such problems by telling the respondent that giving more information before the interview might interfere with it, but by being uncompromising in letting the respondent know what the interview will require of him and any risks or disadvantage that it might involve.

The development of intrinsic motivation in the interview—emotional and intellectual satisfaction in the process itself—may begin with the introduction, but it matures only as the task and relationship acquire meaning for the respondent. The opportunity to talk to a good listener, to find one's opinions of serious interest to another person, to see that person making a real and successful effort to understand rather than to evaluate or criticize—these are experiences which are rare for many people and which are intrinsically satisfying. Thus, the interviewer should create and maintain an atmosphere in which the respondent feels that he is fully understood and in which he is safe to communicate fully without fear of being judged, criticized, or subsequently identified and disadvantaged. At the same time, the interviewer must focus attention on the content of the communication, encouraging the respondent to consider each topic as deeply, fully, and frankly as the interview objectives require.

The interviewer's means for doing these things are not mysterious; they are his elaborations on the primary (prepared) interview questions and in some respects resemble the processes of mutual influence familiar in informal conversation. Richardson and his colleagues (1965) have referred to them as "encouragements, silences, guggles, and interruptions"; Kahn and Cannell (1957) have called them "controlled nondirective probing"; other authors have used still other terms. The specific behaviors proposed for the interviewer by these various authors are more alike than their terminology. The following is a list of some of these behaviors.

Brief expectant pauses.

Brief expressions of understanding and interest:
"I see, um-hm."
"Yes, I understand."

Neutral requests for additional information:
"How do you mean?"
"I'd like to know more of your thinking on that."
"What do you have in mind there?"
"Is there anything else?"
"Can you tell me more about that?"

Echo or near repetition of the respondent's words:
Respondent: "I've taken these treatments for almost six months, and I'm not getting any better."
Interviewer: "You're not getting better?"

Summarizing or reflecting respondent expressions:
Examples would follow respondent statements, stating the interviewer's understanding of a key feeling or meaning. Such summaries often begin with phrases like "You feel that———" or "You mean that———"

Requests for specific kinds of additional information:
"*Why* do you think that is so?"
"*How* did that become clear to you?"
"*When* was that?"

Requests for clarification:
"I'm not clear on that."
"Could you explain what you mean?"

Repetition of a primary question:
Interviewer: "What kind of work do you do?"
Respondent: "I work at the paper mill."
Interviewer: "I see. What kind of work do you do there?"

To these specific forms of supplementing the primary questions in an interview must be added explanations, reassurances, and further information about the interview.

The separation of such interviewing techniques from the formulation of the major questions that present the topics of inquiry is in some degree arbitrary, of course. This distinction between primary and supplementary or "probe" questions is perhaps least useful in the two extreme forms of the interview, that is, when the interviewer is formulating all questions on the spur of the moment in a completely unstructured situation, and when the interviewer is absolutely restricted to reading prepared questions from a script or schedule. But most interviews are well within these extremes and involve some mixture of predetermined questions and spontaneous interactive elaborations on them. In most social research of scale, this functional distinction is emphasized by a division of labor; the people who conduct the interviews have usually not developed the basic questions to be asked. In such circumstances, it becomes essential to develop an understanding of the complementary functions of question formulation and elaboration or probing, so that they may in fact complement each other instead of producing a validity-destroying and unintended competition.

Selection and training of interviewers

Evidence for the importance of interviewer selection and training is compelling and of long standing, although much of it is also indirect. It is clear, for example, that a variety of interviewer characteristics—demographic and attitudinal—can affect the interview product. It is less clear under what circumstances and through what chain of events these effects occur. Race and religion have been shown to inhibit responses, particularly when the respondent holds critical views of a minority to which the interviewer apparently belongs (Hyman et al. 1954; Robinson & Rhode 1946). Differences in age and sex between interviewer and respondent have been shown to reduce the communication of some kinds of data (Benney et al. 1956). Moreover, there is evidence that the social class of interviewers is reflected in the data they obtain on political issues (Katz 1942).

The unwanted relationship between interviewers' attitudes and interview data, which was first described by Rice (1929), has been demonstrated repeatedly (Cahalan et al. 1947; Guest 1947; Ferber & Wales 1952). Stanton and Baker (1942) have demonstrated that similar outcomes may result from an "expectation bias," that is, the interviewer's expectations of the respondent. Studies of interviewer personality and its effects are fewer, but one by Richardson (1965), using TAT (Thematic Apperception Test) measures of personality in relation to interviewer performance, suggests that the effective interviewer enjoys people, seeks friendly relations with them, and has insight into the complex of feeling relationships among widely varying types of people.

It is a plausible assumption that most of the characteristics of interviewers that have been shown to create biased data do so via the interviewers' behavior rather than through the reaction of respondents to some nonbehavioral characteristic. In other words, interviewers make characteristic errors which might be avoided to some extent by appropriate techniques and training (Cannell 1953) and by selection on the basis of intelligence (Guest & Nuckols 1950) and other measurable criteria. [*See* ERRORS, *article on* NONSAMPLING ERRORS.]

As yet there is not enough research on the selection and training of interviewers to warrant making definitive statements. However, it appears that the most important consideration in selection is that the interviewer must be seen by the respondent as being "within range" of communication on the interview topic. This does not imply a matching on education, age, sex, or other characteristics but, rather, suggests that the respondent must perceive the interviewer as having sufficient knowledge and understanding so that effective communication between the two is possible.

Evidence for the effects of interviewer training under varying circumstances is sketchy; opinions are strong—and variable. They range from reliance on brief written instructions (characteristic of much commercial opinion polling and market research) to the proposals of Nadel (1951) and Kluckhohn (1945) that interviewers in anthropological research should be psychoanalyzed as well as intensively trained in more specific ways. There is evidence that training makes a difference, some of it indirect and some direct. Cannell (1953) found that carefully trained interviewers produced results that were relatively free of the class and attitudinal biases reported by other investigators; his research, however, did not include experimental comparisons among interviewers differing in training. Richardson (1965) found that intensive training in field methods produced significant increases in measures of interviewing performance and no significant changes in personality measures (TAT). Individual differences in training effects were large, and the performance of individual interviewers before training was a poor predictor of posttraining performance.

To be effective, a training program should place heavy emphasis on interviewing practice. Role playing, observation by experienced persons, and tape recordings are useful, since they provide the opportunity for immediate feedback, which is a most important aspect of training.

Any article on interviewing in social research written today should be tentative in tone. For a number of reasons, the field is in flux. For one thing, some new ideas have been proposed after a long and relatively static period. Webb, Campbell, and their colleagues (Webb et al. 1966) have argued persuasively for less addiction to interviewing among social research workers and more attention to "unobtrusive measures." Richardson, Dohrenwend, and Klein (1965) have presented the opposing notions of "stressed" versus "unstressed" interviews in terms that invite research on the appropriate use of differing techniques. Cannell (see U.S. Department . . . 1965*a*; 1965*b*), Fowler (1965), and others have questioned the earlier emphasis on avoidance of interviewer influence and have begun research on the interviewer's functions as teacher and reinforcer of appropriate respondent behavior, as well as permissive encourager of conversation.

A second factor that makes for change in interviewing theory and practice is cultural and historical; the violation of privacy for trivial or questionable purposes has brought the rights and roles of interviewers and respondents under some discussion in the United States. It is a poor time for predicting the outcome of a trend that is so new and that may show vastly different forms in different cultures. Nevertheless, it seems likely that increasing sophistication about the reasons for interviewing and the consequences of the interview is a part of technological and industrial development in its contemporary forms. Whether the increasing sophistication of respondents will make the collection of data by interviewing easier, more difficult, or virtually impossible depends on the visible uses and abuses of the interview. In all but its most extreme and indirect forms, the interviewing technique is ultimately dependent upon a societal record of individual protection, respect for the confidentiality of personal data, and relevant and benign use of information in research, industry, and government.

ROBERT L. KAHN AND CHARLES F. CANNELL

[*See* ETHICAL ISSUES IN THE SOCIAL SCIENCES. *See also* Becker 1968; Butler 1968; Kadushin 1968; Lindzey & Thorpe 1968; Mullahy 1968; Powdermaker 1968; Stanton 1968; Wax 1968.]

BIBLIOGRAPHY

ADORNO, THEODOR W. et al. 1950 *The Authoritarian Personality.* American Jewish Committee, Social Studies Series, No. 3. New York: Harper.

ARGYRIS, CHRIS 1964 *Integrating the Individual and the Organization.* New York: Wiley.

ASCH, SOLOMON E. (1952) 1959 *Social Psychology.* Englewood Cliffs, N.J.: Prentice-Hall.

ATKINSON, JOHN W. (editor) 1958 *Motives in Fantasy, Action, and Society: A Method of Assessment and Study.* Princeton, N.J.: Van Nostrand.

BANCROFT, GERTRUDE; and WELCH, EMMETT H. 1946 Recent Experience With Problems of Labor Force Measurement. *Journal of the American Statistical Association* 41:303–312.

BARTLETT, FREDERIC C. (1932) 1950 *Remembering: A Study in Experimental and Social Psychology.* Cambridge Univ. Press.

►BECKER, HOWARD S. 1968 Observation: I. Social Observation and Social Case Studies. Volume 11, pages 232–238 in *International Encyclopedia of the Social Sciences.* Edited by David L. Sills. New York: Macmillan and Free Press.

BEEZER, R. H. 1956 *Research Methods of Interviewing Foreign Informants.* Technical Report, No. 30. Washington, D.C.: George Washington Univ., Human Resources Research Office.

BENNEY, MARK; RIESMAN, DAVID; and STAR, SHIRLEY A. 1956 Age and Sex in the Interview. *American Journal of Sociology* 62:143–152.

BINGHAM, WALTER; and MOORE, BRUCE V. (1931) 1959 *How to Interview.* 4th ed., rev. New York: Harper.

BOOTH, CHARLES et al. (1889–1891) 1902–1903 *Life and Labour of the People in London.* 17 vols. London: Macmillan.

BORING, EDWIN G. (1929) 1950 *A History of Experimental Psychology.* 2d ed. New York: Appleton.

►BUTLER, JOHN 1968 Mental Disorders, Treatment of: III. Client-centered Counseling. Volume 10, pages 178–185 in *International Encyclopedia of the Social Sciences.* Edited by David L. Sills. New York: Macmillan and Free Press.

CAHALAN, DON; TAMULONIS, VALERIE; and VERNER, HELEN W. 1947 Interviewer Bias Involved in Certain Types of Opinion Survey Questions. *International Journal of Opinion and Attitude Research* 1, no. 1: 63–77.

CAMPBELL, DONALD T. 1957 Factors Relevant to Validity of Experiments in Social Settings. *Psychological Bulletin* 54:297–312.

CANNELL, CHARLES F. 1953 A Study of the Effects of Interviewers' Expectations Upon Interviewing Results. Ph.D. dissertation, Ohio State Univ.

COOMBS, CLYDE H. 1964 *A Theory of Data.* New York: Wiley.

COUCH, A.; and KENISTON, K. 1960 Yeasayers and Naysayers: Agreeing Response Set as a Personality Variable. *Journal of Abnormal and Social Psychology* 60: 151–174.

CRONBACH, LEE J. 1946 Response Sets and Test Validity. *Educational and Psychological Measurement* 6: 475–494.

CRONBACH, LEE J. (1949) 1960 *Essentials of Psychological Testing.* 2d ed. New York: Harper.

FERBER, ROBERT; and WALES, HUGH G. 1952 Detection and Correction of Interviewer Bias. *Public Opinion Quarterly* 16:107–127.

FOWLER, F. J. JR. 1965 Education, Interaction, and Interview Performance. Ph.D. dissertation, Univ. of Michigan.

FREEDMAN, RONALD; WHELPTON, PASCAL K.; and CAMPBELL, ARTHUR A. 1959 *Family Planning, Sterility, and Population Growth.* New York: McGraw-Hill.

GORDEN, RAYMOND L. 1954 An Interaction Analysis of the Depth-interview. Ph.D. dissertation, Univ. of Chicago.

GUEST, LESTER L. 1947 A Study of Interviewer Competence. *International Journal of Opinion and Attitude Research* 1, no. 4:17–30.

GUEST, LESTER L.; and NUCKOLS, ROBERT 1950 A Laboratory Experiment in Recording in Public Opinion Interviewing. *International Journal of Opinion and Attitude Research* 4:336–352.

GURIN, GERALD; VEROFF, JOSEPH; and FELD, SHEILA 1960 *Americans View Their Mental Health: A Nationwide Interview Survey.* Joint Commission on Mental Illness and Health, Monograph Series, No. 4. New York: Basic Books.

HANSEN, MORRIS H. et al. 1951 Response Errors in Surveys. *Journal of the American Statistical Association* 46:147–190.

HERZBERG, FREDERICK; MAUSNER, BERNARD; and SNYDERMAN, BARBARA B. 1959 *The Motivation to Work.* New York: Wiley.

HILDUM, DONALD C.; and BROWN, ROGER W. 1956 Verbal Reinforcement and Interviewer Bias. *Journal of Abnormal and Social Psychology* 53:108–111.

HYMAN, HERBERT H. et al. (1954) 1962 *Interviewing in Social Research.* Univ. of Chicago Press.

►KADUSHIN, CHARLES 1968 Reason Analysis. Volume 13, pages 338–343 in *International Encyclopedia of the Social Sciences.* Edited by David L. Sills. New York: Macmillan and Free Press.

KAHN, ROBERT L. 1952 A Comparison of Two Methods of Collecting Data for Social Research: The Fixed-alternative Questionnaire and the Open-ended Interview. Ph.D. dissertation, Univ. of Michigan.

KAHN, ROBERT L.; and CANNELL, CHARLES F. 1957 *The Dynamics of Interviewing: Theory, Technique and Cases.* New York: Wiley.

KAHN, ROBERT L. et al. 1964 *Organizational Stress: Studies in Role Conflict and Ambiguity.* New York: Wiley.

KAPLAN, ABRAHAM 1964 *The Conduct of Inquiry: Methodology for Behavioral Science.* San Francisco: Chandler.

KATONA, GEORGE 1960 *The Powerful Consumer: Psychological Studies of the American Economy.* New York: McGraw-Hill.

KATZ, DANIEL 1942 Do Interviewers Bias Poll Results? *Public Opinion Quarterly* 6:248–268.

KINSEY, ALFRED C. et al. 1948 *Sexual Behavior in the Human Male.* Philadelphia: Saunders.

KLUCKHOHN, CLYDE 1945 The Personal Document in Anthropological Science. Pages 79–173 in Louis R. Gottschalk, Clyde Kluckhohn, and Robert Angell, *The Use of Personal Documents in History, Anthropology and Sociology.* New York: Social Science Research Council.

KRASNER, LEONARD 1958 Studies of the Conditioning of Verbal Behavior. *Psychological Bulletin* 55:148–170.

LAZARSFELD, PAUL F. 1944 The Controversy Over Detailed Interviews—An Offer for Negotiation. *Public Opinion Quarterly* 8:38–60.

LENSKI, G. E.; and LEGGETT, J. C. 1960 Caste, Class and Deference in the Research Interview. *American Journal of Sociology* 65:463–467.

LEWIN, KURT 1947 Frontiers in Group Dynamics. Parts 1–2. *Human Relations* 1:5–41, 143–153. → Part 1: Concept, Method and Reality in Social Science, Social Equilibria and Social Change. Part 2: Channels of Group Life; Social Planning and Action Research.

LIKERT, RENSIS 1961 *New Patterns of Management.* New York: McGraw-Hill.

►LINDZEY, GARDNER; and THORPE, JOSEPH S. 1968 Projective Methods: I. Projective Techniques. Volume 12, pages 561–568 in *International Encyclopedia of the Social Sciences.* Edited by David L. Sills. New York: Macmillan and Free Press.

LITWAK, EUGENE 1956 A Classification of Biased Questions. *American Journal of Sociology* 62:182–186.

MANDLER, GEORGE; and KAPLAN, WARREN 1956 Subjective Evaluation and Reinforcing Effect of a Verbal Stimulus. *Science* 124:582–583.

MERTON, ROBERT K.; FISKE, MARJORIE; and KENDALL, PATRICIA L. (1944) 1956 *The Focused Interview: A Manual of Problems.* Glencoe, Ill.: Free Press.

►MESSICK, SAMUEL 1968 Response Sets. Volume 13, pages 492–496 in *International Encyclopedia of the Social Sciences.* Edited by David L. Sills. New York: Macmillan and Free Press.

METZNER, HELEN; and MANN, FLOYD C. 1951 A Limited Comparison of Two Methods of Data Collection: The Fixed-alternative Questionnaire and the Open-ended Interview. Unpublished manuscript, Univ. of Michigan, Institute for Social Research.

MICHIGAN, UNIVERSITY OF, SURVEY RESEARCH CENTER 1960 *The American Voter,* by Angus Campbell et al. New York: Wiley.

MILGRAM, STANLEY 1965 Some Conditions of Obedience and Disobedience to Authority. Pages 243–262 in Ivan D. Steiner and Morton Fishbein (editors), *Current Studies in Social Psychology.* New York: Holt.

►MULLAHY, PATRICK 1968 Sullivan, Harry Stack: II. Interpersonal Theory. Volume 15, pages 398–406 in *International Encyclopedia of the Social Sciences.* Edited by David L. Sills. New York: Macmillan and Free Press.

NADEL, SIEGFRIED F. 1951 *The Foundations of Social Anthropology.* London: Cohen & West; Glencoe, Ill.: Free Press. → See especially pages 169–176.

PAYNE, STANLEY L. 1951 *The Art of Asking Questions.* Studies in Public Opinion, No. 3. Princeton Univ. Press.

►POWDERMAKER, HORTENSE 1968 Field Work. Volume 5, pages 418–424 in *International Encyclopedia of the Social Sciences.* Edited by David L. Sills. New York: Macmillan and Free Press.

QUAY, HERBERT 1959 The Effect of Verbal Reinforcement on the Recall of Early Memories. *Journal of Abnormal and Social Psychology* 59:254–257.

RICE, STUART A. 1929 Contagious Bias in the Interview: A Methodological Note. *American Journal of Sociology* 35:420–423.

RICHARDSON, STEPHEN A. 1965 A Study of Selected Personality Characteristics of Social Science Field Workers. Pages 328–358 in Stephen A. Richardson, Barbara S. Dohrenwend, and David Klein, *Interviewing: Its Forms and Functions.* New York: Basic Books.

RICHARDSON, STEPHEN A.; DOHRENWEND, BARBARA S.; and KLEIN, DAVID 1965 *Interviewing: Its Forms and Functions.* New York: Basic Books. → See especially pages 64–65 and 277–278.

RICHARDSON, STEPHEN A.; HASTORF, ALBERT H.; and DORNBUSCH, SANFORD M. 1964 Effects of Physical Disability on a Child's Description of Himself. *Child Development* 35:893–907.

RIESMAN, DAVID 1958 Some Observations on the Interviewing in the Teacher Apprehension Study. Pages 266–370 in Paul F. Lazarsfeld and Wagner Thielens, Jr., *The Academic Mind: Social Scientists in a Time of Crisis.* Glencoe, Ill.: Free Press.

ROBINSON, DUANE; and ROHDE, SYLVIA 1946 Two Experiments With an Anti-Semitism Poll. *Journal of Abnormal and Social Psychology* 41:136–144.

SALZINGER, KURT; and PISONI, STEPHANIE 1960 Reinforcement of Verbal Affect Responses of Normal Subjects During an Interview. *Journal of Abnormal and Social Psychology* 60:127–130.

SASLOW, GEORGE et al. 1957 Test–Retest Stability of Interaction Patterns During Interviews Conducted One Week Apart. *Journal of Abnormal and Social Psychology* 54:295–302.

SELLTIZ, CLAIRE et al. (1959) 1964 *Research Methods in Social Relations.* Rev. ed. New York: Holt.

SHAPIRO, SAM; and EBERHART, JOHN C. 1947 Interviewer Differences in an Intensive Interview Survey. *International Journal of Opinion and Attitude Research* 1, no. 2:1–17.

SMITH, M. BREWSTER; BRUNER, JEROME S.; and WHITE, ROBERT W. 1956 *Opinions and Personality.* New York: Wiley.

►STANTON, ALFRED H. 1968 Sullivan, Harry Stack: I. Life and Work. Volume 15, pages 396–398 in *International Encyclopedia of the Social Sciences.* Edited by David L. Sills. New York: Macmillan and Free Press.

STANTON, FRANK; and BAKER, KENNETH H. 1942 Interviewer-bias and the Recall of Incompletely Learned Materials. *Sociometry* 5:123–134.

U.S. DEPARTMENT OF HEALTH, EDUCATION, AND WELFARE, PUBLIC HEALTH SERVICE 1965*a* *Reporting of Hospitalization in Health Interviewing Survey: Methodological Study of Several Factors Affecting Reporting of Hospital Episodes.* Prepared by C. F. Cannell et al. Washington: Government Printing Office.

U.S. DEPARTMENT OF HEALTH, EDUCATION, AND WELFARE, PUBLIC HEALTH SERVICE 1965*b* *Comparison of Hospitalization Reporting.* Prepared by C. F. Cannell et al. Washington: Government Printing Office.

►WAX, ROSALIE HANKEY 1968 Observation: II. Participant Observation. Volume 11, pages 238–241 in *International Encyclopedia of the Social Sciences.* Edited by David L. Sills. New York: Macmillan and Free Press.

WEBB, EUGENE et al. 1966 *Unobtrusive Measures.* Chicago: Rand McNally.

Postscript

The main article, written a decade ago, suggests that both theory and practice of data collection by interview were in a state of flux, partly because of some then recent methodological work and partly because of increasing suspiciousness and sensitivity of respondents. In the intervening years, the problems of interviewing have become more serious and some of the proposed remedies have become more explicit.

Response rates. Obtaining response rates high enough to do justice to probability sample designs has become increasingly difficult and costly. Response rates have dropped throughout the United States, but the greatest decreases have occurred in the central parts of the major metropolitan areas. Research workers seem to agree that the problem exists, although they differ as to its magnitude and urgency. Some cite a severe drop, others a more gradual and moderate tendency.

Virtually no data exist on the causes for the decrease, but three possibilities seem plausible and important. First is the matter of personal safety. The incidence of burglary, robbery, and assault in urban areas has increased, and there has been an attendant increase in publicity and citizen concern over such crimes. Some potential respondents are unwilling to admit survey interviewers to their homes; some apartment houses and multi-family developments maintain guards who allow access only at the request of a tenant. Moreover, the concern for personal safety affects interviewers as well as respondents. Interviewers are reluctant to go to addresses in some areas, especially in the evenings.

A second reason for decreasing rates of response has to do with the changing patterns of participation in the labor force, and with other changes in household composition. Increases in the proportion of women in the labor force, decreasing household size, increasing leisure activities away from home—all add to the difficulty of finding a responsible adult who can provide the required information or set an appointment for an interview with the designated person.

The third factor that seems involved in the declining response rate is the sensitivity or opposition of respondents themselves, after potential respondents have been found and approached. Invasion of privacy has become a major issue, an issue deepened and enlarged by concern about the capacities and potential abuses of computer technology, and by an apparent loss of confidence and trust in public and private institutions.

Attempts to maintain adequate response rates under these conditions have not been wholly successful, and even their partial success has been costly. Interviewers make more calls to find people at home; research organizations devote more funds to converting refusals to completed interviews. Imaginative research on the improvement of response rates is of critical importance.

Response validity. Deficiencies in response rates are viewed with greater alarm than other threats to the quality of survey data because they are more obvious and more readily quantified. Indeed, the rate of response has often been taken as the sole indicator of survey quality. There is, however, a growing sensitivity among researchers to other sources of bias, especially those having to do with response validity. Earlier approaches to this problem (Hyman et al. 1954) emphasized the characteristics of the interviewer and the respondent, separately or in combination. More recent work has focused on the content of the interview itself and the resultant demands on the respondent. Sudman and Bradburn (1974) submitted a large number of studies to complex analyses in an attempt to isolate variables associated with response validity. Their work confirms the importance of the nature and complexity of the task given the respondent; potential bias begins here, rather than with some characteristic of interviewer or respondent that supposedly intrudes regardless of interview content.

The centrality of the interview task is reflected in a new emphasis on cognitive factors in interviewing, including the retrieval of information from memory, and the organization and integration of information in response to cues from the interviewer. (See *Advances in Health Survey Research Methods* 1977.) An increasing number of methodological investigations are designed to investigate the effects of cognitive factors on response validity. Some of these investigations (Cannell et al. 1978) include an enlargement of the interviewer role, especially its instructional aspects. The underlying hypothesis is that direct instruction and positive reinforcement of responses at the appropriate length and detail will improve the validity of the interview data—by increasing the respondent's understanding of the task, acceptance of it, and effort to accomplish it.

Telephone interviewing. The most dramatic development in interviewing methods has been the increased use of the telephone and the telephone system, both for data collection and to a lesser extent for sample selection. The relative economy of telephone interviewing (variously estimated as one-third to one-half the cost of personal interviews) has always been attractive, but the biases involved in the exclusion of households without telephones were formerly quite large. Now more than 90 per cent of the households in the United States are accessible by telephone, and some of them are almost certainly not willing to be interviewed personally. The introduction of the interview by telephone, and the request for cooperation by telephone present their own difficulties. They avoid, however, the problems associated with respondent fears of opening the door to a stranger.

In addition, interviewing by telephone avoids interviewer fears as it does those of respondents, regardless of the time of day or night. Given these facts, along with rising costs and falling response rates, some turn toward telephone interviewing was almost inevitable. Concurrent developments in sampling, coding, and the transfer of data to computer tape have both stimulated and been stimulated by the interest in telephone interviewing. For example, the selection of samples by means of random-digit dialing, the feasibility of having the interviewer enter coded responses at the time of interview, and the use of computer consoles to permit direct data entry (thus eliminating an additional step in coding and key punching) are being investigated as part of the general interest in data collection by telephone.

More relevant to the immediate topic of interviewing, however, are a number of issues of technique that have yet to be investigated. The use of the telephone has thus far not been guided by techniques and methods specific to its strengths and limitations. The tendency has been, predictably enough, to proceed by telephone much as would have been done in person. And yet, communication by telephone differs from face-to-face communication in several significant ways, the most obivous of them having to do with nonverbal stimuli. Many of the usual feedback cues—gestures and facial expressions of approval or disapproval, understanding or puzzlement—are missing in the telephone situation. Research has demonstrated the importance of such feedback as a source of information and motivation. Work is needed on how best to compensate for the loss of nonverbal communication in the telephone interview.

A related difference between the two modes of interviewing has to do with the opening minutes of the interview process—the introduction of the interview and the negotiation, formal or informal, of informed consent. The personal interview begins when the potential respondent opens the door to the interviewer. The acquaintance process begins here, as the respondent evaluates the interviewer's request and the interviewer as a person, not only on the basis of what is said but on the basis of appearance, dress, and behavior. The doorstep contact is followed by the interviewer's entrance into the home, and both participants engage in a process of getting settled and ready to begin work. The interviewer is helped to find a place to sit and write; the printed schedule of questions becomes visible; the pencil is poised. All this may take only a few moments, but it is a time of informal conversation and reassurance during which interviewer and respondent become increasingly comfortable with each other. It is also a time rich in nonverbal communication.

The telephone not only excludes the nonverbal content of this acquaintance process; it tends to exclude much of the verbal exchange as well. The brief chat by which people begin a personal transaction before getting down to business has no spontaneous counterpart in a telephone conversation between strangers.

Researchable questions come quickly to mind. Does the absence of such interpersonal conventions really affect respondents' subsequent behavior in the interview? Are there lingering uncertainties or ambiguities after the brief telephone introductions that are somehow dispelled in the course of personal introductions and exchange of pleasantries? And do any such differences demonstrably diminish the richness and validity of the data? If they do, what inventions of telephone technique can provide the functional equivalent of the subtle craft that personal interviewing has gradually become?

Before 1970, little research had been done on the advantages and disadvantages of interviewing by telephone and even less on the selection of population samples by the random generation of telephone numbers. Within the past few years, however, there has been a surge of research activity along these lines, most of it yet to be published. Statements about the outcomes of such work must therefore be more predictive than integrative. Several such predictions follow, all of them cast in terms of comparison between the telephone and personal modes of interviewing.

(1) Telephone interviewing offers significant economies in the collection of social data.

(2) Telephone interviewing involves fewer limitations—for example, in the duration of the interview and its substantive content—than had been generally supposed.

(3) The question of response rate must await further experience and research. Experience thus far suggests greater variance within either the telephone or the personal mode of interviewing than between the two modes. The concentration of nonresponse in central metropolitan areas, however, seems to be more characteristic of personal-interview studies than of telephone surveys.

(4) The possibility of mixed-mode interviewing is attractive but complicated. Procedures have yet to be developed for the efficient melding of multi-stage areal household samples and single-stage samples generated by random-digit dialing of telephone numbers.

(5) The preservation of privacy and the nature of informed consent have somewhat different meanings in telephone and personal interviews, but the differences are yet to be investigated. Current experience suggests that respondents in telephone surveys feel more free to refuse an interview or to interrupt one already begun, which is certainly one indicator of consent. On the other hand, the presentation of credentials, which is the initial means of making consent informed, is more difficult by telephone.

(6) Such differences in introducing the interviewer and the inquiry exemplify a more general question: what are the differences in interviewing technique implied by the two modes of data collection? Some are obvious; the cue cards and visual material that have become common in personal interviews must be dropped in telephone surveys, and the wording and sequence of questions must be adapted correspondingly. Other implications for interviewing technique are less apparent. It seems likely, however, that the emphasis on quasi-clinical rapport between interviewer and respondent is inappropriate for the telephone. The model of the interview has been called into question by recent methodological research, and the use of the telephone will accelerate the development of alternative models of interviewer–respondent interaction.

The full impact of telephone techniques of sampling and interviewing has yet to be felt. It is the fate of new technologies to be used initially as if they were mere conveniences for performing old tasks; their full potentialities emerge more gradually. The computer, for example, was regarded initially as a conveniently rapid counter–sorter. The optimum use of telephone sampling and data collection is only beginning to be understood.

ROBERT L. KAHN AND
CHARLES F. CANNELL

[*See also* ERRORS, *article on* NONSAMPLING ERRORS.]

ADDITIONAL BIBLIOGRAPHY

A relevant study by Robert M. Groves and Robert L. Kahn is planned for publication by Academic Press as Comparing Telephone and Personal Interview Surveys.

Advances in Health Survey Research Methods: Proceedings of a National Invitational Conference, Airlie House, Airlie, Virginia, May 1–2, 1975. 1977 National Center for Health Services Research, Research Proceedings Series. DHEW Publication No. (HRA) 77–3154. Washington: U.S. Dept. of Health, Education, and Welfare.

CANNELL, CHARLES F.; OKSENBERG, LOIS; and CONVERSE, JEAN M. (editors) 1978 *Experiments in Interviewing Techniques: Field Experiments in Health Reporting, 1971–1977.* National Center for Health Services Research, Research Proceedings Series. DHEW Publication No. (HRA) 78-3204. Washington: U.S. Dept. of Health, Education, and Welfare.

SUDMAN, SEYMOUR; and BRADBURN, NORMAN M. 1974 *Response Effects in Surveys: A Review and Synthesis.* National Opinion Research Center Monographs in Social Research, No. 16. Chicago: Aldine.

INTRACLASS CORRELATION

See MULTIVARIATE ANALYSIS, *article on* CORRELATION METHODS.

INVENTORY CONTROL THEORY

This article was first published in IESS *as "Inventories: II. Inventory Control Theory" with a companion article less relevant to statistics.*

Inventory control theory seeks to establish optimal inventory decision rules for individuals and business firms. Inventory decisions may arise in purchasing or in production. A retailer or wholesaler generally must hold stocks of goods to meet the demands of his customers, and he must therefore decide on the quantities of the goods he will purchase and hold and at what point he will place new orders for each good. A manufacturer generally must hold inventories of finished goods because it is not possible (or economical) to synchronize perfectly his sales and his production process. His inventory decisions are therefore closely related to decisions regarding the scheduling of production. A manufacturer must also hold stocks of semifinished goods and stocks of raw materials. Inventory decisions regarding semifinished goods are, again, closely related to the scheduling of production, while those regarding raw materials are related both to production scheduling and to purchasing.

The optimal decision rules established for various inventory problems are not only of prescriptive value; they may also provide useful insights into observed fluctuations of inventories at the microeconomic and macroeconomic levels.

Historical sketch. In past decades there were occasional periods of intensive interest in inventory theory, sometimes as the aftermath of forced inventory liquidation. For the most part, the literature consisted of a few articles in business journals that had little impact on business behavior and no impact on economic theory. More recently there has been an upsurge of interest which has far surpassed any of its predecessors with respect to the quantity and quality of the work accomplished and with respect to its over-all effect on business behavior and economic theory. Statisticians and economists have become interested in industrial problems concomitantly with increased attention in business to the techniques of advanced management, including operations research and management science. The development of these latter areas has included much detailed attention to inventory theory. [*See* OPERATIONS RESEARCH.]

The earliest attempts at developing inventory theory were primarily concerned with the problem of determining economical lot sizes in purchasing or production (discussed below). Raymond's book (1931) is illustrative of these attempts. During World War II, a useful probabilistic model for controlling stocks was developed. Shortly thereafter a probabilistic version of economical lot-size analysis was developed by Whitin (1953), whose book was the first in English that dealt with probabilistic inventory systems in any detail. Several economists and mathematicians have provided rigorous mathematical analyses of inventory systems, the most noteworthy contributions being an article by Arrow, Harris, and Marschak (1951) and the rather abstract mathematical papers by Dvoretzky, Kiefer, and Wolfowitz (1952). The past few years have given rise to more than ten books on inventory theory, as well as chapters dealing with inventories in almost all of the many books on operations research. At the same time, business firms have been stressing the importance of stock control far more than ever before, as evidenced by the many new corporate vice-presidents in charge of inventory control. Several of the formal mathematical approaches to inventory analysis have been applied in practice by business firms.

From the standpoint of the national economy also, inventory theory has received a considerable amount of attention. One important development was Metzler's formulation (1941) of a business cycle theory in which inventory behavior is the primary causal factor. A business cycle study of much empirical and theoretical interest was published by Abramovitz (1950), who showed that changes in inventory investment constituted a major component of the changes in national income in the five business cycles between the two world wars.

Approaches to inventory theory

There are a wide variety of mathematical analyses of inventory problems. A few different types will be presented here for the purpose of illustration. The mathematical analysis underlying each example will be avoided in order to make the material accessible to readers who are not mathematically oriented.

The newsboy problem. Suppose a newsboy is faced with the problem of determining how many papers to stock when his daily sales vary in a probabilistic manner. He buys papers at a unit cost C and sells them at a unit price P. He can be reimbursed an amount R for each paper not sold. How many papers should he stock to maximize his daily profit? This problem may be simply formulated in terms of the familiar marginal analysis. Let $p(x)$ be the probability that customers will demand x or more papers. If the xth paper is sold, the newsboy makes a marginal profit of $P-C$ on this paper. If it is not sold, the newsboy incurs a marginal loss of $C-R$. Weighting the marginal profit and the marginal loss by their probabilities, $p(x)$ and $1-p(x)$ respectively, one can readily ascertain that the xth paper should be stocked if

$$(1) \quad p(x)(P-C) \geqslant (1-p(x))(C-R).$$

Simple computations show that condition (1) will be satisfied if $p(x) > (C-R)/(P-R)$. That is, the newsboy should continue to add papers to his stock as long as the probability of selling the marginal paper exceeds a known critical ratio. It can readily be seen that the newsboy should stock more papers the higher the profit margin $(P-C)$ and the lower the loss on papers not sold $(C-R)$. One of the principal lessons is that he should not, in general, stock the number of papers that corresponds to average sales. The existence of random demand changes the basic nature of the problem.

Economical lot-size problems. As mentioned before, the earliest inventory problems subjected to mathematical analysis were those involving the determination of economical purchase quantities of goods for inventories. Consider, for example, the case of a retailer who must hold stocks of the goods he sells. There are some inventory costs that decrease as the quantity of inventory goods he orders increases—for example, costs of procurement and costs of receiving. These costs are usually referred to as "ordering costs." Other inventory costs increase with the size of the quantity ordered, for example, costs of holding inventories—interest, depreciation, obsolescence, etc. These costs are usually referred to as "carrying costs." The problem for the retailer is, when faced with a known demand, to purchase a lot (a quantity of the inventory good) that fulfills demand and minimizes the sum of the ordering and carrying costs.

Let Y be the number of units the retailer sells per year (assumed to be sold at a constant rate during the year). At one extreme, he could purchase Y units at the beginning of the year; at the other extreme, he could purchase $Y/365$ units each day of the year. Obviously the first policy would entail very high carrying costs, and the second policy would entail very high ordering costs. If S is ordering costs per order and I is carrying costs per unit per year (both assumed to be constant), it can be shown that minimum costs will be incurred for lots of size Q^* where

$$(2) \qquad Q^* = \sqrt{\frac{2YS}{I}}.$$

This equation indicates that the optimal lot size varies proportionately with the square root of expected sales and the square root of procurement expenses and varies inversely with the square root of unit inventory carrying costs.

The problem of determining economical lot sizes in manufacturing has been subjected to a similar analysis. In these cases, the lot is the amount to be produced rather than purchased, and S is defined as the cost per setup, i.e., the clerical and other costs of preparing the machines for a production run. The identical formula results. Although many restrictive assumptions are made in deriving them, economical lot-size formulas are perhaps the most widely applied mathematical technique of inventory analysis.

Probabilistic lot-size models. Probabilistic lot-size models deal with the problem of determining economical lot sizes when demand is not known with certainty but varies about a given mean in accordance with a known probability distribution. Because of the random variations in demand, it is possible to incur unintended stockouts or shortages. The optimal lot size is the one that minimizes an expected-cost expression which includes procurement costs, carrying costs, and stockout costs. The details of this analysis will not be presented here. The result specifies an optimal reorder-point quantity (a point at which orders will be initiated) as well as an optimal lot size to be ordered. The optimal reorder-point quantity varies directly with the demand level, demand variance, and shortage penalty, and varies inversely with the unit inventory carrying charges and setup costs. Typically, the optimal lot size is higher in the probabilistic case than in the case of certainty, since the fewer the number of orders placed, the smaller is the expected number of stockouts. Hence the introduction of stockout costs makes it worthwhile to buy in larger quantities.

The mathematical analysis underlying the probabilistic lot-size model is presented at widely different levels of generality and sophistication in the literature. The literature also contains solutions to the lot-size problem for cases of certain and probabilistic demands in which there are variations in average demand over time (Hadley & Whitin 1963).

Linear programming. Another technique used for analyzing inventory control problems is linear programming [*see* PROGRAMMING]. Business sales often behave roughly in accordance with a known seasonal pattern. If the fluctuations in sales are met by corresponding fluctuations in production, overtime costs will be incurred. Alternatively, if production is kept relatively constant, the fluctuations in sales may be absorbed by inventory adjustments. Linear programming analysis of the problem makes it possible to determine the production schedule that will meet sales at the minimum combined overtime costs and inventory carrying charges. The approach can be trivially extended to handle other situations in which marginal costs increase as the level of output increases. However, the linear programming approach has not yet been extended to allow for random variations in demand or to include lot-size considerations.

Linear decision rules. The linear decision rule approach, developed at the Carnegie Institute of Technology (Holt et al. 1960), takes into account more types of cost factors than does the linear programming approach. Specifically, the approach minimizes a quadratic cost function including regular payroll costs, costs of overtime and idle time, costs of changing the level of the work force, and costs involved in having either too large or too small an inventory level. The quadratic approximation to costs plays a vital role in two ways. First, the derivatives of this function are linear, making it feasible to solve the equations resulting from setting the first derivatives of the function with respect to work force and production levels equal to zero. Second, when the cost function is quadratic, it is possible to consider only average sales, rather than the probability distribution of sales, for it has been demonstrated that the results are identical. The linear decision rules resulting from the solution of the derivative equations are simple linear expressions that can easily be handled in hand computations. The rules indicate the changes in the level of production and work force that are desirable. The derivative equations need only be solved again when cost conditions change.

Waiting-line theory. The "queuing" or waiting-line approach to the inventory problem has received a considerable amount of attention in recent years (Morse 1958). The level of inventory serves as the queue, which is depleted by customer demands and increased by production or procurement. Mathematical expressions (or "equations of detailed balance") for the rate of change of the probabilities that the queue is at each of its possible levels or "states" are developed. Under long-run, steady-state conditions these state probabilities remain unchanged, i.e., the probability that there are exactly x items in the queue at a random instant of time remains constant. This implies that each of the equations of detailed balance can be set equal to zero, making it possible to solve for the (steady) state probabilities. These state probabilities, combined with the associated costs of each state, can be used to make simple evaluations of the costs of various inventory policies. The assumption required concerning the nature of demands and/or deliveries is typically quite restrictive, so that the approach cannot be applied to a wide range of problems. [*See* QUEUES.]

Inventory theory and the theory of the firm

Classical versions of the economic theory of the firm do not take inventories into account explicitly in any way (see Wolfe 1968). Since inventories are of considerable importance in the actual operation of almost all firms, the theory appears to have serious deficiencies on this score. Only in the case of stationary demand known with certainty and stationary cost conditions can inventories be included in the classical theory. It has been shown that inventory carrying charges and setup costs can be included in the traditional long-run cost curve of the firm, i.e., the envelope of the short-run cost curves, the short-run average cost curves being based on a fixed time between orders, i.e., a fixed lot size (Wagner & Whitin 1958). Some nonstationary inventory situations can be handled by price-discrimination techniques. However, there remain important fundamental differences between inventory theory and classical economic theory. For example, consider the costs included in the linear decision rule example above. Of the several types of costs discussed, only regular payroll costs are taken into consideration by the classical approach, for under stationary demand conditions there would be no overtime, no idle time, no changes in the work force, and no changes in the average inventory level. The very existence of this inventory approach is based on nonstationarity. No long-run equilibrium is ever achieved. A more realistic theory of the firm must allow for some of these nonstationary aspects.

Another aspect of inventory theory that has relevance for the theory of the firm is the existence of economies of scale in most inventory models. Lot-size analysis indicates that inventory costs vary less than proportionately with sales, and the analysis of reorder-point stocks also gives rise to economies of scale because, by the law of large numbers, stocks held as protection against random variations in demand vary less than proportionately with demand. Thus at least two causes of decreasing average costs are established. The arguments for increasing average costs, which are an essential ingredient of classical economic theory (both the theory of the firm and the theory of the economy), have not been convincing, typically being rather vague statements concerning diseconomies of large-scale management or control. Inventory analysis has much to contribute to problems of returns to scale, including problems of vertical and horizontal integration. Few attempts have thus far been made to complete such analyses. (See Bain 1968.)

One of the few attempts to incorporate inventory behavior into the theory of the firm was Boulding's reconstruction of economic theory on the basis of balance sheet considerations (1950). "Preferred asset ratios" played a vital role in his analysis, but he spent little time explaining the basic determinants of these ratios.

Inventory theory and aggregate economics

At the level of aggregate economic analysis, inventory theory is of interest from several standpoints. Its relevance to business cycle theory was mentioned above. In addition, inventory theory can readily be related to Keynesian economics through the three Keynesian motives for holding cash (or goods)—the transactions motive, the precautionary motive, and the speculative motive. According to the transactions motive, it is necessary and desirable to hold some inventories of goods for the purpose of meeting demand. Lot-size analysis provides an approach to determining the quantities that should be held for this purpose in order to minimize the sum of setup costs and inventory carrying charges. The determination of reorder-point quantities involves the precautionary motive. Safety or "cushion" stocks are held to avoid stockouts arising from random sales variations. Finally, inventories may be held for speculative reasons, that is to say, in anticipation of changes in demand or supply conditions. Aggregate levels of stocks held for any of these reasons are of significance in aggregate models of the economy.

Stocks of money have also been subjected to probabilistic inventory analysis. (Here, brokerage fees play the role of setup costs.) For example, the precautionary motive for holding stocks of cash was discussed by Edgeworth in 1888 in connection with the determination of bank reserve ratios.

In a general sense, the behavior of economic aggregates depends upon the behavior of the detailed components of the aggregates, which, in turn, depend quite heavily on inventory considerations. Thus, a better understanding of inventory theory is needed for a more complete theory of aggregate economics.

T. M. WHITIN

BIBLIOGRAPHY

ABRAMOVITZ, MOSES 1950 *Inventories and Business Cycles, With Special Reference to Manufacturers' Inventories.* New York: National Bureau of Economic Research.

ARROW, KENNETH J.; HARRIS, THEODORE; and MARSCHAK, JACOB 1951 Optimal Inventory Policy. *Econometrica* 19:250–272.

ARROW, KENNETH J.; KARLIN, SAMUEL; and SCARF, HERBERT 1958 *Studies in the Mathematical Theory of Inventory and Production.* Stanford Mathematical Studies in the Social Sciences, No. 1. Stanford Univ. Press.

►BAIN, JOE S. 1968 Economies of Scale. Volume 4, pages 491–495 in *International Encyclopedia of the Social Sciences.* Edited by David L. Sills. New York: Macmillan and Free Press.

BOULDING, KENNETH E. 1950 *A Reconstruction of Economics.* New York: Wiley.

BUCHAN, JOSEPH; and KOENIGSBERG, ERNEST 1963 *Scientific Inventory Management.* Englewood Cliffs, N.J.: Prentice-Hall.

DVORETZKY, A.; KIEFER, J.; and WOLFOWITZ, J. 1952 The Inventory Problem: I. Case of Known Distributions of Demand; II. Case of Unknown Distributions of Demand. *Econometrica* 20:187–222, 450–466.

EDGEWORTH, FRANCIS Y. 1888 The Mathematical Theory of Banking. *Journal of the Royal Statistical Society* 51:113–127.

FETTER, ROBERT B.; and DALLECK, WINSTON C. 1961 *Decision Models for Inventory Management.* Homewood, Ill.: Irwin.

HADLEY, GEORGE; and WHITIN, T. M. 1963 *Analysis of Inventory Systems.* Englewood Cliffs, N.J.: Prentice-Hall.

HANSSMANN, FRED 1962 *Operations Research in Production and Inventory Control.* New York: Wiley.

HOLT, CHARLES C. et al. 1960 *Planning Production, Inventories, and Work Force.* Englewood Cliffs, N.J.: Prentice-Hall.

METZLER, LLOYD A. (1941) 1965 The Nature and Stability of Inventory Cycles. Pages 100–129 in American Economic Association, *Readings in Business Cycles.* Homewood, Ill.: Irwin. → First published in Volume 23 of *Review of Economic Statistics.*

MILLS, EDWIN S. 1962 *Price, Output, and Inventory Policy: A Study in the Economics of the Firm and Industry.* New York: Wiley.

MORSE, PHILIP M. 1958 *Queues, Inventories and Maintenance: The Analysis of Operational Systems With Variable Demand and Supply.* New York: Wiley.

RAYMOND, FAIRFIELD E. 1931 *Quantity and Economy in Manufacture.* New York: McGraw-Hill.

STARR, MARTIN; and MILLIER, DAVID W. 1962 *Inventory Control: Theory and Practice.* Englewood Cliffs, N.J.: Prentice-Hall.

WAGNER, HARVEY M. 1962 *Statistical Management of Inventory Systems.* New York: Wiley.

WAGNER, HARVEY M.; and WHITIN, T. M. 1958 Dynamic Problems in the Theory of the Firm. *Naval Research Logistics Quarterly* 5:53–74.

WHITIN, T. M. 1953 *The Theory of Inventory Management.* Princeton Univ. Press.

►WOLFE, J. N. 1968 Firm, Theory of the. Volume 5, pages 455–460 in *International Encyclopedia of the Social Sciences.* Edited by David L. Sills. New York: Macmillan and Free Press.

ITEM ANALYSIS

See PSYCHOMETRICS.

JK

JACKKNIFE TECHNIQUE

See ERRORS, *article on* NONSAMPLING ERRORS.

KEPLER, JOHANNES

► *This article was specially written for this volume.*

Johannes Kepler (1571–1630), the great astronomer, was not, of course, a statistician in the present sense, but statistical ideas appear in his astronomy. In treating observations, Kepler strove for the least possible effects of random and systematic errors ([1609] 1929, p. 209 and ch. 51), demanding ([1609] 1929, p. 311) that a symmetric program of observations be carried out so as to exclude systematic influence. Such requirements, found even in the works of ancient astronomers, constitute part of the prehistory of the design of experiments.

Kepler repeatedly adjusted direct observations, usually following the rule of the arithmetic mean. He noticed the inevitability of errors ([1609] 1929, p. 114). For indirect observations he used iterations to calculate unknowns, starting from one or another set of observations and each time checking end results against redundant observations. To find the actual paths of planets, he deliberately corrupted his data by small arbitrary quantities, possibly selected with regard for their stochastic properties ([1609] 1929, p. 197) thus in effect using the Monte Carlo method. [*See* RANDOM NUMBERS.] Kepler also used the minimax principle; he came to reject the Ptolemaic system of the world because (even apart from physical considerations) the approximately calculated minimax error of its fit exceeded the possible error of Tychonic observations ([1609] 1929, p. 166).

In astronomy proper, Kepler had to find room for random causes for orbital eccentricities, and also for his speculations about the end of the world. He repeatedly returned (1609; 1616; [1618] 1952, p. 932) to the principle that elliptical deviation from circular motion is a corruption arising from random influences; this point of view seems to have been borrowed both by Kant ([1775] 1910, p. 337) and by Laplace ([1796] 1884, p. 504).

Assuming an incorrect form of the third kinematic law, Kepler ([1596] 1936, p. 114) rejected any possibility of a simultaneous return of all the planets to their position at the moment of creation. This meant, according to Kepler, that the end of the world is impossible. But then, in the second edition of his *Mysterium cosmographicum*, bearing in mind the correct form of the third law, he says (1936, p. 145) that such a simultaneous return is possible but that its probability is extremely small because, if two numbers were taken "at random," they would most probably be incommensurable. This argument, actually borrowed from Oresme ([1351] 1966, pp. 247, 422), was the first one in which stochastic considerations were applied to an abstract mathematical notion and also used to prove an important fact in natural science.

Kepler ([1610] 1941, p. 248) considered himself to be the founder of scientific astrology, a science of the general influence of heaven upon earth. He was sure ([1619] 1939, pp. 256, 263) that there exists a connection between remarkable mutual positions of the heavenly bodies (aspects) and meteorological phenomena on the earth. In his time, meteorology was not yet a quantitative science, however, so he could not seek quantitative confirmation.

Another feature of Kepler's astrology is his presumption of divine care for mankind (1610). If one bears in mind that astrology had to do with the general destiny of nations as decided by the prevailing aspects as well as by geographical conditions, and so on, it follows that both in his presumptions and in his aims, Kepler the astrologer resembles the founders of political arithmetic, Graunt and Petty. Completely lacking in Kepler's work, however, are statistical data on populations that were to form the basis of political arithmetic. Such data could not have existed in feudal, subdivided Germany.

Another line of development connects Kepler with Jakob I Bernoulli. Kepler considered it impossible from the standpoint of both religion and scientific astrology to say definitely whether or not a man, absent for many years, is still alive ([1610] 1941, p. 238). On the other hand, Bernoulli, in his *Ars conjectandi,* poses this same problem and, it seems, is quite prepared to weigh corresponding probabilities against each other. This was the beginning of "moral" applications of probability completely lacking in Kepler's work.

O. B. SHEYNIN

[*Related material may be found in the biographies of* GRAUNT; PETTY; *and the* BERNOULLI FAMILY.]

WORKS BY KEPLER

(1596 and 1621) 1936 *Das Weltgeheimnis.* Munich and Berlin: Oldenbourg. → A reprint of the 1923 edition published by Filser. First published in Latin as *Mysterium cosmographicum.*

(1609) 1929 *Neue Astronomie.* Munich: Oldenbourg. → First published in Latin.

(1610) 1941 Tertius interveniens. Volume 4, pages 145–258 in *Johannes Keplers Gesammelte Werke.* Munich: Beck.

(1618) 1952 *Epitome of Copernican Astronomy, Books IV and V.* Great Books of the Western World, Vol. 16. Chicago: Encyclopaedia Brittanica. → First published in Latin.

(1619) 1939 *Weltharmonik.* Munich and Berlin: Oldenbourg. → First published in Latin.

SUPPLEMENTARY BIBLIOGRAPHY

Four Hundred Years: Proceedings of a Conference Held in Honour of Johannes Kepler. 1975 Edited by Arthur Beer and Peter Beer. Vistas in Astronomy, Vol. 18. New York: Pergamon.

KANT, IMMANUEL (1755) 1910 Allgemeine Naturgeschichte und Theorie des Himmels. Volume 1, pages 215–368 in *Gesammelte Schriften.* Berlin: Reimer.

LAPLACE, PIERRE SIMON DE (1796) 1884 Exposition du système du monde. Volume 6, pages 1–509 in *Oeuvres complètes.* Paris: Gauthier-Villars.

ORESME, NICOLAS (1351) 1966 *De proportionibus proportionum* and *Ad pauca respicientes.* Madison: Univ. of Wisconsin Press. → In the original Latin with an English translation, introduction, and critical notes by Edward Grant.

SHEYNIN, O. B. 1973 Mathematical Treatment of Astronomical Observations (Historical Essay). *Archive for History of Exact Sciences* 11:97–126.

SHEYNIN, O. B. 1974 On the Prehistory of the Theory of Probability. *Archive for History of Exact Sciences* 12:97–141.

SHEYNIN, O. B. 1977 Kepler as a Statistician. *Bulletin of the International Statistical Institute* 46:341–354.

KEYNES, JOHN MAYNARD: CONTRIBUTIONS TO STATISTICS

This article was first published in IESS *with a companion article less relevant to statistics.*

Keynes took his degree in mathematics: it was therefore natural that his fellowship thesis should be on a mathematical subject. He chose probability as the topic, and out of this thesis grew *A Treatise on Probability* (1921), his single great contribution to the subject. Of the five parts of this large book, the second attempts to reduce to logical formulas the fundamental theorems of the probability calculus, a mathematical exercise in the tradition of Whitehead and Russell that has had little influence. Another part of the book is historical and bibliographical; the bibliography lists 600 items. Keynes's passion for collecting reveals itself in this admirable compendium, which brought Todhunter's and Laurent's earlier historical treatments of logic up to date. A third feature is a fine critique of some views of probability that were held then and are still popular. Among these are the idea of probability as a subjective degree of belief in a proposition, given the evidence, and the notion of probability as a limiting frequency associated with a certain type of infinite sequence.

The main contribution of Keynes is the argument that probability is a primitive idea—a logical relation between a proposition and the evidence bearing on the truth of the proposition. Thus, with the subjectivists, he held that it is a relation between propositions and evidence; but he supported the frequentists in thinking that it is an objective notion. His conception of probability was pursued with great thoroughness and in a style worthy of attention for its literary merits. For Keynes, the purpose of probability theory is to systematize inference processes. He therefore attempted to formulate certain rules of probability and to develop a calculus. Furthermore, he tried to develop the logical foundations of statistical arguments.

Keynes's viewpoint and the program were novel and important. They have had a great influence on probabilists and statisticians. Unfortunately, they were marred by a serious restriction that Keynes imposed in refusing to admit that all probabilities can be compared. He was prepared to assume only that probabilities are partially ordered. Related to this difficulty is his refusal to recognize that a numerical measure of probability is always appropriate. As Ramsey was later to point out (1923–1928), this refusal to introduce numbers is surprising in view of Keynes's obvious knowledge of Russell's work on the correspondence between order relations and numbers. But without numbers, progress is difficult if not impossible.

In a biographical essay on Ramsey, Keynes (1933, p. 300) later withdrew his objections and admitted the correctness of Ramsey's view of probability as expressed in terms of bets. He also admitted Ramsey's argument that the rules of probability are logical deductions from proper betting behavior, and not primitive axioms. In the hands of Savage, Ramsey's work has led to many interesting developments in what is now often called Bayesian probability, which is having an increasing influence on practical statistics. It is interesting to note that while Keynes was working on the treatise, Jeffreys (1939), also in Cambridge, was developing a similar objective, logical theory. But since he admitted numbers, he made much more progress on the calculus than did Keynes. According to Jeffreys, both he and Keynes were influenced by W. E. Johnson, a lecturer in philosophy at Cambridge.

DENNIS V. LINDLEY

[*Directly related are the articles on* BAYESIAN INFERENCE *and* PROBABILITY.]

WORKS BY KEYNES

(1921) 1952 *A Treatise on Probability.* London: Macmillan. → A paperback edition was published in 1962 by Harper.

(1933) 1951 *Essays in Biography.* New ed. Edited by Geoffrey Keynes. New York: Horizon Press. → A paperback edition was published in 1963 by Norton.

SUPPLEMENTARY BIBLIOGRAPHY

JEFFREYS, HAROLD (1931) 1957 *Scientific Inference.* 2d ed. Cambridge Univ. Press.

JEFFREYS, HAROLD (1939) 1961 *Theory of Probability.* 3d ed. Oxford: Clarendon.

RAMSEY, FRANK P. (1923–1928) 1931 *The Foundations of Mathematics and Other Logical Essays.* New York: Harcourt.

○SAVAGE, LEONARD J. (1954) 1972 *The Foundations of Statistics.* Rev. ed. New York: Dover. → Includes a new preface.

Postscript

The development by Ramsey has been much extended by the work of de Finetti, whose writings are now available in English.

DENNIS V. LINDLEY

ADDITIONAL BIBLIOGRAPHY

DE FINETTI, BRUNO (1970) 1974–1975 *Theory of Probability.* 2 vols. New York and London: Wiley. → First published in Italian.

KING, GREGORY

Gregory King (1648–1712), the son of a mathematician who was also a landscape gardener, was born at Lichfield. He got his early education at home, at the Free School, and as clerk to the antiquary Sir William Dugdale, whose service he entered at the age of 14 and with whom he spent several years traveling in the English counties on heraldic surveys. He was a well-informed topographer and surveyor; he assisted in the production of *Itinerarium Angliae: Or, a Book of the Roads,* 1675; and for a while he supported himself by mapmaking, engraving, and surveying. He was a skilled and well-known genealogist and became successively Rouge Dragon, registrar of the College of Arms, and Lancaster herald. He was a successful practicing accountant; he taught bookkeeping as a young man and in later life became secretary to the comptroller of army accounts and to the commissioners of public accounts. Finally, he was a distinguished political arithmetician. His autobiography (which, however, ends in 1694) was reproduced by James Dalloway as an appendix to *Inquiries Into the Origin and Progress of the Science of Heraldry in England* (1793).

It is as a statistician that King makes his claim to fame as a social scientist although, so far as we know, he did not become interested in political arithmetic until 1695. In that year he published a broadsheet summarizing the rates of duties payable under the Act of 1694, which levied taxes on marriages, births, burials, bachelors, and childless widowers. Possibly King had been involved in designing the statistical inquiries that were essential to the assessment of these taxes: he was certainly interested in the results, which were immensely significant sources of demographic information in an age when the size and trend of the population was a matter of great political interest and much speculation. There is no record of his having published any other work relating to political arithmetic during his lifetime, although his work was well known among his contemporaries and his estimates were freely used and quoted by Charles Davenant.

Unlike most of his contemporary political arithmeticians, King was a scholar rather than a politician. Perhaps this was why he never published his estimates, being content to make them freely available as a basis for economic policy-making or analysis rather than using them to support his own special pleadings. He was primarily interested in finding the exact truth about the dimensions of the national economy, so far as the available data would let him. It is evident from the notes and communications which have survived that he was completely honest about the limitations of his material and amazingly methodical in his use of it, and the more modern scholars have probed his methods and uncovered new sources of his notes, the more they have tended to admire his results. His famous "Scheme of the Income and Expences of the Several Families of England," given in his *Natural and Political Observations and Conclusions Upon the State and Condition of England* (1696), and his international comparisons of national income and expenditure for England, France, and Holland in his *Of the Naval Trade of England A° 1688 and the National Profit Then Arising Thereby* (1697) were based essentially on guesswork, but as explicit statements of the views of a particularly well informed observer they are profoundly revealing. They inspired comparable calculations by Patrick Colquhoun in the early nineteenth century and became bench mark data of immense value to students of long-term growth.

All King's estimates were made with an accountant's meticulous concern for internal consistency, and in this respect his national income estimates were in advance of any calculations made in this field until the mid-twentieth century. It is possible to extract from the national income and balance of payments estimates given in his two tracts, supplemented with additional estimates quoted by Davenant, a complete, articulated set of double-entry social accounts as well as an abundance of detail on the content of national income, output, and expenditure in 1688 and 1695. He also made estimates of the national capital, its content, and its rate of increase through the seventeenth century. His population estimates were based on careful analyses of actual enumerations for particular places, corrected for technical errors and adjusted to a national basis, on assumptions that modern demographers (basing their judgments on the results of nineteenth-century census enumerations) have found to be both consistent and plausible. His schedule of the relation between changes in the price of wheat and deviations from the normal wheat harvest, which was originally published by Davenant and became known as "Gregory King's Law," represents a piece of demand analysis of a kind that we find in no other source until the early twentieth century.

PHYLLIS DEANE

WORKS BY KING

○ *There are manuscripts and calculations by King in the British Museum, the Public Record Office (London), the Bodleian Library (Oxford), and the Library of the Greater London Council.*

1695 *A Scheme of the Rates and Duties Granted to His Majesty Upon Marriages, Births, and Burials, and Upon Batchelors and Widowers, for the Term of Five Years from May 1, 1695.* London: A broadsheet.

(1696) 1936 Natural and Political Observations and Conclusions Upon the State and Condition of England. Pages 12–56 in *Two Tracts by Gregory King.* Edited by George E. Barnett. Baltimore: Johns Hopkins Press. → The manuscript of 1696 was first published in 1802 in George Chalmers' *An Estimate of the Comparative Strength of Great-Britain.*

(1697) 1936 Of the Naval Trade of England A° 1688 and the National Profit Then Arising Thereby. Pages 60–76 in *Two Tracts by Gregory King.* Edited by George E. Barnett. Baltimore: Johns Hopkins Press. → The manuscript of 1697 was first published in 1936.

1793 Some Miscellaneous Notes of the Birth, Education, and Advancement of Gregory King. Appendix 2 in James Dalloway, *Inquiries Into the Origin and Progress of the Science of Heraldry in England.* Gloucester (England): Raikes. → The autobiography covers the years 1648 to 1694.

SUPPLEMENTARY BIBLIOGRAPHY

DEANE, PHYLLIS 1955 The Implications of Early National Income Estimates for the Measurement of Long-term Growth in the United Kingdom. *Economic Development and Cultural Change* 4:3–38.

ˡGLASS, D. V. 1946 Gregory King and the Population of England and Wales at the End of the Seventeenth Century. *Eugenics Review* New Series 38:170–183.

ˡGLASS, D. V. 1950 Gregory King's Estimate of the Population of England and Wales, 1695. *Population Studies* 3:338–374.

REES, J. F. 1932 Gregory King. Volume 8, page 565 in *Encyclopaedia of the Social Sciences.* New York: Macmillan.

YULE, G. UDNY 1915 Crop Production and Price: A Note on Gregory King's Law. *Journal of the Royal Statistical Society* 78:296–298.

Postscript

Further scholarship concerning King has included the publication in facsimile (Laslett 1973) of his *Natural and Political Observations Upon the State and Condition of England,* together with a manuscript notebook now in the Library of the Greater London Council containing King's worksheets for his population estimates. In addition, the two papers by Glass (1946; 1950) have been superseded by Glass (1965).

PHYLLIS DEANE

ADDITIONAL BIBLIOGRAPHY

[1]GLASS, D. V. 1965 Two Papers by Gregory King. Pages 159–220 in D. V. Glass and D. E. C. Eversley (editors), *Population in History: Essays in Historical Demography.* London: Arnold; Chicago: Aldine.

LASLETT, PETER (editor) 1973 *The Earliest Classics: John Graunt and Gregory King.* Westmead (England): Gregg International. → Facsimiles of classic works, with an introduction by the editor.

KŐRÖSY, JÓZSEF

József Kőrösy (1844–1906) was a pioneer Hungarian statistician. He was born in Pest, where his father was a merchant. The family moved to the country to take up farming, but Kőrösy's father died when he was six, and he returned to Pest for his education. Financial difficulties forced him to take a job immediately upon graduating from secondary school. After working first as an insurance clerk, he became a journalist and wrote a column on economics. Although he never regularly attended a university and was almost entirely self-taught, he nevertheless acquired both proficiency in languages and advanced knowledge in his professional specialty.

The articles that Kőrösy wrote on economics revealed his excellent sense for statistics and attracted attention. Therefore, in 1869 when a municipal statistical office was first set up in Pest, Kőrösy was appointed director. Only two years had passed since Károly Keleti had set up the Hungarian national statistical service. Prior to that time there had been no such service provided for Hungary; the Austrian government statistical service in Vienna had been responsible for all data relating to Hungary and had furthermore treated most of these data as confidential. Consequently there had been little opportunity for the development of statistical thought in Hungary.

Kőrösy was a research scholar as well as an official statistician—a rare combination—and he turned the Budapest statistical office into a model research institution. (Budapest was formed in 1873, and Kőrösy's office at that time became the statistical office for the new city.) Data were collected covering nearly every aspect of Budapest life. Publications by the office included both methodological experiments and pragmatic analyses, usually written by Kőrösy. He edited the statistical publications of Budapest and was in charge of publishing the first comparative statistics relating to big cities (for vital statistics, see 1874; for finances, see 1877; and so forth).

From 1883 on, Kőrösy was also a reader at the University of Budapest; he lectured there in the field of demography. He was a member of many statistical organizations and of other scientific societies and the recipient of numerous honors and awards, both at home and abroad.

Kőrösy contributed voluminously to the statistical and demographic literature of his age, and his contributions covered a wide range of subjects. His studies of Budapest included detailed analyses both of the city's population census and of mortality. He also used Budapest data to study patterns of contagious diseases, pauperism, public education, construction, taxation, and—with newly developed indicators—corporation profits.

His most outstanding papers from the methodological point of view were those in which he developed the first "natality" (or fertility) tables. These tables and other investigations showed the relationship between the parents' ages and the viability of the newborn infant, as well as the effect of the ages of spouses and the length of their marriage on fertility and family size.

Kőrösy also made important contributions to the problem of obtaining reliable mortality tables. Although his "individual method" of constructing generation mortality tables was not practicable, his proposals make him a precursor of modern cohort analysis. The individual method is an extremely laborious one: Kőrösy planned to follow the life histories of a certain number of persons by using a separate chart for each individual, a sort of perpetual register from birth to death; the charts would then be used to construct mortality tables. In this way precise and valuable data could be obtained, but the method was far too costly to be the final solution to the problem of compiling mortality rates, as Kőrösy believed it was (Saile 1927, pp. 237–238). Other pioneer research by Kőrösy dealt with the hereditary character of certain illnesses and with the effects of weather, housing conditions, educational level, and income level on morbidity and mortality.

Kőrösy tried many methodological innovations, most of which were successful. For example, his ingenious coefficients of relative intensity of morbidity or mortality enabled him to point out connections between phenomena when the size and distribution of the basic population are unknown. And although he was working at a time when the use of mathematics in statistics was not yet commonplace, the indexes he introduced for measuring association are very similar to those developed later by Pearson and Yule (see Jordan 1927, p. 337; Goodman & Kruskal 1959). He also had a great deal to do with introducing standardized descriptions of the causes of death—here he worked with Bertillon—and was active in attempts to standardize population censuses internationally. Along with Ogle in Britain and Koch in Germany, Kőrösy introduced the standardized death rate, which mitigates the problem of comparing over-all death rates in populations of diverse age distributions (Kőrösy 1892–1893; *Annual Summary* . . . 1883; Hamburg, Statistisches Landesamt 1883).

The statistics collected by Kőrösy's office did much to lower the death rate in Budapest, in part by making clear the need for improved health standards. Battling against false statistics and erroneous methods, he fought for the use of smallpox vaccine (see Westergaard 1932, pp. 253–254).

In doctrine he stood close to Wilhelm Lexis, although he was ahead of Lexis in his knowledge of the laws of demography. Although Kőrösy did not write textbooks on either statistics or demography, his collaborators and students, most notably Gusztáv Thirring, built on the extensive and sound foundations he laid.

LAJOS THIRRING

[*For the historical context of Kőrösy's work, see* SOCIAL RESEARCH, THE EARLY HISTORY OF; VITAL STATISTICS; *and the biography of* LEXIS. *For discussion of the subsequent development of his ideas, see* STATISTICS, DESCRIPTIVE, *article on* ASSOCIATION; *and the biographies of* PEARSON *and* YULE.]

WORKS BY KŐRÖSY

1871 *Pest szabad királyi város az 1870, évben: A népszámlálás és népleirás eredménye.* Budapest: Ráth. → Also published in 1872 in German by the same publisher.

1874 *Welche Unterlagen hat die Statistik zu beschaffen um richtige Mortalitäts-tabellen zu gewinnen?* Berlin: Engel. → Reprinted in section 1 of International Statistical Congress, Ninth, Budapest, 1876, *Programme.*

1877 *Statistique internationale des grandes villes.* Deuxième Section: Finances. Budapest: Ráth; Paris: Guillaumin.

1881 *Projet d'un recensement du monde.* Paris: Guillaumin.

1885 On the Unification of Census Record Tables. *Journal of the Royal Statistical Society* Series A Jubilee Volume: 159–170.

(1887) 1889 *Kritik der Vaccinations-statistik und neue Beiträge zur Frage des Impfschutzes.* Berlin: Puttkammer & Mühlbrecht.

1892*a* Mortalitäts-coëfficient und Mortalitäts-index. Institut international de statistique, *Bulletin* 6, no. 2:305–361.

1892*b* Wissenschaftliche Stellung und Grenzen der Demologie. *Allgemeines statistisches Archiv* 2:397–418.

1892–1893 Einfluss des Alters der Eltern auf die Vitalität ihrer Kinder. Volume 10, pages 262–263 in International Congress of Hygiene and Demography, London, 1891, *Transactions.* London: Eyre & Spottiswoode.

1894 Über den Zusammenhang zwischen Armuth und infectiösen Krankheiten und über die Methode der Intensitätsrechnung. *Zeitschrift für Hygiene und Infectionskrankheiten* 18:505–528.

1896 An Estimate of the Degrees of Legitimate Natality as Derived From a Table of Natality Compiled by the Author From His Observations Made at Budapest. Royal Society of London, *Philosophical Transactions* Series B 186:781–875.

1900 *La statistique des résultats financières des sociétés anonymes.* Paris: Dupont.

1906 Über die Statistik der Ergiebigkeit der Ehen. Institut International de Statistique, *Bulletin* 15, no. 2: 404–416.

SUPPLEMENTARY BIBLIOGRAPHY

Annual Summary of Births, Deaths, and Causes of Death in London and Other Large Cities. 1883 Great Britain, General Register Office, *Weekly Return of Births and Deaths in London and in Twenty-two Other Large Towns of the United Kingdom* 54:i–lv.

GOODMAN, LEO A.; and KRUSKAL, WILLIAM H. 1959 Measures of Association for Cross Classifications: II. Further Discussion and References. *Journal of the American Statistical Association* 54:123–163.

HAMBURG, STATISTISCHES LANDESAMT *Statistik des Hamburgischen Staats* [1883], Heft 12.

J. A. B. 1907 Dr. Joseph Kőrösy.—M. G. Olanesco. *Journal of the Royal Statistical Society* 70:332–333. → The author of the article is J. A. Baines.

JORDAN, CHARLES 1927 Les coefficients d'intensité relative de Kőrösy. Société Hongroise de Statistique, *Revue* 5:332–345.

LAKY, DÉSIRÉ 1939 Les représentants académiciens de la grande époque de la statistique hongroise. Société Hongroise de Statistique, *Journal* 17:365–378.

SAILE, TIVADAR A. 1927 *Influence de Joseph de Kőrösy sur l'évolution de la statistique.* Budapest: Magyar Tudományos Akadémia.

THIRRING, GUSZTÁV 1907 Joseph de Kőrösy. Institut International de Statistique, *Bulletin* 16, no. 1:150–155.

WESTERGAARD, HARALD L. 1932 *Contributions to the History of Statistics.* London: King.

L

LAPLACE, PIERRE SIMON DE

Pierre Simon de Laplace (1749–1827), renowned French mathematician, was born in Beaumont-en-Auge, a village 4 miles west of Pont l'Évêque in Normandy. He was the second of two children of Pierre de Laplace, a syndic of the parish, who owned and farmed a small estate. When Laplace arrived in Paris, barely twenty years old, he had finished his studies and begun his own research. His ability soon impressed d'Alembert, whose disciple he was to become. D'Alembert's patronage secured Laplace a position as a teacher at the École Royale Militaire, where he remained until changes in the organization brought his teaching there to an end. In 1783 he became an artillery inspector (Duveen & Hahn 1957), and in this capacity he made the acquaintance of the young Bonaparte. A mutual respect developed, and from then on Laplace enjoyed Bonaparte's increasingly powerful support. The two became colleagues at the Académie des Sciences; as first consul, Napoleon appointed Laplace minister of the interior, a position he gave up shortly afterward to become chancellor of the Sénat Conservateur. In 1814, however, he turned against Napoleon; he was eventually made a marquis by Louis XVIII, a surprising but not uncommon turn of events in those troubled times.

Laplace achieved distinction not only in mathematics; it was his literary style that won him election to the Académie Française in 1816. His famous sentence on the hidden determinism of natural laws is a good example of his style:

> If there were an intelligence that for a given instant could comprehend all the forces that animate nature and the condition of each being that composes it; if, moreover, this intelligence were sufficiently great to submit these data to analysis, it would create a single formula that would embrace both the movements of the vastest bodies of the universe and those of the smallest atoms: to this intelligence nothing would be uncertain, and the future, as the past, would be present to its eyes. ([1814] 1951, p. 4)

Although Laplace's work has literary elegance, his demonstrations are not always rigorous; often they are even obscure. He frequently wrote "it is clear that . . ." in place of long and difficult calculations.

Laplace presided over the famous Société d'Arcueil, one of the informal scientific societies that flourished in the nineteenth century. It took its name from Laplace's estate, where he and the great chemist Claude Louis Berthollet periodically received their students to discuss scientific questions in an informal atmosphere.

Although our primary concern here is with Laplace's contributions to probability and statistics, it is worth noting that the full scope of his work includes physics (the theory of capillary phenomena, the exact formula for the speed of sound), pure mathematics, and celestial mechanics (he was dubbed a "second Newton"). In probability and its applications Laplace systematized and further extended the scattered researches of his predecessors, bringing the subject to full flower in the third edition (1820) of his great treatise, *Théorie analytique des probabilités* (1812). This is why Todhunter wrote: "On the whole, the Theory of Probability is more indebted to him than to any other mathematician" ([1865] 1949, p. 464).

Laplace was not content to make important discoveries; he also thought it necessary to communicate them to a wide public. To this end he wrote two popular works addressed to the intelligent and educated general reader, *Exposition du système du monde* (1796) and *Essai philosophique sur les probabilités* (1814). Other French mathematicians have continued this practice; thus, Émile Borel thought it useful to write an analogous work, *Le hasard* (1914), which took into consideration the progress made since Laplace.

Generating functions and characteristic functions. Motivated by problems arising, for instance, in the mathematical treatment of games of chance, Laplace, in the epochal *Mémoire sur les suites* (1782), developed the general theory of a powerful power-series technique for solving finite-difference equations, or recurrence relations, which he termed the "method of generating functions" (*calcul des fonctions génératrices; ibid.*, p. 1). In Book 1, Part 1, of his *Théorie analytique des probabilités* (1812) he reproduced this *Mémoire* almost entirely, and in Book 2 he made repeated use of generating functions in solving a great variety of probability problems arising in the mathematical treatment of games of chance. The probability-generating function of a random variable, X—that is, the mean (or expected) value of t^X, where t is a "dummy" real variable—had already been used (without being named) by De Moivre, in studies of games of chance (1730, pp. 191–197; [1718] 1756, pp. 41–43), and by Simpson and Lagrange, in their studies of the distribution of the arithmetic mean of independent observations under various laws of error (Simpson 1756; 1757; Lagrange 1770–1773). However, it is Laplace's extensive discussion of generating functions (1782) and the applications of them in his *Théorie analytique des probabilités* that is the actual source of their widespread use in probability theory, combinatory analysis, and the solution of finite-difference equations and recurrence formulas (see David & Barton 1962; Feller 1950; Fréchet 1940–1943; Jordan 1939; Riordan 1958; Uspensky 1937). The invention of the characteristic function (*fonction caractéristique*) of the distribution of a random variable X (that is, the expected value of e^{itX}, where $i = \sqrt{-1}$) and the associated inversion formula for deducing the distribution function of a random variable from its characteristic function are often attributed to later writers, such as A. L. Cauchy, 1789–1857, and Henri Poincaré, 1854–1912. Laplace, however, introduced this very same (characteristic) function, without assigning it a name, and gave the associated inversion formula for the case of a discrete random variable, in article 21 of Book 1 of his *Théorie analytique des probabilités* (it also appears in *Oeuvres complètes*, vol. 7, pp. 83–84) and then employed such functions systematically in Book 2, Chapter 4 (cf. Molina 1930, arts. 3–4). The technical advantages possessed by the characteristic function permit simplification of many manipulations and proofs, although when X takes only integral values the generating function is usually adequate. [*See* PROBABILITY, *article on* FORMAL PROBABILITY.]

Bayesian inference. Laplace was apparently the first to have stated in a general form what is now called Bayes' theorem (*Théorie analytique* [1812] 1820, book 2, art. 1; Molina 1930, appendix I). Several scholars have deprecated Laplace's originality in this area of "probability of causes," or Bayesian inference. [*See* BAYESIAN INFERENCE *for a discussion of the probability of causes.*] But Laplace did introduce an essential innovation here. Bayes had considered only the case where the a priori probabilities are equal (as an evident hypothesis when one is ignorant of these probabilities); Laplace extended Bayes' theorem to cover the general case where these a priori probabilities are not necessarily equal. What has generally not been stressed enough is that Bayes' theorem—whether generalized or not—is really only an interpretation of a standard relationship for conditional probabilities. [*See* PROBABILITY, *article on* FORMAL PROBABILITY.] It is the interpretation of the formula itself that is important here. A discussion by Molina (1930) clearly establishes Laplace's position and demonstrates previous misunderstandings of it.

Normal distribution. Laplace should have the major credit for discovering and demonstrating the central rôle of the normal distribution in the mathematical theory of probability and for determining its principal mathematical properties [*see* DISTRIBUTIONS, STATISTICAL, *article on* SPECIAL CONTINUOUS DISTRIBUTIONS].

De Moivre in 1733 had shown how to employ integrals of the function $\exp(-t^2)$ to approximate sums of successive terms of the binomial expansion of $(a+b)^n$ when n is large, obtaining what we today call the normal approximation to the binomial probability distribution [*see* DISTRIBUTIONS, STATISTICAL, *article on* APPROXIMATIONS TO DISTRIBUTIONS]. (De Moivre's analysis is readily available in Smith [1929] 1959, pp. 566–575.) Laplace extended this approach in two important directions.

Approximations to the normal integral. He developed (1781) a general method for approximating an arbitrary definite integral by a series expansion in terms of integrals of the function $\exp(-t^2)$ and its derivatives, anticipating by a century the so-called Gram–Charlier Type A series expansion. He then utilized this technique to approximate various discrete and continuous probability distributions arising in games of chance and other problems in the calculus of probabilities. Remarking (1786*a*, p. 305) that the integral of $\exp(-t^2)$ arises so frequently that it would be useful to have a table of its values for a succession of limits of integration, Laplace provided (1785, sec. VI; 1805; [1812] 1820, book 1, art. 27) the now well-known power-series, asymptotic-series, and continued fraction expansions for integrals of the function $\exp(-t^2)$. These many results collectively constitute Laplace's first great contribution to the central role of the normal distribution today, and throughout their development the function $\exp(-t^2)$ seems to have been regarded exclusively as an approximating function —neither De Moivre nor Laplace appears to have regarded it or any corresponding expression as a law of error or even as a probability distribution in its own right.

Central limit theorem. Laplace's second great contribution to the establishment of the leading role of the normal distribution is his discovery and proof of what we today call the (classical) central limit theorem. De Moivre's result of 1733 is a special case of this theorem, as are many of Laplace's results in his papers of 1781, 1785, and 1786, but the theorem itself appears for the first time in the important *Mémoire* (1810, sec. VI), for the case of n independent errors from a common arbitrary symmetric discrete distribution on the interval $(-a, +a)$. Then (*ibid.*, sec. VI), by considering the discrete distribution corresponding to a fine subdivision of his double exponential law of error (1774, sec. V), $\frac{1}{2}m \exp(-m|x|)$, he illustrated the extension of this important theorem to symmetric continuous distributions on the interval $(-\infty, +\infty)$. Gauss (1809, arts. 175–177) had deduced his law of error, $C \exp(-h^2x^2)$, and thence his development of the method of least squares, from the principle of the arithmetic mean. Laplace's central limit theorem provided an alternative justification, or "proof," of this law of error as the limit of the distribution of the sum of n independent random errors as $n \to \infty$ when the relative contribution of each to their sum tends to 0 as n increases, and thus he provided a valid basis for the method of least squares when the individual results involved "are each determined by a very large number of observations, whatever be the laws of facility of the errors of these observations," and hence "a reason for employing it in all cases" (1810, supplement, p. 353). [*See* DISTRIBUTIONS, STATISTICAL, *article on* SPECIAL CONTINUOUS DISTRIBUTIONS; *see also* LINEAR HYPOTHESES, *article on* REGRESSION.]

Laplace's proof had only the limited precision current in his time, but it was later made more rigorous and more general by Aleksandr Mikhailovich Liapunov, Paul Lévy, and others. Laplace's demonstration centers on the hypothesis that one deals with the *sum* of independent random variables, but in applications neither independence nor summation may be appropriate. Other functions of random variables can lead to other distributions; for example, if instead of a sum one deals with the greatest random variable, entirely different limit distributions can result (Fréchet 1927). One of these distributions has found application in flood protection, breaking strength of materials, and other areas (Gumbel 1958).

Estimation. Laplace's writings contain the seeds of ideas that have been carefully studied only in recent years: optimum point estimation, hypothesis testing, and confidence interval estimation. Laplace proposed (see 1774, sec. V) that when estimating a parameter, θ, one use that function $T = T(Y_1, Y_2, \cdots)$ of the observations $Y_1, Y_2, \cdots$ for which the mean (or expected) absolute error of estimation, $E[|T - \theta|]$, is a minimum for the given probability distribution of errors. For the case of three independent identically distributed observations he gave an explicit algorithm for finding such a function to use for estimating the location parameter of their common distribution, $f(y - \theta)$, when this is completely specified except for the value of θ. He subsequently extended (1781, sec. XXX) this procedure (which is the same as Pitman's method of "close estimation"—see Pitman 1939) to cover n independent observations in the one-parameter case. Gauss's reformulation (1821) of the method of least squares in terms of minimum mean-square error of estimation (i.e., min $E[(T - \theta)^2]$) stems directly from this earlier work of Laplace's.

Many modern statisticians see the problem of confidence intervals as one of providing a random interval that contains, with at least a specified probability, some parameter of the distribution sampled. Laplace, however, gave another—somewhat vague—interpretation of the interval (*Oeuvres complètes*, vol. 7, pp. 286–287). Major differences of opinion persist about the interpretation of such intervals. [*See* BAYESIAN INFERENCE; ESTIMATION, *article on* CONFIDENCE INTERVALS AND REGIONS; FIDUCIAL INFERENCE.]

Applications to demography. Laplace was not satisfied merely to describe useful statistical methods; he never stopped applying them, most particularly to demography. In this area he was preceded by the great naturalist Buffon, whose early career was in mathematics, although this is not generally known, and who solved the famous problem of the needle thrown onto a table covered with parallel lines. His "Essai d'arithmétique morale" of 1777 deals with demographic statistics; he notes, for example, the propensity that most people have to use round numbers in stating their age. Laplace, using more general and more precise mathematical methods and ideas, was able to treat demographic problems much more intensively. Buffon had already noted the general preponderance of male over female births; Laplace noted that during the years 1745–1785, 393,386 boys and 377,555 girls were born in Paris, and he proved that given certain natural hypotheses, the probability that the chance of a male birth would be more than $\frac{1}{2}$ is $1 - 1/N$; in this example N is a very large number, greater than 10^{72}. In an anomalous case cited by Buffon—the birth of 203 boys and 213 girls in the village of Vitteaux over a five-year period—the same probability is only about $\frac{1}{3}$. Laplace pointed out that there is no logical contradiction between the two cases, since it is only for large populations that the value of the sought probability is near unity—i.e., that the event is almost certain.

An exact census of the French population would have been a very difficult undertaking in Laplace's time, and he therefore tried to estimate the population indirectly, by a rather curious method. He used in the approximation the rough constancy of the ratio between total population and annual number of births, so that if the numerical value of the ratio is known, one need only multiply it by the number of births to obtain an estimate of the population. The government accepted his proposal that studies be made to determine annual births and total population for thirty *départements*, selecting in each *département* only those parishes whose mayors seemed sufficiently conscientious. On September 22, 1802 (the Republic's New Year's Day), the inhabitants of these parishes were counted, and the number of births in each of the three preceding years was also recorded. From the resulting ratio and an estimate of the total number of yearly births in France, Laplace estimated the population to be 25 million.

Laplace was also interested in statistics having to do with the duration of marriages and with life insurance. In addition, he studied the application of statistics to problems of social order, such as the validity of trial evidence and of court judgments, and the concept of "moral expectation." This concept, introduced by Daniel Bernoulli, is based on the observation that the richer a man is, the less concerned he is about any fixed moderate sum of money. Laplace added to this generalization the qualification "all other things being equal," meaning that the benefit to be calculated depends in general on an infinity of circumstances that are impossible to evaluate and that are relative to the individual who is calculating it (see Georgescu-Roegen 1968). The work that Laplace did on trial evidence and court judgments has been especially controversial; J. L. F. Bertrand, in his caustic way, treated it with complete disdain, but Borel ([1914] 1948, pp. 251–262) felt that Bertrand was too harsh.

MAURICE FRÉCHET

[*Directly related is the entry* PROBABILITY. *Other relevant material may be found in* SOCIAL RESEARCH, THE EARLY HISTORY OF; *and in the biographies of* GAUSS *and* MOIVRE.]

WORKS BY LAPLACE

○1771 Recherches sur le calcul intégral aux différences infiniment petites, aux différences finies. Accademia delle Scienze di Torino, *Memorie* 4:273–375. → Volume 4, dated 1766–1769, was first published under the title *Mélanges de philosophie et de mathématique de la Société Royale de Turin.*

(1774) 1891 Mémoire sur la probabilité des causes par les événements. Volume 8, pages 27–65 in Pierre Simon de Laplace, *Oeuvres complètes.* Paris: Gauthier-Villars. → Includes "Problème III: Déterminer le milieu que l'on doit prendre entre trois observations données d'un même phénomène" on pages 41–48.

(1781) 1893 Mémoire sur les probabilités. Volume 9, pages 383–485 in Pierre Simon de Laplace, *Oeuvres complètes.* Paris: Gauthier-Villars.

(1782) 1894 Mémoire sur les suites. Volume 10, pages 1–89 in Pierre Simon de Laplace, *Oeuvres complètes.* Paris: Gauthier-Villars.

(1785) 1894 Mémoire sur les approximations des formules qui sont fonctions de très grands nombres. Volume 10, pages 209–291 in Pierre Simon de Laplace, *Oeuvres complètes.* Paris: Gauthier-Villars.

(1786*a*) 1894 Suite du mémoire sur les approximations des formules qui sont fonctions de très grands nombres. Volume 10, pages 296–338 in Pierre Simon de Laplace, *Oeuvres complètes.* Paris: Gauthier-Villars.

(1786*b*) 1895 Sur les naissances, les mariages et les morts à Paris, depuis 1771 jusqu'en 1784, et dans toute l'étendue de la France, pendant les années 1781 et 1782. Volume 11, pages 35–46 in Pierre Simon de Laplace, *Oeuvres complètes.* Paris: Gauthier-Villars.

(1796) 1836 *Exposition du système du monde.* 2 vols. 6th ed. Paris: Bachelier.

(1805) 1880 Traité de mécanique céleste. Part 2: Théories particulières des mouvements célestes. Volume 4 in Pierre Simon de Laplace, *Oeuvres complètes.* Paris: Gauthier-Villars.

(1810) 1898 Mémoire sur les approximations des formules qui sont fonctions de très grands nombres et sur leur application aux probabilités. Volume 12, pages 301–345 in Pierre Simon de Laplace, *Oeuvres complètes*. Paris: Gauthier-Villars. → A supplement to the "Mémoire" appears on pages 349–353.

(1812) 1820 *Théorie analytique des probabilités*. 3d ed., rev. Paris: Courcier. → Also published as Volume 1 of *Oeuvres complètes de Laplace*.

(1814) 1951 *A Philosophical Study on Probabilities*. New York: Dover. → First published as *Essai philosophique sur les probabilités*.

Oeuvres complètes de Laplace. 14 vols. Paris: Gauthier-Villars, 1878–1912.

SUPPLEMENTARY BIBLIOGRAPHY

BOREL, ÉMILE (1914) 1948 *Le hasard*. New ed., rev. & enl. Paris: Presses Universitaires de France.

COLBERT-LAPLACE, A. 1929 Letter to Karl Pearson, Dated 16 February 1929. *Biometrika* 21:203–204.

DANTZIG, D. VAN 1955 Laplace, probabiliste et statisticien, et ses précurseurs. *Archives internationales d'histoire des sciences* 8:27–37.

DAVID, F. N. 1965 Some Notes on Laplace. Pages 30–44 in Jerzy Neyman and Lucien M. Le Cam (editors), *Bernoulli, 1713; Bayes, 1763; Laplace, 1813*. New York: Springer.

DAVID, F. N.; and BARTON, D. E. 1962 *Combinatorial Chance*. London: Griffin; New York: Hafner.

DUVEEN, DENIS I.; and HAHN, ROGER 1957 Laplace's Succession to Bézout's Post of Examinateur des Élèves de l'Artillerie. *Isis* 48:416–427.

EISENHART, CHURCHILL 1964 The Meaning of "Least" in Least Squares. *Journal of the Washington Academy of Sciences* 54:24–33.

FELLER, WILLIAM (1950) 1957 *An Introduction to Probability Theory and Its Applications*. Vol. 1. 2d ed. New York: Wiley.

FRÉCHET, MAURICE 1927 Sur la loi de probabilité de l'écart maximum. Polskie Towarzystwo Matematyczne, *Annales: Rocznik* 6:93–122.

FRÉCHET, MAURICE 1940–1943 *Les probabilités associées à un système d'évènements compatibles et dépendants*. Parts 1–2. Paris: Hermann.

GAUSS, CARL FRIEDRICH (1809) 1963 *Theory of Motion of the Heavenly Bodies Moving About the Sun in Conic Sections*. New York: Dover. → First published in Latin.

GAUSS, CARL FRIEDRICH (1821) 1880 Theoria combinationis observationum erroribus minimis obnoxiae. Pars prior. Volume 4, pages 1–26 in *Carl Friedrich Gauss Werke*. Göttingen (Germany): Dieterichsche Universitäts-Druckerei. → A French translation was published in Paris in 1855 under the title *Méthode des moindres carrés: Mémoires sur la combinaison des observations*.

►GEORGESCU-ROEGEN, NICHOLAS 1968 Utility. Volume 16, pages 236–267 in *International Encyclopedia of the Social Sciences*. Edited by David L. Sills. New York: Macmillan and Free Press.

GUMBEL, E. J. 1958 *Statistics of Extremes*. New York: Columbia Univ. Press.

JORDAN, CHARLES (1939) 1947 *Calculus of Finite Differences*. 2d ed. New York: Chelsea.

LAGRANGE, JOSEPH LOUIS (1770–1773) 1868 Mémoire sur l'utilité de la méthode de prendre le milieu entre les résultats de plusieurs observations; dans lequel on examine les avantages de cette méthode par le calcul des probabilités; et où l'on résout différents problèmes relatifs à cette matière. Volume 2, pages 173–234 in Joseph Louis Lagrange, *Oeuvres de Lagrange*. Paris: Gauthier-Villars.

LÉVY, PAUL 1925 *Calcul des probabilités*. Paris: Gauthier-Villars.

MACMAHON, PERCY A. 1915 *Combinatory Analysis*. Vol. 1. Cambridge Univ. Press.

MERRIMAN, MANSFIELD 1877 A List of Writings Relating to the Method of Least Squares, With Historical and Critical Notes. Connecticut Academy of Arts and Sciences, *Transactions* 4:151–232.

MOIVRE, ABRAHAM DE (1718) 1756 *The Doctrine of Chances: Or, a Method of Calculating the Probabilities of Events in Play*. 3d ed. London: Millar.

MOIVRE, ABRAHAM DE 1730 *Miscellanea analytica de seriebus et quadraturis*. . . . London: Tonson & Watts.

MOIVRE, ABRAHAM DE (1733) 1959 A Method of Approximating the Sum of the Terms of the Binomial $(a+b)^n$ Expanded Into a Series, From Whence Are Deduced Some Practical Rules to Estimate the Degree of Assent Which Is to Be Given to Experiments. Volume 2, pages 566–575 in David Eugene Smith, *A Source Book in Mathematics*. New York: Dover. → First published as "Approximatio ad summam terminorum binomii $(a+b)^n$ in seriem expansi."

MOLINA, E. C. 1930 Theory of Probability: Some Comments on Laplace's *Théorie analytique*. American Mathematical Society, *Bulletin* 36:369–392.

NEWMAN, JAMES R. 1956 Commentary on Pierre Simon de Laplace. Volume 2, pages 1316–1324 in James R. Newman (editor), *The World of Mathematics*. New York: Simon & Schuster.

PEARSON, KARL 1929 Laplace: Being Extracts From the Lectures Delivered by Karl Pearson. *Biometrika* 21:202–216.

PITMAN, E. J. G. 1939 The Estimation of the Location and Scale of Parameters of a Continuous Population of Any Given Form. *Biometrika* 30:391–421.

RIORDAN, JOHN 1958 *An Introduction to Combinatorial Analysis*. New York: Wiley.

SIMON, G. A. 1929 Les origines de Laplace: Sa généalogie, ses études. *Biometrika* 21:217–230.

SIMPSON, THOMAS 1756 A Letter to the Right Honourable George, Earl of Macclesfield, President of the Royal Society, on the Advantage of Taking the Mean of a Number of Observations, in Practical Astronomy. Royal Society of London, *Philosophical Transactions* 49:82–93.

SIMPSON, THOMAS 1757 An Attempt to Show the Advantage Arising by Taking the Mean of a Number of Observations in Practical Astronomy. Pages 64–75 in Thomas Simpson, *Miscellaneous Tracts on Some Curious, and Very Interesting Subjects in Mechanics, Physical-astronomy, and Speculative Mathematics*. London: Nourse.

SMITH, DAVID EUGENE (1929) 1959 *A Source Book in Mathematics*. 2 vols. New York: Dover.

TODHUNTER, ISAAC (1865) 1949 *A History of the Mathematical Theory of Probability From the Time of Pascal to That of Laplace*. New York: Chelsea.

USPENSKY, JAMES V. 1937 *Introduction to Mathematical Probability*. New York: McGraw-Hill.

WALKER, HELEN M. 1929 *Studies in the History of Statistical Method, With Special Reference to Certain Educational Problems*. Baltimore: Williams & Wilkins.

WHITTAKER, EDMUND 1949*a* Laplace. *Mathematical Gazette* 33:1–12.

WHITTAKER, EDMUND 1949*b* Laplace. *American Mathematical Monthly* 56:369–372.

Wilson, Edwin B. 1923 First and Second Laws of Error. *Journal of the American Statistical Association* 18:841–851.

Postscript

The material here amplifies the discussion of Laplace's work on estimation and comments on the levels of development and understanding achieved in Laplace's theoretical and applied work.

The growth of Laplace's work over the span of his career accurately reflects (and was the major determinant of) the development of the fields of probability and statistics over this period; indeed his work was primarily responsible for the synthesis of these fields into what we now call mathematical statistics. In the early years of the eighteenth century, the works of Jacques Bernoulli in 1713, Montmort in 1713, and De Moivre in 1718 excited considerable interest in the theory of probability and its potential for application in areas far afield from the games of chance that had provided its early nourishment. [*See the biographies of the* Bernoulli family *and* Moivre.] To the early results of Jacques Bernoulli and De Moivre on laws of large numbers were added mathematical tools for studying the distribution of averages or means (Simpson 1756; Lagrange 1770–1773) and principles akin to maximum likelihood for the determination of the "best average" (Lambert's work in 1760 is discussed by Sheynin 1971; 1972; see also Daniel Bernoulli 1769). In this line of development, Laplace's memoirs of 1774 and 1781 were a novel departure in many respects: in 1774 Laplace discovered Bayes' theorem (perhaps in ignorance of Bayes' earlier essay; see Stigler 1975) and, in conjunction with his introduction of the first loss function, became the first to bring the theorem to bear on a statistical problem involving a model other than the binomial. He proposed that an unknown quantity be estimated by the number that minimizes the expected absolute error, the expectation being taken with respect to the posterior distribution relative to a uniform prior, and proved that this would be accomplished by using the median of the posterior distribution. In 1781, in an anticipation of later proofs of the optimality of least squares estimators, Laplace extended a calculation performed for a particular case in 1774 to show that for any reasonable, symmetric error density, the posterior median approached the sample mean as the scale parameter increased indefinitely.

Meanwhile, the increasing complexity of the problems being attacked in celestial mechanics by scientists such as Euler in 1749 and Mayer in 1750 and the increasing availability of reliable astronomical data produced in turn more difficult and important statistical problems. Methods of combining and solving the inconsistent systems of linear equations that resulted from these investigations were introduced and used by Mayer in 1750 and Boscovich in 1757. (See Sheynin 1972; Eisenhart 1961.) These methods were formulated without reference to the theory of probability, with no explicit mathematical model for the observational procedure, and with no attempt to quantify and minimize the uncertainty of the solutions of the equations. Laplace himself added to this line of development in a series of memoirs in the 1780s on celestial mechanics and in his *Traité de mécanique céleste* (1798–1805), providing mathematical algorithms or examples of the use of a wide variety of methods for solving multiple regression problems, including minimizing the maximum residual, minimizing the sum of absolute residuals, and aggregating the inconsistent equations by linear combination into a collection of consistent equations equal in number to the number of unknowns. (Simple versions of all those methods had been considered by earlier workers; see Stigler 1975.)

The development of nonprobabilistic methods for combining observations reached its zenith in 1805 with Legendre's publication of the method of least squares. But in 1809–1812 Gauss and Laplace separately tied the method of least squares to a probability model, and over the succeeding years, until Laplace's death in 1827, there was an explosive development of statistical models, methods, and techniques. Methods and concepts first enunciated a half century earlier by Simpson, Lagrange, and Laplace were recalled and merged with later techniques of mathematical analysis and the newer probability models to produce in a short period a body of work that is at least superficially similar to the major portion of modern mathematical statistics.

In 1810, Laplace employed his central limit theorem to provide a Bayesian justification for the method of least squares: if one were combining observations, each one of which was itself the mean of a large number of observations, then the least squares estimates would not only maximize the likelihood function but would also minimize the expected posterior absolute error, without an assumption about the particular error distribution or a circular appeal to the principle of the arithmetic mean, such as Gauss had employed in 1809. In 1811, Laplace reconsidered the linear regression problem from a non-Bayesian point of view. He restricted attention to linear unbiased estimators of the coefficients and showed (via a central-limit-type argument, with no assumption about the error distribution) that least squares estimators had the smallest asymptotic variance in this class, and thus both minimized the expected absolute error and maximized the probability that the estimator would be in any symmetric interval about the unknown coefficient. His derivation included the joint limiting distribution of the least squares estimators of two parameters.

In his *Théorie analytique des probabilités* (1812) and its supplements of 1814, 1818, 1820, Laplace expanded his earlier work, with additional material including an asymptotic expansion for posterior distributions via an approach that is the basis of modern work on this subject (Laplace [1812] 1820, book 2, chapter 4, sec. 23), a proof of the joint asymptotic normality of the least squares and least absolute deviations estimators (in the second supplement; see Stigler 1973), and a proof of the asymptotic equivalence of the sum of squared errors and the residual sum of squares (in the third supplement).

By the end of his life, Laplace's statistical understanding had matured to the point where, in an 1827 work on the application of his statistical methods to the problem of determining the moon's tidal effect on the earth's atmosphere (see Stigler 1975), he ingeniously blocked out day-to-day variations by considering first and second differences of barometric readings instead of the readings themselves, and he correctly performed a multiple regression analysis with correlated errors and known correlation structure. His analysis was marred by the fact that he tested the null hypothesis that the regression coefficients were zero using the total sum of squares (rather than the residual sum of squares) to estimate the variance, and when he performed a multiple comparison between quarterly mean diurnal barometric changes, he neglected to take account of the correlation between the quarterly means and the over-all mean and the effect of the multiple comparison upon the significance level. But his level of development was remarkable nonetheless.

A complete catalog of Laplace's contributions to probability and statistics would touch every area of these fields and reveal the pervasiveness of his influence upon the work of the last century and a half. The tools and concepts he developed are still in daily use; it seems fair to say that Laplace was more responsible for the early development of mathematical statistics than was any other person.

STEPHEN M. STIGLER

ADDITIONAL BIBLIOGRAPHY

BERNOULLI, DANIEL (1769) 1970 The Most Probable Choice Between Several Discrepant Observations and the Formation Therefrom of the Most Likely Induction. Translated by C. G. Allen. Volume 1, pages 157–167 in E. S. Pearson and Maurice G. Kendall (editors), *Studies in the History of Statistics and Probability.* London: Griffin; Darien, Conn.: Hafner. → First published in Latin.

CROSLAND, MAURICE P. 1967 *The Society of Arcueil: A View of French Science at the Time of Napoleon I.* Cambridge, Mass.: Harvard Univ. Press. → Includes a great deal of material on Laplace's career and his role in the social structure of French science.

EISENHART, CHURCHILL (1961) 1963 Boscovich and the Combination of Observations. Chapter 7 in Lancelot L. Whyte (editor), *Roger Joseph Boscovich, S.J., F.R.S., 1711–1787: Studies of His Life and Work on the 250th Anniversary of His Birth.* New York: Fordham Univ. Press.

FOURIER, JOSEPH 1829 Historical Eloge of the Marquis de Laplace. *Philosophical Magazine* Second Series 6:370–381. → First published in French.

GILLISPIE, CHARLES C. 1972 Probability and Politics: Laplace, Condorcet, and Turgot. American Philosophical Society, *Proceedings* 116:1–20.

PLACKETT, R. L. 1949 A Historical Note on the Method of Least Squares. *Biometrika* 36:458–460.

SHEYNIN, O. B. 1971 J. H. Lambert's Work on Probability. *Archive for History of Exact Sciences* 7:244–256.

SHEYNIN, O. B. 1972 On the Mathematical Treatment of Observations by L. Euler. *Archive for History of Exact Sciences* 9:45–56.

STIGLER, STEPHEN M. (1973) 1977 Laplace, Fisher, and the Discovery of the Concept of Sufficiency. Volume 2, pages 271–277 in Maurice G. Kendall and R. L. Plackett (editors), *Studies in the History of Statistics and Probability.* London: Griffin; New York: Hafner. →First published in *Biometrika* 60:439–445.

STIGLER, STEPHEN M. 1974 Gergonne's 1815 Paper on the Design and Analysis of Polynomial Regression Experiments. *Historia Mathematica* 1:431–447.

STIGLER, STEPHEN M. 1975 Napoleonic Statistics: The Work of Laplace. Studies in the History of Probability and Statistics, No. 34. *Biometrika* 62:503–517.

LATENT STRUCTURE

A scientist is often interested in quantities that are not directly observable but can be investigated only via observable quantities that are probabilistically connected with those of real interest. Latent structure models relate to one such situation in which the observable or manifest quantities are multivariate multinomial observations, for example, answers by a subject or respondent to dichotomous or trichotomous questions. Models relating polytomous observable variables to unobservable or latent variables go back rather far; some early references are Cournot (1838), Weinberg (1902), Benini (1928), and deMeo (1934). These models typically express the multivariate distribution of the observable variables as a mixture of multivariate distributions, where the distribution of the latent variable is the mixing distribution [*see* DIS-

TRIBUTIONS, STATISTICAL, *article on* MIXTURES OF DISTRIBUTIONS].

Lazarsfeld (1950) first introduced the term *latent structure model* for those models in which the variables distributed according to any of the component multivariate distributions of the mixture are assumed to be stochastically independent. (Thus a latent structure model of a subject's answers to 50 dichotomous questions—the latent class model of this article—assumes that subjects fall into relatively few classes, called latent classes, with the variable that relates the subject to his class being the latent variable. The distribution of this latent variable, that is, the distribution of the subjects among latent classes, is the mixing distribution. Within each class it is assumed that the responses to the 50 dichotomous questions are stochastically independent.) A basic reference for the general form of latent structure models is Anderson (1959).

The present article—restricted to the case of dichotomous questions—emphasizes the problems of identifiability and efficient statistical estimation of the parameters of latent structure models, points out difficulties with methods that have been proposed, and summarizes doubts currently held about the possibility of good estimation.

The simplest of the latent structure models and almost the only one in which the problem of parameter estimation has been carefully addressed is the *latent class model.* In this model, each observation in the sample is a vector $\boldsymbol{x}$ with p two-valued items or coordinates, conveniently coded by writing each either as 0 or as 1. The latent class model postulates that there is a small number m of classes, called *latent classes*, into which potential observations on the population can be classified such that within each class the p coordinates of the vector $\boldsymbol{x}$ are statistically independent. This is not to say that all identical observations in the sample are automatically considered as coming from the same class. Rather, associated with each class is a probability distribution on the 2^p possible vectors $\boldsymbol{x}$, such that the p coordinates of $\boldsymbol{x}$ are (conditionally) independent. An observation vector $\boldsymbol{x}$ thus has a probability distribution that is a mixture of the probability distributions of $\boldsymbol{x}$ associated with each of the latent classes.

An example of the above model comes from the study (Lazarsfeld 1950) of the degree of ethnocentrism of American soldiers during World War II. Because it is not known how to measure ethnocentrism directly, a sample of soldiers was asked the following three questions: Do you believe that our European allies are much superior to us in strategy and fighting morale? Do you believe that the majority of all equipment used by all the allies comes from American lend-lease shipment? Do you believe that neither we nor our allies could win the war if we didn't have each other's help?

Here $p = 3$ and $\boldsymbol{x}$ is the vector of responses to the three questions, with Yes coded as 1 and No coded as 0. A suitable latent class model would postulate that there are two latent classes (so that $m = 2$), such that within each class the answers to the three questions are stochastically independent. Postulating the existence of any more than two latent classes would, as will be seen later, lead to difficulties, since the parameters of such a latent class model could not be consistently estimated. The two latent classes would probably be composed of ethnocentric and nonethnocentric soldiers, respectively. However, this need not be the case, and in fact it may happen that the two latent classes will have no reasonable interpretation, let alone the hoped-for interpretation. This phenomenon of possible noninterpretability is characteristic not only of the latent class model but also of the factor analysis and other mixture-of-distributions models.

The latent class model. Let σ denote a subset (unordered) of the integers $(1, 2, \cdots, p)$, possibly the null subset ϕ. (Other subsets will, for concreteness, be denoted by writing their members in customary numerical order.) Let π_σ denote the probability that for a randomly chosen individual each coordinate of $\boldsymbol{x}$ with index a member of σ is a 1, and define $\pi_\phi = 1$. For example, $\pi_{2,7,19}$ is the probability that the second, seventh, and nineteenth coordinates of $\boldsymbol{x}$ are all 1, forgetting about—or marginally with respect to—the values of the other coordinates of $\boldsymbol{x}$.

Since the order of coordinates is immaterial for such a probability, one is justified in dealing with the 2^p unordered σ's, but a specific order in naming the subset is helpful for exposition. The π_σ's are notationally a more convenient set of parameters than what might be considered the 2^p natural parameters of the multinomial distribution of $\boldsymbol{x}$.

A concise description of the natural parameters of the distribution of $\boldsymbol{x}$ is the following. Let $\bar{\sigma}$ denote that subset of the integers $(1, 2, \cdots, p)$ which is the complement of σ. Let $\pi_{\sigma:\bar{\sigma}}$ denote the probability that for a randomly chosen individual each coordinate of $\boldsymbol{x}$ with index of a member of σ is a 1 and each coordinate of $\boldsymbol{x}$ with index a member of $\bar{\sigma}$ is a 0. The 2^p $\pi_{\sigma:\bar{\sigma}}$'s are the natural parameters of the multinomial distribution of $\boldsymbol{x}$, since they are the probabilities of each of the 2^p possible observation values. For example, in the ethnocentrism case, $\pi_{1,2:3}$ would be the probability that the first two questions are answered Yes, while the third question is answered No. The π_σ's and $\pi_{\sigma:\bar{\sigma}}$'s are related by a nonsingular linear transformation.

Let ν_α be the probability that the observation vector $\boldsymbol{x}$ is a member of the αth latent class, where $\alpha = 1, 2, \cdots, m$ and $\sum \nu_\alpha = 1$. Let $\lambda_{\alpha\sigma}$ be the probability that if $\boldsymbol{x}$ is a vector chosen at random from the αth class, then each coordinate of $\boldsymbol{x}$ with index a member of σ is a 1. Clearly $\pi_\sigma = \sum_\alpha \nu_\alpha \lambda_{\alpha\sigma}$.

Let σ_i denote the ith member of σ, with the members of σ arranged in some order, say numerical. The fundamental independence assumption of the latent class model then says that for each α

$$\lambda_{\alpha\sigma} = \prod_{\sigma_i \epsilon \sigma} \lambda_{\alpha\sigma_i}$$

for all σ. That is, the probability (conditional on $\boldsymbol{x}$ being in the αth latent class) of any given set of coordinates of $\boldsymbol{x}$ being all 1's is the product of the probabilities of each of these coordinates being a 1. Then

$$\pi_\sigma = \sum_{\alpha=1}^{m} \nu_\alpha \prod_{\sigma_i \epsilon \sigma} \lambda_{\alpha\sigma_i}$$

for all σ. These equations are called the *accounting equations* of the latent class model. Thus the $m(p+1)$ parameters of the model are the *latent parameters* $\lambda_{\alpha i}$ and the ν_α, $\alpha = 1, \cdots, m$, $i = 1, \cdots, p$. These completely determine the 2^p *manifest parameters*, the π_σ, via the accounting equations.

Parameter estimation. Suppose that the number of latent classes, m, is known to the investigator. (This assumption is made because it underlies all the theoretical work on the estimation of parameters of the latent class model. In practice m is unknown, but a pragmatic approach is to assume a particular small value of m, proceed with the estimation, see how well the estimated model fits the manifest data, and alter m and begin again if the fit is poor.) Then a central statistical problem is that of estimating the parameters of the model, the ν's and λ's, from a random sample of n vectors $\boldsymbol{x}$. (The typical sample in survey work is a stratified rather than a simple random sample. However, the problem of estimating latent parameters from such samples is much more complicated, and as yet has hardly been touched.)

Let n_σ be the number of vectors in the sample with 1's in each component whose index is a member of σ, and let $p_\sigma = n_\sigma/n$. If the model were simply a multinomial model with parameters the π_σ's, then the p_σ's would be maximum likelihood estimators of the π_σ's. If for each set of 2^p π_σ's there is a unique set of latent parameters, ν_α's and $\lambda_{\alpha i}$'s, $\alpha = 1, \cdots, m$, $i = 1, \cdots, p$, then the ν's and λ's are functions of the π_σ's, and evaluating these functions at the p_σ's as arguments will yield estimators (actually consistent estimators) of the latent parameters. But the "if" in the last sentence is most critical; it is the identifiability condition, common to all models relating distributions of observable random variables to distributions of unobservable random variables. Consequently, most of the work on parameter estimation in latent class analysis is really a by-product of work on finding constructive procedures, that is, procedures that explicitly derive the unique latent parameters as function of the π's, for proving the identifiability of a latent class model associated with a given m and p. With such a constructive procedure available, one can replace the π's by their estimates, the p's, and use the procedure to determine estimates of the ν's and λ's. The following description of estimation procedures based on constructive proofs of identifiability will thus really be a description of the constructive procedure for determining the ν's and λ's from a subset of the π's.

Green's method of estimation. The earliest constructive procedure was given by Green (1951). Let $\boldsymbol{D}_i$ be the $m \times m$ diagonal matrix with $\lambda_{\alpha i}$, $\alpha = 1, \cdots, m$, on the diagonal, and let $\boldsymbol{L}$ be the $(p+1) \times m$ matrix with first row a vector of 1's and jth row $(j = 2, \cdots, p+1)$ the vector of $(\lambda_{1,j-1}, \cdots, \lambda_{m,j-1})$. Let $\boldsymbol{N}$ be the $m \times m$ diagonal matrix with ν_α, $\alpha = 1, \cdots, m$, on the diagonal. For σ a subset of $(1, 2, \cdots, p)$, define $\boldsymbol{D}_\sigma = \prod_{\sigma_j \epsilon \sigma} \boldsymbol{D}_{\sigma_j}$. Form the matrix $\Pi_\sigma = \boldsymbol{L N D}_\sigma \boldsymbol{L}'$, where the prime denotes the matrix transpose. The (i,j)th element of this matrix is

$$\sum_{\alpha=1}^{m} \nu_\alpha \lambda_{\alpha i} \lambda_{\alpha j} \prod_{\sigma_k \epsilon \sigma} \lambda_{\alpha\sigma_k}$$

If $i \neq j$ and $i,j \not\epsilon\, \sigma$ then the (i,j)th element of this matrix is the manifest parameter $\pi_{ij\sigma}$. Otherwise the (i,j)th element of this matrix can formally be defined as a quantity called $\pi_{ij\sigma}$, where the subscript of π may have repeated elements. Since π's with repeated subscripts are not manifest parameters and have no empirical counterpart but are merely formal constructs based on the latent parameters, they are not estimable directly from the n_σ's. However, Green provided some rules for guessing at values of these π's (one rule is given below) so that the matrix Π_σ can be partly estimated and partly guessed at, given data.

Let $\Pi_0 = \boldsymbol{LNL}'$, $\boldsymbol{N}^{\frac{1}{2}}$ be the $m \times m$ diagonal matrix with $\sqrt{\nu_\alpha}$, $\alpha = 1, \cdots, m$, on the diagonal, $\boldsymbol{D} = \sum_{k=1}^{m} \boldsymbol{D}_k$, and $\boldsymbol{A} = \boldsymbol{LN}^{\frac{1}{2}}$. Then $\Pi = \sum_k \Pi_k = \boldsymbol{ADA}'$.

Under the assumptions that $m \leqslant p + 1$, rank $\boldsymbol{A} = m$, and all the diagonal elements of $\boldsymbol{D}$ are different and nonzero, the following procedure determines the matrices $\boldsymbol{L}$ and $\boldsymbol{N}$ of latent parameters.

Factor Π_0 as $\Pi_0 = \boldsymbol{BB}'$ and Π as $\Pi = \boldsymbol{CC}'$. (The matrices $\boldsymbol{B}$ and $\boldsymbol{C}$ are not unique, but any factorization will do.) Let $\boldsymbol{T} = (\boldsymbol{BB}')^{-1}\boldsymbol{B}'\boldsymbol{C}$. A complete principal component analysis of $\boldsymbol{TT}'$ will yield an orthogonal matrix $\boldsymbol{Q}$, and it can be shown that $\boldsymbol{A} = \boldsymbol{BQ}$. Since the first row of $\boldsymbol{L}$ is a vector of 1's, the first row of $\boldsymbol{A}$ is an estimate of the vector $(\sqrt{\nu_1}, \cdots, \sqrt{\nu_m})$, so that $\boldsymbol{N}$ is easily determined. The matrix $\boldsymbol{L}$ is then just $\boldsymbol{AN}^{-\frac{1}{2}}$.

The major shortcoming of this procedure is the problem of how to guess at values of the π's bearing repeated subscripts. No one has yet devised a rule which, when applied to a set of p's, will yield consistent estimators of $\boldsymbol{L}$ and $\boldsymbol{N}$. For example, Green suggests using $p_{ii} = p_i^2 + \max_{j \neq i}(p_{ij} - p_i p_j)$ as a guess at π_{ii}. Yet in the case $m = 2$, $p = 3$ with latent parameters $\nu_1 = \nu_2 = .5$, $\lambda_{11} = .9$, $\lambda_{12} = .2$, $\lambda_{13} = .8$, $\lambda_{21} = .7$, $\lambda_{22} = .9$, $\lambda_{23} = .4$, if $i = 2$, $\max_{j \neq 2}(p_{2j} - p_2 p_j)$ is a consistent estimator of $-.07$, so that p_{22} is a consistent estimator of something smaller than π_2^2. But $\pi_{ii} \geqslant \pi_i^2$, so that p_{22} is not a consistent estimator of π_{22}.

Determinantal method of estimation. A matricial procedure that does not have the above shortcoming, since it involves only estimable π's, was first suggested by Lazarsfeld and Dudman (see Lazarsfeld 1951) and independently by Koopmans (1951), developed by Anderson (1954), and extended by Gibson (1955; 1962) and Madansky (1960). For ease of exposition, the procedure will be described only for the cases treated by Anderson.

● Assume that $p \geqslant 2m - 1$. In that case, $2m - 1$ different items can be selected from the p items (say, the first $2m - 1$) and the following matrices of π's involving only these items formed. Let

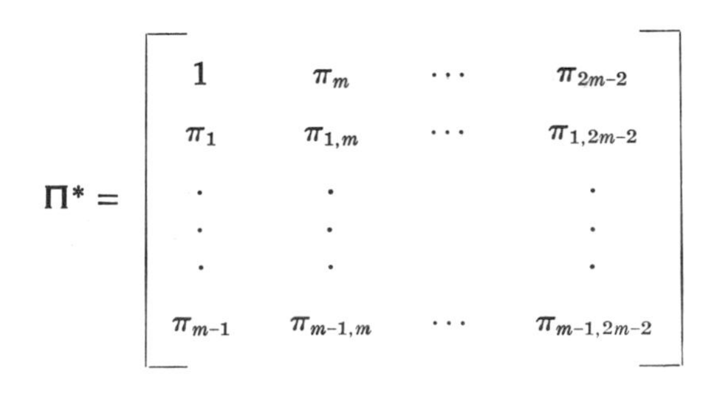

$$\Pi^* = \begin{bmatrix} 1 & \pi_m & \cdots & \pi_{2m-2} \\ \pi_1 & \pi_{1,m} & \cdots & \pi_{1,2m-2} \\ \cdot & \cdot & & \cdot \\ \cdot & \cdot & & \cdot \\ \cdot & \cdot & & \cdot \\ \pi_{m-1} & \pi_{m-1,m} & \cdots & \pi_{m-1,2m-2} \end{bmatrix}$$

and let $\tilde{\Pi}$ be the matrix Π^* with the 1 replaced by π_{2m-1} and all the π's having the additional subscript $2m - 1$. Let Λ_1 be an $m \times m$ matrix with the first row a vector of 1's and the jth row $(j = 2, \cdots, m)$ the vector $(\lambda_{1,j-1}, \cdots, \lambda_{m,j-1})$, and let Λ_2 be an $m \times m$ matrix with first row a vector of 1's and the jth row $(j = 2, \cdots, m)$ the vector $(\lambda_{1,m+j-2}, \cdots, \lambda_{m,m+j-2})$. Let $\boldsymbol{N}$ and $\boldsymbol{D}_{2m-1}$ be defined as above. Then $\Pi^* = \Lambda_1 \boldsymbol{N} \Lambda_2'$ and $\tilde{\Pi} = \Lambda_1 \boldsymbol{N} \boldsymbol{D}_{2m-1} \Lambda_2'$. Thus if the diagonal elements of $\boldsymbol{D}_{2m-1}$ are distinct and if Λ_1, $\boldsymbol{N}$, and Λ_2 are of full rank, then the diagonal elements of $\boldsymbol{D}_{2m-1}$ are the roots θ of the determinantal equation $|\tilde{\Pi} - \theta \Pi^*| = 0$.

If $\boldsymbol{Z}$ is the matrix of characteristic vectors corresponding to the roots $\theta_1, \cdots, \theta_m$, then the columns of $\Pi^* \boldsymbol{Z}$ are proportional to the columns of Λ_1, with the constant of proportionality determined by the condition that the first row of Λ_1 is a vector of 1's. A similar argument using the transposes of $\tilde{\Pi}$ and Π^* yields Λ_2, and $\boldsymbol{N}$ is determined by $\boldsymbol{N} = \Lambda_1^{-1} \Pi^* \Lambda_2'^{-1}$.

○ A difficulty with this procedure is that it depends critically on which $2m - 1$ items are chosen from the p items, on which of these $2m - 1$ are chosen to define Π^*, and on the allocation of the $2m - 2$ items to the rows and columns defining Π^*. That is, it depends critically on the ordering of the items. There are no general rules available for an ordering of the items that will yield relatively efficient estimators of the latent parameters.

The most important shortcoming of this procedure and of its extensions (which involve more of the π's) is that there is no guarantee that when the procedure is used with a set of p's it will produce permissible estimates of the latent parameters, that is, estimates that are real numbers between 0 and 1. In four sampling experiments with $n = 1{,}000$, $m = 3$, and $p = 8$, Anderson and Carleton (1957) found that of 2,240 determinantal equations only 33.7 per cent had all roots between 0 and 1. Madansky (1959) computed the asymptotic variance of the determinantal estimates for the case $m = 2$, $p = 3$, a case in which these estimators, if permissible, are the maximum likelihood estimators of the latent parameters, and found the results presented in Table 1, where n is the sample size. Thus, sample sizes must be greater than 1,116 for the variance of the estimators of all the parameters to be less than 1.

Table 1

Parameter	Value	Asymptotic variance
ν_1	3/4	$1115.42/n$
λ_{11}	1/2	$39.00/n$
λ_{12}	1/3	$60.89/n$
λ_{13}	1/3	$4.96/n$
λ_{21}	1/4	$303.00/n$
λ_{22}	2/3	$611.53/n$
λ_{23}	1/4	$31.00/n$

Rounding error also affects the estimates greatly. The parameters of the multinomial distribution for the above model are given in Table 2.

For a sample of size 40, if one had actually observed the expected number of respondents for each of the response patterns (rounded to the nearest integer), then the sample would have the composition shown in Table 3. Table 4 shows the p_σ's based on these data (π_σ being given for comparison). The determinantal estimates of the latent parameters are given in the third column of Table 5. (The fourth column will be discussed below.)

Partitioning method of estimation. A third estimation procedure (Madansky 1959) looks at the problem in a different light. Since the latent classes are defined as those classes within which the p components of the vector $\boldsymbol{x}$ are statistically independent, one might (at least conceptually) look at all possible assignments of the n observations into m classes and find that assignment for which the usual χ^2 test statistic for independence is smallest. The estimates of the latent parameters would then just be the appropriate proportions based on this best assignment. They would always be permissible. Although for finite samples they would not be identical with minimum χ^2 estimates, they would have the same asymptotic properties and thus be asymptotically equivalent to maximum likelihood estimates.

Madansky (1959) introduced another measure of independence, simpler to compute than χ^2, and found that the asymptotic efficiency of the estimators of the latent parameters from this procedure, in the example described above, is about .91. The obvious shortcoming of this idea is that it is too time consuming to carry out all the possible assignments, even for moderate samples on an electronic computer. In the example described above, for a sample of size 40 it took four hours of computation on the IBM 704 to enumerate and assess all the assignments into two classes. The resulting estimates are shown in the fourth column of Table 5.

Scoring methods. Current activity on estimation procedures for the latent class model (Henry 1964) is directed toward writing computer routines using the scoring procedure described by McHugh (1956) to obtain best asymptotically normal estimates of the latent parameters. The scoring procedure will yield estimators with the same large asymptotic variances as those indicated by the above example of the maximum likelihood estimators' asymptotic variances. Also, the scoring procedure has the same permissibility problem associated with it as did the determinantal approach described above. However, the problem can be alleviated for this procedure by using a set of consistent permissible estimators for initial values in the scoring procedure.

ALBERT MADANSKY

[*See also* SCALING, MULTIDIMENSIONAL. *Directly related are the entries* DISTRIBUTIONS, STATISTICAL, *article on* MIXTURES OF DISTRIBUTIONS; FACTOR ANALYSIS AND PRINCIPAL COMPONENTS; STATISTICAL IDENTIFIABILITY.]

Table 2

$$\pi_{123;\phi} = 10/192$$
$$\pi_{23;1} = 14/192$$
$$\pi_{13;2} = 17/192$$
$$\pi_{12;3} = 22/192$$
$$\pi_{3;12} = 19/192$$
$$\pi_{2;13} = 34/192$$
$$\pi_{1;23} = 35/192$$
$$\pi_{\phi;123} = 41/192$$

Table 3

Response pattern	Number observed
123;ϕ	2
23;1	3
13;2	4
12;3	4
3;12	4
2;13	7
1;23	7
ϕ;123	9

Table 4

σ	p_σ	π_σ
1	.425	.4375
2	.400	.4167
3	.325	.3125
12	.150	.1667
13	.150	.1406
23	.125	.1250
123	.050	.0521

Table 5 — Parameter estimates for two methods*

Parameter	Value	Determinantal estimate	Partitioning estimate
ν_1	.75	.23	.58
λ_{11}	.50	.82	.00
λ_{12}	.33	.23	.43
λ_{13}	.33	.42	.30
λ_{21}	.25	.30	1.00
λ_{22}	.67	.45	.35
λ_{23}	.25	.29	.35

* $n = 40$.

Source: Madansky 1959, p. 21.

BIBLIOGRAPHY

ANDERSON, T. W. 1954 On Estimation of Parameters in Latent Structure Analysis. *Psychometrika* 19:1–10.

ANDERSON, T. W. 1959 Some Scaling Models and Estimation Procedures in the Latent Class Model. Pages 9–38 in Ulf Grenander (editor), *Probability and Statistics*. New York: Wiley.

ANDERSON, T. W.; and CARLETON, R. O. 1957 Sampling Theory and Sampling Experience in Latent Structure Analysis. *Journal of the American Statistical Association* 52:363 only.

BENINI, RODOLFO 1928 Gruppi chiusi e gruppi aperti in alcuni fatti collettivi di combinazioni. International Statistical Institute, *Bulletin* 23, no. 2:362–383.

COURNOT, A. A. 1838 Mémoire sur les applications du calcul des chances à la statistique judiciaire. *Journal de mathématiques pures et appliquées* 3:257–334.

DEMEO, G. 1934 Su di alcuni indici atti a misurare l'attrazione matrimoniale in classificazioni dicotome. Accademia delle Scienze Fisiche e Matematiche, Naples, *Rendiconto* 73:62–77.

GIBSON, W. A. 1955 An Extension of Anderson's Solution for the Latent Structure Equations. *Psychometrika* 20:69–73.

GIBSON, W. A. 1962 Extending Latent Class Solutions to Other Variables. *Psychometrika* 27:73–81.

GREEN, BERT F. JR. 1951 A General Solution for the Latent Class Model of Latent Structure Analysis. *Psychometrika* 16:151–166.

HENRY, NEIL 1964 The Computation of Efficient Estimates in Latent Class Analysis. Unpublished manuscript, Columbia Univ., Bureau of Applied Social Research.

KOOPMANS, T. C. 1951 Identification Problems in Latent Structure Analysis. Cowles Commission Discussion Paper: Statistics, No. 360. Unpublished manuscript.

LAZARSFELD, PAUL F. 1950 The Logical and Mathematical Foundation of Latent Structure Analysis. Pages 362–412 in Samuel A. Stouffer et al., *Measurement and Prediction*. Princeton Univ. Press.

LAZARSFELD, PAUL F. 1951 *The Use of Mathematical Models in the Measurement of Attitudes*. Research Memorandum RM-455. Santa Monica (Calif.): RAND Corporation.

LAZARSFELD, PAUL F. 1959 Latent Structure Analysis. Pages 476–543 in Sigmund Koch (editor), *Psychology: A Study of a Science*. Volume 3: Formulations of the Person and the Social Context. New York: McGraw-Hill.

MCHUGH, RICHARD B. 1956 Efficient Estimation and Local Identification in Latent Class Analysis. *Psychometrika* 21:331–347.

MCHUGH, RICHARD B. 1958 Note on "Efficient Estimation. . . ." *Psychometrika* 23:273–274. → This is a correction to McHugh 1956.

MADANSKY, ALBERT 1959 *Partitioning Methods in Latent Class Analysis*. Paper P-1644. Santa Monica (Calif.): RAND Corporation.

MADANSKY, ALBERT 1960 Determinantal Methods in Latent Class Analysis. *Psychometrika* 25:183–198.

WEINBERG, WILHELM 1902 Beiträge zur Physiologie und Pathologie der Mehrlingsgeburten beim Menschen. *Pflüger's Archiv für die gesamte Physiologie des Menschen und der Tiere* 88:346–430.

Postscript

The long-awaited book by Lazarsfeld and Henry (1968) includes a clear elementary presentation of the latent class model and more general latent structure models. The scoring method for estimation in the latent class model (Henry 1964) is described therein, as is some unpublished work by W. A. Gibson extending Green's method of estimation. Appendix B contains a paper by T. W. Anderson, "Large Sample Distribution Theory for Estimates of the Parameters of a Latent Class Model," which derives the asymptotic distribution of estimates obtained by the determinantal method of estimation.

Instead of using the method of scoring to determine estimates that are asymptotically equivalent to the maximum likelihood estimates, Goodman (1974*a*; 1974*b*) suggests that one apply an iterative procedure to find the roots of the likelihood equation directly. Moreover, such an iterative procedure extends quite readily to handle cases where there are special "restrictions" on the latent parameters, for example, if it is known that $\lambda_{ij} = \lambda_{ik}$ for some i, j, k.

The iterative procedure, in our notation, is the following. Let σ_i^* denote a subset of the p integers $(1, 2, \cdots, p)$ excluding the integer i. Then, for $\alpha = 1, \cdots, m$ and $i = 1, \cdots, p$, the $(t+1)$st estimate of the latent parameters is obtained from the tth estimate by

$$\hat{\nu}_\alpha^{(t+1)} = \sum_\sigma p_\sigma \hat{\nu}_\alpha^{(t)} \hat{\lambda}_{\alpha\sigma}^{(t)} \Big/ \hat{\pi}_\sigma^{(t)},$$

$$\hat{\lambda}_{\alpha i}^{(t+1)} = \sum_{\sigma_i^*} p_{i\sigma_i^*} \hat{\nu}_\alpha^{(t)} \hat{\lambda}_{\alpha i}^{(t)} \hat{\lambda}_{\alpha\sigma_i^*}^{(t)} \Big/ \hat{\pi}_{i\sigma_i^*}^{(t)},$$

where

$$\hat{\pi}_\sigma^{(t)} = \sum_{\alpha=1}^m \hat{\nu}_\alpha^{(t)} \hat{\lambda}_{\alpha\sigma}^{(t)},$$

and

$$\hat{\lambda}_{\alpha\sigma}^{(t)} = \prod_{\sigma_i \in \sigma} \hat{\lambda}_{\alpha\sigma_i}^{(t)}.$$

The initial values of the $\hat{\nu}_\alpha^{(t)}$ and $\hat{\lambda}_{\alpha i}^{(t)}$ are selected so as to satisfy the accounting equations

$$p_\sigma = \sum_{\alpha=1}^m \hat{\nu}_\alpha^{(0)} \prod_{\sigma_i \in \sigma} \hat{\lambda}_{\alpha\sigma_i}^{(0)}$$

and the constraints $1 \geqslant \hat{\nu}_\alpha^{(0)} \geqslant 0$, $1 \geqslant \hat{\lambda}_{\alpha i}^{(0)} \geqslant 0$, $\sum_{\alpha=1}^m \hat{\nu}_\alpha^{(0)} = 1$. Since the likelihood function is not concave in the latent parameters, this iterative procedure may lead to a local rather than global maximizer. A suggested approach is to vary the initial values of the latent parameter estimates and (after the iterative procedure has converged) compare the resulting likelihood function values.

ALBERT MADANSKY

[*See also the biography of* LAZARSFELD.]

ADDITIONAL BIBLIOGRAPHY

GOODMAN, LEO A. 1974*a* The Analysis of Systems of Qualitative Variables When Some of the Variables Are Unobservable: I. A Modified Latent Structure Approach. *American Journal of Sociology* 79:1179–1259.

GOODMAN, LEO A. 1974*b* Exploratory Latent Structure Analysis Using Both Identifiable and Unidentifiable Models. *Biometrika* 61:215–231.

LAZARSFELD, PAUL F.; and HENRY, NEIL W. 1968 *Latent Structure Analysis*. Boston: Houghton Mifflin.

LATIN SQUARES

See EXPERIMENTAL DESIGN.

LAW

See PUBLIC POLICY AND STATISTICS; STATISTICS AS LEGAL EVIDENCE.

LAZARSFELD, PAUL F.

► *This article was specially written for this volume.*

Paul Felix Lazarsfeld was born on February 13, 1901, in Vienna and died on August 30, 1976, in New York City. Lazarsfeld's principal academic contributions were to sociology; he was probably the most important force in what came to be known as American sociology, which invaded Europe in the 1950s and 1960s. In making those contributions, Lazarsfeld not only did a great deal to bring statistical analysis into sociology but also himself made several major contributions to statistical analysis. This brief biography concentrates on those statistical contributions; fuller appreciations of Lazarsfeld's work, and a full bibliography, will be found in a volume in preparation by Merton, Coleman, and Rossi.

Lazarsfeld grew up in Vienna and in 1925 received his PH.D. in mathematics from the University of Vienna. Almost immediately after that, he was attracted to social psychology by Charlotte and Karl Bühler, and he taught as an assistant at their institute at the University of Vienna. While there, he carried out, with Marie Jahoda and Hans Zeisel, research leading to *Die Arbeitslosen von Marienthal* (1933), a study of massive unemployment in an Austrian town. This early research showed the inventiveness in use of qualitative and quantitative data that was to mark Lazarsfeld's work (for example, the comparison of diaries of unemployed men and their busy wives, the measurement of the pace of walking, and changes in the circulation of the political newspaper and sports-oriented newspaper after unemployment began).

In 1933, after spending a short period as an interpreter in France (a country to which he always felt strong emotional ties and to which he frequently returned), Lazarsfeld arrived in the United States as a Rockefeller Foundation fellow. With the rise of the Nazis, he remained in the United States and took employment at the University of Newark. There he began an activity that had in two ways a lasting impact on American social science. First, he established an applied social research center, the prototype for his later Bureau of Applied Social Research at Columbia University and subsequently for applied social research institutes in universities throughout the country. Second, in this new center he began radio audience research, the prototype and forerunner of the field of mass communications research (see, for example, 1940*a*). In 1937 he moved to Princeton University, and then to Columbia in 1940, where he joined the sociology department and established (if the word "established" can be used to describe a long struggle to create a legitimate position for this organizational innovation in the university) the Bureau of Applied Social Research.

From 1940 until his retirement in 1969, he remained at Columbia, where in 1962 he became the first Quetelet Professor of Social Sciences. At the time of his death, he was Distinguished Professor of Social Science at the University of Pittsburgh. During his career, Lazarsfeld received honorary degrees from Columbia, the University of Chicago, the Sorbonne, the University of Vienna, and Yeshiva University. He was a member of the National Academy of Sciences and was the 51st president of the American Sociological Association.

A major contribution of Lazarsfeld to statistical analysis in social science was the idea of a "latent structure" (of attitudes or other unobservable entities) that could be recovered only through its probabilistic connection to observed data. The principle of local independence—that is, at a given point in the structure responses are independent—provided leverage to recover the latent structure from observed responses. [*See* LATENT STRUCTURE; *see also* 1950*a*; 1954*a*; 1959; 1965; 1968.]

A second major development that gained strong impetus from Lazarsfeld was the conversion of sample surveys from instruments for estimating characteristics of the population to instruments for causal analysis. In a seminal paper with Patricia Kendall (1950*b*), he developed his so-called elaboration formula, which made use of partial correlations in novel ways for inferring causal structures among variables. [*See* SAMPLE SURVEYS; SURVEY ANALYSIS; CAUSATION.]

A third major development in which Lazarsfeld was highly influential was the introduction of panel analysis into sample surveys, by repeated interviews of the same sampled persons. Lazarsfeld pioneered the use of panels as part of his continuing effort to identify causal relations among variables from survey data. [*See* PANEL STUDIES; *see also* 1938; 1940*b*; 1941; 1948; 1951.] This first took the form of the 16-fold table, cross-tabulating two dichotomous attributes observed at two points in time; later it took other forms, all aimed at identifying the direction of causality that led to an observed correlation. [*See* SURVEY ANALYSIS, *especially the article on* THE ANALYSIS OF ATTRIBUTE DATA; *see also* 1961; 1972*a*; 1972*b*.]

This brief description of Lazarsfeld's work most directly important to statistics ignores much of his work in sociology. It also omits a central element of Lazarsfeld's importance, for his own publications and their impact constitute only a fraction of his influence on social research. He could not and would not keep his intellectual concerns to himself. It was difficult to be in or near sociology at Columbia and not be drawn into Lazarsfeld's activities. He pulled into his orbit at one time or another colleagues and students of all stripes, ranging from C. Wright Mills to R. Duncan Luce. It was his influence, for example, that led T. W. Anderson to begin exploring the use of Markov chains for modeling attitude change, an approach then novel but now commonplace. [*See* MARKOV CHAINS.] In teaching as well as research, Lazarsfeld was interested in the interaction of his ideas with those of others. For some years, he taught an influential seminar in mathematical sociology with Ernest Nagel. His seminars at the Bureau of Applied Social Research were a source of stimulation for many years.

The pattern of intellectual interaction characteristic of Lazarsfeld had its strongest and longest manifestation in his relation with Robert Merton. For more than twenty years, Lazarsfeld and Merton, in research, writing, and teaching, both together and separately, constituted Columbia sociology. (See 1954*b*; 1972*c*; for Lazarsfeld's own description of the collaboration, see 1975.) More than that, this combination constituted for a period of years the dominant force in American sociology. It was a force that, probably more than any other, brought statistical analysis into sociology. The statistical techniques were sometimes crude but always ingenious, and they pointed the way for the widespread use of statistical methods in sociology today.

JAMES S. COLEMAN

[*For an example of Lazarsfeld's statistical work, see* SURVEY ANALYSIS, *article on* THE ANALYSIS OF ATTRIBUTE DATA.]

WORKS BY LAZARSFELD

(1933) 1970 LAZARSFELD, PAUL F.; JAHODA, MARIE; and ZEISEL, HANS *Marienthal: The Sociography of an Unemployed Community*. Chicago: Aldine. → First published as *Die Arbeitslosen von Marienthal: Ein soziographischer Versuch über die Wirkungen langdauernder Arbeitslosigkeit*. The English edition includes a new preface.

1938 LAZARSFELD, PAUL F.; and FISKE, MARJORIE The "Panel" as a New Tool for Measuring Opinion. *Public Opinion Quarterly* 2:596–612.

1940*a* *Radio and the Printed Page: An Introduction to the Study of Radio and Its Role in the Communication of Ideas*. New York: Duell. → Reprinted in 1971 by Arno.

1940*b* "Panel" Studies. *Public Opinion Quarterly* 4:122–128.

1941 Repeated Interviews as a Tool for Studying Changes in Opinion and Their Causes. American Statistical Association, *Bulletin* 2:3–7.

1948 The Use of Panels in Social Research. American Philosophical Society, *Proceedings* 92:405–410. → Reprinted in Lazarsfeld et al. 1972*d*.

1950*a* The Logical and Mathematical Foundations of Latent Structure Analysis. Pages 362–412 in Samuel A. Stouffer et al., *Measurement and Prediction*. Studies in Social Psychology in World War II, Vol. 4. Princeton Univ. Press.

1950*b* LAZARSFELD, PAUL F.; and KENDALL, PATRICIA L. Problems of Survey Analysis. Pages 133–196 in Paul F. Lazarsfeld and Robert K. Merton (editors), *Continuities in Social Research: Studies in the Scope and Method of "The American Soldier."* Glencoe, Ill.: Free Press.

1950*c* LAZARSFELD, PAUL F.; and MERTON, ROBERT K. (editors) *Continuities in Social Research: Studies in the Scope and Method of "The American Soldier."* Glencoe, Ill.: Free Press.

1951 LAZARSFELD, PAUL F.; ROSENBERG, MORRIS; and THIELENS, WAGNER JR. The Panel Study. Part 2, pages 587–609 in Marie Jahoda, Morton Deutsch, and Stuart W. Cook (editors), *Research Methods in Social Relations, With Special Reference to Prejudice*. New York: Dryden.

1954*a* A Conceptual Introduction to Latent Structure Analysis. Pages 349–387 in Paul F. Lazarsfeld (editor), *Mathematical Thinking in the Social Sciences.* Glencoe, Ill.: Free Press. → Reprinted in 1969 by Russell & Russell.

1954*b* LAZARSFELD, PAUL F.; and MERTON, ROBERT K. Friendship as Social Process: A Substantive and Methodological Analysis. Pages 18–66 in Monroe Berger, Theodore Abel, and Charles Page (editors), *Freedom and Control in Modern Society.* New York: Van Nostrand.

1955 LAZARSFELD, PAUL F.; and ROSENBERG, MORRIS (editors) *The Language of Social Research: A Reader in the Methodology of Social Research.* Glencoe, Ill.: Free Press.

1959 Latent Structure Analysis. Volume 3, pages 476–543 in Sigmund Koch (editor), *Psychology: A Study of the Sciences.* Volume 3: *Formulations of the Person and the Social Context.* New York: McGraw-Hill.

1961 The Algebra of Dichotomous Systems. Pages 111–157 in Herbert Solomon (editor), *Studies in Item Analysis and Prediction.* Stanford Univ. Press. → Reprinted in Lazarsfeld et al. 1972*d*.

1965 LAZARSFELD, PAUL F.; and HENRY, NEIL W. The Application of Latent Structure Analysis to Quantitative Ecological Data. Pages 333–348 in Fred Massarik and Philburn Ratoosh (editors), *Mathematical Explorations in Behavioral Science.* Homewood, Ill.: Dorsey.

1968 LAZARSFELD, PAUL F.; and HENRY, NEIL W. *Latent Structure Analysis.* Boston: Houghton Mifflin.

1972*a* Regression Analysis With Dichotomous Attributes. *Social Science Research* 1:25–34. → Comments appear on subsequent pages, and Lazarsfeld's reply appears on pages 425–427. Reprinted in Lazarsfeld et al. 1972*d*.

1972*b* Mutual Relations Over Time of Two Attributes: A Review and Integration of Various Approaches. Pages 461–480 in Muriel Hammer, K. Salsinger, and S. Sutton (editors), *Psychopathology: Festschrift in Honor of Joseph Zubin.* New York: Wiley.

1972*c* LAZARSFELD, PAUL F.; and MERTON, ROBERT K. A Professional School for Training in Social Research. Pages 361–391 in Paul F. Lazarsfeld, *Qualitative Analysis: Historical and Critical Essays.* Boston: Allyn & Bacon.

1972*d* LAZARSFELD, PAUL F.; PASANELLA, ANN K.; and ROSENBERG, MORRIS (editors) *Continuities in the Language of Social Research.* New York: Free Press.

1975 Working With Merton. Pages 35–66 in Lewis A. Coser (editor), *The Idea of Social Structure: Papers in Honor of Robert K. Merton.* New York: Harcourt.

SUPPLEMENTARY BIBLIOGRAPHY

A collection of papers in honor of Lazarsfeld, edited by Robert K. Merton, James S. Coleman, and Peter H. Rossi, is scheduled for publication by The Free Press as Qualitative and Quantitative Social Research. *A complete bibliography is to be included.*

LEAST SQUARES

See ESTIMATION; LINEAR HYPOTHESES.

LEGAL EVIDENCE

See STATISTICS AS LEGAL EVIDENCE.

LEXIS, WILHELM

Wilhelm Lexis (1837–1914), a German statistician and economist, made major contributions to the theory of statistics and its application, particularly in population research and economic time series. As a mathematician Lexis was deeply skeptical about the state of mathematical economics in his time. His criticism of certain contemporary work in mathematical economics led him to some fundamental observations on economic events and their interdependence.

Lexis was born in Eschweiler, near Aachen, Germany. His studies were widespread, and his interests ranged from law to the natural sciences and mathematics. He graduated from the University of Bonn in 1859, having written a thesis on analytical mechanics; he also obtained a degree in mathematics. For some time he did research in Bunsen's chemical laboratories in Heidelberg. In 1861, Lexis went to Paris to study the social sciences, and his studies led to his first major publication (1870), a treatise on French export policies. This work displays the feature that characterizes his later economic writings: a skepticism toward "pure economics" and toward the application of supposedly descriptive mathematical models which have no reference to economic reality. Even in this early work he insisted that economic theory should be founded on quantitative economic data. In Lexis' view an elaborate general economic equilibrium analysis, whose main problem was to match the unknowns with an equal number of equations, could make no contribution to the understanding or solution of economic problems and therefore should not be taken too seriously. He was one of a number of mathematically trained students of economics who, in the second half of the nineteenth century, became alienated from that discipline; another, more famous one was Max Planck, who, after attempting to read Marshall's work, threw it away and changed his course of study for good (see Schumpeter 1954, pp. 957–958).

Lexis was appointed to the University of Strassburg in 1872. While there, he wrote his introduction to the theory of population (1875). From Strassburg he went to Dorpat in 1874, as professor of geography, ethnology, and statistics, and then to Freiburg in 1876, as professor of economics. His major contributions to statistics were made while he was at Freiburg (1876; 1877; 1879*a*), and he also published papers on economics at this time (1879*b*; 1881; 1882*a*; 1882*b*). After an interlude at the University of Breslau from 1884 to 1887, he was appointed professor of political science at the University of Göttingen.

Lexis' activities in his later years were remarkably diverse. An editor of the *Handwörterbuch der Staatswissenschaften*, the major German economic encyclopedia, he was also an active contributor, and he was director of the first institute of actuarial sciences in Germany. In the 1890s he published and edited several volumes pertaining to education, particularly to the university system (1893; 1901; 1902; 1904*a*; 1904*b*). Lexis' works on population research, economics, and statistics during the subsequent decade bring together and refine some of his earlier arguments (see 1903; 1906*a*; 1906*b*; 1908; 1914). He died in Göttingen during the very first days of World War I.

Statistics

Lexis' major contributions were in statistics. His statistical work originated in problems he encountered in population research (1875; 1879*b*; 1891; 1903), sociology (1877), and economics (1870; 1879*a*; 1908; 1914). In connection with his studies of social mass phenomena (1877) and the time series encountered in several of the social sciences (1879*b*), Lexis came upon problems concerning statistical homogeneity which apparently had been neglected up to then, although, as Bortkiewicz pointed out (1918), Dormoy (1874; 1878) developed similar ideas at about the same time. (Other possible forerunners of Lexis were Bienaymé [1855], Cournot [1843], and R. Campbell [1859].) Lexis credited Dormoy with having anticipated some of his ideas; nevertheless, it was Lexis who more or less independently gave a new direction to the analysis of statistical series and led statisticians in the shift of emphasis from the purely mathematical approach, with which Laplace is associated, to an empirical or inductive approach (see Keynes 1921, pp. 392 ff.). He also initiated the analysis of dispersion and variance in his attempts to develop statistics with which to evaluate qualitative changes in populations over time (see Keynes 1921; Pólya 1919).

Lexis showed that in the universe of social mass phenomena the conditions of statistical homogeneity (random sampling from a stable distribution) are seldom, if ever, fulfilled (1877; 1879*a*; 1879*b*). The underlying probability structure may well differ from one part of the sample to another because of special circumstances related to dispersion in space, time, or other factors. A universe in which individual samples are drawn from potentially different populations is now known as a *Lexis universe* (see, for example, Herdan 1966). To some extent Lexis' work was a reaction to the uncritical assumptions of homogeneity made in statistical work before his time—for instance, by Quetelet. As Keynes (1921) pointed out, Quetelet and others simply asserted, with little evidence, the probabilistic stability from year to year of various social statistics.

Lexis' work centered on the dispersion of observations around their local means and on the behavior of the means and dispersions over time. He devised statistics to measure the degree of stability of such time series and arrived at the useful generalization that these statistics would either confirm statistical homogeneity, indicating a *Bernoulli series*, or diverge from it, indicating a *Poisson series* or a *Lexis series*. (This terminology was developed by C. V. L. Charlier; see A. Fisher 1915, p. 117).

Lexis considered only dichotomous variates (male–female, living–dead, etc.), but the argument he advanced holds equally for numerical variates in the ordinary sense (see also Pólya 1919). In the following, Lexis' ideas are given in a generalized form.

Let x_{ij} $(i = 1, \cdots, n,\ j = 1, \cdots, m)$ be a set of n samples with m observations each, and let the arithmetic mean of the x_{ij} in sample i be $\bar{x}_i$ and the arithmetic mean of the x_{ij} over all n samples be $\bar{x}$. Similarly, let $a_{ij} = E(x_{ij})$, the expectation of x_{ij}, so that $\bar{a}_i = E(\bar{x}_i)$ and $\bar{a} = E(\bar{x})$. Lexis considered the following quadratic forms which measure dispersion in three different senses:

$$s_w^2 = \frac{1}{nm} \sum_i \sum_j (x_{ij} - \bar{x}_i)^2,$$

$$s_b^2 = \frac{1}{n} \sum_i (\bar{x}_i - \bar{x})^2,$$

$$s^2 = \frac{1}{nm} \sum_i \sum_j (x_{ij} - \bar{x})^2,$$

where s_b^2 has rank (degrees of freedom) $r_b = n - 1$, s_w^2 has rank $r_w = n(m-1)$, and s^2 has rank $r = r_b + r_w = nm - 1$. (The subscripts "w" and "b" are used to indicate that s_w^2 comprises within sample dispersion and s_b^2 between sample dispersion.) Furthermore,

$$s^2 = s_b^2 + s_w^2.$$

If the n samples of m objects each are drawn at random from the same population, then the expected value of each observation equals the expected value of the sample mean and the expected value of the mean of all observations, that is,

$$a_{ij} = \bar{a}_i = \bar{a},$$

and statistical homogeneity is present. Repeated independent measurements of a distance and $n \times m$ drawings of balls from an urn with each ball returned after it is drawn are examples of such series. In this case the three quadratic forms, multiplied by appropriate constants, have the same expectations. Specifically, when statistical homogeneity holds,

$$E\left(\frac{n}{r}s^2\right) = E\left(\frac{n}{r_b}s_b^2\right) = E\left(\frac{n}{r_w}s_w^2\right),$$

and the common value is $1/m$ times the variance of the underlying population. A set of samples drawn under such conditions is known as a Bernoulli series (see, e.g., A. Fisher 1915).

It may, however, be the case that statistical homogeneity holds within the samples ($a_{i1} = a_{i2} = \cdots = a_{im}$) but not between samples—that is, the n samples may be random samples from different populations. In this case a Lexis series is generated (a supernormal series, in Lexis' terminology). Such a series is expected when, for example, each set of balls (one sample) is drawn from a different urn. Other examples of such series explain further the importance of the Lexis series: m observations made at time t_0, another m observations made at time t_1, and so forth, up to t_n, will give rise to a *time series* of $m \times n$ observations, where the ith sample (covering one period, t_i) may well come from a single population but where between different periods such changes occurred that statistical homogeneity is no longer preserved—that is, the samples come from different populations.

Similarly, social or economic samples of m observations drawn from n different geographical regions (nations) are likely to come from different statistical populations, although in each region (nation) the m observations of the sample come from the same distribution (*interregional series, international series*). In short, if the over-all dispersion is caused not only by chance variations about a constant but also by trends and other systematic factors varying between samples, then a Lexis series will be generated. (A comprehensive and elementary treatment of Lexis series is given in Pólya 1919.)

We expect in this case that the variance between the samples will contribute relatively more to the over-all variance of the $m \times n$ observations than will the variance within the samples and that the expected value of an observation will equal the sample mean, whereas the sample mean is expected to differ from the mean of all observations. Further, although $a_{ij} = \bar{a}_i$, $\bar{a}_i \neq \bar{a}$ for at least some i, and

$$E\left(\frac{n}{r_w}s_w^2\right) < E\left(\frac{n}{r}s^2\right) < E\left(\frac{n}{r_b}s_b^2\right).$$

Statistical homogeneity is not preserved.

Much less realistic, but a formal complement to the Bernoulli and Lexis series, is the Poisson series (a subnormal series, in Lexis' terminology). The Poisson model was developed as one that would generate higher within sample than between sample variability. In this case the jth observation of each sample is drawn from a fixed population, but the populations differ according to j. In short, $a_{1j} = a_{2j} = \cdots = a_{nj}$, but for a fixed i the a_{ij} are not all equal. Hence $\bar{a}_i = \bar{a}$, and there is no between sample variability coming from the a's. It follows that

$$E\left(\frac{n}{r_b}s_b^2\right) < E\left(\frac{n}{r}s^2\right) < E\left(\frac{n}{r_w}s_w^2\right).$$

Other kinds of models leading to subnormal dispersion have also been considered.

Lexis proposed a statistic, based on the above quadratic forms, to describe the extent to which a given series is homogeneous, supernormal, or subnormal. The statistic, called the Lexis quotient, is

$$L = \frac{s_b^2/r_b}{s^2/r},$$

a monotone increasing function of another statistic, $(s_b^2/r_b)/(s_w^2/r_w)$, which might be used alternatively. A. A. Chuprov showed later (1922) that in the case of statistical homogeneity, $E(L) = 1$ and the variance of L is approximately $2/(n-1)$.

A further elaboration of this leads to significance tests. A first step in that direction is made by adding to and subtracting from the expected value of L its standard error, obtaining $1 \pm \sqrt{2/(n-1)}$. If L lies within the boundaries thus calculated, one may conclude with confidence of approximately two out of three that the statistical mass is homogeneous; if L is significantly larger than 1, one may conclude that the series was drawn from a Lexis universe; and if L is significantly smaller than 1, one may conclude that the series is Poisson.

The relevance and connections of the Lexis series and Lexis' L to the analysis of variance and the chi-square distribution were later shown by many authors. The formal connection between L and the χ^2 statistic of Pearson was elaborated by R. A. Fisher. Fisher showed that in the case of a $2 \times n$

classification the χ^2 statistic is just nL. [*See* COUNTED DATA; *see also* R. A. Fisher 1928; Gebelein & Heite 1951.]

The relation of the L-statistic to the F-statistic is very direct, and we may say that Lexis anticipated the F-statistic (see Coolidge 1921; Rietz 1932; Geiringer 1942*a*; 1942*b*; Gini 1956; Herdan 1966). Whereas in L one compares the variance between the samples to the variance of all $n \times m$ observations, that is,

$$L = \frac{s_b^2/r_b}{s^2/r},$$

in the F-statistic one compares the variance between the samples to the variance within the samples, that is,

$$F = \frac{s_b^2/r_b}{s_w^2/r_w}.$$

Furthermore,

$$L \gtreqless 1 \quad \text{if and only if} \quad F \gtreqless 1.$$

This concurrence is based on the previously stated equality

$$s^2 = s_b^2 + s_w^2.$$

Although the asymptotic distribution of the F-statistic as $m \to \infty$ was established by R. A. Fisher (1925, p. 97) and by W. G. Cochran (1934, p. 178) and generalized by M. G. Madow (1940), the same distribution was established for Lexis' L as early as 1876 by F. R. Helmert, using the method of characteristic functions.

Bortkiewicz extended the application of Lexis' theory of dispersion, and Chuprov (1922) extended Lexis' theory and gave it the most comprehensive treatment. Others influenced by Lexis were J. von Kries, H. Westergaard, and F. Y. Edgeworth, the only Anglo–American scholar closely familiar with statistical work on the Continent at that time (see Keynes 1921).

Economics

Lexis' contributions to economic theory were less appreciated than were his contributions to statistics; many economists, including Schumpeter, largely ignored them. Such a negative assessment of Lexis' work proves to be not entirely justified after closer examination of the reasons that led him to criticize certain aspects of the mathematical and "pure" economics of his contemporaries. His main contribution was a valid criticism of the work done at his time, particularly that of the Austrian school and the Lausanne school. His criticism was informed in part by the outlook of the historical school, which was prevalent in Germany, and accordingly he believed that it was necessary to incorporate in any theory of value and demand the element of time as well as the phenomenon of the recurrence of wants.

Lexis accepted Gossen's analysis of human behavior because Gossen appreciated all the shortcomings of any such theory. The criticisms Lexis made of the Austrian school seem contradictory but are only superficially so: he deplored the lack of mathematics in the work of some authors, especially Carl Menger, and found fault with the application of inappropriate mathematics in the work of others, especially Auspitz and Lieben.

Lexis regarded the concept of utility as being rather vague, since utility cannot be measured. He argued that to say that the utility of a good (set of goods) is equal to, larger than, or less than the utility of some other good permits a partial or complete preordering of utilities. Complementarity and substitution effects imply, however, subadditivity and superadditivity of utilities, which render futile any attempt to aggregate utilities and demand correspondences. Lexis questioned the convexity and continuity assumptions of preference orderings.

The controversy then raging over how to determine total utility given the marginal utility correspondences (a controversy between E. von Böhm-Bawerk and F. von Wieser) was correctly interpreted by Lexis and led him to a discussion of Gossen's other laws, most notably the equalization of marginal utilities. Any such theorem, he believed, must be hedged by a number of qualifications; it is particularly important to consider the time element connected with demand and consumption. Want and satisfaction are both felt and exercised over time. At one and the same time only a limited set of wants can be satisfied. One can eat, drink, sleep, and work, but these activities are to some extent mutually exclusive. Thus, the individual has to decide what sequence to follow in satisfying his set of wants. This sequence will be determined, according to Lexis, by the intensity of wants and by their periodicity, the most fundamental rhythms being the day, the year, and one lifetime. The demand of an individual will be classified and exercised accordingly. Intensive wants will be satisfied first, on a daily, yearly, or other basis, depending on the periodicity of recurrence of wants. Other, less intensive wants will be satisfied after full satisfaction of the intensive wants has been achieved. The implications are far-reaching but perhaps misleading; they have not been accepted by subsequent economists. Individual demand correspondences have to be defined by the period to which they relate, which in turn requires a reformulation of the theory of demand and implies the necessity of defining the demand for (consumption of) each good at different times as quantitatively different. This entails no theoretical difficulties in a general mathematical equilibrium analysis, but it does prevent the theory from having any operational value.

The concept of preferential ordering over time induced Lexis to observe that certain more intensive wants will be saturated whereas other, less intensive wants will be satisfied partially, implying satisfaction of zero marginal utility for the first set of wants and some positive marginal utility for the second. Lexis supported this conclusion by referring to economic reality. However, his statement about the equalization of marginal utilities turns out ultimately to be incorrect if we allow for errors of judgment, evaluations of uncertainties and risks, and diversity of attributes of each good for any individual.

Thus, Lexis' skepticism about the potential of the marginal utility theory in economics was based on the difficulty of measuring utility, the existence of subadditivities and superadditivities in utility correspondences, and the impossibility of aggregating individual preferences. The introduction of the time element into the theory of value and demand adds interesting arguments to general equilibrium analysis which imply, according to Lexis, obvious refutations of the equalization of marginal utilities. As a consequence of his skepticism, Lexis turned, in the rest of his economic work, to a rather dry description of economic events, which failed to be attractive to more speculative minds.

KLAUS-PETER HEISS

WORKS BY LEXIS

1870 *Die französischen Ausfuhrprämien im Zusammenhange mit der Tarifgeschichte und Handelsentwicklung Frankreichs seit der Restauration.* Bonn: Marcus.

1875 *Einleitung in die Theorie der Bevölkerungsstatistik.* Strassburg: Trübner.

1876 Das Geschlechtsverhältniss der Geborenen und die Wahrscheinlichkeitsrechnung. *Jahrbücher für Nationalökonomie und Statistik* 27:209–245.

1877 *Zur Theorie der Massenerscheinungen in der menschlichen Gesellschaft.* Freiburg im Breisgau: Wagner.

(1879*a*) 1942 *Über die Theorie der Stabilität statistischer Reihen (The Theory of the Stability of Statistical Series).* Minneapolis, Minn: WPA. → First published in Volume 32 of *Jahrbücher für Nationalökonomie und Statistik.*

1879*b* Gewerkvereine und Unternehmerverbände in Frankreich. Verein für Socialpolitik, Berlin, *Schriften* 17:1–280.

1881 *Erörterungen über die Währungsfrage.* Leipzig: Duncker & Humblot.

(1882*a*) 1890 Die volkswirthschaftliche Konsumtion. Volume 1, pages 685–722 in *Handbuch der politischen Oekonomie.* 3d ed. Edited by Gustav Schönberg. Tübingen: Laupp.

(1882*b*) 1891 Handel. Volume 2, pages 811–938 in *Handbuch der politischen Oekonomie.* 3d ed. Edited by Gustav Schönberg. Tübingen: Laupp.

1886 Über die Wahrscheinlichkeitsrechnung und deren Anwendung auf die Statistik. *Jahrbücher für Nationalökonomie und Statistik* 47:433–450.

1891 Bevölkerungswesen, II: Bevölkerungswechsel, 1: Allgemeine Theorie des Bevölkerungswechsels. Volume 2, pages 456–463 in *Handwörterbuch der Staatswissenschaften.* Jena: Fischer.

1893 *Die deutschen Universitäten: Für die Universitätsausstellung in Chicago 1893.* 2 vols. Berlin: Asher.

(1895*a*) 1896 *The Present Monetary Situation.* American Economic Association, Economic Studies, Vol. 1, No. 4. New York: Macmillan. → First published as *Der gegenwärtige Stand der Währungsfrage.*

1895*b* Grenznutzen. Volume 1, pages 422–432 in *Handwörterbuch der Staatswissenschaften: Supplementband.* Jena: Fischer.

1901 *Die neuen französischen Universitäten.* Munich: Academischer Verlag.

1902 LEXIS, WILHELM (editor) *Die Reform des höheren Schulwesens in Preussen.* Halle: Waisenhaus.

1903 *Abhandlungen zur Theorie der Bevölkerungs- und Moralstatistik.* Jena: Fischer. → Contains reprints of 1876 and 1879*a*.

1904*a* LEXIS, WILHELM (editor) *Das Unterrichtswesen im Deutschen Reich.* 4 vols. Berlin: Asher.

1904*b* *A General View of the History and Organisation of Public Education in the German Empire.* Berlin: Asher.

1906*a* Das Wesen der Kultur. Pages 1–53 in *Die allgemeinen Grundlagen der Kultur der Gegenwart.* Die Kultur der Gegenwart, vol. 1, part 1. Berlin: Teubner.

1906*b* *Das Handelswesen.* 2 vols. Sammlung Göschen, Vols. 296–297. Berlin: Gruyter. → Volume 1: *Das Handelspersonal und der Warenhandel.* Volume 2: *Die Effektenbörse und die innere Handelspolitik.*

1908 Systematisierung, Richtungen und Methoden der Volkswirtschaftslehre. Volume 1, pages I:1–45 in *Die Entwicklung der deutschen Volkswirtschaftslehre im neunzehnten Jahrhundert.* Leipzig: Duncker & Humblot.

(1910) 1926 *Allgemeine Volkswirtschaftslehre.* 3d ed., rev. Die Kultur der Gegenwart, vol. 2, part 10, section 1. Berlin and Leipzig: Teubner.

(1914) 1929 *Das Kredit- und Bankwesen.* 2d ed. Sammlung Göschen, Vol. 733. Berlin: Gruyter.

SUPPLEMENTARY BIBLIOGRAPHY

BAUER, RAINALD K. 1955 Die Lexissche Dispersionstheorie in ihren Beziehungen zur modernen statistischen Methodenlehre, insbesondere zur Streuungsanalyse (Analysis of Variance). *Mitteilungsblatt für mathematische Statistik und ihre Anwendungsgebiete* 7: 25–45.

BIENAYMÉ, JULES (1855) 1876 Sur un principe que M. Poisson avait cru découvrir et qu'il avait appelé loi des grands nombres. *Journal de la Société de Statistique de Paris* 17:199–204.

BORTKIEWICZ, LADISLAUS VON 1901 Über den Präcisiongrad des Divergenzkoëffizienten. Verband der

Österreichischen und Ungarischen Versicherungs-Techniker, *Mitteilungen* 5:1–3.

BORTKIEWICZ, LADISLAUS VON 1909–1911 Statistique. Part 1, Volume 4, pages 453–490 in *Encyclopédie des sciences mathématiques*. Paris: Gauthier-Villars.

BORTKIEWICZ, LADISLAUS VON 1915 Wilhelm Lexis [Obituary]. International Statistical Institute, *Bulletin* 20, no. 1:328–332.

BORTKIEWICZ, LADISLAUS VON 1917 Wahrscheinlichkeitstheoretische Untersuchungen über die Knabenquote bei Zwillingsgeburten. Berliner Mathematische Gesellschaft, *Sitzungsberichte* 17:8–14.

BORTKIEWICZ, LADISLAUS VON 1918 Der mittlere Fehler des zum Quadrat erhobenen Divergenzkoëffizienten. Deutsche Mathematiker-Vereinigung, *Jahresbericht* 27: 71–126.

BORTKIEWICZ, LADISLAUS VON 1930 Lexis und Dormoy. *Nordic Statistical Journal* 2:37–54.

BORTKIEWICZ, LADISLAUS VON 1931 The Relations Between Stability and Homogeneity. *Annals of Mathematical Statistics* 2:1–22.

CAMPBELL, ROBERT 1859 On the Probability of Uniformity in Statistical Tables. *Philosophical Magazine* 18:359–368.

CHUPROV, ALEKSANDR A. 1905 Die Aufgaben der Theorie der Statistik. *Jahrbuch für Gesetzgebung, Verwaltung und Volkswirtschaft im Deutschen Reich* 29: 421–480. → The author's name is given its German transliteration, Tschuprow.

CHUPROV, ALEKSANDR A. 1922 Ist die normale Stabilität empirisch nachweisbar? *Nordisk statistisk tidskrift* 1:369–393. → The author's name is given its German transliteration, Tschuprow.

COCHRAN, W. G. 1934 The Distribution of Quadratic Forms in a Normal System, With Applications to the Analysis of Covariance. Cambridge Philosophical Society, *Proceedings* 30:178–191.

COOLIDGE, JULIAN L. 1921 The Dispersion of Observations. American Mathematical Society, *Bulletin* 27: 439–442.

COURNOT, ANTOINE AUGUSTIN 1843 *Exposition de la théorie des chances et des probabilités*. Paris: Hachette.

DORMOY, ÉMILE 1874 *Théorie mathématique des paris de courses*. Paris: Gauthier-Villars. → Also published in Volume 3 of *Journal des actuaires français*.

DORMOY, ÉMILE 1878 *Théorie mathématique des assurances sur la vie*. 2 vols. Paris: Gauthier-Villars.

EDGEWORTH, F. Y. 1885 Methods of Statistics. Pages 181–217 in Royal Statistical Society, London, *Jubilee Volume*. London: Stanford.

FISHER, ARNE 1915 *The Mathematical Theory of Probabilities and Its Application to Frequency Curves and Statistical Methods*. Vol. 1. London: Macmillan.

FISHER, R. A. 1925 Applications of "Student's" Distribution. *Metron* 5, no. 3:90–104.

FISHER, R. A. 1928 On a Distribution Yielding the Error Functions of Several Well Known Statistics. Volume 2, pages 805–813 in International Congress of Mathematicians (New Series), Second, Toronto, 1924, *Proceedings*. Univ. of Toronto Press.

GEBELEIN, HAND; and HEITE, H.-J. 1951 *Statistische Urteilsbildung*. Berlin: Springer.

GEIRINGER, HILDA 1942*a* A New Explanation of Nonnormal Dispersion in the Lexis Theory. *Econometrica* 10:53–60.

GEIRINGER, HILDA 1942*b* Observations on Analysis of Variance Theory. *Annals of Mathematical Statistics* 13:350–369.

GINI, C. 1956 Généralisations et applications de la théorie de la dispersion. *Metron* 18, no. 1/2:1–75.

HELMERT, F. R. 1876 Ueber die Wahrscheinlichkeit der Potenzsummen der Beobachtungsfehler und über einige damit im Zusammenhange stehenden Fragen. *Zeitschrift für Mathematik und Physik* 21:192–218.

HERDAN, G. 1966 *The Advanced Theory of Language as Choice and Chance*. New York: Springer.

KEYNES, JOHN MAYNARD (1921) 1952 *A Treatise on Probability*. London: Macmillan. → A paperback edition was published in 1962 by Harper.

KRIES, JOHANNES VON (1886) 1927 *Die Principien der Wahrscheinlichkeitsrechnung: Eine logische Untersuchung*. 2d ed. Tübingen: Mohr.

MADOW, WILLIAM G. 1940 Limiting Distributions of Quadratic and Bilinear Forms. *Annals of Mathematical Statistics* 11:125–146.

►OBERSCHALL, ANTHONY 1965 *Empirical Social Research in Germany, 1848–1914*. Paris: Mouton. → See especially pages 48–51.

PÓLYA, GEORG 1919 Anschauliche und elementare Darstellung der Lexisschen Dispersionstheorie. *Zeitschrift für schweizerische Statistik und Volkswirtschaft* 55: 121–140.

RIETZ, H. L. 1932 On the Lexis Theory and the Analysis of Variance. American Mathematical Society, *Bulletin* 38:731–735.

SCHUMPETER, JOSEPH A. (1954) 1960 *History of Economic Analysis*. Edited by E. B. Schumpeter. New York: Oxford Univ. Press.

VON MISES, RICHARD 1932 Théorie des probabilités: Fondements et applications. Paris, Université de, Institut Henri Poincaré, *Annales* 3:137–190.

LIFE TABLES

The life table (also referred to as the mortality table) is a statistical device used to compute chances of survivorship and death and average remaining years of life, for specific years of age. The concept of the life table is applicable not only to humans (Spiegelman 1957) and other species of life (Haldane 1953; Ciba Foundation 1959) but also to items of industrial equipment (Dublin, Lotka, & Spiegelman [1936] 1949) and other defined aggregates subject to a measurable process of attrition. Life tables can also be developed further for computing the chances of other vital events in human life, such as marriage and remarriage, the birth of children, widowhood, illness and disability, and labor force participation and retirement (Spiegelman 1957); and they enter into a wide variety of annuity and life insurance computations (Hooker & Longley-Cook 1953–1957; Jordan 1952).

The conventional life table

The conventional form of a life table for the general population is illustrated in Table 1. The original data are recorded deaths and the census of population classified according to age (this step is

Table 1 — Life table for white females, United States, 1949–1951 [a]

Year of age	RATE OF MORTALITY PER 1,000: Number dying between ages x and x + 1 among 1,000 living at age x	OF 100,000 BORN ALIVE: Number surviving to exact age x	Number dying between ages x and x + 1	Number of years lived by the cohort between ages x and x + 1	Total number of years lived by the cohort from age x on, until all have died	Average number of years lived after age x per person surviving to exact age x [b]
x	$1{,}000q_x$	l_x	d_x	L_x	T_x	$\mathring{e}_x$
0	23.55	100,000	2,355	97,965	7,203,179	72.03
1	1.89	97,645	185	97,552	7,105,214	72.77
2	1.12	97,460	109	97,406	7,007,662	71.90
3	0.87	97,351	85	97,308	6,910,256	70.98
4	0.69	97,266	67	97,233	6,812,948	70.04
5	0.60	97,199	59	97,169	6,715,715	69.09
6	0.53	97,140	52	97,114	6,618,546	68.13
7	0.48	97,088	46	97,065	6,521,432	67.17
8	0.44	97,042	43	97,020	6,424,367	66.20
9	0.41	96,999	39	96,980	6,327,347	65.23
⋮	⋮	⋮	⋮	⋮	⋮	⋮
100	388.39	294	114	237	566	1.92
101	407.52	180	73	143	329	1.83
102	426.00	107	46	84	186	1.74
103	443.67	61	27	48	102	1.66
104	460.76	34	16	26	54	1.59
⋮	⋮	⋮	⋮	⋮	⋮	⋮

a. Based upon recorded deaths in the United States during the three-year period 1949–1951, recorded births for each year from 1944 through 1951, and the census of population taken April 1, 1950; for details, see U.S. Public Health Service 1959, pp. 149–158.

b. Represents complete expectation of life, or average future lifetime.

Source: U.S. Public Health Service 1954–1955, p. 18.

not shown on the table). From these data were computed the *rates of mortality*, conventionally designated as q_x, for each year of age, x. These rates show the proportion of deaths occurring within the year of age among those who attain that age; the rates are usually shown per thousand ($1{,}000q_x$). For example, Table 1 shows that of every 1,000 who just attained age 0 (the newly born), 23.55 died before reaching their first birthday; similarly, of every 1,000 who attained age six, 0.53 died within that year of age. Typically, mortality rates for a general population start at a high point in the first year of life, fall rapidly to a minimum at about age ten, and then rise with advance in years. The rise is gradual to about age 40, and then becomes increasingly rapid; since the maximum attainable age for human beings is in the neighborhood of 110 years, life tables seldom go beyond that point.

Once one knows the mortality rates at each age of life, it becomes possible to compute the number of *survivors* (column l_x of the life table) and also the number of *deaths* (column d_x). It is usually most convenient to start the population life table with a base (radix) of 100,000 newborn individuals. In the example presented here, where there is a death rate of 23.55 per 1,000 at age 0, among the 100,000 newly born there must be 2,355 deaths in the first year of life. The number of survivors to attain age 1 is then $100{,}000 - 2{,}355 = 97{,}645$. With a mortality rate of 1.89 per 1,000 at age 1, among the 97,645 who attained that age there are

$$97{,}645 \times \frac{1.89}{1{,}000} = 185 \text{ deaths.}$$

The number of survivors to age 2 is then calculated in the same way:

$$97{,}645 - 185 = 97{,}460.$$

This procedure is continued to the end of the life table. Obviously, the number in the survivorship column, l_x, at any attained age is equal to the sum of the deaths in the d_x column for that and all higher ages.

To compute the *expectation of life* ($\mathring{e}_x$), or average future lifetime, for any attained age, it will be assumed that deaths, d_x, are uniformly distributed over the year of age, x. Equivalent to this is the assumption that each of the persons dying lived one-half year after the last birthday. Thus, among the 294 in Table 1 who attained age 100, there were 114 deaths during that year of age, and these individuals lived $\frac{1}{2} \times 114$ years after their last birthday. Similarly, the 73 who died at age 101 lived $1\frac{1}{2}$

years each after attaining age 100, and the 46 who died at age 102 lived $2\frac{1}{2}$ years each after attaining age 100, and so on, to the last death. Altogether, the total number of years of life lived from age 100 on by the 294 who attained that age is $(\frac{1}{2} \times 114) + (1\frac{1}{2} \times 73) + (2\frac{1}{2} \times 46) + (3\frac{1}{2} \times 27) + \cdots = 566$. This is the figure for age 100 in the column headed T_x. Since the 294 who attained age 100 lived a total of 566 years from their 100th birthday until the death of the last survivor, the average remaining lifetime was

$$566 \div 294 = 1.92 \text{ years.}$$

This is more commonly known as the expectation of life, $\mathring{e}_x$; as an average, it is not applicable to any specific individual.

[1]In Table 1 the life table symbols at the head of each column are defined by the terms above them. Reference has already been made to each, except L_x, which denotes the total number of years lived within the year of age by the number, l_x, who attain that age. It has been assumed that each of the persons dying lived only one half year after the last birthday. Accordingly, among the number, l_x, who attain age x, the years of life lived by those dying during that year of age is $\frac{1}{2}d_x$. The years of life lived by the survivors is l_{x+1}, which is equal to $l_x - d_x$. The sum of $\frac{1}{2}d_x$ and $l_x - d_x$ is the total number of years lived within that year of age. Thus,

$$L_x = l_x - \tfrac{1}{2}d_x .$$

Since T_x is the total number of years lived from age x on by those who attain age x, it follows that

$$T_x = L_x + L_{x+1} + L_{x+2} + \cdots$$

and also that

$$T_x = L_x + T_{x+1} .$$

It should be recognized that except for the mortality rates, which represent an actually observed situation, *all other columns of figures in the life table represent a hypothetical situation.* Thus, the survivorship column and the column of life table deaths show only the expected number of survivals and deaths for successive ages, on the assumption that the mortality rates observed during the specified calendar period continue without change over time. The same assumption underlies the column of figures for expectation of life.

Life table formulas. It will be seen from the preceding discussion that the construction of life tables rests upon a small number of elementary assumptions, which can be summarized in the following formulas:

$$l_x - l_{x+1} = d_x ,$$

$$q_x = \frac{d_x}{l_x} .$$

Moreover, it is evident that if p_x denotes the probability of surviving one year after attaining age x, then

$$p_x = \frac{l_{x+1}}{l_x} = \frac{l_x - d_x}{l_x} = 1 - q_x .$$

Similarly, if ${}_np_x$ denotes the probability of surviving n years after attaining age x, then

$${}_np_x = \frac{l_{x+n}}{l_x} = 1 - {}_nq_x ,$$

where ${}_nq_x$ is the probability of dying within n years after attaining age x. Thus,

$${}_nq_x = 1 - {}_np_x = \frac{l_x - l_{x+n}}{l_x} .$$

Another measure of mortality is the "force of mortality." This measure takes into account the fact that mortality varies continually with advance in age. In this sense, the rate of mortality in the brief instant after attaining exact age x will be different from that for the brief instant just before leaving age x to attain exact age $x + 1$. The force of mortality, μ_x, is the annual rate of loss of lives, corresponding to the loss, at any instant of time, per head surviving at that time. In terms of the calculus,

$$\mu_x = -\frac{1}{l_x} \cdot \frac{dl_x}{dx} = -\frac{d \log l_x}{dx} ,$$

where d/dx denotes the derivative of the specified function with respect to x.

The force of mortality at age x may be approximated by

$$\mu_x = \frac{l_{x-1} - l_{x+1}}{2l_x}$$

or, more closely, by

$$\mu_x = \frac{8(l_{x-1} - l_{x+1}) - (l_{x-2} - l_{x+2})}{12l_x} .$$

The relevant approximation formulas have been discussed by Jordan (1952, pp. 19–21).

Life table computation. The first task to be carried out in computing a life table for any specific population is to convert the *central death rate,* m_x—that is, the average annual death rate for persons of a given age—into a *mortality rate,* q_x, such as has already been described. A means of doing this is illustrated as follows. In any specified community, let D_x denote the number of deaths recorded within a calendar year of individuals at age x on last birthday (or average annual deaths for a calendar period). Also, let P_x denote the number of people at age x on last birthday on the middate of the calendar year or period; this is an approximation to the average number living and, therefore, to the number of years of life lived within the year of age. Then the central death rate at age x for the community is

$$m_x = \frac{D_x}{P_x}.$$

The problem is to convert the central death rate, m_x, into a mortality rate, q_x.

In the life table the number of years of life lived during the year of age x is L_x and deaths during age x number d_x, so that the central death rate m_x is

$$m_x = \frac{d_x}{L_x} = \frac{d_x}{l_x - \frac{1}{2}d_x}.$$

Since $d_x = l_x \cdot q_x$,

$$m_x = \frac{l_x \cdot q_x}{l_x - \frac{1}{2}l_x q_x} = \frac{q_x}{1 - \frac{1}{2}q_x}.$$

Solving for q_x yields

$$q_x = \frac{m_x}{1 + \frac{1}{2}m_x}.$$

In terms of the recorded (observed) deaths and population,

$$q_x = \frac{D_x/P_x}{1 + \frac{1}{2}D_x/P_x} = \frac{D_x}{P_x + \frac{1}{2}D_x}.$$

In practice, however, the mortality rates at the very early ages are usually computed on the basis of a population estimated from recorded births and deaths, since census data for this stage of life are usually unreliable. The risk of mortality in infancy is highest in the first month following birth, and decreases rapidly thereafter; accordingly, the assumption of a uniform distribution of deaths is not valid for the first year of age. For the terminal ages of life, the basic data are usually meager and unreliable; various artifices are therefore used to compute these mortality rates. The mortality rates for the broad range of intervening ages are generally subjected to mathematical procedures of interpolation and graduation in order to produce a smooth progression of figures (Spiegelman 1955, p. 72). A complete life table shows the figures in each column for every age of life. An abridged life table shows figures for only selected ages, such as every fifth or tenth year of age.

Life tables directly from census data

Where death data are grossly inadequate or lacking, a life table may be approximated from the age distributions of population in two consecutive censuses, as in the following simplified example.

Assume two censuses, five years apart, with correct reporting of ages and with no migration. Then, clearly, the population at age $x + 5$ in the second census, P''_{x+5}, consists of survivors of the population five years younger at the time of the first census, P'_x. The ratio of P''_{x+5} to P'_x accordingly is a five-year survivorship rate for a population at age x last birthday. Assuming a uniform distribution of population over the year of age, this population is approximately at an average attained age $x + \frac{1}{2}$. Thus,

$${}_5p_{x+\frac{1}{2}} = \frac{P''_{x+5}}{P'_x}.$$

Having arrived at a series of values of ${}_5p_{x+\frac{1}{2}}$ according to age, it is possible to work back to a series of mortality rates, q_x. In using this method, allowance may be made for migration (Mortara 1949).

There is also a method of life table estimation that can be used when a population age distribution is available from only one census (Stolnitz 1956). If there is good reason to believe that the size of the population of a community has been virtually stationary over time and that mortality according to age has remained essentially unchanged over time, then its age distribution is clearly very much like that of the life table column L_x. In other words, the number living, P_x, at age x last birthday is proportionate to L_x. Thus,

$${}_5p_{x+\frac{1}{2}} = \frac{P_{x+5}}{P_x} = \frac{L_{x+5}}{L_x},$$

and q_x may be estimated, as in the case with two consecutive censuses.

Consider now a population that may be regarded as stable, in the sense that it is growing at a constant annual rate, r, and that mortality at each age is also constant over time. This growth results solely from an excess of births over deaths each year; there is no migration. Then, for an interval of five years,

$$P''_x = P'_x(1 + r)^5.$$

Likewise, P''_{x+5} consists of survivors of P'_x as before. It follows that

$${}_5p_{x+\frac{1}{2}} = \frac{P''_{x+5}}{P'_x} = \frac{P''_{x+5}}{P''_x/(1+r)^5} = \frac{P''_{x+5}(1+r)^5}{P''_x},$$

so that use is made of the population at the second census only. Stolnitz generalized this approach by tracing the populations P''_{x+5} and P''_x from their respective births, $x + 5$ and x years previously, namely B_{x+5} and B_x. For this, he introduced survival factors to the same attained age x last birthday, namely S'_x and S''_x, and made use of the five-year survivorship ratio, ${}_5p_{x+\frac{1}{2}}$. Thus,

$$\frac{P''_{x+5}}{P''_x} = \frac{B_{x+5}}{B_x} \cdot \frac{S'_x}{S''_x} \cdot {}_5p_{x+\frac{1}{2}} .$$

Stolnitz shows how the birth ratio and the ratio of survival factors may be estimated from other experiences. With such estimates it becomes possible to compute ${}_5p_{x+\frac{1}{2}}$ from the age distribution of a single census.

Model life tables for developing areas. In the developing areas the problem is to estimate a life table for a population with scanty mortality data or from data gathered in a special survey. Since the mortality rate in infancy or the first few years of life is frequently indicative of the general level of mortality, such a rate may be used as the basis for estimation of life table values. Such an observed mortality rate, with suitable adjustment to enhance its validity, is used as a key to select one of a series of life table mortality rates (q_0, ${}_4q_1$, and ${}_5q_x$ for x at five-year intervals) from 40 theoretical model series (United Nations 1955*a*). These models were derived from a study of the patterns of mortality rates in existing life tables. For refinement, the series of life table mortality rates may be selected by interpolating among the models on the basis of the key rate. Further refinement is possible by computing from the equations used to derive the models. Although these model life tables of the United Nations have been subject to technical criticisms, they are widely used (Gabriel & Ronen 1958; Kurup 1966). A more extensive set of model life tables, prepared at Princeton University, takes into account variations in the patterns of mortality between four broad geographic regions, defined as East, West, North, and South, in addition to variations in the level of mortality within each region (Coale & Demeny 1966).

Life tables directly from death data

As pointed out before, in a population that is virtually stationary, with mortality rates essentially unchanged over time, the age distribution corresponds closely to that in a life table. Only in such a situation is it feasible to cumulate the distribution of deaths according to age, starting with the highest age and noting the total for each age, running back to birth, in order to approximate the survivorship column of the life table. This approach is not applicable in any other situation, since the age distribution of deaths will be influenced by the age distribution of the population. Thus, a population with a large proportion of aged persons will have a large proportion of its deaths at the older ages, irrespective of the level of its mortality rates.

Multiple decrement tables

In a multiple decrement table, the survivorship column of the life table is split, in passing from one age to the next, into two or more component parts, on the basis of changes in status or of newly acquired characteristics (Jordan 1952, pp. 237, 251; Bailey & Haycocks 1946). One example is the case where the survivorship column is split, on the basis of marriage rates according to age, to distinguish those who marry from those who remain single. In another example, shown in Table 2, the survivorship column of the life table is split to show those who become permanently disabled lives apart from those who remain as active lives. This table shows, in addition to the numbers surviving to successive ages as active lives and as perma-

Table 2 — Example of a double decrement table, with decrements by death and by disability

				Of 100,000 born alive[a]						
	RATE OF MORTALITY PER 1,000		DISABILITY RATE PER 1,000[b]	NUMBER SURVIVING TO EXACT AGE x			ACTIVE LIVES DISABLED	NUMBER DYING BETWEEN AGES x AND x + 1		
Year of age, x	Among active lives	Among disabled lives	Among active lives	Total	As active lives	As disabled lives[c]	Between ages x and x + 1	Total	Among active lives	Among disabled lives
15	7.55	267	0.587	66,949	66,949	0	40	511	505	6
16	7.47	254	0.584	66,438	66,404	34	39	509	496	13
17	7.40	241	0.581	65,929	65,869	60	38	507	487	20
18	7.40	229	0.578	65,422	65,344	78	38	506	484	22
19	7.40	217	0.575	64,916	64,822	94	38	504	480	24
⋮	⋮	⋮	⋮	⋮	⋮	⋮	⋮	⋮	⋮	⋮

a. The radix in the source (100,000 at age 10) was changed to 100,000 at birth.

b. Per 1,000 active lives at exact age x.

c. Assuming no lives were disabled before age 15.

Source: Adapted from Hunter et al. 1932, p. 92.

nently disabled lives, the rates of mortality for each of these categories and the rates at which active lives become permanently disabled. The column of life table deaths is also split to show the number of deaths among the permanently disabled separately from that among the active. It is assumed that the number of newly disabled lives in any year of age is uniformly distributed over that year; consequently, they are exposed to the mortality rate of the disabled for an average of one half of a year.

Select tables

The life table has been described in terms of rates of mortality dependent only upon attained age; in describing multiple decrement tables, reference was made to rates of disability and of marriage according to attained age. In select tables, rates of mortality (or other rates) are shown on the basis of both the age at acquisition of a new characteristic and the duration since that acquisition (Jordan 1952, p. 26). This two-way classification constitutes a select table, since some selective process is present at the time of acquisition. For example, mortality rates for permanently disabled lives may be shown not only for the age at which disablement occurred but also separately for each subsequent year of disability. Another example of a select table is the two-way classification of rates of remarriage for widows, in relation to both age at widowhood and years since that event. Such a two-way classification of rates is shown in Table 3. In that table, remarriage rates after the fifth year of widowhood are shown only on the basis of attained age, since duration in this case is only of minor influence upon the rates. The table as a whole is known as a select and ultimate table. That portion showing rates according to duration since widowhood is the select table; the ultimate table is that portion showing rates only according to attained age, since duration is no longer of any importance. Select and ultimate tables are used in life insurance mortality investigations. The choice of the number of durations to be shown for the select period is a matter of study in each experience.

Cohort or generation life tables

In the foregoing account of the conventional life table and the related multiple decrement and select tables, the rates of mortality and other rates of attrition were based upon observations during some specified year or other period. The hypothetical nature of the conventional life table with respect to the time period of observation has already been indicated. A realistic picture of the mortality and survivorship experience of a cohort traced from birth is obtained by observing these events each year in a generation born at the latest 100 years ago. In that way, a record would be obtained of the number surviving to successive ages in successive years and also of the corresponding number of deaths at each age; the mortality rates according to age in successive years may then be computed. After the last death, it would be possible to compute the average length of life of the generation and the average years of life remaining after each age. Such a table is called a generation life table, since it reflects the actual mortality rates of a cohort as it ages in successive years. The table derived from mortality rates for a calendar year or period is called a current life table (Dublin, Lotka, & Spiegelman [1936] 1949, p. 174; Jacobson 1964). Thus, the expectation of life computed from mortality rates observed in 1850 will understate the average length of life of the generation born that year, because of the reductions in mortality since then. In general, with the trend toward lower mortality, the expectation of life at birth computed from a current life table understates the average length of life of a newly born generation.

Further applications

In addition to the applications of the life table that are mentioned in the opening paragraph of this article, increasing use is being made of it as

Table 3 — Example of select and ultimate table, showing probabilities of remarriage during widowhood

	YEARS ELAPSED SINCE HUSBAND'S DEATH						
	SELECT TABLE					ULTIMATE TABLE	
Age at widowhood	0	1	2	3	4	5 or more	Attained age
35	0.0201	0.0490	0.0386	0.0376	0.0230	0.0163	40
36	0.0184	0.0449	0.0354	0.0345	0.0211	0.0149	41
37	0.0169	0.0412	0.0324	0.0316	0.0193	0.0137	42
38	0.0155	0.0377	0.0297	0.0290	0.0177	0.0126	43
39	0.0142	0.0345	0.0272	0.0266	0.0162	0.0115	44

Source: Adapted from Myers 1949, p. 73.

an analytic tool in social and economic problems. Several interesting and important examples may be cited in the field of demography (for references to examples, see Spiegelman 1955; 1957). John Durand (1960) made use of the United Nations model life tables, cited previously, as an adjunct in arriving at estimates of expectation of life at birth for the western Roman Empire. The life table is fundamental in the stable-population theory developed by A. J. Lotka (Dublin & Lotka 1925) and also in Lotka's work on the structure of a growing population (1931). In the field of education E. G. Stockwell and C. B. Nam (1963) prepared school life tables to show the joint effects of death and school dropouts on school attendance patterns. B. C. Churchill (1955) studied the mortality and survival of manufacturing, wholesale trade, and retail trade firms in the United States; in similar fashion A. J. Jaffe (1961) has used data from the censuses of manufactures in Puerto Rico to prepare survival curves according to the age of the establishment.

MORTIMER SPIEGELMAN

[*See also* DEMOGRAPHY; VITAL STATISTICS; *the biography of* GRAUNT; Grauman 1968; Mayer 1968; Spengler 1968.]

BIBLIOGRAPHY

A very elementary account of the essentials of the life table is given in Dublin, Lotka, & Spiegelman [1936] 1949. *A wholly nontechnical account of the life table, including double decrement and select tables, with brief descriptions of applications, will be found in* Spiegelman 1957. *The beginner in graduate study who has a nonmathematical background but a sense of arithmetic will find the chapter on the life table in* Barclay 1958 *a good introduction to the subject. A corresponding account of the life table, with some further development, is contained in* Pressat 1961. U.S. Bureau of the Census 1951 *provides step-by-step directions for elementary life table construction, as well as exercises for the beginning student. More technical is the exposition of the life table in* Benjamin 1959. *The student with a background in the calculus seeking a more comprehensive understanding of the life table, double decrement tables, and select tables may start with* Hooker & Longley-Cook 1953–1957; Jordan 1952. *The theoretical aspects of double and higher-order decrement tables are discussed in* Bailey & Haycocks 1946. *A firm understanding of the techniques of life table construction requires a good background in the means for estimating the exposed-to risk, as given in* Gershenson 1961, *and also for a grasp of the elements of graduation and interpolation, as given in* Miller 1946. *The principal techniques used in the construction of life tables are described in* Spiegelman 1955, *which also treats, in detail, the special situations at the early ages, where the assumption of a uniform distribution of deaths over the year of age is not applicable, and at extreme old age, where artifices are used to complete the column of mortality rates.*

BAILEY, WALTER G.; and HAYCOCKS, HERBERT W. 1946 *Some Theoretical Aspects of Multiple Decrement Tables.* Edinburgh: Constable.

BARCLAY, GEORGE W. 1958 *Techniques of Population Analysis.* New York: Wiley.

BENJAMIN, BERNARD (1959) 1960 *Elements of Vital Statistics.* London: Allen & Unwin; Chicago: Quadrangle Books.

BRASS, WILLIAM 1963 The Construction of Life Tables From Child Survivorship Ratios. Volume 1, pages 294–301 in International Population Conference, New York, 1961, *Proceedings.* London: International Union for the Scientific Study of Population.

CHIANG, CHIN L. 1960 A Stochastic Study of the Life Table and Its Applications: 2. Sample Variance of the Observed Expectation of Life and Other Biometric Functions. *Human Biology* 32:221–238.

CHURCHILL, BETTY C. 1955 Age and Life Expectancy of Business Firms. *Survey of Current Business* 35, no. 12:15–19, 24.

CIBA FOUNDATION 1959 *Colloquia on Aging.* Volume 5: The Lifespan of Animals. Boston: Little.

COALE, ANSLEY J.; and DEMENY, PAUL 1966 *Regional Model Life Tables and Stable Populations.* Princeton Univ. Press.

DUBLIN, LOUIS I.; and LOTKA, ALFRED J. 1925 On the True Rate of Natural Increase. *Journal of the American Statistical Association* 20:305–339.

DUBLIN, LOUIS I.; LOTKA, ALFRED J.; and SPIEGELMAN, M. (1936) 1949 *Length of Life.* Rev. ed. New York: Ronald Press. → The 1936 edition was written by Dublin and Lotka only; citations in the text refer to the 1949 edition.

DURAND, JOHN D. 1960 Mortality Estimates From Roman Tombstone Inscriptions. *American Journal of Sociology* 65:365–373.

GABRIEL, K. R.; and RONEN, ILANA 1958 Estimates of Mortality From Infant Mortality Rates. *Population Studies* 12:164–169.

GERSHENSON, HARRY 1961 *Measurement of Mortality.* Chicago: The Society of Actuaries.

►GRAUMAN, JOHN V. 1968 Population: VI. Population Growth. Volume 12, pages 376–381 in *International Encyclopedia of the Social Sciences.* Edited by David L. Sills. New York: Macmillan and Free Press.

GREVILLE, T. N. E. 1966 *Methodology of the National, Regional, and State Life Tables for the United States: 1959–61.* Washington: National Center for Health Statistics.

HALDANE, J. B. S. 1953 Some Animal Life Tables. Institute of Actuaries, London, *Journal* 79:83–89.

HOOKER, PERCY F.; and LONGLEY-COOK, L. H. 1953–1957 *Life and Other Contingencies.* 2 vols. Cambridge Univ. Press.

HUNTER, ARTHUR et al. 1932 *Disability Benefits in Life Insurance Policies.* 2d ed. Actuarial Studies, No. 5. Chicago: Actuarial Society of America.

JACOBSON, P. H. 1964 Cohort Survival for Generations Since 1840. *Milbank Memorial Fund Quarterly* 42: 36–53.

JAFFE, A. J. 1961 The Calculation of Death Rates for Establishments With Supplementary Notes on the Calculation of Birth Rates. *Estadistica: Journal of the Inter-American Statistical Institute* [1961]:513–526.

JONES, J. P. 1962 *Remarriage Tables Based on Experience Under OASDI and U.S. Employees Compensation Systems.* Actuarial Study No. 55. Washington: U.S. Social Security Administration.

[2]JORDAN, CHESTER W. 1952 *Society of Actuaries' Textbook on Life Contingencies.* Chicago: The Society of Actuaries.

KEYFITZ, NATHAN 1966 A Life Table That Agrees With the Data. *Journal of the American Statistical Association* 61:305–312.

KURUP, R. S. 1965 A Revision of Model Life Tables. Unpublished manuscript. → Paper presented at the second World Population Conference.

LOTKA, ALFRED J. 1931 The Structure of a Growing Population. *Human Biology* 3:459–493.

►MAYER, KURT B. 1968 Population: IV. Population Composition. Volume 12, pages 362–370 in *International Encyclopedia of the Social Sciences*. Edited by David L. Sills. New York: Macmillan and Free Press.

MILLER, MORTON D. 1946 *Elements of Graduation*. Chicago: Actuarial Society of America.

MORTARA, GIORGIO 1949 *Methods of Using Census Statistics for the Calculation of Life Tables and Other Demographic Measures*. Population Studies, No. 7. Lake Success, N.Y.: United Nations, Department of Social Affairs.

MYERS, ROBERT J. 1949 Further Remarriage Experience. Casualty Actuarial Society, *Proceedings* 36:73–104.

PRESSAT, ROLAND 1961 *L'analyse démographique*. Paris: Presses Universitaires de France.

SIRKEN, MONROE G. (1964) 1966 *Comparison of Two Methods of Constructing Abridged Life Tables*. Rev. ed. Series 2, No. 4. Washington: National Center for Health Statistics.

►SPENGLER, JOSEPH J. 1968 Lotka, Alfred J. Volume 9, pages 475–476 in *International Encyclopedia of the Social Sciences*. Edited by David L. Sills. New York: Macmillan and Free Press.

[3]SPIEGELMAN, MORTIMER 1955 *Introduction to Demography*. Chicago: The Society of Actuaries. → See the references in Chapter 5 for materials on life tables.

SPIEGELMAN, MORTIMER 1957 The Versatility of the Life Table. *American Journal of Public Health* 47:297–304. → Contains a list of references.

STOCKWELL, EDWARD G.; and NAM, CHARLES B. 1963 Illustrative Tables of School Life. *Journal of the American Statistical Association* 58:1113–1124.

STOLNITZ, GEORGE J. 1956 *Life Tables From Limited Data: A Demographic Approach*. Princeton Univ., Office of Population Research.

UNITED NATIONS, DEPARTMENT OF SOCIAL AFFAIRS 1955*a* *Age and Sex Patterns of Mortality: Model Life Tables for Under-developed Countries*. Population Studies, No. 22. New York: United Nations.

UNITED NATIONS, DEPARTMENT OF SOCIAL AFFAIRS 1955*b* *Methods of Appraisal of Basic Data for Population Estimates*. New York: United Nations.

U.S. BUREAU OF THE CENSUS 1951 *Handbook of Statistical Methods for Demographers*. Washington: Government Printing Office.

U.S. PUBLIC HEALTH SERVICE 1954–1955 [Life Tables for 1949–1951.] U.S. National Office of Vital Statistics, *Vital Statistics: Special Reports* 41, no. 1; no. 2.

U.S. PUBLIC HEALTH SERVICE 1959 [Life Tables for 1949–1951.] U.S. National Office of Vital Statistics, *Vital Statistics: Special Reports* 41, no. 5:149–158.

U.S. PUBLIC HEALTH SERVICE 1961 *Guide to United States Life Tables, 1900–1959*. Bibliography Series, No. 42. Washington: Government Printing Office.

Postscript

[1]The assumption that deaths are uniformly distributed over the year of age is not appropriate for the first year of life because in that year the distribution of deaths by age is markedly nonuniform. Accordingly, the first entry in the L_x column of Table 1 is not equal to $l_0 - \frac{1}{2} d_0$. The method of calculating this entry is described by Spiegelman (1968, chapter 5).

A useful technique is the subdivision of the d_x column of the life table into components corresponding to major groups of causes of death. As a by-product one can calculate the probability that a person at a given age will eventually die from a given cause or the gain in expectation of life at a given age that would result from the complete elimination of a given cause of death. Such results for the United States in 1959–1961 have been published by the U.S. National Center for Health Statistics (1968). Preston et al. (1972) have compiled similar data for many countries at various dates and have shown that important conclusions can be drawn from international comparisons of such data.

EDWARD B. PERRIN

ADDITIONAL BIBLIOGRAPHY

[2]JORDAN, CHESTER W. (1952) 1967 *Society of Actuaries' Textbook on Life Contingencies*. 2d ed. Chicago: The Society of Actuaries.

KEYFITZ, NATHAN 1968 *Introduction to the Mathematics of Population*. Reading, Mass.: Addison-Wesley.

PRESTON, SAMUEL H.; KEYFITZ, NATHAN; and SCHOEN, ROBERT 1972 *Causes of Death: Life Tables for National Populations*. New York: Seminar Press.

SHRYOCK, HENRY S. et al. 1971 *The Methods and Materials of Demography*. 2 vols. in 1. Washington: U.S. Bureau of the Census.

[3]SPIEGELMAN, MORTIMER (1955) 1968 *Introduction to Demography*. Rev. ed. Cambridge, Mass.: Harvard Univ. Press.

U.S. NATIONAL CENTER FOR HEALTH STATISTICS 1968 *Life Tables, 1959–1961, Report 6: United States Life Tables by Causes of Death, 1959–1961*. Washington: The Center.

LIFE TESTING

See under QUALITY CONTROL, STATISTICAL.

LIKELIHOOD

The *likelihood function* is important in nearly every part of statistical inference, but concern here is with just the *likelihood principle*, a very general and problematic concept of statistical evidence. [*For discussion of other roles of the likelihood function, see* ESTIMATION; HYPOTHESIS TESTING; SUFFICIENCY.]

The likelihood function is defined in terms of the probability law (or density function) assumed to represent a sampling or experimental situation: When the observation variables are fixed at the values actually observed, the resulting function of the unknown parameter(s) is the likelihood function. (More precisely, two such functions identical except for a constant factor are considered equivalent representations of the same likelihood function.)

The likelihood principle may be stated in two parts: (1) the likelihood function, determined by the sample observed in any given case, represents fully the evidence about parameter values available in those observations (this is the *likelihood axiom*); and (2) the evidence supporting one parameter value (or point) as against another is given by relative values of the likelihood function (likelihood ratios).

For example, suppose that a random sample of ten patients suffering from migraine are treated by an experimental drug and that four of them report relief. The sampling is binomial, and the investigator is interested in the unknown proportion, p, in the population of potential patients, who would report relief. The likelihood function determined by the sample is a function of p,

$$\binom{10}{4} p^4(1-p)^6, \qquad 0 \leqslant p \leqslant 1,$$

whose graph is shown in Figure 1. This likelihood function has a maximum at $p = .4$ and becomes very small, approaching 0, as p approaches 0 or 1. Hence, according to the likelihood principle, values of p very near .4 are supported by the evidence in this sample, as against values of p very near 0 or 1, with very great strength, since the corresponding likelihood ratios $(.4)^4(.6)^6/p^4(1-p)^6$ are very large.

A different rule for sampling patients would be to treat and observe them one at a time until just four had reported relief. A possible outcome would be that just ten would be observed, with six reporting no relief and of course four reporting relief. The probability of that observed outcome is

$$\binom{9}{3} p^4(1-p)^6, \qquad 0 \leqslant p \leqslant 1.$$

This function of p differs from the previous one by only a constant factor and hence is considered to be an alternative, equivalent representation of the *same* likelihood function. The likelihood principle asserts that therefore the evidence about p in the two cases is the same, notwithstanding other differences in the two probability laws, which appear for other possible samples.

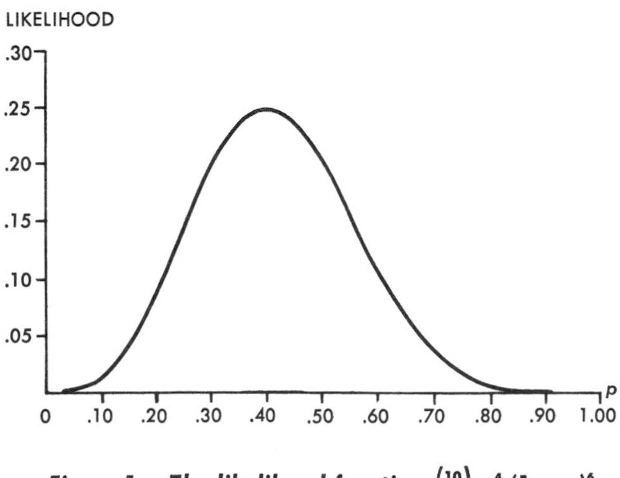

Figure 1 — The likelihood function $\binom{10}{4} p^4(1-p)^6$

Relation to other statistical theory. The likelihood principle is incompatible with the main body of modern statistical theory and practice, notably the Neyman–Pearson theory of hypothesis testing and of confidence intervals, and incompatible in general even with such well-known concepts as standard error of an estimate and significance level [*see* ESTIMATION; HYPOTHESIS TESTING].

To illustrate this incompatibility, observe that in the example two distinct sampling rules gave the same likelihood function, and hence the same evidence under the likelihood principle. On the other hand, different determinations of a lower 95 per cent confidence limit for p are required under the respective sampling rules, and the two confidence limits obtained are different. The likelihood principle, however, is given full formal justification and interpretation within Bayesian inference theories and much interest in the principle stems from recently renewed interest and developments in such theories [*see* BAYESIAN INFERENCE].

Finally, on grounds independent of the crucial and controversial Bayesian concepts of prior or personal probability, interest and support for the likelihood principle arises because most standard statistical theory fails to include (and clearly implicitly excludes) any precise general concept of evidence in an observed sample, while several concepts of evidence that many statisticians consider appropriate have been found on analysis to entail the likelihood axiom. Some of these concepts have become part of a more or less coherent widespread body of theory and practice in which the Neyman–Pearson approach is complemented by concepts of evidence often left implicit. Such concepts also appear as basic in some of Fisher's theories. When formulated as axioms and analyzed, these concepts have been discovered to be equivalent to the likelihood axiom and hence basically incompatible with, rather than possible complements to, the Neyman–Pearson theory.

General concepts of statistical evidence. The central one of these concepts is that of *conditionality* (or *ancillarity*), a concept that appeared first in rather special technical contexts. Another somewhat similar concept of evidence, which can be illustrated more simply here and which also entails the major part of the likelihood axiom, is the *censoring axiom:* Suppose that after interpretation of the outcome described with the second sampling rule of the example, it is discovered that the reserve supply of the experimental drug had been accidentally destroyed and is irreplaceable and that no more than ten patients could have been treated with the supply on hand for the experiment. Is the interpretation of the outcome to be changed? In the hypothetical possible case that seven or more patients reported no relief before a fourth reported relief, the sampling plan could not have been carried through even to the necessary eleventh patient: The change of conditions makes unavailable ("censored") the information whether if an outcome were to include more than six patients reporting no relief, that number would be seven, or eight, or any specific larger number. But in fact the outcome actually observed was a physical event unaffected, except in a hypothetical sense, by the differences between intended and realizable sampling plans. Many statisticians consider such a hypothetical distinction irrelevant to the evidence in an outcome. It follows readily from the general formulation of such a concept that the evidence in the observed outcome is characterized by just the function $\binom{9}{3}p^4(1-p)^6$, $0 \leqslant p \leqslant 1$. More generally this censoring concept is seen to be the likelihood axiom, slightly weakened by disallowance of an arbitrary constant factor; the qualification is removable with adoption of another very weak "sufficiency" concept concerning evidence (see Birnbaum 1961; 1962; Pratt 1961).

Interpreting evidence. The only method proposed for interpreting evidence just through likelihood functions, apart from Bayesian methods, is that stated as part (2) of the likelihood principle above. The briefness and informality of these statements and interpretation are typical of those given by their originator, R. A. Fisher (1925; 1956), and their leading proponent, G. A. Barnard (1947; 1949; 1962). "Likelihood ratio" appears in such interpretations as a primitive term concerning statistical evidence, associated in each case with a nonnegative numerical value, with larger values representing qualitatively greater support for one parameter point against the other, and with unity (representing "no evidence") the only distinguished point on the scale. But likelihood ratio here is not subject to definition or interpretation in terms of other independently defined extramathematical concepts.

Only in the simplest case, where the parameter space has but two points (a case rare in practice but of real theoretical interest), are such interpretations of likelihood functions clearly plausible; and in this case they appear to many to be far superior to more standard methods, for example, significance tests. In such cases the likelihood function is represented by a single likelihood ratio.

In the principal case of larger parameter spaces, such interpretations can be seriously misleading with high probability and are considered unacceptable by most statisticians (see Stein 1962; Armitage 1963). Thus progress in clarifying the important problem of an adequate non-Bayesian concept of statistical evidence leaves the problem not only unresolved but in a positively anomalous state.

Another type of argument supporting the likelihood principle on non-Bayesian grounds is based upon axioms characterizing rational decision making in situations of uncertainty, rather than concepts of statistical evidence (see, for example, Cornfield 1966; Luce & Raiffa 1957).

Likelihoods in form of normal densities. Attention is sometimes focused on cases where likelihood functions have the form of normal density functions. This form occurs in the very simple and familiar problem of inferences about the mean of a normal distribution with known variance (with ordinary sampling). Hence adoption of the likelihood axiom warrants and invites identification with this familiar problem of all other cases where such likelihood functions occur. In particular, the maximum likelihood estimator in any such case is thus related to the classical estimator of the normal mean (the sample mean), and the curvature of the likelihood function (or of its logarithm) at its maximum is thus related to the variance of that classical estimator. In similar vein, transformations of parameters have been considered that tend to give likelihood functions a normal density shape (in problems with several parameters as well as those with only one). (See, for example, Anscombe 1964.)

Likelihoods in nonparametric problems. In nonparametric problems, there is no finite set of parameters that can be taken as the arguments of a likelihood function, and it may not be obvious that the likelihood axiom has meaning. However, "nonparametric" is a sometimes misleading name for a very broad mathematical model that includes all specific parametric families of laws among those allowed; hence in principle it is simple to imagine (although in practice formidably awkward to represent) the extremely inclusive parameter space containing a point representing each (absolutely) continuous law, and for each pair of such points a likelihood ratio (determined as usual from the observed sample).

ALLAN BIRNBAUM

[*Other relevant material may be found in* DISTRIBUTIONS, STATISTICAL, *articles on* SPECIAL CONTINUOUS DISTRIBUTIONS *and* SPECIAL DISCRETE DISTRIBUTIONS.]

BIBLIOGRAPHY

ANSCOMBE, F. J. 1964 Normal Likelihood Function. Institute of Statistical Mathematics, *Annals* 26:1–19.

ARMITAGE, PETER 1963 Sequential Medical Trials: Some Comments on F. J. Anscombe's Paper. *Journal of the American Statistical Association* 58:384–387.

BARNARD, G. A. 1947 [Review of] *Sequential Analysis* by Abraham Wald. *Journal of the American Statistical Association* 42:658–664.

BARNARD, G. A. 1949 Statistical Inference. *Journal of the Royal Statistical Society* Series B 11:116–149.

BARNARD, G. A.; JENKINS, G. M.; and WINSTEN, C. B. 1962 Likelihood Inference and Time Series. *Journal of the Royal Statistical Society* Series A 125:321–375. → Includes 20 pages of discussion.

BIRNBAUM, ALLAN 1961 On the Foundations of Statistical Inference: I. Binary Experiments. *Annals of Mathematical Statistics* 32:414–435.

BIRNBAUM, ALLAN 1962 On the Foundations of Statistical Inference. *Journal of the American Statistical Association* 57:269–326. → Includes 20 pages of discussion. See especially John W. Pratt's comments on pages 314–315.

CORNFIELD, JEROME 1966 Sequential Trials, Sequential Analysis and the Likelihood Principle. *American Statistician* 20:18–23.

COX, D. R. 1958 Some Problems Connected With Statistical Inference. *Annals of Mathematical Statistics* 29:357–372.

FISHER, R. A. (1925) 1950 Theory of Statistical Estimation. Pages 11.699*a*–11.725 in R. A. Fisher, *Contributions to Mathematical Statistics*. New York: Wiley. → First published in Volume 22, Part 5 of the Cambridge Philosophical Society, *Proceedings*.

FISHER, R. A. (1956) 1959 *Statistical Methods and Scientific Inference*. 2d ed., rev. New York: Hafner; London: Oliver & Boyd.

LUCE, R. DUNCAN; and RAIFFA, HOWARD 1957 *Games and Decisions: Introduction and Critical Survey*. A Study of the Behavioral Models Project, Bureau of Applied Social Research, Columbia University. New York: Wiley.

PRATT, JOHN W. 1961 [Review of] *Testing Statistical Hypotheses* by E. L. Lehmann. *Journal of the American Statistical Association* 56:163–167.

STEIN, CHARLES M. 1962 A Remark on the Likelihood Principle. *Journal of the Royal Statistical Society* Series A 125:565–573. → Includes five pages of comments by G. A. Barnard.

Postscript

In the years since the main article was published there seem to have been no major developments or changes in viewpoint about the likelihood concept. The new references listed here provide appropriate additional elaboration on theoretical viewpoints and problems concerning the likelihood concept and on questions arising in its various specific applications.

ALLAN BIRNBAUM

ADDITIONAL BIBLIOGRAPHY

BIRNBAUM, ALLAN 1961 Concepts of Statistical Evidence. Pages 112–143 in Sidney Morgenbesser et al. (editors), *Philosophy, Science, and Method: Essays in Honor of Ernest Nagel*. New York: St. Martin's.

BIRNBAUM, ALLAN 1972 More on Concepts of Statistical Evidence. *Journal of the American Statistical Association* 67:858–861.

COX, D. R.; and HINKLEY, D. V. 1974 *Theoretical Statistics*. London: Chapman & Hall. → Chapter 2 and other parts indexed under "likelihood" give theoretical exposition, numerous examples, and further references.

EDWARDS, A. W. F. 1972 *Likelihood: An Account of the Statistical Concept of Likelihood and Its Application to Scientific Inference*. Cambridge Univ. Press. → Reviewed by A. P. Dempster, *Science* 177 (1972): 878–879; by B. M. Hill, *Journal of the American Statistical Association* 68 (1973):487–489; and by Ian Hacking, *British Journal for the Philosophy of Science* 23 (1972):132–137.

EDWARDS, A. W. F. 1974 The History of Likelihood. *International Statistical Review* 42:9–15.

HACKING, IAN 1965 *Logic of Statistical Inference*. Cambridge Univ. Press. → Reviewed by G. A. Barnard, *British Journal for the Philosophy of Science* 23 (1972):123–132; and by A. P. Dempster, *Journal of the American Statistical Association* 61 (1966):1233–1235.

HACKING, IAN 1971 Jacques Bernoulli's *Art of Conjecturing*. *British Journal for the Philosophy of Science* 22:209–229.

KALBFLEISCH, J. D. 1971 Likelihood Methods of Prediction. Pages 378–392 in Symposium on the Foundations of Statistical Inference, University of Waterloo, 1970, *Foundations of Statistical Inference: Proceedings*. Edited by V. P. Godambe and D. A. Sprott. Toronto: Holt. → Includes discussion.

KALBFLEISCH, J. D. 1975 Sufficiency and Conditionality. *Biometrika* 62:251–259. → Discussion and a reply by Kalbfleisch on pages 260–268.

SYMPOSIUM ON THE FOUNDATIONS OF STATISTICAL INFERENCE, UNIVERSITY OF WATERLOO, *1970* 1971 *Foundations of Statistical Inference: Proceedings*. Edited by V. P. Godambe and D. A. Sprott. Toronto: Holt. → Papers by the following authors, and discussion thereof, bear on the likelihood approach: G. A. Barnard, J. Hájek, J. S. Williams, A. Plante, and J. D. Kalbfleisch.

LIKELIHOOD RATIO

See HYPOTHESIS TESTING; LIKELIHOOD.

LINEAR HYPOTHESES

I. REGRESSION — *Evan J. Williams*
II. ANALYSIS OF VARIANCE — *Julian C. Stanley*
III. MULTIPLE COMPARISONS — *Peter Nemenyi*

I REGRESSION

Regression analysis, as it is presented in this article, is an important and general statistical tool. It is applicable to situations in which one observed variable has an expected value that is assumed to be a function of other variables; the function usually has a specified form with unspecified parameters. For example, an investigator might assume that under appropriate circumstances the expected score on an examination is a linear function of the length of training period. Here there are two parameters, slope and intercept of the line. The techniques of regression analysis may be classified into two kinds: (1) testing the concordance of the observations with the assumed model, usually in the framework of some broader model, and (2) carrying out estimation, or other sorts of inferences, about the parameters when the model is assumed to be correct. This area of statistics is sometimes known as "least squares," and in older publications it was called "the theory of errors."

In the regression relations discussed in this article only one variable is regarded as random; the others are either fixed by the investigator (where experimental control is possible) or selected in some way from among the possible values. The relation between the expected value of the random variable (called the dependent variable, the predictand, or the regressand) and the nonrandom variables (called regression variables, independent variables, predictors, or regressors) is known as a regression relation. Thus, if a random variable Y, depending on a variable x, varies at random about a linear function of x, we can write

$$Y = \beta_0 + \beta_1 x + e,$$

which expresses a linear regression relation. The parameters β_0 and β_1 are the regression coefficients or parameters, and e is a random variable with expected value zero. Usually the e's corresponding to different values of Y are assumed to be uncorrelated and to have the same variance. If η denotes the expected value of Y, the basic relation may be expressed alternatively as

$$E(Y) = \eta = \beta_0 + \beta_1 x.$$

The parameters in the relation will be either unknown or given by theory; observations of Y for different values of x provide the means of estimating these parameters or testing the concordance of the simple linear structure with the data.

Linear models, linear hypotheses. A regression relation that is linear in the unknown parameters is known as a linear model, and the assertion of such a model as a basis for inference is the assertion of a linear hypothesis. Often the term "linear hypothesis" refers to a restriction on the linear model (for example, specifying that a parameter has the value 7 or that two parameters are equal) that is to be tested. The importance of the linear model lies in its ease of application and understanding; there is a well-developed body of theory and techniques for the statistical treatment of linear models, in particular for the estimation of their parameters and the testing of hypotheses about them.

Needless to say, the description of a phenomenon by means of a linear model is usually a matter of convenience; the model is accepted until some more elaborate one is required. Nevertheless, the linear model has a wide range of applicability and is of great value in elucidating relationships, especially in the early stages of an investigation. Often a linear model is applicable only after transformations of the independent variables (like x in the above example), the dependent variable (Y, above), or both [*see* STATISTICAL ANALYSIS, SPECIAL PROBLEMS OF, *article on* TRANSFORMATIONS OF DATA].

In its most general form, regression analysis includes a number of other statistical techniques as special cases. For instance, it is not necessary that the x's be defined as metric variables. If the values of the observations on Y are classified into a number of groups—say, p—then the regression relation is written $E(Y) = \beta_1 x_1 + \beta_2 x_2 + \cdots + \beta_p x_p$, and x_i may be taken to be 1 for all observations in the ith group and 0 for all the others. The p x-variables will then specify the different groups, and the regression relation will define the mean value of Y for each group. In the simplest case, with two groups,

$$E(Y) = \beta_1 x_1 + \beta_2 x_2,$$

where $x_1 = 1$ and $x_2 = 0$ for the first group, and vice versa for the second.

The estimation of the population mean from a sample is a special case, since the model is then just

$$E(Y) = \beta_0,$$

β_0 being the mean of the population.

This treatment of the comparison of different groups is somewhat artificial, although it is important to note that it falls under the regression rubric. Such comparisons are generally carried out by means of the technique known as the analysis of variance [*see* LINEAR HYPOTHESES, *article on* ANALYSIS OF VARIANCE].

When a regressor is not measured quantitatively but is given only as a ranking (for example, order in time of archeological specimens, social position of occupation), it may still provide a regression relation suitable for estimation or prediction. The simplest way to include such a variable in a relation is to replace the qualitative values (rankings) by arbitrary numerical scores, equally spaced (see, for example, Strodtbeck et al. 1957). More refined methods would use scores spaced according to some measure of "distance" between successive rankings; thus, in some instances the scores have been chosen so that their frequency distribution approximates a grouped normal distribution. Since any method of scoring is arbitrary, the method that is used must be judged by the relations based on it as well as by its theoretical cogency. Simple scoring systems, which can be easily understood, are usually to be preferred.

When both the dependent variable and the regression variable are qualitative, each may be replaced by arbitrary scores as indicated above. Alternative methods determine scores for the dependent variable that are most highly correlated (formally) with the regressor scores or, if the regressor scores for any set of data are open to choice, choose scores for both variables so that the correlation is maximized. The calculation and interpretation of the regression relations for such situations have been discussed by Yates (1948) and by Williams (1952).

Regression, correlation, functional relation. The regression relation is a one-way relation between variables in which the expected value of one random variable is related to nonrandom values of the other variables. It is to be distinguished from other types of statistical relations, in particular from correlation and functional relationships. [*See* MULTIVARIATE ANALYSIS, *article on* CORRELATION METHODS.] Correlation is a relation between two or more random variables and may be described in terms of the amount of variation in one of the variables associated with variation in the other variable or variables. The functional relation, by contrast, is a relation between the *expected values* of random variables. If quantities related by some physical law are subject to errors of measurement, the functional relation between expected values, rather than the regression relation, is what the investigator generally wants to determine.

Although the regression relation relates a random variable to other, nonrandom variables, in many situations it will apply also when the regression variables are random; then the regression, conditional on the observed values of the random regression variables, is determined. Here the *expected value* of one random variable is related to the *observed values* of the other random variables. For a discussion of the fitting of regression lines when the regression variables are subject to error, see Madansky (1959). When more than one variable is to be considered as random, the problem is usually thought of as one of multivariate analysis [*see* MULTIVARIATE ANALYSIS].

History. The method of least squares, on which most methods of estimation for linear models are based, was apparently first published by Adrien Legendre (1805), but the first treatment along the lines now familiar was given by Carl Friedrich Gauss (1821, see in 1855). Gauss showed that the method gives estimators of the unknown parameters with minimum variance among unbiased linear estimators. This basic result is sometimes known as the Gauss–Markov theorem, and the least squares estimators as Gauss–Markov estimators.

The term "regression" was first used by Francis Galton, who applied it to certain relations in the theory of heredity, but the term is now applied to relationships in general and to nonlinear as well as to linear relationships.

The linearity of linear hypotheses rests in the way the parameters appear; the x's may be highly nonlinear functions of underlying nonrandom variables. For example,

$$\eta = \beta_0 + \beta_1 x_1 + \beta_2 x_1^2 + \beta_3 x_1^3$$

and

$$\eta = \beta_1 e^{x_1} + \beta_2 \tan x_1$$

both fall squarely under the linear hypothesis model, whereas

$$\eta = \beta_1 e^{\beta_2 x_1}$$

does not fit that model.

There is now a vast literature dealing with the linear model, and the subject is also treated in most statistical textbooks.

Application in the social sciences. There has been a good deal of discussion about the type of model that should be used to describe relations between variables in the social sciences, particularly in economics. Linear regression models have often been considered inadequate for complex economic phenomena, and more complicated models have been developed. Recent work, however, indicates that ordinary linear regression methods have a wider scope than had been supposed. For example, there has been much discussion about how to treat data correlated in time, for which the residuals from the regression equation (the e's) show autocorrelation. This autocorrelation may be the result of autocorrelation in the variables not included in the model. Geary (1963) suggests that in such circumstances the inclusion of additional regression variables may effectively eliminate the autocorrelation among the residuals, so that standard methods may be applied.

Further discussion of the applicability of regression methods to economic data is given by Ezekiel and Fox (1930, chapters 20 and 24) and also by Wold and Juréen (see in Wold 1953).

Investigators should be encouraged to employ the simple methods of regression analysis as a first step before turning to more elaborate techniques. Despite the relative simplicity of its ideas, it is a powerful technique for elucidating relations, and its results are easily understood and applied. More elaborate techniques, by contrast, do not always provide a readily comprehensible interpretation.

Assumptions in regression analysis. A regression model may be expressed in the following way:

$$E(Y) = \eta = \beta_0 + \beta_1 x_1 + \cdots + \beta_p x_p,$$
$$Y = \eta + e,$$

where Y is the random variable, η is its expected value, the x's are known variables, the β's are unknown coefficients, and e is a random error or deviation with zero mean. In the notation for variables, either fixed or random, subscripts are used only to distinguish the different variables but not to distinguish different observations of the same variable. The context generally makes the meaning clear. Thus, the above expression is an abbreviated form of

$$E(Y_j) = \eta_j = \beta_0 + \beta_1 x_{1j} + \beta_2 x_{2j} + \cdots + \beta_p x_{pj}, \quad j = 1, 2, \cdots, n;$$

$$Y_j = \eta_j + e_j.$$

This model is perfectly general; however, in estimating the coefficients, it is usually assumed that the e_j are mutually uncorrelated and are of equal variance (homoscedastic).

If there is no regressor variable that is identically one (as in the two-sample situation described earlier), the β_0 term might well be omitted. This is primarily a matter of notational convention.

The additional assumption that the errors are normally distributed is convenient and simplifies the theory. It can be shown that, on this assumption, the linear estimators given by least squares are in fact the maximum likelihood (m.l.) estimators of the parameters. In addition, the residual sum of squares $\sum(Y - \hat{\eta})^2$ (see below) is the basis for the m.l. estimator of the error variance (σ^2).

Apart from the theoretical advantages of the assumption of normality of the e's, there are the practical advantages that efficient methods of estimation and suitable tests of significance are relatively easy to apply and that the test statistics have well-known properties and are extensively tabulated. The normality assumption is often reasonable in applications, since even appreciable departures from it do not as a rule seriously invalidate regression analyses based upon normality [*see* ERRORS, *article on* EFFECTS OF ERRORS IN STATISTICAL ASSUMPTIONS].

Some departures from assumptions may be expected in certain situations. For example, if some of the measurements of Y are much larger than others, the associated errors, either errors of measurement or errors resulting from uncontrolled random effects, may well be correspondingly larger, so that the variances of the errors will be heterogeneous (the errors are heteroscedastic). Again, with annual data it is to be expected that errors may arise from unobserved factors whose influence from year to year will be associated, so the errors will not be independent (but see Geary 1963). It is often possible in particular cases to transform the data so that they conform more closely to the assumptions; for instance, a logarithmic or square-root transformation of Y will often give a variable whose errors have approximately constant variance [*see* STATISTICAL ANALYSIS, SPECIAL PROBLEMS OF, *article on* TRANSFORMATIONS OF DATA]. This will amount to replacing the linear model for Y with a linear model for the transformed variable. In practice this often gives a satisfactory representation of the data, any departure from the model being attributed to error.

The method of least squares determines, for the parameters β_i in the regression equation, estimators that minimize the sum of squares of deviations of the Y-values from the values given by the equation. This sum of squares is

$$\sum(Y - \eta)^2 = \sum(Y - \beta_0 - \beta_1 x_1 - \cdots - \beta_p x_p)^2.$$

In the following discussion the estimated β's are denoted by b's, and the corresponding estimator of η is denoted by $\hat{\eta}$, so that

$$\hat{\eta} = b_0 + b_1 x_1 + \cdots + b_p x_p$$

and the minimized sum of squared deviations is $\sum(Y - \hat{\eta})^2$.

The method has the twofold merit of minimizing not only the sum of squares of deviations but also the variance of the estimators b_i (among unbiased linear estimators). Thus, for most practical purposes the method of least squares gives estimators with satisfactory properties. Sometimes such estimators are not appropriate—for example, when errors in one direction are more serious than those in the other—but those cases are usually apparent to the investigator.

The method of least squares applies equally well when the errors are heteroscedastic or even correlated, provided the covariance structure of the errors is known (apart from a constant of proportionality, which may be estimated from the data). The method can be generalized to take account of the general correlation structure or, equivalently, a linear transformation of the observations may be used to reduce the problem to the simpler case of uncorrelated homoscedastic errors. (Details may be found in Rao 1965, chapter 4.)

When the correlation structure is unknown, the method of least squares may still be applied. If the data are analyzed as though the errors are uncorrelated and homoscedastic, the estimators of the parameters will be unbiased, although they will be less precise than if based on the correct model.

On the other hand, if the assumed linear model is incorrect—for example, if the relation is quadratic in one of the variables but only a linear model is fitted—then the estimators are liable to serious bias.

Since the form of the underlying model is almost always unknown there is usually a corresponding risk of bias. This problem has been studied in various contexts, but there is still much to be done; see Box and Wilson (1951), Box and Andersen (1955), and Plackett (1960, chapter 2).

Simple linear regression

In the simple linear regression model the expected value of Y is a linear function of a single variable, x_1:

$$E(Y) = \eta = \beta_0 + \beta_1 x_1.$$

The parameter β_0 is the intercept, and the parameter β_1 is the slope, of the regression line. This model is a satisfactory one in many cases, even if a number of variables affect the expected value of Y, for one of these may have a predominating influence, and although the omission of variables from the relation will lead to some bias, this may not be important in view of the increased simplicity of the model.

In studying the relation between two variables it is almost always desirable to plot a scatter diagram of the points representing the observations. The x-axis, or abscissa, is usually used for the regression variable and the y-axis, or ordinate, for the random variable. If the regression relation is linear the points should show a tendency to fall near a straight line, though if the variation is large this tendency may well be masked. Although for some purposes a line drawn "by eye" is adequate to represent the regression, in general such a line is not sufficiently accurate. There is always the risk of bias in both the position and the slope of the line. Because there is a tendency for the deviations from the line in both the x and y directions to be taken into account in determining the fit, lines fitted by eye are often affected by the scales of measurement used for the two axes. Since Y is the random variable, only the deviations in the y direction should be taken into account in determining the fit of the line. Often the investigator, knowing that there may be error in x_1, may attempt to take it into account. It should be understood that this procedure will give an estimate not of the regression relation but of underlying structure, which often differs from the regression relation. Another and more serious shortcoming of lines drawn by eye is that they do not provide an estimate of the variance about the line, and such an estimate is almost always required.

The method of least squares is commonly used when an arithmetical method of fitting is required, because of its useful properties and its relative ease of application. The equations for the least squares estimators, b_i, based on n pairs of observations (x_1, Y), are as follows:

$$b_1 = \sum Y(x_1 - \bar{x}_1) / \sum (x_1 - \bar{x}_1)^2,$$
$$b_0 = \bar{Y} - b_1\bar{x}_1.$$

Here the summation is over the observed values, $\bar{x}_1 = \sum x_1/n$, and $\bar{Y} = \sum Y/n$. (Note that the observations on x_1 need not be all different, although they must not all be the same.) The estimated regression function is

$$\hat{\eta} = b_0 + b_1 x_1.$$

The minimized sum of squares of deviations is

$$\begin{aligned}\sum(Y - \hat{\eta})^2 &= \sum(Y - \bar{Y})^2 - b_1^2\sum(x_1 - \bar{x}_1)^2 \\ &= \sum(Y - \bar{Y})^2 - b_1\sum Y(x_1 - \bar{x}_1).\end{aligned}$$

The standard errors (estimated standard deviations) of the estimators may be derived from the minimized sum of squares of deviations. Two independent linear parameters have been fitted, and it may readily be shown that the expected value of this minimized sum of squares is $(n-2)\sigma^2$, where σ^2 is the common variance of the residual errors. Consequently, an unbiased estimator of σ^2 is given by

$$s^2 = \sum(Y - \hat{\eta})^2/(n-2),$$

and this is the conventional estimator of σ^2. (The m.l. estimator of σ^2 is $\sum(Y - \hat{\eta})^2/n$.) The sum of squares for deviations is said to have $n - 2$ degrees of freedom, representing the number of linearly independent quantities on which it is based.

The estimated variances of the estimators are est. var $(b_1) = s^2/\sum(x_1 - \bar{x}_1)^2$ and est. var $(b_0) = s^2\sum x_1^2/[n\sum(x_1 - \bar{x}_1)^2]$, and the estimated covariance is est. cov $(b_0, b_1) = -s^2\bar{x}_1/\sum(x_1 - \bar{x}_1)^2 = -\bar{x}_1$ var (b_1).

Separate confidence limits for the parameters β_0 and β_1 may be determined from the estimators and their standard errors, using Student's *t*-distribution [*see* ESTIMATION, *article on* CONFIDENCE INTERVALS AND REGIONS]. If $t_{\alpha;n-2}$ denotes the α-level of this distribution for $n-2$ degrees of freedom, the $1 - \alpha$ confidence limits for β_1 are $b_1 \pm t_{\alpha;n-2}\, s/\sqrt{\sum(x_1 - \bar{x}_1)^2}$. Confidence limits for the intercept, β_0, may be determined in a similar way but are not usually of interest. In a few cases it may be necessary to determine whether the estimator b_0 is in agreement with some theoretical value of the intercept. Thus, in some situations it is reasonable to expect the regression line to pass through the origin, so that $\beta_0 = 0$. It will then be necessary to test the significance of the departure of b_0 from zero or, equivalently, to determine whether the confidence limits for β_0 include zero.

When it is assumed that $\beta_0 = 0$ and there is no need to test this hypothesis, then the regression has only one unknown parameter; in such a case the sum of squares for deviations from the regression line, used to estimate the residual variance, will have $n - 1$ degrees of freedom.

When the parameters β_0 and β_1 are both of interest, a joint confidence statement about them may be useful. The joint confidence region is usually an ellipse centered at (b_0, b_1) and containing all values of (β_0, β_1) from which (b_0, b_1) does not differ with statistical significance as measured by an *F*-test (see the section on significance testing, below). [*The question of joint confidence regions is discussed further in* LINEAR HYPOTHESES, *article on* MULTIPLE COMPARISONS.]

Choice of experimental values. The formula for the variance of the regression coefficient b_1 shows that it is the more accurately determined the larger is $\sum(x_1 - \bar{x}_1)^2$, the sum of squares of the values of x_1 about their mean. This is in accordance with common sense, since a greater spread of experimental values will magnify the regression effect yet will in general leave the error component unaltered. If accurate estimation of β_1 were the only criterion, the optimum allocation of experimental points would be in equal numbers at the extreme ends of the possible range. However, the assumption that a regression is linear, although satisfactory over most of the possible range, is often likely to fail near the ends of the range; for this and other reasons it may be desirable to check the linearity of the regression, and to do so points other than the two extreme values must be observed. In practice, where little is known about the form of the regression relation it is usually desirable to take points distributed uniformly throughout the range. If the experimental points are equally spaced, this will facilitate the fitting of quadratic or higher degree polynomials, using tabulated orthogonal polynomials as described below.

Confidence limits for the regression line. The estimated regression function is

$$\begin{aligned}\hat{\eta} &= b_0 + b_1x_1\\ &= \bar{Y} + b_1(x_1 - \bar{x}_1),\end{aligned}$$

and corresponding to any specified value, x_{1*}, of x_1, the variance of $\hat{\eta}$ is estimated as

$$\text{est. var}\,(\hat{\eta}) = s^2\left(\frac{1}{n} + \frac{(x_{1*} - \bar{x}_1)^2}{\sum(x_1 - \bar{x}_1)^2}\right).$$

Thus, for any specified value of x_1, confidence limits for η can be determined according to the formula

$$Y_L = \hat{\eta} \pm t_{\alpha;n-2}\, s\sqrt{\left(\frac{1}{n} + \frac{(x_{1*} - \bar{x}_1)^2}{\sum(x_1 - \bar{x}_1)^2}\right)}.$$

The locus of these limits consists of the two branches of a hyperbola, lying on either side of the fitted regression line; this locus defines what may be described as a confidence curve. A typical regression line fitted to a set of points is shown in Figure 1 with the 95 per cent confidence curve shown as the two inside upper and lower curves, Y_L.

The above limits are appropriate for the estimated value of η corresponding to a *given* value of x_1. They do not, however, set limits to the whole line. Such limits are given by a method developed by Working and Hotelling, as described, for example, by Kendall and Stuart ([1961] 1973). [*See also* LINEAR HYPOTHESES, *article on* MULTIPLE COMPARISONS.] As might be expected, these limits lie outside the corresponding limits for the same probability for a single value of x_1. The limits may be regarded as arising from the envelope of all lines whose parameters fall within a suitable confidence region. These limits are given by

$$Y_{WH} = \hat{\eta} \pm s\sqrt{2F_{1-\alpha;2,n-2}\left(\frac{1}{n} + \frac{(x_{1*} - \bar{x}_1)^2}{\sum(x_1 - \bar{x}_1)^2}\right)},$$

where $F_{1-\alpha;2,n-2}$ is the tabulated value for the F-distribution with 2 and $n-2$ degrees of freedom at confidence level $1-\alpha$. These limits, for a 95 per cent confidence level, are shown as a pair of broken lines in Figure 1.

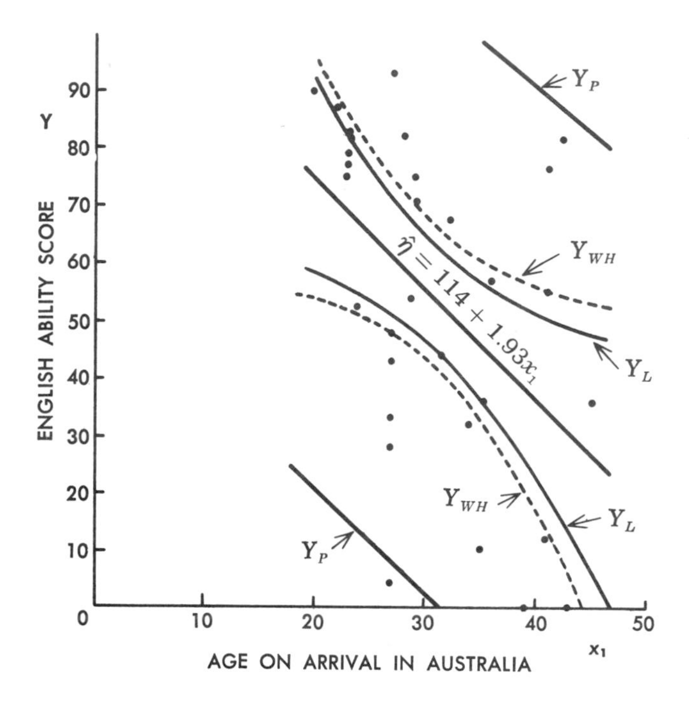

Figure 1 — Regression line and associated 95 per cent confidence regions*

* The Y_P curves, although they appear straight in the figure, are hyperbolas like the other Y curves.

Source of data: Martin, Jean I., 1965, *Refugee Settlers: A Study of Displaced Persons in Australia.* Canberra: Australian National University.

The user of the confidence limits must be clear about which type of limits he requires. If he is interested in the limits on the estimated η for a particular value x_{1*} or in only one pair of limits at a time, the inner limits, Y_L, will be appropriate, but if he is interested in limits for many values of x_{1*} (some of which may not be envisaged when the calculations are being made), the Working–Hotelling limits, Y_{WH}, will be needed.

Application of the regression equation. The regression equation is usually determined not only to provide an empirical law relating variables but also as a means of making future estimates or predictions. Thus, in studies of demand, regression relations of demand on price and other factors enable demand to be predicted for future occasions when one or more of these factors is varied. Such prediction is provided directly by the regression equation. It should be noted, however, that the standard error of prediction will be greater than the standard error of the estimated points ($\hat{\eta}$) on the regression line. This is because a future observation will vary about its regression value with variance equal to the variance of individual values about the regression in the population. When standard errors are being quoted, it is important to distinguish between the standard error of the point $\hat{\eta}$ on the regression line and the standard error of prediction. The estimated variance of prediction is

$$\text{est. var } (\hat{\eta}_P) = s^2\left(1 + \frac{1}{n} + \frac{(x_{1*} - \bar{x}_1)^2}{\sum(x_1 - \bar{x}_1)^2}\right).$$

The outside upper and lower curves in Figure 1 are confidence limits for prediction, Y_P, based on this variance. Clearly, for making predictions of this sort there is little point in determining the regression line with great accuracy. The major part of the error in such cases will be the variance of individual values.

The formula for the standard error of $\hat{\eta}$ or $\hat{\eta}_P$ shows that the error of estimation increases as the x_1-value departs from the mean of the sample, so that when the deviation from the mean is large the variance of estimate can be so great as to make the estimate worthless. This is one reason why investigators should be discouraged from attempting to draw inferences beyond the range of the observed values of x_1. The other reason is that the assumed linear regression, even though satisfactory within the observed range, may not hold true outside this range.

Inverse estimation. In many situations the investigator is primarily interested in determining the value of x_1 corresponding to a given level or value, η_*. Thus, although it is still appropriate to determine the regression of the random variable Y on the fixed variable x_1, the inference has to be carried out in reverse. For example, if a drug that affects the reaction time of individuals is being tested at different levels, the reaction time Y will be a random variable with regression on the dose level x_1. However, the purpose of the investigation may be to determine a dose level that will lead to a given time of reaction on the average. The experimental doses, being fixed, cannot be treated as random, so that it is inappropriate to determine a regression of x_1 on Y, and such a pseudo regression would give spurious results. In such situations the value of x_1 corresponding to a given value of η has to be estimated from the regression of Y on x_1.

The regression equation can be rearranged to give an estimator of x_1 corresponding to a given value, η_*,

$$\hat{X}_* = (\eta_* - b_0)/b_1 .$$

The approximate estimated variance of the estimator is

$$\text{est. var}\,(\hat{X}_*) \cong \frac{s^2}{b_1^2}\left(\frac{1}{n} + \frac{(\hat{X}_* - \bar{x}_1)^2}{\sum(x_1 - \bar{x}_1)^2}\right)$$

$$= \frac{s^2}{b_1^2}\left(\frac{1}{n} + \frac{(\eta_* - \bar{Y})^2}{b_1^2\sum(x_1 - \bar{x}_1)^2}\right)$$

A more precise method of treating such a problem is to determine confidence limits for η given x_1 and to determine from these, by rearranging the equation, confidence limits for x_1. For the regression shown in Figure 1, the 95 per cent confidence curves (the inner curves, Y_L, on either side of the line) will in this way give confidence limits for x_1 corresponding to a given value of η. The point at which the horizontal line $Y = \eta$ cuts the regression line gives the estimate of x_1; the points at which the line cuts the upper and lower curves give, respectively, lower and upper confidence limits for x_1. This may be demonstrated by an extension of the reasoning leading to confidence limits. [*See* ESTIMATION, *article on* CONFIDENCE INTERVALS AND REGIONS.]

Sometimes, rather than a hypothetical regression value, η_*, a single observed value, y_* (not in the basic sample), is given, and limits are required for the value of x_1 that could be associated with such a value. The estimator $\hat{X}_*$ is given by

$$\hat{X}_* = (y_* - b_0)/b_1 ,$$

and its approximate estimated variance (which must take into account the variation between responses on Y to a given value of x_1) is

$$\text{est. var}\,(\hat{X}_*) \cong \frac{s^2}{b_1^2}\left(1 + \frac{1}{n} + \frac{(\hat{X}_* - \bar{x}_1)^2}{\sum(x_1 - \bar{x}_1)^2}\right).$$

Using this augmented variance, confidence limits on x_1 corresponding to a given y_* may be found. For more precise determination of the confidence limits for prediction, the locus of limits for y_* given x_1 may be inverted to give limits for x_1 given y_*. In Figure 1, the outer curves are these loci (for the 95 per cent confidence level); the 95 per cent limits for x_1 will be given by the intersection of the line $Y = y_*$ with these confidence curves for prediction.

Multiple regression

In many situations where a single regression variable is not adequate to represent the variation in the random variable Y, a multiple regression is appropriate. In other situations there may be only one regression variable, but the assumed relation, rather than being linear, is a quadratic or a polynomial of higher degree. Since both multiple linear regression and polynomial regression relations are linear in the unknown parameters, the same techniques are applicable to both; in fact, polynomial regression is a special case of multiple regression. The number of variables to include in a multiple regression, or the degree of polynomial to be applied, is to some extent a matter of judgment and convenience, although it must be remembered that a regression equation containing a large number of variables is usually inconvenient to use as well as difficult to calculate. With the use of electronic computers, however, there is greater scope for increasing the number of regression variables, since the computations are routine.

Consider the multiple regression equation

$$E(Y) = \eta = \beta_0 + \beta_1 x_1 + \cdots + \beta_p x_p ,$$

with p regression variables and a constant term. The estimation of these $p + 1$ unknown parameters can be systematically carried out if β_0 is also regarded as a regression coefficient corresponding to a regression variable x_0 that is always unity. As in simple regression, the method of least squares provides unbiased linear estimators of the coefficients with minimum variance and also provides estimators of the standard errors of these coefficients. The quantities required for determining the estimators are the sums of squares and products of the x-values, the sums of products of the observed Y with each of the x-values, and the sum of squares of Y. The method of least squares gives a set of linear equations for the b's, called the *normal equations*:

$$\begin{aligned}
b_0 t_{00} + b_1 t_{01} + \cdots + b_p t_{0p} &= u_0 \\
b_0 t_{10} + b_1 t_{11} + \cdots + b_p t_{1p} &= u_1 \\
\vdots \qquad \vdots \qquad\quad & \;\;\vdots \qquad \vdots \\
b_0 t_{p0} + b_1 t_{p1} + \cdots + b_p t_{pp} &= u_p
\end{aligned}$$

where $t_{hi} = t_{ih} = \sum x_h x_i$ and $u_i = \sum Y x_i$. These equations can be written in matrix form as

$$\boldsymbol{Tb} = \boldsymbol{u},$$

where $\boldsymbol{T} = (t_{hi})$ and $\boldsymbol{u}$ is the vector of the u_i. The solution requires the inversion of the matrix $\boldsymbol{T}$, the inverse matrix being denoted by $\boldsymbol{T}^{-1}$ (with typical element t^{hi}). The solution may be written in matrix form as

$$\boldsymbol{b} = \boldsymbol{T}^{-1}\boldsymbol{u}$$

or in extended form as

$$b_0 = t^{00}u_0 + t^{01}u_1 + \cdots + t^{0p}u_p,$$

and so forth.

The variance of b_i is $t^{ii}\sigma^2$, and the covariance of b_i and b_j is $t^{ij}\sigma^2$. It should be remarked that in the special case of "regression through the origin"—that is, when the constant term β_0 is assumed to be zero—the first equation and the first term of each other equation are omitted; the constant regressor x_0 and its coefficient β_0 thus have the same status as any other regression variable.

When the constant term is included, computational labor may be reduced and arithmetical accuracy increased if the sums of squares and products are taken about the means. That is, the t_{hi} and u_i are replaced by

$$t'_{hi} = t_{hi} - \sum x_h \sum x_i / n = \sum (x_h - \bar{x}_h)(x_i - \bar{x}_i)$$

and

$$u'_i = u_i - \sum Y \sum x_i / n,$$

respectively. All the sums of products with zero subscripts then vanish, and the sums of squares are reduced in magnitude. The constant term has to be estimated separately; it is given by

$$\begin{aligned} b_0 &= \bar{Y} - b_1\bar{x}_1 - b_2\bar{x}_2 - \cdots - b_p\bar{x}_p \\ &= (\sum Y - b_1\sum x_1 - b_2\sum x_2 - \cdots - b_p\sum x_p)/n. \end{aligned}$$

The computational aspects of matrix inversion and the determination of the regression coefficients are dealt with in many statistical texts, including Williams (1959); in addition, many programs for matrix inversion are available for electronic computers.

Effect of heteroscedasticity. When the error variance of the dependent variable Y is different in different parts of its range (or, strictly, of the range of its expected value, η), estimators of regression coefficients ignoring the heteroscedasticity will be unbiased but of reduced accuracy, as already mentioned. The calculation of improved estimators may then sometimes be necessary.

There are some problems in taking heteroscedasticity into account. Among them is the problem of specification: defining the relation between expected value and variance. Often, with adequate data, the estimated ($\hat{\eta}$) values from the usual unweighted regression line can be grouped and the mean squared deviation from these values for each group used as a rough measure of the variance. The regression can then be refitted, each value being given a weight inversely proportional to the estimated variance. Two iterations of this method are likely to give estimates of about the accuracy practically attainable. If an empirical relation between expected value and error variance can be deduced, this simplifies the problem somewhat; however, the weight for each observation has to be determined from a provisionally fitted relation, so iteration is still required.

To calculate a weighted regression, each observation Y_j ($j = 1, 2, \cdots, n$) is given a weight w_j instead of unit weight as in the standard calculation. These weights will be the reciprocals of the estimated variances of each value. Then, if weighted quantities are distinguished by the subscript w,

$$t_{whi} = \sum w x_h x_i,$$
$$u_{wi} = \sum w Y x_i,$$

and the normal equations are

$$b_{w0}t_{w00} + b_{w1}t_{w01} + \cdots + b_{wp}t_{w0p} = u_{w0}$$

and so on, or in matrix form

$$\boldsymbol{T}_w \boldsymbol{b}_w = \boldsymbol{u}_w.$$

The solution is

$$\boldsymbol{b}_w = \boldsymbol{T}_w^{-1}\boldsymbol{u}_w,$$

and the variances of the estimators b_{wi} are approximately $t_w^{ii}\sigma^2$.

When the weights are estimated from the data, as in the iterative method just described, some allowance has to be made in an exact analysis for errors in the weights. This inaccuracy will somewhat reduce the precision of the estimators. However, for most practical purposes, and provided that the number of observations in each group for which weights are estimated is not too small, the errors in the weights may be ignored. (For further discussion of this question see Cochran & Carroll 1953.)

Estimability of the coefficients. It is intuitively clear in a general way that the $p + 1$ regression variables included in a regression equation should not be too nearly linearly dependent on one another, for then it might be expected that these regression variables could be approximately expressed in terms of a smaller number.

More precisely, in order that meaningful estimators of the regression coefficients exist, it is necessary that the variables be linearly independent (or, equivalently, $\boldsymbol{T}$ must be nonsingular). That is, no one variable should be expressible as a linear combination of the others or, expressed symmetrically, no linear combination of the variables vanishes unless all coefficients are zero. Clearly, if only $p - r$ of the variables are linearly independent, then the regression relation may be represented as a regression on these $p - r$, together with arbitrary multiples of the vanishing linear combinations. From the practical point of view, this lack of estimability will cause no problems, provided that the regression on a set of $p - r$ linearly independent variables is calculated. Estimation from the equation will be unaffected, but for testing the significance of the regression it must be noted that the regression sum of squares has not $p + 1$ but $p - r$ degrees of freedom, and the residual has $n - p + r$.

However, if the lack of estimability is ignored, the calculations to determine the $p + 1$ coefficients either will fail (since the matrix $\boldsymbol{T}$, being singular, has no inverse) or will give misleading results (if an approximate value of $\boldsymbol{T}$, having an inverse, is used in calculation and the lack of estimability is obscured).

When the regression variables, although linearly independent, are barely so (in the sense that the matrix $\boldsymbol{T}$, although of rank $p + 1$, is "almost singular," having a small but nonvanishing determinant), the regression coefficients will be estimable but will have large standard errors. In typical cases, many of the estimated coefficients will not differ with statistical significance from zero; this merely reflects the fact that the corresponding regression variable may be omitted from the equation and the remaining coefficients adjusted without significant worsening of the fit.

In this situation, as in the case of linear dependence, these effects are not usually important in practice; however, they may suggest the advisability of reducing the number of regression variables included in the equation. [*For further discussion, see* STATISTICAL IDENTIFIABILITY.]

Conditions on the coefficients. Sometimes the regression coefficients β_i are assumed to satisfy some conditions based on theory. Provided these conditions are expressible as linear equations in the coefficients, the method of least squares carries through and leads, as before, to unbiased estimators satisfying the conditions and with minimum variance among linear estimators. It will be clear that with $p + 1$ regression coefficients subject to $r + 1$ independent linear restrictions, $r + 1$ of the coefficients may be eliminated, so that the restricted regression is equivalent to one with $p - r$ coefficients. Thus, in principle there is a choice between expressing the model in terms of $p - r$ unrestricted coefficients or $p + 1$ restricted ones; often the latter has advantages of symmetry and interpretability.

A simple example of restricted regression is one in which η is a weighted average of the x's but with unknown weights, $\beta_1, \cdots, \beta_p$. Here the side conditions would be $\beta_0 = 0$, $\beta_1 + \cdots + \beta_p = 1$.

As the introduction of side conditions effectively reduces the number of linearly independent coefficients, such conditions are useful in restoring estimability when the coefficients are nonestimable. In many problems these side conditions may be chosen to have practical significance. For example, where an over-all mean and a number of treatment "effects" are being estimated, it is conventional to specify the effects so that their mean vanishes; with this specification they represent deviations from the over-all mean.

When a restricted regression is being estimated, it will often be possible and of interest to estimate the unrestricted regression as well, in order to see the effect of the restrictions and to test whether the data are concordant with the conditions assumed. The test of significance consists of comparing the $(p + 1)$-variable (unrestricted) regression with the $(p - r)$-variable (restricted) regression, in the manner described in the section on significance testing. This test of concordance is independent of the test of significance of any of the restricted coefficients.

Further details and examples of restricted regression are given by Rao (1965, p. 189) and Williams (1959, pp. 49–58). In the remainder of this article, the notation will presume unrestricted regression.

Missing values. When observations on some of the variables are missing, the simplest and usually the only practicable procedure is to ignore the corresponding values of the other variables—that is, to work only with complete sets of observations. However, it is sometimes possible to make use of the incomplete data, provided some additional assumptions are made. Methods have been developed under the assumption that (a) the missing values are in some sense randomly deleted, or the assumption that (b) the variables are all random and follow a multivariate normal distribution. Assumption (b) is treated by Anderson (1957) and Rao (1952, pp. 161–165). It is sometimes found, after the least squares equations for the constants in a regression relation have been set up, that some of the values of the dependent variable are unreliable or missing altogether. Rather than recalculate the equations it is often more convenient to replace the missing value by the value expected from the regression relation. This substitution conserves the form of the estimating equations, usually with little disturbance to the significance tests or the variances of the estimators.

The techniques of "fitting missing values" have

been most fully developed for experiments designed in such a way that the estimators of various constants are either uncorrelated or have a symmetric pattern of correlations and the estimating equations have a symmetry of form that simplifies their solution. Missing values in such experiments destroy the symmetry and make estimation more difficult; it is therefore a great practical convenience to replace the missing values. Details of the method applied to designed experiments will be found in Cochran and Cox (1950). For applications to general regression models see Kruskal (1961).

The technique is itself an application of the method of least squares. To replace a missing value Y_j, a value $\hat{\eta}_j$ is chosen so as to minimize its contribution to the residual sum of squares. Thus, the estimate is equivalent to the one that would have been obtained by a fresh analysis; the calculation is simplified by the fact that estimates for only one or a few values are being calculated. The degrees of freedom for the residual sum of squares are reduced by the number of values thus fitted. For most practical purposes it is then sufficiently accurate to treat the fitted values as though they were original observations. The exact analysis is described by Yates (1933) and, in general terms, by Kruskal (1961).

Significance testing. In order to determine the standard errors of the regression coefficients and to test their significance, it is necessary to estimate the residual variance, σ^2. The sum of squares of deviations, $\sum(Y-\hat{\eta})^2$, which may readily be shown to satisfy

$$\begin{aligned}\sum(Y-\hat{\eta})^2 &= \sum Y^2 - \boldsymbol{u}'\boldsymbol{T}^{-1}\boldsymbol{u}\\ &= \sum Y^2 - \boldsymbol{b}'\boldsymbol{u},\end{aligned}$$

is found under $p+1$ constraints and so may be said to have $n-p-1$ degrees of freedom; if the model assumed is correct, so that the deviations are purely random, the expected value of the sum of squares is $(n-p-1)\sigma^2$. Accordingly, the *residual mean square*,

$$s^2 = \sum(Y-\hat{\eta})^2/(n-p-1),$$

is an unbiased estimator of the residual variance. The variances of the regression coefficients are estimated by

$$\text{est. var}\,(b_i) = t^{ii}s^2,$$

and the standard errors are the square roots of these quantities. The inverse matrix thus is used both in the calculation of the estimators and in the determination of their standard errors. From the off-diagonal elements t^{hi} of the inverse matrix are derived the estimated covariances between the estimators,

$$\text{est. cov}\,(b_h, b_i) = t^{hi}s^2.$$

The splitting of the total sum of squares of Y into two parts, a part associated with the regression effects and a residual part independent of them, is a particular example of what is known as the analysis of variance [*see* LINEAR HYPOTHESES, *article on* ANALYSIS OF VARIANCE].

Testing for regression effects. The regression sum of squares, being based on $p+1$ estimated quantities, will have $p+1$ degrees of freedom. When regression effects are nonexistent, the expected value of each part is proportional to its degrees of freedom. Accordingly, it is often convenient and informative to present these two parts, and their corresponding mean squares, in an analysis-of-variance table, such as Table 1.

In the table, the final column gives the expected values of the two mean squares; it shows that real regression effects inflate the regression sum of squares but not the residual sum of squares. This fact provides the basis for tests of significance of a calculated regression, since large values of the ratio of regression mean square to residual mean square give evidence for the existence of a regression relation.

Significance of a single coefficient. The question may arise whether one or more of the regression variables contribute to the relation anything that is not already provided by the other variables. In such circumstances the relevant hypothesis to be examined is that the β's corresponding to these variables are zero. A more general hypothesis that

Table 1 — Analysis-of-variance table for testing regression effects

Source	Degrees of freedom	Sum of squares	Mean square	Expected mean square
Regression	$p+1$	$\sum b_i u_i$	$\dfrac{\sum b_i u_i}{p+1}$	$\sigma^2 + \dfrac{\boldsymbol{\beta}'\boldsymbol{T}\boldsymbol{\beta}}{p+1}$
Residual	$n-p-1$	$\sum Y^2 - \sum b_i u_i = (n-p-1)s^2$	s^2	σ^2
Total	n	$\sum Y^2$		

may sometimes need to be tested is that certain of the β's take assigned values.

The simplest test is that of the statistical significance of a single coefficient—say, b_i. The test will be of its departure from zero, if the contribution of x_i to the regression is in question. More generally, when β_i is specified, as, say, β_i^*, it will be relevant to test the significance of departure of b_i from β_i^*. The significance test in either case is the same; the squared difference between estimated and hypothesized values is compared with the estimated variance of that difference, which is s^2t^{ii}.

The ratio $F = (b_i - \beta_i^*)^2/(s^2t^{ii})$ has the F-distribution with 1 and $n-p-1$ degrees of freedom if the difference is in fact due to sampling fluctuations alone; in this case, the F-statistic is just the square of the usual t-statistic. When β_i differs from β_i^* the F-statistic will tend to be larger, so that a right-tail test is indicated.

Testing several coefficients. To test a number of regression variables—or, more precisely, their regression coefficients—the method of least squares is equivalent to fitting a regression with and without the variables in question and testing the difference in the regression sums of squares against the estimated error variance. To choose a specific example, suppose the last q coefficients in a p-variable regression are to be tested. If the symbol S^2 is used to stand for sum of squares, the sum of squares for regression on all p variables may be written

$$S_p^2 = \boldsymbol{u}_p'\boldsymbol{T}_p^{-1}\boldsymbol{u}_p,$$

with $p+1$ degrees of freedom, and the corresponding sum of squares on the first $p-q$ variables as

$$S_{p-q}^2 = \boldsymbol{u}_{p-q}'\boldsymbol{T}_{p-q}^{-1}\boldsymbol{u}_{p-q}$$

with $p-q+1$ degrees of freedom. The difference, a sum of squares with q degrees of freedom, provides a criterion for testing the significance of the q regression coefficients. The ratio

$$F = (S_p^2 - S_{p-q}^2)/(qs^2)$$

has, under the null hypothesis that the last q coefficients are zero, the F-distribution with q and $n-p-1$ degrees of freedom. This simultaneous test of q coefficients may also be adapted to testing the departure of the q coefficients from theoretical values, not necessarily zero.

The significance test may be conveniently set out as in Table 2, where only the mean squares required for the significance test appear in the last column.

When $q = 1$, this test reduces to the test for a single regression coefficient, and the F-ratio

$$F = (S_p^2 - S_{p-1}^2)/s^2$$

is then identical with the F-ratio given above for making such a test.

Linear combinations of coefficients. Sometimes it is necessary to test the significance of one or more linear combinations of the coefficients—that is, to test hypotheses about linear combinations of the β's. A common example is the comparison of two coefficients, β_1 and β_2, say, for which the comparison $b_1 - b_2$ is relevant. The F-test applies to such comparisons also. Thus, for the difference $b_1 - b_2$, the estimated variance is $s^2(t^{11} - 2t^{12} + t^{22})$, and $F = (b_1 - b_2)^2/[s^2(t^{11} - 2t^{12} + t^{22})]$, with 1 and $n-p-1$ degrees of freedom.

In general, to test the departure from zero of k linear combinations of regression coefficients the procedure is as follows. Let the linear combinations (expressed in matrix notation) be

$$\boldsymbol{\Gamma}'\boldsymbol{b},$$

where $\boldsymbol{\Gamma}$ is a $(p+1)\times k$ matrix of known constants. Then the estimated covariance matrix of these linear combinations is

$$s^2\boldsymbol{\Gamma}'\boldsymbol{T}^{-1}\boldsymbol{\Gamma},$$

and the F-ratio is

$$F = \boldsymbol{b}'\boldsymbol{\Gamma}(\boldsymbol{\Gamma}'\boldsymbol{T}^{-1}\boldsymbol{\Gamma})^{-1}\boldsymbol{\Gamma}'\boldsymbol{b}/ks^2,$$

Table 2 — Analysis-of-variance table for testing several regression coefficients

Source	Degrees of freedom	Sum of squares	Mean square
Regression on $p-q$ variables	$p-q+1$	S_{p-q}^2	...
Additional q variables	q	$S_p^2 - S_{p-q}^2$	$(S_p^2 - S_{p-q}^2)/q$
Regression on all p variables	$p+1$	S_p^2	...
Residual	$n-p-1$	$(n-p-1)s^2$	s^2
Total	n	$\sum Y^2$	

with k and $n-p-1$ degrees of freedom. Of course, this test can also be adapted to testing the departure of these linear combinations from preassigned values other than zero.

When the population coefficients β_i are in fact nonzero, the expected value of the regression mean square in the analysis of variance shown in Table 1 will be larger than σ^2 by a term that depends on both the magnitude of the coefficients and the accuracy with which they are estimated (see, for example, the last column of Table 1). Clearly, the greater this term, called the *noncentrality*, the greater the probability that the null hypothesis will be rejected at the adopted significance level. The F-test has certain optimum properties, but other tests may be preferred in special circumstances.

Multivariate analogues

Although hitherto only the regression of a single dependent variable Y on one or more regressors x_i has been discussed, it will be realized that often the simultaneous regressions of a number of random variables on the same regressors will be of importance. For instance, in a sociological study of immigrants the regressions of annual income and size of family on age, educational level, and period of residence in the country may be determined; here there are two dependent variables and three regressors.

Often the relations among the different dependent variables will also be of interest, or various linear combinations of the variables, rather than the original variables themselves, may be studied. The linear combination that is most highly correlated with the regressors may sometimes be relevant to the investigation, but the linear compounds will usually be chosen for their practical relevance rather than their statistical properties. [*For further discussion of multivariate analogues, see* MULTIVARIATE ANALYSIS, *especially the general article,* OVERVIEW, *and the article on* CLASSIFICATION AND DISCRIMINATION.]

Polynomial regression

When the relation between two variables, x_1 and Y, appears to be curvilinear, it is natural to fit some form of smooth curve to the data. For some purposes a freehand curve is adequate to represent the relation, but if the curve is to be used for prediction or estimation and standard errors are required, some mathematical method of fitting, such as the method of least squares, must be used. The freehand fitting of a curvilinear relation has all the disadvantages of freehand fitting of a straight line, with the added disadvantage that it is more difficult to distinguish real trends from random fluctuations.

The polynomial form is

$$E(Y) = \eta = \beta_0 + \beta_1 x_1 + \beta_2 x_1^2 + \cdots + \beta_p x_1^p.$$

Being a linear model, it has the advantages of simplicity, flexibility, and relative ease of calculation. It is for such reasons, not because it necessarily represents the theoretical form of the relation, that a polynomial regression is often fitted to data.

Orthogonal polynomials. The computations in polynomial regression are exactly the same as those in multiple regression, except that some simplification of the arithmetic may be introduced if the same values of x_1 are used repeatedly. Then instead of using the powers of x_1 as the regression variables, these are replaced by orthogonal polynomials of successively increasing degree, so defined that the sum of products of any pair of them, over their chosen values, is zero.

This procedure has the twofold advantage that, first, all the off-diagonal elements of the matrix $\boldsymbol{T}$ are zero, so the calculation of regression coefficients and their standard errors is much simplified, and, second, the regression coefficient on each polynomial and the corresponding sum of squares can be independently determined.

Because it is common for investigators to use data with values of the independent variables equally spaced, the orthogonal polynomials for this particular case have been extensively tabulated. Fisher and Yates (1938) tabulate these orthogonal polynomials up to those of fifth degree, for numbers of equally spaced points up to 75. However, if the data are not equally spaced the tabulated polynomials are not applicable, and the regression must be calculated directly.

Testing adequacy of fit. The question of what degree of polynomial is appropriate to fit to a set of data is discussed below (see "Considerations in regression model choice"). If for each value of x_1 there is an array of values for Y, the variation in the data can be analyzed into parts between and within arrays by the techniques of analysis of variance [*see* LINEAR HYPOTHESES, *article on* ANALYSIS OF VARIANCE]. The sum of squares between arrays can be further analyzed into that part accounted for by regression and that part not so accounted for (deviation from regression). The adequacy of a polynomial fitted to the data is indicated by nonsignificant deviation from regression.

When there is but one observation of Y for each value of x_1, such an analysis is not possible. To test the adequacy of a pth-degree polynomial regression, a common though not strictly defensible procedure is to fit a polynomial of degree $p+1$ and test whether the coefficient b_{p+1} of x_1^{p+1} is significant. Anderson (1962) has treated this problem as a multiple decision problem and has provided optimal procedures that can readily be applied.

Estimation of maxima. Sometimes a polynomial regression is fitted in order to estimate the value of x_1 that yields a maximum value of η. A detailed discussion of the estimation of maxima is given by Hotelling (1941). To give an idea of the methods that are used, consider a quadratic regression of the form

$$\hat{\eta} = b_0 + b_1x_1 + b_2x_1^2.$$

A maximum (or minimum) value of $\hat{\eta}$ occurs at the point $x_m = -b_1/2b_2$, and this value is taken as the estimated position of the maximum. Confidence limits for the position can be determined by means of the following device. If the position of the maximum of the true regression curve is ξ, then $\xi = -\beta_1/2\beta_2$, so that $\beta_1 + 2\beta_2\xi = 0$. Consequently the quantity

$$b_1 + 2b_2\xi$$

is distributed with mean zero and estimated variance

$$s^2(t^{11} + 4t^{12}\xi + 4t^{22}\xi^2).$$

The confidence limits for ξ with confidence coefficient $1-\alpha$ are given by the roots of the equation

$$\frac{(b_1 + 2b_2\xi)^2}{s^2(t^{11} + 4t^{12}\xi + 4t^{22}\xi^2)} = F_{\alpha;1,n-3},$$

where $F_{\alpha;1,n-3}$ is the α-point of the F-distribution with 1 and $n-3$ degrees of freedom, abbreviated below as F_α. The solution of this equation may be simplified by writing

$$g_{11} = \frac{F_\alpha s^2 t^{11}}{b_1^2}, \qquad g_{12} = \frac{F_\alpha s^2 t^{12}}{b_1 b_2}, \qquad g_{22} = \frac{F_\alpha s^2 t^{22}}{b_2^2},$$

so that the confidence limits become

$$X_L = x_m \frac{(1-g_{12}) \pm \sqrt{(1-g_{12})^2 - (1-g_{11})(1-g_{22})}}{1-g_{22}}.$$

Note that these limits are not, in general, symmetrically placed about the estimated value $-b_1/2b_2$, since allowance is made for the skewness of the distribution of the ratio. Note also that the limits will include infinite values and will therefore not be of practical use, unless b_2 is significant at the α-level. In terms of the g-values, this means that g_{22} must not exceed 1.

When the regression model is a polynomial in two or more variables, investigation of maxima and other aspects of shape becomes more complex. [*A discussion of this problem appears in* EXPERIMENTAL DESIGN, *article on* RESPONSE SURFACES.]

Nonlinear models

In a nonlinear model the regression function is nonlinear in one or more of the parameters. Familiar examples are the exponential regression,

$$\eta = \beta_0 + \beta_1 e^{\beta_2 x_1},$$

and the logistic curve,

$$\eta = \frac{\beta_1}{1 + e^{-\beta_2 x_1}},$$

β_2 being the nonlinear parameter in each example. Such nonlinear models usually originate from theoretical considerations but nevertheless are often useful for applying to observational data.

Sometimes the model can be reduced to a linear form by a transformation of variables (and a corresponding change in the specification of the errors). The exponential regression with $\beta_0 = 0$ may thus be reduced by taking logarithms of the dependent variable and assuming that the errors of the logarithms, rather than the errors of the original values, are distributed about zero. If $Z = \log_e Y$ and $E(Z) = \xi$, the exponential model with $\beta_0 = 0$ reduces to $\xi = \log_e \beta_1 + \beta_2 x_1$, a linear model.

The general models shown above cannot be reduced to linear models in this way. For nonlinear models generally, the nonlinear parameters must be estimated by successive approximation. The following method is straightforward and of general applicability.

Suppose the model is

$$\eta = \beta_0 + \beta_1 f(x_1, \beta_2)$$

where $f(x_1, \beta_2)$ is a nonlinear function of β_2, and the estimated regression, determined by least squares, is

$$\hat{\eta} = b_0 + b_1 f(x_1, c).$$

If c_0 is a trial value of c (estimated by graphical or other means), the values of $f(x_1, c)$ and its first derivative with respect to c (denoted, for brevity, by f and f', respectively) are calculated for each value of x_1, with $c = c_0$. The regression of Y on f and f' is then determined in the usual way, yielding the regression equation

$$\hat{\eta} = b_0 + b_1 f + b_2 f'.$$

A first adjustment to c_0 is given by b_2/b_1, giving the new approximation

$$c_1 = c_0 + b_2/b_1.$$

The process of recalculating the regression on f and f' and determining successive approximations to c can be continued until the required accuracy is attained (for further details see Williams 1959).

The method is an adaptation of the *delta method*, which utilizes the principle of *propagation of error*. If a small change, $\delta\beta_2$, is made in a parameter β_2, the corresponding change in a function $f(\beta_2)$ is, to a first approximation, $f'(\beta_2)\delta\beta_2$. The use of this method allows the replacement of the nonlinear equations for the parameters by approximate linear equations for the adjustments. For a regression relation of the form

$$\eta = \beta_0 + \beta_1 e^{\beta_2 x_1},$$

Stevens (1951) provides a table to facilitate the calculation of the nonlinear parameter by a method similar to that described above, and Pimentel Gomes (1953) provides tables from which, with a few preliminary calculations, the least squares estimate of the nonlinear parameter can be read off easily.

Considerations in regression model choice

In deciding which of several alternative models shall be used to interpret a relationship, a number of factors must be taken into account. Other things being equal, the model which represents the predictands most closely (where "closeness" is measured in terms of some criterion such as minimum mean square error among linear estimators) will be used. However, questions of convenience and simplicity should also be considered. A regression equation that includes a large number of regression variables is not convenient to use, and an equation with fewer variables may be only slightly less accurate. In deciding between alternative models, the residual variance is therefore not the only factor to take into account.

In polynomial regression particularly, the assumed polynomial form of the model is usually chosen for convenience, so that a polynomial of given degree is not assumed to be the true regression model. Because of this, the testing of individual polynomial coefficients is little more than a guide in deciding on the degree of polynomial to be fitted. Of far more importance is a decision on what degree of variability about the regression model is acceptable, and this decision will be based on practical rather than merely statistical considerations.

Besides the question of including additional variables in a regression, for which significance tests have already been described, there is also the question of alternative regression variables. The alternatives for a regression relation could be different variables or different functions of the same variable—for instance, x_1 and $\log x_1$.

For comparison of two or more individual variables as predictors, a test devised by Hotelling (1940) is suitable, although not strictly accurate. It is based on the correlations between Y and the different predictors and of the predictors among themselves. For comparing two regression variables x_1 and x_2, the test statistic is

$$F = \frac{(u'_1/\sqrt{t'_{11}} - u'_2/\sqrt{t'_{22}})^2}{2s^2(1 - t'_{12}/\sqrt{t'_{11}t'_{22}})},$$

which is distributed approximately as F with 1 and $n-3$ degrees of freedom. Here, as before,

$$u'_i = \sum Y(x_i - \bar{x}_i),$$
$$t'_{hi} = \sum (x_h - \bar{x}_h)(x_i - \bar{x}_i),$$

and s^2 is the mean square of residuals from the regression of Y on x_1 and x_2, with $n-3$ degrees of freedom.

EVAN J. WILLIAMS

BIBLIOGRAPHY

ANDERSON, T. W. 1957 Maximum Likelihood Estimates for a Multivariate Normal Distribution When Some Observations Are Missing. *Journal of the American Statistical Association* 52:200–203.

ANDERSON, T. W. 1962 The Choice of the Degree of a Polynomial Regression as a Multiple Decision Problem. *Annals of Mathematical Statistics* 33:255–265.

BOX, GEORGE E. P.; and ANDERSEN, S. L. 1955 Permutation Theory in the Derivation of Robust Criteria and the Study of Departures From Assumption. *Journal of the Royal Statistical Society* Series B 17:1–26.

BOX, GEORGE E. P.; and WILSON, K. B. 1951 On the Experimental Attainment of Optimum Conditions. *Journal of the Royal Statistical Society* Series B 13:1–45. → Contains seven pages of discussion.

COCHRAN, WILLIAM G.; and CARROLL, SARAH P. 1953 A Sampling Investigation of the Efficiency of Weighting Inversely as the Estimated Variance. *Biometrics* 9: 447–459.

COCHRAN, WILLIAM G.; and COX, GERTRUDE M. (1950) 1957 *Experimental Designs*. 2d ed. New York: Wiley.

EZEKIEL, MORDECAI; and FOX, KARL A. (1930) 1961 *Methods of Correlation and Regression Analysis: Linear and Curvilinear*. New York: Wiley.

FISHER, R. A.; and YATES, FRANK (1938) 1963 *Statistical Tables for Biological, Agricultural and Medical Research*. 6th ed., rev. & enl. Edinburgh: Oliver & Boyd; New York: Hafner.

GAUSS, CARL F. 1855 *Méthode des moindres carrés: Mémoires sur la combinaison des observations*. Translated by J. Bertrand. Paris: Mallet-Bachelier. → An authorized translation of Carl Friedrich Gauss's works on least squares.

GEARY, R. C. 1963 Some Remarks About Relations Between Stochastic Variables: A Discussion Document. Institut International de Statistique, *Revue* 31:163–181.

HOTELLING, HAROLD 1940 The Selection of Variates for

Use in Prediction With Some Comments on the General Problem of Nuisance Parameters. *Annals of Mathematical Statistics* 11:271–283.

HOTELLING, HAROLD 1941 Experimental Determination of the Maximum of a Function. *Annals of Mathematical Statistics* 12:20–45.

○KENDALL, MAURICE G.; and STUART, ALAN (1961) 1973 The General Theory of Regression. Volume 2, chapter 28, pages 361–390 in Maurice G. Kendall and Alan Stuart, *The Advanced Theory of Statistics*. Volume 2: *Inference and Relationship*. 3d ed. London: Griffin; New York: Hafner. → Early editions of Volume 2, first published in 1946, were written by Kendall alone. Stuart became a joint author on later, renumbered editions in the three-volume set.

KRUSKAL, WILLIAM H. 1961 The Coordinate-free Approach to Gauss–Markov Estimation and Its Application to Missing and Extra Observations. Volume 1, pages 435–451 in Symposium on Mathematical Statistics and Probability, Fourth, Berkeley, *Proceedings*. Berkeley and Los Angeles: Univ. of California Press.

LEGENDRE, ADRIEN M. (1805) 1959 On a Method of Least Squares. Volume 2, pages 576–579 in David Eugene Smith, *A Source Book in Mathematics*. New York: Dover. → First published as "Sur la méthode des moindres carrés" in Legendre's *Nouvelles méthodes pour la détermination des orbites des comètes*.

MADANSKY, ALBERT 1959 The Fitting of Straight Lines When Both Variables Are Subject to Error. *Journal of the American Statistical Association* 54:173–205.

PIMENTEL GOMES, FREDERICO 1953 The Use of Mitscherlich's Regression Law in the Analysis of Experiments With Fertilizers. *Biometrics* 9:498–516.

PLACKETT, R. L. 1960 *Principles of Regression Analysis*. Oxford: Clarendon.

RAO, C. RADHAKRISHNA 1952 *Advanced Statistical Methods in Biometric Research*. New York: Wiley.

RAO, C. RADHAKRISHNA 1965 *Linear Statistical Inference and Its Applications*. New York: Wiley.

STEVENS, W. L. 1951 Asymptotic Regression. *Biometrics* 7:247–267.

STRODTBECK, FRED L.; MCDONALD, MARGARET R.; and ROSEN, BERNARD C. 1957 Evaluation of Occupations: A Reflection of Jewish and Italian Mobility Differences. *American Sociological Review* 22:546–553.

WILLIAMS, EVAN J. 1952 Use of Scores for the Analysis of Association in Contingency Tables. *Biometrika* 39: 274–289.

WILLIAMS, EVAN J. 1959 *Regression Analysis*. New York: Wiley.

WOLD, HERMAN 1953 *Demand Analysis: A Study in Econometrics*. New York: Wiley.

YATES, FRANK 1933 The Analysis of Replicated Experiments When the Field Results Are Incomplete. *Empire Journal of Experimental Agriculture* 1:129–142.

YATES, FRANK 1948 The Analysis of Contingency Tables With Groupings Based on Quantitative Characters. *Biometrika* 35:176–181.

Postscript

Pitfalls in attempting to assess relative effects of regression variables. Researchers often feel a desire to assess the separate effects of the different (independent) variables in a regression relation. With two or more regression variables under consideration, there is a natural interest in such a partitioning to exhibit the relative importance of the contribution of each variable to the relation.

Concepts of relative importance are generally without meaning unless there is a specific "natural" ordering of the regression variables. Such an ordering is exemplified by the successive terms of a polynomial representing a curvilinear regression; the successive terms usually fitted in such a regression are increasing powers of an underlying variable. In some other contexts a similar ordering may be appropriate; for example, in a study of the effects of environmental and social factors on learning ability, it may be natural to include an environmental variable before a social variable in the regression relation.

For such an ordered set of variables, $x_1, x_2, \cdots$, it may be found, for example, that the regression on x_1 accounts for 42 per cent of the variability of Y; of the remaining 58 per cent, x_2 could account for a fraction 55 per cent, making up $58 \times 55/100$ or 32 per cent of the variability in Y. In this example the successive contributions would be 42, 32, totaling 74 per cent and leaving 26 per cent as residual variability still unaccounted for.

Such a partition of the total variability may have practical significance in certain contexts. It should be emphasized, however, that it applies only to a particular ordering of regression variables. If the ordering is changed, the partition will be changed. That this is so can be seen in a general way because the effects of different regression variables are nearly always intertwined, and there is no unique allocation of these effects among the variables. Only if the vectors of the regression variables are orthogonal is the partition unchanged as order changes.

In discussing the general situation where there is no inherent ordering of variables, two other points should be made. First, partitioning the separate effects need have nothing to do with causality, a different and more difficult philosophical concept. [*See* CAUSATION.] Second, the discussion relates to inherent relationships, properties of the population under study; thus we are not here concerned with the statistical significance, for example, of the estimated relationship, a property of the observed sample. Of course, the features of the population are reflected (often, alas, imperfectly) in those of the sample, so that the remarks here are relevant to the interpretation of the data of a sample as much as to the population from which they are drawn.

Consider first, for simplicity, a regression that is linear in two variables; the conclusions drawn will apply also to more complicated examples. I begin with the case in which the regression and dependent variables are jointly random. The reason is that then I can sensibly discuss both separate and joint regressions. Later I shall return briefly to the case of primary interest here, that of fixed regression variables.

In order to compare different regression relations for the same population or sample, note that such regressions exist only when x_1, x_2, and Y are jointly distributed *random variables*. (I here ignore trivial cases such as zero regressions, which are in any case of no interest for this discussion.) If the regression of Y on x_1 and x_2 exists (that is, if the expected value of Y for specified values of x_1 and x_2 is determinate), then there also exists a regression of Y on x_1 alone, the expectation of the double regression function taken over the distribution of x_2; similarly a regression of Y on x_2 exists.

If x_1 and x_2 are *uncorrelated*, then the contributions of x_1 and x_2 to the regression, as measured by the variances of the two simple regression functions, are additive; their sum is equal to the variance of the double regression function. Under the very strong assumption of no correlation between x_1 and x_2, then, the separate contributions to the regression are identifiable and acceptable as reasonable measures of the importance of the two regressors.

By extension from this highly special situation, the view has developed that effects in general can be allocated to different regression variables, but in fact no meaningful allocation can be made, except in the simple situation of noncorrelation just described. Unfortunately, noncorrelation seldom holds, so that any attempt at allocation is generally invalid.

First, as already pointed out, only when the regression variables are random variables will both multiple and simple regressions exist. Second, even if the regression variables are random, they will usually be correlated, Even if these problems are ignored in the practical interpretation of sample data, the regression variables will in general be correlated in the sample, thus rendering impossible an additive interpretation.

Example to illustrate problem of partitioning effects. I shall show that it is possible for the sum of squares for double regression on x_1 and x_2 to be large even when the sums of squares for the simple regressions on x_1 and x_2 are both small, so that a partitioning of the double regression effect to the individual variables can have no meaning.

For the regression of the dependent variable Y on regression variables x_1 and x_2, I adapt the notation of the main article and denote by S_2 the sum of squares for the double regression, and by $S_1(1)$ and $S_1(2)$ the sums of squares for the two simple regressions on x_1 and x_2 respectively. Clearly the sum of squares for regression cannot be decreased by the inclusion of an additional regression variable; it therefore follows that the minimum possible value of S_2 is the greater of $S_1(1)$ and $S_1(2)$. What is at first sight surprising is that the maximum value of S_2 need bear no relation to the simple regression sums of squares, and is in fact equal to the total sum of squares $\sum Y^2$. This result holds with one exception: if $S_1(1)$ and $S_1(2)$ both vanish, then so does S_2. (This situation is much like that of projecting a vector $\mathbf{Y}$ onto a plane defined by vectors $\boldsymbol{x}_1$ and $\boldsymbol{x}_2$. The squared length of the projection is at most that of $\mathbf{Y}$; unless $\boldsymbol{x}_1$ and $\boldsymbol{x}_2$ are orthogonal, the squared length of the projection onto the plane bears no special relationship to the two separate projections onto the lines defined by $\boldsymbol{x}_1$ and $\boldsymbol{x}_2$.)

As a simple yet realistic example, consider a random variable V that is a linear function of Y but whose measurement is affected by random errors U uncorrelated with Y; let x_1 be the measured value of $V + U$. Suppose now that a variable that is a linear function of the errors U can be observed; call this x_2. Then it is not difficult to see that, although the association between Y and x_1 will be less than perfect (so that for a sample, $S_1(1)$ will be less than $\sum Y^2$, and may be quite small), the incorporation of x_2 into the regression will provide a correction for the errors of measurement; the double regression function will be just a linear function of V, identical with Y, and $S_2 = \sum Y^2$.

This example may be described in both population and sample terms. It is apparent that x_2 exerts its effect, not through its relation with Y (since it is not so related), but through its relation with x_1, to which it acts as a correction. It would be absurd to attribute to x_2 an "effect on Y." Yet the inclusion of x_2 in the regression has the effect of making the regression an identity, and of increasing the regression sum of squares to its maximum possible value.

Thus, efforts to allocate regression effects to individual regression variables are fruitless, and add nothing to the interpretation of the regression relation.

Implications for scientific method. The results demonstrated with two regression variables apply also in more general situations; they do not depend on the form specified for the regression function, and they certainly have nothing to do with the techniques of fitting regressions. They show that, when a relation is established between one regression variable and others, no general meaning can be attached to the contribution of one of the variables to the relation.

The above conclusions for random regression variables are true a fortiori for fixed regression variables. Since only one regression function exists in this situation, there is no possibility of studying the contribution of individual variables to the relation. In a strictly mathematical sense, this is obvious; if the relation is

$$\eta = f(x_1, x_2, \cdots, x_p),$$

then the relation between η and x_1 has no meaning except in trivial cases (for example, when η depends only on x_1 and not on the other x's).

The techniques of statistical analysis for regression relations are seductive in providing "sums of squares for effects of given variables" and so suggesting that effects can in fact be attributed to the different variables. The above discussion shows that this concept is illusory. In general the only realistic interpretation of a regression relation is that the dependent variable is subject to the combined effect of a number of variables. Any attempt to take the interpretation further can lead only to misinterpretation and confusion.

And yet, there is an important sense, still awaiting careful analysis, in which one might write of separate regressions. A formula like that displayed above is a central part of the specification of a model; no model is ever perfectly correct, and one often adopts a simpler model that fits observations not so well as a more complex one. These comments can lead rapidly to fundamental questions of philosophy; I do not continue now along that path, but it is important to know that it exists.

The role of the regression function. One must keep in mind the positive role played by the regression function itself. Corresponding to the sample, and to a set of values of the regression variables, the regression function provides the unbiased estimator of the dependent variable that is of minimum variance in the class of linear estimators. In the practical applications of regression analysis, such an estimator, and estimators derived from it, are of primary interest. Partitioning of effects among the terms in the regression function would be of secondary importance even if it were feasible.

For example, if in a number of cities the health rating (Y) were believed to depend on two factors, air purity (x_1) and water purity (x_2), measured in some suitable units, it might be assumed that there existed a regression, $EY = \beta_0 + \beta_1 x_1 + \beta_2 x_2$. City administrations might well ask about the relative importance of x_1 and x_2 in this regression function in order to decide where to put their efforts toward decreasing pollution. For any practical health program, however, a means would be sought to optimize expected health ratings for given outlays on air and water purification. If C_1 and C_2 are the costs in dollars of increasing x_1 and x_2, respectively, by one unit, and if steps taken to improve x_1 result in a reduction of x_2 by an amount q_2, whereas steps taken to improve x_2 result in a reduction of x_1 by an amount q_1, then a program allocating funds f_1 and f_2 ($f_1 + f_2$ fixed) to improving x_1 and x_2 repectively, will lead to an expected improvement in health rating of

$$\frac{f_1}{C_1}(\beta_1 - \beta_2 q_2) + \frac{f_2}{C_2}(\beta_2 - \beta_1 q_1).$$

Because of linearity, the most economic program is then to make all funds available to improve air purity (x_1) if $(\beta_1 - \beta_2 q_2)/C_1 > (\beta_2 - \beta_1 q_1)/C_2$; and to improve water purity (x_2) if $(\beta_1 - \beta_2 q_2)/C_1 < (\beta_2 - \beta_1 q_1)/C_2$.

Note that q_1 and q_2 may be either positive or negative, depending on circumstances; if these factors are unknown but believed to be small, a satisfactory approximation will be provided by the rule: improve air purity (x_1) if $\beta_1/C_1 > \beta_2/C_2$; improve water purity ($x_2$) if $\beta_1/C_1 < \beta_2/C_2$.

Of course, the model might include an interaction term, perhaps one in the product of x_1 and x_2, and that would make matters more complex. In another, perhaps more fundamental direction, it is possible that an analysis that takes into account uncertainties in the coefficients and costs might lead to a *mixed* strategy, one in which funds are put into improvement of both air and water quality. In any case—and this is the moral—the allocation is not based on any attempted partitioning of the two kinds of effect.

Ridge regression. A method known as ridge regression has been advocated (see especially Hoerl & Kennard 1970) as a means of overcoming some of the supposed shortcomings of the standard method of least squares for estimating regression coefficients. Ridge regression is motivated by the fact that, when regression variables are closely associated, the least-squares estimators of the regression coefficients are subject to rather large standard errors; as a result, these estimators, even though free from bias, may differ substantially from values deemed reasonable on theoretical or other grounds. Ridge regression replaces the least-squares estimators with biased estimators that tend to have smaller mean square error; the bias introduced is thus compensated by reduction in variance.

To calculate a ridge regression, the regression variables are first scaled so that the matrix $\mathbf{X'X} = \mathbf{T}$

is a "correlation matrix" with unit diagonal elements; then $\boldsymbol{T}$ is replaced in the least-squares equations by $\boldsymbol{T} + k\boldsymbol{I}$, where $\boldsymbol{I}$ is the unit matrix. If $\boldsymbol{T}$ is "almost singular," $\boldsymbol{T} + k\boldsymbol{I}$ will be less so; in any case, the elements of its inverse will be reduced if k is sufficiently large. (Values of k between zero and unity are advocated in practice.) If $\boldsymbol{u}$ is defined as in the main article, the ridge estimators $\boldsymbol{b}^*$ are then given by the equations

$$(\boldsymbol{T} + k\boldsymbol{I})\boldsymbol{b}^* = \boldsymbol{u},$$

so that

$$\boldsymbol{b}^* = (\boldsymbol{T} + k\boldsymbol{I})^{-1}\boldsymbol{u}.$$

The ridge estimator $\boldsymbol{b}^*$ depends on k; as k is increased, $\boldsymbol{b}^*$ will become more biased and will ultimately approach zero; the variance of $\boldsymbol{b}^*$ will, however, decrease. The central question of ridge regression is how to choose k to achieve a satisfactory compromise between bias and variance.

A commonly used measure of adequacy of an estimator is the mean square error (the sum of the variance and squared bias summed for all the components of $\boldsymbol{b}^*$). The value of k for which the mean square error is a minimum is described as optimal. Hemmerle (1975) presents a method for determining this optimal value of k.

In applications, however, it is important to decide just how much bias can be tolerated in return for a reduction in variance; the mean square error, which gives equal weight to the variance and squared bias for each component of $\boldsymbol{b}^*$, is not necessarily appropriate in all situations.

A generalization of ridge regression is to add to $\boldsymbol{T}$ in the normal equations, not a multiple of the unit matrix, but some other diagonal matrix, or even some arbitrary positive definite symmetric matrix. The use of any particular added matrix must be justified in any case.

Space does not permit a detailed discussion of ridge regression. There appears to be little place for it in the frequent situation where theory does not provide a priori values for regression coefficients. The following points should, however, be noted.

If prior information about a regression coefficient, for a strictly comparable situation, is in fact available, this information may be incorporated to produce a combined estimator from the present and the previous data. Ridge regression, as usually applied, does not utilize such prior information, and tends to bias estimators toward zero; it may thus give estimators that are farther than the least-squares estimator from the value sought. A possible modification of the ridge technique would be to work with the difference of the present regression coefficient from the previous one; when this difference is biased toward zero, the practical effect would be to bias the regression coefficient toward the previous one. But the usual approach, which is more straightforward, is to average the present and the previous regression coefficients, weighting each inversely as its variance in the usual way, to give a combined regression coefficient with minimum variance.

The choice of added matrix (that is, $k\boldsymbol{I}$ in the original method) is arbitrary; clearly the introduction of an arbitrary element in the analysis of data, even inaccurate data, is in general unsatisfactory.

Finally, in many practical situations, one is interested not in the regression coefficients as such, but in the regression *function* (estimator of the dependent variable) in the region covered by the observed values of the regression variables. Now, it is well known that any estimated quantity, such as a regression function, is relatively accurately estimated for points inside the region of observations, but may be inaccurately estimated outside this region. The remedy for this inaccuracy is not to manipulate the existing data (since the variances of estimators by least squares are already the minimum possible among those of unbiased linear estimators), but to extend the region of observation to the region of interest.

If the region of possible observation is such that the regression coefficients cannot be estimated accurately, this suggests that the values of the regression coefficients themselves may not be of great practical interest. It should be emphasized that the regression function will still be accurately estimated within the region of observation even when its components, the regression coefficients, are not. Methods, such as ridge regression, for improving estimators may be found not to improve the estimation within the region of interest.

Path coefficients. In some complex situations, the relation between two variables somewhat separated but belonging to the same system may be inferred from the relations among successive variables forming a "path" between the two variables under consideration. In such situations, all possible paths connecting the two variables via related variables need to be taken into account in assessing the relationship. Sewall Wright (1921) gave the first general description of the method, with particular application to genetic examples. The numerical factors linking successive variables, and representing the strength of association between them, he called *path coefficients;* they are in fact standardized partial regression coefficients.

The method has great potential application in those sciences in which the variables under study are linked in complicated ways. [*For a discussion of path analysis, see* SURVEY ANALYSIS.]

EVAN J. WILLIAMS

[*For a discussion of stepwise procedures for estimating regression functions, see* MULTIVARIATE ANALYSIS, *article on* CORRELATION METHODS.]

ADDITIONAL BIBLIOGRAPHY

DANIEL, CUTHBERT; and WOOD, FRED S. 1971 *Fitting Equations to Data: A Computer Analysis of Multifactor Data for Scientists and Engineers.* New York: Wiley.

DRAPER, NORMAN R.; and SMITH, H. 1966 *Applied Regression Analysis.* New York: Wiley.

DUNCAN, OTIS DUDLEY 1966 Path Analysis: Sociological Examples. *American Journal of Sociology* 72:1–16.

HEMMERLE, WILLIAM J. 1975 An Explicit Solution for Generalized Ridge Regression. *Technometrics* 17:309–314.

HOCKING, R. R. 1976 The Analysis and Selection of Variables in Linear Regression. *Biometrics* 32:1–49.

HOERL, ARTHUR E.; and KENNARD, ROBERT W. 1970 Ridge Regression: Biased Estimation for Nonorthogonal Problems. *Technometrics* 12:55–67.

LI, CHING CHUN 1968 Fisher, Wright, and Path Coefficients. *Biometrics* 24:471–483.

LI, CHING CHUN 1975 *Path Analysis: A Primer.* Pacific Grove, Calif.: Boxwood.

MARQUARDT, DONALD W. 1970 Generalized Inverses, Ridge Regression, Biased Linear Estimation, and Nonlinear Estimation. *Technometrics* 12:591–612.

NELDER, JOHN A. 1972 Discussion on the paper by Professor Lindley and Dr. Smith. *Journal of the Royal Statistical Society* Series B, pt. 1, 34:18–20.

SPJØTVOLL, EMIL 1972 Multiple Comparison of Regression Functions. *Annals of Mathematical Statistics* 43:1076–1088.

TUKEY, JOHN W. (1954) 1964 Causation, Regression, and Path Analysis. Pages 35–66 in Oscar Kempthorne et al. (editors), *Statistics and Mathematics in Biology.* Reprint ed. New York: Hafner. → Papers presented at the biostatistics conference held at Iowa State College, June–July 1952.

WRIGHT, SEWALL 1921 Correlation and Causation. *Journal of Agricultural Research* 20:557–585.

WRIGHT, SEWALL 1960 Path Coefficients and Path Regressions: Alternative or Complementary Concepts? *Biometrics* 16:189–202.

II
ANALYSIS OF VARIANCE

Analysis of variance is a body of statistical procedures for analyzing observational data that may be regarded as satisfying certain broad assumptions about the structure of means, variances, and distributional form. The basic notion of analysis of variance (or ANOVA) is that of comparing and dissecting empirical dispersions in the data in order to understand underlying central values and dispersions.

This basic notion was early noted and developed in special cases by Lexis and von Bortkiewicz [*see* LEXIS; BORTKIEWICZ]. Not until the pioneering work of R. A. Fisher (1925; 1935), however, were the fundamental principles of analysis of variance and its most important techniques worked out and made public [*see* FISHER, R. A.]. Early applications of analysis of variance were primarily in agriculture and biology. The methodology is now used in every field of science and is one of the most important statistical areas for the social sciences. (For further historical material see Sampford 1964.)

Much basic material of analysis of variance may usefully be regarded as a special development of regression analysis [*see* LINEAR HYPOTHESES, *article on* REGRESSION]. Analysis of variance extends, however, to techniques and models that do not strictly fall under the regression rubric.

In analysis of variance all the standard general theories of statistics, such as point and set estimation and hypothesis testing, come into play. In the past there has sometimes been overemphasis on testing hypotheses.

One-factor analysis of variance

A simple experiment will now be described as an example of ANOVA. Suppose that the publisher of a junior-high-school textbook is considering styles of printing type for a new edition; there are three styles to investigate, and the same chapter of the book has been prepared in each of the three styles for the experiment. Junior-high-school pupils are to be chosen at random from an appropriate large population of such pupils, randomly assigned to read the chapter in one of the three styles, and then given a test that results in a reading-comprehension score for each pupil.

Suppose that the experiment is set up so that P_1, P_2, and P_3 pupils (where $P_1 = P_2 = P_3 = P$) read the chapter in styles 1, 2, and 3, respectively, and that X_{ps} denotes the comprehension score of the pth pupil reading style s. (Here $s = 1, 2, 3$; in general, $s = 1, 2, \cdots, S$.) There is a hypothetical mean, or expected, value of X_{ps}, μ_s, but X_{ps} differs from μ_s because, first, the pupils are chosen randomly from a population of pupils with different inherent means and, second, a given pupil, on hypothetical repetitions of the experiment, would not always obtain the same score. This is expressed by writing

$$X_{ps} = \mu_s + e_{ps}. \qquad (1)$$

Then the assumptions are made that the e_{ps} are all independent, that they are all normally distributed, and that they have a common (usually unknown) variance, σ^2. By definition, the expectation of e_{ps} is zero.

Because differences among the pupils reading a particular style of type are thrown into the random "error" terms (e_{ps}), μ_s, the expectation of X_{ps}, does not depend on p. It is convenient to rewrite (1) as

$$X_{ps} = \mu + (\mu_s - \mu) + e_{ps},$$

where $\mu = (\sum \mu_s)/S$, the average of the μ_s. For simplicity, set $\alpha_s = \mu_s - \mu$ (so that $\alpha_1 + \alpha_2 + \cdots + \alpha_S = 0$) and write the structural equation finally in the conventional form

$$(2) \qquad X_{ps} = \mu + \alpha_s + e_{ps}.$$

Here α_s is the differential effect on comprehension scores of style s for the relevant population of pupils. The unknowns are μ, the α_s, and σ^2.

Note that this structure falls under the linear regression hypothesis with coefficients 0 or 1. For example, if $E(X_{ps})$ represents the expected value of X_{ps},

$$E(X_{p1}) = 1 \cdot \mu + 1 \cdot \alpha_1 + 0 \cdot \alpha_2 + 0 \cdot \alpha_3 + \cdots + 0 \cdot \alpha_S,$$
$$E(X_{p2}) = 1 \cdot \mu + 0 \cdot \alpha_1 + 1 \cdot \alpha_2 + 0 \cdot \alpha_3 + \cdots + 0 \cdot \alpha_S.$$

Consider how this illustrative experiment might be conducted. After defining the population to which he wishes to generalize his findings, the experimenter would use a table of random numbers to choose pupils to read the chapter printed in the different styles. (Actually, he would probably have to sample intact school classes rather than individual pupils, so the observations analyzed might be class means instead of individual scores, but this does not change the analysis in principle.) After the three groups have read the same chapter under conditions that differ only in style of type, a single test covering comprehension of the material in the chapter would be administered to all pupils.

The experimenter's attention would be focused on differences between average scores of the three style groups (that is, $\bar{X}_{\cdot 1}$ versus $\bar{X}_{\cdot 2}$, $\bar{X}_{\cdot 2}$ versus $\bar{X}_{\cdot 3}$, and $\bar{X}_{\cdot 1}$ versus $\bar{X}_{\cdot 3}$) relative to the variability of the test scores within these groups. He estimates the μ_s via the $\bar{X}_{\cdot s}$, and he attempts to determine which of the three averages, if any, differ with statistical significance from the others. Eventually he hopes to help the publisher decide which style of type to use for his new edition.

ANOVA of random numbers—an example. An imaginary experiment of the kind outlined above will be analyzed here to illustrate how ANOVA is applied. Suppose that the three P_s are each 20, that in fact the μ_s are all exactly equal to 0, and that $\sigma^2 = 1$ (setting $\mu_s = 0$ is just a convenience corresponding to a conventional origin for the comprehension-score scale).

Sixty random normal deviates, with mean 0 and variance 1, were chosen by use of an appropriate table (RAND Corporation 1955). They are listed in Table 1, where the second column from the left should be disregarded for the moment—it will be used later, in a modified example. From the "data" of Table 1 the usual estimates of the μ_s are just the column averages, $\bar{X}_{\cdot 1} = -0.09$, $\bar{X}_{\cdot 2} = 0.10$, and $\bar{X}_{\cdot 3} = 0.08$. The estimate of μ is the over-all mean, $\bar{X}_{\cdot\cdot} = 0.03$, and the estimates of the α_s are $-0.09 - 0.03 = -0.12$, $0.10 - 0.03 = 0.07$, and $0.08 - 0.03 = 0.05$. Note that these add to zero, as required. In ANOVA, for this case, two quantities are compared. The first is the dispersion of the three μ_s estimates—that is, the sum of the $(\bar{X}_{\cdot s} - \bar{X}_{\cdot\cdot})^2$, conveniently multiplied by 20, the common sample size. This is called the between-styles dispersion or sum of squares. Here it is 0.4466. (These calculations, as well as those below, are made with the raw data of Table 1, not with the rounded means appearing there.) The second quantity is the within-sample dispersion, the sum of the three quantities $\sum_p (X_{ps} - \bar{X}_{\cdot s})^2$.

Table 1 — Data for hypothetical experiment; 60 random normal deviates

	X_{p1}	$X_{p1} + 1$*	X_{p2}	X_{p3}
	0.477	1.477	−0.987	1.158
	−0.017	0.983	2.313	0.879
	0.508	1.508	0.016	0.068
	−0.512	0.488	0.483	1.116
	−0.188	0.812	0.157	0.272
	−1.073	−0.073	1.107	−0.396
	−0.412	0.588	−0.023	−0.983
	1.201	2.201	0.898	−0.267
	−0.676	0.324	−1.404	0.327
	−1.012	−0.012	−0.080	0.929
	0.997	1.997	−1.258	−0.603
	−0.127	0.873	−0.017	0.493
	1.178	2.178	1.607	−1.243
	−1.507	−0.507	0.005	−0.145
	1.010	2.010	0.163	1.334
	−0.528	0.472	−0.771	−0.906
	−0.139	0.861	0.485	−1.633
	0.621	1.621	0.147	0.424
	−2.078	−1.078	−1.764	−0.433
	0.485	1.485	0.986	1.245
Mean	−0.09	0.91	0.10	0.08
Variance	0.83	0.83	1.03	0.78

* This column was obtained by adding 1 to each deviate of the first column.

This is called the within-style dispersion or sum of squares. Here it is 50.1253.

This comparison corresponds to the decomposition

$$X_{ps} - \bar{X}.. = (\bar{X}._s - \bar{X}..) + (X_{ps} - \bar{X}._s)$$

and to the sum-of-squares identity

$$\begin{aligned}\sum_p \sum_s (X_{ps} - \bar{X}..)^2 &= \sum_p \sum_s (\bar{X}._s - \bar{X}..)^2 + \sum_p \sum_s (X_{ps} - \bar{X}._s)^2 \\ &= 20 \sum_s (\bar{X}._s - \bar{X}..)^2 + \sum_p \sum_s (X_{ps} - \bar{X}._s)^2,\end{aligned}$$

which shows how the factor of 20 arises. Such identities in sums of squares are basic in most elementary expositions of ANOVA.

The fundamental notion is that the within-style dispersion, divided by its so-called *degrees of freedom* (here, degrees of freedom for error), unbiasedly estimates σ^2. Here the degrees of freedom for error are 57 (equals 60 [for the total number of observations] minus 3 [for the number of μ_s estimated]). On the other hand, the between-styles dispersion, divided by its degrees of freedom (here 2), estimates σ^2 unbiasedly if and only if the μ_s are equal; otherwise the estimate will tend to be larger than σ^2. Furthermore, the between-styles and within-style dispersions are statistically independent. Hence, it is natural to look at the ratio of the two dispersions, each divided by its degrees of freedom. The result is the F-statistic, here

$$\frac{0.4466/2}{50.1253/57} = 0.25.$$

In repeated trials with the null hypothesis (that there are no differences between the μ_s) true, the F-statistic follows an F-distribution with (in this case) 2 and 57 degrees of freedom [*see* DISTRIBUTIONS, STATISTICAL, *article on* SPECIAL CONTINUOUS DISTRIBUTIONS]. Level of significance is denoted by "α" (which should not be confused with the totally unrelated "α_s," denoting style effect; the notational similarity stems from the juxtaposition of two terminological traditions and the finite number of Greek letters). The F-test at level of significance α of the null hypothesis that the styles are equivalent rejects that hypothesis when the F-statistic is too large, greater than its 100α percentage point, here $F_{\alpha;2,57}$. If $\alpha = 0.05$, which is a conventional level, then $F_{.05;2,57} = 3.16$, so 0.25 is much smaller than the cutoff point, and the null hypothesis is, of course, not rejected. This is consonant with the fact that the null hypothesis *is* true in the imaginary experiment under discussion.

Table 2 summarizes the above discussion in both algebraic and numerical form. The algebraic form is for S styles with P_s students at the sth style.

To reiterate, in an analysis of variance each kind of effect (treatment, factor, and others to be discussed later) is represented by two basic numbers. The first is the so-called *sum of squares* (*SS*), corresponding to the effect; it is random, depending upon the particular sample, and has two fundamental properties: (*a*) If the effect in question is wholly *absent*, its sum of squares behaves probabilistically like a sum of squared independent normal deviates with zero means. (*b*) If the effect in question is *present*, its sum of squares tends to

Table 2 — Analysis-of-variance table for one-factor experiment

(a) ANOVA of 60 random normal deviates

Source of variation	*df*	*SS*	*MS*	*F*	*Tabled* $F_{.05;2,57}$
Between styles	$3 - 1 = 2$	0.4466	0.2233	0.25	3.16
Within styles	$60 - 3 = 57$	50.1253	0.8794*		
Total	$60 - 1 = 59$	50.5719			

* Actually, σ^2 here is known to be 1.

(b) ANOVA of general one-factor experiment with S treatments

Source of variation	*df*	*MS*	*EMS*
Between treatments	$S - 1$	$\sum_{s=1}^{S} P_s(\bar{X}._s - \bar{X}..)^2/(S-1)$	$\sigma^2 + \sum_{s=1}^{S} P_s\alpha_s^2/(S-1)$
Within treatments	$P_+ - S$*	$\sum_{s=1}^{S}\sum_{p=1}^{P_s}(X_{ps} - \bar{X}._s)^2/(P_+ - S)$*	σ^2
Total	$P_+ - 1$*	$\sum_{s=1}^{S}\sum_{p=1}^{P_s}(X_{ps} - \bar{X}..)^2/(P_+ - 1)$*	

* Here P_+ is used for $\sum_{s=1}^{S} P_s$ for convenience.

be relatively large; in fact, it behaves probabilistically like a sum of squared independent normal deviates with *not* all means zero.

The second number is the so-called *degrees of freedom* (*df*). This quantity is not random but depends only on the structure of the experimental design. The *df* is the number of independent normal deviates in the description of sums of squares just given.

A third (derived) number is the so-called *mean square* (*MS*), which is computed by dividing the sum of squares by the degrees of freedom. When an effect is wholly absent, its mean square is an unbiased estimator of underlying variance, σ^2. When an effect is present, its mean square has an expectation greater than σ^2.

In the example considered here, each observation is regarded as the sum of (*a*) a grand mean, (*b*) a printing-style effect, and (*c*) error. It is conventional in analysis-of-variance tables not to have a line corresponding to the grand mean and to work with sample residuals centered on it; that convention is followed here. Printing-style effect and error differ in that the latter is assumed to be wholly random, whereas the former is not random but may be zero. The mean square for error estimates underlying variance unbiasedly and is a yardstick for judging other mean squares.

In the standard simple designs to which ANOVA is applied, it is customary to define effects so that the several sums of squares are statistically independent, from which additivity both of sums of squares and of degrees of freedom follows [*see* PROBABILITY, *article on* FORMAL PROBABILITY]. In the example, $SS_{\text{between}} + SS_{\text{within}} = SS_{\text{total}}$, and $df_b + df_w = df_{\text{total}}$. (Here, and often below, the subscripts "*b*" and "*w*" are used to stand for "between" and "within," respectively.) This additivity is computationally useful, either to save arithmetic or to verify it.

Analysis-of-variance tables, which, like Table 2, are convenient and compact summaries of both the relevant formulas and the computed numbers, usually also show *expected mean squares* (*EMS*), the average value of the mean squares over a (conceptually) infinite number of experiments. In fixed-effects models (such as the model of the example) these are always of the form σ^2 (the underlying variance) plus an additional term that is zero when the relevant effect is absent and positive when it is present. The additional term is a convenient measure of the magnitude of the effect.

Expected mean squares, such as those given by the two formulas in Table 2, provide a necessary condition for the *F*-statistic to have an *F*-distribution when the null hypothesis is true. (Other conditions, such as independence, must also be met.) Note that if the population mean of the *s*th treatment, μ_s, is the same for all treatments (that is, if $\alpha_s = 0$ for all *s*) then the expected value of MS_b will be σ^2, the same as the expected value of MS_w. If the null hypothesis is true, the average value of the *F* from a huge number of identical experiments employing fresh, randomly sampled experimental units will be $(P_+ - S)/(P_+ - S - 2)$, which is very nearly 1 when, as is usually the case, the total number of experimental units, P_+, is large compared with *S*. Expected mean squares become particularly important in analyses based on models of a nature somewhat different from the one illustrated in Tables 1 and 2, because in those cases it is not always easy to determine which mean square should be used as the denominator of *F* (see the discussion of some of these other models, below).

The simplest *t*-tests. It is worth digressing to show how the familiar one-sample and two-sample *t*-tests (or Student tests) fall under the analysis-of-variance rubric, at least for the symmetrical two-tail versions of these tests.

Single-sample t-test. In the single-sample *t*-test context, one considers a random sample, X_1, X_2, $\cdots$, X_P, of independent normal observations with the same unknown mean, μ, and the same unknown variance, σ^2. Another way of expressing this is to write

$$X_p = \mu + e_p, \qquad p = 1, \cdots, P,$$

where the e_p are independent normal random variables, with mean 0 and common variance σ^2. The usual estimator of μ is $\bar{X}.$, the average of the X_p, and this suggests the decomposition into average and deviation from average,

$$X_p = \bar{X}. + (X_p - \bar{X}.),$$

from which one obtains the sum-of-squares identity

$$\begin{aligned}\textstyle\sum X_p^2 &= \textstyle\sum (X_p - \bar{X}.)^2 + \sum \bar{X}.^2 + 2\bar{X}.\sum (X_p - \bar{X}.)\\ &= \textstyle\sum (X_p - \bar{X}.)^2 + P\bar{X}.^2\end{aligned}$$

(since $\sum (X_p - \bar{X}.) = 0$), a familiar algebraic relationship. Since the usual unbiased estimator of σ^2 is $s^2 = \sum (X_p - \bar{X}.)^2/(P-1)$, the sum-of-squares identity may be written

$$\textstyle\sum X_p^2 = (P-1)s^2 + P\bar{X}.^2 .$$

Ordinarily the analysis-of-variance table is not written out for this simple case; it is, however, the one shown in Table 3. In Table 3 the total row is the actual total including all observations; it is of the essence that the row for mean is separated out.

Table 3

Effect	*df*	*SS*	*EMS*
Mean	1	$P\bar{X}_.^2$	$\sigma^2 + P\mu^2$
Error	$P-1$	$\sum(X_p - \bar{X}.)^2$	σ^2
Total	P	$\sum X_p^2$	

The F-statistic for testing that $\mu = 0$ is the ratio of the mean squares for mean and error,

$$\frac{P\bar{X}_.^2}{\sum(X_p - \bar{X}.)^2/(P-1)} = \frac{P\bar{X}_.^2}{s^2},$$

which, under the null hypothesis, has an F-distribution with 1 and $P-1$ degrees of freedom. Notice that the above F-statistic is the square of

$$\frac{\bar{X}.}{s/\sqrt{P}},$$

which is the ordinary t-statistic (or Student statistic) for testing $\mu = 0$. If a symmetrical two-tail test is wanted, it is immaterial whether one deals with the t-statistic or its square. On the other hand, for a one-tail test the t-statistic would be referred to the t-distribution with $P-1$ degrees of freedom [*see* DISTRIBUTIONS, STATISTICAL, *article on* SPECIAL CONTINUOUS DISTRIBUTIONS].

It is important to note that a confidence interval for μ may readily be established from the above discussion [*see* ESTIMATION, *article on* CONFIDENCE INTERVALS AND REGIONS]. The symmetrical form is

$$\bar{X}. - \sqrt{F_{\alpha;1,P-1}}\, s/\sqrt{P} \leqslant \mu \leqslant \bar{X}. + \sqrt{F_{\alpha;1,P-1}}\, s/\sqrt{P}.$$

Alternatively, $F_{\alpha;1,P-1}$ can be replaced by the upper $100(\alpha/2)$ per cent point for the t-distribution with $P-1$ degrees of freedom, $t_{\alpha/2;P-1}$.

Suppose, for example, that from a normally distributed population there has been drawn a random sample of 25 observations for which the sample mean, $\bar{x}.$, is 34.213 and the sample variance, s^2, is 49.000. What is the population mean, μ? The usual point estimate from this sample is 34.213. How different from μ is this value likely to be? For $\alpha = .05$, a 95 per cent confidence interval is constructed by looking up $t_{.025;24} = 2.064$ in a table (for instance, McNemar [1949] 1962, p. 430) and substituting in the formula

$$\text{Confidence}\left[34.213 - 2.064\left(\frac{\sqrt{49.000}}{\sqrt{25}}\right) \leqslant \mu \leqslant 34.213 + 2.064\left(\frac{\sqrt{49.000}}{\sqrt{25}}\right)\right] = 0.95.$$

Thus,

$$\text{Confidence}\,[31.32 \leqslant \mu \leqslant 37.10] = 0.95.$$

This result means that if an infinite number of samples, each of size $P = 25$, were drawn randomly from a normally distributed population and a confidence interval for each sample were set up in the above way, only 5 per cent of the intervals would fail to cover the mean of the population (which is a certain fixed value).

Similarly, from this one sample the unbiased point estimate of σ^2 is the value of s^2, 49.000. Brownlee ([1960] 1965, page 282) shows how to find confidence intervals for σ^2 [*see also* VARIANCES, STATISTICAL STUDY OF].

Is it "reasonable" to suppose that the mean of the population from which this sample was randomly chosen is as large as, say, 40? No, because that number does not lie within even the 99 per cent confidence interval. Therefore it would be unreasonable to conclude that the sample was drawn from a population with a mean as great as 40. The relevant test of statistical significance is

$$t = \frac{34.213 - 40.000}{7/5} = \frac{-5.787(5)}{7} = -4.134,$$

the absolute magnitude of which lies beyond the 0.9995 percentile point (3.745) in the tabled t-distribution for 24 degrees of freedom. Therefore, the difference is statistically significant beyond the $0.0005 + 0.0005 = 0.001$ level. The null hypothesis being tested was H_0: $\mu = 40$, against the alternative hypothesis H_a: $\mu \neq 40$. Just as the confidence interval indicated that it is unreasonable to suppose the mean to be equal to 40, this test also shows that 40 will lie outside the 99 per cent confidence interval; however, of the two procedures, the confidence interval gives more information than the significance test.

Two-sample t-test. In the two sample t-test context, there are two random samples from normal distributions assumed to have the same variance, σ^2, and to have means μ_1 and μ_2. Call the observations in the first sample $X_{11}, \cdots, X_{P_1 1}$ and the observations in the second sample $X_{12}, \cdots, X_{P_2 2}$. The most usual null hypothesis is $\mu_1 = \mu_2$, and for that the t-statistic is

$$\frac{\bar{X}._1 - \bar{X}._2}{s\sqrt{(1/P_1) + (1/P_2)}},$$

where the P's are the sample sizes, the $\bar{X}$'s are the sample means, and s^2 is the estimate of σ^2 based on the pooled within-sample sum of squares,

$$s^2 = \frac{1}{P_1 + P_2 - 2}\left\{\sum_p (X_{p1} - \bar{X}._1)^2 + \sum_p (X_{p2} - \bar{X}._2)^2\right\}.$$

Here $P_1 + P_2 - 2$ is the number of degrees of free-

Table 4

Effect	df	SS	EMS*
Style	1	$\sum P_s(\bar{X}._s - \bar{X}..)^2 = \dfrac{(\bar{X}._1 - \bar{X}._2)^2}{(1/P_1)+(1/P_2)}$	$\sigma^2 + \dfrac{(\mu_1 - \mu_2)^2}{(1/P_1)+(1/P_2)}$
Error	$P_1 + P_2 - 2$	$(P_1 + P_2 - 2)s^2$	σ^2
Total	$P_1 + P_2 - 1$	$\sum\sum (X_{ps} - \bar{X}..)^2$	

* Note that the expected mean square for style is σ^2 plus what is obtained by formal substitution for the random variables ($\bar{X}._1$, $\bar{X}._2$) in the sum of squares of their respective expectations (divided by *df*, which here is 1). This relationship is a perfectly general one in the analysis-of-variance model now under discussion, but it must be changed for other models that will be mentioned later.

dom for error, the total number of observations less the number of estimated means ($\bar{X}._1$ and $\bar{X}._2$ estimate μ_1 and μ_2, respectively). Under the null hypothesis, the *t*-statistic has the *t*-distribution with $P_1 + P_2 - 2$ degrees of freedom.

The basic decomposition is

$$X_{ps} - \bar{X}.. = (\bar{X}._s - \bar{X}..) + (X_{ps} - \bar{X}._s),$$

leading to the sum-of-squares decomposition

$$\sum_p\sum_s(X_{ps} - \bar{X}..)^2 = \sum_s P_s(\bar{X}._s - \bar{X}..)^2 + \sum_p\sum_s(X_{ps} - \bar{X}._s)^2.$$

Since *s* has only the values 1 and 2,

$$\bar{X}.. = \frac{P_1}{P_1 + P_2}\bar{X}._1 + \frac{P_2}{P_1 + P_2}\bar{X}._2,$$

and therefore

$$\sum_s P_s(\bar{X}._s - \bar{X}..)^2 = \frac{P_1P_2}{P_1 + P_2}(\bar{X}._1 - \bar{X}._2)^2$$

$$= \frac{(\bar{X}._1 - \bar{X}._2)^2}{(P_1 + P_2)/P_1P_2}$$

$$= \frac{(\bar{X}._1 - \bar{X}._2)^2}{(1/P_1) + (1/P_2)}.$$

The analysis-of-variance table may be written as in Table 4. The *F*-statistic for the null hypothesis that $\mu_1 = \mu_2$ is

$$\frac{(\bar{X}._1 - \bar{X}._2)^2/(P_1^{-1} + P_2^{-1})}{s^2} = \frac{(\bar{X}._1 - \bar{X}._2)^2}{s^2[(1/P_1) + (1/P_2)]},$$

and this is exactly the square of the *t*-statistic for the two-sample problem.

Note that the two-sample problem as it is analyzed here is only a special case (with $S = 2$) of the *S*-sample problem presented earlier.

The numerical example continued. Returning to the numerical example of Table 1, add 1 to every number in the leftmost column to obtain the second column and consider the numbers in the second column as the observations for style 1. Now $\mu_1 = 1$ and $\mu_2 = \mu_3 = 0$. What happens to the analysis of variance and the *F*-test? Table 5 shows the result; the *F*-statistic is 5.41, which is of high statistical significance since $F_{.01;2,57} = 5.07$. Thus, one would correctly reject the null hypothesis of equality among the three μ_s.

The actual value of μ is $\frac{1}{3} \cong 0.33$, and that of α_1 is $\frac{2}{3} \cong 0.67$. The estimate of μ is 0.36, and that of α_1 is 0.55.

With three styles, one can consider many contrasts—for example, style 1 versus style 2, style 1 versus style 3, style 2 versus style 3, $\frac{1}{2}$(style 1 + style 2) versus style 3. There are special methods for dealing with several contrasts simultaneously [*see* LINEAR HYPOTHESES, *article on* MULTIPLE COMPARISONS].

ANOVA with more than one factor

In the illustrative example being considered here, suppose that the publisher had been interested not only in style of type but also in a second factor, such as the tint of the printing ink (t). If he had three styles and four tints, a complete "crossed" *factorial design* would require $3 \times 4 = 12$ experimental conditions ($s_1t_1, s_1t_2, \cdots, s_3t_4$). From $12P$ experimental units he would assign P units at random to each of the 12 conditions, conduct his experiment, and obtain outcome measures to analyze. The total variation between the $12P$ outcome measures can be partitioned into four sources rather than into the two found with one factor. The sources of variation are the following: be-

Table 5 — One-factor ANOVA of 60 transformed random normal deviates

Source of variation	df	SS	MS	EMS	F
Between styles	2	8.9246	4.4623	$\sigma^2 + 20\sum_{s=1}^{3}\alpha_s^2/2$	5.41
Within styles	57	50.1253	0.8794	σ^2	
Total	59	59.0499			

Table 6 — ANOVA of a complete, crossed-classification, two-factor factorial design with P experimental units for each factor-level combination

Source of variation	df	SS	EMS
Between styles	$S-1$	$PT\sum_{s=1}^{S}(\bar{X}_{\cdot s \cdot} - \bar{X}_{\cdots})^2$	$\sigma^2 + PT\sum_s \alpha_s^2/(S-1)$
Between tints	$T-1$	$PS\sum_{t=1}^{T}(\bar{X}_{\cdot\cdot t} - \bar{X}_{\cdots})^2$	$\sigma^2 + PS\sum_t \beta_t^2/(T-1)$
Styles × tints (interaction)	$(S-1)(T-1)$	$P\sum_{s=1}^{S}\sum_{t=1}^{T}(\bar{X}_{\cdot st} - \bar{X}_{\cdot s \cdot} - \bar{X}_{\cdot\cdot t} + \bar{X}_{\cdots})^2$	$\sigma^2 + P\sum_s\sum_t \gamma_{st}^2/(S-1)(T-1)$
Within style–tint combinations	$ST(P-1)$	$\sum_{p=1}^{P}\sum_{s=1}^{S}\sum_{t=1}^{T}(X_{pst} - \bar{X}_{\cdot st})^2$	σ^2
Total	$PST-1$	$\sum_{p=1}^{P}\sum_{s=1}^{S}\sum_{t=1}^{T}(X_{pst} - \bar{X}_{\cdots})^2$	

tween styles, between tints, *interaction* of styles with tints, and within style–tint combinations (error).

The usual model for the two-factor crossed design is

$$X_{pst} = \mu + \alpha_s + \beta_t + \gamma_{st} + e_{pst},$$

where $\sum_s \alpha_s = \sum_t \beta_t = \sum_s \gamma_{st} = \sum_t \gamma_{st} = 0$, and the e_{pst} are independent normally distributed random variables with mean 0 and equal variance σ^2 for each st combination. The analysis-of-variance procedure for this design appears in Table 6. The α_s and β_t represent *main effects* of the styles and tints; the γ_{st} denote (two-factor) *interactions*.

Interaction. The two-factor design introduces interaction, a concept not relevant in one-factor experiments. It might be found, for example, that, although in general s_1 is an ineffective style and t_3 is an ineffective tint, the particular combination s_1t_3 produces rather good results. It is then said that style interacts with tint to produce nonadditive effects; if the effects were additive, an ineffective style combined with an ineffective tint would produce an ineffective combination.

Interaction is zero if $E(X_{pst}) = \mu + \alpha_s + \beta_t$ for every st, because under this condition the population mean of the stth combination is the population grand mean plus the sum of the effects of the sth style and the tth tint. Then the interaction effect, γ_{st}, is zero for every combination. Table 7 contains hypothetical data showing population means, $\bar{\mu}_{st}$, for zero interaction (Lubin 1961 discusses types of interaction). Note that for every cell of Table 7, $\bar{\mu}_{st} - (\bar{\mu}_{s\cdot} - \mu) - (\bar{\mu}_{\cdot t} - \mu) = \mu = 3$. (Here $\bar{\mu}_{\cdot\cdot}$ is written as μ for simplicity.) For example, for tint 1 and style 1,

$$3-(5-3)-(1-3)=3.$$

One tests for interaction by computing $F = MS_{\text{styles}\times\text{tints}}/MS_{\text{within style–tint}}$, comparing this F with the F's tabled at various significance levels for $(S-1)(T-1)$ and $ST(P-1)$ degrees of freedom.

If there were but one subject reading with each style–tint combination (that is, if there were no replication), further assumptions would have to be made to permit testing of hypotheses about main effects. In particular, it is commonly then assumed that the style × tint interaction is zero, so that the expected mean square for interaction in Table 6 reduces to the underlying variance, and the $MS_{\text{styles}\times\text{tints}}$ may be used in the denominator of the F's for testing main effects. No test of the assumption of additivity is possible through $MS_{\text{within style–tint}}$, because this quantity cannot be calculated. However, Tukey (1949; see also Winer 1962, pp. 216–220) has provided a one-degree-of-freedom test for interaction, or nonadditivity, of a special kind that can be used for testing the hypothesis of no interaction for these unreplicated experiments of the fixed-effects kind. (See Scheffé 1959, pp. 129–134.)

The factorial design may be extended to three or more factors. With three factors there are four sums of squares for interactions: one for the three-factor interaction (sometimes called a second-

Table 7 — Zero interaction of two factors (hypothetical population means $\bar{\mu}_{st}$)

Style \ Tint	1	2	3	4	Row means ($\bar{\mu}_{s\cdot}$)
1	3	4	5	8	5
2	0	1	2	5	2
3	0	1	2	5	2
Column means ($\bar{\mu}_{\cdot t}$)	1	2	3	6	$3 = \mu$

order interaction, because a one-factor "interaction" is a main effect) and one each for the three two-factor (that is, first-order) interactions. If the three factors are A, B, and C, their interactions might be represented as $A \times B \times C$, $A \times B$, $A \times C$, and $B \times C$. For example, a style of type that for the experiment as a whole yields excellent comprehension may, when combined with a generally effective size of type and a tint of paper that has over-all facilitative effect, yield rather poor results. One three-factor factorial experiment permits testing of the hypothesis that there is a no second-order interaction and permits the magnitude of such interaction to be estimated, whereas three one-factor experiments or a two-factor experiment and a one-factor experiment do not. Usually, three-factor nonadditivity is difficult to explain substantively.

A large number of more complex designs, most of them more or less incomplete in some respect as compared with factorial designs of the kind discussed above, have been proposed. [*See* EXPERIMENTAL DESIGN; *see also* Winer 1962; Fisher 1935.]

The analysis of covariance

Suppose that the publisher in the earlier, style-of-type example had known reading-test scores for his 60 pupils prior to the experiment. He could have used these antecedent scores in the analysis of the comprehension scores to reduce the magnitude of the mean square within styles, which, as the estimate of underlying variance, is the denominator of the computed F. At the same time he would adjust the subsequent style means to account for initial differences between reading-test-score means in the three groups. One way of carrying out this more refined analysis would be to perform an analysis of variance of the differences between final comprehension scores and initial reading scores—say, $X_{ps} - Y_{ps}$. A better prediction of the outcome measure, X_{ps}, might be secured by computing $\alpha + \beta Y_{ps}$, where α and β are constants to be estimated.

By a statistical procedure called the *analysis of covariance* one or more antecedent variables may be used to reduce the magnitude of the sum of squares within styles and also to adjust the observed style means for differences between groups in average initial reading scores. If $\beta \neq 0$, then the adjusted sum of squares within treatments (which provides the denominator of the F-ratio) will be less than the unadjusted SS_w of Table 2, thereby tending to increase the magnitude of F. For each independent antecedent variable one uses, one degree of freedom is lost for SS_w and none for SS_b; the loss of degrees of freedom for SS_w will usually be more than compensated for by the decrease in its magnitude.

A principal statistical condition needed for the usual analysis of covariance is that the regression of outcome scores on antecedent scores is the same for every style, because one computes a single within-style regression coefficient to use in adjusting the within-style sum of squares. Homogeneity of regression can be tested statistically; see Winer (1962, chapter 11). Some procedures to adopt in the case of heterogeneity of regression are given in Brownlee (1960).

The regression model chosen must be appropriate for the data if the use of one or more antecedent variables is to reduce MS_w appreciably. Usually the regression of outcome measures on antecedent measures is assumed to be linear.

The analysis of covariance can be extended to more than one antecedent variable and to more complex designs. (For further details see Cochran 1957; Smith 1957; Winer 1962; McNemar 1949.)

Models—fixed, finite, random, and mixed

In the example, the publisher's "target population" of styles of print consisted of just those 3 styles that he tried out, so he exhausted the population of styles of interest to him. Suppose that, instead, he had been considering 39 different styles and had drawn *at random* from these 39 the 3 styles he used in the experiment. His intention is to determine from the experiment based on these 3 styles whether it would make any difference which one of the 39 styles he used for the textbook (of course, in practice a larger sample of styles would be drawn). If the styles did seem to differ in effectiveness, he would estimate from his experimental data involving only 3 styles the variance of the 39 population means of the styles. Then he might perform further experiments to find the most effective styles.

Finite-effects models. Thus far in this article the model assumed has been the *fixed-effects* model, in which one uses in the experiment itself all the styles of type to which one wishes to generalize. The 3-out-of-39 experiment mentioned above illustrates a *finite-effects* model, with only a small percentage (8 per cent, in the example given) of the styles drawn at random for the experiment but where one has the intention of testing the null hypothesis

$$H_0: \mu_1 = \mu_2 = \cdots = \mu_{39}$$

against all alternative hypotheses and estimating the "variance," $\sigma^2_{\text{style}} = \sum_{s=1}^{39} (\mu_s - \mu)^2/(39-1)$ from

$MS_b = 20 \sum_{s=1}^{3}(\bar{X}_{.s} - \bar{X}_{..})^2/(3-1)$ and $MS_w = \sum_{s=1}^{3}\sum_{p=1}^{20}(X_{ps} - \bar{X}_{.s})^2/[3(20-1)]$.

Random-effects models. If the number of "levels" of the factor is very large, so that the number of levels drawn randomly for the experiment is a negligible percentage of the total number, then one has a *random-effects* model, sometimes called a components-of-variance model or Model II. This model would apply if, for example, one drew 20 raters at random from an actual or hypothetical population of 100,000 raters and used those 20 to rate each of 25 subjects who had been chosen at random from a population of half a million. (Strictly speaking, the number of raters and the number of subjects in the respective populations would have to be infinite to produce the random-effects model, but for practical purposes 100,000/20 and 500,000/25 are sufficiently large.) If every rater rated every subject on one trait (say, gregariousness) there would be $20 \times 25 = 500$ ratings, one for each experimental combination—that is, one for each rater–subject combination.

This, then, would be a two-factor design without replication, that is, with just one rating per rater–subject combination. (Even if the experimenter had used available raters and subjects rather than drawing them randomly from any populations, he would probably want to generalize to other raters and subjects "like" them; see Cornfield & Tukey 1956, p. 913.)

The usual model for an experiment thus conceptualized is

$$X_{rs} = \mu + a_r + b_s + e_{rs},$$

where μ is a grand mean, the a's are the (random) rater effects, the b's are (random) subject effects, and the e's combine interaction and inherent measurement error. The $20 + 25 + (20 \times 25)$ random variables are supposed to be independent and assumed to have variances as follows:

$$\text{var } a_r = \sigma_r^2, \qquad \text{var } b_s = \sigma_s^2, \qquad \text{var } e_{rs} = \sigma_e^2.$$

For F-testing purposes, a, b, and e are supposed to be normally distributed.

The analysis-of-variance table in such a case is similar to those presented earlier, except that the expected mean square column is changed to the one shown in Table 8.

Table 8

Effect	EMS
Rater	$\sigma_e^2 + 25\,\sigma_r^2$
Subject	$\sigma_e^2 + 20\,\sigma_s^2$
Error	σ_e^2

The F-statistic for testing the hypothesis that the main effect of subjects is absent (that $\sigma_s^2 = 0$) is MS_s/MS_{error}, where

$$MS_{\text{error}} = \frac{1}{19 \times 24}\sum\sum(X_{rs} - \bar{X}_{r.} - \bar{X}_{.s} + \bar{X}_{..})^2.$$

Under the null hypothesis that $\sigma_s^2 = 0$, the F-statistic has an F-distribution with 24 and 19×24 degrees of freedom. (A similar F-statistic is used for testing $\sigma_r^2 = 0$.) An unbiased estimator of σ_s^2 is

$$\frac{1}{20}(MS_s - MS_{\text{error}})$$

with a similar estimator for σ_r^2. A serious difficulty with these estimators is that they may take negative values; perhaps the best resolution of that difficulty is to enlarge the model. See Nelder (1954), and for another approach and a bibliography, see Thompson (1962).

Note that here it appears impossible to separate random interaction from inherent variability, both of which contribute to σ_e^2, the variance of the e's; in the random-effects model, however, this does not jeopardize significance tests for main effects.

In more complex Model II situations, the F-tests used are inherently different from their Model I analogues; in particular, sample components of variance are often most reasonably compared, not with the "bottom" estimator of σ^2, but with some other—usually an interaction—component of variance. (See Hays 1963, pp. 356–489; Brownlee [1960] 1965, pp. 309–396, 467–529.)

Mixed models. If all the levels of one factor are used in an experiment while a random sample of the levels of another factor is used, a *mixed model* results. Mixed models present special problems of analysis that have been discussed by Scheffé (1959, pp. 261–290) and by Mood and Graybill (1963).

Other topics in ANOVA

Robustness of ANOVA. Fixed-effects models are better understood than the other models and therefore, *where appropriate*, can be used with considerable confidence. Fixed-effects ANOVA seems "robust" for type I errors to departures from certain mathematical assumptions underlying the F-test, provided that the number of experimental units is the same for each experimental combination. Two of these assumptions are that the e's are normally distributed and that they have common variance σ^2 for every one of the experimental combinations. In particular, the common-variance assumption can be relaxed without greatly affecting the probability values for computed F's. If the number of experimental units does not vary from one factor-level combination to another, then it may be unnecessary to test for heterogeneity of variances preliminary to performing an ANOVA, because ANOVA is robust to such heterogeneity. (In fact, it may be unwise to make such a test, because the usual test for heterogeneity of variance is more sensitive to nonnormality than is ANOVA.) For further discussion of this point see Lindquist (1953, pp. 78–86), Winer (1962, pp. 239–241), Brownlee ([1960] 1965, chapter 9), and Glass (1966). Brownlee (1960) and others have provided the finite-model expected mean squares for the complete three-factor factorial design, from which one can readily determine expected mean squares for three-factor fixed, mixed, and random models.

Analysis-of-variance F's are unaffected by linear transformation of the observations—that is, by changes in the X_{ps} of the form $a + bX_{ps}$, where a and b are constants ($b \neq 0$). Multiplying every observation by b multiplies every mean square by b^2. Adding a to every observation does not change the mean squares. Thus, if observations are two-decimal numbers running from, say, -1.22 upward, one could, to simplify calculations, drop the decimal (multiply each number by 100) and then add 122 to each observation. The lowest observation would become $100(-1.22) + 122 = 0$. Each mean square would become $100^2 = 10{,}000$ times as large as for the decimal fractions. With the increasing availability of high-speed digital computers, coding of data is becoming less important than it was formerly.

A brief classification of factors. The ANOVA "factors" considered thus far are style of printing type, tint of ink, rater, and subject. Styles differ from each other qualitatively, as do raters and subjects. Tint of ink might vary more quantitatively than do styles, raters, and subjects—as would, for example, size of printing type or temperature in a classroom. Thus, one basis for classifying factors is whether or not their levels are ordered and, if they are, whether meaningful numbers can be associated with the factor levels.

Another basis for classification is whether the variable is manipulated by the experimenter. In order to conduct a "true" experiment, one must assign his experimental units in some (simple or restrictive) random fashion to the levels of at least one manipulated factor. ANOVA may be applied to other types of data, such as the scores of Englishmen versus Americans on a certain test, but this is an associational study, not a stimulus–response experiment. Obviously, nationality is not an independent variable in the same sense that printing type is. The direct "causal" inference possible from a well-conducted style-of-type experiment differs from the associational information obtained from the comparison of Englishmen's scores with those of Americans (see Stanley 1961; 1965; 1966; Campbell & Stanley 1963). Some variables, such as national origin, are impossible to manipulate in meaningful ways, whereas others, such as "enrolls for Latin versus does not enroll for Latin," can in principle be manipulated, even though they usually are not.

Experimenters use nonmanipulated, classification variables for two chief reasons. First, they may wish to use a factor explicitly in a design in order to isolate the sum of squares for the main effect of that factor so that it will not inflate the estimate of underlying variance—that is, so it will not make the denominator mean square of F unnecessarily large. For example, if the experimental units available for experimentation are children in grades seven, eight, and nine, and if IQ scores are available, it is wise in studying the three styles of type to use the three (ordered) grades as one fixed-effects factor and a number of ordered IQ levels—say, four—as another fixed-effects factor. If the experimenter suspects that girls and boys may react differently to the styles, he will probably use this two-level, unordered classification (girls versus boys) as the third factor. This would produce $3 \times 4 \times 2 \times 3 = 72$ experimental combinations, so with at least 2 children per combination he needs not less than 144 children.

Probably most children in the higher grades read better, regardless of style, than do most children in the lower grades, and children with high IQ's tend to read better than children with lower IQ's, so the main effects of grade and of IQ should be large. Therefore, the variation within grade–IQ–sex–style groups should be considerably less than within styles alone.

A second reason for using such stratifying or leveling variables is to study their interactions with the manipulated variable. Ninth graders might do relatively better with one style of type and seventh graders relatively better with another style, for example. If so, the experimenter might decide to recommend one style of type for ninth graders and another for seventh graders. With the above design one can isolate and examine one four-factor interaction, four three-factor interactions, six two-factor interactions, and four main effects, a total of $2^4 - 1 = 15$ sources of variation across conditions. In the fixed-effects model all of these are tested against the variation within the experimental combinations, pooled from all combinations. Testing 15 sources of variation instead of 1 will tend to cause more apparently significant *F*'s at a given tabled significance level than would be expected under the null hypothesis. For any one of the significance tests, given that the null hypothesis is true, one expects 5 spurious rejections of the true null hypothesis out of 100 tests; thus, if an analyst keeps making *F*-tests within an experiment, he has more than a .05 probability of securing at least one statistically significant *F*, even if no actual effects exist. There are systematic ways to guard against this (see, for example, Pearson & Hartley 1954, pp. 39–40). At least, one should be suspicious of higher-order interactions that seem to be significant at or near the .05 level. Many an experimenter utilizing a complex design has worked extremely hard trying to interpret a spuriously significant high-order interaction and in the process has introduced his fantasies into the journal literature.

Studies in which researchers do not manipulate any variables are common and important in the social sciences. These include opinion surveys, studies of variables related to injury in automobile accidents, and studies of the Hiroshima and Nagasaki survivors. ANOVA proves useful in many such investigations. [*See* Campbell & Stanley 1963; Lindzey 1954; *see also* EXPERIMENTAL DESIGN, *article on* QUASI-EXPERIMENTAL DESIGN.]

"Nesting" and repeated measurements. Many studies and experiments in the social sciences involve one or more factors whose levels do not "cross" the levels of certain other factors. Usually these occur in conjunction with repeated measurements taken on the same individuals. For example, if one classification is school and another is teacher within school, where each teacher teaches two classes within her school with different methods, then teachers are said to be "nested" within schools. Schools can interact with methods (a given method may work relatively better in one school than in another) and teachers can interact with methods *within schools* (a method that works relatively better for one teacher does not necessarily produce better results for another teacher in the same school), but schools cannot interact with teachers, because teachers do not "cross" schools—that is, the same teacher does not teach at more than one school.

This does not mean that a given teacher might not be more effective in another school but merely that the experiment provides no evidence on that point. One could, somewhat inconveniently, devise an experiment in which teachers did cross schools, teaching some classes in one school and some in another. But an experimenter could not, for example, have boys cross from delinquency to nondelinquency and vice versa, because delinquency–nondelinquency is a personal rather than an environmental characteristic. (For further discussion of nested designs see Brownlee [1960] 1965, chapters 13 and 15.)

If the order of repeated measurements on each individual is randomized, as when each person undergoes several treatments successively in random order, there is more likelihood that ANOVA will be appropriate than when the order cannot be randomized, as occurs, for instance, when the learning process is studied over a series of trials. Complications occur also if the successive treatments have differential residual effects; taking a difficult test first may discourage one person in his work on the easier test that follows but make another person try harder. These residual effects seem likely to be of less importance if enough time occurs between successive treatment levels for some of the immediate influence of the treatment to dissipate. Human beings cannot have their memories erased like calculating machines, however, so repeated-measurement designs, although they usually reduce certain error terms because intraindividual variability tends to be less than interindividual variability, should not be used indiscriminately when analogous designs without repeated measurements are experimentally and financially feasible. (For further discussion see Winer 1962; Hays 1963, pp. 455–456; Campbell & Stanley 1963.)

Missing observations. For two factors with levels $s = 1, 2, \cdots, S$ and $t = 1, 2, \cdots, T$ in the experiment, such that the number of experimental units for the *st*th experimental combination is n_{st}, one usually designs the experiment so that $n_{st} = n$, a constant for all *st*. A few missing observations at the end of the experiment do not rule out a slightly adjusted simple ANOVA, if they were not caused differentially by the treatments. If, for example, one treatment was to administer a severe shock on several occasions, and the other was to give ice cream each time, it would not be surprising to find that fewer shocked than fed experimental subjects come for the final session. The outcome measure might be arithmetical-reasoning score; but if only the more shock-resistant subjects take the final test, comparison of the two treatments may be biased. There would be even more difficulty with, say, a male–female by shocked–fed design, because shocking might drive away more women than men (or vice versa).

When attrition is not caused differentially by the factors one may, for one-factor ANOVA, perform the usual analysis. For two or more factors, adjustments in the analysis are required to compensate for the few missing observations. (See Winer 1962, pp. 281–283, for example, for appropriate techniques.)

The power of the *F*-test. There are two kinds of errors that one can make when testing a null hypothesis against alternative hypotheses: one can reject the null hypothesis when in fact it is true, or one can fail to reject the null hypothesis when in fact it is false. Rejecting a true null hypothesis is called an "error of the first kind," or a "type I error." Failing to reject an untrue null hypothesis is called an "error of the second kind" or "type II error." The probability of making an error of the first kind is called the *size* of the significance test and is usually signified by α. The probability of making an error of the second kind is usually signified by β. The quantity $1-\beta$ is called the *power* of the significance test.

If there is no limitation on the number of experimental units available one can fix both α and β at any desired levels prior to the experiment. To do this some prior estimate of σ^2 is required, and it is also necessary to state what nonnull difference among the factor-level means is considered large enough to be worth detecting. This latter requirement is quite troublesome in many social science experiments, because a good scale of value (such as dollars) is seldom available. For example, how much is a one-point difference between the mean of style 1 and style 2 on a reading-comprehension test worth educationally? Intelligence quotients and averages of college grades are quasi-utility scales, although one seldom thinks of them in just that way. How much is a real increase in IQ from 65 to 70 worth? How much more utility for the college does a grade-point average of 2.75 (where $C=2$ and $B=3$) have than a grade-point average of 2.50? (For further discussion of this topic see Chernoff & Moses 1959.)

In the hypothetical printing-styles example (Tables 1 and 5) it is known that $\sigma^2=1$ and that the population mean of style 1 is one point greater than the population means of styles 2 and 3, so with this information it is simple to enter Winer's Table B.11 (1962, p. 657) with, for example, $\alpha=.05$ and $\beta=.10$ and to find that for each of the three styles $P=20$ experimental units are needed.

In actual experiments, where σ^2 and the $\sum_{s=1}^{S} P_s \alpha_s^2$ of interest to the experimenter are usually not known, the situation is more difficult (see Brownlee [1960] 1965, pp. 97–111; McNemar [1949] 1962, pp. 63–69; Hays 1963; and especially Scheffé 1959, pp. 38–42, 62–65, 437–455).

Alternatives to analysis of variance

If one conducted an experiment to determine how well ten-year-old boys add two-digit numbers at five equally spaced atmospheric temperatures, he could use the techniques of regression analysis to determine the equation for the line that best fits the five means (in the sense of minimum squared discrepancies). This line might be of the simple form $\alpha+\beta T$ (that is, straight with slope β and intercept α) or it might be based on some other function of T. [*See* Winer 1962 *for further discussion of trend analysis; see also* LINEAR HYPOTHESES, *article on* REGRESSION.]

The symmetrical two-tail t-test is a special case of the F-test; $t_{df}^2 = F_{1,df}$. Likewise, the unit normal deviate (z), called "critical ratio" in old statistics textbooks when used for testing significance, is a special case of F: $z^2=F_{1,\infty}$. The F-distribution is closely related to the chi-square distribution. [*For further discussion of these relationships, see* DISTRIBUTIONS, STATISTICAL, *article on* SPECIAL CONTINUOUS DISTRIBUTIONS.]

For speed and computational ease, or when assumptions of ANOVA are violated so badly that results would seem dubious even if the data were transformed, there are other procedures available (see Winer 1962). Some of these procedures involve consecutive, untied ranks, whose means and variances are parameters dependent only on the number of ranks; an important example is the Kruskal–Wallis analysis of variance for ranks (Winer 1962, pp. 622–623). Other procedures employ the binomial expansion $(p+q)^n$ or the chi-square approximation to it for "sign tests." Still others involve dichotomizing the values for each treatment at the median and computing χ^2. Range tests may be used also. [*See* Winer 1962, p. 77; McNemar (1949) 1962, chapter 19. *Some of these procedures are discussed in* NONPARAMETRIC STATISTICS.]

When the normal assumption is reasonable, there are often available testing and other procedures that are competitive with the F-test. The latter has factotum utility, and it has optimal properties when the alternatives of interest are symmetrically arranged relative to the null hypothesis. But when the alternatives are asymmetrically arranged, or in other special circumstances, competitors to F procedures may be preferable. Particularly worthy of mention are Studentized range tests (see Scheffé 1959, pp. 82–83) and half-normal plotting (see Daniel 1959).

Special procedures are useful when the alternatives specify an ordering. For example, in the style-of-type example it might be known before the experiment that if there *is* any difference between the styles, style 1 is better than style 2, and style 2 better than style 3 (see Bartholomew 1961; Chacko 1963).

It is also important to mention here the desirability of examining residuals (observations less the estimates of their expectations) as a check on the model and as a source of suggestions toward useful modifications. [*See* STATISTICAL ANALYSIS, SPECIAL PROBLEMS OF, *article on* TRANSFORMATIONS OF DATA; *see also* Anscombe & Tukey 1963. *Often an observed value appears to be so distant from the other values that the experimenter is tempted to discard it before performing an ANOVA. For a discussion of procedures in such cases, see* STATISTICAL ANALYSIS, SPECIAL PROBLEMS OF, *article on* OUTLIERS.]

Multivariate analysis of variance. The analysis of variance is multivariate in the independent variables (the factors) but univariate in the dependent variables (the outcome measures). S. N. Roy (for example, see Roy & Gnanadesikan 1959) and others have developed a multivariate analysis of variance (MANOVA), multivariate with respect to both independent and dependent variables, of which ANOVA is a special case. A few social scientists (for example, Rodwan 1964; Bock 1963) have used MANOVA, but as yet it has not been used widely by workers in these disciplines.

JULIAN C. STANLEY

BIBLIOGRAPHY

ANSCOMBE, F. J.; and TUKEY, JOHN W. 1963 The Examination and Analysis of Residuals. *Technometrics* 5:141–160.

BARTHOLOMEW, D. J. 1961 Ordered Tests in the Analysis of Variance. *Biometrika* 48:325–332.

BOCK, R. DARRELL 1963 Programming Univariate and Multivariate Analysis of Variance. *Technometrics* 5: 95–117.

BROWNLEE, KENNETH A. (1960) 1965 *Statistical Theory and Methodology in Science and Engineering.* 2d ed. New York: Wiley.

CAMPBELL, DONALD T.; and STANLEY, JULIAN C. 1963 Experimental and Quasi-experimental Designs for Research on Teaching. Pages 171–246 in Nathaniel L. Gage (editor), *Handbook of Research on Teaching.* Chicago: Rand McNally. → Republished in 1966 as a separate monograph titled *Experimental and Quasi-experimental Designs for Research.*

CHACKO, V. J. 1963 Testing Homogeneity Against Ordered Alternatives. *Annals of Mathematical Statistics* 34:945–956.

CHERNOFF, HERMAN; and MOSES, LINCOLN E. 1959 *Elementary Decision Theory.* New York: Wiley.

COCHRAN, WILLIAM G. 1957 Analysis of Covariance: Its Nature and Uses. *Biometrics* 13:261–281.

CORNFIELD, JEROME; and TUKEY, JOHN W. 1956 Average Values of Mean Squares in Factorials. *Annals of Mathematical Statistics* 27:907–949.

DANIEL, CUTHBERT 1959 Use of Half-normal Plots in Interpreting Factorial Two-level Experiments. *Technometrics* 1:311–341.

FISHER, R. A. (1925) 1958 *Statistical Methods for Research Workers.* 13th ed. New York: Hafner. → Previous editions were also published by Oliver & Boyd.

FISHER, R. A. (1935) 1960 *The Design of Experiments.* 7th ed. London: Oliver & Boyd; New York: Hafner.

GLASS, GENE V. 1966 Testing Homogeneity of Variances. *American Educational Research Journal* 3:187–190.

[GOSSET, WILLIAM S.] (1908) 1943 The Probable Error of a Mean. Pages 11–34 in William S. Gosset, *"Student's" Collected Papers.* London: University College, Biometrika Office. → First published in Volume 6 of *Biometrika.*

[1]HAYS, WILLIAM L. 1963 *Statistics for Psychologists.* New York: Holt.

LINDQUIST, EVERET F. 1953 *Design and Analysis of Experiments in Psychology and Education.* Boston: Houghton Mifflin.

[2]LINDZEY, GARDNER (editor) (1954) 1959 *Handbook of Social Psychology.* 2 vols. Cambridge, Mass.: Addison-Wesley. → Volume 1: *Theory and Method.* Volume 2: *Special Fields and Applications.*

LUBIN, ARDIE 1961 The Interpretation of Significant Interaction. *Educational and Psychological Measurement* 21:807–817.

MCLEAN, LESLIE D. 1967 Some Important Principles for the Use of Incomplete Designs in Behavioral Research. Chapter 4 in Julian C. Stanley (editor), *Improving Experimental Design and Statistical Analysis.* Chicago: Rand McNally.

[3]MCNEMAR, QUINN (1949) 1962 *Psychological Statistics.* 3d ed. New York: Wiley.

MOOD, ALEXANDER M.; and GRAYBILL, FRANKLIN A. 1963 *Introduction to the Theory of Statistics.* 2d ed. New York: McGraw-Hill. → The first edition was published in 1950.

NELDER, J. A. 1954 The Interpretation of Negative Components of Variance. *Biometrika* 41:544–548.

PEARSON, EGON S.; and HARTLEY, H. O. (editors) (1954) 1966 *Biometrika Tables for Statisticians.* Volume 1. 3d ed. Cambridge Univ. Press. → A revision of *Tables for Statisticians and Biometricians* (1914), edited by Karl Pearson.

RAND CORPORATION 1955 *A Million Random Digits With 100,000 Normal Deviates.* Glencoe, Ill.: Free Press.

RODWAN, ALBERT S. 1964 An Empirical Validation of the Concept of Coherence. *Journal of Experimental Psychology* 68:167–170.

ROY, S. N.; and GNANADESIKAN, R. 1959 Some Contributions to ANOVA in One or More Dimensions: I and II. *Annals of Mathematical Statistics* 30:304–317, 318–340.

SAMPFORD, MICHAEL R. (editor) 1964 In Memoriam Ronald Aylmer Fisher, 1890–1962. *Biometrics* 20, no. 2:237–373.

SCHEFFÉ, HENRY 1959 *The Analysis of Variance.* New York: Wiley.

SMITH, H. FAIRFIELD 1957 Interpretation of Adjusted Treatment Means and Regressions in Analysis of Covariance. *Biometrics* 13:282–308.

STANLEY, JULIAN C. 1961 Studying Status vs. Manipulating Variables. Phi Delta Kappa Symposium on Educational Research, *Annual Phi Delta Kappa Symposium on Educational Research:* [*Proceedings*] 2:173–208. → Published in Bloomington, Indiana.

STANLEY, JULIAN C. 1965 Quasi-experimentation. *School Review* 73:197–205.

STANLEY, JULIAN C. 1966 A Common Class of Pseudo-experiments. *American Educational Research Journal* 3:79–87.

THOMPSON, W. A. JR. 1962 The Problem of Negative Estimates of Variance Components. *Annals of Mathematical Statistics* 33:273–289.

TUKEY, JOHN W. 1949 One Degree of Freedom for Non-additivity. *Biometrics* 5:232–242.

[4]WINER, BEN J. 1962 *Statistical Principles in Experimental Design.* New York: McGraw-Hill.

ADDITIONAL BIBLIOGRAPHY

BRACHT, GLENN H.; and GLASS, GENE V. 1968 The External Validity of Experiments. *American Educational Research Journal* 5:437–474.

GLASS, GENE V.; and STANLEY, JULIAN C. 1971 *Statistical Methods in Education and Psychology.* Englewood Cliffs, N.J.: Prentice-Hall.

[1]HAYS, WILLIAM L. (1963) 1973 *Statistics for the Social Sciences.* 2d ed. New York: Holt. → First published as *Statistics for Psychologists.*

[2]LINDZEY, GARDNER; and ARONSON, ELLIOT (editors) 1968 *Handbook of Social Psychology.* 5 vols. 2d ed. Reading, Mass.: Addison-Wesley. → Volume 1: *Systematic Positions.* Volume 2: *Research Methods.* Volume 3: *The Individual in a Social Context.* Volume 4: *Group Psychology and Phenomena of Interaction.* Volume 5: *Applied Social Psychology.*

[3]MCNEMAR, QUINN (1949) 1969 *Psychological Statistics.* 4th ed. New York: Wiley.

PEARCE, STANLEY C. 1965 *Biological Statistics: An Introduction.* New York: McGraw-Hill. → See especially Chapter 9, "Interactions and Confounding."

STANLEY, JULIAN C. 1971 Design of Controlled Experiments in Education. Volume 3, pages 474–483 in *The Encyclopedia of Education.* Edited by Lee C. Deighton. New York: Macmillan and Free Press.

STEGER, JOSEPH A. (editor) 1971 *Readings in Statistics for the Behavioral Scientist.* New York: Holt.

WALKER, HELEN M. 1940 Degrees of Freedom. *Journal of Educational Psychology* 31:253–269. → Reprinted in Steger (1971, pp. 349–363).

WALSTER, ELAINE; CLEARY, T. ANNE; and CLIFFORD, MARGARET M. 1971 The Effect of Race and Sex on College Admission. *Sociology of Education* 44:237–244.

[4]WINER, BEN J. (1962) 1971 *Statistical Principles in Experimental Design.* 2d ed. New York: McGraw-Hill.

III
MULTIPLE COMPARISONS

Multiple comparison methods deal with a dilemma arising in statistical analysis: On the one hand, it would be unfortunate not to analyze the data thoroughly in all its aspects; on the other hand, performing several significance tests, or constructing several confidence intervals, for the same data compounds the error rates (significance levels), and it is often difficult to compute the over-all error probability.

Multiple comparison and related methods are designed to give simple over-all error probabilities for analyses that examine several aspects of the data simultaneously. For example, some simultaneous tests examine all differences between several treatment means.

Cronbach (1949, especially pp. 399–403) describes the problem of inflation of error probabilities in multiple comparisons. The solutions now available are, for the most part, of a later date (see Ryan 1959; Miller 1966). Miller's book provides a comprehensive treatment of the major aspects of multiple comparisons.

Normal means—confidence regions, tests

1. Simultaneous limits for several means. As a simple example of a situation in which multiple comparison methods might be applied, suppose that independent random samples are drawn from three normal populations with unknown means, μ_1, μ_2, μ_3, but *known* variances, $\sigma_1^2, \sigma_2^2, \sigma_3^2$. If only the first sample were available, a 99 per cent confidence interval could be constructed for μ_1:

$$(1)\quad \bar{X}_1 - 2.58\sigma_1/\sqrt{n_1} < \mu_1 < \bar{X}_1 + 2.58\sigma_1/\sqrt{n_1},$$

where $\bar{X}_1$ is the sample mean, and n_1 the size, of the first sample. In hypothetical repetitions of the procedure, the confidence interval covers, or includes, the true value of μ_1 99 per cent of the time in the long run. [*See* ESTIMATION, *article on* CONFIDENCE INTERVALS AND REGIONS.]

If all three samples are used, three statements like (1) can be made, successively replacing the subscript "1" by "2" and "3." The probability that all three statements *together* are true, however, is not .99 but .99 × .99 × .99, or .9703.

In a coordinate system with three axes marked μ_1, μ_2, and μ_3, the three intervals together define a 97 per cent (approximately) confidence box. This confidence box is shown in Figure 1. In order to obtain a 99 per cent confidence box—that is, to have all three statements hold simultaneously with probability .99—the confidence levels for the three individual statements must be increased. One method would be to make each individual confidence level equal to .9967, the cube root of .99.

The simple two-tail test of the null hypothesis (H_0) $\mu_1 = 0$ rejects it (at significance level .01) if

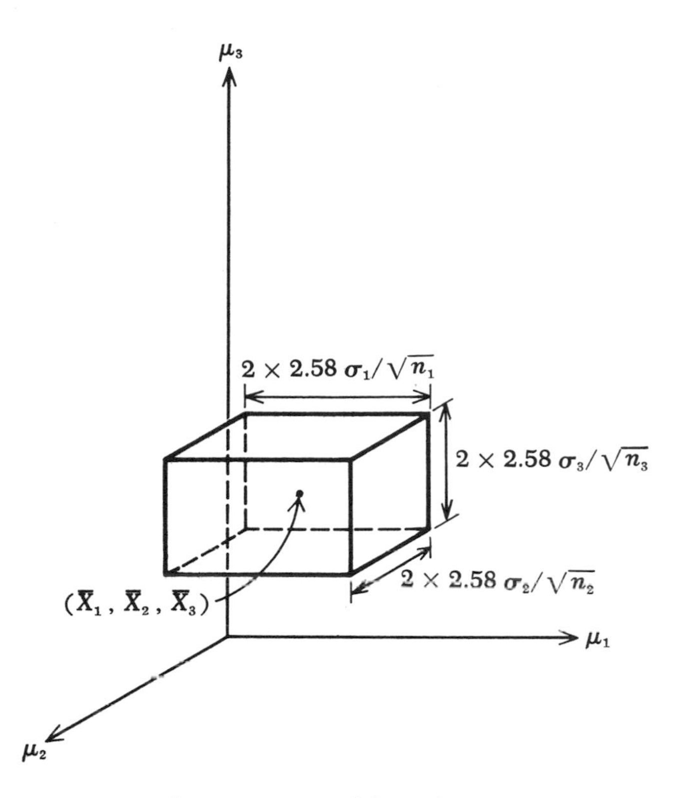

Figure 1 — A confidence box

the value 0 is not caught inside the confidence interval (1). It is natural to think of extending this test to the composite null hypothesis $\mu_1 = 0$ *and* $\mu_2 = 0$ *and* $\mu_3 = 0$ by rejecting the composite hypothesis if the point (0,0,0) is outside the confidence box corresponding to (1). The significance level of this procedure, however, is not .01 but $1 - .9703$, almost .03. In order to reduce the significance level to .01, "2.58" in (1) must be replaced by a higher number. If this is done symmetrically, the significance level for each of the three individual statements like (1) must be .0033. In this argument any hypothetical values of the means, $\mu_1^*, \mu_2^*, \mu_3^*$, may be used in place of 0,0,0 to specify the null hypothesis; the point $(\mu_1^*, \mu_2^*, \mu_3^*)$ then takes the place of (0,0,0).

The same principles can be applied just as easily to the case where the three variances are not known but are estimated from the respective samples, in which case 1 per cent points of Student's *t*-distribution take the place of 2.58. Of course, any other significance levels may also be used instead of 1 per cent.

Pooled estimate of variance. The problem considered so far is atypically simple because the three intervals are statistically *independent*, so that probabilities can simply be multiplied. This is no longer true if the variances are unknown but are assumed to be equal and are estimated by a single *pooled* estimate of variance, $\hat{\sigma}^2$, which is the sum of the three within-sample sums of squares divided by $n_1 + n_2 + n_3 - 3$. This is equal to the mean square used in the denominator of an analysis-of-variance F [*see* LINEAR HYPOTHESES, *article on* ANALYSIS OF VARIANCE]. The conditions

$$\bar{X}_i - M\hat{\sigma}/\sqrt{n_i} < \mu_i < \bar{X}_i + M\hat{\sigma}/\sqrt{n_i}, \quad i = 1,2,3$$

(where M is a constant to be chosen), use the same $\hat{\sigma}$ and hence are *not* statistically independent. Thus, the probability that all three hold simultaneously is not the product of the three separate probabilities, although this is still a surprisingly good approximation, adequate for most purposes.

Critical values, M_α, have, however, been computed for $\alpha = .05$ and .01 and for any number of degrees of freedom $(n_1 + n_2 + n_3 - 3)$ of $\hat{\sigma}^2$. If M_α is substituted for M in the three intervals, the probability that all three conditions simultaneously hold is $1 - \alpha$ (Tukey 1953).

Exactly the same principles described for the problem of estimating, or testing, three population means also apply to k means. A table providing critical values M_α for $k = 2, 3, \cdots, 10$ and for various numbers of degrees of freedom, $N - k$, has been computed by Pillai and Ramachandran (1954). Part of the table is reproduced in Miller (1966). The square of M_α was tabulated earlier by Nair (1948*a*) for use in another context (see Section 7, below). This table is reproduced in Pearson and Hartley ([1954] 1966, table 19).

Notation. In the following exposition, "$\bar{X}_i$" and "μ_i" represent sample and population means, respectively ($i = 1, \cdots, k$), "σ^2" the population variance, generally assumed to be common to all k populations, "$\hat{\sigma}^2$" the pooled sample estimate of σ^2, and "*SE*" the estimated standard error of a statistic (*SE* will depend on $\hat{\sigma}^2$, on the particular statistic, and on the sample sizes involved). The symbol "$\sum$" always denotes summation over i, from 1 to k, unless otherwise specified; N denotes $\sum n_i$, the total sample size, and "*ddf*" stands for "denominator degrees of freedom," the degrees of freedom of $\hat{\sigma}^2$.

2. Treatments versus control (Dunnett). Many studies are concerned with the *difference* between means rather than with the means themselves. For example, sample 1 may consist of *controls* (that is, observations taken under standard conditions) to be used for comparison with samples $2, 3, \cdots, k$ (taken under different treatments or nonstandard conditions), for the purpose of estimating the treatment effects, $\mu_2 - \mu_1, \cdots, \mu_k - \mu_1$. For $k = 3, 4, \cdots, 10$, for any number of denominator degrees of freedom, $N - k$, greater than 4, and for $\alpha = .05$ and .01, Dunnett (1955; also in

Miller 1966) has tabulated critical values D_α such that with probability approximately equal to $1-\alpha$, all $k-1$ statements

$$|(\bar{X}_i - \bar{X}_1) - (\mu_i - \mu_1)| < D_\alpha SE, \quad i = 2, 3, \cdots, k,$$

will be simultaneously true—that is, all $k-1$ effects $\mu_i - \mu_1$ will be covered by confidence intervals centered at $\bar{X}_i - \bar{X}_1$ with half-lengths $D_\alpha SE$, where $SE = \sqrt{(1/n_i) + (1/n_1)}\,\hat{\sigma}$.

The over-all probability is exactly $1-\alpha$ if all k sample sizes are equal. It is not the product of $k-1$ probabilities (obtained from Student's t-distribution) of the separate confidence statements, because these are not statistically independent; dependence comes not only from the common estimator of σ in all statements but also from the correlation ($\rho \cong .5$ for sample sizes roughly the same) between any two differences $\bar{X}_i - \bar{X}_1$ with $\bar{X}_1$ in common. Surprisingly enough, the product rule gives a close approximation just the same.

Viewed as restrictions on the point (μ_1, μ_2, μ_3) in three-space, the two (pairs of) inequalities for $k = 3$ define a confidence region that is the intersection of the slab bounded by two parallel planes, $\mu_2 - \mu_1 = \bar{X}_2 - \bar{X}_1 \pm D_\alpha SE$, and another slab at 45° to the first slab. This is illustrated in Figure 2, where for simplicity all n_i are assumed to be equal. The region is a prism that is infinite in length, is parallel to the 45° line $\mu_1 = \mu_2 = \mu_3$, and has a rhombus as its cross section.

Dunnett's significance test rejects the null hypothesis, H_0: $\mu_2 = \cdots = \mu_k = \mu_1$, in favor of the alternative hypothesis that one or more of the μ_i differ from μ_1 if the $k-1$ confidence intervals do not all contain the value 0 or, equivalently, if

$$(2) \qquad |t_{i1}| = |\bar{X}_i - \bar{X}_1|/SE \geqslant D_\alpha$$

for any i ($i = 2, \cdots, k$). If the null hypothesis is of the less trivial form $\mu_i - \mu_1 = d_{i1}$, where the d_{i1} are any specified constants, then d_{i1} is subtracted from the differences of sample means in the numerators of t_{i1}.

The probability of rejecting H_0 if it is true, called the *error rate experimentwise*, is exactly the stated α if all sample sizes are equal, and is approximately α for unequal n_i, provided the inequality is not gross. Dunnett (1955) showed that a design using equal n_i, $i = 2, \cdots, k$, but with n_1 larger in about the proportion $\sqrt{k-1} : 1$ is most efficient. Unfortunately this leads to true error rates exceeding the stated α if Dunnett's table is used, and it is then safer to substitute a Bonferroni t-statistic for Dunnett's D_α if k is as big as 6 or 10 (for Bonferroni t,

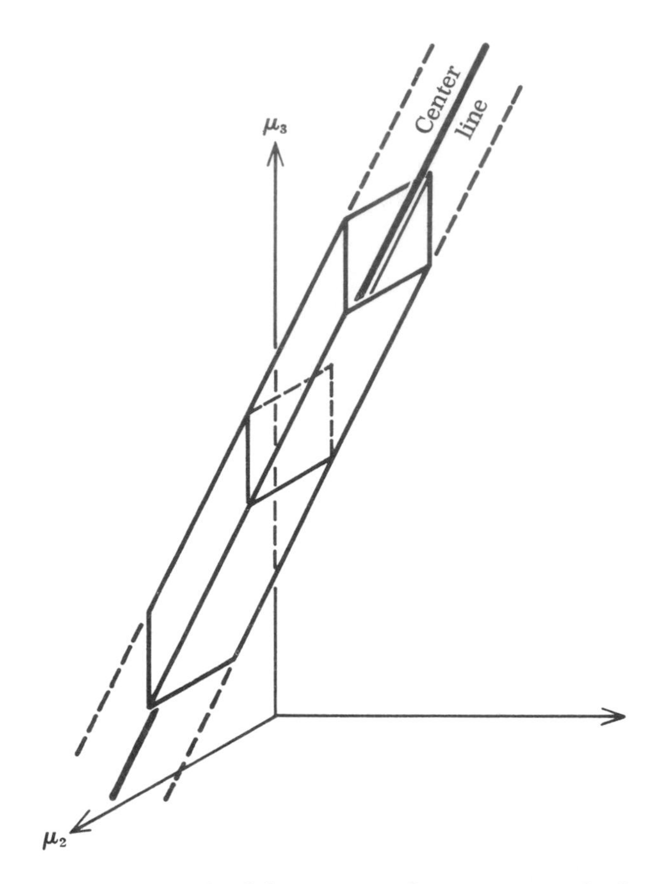

Figure 2 — Confidence region for Dunnett method

see Section 14, below; see also Miller 1966, table 2).

Simultaneous one-tail tests are of the same form as (2), above, except that the absolute-value signs are removed and an appropriate smaller critical value, D_α, also tabulated in Dunnett (1955), is used. The corresponding confidence intervals are one-sided, extending to infinity on the other side.

13. All differences—Tukey method. In order to compare several means with one another rather than only with a single control, a method of Tukey's (1953) is suitable. It provides simultaneous confidence intervals (or significance tests, if desired) for all $\binom{k}{2} = \frac{1}{2}k(k-1)$ differences, $\mu_i - \mu_j$, among k means.

A constant, T_α, is chosen so that the probability is at least $1-\alpha$ that all $\binom{k}{2}$ statements

$$|(\bar{X}_i - \bar{X}_j) - (\mu_i - \mu_j)| < T_\alpha SE,$$

or, equivalently,

$$|t_{ij}| = |(\bar{X}_i - \bar{X}_j) - (\mu_i - \mu_j)|/SE < T_\alpha,$$

will be simultaneously true. Here SE is equal to $\sqrt{(1/n_i) + (1/n_j)}\,\hat{\sigma}$. The probability is exactly $1-\alpha$ if the sample sizes are equal (Tukey 1953; Kurtz 1956; Kramer 1956).

Simultaneous confidence intervals for all the differences, $\mu_i - \mu_j$, are centered at $\bar{X}_i - \bar{X}_j$ with half-lengths $T_\alpha SE$. In a significance test of the null hypothesis, H_0, that the differences, $\mu_i - \mu_j$, have any specified (mutually consistent) values, d_{ij} (often 0), one substitutes d_{ij} for $\mu_i - \mu_j$ in the t-ratios and rejects H_0 if the largest ratio is not less than T_α.

The constant, T_α, is $R_\alpha/\sqrt{2} = .707R_\alpha$, where R_α is the upper α-point in the distribution of the Studentized range. Table 29 of Pearson and Hartley ([1954] 1966) shows R_α for $\alpha = .1$, .05, and .01, for values of k up to 20, and for any number of *ddf*. Briefer tables are found in Vianelli (1959) and in a number of textbooks—for example, Winer (1962). More extensive tables prepared by Harter (1960) can also be found in Miller (1966).

Geometrically, Tukey's $(1 - \alpha)$-confidence region can be obtained, for $k = 3$, by widening and thickening Dunnett's prism (Figure 2) in the proportion $T_\alpha : D_\alpha$ and then removing a pair of triangular prisms by intersection with a third slab. The cross section is hexagonal.

Tukey's multiple comparisons are frequently used after an F-test rejects H_0 but may also be used in place of F.

Simplified multiple t-tests. Simplified multiple t-tests, which were developed by Tukey, use the sum of sample ranges in place of σ and a critical value, T_α, adjusted accordingly. (See Kurtz et al. 1965.)

4. **One outlying mean (slippage).** In comparing k populations it may be desirable to find out whether one of them (which one is not specified in advance) is outstanding (has "slipped") relative to the others. Then using k independent treatment samples one may examine the differences, $\bar{X}_i - \bar{X}$, where $\bar{X} = \sum_{i=1}^{k} \sum_{j=1}^{n_i} X_{ij}/N = \sum n_i \bar{X}_i/N$. Let $\mu = \sum n_i \mu_i/N$.

Halperin provided critical values H_α such that with probability approximately $1 - \alpha$,

$$|(\bar{X}_i - \bar{X}) - (\mu_i - \mu)| < H_\alpha \sqrt{\frac{k}{k-1}\left(\frac{1}{n_i} - \frac{1}{N}\right)}\,\hat{\sigma}$$

simultaneously for $i = 1, \cdots, k$ (Halperin et al. 1955). The probability is exactly $1 - \alpha$ in the case of equal n_i. This provides two-sided tests for the null hypothesis that all $\mu_i = \mu$ and simultaneous confidence intervals for all the $\mu_i - \mu$, in the usual way. In case the table is not at hand, a good approximation to the right-hand side of the inequality is (upper $(\alpha/2k)$-point of Student's t) $\times$ $\sqrt{(1/n_i) - (1/N)}\,\hat{\sigma}$.

Critical values for the corresponding one-sided test, to ascertain whether one of the means has slipped in a specified direction (for example, whether it has slipped down), were first computed by Nair (1952). David (1962*a*; 1962*b*) provides improved tables. A refinement of Nair's test and of Halperin's is presented by Quesenberry and David (1961). In Pearson and Hartley ([1954] 1966), tables 26*a* and 26*b* (and the explanation on p. 51) pertain to these methods, whereas table 26 is Nair's statistic.

5. **Contrasts—Scheffé method.** A *contrast* in k population means is a linear combination, $\sum c_i \mu_i$, with coefficients adding up to zero, $\sum c_i = 0$. This is always equal to a multiple of the difference between weighted averages of two sets of means—that is, constant $\times$ $(\sum_h a_h \mu_h - \sum_j b_j \mu_j)$ with summations running over two subsets of the subscripts $(1, \cdots, k)$ having no subscript in common and with $\sum_h a_h = 1$, $\sum_j b_j = 1$. The simple differences, $\mu_h - \mu_j$, are special contrasts. Some other examples include contrasts representing a difference between two *groups* of means (for example, $\frac{1}{3}[\mu_2 + \mu_3 + \mu_5] - \frac{1}{2}[\mu_1 + \mu_4]$) or slippage of one mean (for example, $\mu_2 - \bar{\mu}$, since this is equal to $\{[k-1]/k\}\mu_2 - [1/k][\mu_1 + \mu_3 + \mu_4 + \cdots + \mu_k]$), or trend (for example, $-3\mu_1 - \mu_2 + \mu_3 + 3\mu_4$).

In an exploratory study to compare k means when little is known to suggest a specific pattern of differences in advance, any and all striking contrasts revealed by the data will be of interest. Also, when looking for slippage or simple differences one may wish to take account of some other, unanticipated, pattern displayed by the data.

Any of the systems of multiple comparisons discussed in sections 1–4 can be adapted to obtain tests, or simultaneous intervals, for all contrasts. For example, the $k - 1$ simultaneous conditions $|(\bar{X}_i - \bar{X}_1) - (\mu_i - \mu_1)| < D_\alpha \sqrt{(1/n_i) + (1/n_1)}\,\hat{\sigma}$, where D_α represents the critical value of the Dunnett statistic as defined in Section 2, above, imply that every contrast, $\sum c_i \mu_i$, falls into an interval of half-length $D_\alpha \sum_2^k (|c_i| \sqrt{2/n_i})\,\hat{\sigma}$, centered at $\sum c_i \bar{X}_i$, in the case of equal sample sizes.

The following method, developed by Scheffé, however, is more efficient for all-contrasts analyses, because it yields shorter intervals for most contrasts. Scheffé proved that

$$\sqrt{(k-1)F} = \max[\textstyle\sum c_i(\bar{X}_i - \mu_i)/SE],$$

the largest of all the (infinitely many) Studentized contrasts, where F is the analysis-of-variance F-ratio for testing equality of all the μ_i, and where

$SE = \sqrt{\sum(c_i^2/n_i)}\,\hat{\sigma}$. Thus,

$$\begin{aligned} 1-\alpha &= Pr\{F < F_\alpha\} \\ &= Pr\{all \text{ Studentized contrasts} < \sqrt{(k-1)F_\alpha}\} \\ &= Pr\left\{\frac{\sum c_i(\bar{X}_i - \mu_i)}{SE} < \sqrt{(k-1)F_\alpha} \right. \\ &\qquad\qquad \left. \text{for } all \text{ sets of } c_i \text{ with } \sum c_i = 0\right\}. \end{aligned}$$

Simultaneous confidence intervals for all contrasts, $\sum c_i\mu_i$, are centered at $\sum c_i\bar{X}_i$ and have half-lengths $SE\sqrt{(k-1)F_\alpha}$. The confidence level is *exactly* the stated $1-\alpha$, regardless of whether sample sizes are equal.

For $k = 3$, any particular interval can be depicted in (μ_1, μ_2, μ_3)-space by a pair of parallel planes equidistant from the line given by $\mu_1 - \bar{X}_1 = \mu_2 - \bar{X}_2 = \mu_3 - \bar{X}_3$ through the point $(\bar{X}_1, \bar{X}_2, \bar{X}_3)$. Together these planes constitute all the tangent planes of the cylinder (in the "variables" μ_1, μ_2, μ_3),

$$\sum n_i[(\bar{X}_i - \bar{X}) - (\mu_i - \mu)]^2 = (3-1)\hat{\sigma}^2 F_\alpha,$$

where F_α has degrees of freedom $3-1$ and $n-3$. This cylinder, like the prism of Figure 2, is infinite in length and equally inclined to the coordinate axes. (As in the case of the regions for Dunnett's and Tukey's procedures, the addition of the same constant to each of the coordinates X_1, X_2, X_3 of a point on the surface will move this point along the surface.) See Figure 3.

Significance test. A value of $F \geqslant F_\alpha$ implies $\sum c_i(\bar{X}_i - \mu_i) \geqslant SE\sqrt{(k-1)F_\alpha}$ for at least one contrast (namely, at least for the maximum Studentized contrast). Scheffé's multiple comparison test declares $\sum c_i\bar{X}_i$ to be statistically significant—that is, $\sum c_i\mu_i$ different from zero—for all those contrasts for which the inequality is true. Thus, one may test every contrast of interest, or every contrast that looks promising, and incur a risk of just α of falsely declaring any $\sum c_i\mu_i$ whatsoever to be different from zero; in other words, the probability of making no false statement of the form $\sum c_i\mu_i \neq 0$ is $1-\alpha$, the probability of making one or more such statements is α. Of course, the Scheffé approach gives a larger confidence interval (or decreased power) than the analogous procedure if only a single contrast is of interest.

General linear combinations. Simultaneous confidence intervals, or tests, can also be obtained for all possible linear combinations, $\sum c_i\mu_i$, with the restriction $\sum c_i = 0$ lifted. Then Scheffé's confidence and significance statements for contrasts remain applicable, except that $(k-1)F_\alpha$ is changed to kF_α and the numerator degrees of freedom of F are changed from $k-1$ to k. (See Miller 1966, chapter 2, sec. 2).

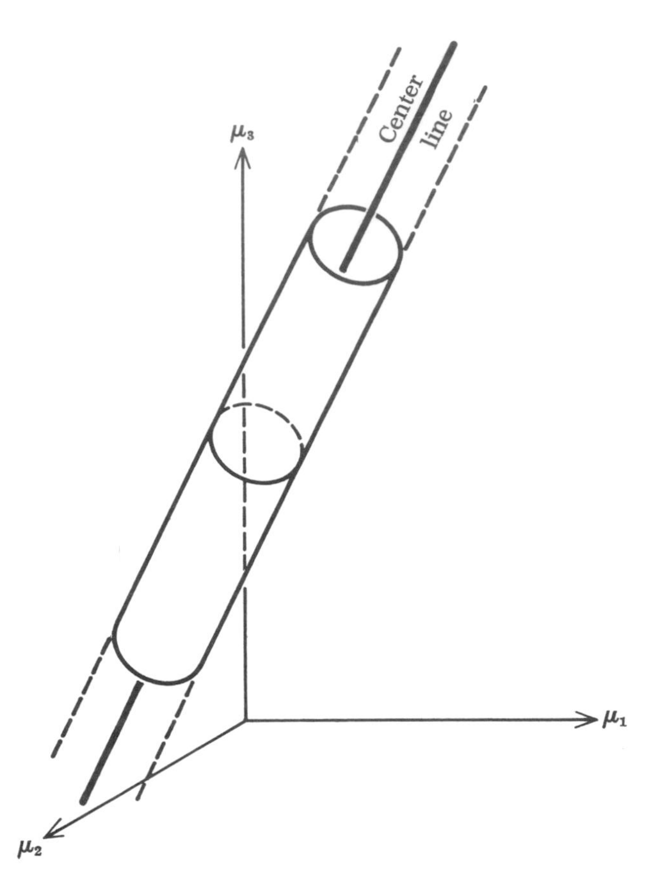

Figure 3 — Confidence region for all contrasts (Scheffé)

A confidence region for all (standardized) linear combinations consists of the ellipsoid in the k-dimensional space with axes labeled $\mu_1, \mu_2, \cdots, \mu_k$, $\sum n_i(\bar{X}_i - \mu_i)^2 < kF_\alpha\hat{\sigma}^2$. For $k = 3$, any particular interval can be depicted in (μ_1, μ_2, μ_3)-space by a pair of parallel planes equidistant from the point $(\bar{X}_1, \bar{X}_2, \bar{X}_3)$. Together these planes constitute all the tangent planes of the confidence ellipsoid (in the "variables" μ_1, μ_2, μ_3).

Tukey (1953) and Miller (1966) also discuss the generalization of the application of intervals based on the Studentized range (referred to in Section 3, above) to take care of all linear combinations. Simultaneous intervals for all linear combinations can also be based on the Studentized maximum modulus (Section 1); half-lengths become $M_\alpha\sqrt{1/n}\hat{\sigma} \cdot \sum|c_i|$ (Tukey 1953).

All of these methods dealing with contrasts and general linear combinations are described in Miller (1966).

Further discussion of normal populations

6. Newman–Keuls and Duncan procedures. The Newman–Keuls procedure is a multiple comparison test for all differences. It does not provide a confidence region. The sample means are arranged and renumbered in order of magnitude, so that $\bar{X}_1 < \bar{X}_2 < \cdots < \bar{X}_k$. The first step is the same as Tukey's test; the null hypothesis is rejected or accepted according as $\bar{X}_k - \bar{X}_1$, the range of the sample means, is $\geqslant$ or $< T_{\alpha;k} SE$, where $T_{\alpha;k}$ is the upper α-point of Tukey's statistic for k means and $N - k$ *ddf*.

Accepting H_0 means that there is not enough evidence to establish differences between any of the population means, and the analysis is complete (all k means are then called "homogeneous"). On the other hand, if the null hypothesis is rejected, so that μ_k, the population mean corresponding to the largest sample mean, is declared to be different from μ_1, the population mean corresponding to the smallest sample mean, the next step is to test $\bar{X}_{k-1} - \bar{X}_1$ and $\bar{X}_k - \bar{X}_2$ similarly, but with $T_{\alpha;k-1}$ in place of $T_{\alpha;k}$ (the original pooled variance estimator $\hat{\sigma}^2$ and $N - k$ *ddf* are used throughout). A subrange of means that is not found statistically significant is called homogeneous. As long as a subrange is statistically significant, the two subranges obtained by removing in one case its largest and in the other case its smallest $\bar{X}_i$ are tested, using a critical value $T_{\alpha;h}$, where h is only the number of means left in the new subranges—but testing is limited by the rule that every subrange contained in a homogeneous range of means is not tested but is automatically declared to be homogeneous. The result of the whole procedure is to group the means into homogeneous sets, which may also be represented diagrammatically by connecting lines, as in the example presented in Section 10, below.

Critics of the Newman–Keuls method object that the error probabilities, such as that of falsely declaring $\mu_2 \neq \mu_5$, are not even known in this test; its supporters, however, argue that power should not be wasted by judging subranges by the same stringent criterion used for the full range of all k sample means.

Duncan (1955) goes a step further, arguing that even $T_{\alpha;h}$ is too stringent a criterion because the $\frac{1}{2}h(h-1)$ differences between h means have only $h - 1$ degrees of freedom. He concludes that $T_{\gamma;h}$ should be used instead, where $1 - \gamma = (1 - \alpha)^{h-1}$. This further increases the power—and the effective type I error probability. For a study of error rates of Tukey, Newman–Keuls, Duncan, and Student tests, see Harter (1957).

7. General Model I design. The F-test in the one-way analysis of variance and the multiple comparison methods already discussed are based on the fact that *ddf* times $\hat{\sigma}^2/\sigma^2$ has a chi-square distribution and is independent of the sample means. This condition is also satisfied by the residual variance used in randomized blocks, factorial designs, Latin squares, and all Model I designs. Therefore, all these designs permit the use of the methods, and tables, of sections 1–6, to compare the means defined by any one factor, provided that these are independent.

In certain instances of nonparametric multiple comparisons and in certain instances of multiple comparisons of interactions in balanced factorial designs, where the (adjusted or transformed) observations are not independent but equicorrelated, the multiple comparison methods of sections 2–6 still apply: The use of the adjusted error variance, $(1 - \rho)\hat{\sigma}^2$, to compute standard errors fully compensates for the effect of equal correlations (see Tukey 1953; Scheffé 1953; Miller 1966, pp. 41–42, 46–47). Scheffé's method can also be adapted for use with unequal correlations (see Miller 1966, p. 53).

When several factors, and perhaps some interactions, are t-tested in the same experiment, the question arises whether extra adjustment should not be made for the resulting additional compounding of error probabilities. One method open to an experimenter willing to sacrifice power for strict experimentwise control of type I error is the conservative one of using error rates per t-test of α/(number of t-tests contemplated), that is, using Bonferroni t-statistics (see Section 14). For experimentwise control of error rates in the special case of a 2^r factorial design, Nair (1948*a*) has tabulated percentage points of the largest of r independent χ^2's with one degree of freedom, divided by an independent variance estimator (Pearson & Hartley [1954] 1966, table 19). The statistic is equal to the square of the Studentized maximum modulus introduced in Section 1.

8. An example—juxtaposition of methods. Three competing theories about how hostility evoked in people by willfully imposed frustration may be diminished led Rothaus and Worchel (1964) to goad 192 experimental subjects into hostility by unfair administration of a test of coordination and then to apply the following "treatments" to four groups, each composed of 48 subjects: (1) no treatment (control); (2) fair readministration of the test, seemingly as a result of a grievance procedure (instrumental communication); (3) an opportunity for verbal expression of hostility (catharsis); (4) conversation to the effect that the test was unfair and the result therefore not indicative of failure on the subjects' part (ego support).

Table 1 — Analysis of variance of hostility scores

Source	df	Mean square	F-ratio
4 treatments	3	369.77	3.38*
3 subgroups	2	151.31	1.38
2 sexes	1	41.14	0.38
2 BIHS levels	1	2.68	0.02
All the interactions, none of them statistically significant	40		
4 replications (nested)	144	$109.54 = \hat{\sigma}^2$	

* Denotes statistical significance at the 5 per cent level.

After treatment all subjects were given another—fair—test of coordination. Each treatment group was subdivided into three subgroups, a different experimenter working with each subgroup. All subjects had been given Behavioral Items for Hostility Scales (BIHS) three weeks before the experiment.

The experimental plan was factorial: 4 treatments $\times$ 3 subgroups $\times$ 2 sexes $\times$ 2 BIHS score groups (high versus low) $\times$ 4 replications. The study variable, X, was hostility measured on the Social Sensitivity Scale at the end of the experiment.

The sample means (unordered) for the four treatment groups were $\bar{x}_1 = 47.08$, $\bar{x}_2 = 42.00$, $\bar{x}_3 = 48.53$, $\bar{x}_4 = 45.40$.

In fact, the numbers in Table 1 reflect an analysis of covariance. The mean squares shown are adjusted mean squares, the sample means are adjusted means, and $\hat{\sigma}^2$ has 143 *df*. But for the sake of simplicity of interpretation the data will be treated as if they had come from a $4 \times 3 \times 2 \times 2$ factorial analysis of variance. The estimated standard error for differences between two means, *SE*, is $\sqrt{[(1/48) + (1/48)] \times 109.54} = 2.136$.

Dunnett comparisons. The Dunnett method, with $\alpha = .05$, would be applied to the data of the experiment, as analyzed in Table 2. As indicated in Table 2, the one-tail test in the direction of the theory (H_1) under study declares μ_2 to be less than μ_1. Thus, the conclusion, if the one-sided Dunnett test and the 5 per cent significance level are adopted, is that instrumental communication reduces hostility but that the evidence does not confirm any reduction due to ego support or catharsis. If the two-tail test had been chosen, allowing for a possible *increase* in hostility due to treatment, the conclusion would be that there is insufficient evidence to reject.

All pairs—Tukey and Scheffé methods. A comparison of all possible pairs of means by the methods of Tukey and Scheffé is shown in Table 3. The tests of Tukey and Scheffé in this case both discount $\bar{x}_1 - \bar{x}_3$ but declare $\bar{x}_2 - \bar{x}_3$ "significant." The conclusion is that instrumental communication leaves the mean hostility of frustrated subjects lower than ego support does, but no other difference is established; specifically, neither test would conclude that instrumental communication actually reduces hostility as compared with no treatment or that ego support increases (or reduces) it.

In addition to the simple differences, the data suggest testing a contrast related to the alternate hypothesis $\mu_2 < \mu_4 < \mu_1 < \mu_3$, for example, the contrast $-3\mu_2 - \mu_4 + \mu_1 + 3\mu_3$. (It is legitimate, for these procedures, to choose such a contrast after inspecting the data.) For the present example, $-3\bar{x}_2 - \bar{x}_4 + \bar{x}_1 + 3\bar{x}_3 = 21.27$; the *SE* for a Scheffé test is $\sqrt{(3^2 + 1^2 + 1^2 + 3^2)/48}\hat{\sigma} = \sqrt{10} \times 2.136 = 6.755$, and $t = 21.27/6.755 = 3.15$, statistically significant at the 5 per cent level ($3.15 > 2.80$). The conclusion is that $\mu_2 \leqslant \mu_4 \leqslant \mu_1 \leqslant \mu_3$ with at least one strict inequality holding. A Scheffé 95 per cent confidence interval for $-3\mu_2 - \mu_4 + \mu_1 + 3\mu_3$ is (1.36, 40.18). A Tukey test would not find this contrast statistically significant. For this analysis 2.136 is multiplied by $\frac{1}{2}(|-3| + |-1| + |1| + |3|) = 4$, instead of by $\sqrt{10}$, yielding 8.544. Thus, in this case $t = 21.27/8.544 = 2.49$, which is less than 2.60, and a confidence interval is $(-0.94, +43.48)$.

The *SE* for individual $\bar{x}_i$, also used in slippage statistics, is $\sqrt{(1/48)\,\hat{\sigma}^2} = 2.136/\sqrt{2} = 1.510$. For $k = 4$, $M_{.05} = 2.50$, and simultaneous confidence intervals for the four μ_i are centered at the $\bar{x}_i$ and have half-lengths $2.50 \times 1.51 = 3.78$.

The 5 per cent critical value tabulated by Halperin et al. for two-sided slippage tests is 2.23; thus $(\bar{x}_2 - \bar{x})/1.51 = 2.48$ is statistically signi-

Table 2 — Dunnett comparisons of control with three treatments, $\alpha = .05$

			DUNNETT METHOD: TWO-SIDED		DUNNETT METHOD: ONE-SIDED	
Pair	$\bar{x}_i - \bar{x}_j$	t-ratio: $\frac{\bar{x}_i - \bar{x}_j}{2.136}$	Test: $D_\alpha = 2.40$	Confidence interval (half-length $= 2.136D_\alpha = 5.13$)	Test: $D_\alpha = 2.08$	Confidence interval (lower length $= 2.136D_\alpha = 4.44$)
(1)–(2)	5.08	2.38	(near significance)	(−0.05,10.21)	*	(0.64,∞)
(1)–(3)	−1.45	−0.68	—	(−6.58,3.68)	—	(−5.89,∞)
(1)–(4)	1.68	0.79	—	(−3.45,6.81)	—	(−2.76,∞)

* Statistically significant at the 5 per cent level; all other comparisons do not reach statistical significance at the 5 per cent level.

Table 3 — All pairs, by Tukey and by Scheffé method, $\alpha = .05$

			TUKEY METHOD		SCHEFFÉ METHOD	
Pair	$\bar{x}_i - \bar{x}_j$	t-ratio: $\frac{\bar{x}_i - \bar{x}_j}{2.136}$	Test: $T_\alpha = 2.60$	Confidence interval (half-length = $2.136T_\alpha = 5.55$)	Test: $\sqrt{3F_\alpha} = 2.80$	Confidence interval (half-length = $2.136\sqrt{3F_\alpha} = 5.98$)
(1)–(2)	5.08	2.38	—	(−0.47,10.63)	—	(−0.90,11.06)
(1)–(3)	−1.45	−0.68	—	(−7.00,4.10)	—	(−7.43,4.53)
(1)–(4)	1.68	0.79	—	(−3.87,7.23)	—	(−4.30,7.66)
(2)–(3)	−6.53	−3.06	*	(−12.08,−0.98)	*	(−12.51,−0.55)
(2)–(4)	−3.40	−1.59	—	(−8.95,2.15)	—	(−9.38,2.58)
(3)–(4)	3.13	1.47	—	(−2.42,8.68)	—	(−2.85,9.11)

* Statistically significant at the 5 per cent level; all other pairs do not reach statistical significance at the 5 per cent level.

ficant, whereas the other three t-ratios for slippage are not. The conclusion of this test would be that mean hostility after instrumental communication is low compared with that after other treatments; no other treatment can be singled out as leaving hostility either low or high compared with that after other treatments.

An example of Newman–Keuls and Duncan tests is given in Section 10, below.

Other multiple comparison methods

9. Nonparametric multiple comparisons. The multiple comparison approach has been articulated with nonparametric (or distribution-free) methods in several ways [*for background see* NONPARAMETRIC STATISTICS].

For example, one of the simplest nonparametric tests is the sign test. Suppose that an experiment concerning techniques for teaching reading deals with school classes and that each class is divided in half at random. One half is taught by method 1, the other by method 2, the methods being allocated at random. Suppose further that improvement in average score on a reading test after two months is the basic observation but that one chooses to consider only whether the pupils taught by method 1 gain more than the pupils taught by method 2, or vice versa, and not the magnitude of the difference. If C is the number of classes for which the pupils taught with method 1 have a larger average gain than those taught with method 2, then the (two-sided) sign test rejects the null hypothesis of equal method effect when the absolute value of C is larger than a critical value. The critical value comes simply from a symmetrical binomial distribution.

Suppose now that there are k teaching methods, where k might be 3 or 4, and the classes are each divided at random into k groups and assigned to methods. Let C_{ij} $(i \neq j)$ be the number of classes for which the average gain in reading-test score for the group taught by method i is greater than that for the group taught by method j. Each C_{ij} taken separately has (under the null hypothesis that the corresponding two methods are equally effective) a symmetric binomial distribution which is approximated asymptotically by $\frac{1}{2}n + z\frac{1}{2}\sqrt{n} + \frac{1}{2}$, where n is the number of classes, z is a standard normal variable, and $\frac{1}{2}$ is a continuity correction. But to test for the equality of all k methods, the largest $|C_{ij}|$ should be used. The critical values of this statistic may be approximated by $T_\alpha\frac{1}{2}\sqrt{n} + \frac{1}{2}$, where T_α is the upper α point for Tukey's statistic with k groups and $ddf = \infty$.

The same procedure is feasible for other two-sample test statistics—for example, rank sums. An analogous method works for comparing $k - 1$ treatments with a control; in the teaching-method experiment, if method 1 were the control, this would mean using as the test statistic the maximum over $j \neq 1$ of $|C_{1j}|$ (or of C_{1j} in the one-sided case). For a discussion of this material, see Steel (1959).

Joint nonparametric confidence intervals may sometimes be obtained in a similar way. Given a confidence interval estimation procedure related to any two-sample test statistic with critical value S_α (see Moses 1953; 1965), the same procedure with S_α replaced by its multiple comparison analogue C_α yields confidence intervals with a joint confidence level of $1 - \alpha$.

A second class of nonparametric multiple comparison tests arises by analogy with normal theory analysis of variance for the one-way classification and other simple designs [*see* LINEAR HYPOTHESES, *article on* ANALYSIS OF VARIANCE]. The procedures start by transforming the observations into ranks or other kinds of simplified scores (except that the so-called permutation tests leave the observations unaltered). The analysis is conditional on the totality of scores and uses as its null distribution that obtained from random allocations of the observed scores to treatments. The test statistic may be the ordinary F-ratio on the scores, but modified so that the denominator is the exact over-all variance of the given scores. This statistic's null distribution is approximately F, with $k-1$ and ∞ as degrees of freedom (where k is the number of treatments), or, equivalently, $k-1$ times the F-test statistic has as approximate null distribution the chi-square distribution with $k-1$ degrees of freedom. Similar adaptations hold for the Tukey test statistic and others. The approach may also be extended to randomized block designs; in another direction, the approach may be extended to compare dispersion, rather than location. Discussions of this material are given by Nemenyi (1963) and Miller (1966, chapter 2, sec. 1.4, and chapter 4, sec. 7.5).

A difficulty with these test procedures is that confidence sets cannot generally be obtained in a straightforward way.

A third nonparametric approach to multiple comparisons is described by Walsh (1965, pp. 535–536). The basic notion applies when there are a number of observations for each treatment or treatment combination. Such a set of observations is divided into several subsets; the average of each subset is taken. These averages are then treated by normal theory procedures of the kind discussed earlier.

For convenient reference, a few 5 per cent and 1 per cent critical points of multiple comparison statistics with $ddf = \infty$ are listed in Table 4.

10. An example. As an illustration of some distribution-free multiple comparison methods, consider the following data from Vaughan, Sjoberg, and Smith (1966), who sent questionnaires to a sample of scientists listed in *American Men of Science* in order to compare scientists in four different fields with respect to the role that traditional religion plays in their lives. Table 5 summarizes responses to the question about frequency of church attendance and shows some of the calculations.

Using the data of Table 5, illustrative significance tests of the null hypothesis of four identical population distributions, against various alternatives, will be performed at the 1 per cent level.

The method of Yates (1948) begins by assigning ascending numerical scores, u, to the four ordered categories; arithmetically convenient scores, as shown in the last column of Table 5, are -1, 0, 1, and 2. Sample totals of scores are calculated—for example, $T_1 = 44(-1)+38(0)+52(1)+33(2) = 74$, and the average score for sample i is $\bar{u}_i = T_i/n_i$. From the combined sample (margin) Yates computes an average score, $\bar{u} = T/N = .210$, and the variance of scores,

$$\sigma^2 = \frac{N}{N-1}(\text{average square} - \bar{u}^2),$$

giving $(638/637)\{[247(1) + 108(0) + 185(1) + 98(4)]/638 - .210^2\} = 1.2494$.

Yates then computes a variance between means, $T_1^2/n_1 + \cdots + T_4^2/n_4 - T^2/N = 17.05$, and the critical ratio used is either $F = (17.05/3)/1.2494$ or $\chi^2 = 17.05/1.2494 = 13.7$. The second of these is referred to a table of chi-square with 3 df and found significant at the 1 per cent level (in fact, $P = .0034$).

It follows that some contrasts must be statistically significant. The almost linear progression of the sample mean scores suggests calculating $3\bar{u}_1 + \bar{u}_2 - \bar{u}_3 - 3\bar{u}_4 = 1.438$. For the denominator, $(3^2/167 + 1^2/159 + 1^2/155 + 3^2/157)\sigma^2 = .1240 \times 1.2494 = .1549$, so that $\chi^2 = 1.438^2/.1549 = 13.35$, or its square root, $z = 1.438/\sqrt{.1549} = 3.65$. (This comes close to the value $\sqrt{13.7} = 3.70$ of the largest standardized contrast—see Section 5.) When 3.65

Table 4 — Selected 5 per cent and 1 per cent critical points of multiple comparison statistics with ddf $= \infty$

	DUNNETT		TUKEY	NAIR–HALPERIN		SCHEFFÉ	DUNCAN
	$k-1$ versus one		all pairs	outlier tests			
k	one-tail	two-tail	range/$\sqrt{2}$	one-tail	two-tail	$\sqrt{\chi^2_{k-1}}$	
			5 per cent level				
2	1.64	1.96	1.96	1.39	1.39	1.96	1.96
3	1.92	2.21	2.34	1.74	1.91	2.45	2.06
4	2.06	2.35	2.57	1.94	2.14	2.80	2.13
5	2.16	2.44	2.73	2.08	2.28	3.08	2.18
6	2.23	2.51	2.85	2.18	2.39	3.23	2.23
			1 per cent level				
2	2.33	2.58	2.58	1.82	1.82	2.58	2.58
3	2.56	2.79	2.91	2.22	2.38	3.03	2.68
4	2.68	2.92	3.11	2.43	2.61	3.37	2.76
5	2.77	3.00	3.25	2.57	2.76	3.64	2.81
6	2.84	3.06	3.36	2.68	2.87	3.88	2.86

Table 5 — Frequency of church attendance of scientists in four different fields

Church attendance	(1) Chemical engineers	(2) Physicists	(3) Zoologists	(4) Geologists	Combined sample	Score u
Never	44	65	66	72	247	−1
Not often	38	19	21	30	108	0
Often	52	46	49	38	185	1
Very often	33	29	19	17	98	2
Sample size, n_i	167	159	155	157	$N = 638$	
$T_i = \sum \text{Frequency} \cdot u$	74	39	21	0*	$T = 134$	
$\bar{u}_i = T_i/n_i$	0.443	0.245	0.136	0.000	$\bar{u} = 0.210$	
$1/n_i$	0.005988	0.006289	0.006452	0.006369	$1/N = 0.001567$	

* It is purely accidental that T_4 exactly equals 0.

Source: Vaughan et al. 1966.

is referred to the Scheffé table (in Table 4, above) for $k = 4$, or when 13.35 is referred to a table of chi-square with 3 *df*, each is found to be statistically significant (in fact, $P = .0040$). The conclusion that can be drawn from this one-sided test for trend is that the population mean scores are ordered $\tilde{\mu}_1 \geqslant \tilde{\mu}_2 \geqslant \mu_3 \geqslant \tilde{\mu}_4$ with at least one strict inequality holding. Had a trend in this particular order been predicted ahead of time and postulated as the sole alternative hypothesis to be considered, $z = 3.65$ could have been judged by the normal table, yielding $P = .00013$. The two-tail version of this test is Yates's one-degree-of-freedom chi-square for trend (1948).

Another contrast that may be tested is the simple difference $\bar{u}_1 - \bar{u}_4 = .443 - .000$. Here $SE = \sqrt{[(1/167) + (1/157)] \times 1.2494} = .1243$, and $z_{14} = .443/.1243 = 3.57$. Because it is greater than 3.37, this contrast is statistically significant. Similarly, $z_{13} = (.443 - .136)/.1239 = 2.48$, but this is not significant at the 1 per cent level, and the other simple differences are still smaller.

If Tukey's test had been adopted instead of Scheffé's, the same ratios would be compared with the critical value 3.11 ($k = 4$, $\alpha = .01$). The conclusions would be the same in the present case. Tukey's method could also be used to test other contrasts.

In the present example, the Newman–Keuls procedure would also have led to the same conclusions about simple differences: $z_{14} = 3.57$ is called significant because it is greater than 3.11; then z_{13} (which equals 2.48) and z_{24} (which is still smaller) are compared with 2.91 and found "not significant," and the procedure ends. The conclusions may be summarized as follows:

$$\underline{.443 \quad .245 \quad .136} \quad .000,$$

where the absence of a line connecting $\bar{u}_1$ with $\bar{u}_4$ signifies that $\tilde{\mu}_1$ and $\tilde{\mu}_4$ are declared unequal. It may be argued that a conclusion of the form "*A*, *B*, and *C* homogeneous, *B*, *C*, and *D* homogeneous, but *A*, *B*, *C*, and *D* not homogeneous" is self-contradictory. This is not necessarily the case if the interpretation is the usual one that *A*, *B*, and *C may* be equal (not enough evidence to prove them unequal) and *B*, *C*, and *D* may be equal, but *A* and *D* are not equal.

In Duncan's procedure the critical value 3.11 used in the first stage would be replaced by 2.76 (see Table 5), and the critical value 2.91 used at the second stage ($k = 3$) would be replaced by 2.68. Since $3.57 > 2.76$ but $2.48 < 2.68$, Duncan's test leads to the same conclusion in the present example as the Newman–Keuls procedure.

A Halperin outlier test would use max $|\bar{u}_i - \bar{u}|$, in this case $.443 - 2.10 = .233$, divide it by

$$\sqrt{4/3[(1/167) - (1/638)]\,1.2494} = .08582,$$

and compare the resulting ratio, 2.72, with the critical value, 2.61 ($k = 4$, 1 per cent level). The next largest ratio is $(.210 - .000)/.08944 = 2.35$. The conclusion is that chemical engineers tend to report more frequent church attendance than the other groups, but nothing can be said about geologists. If the outlier contrasts had been tested as part of a Scheffé test for all contrasts, none of them would have been found significant at the 1 per cent level (critical value 3.37) or even at the 5 per cent level.

What would happen if unequally spaced scores had been used instead of −1, 0, 1, 2 to quantify the four degrees of religious loyalty? In fact, Vaughan and his associates described the ordered categories not verbally but as frequency of church attendance per month grouped into 0, 1, 2–4, 5+. Although we do not know whether frequency of church attendance is a linear measure of the importance of religion in a person's life, the scores (0, 1, 3, 6) could reasonably have been assigned. In the present case this would lead to essentially the same conclusions that the other scoring led to: The mean scores become 2.35, 2.08, 1.82, and 1.57, Yates's χ^2 changes from 13.7 to 12.5, the standardized contrast for trend changes from $\sqrt{13.35}$ to very nearly $\sqrt{12.5}$, z_{14} changes from 3.57 to 3.36, and z_{13} and z_{24} again have values too small for statistical significance by Tukey's criterion or by Newman–Keuls'.

A fundamentally different assignment of scores —for example, 1, 0, 0, 1—would be used to test for differences in spread. It yields sample means, $\bar{u}_i$, of .461, .591, .548, .576, $\bar{u} = 0.541$ and a variance, σ^2, of .2484. Yates's analysis-of-variance χ^2 is 1.449/0.2484, that is, only 5.83, so $P = .12$. Thus, no contrast is called significant in a Scheffé test (or, it turns out, in any other multiple comparison test at the 1 per cent significance level). In the present example these tests for spread are unreliable, because the presence of sample location differences, noted above, can vitiate the results of the test for differences in spread.

Throughout the numerical calculations in this section, the continuity correction has been neglected. In the case of unequal sample sizes it is difficult to determine what continuity correction would yield the most accurate results, and the effect of the adjustment would be slight anyway. When sample sizes are equal, the use of $|T_i - T_j| - \frac{1}{2}$ in place of $|T_i - T_j|$ is recommended, as it frequently (although not invariably) improves the fit of the asymptotic approximation used.

11. Comparisons for differences in scale. Standand multiple comparisons of variances of k normal populations, by Cochran (1941), David (1952), and others, utilize ratios of the $\hat{\sigma}_i^2$. These methods should be used with caution, because they are ultrasensitive to slight nonnormality.

Distribution-free multiple comparison tests for scale differences are also available. Any rank test may be used with a Siegel–Tukey reranking [*see* NONPARAMETRIC STATISTICS, *article on* RANKING METHODS]. Such methods, too, require caution, because—especially in a joint ranking of all k samples—any sizable location differences may masquerade as differences in scale (Moses 1963).

Safer methods—but with efficiencies of only about 50 per cent for normal distributions—are adaptations of some tests by Moses (1963). In these tests a small integer, s, such as 2 or 3, is chosen, and each sample is randomly subdivided into subgroups of s observations. Let y be the range or variance of a subgroup. Then any multiple comparison tests may be applied to the k samples of y's (or $\log y$'s), at the sacrifice of between-subgroups information. The effective sample sizes have been reduced to $[n_i/s]$; if these are small (about 6), either a nonparametric test or, at any rate, $\log y$'s should be used. (Some nonparametric multiple comparison tests, such as the median test, have no power—that is, they cannot possibly reject the null hypothesis—at significance levels such as .05 with small samples. But rank tests can be used with several samples as small as 4 or 5.)

12. Multiple comparisons of proportions. A simultaneous test for all differences between k proportions $p_1, \cdots, p_k$, based on large samples, can be obtained by comparing

$$\text{Max} \frac{X_i/n_i - X_j/n_j}{\sqrt{\frac{N}{N-1}\left(\frac{1}{n_i} + \frac{1}{n_j}\right)\frac{X}{N}\left(1 - \frac{X}{N}\right)}}$$

with a critical value of Tukey's statistic (Section 3), where X_i, $i = 1, \cdots, k$, denotes the number of "successes" in sample i and $X = \sum X_i$. Analogous asymptotic tests can be used for comparison of several treatments with a control and other forms of multiple comparisons. If X/N is small, the sample sizes must be very large for this asymptotic approximation to be adequate. (For a similar method see Ryan 1960.)

Small-sample multiple comparison tests of proportions may be carried out by transforming the counts into normal variables with known equal variances and then applying any test of sections 1–7 to these standardized variables (using ∞ *ddf*). [*See* STATISTICAL ANALYSIS, SPECIAL PROBLEMS OF, *article on* TRANSFORMATIONS OF DATA; *see also* Siotani & Ozawa 1958.]

A $(1 - \alpha)$-confidence region for k population proportions is composed of a $\sqrt[k]{1 - \alpha}$-confidence interval for each of them. Simultaneous confidence intervals for a set of differences of proportions may be approximated by using Bonferroni's inequality (see Section 14). For a discussion of confidence regions for multinomial proportions, see Goodman (1965).

Some discussion of multiple comparisons of proportions can be found in Walsh (1965, for example, pp. 536–537).

13. Selection and ranking. The approach called selection or ranking *assumes* a difference between populations and seeks to select the population(s) with the highest mean—or variance or proportion —or to arrange all k populations in order [*see* SCREENING AND SELECTION; *see also* Bechhofer

1958]. Bechhofer, Kiefer, and Sobel (1967) have written a monograph on the subject.

Error rates, choice of method, history

14. Error rates and choice of method. In a significance test comparing two populations, the significance level is defined as

$$a = \frac{\text{number of type I errors}}{\text{number of comparisons when } H_0 \text{ is true}}$$

in repeated use of the same criterion. This is termed the *error rate per comparison.* The corresponding confidence level for confidence intervals is $1 - a$.

For analyses of k-sample experiments one may instead define the *error rate per experiment,*

$$a' = \frac{\text{number of type I errors}}{\text{number of experiments analyzed, when } H_0 \text{ is true}}.$$

This is related to what Miller (1966) terms the "expected error rate." For m (computed or implied) comparisons per experiment, $a' = ma$; $a = a'/m$ (see Stanley 1957).

Standard multiple comparison tests specify an *error rate experimentwise* (or "familywise"):

$$\alpha = \frac{\text{number of experiments, when } H_0 \text{ is true, leading to any type I errors,}}{\text{number of experiments analyzed, when } H_0 \text{ is true}}.$$

Miller refers to this as the "probability of a non-zero family error rate" or "probability error rate."

[2]The only difference between a' and α is that α counts multiple rejections in a single experiment as only one error whereas a' counts them as more than one. Hence, $\alpha \leqslant a'$; this is termed *Bonferroni's inequality.*

On the other hand, it is also true that unless a' is large, α is almost equal to a', so that α and a' may be used interchangeably for practical purposes. For example, D_α for 6 treatments and a control, M_α for $k = 6$, and T_α for $k = 4$ ($\binom{4}{2} = 6$), are all approximately equal to the two-tailed critical value $|t|_{\alpha/6} = t_{\alpha/12}$ of Student's t. More generally, m individual comparisons may safely be made using any statistic at significance level α/m per comparison when it is desired to avoid error rates greater than α experimentwise; this procedure may be applied to comparisons of several correlation coefficients or other quantities for which multiple comparison tables are not available. Only when α is about .10 or more, or when m is very big, does this lead to serious waste. Then a' grossly overstates α, power is lost, and confidence intervals are unnecessarily long (see Stanley 1957; Ryan 1959; Dunn 1961).

Some authors refer to (α/m)-points as Bonferroni statistics and to their use in multiple comparisons as the Bonferroni method. Table 2 in Miller (1966) shows Bonferroni t-statistics, $(.05/2m)$-points of Student's t for various m and various numbers of *ddf*.

Bonferroni's second inequality (see Halperin et al. 1955, p. 191) may sometimes be used to obtain an upper limit for the discrepancy $a' - \alpha$ and a second approximation to critical values for error rates α experimentwise. This works best in the case of slippage statistics and was used by Halperin and his associates (1955), Doornbos and Prins (1958), Thompson and Willke (1963), and others.

The choice between "experimentwise" and "per comparison" is largely a matter of taste. An experimenter should make it consciously, aware of the implications: A given error probability, a, per comparison implies that the risk of at least one type I error in the analysis is much greater than a; indeed, about $a \times m$ such errors will probably occur.

Perhaps analyses reporting error rates experimentwise are generally the most honest, or transparent. However, too dogmatic an application of this principle would lead to all sorts of difficulties. Should not the researcher who in the course of his career analyzes 86 experiments involving 1,729 relevant contrasts control the error rate *lifetimewise*? If he does not, he is almost bound to make a false positive inference sooner or later.

Sterling (1959) discusses the related problem of concentration of type I errors in the literature that result from the habit of selecting significant findings for publication [*see* FALLACIES, STATISTICAL, *for further discussion of this problem*].

There is another context in which the problem of choosing error rates arises: If an experimenter laboriously sets up expensive apparatus for an experiment to compare two treatments or conditions in which he is especially interested, he often feels that it would be unfortunate to pass up the opportunity to obtain additional data of secondary interest at practically no extra cost or trouble; so he makes observations on populations 3, 4, $\cdots$, k as well. It is then possible that the results are such that a two-sample test on the data of primary interest would have shown statistical significance, but no "significant differences" are found in a multiple comparison test. If the bonus observations thus drown out, so to speak, the significant difference, was the experimenter wrong to read them? He was not—the opportunity to obtain extra information should not be wasted, but the analysis should be planned ahead of time with the experimenter's interests and priorities in mind. He could decide to analyze his primary and subsidiary results as if they had come from separate experiments, or he could conduct multiple comparisons with an overall error rate enlarged to avoid undue loss of power, or he could use a method of analysis which subdivides α, allocating a certain (large) part to the primary comparison and the rest to "data snooping" among the extra observations (Miller 1966, chapter 2, sec. 2.3).

Whenever it is decided to specify error rates experimentwise, a choice between different systems of multiple comparisons (different shapes of confidence regions) remains to be made. In order to study simple differences or slippage only, one of the methods of sections 2–4 above (or a nonparametric version of them) is best—that is, yields the shortest confidence intervals and most powerful tests, provided the n_i are (nearly) equal. But Scheffé's approach (see section 5) is better if a variety of contrasts may receive attention.

When sample sizes are grossly unequal, probability statements based on existing Tukey or Dunnett tables, computed for equal n's, become too inaccurate. Pending the appearance of appropriate new tables, it is better to use Scheffé's method, which furnishes exact probabilities. The Bonferroni statistics discussed above offer an alternative solution, preferable whenever attention is strictly limited to a few contrasts chosen ahead of time. Miller (1966, especially chapter 2, secs. 2.3 and 3.3) discusses these questions in some detail.

15. History of multiple comparisons. An early, isolated example of a multiple comparison method was one developed by Working and Hotelling (1929) to obtain a confidence belt for a regression line (see Miller 1966, chapter 3; Kerrich 1955). This region also corresponds to simultaneous confidence intervals for the intercept and slope [*see* LINEAR HYPOTHESES, *article on* REGRESSION]. Hotelling (1927) had already developed the idea of simultaneous confidence interval estimation earlier in connection with the fitting of logistic curves to population time series. In his famous paper introducing the T^2-statistic, Hotelling (1931) also introduced the idea of simultaneous tests and a confidence ellipsoid for the components of a multivariate normal mean.

The systematic development of multiple comparison methods and theory began later, in connection with the problem of comparing several normal means. The usual method had been the analysis-of-variance F-test, sometimes accompanied by t-tests at a stated significance level, a (usually 5 per cent), per comparison.

Fisher, in the 1935 edition of *The Design of Experiments*, pointed out the problem of inflation of error probabilities in such multiple t-tests and recommended the use of t-tests at a stated level a' per experiment. Pearson and Chandra Sekar further discussed the problem (1936). Newman (1939), acting on an informal suggestion by Student, described a test for all differences based on tables of the Studentized range and furnished a table of approximate 5 per cent and 1 per cent points. Keuls formulated Newman's test more clearly much later (Keuls 1952).

Nair made two contributions in 1948, the one-sided test for slippage of means and a table for simultaneous F-tests in a 2^r factorial design. Also in the late 1940s, Duncan and Tukey experimented with various tests for normal means which were forerunners of the multiple comparison tests now associated with their names.

The standard methods for multiple comparisons of normal means were developed between 1952 and 1955 by Tukey, Scheffé, Dunnett, and Duncan. Tukey wrote a comprehensive volume on the subject which was widely circulated in duplicated form and extensively quoted but which has not been published (1953). The form of Tukey's method described in Section 3 for unequal n's was given independently by Kurtz and by Kramer in 1956. Also in the early and middle 1950s, some multiple comparison methods for normal *variances* were published, by Hartley, David, Truax, Krishnaiah, and others. Cochran's slippage test for normal variances was published, for use as a substitute for Bartlett's test for homogeneity of variances, as early as 1941 (see Cochran 1941).

Selection and ranking procedures for means, variances, and proportions have been developed since 1953 by Bechhofer and others.

An easy, distribution-free slippage test was proposed by Mosteller in 1948—simply count the number of observations in the most extreme sample lying beyond the most extreme value of all the other samples and refer to a table by Mosteller and Tukey (1950). Other distribution-free multiple comparison methods—although some of them can be viewed as applications of S. N. Roy's work of 1953—did not begin to appear until after 1958.

The most important applications of the very general methodology developed by the school of Roy and Bose since 1953 have been *multivariate* multiple comparison tests and confidence regions. Such work by Roy, Bose, Gnanadesikan, Krishnaiah,

Gabriel, and others is generally recognizable by the word "simultaneous" in the title—for example, SMANOVA, that is, simultaneous multivariate analysis of variance (see Miller 1966, chapter 5).

Another recent development is the appearance of some Bayesian techniques for multiple comparisons. These are discussed by Duncan in the May 1965 issue of *Technometrics*, an issue which is devoted to articles on multiple comparison methods and theory and reflects a cross section of current trends in this field.

PETER NEMENYI

BIBLIOGRAPHY

The only comprehensive source for the subject of multiple comparisons to date is Miller 1966. *Multiple comparisons of normal means (and variances) are summarized by a number of authors, notably David* 1962*a and* 1962*b*. *Several textbooks on statistics—e.g.,* Winer 1962*—also cover some of this ground. Many of the relevant tables, for normal means and variances, can also be found in* David 1962*a and* 1962*b*; Vianelli 1959; *and* Pearson & Hartley 1954; *these volumes also provide explanations of the derivation and use of the tables.*

BECHHOFER, R. E. 1958 A Sequential Multiple-decision Procedure for Selecting the Best One of Several Normal Populations With a Common Unknown Variance, and Its Use With Various Experimental Designs. *Biometrics* 14:408–429.

BECHHOFER, R. E.; KIEFER, J.; and SOBEL, M. 1967 Sequential Ranking Procedures. Unpublished manuscript. → Projected for publication by the University of Chicago Press in association with the Institute of Mathematical Statistics.

COCHRAN, W. G. 1941 The Distribution of the Largest of a Set of Estimated Variances as a Fraction of Their Total. *Annals of Eugenics* 11:47–52.

CRONBACH, LEE J. 1949 Statistical Methods Applied to Rorschach Scores: A Review. *Psychological Bulletin* 46:393–429.

DAVID, H. A. 1952 Upper 5 and 1% Points of the Maximum *F*-ratio. *Biometrika* 39:422–424.

DAVID, H. A. 1962*a* Multiple Decisions and Multiple Comparisons. Pages 144–162 in Ahmed E. Sarhan and Bernard G. Greenberg (editors), *Contributions to Order Statistics*. New York: Wiley.

DAVID, H. A. 1962*b* Order Statistics in Shortcut Tests. Pages 94–128 in Ahmed E. Sarhan and Bernard G. Greenberg (editors), *Contributions to Order Statistics*. New York: Wiley.

DOORNBOS, R.; and PRINS, H. J. 1958 On Slippage Tests. Part 3: Two Distribution-free Slippage Tests and Two Tables. *Indagationes mathematicae* 20:438–447.

DUNCAN, DAVID B. 1955 Multiple Range and Multiple *F* Tests. *Biometrics* 11:1–42.

DUNCAN, DAVID B. 1965 A Bayesian Approach to Multiple Comparisons. *Technometrics* 7:171–222.

DUNN, OLIVE J. 1961 Multiple Comparisons Among Means. *Journal of the American Statistical Association* 56:52–64.

DUNNETT, CHARLES W. 1955 A Multiple Comparison Procedure for Comparing Several Treatments With a Control. *Journal of the American Statistical Association* 50:1096–1121.

FISHER, R. A. (1935) 1960 *The Design of Experiments*. 7th ed. London: Oliver & Boyd; New York: Hafner.

FISHER, R. A.; and YATES, FRANK (1938) 1963 *Statistical Tables for Biological, Agricultural, and Medical Research*. 6th ed., rev. & enl. Edinburgh: Oliver & Boyd; New York: Hafner.

GABRIEL, K. R. 1966 Simultaneous Test Procedures for Multiple Comparisons on Categorical Data. *Journal of the American Statistical Association* 61:1081–1096.

GOODMAN, LEO A. 1965 On Simultaneous Confidence Intervals for Multinomial Proportions. *Technometrics* 7:247–252.

HALPERIN, M.; GREENHOUSE, S.; CORNFIELD, J.; and ZALOKAR, J. 1955 Tables of Percentage Points for the Studentized Maximum Absolute Deviate in Normal Samples. *Journal of the American Statistical Association* 50:185–195.

HARTER, H. LEON 1957 Error Rates and Sample Sizes for Range Tests in Multiple Comparisons. *Biometrics* 13:511–536.

HARTER, H. LEON 1960 Tables of Range and Studentized Range. *Annals of Mathematical Statistics* 31:1122–1147.

HARTLEY, H. O. 1950 The Maximum *F*-ratio as a Short-cut Test for Heterogeneity of Variance. *Biometrika* 37:308–312.

HOTELLING, HAROLD 1927 Differential Equations Subject to Error, and Population Estimates. *Journal of the American Statistical Association* 22:283–314.

HOTELLING, HAROLD 1931 The Generalization of Student's Ratio. *Annals of Mathematical Statistics* 2:360–378.

KERRICH, J. E. 1955 Confidence Intervals Associated With a Straight Line Fitted by Least Squares. *Statistica neerlandica* 9:125–129.

KEULS, M. 1952 The Use of "Studentized Range" in Connection With an Analysis of Variance. *Euphytica* 1:112–122.

KRAMER, CLYDE Y. 1956 Extension of Multiple Range Tests to Group Means With Unequal Number of Replications. *Biometrics* 12:307–310.

KRAMER, CLYDE Y. 1957 Extension of Multiple Range Tests to Group Correlated Adjusted Means. *Biometrics* 13:13–18.

KRISHNAIAH, P. R. 1965*a* On a Multivariate Generalization of the Simultaneous Analysis of Variance Test. Institute of Statistical Mathematics (Tokyo), *Annals* 17, no. 2:167–173.

KRISHNAIAH, P. R. 1965*b* Simultaneous Tests for the Equality of Variance Against Certain Alternatives. *Australian Journal of Statistics* 7:105–109.

KURTZ, T. E. 1956 An Extension of a Method of Making Multiple Comparisons (Preliminary Report). *Annals of Mathematical Statistics* 27:547 only.

KURTZ, T. E.; LINK, R. F.; TUKEY, J. W.; and WALLACE, D. L. 1965 Short-cut Multiple Comparisons for Balanced Single and Double Classifications. Part 1: Results. *Technometrics* 7:95–169.

MCHUGH, RICHARD B.; and ELLIS, DOUGLAS S. 1955 The "Post Mortem" Testing of Experimental Comparisons. *Psychological Bulletin* 52:425–428.

MILLER, RUPERT G. 1966 *Simultaneous Statistical Inference*. New York: McGraw-Hill.

MOSES, LINCOLN E. 1953 Nonparametric Methods. Pages 426–450 in Helen M. Walker and Joseph Lev, *Statistical Inference*. New York: Holt.

MOSES, LINCOLN E. 1963 Rank Tests of Dispersion. *Annals of Mathematical Statistics* 34:973–983.

MOSES, LINCOLN E. 1965 Confidence Limits From Rank Tests (Reply to a Query). *Technometrics* 7:257–260.

MOSTELLER, FREDERICK W.; and TUKEY, JOHN W. 1950 Significance Levels for a *k*-sample Slippage Test. *Annals of Mathematical Statistics* 21:120–123.

NAIR, K. R. 1948*a* The Studentized Form of the Extreme Mean Square Test in the Analysis of Variance. *Biometrika* 35:16–31.

NAIR, K. R. 1948*b* The Distribution of the Extreme Deviate From the Sample Mean and Its Studentized Form. *Biometrika* 35:118–144.

NAIR, K. R. 1952 Tables of Percentage Points of the "Studentized" Extreme Deviate From the Sample Mean. *Biometrika* 39:189–191.

NEMENYI, PETER 1963 Distribution-free Multiple Comparisons. Ph.D. dissertation, Princeton Univ.

NEWMAN, D. 1939 The Distribution of the Range in Samples From a Normal Population, Expressed in Terms of an Independent Estimate of Standard Deviation. *Biometrika* 31:20–30.

PEARSON, EGON S.; and CHANDRA SEKAR, C. 1936 The Efficiency of Statistical Tools and a Criterion for the Rejection of Outlying Observations. *Biometrika* 28: 308–320.

PEARSON, EGON S.; and HARTLEY, H. O. (editors) (1954) 1966 *Biometrika Tables for Statisticians*. Vol. 1. 3d ed. Cambridge Univ. Press.

PILLAI, K. C. S.; and RAMACHANDRAN, K. V. 1954 On the Distribution of the Ratio of the *i*th Observation in an Ordered Sample From a Normal Population to an Independent Estimate of the Standard Deviation. *Annals of Mathematical Statistics* 25:565–572.

QUESENBERRY, C. P.; and DAVID, H. A. 1961 Some Tests for Outliers. *Biometrika* 48:379–390.

ROESSLER, R. G. 1946 Testing the Significance of Observations Compared With a Control. American Society for Horticultural Science, *Proceedings* 47:249–251.

ROTHAUS, PAUL; and WORCHEL, PHILIP 1964 Ego Support, Communication, Catharsis, and Hostility. *Journal of Personality* 32:296–312.

ROY, S. N.; and BOSE, R. C. 1953 Simultaneous Confidence Interval Estimation. *Annals of Mathematical Statistics* 24:513–536.

ROY, S. N.; and GNANADESIKAN, R. 1957 Further Contributions to Multivariate Confidence Bounds. *Biometrika* 44:399–410.

RYAN, THOMAS A. 1959 Multiple Comparisons in Psychological Research. *Psychological Bulletin* 56:26–47.

RYAN, THOMAS A. 1960 Significance Tests for Multiple Comparisons of Proportions, Variances, and Other Statistics. *Psychological Bulletin* 57:318–328.

SCHEFFÉ, HENRY 1953 A Method for Judging All Contrasts in the Analysis of Variance. *Biometrika* 40:87–104.

SIOTANI, M.; and OZAWA, MASARU 1958 Tables for Testing the Homogeneity of *k* Independent Binomial Experiments on a Certain Event Based on the Range. Institute of Statistical Mathematics (Tokyo), *Annals* 10:47–63.

STANLEY, JULIAN C. 1957 Additional "Post Mortem" Tests of Experimental Comparisons. *Psychological Bulletin* 54:128–130.

STEEL, ROBERT G. D. 1959 A Multiple Comparison Sign Test: Treatments vs. Control. *Journal of the American Statistical Association* 54:767–775.

STERLING, THEODORE D. 1959 Publication Decisions and Their Possible Effects on Inferences Drawn From Tests of Significance—or Vice Versa. *Journal of the American Statistical Association* 54:30–34.

THOMPSON, W. A. JR.; and WILLKE, T. A. 1963 On an Extreme Rank Sum Test for Outliers. *Biometrika* 50: 375–383.

TRUAX, DONALD R. 1953 An Optimum Slippage Test for the Variances of *k* Normal Populations. *Annals of Mathematical Statistics* 24:669–674.

TUKEY, J. W. 1953 The Problem of Multiple Comparisons. Unpublished manuscript, Princeton Univ.

VAUGHAN, TED R.; SJOBERG, G.; and SMITH, D. H. 1966 Religious Orientations of American Natural Scientists. *Social Forces* 44:519–526.

VIANELLI, SILVIO 1959 *Prontuari per calcoli statistici: Tavole numeriche e complementi*. Palermo: Abbaco.

WALSH, JOHN E. 1965 *Handbook of Nonparametric Statistics*. Volume 2: Results for Two and Several Sample Problems, Symmetry, and Extremes. Princeton, N.J.: Van Nostrand.

WINER, B. J. 1962 *Statistical Principles in Experimental Design*. New York: McGraw-Hill.

WORKING, HOLBROOK; and HOTELLING, HAROLD 1929 Application of the Theory of Error to the Interpretation of Trends. *Journal of the American Statistical Association* 24 (March Supplement):73–85.

YATES, FRANK 1948 The Analysis of Contingency Tables With Groupings Based on Quantitative Characters. *Biometrika* 35:176–181.

Postscript

Miller (1976) provides an extensive bibliography of publications in the field of multiple comparisons that have appeared in the decade 1966–1976 and a summary of some of these new developments.

[1]Spjøtvoll and Stoline (1973) proposed a new modification of the Tukey Studentized range procedure for the case of unequal sample sizes. The form described in the main article, using the same standard error ($\sqrt{(1/n_i + 1/n_j)}\hat{\sigma}$) as in the Student two-sample t, is itself a modification, proposed independently by Kurtz (1956) and Kramer (1956). The form originally suggested by Tukey for unequal n's uses the average of all the $(1/n_i)$ in place of $1/n_i$ and $1/n_j$; both methods only approximate the stated confidence level $1-\alpha$. Spjøtvoll and Stoline use the larger of $1/n_i$ and $1/n_j$ in place of both and show that a further slight modification, use of the Studentized augmented range in place of the Studentized range, will make the experimentwise error rate exactly equal to the stated $1-\alpha$. (Tables for this apparently are not available to date.) Hochberg (1975; 1976) modified the method by using tables of the Studentized maximum modulus of $\binom{k}{2}$ means with the standard error of Kramer (or Student); in

the case of ddf = ∞ (known variance) this is the same as the method of Šidák (1967) mentioned below.

Petrinovich and Hardyck (1969), Ury (1976), Einot and Gabriel (1975), Games (1976), the above-mentioned authors, and others have undertaken studies of experimentwise confidence levels (or error rates) or lengths of confidence intervals under these methods using the modified Tukey procedures compared with Tukey, Scheffé, Bonferroni, and Duncan's multiple comparisions. Which one of these procedures—Scheffé, Tukey, the modifications, or Bonferroni—will be a little more satisfactory than the others depends on k, the n's, the ratio of n's, and the number of degrees of freedom with which the variance is estimated. Some of the papers offer complicated rules for choosing among procedures.

[2]A discovery of some theoretical interest is a refinement of the Bonferroni inequality by Šidák (1967). Bonferroni's inequality says that the use of t statistics (or any statistics) at confidence levels $1 - \alpha/m$ per comparison in m comparisons guarantees that the experimentwise confidence level will be $\geqslant 1 - \alpha$. So does the use of t statistics at confidence levels $(1 - \alpha)^{1/m}$. This is what Šidák proved; it is remarkable because the formula makes the experimentwise confidence level exactly equal to $1 - \alpha$ in the case of m *independent* comparisons, and the inequality says that positive and/or negative correlations alike will all increase the experimentwise confidence level. Šidák's modification makes for slightly more powerful tests and shorter confidence intervals than Bonferroni's inequality.

For a Bayesian treatment of multiple comparisons, see Duncan (1975). There has also been work in multiple comparison tests for interactions, for example, Bradu and Gabriel (1974), and in setting confidence bands around regression lines and response surfaces, for example, Bohrer (1969).

Comment. Whenever a survey or experiment yields more than two groups of observations, and especially when more than one factor is varied, there is an opportunity to make more than one comparison. Indeed the comparisons of potential interest to the researcher may be numerous. When making multiple comparisons, it is important to recognize that the opportunity to reach false conclusions is also multiple, and the use of tables computed to limit the risk of error in one comparison to a standard α will lead to a much higher error risk per experiment. Tables of adjusted critical values have been prepared that can be substituted for the standard critical values in order to reduce the error risk to 5 per cent or 1 per cent experimentwise, provided the set of comparisons contemplated is of a certain form. The use of such tables may be convenient and sensible under certain circumstances, but their value has perhaps been exaggerated in some literature. There is no justification for the dogma that error risks must always be expressed per this or per that, for the practicing statistician must be aware that each additional day of a professional life entails additional risks of Type 1, Type 2, and other errors.

On the other hand a good argument can be made that it is inappropriate just blandly to declare a long list of observed differences in an experiment "significant" at the 5 per cent level without pointing out that some of them would be almost certain to qualify just by chance alone. When a study provides a potential for many comparisons including just a few bearing on the question of paramount interest to the investigator, the following procedure may be a reasonable compromise. Identify the comparisons of paramount interest ahead of time (before looking at data). If there are m of these, allow $\alpha = 5$ per cent (or 1 per cent) error probability to be divided among them (α/m each) so that any positive statements about them are made with $1 - \alpha$ confidence. Then, test any other striking differences when the data are in, using an ordinary table to allow an error risk of .05 or .01 per comparison. Any positive results (that is, rejections of null hypotheses) obtained in the final step should be treated as only suggestive and possibly worth following up in subsequent research.

PETER NEMENYI

ADDITIONAL BIBLIOGRAPHY

BOHRER, ROBERT 1969 On One-sided Confidence Bounds for Response Surfaces. International Statistical Institute, *Bulletin* 43, no. 2:255–257.

BRADU, DAN; and GABRIEL, K. R. 1974 Simultaneous Statistical Inference on Interactions in Two-way Analysis of Variance. *Journal of the American Statistical Association* 69:428–436.

DUNCAN, DAVID B. 1975 *t* Tests and Intervals for Comparisons Suggested by the Data. *Biometrics* 31:339–359.

EINOT, ISRAEL; and GABRIEL, K. R. 1975 A Study of the Powers of Several Methods of Multiple Comparisons. *Journal of the American Statistical Association* 70: 574–583.

GAMES, P. S.; and HOWELL, J. F. 1976 Pairwise Multiple Comparison Procedures With Unequal *N*'s and/or Variances: A Monte Carlo Study. *Journal of Educational Statistics* 1:113–115.

HOCHBERG, YOSEF 1975 An Extension of the *T*-method to General Unbalanced Models of Fixed Effects. *Journal of the Royal Statistical Society* Series B 37:426–433.

HOCHBERG, YOSEF 1976 A Modification of the *T*-method of Multiple Comparisons for a One-way Layout With Unequal Variances. *Journal of the American Statistical Association* 71:200–203.

MILLER, ROBERT G. JR. 1976 Advances in Multiple Comparisons in the Last Decade. Stanford Univ., Div. of Biostatistics, Technical Report No. 26.

PETRINOVICH, LEWIS F.; and HARDYCK, CURTIS D. 1969 Error Rates for Multiple Comparison Methods: Some Evidence Concerning the Frequency of Erroneous Conclusions. *Psychological Bulletin* 71:43–54.

ŠIDÁK, ZBYNĚK 1967 Rectangular Confidence Regions for the Means of Multivariate Normal Distributions. *Journal of the American Statistical Association* 62: 626–633.

SPJØTVOLL, EMIL; and STOLINE, MICHAEL R. 1973 An Extension of the *T*-method of Multiple Comparisons to Include the Cases With Unequal Sample Sizes. *Journal of the American Statistical Association* 68: 975–978.

URY, HANS K. 1976 A Comparison of Four Procedures for Multiple Comparisons Among Means (Pairwise Contrasts) for Arbitrary Sample Sizes. *Technometrics* 18:89–97.

LINEAR PROGRAMMING

See OPERATIONS RESEARCH; PROGRAMMING.

LINEAR REGRESSION

See under LINEAR HYPOTHESES.

LOCATION AND DISPERSION

See under STATISTICS, DESCRIPTIVE.

LOCATION THEORY

See GEOGRAPHY, STATISTICAL.

LOG-LINEAR MODELS

See COUNTED DATA; SOCIAL MOBILITY; SURVEY ANALYSIS.

LONGITUDINAL STUDIES

See PANEL STUDIES; TIME SERIES.

M

MAHALANOBIS, P. C.

► *This article was specially written for this volume.*

Prasanta Chandra Mahalanobis (1893–1972) was born in a well-known family of Brahmos (a protestant theist movement within the fold of Hinduism) in Calcutta and had his early education in Brahmo Boys' School. After earning a B.SC. with honors in physics in 1912 from Presidency College, Calcutta, he went to Cambridge, where he took part I of the mathematics tripos in 1914 and part II of the physics tripos in 1915. Awarded a senior scholarship by King's College, Cambridge, he intended to work with the physicist C. T. R. Wilson at the Cavendish Laboratory, but upon returning to India for a short vacation he found so much to hold his interest that he remained there. Shortly after his return, Mahalanobis fell in love with the 16-year-old Nirmalkumari (or Rani as she is popularly known), and they were married in 1923.

Phenomenal activity. Mahalanobis was offered a teaching position in the physics department of Presidency College at Calcutta soon after his return from Cambridge. He joined the Indian Educational Service (IES) in 1915, held the post of professor of physics for a long time, and later became principal of Presidency College. He retired in 1948. During his life he held several distinguished posts, many of them simultaneously. He was the chief executive of the Indian Statistical Institute as secretary and director continuously from the founding of the institute in 1931; he was honorary statistical advisor to the government of India from 1949; and he was associated with the work of the Planning Commission as a member (1955–1967). He also served as head of the department of statistics of Calcutta University (1941–1945) and statistical advisor to the government of Bengal (1945–1948). He was a member of the U.N. Statistical Subcommission on Sampling from the time it was formed in 1946 and its chairman from 1954 to 1958; he was general secretary of the Indian Science Congress (1945–1948) and later its treasurer (1952–1955) and president (1949–1950); and he was president of the Indian National Science Academy (1957–1958) and editor of *Sankhyā: The Indian Journal of Statistics,* from its foundation in 1933.

An original thinker, Mahalanobis was suspicious of accepted knowledge, chose challenging problems irrespective of subject matter, and approached them in unconventional ways. According to him, "the value of science to society lies in its unorthodoxy and ability to challenge accepted concepts and theories" (1960, p. 112).

When Mahalanobis founded the Indian Statistical Institute in 1931, it employed one part-time computing assistant and had an annual budget of $50. At the time of his death, forty years later, the institute had nearly 2,000 employees and an annual budget of $2.5 million. The development of the institute as a teaching and research organization of international importance and recognition was itself a Herculean task.

Scientific contributions. While in Cambridge, Mahalanobis came across volumes of *Biometrika,* edited by Karl Pearson. He got so interested in them that he took a complete set back to India. He started reading them on the boat during his journey and discovered that statistics was a new discipline capable of wide application. Upon arrival in India, he looked for problems to which he could apply statistics. The resulting highly interesting problems in meteorology and anthropology marked the turning point in his career from a physicist to a statistician.

Mahalanobis distance. The first opportunity to use statistical methods came to Mahalanobis when N. Anandale (then director of the Zoological Survey of India) asked him to analyze anthropometric measurements taken on Anglo-Indians (of mixed British and Indian parentage) in Calcutta. This study led to Mahalanobis's first scientific paper (1922), and it was followed by other anthropometric investigations leading to formulation of the D^2 statistic (1930; 1936), often known in the statistical literature as *Mahalanobis distance* and widely used in taxonomic classification (1940*b*). [*See* MULTIVARIATE ANALYSIS, *article on* CLASSIFICATION AND DISCRIMINATION; *see also* CLUSTERING.]

Suppose each individual of a population is characterized by p measurements, perhaps anthropological, but the approach is generally applicable. The means (averages) of the p measurements can be represented as a point in a p dimensional space. Corresponding to k given populations, we have a configuration of k points in the p space. If $p = 2$, the points can be represented on a two-dimensional chart, and the affinities between populations as measured by nearness of mean values can be graphically examined. We may find that some populations are close to each other in the mean values of the characters studied, thus forming a cluster. The entire configuration of points may then be described in terms of distinct clusters and relations between clusters. We can also determine the groups forming clusters at different levels of mutual distances. Such a description may help in drawing inferences on interrelations between populations and speculating on their origins.

As p grows larger than 2, however, graphical examination becomes difficult or impossible. How might one sensibly measure the difference between the mean vectors of, for example, populations 1 and 2? Call these vectors $\boldsymbol{\delta}_1$ and $\boldsymbol{\delta}_2$. A direct approach—one that had been used by anthropologists—was to work with the squared length of the column vector $\boldsymbol{\delta}_1 - \boldsymbol{\delta}_2$,

$$(\boldsymbol{\delta}_1 - \boldsymbol{\delta}_2)'(\boldsymbol{\delta}_1 - \boldsymbol{\delta}_2),$$

but this approach pays no attention to internal population covariation among the coordinates, nor to heterogeneity of variances. A related disadvantage of this simple sum-of-squares distance is its noninvariance under linear transformations of the coordinate system.

Mahalanobis, perhaps motivated by his studies of mathematical physics, suggested a more useful kind of distance, providing that the dispersion (variance–covariance) matrix of the two populations is the same, say $\boldsymbol{\Lambda}$. The Mahalanobis distance, traditionally given in squared form and called D^2, is

$$D^2 = (\boldsymbol{\delta}_1 - \boldsymbol{\delta}_2)'\boldsymbol{\Lambda}^{-1}(\boldsymbol{\delta}_1 - \boldsymbol{\delta}_2),$$

so that D^2 is in a sense the standardized squared difference between $\boldsymbol{\delta}_1$ and $\boldsymbol{\delta}_2$, generalizing the univariate $(\mu_1 - \mu_2)^2/\sigma^2$ where σ^2 is the common variance and μ_1 and μ_2 are mean values.

If $\boldsymbol{\Lambda}$ is a common dispersion matrix for all k populations, one may then examine the $k(k-1)/2$ different D^2's and base on them various statistical analyses, including cluster analysis. One large-scale anthropometric cluster analysis was published by Mahalanobis and others in 1949. These developments were closely related to research by R. A. Fisher and by Harold Hotelling. The approach is especially useful when the clusters of points for each population are, roughly speaking, ellipsoidal and with about the same shape from population to population. [*See the biographies of* FISHER *and* HOTELLING.]

Fuller and more precise statements of the above exposition will be found in Mahalanobis's writings listed in the bibliography and in chapter 9 of my monograph *Advanced Statistical Methods in Biometric Research* (1952).

Mahalanobis argued that inferences based on distances among populations might depend on the particular measurements chosen for study. The configuration may change, and even the order relations between distances may be disturbed, if one set of measurements is replaced by another set. Mahalanobis therefore laid down an important axiom for the validity of cluster analysis (1936) called *dimensional convergence* of D^2.

When a comparison between two real populations is made, one should ideally consider all possible (relevant) measurements, typically infinite in number. Consequently, cluster analysis of a given set of populations should ideally be based on distances computed on an infinite number of measurements. If D_p^2 and D_∞^2 denote Mahalanobis distances between two populations based on p characters and all possible characters respectively, then it can be shown under some conditions that

$$D_p^2 \rightarrow D_\infty^2 \text{ as } p \rightarrow \infty$$

(naturally after making that expression precise; see Rao 1954). Since in practice one can study only a finite number p of characteristics, D_p^2 should be a good approximation to D_∞^2 if cluster analysis is to be stable. Mahalanobis showed that stability, in important senses, was possible if and only if D^2_∞ is finite.

Meteorological research. Sir Gilbert Walker (then director general of observatories in India) referred to Mahalanobis some meteorological problems for statistical study. This resulted in two memoirs and a note on upper air variables (1923). Correcting meteorological data for errors of observation, Mahalanobis established by purely statistical methods that the region of highest control for changes in weather conditions on the surface on the earth is located about four kilometers above sea level (a result rediscovered later by Franz Bauer in Germany from physical considerations).

Early examples of operations research. In 1922 a disastrous flood occurred in North Bengal. An expert government committee of engineers was about to recommend the construction of expensive retarding basins to hold up the flood water, when the question was referred to Mahalanobis for examination. A statistical study of rainfall and floods extending over a period of fifty years showed that the proposed retarding basins would be of no value in controlling floods in North Bengal. The real need was improvement of rapid drainage and not holding up the flood water. Specific remedies were recommended, many of which were implemented and proved effective (1927).

A similar question of flood control in Orissa was referred to Mahalanobis, after a severe flood of the Brahmini River in 1926. An expert committee of engineers were of the opinion that the bed of the Brahmini had risen, and they recommended increasing the height of river embankments by several feet. The statistical study covering a period of about sixty years showed that no change had occurred in the river bed and that construction of dams for holding excessive flood water in the upper reaches of the river would provide an effective flood control (1931; 1940*c*). Mahalanobis pointed out that dams could be used for the generation of electric power needed for the economic development of the region. He also gave first calculations for a multipurpose (flood control, irrigation, and power) scheme for the Mahanadi system in Orissa, which formed the basis of the Hirakud hydroelectric project inaugurated about thirty years later in 1957.

Large-scale sample surveys. Large-scale sample survey techniques as practiced today owe much to the pioneering work of Mahalanobis in the 1940s and 1950s. He saw the need for sample surveys in collecting information, especially in developing countries, where official statistical systems are poor and data are treated as an integral part of the administrative system regulated by the principle of authority. A sample survey, properly conducted, would provide a wealth of data, useful for planning and policy purposes, expeditiously, economically, and with a reasonable degree of accuracy, and at the same time ensure objectivity of data. [*See also* SAMPLE SURVEYS.]

The methodology of large-scale sample surveys was developed during 1937–1944 in connection with the numerous surveys planned and executed by the Indian Statistical Institute. The survey topics included consumer expenditure, tea-drinking habits, public opinion and public preferences, acreage under a crop, crop yields, velocity of circulation of rupee coins, and incidence of plant diseases. The basic results on large-scale sample surveys were published in 1944 and also presented at a meeting of the Royal Statistical Society (1961*a*).

The *Philosophical Transactions* memoir on sample surveys (1944) is a classic in many ways, touching on fundamental problems: what is randomness? what constitutes a random sample? can different levels of randomness be identified? It gives the basic theory of sample surveys and estimation procedures.

In the 1944 paper, Mahalanobis described a variety of designs now in common use, such as simple random sampling with or without replacement, stratified, systematic, and cluster sampling. He was also familiar with multistage and multiphase sampling, and with ratio and regression methods of estimation. He was, in a sense, conscious of selection with probability proportional to size (area); he pointed out that selection of fields on the basis of cumulative totals of the areas of millions of fields was difficult.

Mahalanobis made three notable contributions to sample survey techniques: pilot surveys, concepts of optimum survey design, and interpenetrating network of samples.

A pilot survey provides basic information on operational costs and the variability of characters, which are two important factors in designing an optimum survey. It gives an opportunity to test the suitability of certain schedules or questionnaires to be used in the survey. A pilot survey can also be used to construct a suitable frame for sampling of units.

From the beginning, Mahalanobis was clear about the principles of good sample design. He wrote (1940*a*, p. 513) that

> from the statistical point of view our aim is to evolve a sampling technique which will give, for any given total expenditure, the highest possible accuracy in the

final estimate and for given precision in the estimates, the minimum possible total expenditure. . . .

For this it is necessary to determine three things, viz., (a) what is the best size of the sample units?, (b) what is the total number of sample units which should be used to obtain the desired degree of accuracy in the final estimates?, and (c) what is the best way of distributing the sampling units among different districts, regions, or zones covered by the survey?

Mahalanobis constructed appropriate variance and cost functions in a variety of situations and used them in designing actual surveys.

As a physicist, Mahalanobis was aware of instrumental errors and personal bias in taking measurements, and consequently he stressed the need for repeating measurements with different instruments, different observers, and under different conditions. He maintained that a statistical survey was like a scientific experiment and that the planning of a survey required the same discipline and rigor as do other investigations. He advocated built-in cross-checks to validate survey results. For this purpose, he developed the concept of interpenetrating network of samples (i.p.n.s.).

A simple design using i.p.n.s. is as follows. Suppose that a given area is divided into four strata and that we have four investigators for the field work. The normal practice is to assign one stratum to each investigator to cover all units (randomly) chosen from that particular stratum. With i.p.n.s., however, each investigator works in all the strata and covers a random quarter of its units. Thus the i.p.n.s. design provides four independent (parallel) estimates of the characteristic under study for the region as a whole, corresponding to the four different investigators. The validity of the survey will be in doubt if the four estimates differ widely. In such a case, it may be possible, by further data analysis, to take the differences properly into account when reporting the final estimate. Such a comparison or critical study would not have been possible if the four different strata had been assigned to four different investigators, because stratum and investigator differences would have been confounded.

Fractile graphical analysis. Fractile graphical analysis is an important generalization of the method and use of concentration (or Lorenz) curves [*see* GRAPHIC PRESENTATION]. A Lorenz curve for wealth in a population tells, for example, that the least wealthy 50 per cent of the population owns 10 per cent of the wealth. (If wealth were equally distributed, the Lorenz curve would be a straight line.) The comparison of Lorenz curves for two or more populations is a graphical way to compare their distributions of wealth, income, numbers of acres owned, frequency of use of library books, and so on.

One of Mahalanobis's contributions in this domain was to stress the extension of the Lorenz curve idea to two variables (1958*b*; 1958*c*). Thus one can consider, for example, both wealth and consumption for families, and draw a curve from which it may be read that the least wealthy 50 per cent of the families consume 27 per cent of total consumption, or a certain quantity per family on the average. Or, treating the variables in the other direction, one might find, perhaps, that the 20 per cent least-consuming families account for 15 per cent of the wealth or a certain value per family. (The numbers in these examples are hypothetical and only for illustration.) Such bivariate generalized Lorenz curves can, of course, also be usefully compared across populations.

Economic planning. For Mahalanobis, statistics—and science generally—were meaningful only as they helped in understanding the problems of the real world and particularly the problems of poverty in India. This led him to develop and devour statistics relating to the Indian economy with a passion displayed by few economists. He never studied traditional economic theory, and he believed that his lack of formal education in economics enabled him to work on economic problems of the country with a sense of realism. He said (1961*b*, p. 103), "The sophisticated economic theories which may be appropriate for advanced countries acted for a long time as thought barriers to economic progress in India."

Soon after independence, in 1949, Mahalanobis was appointed by Jawaharlal Nehru as honorary statistical advisor to the cabinet—a post that he held until his death. He believed that, before any worthwhile thinking about the economic problems of India could be done, it was necessary to have the basic national income statistics. On his advice, the government of India appointed in 1949 the National Income Committee "to prepare a report on the national income and related estimates, to suggest measures for improving the quality of available data and for the collection of further essential statistics and to recommend ways and means of promoting research in the field of national income." Mahalanobis was appointed chairman of this committee. The committee's reports laid the foundations of systematic and continuous work on national income in India.

The first major task was development of a national statistical system for collection and tabulation of official data at the state and federal levels. To this end, Mahalanobis advised the government of India to set up the central statistical unit that became the Central Statistical Organization (CSO). Simultaneously, plans were made to establish statistical bureaus in all the states to ensure efficient collection and reporting of data at the state level. But the work of the CSO and the state bureaus needed to be supplemented by extensive field surveys covering the entire country. For this, Mahalanobis recommended the establishment of the independent National Sample Survey (NSS) organization to operate with its own field staff and technical management.

Beginning in 1950, the NSS started turning out vast amounts of data relating to many important but dark areas of the Indian economy, for example, consumer expenditure, distribution and land holdings, unemployment, economics of cultivation, vital statistics, household enterprises, and amenities like postal, educational, and health services available in rural areas. Soon Mahalanobis was called upon by Jawaharlal Nehru to help in government planning. Setting himself the twin objectives of doubling the national income and reducing unemployment considerably over a period of twenty years, he produced what are known as two- and four-sector models for economic development with investment in various sectors of the economy as the center piece. In his 1950 presidential address "Why Statistics?", delivered to the Indian Science Congress at Poona, Mahalanobis first expounded his ideas about using capital–output ratio to derive estimates of the level of investment required for a stipulated increase in income. Subsequently he developed independently a forward-looking Harrod–Domar type of model for economic growth (1953; 1955*a*; 1955*b*).

Mahalanobis's association with the Planning Commission, and his involvement in the formulation of its Second Five Year Plan brought him into contact with problems of wide national and international importance. He wrote extensively on such subjects as (1) the priority of basic industries, (2) the role of scientific research, technical manpower, and education in economic development, (3) the industrialization of poorer countries and world peace (1958*a*), and (4) labor and unemployment.

Honors and awards. Mahalanobis was elected as a fellow of the Royal Society of London in 1945, of the Econometric Society, U.S.A., in 1951, and of the Pakistan Statistical Association in 1961; as an honorary fellow of the Royal Statistical Society, U.K., in 1954 and of King's College, Cambridge, in 1959; as honorary president of the International Statistical Institute in 1957; and as a foreign member of the U.S.S.R. Academy of Science in 1958. He received the Weldon medal from Oxford University in 1944, a gold medal from the Czechoslovak Academy of Sciences in 1963, the Sir Deviprasad Sarvadhikari gold medal in 1957, the Durgaprasad Khaitan gold medal in 1961, and the Srinivasa Ramanujam gold medal in 1968. He received honorary doctorates from Calcutta, Delhi, Stockholm, and Sofia universities and one of the highest civilian awards, Padmavibhushan, from the government of India.

Life with a mission. The 79 years of Mahalanobis's life were full of activity. His contributions were massive on the academic side as the builder of the Indian Statistical Institute, founder and editor of *Sankhyā*, organizer of the Indian Statistical Systems, pioneer in the applications of statistical techniques to practical problems, promoter of the statistical quality control movement for improvement of industrial products, architect of the Indian Second Five Year Plan, and so on.

Statistical science was a virgin field and practically unknown in India before the 1920s. Developing statistics was like exploring a new territory; it needed a pioneer and adventurer like Mahalanobis, with his indomitable courage and tenacity to fight all opposition, clear all obstacles, and open new paths of knowledge for the advancement of science and society. (See Rao 1973.)

The Mahalanobis era in statistics, which started in the early 1920s, ended with his death in 1972. It will be remembered as a golden period of statistics, marked by an intensive development of a new (key) technology (1965), and its application for the welfare of mankind.

C. RADHAKRISHNA RAO

WORKS BY MAHALANOBIS

1922 Anthropological Observations on the Anglo-Indians of Calcutta: I. Analysis of the Male Stature. *Records of the Indian Museum* 23: 1–96.

1923 Correlation of Upper Air Variables. *Nature* 112: 323–324.

1927 Report on Rainfall and Floods in North Bengal, 1870–1922. 2 vols. Submitted to the government of Bengal. → Volume 1: text. Volume 2: 28 maps.

1930 On Tests and Measures of Group Divergence: I. Theoretical Formulae. *Journal of the Proceedings of the Asiatic Society of Bengal* New Series 26: 541–588.

1931 Statistical Study of the Level of the Rivers of Orissa and the Rainfall in the Catchment Areas During the Period 1868–1928. Report submitted to the government of Bihar and Orissa.

1936 On Generalised Distance in Statistics. National Institute of Science, India, *Proceedings* 2: 49–55.

1940*a* A Sample Survey of the Acreage Under Jute in Bengal. *Sankhyā* 4: 511–530.

1940*b* The Application of Statistical Methods in Physical Anthropometry. *Sankhyā* 4: 594–598.

1940*c* Rain Storms and River Floods in Orissa. *Sankhyā* 5: 1–20.

1944 On Large Scale Sample Surveys. Royal Society of London, *Philosophical Transactions* Series B 231: 329–451.

1949 MAHALANOBIS, P. C.; MAJUMDAR, D. N.; and RAO, C. RADHAKRISHNA Anthropometric Survey of the United Provinces, 1941: A Statistical Study. *Sankhyā* 9:89–324.

1953 Some Observations on the Process of Growth of National Income. *Sankhyā* 12:307–312.

1955*a* Draft Plan-frame for the Second Five Year Plan, 1956/57–1960/61: Recommendations for the Formulation of the Second Five Year Plan. *Sankhyā* 16:63–90.

1955*b* The Approach of Operational Research to Planning in India. *Sankhyā* 16:3–62.

1958*a* Industrialisation of Underdeveloped Countries: A Means to Peace. *Bulletin for Atomic Scientists* 15:12–17. → Also in *Sankhyā* 22 (1960):173–182.

1958*b* A Method of Fractile Graphical Analysis With Some Surmises of Results. Bose Research Institute, *Transactions* 22:223–230.

1958*c* A Method of Fractile Graphical Analysis. *Econometrica* 28:325–351. → Also in *Sankhyā* Series A 23 (1961):41–64.

1960 A Note on Problems of Scientific Personnel. *Science and Culture* 27:101–128.

1961*a* *Experiments in Statistical Sampling in the Indian Statistical Institute.* Calcutta: Asia Publishing House and Statistical Publishing Society. → Also in *Journal of the Royal Statistical Society* Series A 109:325–378.

1961*b* Statistics for Economic Development. Operations Research Society of Japan, *Journal* 3:98–112.

1965 Statistics as a Key Technology. *American Statistician* 19, no. 2:43–46.

SUPPLEMENTARY BIBLIOGRAPHY

Series A and B of Sankhyā 35 (1973) *are dedicated to the memory of Mahalanobis. The December issues of both series carry supplements with articles by his close associates describing aspects of his life and work.*

RAO, C. RADHAKRISHNA 1952 *Advanced Statistical Methods in Biometric Research.* New York: Hafner. → Reprinted with corrections in 1970 and 1974.

RAO, C. RADHAKRISHNA 1954 On the Use and Interpretation of Distance Functions in Statistics. International Statistical Institute, *Bulletin* 34, no. 2:90–97.

RAO, C. RADHAKRISHNA 1973 Prasanta Chandra Mahalanobis, 1893–1972. Royal Society of London, *Biographical Memoirs of the Fellows* 19:455–492.

MAPS

See GEOGRAPHY, STATISTICAL; GRAPHIC PRESENTATION.

MARKOV CHAINS

A Markov chain is a chance process having the special property that one can predict its future just as accurately from a knowledge of the present state of affairs as from a knowledge of the present together with the entire past history.

The theory of social mobility illustrates the idea. Considering only eldest sons, note the status of successive generations of a particular male line in a society divided into three classes: upper, middle, and lower. If we assume the movement of a family among the three social classes is a chance process —governed by probabilistic laws rather than deterministic ones—several possibilities present themselves. An independent process represents a perfectly mobile society: the probability that a man is in a particular class depends in no way on the class of his father. At the other extreme is a society in which the probability that a man is in a particular class depends on the class of his father, that of his grandfather, that of his great-grandfather, etc.

A Markov chain is a process of intermediate complexity: the probability that a man is in a given class may depend on the class of his father, but it does not further depend on the classes of his earlier antecedents. For instance, while upper-class and middle-class fathers may have different probabilities of producing sons of a given class, an upper-class father whose own father was also upper class must have the same probability of producing a son of a given class as does an upper-class father whose father was, say, lower class. (This example will be used to illustrate each concept introduced below.)

No chance process encountered in applications is truly independent—in particular, no society is perfectly mobile. While in the same way no natural process exactly satisfies the Markov chain condition, many of them come close enough to make a Markov chain model useful.

Formal definitions. For an exact formulation of the Markov chain concept, introduced in 1907 by the Russian mathematician A. A. Markov, imagine a system (family, society, person, organism) that passes with each unit of time (minute, hour, generation) from one to another of the s *states* $E_1, E_2, \cdots, E_s$. (Upper class, E_1, middle class, E_2, and lower class, E_3, are the states in the social mobility example.) Assume that a chance process governs the evolution of the system; the chance process is the collection of probability laws describing the way in which the system changes with time. The system is that which undergoes change; a particular analysis may involve many systems of the same kind (many families, or many societies, etc.), all obeying the same process or set of probability laws.

Since the system passes through various states in sequence, time moves in jumps, rather than continuously; hence the integers 1, 2, 3, $\cdots$ provide a natural time index. Denote by $P(E_k|E_i, E_j)$ the conditional probability that at time $n+2$ the system is in state E_k, given that at times n and $n+1$ it was in states E_i and E_j in that order; and similarly for longer or shorter conditioning sequences of states. We make the usual assumption that the conditional probabilities just defined do not depend on n (that is, the conditional probabilities do not change as time passes). The process is a Markov chain if $P(E_k|E_i, E_j) = P(E_k|E_j)$, $P(E_l|E_i, E_j, E_k) = P(E_l|E_k)$, $P(E_m|E_i, E_j, E_k, E_l) = P(E_m|E_l)$, etc.

The *transition probabilities* $p_{ij} = P(E_j|E_i)$ determine the fundamental properties of the Markov chain; they form an s by s *transition matrix* $\boldsymbol{P} = (p_{ij})$, the basic datum of the process. (A matrix such as $\boldsymbol{P}$ that has only nonnegative entries and has rows summing to 1 is called a stochastic matrix.)

An independent process can be considered a special sort of Markov chain for which $p_{ij} = p_j$ (that is, the p_{ij} do not depend on i): the rows of $\boldsymbol{P}$ are all the same in this case. The *second-order transition probabilities* provide a second example of information contained in $\boldsymbol{P}$. If the system is presently in state E_i, then the conditional probability that it will pass to E_j and then to E_k in the next two steps is $P(E_j|E_i)P(E_k|E_i, E_j) = p_{ij}p_{jk}$; hence the conditional probability that the system will occupy E_k two time units later is

$$p_{ik}^{(2)} = \sum_{j=1}^{s} p_{ij}p_{jk}.$$

(Summing over the index j accounts for all the possible intermediate states.) But this second-order transition probability $p_{ik}^{(2)}$ is just the (i,k)th entry in $\boldsymbol{P}^2$ (the matrix $\boldsymbol{P}$ times itself in the sense of matrix multiplication).

Social mobility in England and Wales is approximately described by a Markov chain with transition matrix shown in Table 1. (These numbers must, of course, be arrived at empirically. The problem of estimation of transition probabilities is discussed later.) For example, a middle-class man has chance .054 of producing a son (recall that only eldest sons are considered here) who enters the upper class. Now the chance that a man in the upper class has a middle-class grandson is $p_{12}^{(2)} = (.448 \times .484) + (.484 \times .699) + (.068 \times .503) = .589$, while the chance that a man in the middle class has a middle-class grandson is $p_{22}^{(2)} = .639$.

Table 1

	E_1	E_2	E_3
E_1 *(upper class)*	.448	.484	.068
E_2 *(middle class)*	.054	.699	.247
E_3 *(lower class)*	.011	.503	.486

Notice that these last two probabilities are different, which raises a point often misunderstood. The Markov chain definition prescribes that an upper-class man with upper-class antecedents must have the same chance of producing a middle-class son as an upper-class man with lower-class antecedents has. But influence—stochastic influence, so to speak—of a man on his grandson, which does exist if $p_{12}^{(2)} \neq p_{22}^{(2)}$, is entirely consistent with the definition, which requires only that the grandfather exert this influence exclusively through the intermediate generation (the father).

To describe natural phenomena by Markov chains requires idealization; the social mobility example carried through here makes this obvious. Since Markov chains allow for dependence, however, they can with satisfactory accuracy account for the evolution of diverse social, psychological, biological, and physical systems for which an independent chance process would make too crude a model. On the other hand, Markov chains have simple enough structure to be mathematically tractable.

As another example of an approximate Markov chain, consider sociological panel studies. A potential voter is asked his party preference each month for six months preceding an election; his answer places him in state E_1 (Republican), E_2 (Democrat), or E_3 (undecided). The voter's progress among these states (approximately) obeys the Markov rule if, in predicting his August preference on the basis of his previous ones, one can without (essential) loss disregard them all except the most recent one, namely that for July (and similarly for the other predictions). In learning theory, Markov models have proved fruitful for describing organisms that change state from trial to trial in response to reinforcement in a learning process. (See Bibliography for applications in such areas as industrial inspection, industrial mobility, sickness and accident statistics, and economics.)

Mathematical analysis. The transition matrix $\boldsymbol{P}$ determines many of the characteristics of a Markov chain. We have seen that the (i,j)th entry $p_{ij}^{(2)}$ of $\boldsymbol{P}^2$ is the conditional probability, given that the system is in E_i, that it will be in E_j two steps later. In the same way, the nth *order transition probability* $p_{ij}^{(n)}$, the conditional probability that the system will occupy E_j after n steps if it is now in E_i, is the (i,j)th entry in $\boldsymbol{P}^n$. But $\boldsymbol{P}$ alone does not determine the absolute probability $a_i^{(n)}$ that at time n the system is in state E_i. For this we need $\boldsymbol{P}$ and the *initial probabilities* $a_i^{(1)}$.

The probability that the system occupies states E_i and E_j respectively at times n and $n+1$ is $a_i^{(n)}p_{ij}$; adding over the states possible at time n, we conclude that $\sum_i a_i^{(n)} p_{ij} = a_j^{(n+1)}$, or, if $\boldsymbol{a}^{(n)}$ denotes the row vector $(a_1^{(n)}, \cdots, a_s^{(n)})$, $\boldsymbol{a}^{(n)}\boldsymbol{P} = \boldsymbol{a}^{(n+1)}$. More generally, the mth order transition probabilities link the absolute probabilities for time $n+m$ to those for time n: $\sum_i a_i^{(n)} p_{ij}^{(m)} = a_j^{(n+m)}$ or $\boldsymbol{a}^{(n)}\boldsymbol{P}^m = \boldsymbol{a}^{(n+m)}$. Thus the absolute probabilities $a_i^{(n)}$ are completely determined, via the relation $\boldsymbol{a}^{(n)} = \boldsymbol{a}^{(1)}\boldsymbol{P}^{n-1}$, by the transition probabilities p_{ij} and the initial probabilities $a_i^{(1)}$. In other words, $\boldsymbol{P}$ and $\boldsymbol{a}^{(1)}$ completely specify the probability laws of the Markov chain.

Chains with all $p_{ij} > 0$. The most important mathematical results about Markov chains concern the stability of the $a_i^{(n)}$ and the behavior of $p_{ij}^{(n)}$ for large n. Although some of the p_{ij} may be zero, suppose they are all strictly greater than zero. (Later this restriction will be lifted and other chains considered.) One expects that for large n, $p_{ij}^{(n)}$ should not depend much on i—the effect of the initial state should wear off. This is indeed true: there exist positive numbers $p_1, \cdots, p_s$ such that $p_{ij}^{(n)}$ is close to p_j for large n ($\lim_{n\to\infty} p_{ij}^{(n)} = p_j$ for all i and j). It follows that $a_i^{(n)} = \sum_k a_k^{(1)} p_{ki}^{(n-1)}$ approaches $\sum_k a_k^{(1)} p_i = p_i$ as n becomes large, so that the absolute probabilities $a_i^{(n)}$ stabilize near the p_i, no matter what the initial probabilities $a_i^{(1)}$ are.

The numbers p_i have several important mathematical properties, which in turn have probability interpretations. Clearly they sum to one: $\sum_i p_i = 1$. Moreover, if n is large, $\sum_i p_{ki}^{(n)} p_{ij}$ is near $\sum_i p_i p_{ij}$, while $p_{kj}^{(n+1)}$ is near p_j. Since $\sum_i p_{ki}^{(n)} p_{ij} = p_{kj}^{(n+1)}$, we have $\sum_i p_i p_{ij} = p_j$, or, with $\boldsymbol{p} = (p_1, \cdots, p_s)$, $\boldsymbol{pP} = \boldsymbol{p}$. Therefore the p_j solve the system

$$(1)\qquad \sum_{i=1}^{s} p_i p_{ij} = p_j, \qquad j = 1, 2, \cdots, s, \qquad \sum_{i=1}^{s} p_i = 1$$

of $s+1$ equations in s unknowns. Since it turns out that this system has but one solution, the limits p_j can be found by solving it, and this is the method used in practice. The vector $\boldsymbol{p}$ for the social mobility example is (.067, .624, .309). Whatever a man's class, there is probability near .309 that a descendant a great many generations later is in the lower class.

The quantities p_j connect up with the absolute probabilities $a_j^{(n)}$. If $\boldsymbol{a}^{(n)} = \boldsymbol{p}$, then $\boldsymbol{a}^{(n+1)} = \boldsymbol{a}^{(n)}\boldsymbol{P} = \boldsymbol{pP} = \boldsymbol{p}$, and similarly $\boldsymbol{a}^{(m)} = \boldsymbol{p}$ for all m beyond n. In other words, if the system has at a given time probability exactly p_j of being in state E_j (for each j) then the same holds ever after. For this reason, the p_j are called a set of *stationary probabilities;* since the solution to (1) is unique, they form the only such set.

Thus a system evolving according to the laws of the Markov chain has long-range probability approximately p_j of being in E_j; and if the probability of being in E_j at a particular time is exactly p_j, this relationship is preserved in the future. The law of large numbers gives a complementary way of interpreting these facts. Imagine a large number of systems evolving independently of one another, each according to the laws of a common Markov chain. No matter what the initial states of the many systems may be, after a long time the numbers of systems in the various states will become approximately proportional to the p_j. And once this state of affairs is achieved, it will persist, except for fluctuations that are proportionately small if the number of systems involved is large. In the social mobility example, the proportions in the three classes of a large number of family lines will eventually be nearly 067:624:309.

A system setting out from state E_i is certain to return again to E_i after one or more steps; the expected value of the number of steps until first return turns out to be $1/p_i$, a result that agrees qualitatively with intuition: the lower the probability of a state the longer the average time between successive passages through it. There exist also formulas for the expected number of steps to reach E_j for a system starting in E_i. In the example from mobility theory, the mean time to pass from the lower class to the upper class is 26.5 generations, while the mean for the return trip is but 5.6 generations (which should be a lesson to us all).

Other Markov chains. In the analysis sketched above, it was assumed that the p_{ij} were all strictly greater than zero. More generally, the same results hold even if some of the p_{ij} equal zero, provided the Markov chain has the property that the states all communicate (for each i and j there is an n for which $p_{ij}^{(n)} > 0$) and provided also that there is no periodicity (no integer m exceeding 1 exists such that a passage from E_i back to E_i is possible in n steps only if n is divisible by m). Such a chain is called ergodic. The results can also be reformulated to cover nonergodic chains.

The theory extends in still other useful ways. (a) A chance process is a Markov chain of second order if the probability distribution of future states depends only on the two most recently visited states: $P(E_l|E_i, E_j, E_k) = P(E_l|E_j, E_k)$, $P(E_m|E_i, E_j, E_k, E_l) = P(E_m|E_k, E_l)$, etc. These processes contain ordinary Markov chains as a special case and make possible a more precise description of some phenomena. The properties derived above carry over to chains of second (and higher) order. (b) If, for instance, the system is a population and the state is its size, there are then infinitely many possible states. The theory generalizes to cover such examples, but the mathematics becomes more difficult; for example (1) becomes a system of infinitely many equations in infinitely many unknowns. (*c*) Continuous time, rather than time that goes in jumps, is appropriate for the description of some systems—for example, populations that change state (size) because of births and deaths that can occur at any instant of time. Such processes in principle admit an approximate description within the discrete-time theory: one observes the system periodically (every year or minute or microsecond) and ignores the state occupied at other time points. However, since such an analysis is often unnatural, an extensive theory of continuous-time processes has grown up.

Statistical analysis. One may believe a given process to be a Markov chain—or approximately a Markov chain—because of his knowledge of the underlying mechanism. Or he may hypothesize that it is a Markov chain, hoping to check this assumption against actual data. Under the Markov assumption, he may want to draw from actual data conclusions about the transition probabilities p_{ij}, which are the basic parameters of the model. In other words, statistical problems of estimation and hypothesis testing arise for Markov chains just as they do for independent processes [*see* ESTIMATION; HYPOTHESIS TESTING].

Suppose one observes n systems (governed by a common Markov chain), of which n_i start in state E_i, and follows each of them through one transition. If n_{ij} is the number that step from E_i to E_j, so that $\sum_j n_{ij} = n_i$, then the log-likelihood function is $\sum_{i,j} n_{ij} \log p_{ij}$; maximizing this function with the s constraints $\sum_j p_{ij} = 1$ gives the natural ratios n_{ij}/n_i as the maximum likelihood estimators of the p_{ij}. The numerical matrix given for the social mobility example was obtained in this way from a sample of some 3,500 father–son pairs in England and Wales. If the n_i are large, the estimators are approximately normally distributed about their mean values p_{ij}. By seeing how far the n_{ij}/n_i differ from putative transition probabilities p_{ij}, one can test the null hypothesis that these p_{ij} are the true parameter values. The statistic appropriate for this problem is $\sum_{i,j}(n_{ij} - n_i p_{ij})^2/n_i p_{ij}$; it has approximately a chi-square distribution with $s(s-1)$ degrees of freedom if the n_i are large.

Other tests are possible if the systems under observation are traced for more than one step. Suppose n_{ijk} is the number of the systems that start in E_i, step to E_j, and then step to E_k. If the process is a Markov chain, then (*a*) the chance that the second step carries the system from E_j to E_k does not depend on the initial state E_i and (*b*) the transition probabilities for the second step are the same as those for the first.

Let a dot indicate an index that has been summed out; for example, $n_{\cdot ij} = \sum_k n_{kij}$ is the number of systems that are in states E_i and E_j at times 2 and 3, respectively (with the state at time 1 completely unspecified). Assuming (*a*), one can test (*b*) by comparing the estimators $n_{ij\cdot}/n_{i\cdot\cdot}$ of the transition probabilities for the first step with the estimators $n_{\cdot ij}/n_{\cdot i\cdot}$ of the transition probabilities for the second step. And one can test (*a*) itself by comparing the ratios $n_{\cdot jk}/n_{\cdot j\cdot}$ (the estimated second-step transition probabilities) with the ratios $n_{ijk}/n_{\cdot jk}$ (the estimated second-step transition probabilities, allowing for possible further influence of the initial state); if (*a*) holds, $n_{ijk}/n_{ij\cdot}$ should, independently of i, be near $n_{\cdot jk}/n_{\cdot j\cdot}$; thus one can check on the Markov chain assumption itself.

This statistical analysis, based on following a large number of independently evolving systems through a small number of steps, really falls under the classical chi-square and maximum likelihood theory. There is also a statistical theory for the opposite case, following a small number of systems (perhaps just one) through many steps. Although the estimates and tests for this case have forms similar to those for the first case, the derivation of their asymptotic properties is more involved.

The statistical analysis of Markov chains may be extended in various directions. For example, a mode of analysis in which *times of stay* in the states are considered, together with transition probabilities, is discussed by Weiss and Zelen (1965).

PATRICK BILLINGSLEY

[*Other relevant material may be found in* PANEL STUDIES; SOCIAL MOBILITY.]

BIBLIOGRAPHY

ANDERSON, T. W.; and GOODMAN, LEO A. 1957 Statistical Inference About Markov Chains. *Annals of Mathematical Statistics* 28:89–110. → A detailed treatment of the problem of many short samples.

BILLINGSLEY, PATRICK 1961 Statistical Methods in Markov Chains. *Annals of Mathematical Statistics* 32: 12–40. → A review paper covering the problem of one long sample from a Markov chain. Contains a large bibliography.

BLUMEN, ISADORE; KOGAN, MARVIN; and MCCARTHY, PHILIP 1955 *The Industrial Mobility of Labor as a Probability Process.* Cornell Studies in Industrial and Labor Relations, Vol. 6. Ithaca, N.Y.: Cornell Univ. Press. → A detailed empirical study of mobility problems.

FELLER, WILLIAM 1950–1966 *An Introduction to Probability Theory and Its Applications.* 2 vols. New York: Wiley. → A classic text covering continuous-time processes as well as chains with infinitely many state

Excellent examples. The second edition of the first volume was published in 1957.

GLASS, D. V.; and HALL, J. R. 1954 Social Mobility in Great Britain: A Study in Inter-generation Changes in Status. In D. V. Glass (editor), *Social Mobility in Britain*. London: Routledge. → This is the source of the original data for the social mobility example.

GOODMAN, LEO A. 1961 Statistical Methods for the Mover–Stayer Model. *Journal of the American Statistical Association* 56:841–868.

GOODMAN, LEO A. 1962 Statistical Methods for Analyzing Processes of Change. *American Journal of Sociology* 68:57–78. → This and the preceding paper treat modifications of the Markov model for mobility problems.

JAFFE, JOSEPH; CASSOTTA, LOUIS; and FELDSTEIN, STANLEY 1964 Markovian Model of Time Patterns of Speech. *Science* 144:884–886.

KEMENY, JOHN G.; and SNELL, J. LAURIE 1960 *Finite Markov Chains*. Princeton, N.J.: Van Nostrand. → Contains both theory and examples. This is the source of the figures for the social mobility example.

PRAIS, S. J. 1955 Measuring Social Mobility. *Journal of the Royal Statistical Society* Series A 118:56–66. → This is the source of the analysis of the social mobility example.

WEISS, GEORGE H.; and ZELEN, MARVIN 1965 A Semi-Markov Model for Clinical Trials. *Journal of Applied Probability* 2:269–285.

MATCHED CONTROLS

See EXPERIMENTAL DESIGN, *article on* QUASI-EXPERIMENTAL DESIGN.

MATHEMATICAL MODELS

See MODELS, MATHEMATICAL.

MATHEMATICAL STATISTICS

See STATISTICS.

MATHEMATICS

The history of mathematics, and to some extent its content, can be thought of as involving three major phases. Ancient mathematics, covering the period from the earliest written records through the first few centuries A.D., culminated in Euclidean geometry, the elementary theory of numbers, and ordinary algebra. Equally important, this phase saw the evolution and partial clarification of axiomatic systems and deductive proofs. The next major phase, classical mathematics, began more than 1,000 years later, with the Cartesian fusion of geometry and algebra and the use of limiting processes in the calculus. From these evolved, during the eighteenth and nineteenth centuries, the several aspects of classical analysis. Other contributions of this phase include non-Euclidean geometries, the beginnings of probability theory, vector spaces and matrix theory, and a deeper development of the theory of numbers. About a hundred years ago the third and most abstract and demanding phase, known as modern mathematics, began to evolve and become separate from the classical period. This phase has been concerned with the isolation of several recurrent structures of analysis worthy of independent study —these include abstract algebraic systems (for example, groups, rings, and fields), topological spaces, symbolic logic, and functional analysis (Hilbert and Banach spaces, for example)—and various fusions of these systems (for example, algebraic geometry and topological groups). The rate of growth of mathematics has been so great that today most mathematicians are familiar in detail with the major developments of only a few branches of the subject.

Our purpose is to give some hint of these topics. The reader interested in a somewhat more detailed treatment will find the best single source to be *Mathematics: Its Content, Methods, and Meaning*, the translation of a Russian work (Akademiia Nauk S.S.S.R. 1956). Other general works are Courant and Robbins (1941), Friedman (1966), and Newman (1956). More specific references are given where appropriate. We do not here discuss probability, mathematical statistics, or computation, even though they are especially important mathematical disciplines for the social sciences, because they are covered in separate articles in the encyclopedia.

Ancient mathematics

The history of ancient mathematics divides naturally into three periods. In the first period, the pre-Hellenic age, the beginnings of systematic mathematics took place in ancient Egypt and in Mesopotamia. Contrary to much popular opinion, the mathematical developments in Mesopotamia were deeper and more substantial than those in Egypt. The Babylonians developed elementary arithmetic and algebra, particularly the computational aspects of algebra, to a surprising degree. For example, they were able to solve the general quadratic equation, $ax^2 + bx + c = 0$. An authoritative and readable account of Babylonian mathematics as well as of Greek mathematics is presented by Neugebauer (1951).

The second period of ancient mathematics was the early Greek, or Hellenic, age. The fundamentally new step taken by the Greeks was to introduce the concept of a mathematical proof. These developments began around 600 B.C. with Thales,

Pythagoras, and others, and reached their high points a little more than a century later in the work of Eudoxus, who is responsible for the theory of proportions, which in antiquity held the place now held by the modern theory of real numbers.

The third period is the Hellenistic age, which extended from the third century B.C. to the sixth century A.D. The early part of this period, sometimes called the golden age of ancient mathematics, encompassed Euclid's *Elements* (about 300 B.C.), which is the most important textbook ever written in mathematics, the work on conics by Apollonius (about 250 B.C.), and above all the extensive and profound work of Archimedes on metric geometry and mathematical physics (Archimedes died in 212 B.C.). The second most important systematic treatise of ancient mathematics, after Euclid's *Elements*, is Ptolemy's *Almagest* (about A.D. 150). Ptolemy systematized and extended Greek mathematical astronomy and its mathematical methods. The mathematical sophistication of Archimedes and the richness of applied mathematics evidenced by the *Almagest* were not equaled until the latter part of the seventeenth century.

Classical analysis

The intertwined and rapid growth of mathematics and physics during the seventeenth, eighteenth, and nineteenth centuries centered in a major way on what is now called classical analysis: the calculus of Newton and Leibniz, differential and integral equations and the special functions that are their solutions, infinite series and products, functions of a complex variable, extremum problems, and the theory of transforms. At the basis of all this are two major ideas, *function* and *limit*. The first evolved slowly, beginning with the correspondence, established in the Cartesian fusion of the two best-developed areas of ancient mathematics, between algebraic expressions and simple geometric curves and surfaces, until we now have the present, very simple definition of the term "function." A set f of points in the plane (ordered pairs of numbers) of the form (x, y) is called a function if at most one y is associated with each x. If (x, y) is a member of f, it is customary to write $y = f(x)$; x is sometimes called the independent variable and y the dependent variable, but no causal meaning should be read into this terminology.

The notion and notation may be generalized to more than one independent variable; if g is a set of ordered triples (x, y, z) with at most one z associated with each pair (x, y), then $z = g(x, y)$ is called a function of two arguments. Since the most general notion of function can relate any two sets of objects, not just sets of numbers, it is sometimes desirable to emphasize the numerical character of the function. Then f is said to be a real-valued function of a real variable; here the term "real" refers to real numbers (in contrast to complex numbers, which will be discussed later).

Although a real-valued function has been defined as a set of ordered pairs of numbers, (x, y), where the domain of x is is an unspecified set of numbers, the subsequent discussion of functions is mostly confined to the familiar case in which the domain of x is an interval of numbers. Even when the discussion applies more generally, it is helpful to keep the interval case in mind.

A desire to understand limits was apparent in Greek mathematics, but a correct definition of the concept eluded the Greeks. A fully satisfactory definition, which was not evolved until the nineteenth century (by Augustin Louis Cauchy), is the following: b is the limit of f at a if and only if for every positive number ϵ there is a positive number δ such that, when the absolute value of $x - a$ is less than δ and greater than 0 (that is, $0 < |x - a| < \delta$), the absolute value of $f(x) - b$ is less than ϵ (that is, $|f(x) - b| < \epsilon$). In other words, b is the limit of f at a if x can be chosen sufficiently close to a (but not equal to a) to force $f(x)$ to be as close to b as desired. Symbolically, this is written $\lim_{x\to a} f(x) = b$. The limit of f at a may exist even though $f(a)$ is not defined; moreover, when $f(a)$ is defined, b may or may not equal $f(a)$. If it does—that is, if $f(x)$ is "near" $f(a)$ whenever x is "near" a—then f is said to be continuous at a. If f is continuous at each a in an interval, f is said to be continuous over that interval.

The calculus. The calculus defines two new concepts, the derivative and the integral, in terms of function and limit. They and their surprising relationship serve as the basis of the rest of mathematical analysis.

The derivative. The first definition arises as the answer to the question "Given a function f, what is its slope (or, equivalently, its direction or rate of change) at any point x?" For example, suppose that $y = f(x)$ represents the distance, y, that a particle has moved in x units of time; then what is the rate of change of distance—the instantaneous velocity—at time x? If h is a short period of time, then an approximate answer is the distance traversed between x and $x + h$, that is, $f(x + h) - f(x)$, divided by the time, h, taken to

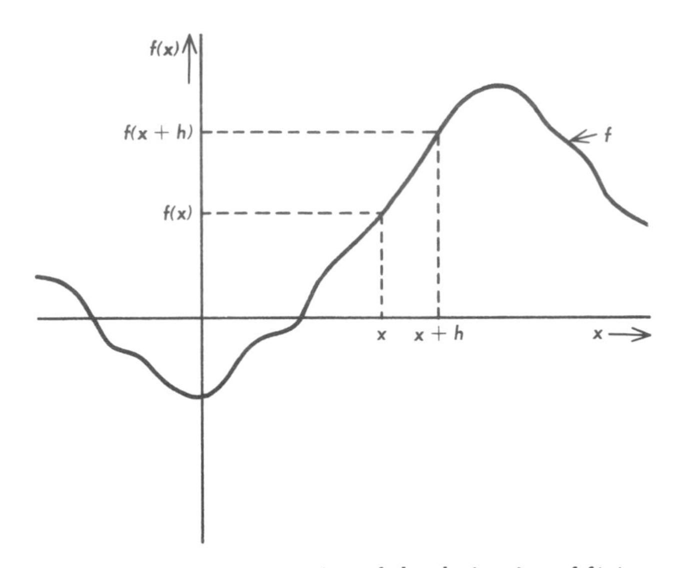

Figure 1 — Approximation of the derivative of f(x)

travel that distance (see Figure 1). The approximation is better the smaller the value of h, which suggests the definition of the rate of change of f at x as the limit of this ratio as h approaches 0, that is,

$$\lim_{h \to 0} \frac{f(x+h) - f(x)}{h}.$$

This limit, if it exists, is denoted by $f'(x)$ (or by $df(x)/dx$ or by dy/dx) and is called the *derivative* of f at x. If $f'(x)$ exists, then f can be shown to be continuous at x, but the converse is not true in general.

One of the earliest and most important applications in the social sciences of the concept of a derivative has been to the mathematics of marginal concepts in economics. For example, let x represent output, $C(x)$ the cost of output x, and $R(x)$ the revenue derived from output x; then $C'(x)$ and $R'(x)$ (or $dC(x)/dx$ and $dR(x)/dx$) are the marginal cost and marginal revenue, respectively. Marginal utility, marginal rate of substitution, and other marginal concepts are defined in a similar fashion. Many of the fundamental assumptions of economic theory receive precise formulation in terms of these marginal concepts.

The integral. The second concept in the calculus arises as the answer to the question "What is the area between the graph of a function f and the line $y = 0$ (the horizontal axis, or abscissa, of the coordinate system) over the interval from a to b?" (Regions below the abscissa are treated as negative areas to be subtracted from the positive ones above the abscissa; see Figure 2.) The solution, which will not be stated precisely, involves the following steps: the abscissa is partitioned into a finite number of intervals; using the height of the function at some value within each interval, the function is approximated by the resulting step function; the area under the step function is calculated as the sum of the areas of the rectangles of which it is composed; and, finally, the limit of this sum is calculated as the widths of the intervals approach zero (and, therefore, as their number approaches infinity). When this limit exists, it is called the *Riemann integral* of f from a to b and is symbolized as $\int_a^b f(x)\,dx$. It can be shown that the Riemann integral exists if f is continuous over the interval; it also exists for some discontinuous functions. For more advanced work, the concept of the length of an interval is generalized to the concept of the *Lebesgue measure* of a set, and the Riemann integral is generalized to the *Lebesgue integral.* Roughly, the vertical columns used to approximate the area in the Riemann integral are replaced in the Lebesgue integral by horizontal slabs.

Although the interpretation of the integral as an extension of the elementary concept of area is important, even more important is its relation (called the fundamental theorem of the calculus) to the derivative: Consider $F(x) = \int_a^x f(u)\,du$ as a function of the upper limit, x, of the interval over which the integral is computed; it can then be proved that the derivative of this function, $F'(x)$, exists and is equal to $f(x)$. Put another way, the rate of change at x of the area generated by f is equal to the value of f at x; or put still another way, the operation of taking the derivative undoes the operation of integration. This fact plays a crucial role in the solution of many problems of classical applied mathematics that are formulated in terms of derivatives of functions.

Introductions to the calculus and elementary parts of analysis are Apostol (1961–1962) and Bartle (1964).

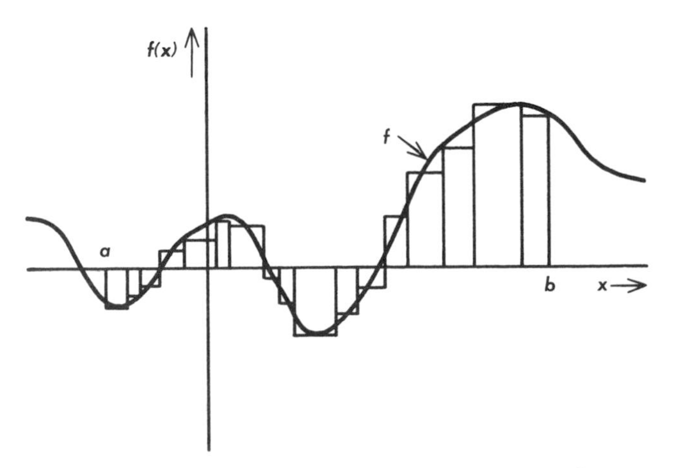

Figure 2 — Approximation of the integral of f(x)

Implicit definitions of functions. An algebraic equation such as $2x^2 - 5x - 3 = 0$ implicitly defines two numbers (namely, the two values of x, 3 and $-\frac{1}{2}$) for which the equality holds. Other algebraic equations implicitly define sets of numbers for which they hold.

A functional equation is an equality stated in terms of an unknown function; it implicitly defines those functions (as in the algebraic case, there may be more than one) that render the equality true.

Ordinary differential equations. Suppose it is postulated that the amount of interest (that is, the rate of change of money at time t) is proportional to (that is, is a constant fraction, k, of) the amount, $f(t)$, of money that has been saved. (This is the case of continuous compound interest.) Then f satisfies the equation $f'(t) = kf(t)$. This is a simple example of an *ordinary differential equation*, the solution of which is any function having the property that its derivative is k times the function. The solutions are $f(t) = f(0) \exp(kt)$, where $f(0)$ denotes the initial amount of money at time $t = 0$. Another simple economic example is the differential equation that arises from the assumption that marginal cost always equals average cost (that is, $dC(x)/dx = C(x)/x$) which has the solution that average cost is constant, that is, that $C(x) = kx$ for some constant, k.

Some laws of classical physics are formulated as second-order, linear, ordinary differential equations of the form

$$f''(t) + P(t)f'(t) + Q(t)f(t) = R(t),$$

where f'' is the derivative of f' (f'' is called the second derivative of f) and P, Q, and R are given functions. If, for example, f denotes distance, then this differential equation asserts that at each time t, a linear relation holds among distance, velocity, and acceleration. A vast literature is concerned with the solutions to this class of equations for different restrictions on P, Q, and R; most of the famous special functions used in physics—Bessel, hypergeometric, Hankel, gamma, and so on—are solutions to such differential equations (see Coddington 1961).

Partial differential equations. Many physical problems require differential equations a good deal more complicated than those just mentioned. For example, suppose that there is a flow of heat along one dimension, x. Let $f(x,t)$ denote the temperature at position x at time t. With t fixed, one can find the rate of change (the derivative) of temperature with changes in x; denote this by $\partial f(x,t)/\partial x$ and its second derivative with respect to x by $\partial^2 f(x,t)/\partial x^2$. These are called partial derivatives. Similarly, holding x fixed, the derivative with respect to t is denoted by $\partial f(x,t)/\partial t$. According to classical physics, temperature changes due to conduction in a homogeneous one-dimensional medium satisfy the following *partial differential equation*:

$$\frac{\partial f(x,t)}{\partial t} = \frac{k}{\rho\sigma}\frac{\partial^2 f(x,t)}{\partial x^2},$$

where k is the thermal conductivity, ρ the density, and σ the specific heat of the medium. Problems involving two or more independent variables (usually, time and some or all of the three space coordinates)—fluid flow, heat dissipation, elasticity, electromagnetism, and so on—lead to partial differential equations. Their solution is often very complex and requires the specification of the unknown function along a boundary of the space. This requirement is called a boundary condition. (See Akademiia Nauk S.S.S.R. [1956] 1964, chapter 6.)

Integral equations. Some physical problems lead to *integral equations*. In one type, functions g and K of one and two variables, respectively, and a constant, λ, are given, and the problem is to find those functions, f, for which

$$f(x) = g(x) + \lambda \int_a^b K(x,y)f(y)\, dy.$$

This equation is called Fredholm's linear integral equation or the inhomogeneous linear integral equation. Basically, it asserts that the value of some quantity f at a point x is equal to an impressed value, $g(x)$, plus a weighted average of its value at all other points. Integral equations arise in empirical contexts for which it is postulated that the value of a function at a point depends on the behavior of the function over a large region of its domain. Thus, in the example just considered the value of f at x depends on the integrand $K(x,y)f(y)$ integrated over the interval (a,b). There is a large body of literature dealing with the solution of various types of integral equations, especially those of interest in physics and probability theory.

Functional equations. Although both differential and integral equations (and mixtures of the two, called integrodifferential equations) are examples of functional equations, that term is often restricted to equations that involve only the unknown function, not its derivatives or integrals. A simple, well-known example is $f(xy) = f(x) + f(y)$, which implicitly defines those functions that transform multiplication into addition. If f is required to be continuous, then the solutions are $K\int_1^x dz/z$, where K is a positive constant; this integral is called the natural logarithm. The choice of K is usually referred to as the selection of the base of the logarithm.

Difference equations are functional equations of special importance in the social sciences. They arise both in the study of discrete stochastic processes (in learning theory, for example) and as discrete analogues of differential equations. Here the unknown function is defined only on the integers (or, equivalently, on any equidistant set of points), not on all of the real numbers, and so the function is written $f_n = f(n)$, where n is an integer. The equation states a relation among values of the unknown function for several successive integers. For example, the second-order, linear difference equation—the analogue of the second-order, linear, ordinary differential equation, described above—is of the form

$$f_{n+2} + P_n f_{n+1} + Q_n f_n = R_n .$$

In some probabilistic models of the learning process it is postulated (or derived from more primitive assumptions) that the probability of a particular response on trial $n + 1$, denoted by p_{n+1}, is some function of p_n and of the actual events that occurred on trial n. The simplest such assumption is the linear one, that is, $p_{n+1} = \alpha p_n + \beta$, where α and β are parameters that depend upon the events that actually occur. If there is a run of trials during which the same events occur, so that α and β are constant, then the solution to the above first-order, linear difference equation is

$$p_k = \alpha^k p_0 + \frac{(1-\alpha^k)}{1-\alpha}\beta.$$

When different events occur on different trials, the equation to be solved becomes considerably more complex. An introduction to difference equations is Goldberg (1958).

Given a functional equation—in the most general sense—the answer to the question of whether a solution exists is not usually obvious. Exhibiting a solution, of course, answers the question affirmatively, but often the existence of a solution can be proved before one is found. Such a result is known as an *existence theorem.* If a solution exists, it is also not usually obvious whether it is unique and, if it is not unique, how two different solutions relate to one another. A statement of the nature of the nonuniqueness of the solutions is known, somewhat inappropriately, as a *uniqueness theorem.* Some rather general existence and uniqueness theorems are available for differential and integral equations, but in less well understood cases considerable care is needed to discover just how restrictive the equation is.

A general work on functional equations is Aczél (1966).

Three other areas of classical analysis. Three other branches of classical analysis will be briefly discussed.

Extremum problems. For what values of its argument does a function assume its maximum or its minimum value? This type of problem arises in theoretical and applied physics and in the social sciences. In its simplest form, a real-valued function f is defined over some interval of the real numbers, and the problem is to find those x_0 for which $f(x_0)$ is a maximum or a minimum. If f is differentiable and if x_0 is not one of the end points of the interval, a necessary condition is that $f'(x_0) = 0$; moreover, x_0 is a local maximum if $f''(x_0) < 0$ and a local minimum if $f''(x_0) > 0$. (These statements should be intuitively clear for graphs of simple functions.) From these results it is easy to find, for example, which rectangle has the maximum area when the perimeter is held constant: it is the square whose sides are each equal to a quarter of the perimeter.

A much more difficult and interesting problem—the subject of the *calculus of variations*—is to find which function (or functions) f of a given family of functions causes a given function F of f (known as a functional) to assume its maximum or minimum value. For example, let f be a continuous function that passes through two fixed points in the plane, and let $F(f)$ be the surface area of the body that is generated by rotating f about the abscissa. A question that may be asked is "For which f (or f's) is $F(f)$ a minimum?" A major tool in the solution of this problem is a second-order, ordinary differential equation, known as Euler's equation, that f must necessarily satisfy (just as the solution x_0 to the simpler problem necessarily satisfies $f'(x_0) = 0$). (See Akademiia Nauk S.S.S.R. [1956] 1964, chapter 8.)

Within the past twenty years new classes of extremum problems have been posed and partially solved; they are mainly of concern in the social sciences, and they go under the names of linear, nonlinear, and dynamic programming. An example of a linear programming problem is the following diet problem. Each of several foodstuffs, $f_1, f_2, \cdots, f_k$, contains known amounts of various nutritional components, such as vitamins and proteins. Let f_{ij} be the amount of component j in food f_i, $j = 1, 2, \cdots, n$, and let a_j be the minimum amount of component j acceptable in the diet. If x_i is the amount of food f_i in the diet, the diet will be acceptable only if the following n inequalities are fulfilled:

$$x_1 f_{1j} + x_2 f_{2j} + \cdots + x_k f_{kj} \geqslant a_j, \qquad j = 1, 2, \cdots, n.$$

If p_i denotes the price of food f_i, the problem is to choose the x_i so as to minimize the cost,

$$x_1 p_1 + x_2 p_2 + \cdots + x_k p_k,$$

while fulfilling the above linear inequalities. [*See* PROGRAMMING.]

Functions of a complex variable. One of the most beautiful subfields of analysis is the theory of functions of a complex variable, which was developed in the nineteenth century, starting with the work of Cauchy. It has been significant in the growth of several two-dimensional, continuous physical theories, including parts of electromagnetism, hydrodynamics, and acoustics, but so far its applications in the social sciences have been mainly restricted to mathematical statistics, as in the concept of the characteristic function of a probability distribution. A complex number, z, is of the form $z = x + iy$, where x and y are real numbers and $i = \sqrt{-1}$. Sums and products are defined in such a way that the resulting arithmetic reduces to that of the ordinary numbers when $y = 0$. Because a point (x,y) in the plane can be (usefully) identified with the complex number $x + iy$, functions from the plane into the plane can be interpreted as complex-valued functions of a complex variable. If the derivative of such a function exists at all points of a region, derivatives of all orders exist and the function can be expressed as a convergent *power series* of the form $a_0 + a_1 z + a_2 z^2 + \cdots$ for some circle of z's within that region. It is clear from this result that the mere supposition that the derivative exists is a much stronger condition for complex-valued functions than for ordinary numerical functions. Such functions, which are called *analytic*, are very strongly constrained—among other things, specifying an analytic function over a small region determines it completely—and this fact has been effectively exploited to solve many two-dimensional problems of theoretical and practical interest. Interestingly, the theory cannot be neatly generalized beyond two dimensions. An introductory work on functions of a complex variable is Cartan (1961).

Integral transforms. Suppose that f is any continuous, real-valued function defined over an interval from a to b and that K is a fixed, continuous, real-valued function of two variables, the first of which is also on the interval from a to b; then $I(f,y) = \int_a^b K(x,y) f(x)\, dx$ is called an *integral transform* of f. If K satisfies certain restrictions, knowing I is equivalent to knowing f. Nevertheless, if K is carefully chosen, I may have convenient properties not possessed by f. For example, if $a = 0$, $b = \infty$, and $K(x,y) = e^{-xy}$, then I, which is then known as the *Laplace transform* and which is closely related to the moment-generating function of statistics, has the property that it converts certain integrals (convolutions) of two functions into multiplications of their transforms. In statistics such a convolution represents the distribution of the sum of two independent random variables. Another well-known and important example is the *Fourier transform*, which is used widely in statistics, and to a lesser extent in probabilistic models of behavior, to obtain a probability distribution from its characteristic function.

Theory of numbers

Despite several intellectual crises that led mathematicians to introduce new types of numbers into mathematics, it was not until about a hundred years ago that numbers were treated as being something other than intuitively understood. The natural numbers, $1, 2, 3, \cdots$, and their ratios, the positive rationals, are ancient concepts. The Greeks first noted their incompleteness when they showed that they are inadequate to represent $\sqrt{2}$, the length of the diagonal of a square whose side is of length 1. Certain irrational numbers had to be added, and later 0, negative numbers, and complex numbers were added so that certain classes of equations would all have solutions. To clarify this patchwork and to understand the uniqueness of the additions, nineteenth-century mathematicians undertook the axiomatization of various aspects of the number system. Perhaps the most subtle step was the definition of irrational numbers in terms of sets of rational numbers (roughly, the set of all rationals less than the irrational to be defined).

The axiomatization of numbers is not really the mainstream of the "theory of numbers." When one sees a book or course with that title, it usually refers to the study of properties of the natural numbers, mainly the prime numbers. Recall that an integer is prime if it is divisible only by 1 and itself; the first few primes are 3, 5, 7, 11, and 13. In addition to the many results that can be proved directly (some of which were known to the ancients), such as that every integer can be represented uniquely as the product of powers of primes and that there are infinitely many primes, other results have depended upon the application of deep results from analysis. For example, parts of the theory of functions of a complex variable were used to show that the number of primes not larger than n divided by the number $n/\ln n$, where $\ln n$ is the natural logarithm of n, that is, $\int_1^n dx/x$, is a ratio that approaches 1 as n becomes large. Not only has this work greatly increased the depth of understanding of integers, but it has fed back into analysis and was one of the factors leading to the development of parts of contemporary abstract algebra.

Many applications of mathematics (for example, in statistics) involve counting the number of distinct events or objects that satisfy certain conditions; often these counting problems are quite difficult. Theorems providing explicit formulas or recursion schemes are called combinatorial theorems. One of the earliest important examples was the binomial theorem for the expansion of $(a+b)^n$, which is now part of every elementary algebra course. [*See* PROBABILITY, *article on* FORMAL PROBABILITY.]

A general introduction to the theory of numbers is Ore (1948).

Algebra

Classically, algebra was the theory of solving equations expressed in terms of the four arithmetical operations—addition, subtraction, multiplication, and division. The linear and quadratic equations of elementary algebra are familiar examples. Historically, the expression of mathematical problems in the form of equations, using letters to stand for the unknown numbers, was a major step in clarifying and simplifying the mathematical nature of many kinds of problems. Perhaps the most important consequence of the introduction of letters and the use of equations was the extension of routine methods of calculation to quite complicated settings. The introduction of algebraic equations probably ranks in importance in the history of ideas with the earlier invention, probably first by the Babylonians, of the place-value system of notation for numbers; such a system was needed to develop simple algorithms for performing arithmetical computations.

The general theory of algebraic equations, the elementary parts of which are studied in high school, has a long and distinguished history in mathematics. The proof by Niels Henrik Abel in 1824 that solutions of an algebraic equation of degree five or greater, where the degree is the highest exponent of any term in the equation, cannot be expressed in terms of radicals (that is, expressions definable in terms of square roots) was one of the most important mathematical results of the first half of the nineteenth century. Another result of basic importance is the fundamental theorem of algebra, which was first proved in the eighteenth century but which was proved rigorously only in the last half of the nineteenth century. This theorem asserts that every algebraic equation always has at least one root that is a real or a complex number. Also of great significance were the proofs that not all numbers are roots of algebraic equations; numbers that are not such roots are called transcendental numbers. The most famous proofs of this sort are Charles Hermite's (in 1873) that e is transcendental and F. Lindemann's (in 1882) that π is transcendental.

Orderings. Much of the work in algebra during the present century has been devoted to generalized mathematical systems that are characterized not in terms of the four fundamental arithmetical operations but in terms of generalizations of these operations and of the familiar ordering relations of "less than" and "greater than."

In a number of the social sciences the theory of binary relations has received extensive application. From an algebraic standpoint a *binary relation structure* may be characterized as consisting of a set A and a set R of ordered pairs (x, y), where x and y are both elements of A. Such an R is called a binary relation on A. A relation R is said to be a *partial ordering* of A when it is reflexive, antisymmetric, and transitive—that is, when it satisfies the following three properties: *reflexive*: for every x in A, xRx; *antisymmetric*: for every x and y in A, if xRy and yRx, then $x=y$; *transitive*: for every x, y, and z in A, if xRy and yRz, then xRz. If R is also connected in A (that is, if for any two elements x and y in A with $x \neq y$, either xRy or yRx) then R is said to be a *complete* or *simple ordering* or, sometimes, a *linear ordering* of A. The concept of a complete ordering is a direct abstraction of the order properties of "$\leqslant$" with respect to the real numbers. A familiar use of the concept of an ordering relation is in utility theory, particularly in the classical theory of demand in economics, in which it is assumed that each individual has an ordering relation over the set of commodity bundles or, more generally, over the set of alternatives with which he is presented. The general concept of ordering relations also has far-ranging applications in the theory of measurement within psychology and sociology, and more general binary relations have been extensively applied in anthropology in the study of kinship systems.

Partial orderings can be extended in another direction by imposing additional conditions to obtain *lattices*, which have also been used in the social sciences. In a different direction, but still within the framework of binary relations, is the *theory of graphs*, in which no restrictions are placed on the binary relation, R. Applications of graph theory have been made to social-psychological and sociological problems, especially to provide a mathematical method for representing various kinds of relationships between persons.

Groups, rings, and fields. Another direction of generalization of classical algebra has been to what are called groups, rings, and fields. A *group* is a set A together with a binary operation, $\circ$, satisfying the following axioms. First, the operation $\circ$ is associative, that is, for x, y, and z in A, $x \circ (y \circ z) = (x \circ y) \circ z$. Second, there is an element e, called the identity, of the set A such that for every x in A, $x \circ e = e \circ x = x$. And, finally, for each element x of A there is an inverse element x^{-1} such that $x \circ x^{-1} = e$. It is obvious that if A is taken as the set of integers, $\circ$ as the operation of addition, e as the number 0, and the inverse of x as the negative of x, then the set of integers is a group under the binary operation of addition. The theory of groups has had profound ramifications in other parts of mathematics and in the sciences, ranging from the theory of algebraic equations to geometry and physics. The reason for the fundamental importance of group theory is perhaps best summarized by stating that a group is the appropriate way to formulate the very important concept of symmetry. In the range of applications of group theory just mentioned, the underlying thread is the concept of symmetry, whether it is in the symmetry of the roots of an equation or the symmetry properties of the fundamental particles of physics. As a simple example, consider the finite group of rotations 90°, 180°, 270°, and 360°. A square does not change its apparent orientation under such a rotation about its center, but an equilateral triangle does. This group of rotations is the symmetry group of rotations for a square but not, of course, for an equilateral triangle. Although the methods and results of group theory have not yet had special applications of depth in the social sciences, they are important to many of the general mathematical results that have been applied.

The theories of rings and fields represent rather direct generalization of arithmetical properties of the number system. The theory of groups is fundamentally a generalization of the concept of a single binary operation, such as addition or multiplication, whereas rings and fields are algebraic systems that have two fundamental operations. The most familiar example of a field or of a ring is the set of rational numbers or of real numbers with respect to the operations of addition and multiplication.

Boolean algebras. Algebraic aspects of the theory of sets have been studied under the heading of Boolean algebras. The concept of an algebra of sets, that is, a collection of sets closed under union and complementation, is fundamental in the modern theory of probability, where events are interpreted as sets of possible outcomes and numerical probabilities are assigned to events. [*See* PROBABILITY, *article on* FORMAL PROBABILITY.]

Isomorphism and homomorphism. It should be mentioned that certain very general mathematical concepts find their most natural definition and application in modern algebra. One of the most important concepts is that of the isomorphism of two mathematical systems. An *isomorphism* is a one-to-one mapping of a system A onto a system B in which the operations and relations of A are preserved under the mapping and have the same structure as the operations and relations of system B. If the mapping is not one-to-one but the operations and relations are preserved, then it is called a *homomorphism*. A well-known application of the concept of isomorphism in the social sciences is in theories of fundamental measurement in which one shows that an appropriate algebra of empirical operations is isomorphic to some numerical algebra. It is this isomorphism that permits the direct application of computational methods to the results of measurement.

Introductory works on algebra, both for this and for the next section, are Birkhoff and MacLane (1941) and Mostow, Sampson, and Meyer (1963).

Vector spaces and matrix algebra

Linear algebra is one of the most important generalizations of classical elementary algebra. The objects to which the operations of addition and multiplication are applied are now matrices, vectors of an n-dimensional space, and linear transformations (an $n \times n$ matrix is a particular representation of a linear transformation in n-dimensional space). More particularly, linear algebra arises as a generalization of the linear equations so familiar in elementary algebra, and historically one of the most important tasks of linear algebra has been to find solutions of systems of linear equations. As many research workers in the social sciences know, the numerical solution of linear equations can be an extremely laborious and difficult affair when the number of equations is large. The set of coefficients of a system of linear equations gives rise to the concept of a rectangular array of numbers, which is precisely what a matrix is. An algebra of matrices in terms of addition and multiplication may be constructed; the distinguishing feature of this algebra, as compared with the algebra of the real numbers, is that multiplication is not commutative—that is, ***AB*** is not usually equal to ***BA***, and the product of two nonzero matrices can be zero.

The intuitive geometric concept of a vector may be represented by a column or row of n numbers, and an algebra of vectors, which bears a close resemblance to the algebra of numbers, may be constructed. Simple (linear) transformations of vectors, such as rotations and stretches of the coordinate system in space, can be interpreted as multiplication by matrices. The interaction between the geometrical intuitions about n-dimensional space and the algebraic techniques of calculation provided by linear algebra and the theory of matrices have made them powerful tools in the application of mathematics to many parts of science. These applications have been particularly prominent in statistics (for example, in factor analysis), as well as in economics, where it is often useful to treat n-dimensional bundles of commodities as vectors.

Topology and abstract spaces

Intuitively, a topological transformation of a geometrical figure or object is a deformation that introduces neither breaks nor fusions in the object. Put more exactly, a topological transformation is one that is one-to-one, is continuous, and has a continuous inverse. If one starts with a circle—perhaps the best example of a simple closed curve—one can deform it topologically into an ellipse or into the shape of a crescent, but one cannot deform it topologically into a figure eight, for example, because then two distinct points of the circle are fused as the intersection point of the eight. Also, one cannot deform it into a straight line segment, because to do so would introduce a break in the circle. Many familiar qualitative geometrical properties are topological invariants in the sense that they are not altered (are invariant) under topological transformations. Examples are the property of being inside or outside a closed figure in the plane; the property of a surface being closed, such as the surface of a sphere or an ellipsoid; or the property of the dimension of an object. For example, the surface of a sphere cannot be topologically transformed into a one-dimensional curve or a three-dimensional sphere. We shall not attempt here to give an exact definition of continuity as it is used in topology; we simply remark that it is a reasonable generalization of the concept of continuity used in analysis.

Topological methods and results have far-reaching applications in many branches of mathematics, but as yet the methods themselves have not been directly applied in those parts of the social sciences concerned extensively with empirical data. The most direct applications have been in economics, where topological fixed-point theorems have been of great importance in investigating the conditions guaranteeing the existence of a stable equilibrium in a competitive economy. The classical example of a fixed-point theorem—first proved by L. E. J. Brouwer, at the beginning of this century—states that for every topological mapping of an n-dimensional sphere into itself there is always at least one point that maps into itself, that is, remains fixed. Familiar examples of such mappings are rotations in two or three dimensions for which the center of the rotation is the fixed point of the transformation.

Topological space. As a typical example of abstraction in modern mathematics, the initial concept of a topological transformation of familiar geometrical figures has led to the general abstract notion of a topological space. Roughly speaking, a *topological space* consists of a set, X, and a family, $\mathcal{T}$, of subsets of X, called *open sets*, for which the following four conditions are satisfied: the empty set is in $\mathcal{T}$; X is in $\mathcal{T}$; the union of arbitrarily many sets each of which is in $\mathcal{T}$ is also in $\mathcal{T}$; and the intersection of any finite number of sets from $\mathcal{T}$ is also in $\mathcal{T}$. The concept of an open set is a generalization of the notion of an open interval of real numbers (an interval that does not include its end points). For example, the natural topology of the real line is the family of open intervals together with the sets that are formed from arbitrary unions and finite intersections of open intervals. Generally speaking, the notion of open set is used to express the idea of continuity. The important thing about a continuous function is that it does not jumble neighboring points too much, and this requirement may be expressed by requiring of a topological transformation that open sets be mapped into open sets and that the inverse of an open set be an open set.

Metric space. Other kinds of abstract spaces have come into prominence in the development of topology. Perhaps the most important is the concept of a metric space. A set, X, together with a distance function, d, that maps pairs of points into real numbers is called a *metric space* if d satisfies the following conditions: $d(x, y) = 0$ if and only if $x = y$, that is, the distance between x and y is 0 if and only if x and y are the same point; $d(x, y) \geqslant 0$, which asserts that distance is a nonnegative real number; $d(x, y) = d(y, x)$, that is, distance is symmetric; and, finally, $d(x, y) + d(y, z) \geqslant d(x, z)$, which is known as the triangle inequality. The concept of a metric space has had important applications in many parts of mathematics and is a fundamental concept in modern mathematics. It has been applied in recent work in scaling theory in psychology and sociology, particularly to the problems of multidimensional scaling, and also in certain areas of mathematical economics [*see* SCALING]. It is clear that the notion of a metric space generalizes, in a very natural way, the concept of distance in Euclidean space.

A typical metric problem raised in the social sciences is this: Given data in the form of "distances" among a finite set of points, what is the smallest dimensional Euclidean space within which the points can be embedded so that these distances equal the Euclidean or some other preassigned metric of that space? Recently this problem has been effectively generalized by permitting certain transformations of the "distances" that preserve their metric property. Little has yet been done about embeddings in non-Euclidean spaces.

An introductory work on topology is Hocking and Young (1961).

Foundations

As was remarked above, the concept of a rigorous mathematical proof originated in ancient Greek mathematics. The modern formal axiomatic method, characteristic of twentieth-century mathematical research and one of the most important topics to be clarified in modern research on foundations of mathematics, is conceptually very close to the approach followed in Euclid's *Elements*. The main difference is that the primitive concepts of the theory are now treated as undefined or meaningless. All that is assumed about them must be formally expressed in the axioms. In contrast, in the *Elements* primitive concepts such as those of point and line are given an interpretation or meaning from the very beginning. This modern conception originated with David Hilbert, who provided the first complete, modern axiomatization of geometry in 1889. It is customary to say that the concepts of the theory are implicitly defined by the axioms. What is not recognized often enough is that the collection of axioms together *explicitly* defines the theory embodied in the concepts. Thus, in slightly more exact phrasing, the axioms of Euclidean geometry define the theory of Euclidean geometry by defining the phrase "is a model of Euclidean geometry." In the same fashion, the axioms of group theory define the theory of groups by specifying what kinds of objects are called groups or, in other words, what kinds of objects are models of the theory of groups (here we are using the term "model" in the logical or mathematical sense).

A more particular aim of foundational research has been to provide a set of axioms that would serve as a basis for the main body of mathematics. At least three major positions on the foundations of mathematics have been enunciated in the twentieth century; they differ in their conception of the nature of mathematical objects.

Intuitionism. Intuitionism holds that in the most fundamental sense mathematical objects are themselves thoughts or ideas. The intuitionist holds that one can never be certain that he has correctly expressed the mathematics when it is formalized as a mathematical theory. As part of this thesis, the classical logic of Aristotle, in particular the law of excluded middle, has been challenged by Brouwer and other intuitionists because it permits the derivation of purely existential, nonconstructive statements about mathematical objects. In particular the validity of classical *reductio ad absurdum* proofs depends upon this logical law. Although intuitionists express themselves in a way which suggests a psychological analysis of mathematics, it should be emphasized that their conception of mathematical objects as thoughts has not been seriously explored by any intuitionists from the standpoint of scientific psychology.

Platonism. A second view of mathematics, the Platonistic one, is that mathematical objects are abstract objects that exist independently of human thought or activity. Those who hold that set theory or logic itself provides an appropriate foundation for mathematics (adherents of logicism) usually adopt some form of Platonism in their basic attitude. From the standpoint of working mathematics, set theory—and thus Platonism—has been the most influential conception of mathematics in this century. Set theory itself originated in the late nineteenth century with the revolutionary work of Georg Cantor. Its foundations were called into question by Bertrand Russell's discovery of a simple paradox which arises in considering the set of all objects that are not members of themselves. If it is supposed that to every property there corresponds the set of objects having this property, then a contradiction within classical logic may easily be derived by considering the set whose members are those and only those sets that are not members of themselves. An apparently satisfactory foundation for set theory, which avoids this and related paradoxes, was formulated in 1908 by Ernst Zermelo, and with suitable technical extensions it provides a satisfactory basis for most of the mathematics published in this century.

Formalism. The third influential position on the foundation of mathematics, called formalism, was developed by Hilbert and others. This view is that the primary mathematical objects are the symbols in which mathematics is written. This carries to the extreme the development of the axiomatic method begun by the Greeks. Under the formalist account the interpretation and use of mathematics must then be given from outside pure mathematics. From a psychological or behavioral standpoint, there is much that is appealing about formalism, but again little effort has yet been made to relate the detailed results and methods of formalism to theoretical or experimental work in scientific psychology.

Relevance of research on foundations. In view of the high degree of agreement about the validity of most published pieces of mathematics, the skeptical social scientist may question the real relevance of these varying views about the foundations of mathematics to working mathematics itself. There is a highly invariant content of mathematics recognized by almost all mathematicians, including those concerned with the foundations of mathematics, and this invariant content is essentially untouched by radically different philosophical views about the nature of mathematical objects. A reasonable conjecture is that future research in the foundations of mathematics will attempt to capture this invariant content by concentrating on the character of mathematical thinking rather than on the nature of mathematical objects.

One other important aspect of foundational research in the twentieth century is the fundamental work on mathematical logic, in particular the attempt by Gottlob Frege, A. N. Whitehead, Bertrand Russell, and others to reduce all of mathematics to purely logical assumptions. These efforts have led to great clarification of the nature of mathematics itself and to vastly increased standards of precision in talking about mathematical proofs and the structure of mathematical systems. Of major importance were the deep results of Kurt Gödel (1931) on the logical limitations of any formal system rich enough to express elementary number theory. His results show that any such formal system must be essentially incomplete in the sense that not all true sentences of the theory can be proved as theorems.

An introductory work on foundations is Kneebone (1963).

Mathematics applied to social sciences

Applications of mathematics to specific social science problems are described, and detailed references are given, elsewhere in this encyclopedia. That material is not repeated here; several reasonably general references are Allen (1938), Coleman (1964), Kemeny and Snell (1962), Luce (1964), Luce, Bush, and Galanter (1963–1965), Samuelson (1947). Suffice it to say that these applications involve only fragments of the whole of mathematics, and they have not been as successful as those in the physical sciences. The reasons are many, among them these: the effort so far expended is much less; the basic empirical concepts and variables have not been isolated and purified to the same degree; mathematics grew up with and was to some extent molded by the needs of physics, and so it may very well be less suited to social science problems if these problems are of a basically different character from those of physics; a typical social science problem appears to involve more variables than one is accustomed to handling in physics; and, finally, social scientists are generally not extensively trained in mathematics.

A social scientist who attempts to formulate and solve a scientific problem in mathematical terms is often disappointed with the mathematics he can find. This may happen simply because a mathematical system appropriate to his problem does not seem to have been invented, or, as is more common, the definite and often quite complex mathematical system that he happens to want to understand in depth has not been investigated in any detail. In this century especially, mathematicians have tended to focus on very general classes of systems, and the theorems concern properties that are true of all or of large subclasses of them; however, these results do not usually provide much detailed information about any particular member of the class.

As an example, the axioms of group theory are not categorical—that is, two groups need not be isomorphic. Therefore, theorems about groups in general tell one little about the specific properties of a particular group. But this is what is of interest when a particular group is used to represent an empirical structure, as in modern particle physics. When this happens, it is necessary for the applied mathematician to carry out considerable mathematical analysis to achieve the understanding he needs to answer scientifically interesting questions.

We have already discussed two parts of mathematics in which highly specific systems have been explored in depth: classical analysis and matrix algebra. A primary motivation for this detailed work was the needs of physical science. In fortunate instances, a problem may be formulated in terms of one of these systems, in which case specific results can sometimes be extracted from the existing literature. Examples where this has been done are in the application of matrix algebra to factor analysis and of Markov chains (a part of probability theory) to several areas, including learning, social interaction, and social structure [*see* FACTOR ANALYSIS AND PRINCIPAL COMPONENTS; MARKOV CHAINS].

Theory as detailed as this, however, is not typical of contemporary mathematics. We have in mind such active areas as associative and nonassociative algebras, homological algebra, group theory, topological groups, algebraic topology, rings, manifolds, and functional analysis.

The generality of contemporary mathematics can be seductive in that it invites sophistic treatments of scientific problems. It is often not difficult to find some general branch of mathematics within which to cast a specific social or behavioral problem without, however, actually capturing in detail the various constraints of the problem. Without these constraints few explicit results and predictions can be proved. Nevertheless, the real emptiness of such endeavors can be shrouded for the unwary in the impressive symbolism and ringing terms of whatever mathematics it is that is not being seriously used.

If the growth of the social sciences parallels at all that of the physical sciences, they will study in detail various systems, which, although of peripheral mathematical interest, are of substantive interest. Indeed, some examples already exist, including these: (1) Just as classes of maximum and minimum problems have been formulated and solved in the physical sciences, other classes have arisen in the social sciences, such as linear, nonlinear, and dynamic programming, game theory, and statistical decision theory. (2) Various mathematical structures that may correspond to (parts of) empirical structures have been investigated, for example, aspects of the theory of relations and the closely related theory of graphs, matrix algebra, and concatenation algebras, which arose in the study of grammar and syntax. (3) Underlying the success of much physical theory is the fact that many variables can be represented numerically. The theories that account for this in physics are not suitable for the social sciences, but alternative possibilities are under active development, particularly in terms of theories of fundamental and derived measurement. The mathematics is reasonably involved, although for the most part the proofs are self-contained. (4) Although the theory of stochastic processes is a well-developed part of probability theory, a number of the processes that have found applications in the social sciences had not previously been studied by probabilists; their properties have been partially worked out in the social science literature. Among the most prominent examples are the nonstationary processes that have arisen in learning theory. Some of these postulate that on each trial one of several operators Q_i transforms a response probability into the corresponding probability on the next trial. Two special cases have been most adequately studied. One assumes that the Q_i are linear operators and the other assumes that the operators commute with one another—that is, $Q_iQ_j = Q_jQ_i$. (See Ross 1968.)

As increasing use is made of mathematics in the social sciences, one may anticipate the investigation of very specific mathematical systems and, ultimately, the isolation of interesting abstract properties from these systems for further study and generalization as pure mathematics.

R. DUNCAN LUCE AND PATRICK SUPPES

BIBLIOGRAPHY

ACZÉL, J. 1966 *Lectures on Functional Equations and Their Applications.* New York: Academic Press.

AKADEMIIA NAUK S.S.S.R., MATEMATICHESKII INSTITUT (1956) 1964 *Mathematics: Its Content, Methods, and Meaning.* Edited by A. D. Aleksandrov, A. N. Kolmogorov, and M. A. Laurent'ev. 3 vols. Cambridge, Mass.: M.I.T. Press. → First published in Russian.

ALLEN, R. G. D. (1938) 1962 *Mathematical Analysis for Economists.* London: Macmillan.

APOSTOL, TOM M. 1961–1962 *Calculus.* 2 vols. New York: Blaisdell.

BARTLE, ROBERT G. 1964 *The Elements of Real Analysis.* New York: Wiley.

BIRKHOFF, GARRETT; and MACLANE, SAUNDERS (1941) 1965 *A Survey of Modern Algebra.* 3d ed. New York: Macmillan.

CARTAN, HENRI (1961) 1963 *Elementary Theory of Analytic Functions of One or Several Complex Variables.* Reading, Mass.: Addison-Wesley. → First published in French.

CODDINGTON, EARL A. (1961) 1964 *An Introduction to Ordinary Differential Equations.* Englewood Cliffs, N.J.: Prentice-Hall.

COLEMAN, JAMES S. 1964 *Introduction to Mathematical Sociology.* New York: Free Press.

COURANT, RICHARD; and ROBBINS, HERBERT (1941) 1961 *What Is Mathematics? An Elementary Approach to Ideas and Methods.* Oxford Univ. Press.

FRIEDMAN, BERNARD 1966 What Are Mathematicians Doing? *Science* 154:357–362.

GÖDEL, KURT (1931) 1965 On Formally Undecidable Propositions of the *Principia mathematica* and Related Systems. I. Pages 4–38 in Martin Davis (editor), *The Undecidable: Basic Papers on Undecidable Propositions, Unsolvable Problems and Computable Functions.* Hewlett, N.Y.: Raven. → First published in German in Volume 38 of the *Monatshefte für Mathematik und Physik.*

GOLDBERG, SAMUEL 1958 *Introduction to Difference Equations: With Illustrative Examples From Economics, Psychology, and Sociology.* New York: Wiley. → A paperback edition was published in 1961.

HOCKING, JOHN G.; and YOUNG, GAIL S. 1961 *Topology.* Reading, Mass.: Addison-Wesley.

KEMENY, JOHN G.; and SNELL, J. LAURIE 1962 *Mathematical Models in the Social Sciences.* Boston: Ginn.

KNEEBONE, G. T. 1963 *Mathematical Logic and the Foundations of Mathematics: An Introductory Survey.* New York: Van Nostrand.

LUCE, R. DUNCAN 1964 The Mathematics Used in Mathematical Psychology. *American Mathematical Monthly* 71:364–378.

LUCE, R. DUNCAN; BUSH, ROBERT R.; and GALANTER, EUGENE (editors) 1963–1965 *Handbook of Mathematical Psychology.* 3 vols. New York: Wiley.

MOSTOW, GEORGE; SAMPSON, JOSEPH H.; and MEYER, JEAN-PIERRE 1963 *Fundamental Structures of Algebra.* New York: McGraw-Hill.

NEUGEBAUER, OTTO (1951) 1957 *The Exact Sciences in Antiquity.* 2d ed. Providence, R.I.: Brown Univ. Press.

NEWMAN, JAMES R. (editor) 1956 *The World of Mathematics.* 4 vols. New York: Simon & Schuster.

ORE, ØYSTEIN 1948 *Number Theory and Its History.* New York: McGraw-Hill.

►ROSS, LEONARD E. 1968 Learning Theory. Volume 9, pages 189–197 in *International Encyclopedia of the Social Sciences.* Edited by David L. Sills. New York: Macmillan and Free Press.

SAMUELSON, PAUL A. (1947) 1958 *Foundations of Economic Analysis.* Harvard Economic Studies, Vol. 80. Cambridge, Mass.: Harvard Univ. Press.

Postscript

Because the values that guide and the attitudes that are expressed by pure mathematicians are often poorly understood by nonmathematicians, the lucid article (1972) by Adler in the *New Yorker* on these matters is most instructive.

At a substantive level, the Committee on Support of Research in the Mathematical Sciences issued in 1969 a collection of essays, *The Mathematical Sciences,* "intended not only for the nonmathematical scientist but also for the scientifically oriented layman." These essays are designed to give some sense of the state of contemporary mathematics—pure and applied. One topic omitted is the development of nonstandard analysis (Davis & Hersh 1972), which makes rigorous the concept of an infinitesimal. We mention this explicitly because it is beginning to play a role in some work in economics and the theory of measurement.

R. DUNCAN LUCE AND PATRICK SUPPES

ADDITIONAL BIBLIOGRAPHY

ADLER, ALFRED 1972 Reflections: Mathematics and Creativity. *New Yorker* Feb. 19:39–45.

BROWDER, FELIX E. 1976 Does Pure Mathematics Have a Relation to the Sciences? *American Scientist* 64:542–549.

DAVIS, MARTIN; and HERSH, REUBEN 1972 Nonstandard Analysis. *Scientific American* 226, June:78–86.

The Mathematical Sciences: A Collection of Essays. 1969 Edited by the National Research Council's Committee on Support of Research in the Mathematical Sciences with the collaboration of George A. W. Boehm. Cambridge, Mass.: M.I.T. Press.

RAPOPORT, ANATOL 1976 Directions in Mathematical Psychology. Parts 1 and 2. *American Mathematical Monthly* 83:85–106, 153–172.

MAXIMUM LIKELIHOOD
See ESTIMATION.

MEAN VALUES
See STATISTICS, DESCRIPTIVE, *article on* LOCATION AND DISPERSION.

MEASUREMENT
See ECONOMETRICS; EVALUATION RESEARCH; PANEL STUDIES; PSYCHOMETRICS; SOCIOMETRY; STATISTICS; STATISTICS, DESCRIPTIVE, *article on* LOCATION AND DISPERSION; SURVEY ANALYSIS.

MINIMAX
See DECISION THEORY.

MINIMUM REGRET
See DECISION THEORY.

MISES, RICHARD VON
See VON MISES, RICHARD.

MIXTURES OF DISTRIBUTIONS
See under DISTRIBUTIONS, STATISTICAL.

MOBILITY
See SOCIAL MOBILITY.

MODELS, MATHEMATICAL

Although mathematical models are applied in many areas of the social sciences, this article is limited to mathematical models of individual behavior. For applications of mathematical models in econometrics, see ECONOMETRIC MODELS, AGGREGATE. *Other articles discussing modeling in general include* PROBABILITY, SIMULATION, *and* SIMULTANEOUS EQUATION ESTIMATION. *Specific models are discussed in various articles dealing with substantive topics.*

Theories of behavior that have been developed and presented verbally, such as those of Hull or Tolman or Freud, have attempted to describe and predict behavior under any and all circumstances. Mathematical models of individual behavior, by contrast, have been much less ambitious: their goal has been a precise description of the data obtained from restricted classes of behavioral experiments concerned with simple and discrimination learning; with detection, recognition, and discrimination of simple physical stimuli; with the patterns of preference exhibited among outcomes; and so on. Models that embody very specific mathematical assumptions, which are at best approximations applicable to highly limited situations, have been analyzed exhaustively and applied to every conceivable aspect of available data. From this work broader classes of models, based on weaker assumptions and thus providing more general predictions, have evolved in the past few years. The successes of the special models have stimulated, and their failures have demanded, these generalizations. The number and variety of experiments to which these mathematical models have been applied have also grown, but not as rapidly as the catalogue of models.

Most of the models so far developed are restricted to experiments having discrete trials. Each trial is composed of three types of events: the presentation of a stimulus configuration selected by the experimenter from a limited set of possible presentations; the subject's selection of a response from a specified set of possible responses; and the experimenter's feedback of information, rewards, and punishments to the subject. Primarily because the response set is fixed and feedback is used, these are called choice experiments (Bush et al. 1963). Most psychophysical and preference experiments, as well as many learning experiments, are of this type. Among the exceptions are the experiments without trials e.g., vigilance experiments and the operant conditioning methods of Skinner. Currently, models for these experiments are beginning to be developed.

Measures. With attention confined to choice experiments, three broad classes of variables necessarily arise—those concerned with stimuli, with responses, and with outcomes. The response variables are, of course, assumed to depend upon the (experimentally) independent stimuli and upon the outcome variables, and each model is nothing more or less than an explicit conjecture about the nature of this dependency. Usually such conjectures are stated in terms of some measures, often numerical ones, that are associated with the variables. Three quite different types of measures are used: physical, probabilistic, and psychological. The first two are objective and descriptive; they can be introduced and used without reference to any psychological theory, and so they are especially popular with atheoretical experimentalists, even though the choice of a measure usually reflects a theoretical attitude about what is and is not psychologically relevant. Although we often use physical measures to characterize the events for which probabilities are defined, this is only a labeling function which makes little or no use of the powerful mathematical structure embodied in many physical measures. The psychological measures are constructs within some specifiable psychological theory, and their calculation in terms of observables is possible only within the terms of that theory. Examples of each type of measure should clarify the meaning.

Physical measures. In experimental reports, the stimuli and outcomes are usually described in terms of standard physical measures: intensity, frequency, size, weight, time, chemical composition, amount, etc. Certain standard response measures are physical. The most ubiquitous is response latency (or reaction time), and it has received the attention of some mathematical theorists (McGill 1963). In addition, force of response, magnitude of displacement, speed of running, etc., can sometimes be recorded. Each of these is unique to certain experimental realizations, and so they have not been much studied by theorists.

Probability measures. The stimulus presentations, the responses, and the outcomes can each be thought of as a sequence of selections of elements from known sets of elements, i.e., as a schedule over trials. It is not usual to work with the specific schedules that have occurred but, rather, with the probability rules that were used to generate them. For the stimulus presentations and the outcomes, the rules are selected by the experimenter, and so there is no question about what they are. Not only are the rules not known for the responses, but even their general form is not certain. Each response theory is, in fact, a hypothesis about the form of these rules, and certain relative frequencies of responses are used to estimate the postulated conditional response probabilities.

Often the schedules for stimulus presentations are simple random ones in the sense that the probability of a stimulus' being presented is independent of the trial number and of the previous history of the experiment; but sometimes more complex contingent schedules are used in which various conditional probabilities must be specified. Most outcome schedules are to some degree contingent, usually on the immediately preceding presentation and response, but sometimes the dependencies reach further back into the past. Again, conditional probabilities are the measures used to summarize the schedule. [*See* PROBABILITY.]

Psychological measures. Most psychological models attempt to state how either a physical measure or a probability measure of the response depends upon measures of the experimental independent variables, but in addition they usually include unknown free parameters—that is, numerical constants whose values are specified neither by the experimental conditions nor by independent measurements on the subject. Such parameters must, therefore, be estimated from the data that have been collected to test the adequacy of the theory, which thereby reduces to some degree the stringency of the test. It is quite common for current psychological models to involve only probability measures and unknown numerical parameters, but not any physical measures. When the numerical parameters are estimated from different sets of data obtained by varying some independent variables under the experimenter's control, it is often found that the parameters vary with some variables and not with others. In other words, the parameters are actually functions of some of the experimental variables, and so they can be, and often are, viewed as psychological measures (relative to the model within which they appear) of the variables that affect them. Theories are sometimes then provided for this dependence, although so far this has been the exception rather than the rule.

The theory of signal detectability, for example, involves two parameters: the magnitude, d', of the psychological difference between two stimuli; and a response criterion, c, which depends upon the outcomes and the presentation schedule. Theories for the dependence of d' and c upon physical measures have been suggested (Luce 1963; Swets 1964). Most learning theories for experiments with only one presentation simply involve the conditional outcome probabilities and one or more free parameters. Little is known about the dependence of these parameters upon experimentally manipulable variables. In certain scaling theories, numerical parameters are assigned to the response alternatives and are interpreted as measures of response strength (Luce & Galanter 1963). In some models these parameters are factored into two terms, one of which is assumed to measure the contribution of the stimulus to response strength and the other of which is the contribution due to the outcome structure.

The phrasing of psychological models in terms only of probability measures and parameters (psychological measures) has proved to be an effective research strategy. Nonetheless, it appears important to devise theories that relate psychological measures to the physical and probability measures that describe the experiments. The most extensive mathematical models of this type can be found in audition and vision (Hurvich et al. 1965; Zwislocki 1965). The various theories of utility are, in part, attempts to relate the psychological measure called utility to physical measures of outcomes, such as amounts of money, and probability measures of their schedules, such as probabilities governing gambles (Luce & Suppes 1965). In spite of the fact that it is clear that the utilities of outcomes must be related to learning parameters, little is known about this relation. [*See* GAME THEORY; *see also* Devereux 1968; Georgescu-Roegen 1968.]

The nature of the models. The construction of a mathematical model involves decisions on at least two levels. There is, first, the over-all perspective about what is and is not important and about the best way to secure the relevant facts. Usually this is little discussed in the presentation of a model, mainly because it is so difficult to make the discussion coherent and convincing. Nonetheless, this is what we shall attempt to deal with in this section. In the following section we turn to the second level of decision: the specific assumptions made.

Probability vs. determinism. One of the most basic decisions is whether to treat the behavior as if it arises from some sort of probabilistic mechanism, in which case detailed, exact predictions are not possible, or whether to treat it as deterministic, in which case each specific response is susceptible to exact prediction. If the latter decision is made, one is forced to provide some account of the observed inconsistencies of responses before it is possible to test the adequacy of the model. Usually one falls back on either the idea of errors of measurement or on the idea of systematic changes with time (or experience), but in practice it has not been easy to make effective use of either idea, and most workers have been content to develop probability models. It should be pointed out that, as far as the model is concerned, it is immaterial whether the model builder believes the behavior to be inherently probabilistic, or its determinants to be too complex to give a detailed analysis, or that there are uncontrolled factors which lead to experimental errors.

Static vs. dynamic models. A second decision is whether the model shall be dynamic or static. (We use these terms in the way they are used in physics; static models characterize systems which do not change with time or systems which have reached equilibrium in time, whereas dynamic models are concerned with time changes.) Some dynamic models, especially those for learning, state how conditional response probabilities change with experience. Usually these models are not very helpful in telling us what would happen if, for example, we substituted a different but closely related set of response alternatives or outcomes. In static models the constraints embodied in the model concern the relations among response probabilities in several different, but related, choice situations. The utility models for the study of preference are typical of this class.

The main characteristic of the existing dynamic models is that the probabilities are functions of a discrete time parameter. Such processes are called stochastic, and they can be thought of as generating branching processes through the fanning out of new possibilities on each trial (Snell 1965). Each individual in an experiment traces out one path of the over-all tree, and we attempt to infer from a small but, it is hoped, typical sample of these paths something about the probabilities that supposedly underlie the process. Usually, if enough time is allowed to pass, such a process settles down —becomes asymptotic—in a statistical sense. This is one way to arrive at a static model; and when we state a static model, we implicitly assume that it describes (approximately) the asymptotic behavior of the (unknown) dynamic process governing the organisms.

Psychological vs. mathematical assumptions. Another distinction is that between psychological and formal mathematical assumptions. This is by no means a sharp one, if for no other reason than that the psychological assumptions of a mathematical model are ultimately cast in formal terms and that psychological rationales can always be evolved for formal axioms. Roughly, however, the distinction is between a structure built up from elementary principles and a postulated constraint concerning observable behavior. Perhaps the simplest example of the latter is the axiom of transitivity of preferences; if a is preferred to b and b is preferred to c, then a will be preferred to c. This is not usually derived from more basic psychological postulates but, rather, is simply asserted on the grounds that it is (approximately) true in fact. A somewhat more complex, but essentially similar, example is the so-called choice axiom which postulates how choice probabilities change when the set of possible choices is either reduced or augmented (Luce 1959). Again, no rationale was originally given except plausibility; later, psychological mechanisms were proposed from which it derives as a consequence.

The most familiar example of a mathematical model which is generally viewed as more psychological and less formal is stimulus sampling theory. In this theory it is supposed that an organism is exposed to a set of stimulus "elements" from which one or more are sampled on a trial and that these elements may become "conditioned" to the performed response, depending upon the outcome that follows the response (Atkinson & Estes 1963). The concepts of sampling and conditioning are interpreted as elementary psychological processes from which the observed properties of the choice behavior are to be derived. Lying somewhere between the two extremes just cited are, for example, the linear operator learning models (Bush & Mosteller 1955; Sternberg 1963). The trial-by-trial changes in response probabilities are assumed to be linear, mainly because of certain formal considerations; the choice of the limit points of the operators in specific applications is, however, usually based upon psychological considerations; and the resulting mathematical structure is not evaluated directly but, rather, in terms of its ability to account for the observed choice behavior as summarized in such observables as the mean learning curve, the sequential dependencies among responses, and the like.

Recurrent theoretical themes. Beyond a doubt, the most recurrent theme in models is independence. Indeed, one can fairly doubt whether a serious theory exists if it does not include statements to the effect that certain measures which contribute to the response are in some way independent of other measures which contribute to the same response. Of course, independence assumes different mathematical forms and therefore has different names, depending upon the problem, but one should not lose sight of the common underlying intuition which, in a sense, may be simply equivalent to what we mean when we say that a model helps to simplify and to provide understanding of some behavior.

Statistical independence. In quite a few models simple statistical independence is invoked. For example, two chance events, A and B, are said to be independent when the conditional probability of A, given B, is equal to the unconditional probability of A; equivalently, the probability of the joint event AB is the product of the separate probabilities of A and B.

A very simple substantive use of this notion is contained in the choice axiom which says, in effect, that altering the membership of a choice set does not affect the relative probabilities of choice of two alternatives (Luce 1959). More complex notions of independence are invoked whenever the behavior is assumed to be described by a stochastic process.

Each such process states that some, but not all, of the past is relevant in understanding the future: some probabilities are independent of some earlier events. For example, in the "operator models" of learning, it is assumed that the process is "path independent" in the sense that it is sufficient to know the existing choice probability and what has happened on that trial in order to calculate the choice probability on the next trial (Bush & Mosteller 1955). In the "Markovian" learning models, the organism is always in one of a finite number of states which control the choice probabilities, and the probabilities of transition from one state to another are independent of time, i.e., trials (Atkinson & Estes 1963). Again, the major assumption of the model is a rather strong one about independence of past history. [*See* MARKOV CHAINS.]

Additivity and linearity. Still another form of independence is known as additivity. If r is a response measure that depends upon two different variables assuming values in sets A_1 and A_2, then we say that the measure is additive (over the independent variables) if there exists a numerical measure r_1 on A_1 and r_2 on A_2 such that for x_1 in A_1 and x_2 in A_2, $r(x_1,x_2) = r_1(x_1) + r_2(x_2)$. This assumption for particular experimental measures r is frequently postulated in the models of analysis of variance as well as derived from certain theories of fundamental measurement. A special case of additivity known as linearity is very important. Here there is but one variable (that is, $A_1 = A_2 = A$); any two values of that variable, x and x' in A, combine through some physical operation to form a third value of that variable, denoted $x * x'$; and there is a single measure r on A (that is, $r_1 = r_2 = r$) such that $r(x * x') = r(x) + r(x')$. Such a requirement captures the superposition principle and leads to models of a very simple sort. These linear models have played an especially important role in the study of learning, where it is postulated that the choice probability on one trial, p_n, can be expressed linearly in terms of the probability, p_{n-1}, on the preceding trial. Other models also postulate linear transformations, but not necessarily on the response probability itself. In the "beta" model, the quantity $p_n/(1 - p_n)$ is assumed to be transformed linearly; this quantity is interpreted as a measure of response strength (Luce 1959).

Commutativity. The "beta" model exhibits another property that is of considerable importance, namely, commutativity. The essence of commutativity is that the order in which the operators are applied does not matter; that is, if A and B are operators, then the composite operator AB (apply B first and then A) is the same as the operator BA. Again, there is a notion of independence—independence of the order of application. It is an extremely powerful property that permits one to derive a considerable number of properties of the resulting process; however, it is generally viewed with suspicion, since it requires the distant past to have exactly the same effect as the recent past. A commutative model fails to forget gradually.

Nature of the predictions. As would be expected, models are used to make a variety of predictions. Perhaps the most general sorts of predictions involve broad classes of models. For example, probabilistic reinforcement schedules for a certain class of distance-diminishing models, i.e., ones that require the behavior of two subjects to become increasingly similar when they are identically reinforced, can be shown to be ergodic, which means that these models exhibit the asymptotic properties that are commonly taken for granted. A second example is the combining-of-classes theorem, which asserts that if the theoretical descriptions of behavior are to be independent of the grouping of responses into classes, then only the linear learning models are appropriate.

At a somewhat more detailed level, but still encompassing several different models, are predictions such as the mean learning curve, response operating characteristics, and stochastic transitivity of successive choices among pairs of alternatives. Sometimes it is not realized that conceptually quite different models, which make some radically different predictions, may nonetheless agree completely on other features of the data, often on ones that are ordinarily reported in experimental studies. Perhaps the best example of this phenomenon arises in the analysis of experiments in which subjects learn arbitrary associations between verbal stimuli and responses. A linear incremental model, of the sort described above, predicts exactly the same mean learning curve as does a model that postulates that the arbitrary association is acquired on an all-or-none basis. On the face of it, this result seems paradoxical. It is not, because in the latter model, different subjects acquire the association on different trials, and averaging over subjects thereby leads to a smooth mean curve that happens to be identical with the one predicted by the linear model. Actually, a wide variety of models predict the same mean learning curve for many probabilistic schedules of reinforcement, and so one must turn to finer-grained features of the data to distinguish among the models. Among these differential predictions are the distribution of runs of the same response, the expected number of such runs, the variance of the number of successes in a fixed block of trials, the mean number of total errors, the mean trial of last error, etc. [*See* STATISTICAL IDENTIFIABILITY.]

The classical topic of individual differences raises issues of a different sort. For the kinds of predictions discussed above it is customary to pool individual data and to analyze them as if they were entirely homogeneous. Often, in treating learning data this way, it is argued that the structural conditions of the experiment are sufficiently more important determinants of behavior than are individual differences so that the latter may be ignored without serious distortion. For many experiments to which models have been applied with considerable success, simple tests of this hypothesis of homogeneity are not easily made. For example, when a group of 30 or 40 subjects is run on 12 to 15 paired-associate items, it is not useful to analyze each subject item because of the large relative variability which accompanies a small number of observations. On the other hand, in some psychophysical experiments in which each subject is run for thousands of trials under constant conditions of presentation and reinforcement, it is possible to treat in detail the data of individuals. The final justification for using group data, on the assumption of identical subjects, is the fact that for ergodic processes, which most models are, the predictions for data averaged over subjects are the same as those for the data of an individual averaged over trials.

Another issue, which relates to group versus individual data, is parameter invariance. One way of asking if a group of individuals is homogeneous is to ask whether, within sampling error, the parameters for individuals are identical. Thus far, however, more experimental attention has been devoted to the question of parameter invariance for sets of group data collected under different experimental conditions. For instance, the parameters of most learning models should be independent of the particular reinforcement schedule adopted by the experimenter. Although in many cases a reasonable degree of parameter invariance has been obtained for different schedules, it is fair to say that the results have not been wholly satisfactory.

For a detailed discussion of the topics of this section, see Sternberg (1963) and Atkinson and Estes (1963).

Model testing. Most of the mathematical models used to analyze psychological data require that at least one parameter, and often more, be estimated from the data before the adequacy of the model can be evaluated. In principle, it might be desirable to use maximum-likelihood methods for estimation. Perhaps the central difficulty which prevents our using such estimators is that the observable random variables, such as the presentation, response, and outcome random variables, form chains of infinite order. This means that their probabilities on any trial depend on what actually happened in all preceding trials. When that is so, it is almost always impractical to obtain a useful maximum-likelihood estimator of a parameter. In the face of such difficulties, less desirable methods of estimation have perforce been used. Theoretical expressions showing the dependency on the unknown parameter of, for example, the mean number of total errors, the mean trial of first success, and the mean number of runs, have been equated to data statistics to estimate the parameters. The classical methods of moments and of least squares have sometimes been applied successfully. And, in certain cases, maximum-likelihood estimators can be approximated by pseudo-maximum-likelihood ones that use only a limited portion of the immediate past. For processes that are approximately stationary, a small part of the past sometimes provides a very good approximation to the full chain of infinite order, and then pseudo-maximum-likelihood estimates can be good approximations to the exact ones. Because of mathematical complexities in applying even these simplified techniques, Monte Carlo and other numerical methods are frequently used. [*See* ESTIMATION.]

Once the parameters have been estimated, the number of predictions that can be derived is, in principle, enormous: the values of the parameters of the model, together with the initial conditions and the outcome schedule, uniquely determine the probability of all possible combinations of events. In a sense, the investigator is faced with a plethora of riches, and his problem is to decide what predictions are the most significant from the standpoint of providing telling tests of a model. In more classical statistical terms, what can be said about the goodness of fit of the model?

Just as with estimation, it might be desirable to evaluate goodness of fit by a likelihood ratio test. But, a fortiori, this is not practical when maximum-likelihood estimators themselves are not feasible. Rather, a combination of minimum chi-square techniques for both estimation and testing goodness of fit have come to be widely used in recent years. No single statistic, however, serves as a satisfactory over-all evaluation of a model, and so the report usually summarizes its successes and failures on a rather extensive list of measures of fit.

A model is never rejected outright because it does not fit a particular set of data, but it may disappear from the scene or be rejected in favor of another model that fits the data more adequately. Thus, the classical statistical procedure of accepting or rejecting a hypothesis—or model—is in fact seldom directly invoked in research on mathematical models; rather, the strong and weak points of the model are brought out, and new models are sought that do not have the discovered weaknesses. [*See* GOODNESS OF FIT; *more detail on these topics can be found in* Bush 1963].

Impact on psychology. Although the study of mathematical models has come to be a subject in its own right within psychology, it is also pertinent to ask in what ways their development has had an impact on general experimental psychology.

For one, it has almost certainly raised the standards of systematic experimentation: the application of a model to data prompts a number of detailed questions frequently ignored in the past. A model permits one to squeeze more information out of the data than is done by the classical technique of comparing experimental and control groups and rejecting the null hypothesis whenever the difference between the two groups is sufficiently large. A successful test of a mathematical model often requires much larger experiments than has been customary. It is no longer unusual for a quantitative experiment to consist of 100,000 responses and an equal number of outcomes. In addition to these methodological effects on experimentation and on data analysis, there have been substantive ones. Of these we mention a few of the more salient ones.

Probability matching. A well-known finding, which dates back to Humphreys (1939), is that of probability matching. If either one of two responses is rewarded on each trial, then in many situations organisms tend to respond with probabilities equal to the reward probabilities rather than to choose the more often rewarded response almost all of the time. Since Humphreys' original experiment, many similar ones have been performed on both human and animal subjects to discover the extent and nature of the phenomenon, and a great deal of effort has been expended on theoretical analyses of the results. Estes (1964) has given an extensive review of both the experimental and the theoretical literature. Perhaps the most important contribution of mathematical models to this problem was to provide sets of simple general assumptions about behavior which, coupled with the specification of the experimenter's schedule of outcomes, predict probability matching. As noted above, investigators have not been content with just predicting the mean asymptotic values but have dealt in detail with the relation between predicted and observed conditional expectations, run distributions, variances, etc. Although this experimental paradigm for probability learning did not originate in mathematical psychology, its thorough exploration and the resulting interpretations of the learning process have been strongly promoted by the many predictions made possible by models for this paradigm.

The all-or-none model. A second substantive issue to which a number of investigators have addressed mathematical models is whether or not simple learning is of an all-or-none character. As noted earlier, the linear model assumes learning to be incremental in the sense that whenever a stimulus is presented, a response made, and an outcome given, the association reinforced by the outcome is thereby made somewhat more likely to occur. In contrast, the simple all-or-none model postulates that the subject is either completely conditioned to make the correct response, or he is not so conditioned. No intermediate states exist, and until the correct conditioning association is established on an all-or-none basis, his responses are determined by a constant guessing probability. This means that learning curves for individual subjects are flat until conditioning occurs, at which point they exhibit a strong discontinuity. The problem of discriminating the two models must be approached with some care since, for instance, the mean learning curve obtained by averaging data over subjects, or over subjects and a list of items as well, is much the same for the two models. On the other hand, analyses of such statistics as the variance of total errors, the probability of an error before the last error, and the distribution of last errors exhibit sharp differences between the models. For paired-associates learning, the all-or-none model is definitely more adequate than the linear incremental model (Atkinson & Estes 1963). Of course, the issue of all-or-none versus incremental learning is not special to mathematical psychology; however, the application of formal models has raised detailed questions of data analysis and posed additional theoretical problems not raised, let alone answered, by previous approaches to the problem.

Reward and punishment. The classic psychological question of the relative effects of reward and punishment (or nonreward) has also arisen in work on models, and it has been partially answered. In some models, such as the linear one, there are two rate parameters, one of which represents the effect of reward on a single trial and the other of which represents the effect of nonreward. Their estimated values provide comparable measures of the effects of these two events for those data from which they are estimated. For example, Bush and Mosteller (1955) found that a trial on which a dog avoided shock (reward) in an avoidance training experiment produced about the same change in response probabilities as three trials of nonavoidance (punishment). No general law has emerged, however. The relative effects of reward and nonreward seem to vary from one experiment to another and to depend on a number of experimental variables.

When using a model to estimate the relative effects of different events, the results must be interpreted with some care. The measures are meaningful only in terms of the model in which they are defined. A different model with corresponding reward and nonreward parameters may lead to the opposite conclusion. Thus, one must decide which model best accounts for the data and use it for measuring the relative effects of the two events. Very delicate issues of parameter estimation arise, and examples exist where opposite conclusions have been drawn, depending on the estimators used. The alternative is to devise more nonparametric methods of inference which make weaker assumptions about the learning process. A detailed discussion of these problems is given by Sternberg (1963, pp. 109–116). (See Pliskoff & Ferster 1968.)

Homogenizing a group. If one wishes to obtain a homogeneous group of subjects after a particular experimental treatment, should all subjects be run for a fixed number of trials, or should each subject be run until he meets a specific performance criterion? Typically it is assumed by those who use such a criterion that individual subjects differ; that, for example, some are fast learners and some are slow. It is further assumed that all subjects will achieve the same performance level if each is run to a criterion such as ten successive successes. Now it is clear that for identical subjects, it is simpler to run them all for the same number of trials and perhaps use a group performance criterion. It is, however, less obvious whether it would be better to do this than to run each to a criterion. An analysis of stochastic learning models has shown that running each of identical subjects to a criterion introduces appreciable variance in the terminal performance levels. One can study individual differences only in terms of a model and assumptions about the distributions of the model parameters. When this is done, it becomes evident that very large individual differences must exist to justify using the criterion method of homogenizing a group of subjects.

Psychophysics. The final example is selected from psychophysics. With the advent of signal detection theory it became increasingly apparent that the classical methods for measuring sensory thresholds are inherently ambiguous, that they depend not only, as they are supposed to, on sensitivity but also on response biases (Luce 1963; Swets 1964). Consider a detection experiment in which the stimulus is presented only on a proportion π of the trials. Let $p(Y|s)$ and $p(Y|n)$ be the probabilities of a "Yes" response to the stimulus and to no stimulus respectively. If the experiment is run several times with different values of π between 0 and 1, then $p(Y|n)$, as well as $p(Y|s)$, which is a classical threshold measure, varies systematically from 0 to 1. The data points appear to fall on a smooth, convex curve, which shows the relation, for the subject, between correct responses to stimuli and incorrect responses to no-stimulus trials (false alarms). Its curvature, in effect, characterizes the subject's sensitivity, and the location of the data point along the curve represents the amount of bias, i.e., his over-all tendency to say "Yes," which varies with π, with the payoffs used, and with instructions. Several conceptually different theories, which are currently being tested, account for such curves; it is clear that any new theory will be seriously entertained only if it admits to some such partition of the response behavior into sensory and bias components. This point of view is, of course, applicable to any two-stimulus–two-response experiment, and often it alters significantly the qualitative interpretation of data. [*See* PSYCHOPHYSICS; *see also* Jerison 1968.]

Although one cannot be certain about what will happen next in the application of mathematical models to problems of individual behavior, certain trends seem clear. (1) The ties that have been established between mathematical theorists and experimentalists appear firm and productive; they probably will be strengthened. (2) The general level of mathematical sophistication in psychology can be expected to increase in response to the increasing numbers of experimental studies that stem from mathematical theories. (3) The major applications will continue to center around well-defined psychological issues for which there are accepted experimental paradigms and a considerable body of data. One relatively untapped area is operant (instrumental) conditioning. (4) Along with models for explicit paradigms, abstract principles (axioms) of behavior that have wide potential applicability are being isolated and refined, and attempts are being made to explore general qualitative properties of whole classes of models. (5) Even though the most successful models to date are probabilistic, the analysis of symbolic and conceptual processes seems better handled by other mathematical techniques, and so more nonprobabilistic models can be anticipated.

ROBERT R. BUSH, R. DUNCAN LUCE, AND PATRICK SUPPES

[*See also* DECISION MAKING, *article on* PSYCHOLOGICAL ASPECTS; SIMULATION, *article on* INDIVIDUAL BEHAVIOR. *Other relevant material may be found in* MATHEMATICS; PROBABILITY; PSYCHOPHYSICS; Jerison 1968; Maron 1968; Ross 1968; Torgerson 1968.]

BIBLIOGRAPHY

ATKINSON, RICHARD C.; and ESTES, WILLIAM K. 1963 Stimulus Sampling Theory. Volume 2, pages 121–268 in R. Duncan Luce, Robert R. Bush, and Eugene Galanter (editors), *Handbook of Mathematical Psychology*. New York: Wiley.

BUSH, ROBERT R. 1963 Estimation and Evaluation. Volume 1, pages 429–469 in R. Duncan Luce, Robert R. Bush, and Eugene Galanter (editors), *Handbook of Mathematical Psychology*. New York: Wiley.

BUSH, ROBERT R.; GALANTER, EUGENE; and LUCE, R. DUNCAN 1963 Characterization and Classification of Choice Experiments. Volume 1, pages 77–102 in R. Duncan Luce, Robert R. Bush, and Eugene Galanter (editors), *Handbook of Mathematical Psychology*. New York: Wiley.

BUSH, ROBERT R.; and MOSTELLER, FREDERICK 1955 *Stochastic Models for Learning*. New York: Wiley.

►DEVEREUX, EDWARD C. JR. 1968 Gambling. Volume 6, pages 53–62 in *International Encyclopedia of the Social Sciences*. Edited by David L. Sills. New York: Macmillan and Free Press.

ESTES, WILLIAM K. 1964 Probability Learning. Pages 89–128 in Symposium on the Psychology of Human Learning, University of Michigan, 1962, *Categories of Human Learning*. Edited by Arthur W. Melton. New York: Academic Press.

►GEORGESCU-ROEGEN, NICHOLAS 1968 Utility. Volume 16, pages 236–267 in *International Encyclopedia of the Social Sciences*. Edited by David L. Sills. New York: Macmillan and Free Press.

HUMPHREYS, LLOYD G. 1939 Acquisition and Extinction of Verbal Expectations in a Situation Analogous to Conditioning. *Journal of Experimental Psychology* 25: 294–301.

HURVICH, LEO M.; JAMESON, DOROTHEA; and KRANTZ, DAVID H. 1965 Theoretical Treatments of Selected Visual Problems. Volume 3, pages 99–160 in R. Duncan Luce, Robert R. Bush, and Eugene Galanter (editors), *Handbook of Mathematical Psychology*. New York: Wiley.

►JERISON, HARRY J. 1968 Attention. Volume 1, pages 444–449 in *International Encyclopedia of the Social Sciences*. Edited by David L. Sills. New York: Macmillan and Free Press.

LUCE, R. DUNCAN 1959 *Individual Choice Behavior*. New York: Wiley.

LUCE, R. DUNCAN 1963 Detection and Recognition. Volume 1, pages 103–190 in R. Duncan Luce, Robert R. Bush, and Eugene Galanter (editors), *Handbook of Mathematical Psychology*. New York: Wiley.

LUCE, R. DUNCAN; and GALANTER, EUGENE 1963 Psychophysical Scaling. Volume 1, pages 245–308 in R. Duncan Luce, Robert R. Bush, and Eugene Galanter (editors), *Handbook of Mathematical Psychology*. New York: Wiley.

LUCE, R. DUNCAN; and SUPPES, PATRICK 1965 Preference, Utility, and Subjective Probability. Volume 3, pages 249–410 in R. Duncan Luce, Robert R. Bush, and Eugene Galanter (editors), *Handbook of Mathematical Psychology*. New York: Wiley.

MCGILL, WILLIAM J. 1963 Stochastic Latency Mechanisms. Volume 1, pages 309–360 in R. Duncan Luce, Robert R. Bush, and Eugene Galanter (editors), *Handbook of Mathematical Psychology*. New York: Wiley.

►MARON, M. E. 1968 Cybernetics. Volume 4, pages 3–6 in *International Encyclopedia of the Social Sciences*. Edited by David L. Sills. New York: Macmillan and Free Press.

►PLISKOFF, STANLEY S.; and FERSTER, CHARLES B. 1968 Learning: IV. Reinforcement. Volume 9, pages 135–143 in *International Encyclopedia of the Social Sciences*. Edited by David L. Sills. New York: Macmillan and Free Press.

►ROSS, LEONARD E. 1968 Learning Theory. Volume 9, pages 189–197 in *International Encyclopedia of the Social Sciences*. Edited by David L. Sills. New York: Macmillan and Free Press.

SNELL, J. LAURIE 1965 Stochastic Processes. Volume 3, pages 411–486 in R. Duncan Luce, Robert R. Bush, and Eugene Galanter (editors), *Handbook of Mathematical Psychology*. New York: Wiley.

STERNBERG, SAUL 1963 Stochastic Learning Theory. Volume 2, pages 1–120 in R. Duncan Luce, Robert R. Bush, and Eugene Galanter (editors), *Handbook of Mathematical Psychology*. New York: Wiley.

SWETS, JOHN A. (editor) 1964 *Signal Detection and Recognition by Human Observers: Contemporary Readings*. New York: Wiley.

►TORGERSON, WARREN S. 1968 Scaling. Volume 14, pages 25–39 in *International Encyclopedia of the Social Sciences*. Edited by David L. Sills. New York: Macmillan and Free Press.

ZWISLOCKI, JOZEF 1965 Analysis of Some Auditory Characteristics. Volume 3, pages 1–98 in R. Duncan Luce, Robert R. Bush, and Eugene Galanter (editors), *Handbook of Mathematical Psychology*. New York: Wiley.

Postscript

In an attempt to provide clear examples of the impact of mathematical models on psychological research, Krantz et al. (1974) had various authors exposit topics they judged important. Volume 1, *Learning, Memory, and Thinking*, suggests—and an examination of recent volumes of *Cognitive Psychology* and the *Journal of Mathematical Psychology* confirms—a shift away from operator and Markov models of learning to less mathematical, more computer-oriented models of memory and information processing. Although little is excluded by the latter term, it does suggest a distinctive theoretical attitude, different from the earlier impact of behaviorism on the learning models. The characteristic uses of reaction-time measures in the study of short-term memory are summarized by Sternberg (1969).

Volume 2, *Measurement, Psychophysics, and Neural Information Processing,* illustrates two more trends: an increased focus on questions of measurement and new, more global concepts in psychophysics. To be sure, scaling has long been a staple of psychometrics, and test theory (Lord & Novick 1968) is of great importance in the analysis of psychological tests. But new developments in multidimensional scaling [*see* SCALING, MULTIDIMENSIONAL], functional measurement, and the theoretical foundations of measurement (see also Krantz et al. 1971) have captured much attention. Psychophysics, long a heavy user of mathematical models, has shown a tendency toward models of more comprehensive scope, particularly those that attempt some sort of rapprochement with neurophysiology.

Three topics that are not covered in these volumes and that are generally conceded to be of current importance are perception, preference, and psycholinguistics. No up-to-date surveys exist. Some perceptual work (for example, the analysis of Mach bands, spatial frequencies, context effects) blends smoothly into psychophysical modeling, but other work such as the geometry of visual perception is of a rather different character (see Indow 1974 for a list of references). Considerable modeling, both algebraic and probabilistic, continues in the area of preferential choice; much of it is sophisticated and imaginative but the field does not seem to have jelled conceptually, even though some of the models are finding application in economics. A healthy interplay between psycholinguistics, a largely empirical field, and formal linguistics has developed. In particular, sophisticated work on the learning and the learnability of language is under way, as well as on the semantics of natural language.

R. DUNCAN LUCE AND PATRICK SUPPES

ADDITIONAL BIBLIOGRAPHY

INDOW, TAROW 1974 Geometry of Frameless Binocular Perceptual Space. *Psychologia* 17:50–63.

KRANTZ, DAVID H. et al. (editors) 1971 *Foundations of Measurement.* Volume 1: *Additive and Polynomial Representations.* New York: Academic Press.

KRANTZ, DAVID H. et al. (editors) 1974 *Contemporary Developments in Mathematical Psychology.* 2 vols. San Francisco: Freeman. → Volume 1: *Learning, Memory, and Thinking.* Volume 2: *Measurement, Psychophysics, and Neural Information Processing.*

LORD, FREDERIC M.; and NOVICK, MELVIN R. 1968 *Statistical Theories of Mental Test Scores.* Reading, Mass.: Addison-Wesley.

STERNBERG, SAUL 1969 Memory-scanning: Mental Processes Revealed by Reaction-time Experiments. *American Scientist* 57:421–457.

MOIVRE, ABRAHAM DE

Abraham de Moivre (1667–1754) was born of French Protestant parents named Moivre. (He was also known as Demoivre; as part of the return address of a letter to Johann Bernoulli he himself wrote his name as deMoivre.) He studied mathematics and physics in Paris under Ozanam, and emigrated to England when he was 21 to escape religious persecution (Walker 1934). Although de Moivre was a mathematical genius of outstanding analytical power and was in contact by correspondence and in person (at the Royal Society) with many of the leading mathematicians of the day, he never succeeded in obtaining a university appointment. Instead, he had to live by tutoring noblemen's sons and by advising gamblers and speculators who dealt in annuities, which were a popular form of investment in the first half of the eighteenth century (Walford 1871). This misfortune for de Moivre is posterity's gain, for the problems he met in his consulting practice and his successful solution of them provided the material for his two great textbooks. In fact, during his last years de Moivre must have relied heavily on the sales of the later editions of his book on annuity calculations.

De Moivre's practical text on probability first appeared in 1718 as a translation and revision of his Latin article of 1711. It was dedicated to Isaac Newton, who is accorded the author's thanks for his writings and conversations. In its final form, published in 1756, this book is notable for its original treatment of the following topics, all of which play a central role in the modern theory of probability:

(1) The general laws of addition (David & Barton 1962, chapter 2) and multiplication of probabilities (Montucla [1758] 1802, part 5, book 1, chapter 39);

(2) The binomial distribution law (Cantor 1898, chapter 96);

(3) Probability-generating functions (Seal 1949*a*);

(4) Difference equations involving probabilities and their solution by means of recurring series (Czuber 1900);

(5) New and general solutions of problems on the duration of play, or "gambler's ruin" (Todhunter 1865, chapter 9);

(6) The limiting form of the binomial term

$$\binom{n}{x} p^x(1-p)^{n-x}, \qquad x = 0, 1, 2, \cdots, n; \quad 0 < p < 1,$$

when (*a*) $n \to \infty$ with np remaining finite, and (*b*) $n \to \infty$ and $np \to \infty$. In case (*a*) only the term with $x = 0$ was considered (David 1962). In case (*b*) the result

$$\{2\pi np(1-p)\}^{-\frac{1}{2}} \exp[-(x-np)^2/2np(1-p)],$$

namely, the ordinate of the normal distribution, was obtained explicitly (in different notation). Included in this book were the trigonometrical theorem that goes by de Moivre's name and his approximation to the logarithm of a factorial which was improved by Stirling's discovery in the same year that the value of the series contained therein was 2π.

Some of the mathematical derivations of this probability text were published for a wider circle of readers in Latin (1730).

De Moivre's other textbook laid the foundations of the mathematics of life contingencies (Saar 1923). Although the first edition sold slowly, Thomas Simpson's plagiaristic text of 1742 spurred de Moivre to a complete revision published in the following year (Young 1908). The success of this edition is indicated by the two further editions, with minor changes, that followed within nine years. In 1756 a final, thoroughly revised edition was printed as the last section of the third edition of *The Doctrine of Chances*. In an appendix, reference is made to a paper published in 1755 by James Dodson, the father of scientific life insurance (Ogborn 1962) and possibly the "friend" who edited the posthumous edition.

The originality of this life contingency textbook is attested by its inclusion of the following:

(*a*) The recursion formula for calculating a life annuity at age x, given that at age $x + 1$ (though it is doubtful whether the author envisaged the calculation of the whole set of annuity values by starting at the "oldest age" [Young 1908]);

(*b*) General relations for survivorship and reversionary annuities in terms of single and joint life annuities;

(*c*) Use of the calculus to obtain the value of a continuous annuity-certain;

(*d*) A law of mortality, namely, that of uniform decrements in the number of survivors, which was in substantial agreement with the Breslau table, published by his friend Halley in 1693; and as a result:

(*e*) Easily computable values of single and joint life annuities for limited terms or for life;

(*f*) Expressions for the computation of complex survivorship probabilities;

(*g*) The value of a life annuity with a proportionate payment in the year of death.

All these results originated with de Moivre himself.

○ The earliest published works of de Moivre in the 1690s were influenced by Newton's method of fluxions and theory of series (Cantor 1898, chapter 86), and his interest in probability dates only from the first edition of Montmort's *Essay* (1708). Perhaps his most important contribution, first printed in 1733 as a supplement to the *Miscellanea analytica*, was his improvement of the wide limits obtained by Bernoulli (1713) in his statement of the law of large numbers. For this purpose de Moivre utilized the result mentioned in (6) to obtain the sum of the binomial probabilities from $x = np - \ell$ to $x = np + \ell$ with $p = \frac{1}{2}$ and $\ell = k(n/4)^{\frac{1}{2}}$, $k = 1,2,3$, by approximate quadrature (by ordinate summation and the three-eighths rule) of the normal ordinates. While this constitutes the first tabulation of the normal areas at one, two, and three standard deviations from the mean, there is no evidence that de Moivre thought in terms of a continuous probability distribution (Seal 1954; 1957). Nevertheless, this work is clearly the basis for the subsequent demonstration by Laplace (1812, pp. 275–284) that the binomial tends to the normal when n is large (Pearson 1924).

It may be added that de Moivre was a pure mathematician little interested in the practical applications of his theory. Although he wrote on life contingencies, only in the final Appendix of the posthumous edition of 1756 is there a brief reference to mortality data later than those of the Breslau table. Actually, the early and middle years of the eighteenth century saw the publication of several collections of mortality statistics that would have been fertile ground for the application of de Moivre's improved version of Bernoulli's theorem (Seal 1949*b*). These data had led to a widespread belief in the divine regularity of demographic ratios, and a few paragraphs in the 1738 and 1756 editions of the *Doctrine* refer to a connection between this belief and Bernoulli's theorem. Unfortunately, the topic was not pursued by de Moivre or his contemporaries (Westergaard 1932, chapters 7, 10) and cannot be regarded as indicating that de Moivre was interested in theology (Walker 1929) or that he influenced the demographers of the eighteenth and early nineteenth centuries (Pearson 1926).

HILARY L. SEAL

[*For the historical context of de Moivre's work, see the article on the* BERNOULLI FAMILY. *For discussion of the subsequent development of de Moivre's ideas, see* DISTRIBUTIONS, STATISTICAL; LIFE TABLES; PROBABILITY; *and the biography of* LAPLACE.]

WORKS BY DE MOIVRE

In many reference works, de Moivre is alphabetized under D; however, in line with the cataloguing practice of major libraries, we have listed him under M. Consistent with this, we have used a lower-case "d" for the particle.

1711 De mensura sortis seu, de probabilitate eventuum in ludis a casu fortuito pendentibus. Royal Society of London, *Philosophical Transactions* 27:213–264. → Reprinted by Kraus (New York) in 1963.

○(1718) 1756 *The Doctrine of Chances: Or, a Method of Calculating the Probabilities of Events in Play.* 3d ed. London: Millar. → Facsimiles of both the second (1738) and the third (1756) editions were published by Chelsea (New York) and Cass (London) in 1967.

(1725) 1752 *Annuities Upon Lives: Or, the Valuation of Annuities Upon Any Number of Lives, as Also, of Reversions.* 4th ed. London: Millar. → The Appendix concerns the expectations of life and the probabilities of survivorship.

1730 *Miscellanea analytica de seriebus et quadraturis. . . .* London: Tonson & Watts.

SUPPLEMENTARY BIBLIOGRAPHY

BERNOULLI, JAKOB (1713) 1899 *Wahrscheinlichkeitsrechnung: (Ars conjectandi).* 2 vols. Leipzig: Engelmann. → First published posthumously in Latin.

CANTOR, MORITZ 1898 *Vorlesungen über Geschichte der Mathematik.* Volume 3: Von 1668–1758. Leipzig: Teubner.

CZUBER, EMANUEL (1900) 1906 Calcul des probabilités. Section 1, volume 4, pages 1–46 in *Encyclopédie des sciences mathématiques.* Paris: Gauthier-Villars. → First published in German in *Encyklopädie der mathematischen Wissenschaften.*

DAVID, F. N. 1962 *Games, Gods and Gambling: The Origins and History of Probability and Statistical Ideas From the Earliest Times to the Newtonian Era.* New York: Hafner.

DAVID, F. N.; and BARTON, D. E. 1962 *Combinatorial Chance.* New York: Hafner.

LAPLACE, PIERRE SIMON DE (1812) 1820 *Théorie analytique des probabilités.* 3d ed., rev. Paris: Courcier.

[MONTMORT, PIERRE RÉMOND DE] (1708) 1713 *Essay d'analyse sur les jeux de hazard.* 2d ed. Paris: Quillau. → First published anonymously.

MONTUCLA, JEAN É. (1758) 1802 *Histoire des mathématiques dans laquelle on rend compte de leurs progrès depuis leur origine jusqu'à nos jours. . . .* Paris: Agasse.

OGBORN, MAURICE EDWARD 1962 *Equitable Assurances: The Story of Life Assurance in the Experience of the Equitable Life Assurance Society, 1762–1962.* London: Allen & Unwin.

PEARSON, KARL 1924 Historical Note on the Origin of the Normal Curve of Errors. *Biometrika* 16:402–404.

PEARSON, KARL 1926 Abraham de Moivre. *Nature* 117: 551–552.

SAAR, J. DU 1923 De beteekenis van De Moivre's werk over lijfrenten voor de ontwikkeling van de verzekeringswetenschap. *Verzekerings archief* 4:28–45.

SEAL, HILARY L. 1949*a* The Historical Development of the Use of Generating Functions in Probability Theory. Vereinigung schweizerischer Versicherungsmathematiker, *Mitteilungen* 49:209–228.

SEAL, HILARY L. 1949*b* Mortality Data and the Binomial Probability Law. *Skandinavisk aktuarietidskrift* 32: 188–216.

SEAL, HILARY L. 1954 A Budget of Paradoxes. *Journal of the Institute of Actuaries Students' Society* 13: 60–65.

SEAL, HILARY L. 1957 A Correction. *Journal of the Institute of Actuaries Students' Society* 14:210–211.

SIMPSON, THOMAS (1742) 1775 *The Doctrine of Annuities and Reversions, Deduced From General and Evident Principles. . . .* 2d ed. London: Printed for J. Nourse.

TODHUNTER, ISAAC (1865) 1949 *A History of the Mathematical Theory of Probability From the Time of Pascal to That of Laplace.* New York: Chelsea.

WALFORD, CORNELIUS 1871 *The Insurance Cyclopaedia.* Volume 1. London: Layton. → See especially pages 98–169 on "Annuities."

WALKER, HELEN M. 1929 *Studies in the History of Statistical Method: With Special Reference to Certain Educational Problems.* Baltimore: Williams & Wilkins.

WALKER, HELEN M. 1934 Abraham de Moivre. *Scripta mathematica* 2:316–333.

WESTERGAARD, HARALD L. 1932 *Contributions to the History of Statistics.* London: King.

YOUNG, T. E. 1908 Historical Notes Relating to the Discovery of the Formula $a_x = vp_x(1 + a_{x+1})$: And to the Introduction of the Calculus in the Solution of Actuarial Problems. *Journal of the Institute of Actuaries* 42:188–205.

Postscript

De Moivre's quadrature of the normal curve extended from zero to ℓ and approximated the sum of bionomial terms starting from $x = np$ and going to $x = np + \ell(npq)^{\frac{1}{2}}$. He doubled the result to obtain the sum of terms extending over $x = np \pm \ell(npq)^{\frac{1}{2}}$ and Czuber (1898) incorrectly saw this as an error because of the inclusion of the middle binomial term twice.

A definitive review of de Moivre's life and work is that of Schneider (1968). An interesting search for the remaining copies of the (second) supplement to the *Miscellanea analytica* is described by Daw and Pearson (1972).

HILARY L. SEAL

ADDITIONAL BIBLIOGRAPHY

CZUBER, EMANUEL 1898 Die Entwicklung der Wahrscheinlichkeitstheorie und ihrer Anwendungen. *Deutschen Mathematiker-Vereinigung* 7, no. 2: 1–271.

DAW, R. H.; and PEARSON, E. S. (1972) 1977 Abraham de Moivre's 1733 Derivation of the Normal Curve: A Bibliographical Note. Volume 2, pages 63–66 in Maurice G. Kendall and R. L. Plackett (editors), *Studies in the History of Statistics and Probability*. London: Griffin; New York: Macmillan. → First published in *Biometrika* 59: 677–680.

SCHNEIDER, IVO 1968 Der Mathematiker Abraham de Moivre, 1667–1754. *Archive for History of Exact Sciences* 5: 177–317.

MONTE CARLO METHODS

See RANDOM NUMBERS; SIMULATION.

MORBIDITY

See EPIDEMIOLOGY.

MULTIDIMENSIONAL SCALING

See SCALING, MULTIDIMENSIONAL.

MULTIPLE COMPARISONS

See under LINEAR HYPOTHESES.

MULTIPLE CORRELATION

See LINEAR HYPOTHESES, *article on* REGRESSION; MULTIVARIATE ANALYSIS, *article on* CORRELATION METHODS; STATISTICS, DESCRIPTIVE, *article on* ASSOCIATION.

MULTIVARIATE ANALYSIS

I.	OVERVIEW	*Ralph A. Bradley*
II.	CORRELATION METHODS	*Robert F. Tate*
III.	CLASSIFICATION AND DISCRIMINATION	*T. W. Anderson*

I OVERVIEW

Multivariate analysis in statistics is devoted to the summarization, representation, and interpretation of data when more than one characteristic of each sample unit is measured. Almost all data-collection processes yield multivariate data. The medical diagnostician examines pulse rate, blood pressure, hemoglobin, temperature, and so forth; the educator observes for individuals such quantities as intelligence scores, quantitative aptitudes, and class grades; the economist may consider at points in time indexes and measures such as per-capita personal income, the gross national product, employment, and the Dow-Jones average. Problems using these data are multivariate because inevitably the measures are interrelated and because investigations involve inquiry into the nature of such interrelationships and their uses in prediction, estimation, and methods of classification. Thus, multivariate analysis deals with samples in which for each unit examined there are observations on two or more stochastically related measurements. Most of multivariate analysis deals with estimation, confidence sets, and hypothesis testing for means, variances, covariances, correlation coefficients, and related, more complex population characteristics.

Only a sketch of the history of multivariate analysis is given here. The procedures of multivariate analysis that have been studied most are based on the multivariate normal distribution discussed below.

Robert Adrian considered the bivariate normal distribution early in the nineteenth century, and Francis Galton understood the nature of correlation near the end of that century. Karl Pearson made important contributions to correlation, including multiple correlation, and to regression analysis early in the present century. G. U. Yule and others considered measures of association in contingency tables, and thus began multivariate developments for counted data. The pioneering work of "Student" (W. S. Gosset) on small-sample distributions led to R. A. Fisher's distributions of simple and multiple correlation coefficients. J. Wishart derived the joint distribution of sample variances and covariances for small multivariate normal samples. Harold Hotelling generalized the Student t-statistic and t-distribution for the multivariate problem. S. S. Wilks provided procedures for additional tests of hypotheses on means, variances, and covariances. Classification problems were given initial consideration by Pearson, Fisher, and P. C. Mahalanobis through measures of racial likeness, generalized distance, and discriminant functions, with some results similar to the work of Hotelling. Both Hotelling and Maurice Bartlett made initial studies of canonical correlations, intercorrelations between two sets of variates. More recent research by S. N. Roy, P. L. Hsu, Meyer Girshick, D. N. Nanda, and others has dealt with the distributions of certain characteristic roots and vectors as they relate to multivariate problems, notably to canonical correlations and multivariate analysis of variance. Much attention has also been given to the reduction of multivariate data and its interpretation through many papers on factor analysis and principal components. [*For further discussion of the history of these special areas of multivariate analysis and of their present-day applications, see* COUNTED DATA; DISTRIBUTIONS, STATISTICAL, *article on* SPECIAL

CONTINUOUS DISTRIBUTIONS; FACTOR ANALYSIS AND PRINCIPAL COMPONENTS; MULTIVARIATE ANALYSIS, *articles on* CORRELATION METHODS *and* CLASSIFICATION AND DISCRIMINATION; STATISTICS, DESCRIPTIVE, *article on* ASSOCIATION; *and the biographies of* FISHER; GALTON; GIRSHICK; GOSSET; PEARSON; WILKS; YULE.]

Basic multivariate distributions

Scientific progress is made through the development of more and more precise and realistic representations of natural phenomena. Thus, science, and to an increasing extent social science, uses mathematics and mathematical models for improved understanding, such mathematical models being subject to adoption or rejection on the basis of observation [*see* MODELS, MATHEMATICAL]. In particular, stochastic models become necessary as the inherent variability in nature becomes understood.

The multivariate normal distribution provides the stochastic model on which the main theory of multivariate analysis is based. The model has sufficient generality to represent adequately many experimental and observational situations while retaining relative simplicity of mathematical structure. The possibility of applying the model to transforms of observations increases its scope [*see* STATISTICAL ANALYSIS, SPECIAL PROBLEMS OF, *article on* TRANSFORMATIONS OF DATA]. The large-sample theory of probability and the multivariate central limit theorem add importance to the study of the multivariate normal distribution as it relates to derived distributions. Inquiry and judgment about the use of any model must be the responsibility of the investigator, perhaps in consultation with a statistician. There is still a great deal to be learned about the sensitivity of the multivariate model to departures from that distributional assumption. [*See* ERRORS, *article on* EFFECTS OF ERRORS IN STATISTICAL ASSUMPTIONS.]

The multivariate normal distribution. Suppose that the characteristics or variates to be measured on each element of a sample from a population, conceptual or real, obey the probability law described through the multivariate normal probability density function. If these variates are p in number and are designated by $X_1, \cdots, X_p$, the multivariate normal density contains p parameters, or population characteristics, $\mu_1, \cdots, \mu_p$, representing, respectively, the means or expected values of the variates, and $\frac{1}{2}p(p+1)$ parameters σ_{ij}, $i, j = 1, \cdots, p$, $\sigma_{ji} = \sigma_{ij}$, representing variances and covariances of the variates. Here σ_{ii} is the variance of X_i (corresponding to the variance σ^2 of a variate X in the univariate case) and $\sigma_{ij} = \sigma_{ji}$ is the covariance of X_i and X_j. The correlation coefficient between X_i and X_j is $\rho_{ij} = \sigma_{ij}/\sqrt{\sigma_{ii}\sigma_{jj}}$.

The multivariate normal probability density function provides the probability density for the variates $X_1, \cdots, X_p$ at each point $x_1, \cdots, x_p$ in the sample or observation space. Its specific mathematical form is

$$f(x_1, \cdots, x_p) = (2\pi)^{-p/2}|\boldsymbol{\Sigma}|^{-\frac{1}{2}} \exp[-\tfrac{1}{2}(\boldsymbol{x}-\boldsymbol{\mu})'\boldsymbol{\Sigma}^{-1}(\boldsymbol{x}-\boldsymbol{\mu})],$$

$-\infty < x_i < \infty$, $i = 1, \cdots, p$. [*For the explicit form of this density in the bivariate case* ($p = 2$), *see* MULTIVARIATE ANALYSIS, *article on* CORRELATION METHODS.]

(Vector and matrix notation and an understanding of elementary aspects of matrix algebra are important for any real understanding or application of multivariate analysis. Thus, $\boldsymbol{x}'$ is the vector $(x_1, \cdots, x_p)$, $\boldsymbol{\mu}'$ is the vector $(\mu_1, \cdots, \mu_p)$, and $(\boldsymbol{x}-\boldsymbol{\mu})'$ is the vector $(x_1-\mu_1, \cdots, x_p-\mu_p)$. Also, $\boldsymbol{\Sigma}$ is the $p \times p$, symmetric matrix which has elements σ_{ij}, $\boldsymbol{\Sigma} = [\sigma_{ij}]$, $|\boldsymbol{\Sigma}|$ is the determinant of $\boldsymbol{\Sigma}$ and $\boldsymbol{\Sigma}^{-1}$ is its inverse. The prime indicates "transpose," and thus $(\boldsymbol{x}-\boldsymbol{\mu})'$ is the transpose of $(\boldsymbol{x}-\boldsymbol{\mu})$, a column vector.)

Comparison of $f(x_1, \cdots, x_p)$ with $f(x)$, the univariate normal probability density function, may assist understanding; for a univariate normal variate X with mean μ and variance σ^2,

$$f(x) = (2\pi)^{-\frac{1}{2}}(\sigma^2)^{-\frac{1}{2}} \exp\{-\tfrac{1}{2}(x-\mu)(\sigma^2)^{-1}(x-\mu)\} = \frac{1}{\sqrt{2\pi}\sigma} e^{-(x-\mu)^2/2\sigma^2},$$

where $-\infty < x < \infty$.

The multivariate normal density may be characterized in various ways. One direct method begins with p independent, univariate normal variables, $U_1, \cdots, U_p$, each with zero mean and unit variance. From the independence assumption, their joint density is the product

$$(2\pi)^{-p/2}\exp\{-\tfrac{1}{2}(u_1^2 + \cdots + u_p^2)\},$$

a very special case of the multivariate normal probability density function. If variates $X_1, \cdots, X_p$ are linearly related to $U_1, \cdots, U_p$ so that $\boldsymbol{X} = \boldsymbol{AU} + \boldsymbol{\mu}$, in matrix notation, with $\boldsymbol{X}$, $\boldsymbol{U}$, and $\boldsymbol{\mu}$ being column vectors and $\boldsymbol{A}$ being a $p \times p$ nonsingular matrix of constants a_{ij}, then

$$X_i = a_{i1}U_1 + \cdots + a_{ip}U_p + \mu_i, \qquad i = 1, \cdots, p.$$

Clearly, the mean of X_i is $E(X_i) = \mu_i$, where μ_i is a known constant and E represents "expectation."

The variance of X_i is

$$\text{var}(X_i) = \sum_{k=1}^{p} a_{ik}^2 = \sigma_{ii},$$

and the covariance of X_i and X_j, $i \neq j$, is

$$\text{cov}(X_i, X_j) = \sum_{k=1}^{p} a_{ik} a_{jk} = \sigma_{ij}.$$

Standard density function manipulations then yield the joint density function of $X_1, \cdots, X_p$ as that already given as the general p-variate normal density. If the matrix $\mathbf{A}$ is singular, the results for $E(X_i)$, $\text{var}(X_i)$, and $\text{cov}(X_i, X_j)$ still hold and $X_1, \cdots, X_p$ are said to have a singular multivariate normal distribution; although the joint density function cannot be written, the concept is useful.

A second characterization of the p-variate normal distribution is the following: $X_1, \cdots, X_p$ have a p-variate normal distribution if and only if $\sum_{i=1}^{p} a_i X_i$ is univariate normal for all choices of the coefficients a_i, that is, if and only if all linear combinations of the X_i are univariate normal.

The multivariate normal cumulative distribution function represents the probability of the joint occurrence of the events $X_1 \leqslant x_1, \cdots, X_p \leqslant x_p$ and may be written

$$P(X_1 \leqslant x_1, \cdots, X_p \leqslant x_p) = F(x_1, \cdots, x_p)$$
$$= \int_{-\infty}^{x_1} \cdots \int_{-\infty}^{x_p} f(u_1, \cdots, u_p) du_1 \cdots du_p,$$

indicating that probabilities that observations fall into regions of the p-dimensional variate space may be obtained by integration. Tables of $F(x_1, \cdots, x_p)$ are available for $p = 2, 3$ (see Greenwood & Hartley 1962).

Some basic properties of the p-variate normal distribution in terms of $\mathbf{X} = (X_1, \cdots, X_p)$ are the following.

(*a*) Any subset of the X_i has a multivariate normal distribution. In fact, any set of q linear combinations of the X_i has a q-variate normal distribution, a result following directly from the linear combination characterization, $q \leqslant p$.

(*b*) The conditional distribution of q of the X_i, given the $p - q$ others, is q-variate normal.

(*c*) If $\sigma_{ij} = 0$. $i \neq j$, then X_i and X_j are independent.

(*d*) The expectation and variance of $\sum_{i=1}^{n} a_i X_i$ are $\sum_{i=1}^{p} a_i \mu_i$ and $\sum_{i=1}^{p} \sum_{j=1}^{p} a_i a_j \sigma_{ij}$.

(*e*) The covariance of $\sum_{i=1}^{n} a_i X_i$ and $\sum_{i=1}^{n} b_i X_i$ is $\sum_{i=1}^{p} \sum_{j=1}^{p} a_i b_j \sigma_{ij}$.

A cautionary note is that $X_1, \cdots, X_p$ may be separately (marginally) univariate normal while the joint distribution may be very nonnormal.

The geometric properties of the p-dimensional surface defined by $y = f(x_1, \cdots, x_p)$ are interesting. Contours of the surface are p-dimensional ellipsoids. All inflection points of the surface occur at constant y and hence fall on the same horizontal ellipsoidal cross section. Any vertical cross section of the surface leads to a subsurface that is normal or multivariate normal in form and is capable of representation as a normal probability density surface except for a proportionality constant.

Characteristic and moment-generating functions yield additional methods of description of random variables [*see* DISTRIBUTIONS, STATISTICAL, *article on* SPECIAL CONTINUOUS DISTRIBUTIONS]. For the multivariate normal distribution, the moment-generating function is

$$M(t_1, \cdots, t_p)$$
$$= \int_{-\infty}^{\infty} \cdots \int_{-\infty}^{\infty} \exp(t_1 x_1 + \cdots + t_p x_p) \cdot f(x_1, \cdots, x_p)\, dx_1 \cdots dx_p$$
$$= \exp(\mathbf{t}'\boldsymbol{\mu} + \tfrac{1}{2}\mathbf{t}'\boldsymbol{\Sigma}\mathbf{t}),$$

where $\mathbf{t}' = (t_1, \cdots, t_p)$. The moment-generating function may describe either the nonsingular or singular p-variate normal distribution. Note that, from its definition, the matrix $\boldsymbol{\Sigma}$ may be shown to be nonnegative definite. When $\boldsymbol{\Sigma}$ is positive definite the multivariate density may be specified as $f(x_1, \cdots, x_p)$. When $\boldsymbol{\Sigma}$ is singular, $\boldsymbol{\Sigma}^{-1}$ does not exist, and the density may not be given. However, $M(t_1, \cdots, t_p)$ may still be given and can thus describe the singular multivariate normal distribution. To say that $\mathbf{X}$ has a singular distribution is to say that $\mathbf{X}$ lies in some hyperplane of dimension less than p.

The multivariate normal sample. Table 1 illustrates a multivariate sample with $p = 4$ and sample size $N = 10$; the data here are head measurements.

Table 1 — Measurements taken on first and second adult sons in a sample of ten families

HEAD LENGTH		HEAD BREADTH	
First son	Second son	First son	Second son
X_1	X_2	X_3	X_4
191	179	155	145
195	201	149	152
181	185	148	149
183	188	153	149
176	171	144	142
208	192	157	152
189	190	150	149
197	189	159	152
188	197	152	159
192	187	150	151

Source: Based on original data by G. P. Frets, presented in Rao 1952, table 7b.2β.

One can anticipate covariance or correlation between head length and head breadth and between head measurements of first and second sons. Hence, for most purposes it will be important to treat the data as a single multivariate sample rather than as several univariate samples.

General notation for a multivariate sample is developed in terms of the variates $X_{1\alpha}, \cdots, X_{p\alpha}$ representing the p observation variates for the αth sample unit (for example, the αth family in the sample), $\alpha = 1, \cdots, N$. In a parallel way $x_{i\alpha}$ may be regarded as the realization of $X_{i\alpha}$ in a particular set of sample data. For multivariate normal procedures, standard data summarization involves calculation of the sample means, $\bar{x}_i = \sum_{\alpha=1}^{N} x_{i\alpha}/N$, $i = 1, \cdots, p$, and the sample variances and covariances,

$$s_{ij} = \sum_{\alpha=1}^{N} (x_{i\alpha} - \bar{x}_i)(x_{j\alpha} - \bar{x}_j)/(N-1)$$

$$= \left[\sum_{\alpha=1}^{N} (x_{i\alpha}x_{j\alpha}) - N\bar{x}_i\bar{x}_j\right] \Big/ (N-1), \quad i, j = 1, \cdots, p, \quad s_{ij} = s_{ji}.$$

Sample correlation coefficients may be computed from $r_{ij} = s_{ij}/\sqrt{s_{ii}s_{jj}}$. For the data of Table 1, the sample values of the statistics are given in Table 2.

Table 2 — Sample statistics for measurements taken on sons

MEANS, $\bar{x}_i$

Variate	1	2	3	4
	190.0	187.9	151.7	150.0

VARIANCES AND COVARIANCES, s_{ij}

Variate j \ Variate i	1	2	3	4
1	81.56	42.00	29.56	18.67
2		72.32	11.86	33.11
3			20.01	7.78
4				20.67

CORRELATIONS, r_{ij}

Variate j \ Variate i	1	2	3	4
1	1	.55	.73	.46
2		1	.31	.86
3			1	.38
4				1

The required assumptions for the simpler multivariate normal procedures are that the observation vectors $(X_{1\alpha}, \cdots, X_{p\alpha})$ are independent in probability and that each such observation vector consists of p variates following the same multivariate normal law—that is, having the same probability density $f(x_1, \cdots, x_p)$ with the same parameters, elements of $\boldsymbol{\mu}$ and $\boldsymbol{\Sigma}$. The joint density for the $p \times N$ random variables $X_{i\alpha}$ is, by the independence assumption, just the product of N p-variate normal densities, each having the same μ's and σ's. The joint density may be expressed in terms of $\boldsymbol{\mu}$ and $\boldsymbol{\Sigma}$ and $\bar{\boldsymbol{x}}$ and $\boldsymbol{s}$, where $\boldsymbol{s}$ is the symmetric $p \times p$ matrix with elements s_{ij} and $\bar{\boldsymbol{x}}' = (\bar{x}_1, \cdots, \bar{x}_p)$.

Elements of $\mathbf{S}$, the matrix of random variables corresponding to $\boldsymbol{s}$, and of the vector $\bar{\mathbf{X}}$ constitute a set of sufficient statistics for the parameters in $\boldsymbol{\Sigma}$ and $\boldsymbol{\mu}$ [*see* SUFFICIENCY]. Furthermore, it may be shown that $\mathbf{S}$ and $\bar{\mathbf{X}}$ are independent.

Basic derived distributions. The distribution of the vector of sample means, $\bar{\mathbf{X}} = (\bar{X}_1, \cdots, \bar{X}_p)$, is readily described for the random sampling under discussion. That distribution is again p-variate normal with the same mean vector, $\boldsymbol{\mu}$, as in the underlying population but with covariance matrix $N^{-1}\boldsymbol{\Sigma}$. There is complete analogy here with the univariate case.

The joint probability density function of the sample variances and covariances, S_{ij}, has been named the Wishart distribution after its developer. This density is

$$h(s_{11}, s_{12}, \cdots, s_{pp}) = \frac{[\frac{1}{2}(N-1)]^{\frac{1}{2}p(N-1)}|\boldsymbol{s}|^{\frac{1}{2}(N-p-2)}\exp[-\frac{1}{2}(N-1)\operatorname{tr}\boldsymbol{\Sigma}^{-1}\boldsymbol{s}]}{\pi^{p(p-1)/4}|\boldsymbol{\Sigma}|^{\frac{1}{2}(N-1)}\prod_{i=1}^{p}\Gamma[\frac{1}{2}(N-i)]},$$

where $-\infty < s_{ij} < \infty$, $i < j$, $0 \leqslant s_{ii} < \infty$, and $i, j = 1, \cdots, p$, and the matrix $\boldsymbol{s}$ is positive definite.

The Wishart density is a generalization of the chi-square density with $N-1$ degrees of freedom for $(N-1)S^2/\sigma^2$ in the univariate case, in which S^2 is the sample variance based on N independent observations from a univariate normal population. Anderson (1958, sec. 14.3) has a note on the noncentral Wishart distribution, a generalization of the noncentral chi-square distribution.

Procedures on means, variances, covariances

Many of the simpler multivariate statistical procedures were developed as extensions of useful univariate methods dealing with tests of hypotheses and related confidence intervals for means, variances, and covariances. Small-sample distributions of important statistics of multivariate analysis have been found; almost invariably the starting point in the derivations is the joint probability density of sample means and sample variances and covariances, the product of a multivariate normal density and a Wishart density, or one of these densities separately.

Inferences on means, dispersion known. If $\boldsymbol{\mu}^*$ is a p-element column vector of given constants and if the elements of $\boldsymbol{\Sigma}$ are known, it was shown long ago, perhaps first by Karl Pearson, that when $\boldsymbol{\mu}^* = \boldsymbol{\mu}$, $Q(\bar{\mathbf{X}}) = N(\bar{\mathbf{X}} - \boldsymbol{\mu}^*)'\boldsymbol{\Sigma}^{-1}(\bar{\mathbf{X}} - \boldsymbol{\mu}^*)$ has the central chi-square distribution with p degrees of freedom [*see* DISTRIBUTIONS, STATISTICAL, *article on* SPECIAL CONTINUOUS DISTRIBUTIONS]. It was later shown more generally that $Q(\bar{\mathbf{X}})$ has the noncentral chi-square distribution with p degrees of freedom and noncentrality parameter $\tau^2 = N(\boldsymbol{\mu}^* - \boldsymbol{\mu})'\boldsymbol{\Sigma}^{-1}(\boldsymbol{\mu}^* - \boldsymbol{\mu})$ when $\boldsymbol{\mu}^* \neq \boldsymbol{\mu}$. (The symbol τ^2 is consistent with the notation of Anderson 1958, sec. 5.4.)

A null hypothesis, H_{01}: $\boldsymbol{\mu} = \boldsymbol{\mu}^*$, specifying the means of the multivariate normal density when $\boldsymbol{\Sigma}$ is known and when the alternative hypothesis is general, $\boldsymbol{\mu} \neq \boldsymbol{\mu}^*$, may be of interest in some experimental situations. With significance level α the critical region of the sample space, the region of rejection of the hypothesis H_{01}, is that region where $Q(\bar{\mathbf{x}}) \geqslant \chi^2_{p;\alpha}$ ($\chi^2_{p;\alpha}$ being the tabular value of a chi-square variate χ^2_p with p degrees of freedom such that $P\{\chi^2_p \geqslant \chi^2_{p;\alpha}\} = \alpha$) [*see* HYPOTHESIS TESTING]. The power of this test may be computed when H_{01} is false, that is, when $\boldsymbol{\mu} \neq \boldsymbol{\mu}^*$, by evaluation of the probability, $P\{\chi'^2_p \geqslant \chi^2_{p;\alpha}\}$, where χ'^2_p is a noncentral chi-square variate with p degrees of freedom and noncentrality τ^2.

When the alternative hypotheses are one-sided, in the sense that each component of $\boldsymbol{\mu}$ is taken to be greater than or equal to the corresponding component of $\boldsymbol{\mu}^*$, the problem is more difficult. First steps have been taken toward the solution of this problem (see Kudô 1963; Nüesch 1966).

Since $\boldsymbol{\mu}$ is unknown, it is estimated by $\bar{\mathbf{x}}$. Corresponding to the test given above, the confidence region with confidence coefficient $1 - \alpha$ for the μ_i consists of all values $\boldsymbol{\mu}^*$ for which the inequality $Q(\bar{\mathbf{x}}) \leqslant \chi^2_{p;\alpha}$ holds [*see* ESTIMATION, *article on* CONFIDENCE INTERVALS AND REGIONS]. This confidence region is the surface and interior of an ellipsoid centered at the point whose coordinates are the elements of $\bar{\mathbf{x}}$ in the p-dimensional parameter space of the elements of $\boldsymbol{\mu}$.

Paired sample problems may also be handled. Let $Y_1, \cdots, Y_{2p}$ be $2p$ variates with means $\xi_1, \cdots, \xi_{2p}$ having a multivariate normal density, and let $y_{j\alpha}$, $j = 1, \cdots, 2p$, $\alpha = 1, \cdots, N$, be independent multivariate observations from this multivariate normal population. Suppose that Y_i and Y_{p+i}, $i = 1, \cdots, p$, are paired variates. Then $X_i = Y_i - Y_{p+i}$, $i = 1, \cdots, p$, make a set of multivariate normal variates with parameters that again may be designated as the elements of $\boldsymbol{\mu}$ and $\boldsymbol{\Sigma}$, $\mu_i = \xi_i - \xi_{p+i}$. Similarly, take $x_{i\alpha} = y_{i\alpha} - y_{p+i,\alpha}$ and $\bar{x}_i = \bar{y}_i - \bar{y}_{p+i}$. Inferences on the means, μ_i, of the difference variates, X_i, when $\boldsymbol{\Sigma}$ is known may be made on the same basis as above for the simple sample. In the paired situation it will often be appropriate to take $\boldsymbol{\mu}^* = \mathbf{0}$, that is, H_{01} : $\boldsymbol{\mu} = \mathbf{0}$. Here $\mathbf{0}$ denotes a vector of 0's. For example, in Table 1 the data can be paired through the association of first and second sons in a family; a pertinent inquiry may relate to the equalities of both mean head lengths and mean head breadths of first and second sons. For association with this paragraph, columns in Table 1 should have variate headings Y_1, Y_3, Y_2, and Y_4, indicating that $p = 2$; then $X_1 = Y_1 - Y_3$ measures difference in head lengths of first and second sons and $X_2 = Y_2 - Y_4$ measures difference in head breadths.

There are also nonpaired versions of these procedures. In a table similar to Table 1 the designations "first son" and "second son" might be replaced by "adult male American Indian" and "adult male Eskimo." Then the data could be considered to consist of ten bivariate observations taken at random from each of the two indicated populations with no basis for the pairing of the observation vectors. Anthropological study might require comparisons of mean head lengths and mean head breadths for the two racial groups. The procedures of this section may be adapted to this problem. Suppose that $X^{(1)}_1, \cdots, X^{(1)}_p$ and $X^{(2)}_1, \cdots, X^{(2)}_p$ are the p variates for the two populations, the two sets of variates being stochastically independent of each other and having multivariate normal distributions with common dispersion matrix $\boldsymbol{\Sigma}^*$ but with means $\mu^{(1)}_1, \cdots, \mu^{(1)}_p$ and $\mu^{(2)}_1, \cdots, \mu^{(2)}_p$, respectively. The corresponding sample means are $\bar{x}^{(1)}_1, \cdots, \bar{x}^{(1)}_p$ and $\bar{x}^{(2)}_1, \cdots, \bar{x}^{(2)}_p$ based respectively on samples of independent observations of sizes N_1 and N_2 from the two populations. Definition of $\boldsymbol{\mu} = \boldsymbol{\mu}^{(1)} - \boldsymbol{\mu}^{(2)}$, $\bar{\mathbf{x}} = \bar{\mathbf{x}}^{(1)} - \bar{\mathbf{x}}^{(2)}$, and $N\boldsymbol{\Sigma}^{-1} = [N_1N_2/(N_1 + N_2)]\boldsymbol{\Sigma}^{*-1}$ permits association and use of $Q(\bar{\mathbf{x}})$ and its properties for this two-sample problem. If the dispersion matrices of the two populations are known but different, a slight modification of the procedure is readily available.

Jackson and Bradley (1961) have extended these methods to sequential multivariate analysis [*see* SEQUENTIAL ANALYSIS].

Generalized Student procedures. In the preceding section it was assumed that $\boldsymbol{\Sigma}$ was known, but in most applications this is not the case. Rather, $\boldsymbol{\Sigma}$ must be estimated from the data, and the generalized Student statistic or Hotelling's T^2, $T^2(\overline{\mathbf{X}},\mathbf{S}) = N(\overline{\mathbf{X}} - \boldsymbol{\mu}^*)'\mathbf{S}^{-1}(\overline{\mathbf{X}} - \boldsymbol{\mu}^*)$, comparable to $Q(\overline{\mathbf{X}})$, is almost always used. (For procedures that are not based on T^2 see Šidák 1967.) It has been shown that $F(\overline{\mathbf{X}},\mathbf{S}) = (N-p)T^2(\overline{\mathbf{X}},\mathbf{S})/p(N-1)$ has the variance-ratio or F-distribution with p and $N-p$ degrees of freedom [*see* DISTRIBUTIONS, STATISTICAL, *article on* SPECIAL CONTINUOUS DISTRIBUTIONS]. The F-distribution is central when $\boldsymbol{\mu} = \boldsymbol{\mu}^*$, that is, when the mean vector of the multivariate normal population is equal to the constant vector $\boldsymbol{\mu}^*$, and is noncentral otherwise with noncentrality parameter τ^2 already defined.

The hypothesis $H_{01}: \boldsymbol{\mu} = \boldsymbol{\mu}^*$ is of interest, as before. The statistic $F(\overline{\mathbf{X}},\mathbf{S})$ takes the role of $Q(\overline{\mathbf{X}})$, and $F_{p,N-p;\alpha}$ takes the role of $\chi^2_{p;\alpha}$, where $F_{p,N-p;\alpha}$ is the tabular value of the variance-ratio variate $F_{p,N-p}$ with p and $N-p$ degrees of freedom such that $P\{F_{p,N-p} \geqslant F_{p,N-p;\alpha}\} = \alpha$. The confidence region for the elements of $\boldsymbol{\mu}$ consists of all values $\boldsymbol{\mu}^*$ for which the inequality $F(\bar{\mathbf{x}},\mathbf{s}) \leqslant F_{p,N-p;\alpha}$ holds; the region is again an ellipsoid centered at $\bar{\mathbf{x}}$ in the p-dimensional parameter space, and the confidence coefficient is $1-\alpha$.

Visualization of the confidence region for the elements of $\boldsymbol{\mu}$ is often difficult. When $p = 2$, the ellipsoid becomes a simple ellipse and may be plotted (see Figure 1). When $p > 2$, two-dimensional elliptical cross sections of the ellipsoid may be plotted, and parallel tangent planes to the ellipsoid may be found that yield crude bounds on the various parameters. One or more linear contrasts among the elements of $\boldsymbol{\mu}$ may be of special interest, and then the dimensionality of the whole problem, including the confidence region, is reduced. Some of the problems of multiple comparisons arise when linear contrasts are used [*see* LINEAR HYPOTHESES, *article on* MULTIPLE COMPARISONS].

For the simple one-sample problem, $\mathbf{s} = [s_{ij}]$ is computed as shown in Table 2. For the paired sample problem, $\mathbf{s}$ in $F(\bar{\mathbf{x}},\mathbf{s})$ is the sample variance–covariance matrix computed from the derived multivariate sample of differences, and $\bar{\mathbf{x}}$ is the sample vector of mean differences, as before. For the unpaired two-sample problem, it is necessary to replace $N\mathbf{s}^{-1}$ in $F(\bar{\mathbf{x}},\mathbf{s})$, just as it was necessary to replace $N\boldsymbol{\Sigma}^{-1}$ when $\boldsymbol{\Sigma}$ was known. Each population has the dispersion matrix $\boldsymbol{\Sigma}^*$, and two sample dispersion matrices $\mathbf{s}^*_{(1)}$ and $\mathbf{s}^*_{(2)}$ may be computed, one for each multivariate sample, to estimate $\boldsymbol{\Sigma}^*$. A "pooled" estimate of the dispersion matrix $\boldsymbol{\Sigma}^*$ is $\mathbf{s}^* = [(N_1-1)\mathbf{s}^*_{(1)} + (N_2-1)\mathbf{s}^*_{(2)}]/(N_1+N_2-2)$, the multivariate generalization of the pooled estimate of variance often used in univariate statistics. For the two-sample problem, $N\mathbf{s}^{-1}$ in $F(\bar{\mathbf{x}},\mathbf{s})$ is replaced by $[N_1N_2/(N_1+N_2)]\mathbf{s}^{*-1}$. All of the assumptions about the populations and about the samples discussed in the preceding section apply for the corresponding generalized Student procedures.

An application of the generalized Student procedures for paired samples may be made for the data in Table 1. The bivariate ($p = 2$) sample of paired differences (in Table 1, column 1 minus column 2, column 3 minus column 4) is exhibited in Table 3. The sample mean differences and sample variances and covariance of the difference data are given in Table 4, along with the elements s^{ij} of $\mathbf{s}^{-1}$. The column headings and statistics in Table 3 and Table 4 have the arguments d simply to distinguish them from the symbols in tables 1 and 2.

Table 3 — Difference data on head measurements, first adult son minus second adult son

HEAD-LENGTH DIFFERENCE $X_1(d)$	HEAD-BREADTH DIFFERENCE $X_2(d)$
12	10
−6	−3
−4	−1
−5	4
5	2
16	5
−1	1
8	7
−9	−7
5	−1

Table 4 — Sample statistics for measurement differences on sons

MEANS, $\bar{x}_i(d)$

Variate	1	2
	2.1	1.7

VARIANCES AND COVARIANCES, $s_{ij}(d)$

Variate *j* \ Variate *i*	1	2
1	69.88	32.14
2		25.12

ELEMENTS OF $\mathbf{s}^{-1}(d)$, $s^{ij}(d)$

Variate *j* \ Variate *i*	1	2
1	.0348	−.0445
2		.0967

For a comparison of first and second sons, it may be appropriate to take $\boldsymbol{\mu}^* = \mathbf{0}$ and compute

$$T^2(\bar{\boldsymbol{x}},\boldsymbol{s}) = 10(2.1,1.7)\begin{pmatrix} .0348 & -.0445 \\ -.0445 & .0967 \end{pmatrix}\begin{pmatrix} 2.1 \\ 1.7 \end{pmatrix}$$

$$= (21,17)\begin{pmatrix} -.00257 \\ .07094 \end{pmatrix}$$

$$= 1.152.$$

$$F(\bar{\boldsymbol{x}},\boldsymbol{s}) = 8(1.152)/2(9) = .512.$$

If a significance level $\alpha = .10$ is chosen, then $F_{2,8;.10} = 3.11$ and the differences between paired means are not statistically significant; indeed, they are less than ordinary variation would lead one to expect. (For some sets of data this sort of result should lead to re-examination of possible biases or nonindependence in the data-collection process.)

To find those values $\boldsymbol{\mu}^*$ in the confidence region for $\boldsymbol{\mu}$, $\boldsymbol{\mu}^*$ must be replaced in $T^2(\bar{\boldsymbol{x}},\boldsymbol{s})$; thus,

$$T^2(\bar{\boldsymbol{x}},\boldsymbol{s})$$
$$= 10(2.1-\mu_1^*, 1.7-\mu_2^*)\begin{pmatrix} .0348 & -.0445 \\ -.0445 & .0967 \end{pmatrix}\begin{pmatrix} 2.1-\mu_1^* \\ 1.7-\mu_2^* \end{pmatrix}$$
$$= .348(\mu_1^*-2.1)^2 - .890(\mu_1^*-2.1)(\mu_2^*-1.7) + .967(\mu_2^*-1.7)^2.$$

The corresponding $F(\bar{\boldsymbol{x}},\boldsymbol{s}) = \frac{4}{9}T^2(\bar{\boldsymbol{x}},\boldsymbol{s})$. The confidence region on μ_1 and μ_2 with confidence coefficient $1-\alpha$ consists of those points in the (μ_1,μ_2)-space inside or on the ellipse described by

$$.155(\mu_1^*-2.1)^2 - .396(\mu_1^*-2.1)(\mu_2^*-1.7) + 4.30(\mu_2^*-1.7)^2 = F_{2,8;\alpha}.$$

This ellipse is plotted in Figure 1 for $\alpha = .05$, .10, .25, $F_{2,8;\alpha} = 4.46$, 3.11, 1.66, for clearer insight into the nature of the region.

A number of variants of the generalized Student procedure have been developed, and other variants are bound to be developed in the future. For example, one may wish to test null hypotheses specifying relationships between the coordinates of $\boldsymbol{\mu}$ (see Anderson 1958, sec. 5.3.5). Again, one may wish to test that certain coordinates of $\boldsymbol{\mu}$ have given values, knowing the values of the other coordinates. For another sort of variant, recall that it was assumed for the two-sample application that the dispersion matrices for the two parent populations were identical. If this assumption is untenable, then a multivariate analogue of the Behrens–Fisher

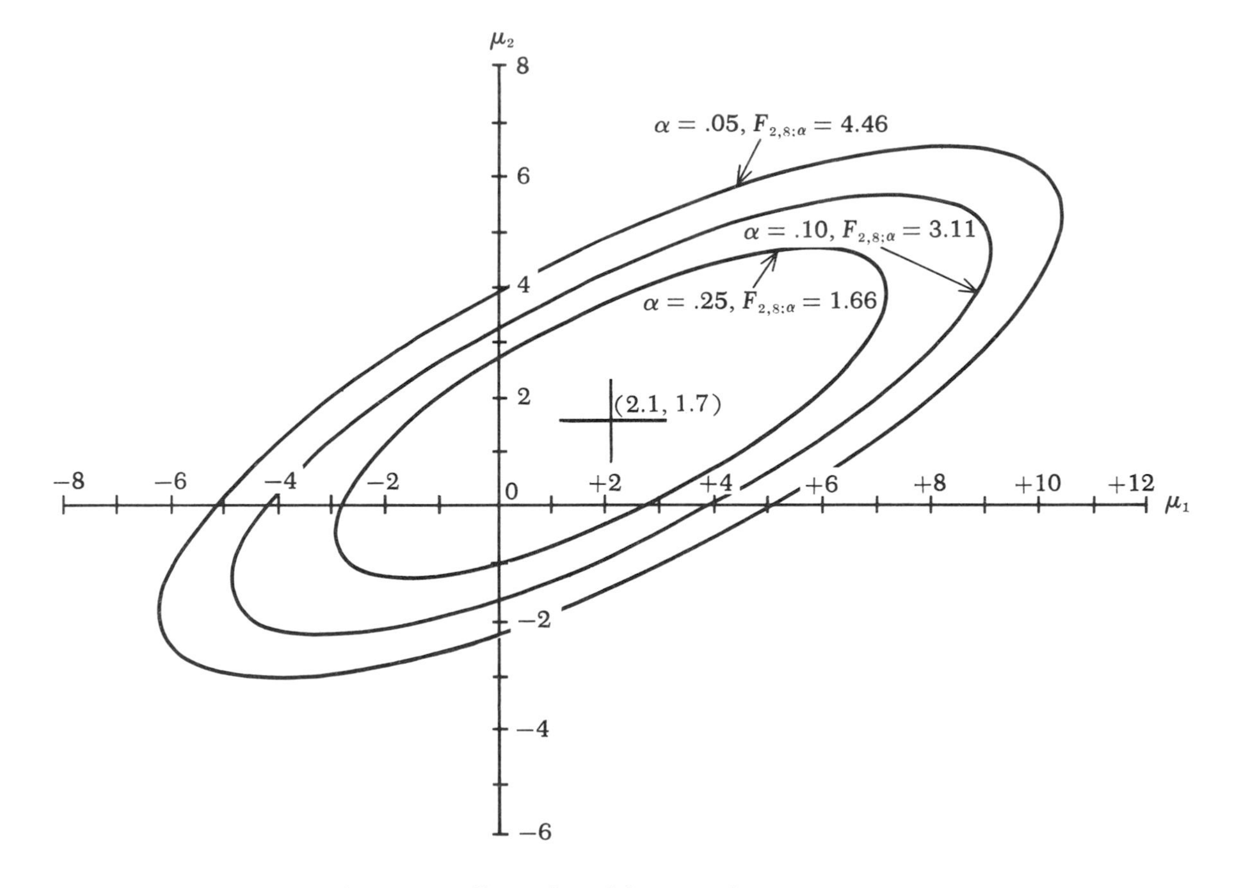

Figure 1 — Elliptical confidence regions on μ_1, μ_2:
$.155(\mu_1-2.1)^2 - .396(\mu_1-2.1)(\mu_2-1.7) + .430(\mu_2-1.7)^2 = F_{2,8;\alpha}$

problem must be considered (Anderson 1958, sec. 5.6). Sequential extensions of the generalized Student procedures have been given by Jackson and Bradley (1961).

Generalized variances. Tests of hypotheses and confidence intervals on variances are conducted easily in univariate cases through the use of the chi-square and variance-ratio distributions. The situation is much more difficult in multivariate analysis.

For the multivariate one-sample problem, hypotheses and confidence regions for elements of the dispersion matrix, $\mathbf{\Sigma}$, may be considered. A first possible hypothesis is H_{02}: $\mathbf{\Sigma} = \mathbf{\Sigma}^*$, a null hypothesis specifying all of the elements of $\mathbf{\Sigma}$. (This hypothesis is of limited interest per se, except when $\mathbf{\Sigma}^* = \mathbf{I}$ or as an introduction to procedures on multivariate linear hypotheses.) It is clear that a test statistic should depend on the elements S_{ij} of $\mathbf{S}$; it is not clear what function of these elements might be appropriate.

The statistic $|\mathbf{S}|$ has been called the generalized sample variance, and $|\mathbf{\Sigma}|$ has been called the generalized variance. The test statistic $|\mathbf{S}|/|\mathbf{\Sigma}^*|$ was proposed by Wilks, who examined its distribution; simple, exact, small-sample distributions are known only when $p = 1,2$. An asymptotic or limiting distribution is available for large N; the statistic $\sqrt{N-1}\,[(|\mathbf{S}|/|\mathbf{\Sigma}|) - 1]/\sqrt{2p}$ has the limiting univariate normal density with zero mean and unit variance. It is clear that when $\mathbf{\Sigma} = \mathbf{\Sigma}^*$ under H_{02}, $\mathbf{S}$ estimates $\mathbf{\Sigma}^*$, and the ratio $|\mathbf{S}|/|\mathbf{\Sigma}^*|$ should be near unity; it is not clear that the ratio may not be near unity when $\mathbf{S} \neq \mathbf{\Sigma}^*$. However, values of $|\mathbf{S}|$ that differ substantially from $|\mathbf{\Sigma}^*|$ should lead to rejection of H_{02} (see Anderson 1958, sec. 7.5).

Wilks's use of generalized variances is only one possible generalization of univariate procedures. Other comparisons of $\mathbf{S}$ and $\mathbf{\Sigma}^*$ are possible. In nondegenerate cases, $\mathbf{\Sigma}^*$ is nonsingular, and the product matrix $\mathbf{S\Sigma}^{*-1}$ should be approximately an identity matrix. All of the characteristic roots from the determinantal equation $|\mathbf{S\Sigma}^{*-1} - \lambda\mathbf{I}| = 0$, where $\mathbf{I}$ is the $p \times p$ identity matrix, should be near unity; the trace, $\operatorname{tr}\mathbf{S\Sigma}^{*-1}$, should be near p. Roy (1957, sec. 14.9) places major emphasis on the largest and smallest roots of $\mathbf{S}$ and $\mathbf{\Sigma}$ and gives approximate confidence bounds on the roots of the latter in terms of those of the former. A test of H_{02} may be devised with the hypothesis being rejected when the corresponding roots of $\mathbf{\Sigma}^*$ fail to fall within the confidence bounds. These and other similar considerations have led to extensive study of the distributions of roots of determinantal equations. Complete and exact solutions to these multivariate problems are not available.

Suppose that two independent multivariate normal populations have dispersion matrices $\mathbf{\Sigma}_{(1)}$ and $\mathbf{\Sigma}_{(2)}$, and samples of independent observation vectors of sizes N_1 and N_2 yield, respectively, sample dispersion matrices $\mathbf{S}_{(1)}$ and $\mathbf{S}_{(2)}$. The hypothesis of interest is H_{03}: $\mathbf{\Sigma}_{(1)} = \mathbf{\Sigma}_{(2)}$. In the univariate case ($p = 1$), the statistic $F = |\mathbf{S}_{(1)}|/|\mathbf{S}_{(2)}| = S_{11}^{(1)}/S_{11}^{(2)}$ is the simple variance ratio and, under H_{03}, has the F-distribution with $N_1 - 1$ and $N_2 - 1$ degrees of freedom. The general likelihood ratio criterion for testing H_{03} is, with minor adjustment,

$$\lambda = |\mathbf{S}_{(1)}|^{\frac{1}{2}(N_1-1)}|\mathbf{S}_{(2)}|^{\frac{1}{2}(N_2-1)}/|\mathbf{S}|^{\frac{1}{2}(N_1+N_2-2)},$$

where

$$\mathbf{S} = \frac{(N_1-1)\mathbf{S}_{(1)} + (N_2-1)\mathbf{S}_{(2)}}{N_1+N_2-2}.$$

If $p = 1$, then λ is a monotone function of F. By asymptotic theory for large N_1 and N_2, $2\log_e\lambda$ may be taken to have the central chi-square distribution with $\frac{1}{2}p(p+1)$ degrees of freedom under H_{03}. Anderson (1958, secs. 10.2, 10.4–10.6) discusses these problems further.

Roy (1957, sec. 14.10) prefers again to consider characteristic roots and develops test procedures and confidence procedures based on the largest and smallest roots of $\mathbf{S}_{(1)}\mathbf{S}_{(2)}^{-1}$. Heck (1960) has provided some charts of upper percentage points of the distribution of the largest characteristic root.

Multivariate analysis of variance. Multivariate analysis of variance bears the same relationship to the problems of generalized variances as does univariate analysis of variance to simple variances. An understanding of the basic principles of the analysis of variance is necessary to consider the multivariate generalization. The theory of general linear hypotheses is pertinent, and concepts of experimental design carry over to the multivariate case. [*See* EXPERIMENTAL DESIGN, *article on* THE DESIGN OF EXPERIMENTS; LINEAR HYPOTHESES, *article on* ANALYSIS OF VARIANCE.]

Consider the univariate randomized block design with v treatments and b blocks. A response, $X_{\gamma\delta}$, on treatment γ in block δ, $\gamma = 1, \cdots, v$, $\delta = 1, \cdots, b$, is expressed in the fixed-effects model (Model I) as the linear function $X_{\gamma\delta} = \mu + \tau_\gamma + \beta_\delta + \epsilon_{\gamma\delta}$, where μ is the over-all mean level of response, τ_γ is the modifying effect of treatment γ ($\sum_{\gamma=1}^{v}\tau_\gamma = 0$), β_δ is the special influence of block δ ($\sum_{\delta=1}^{b}\beta_\delta = 0$), and $\epsilon_{\gamma\delta}$ is a random error such that the set of vb errors are independent univariate normal variates with zero means and equal variances, σ^2. The multivariate generalization of this model replaces the scalar variate $X_{\gamma\delta}$ with a p-variate column vector $\mathbf{X}_{\gamma\delta}$ with elements $X_{\gamma\delta i}$, $i = 1, \cdots, p$, consisting of responses on each of p variates for treatment γ in block δ. Similarly, the scalars μ, τ_γ, β_δ, and $\epsilon_{\gamma\delta}$ are replaced by p-element column vectors, and the vectors $\boldsymbol{\epsilon}_{\gamma\delta}$ constitute a set of vb independent multivariate normal vector variates with zero means and common dispersion matrices, $\boldsymbol{\Sigma}$.

In univariate analysis of variance, treatment and error mean squares are calculated. If these are S_τ^2 and S_ω^2, their forms are

$$S_\tau^2 = b \sum_{\gamma=1}^{v} (\bar{X}_{\gamma\cdot} - \bar{X}_{\cdot\cdot})^2/(v-1)$$

and

$$S_\omega^2 = \sum_{\gamma=1}^{v}\sum_{\delta=1}^{b} (X_{\gamma\delta} - \bar{X}_{\gamma\cdot} - \bar{X}_{\cdot\delta} + \bar{X}_{\cdot\cdot})^2/(v-1)(b-1),$$

where $\bar{X}_{\gamma\cdot} = \sum_{\delta=1}^{b} X_{\gamma\delta}/b$, $\bar{X}_{\cdot\delta} = \sum_{\gamma=1}^{v} X_{\gamma\delta}/v$, and $\bar{X}_{\cdot\cdot} = \sum_{\gamma=1}^{v}\sum_{\delta=1}^{b} X_{\gamma\delta}/vb$. The test of treatment equality is the test of the hypothesis $H_{04} : \tau_1 = \cdots = \tau_v \; (=0)$; the statistic used is $F = S_\tau^2/S_\omega^2$, distributed as F with $v-1$ and $(v-1)(b-1)$ degrees of freedom under H_{04} with large values of F statistically significant. When H_{04} is true, both S_τ^2 and S_ω^2 provide unbiased estimates of σ^2 and are independent in probability, whereas when H_{04} is false, S_ω^2 still gives an unbiased estimate of σ^2, but S_τ^2 tends to be larger.

The multivariate generalization of analysis of variance involves comparison of $p \times p$ dispersion matrices $\mathbf{S}_\tau$ and $\mathbf{S}_\omega$, the elements of which correspond to S_τ^2 and S_ω^2:

$$S_{\tau ij} = \frac{b \sum_{\gamma=1}^{v} (\bar{X}_{\gamma\cdot i} - \bar{X}_{\cdot\cdot i})(\bar{X}_{\gamma\cdot j} - \bar{X}_{\cdot\cdot j})}{(v-1)},$$

$$S_{\omega ij} = \frac{\sum_{\gamma=1}^{v}\sum_{\delta=1}^{b} (X_{\gamma\delta i} - \bar{X}_{\gamma\cdot i} - \bar{X}_{\cdot\delta i} + \bar{X}_{\cdot\cdot i})(X_{\gamma\delta j} - \bar{X}_{\gamma\cdot j} - \bar{X}_{\cdot\delta j} + \bar{X}_{\cdot\cdot j})}{(v-1)(b-1)},$$

for $i, j = 1, \cdots, p$. It can be shown that $\mathbf{S}_\tau$ and $\mathbf{S}_\omega$ have independent Wishart distributions with $v-1$ and $(v-1)(b-1)$ degrees of freedom and identical dispersion matrices, $\boldsymbol{\Sigma}$, under H_{04}. Thus, the multivariate analysis-of-variance problem is reduced again to the problem of comparing two dispersion matrices, $\mathbf{S}_\tau$ and $\mathbf{S}_\omega$, like $\mathbf{S}_{(1)}$ and $\mathbf{S}_{(2)}$ of the preceding section. This is the general situation in multivariate analysis of variance, even though this illustration is for a particular experimental design.

Wilks (1932*a*; 1935) recommended use of the statistic $|\mathbf{S}_\omega|/|\mathbf{S}_\omega + \mathbf{S}_\tau|$, Roy (1953) considered the largest root of $\mathbf{S}_\tau\mathbf{S}_\omega^{-1}$, and Lawley (1938) suggested $\operatorname{tr}(\mathbf{S}_\tau\mathbf{S}_\omega^{-1})$. These statistics correspond roughly to criteria on the product of characteristic roots, the largest root, and the sum of the roots, respectively. They lead to equivalent tests in the univariate case (where only one root exists), but the tests are not equivalent in the multivariate case.

Pillai (1964; 1965) has tables and references on the distribution of the largest root. A paper by Smith, Gnanadesikan, and Hughes (1962) is recommended as an elementary expository summary with a realistic example.

Other procedures. Other, more specialized statistical procedures have been developed for means, variances, and covariances for multivariate normal populations, particularly tests of special hypotheses.

Many models based on the univariate normal distribution may be regarded as special cases of multivariate normal models. In particular, it is often assumed that observations are independent in probability and have homogeneous variances, σ^2. A test of such assumptions may sometimes be made if the sample is regarded as N observation vectors from a p-variate multivariate normal population with special dispersion matrix under a null hypothesis H_{05}: $\boldsymbol{\Sigma} = \sigma^2\boldsymbol{I}$, where $\boldsymbol{I}$ is the $p \times p$ identity matrix and σ^2 is the unknown common variance. This test and a generalization of it are discussed by Anderson (1958, sec. 10.7). See also Wilks (1962, problem 18.21).

Wilks (1946; 1962, problem 18.22) developed a series of tests on means, variances, and covariances for multivariate normal populations. He considered three hypotheses,

$$H_{06} : \mu_i = \mu, \quad \sigma_{ii} = \sigma^2, \quad \sigma_{ij} = \rho\sigma^2, \quad i \neq j, \quad i, j = 1, \cdots, p;$$

$$H_{07} : \sigma_{ii} = \sigma^2, \quad \sigma_{ij} = \rho\sigma^2, \quad i \neq j, \quad i, j = 1, \cdots, p;$$

$$H_{08} : \mu_i = \mu, \quad \text{given that } \sigma_{ii} = \sigma^2, \quad \sigma_{ij} = \rho\sigma^2, \quad i \neq j, \quad i, j = 1, \cdots, p.$$

H_{06} implies equality of means, equality of variances, and equality of covariances; H_{07} makes no assumption about the means but implies equality of variances and equality of covariances; H_{08} is a hypothesis about equality of means given the special dispersion matrix, $\boldsymbol{\Sigma}$, specified through equality of its diagonal elements and equality of its nondiagonal elements. In these hypotheses ρ is the intraclass correlation, which has been considered in various contexts by other authors [*see* MULTIVARIATE ANALYSIS, *article on* CORRELATION METHODS]. Wilks showed that the test of H_{08} leads to the usual, univariate, analysis-of-variance test for treatments in a two-way classification. For H_{06} and H_{07}, likelihood ratio tests were devised and moments of the test statistics were obtained with exact distributions in special cases and asymptotic ones otherwise.

Other topics of multivariate analysis

This general discussion of multivariate analysis would not be complete without mention of basic concepts of other major topics discussed elsewhere in this encyclopedia.

Discriminant functions. Classification problems are encountered in many contexts [*see* MULTIVARIATE ANALYSIS, *article on* CLASSIFICATION AND DISCRIMINATION]. Several populations are known to exist, and information on their characteristics is available, perhaps from samples of individuals or items identified with the populations. A particular individual or item of unknown population is to be classified into one of the several populations on the basis of its particular characteristics. This and related problems were considered by early workers in the field and more recently in the context of statistical decision theory, which seems particularly appropriate for this subject [*see* DECISION THEORY].

Correlation. The simple product-moment correlation coefficient between variates X_i and X_j was defined above as ρ_{ij}, with similarly defined sample correlation, r_{ij} [*see* MULTIVARIATE ANALYSIS, *article on* CORRELATION METHODS]. In the bivariate case ($p = 2$) the exact small sample distributions of r_{12} based on the bivariate normal model were developed by Fisher and Hotelling. The multiple correlation between X_1, say, and the set $X_2, \cdots, X_p$ may be defined as the maximum simple correlation between X_1 and a linear function $\beta_2 X_2 + \cdots + \beta_p X_p$, maximum through choice of $\beta_2, \cdots, \beta_p$.

Partial correlations have been developed as correlations in conditional distributions.

Canonical correlations extend the notion of multiple correlation to two groups of variates. If the variate vector, $\mathbf{X}$, is subdivided so that

$$\mathbf{X} = \begin{bmatrix} \mathbf{X}_{(s)} \\ \mathbf{X}_{(t)} \end{bmatrix},$$

$\mathbf{X}_{(s)}$ being the column vector with elements $X_1, \cdots, X_s$ and $\mathbf{X}_{(t)}$ being the column vector with elements $X_{s+1}, \cdots, X_p$ ($p = s + t$), the largest canonical correlation is the maximum simple correlation between two linear functions, $Y_{(s)} = \sum_{\alpha=1}^{s} \beta_\alpha X_\alpha$ and $Y_{(t)} = \sum_{\alpha=s+1}^{p} \beta_\alpha X_\alpha$. The second largest canonical correlation is the maximum simple correlation between two new linear functions, $Y'_{(s)}$ and $Y'_{(t)}$, similar to $Y_{(s)}$ and $Y_{(t)}$ but uncorrelated with $Y_{(s)}$ and $Y_{(t)}$, and so on. Distribution theory and related problems are given by Anderson (1958, chapter 12) and Wilks (1962, sec. 18.9).

The theory of rank correlation is well developed in the bivariate case [*see* NONPARAMETRIC STATISTICS, *article on* RANKING METHODS]. Tetrachoric and biserial correlation coefficients have been considered for special situations.

Principal components. The problem of principal components and factor analysis is a problem in the reduction of the number of variates and in their interpretation [*see* FACTOR ANALYSIS AND PRINCIPAL COMPONENTS]. The method of principal components considers uncorrelated linear functions of the p original variates with a view to expressing major characteristic variation in terms of a reduced set of new variates. Hotelling has been responsible for much of the development of principal components, and the somewhat parallel treatments of factor analysis have been developed more by psychometricians than by statisticians. References for principal components are Anderson (1958, chapter 11), Wilks (1962, sec. 18.6), and Kendall (1957, chapter 2). Kendall (1957, chapter 3) gives an expository account of factor analysis.

Counted data. The multinomial distribution plays an important role in analysis when multivariate data consist of counts of the number of individuals or items in a sample that have specified categorical characteristics. The multivariate analysis of counted data follows consideration of contingency tables and relationships between the probability parameters of the multinomial distribution. Much has been done on tests of independence in such tables, and recently investigators have developed more systematically analogues of standard multivariate techniques for contingency tables [*see* COUNTED DATA].

Nonparametric statistics. There has been a paucity of multivariate techniques in nonparametric statistics. Except for work on rank correlation, only a few isolated multivariate methods have been developed—for example, bivariate sign tests. The difficulty appears to be that adequate models for multivariate nonparametric methods must contain measures of association (or of nonindependence) that sharply limit the application of the permutation techniques of nonparametric statistics [*see* NONPARAMETRIC STATISTICS].

Missing values. Only limited results are available in multivariate analysis when some observations are missing from observation vectors. Wilks (1932*b*), in considering the bivariate normal distribution with missing observations, provided several methods of parameter estimation and compared them. Maximum likelihood estimation was somewhat complicated, but two *ad hoc* methods proved simpler and yielded exact forms of sampling distributions. Basically, one may obtain estimates of means and variances through weighted averages of means and variances of the available data and estimate correlations from the available data on pairs of variates. If only a few observations are missing, usual analyses should not be much affected; if many observations are missing, little advice may be given except to suggest the use of maximum likelihood techniques and computers for the special situation. It is clearly inappropriate to treat missing observations as zero observations—as has sometimes been done.

Some useful references are Anderson (1957), Buck (1960), Nicholson (1957), and Matthai (1951).

Other multivariate results. In a general discussion of multivariate analysis, it is not possible to consider all areas where multivariate data may arise or all theoretical results of probability and statistics that may be pertinent to multivariate analysis. Many of the theorems of probability admit of multivariate extensions; results in stochastic processes, the theory of games, decision theory, and so on, may have important, although perhaps not implemented, multivariate generalizations.

RALPH A. BRADLEY

BIBLIOGRAPHY

○ *Multivariate analysis is complex in theory, in application, and in interpretation. Basic works should be consulted, and examples of applications in various subject areas should be examined critically. The theory of multivariate analysis is well presented in* Anderson 1958; *its excellent bibliography and reference notations by section make it a good guide to works in the field. Among books on mathematical statistics, other major works are* Rao 1952; Kendall & Stuart (1961) 1973; Roy 1957; Wilks 1946; 1962. Greenwood & Hartley 1962 *gives references to tables. Books more related to the social sciences are* Cooley & Lohnes 1962; Talbot & Mulhall 1962. *Papers that are largely expository and bibliographical are* Tukey 1949; Bartlett 1947; Wishart 1955; Feraud 1942; *and* Smith, Gnanadesikan, & Hughes 1962. *Some applications in the social sciences are given in* Tyler 1952; Rao & Slater 1949; Tintner 1946; Kendall 1957.

ANDERSON, T. W. 1957 Maximum Likelihood Estimates for a Multivariate Normal Distribution When Some Observations Are Missing. *Journal of the American Statistical Association* 52:200–203.

ANDERSON, T. W. 1958 *An Introduction to Multivariate Statistical Analysis.* New York: Wiley.

BARTLETT, M. S. 1947 Multivariate Analysis. *Journal of the Royal Statistical Society* Series B 9 (Supplement): 176–190. → A discussion of Bartlett's paper appears on pages 190–197.

BUCK, S. F. 1960 A Method of Estimation of Missing Values in Multivariate Data Suitable for Use With an Electronic Computer. *Journal of the Royal Statistical Society* Series B 22:302–306.

COOLEY, WILLIAM W.; and LOHNES, PAUL R. 1962 *Multivariate Procedures for the Behavioral Sciences.* New York: Wiley.

FERAUD, L. 1942 Problème d'analyse statistique à plusieurs variables. Lyon, Université de, *Annales* 3d Series, Section A 5:41–53.

GREENWOOD, J. ARTHUR; and HARTLEY, H. O. 1962 *Guide to Tables in Mathematical Statistics.* Princeton Univ. Press. → A sequel to the guides to mathematical tables produced by and for the Committee on Mathematical Tables and Aids to Computation of the National Academy of Sciences–National Research Council of the United States.

HECK, D. L. 1960 Charts of Some Upper Percentage Points of the Distribution of the Largest Characteristic Root. *Annals of Mathematical Statistics* 31:625–642.

JACKSON, J. EDWARD; and BRADLEY, RALPH A. 1961 Sequential χ^2- and T^2-tests. *Annals of Mathematical Statistics* 32:1063–1077.

○KENDALL, MAURICE G. 1957 *A Course in Multivariate Analysis.* London: Griffin; New York: Hafner. → Minor corrections were made in the second printing in 1961.

○KENDALL, MAURICE G.; and STUART, ALAN (1961) 1973 *The Advanced Theory of Statistics.* Volume 2: *Inference and Relationship.* 3d ed. London: Griffin; New York: Hafner. → Early editions of Volume 2, first published in 1946, were written by Kendall alone. Stuart became a joint author on later, renumbered editions in the three-volume set.

KUDÔ, AKIO 1963 A Multivariate Analogue of the One-sided Test. *Biometrika* 50:403–418.

LAWLEY, D. N. 1938 Generalization of Fisher's *z* Test. *Biometrika* 30:180–187.

MATTHAI, ABRAHAM 1951 Estimation of Parameters From Incomplete Data With Application to Design of Sample Surveys. *Sankhyā* 11:145–152.

MORRISON, DONALD F. 1967 *Multivariate Statistical Methods.* New York: McGraw-Hill. → Written for investigators in the life and behavioral sciences.

NICHOLSON, GEORGE E. JR. 1957 Estimation of Parameters From Incomplete Multivariate Samples. *Journal of the American Statistical Association* 52:523–526.

NÜESCH, PETER E. 1966 On the Problem of Testing Location in Multivariate Populations for Restricted Alternatives. *Annals of Mathematical Statistics* 37:113–119.

PILLAI, K. C. SREEDHARAN 1964 On the Distribution of the Largest of Seven Roots of a Matrix in Multivariate Analysis. *Biometrika* 51:270–275.

PILLAI, K. C. SREEDHARAN 1965 On the Distribution of the Largest Characteristic Root of a Matrix in Multivariate Analysis. *Biometrika* 52:405–414.

RAO, C. RADHAKRISHNA 1952 *Advanced Statistical Methods in Biometric Research.* New York: Wiley.

RAO, C. RADHAKRISHNA; and SLATER, PATRICK 1949 Multivariate Analysis Applied to Differences Between Neurotic Groups. *British Journal of Psychology* Statistical Section 2:17–29. → See also "Correspondence," page 124.

ROY, S. N. 1953 On a Heuristic Method of Test Construction and Its Use in Multivariate Analysis. *Annals of Mathematical Statistics* 24:220–238.

ROY, S. N. 1957 *Some Aspects of Multivariate Analysis.* New York: Wiley.

ŠIDÁK, ZBYNĚK 1967 Rectangular Confidence Regions for the Means of Multivariate Normal Distributions. *Journal of the American Statistical Association* 62: 626–633.

SMITH, H.; GNANADESIKAN, R.; and HUGHES, J. B. 1962 Multivariate Analysis of Variance (MANOVA). *Biometrics* 18:22–41.

TALBOT, P. AMAURY; and MULHALL, H. 1962 *The Physical Anthropology of Southern Nigeria: A Biometric Study in Statistical Method.* Cambridge Univ. Press.

TINTNER, GERHARD 1946 Some Applications of Multivariate Analysis to Economic Data. *Journal of the American Statistical Association* 41:472–500.

TUKEY, JOHN W. 1949 Dyadic ANOVA: An Analysis of Variance for Vectors. *Human Biology* 21:65–110.

TYLER, FRED T. 1952 Some Examples of Multivariate Analysis in Educational and Psychological Research. *Psychometrika* 17:289–296.

WILKS, S. S. 1932*a* Certain Generalizations in the Analysis of Variance. *Biometrika* 24:471–494.

WILKS, S. S. 1932*b* Moments and Distributions of Estimates of Population Parameters From Fragmentary Samples. *Annals of Mathematical Statistics* 3:163–195.

WILKS, S. S. 1935 On the Independence of *k* Sets of Normally Distributed Statistical Variables. *Econometrica* 3:309–326.

WILKS, S. S. 1946 Sample Criteria for Testing Equality of Means, Equality of Variances, and Equality of Covariances in a Normal Multivariate Distribution. *Annals of Mathematical Statistics* 17:257–281.

WILKS, S. S. 1962 *Mathematical Statistics.* New York: Wiley. → An earlier version of some of this material was issued in 1943.

WISHART, JOHN 1955 Multivariate Analysis. *Applied Statistics* 4:103–116.

Postscript

Continuing research and new developments have occurred since publication of the main article. The anticipated bibliography of multivariate analysis has been completed by Anderson et al. (1972). The complexity of multivariate analysis in application and interpretation, noted in the introductory paragraph to the main bibliography, is emphasized by Kowalski (1972) in regard to anthropometric research. His comments can be applied to fields other than anthropology, however, and they indicate the need to focus research on the real applied problems. New books are available; Dempster (1969) and Van de Geer (1971) deal respectively with theory and with subject matter application.

Some new direction in multivariate analysis has developed in cluster analysis [*see* CLUSTERING]—a continuation of work in classification with new emphases. A review article by Hartigan (1973) provides a basic bibliography as does the text by Jardine and Sibson (1971). A useful new text by Hartigan (1975) has appeared. Aspects of the work of Joseph B. Kruskal (1964*a*; 1964*b*) on multidimensional scaling [*see* SCALING, MULTIDIMENSIONAL] and of Tukey (1977) on data analysis [*see* DATA ANALYSIS, EXLORATORY] relate to clustering. At the present time, work on cluster analysis is generally limited to the provision of algorithms for the formation of clusters, and much needs to be done on the probabilistic aspects of procedures devised.

RALPH A. BRADLEY

ADDITIONAL BIBLIOGRAPHY

ANDERSON, T. W.; DAS GUPTA, SOMESH; and STYAN, GEORGE P. H. 1972 *A Bibliography of Multivariate Statistical Analysis.* Edinburgh: Oliver & Boyd; New YORK: Halsted. → Reprinted by Krieger in 1977.

DEMPSTER, ARTHUR P. 1969 *Elements of Continuous Multivariate Analysis.* Reading, Mass.: Addison-Wesley.

HARTIGAN, JOHN A. 1973 Clustering. *Annual Review of Biophysics and Bioengineering* 2:81–101.

HARTIGAN, JOHN A. 1975 *Clustering Algorithms.* New York: Wiley.

JARDINE, NICHOLAS; and SIBSON, ROBIN 1971 *Mathematical Taxonomy.* New York: Wiley.

KOWALSKI, CHARLES J. 1972 A Commentary on the Use of Multivariate Statistical Methods in Anthropometric Research. *Journal of Physical Anthropology* 36:119–131.

KRUSKAL, JOSEPH B. 1964*a* Multidimensional Scaling by Optimizing Goodness of Fit to a Nonmetric Hypothesis. *Psychometrika* 29:1–27.

KRUSKAL, JOSEPH B. 1964*b* Nonmetric Multidimensional Scaling: A Numerical Method. *Psychometrika* 29:115–129.

TUKEY, JOHN W. (1970) 1977 *Exploratory Data Analysis.* Reading, Mass.: Addison-Wesley. → First published in a mimeographed "limited preliminary edition."

VAN DE GEER, JOHN P. 1971 *Introduction to Multivariate Analysis for the Social Sciences.* San Francisco: Freeman. → Based on the author's *Inleiding in de Multivariate Analyse.* Common library practice is to alphabetize his works under his Dutch name: Geer, Johannes Petrus, van de.

II
CORRELATION METHODS

● *This article replaces two articles in* IESS, *one by the late Harold Hotelling and the other by Robert F. Tate. The present article combines material from the former with an expanded version of the latter.*

The term "correlation" is used in a variety of contexts to indicate the degree of interrelation between two or more characteristics. One reads, for example, of the correlation between intelligence and wealth, between illiteracy and prejudice, and so on. When used in this sense the term is not sufficiently operational for scientific work. One must instead speak of correlation between numerical measures of characteristics—in short, of correlation between variables.

If statistical inference is to be used, the variables must be random variables, and for them a probability model (strictly, one of a parametric family of models) must be specified. For two random variables, X and Y, such a model will describe the probabilities (or probability densities) with which (X, Y) takes values (x, y); that is, it describes probabilities in the (X, Y) population. One of the parameters of this population is the correlation coefficient; the available information concerning it is often in the form of a simple random sample, $(X_1, Y_1), \cdots, (X_n, Y_n)$, or perhaps in the form of a sample with more complex structure. Correlation theory is thus concerned with the use of samples to estimate, test hypotheses, or carry out other procedures concerning population correlations. Even at this initial stage confusion occasionally sets in. In the early days the habit of confusing labels, and perhaps meanings, between population parameters and sample estimates became ingrained in the writers of the period, and was transmitted to future generations of students.

The so-called Pearson product–moment correlation coefficient—usually denoted by ρ in the population and r in the sample, and usually termed simply the correlation coefficient—is the most frequently encountered, and the main purpose of this article is to survey the situations in which it is employed. Other sorts of correlation include rank correlation, serial correlation, and intraclass correlation. [*For a discussion of rank correlation, see* NONPARAMETRIC STATISTICS, *article on* RANKING METHODS; *for serial correlation, see* TIMES SERIES. *Intraclass correlation is touched on briefly later in this article.*]

The two traditional models for correlation theory are the linear regression model, discussed later in this article [*see also* Binder 1959; LINEAR HYPOTHESES, *article on* REGRESSION] and the joint normal model, which plays a central role in the theory.

For any two random variables, X and Y, it will follow from the definition of ρ that if the variables are independent, they are uncorrelated; hence, to conclude that the hypothesis of zero correlation is false is to assert dependence for X and Y. In the other direction, if X and Y follow a bivariate normal law and are uncorrelated, then they are independent, but this conclusion does not hold in general—the assumption of normality (or some other, similar restriction) is essential; it is even possible that X and Y are uncorrelated and also perfectly related by a (nonlinear) function. If the probability law for X, Y is only approximately bivariate normal, conventional normal theory can still be applied; in fact, considerable departure from normality may be tolerated (Gayen 1951). For large random samples, r itself is in any case approximately normal with mean ρ and with a standard deviation that depends on the joint probability distribution of X, Y.

Research in the theory of correlation can be divided into four historical phases. In the latter part of the nineteenth century. Galton and others realized the value of correlation in their work but could deal with it only in a vague, descriptive way [*see the biography of* GALTON]. About the turn of the century Karl Pearson, Edgeworth, and Yule developed some theory and systematized the use of correlation [*see the biographies of* EDGEWORTH; PEARSON; YULE]. From about 1915 to 1928, R. A. Fisher placed the theory of correlation on a more rigorous footing by deriving exact probability laws and methods of estimation and testing [*see the biography of* FISHER]. During this period Sewall Wright (1918) introduced the method of path analysis, a consideration of which throws much light on the place of correlation methods in the study of causation. Unfortunately, his efforts appear to have been largely unappreciated by social scientists until the 1960s. Finally, in the 1930s first Hotelling and then Wilks, Maurice Kendall, and others, spurred on by psychologists, particularly Spearman and Thurstone, developed principal component analysis (closely related to factor analysis) and canonical correlation. [*See* FACTOR ANALYSIS AND PRINCIPAL COMPONENTS *and the biographies of* HOTELLING; SPEARMAN; WILKS; *see also* Adkins 1968.]

Correlation methods have long been recognized as scientifically important, especially in situations involving many variables. The main value of correlation lies in suggesting lines along which further research can be directed in a search for possible cause-and-effect relations in complex situations, but it can also be used to elucidate causal relationships that have already been assumed, as in the case of path analysis [*see* CAUSATION; SURVEY ANALYSIS].

This article will deal with simple correlation between X and Y, the relation of correlation to regression theory, and some correlation models. Multiple and partial correlation will then be discussed, first in population terms and then in a sample setting. A so-called stepwise regression procedure will be presented as an illustration of the role played by sample partial and multiple correlations. Sets of variables and specialized methods for dealing with them will then be taken up. The final two sections will deal with specialized topics in correlation and path analysis.

Simple correlation

For two jointly distributed random variables, X and Y, denote their population standard deviations by σ_X and σ_Y and their population covariance by σ_{XY}. The correlation coefficient is then defined as $\rho_{XY} = \sigma_{XY}/\sigma_X\sigma_Y$. (Both standard deviations are positive, except for the uninteresting case in which one or both variables are constant; then the correlation coefficient is undefined.) As Feller (1950, p. 186n) has remarked, this definition would lead a physicist to regard ρ_{XY} as "dimensionless covariance."

Some elementary properties of the correlation coefficient follow:

(1) $-1 \leqslant \rho_{XY} \leqslant +1$.

(2) ρ_{XY} is unchanged if constants are added to X or Y, or if X and Y are multiplied by constants with the same algebraic sign; if the signs are different, the sign of ρ_{XY} will change.

(3) ρ_{XY} is ± 1 if and only if a perfect linear relation, $Y = a + bX$, exists (with $b \neq 0$ and $\rho = -1$ for $b < 0$, and $+1$ for $b > 0$).

(4) The variance of a linear function $cX + dY$ in terms of ρ_{XY} is $\sigma^2_{cX+dY} = c^2\sigma^2_X + 2cd\rho_{XY}\sigma_X\sigma_Y + d^2\sigma^2_Y$ (Note in particular that variances are additive in the presence of zero correlation.)

The usual estimator for ρ_{XY}, based on a random sample, $(X_1, Y_1), \cdots, (X_n, Y_n)$, is

$$r_{XY} = \frac{\sum(X_i - \bar{X})(Y_i - \bar{Y})}{[\sum(X_i - \bar{X})^2]^{1/2}[\sum(Y_i - \bar{Y})^2]^{1/2}} = \frac{s_{XY}}{s_X s_Y}$$

with $\bar{X}$, $\bar{Y}$ as the sample means and s_X, s_Y, s_{XY} as the sample standard deviations and covariance. Regardless of the specific distributional assumptions adopted, r_{XY} can be used to estimate ρ_{XY} and will have some desirable properties providing the sample is random: r_{XY} lies between -1 and $+1$ and has approximately ρ_{XY} for its expectation; r_{XY} is a consistent estimator of ρ_{XY}, that is, if the sample size is increased indefinitely, $Pr(|r_{XY} - \rho_{XY}| < \epsilon)$ approaches 1, no matter how small a positive constant ϵ is chosen.

Normal model. If the joint probability law is bivariate normal, that is, if probability is interpreted as volume under the surface

$$f(x, y) = \frac{1}{2\pi\sigma_X\sigma_Y(1 - \rho^2_{XY})^{1/2}} \exp\left\{-\frac{1}{2(1 - \rho^2_{XY})}\left[\left(\frac{x - \mu_X}{\sigma_X}\right)^2 - 2\rho_{XY}\left(\frac{x - \mu_X}{\sigma_X}\right)\left(\frac{y - \mu_Y}{\sigma_Y}\right) + \left(\frac{y - \mu_Y}{\sigma_Y}\right)^2\right]\right\},$$

then $f(x, y)$ factors into an expression in x times an expression in y (the condition defining independence of X, Y) if and only if $\rho_{XY} = 0$.

Under normality, r_{XY} is the maximum likelihood estimator of ρ_{XY}. Further, the probability law of r_{XY} has been derived (Fisher 1915) and tabulated (David 1938). The test statistic calculated as $(n - 2)^{1/2} r_{XY}/(1 - r^2_{XY})^{1/2}$ can be referred to the t table with $n - 2$ degrees of freedom to test H : $\rho_{XY} = 0$. In addition, charts (David 1938), which have been reproduced in many books, are available for the determination of confidence intervals. The variable $z = \tanh^{-1} r_{XY}$ is known (Fisher 1925, pp. 197 ff.) to have an approximate normal law with expectation $\tanh^{-1}\rho_{XY}$ and standard deviation $1/\sqrt{n-3}$; thus, the z transformation is especially useful, for example, in testing whether two (X, Y) populations have the same correlation. Also, it has the advantage of stabilizing variances; that is, the approximate variance of z depends on n but not on ρ_{XY}. An alternate expression for this transformation is $z = \frac{1}{2} \log_e [(1+r)/(1-r)]$. The quantity r_{XY} itself, though approximately normal with expectation ρ_{XY} and standard deviation given by $(1 - \rho^2_{XY})/\sqrt{n}$ for very large n, will still be far from normal for moderate n when ρ_{XY} is not near zero.

The distribution of r, derived by Fisher (1915), is complicated, so when Fisher introduced his z transformation the ramifications of his work on the exact distribution lost some of their relevance. In a lengthy paper Hotelling (1953) carefully reinvestigated questions concerning series expansions in the expression for the probability density, introduced and applied a new method for calculation of moments of r_{XY} and z, and proposed two improvements over the z transformation, one of which is

$$z^* = z - \frac{3z + r_{XY}}{4n}.$$

The variance-stabilizing properties of this transformation turn out to be superior to those of z, since

$$\sigma^2_z = \frac{1}{n-1} + \frac{4 - \rho^2_{XY}}{2(n-1)^2} + O\left(\frac{1}{n^3}\right),$$

$$\sigma^2_{z^*} = \frac{1}{n-1} + O\left(\frac{1}{n^3}\right).$$

The work of Hotelling (1953) is of a special nature: it constitutes a painstaking but necessary effort to refine a limiting result by selecting an improved variant of that result for use with small samples. Similar work was published by Kraemer (1973). Working under the assumption of bivariate normality, she showed for each function $\rho'(n, \rho_{XY})$ of a family she specified that

$$\frac{\sqrt{n-2}\,(r_{XY} - \rho')}{\sqrt{1 - r^2_{XY}}}$$

is approximately distributed as Student's t with $n - 2$ degrees of freedom. For the case $\rho'(n, 0) = 0$, Student's distribution is exact, as was mentioned before.

The sequential-probability-ratio procedure of

Wald was shown by Choi (1971) to apply to the test situation $H_0 : \rho_{XY} = \rho_0$, $H_1 : \rho_{XY} = \rho_1$, and the average-sample-number function was provided. [*See* SEQUENTIAL ANALYSIS.]

Even in the bivariate normal case that was just discussed, r_{XY} does not have population mean exactly ρ_{XY}, but the slight discrepancy can be greatly reduced (Olkin & Pratt 1958) by using the quantity $r_{XY}[1 + (1 - r_{XY}^2)/2(n - 3)]$ instead as an estimator for ρ_{XY}. The price to be paid for this refinement is that the new distribution must be worked out.

Biserial and point-biserial correlations. If one variable, say Y, is dichotomized at some unknown point ω, then the data from the (X, Y) population appear in the form of a sample from an (X, Z) population, with Z one or zero according to whether $Y \geqslant \omega$ or $Y < \omega$. If ρ_{XY} and ω are of interest, they can be estimated by r_b (biserial r) and ω_b, or by maximum likelihood estimators $\hat{\rho}_{XY}$ and $\hat{\omega}$ (Tate 1955*a*; 1955*b*). The latter estimators are jointly normal for large n, and tables of standard deviations are available (Prince & Tate 1966). If ρ_{XZ} is desired, it can be estimated by r_{XZ}, usually called point-biserial r. If, however, the assumption of underlying bivariate normality is correct, then $\rho_{XZ} < \sqrt{2/\pi}\rho_{XY}$, so r_{XZ} would be a bad estimator of ρ_{XY}. Tate (1955*b*) gives an expository discussion of both models. The value of r_{XZ} is doubtful, even when it is correctly applied. It is often better to focus attention on some aspect of the model other than ρ_{XZ}, and to use more conventional methods that deal with data in the form of class frequencies.

An intermediate case between an ordinary correlation model and a biserial model was considered by Mayer (1973). He defined a model in which the underlying distribution is bivariate normal, and one variable is directly observable, while only a monotone function of the other variable is observable. He then introduced the notions of monotone regression and monotone correlation coefficient.

Tetrachoric correlation. If in the bivariate normal case both X and Y are observable only in dichotomized form, the sample values can be arranged in a 2×2 table, and one can calculate r_t, the so-called tetrachoric r. Unfortunately, the tetrachoric model is not amenable to the same type of simple mathematical treatment as is the biserial model. The statistic r_t is now mainly of historical interest (see Kendall & Stuart 1958–1966).

Relation of correlation to regression. The notion of regression is appropriate in a situation in which one needs to predict Y, or to estimate the conditional population mean of Y, for given X [*see* LINEAR HYPOTHESES, *article on* REGRESSION]. The discussion given here will be sufficiently general to bring out the meanings of the correlation coefficient and the correlation ratio in regression analysis and to indicate connections between them. The reader should keep several facts in mind: (1) predictions are described by regression relations, whereas their accuracy is measured by correlation; (2) assumptions of bivariate normality are not required in order to introduce the notion of regression and to carry its development quite far; but (3) when hypotheses are tested, confidence intervals are constructed, or other formal methods of statistical inference are used, an assumption of at least approximate normality for the conditional distribution of Y will be essential.

A prediction of Y from X, $\phi(X)$, is judged "best" (in the sense of least squares), quite apart from assumptions of normality, if it makes the mean-square error, $E(Y - \phi(X))^2$, a minimum. It turns out that for "best" prediction, $\phi(X)$ must be $\mu_{Y|X}$, the mean of the conditional probability law (also often referred to as the regression function) for Y given X, but that if only straight lines are allowed as candidates, the "best" such gives the prediction $A + BX$, with $A = \mu_Y - \mu_X(\rho_{XY}\sigma_Y/\sigma_X)$ and $B = \rho_{XY}\sigma_Y/\sigma_X$. The basic quantities of interest, $\mu_{Y|X}$ and $A+BX$, lead to the following decomposition of $Y-\mu_Y$:

$$Y-\mu_Y = (Y-\mu_{Y|X}) + (A+BX-\mu_Y) + (\mu_{Y|X}-A-BX).$$

It can readily be shown that the right-hand terms are uncorrelated and that therefore, by squaring and taking expected values, one obtains the basic relation

$$\sigma_Y^2 = E(Y-\mu_{Y|X})^2 + E(A+BX-\mu_Y)^2 + E(\mu_{Y|X}-A-BX)^2.$$

These terms may be conveniently interpreted as portions of the variation of Y: the first is the variation unexplained by X, and the sum of the second and third is the variation explained by the "best" prediction, since the second term is the amount explained by the "best" linear prediction. (It should be understood that the use of the word "explained" implies nothing about causation.) The quantity

$$\eta_{XY}^2 = [\sigma_Y^2 - E(Y-\mu_{Y|X})^2]/\sigma_Y^2,$$

the squared correlation ratio for Y on X, is the proportion of variation of Y explained by regression on X. Since it can be shown that $E(A+BX-\mu_Y)^2 = \rho_{XY}^2\sigma_Y^2$, the basic relation may be rewritten as

$$\sigma_Y^2 = E(Y-\mu_{Y|X})^2 + \rho_{XY}^2\sigma_Y^2 + (\eta_{XY}^2 - \rho_{XY}^2)\sigma_Y^2.$$

If the regression is linear, then $\mu_{Y|X} = A+BX$, the third term drops out of the decomposition of $Y-\mu_Y$, and η^2_{XY}, the proposition of explained variation, coincides with ρ^2_{XY}. If in addition $\sigma^2_{Y|X}$, the variance of the conditional distribution for Y given X, is constant over values of X, then this variance coincides with $E(Y-\mu_{Y|X})^2$, and

$$\rho^2_{XY} = (\sigma^2_Y - \sigma^2_{Y|X})/\sigma^2_Y.$$

When X, Y have a bivariate normal distribution, both conditions are met, and hence this last relation is satisfied. Another common way of writing the expression is $\sigma^2_{Y|X} = \sigma^2_Y(1-\rho^2_{XY})$. In any event one can see from the basic relation, and conditions of nonnegativity for mean squares, that $0 \leqslant \rho^2_{XY} \leqslant \eta^2_{XY} \leqslant 1$, with $\rho^2_{XY} = \eta^2_{XY}$ if and only if the regression is linear; when that linearity of regression holds, their common value is zero if and only if the regression is actually constant, and both quantities equal one if and only if the point (X, Y) is constrained to lie on a straight line (except that for horizontal or vertical lines, ρ_{XY} is undefined). It should be noted that in general $\eta^2_{YX} \neq \eta^2_{XY}$, whereas ρ_{XY} is symmetric: $\rho_{XY} = \rho_{YX}$. Traditional terms, now rarely used, are "coefficient of determination" for ρ^2_{XY}, "coefficient of nondetermination" for $1-\rho^2_{XY}$, and "coefficient of alienation" for $(1-\rho^2_{XY})^{1/2}$.

The use of data to predict Y from X by fitting a sample regression line or curve evidently involves two types of error, the error in estimating the true regression curve by a sample curve and the inherent sampling variability of Y (which cannot be reduced by statistical analysis) about the true regression curve. Kruskal (1958) gives a concise summary of the above material, together with further interpretive remarks, and Tate (1966) extends these ideas to the case of three or more variables and the consequent consideration of generalized variances.

It cannot be emphasized too strongly that the correlation coefficient is a measure of the degree of linear relationship. It is frequently the case that for variables Y and X, the regression of Y on X is linear, or at least approximately linear, for those values of X that are of interest or are likely to be encountered. One reason for this is that many processes in nature proceed smoothly and hence can be described by continuous, even differentiable functions. If the regression of Y on X is nonlinear, then one can in any event measure the degree of relationship by a correlation ratio. It may sometimes be desirable to give two measures (that is, to give estimates of both ρ^2_{XY} and $\eta^2_{YX} - \rho^2_{XY}$), one for the degree of linear relationship and one for the degree of additional nonlinear relationship. In the case of nonlinear relationship, however, the problem of estimating $\mu_{Y|X}$ will be complicated unless for each specific value $X = x$ a whole array of Y observations is available, or unless some specific nonlinear functional form is assumed for $\mu_{Y|X}$. In view of the advantages of using normal theory, it is best whenever possible to make the regression approximately linear by a suitable change of variable. [*See* STATISTICAL ANALYSIS, SPECIAL PROBLEMS OF, *article on* TRANSFORMATIONS OF DATA.]

When X, Y follow a bivariate normal law, one has not only linear regression for Y on X but also normality for the conditional law of Y given X and for the marginal law of X. If the conditions of the bivariate normal model are relaxed in order to allow X to have some type of law other than normal, while the remaining properties just mentioned are present, some interesting results can be obtained. It is known, for example (Tate 1966), that for large n, and for random samples, r_{XY} is approximately normal with expectation ρ_{XY} and standard deviation $(1-\rho^2_{XY})(1+\frac{1}{4}\gamma\rho^2_{XY})^{1/2}/\sqrt{n}$, with γ denoting the coefficient of excess (kurtosis minus 3) for the X population. The transformation analogous to Fisher's z, which stabilizes the variance to $O(1/n)$ for this case, is given by $\phi(r) = 2\tanh^{-1} r[(\gamma+4)/(r^2\gamma+4)]^{1/2}/(\gamma+4)^{1/2}$. (For a general treatment of aspects of this case, see Gayen 1951.)

It is an important fact that there is value in r_{XY} even if there exists no population counterpart for it. This arises in the following way. Let x be a fixed variable subject to selection by the experimenter and let Y have a normal law with mean $A+Bx$ and standard deviation $\sigma_{Y|x}$. This is called the linear regression model. The usual mathematical theory developed for this model requires that $\sigma_{Y|x}$ actually not depend on x, although slight deviations from constancy are not serious. If a definite dependence on x exists, it can sometimes be removed by an appropriate transformation of Y [*see* STATISTICAL ANALYSIS, SPECIAL PROBLEMS OF, *article on* TRANSFORMATIONS OF DATA]. Note that the nonrandom character of x here is stressed by use of a lowercase letter.

The quantities A and B can be estimated, as before, by least squares, and r_{xY} enters as r_{XY} did previously. Distribution theory for r_{xY} is, of course, not the same as in the bivariate normal case, since r_{xY} is only formally the same as r_{XY}. In other words, it is important to take into consideration in any given case whether (X, Y) actually has a bivariate distribution or whether $X = x$ behaves as an index for possible Y distributions.

A test of $B = 0$ may be carried out by using

$(n-2)^{1/2} r_{xY}/(1-r_{xY}^2)^{1/2}$, the same statistic that was employed to test $\rho_{XY}=0$ in the bivariate normal model, and it once more is found to have the Student-t distribution with $n-2$ degrees of freedom when $B=0$. More generally, an examination of the derivations for results in the bivariate normal model shows that often at a critical point a conditional distribution of some statistic for fixed $(x_1, \cdots, x_n)$ does not depend on $(x_1, \cdots, x_n)$, and hence is the same as the unconditional distribution. Such results carry over immediately to the linear regression model; the above is one example.

Pitfalls in correlation methods. Two common errors in correlation methods have already been mentioned: concluding that the presence of correlation implies causation, and assuming that no relation between variables is present if correlation is lacking. The literature contains many illustrations, some humorous, of the first type of error; for example, in connection with the high correlation between the number of children and the number of storks' nests in towns of northwestern Europe (Wallis & Roberts 1956, p. 79), the source of correlation presumably is some factor such as economic status or size of house. As an artificial, but mildly surprising, example of the second type of error one should consider the fact that for a standard normal variable X, Y and X are uncorrelated if $Y = X^2$.

A different type of error arises when one tries to control some unwanted condition or source of variation by introducing additional variables. If $U = X/Z$ and $V = Y/Z$, then it is entirely possible that ρ_{UV} will differ greatly from ρ_{XY}. For example, let X equal the number of crimes in an area, Y the number of divorced people in that area, and Z the population in the same area. Thus, ρ_{XY} is presumably near zero, but ρ_{UV} is positive. The difficulty here is clear, but similar difficulties can enter data analysis in insidious ways. Using percentages instead of initial observations can also produce gross misunderstanding. As a very simple example, consider $U = X/(X+Y)$ and $V = Y/(X+Y)$, and the fact that $\rho_{UV} = -1$ even if X and Y are independent. Of course, if additional variables, say, Z and W, are involved, the magnitude of the correlation between $X/(X+Y+Z+W)$ and $Y/(X+Y+Z+W)$ will not be so great. The adjective traditionally applied to this type of correlation is "spurious," though "artificially induced" would be better. Such a correlation can in certain circumstances be useful; for instance, the idea of so-called part–whole correlation (see McNemar [1949] 1962, chapter 10) deserves consideration in certain situations. If, for example, a test score T is made up of scores on separate questions or subtests, say $T_1 + T_2 + \cdots + T_m$, a high correlation r_{TT_i} could not be ascribed wholly to spuriousness. It is altogether possible that T_i would serve as well as T for the purpose at hand. Neyman (1952, chapter 3, pt. 3) discusses "spurious" correlations, and gives several detailed examples.

Multiple and partial correlation

If more than two variables are observed for each individual, say, $p+1$ variables $X_0, X_1, \cdots, X_p$ (it is convenient for some purposes to begin the subscript sequence with zero, especially when one variable, X_0, plays a different role from the others), there are more possibilities to be considered for correlation relationships: simple correlations ρ_{ij} $(i, j = 0, 1, \cdots, p)$, multiple correlations between any variable and a set of the others, and partial correlations between any two variables with the linear effects of all or some of the others removed. (In this section capital letters will be omitted in all subscripts. Also, for the purpose of explaining the meaning of multiple and partial correlation coefficients, and their relations to regression theory, population parameters will be used throughout. All sampling theory will be reserved for the following section, in which partial correlation will be illustrated and a stepwise regression procedure will be explained.)

Multiple correlation. The multiple correlation between X_0 and the set $(X_1, \cdots, X_p)$, denoted by $\rho_{0\cdot12\ldots p}$, is defined to be the largest simple correlation obtainable between X_0 and $a_1X_1 + \cdots + a_pX_p$, as the a_i coefficients vary. (Note that I have departed from the custom of using "R" for multiple correlation in the interests of uniformity with other modern notation.) The multiple correlation possesses the following properties: $\rho_{0\cdot12\ldots p}$ is nonnegative and is at least as large as the absolute value of any simple correlation; if additional variables, X_{p+1}, $X_{p+2}, \cdots$, are included, the multiple correlation cannot decrease. It thus follows that if $\rho_{0\cdot12\ldots p} = 0$, all ρ_{0j} are zero. Also, if $\rho_{0\cdot12\ldots p} = 1$, then a perfect linear relationship, $X_0 = a_0 + a_1X_1 + \cdots + a_pX_p$, exists for some $a_0, a_1, \cdots, a_p$. A variety of multiple correlations may be of interest in any particular setting; the subscript notation may be used to keep track of them. For example, $\rho_{5\cdot24}$ is the multiple correlation between X_5 and the set $\{X_2, X_4\}$.

Regression relationships, in which X_0 is predicted by $X_1, \cdots, X_p$, are analogous to those for simple correlation; for example, when regression is linear and conditional variances are constant, $\rho^2_{0\cdot12\ldots p}$ is the portion of σ_0^2 that is explained by regression, namely, $1 - (\sigma^2_{0\cdot12\ldots p}/\sigma_0^2)$, with $\sigma^2_{0\cdot12\ldots p}$ denoting the mean-square difference between X_0 and its "best" prediction based on $X_1, X_2, \cdots, X_p$: $\mu_{0\cdot 12\ldots p} = B_0 + B_1X_1 + \cdots + B_pX_p$. The coefficients, $B_1, B_2, \cdots, B_p$, were traditionally known as partial regression coefficients but are now usually termed simply regression coefficients, as in the bivariate case. Each coefficient gives the change in $\mu_{0\cdot 12\ldots p}$ per unit change in the variable associated with that coefficient.

Partial correlation. The coefficient of partial correlation $\rho_{01\cdot 2}$ is, roughly speaking, what ρ_{01} would be if the linear effect of X_2 were removed. One can measure "X_0 with the linear effect of X_2 removed" and "X_1 with the linear effect of X_2 removed" by subtracting "best" linear predictions $A_0 + B_2X_2$ and $A_0' + B_2'X_2$, and obtaining the residuals $X_{0\cdot 2} = X_0 - A_0 - B_2X_2$ and $X_{1\cdot 2} = X_1 - A_0' - B_2'X_2$ respectively. Then $\rho_{01\cdot 2}$ is defined as the simple correlation between $X_{0\cdot 2}$ and $X_{1\cdot 2}$. In the same way, the effect of more than one additional variable can be removed, and one may consider $\rho_{01\cdot 23\ldots p}$. Partials between any two other variables, with the linear effects of the remaining $p - 1$ removed, are similarly defined by rearrangement of subscripts. For the case of three variables, $\rho_{01\cdot 2}$ can be expressed in terms of simple correlations as

$$\rho_{01\cdot 2} = \frac{\rho_{01} - \rho_{02}\rho_{12}}{[(1 - \rho_{02}^2)(1 - \rho_{12}^2)]^{1/2}}.$$

Alternatively, if the joint probability law is normal, $\rho_{01\cdot 2}$ is the simple correlation between X_0 and X_1, calculated from the conditional law for X_0 and X_1 given X_2, but this characterization is not true in general. Also, since $\rho_{01\cdot 2}$ is the ordinary correlation between the residuals defined above, $\rho_{01\cdot 2}^2$ may be characterized in terms of the unexplained variance in one residual after linear prediction from the other, namely, $1 - (\sigma_{0\cdot 12}^2/\sigma_{0\cdot 2}^2)$.

To see an important relation between multiple and partial correlation, think of the variables $X_1, X_2, \cdots, X_p$ as being introduced one at a time and producing increases in multiple correlation with X_0. Then

$$\rho_{0\cdot 12\ldots p}^2 = 1 - (1 - \rho_{01}^2)(1 - \rho_{02\cdot 1}^2) \cdots (1 - \rho_{0p\cdot 12\ldots p-1}^2).$$

From this it follows that

$$1 - \rho_{0\cdot 12\ldots p}^2 = (1 - \rho_{0\cdot 12\ldots p-1}^2)(1 - \rho_{0p\cdot 12\ldots p-1}^2),$$

which yields a recursion relation that allows the correction of a multiple correlation when an additional variable is included or excluded.

Many expressions and statements analogous to the foregoing relationships can of course be obtained by selection and rearrangement of subscripts.

Sampling theory for multiple and partial correlation. Since all parameters introduced in the preceding section are actually only simple correlations between appropriate pairs of random variables, one can construct estimators for all parameters, from random samples, by calculating the corresponding functions of sample moments. Thus, for example, $\rho_{0\cdot 12\ldots p}$ is estimated by $r_{0\cdot 12\ldots p}$, which is calculated from the observation pairs $(X_{0i}, \hat{B}_0 + \hat{B}_1X_{1i} + \cdots + \hat{B}_pX_{pi})$; $\rho_{0\cdot 12}$ is estimated by $r_{0\cdot 12}$, which is calculated from observation pairs $(X_{0i} - \hat{A}_0 - \hat{B}_2X_{2i}, X_{1i} - \hat{A}_0' - \hat{B}_2'X_{2i})$; and so on.

Under the joint normal model, the hypothesis that $\rho_{0\cdot 12\ldots p} = 0$ can be tested by referring $[(n - p - 1)/p]\, r_{0\cdot 12\ldots p}^2/(1 - r_{0\cdot 12\ldots p}^2)$ to an F table with p and $n - p - 1$ degrees of freedom (Fisher 1928). Also, $r_{0\cdot 12\ldots p}$, like r_{XY}, is approximately normal for large n with expectation $\rho_{0\cdot 12\ldots p}$ and standard deviation $(1 - \rho_{0\cdot 12\ldots p}^2)/\sqrt{n}$, provided $\rho_{0\cdot 12\ldots p} \neq 0$; if $\rho_{0\cdot 12\ldots p} = 0$, then $nr_{0\cdot 12\ldots p}^2$ has approximately a chi-square law with p degrees of freedom. Fisher's z transformation applies as before except when $\rho_{0\cdot 12\ldots p}$ is zero (Hotelling 1953). Note that $\rho_{0\cdot 12\ldots p}$ and $r_{0\cdot 12\ldots p}$ do not quite reduce to simple correlations if $p = 1$; instead, one finds that $\rho_{0\cdot 1} = |\rho_{01}|$ and $r_{0\cdot 1} = |r_{01}|$. Statements made earlier in reference to simple correlations of biserial data carry over to multiple correlation (Hannan & Tate 1965), and the same tables (Prince & Tate 1966) are applicable.

In connection with partial correlation it was shown by Fisher (1928) that if the multivariate normal model is assumed, many results for $r_{01\cdot 23\ldots p}$ can be obtained from those for r_{01} by replacing $n - 2$ by $n - p - 1$; for example,

$$(n - p - 1)^{1/2} r_{01\cdot 23\ldots p}/(1 - r_{01\cdot 23\ldots p}^2)^{1/2}$$

can be referred to the t table with $n - p - 1$ degrees of freedom as a test of $H: \rho_{01\cdot 23\ldots p} = 0$. Note that this is equivalent to replacing n by (n − the number of variables fixed).

Example of the use of multiple and partial correlation. Consider a set of observations in which X_0 represents grade-point average (GPA), X_1 represents IQ, X_2 represents hours of study per week, and the relationship is sought between X_0 and X_1, with the linear effect of X_2 removed (Keeping 1962, p. 363). Results based on a random sample of 450 schoolchildren showed that $r_{0\cdot 12} = 0.82$, $r_{01} = 0.60$, $r_{02} = 0.32$, $r_{12} = -0.35$, and $r_{01\cdot 2} = 0.80$. The positive correlation between X_0 and X_1, together with the negative correlation between X_1 and X_2 (a more intelligent student need not study so long), obscured somewhat the strength of the relationship between X_0 and X_1. It should perhaps be mentioned that the relation $1 - r_{0\cdot 12}^2 = (1 - r_{02}^2)(1 - r_{01\cdot 2}^2)$ makes it clear that $r_{0\cdot 12} \geqslant r_{01\cdot 2}$, with equality if and only if $r_{02} = 0$. It is true in general, for parameters or sample estimators, that a multiple correlation between a given variable and others is at least as large in magnitude as any simple or partial correlation between that variable and any of the others. In Chapters 12 and 13 of their book, Yule and Kendall (1950) give many examples, along with interpretation and practical advice.

Stepwise regression. Elaborate and useful computational schemes are available for including or excluding variables in a correlation model. One viewpoint (see Ezekiel & Fox [1930] 1959, appendix 2) is that one should generally start with the largest feasible number of independent variables and then subtract one at a time those that make a negligible contribution to the prediction of X_0. Another, diametrically opposed, notion is that one should start with the best single predictor (in the sense of correlation) and then add others, one at a time, until further additions make no substantial improvement. This last approach is known as stepwise regression. The following identity is basic:

$$SS_0 = SS_{0\cdot 12\ldots j} + SS_{12\ldots j}, \qquad j = 1, 2, \cdots, p.$$

The left member denotes $\sum(X_{0i} - \bar{X}_0)^2$, the total variation of X_0, while the first term on the right stands for the sum of squares of residuals for X_0 on $X_1, \cdots, X_j$, the variation not explained by regression on $(X_1, \cdots, X_j)$. The last term represents $\sum(\hat{X}_{0i} - \bar{X}_0)^2$, the variation explained by regression on $(X_1, \cdots, X_j)$, with $\hat{X}_{0i}$ denoting an X_{0i} predicted from $X_{1i}, \cdots, X_{ji}$. Note that in the identity the terms are, when divided by n, estimators for the parameters σ_0^2, $\sigma_{0\cdot 12\ldots j}^2$, and $\sigma^2 - \sigma_{0\cdot 12\ldots j}^2$, which were mentioned earlier in the section on multiple correlation. We begin with X_0 and p available independent variables, $X_1, \cdots, X_p$.

Step 1. Choose the independent variable having the largest squared sample correlation with X_0. Rename this variable X_1, and use the statistic

$$(n - 2)SS_1/SS_{0\cdot 1} = (n - 2)r_{01}^2/(1 - r_{01}^2)$$

to perform an F test (with 1, $n - 2$ degrees of freedom) for $H : B_1 = 0$. [*See* SIGNIFICANCE, TESTS OF.] If H is accepted, stop the procedure, and look for another entire set of independent variables. If H is rejected, proceed to step 2.

Step 2. Choose the remaining variable having the largest squared partial correlation with X_0 when the linear effect of X_1 is removed. Rename this variable X_2, and use the statistic

$$(n-3)(SS_{12} - SS_1)/SS_{0\cdot 12} = (n-3)r_{02\cdot 1}^2/(1 - r_{02\cdot 1}^2)$$

to perform an F test (with 1, $n-3$ degrees of freedom) for $H : B_2 = 0$.

Proceed in this way until the last step in which a new variable is accepted (that is, one step before the H in question is accepted). Call this step j. Then, the last statistic was

$$(n - j - 1)(SS_{12\ldots j} - SS_{12\ldots j-1})/SS_{0\cdot 12\ldots j} = (n - j - 1)r_{0j\cdot 12\ldots j-1}^2/(1 - r_{0j\cdot 12\ldots j-1}^2).$$

Now, calculate the regression formula for X_0 on $X_1, \cdots, X_j$, and report the value of $r_{0\cdot 12\ldots j} = (SS_{12\ldots j}/SS_0)^{1/2}$ as a measure of its worth. If partial correlations themselves are desired, they can be calculated as

$$r_{0j\cdot 12\ldots j-1} = \pm\, [(SS_{12\ldots j} - SS_{12\ldots j-1})/(SS_0 - SS_{12\ldots j-1})]^{1/2},$$

with the sign chosen to be the same as that of $\hat{B}_j$.

These stepwise procedures are still far from fully understood; for example, there are no clear rules for the choice of significance level in the sequence of F tests, one at each step.

The stepwise scheme is an excellent example of a situation in which sample partial correlations play a role in calculation and interpretation, even though their population counterparts may not exist in the model. There is no guarantee, however, that a set of independent variables produced by a stepwise scheme is the set of that particular size that explains the maximum of the total variation SS_0. Note that interpretations of SS_0 as total variation, $SS_{0\cdot 12\ldots j}$ as residual variation, and $SS_{12\ldots j}$ as variation explained by $X_1, \cdots, X_j$ hold at all stages, including the calculation of $r_{0j\cdot 12\ldots j-1}^2$ as the ratio of additional variation explained by the addition of x_j to the variation left unexplained before the addition of x_j. For more on stepwise regression, and other regression procedures, see Neter and Wasserman (1974).

Correlation with sets of variables

Some problems, notably in biology and anthropology, involve so many variables that means, standard deviations, and correlations pose an initial obstacle to sheer comprehension. To solve such problems, it is necessary to characterize the information given by a set of variables or to characterize the relationship between two sets of variables. Advice of an elementary nature is given by Yule and Kendall (1950, chapter 13) in relation to economy in the number of variables to be considered. It must, of course, be a number large enough to give a reasonable picture of the situation yet small enough to be reasonably comprehensible. Methods of data analysis need not rely on statistical tests. Hills (1969) shows that a combination of graphical methods, common sense, and scientific guesswork can be fruitful in the study of a large set of variables. There are also less elementary techniques for problems involving large sets of variables, techniques that were treated in depth about forty years ago. These include canonical correlation, principal components, and factor analysis. Modern computing has brought these methods within the purview of all research workers; yet they have not been widely studied.

The preeminent contributor to the problems of many variables was Hotelling (1933; 1935; 1936; 1968). His original papers deserve to be read not only for content, but for the pedagogical skill displayed in his style. The 1968 paper is also to be recommended for its historical interest.

Canonical correlation. Consider a situation in which a research worker wishes to study the relationship between two sets of variables, $(Y_1, Y_2, \cdots, Y_k)$ and $(X_1, X_2, \cdots, X_p)$. The purpose of canonical correlation theory (Hotelling 1936) is to replace these sets by new (and smaller) sets, at the same time preserving the correlation structure as much as possible. The method is as follows: two linear combinations, one from each set of variables, are so constructed as to have maximum simple correlation with each other. These linear combinations, denoted by U_1 and V_1, are called the first pair of canonical variables; their correlation, ρ_1, is the first canonical correlation. The process is continued by the construction of further pairs of linear combinations, with the provision that each new canonical variable be uncorrelated with all previous ones. If $k \leqslant p$, the process will terminate with $U_1, U_2, \cdots, U_k; V_1, V_2, \cdots, V_k$; and canonical correlations $\rho_1, \rho_2, \cdots, \rho_k$. If $k = 1$, the resulting single canonical correlation is the multiple correlation for Y_1 on $X_1, X_2, \cdots, X_p$. Since $\rho_1 \geqslant \rho_2 \geqslant \cdots \geqslant \rho_k \geqslant 0$, and since many canonical correlations may be small, it is clear that the canonical pairs worth preserving may be few. A point not to be overlooked is that when these canonical variables are found, they can be used as independent variables in regression schemes, since they are in a sense representative of a large body of information.

The usual model specifies random sampling from a joint normal population in $p + k$ variables; estimation of canonical correlations can be carried out with a sample by a scheme that parallels the one for the construction of $\rho_1, \cdots, \rho_k$. The joint probability distribution for sample canonical correlations is known both in exact form and in approximate form for large n.

Before canonical correlations are estimated, it may be wise to carry out an initial test for possible complete lack of correlation between the two sets of variables. The hypothesis that $\rho_1 = \rho_2 = \cdots = \rho_k = 0$, or equivalently that all correlations between an X_i and a Y_j are zero, may be tested by a procedure of Wilks (see Tate 1966). The hypothesis being tested can be rewritten as $1-(1-\rho_1^2)(1-\rho_2^2)\cdots(1-\rho_k^2) = 0$, a form analogous to that of other, related tests. Of the various tests available for this hypothesis, the best will be the one with highest power against the alternative hypothesis of interest. (See Anderson 1958, sec. 14.2; Hotelling 1936.) Jennrich (1970) gives a test for the equality of two correlation matrices.

Recalling that a multiple correlation between X_0 and a number of other variables cannot be decreased by the inclusion of more variables, one can see that it is more or less reasonable that the first so many canonical correlations cannot be decreased by the addition of more variables to either or both sets. This was, in fact, proved by Chen (1971).

Principal components and factor analysis. One of the central needs arising in the application of correlation is that of holding the variables considered down to a manageable number. Mentioned above in connection with canonical correlation, this need is the guiding principle for principal components analysis. The procedure is to consider a single set of variables $X_1, \cdots, X_p$, to postulate a proper measure of total variation for these variables, and then to generate linear (for simplicity and mathematical tractability) functions of them that successively account for decreasing amounts of variation, and are such that each is orthogonal to those preceding it (that is, has coefficient vector at right angles to those preceding it). These linear functions are called principal components. The hope is that a few such functions will account for most of the total variation. The rationale for this procedure is that greater weight in the first few components will be assigned to the variables that vary more, and hence have more ability to discriminate among experimental subjects. Such variables essentially, then, characterize the variable set in this respect. If a new variable X_{p+1} is added to the system, the whole procedure must be repeated. One would expect, however, that if X_{p+1} is highly correlated with original variables that have low weights in the first few components, then it too would have low weights in those components.

In a specific situation a decision must be made regarding commensurability. If the variables $X_1, \cdots, X_p$ are measured in the same units, and can thus reasonably be compared and added, then tr $\mathbf{\Sigma}$ (the trace of their covariance matrix $\mathbf{\Sigma}$) is adopted as the measure of total variation. As principal components are generated, the roots of $|\mathbf{\Sigma}-\lambda \mathbf{I}| = 0$,

$\lambda_1 \geqslant \lambda_2 \geqslant \cdots \geqslant \lambda_p$, become the portions of variation. If $X_1, \cdots, X_p$ are distributed as multivariate normal, principal components will, moreover, be uncorrelated, as orthogonality and zero correlation are equivalent in this case; furthermore, $\lambda_1, \cdots, \lambda_p$ are the variances of the principal components.

If the variables are not measured in the same units, then standardized variables are used instead, and the procedure calls for the correlation matrix $\boldsymbol{P}$ in place of $\boldsymbol{\Sigma}$. The total variation becomes the number of variables, p, which is the trace of $\boldsymbol{P}$, and the principal components in the standardized case, while still orthogonal, are not uncorrelated, even in the multivariate normal case. Sampling in either case is carried out with a random sample of n observations on each of the p variables, and estimates for all quantities are found from the sample covariance matrix or sample correlation matrix, as the case may be, by a procedure that parallels the above.

Factor analysis, which is of vast importance in psychological testing, among other applications, utilizes a similar idea, except that the number of linear compounds to be considered is prescribed by the model. For a discussion of connections between these two methods, see Kendall (1957); for a discussion of principal components, see Hotelling (1933). [*See also* SCALING, MULTIDIMENSIONAL; FACTOR ANALYSIS AND PRINCIPAL COMPONENTS.]

Special topics in correlation theory

Intraclass correlation. In the discussion of sampling from an (X, Y) population and the consequent use of the sample to estimate ρ_{XY}, there has been no question as to the separate identification of the X and Y for each observation. As an example of a situation in which the identification of X and Y is *not* clear, consider the correlation between the weights of identical twins at, say, age five. Here there is, in effect, only one class, that of pairs of weights of twins. Any establishment of two classes—for example, by considering X the weight of the taller twin and Y the weight of the shorter twin—would be arbitrary and seldom helpful. The population of weight pairs has a correlation coefficient, and this gives the intraclass correlation, the correlation coefficient between the two weights in random order. The method for handling this situation works well with data involving triplets (one is still, however, interested in correlation for weights in the same family) or any number of children. Consider n observations (families) on k-tuplets, with $k \geqslant 2$. The method consists essentially in the averaging of products of deviations over all possible $k(k-1)$ pairs of children. If X_{ij} represents the weight of the jth child in the ith family, then the intraclass correlation, r, is given by $nk(k-1)s^2r = nk^2s_m^2 - nks^2$, with $s^2 = \sum\sum(X_{ij} - \bar{X}_i)^2/nk$, the within-families sample variance, and $s_m^2 = \sum(\bar{X}_i - \bar{X})^2/n$, the between-families sample variance. Thus,

$$r = \frac{k(s_m^2/s^2) - 1}{k - 1}.$$

[*Note that in* IESS *the expression for* s^2 *is given incorrectly in "Multivariate Analysis: II. Correlation (1)."*]

It is clear that $r \geqslant -1/(k-1)$ and that for a single family $(n = 1)$, $r = -1/(k-1)$. Intraclass correlation is closely related to components of variance models in the analysis of variance [*see* LINEAR HYPOTHESES, *article on* ANALYSIS OF VARIANCE].

Attenuation. Observations on random variables are frequently subject to measurement errors or, at any rate, are observable only in combination with other random variables, so that in attempting to observe U, V one in fact observes $X = U + E$, $Y = V + F$. Previous methods lead to information about ρ_{XY}, when what is actually desired is information about ρ_{UV}. If E and F are assumed to be uncorrelated with U, V, and each other, then the relation between ρ_{UV} and ρ_{XY} is given by

$$\rho_{XY} = \frac{\operatorname{cov}(U + E, V + F)}{\{(\sigma_U^2 + \sigma_E^2)(\sigma_V^2 + \sigma_F^2)\}^{1/2}} = \frac{\rho_{UV}}{\{(1 + \sigma_E^2/\sigma_U^2)(1 + \sigma_F^2/\sigma_V^2)\}^{1/2}},$$

which shows that $\rho_{XY} \leqslant \rho_{UV}$, with equality occurring only in the trivial case in which both E and F are constant. The coefficient ρ_{UV} is said to be attenuated by the effect of E and F. Correction for attenuation consists of applying to the above relation known or assumed information relative to ρ_{XY}, (σ_E/σ_U), and (σ_F/σ_V) in order to estimate ρ_{UV}. The case of attenuation with multiple correlation is covered in an informative article by Cochran (1970). Simultaneous correction for attenuation of all pairwise correlations among p variables is also considered in an article by Bock and Petersen (1975).

Path analysis

The literature on causation and its connections with correlation is now extensive. Blalock (1971) has edited an excellent collection of papers on this subject, and on causal models in general. By employing statistical inference, graphical techniques, some experience with the type of data involved, and of course ordinary common sense, one can often give convincing demonstrations of causal relationships. Dempster (1971) gives a modern, rather deep treatment of this subject. Nevertheless, it is not possible to give a proof of a causal relationship by an appeal to data, no matter how analyzed. [*See also* CAUSATION.]

With these considerations in mind one can approach the important technique known as path analysis, and at the same time avoid some of the unfruitful, rather tangential discussions that have arisen in connection with it. Some care needs to be exercised in treating path analysis, since certain assumptions that are needed are rarely stated and since the traditional notation may leave the reader in doubt about certain implicit understandings among practitioners of the methods. [*Path analysis is also discussed in* SURVEY ANALYSIS. *The reader should note that terminology varies. For example, what are here called nonresidual ultimate factors and other measured factors may elsewhere be called exogenous variables and endogenous variables. The basic distinctions between measured and unmeasured variables and between variables whose variation is to be accounted for and those that are used solely to do that accounting are, however, standard.*]

Path analysis was introduced by Sewall Wright (1918) in order to interpret and examine relationships between a set of variables by combining what is known of their intercorrelations with what is assumed about causal connections among them. The remainder of this section deals with Wright's method, his assumptions (and some others), and the central concept of the path diagram. It will be seen that the calculations involved in path analysis are based on fairly elementary statistical considerations but that, as a method for presenting and describing models and interpreting the results, it has much to recommend it.

Assumptions underlying the method. The following assumptions are basic:

(1) We deal with a system of interrelated variables. A variable that is a measurement on a specific characteristic is called a factor; others are known as residual variables.

(2) A factor is usually assumed to be a linear function of other factors, and of a residual variable; a factor not so defined is termed an ultimate factor. Each linear relationship contains one residual variable as an independent variable, and this is the only way in which residual variables appear—a separate residual variable for each linear function.

(3) Ultimate factors are assumed in general to be correlated, while residual variables may be correlated with each other, but are uncorrelated with other independent variables of their own linear relations.

Certain other assumptions should be added, following suggestions of Moran (1961):

(4) All variables involved in the system must have some joint probability distribution with second moments (so that all variances, covariances, and correlations mentioned will have meaning).

(5) The conditional expectations of factors given previous factors are linear.

Method and diagram. Consider an example with three random variables measured for university undergraduates. (All variables are standardized. Standardization means zero expectation and unit standard deviation. In practice, standardization is usually obtained by subtracting the observed average and dividing by the observed standard deviation. Such standardization leads to complications in the distribution theory that will not be discussed here.) The three random variables are X_1 = IQ, X_2 = GPA (grade-point average throughout a student's undergraduate career), and X_3 = GRE (graduate record examination) verbal score. Let the uncorrelated residual variables be R_u and R_v. Assume the causal structure

$$(1)\qquad \begin{aligned} X_3 &= p_{31}X_1 + p_{32}X_2 + p_{3u}R_u,\\ X_2 &= p_{21}X_1 + p_{2v}R_v. \end{aligned}$$

These causal assumptions indicate that GPA contributes directly to a student's GRE score, while IQ contributes both directly and indirectly via GPA. The factors here are, of course, X_1, X_2, and X_3; X_1 is the only ultimate factor. The weights p_{ij} are the path coefficients, which are usually unknown constants to be estimated from the data.

It follows (see Wright 1960) from basic operations on correlations that for any of the assumed causal structural equations, $X_o = \sum_{i=1}^{m} p_{oi}X_i + p_{ou}R_u$, and any factor, say, X_q,

$$(2)\qquad \begin{aligned} \rho_{oq} &= \sum_{i=1}^{m} p_{oi}\rho_{iq} + p_{ou}\rho_{ou},\\ \rho_{oo} &= 1 = \sum_{i=1}^{m} p_{oi}\rho_{oi} + p_{ou}^2. \end{aligned}$$

(The subscripts "o" and "q" are dummies for which one can substitute 1, 2, 3, and so on, up to the number of factors. The upper limit of summation, m, depends on the equation. In the example, when o is 3, $m = 2$; when o is 2, $m = 1$; and when o is 1, there is no summation.) These relations are the cornerstones of Wright's method. The first one gives ρ_{oq} as a linear function of correlations between all variables antecedent to both X_o and X_q. For our example equations (2) become

$$(3)\qquad \begin{aligned} \rho_{31} &= p_{31}\rho_{11} + p_{32}\rho_{21} + p_{3u}\rho_{u1},\\ \rho_{32} &= p_{31}\rho_{12} + p_{32}\rho_{22} + p_{3u}\rho_{u2},\\ \rho_{21} &= p_{21}\rho_{11} + p_{2v}\rho_{v1}. \end{aligned}$$

The main assumption (the third mentioned above) for path analysis states that $\rho_{u1} = \rho_{u2} = \rho_{v1} = 0$. Also, $\rho_{11} = \rho_{22} = 1$. These facts, and two applications of the second relation of equations (2), produce

$$\begin{aligned} \rho_{31} &= p_{31} + p_{32}\rho_{21}, \\ \rho_{32} &= p_{31}\rho_{12} + p_{32}, \\ \rho_{21} &= p_{21}, \\ 1 - p_{31}\rho_{31} - p_{32}\rho_{32} &= p_{3u}^2, \\ 1 - p_{21}\rho_{21} &= p_{2v}^2, \end{aligned} \tag{4}$$

for our example. In practice, the correlations are observed, substituted in the equations, and the equations solved. To illustrate, reasonable hypothetical values of the correlations are $\rho_{12} = .35$, $\rho_{13} = .60$, $\rho_{23} = .40$; with these values the solution of equations (4) is $p_{21} = .35$, $p_{32} = .22$, $p_{31} = .52$, $p_{3u} = .77$, $p_{2v} = .94$, and structure (1) is

$$\begin{aligned} X_3 &= .52X_1 + .22X_2 + .77R_u, \\ X_2 &= .35X_1 + .94R_v. \end{aligned}$$

The information on linear structure can in general be expressed by a convenient diagram in which each variable dependent on others has a straight arrow pointing to it from each of those others, and those variables not of the dependent type are connected pairwise by curved double-pointed arrows in all cases for which correlation is deemed possible. The diagram should also contain the path coefficient for each elementary path (from one variable directly to another) and the correlations between all pairs of ultimate factors and/or residuals. Our example is represented by Figure 1, in which the coefficients obtained above have been inserted. Note that in the subscripts for a path coefficient, the variable corresponding to the tip of the arrow appears first.

It is possible to decompose a correlation between any two variables of a system into contributions along separate paths by making repeated application of the first relation in equations (2). Thus, for structure (1), $\rho_{31} = p_{31} + p_{32}p_{21}$, with p_{31} (of magnitude .52 in the example) as the (direct) contribution over the only elementary path, and $p_{32}p_{21}$ (of magnitude .08 in the example) as the (indirect) contribution over the only compound path. The rule for reading contributions directly from the diagram is that one adds the products of elementary path coefficients for the chain of elementary paths composing (compound) paths of the following type: the path connects the two variables; it can be formed by tracing backward through one or more arrows from one variable and then tracing forward to the other variable without turning back; it passes through no variable twice and cannot pass through more than one pair of variables connected by a double arrow. If we consider a variable, X_i, that is determined by certain other variables, including a residual variable R_u, then from the second relation in equations (2), with the aid of the first relation, we can obtain

$$p_{iu}^2 = 1 - \sum_j p_{ij}^2 - 2\sum_{j>k} p_{ij}p_{ik}\rho_{jk},$$

with j and k ranging over the determining variables other than u. Repetition of this procedure by further substitution for ρ_{jk} leads us eventually to a form in which all ρ_{jk} that appear correspond to ultimate and/or residual variables. In this way we obtain equations for the coefficients associated with the residual paths. It was unnecessary, however, to go this far in order to solve equations (4).

The three-factor example of path analysis discussed in this article constitutes what is known as a recursive system (an ordered collection of variables, for which each is a linear function of those preceding); it amounts to an informative method of presenting and interpreting simultaneous regression procedures. Examples of greater complexity can be found in the work of Wright (1921; 1934) and Duncan (1966, reprinted as Chapter 7 of Blalock 1971).

An additional approach has not been touched on here: path regression analysis. It is, in effect, a method of dealing with problems similiar to those already discussed, but with nonstandardized variables. The basic paper is by Tukey (1954). See also the subsequent commentary by Wright (1960).

It will readily become apparent to a reader of the path analysis literature that the path diagram is an integral part of the whole procedure, and that path analysis is more of an art than a science. In conclusion it should be mentioned that the method need not actually concern itself with causal relationships; what is essential is that linear relationships be postulated [*see* SIMULTANEOUS EQUATION ESTIMATION].

ROBERT F. TATE

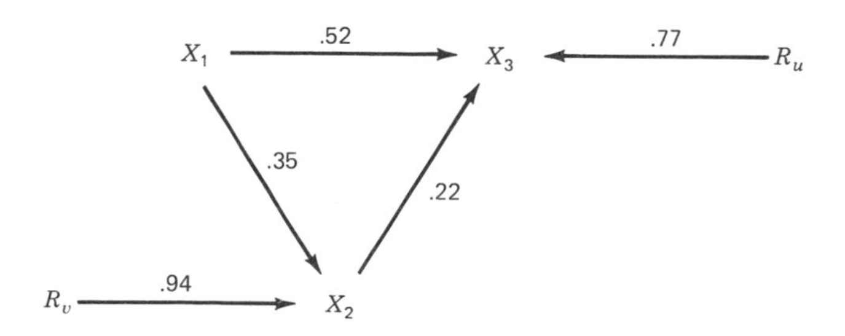

Figure 1 – Path diagram showing structure for X_3 (GRE verbal score), as determined by X_1 (IQ), X_2 (GPA), and two residuals

BIBLIOGRAPHY

In many fields of application there are books describing correlation methods and, just as important, acquainting the reader with special types of data in those fields. Some examples are McNemar 1949 *in psychology,* Croxton, Cowden, & Klein 1939 *in economics and sociology, and* Johnson 1949 *in education. Mathematical treatments on several levels are also available. An excellent elementary work is* Wallis & Roberts 1956, *which requires very little knowledge of mathematics yet presents statistical concepts carefully and fully. Readers with some background in mathematics should find* Anderson & Bancroft 1952, Yule & Kendall 1950, *and* Neter & Wasserman 1974, *at an intermediate level, and* Kendall & Stuart 1958–1966 *and* Morrison 1967, *at a more advanced level, quite useful.*

ADKINS, DOROTHY C. 1968 Thurstone, L. L. Volume 16, pages 22–24 in *International Encyclopedia of the Social Sciences.* Edited by David L. Sills. New York: Macmillan and Free Press.

ANDERSON, RICHARD L.; and BANCROFT, T. A. 1952 *Statistical Theory in Research.* New York: McGraw-Hill.

ANDERSON, T. W. 1958 *An Introduction to Multivariate Statistical Analysis.* New York: Wiley.

BINDER, ARNOLD 1959 Considerations of the Place of Assumptions in Correlational Analysis. *American Psychologist* 14:504–510.

BLALOCK, HUBERT W. JR. (editor) 1971 *Causal Models in the Social Sciences.* Chicago: Aldine-Atherton.

BOCK, R. DARRELL; and PETERSEN, ANNE C. 1975 A Multivariate Correction for Attenuation. *Biometrika* 62:673–678.

CHEN, C. W. 1971 On Some Problems in Canonical Correlation Analysis. *Biometrika* 58:399–400.

CHOI, SUNG C. 1971 Sequential Test for Correlation Coefficients. *Journal of the American Statistical Association* 66:575–576.

COCHRAN, WILLIAM G. 1970 Some Effects of Errors of Measurement on Multiple Correlation. *Journal of the American Statistical Association* 65:22–34.

CROXTON, FREDERICK E.; COWDEN, DUDLEY J.; and KLEIN, SIDNEY (1939) 1967 *Applied General Statistics.* 3d ed. Englewood Cliffs, N.J.: Prentice-Hall. → Klein became a joint author on the third edition.

DAVID, F. N. 1938 *Tables of the Ordinates and Probability Integral of the Distribution of the Correlation Coefficient in Small Samples.* London: University College, Biometrika Office.

DEMPSTER, A. P. 1971 An Overview of Multivariate Data Analysis. *Journal of Multivariate Analysis* 1: 316–346.

DUNCAN, OTIS DUDLEY 1966 Path Analysis: Sociological Examples. *American Journal of Sociology* 72:1–16.

EZEKIEL, MORDECAI; and FOX, KARL A. (1930) 1959 *Methods of Correlation and Regression Analysis: Linear and Curvilinear.* 3d ed. New York: Wiley.

FELLER, WILLIAM (1950) 1968 *An Introduction to Probability Theory and Its Applications.* Volume 1. 3d ed. New York: Wiley.

FISHER, R. A. 1915 Frequency Distribution of the Values of the Correlation Coefficient in Samples From an Indefinitely Large Population. *Biometrika* 10:507–521.

FISHER, R. A. (1925) 1970 *Statistical Methods for Research Workers.* 14th ed. New York: Hafner.

FISHER, R. A. 1928 On a Distribution Yielding the Error Functions of Several Well Known Statistics. Volume 2, pages 805–813 in International Congress of Mathematicians, Second (New Series), Toronto, 1924, *Proceedings.* Univ. of Toronto Press.

GAYEN, A. K. 1951 The Frequency Distribution of the Product–Moment Correlation Coefficient in Random Samples of Any Size Drawn From Non-normal Universes. *Biometrika* 38:219–247.

HANNAN, J. F.; and TATE, ROBERT F. 1965 Estimation of the Parameters for a Multivariate Normal Distribution When One Variable Is Dichotomized. *Biometrika* 52:664–668.

HILLS, MICHAEL 1969 On Looking at Large Correlation Matrices. *Biometrika* 56:249–253.

HOTELLING, HAROLD 1933 Analysis of a Complex of Statistical Variables Into Principal Components. *Journal of Educational Psychology* 24:417–441, 498–520.

HOTELLING, HAROLD 1935 The Most Predictable Criterion. *Journal of Educational Psychology* 26:139–142.

HOTELLING, HAROLD 1936 Relations Between Two Sets of Variates. *Biometrika* 28:321–377.

HOTELLING, HAROLD 1953 New Light on the Correlation Coefficient and Its Transforms. *Journal of the Royal Statistical Society* Series B 15:193–225.

HOTELLING, HAROLD 1968 Multivariate Analysis: III. Correlation (2). Volume 10, pages 545–552 in *International Encyclopedia of the Social Sciences.* Edited by David L. Sills. New York: Macmillan and Free Press.

JENNRICH, ROBERT I. 1970 An Asymptotic χ^2 Test for the Equality of Two Correlation Matrices. *Journal of the American Statistical Association* 65:904–912.

JOHNSON, PALMER O. 1949 *Statistical Methods in Research.* New York: Prentice-Hall.

KEEPING, E. S. 1962 *Introduction to Statistical Inference.* Princeton: Van Nostrand.

KENDALL, MAURICE G. 1957 *A Course in Multivariate Analysis.* London: Griffin; New York: Hafner. → Minor corrections were made in the second printing in 1961.

KENDALL, MAURICE G.; and STUART, ALAN (1958–1966) 1973–1977 *The Advanced Theory of Statistics.* 3 vols. London: Griffin; New York: Macmillan. → Volume 1: *Distribution Theory,* 4th ed., 1977. Volume 2: *Inference and Relationship,* 3d ed., 1973. Volume 3: *Design and Analysis, and Time-series,* 3d ed., 1976. Early editions of Volumes 1 and 2, first published in 1943 and 1946, were written by Kendall alone. Stuart became a joint author on later, renumbered editions in the three-volume set.

KRAEMER, HELENA C. 1973 Improved Approximation to the Non-null Distribution of the Correlation Coefficient. *Journal of the American Statistical Association* 68:1004–1008.

KRUSKAL, WILLIAM H. 1958 Ordinal Measures of Association. *Journal of the American Statistical Association* 53:814–861.

MAYER, LAWRENCE S. 1973 Estimating a Correlation Coefficient When One Variable Is Not Directly Observed. *Journal of the American Statistical Association* 68:420–421.

MCNEMAR, QUINN (1949) 1962 *Psychological Statistics.* 3d ed. New York: Wiley.

MORAN, P. A. P. 1961 Path Coefficients Reconsidered. *Australian Journal of Statistics* 3:87–93.

MORRISON, DONALD F. 1967 *Multivariate Statistical Methods.* New York: McGraw-Hill.

NETER, JOHN; and WASSERMAN, WILLIAM 1974 *Applied Linear Statistical Models: Regression, Analysis of Variance, and Experimental Designs.* Homewood, Ill.: Irwin.

NEYMAN, JERZY (1938) 1952 *Lectures and Conferences on Mathematical Statistics and Probability*. 2d ed., rev. & enl. Washington: U.S. Department of Agriculture. → The first edition was published in mimeographed form.

OLKIN, INGRAM; and PRATT, JOHN W. 1958 Unbiased Estimation of Certain Correlation Coefficients. *Annals of Mathematical Statistics* 29:201–211.

PRINCE, BENJAMIN M.; and TATE, ROBERT F. 1966 The Accuracy of Maximum Lilkelihood Estimates of Correlation for a Biserial Model. *Psychometrika* 31:85–92.

TATE, ROBERT F. 1955*a* The Theory of Correlation Between Two Continuous Variables When One Is Dichotomized. *Biometrika* 42:205–216.

TATE, ROBERT F. 1955*b* Applications of Correlation Models for Biserial Data. *Journal of the American Statistical Association* 50:1078–1095.

TATE, ROBERT F. 1966 Conditional-normal Regression Models. *Journal of the American Statistical Association* 61:477–489.

TATE, ROBERT F. 1968 Multivariate Analysis: II. Correlation (1). Volume 10, pages 537–545 in *International Encyclopedia of the Social Sciences*. Edited by David L. Sills. New York: Macmillan and Free Press.

TUKEY, JOHN W. (1954) 1964 Causation, Regression, and Path Analysis. Pages 35–66 in Oscar Kempthorne et al. (editors), *Statistics and Mathematics in Biology*. Reprint ed. New York: Hafner. → Papers presented at the biostatistics conference held at Iowa State College, June–July 1952.

WALLIS, W. ALLEN; and ROBERTS, HARRY V. 1956 *Statistics: A New Approach*. Glencoe, Ill.: Free Press. → A revised and abridged paperback edition of the first section was published in 1962 by Collier as *The Nature of Statistics*.

WRIGHT, SEWALL 1918 On the Nature of Size Factors. *Genetics* 3:367–374.

WRIGHT, SEWALL 1921 Correlation and Causation. *Journal of Agricultural Research* 20:557–585.

WRIGHT, SEWALL 1934 The Method of Path Coefficients. *Annals of Mathematical Statistics* 5:161–215.

WRIGHT, SEWALL 1960 Path Coefficients and Path Regression: Alternative or Complementary Concepts? *Biometrics* 16:189–202.

YULE, G. UDNY; and KENDALL, MAURICE G. 1950 *An Introduction to the Theory of Statistics*. 14th ed., rev. & enl. London: Griffin; New York: Hafner. → Yule was the sole author of the first edition (1911). Kendall, who became a joint author on the eleventh edition (1937), revised the current edition and added new material to the 1965 printing.

III

CLASSIFICATION AND DISCRIMINATION

This article is numbered IV in IESS.

Classification is the identification of the category or group to which an individual or object belongs on the basis of its observed characteristics. When the characteristics are a number of numerical measurements, the assignment to groups is called by some statisticians *discrimination*, and the combination of measurements used is called a *discriminant function*. The problem of classification arises when the investigator cannot associate the individual directly with a category but must infer the category from the individual's measurements, responses, or other characteristics. In many cases it can be assumed that there are a finite number of populations from which the individual may have come and that each population is described by a statistical distribution of the characteristics of individuals. The individual to be classified is considered as a random observation from one of the populations. The question is, Given an individual with certain measurements, from which population did he arise?

R. A. Fisher (1936), who first developed the linear discriminant function in terms of the analysis of variance, gave as an example the assigning of iris plants to one of two species on the basis of the lengths and widths of the sepals and petals. Indian men have been classified into three castes on the basis of stature, sitting height, and nasal depth and height (Rao 1948). Six measurements on a skull found in England were used to determine whether it belonged to the Bronze Age or the Iron Age (Rao 1952). Scores on a battery of tests in a college entrance examination may be used to classify a prospective student into the population of students with potentialities of completing college successfully or into the population of students lacking such potentialities. (In this example the classification into populations implies the prediction of future performance.) Medical diagnosis may be considered as classification into populations of disease.

The problem of classification was formulated as part of statistical decision theory by Wald (1944) and von Mises (1945). [*See* DECISION THEORY.] There are a number of hypotheses; each hypothesis is that the distribution of the observation is a given one. One of these hypotheses must be accepted and the others rejected. If only two populations are admitted, the problem is the elementary one of testing one hypothesis of a specified distribution against another, although usually in hypothesis testing one of the two hypotheses, the null hypothesis, is singled out for special emphasis [*see* HYPOTHESIS TESTING]. If a priori probabilities of the individual belonging to the populations are known, the Bayesian approach is available [*see* BAYESIAN INFERENCE]. In this article it is assumed throughout that the populations have been determined. (Sometimes the word *classification* is used for the setting up of categories, for example, in taxonomy or typology.) [*See* CLUSTERING; *see also* Tiryakian 1968.]

The characteristics can be numerical measurements (continuous variables), attributes (discrete variables), or both. Here the case of numerical measurements with probability density functions will be treated, but the case of attributes with frequency functions is treated similarly. The theory applies when only one measurement is available ($p = 1$) as well as when several are ($p \geqslant 2$). The classification function based on the approach of statistical decision theory and the Bayesian approach automatically takes into account any correlation between variables. (Karl Pearson's coefficient of racial likeness, introduced in a paper by M. L. Tildesley [1921] and used as a basis of classification, suffered from its neglect of correlation between measurements.)

Classification for two populations

Suppose that an individual with certain measurements $(x_1, \cdots, x_p)$ has been drawn from one of two populations, π_1 and π_2. The properties of these two populations are specified by given probability density functions (or frequency functions), $p_1(x_1, \cdots, x_p)$ and $p_2(x_1, \cdots, x_p)$, respectively. (Each infinite population is an idealization of the population of all possible observations.) The goal is to define a procedure for classifying this individual as coming from π_1 or π_2. The set of measurements $x_1, \cdots, x_p$ can be presented as a point in a p-dimensional space. The space is to be divided into two regions, R_1 and R_2. If the point corresponding to an individual falls in R_1 the individual will be classified as drawn from π_1, and if the point falls in R_2 the individual will be classified as drawn from π_2.

Standards for classification. The two regions are to be selected so that on the average the bad effects of misclassification are minimized. In following a given classification procedure, the statistician can make two kinds of errors: If the individual is actually from π_1 the statistician may classify him as coming from π_2, or if he is from π_2 the statistician may classify him as coming from π_1. As shown in Table 1, the relative undesirability of these two kinds of misclassification are $C(2|1)$, the "cost" of misclassifying an individual from π_1 as coming from π_2, and $C(1|2)$, the cost of misclassifying an individual from π_2 as coming from π_1. These costs may be measured in any consistent units; it is only the ratio of the two costs that is important. While the statistician may not know the costs in each case, he will often have at least a rough idea of them. In practice the costs are often taken as equal.

Table 1 — Costs of correct and incorrect classification

		Population	
		π_1	π_2
Statistician's decision	π_1	0	C(1|2)
	π_2	C(2|1)	0

In the example mentioned earlier of classifying prospective students, one "cost of misclassification" is a measure of the undesirability of starting a student through college when he will not be able to finish and the other is a measure of the undesirability of refusing to admit a student who can complete his course. In the case of medical diagnosis with respect to a specified disease, one cost of misclassification is the serious effect on the patient's health of the disease going undetected and the other cost is the discomfort and waste of treating a healthy person.

If the observation is drawn from π_1, the probability of correct classification, $P(1|1,R)$, is the probability of falling into R_1, and the probability of misclassification, $P(2|1,R) = 1 - P(1|1,R)$, is the probability of falling into R_2. (In each of these expressions R is used to denote the particular classification rule.) For instance,

$$(1) \quad P(1|1,R) = \int_{R_1} p_1(x_1, \cdots, x_p)\, dx_1 \cdots dx_p .$$

The integral in (1) effectively stands for the sum of the probabilities of measurements from π_1 in R_1. Similarly, if the observation is from π_2, the probability of correct classification is $P(2|2,R)$, the integral of $p_2(x_1, \cdots, x_p)$ over R_2, and the probability of misclassification is $P(1|2,R)$. If the observation is drawn from π_1, there is a cost or loss when the observation is incorrectly classified as coming from π_2; the expected loss, or risk, is the product of the cost of a mistake times the probability of making it, $r(1,R) = C(2|1)P(2|1,R)$. Similarly, when the observation is from π_2, the expected loss due to misclassification is $r(2,R) = C(1|2)P(1|2,R)$.

In many cases there are a priori probabilities of drawing an observation from one or the other population, perhaps known from relative abundances. Suppose that the a priori probability of drawing from π_1 is q_1 and from π_2 is q_2. Then the expected loss due to misclassification is the sum of the products of the probability of drawing from each population times the expected loss for that population:

$$(2) \quad \begin{aligned} &q_1 r(1,R) + q_2 r(2,R) \\ &\quad = q_1 C(2|1)P(2|1,R) + q_2 C(1|2)P(1|2,R). \end{aligned}$$

The regions, R_1 and R_2, should be chosen to minimize this expected loss.

If one does not have a priori probabilities of drawing from π_1 and π_2, he cannot write down (2). Then a procedure R must be characterized by the two risks $r(1,R)$ and $r(2,R)$. A procedure R is said to be at least as good as a procedure R^* if $r(1,R) \leqslant r(1,R^*)$ and $r(2,R) \leqslant r(2,R^*)$, and R is better than R^* if at least one inequality is strict. A class of procedures may then be sought so that for every procedure outside the class there is a better one in the class (called a complete class). The smallest such class contains only *admissible* procedures; that is, no procedure out of the class is better than one in the class. As far as the expected costs of misclassification go, the investigator can restrict his choice of a procedure to a complete class and in particular to the class of admissible procedures if it is available.

Usually a complete class consists of more than one procedure. To determine a single procedure as optimum, some statisticians advocate the *minimax principle*. For a given procedure, R, the less desirable case is to have a drawing from the population with the greater risk. A conservative principle to follow is to choose the procedure so as to minimize the maximum risk [*see* DECISION THEORY].

Classification into one of two populations

Known probability distributions. Consider first the case of two populations when a priori probabilities of drawing from π_1 and π_2 are known; then joint probabilities of drawing from a given population and observing a set of variables within given ranges can be defined. The probability that an observation comes from π_1 and that the ith variate is between x_i and $x_i + dx_i$ $(i = 1, \cdots, p)$ is approximately $q_1 p_1(x_1, \cdots, x_p)\, dx_1 \cdots dx_p$. Similarly, the probability of drawing from π_2 and obtaining an observation with the ith variate falling between x_i and $x_i + dx_i$ $(i = 1, \cdots, p)$ is approximately $q_2 p_2(x_1, \cdots, x_p)\, dx_1 \cdots dx_p$. For an actual observation $x_1, \cdots, x_p$, the conditional probability that it comes from π_1 is

$$(3) \qquad \frac{q_1 p_1(x_1, \cdots, x_p)}{q_1 p_1(x_1, \cdots, x_p) + q_2 p_2(x_1, \cdots, x_p)},$$

and the conditional probability that it comes from π_2 is

$$(4) \qquad \frac{q_2 p_2(x_1, \cdots, x_p)}{q_1 p_1(x_1, \cdots, x_p) + q_2 p_2(x_1, \cdots, x_p)}.$$

The conditional expected loss if the observation is classified into π_2 is $C(2|1)$ times (3), and the conditional expected loss if the observation is classified into π_1 is $C(1|2)$ times (4). Minimization of the conditional expected loss is equivalent to the rule

$$(5) \qquad \begin{aligned} R_1 &: C(2|1) q_1 p_1(x_1, \cdots, x_p) > C(1|2) q_2 p_2(x_1, \cdots, x_p), \\ R_2 &: C(2|1) q_1 p_1(x_1, \cdots, x_p) < C(1|2) q_2 p_2(x_1, \cdots, x_p). \end{aligned}$$

(The case of equality in (5) can be neglected if the density functions are such that the probability of equality is zero; if equality in (5) may occur with positive probability, then when such an observation occurs it may be classified as from π_1 with an arbitrary probability and from π_2 with the complementary probability.) Inequalities (5) may also be written

$$(6) \qquad \begin{aligned} R_1 &: \frac{p_1(x_1, \cdots, x_p)}{p_2(x_1, \cdots, x_p)} > k, \\ R_2 &: \frac{p_1(x_1, \cdots, x_p)}{p_2(x_1, \cdots, x_p)} < k, \end{aligned}$$

where $k = [C(1|2)q_2]/[C(2|1)q_1]$. This is the Bayes solution. These results were first obtained in this way by Welch (1939) for the case of equal costs of misclassification.

These inequalities seem intuitively reasonable. If the probability of drawing from π_1 is decreased or if the cost of misclassifying into π_2 is decreased, the inequality in (6) for R_1 is satisfied by fewer points. Since the regions depend on q_1 and q_2, the expected loss does also. The curve A in Figure 1 indicates how the expected loss may vary with q_1 (and $q_2 = 1 - q_1$).

It may very well happen that the statistician errs in assigning his a priori probabilities. (The probabilities might be estimated from a sample of individuals whose populations of origin are known or can be identified by means other than the measurements for classification; for example, disease categories might be identified by subsequent autopsy.) Suppose that the statistician uses $\bar{q}_1$ and $\bar{q}_2$ $(= 1 - \bar{q}_1)$ when q_1 and q_2 $(= 1 - q_1)$ are the actual probabilities of drawing from π_1 and π_2, respec-

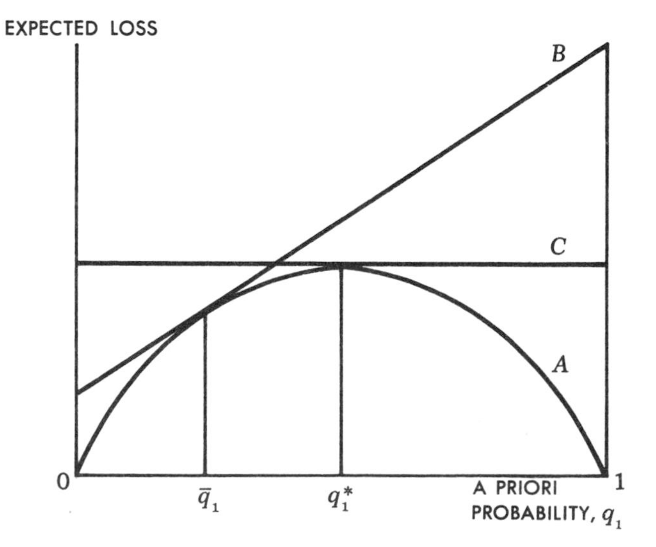

Figure 1 — Expected loss as a function of the a priori probability q_1 for three procedures

tively. Then the actual expected loss is

$$q_1C(2|1)P(2|1,\bar{R}) + (1-q_1)C(1|2)P(1|2,\bar{R}),$$

where $\bar{R}_1$ and $\bar{R}_2$ are based on $\bar{q}_1$ and $\bar{q}_2$. Given the regions $\bar{R}_1$ and $\bar{R}_2$, this is a linear function of q_1 graphed as the line B in Figure 1, a line that touches A at $q_1 = \bar{q}_1$. The line cannot go below A because the best regions are defined by (6). From the graph it is clear that a small error in q_1 is not very important.

When the statistician cannot assign a priori probabilities to the two populations, he uses the fact that the class of Bayes solutions (6) is identical (in most cases) to the class of admissible solutions. A complete class of procedures is given by (6) with k ranging from 0 to ∞. (If the probability that the ratio is equal to k is positive a complete class would have to include procedures that randomize between the two classifications when the value of the ratio is k.)

The minimax procedure is one of the admissible procedures. Since R_2 increases as k increases, and hence $r(1,R)$ increases as k increases, and at the same time $r(2,R)$ decreases, the choice of k giving the minimax solution is the one for which $r(1,R) = r(2,R)$. This is then the average loss, for it is immaterial which population is drawn from. The graph of the risk against a priori probability q_1 is, therefore, a horizontal line (labeled C in Figure 1). Since there is one value of q_1, say q_1^*, such that $k = [C(1|2)(1-q_1)]/[C(2|1)q_1]$, the line C must touch A.

Two known multivariate normal populations. An important example of the general theory is that in which the populations have multivariate normal distributions with the same set of variances and correlations but with different sets of means. [*See* MULTIVARIATE ANALYSIS: OVERVIEW.]

Suppose that $x_1, \cdots, x_p$ have a joint normal distribution with means in π_1 of $Ex_i = \mu_i^{(1)}$ and in π_2 of $Ex_i = \mu_i^{(2)}$. Let the common set of variances and correlations be $\sigma_1^2, \cdots, \sigma_p^2, \rho_{12}, \rho_{13}, \cdots, \rho_{p-1,p}$. It is convenient to write (6) as

$$R_1\colon \ln \frac{p_1(x_1, \cdots, x_p)}{p_2(x_1, \cdots, x_p)} > \ln k,$$

$$R_2\colon \ln \frac{p_1(x_1, \cdots, x_p)}{p_2(x_1, \cdots, x_p)} < \ln k,$$

where "ln" denotes the natural logarithm. In this particular case

$$(7)\quad \ln \frac{p_1(x_1, \cdots, x_p)}{p_2(x_1, \cdots, x_p)} = \sum_{i=1}^{p}\lambda_i x_i - \sum_{i=1}^{p}\lambda_i(\mu_i^{(1)} + \mu_i^{(2)})/2,$$

where $\lambda_1, \cdots, \lambda_p$ form the solution of the linear equations

$$\sum_{j=1}^{p}\sigma_i\sigma_j\rho_{ij}\lambda_j = \mu_i^{(1)} - \mu_i^{(2)}, \qquad i = 1, \cdots, p.$$

The first term on the right side of (7) is the well-known *linear discriminant function* obtained by Fisher (1936) by choosing that linear function for which the difference in expected values for the two populations relative to the standard deviation is a maximum. The second term is a constant consisting of the average discriminant function at the two population means. The regions are given by

$$(8)\quad \begin{aligned} R_1\colon \sum_{i=1}^{p}\lambda_i x_i &> \sum_{i=1}^{p}\lambda_i(\mu_i^{(1)} + \mu_i^{(2)})/2 + \ln k,\\ R_2\colon \sum_{i=1}^{p}\lambda_i x_i &< \sum_{i=1}^{p}\lambda_i(\mu_i^{(1)} + \mu_i^{(2)})/2 + \ln k. \end{aligned}$$

If a priori probabilities are assigned, then k is $[C(1|2)q_2]/[C(2|1)q_1]$. In particular, if $k = 1$ (for example, if $C(1|2) = C(2|1)$ and $q_1 = q_2 = \frac{1}{2}$), $\ln k = 0$, and the procedure is to compare the discriminant function of the observations with the discriminant function of the averages of the respective means.

If a priori probabilities are not known, the same class of procedures (8) is used as the admissible class. Suppose the aim is to find $\ln k = c$, say, so that the expected loss when the observation is from π_1 is equal to the expected loss when the observation is from π_2. The probabilities of misclassification can be computed from the distribution of

$$U = \sum_{i=1}^{p}\lambda_i x_i - \sum_{i=1}^{p}\lambda_i(\mu_i^{(1)} + \mu_i^{(2)})/2$$

when $x_1, \cdots, x_p$ are from π_1 and when $x_1, \cdots, x_p$ are from π_2. Let Δ^2 be the Mahalanobis measure of distance between π_1 and π_2,

$$\Delta^2 = \sum_{i=1}^{p}\lambda_i(\mu_i^{(1)} - \mu_i^{(2)}).$$

The distribution of U is normal with variance Δ^2. If the observation is from π_1 the mean of U is $\frac{1}{2}\Delta^2$; if the observation is from π_2 the mean is $-\frac{1}{2}\Delta^2$.

The probability of misclassification if the observation is from π_1 is

$$\begin{aligned} P(2|1,R) &= Pr\{U \leqslant c|\pi_1\}\\ &= Pr\left\{\frac{U - \frac{1}{2}\Delta^2}{\Delta} \leqslant \frac{c - \frac{1}{2}\Delta^2}{\Delta} \,\middle|\, \pi_1\right\}\\ &= \Phi\left(\frac{c - \frac{1}{2}\Delta^2}{\Delta}\right), \end{aligned}$$

where $\Phi(z)$ is the probability that a normal deviate with mean 0 and variance 1 is less than z. The probability of misclassification if the observation is from π_2 is

$$P(1|2,R) = Pr\{c \leqslant U|\pi_2\}$$
$$= Pr\left\{\frac{c+\frac{1}{2}\Delta^2}{\Delta} \leqslant \frac{U+\frac{1}{2}\Delta^2}{\Delta}\,\Big|\,\pi_2\right\}$$
$$= 1-\Phi\left(\frac{c+\frac{1}{2}\Delta^2}{\Delta}\right) = \Phi\left(\frac{-c-\frac{1}{2}\Delta^2}{\Delta}\right).$$

Figure 2 indicates the two probabilities as the shaded portion in the tails. The aim is to choose c so that

$$r(2,R) = C(1|2)\left[1-\Phi\left(\frac{c+\frac{1}{2}\Delta^2}{\Delta}\right)\right]$$
$$= C(2|1)\Phi\left(\frac{c-\frac{1}{2}\Delta^2}{\Delta}\right) = r(1,R).$$

If the costs of misclassification are equal, $c = 0$ and the common probability of misclassification is $\Phi(\frac{1}{2}\Delta)$. In case the costs of misclassification are unequal, c can be determined to sufficient accuracy by a trial-and-error method with the normal tables.

If the set of variances and correlations in one population is not the same as the set in the other population, the general theory can be applied, but $\ln[p_1(x_1, \cdots, x_p)/p_2(x_1, \cdots, x_p)]$ is a quadratic, not a linear, function of $x_1, \cdots, x_p$. Anderson and Bahadur (1962) treat linear functions for this case.

Classification with estimated parameters. In most applications of the theory the populations are not known but must be inferred from samples, one from each population.

Two multivariate normal populations. Consider now the case in which there are available random samples from two normal populations and in which the aim is to use that information in classifying another observation as coming from one of the two populations. Suppose the sample $(x_{1\gamma}^{(1)}, \cdots, x_{p\gamma}^{(1)})(\gamma = 1, \cdots, N^{(1)})$ is from π_1 and the sample $(x_{1\gamma}^{(2)}, \cdots, x_{p\gamma}^{(2)})(\gamma = 1, \cdots, N^{(2)})$ from π_2. Then $\mu_i^{(1)}$ can be estimated by the mean of the ith variate of the first sample $\bar{x}_i^{(1)}$ and $\mu_i^{(2)}$ by the mean of the second sample $\bar{x}_i^{(2)}$. The usual estimate of $\sigma_i\sigma_j\rho_{ij}$

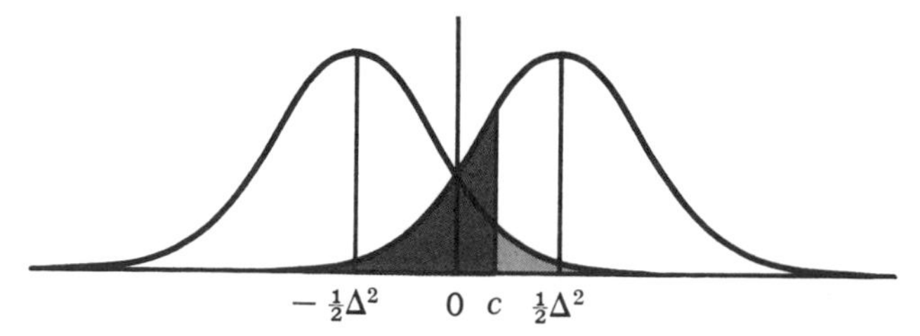

Figure 2 — Probabilities of misclassification as shaded areas under normal densities with means $\pm\frac{1}{2}\Delta^2$ and variance Δ^2

based on the two samples is

$$s_{ij} = \frac{\sum_{\alpha=1}^{N^{(1)}}(x_{i\alpha}^{(1)}-\bar{x}_i^{(1)})(x_{j\alpha}^{(1)}-\bar{x}_j^{(1)}) + \sum_{\alpha=1}^{N^{(2)}}(x_{i\alpha}^{(2)}-\bar{x}_i^{(2)})(x_{j\alpha}^{(2)}-\bar{x}_j^{(2)})}{N^{(1)}+N^{(2)}-2}.$$

These estimates may then be substituted into the definition of U, to obtain a new linear function of $x_1, \cdots, x_p$ depending on these estimates. The classification function is

$$\sum_{i=1}^{p} l_i x_i - \sum_{i=1}^{p} l_i(\bar{x}_i^{(1)}+\bar{x}_i^{(2)})/2,$$

where the coefficients $l_1, \cdots, l_p$ are the solution to

$$\sum_{j=1}^{p} s_{ij}l_j = \bar{x}_i^{(1)} - \bar{x}_i^{(2)}, \qquad i = 1, \cdots, p.$$

Since there are now sampling variations in the estimates of parameters, it is no longer possible to state that this procedure is best in either of the senses used earlier, but it seems to be a reasonable procedure. (A result of Das Gupta [1965] shows that when $N^{(1)} = N^{(2)}$ and the costs of misclassification are equal, the procedure with $c = 0$ is minimax and admissible.)

The exact distributions of the classification statistic based on estimated coefficients cannot be given explicitly; however, the distribution can be indicated as an integral (with respect to three variables). It can be shown that as the sample sizes increase, the distributions of this statistic approach those of the statistic used when the parameters are known. Thus for sufficiently large samples one can proceed exactly as if the parameters were known. Asymptotic expansions of the distributions are available (Bowker & Sitgreaves 1961).

A mnemonic device for the computation of the discriminant function (Fisher 1938) is the introduction of the dummy variate, y, which is equal to a constant (say, 1) when the observation is from π_1 and is equal to another constant (say, 0) when the observation is from π_2. Then (formally) the regression of this dummy variate, y, on the observed variates $x_1, \cdots, x_p$ over the two samples gives a linear function proportional to the discriminant function. In a sense this linear function is a predictor of the dummy variate, y.

In practice the investigator might not be certain that the two populations differ. To test the null hypothesis that $\mu_i^{(1)} = \mu_i^{(2)}$, $i = 1, \cdots, p$, he can use the discriminant function of the difference in sample means

$$\sum_{i=1}^{p} l_i(\bar{x}_i^{(1)} - \bar{x}_i^{(2)}) = 2\left[\sum_{i=1}^{p} l_i\bar{x}_i^{(1)} - \sum_{i=1}^{p} l_i(\bar{x}_i^{(1)}+\bar{x}_i^{(2)})/2\right],$$

which is $(N^{(1)}+N^{(2)})/(N^{(1)}N^{(2)})$ times Hotelling's generalized T^2. The T^2-test may thus be considered as part of discriminant analysis. [*See* MULTIVARIATE ANALYSIS: OVERVIEW.]

Classification for several populations

So far, classification into one of only two groups has been discussed; consider now the problem of classifying an observation into one of several groups. Let $\pi_1, \cdots, \pi_m$ be m populations with density functions $p_1(x_1, \cdots, x_p), \cdots, p_m(x_1, \cdots, x_p)$, respectively. The aim is to divide the space of observations into m mutually exclusive and exhaustive regions $R_1, \cdots, R_m$. If an observation falls into R_g it will be considered to have come from π_g. Let the cost of classifying an observation from π_g as coming from π_h be $C(h|g)$. The probability of this misclassification is

$$P(h|g, R) = \int_{R_h} p_g(x_1, \cdots, x_p)\,dx_1 \cdots dx_p.$$

If the observation is from π_g, the expected loss or risk is

$$r(g, R) = \sum_{\substack{h=1\\h\neq g}}^{m} C(h|g)P(h|g, R).$$

Given a priori probabilities of the populations, $q_1, \cdots, q_m$, the expected loss is

$$\sum_{g=1}^{m} q_g r(g, R) = \sum_{g=1}^{m} q_g\left[\sum_{\substack{h=1\\h\neq g}}^{m} C(h|g)P(h|g, R)\right];$$

$R_1, \cdots, R_m$ are to be chosen to make this a minimum.

Using a priori probabilities for the populations, one can define the conditional probability that an observation comes from a specified population, given the values of observed variates, $x_1, \cdots, x_p$. The conditional probability of the observation coming from π_g is

$$\frac{q_g p_g(x_1, \cdots, x_p)}{q_1 p_1(x_1, \cdots, x_p) + \cdots + q_m p_m(x_1, \cdots, x_p)}.$$

If the observation is classified as from π_h, the expected loss is

$$\sum_{\substack{g=1\\g\neq h}}^{m} \frac{q_g p_g(x)}{\sum_{k=1}^{m} q_k p_k(x)} C(h|g), \tag{9}$$

where x stands for the set $x_1, \cdots, x_p$. The expected loss is minimized at this point if h is chosen to minimize (9). The regions are

$$R_k\colon \sum_{\substack{g=1\\g\neq k}}^{m} q_g p_g(x)C(k|g) < \sum_{\substack{g=1\\g\neq h}}^{m} q_g p_g(x)C(h|g), \qquad h = 1, \cdots, m, \quad h\neq k.$$

If $C(h|g) = 1$ for all g and h $(g\neq h)$, then $x_1, \cdots, x_p$ is in R_k if

$$R_k\colon q_h p_h(x) < q_k p_k(x), \qquad h\neq k. \tag{10}$$

In this case the point $x_1, \cdots, x_p$ is in R_k if k is the index for which $q_g p_g(x)$ is a maximum, that is, π_k is the most probable population, given the observation. If equalities can occur with positive probability so that there is not a unique maximum, then any maximizing population may be chosen without affecting the expected loss.

If a priori probabilities are not given, an unconditional expected loss for a classification procedure cannot be defined. Then one must consider the risks $r(g, R)$ over all values of g and ask for the admissible procedures; the form is (10) when $C(h|g) = 1$ for all g and h $(g\neq h)$. The minimax solution is (10) when $q_1, \cdots, q_m$ are found so that

$$r(1, R) = \cdots = r(m, R). \tag{11}$$

This number is the expected loss. (The theory was first given for the case of equal costs of misclassification by von Mises [1945].)

Several multivariate normal populations. As an example of the theory, consider the case of m multivariate normal populations with the same set of variances and correlations. Let the mean of x_j in π_g be $\mu_j^{(g)}$. Then

$$\ln\frac{p_g(x_1, \cdots, x_p)}{p_h(x_1, \cdots, x_p)} = \sum_{i=1}^{p}\lambda_i^{(g,h)}x_i - \sum_{i=1}^{p}\lambda_i^{(g,h)}(\mu_i^{(g)}+\mu_i^{(h)})/2, \tag{12}$$

where $\lambda_1^{(g,h)}, \cdots, \lambda_p^{(g,h)}$ are the solution to

$$\sum_{j=1}^{p}\sigma_i\sigma_j\rho_{ij}\lambda_j^{(g,h)} = \mu_i^{(g)} - \mu_i^{(h)}, \qquad i = 1, \cdots, p.$$

For the sake of simplicity, assume that the costs of misclassification are equal. If a priori probabilities, $q_1, \cdots, q_m$, are known, the regions are defined by

$$R_g\colon u_{gh}(x_1, \cdots, x_p) > \ln\frac{q_h}{q_g} = \ln q_h - \ln q_g, \qquad h = 1, \cdots, m, \quad h\neq g, \tag{13}$$

where $u_{gh}(x_1, \cdots, x_p)$ is (12). If a priori probabilities are not known, the admissible procedures are given by (13), with $\ln q_h$ replaced by suitable constants c_h. The minimax procedure is (13), for which (11) holds. To determine the constants c_h, use the fact that if the observation is from π_g, $u_{gh}(x_1, \cdots, x_p)$, $h = 1, \cdots, m$ and $h\neq g$, have a joint normal distribution with means

$$(14)\qquad Eu_{gh}(x_1, \cdots, x_p) = \sum_{i=1}^{p} \lambda_i^{(gh)}(\mu_i^{(g)} - \mu_i^{(h)})/2.$$

The variance of $u_{gh}(x_1, \cdots, x_p)$ is twice (14), and the covariance between the variables $u_{gh}(x_1, \cdots, x_p)$ and $u_{gk}(x_1, \cdots, x_p)$ is

$$\sum_{i=1}^{p} \lambda_i^{(g,h)}(\mu_i^{(g)} - \mu_i^{(k)}) = \sum_{i=1}^{p} \lambda_i^{(g,k)}(\mu_i^{(g)} - \mu_i^{(h)}).$$

From these one can determine $P(h|g,R)$ for any set of constants $c_1, \cdots, c_m$.

This procedure divides the space by means of hyperplanes. If $p = 2$ and $m = 3$, the division is by half-lines, as in Figure 3.

If the populations are unknown, the parameters may be estimated from samples, one from each population. If the samples are large enough, the above procedures can be used as if the parameters were known.

An example of classification into three populations has been given in Anderson (1958).

The problem of classification when $(x_1, \cdots, x_p)$ are continuous variables with density functions has been treated here. The same solutions are obtained when the variables are discrete, that is, take on a finite or countable number of values. Then $p_1(x_1, \cdots, x_p)$, $p_2(x_1, \cdots, x_p)$, and so on are the respective probabilities (or frequency functions) of $(x_1, \cdots, x_p)$ in π_1, π_2, and so on. (See Birnbaum & Maxwell 1960; Cochran & Hopkins 1961.) In this case randomized procedures are essential.

For other expositions see Anderson (1951) and Brown (1950). For further examples see Mosteller and Wallace (1964) and Smith (1947).

T. W. ANDERSON

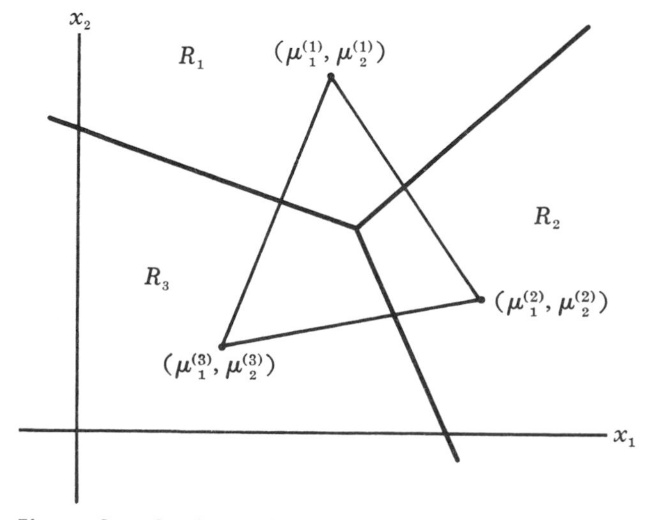

Figure 3 — Regions of classification into one of three multivariate populations

[*Directly related are the entries* CLUSTERING; SCREENING AND SELECTION.]

BIBLIOGRAPHY

ANDERSON, T. W. 1951 Classification by Multivariate Analysis. *Psychometrika* 16:31–50.

ANDERSON, T. W. 1958 *An Introduction to Multivariate Statistical Analysis.* New York: Wiley.

ANDERSON, T. W.; and BAHADUR, R. R. 1962 Classification Into Two Multivariate Normal Distributions With Different Covariance Matrices. *Annals of Mathematical Statistics* 33:420–431.

BIRNBAUM, A.; and MAXWELL, A. E. 1960 Classification Procedures Based on Bayes's Formula. *Applied Statistics* 9:152–169.

BOWKER, ALBERT H.; and SITGREAVES, ROSEDITH 1961 An Asymptotic Expansion for the Distribution Function of the W-classification Statistic. Pages 293–310 in Herbert Solomon (editor), *Studies in Item Analysis and Prediction.* Stanford Univ. Press.

BROWN, GEORGE W. 1950 Basic Principles for Construction and Application of Discriminators. *Journal of Clinical Psychology* 6:58–60.

COCHRAN, WILLIAM G.; and HOPKINS, CARL E. 1961 Some Classification Problems With Multivariate Qualitative Data. *Biometrics* 17:10–32.

DAS GUPTA, S. 1965 Optimum Classification Rules for Classification Into Two Multivariate Normal Populations. *Annals of Mathematical Statistics* 36:1174–1184.

FISHER, R. A. 1936 The Use of Multiple Measurements in Taxonomic Problems. *Annals of Eugenics* 7:179–188.

FISHER, R. A. 1938 The Statistical Utilization of Multiple Measurements. *Annals of Eugenics* 8:376–386.

MOSTELLER, FREDERICK; and WALLACE, DAVID L. 1964 *Inference and Disputed Authorship:* The Federalist. Reading, Mass.: Addison-Wesley.

RAO, C. RADHAKRISHNA 1948 The Utilization of Multiple Measurements in Problems of Biological Classification. *Journal of the Royal Statistical Society* Series B 10:159–193.

RAO, C. RADHAKRISHNA 1952 *Advanced Statistical Methods in Biometric Research.* New York: Wiley.

SMITH, CEDRIC A. B. 1947 Some Examples of Discrimination. *Annals of Eugenics* 13:272–282.

TILDESLEY, M. L. 1921 A First Study of the Burmese Skull. *Biometrika* 13:176–262.

►TIRYAKIAN, EDWARD A. 1968 Typologies. Volume 16, pages 177–186 in *International Encyclopedia of the Social Sciences.* Edited by David L. Sills. New York: Macmillan and Free Press.

VON MISES, RICHARD 1945 On the Classification of Observation Data Into Distinct Groups. *Annals of Mathematical Statistics* 16:68–73.

WALD, ABRAHAM 1944 On a Statistical Problem Arising in the Classification of an Individual Into One of Two Groups. *Annals of Mathematical Statistics* 15:145–162.

WELCH, B. L. 1939 Note on Discriminant Functions. *Biometrika* 31:218–220.

Postscript

Classification is treated in the main article as an area of statistical theory and methodology; however, several of the ideas and procedures of classification occur in other disciplines, often expressed in different terminologies and sometimes in apparent ignorance of the treatment in statistics. In the fields of communications engineering, information processing, and psychophysics, *pattern recognition* and *signal detection* are the terms used to represent the area of classification. The determination of spoken sounds as vowels, consonants, and syllables is the recognition of patterns or the detection of signals. *Taxonomy* is a term used particularly in biology to include the definition of classes or populations (called *classification* by taxonomists) and the assignment of new specimens to these classes (called *identification* by taxonomists and *classification* in this article). The determination of the classes is often termed *clustering*. [*See* CLUSTERING.] The growing interest and activity in this broad area is evidenced by the establishment of the Classification Society, which periodically publishes a review of new literature in the field.

In addition to the applications of classification and discrimination given in the bibliography following the main article, there are many more in the works listed by Cacoullos and Styan (1973) and by Anderson et al. (1972, parts 1 & 2B, sec. I.2).

In the main article, explicit methods are given for multivariate normal distributions, differing with respect to means but not with respect to variances and correlations; if each of a set of populations is approximately normal and the differences in means are large relative to the differences in standard deviations and correlations, the procedures and their evaluations as described are roughly valid. When the model does not hold approximately (for example, if discrete variables occur or if the differences in standard deviations are large relative to the differences in means), the procedures should be modified appropriately. In the papers listed in Cacoullos and Styan (1973) and Anderson et al. (1972), many other methods are given and other aspects investigated.

T. W. ANDERSON

ADDITIONAL BIBLIOGRAPHY

ANDERSON, T. W.; DAS GUPTA, SOMESH; and STYAN, GEORGE P. H. 1972 *A Bibliography of Multivariate Statistical Analysis*. Edinburgh: Oliver & Boyd; New York: Halsted. → Reprinted by Krieger in 1977.

CACOULLOS, T.; and STYAN, GEORGE P. H. 1973 A Bibliography of Discriminant Analysis. Pages 375–434 in NATO Advanced Study Institute on Discriminant Analysis and Applications, Athens, 1972, *Discriminant Analysis and Applications*. Edited by T. Cacoullos. New York: Academic Press. → A comprehensive listing of books, journals, and papers dealing with classification and discrimination.

NATO ADVANCED STUDY INSTITUTE ON DISCRIMINANT ANALYSIS AND APPLICATIONS, ATHENS, 1972 1973 *Discriminant Analysis and Applications*. Edited by T. Cacoullos. New York: Academic Press.

N

NEUMANN, JOHN VON

See VON NEUMANN, JOHN.

NONCENTRAL DISTRIBUTIONS

See DISTRIBUTIONS, STATISTICAL, *article on* SPECIAL CONTINUOUS DISTRIBUTIONS.

NONNORMALITY

See ERRORS, *article on* EFFECTS OF ERRORS IN STATISTICAL ASSUMPTIONS.

NONPARAMETRIC STATISTICS

The articles under this heading are to be regarded not as a handbook of nonparametric procedures but as an introduction that stresses principles through discussion of some important examples. The first article deals mainly with nonparametric inferences about measures of location for one and two populations. The article on ranking methods presents further examples of nonparametric methds, all involving ranking of the observations. Other related topics are treated in the articles on order statistics and on runs.

I. THE FIELD	*I. Richard Savage*
II. ORDER STATISTICS	*Bernard G. Greenberg*
III. RUNS	*Peter G. Moore*
IV. RANKING METHODS	*Herbert A. David*

I
THE FIELD

Nonparametric, or distribution-free, statistical methods are based on explicitly weaker assumptions than such classical parametric procedures as Student's *t*-test, analysis of variance, and standard techniques connected with the correlation coefficient. Examples of nonparametric methods are the sign test, the Wilcoxon tests, and certain confidence interval methods for population quantiles (median, quartile, etc.).

The basic distinction is that parametric procedures make demanding, relatively narrow assumptions about the probability distributions from which empirical observations arise; the most common such distributional assumption is that of normality. Nonparametric methods, in contrast, do not make specific distributional assumptions of this kind.

The dividing line is, of course, not a sharp one; some parametric methods are so robust against errors in distributional assumptions as to be almost nonparametric in practice, while most nonparametric methods are only distribution free for some of their characteristics. For example, most nonparametric hypothesis tests are distribution free under the null hypothesis but distribution dependent with regard to power [*see* ERRORS, *article on* EFFECTS OF ERRORS IN STATISTICAL ASSUMPTIONS].

The underlying motivation for the use of nonparametric methods is the reluctance to make the traditional parametric assumptions, in particular (but not only) the assumption of normality. The relaxation of assumptions is paid for by decreased sharpness of inference when the narrower parametric assumptions are in fact true and by less flexibility in mode of inference. An advantage of some, but by no means all, nonparametric methods is that they are easy and quick to apply; some authors call them "rough and ready."

Nonparametric procedures are, of course, concerned with parameters, that is, numerical characteristics of distributions, but usually with parameters that have desirable invariance properties under modifications in scale of measurement. For example, the median is an important parameter for nonparametric analysis; if X is a random variable with median m and if f is a strictly increasing function, then the median of $f(X)$ is $f(m)$.

History. Many of the first statistical techniques were applied only to massive quantities of data, so massive that the computed statistics had effectively known normal distributions, usually because the statistics were based on sample moments. Later, the need for methods appropriate to small samples became clear, and the exact distributions of many statistics were derived for specific assumed forms (usually normal) of the underlying distribution. This work, begun by William S. Gosset ("Student") and R. A. Fisher, led to the standard statistical tests: for instance, t, chi-square, and F.

Somewhat later, although there was substantial overlap, procedures were developed having exact properties without the need for special distributional assumptions. Procedures were also developed that simplified numerical analysis of data; much of the motivation for all this work was a desire for procedures that could be applied to data given in the form of ranks or comparisons. These developments often arose from the work of social scientists dealing with data that clearly did not come from distributions of standard form; for example, rank sum tests were proposed by G. Deuchler and Leon Festinger (psychologists), and one form of analysis of variance by ranks was proposed by Milton Friedman (an economist).

When dealing with data that clearly do not arise from a standard distribution, an alternative to the use of nonparametric methods is the use of transformations, that is, application to each observation of some common function, chosen so as to make the resulting transformed data appear nearly normal or nearly following some other standard distributional form [*see* STATISTICAL ANALYSIS, SPECIAL PROBLEMS OF, *article on* TRANSFORMATIONS OF DATA].

In the first section below, the nonparametric analysis of experiments generating a single sample is examined. Artificial data are used in order to concentrate attention on the formal aspects of the statistical analysis. The second section is a less detailed discussion of two sample problems, and the last mentions some additional important nonparametric problems.

One-sample problems

A single sample of seven numerical observations will be used to illustrate the application of nonparametric procedures. Suppose the observations are (A) -1.96; (B) $-.77$; (C) $-.59$; (D) $+1.21$; (E) $+.75$; (F) $+4.79$; and (G) $+6.95$. These data might be the scores achieved by seven people on a test, the per cent change in prices of seven stocks between the two points of time, the ratings of seven communities on some sociological scale, and so forth.

There are several kinds of statistical inferences for each of several aspects of the parent population. Also, for each kind of inference about each aspect of the population there exist several nonparametric techniques. The selection of a kind of inference about an aspect by a particular technique is guided by the experimenter's interests; the available resources for analysis; the relative costs; the basic probability structure of the data; the criteria of optimality, such as power of tests or length of confidence intervals; and the sensitivity (robustness) of the technique when the underlying assumptions are not satisfied perfectly.

Point estimation of a mean. Assume the data represent a random sample. As a first problem, the mean (a location parameter) of the population is the aspect of interest; the form of inference desired is a point estimate, and the technique of choice is the sample mean,

$$\bar{x} = \frac{(-1.96) + (-.77) + \cdots + (+6.95)}{7} = 1.483.$$

The following are some justifications for the use of $\bar{x}$: (a) If the sample is from a normal population, then $\bar{x}$ is the maximum likelihood estimator of the population mean, θ. (b) If the sample is from a population with finite mean and finite variance, then $\bar{x}$ is the Gauss–Markov estimator; that is, among linear functions of the observations that have the mean as expected value, $\bar{x}$ has smallest variance. (c) The least squares value is $\bar{x}$; that is, $\bar{x}$ is the value of y that minimizes $(-1.96 - y)^2 + \cdots + (6.95 - y)^2$. (This last holds whether or not the sample is random.)

Result a is parametric since a functional form is selected for the population; in contrast, b is nonparametric. The least squares result, c, is not dependent on probability considerations.

Point estimation of a median. The sample median, $+.75$, is sometimes used as a point estimator of the population mean. (When the sampled population is symmetric, the population mean and median are equal.)

For some populations (two-tailed exponential, for example) the sample median is the maximum likelihood estimator of the population mean. The sample median minimizes the mean absolute deviation, $|-1.96-y| + \cdots + |6.95-y|$.

Parametric confidence intervals for a mean. If it can be assumed that the data are from a normal population (mean θ and variance σ^2 unknown), to form the conventional two-sided confidence interval with confidence level $1-\alpha$ one first computes $\bar{x}$ and s^2 (the sample variance, with divisor $n-1$, where n is sample size) and then forms the interval $\bar{x} \pm t_{n-1,\alpha/2}s/n^{\frac{1}{2}}$, where $t_{n-1,\alpha/2}$ is the upper $\alpha/2$ quantile of the t-distribution with $n-1$ degrees of freedom. For the present data, $\bar{x} = 1.483$, $s^2 = 10.44$, and the conventional 95 per cent confidence interval is $(-1.508, 4.474)$.

Nonparametric confidence intervals for a median. In a sample of size n let $x_{(1)} < x_{(2)} < \cdots < x_{(n)}$ be the ordered observations, the order statistics. (It will be assumed throughout that no ties occur between observations, that is, that the observations come from a continuous population.) If Med is the population median, then the probability of an observation's being less (greater) than Med is $\frac{1}{2}$. The probability that all the observations are less (greater) than Med is 2^{-n}. The event $x_{(i)} < \text{Med} < x_{(n-i+1)}$ occurs when at least i of the observations are less than Med and at least i of the observations are greater than Med. Hence, $x_{(i)} < \text{Med} < x_{(n-i+1)}$ has the same probability as does obtaining at least i heads and at least i tails in n tosses of a fair coin. From the binomial distribution, one obtains

$$Pr(x_{(i)} < \text{Med} < x_{(n-i+1)}) = \sum_{x=i}^{n-i} \binom{n}{x} 2^{-n} = 1 - \sum_{x=0}^{i-1} \binom{n}{x} 2^{-n+1}.$$

In words, $(x_{(i)}, x_{(n-i+1)})$ is a confidence interval for Med with confidence level given by the above formula; for example, $(x_{(1)}, x_{(n)})$ is a confidence interval for Med with confidence level $1-2^{-n+1}$. Thus, for the present data with $i=1$, $(-1.96, 6.95)$ is a confidence interval with confidence level $= 63/64 = .984$ and $(-.77, 4.79)$ has confidence level $= 7/8 = .875$.

Confidence intervals for the mean of a normal population are available at any desired confidence level. For the above nonparametric procedure, particularly for very small sample sizes, there are sharp restrictions on the available confidence levels. The restriction occurs because nonparametric procedures are often based on random variables with discrete distributions even though the underlying random variables have continuous distributions.

Comparisons of confidence intervals. The confidence statement appears more definitive the shorter the confidence interval. One relevant criterion is the expected value of the squared length of the confidence interval. The value of expected squared length for the confidence interval based on the t-distribution is $4n^{-1}\sigma^2 t^2_{n-1,\alpha/2}$. If the normality assumption is dropped, then the t-distribution confidence interval will, in general, no longer have the desired confidence level, although its expected squared length remains the same. (For some specific nonnormal assumptions—uniform distributions, for example—confidence intervals at the desired confidence level may be obtained with much smaller expected squared lengths.) The expected squared length of the order statistic confidence interval, assuming only symmetry of the distribution, is $2[Ex^2_{(i)} - E(x_{(i)}x_{(n-i+1)})]$. If normality actually holds, the confidence interval based on the t-distribution has smaller expected squared length: with $n=7$ the ratio of expected squared lengths is .796 for confidence level .984 and .691 for confidence level .875. A general approximate result is that to obtain equal expected squared lengths one needs a sample size for the order statistics interval of about $\frac{1}{2}\pi \cong 1.57$ times the sample size for the t-distribution interval.

Tests of hypotheses about a mean. Aside from the estimation of a parameter, one often wishes to test hypotheses about it. If the data are from a normal population, then the best procedure, uniformly most powerful unbiased of similar tests, is based on $t = (\bar{x} - \theta_0)n^{\frac{1}{2}}/s$, where θ_0 is the hypothetical value for the population mean. Consider testing the null hypothesis $\theta_0 = -.5$ on the basis of the specific sample of seven; the resulting value of the t-statistic is 1.21, which is statistically significant at the .1575 level (two-sided). For these data and this test statistic the null hypothesis would be rejected if the required significance level exceeds .1575, but if the significance level is smaller, the null hypothesis would not be rejected. The power of this test depends on the quantity $(\theta - \theta_0)^2/\sigma^2$, where θ is the population mean [*see* HYPOTHESIS TESTING].

Sign test for a median. Perhaps the simplest nonparametric test comparable to the t-test is the sign test. It is easily applied, which makes it useful for the preliminary analysis of data and for the analysis of data of passing interest. The null hypothesis for the sign test specifies the population median, Med_0. In the example, the null hypothesis is that the median is $-.5$. The test statistic is the number of observations greater than Med_0. In the example, the value of the statistic is 4. (The term "sign test" arises because one counts the number of differences, observation minus Med_0, that are positive.) If one rejects the null hypothesis whenever the number of positive signs is less than i or greater than $n-i$, $i<n/2$, the significance level will be

$$\sum_{j=0}^{i-1} \binom{n}{j} 2^{-n+1}.$$

In this example the possible significance levels are 0 ($i=0$), $1/64$ ($i=1$), $1/8$ ($i=2$), $29/64$ ($i=3$), 1 ($i=4$). For these data, the sample significance level is 1. The sign test leads to exact levels of significance when the observations are independent and each has probability $\frac{1}{2}$ of exceeding Med_0. It is not necessary to have each observation from the same population. One needs only to compare each observation with the hypothetical median. The test can be applied even when quantitative measurements cannot be made, as long as the signs of the comparisons are available.

The power of the sign test depends on p, the probability that an observation exceeds Med_0. When $p \neq \frac{1}{2}$, the power of the sign test approaches 1 as the sample size increases. Tests having this property are said to be consistent. All reasonable tests, such as those discussed below, will have such consistency properties. When $p \neq \frac{1}{2}$, the power of the sign test is always greater than the significance level. Tests having this property are said to be unbiased.

The sample significance level can be found from a binomial table with $p=\frac{1}{2}$, and the power from a general binomial table. If z is the number of positive results in an experiment with moderately large n, one can approximate the significance level by computing $t'=(2z-n)/n^{\frac{1}{2}}$ and referring t' to a normal table. A somewhat better approximation can be obtained by replacing z with $z+1$ in computing t'; this is called a continuity correction and often arises when a discrete distribution is approximated by a continuous distribution.

The sign test and the nonparametric confidence intervals described above are related in the same manner as the t-test and confidence intervals based on the t-distribution.

The sign procedure is easily put into a sequential form; experiments in this form often save money and time [*see* SEQUENTIAL ANALYSIS].

Conventionalized data—signed ranks. One can think of the sign statistic as the result of replacing the observations by certain conventional numbers, 1 for a positive difference and 0 otherwise, and then proceeding with the analysis of the modified data. A more interesting example is to replace the observations by their signed ranks, that is, to replace the observation whose difference from Med_0 is smallest in absolute value by $+1$ or -1, as that difference is positive or negative, etc. Thus for the present data with $\text{Med}_0=-.5$, the signed ranks are -4, -2, -1, $+5$, $+3$, $+6$, and $+7$.

Signed-rank test or Wilcoxon test for a median. The one-sample signed-rank Wilcoxon statistic, W, is the sum of the positive signed ranks. In the example, $W=21$. The exact null distribution of W can be found when the observations are mutually independent and it is equally likely that the jth smallest absolute value has a negative or positive sign, for example, when the observations come from populations that are symmetrical about the median specified by the null hypothesis.

When the null hypothesis is true, the probability that the Wilcoxon statistic is exactly equal to w is found by counting the number of possible samples that yield w as the value of the statistic and then dividing the count by 2^n. When $n=7$, the largest value for W, 28, is yielded by the one sample having all positive ranks. In that case $Pr(W=28)=2^{-7}$. The distribution of W under the null hypothesis is symmetric around the quantity $n(n+1)/4$, so that $Pr(W=w)=Pr(W=\frac{1}{2}n(n+1)-w)$. Thus for $n=7$, $Pr(W=21)=Pr(W=7)$. A value of 7 for W can be obtained when the positive ranks are (7), (1,6), (2,5), (3,4), (1,2,4)—each parenthesis represents a sample. Thus $Pr(W=21)=Pr(W=7)=5/2^7$. By enumeration, the probability that $W \geqslant 21$ or $W \leqslant 7$ is $36/128=.281$. The present data are therefore statistically significant at the .281 level when using the Wilcoxon test (two-sided). The sign test has about $n/2$ possible significance levels, and the Wilcoxon test has about $n(n+1)/4$ possible significance levels. For small samples, tables of the exact null distribution of W are available. Under the null hypothesis the mean of W is given by $n(n+1)/4$ and the variance of W is given by $n(n+1)(2n+1)/24$. The standardized variable that is derived from W by setting $t'=(W-EW)/(\text{var}\, W)^{\frac{1}{2}}$ has approximately a normal distribution. Most statements about the power of the W test are based on very small samples, Monte Carlo sampling results, or large-sample theory. For small samples, W is easy to compute. As the sample size increases, ranking the data becomes more difficult than computing the t-statistic.

W can be computed by making certain comparisons, without detailed measurements. Denote by m' (n') the number of negative (positive) observations. Let $u_1, \cdots, u_{n'}$ be the values of the positive and $v_1, \cdots, v_{m'}$ be the absolute values of the negative observations. Let $\Delta_{ij}=\Delta(u_i, v_j)$ be equal to 1 if $u_i > v_j$ and equal to 0 otherwise. Define

$$S=\textstyle\sum_i \sum_j \Delta_{ij}.$$

Then

$$W = S + n'(n' + 1)/2.$$

If $m' > 0$ and $n' > 0$, the statistic $S/m'n'$ is an unbiased estimator of the probability that a positive observation will be larger than the absolute value of a negative observation. Hence S is of independent interest. The relationship between S and W involves the random quantity n', so that inferences drawn from S and W need not be the same.

○ *Wilcoxon statistic confidence intervals.* Given a population distribution symmetric about the median, the Wilcoxon test generates confidence intervals for the population median, Med. In the example, the significance level $^{10}/_{128} = .0781$ corresponds to rejecting the null hypothesis when $W \geqslant 25$ or $W \leqslant 3$. Thus the corresponding confidence interval at confidence level $1 - {}^{10}/_{128} = .9219$ consists of values of Med that will make $4 \leqslant W \leqslant 24$. An examination of the original data (the signed ranks are not now sufficient) yields the interval $(-.77, 4.08)$. An examination of some trial values of Med will help in understanding this result. Thus if Med = 4.2, one sees that F has rank 1, G has rank 2, no other observation has positive rank, and $W = 3$, which means this null hypothesis would be rejected. If Med = 4, then F has rank 1, G has rank 3, no other observation has positive rank, and $W = 4$, which means this null hypothesis would be accepted.

Permutation test for a location parameter. The final test considered here for a location parameter (like the median or mean) is the so-called permutation test on the original data. Under the null hypothesis, the observations are mutually independent and come from populations symmetric about the median Med_0. This includes both the cases where the signed ranks are the basic observations and the cases where the signs are the basic observations (scoring +1 for a positive observation and −1 for a negative observation). Given the absolute values of the observations minus Med_0, under the null hypothesis there are 2^n equally likely possible assignments of signs to the absolute values. The nonparametric statistic to be considered is the total, T, of the positive deviations from Med_0. One works with the conditional distribution of T, given the absolute values of the observed differences. Using $\text{Med}_0 = -.5$, for the present data $T = 15.70$. There are 13 configurations of the signs that will give a value of T at least this large and another 13 configurations that will give a value of $17.52 - 15.70 = 1.82$ or smaller, the latter being the lower tail of the symmetrical distribution of T. Thus the significance level of the permutation test on the present data is $^{26}/_{128} = .2031$. With the test, each multiple of $^{1}/_{64}$ is a possible significance level. The computations for this procedure are prohibitive in cost if the sample size is not very small. A partial solution is to use some of the assignments of signs (say a random sample) to estimate the conditional distribution of T. The ordinary t-statistic and T are monotone functions of each other; thus a significance test based on t is equivalent to one based on T. The distribution of T for the permutation test is approximately normal for large values of n. Actually, the *randomization* that occurs in the design of experiments yields the nonparametric structure discussed here.

The power of the permutation test is approximately the same as that of the t-test when the data are from a normal population.

○ Finally, one can construct confidence intervals for Med_0 using the permutation procedure. The confidence interval with level $^{122}/_{128} = .9531$ is $(-1.365, 5.87)$ for the above example.

Tests for a location parameter compared. All the tests are consistent; for a particular alternative and large sample size, each will have power near 1. To compare the large sample power of the tests, a sequence of alternatives approaching the hypothesized null value, Med_0, is usually introduced. Let Med_N be the alternative at the Nth step of this sequence, and assume that $\text{Med}_N - \text{Med}_0 = c/N^{\frac{1}{2}}$. where c is a constant. As N increases, the alternative hypothesis approaches the null hypothesis and larger sample sizes will be required to obtain a desired power for a specified level of significance. The efficiency of test II compared to test I is defined as the ratio $E_{\text{I,II}}(N) = n_{\text{I}}(N)/n_{\text{II}}(N)$, where n_{I} and n_{II} are the sample sizes yielding the desired power for test I and test II. For large N, $E_{\text{I,II}}(N)$ is usually almost independent of test size, power, and the constant c; that is, as N grows, $E_{\text{I,II}}(N)$ usually has a limit, $E_{\text{I,II}}$, called the Pitman efficiency of test II compared to test I. The Pitman efficiency will depend on the tests under consideration and the family of possible distributions. Consider the two experiments, I and II, to test the hypothesized null value, Med_0. They have sample sizes $n_{\text{I}}(N)$ and $n_{\text{II}}(N)$ and costs per observation γ_{I} and γ_{II}. Then the cost of experiment I divided by the cost of experiment II is $\gamma_{\text{I}} n_{\text{I}}(N)/\gamma_{\text{II}} n_{\text{II}}(N) = (\gamma_{\text{I}}/\gamma_{\text{II}})E_{\text{I,II}}$. When this ratio is > 1 (< 1), experiment II costs less (more) than experiment I and the two experiments have the same power functions. The γ associated with the sign test can be much smaller than the γ's associated with the Wilcoxon and permutation tests.

For normal alternatives, where the parameter is the mean, one has

$$E_{t\text{-test, permutation}} = 1,$$
$$E_{t\text{-test, Wilcoxon test}} = 3/\pi = .955,$$
$$E_{t\text{-test, sign test}} = 2/\pi = .637.$$

○ **Tolerance intervals.** Tolerance intervals resemble confidence intervals, but the former permit an interpretation in terms of hypothetical future observations. One nonparametric tolerance interval method, based on the order statistics of a random sample of size n, provides the assertion that at least 90 per cent (say) of the underlying population lies in the interval from $x_{(i)}$ to $x_{(n-i+1)}$. This assertion is made in the confidence sense; that is, its probability of being correct (before the observations are taken) is, say, .95. The relationship between i, the confidence level (here .95), and the coverage level (here .90) is in terms of the incomplete beta function. Special tables and graphs are available (Walsh 1962–1968).

Another form of nonparametric tolerance interval interprets the interval from $x_{(i)}$ to $x_{(n-i+1)}$ as giving a random interval whose probability under the underlying distribution is approximately given by $(n-2i+1)/(n+1)$. In this context the word "approximately" is used to mean the following: the probability measure of the random interval $[x_{(i)}, x_{(n-i+1)}]$ is itself random; its expectation is $(n-2i+1)/(n+1)$. The probability (before any of the observations are made) that a future observation will lie in a tolerance interval of this form is $(n-2i+1)/(n+1)$.

For the data set, with $i = 1$, the interval $[-1.96, 6.95]$ is obtained. This interval is expected (in the above sense) to contain $[(7-2+1)/(7+1)] = 75$ per cent of the underlying population.

Other one-sample tests. Goodness of fit procedures are used to test the hypothesis that the data came from a particular population or to form confidence belts for the whole distribution function. [*These are discussed in* GOODNESS OF FIT.]

Tests of the randomness of a sample (that is, of the assumption that the observations are independent and drawn from the same distribution) can be carried out by counting the number of runs above and below the sample median. [*Such techniques are discussed in* NONPARAMETRIC STATISTICS, *article on* RUNS.]

The statistic based on the number of runs, the Wald–Wolfowitz statistic, was originally proposed as a test of goodness of fit. It is not, however, recommended for this purpose, because other, more powerful tests are available.

Tied observations. The discussion so far has presumed that no pairs of observations have the same value and that no observation equals Med_0. In practice such ties will occur. If ties are not extensive their consequence will be negligible. If exact results are required, the analysis can in principle be carried out conditionally on the observed pattern of ties.

Two-sample problems

Experiments involving two samples allow comparisons between the sampled populations without the necessity of using matched pairs of observations. Comparisons arise naturally when considering the relative advantages of two experimental conditions or two treatments, for example, a treatment and a control, or when absolute standards are unknown or are not available.

Most of the one-sample procedures have two-sample analogues (the exceptions are tests of randomness and tolerance intervals). With two samples, central interest has been focused on the difference between location parameters of the two populations. To be specific, let $x_1, \cdots, x_m$ be the observed values in the first sample and $y_1, \cdots, y_n$ be the observed values in the second sample. Let M_x and M_y be the corresponding location parameters with $\Delta = M_y - M_x$.

Estimation of difference of two means. As a point estimator of Δ when M_x and M_y represent the population means, one often uses the differences in the sample averages, $\bar{y} - \bar{x}$. This is the maximum likelihood estimator of Δ when all of the observations are independent and come from normal populations; it is also the Gauss–Markov estimator whenever the populations have finite variances, and it is the least squares estimator. With the normal assumption, and also assuming equal variances, confidence intervals for Δ can be obtained, utilizing the t-distribution. As in the normal case, nonparametric confidence intervals for Δ will not have the prescribed confidence level unless the two populations differ in location parameter only.

Brown–Mood procedures—two medians. The analogue of the confidence procedure based on signs, the Brown–Mood procedure, is constructed in the following manner: Let $w_1, \cdots, w_{m+n}$ be all of the observations arranged in increasing order, that is, w_1 is the smallest observation in both samples; w_2 is the second smallest observation in both samples; etc. Denote by w^* the median of the combined sample. (It will be assumed that $m+n$ is odd.) Let m^* be the number of w's greater than w^*, from the x-population. When the two populations are the same, m^* has a hypergeometric distribution and a nonparametric test of the hypothesis that the two populations are the same, specifically that $\Delta = 0$, is based on the distribution of m^*. To obtain confidence intervals, replace the x-sample with $x'_i = x_i + \Delta$ $(i = 1, \cdots, m)$; form w', the analogue of the w sequence; compute the median of the w' sequence and call it w'^*; compute the number of observations from the x' sequence above w'^* and call it m'^*; see if one would accept the null hypothesis of no difference between the x' and y populations; if one accepts the null hypothesis, then Δ is in the confidence interval, and if one rejects the null hypothesis, then Δ is not in the confidence interval.

Wilcoxon two-sample procedure. The analogue to the one-sample Wilcoxon procedure is to assign ranks (a set of conventional numbers) to the w's, that is, w_1 is given rank 1, w_2 is given rank 2, etc. The test statistic is the sum of the ranks of those w's from the x-population. When the two populations are identical, the distribution of this test statistic will not depend on the underlying common distribution. The test based on this statistic is called the Wilcoxon test or Mann–Whitney test.

The Wilcoxon test can be used if it is possible to compare each observation from the x-population with each observation from the y-population. The Mann–Whitney version of the Wilcoxon statistic (a linear function of the Wilcoxon statistic) is the number of times an observation from the y-population exceeds an observation from the x-population. When this number is divided by mn, it becomes an unbiased estimator of $Pr(Y > X)$, the probability that a randomly selected y will be larger than a randomly selected x. This parameter has many interpretations and uses, for instance, if stresses and strains are brought together by random selection, it is the probability that the stress will exceed the strain (the system will function).

Permutation procedure. The permutation procedure is based on the conditional distribution of the sum of the observations in the x-sample given the w sequence, when each selection of m of the values from the w sequence is considered equally likely. There are $\binom{m+n}{m}$ such possible selections. The sum of the observations in the x-sample is an increasing function of the usual t-statistic. The importance of the permutation procedure is that it often can be made to mimic the optimal procedures for the parametric situation while retaining exact nonparametric properties.

Comparisons of scale parameters. To compare spread or scale parameters, one can rank the observations in terms of their distances from w^*, the combined sample median. The sum of ranks corresponding to the x-sample is a useful statistic [*see* NONPARAMETRIC STATISTICS, *article on* RANKING METHODS].

When the null hypothesis does not include the assumption that both populations have the same median, the observations in each sample can be replaced by their deviations from their medians, and then the w-sequence can be formed from the deviations in both samples. This ranking procedure will not be exactly nonparametric, but it yields results with significance and confidence levels near the nominal levels.

Other problems

Invariance, sufficiency, and completeness. Such concepts as invariant statistics, sufficient statistics, and complete statistics arise in nonparametric theory. In the two-sample problem, the ranks of the first sample considered as an entity have these properties when only comparisons are obtainable between the observations of the x-samples and y-samples or if the measurement scheme is unique only up to monotone transformations, that is, if there is no compelling reason why particular scales of measurement should be used. For the two-sample problem, confidence intervals based on ranks are invariant under monotone transformations. If a nonparametric confidence interval has been formed for the difference between location parameters, the interval with end points cubed will be the confidence interval that would have been obtained if the original observations had been replaced by their cubes. Maximum likelihood estimation procedures have a similar property.

Multivariate analysis. Although correlation techniques have not been described in this article, Spearman's rank correlation procedure acted as a stimulus for much research in nonparametric analysis. Problems such as partial correlation and multiple correlation have not received adequate nonparametric treatment. For these more complicated problems, the most useful available techniques are associated with the analysis of multidimensional contingency tables. Multivariate tolerance sets are also available. [*See* MULTIVARIATE ANALYSIS, *article on* CORRELATION METHODS; NONPARAMETRIC STATISTICS, *article on* RANKING METHODS; STATISTICS, DESCRIPTIVE, *article on* ASSOCIATION.]

Semantic confusions. In the nonparametric area a number of semantic confusions and multiple uses of the same term have arisen. For example, "rank correlation" has been used by some to refer to the Spearman rank correlation coefficient, by others to refer to both the Spearman coefficient and to the so-called Kendall tau (another measure of bivariate association that is analogous to the Mann–Whitney statistic), and by still others to mean any correlationlike coefficient that depends only on order relationships.

Sometimes the same procedure, described by two superficially different but really equivalent statistics, is erroneously regarded as two separate procedures. An example is confusion between the Mann–Whitney and the Wilcoxon forms of the same test.

Sometimes writers fail to recognize that a word (for example, "runs") describes a variety of different objects.

Analysis of variance analogues. A substantial body of nonparametric procedures has been proposed for the analysis of experiments involving several populations. Analogues have been devised for the simplest standard analysis of variance procedures. One such analogue involves the ranking of all of the measurements in the several samples and then performing an analysis of variance of rank data (Kruskal–Wallis procedure). Another procedure involves each of several judges ranking all populations (the Friedman–Kendall technique).

Decision theory. Decision problems other than estimation and testing have not received extensive nonparametric attention. One result, however, is the Mosteller procedure of deciding whether or not one of several populations has shifted, say to the right. It can be thought of as a multiple-decision problem, since if one rejects the null hypothesis (that the populations are the same) it is natural to decide that the "largest" sample comes from the "largest" population.

I. RICHARD SAVAGE

[*Other relevant material may be found in* ERRORS, *article on* EFFECTS OF ERRORS IN STATISTICAL ASSUMPTIONS; PROBABILITY. *Information on the probability distributions discussed in this article may be found in* DISTRIBUTIONS, STATISTICAL.]

BIBLIOGRAPHY

○ *The extensive literature of nonparametric statistics is indexed in the bibliography of* Savage 1962. *In this index the classification scheme roughly parallels the structure of this article. The index includes citations made to earlier works. Named procedures in this article can be examined in detail by examining the corresponding author's articles listed in the index. Detailed information for applying many nonparametric procedures has been given in* Siegel 1956 *and* Walsh 1962–1968. *The advanced mathematical theory of nonparametric statistics has been outlined in* Fraser 1957 *and* Lehmann 1959. Noether 1967 *is an intermediate-level text, particularly useful for its treatment of ties.*

FRASER, DONALD A. S. 1957 *Nonparametric Methods in Statistics.* New York: Wiley.

LEHMANN, ERICH L. 1959 *Testing Statistical Hypotheses.* New York: Wiley.

NOETHER, GOTTFRIED E. 1967 *Elements of Nonparametric Statistics.* New York: Wiley.

SAVAGE, I. RICHARD 1957 Nonparametric Statistics. *Journal of the American Statistical Association* 52:331–344. → A review of Siegel 1956.

SAVAGE, I. RICHARD 1962 *Bibliography of Nonparametric Statistics.* Cambridge, Mass.: Harvard Univ. Press.

SIEGEL, SIDNEY 1956 *Nonparametric Statistics for the Behavioral Sciences.* New York: McGraw-Hill.

○WALSH, JOHN E. 1962–1968 *Handbook of Nonparametric Statistics.* 3 vols. Princeton, N.J.: Van Nostrand. → Volume 1: *Investigation of Randomness, Moments, Percentiles, and Distributions.* Volume 2: *Results for Two and Several Sample Problems, Symmetry and Extremes.* Volume 3: *Analysis of Variance.*

Postscript

Hollander and Wolfe (1973) is an improved and modernized version of Siegel (1956). An introduction to nonparametric theory is given by Hájek (1969) and to the users of nonparametric statistics by Lehmann (1975). The mathematical theory of nonparametric multivariate procedures is given by Puri and Sen (1971), but this material is not easily applied. Much current effort in the development of statistical methods and theory involves studying the foundations, in particular Bayesian ideas. Savage (1969) describes Bayesian aspects of several nonparametric topics [*see also* BAYESIAN INFERENCE]. The highly theoretical paper by Ferguson (1973) promises a large body of nonparametric Bayesian material.

I. RICHARD SAVAGE

ADDITIONAL BIBLIOGRAPHY

FERGUSON, THOMAS S. 1973 A Bayesian Analysis of Some Nonparametric Problems. *Annals of Statistics* 1:209–230.

HÁJEK, JAROSLAV 1969 *A Course in Nonparametric Statistics.* San Francisco: Holden-Day.

HOLLANDER, MYLES; and WOLFE, DOUGLAS A. 1973 *Nonparametric Statistical Methods.* New York: Wiley.

LEHMANN, E. L. 1975 *Nonparametrics: Statistical Methods Based on Ranks.* San Francisco: Holden-Day.

PURI, MADAN LAL; and SEN, PRANAB KUMAR 1971 *Nonparametric Methods in Multivariate Analysis.* New York: Wiley.

SAVAGE, I. RICHARD 1969 Nonparametric Statistics: A Personal Review. *Sankhyā* Series A 31:107–144.

II

ORDER STATISTICS

Order statistics is a branch of statistics that considers the rank of an observation in a sample as well as its algebraic magnitude. Applications arise in all parts of statistics, from broadly distribution-free (nonparametric) problems to those in which a specific form for the parent population is assumed.

Order statistics methods are particularly useful when a complete sample is unavailable. For example, suppose a biologist is studying treated animals and observes the time of survival from treatment to death. He obtains his observations automatically as order statistics and might end the experiment after a fixed time has elapsed or after a certain number (or proportion) of the animals have died. In such cases estimation of, and testing of hypotheses about, parameters describing survival time can be successfully handled by order statistics.

Early literature in the field was on the use of order statistics for complete samples, usually of small size. Interest in order statistics can be traced back to the use of the median and the range as estimators of location and scale. This method of estimating location and scale has been generalized into linear functions of order statistics for use in censored small samples and as easily computed, but somewhat inefficient, estimators in complete samples. [*See* STATISTICS, DESCRIPTIVE, *article on* LOCATION AND DISPERSION.]

Applications of order statistics also include methods of studying extreme observations in a sample. This theory of extreme values is helpful in the analysis of the statistical aspects of rainfall and floods, in fatigue testing, in the analysis of injury experience of industrial personnel, and in the analysis of oldest ages among survivors. Order statistics are of fundamental importance in screening the extreme individuals in a sample either in a specific selection procedure or in judging whether or not these extreme observations are contaminated and should be rejected as outliers [*see* SCREENING AND SELECTION; STATISTICAL ANALYSIS, SPECIAL PROBLEMS OF, *article on* OUTLIERS]. The range, the "Studentized range," and other functions of order statistics have been incorporated into analysis of variance itself [*see* LINEAR HYPOTHESES, *article on* MULTIPLE COMPARISONS].

There were early and isolated investigations concerning order statistics, such as computations of the relative efficiency of the median versus the mean in samples from the normal distribution (for example, Encke as early as 1834), but the first systematic development of sampling theory in order statistics occurred in 1902, when Pearson considered the "Galton difference problem." He found the expected value of the difference between the rth and $(r+1)$st order statistics in a sample of size n. Daniell (1920) derived the expected value of the rth order statistic and of products of order statistics. He also considered linear estimators for measures of location and scale in the normal distribution. In 1921, Bortkiewicz considered the distribution theory of the sample range, and in 1925, L. H. C. Tippett found the mean value of the sample range and tabulated the cumulative distribution function (cdf) of the largest order statistic for a sample ($n \leqslant 1{,}000$) from the normal distribution.

Definitions and distribution theory

Suppose X is a continuous random variable with cdf $F(x)$ and probability density function $f(x)$. (For discrete random variables the work of Soliman Hasan Abdel-Aty in 1954 and Irving W. Burr in 1955 may be helpful; see the guide at the beginning of the bibliography.) If the elements of a random sample, $X_1, X_2, \cdots, X_n$, are rearranged in order of ascending magnitude such that

$$X_{(1)} \leqslant \cdots \leqslant X_{(n)},$$

then $X_{(r)}$ is called the rth order statistic in the sample of n. Arrangement from least to greatest is always possible, since the probability that two or more X's are equal is zero. In actual samples ties may occur because of rounding or because of insensitive measuring devices; there are special rules designed to handle such cases for each procedure (see, for example, Kendall 1948, chapter 3).

Exact distribution of $X_{(r)}$. The probability element, $\phi(x_{(r)})\,dx_{(r)}$, for the rth order statistic is

$$\frac{n!}{(r-1)!(n-r)!}\cdot [F(x_{(r)})]^{r-1}[1-F(x_{(r)})]^{n-r} f(x_{(r)})\,dx_{(r)}.$$

The heuristic argument for the above expression can be described in terms of Figure 1, showing the x-axis. To say that the rth order statistic lies in a small interval, $(x_{(r)}, x_{(r)}+dx_{(r)})$, is to say that $r-1$ unordered observations are less than $x_{(r)}$, that $n-r$ observations are greater than $x_{(r)}+dx_{(r)}$, and that one observation is in the small interval. These probabilities correspond to the three factors above just to the left of $dx_{(r)}$. The factorial factor at their left allows for rearrangements of the unordered observations.

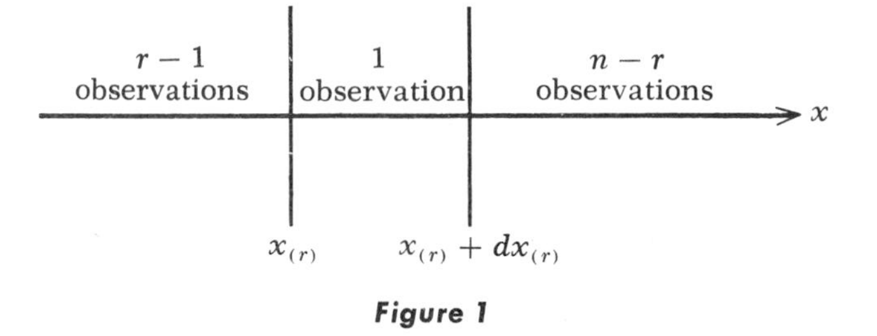

Figure 1

Similar expressions exist for the joint distribution of the ith and jth order statistics, $i < j$. This can lead to the cdf of the sample range and other useful measures.

Limit distributions. Limit distributions will be discussed first for extreme order statistics and then for proportionately defined order statistics.

Extreme order statistics. Limiting (large sample, asymptotic) distributions for the extreme order statistics—the largest and the smallest—have been much studied because of their applicability to the analysis of floods, strength of materials, and so on. For brevity, only the largest order statistic, $X_{(n)}$, will be considered here.

The large sample behavior of $X_{(n)}$ clearly depends on the behavior of $F(x)$ for large values of x, and since $X_{(n)}$ will in general become large and highly variable as n grows, an appropriate kind of centering and scaling must be introduced.

Perhaps the most important case is that in which $F(x)$ approaches unity at an *exponential rate* when x grows but in which $F(x)$ never attains the value unity. More precisely, the case considered is that in which, for some $a > 0$, $[1 - F(x)]/e^{-ax}$ has the limit zero as x grows but in which $1 - F(x)$ never actually attains the value zero. If a centering sequence, $\{u_n\}$, is then defined by

$$F(u_n) = (n-1)/n,$$

and a scaling sequence, $\{\alpha_n\}$, is defined by

$$\alpha_n = nf(u_n),$$

where f is the probability density function associated with F, then the basic result is that $Y = \alpha_n(X_{(n)} - u_n)$ has as its limit distribution the *double exponential,* with cumulative distribution function $\exp(-e^{-y})$.

There are two other kinds of such limit distributions, depending on the structure of F for large x; a detailed discussion was given by E. J. Gumbel in 1958. One of these is, in effect, the distribution studied by W. Weibull in 1949 to investigate the breaking strengths of metals.

The three kinds of limit distributions are intertransformable by changes of variable. Since the limit distributions are relatively simple, large sample estimation is relatively easy; in fact, graphical methods have been worked out and were explained by Gumbel; an exposition of procedural steps is given in Botts (1957).

Proportionately defined order statistics. A second kind of limit distribution is for the rth order statistic where one thinks of r as growing proportionally to n. For any number λ, $0 < \lambda < 1$, the *λ-quantile of the population* is defined as that value of x, ξ_λ, such that

$$F(\xi_\lambda) = \int_{-\infty}^{\xi_\lambda} f(x)\,dx = \lambda.$$

For instance, if $\lambda = .5$, $x_{.50}$ is the median, or the 50th *percentile* (a percentile is a quantile expressed on the base of 100). A λ-quantile, like the median, may sometimes be indeterminate in an interval where $f(x) = 0$. In such a case every ξ in the interval that satisfies $F(\xi_\lambda) = \lambda$ can be taken as the λ-quantile.

The *λ-quantile of a sample* is defined by

$$z_\lambda = \begin{cases} X_{([n\lambda]+1)} & \text{if } n\lambda \text{ is not an integer, where } [n\lambda] \text{ is the largest integer} < n\lambda, \\ \tfrac{1}{2}\{X_{(n\lambda)} + X_{(n\lambda+1)}\} & \text{if } n\lambda \text{ is an integer; this is an arbitrary definition for the indeterminate case, since } z_\lambda \text{ can be any value in the interval } (X_{(n\lambda)}, X_{(n\lambda+1)}). \end{cases}$$

N. V. Smirnoff showed in 1935 that if $r/n = \lambda$ as $n \to \infty$, and if $f(x)$ is continuous and positive at the λ-quantile, ξ_λ, then $\sqrt{n}(X_{(r)} - \xi_\lambda)$ is asymptotically normal with mean zero and variance $\lambda(1-\lambda)/[f(\xi_\lambda)]^2$. This result in the special case of the median ($\lambda = \frac{1}{2}$) was studied by Encke (1834).

The joint limit distribution of two order statistics was also given by Smirnoff. Frederick Mosteller in 1946 extended this to a set of k order statistics with a normal k-variate limit distribution.

The normalized sample quantile given by transforming to $\sqrt{n}\,(z_\lambda - \xi_\lambda)$ has the same asymptotic distribution as $\sqrt{n}(X_{(r)} - \xi_\lambda)$, discussed above. For sampling from a normal distribution with parameters μ and σ, the median of a sample ($z_{.50}$) of size n is asymptotically normal with mean μ and standard deviation $\sigma\sqrt{\pi/2n}$. The sample average has mean μ and standard deviation $\sigma/\sqrt{n}$, so the ratio of standard deviations is $\sqrt{\pi/2}$. In other words, for normal sampling it takes about 100 $\sqrt{\pi/2}$ per cent more observations to obtain the same precision with the sample median as with the sample average. For samples from other distributions, this relationship may be reversed.

Nonparametric procedures for quantiles

Although the distribution of the $X_{(i)}$, the order statistics, depends upon the underlying distribution with cumulative distribution function F, it is important to observe that so long as F is continuous, the distribution of the $F(X_{(i)})$ does *not* depend upon F. A great many nonparametric procedures are based on this fact, and some of them will now be briefly described. (See Mood & Graybill [1950] 1963.)

Confidence limits for quantiles. A simple confidence interval for the quantile at confidence level α is obtained by choosing integers $0 \leqslant r < s \leqslant n$ such that

$$I_\lambda(r, n-r+1) - I_\lambda(s, n-s+1) = 1-\alpha,$$

where I is the incomplete beta distribution [*see* DISTRIBUTIONS, STATISTICAL, *article on* SPECIAL CONTINUOUS DISTRIBUTIONS]. The consequent confidence interval is just $X_{(r)} \leqslant \xi_\lambda \leqslant X_{(s)}$. If the order statistics are symmetrically placed $(s = n-r+1)$, further simplification is achieved. If, further, the quantile is the median, the interval is particularly easy to use; its testing analogue is called the median test. Binomial tail summations may equivalently be used for the incomplete beta distribution values.

An example. Given a sample of 30 observations from any continuous distribution, $1-\alpha$ symmetric confidence intervals for the population median are as shown in Table 1.

Table 1 — Symmetric confidence intervals for the population median, based on a sample of 30 observations

Order statistics	Confidence coefficient
$X_{(8)}$ and $X_{(23)}$	.995
$X_{(10)}$ and $X_{(21)}$	.95
$X_{(11)}$ and $X_{(20)}$	.90

For the quartiles ($\lambda = .25, .50, .75$), with n going from 2 to 1,000, and for several values of α, section 12.3 of Owen (1962) provides convenient tables. Noether (1949) gives an elementary discussion of these procedures and the ones for tolerance intervals.

Tolerance intervals. Nonparametric *tolerance intervals* also may be based on the concept of choosing integers determined by the incomplete beta distribution. A tolerance interval says that at least 100β per cent of the probability distribution described by F is contained in a random interval, so chosen that the statement is true with probability $1-\alpha$. If the ends of the interval are $X_{(r)}$ and $X_{(s)}$, with $r < s \leqslant n$, and if β is given, then α may be computed in terms of the incomplete beta distribution; similarly, if α is given, β may be computed. This was discussed further by R. B. Murphy in 1948.

Furthermore, many nonparametric tests of hypotheses are based squarely on the above fundamental property of the $F(X_{(i)})$. [*Many of these tests are discussed elsewhere in* NONPARAMETRIC STATISTICS; *see also* GOODNESS OF FIT.]

Linear systematic statistics

Mean and standard deviation. A linear systematic statistic is a linear combination of the sample order statistics. This article deals with the use of linear systematic statistics to estimate location parameters (μ) and scale parameters (σ) from a random sample with underlying distribution of known form up to unknown μ and/or σ. The distributions for which linear systematic statistics have been considered include the normal, exponential, rectangular, gamma, logistic, and extreme-value distributions and others of only theoretical interest, such as the right triangular. The choice of the coefficients of a linear systematic statistic should be optimal in some sense—for example, in terms of bias, sampling variance, and computational convenience.

The expectation of $X_{(r)}$ can be expressed as a linear function of μ and σ, and the variance and covariance of the $X_{(r)}$ can be computed up to a scalar constant. Using generalized least squares, E. H. Lloyd in 1952 showed how to find the minimum variance unbiased estimators among all linear combinations. Linear combinations can also be constructed if only a portion of the sample order statistics are used, either from necessity or for convenience (the use of only certain order statistics corresponds to the requirement that certain coefficients of the linear combination be zero). The following general notation for linear systematic statistics will be used:

$$\mu^* = \sum_{i=1}^{n} a_{1i} X_{(i)},$$

$$\sigma^* = \sum_{i=1}^{n} a_{2i} X_{(i)},$$

where the a's are the coefficients.

For the normal distribution Daniel Teichroew in 1956 calculated with ten-decimal precision the expected values of the order statistics and of the products of pairs of order statistics for samples of size 20 and under. These were used by Sarhan and Greenberg in 1956 to compute variances and covariances of the order statistics and to derive tables of optimal coefficients a_{1i} and a_{2i}.

Censored samples. For complete samples this procedure does not represent much of an achievement, although the loss in efficiency in comparison to optimal unbiased minimum variance estimators is often negligible. When some of the sample observations are censored, however, the tables become almost indispensable. [*See* STATISTICAL ANALYSIS, SPECIAL PROBLEMS OF, *article on* TRUNCATION AND CENSORSHIP.] Observations may be censored, usually at the extremes, because the errors of measurement at the extremes are greater than those in the central portion of the distribution or because it is difficult and/or costly to determine their exact magnitude. If censorship is by a fixed point on the abscissa (regardless of whether this point is known to the investigator), the censorship is referred to as Type I. If a certain predetermined percentage of the observations is censored, the censorship is referred to as trimming or as Type II censoring. (F. N. David and Norman L. Johnson in 1956 used the terms "type B" and "type A" to denote Type I and Type II censoring. Another sort of censoring is known as Winsorization, discussed below in relation to the rejection of outlying observations.) Censorship and trimming may be on one or both ends of the ordered sample.

If r_1 and r_2 observations are censored from the left-hand and right-hand sides, respectively, estimates of the parameters can be based upon the values of the $n - r_1 - r_2$ observations remaining after censoring, together with the corresponding coefficients a'_{1i} and a'_{2i}. Tables of the coefficients for the best linear estimators under Type II censoring are given by Sarhan and Greenberg (1962). They also include tables of variances (and covariances) of the estimators and their efficiencies relative to the best linear estimators using complete samples. These tables are valuable because such a use of order statistics gives the best possible estimation among the class of linear statistics for samples up to size 20. Another use of the tables is for the rapid appraisal of patterns of relative information in each order statistic.

An example. Students made measurements (shown in Table 2) of strontium-90 concentration in samples of a test milk in which 9.22 micromicrocuries per liter was the known correct value. One observation was censored at each side because of suspected unreliability. The resulting estimates are $\mu^* = 9.27$ (with variance $.1043\sigma^2$) and $\sigma^* = 2.05$ (with variance $.0824\sigma^2$). If the two censored observations had been trimmed from one side rather than symmetrically, the efficiency in estimating the mean decreases from 95.9 per cent to 93.0 per cent, but that for estimating the standard deviation increases slightly, from 69.9 per cent to 70.9 per cent.

Table 2 — Measurements of strontium-90 concentration

Ordered observations		a'_{1i}	a'_{2i}
1.	—	0	0
2.	7.1	.1884	−.4034
3.	8.2	.1036	−.1074
4.	8.4	.1040	−.0616
5.	9.1	.1041	−.0201
6.	9.8	.1041	.0201
7.	9.9	.1040	.0616
8.	10.5	.1036	.1074
9.	11.3	.1884	.4034
10.	—	0	0

Other linear systematic statistics. In samples larger than size 20, the calculations for optimal linear systematic statistics are cumbersome, and tables of coefficients are not available. Alternative nearly best and nearly unbiased estimators were proposed by A. K. Gupta in 1952, Sarhan and Greenberg in 1956, Gunnar Blom in 1958, and Sarndal in 1962. These alternatives have coefficients that are easier to compute than a'_{1i} and a'_{2i}.

Estimators from less than available sample

The linear systematic statistics discussed above were based on all the order statistics, or on all those available after censorship. Optimality was stressed in terms of minimum variance. The present section considers simpler linear systematic statistics based on less than all available order statistics. They are simpler in the sense that many of the coefficients are zero and/or the nonzero coefficients take only a few, convenient values.

Small samples. Simplified estimators in small samples will be considered for several different distributions.

Normal distribution. Wilfrid J. Dixon in 1957 suggested simple estimators of the mean of a normal distribution that are highly efficient for small samples. Two of these were:

(1) The mean of two optimally selected order statistics;

(2) $\bar{X}_{(2,n-1)} = \{1/(n-2)\}\sum_{i=2}^{n-1} X_{(i)}$, that is, the mean of all order statistics except the largest and smallest.

Estimator (1) is, more specifically, the average of the order statistics with indices $([.27n]+1)$ and $([.73n]+1)$, where $[k]$ is the largest integer in k. Its asymptotic efficiency, relative to the arithmetic mean of the sample, is 81.0 per cent, and in small samples its efficiency, measured in the same way, is never below this figure and rapidly approaches 100 per cent as n declines toward 2. (Asymptotic estimation is further discussed below.) When n is a multiple of 100, computation of the 27th and 73d percentiles runs into the same problem of indeterminacy as does computation of the median of an even-numbered sample. As explained in the definition of quantiles above, an arbitrary working rule for the 27th percentile is $\frac{1}{2}[X_{(.27n)} + X_{(.27n+1)}]$.

The 27th and 73d percentiles as used here are relevant to the problem of optimum groupings discussed below for large samples in the univariate and bivariate normal distributions. Their application to a problem in educational psychology was originally pointed out by Kelley (1939), seeking to select the upper and lower group of persons with a normally distributed score for the validation of test items. He recommended selection of 27 per cent at each extreme to be reasonably certain that the groups were different in respect to the trait. Cureton (1957), using the same method, and David R. Cox in 1957 noted that one-third at each extreme is optimal when the distribution is rectangular.

The second estimator has higher efficiency than the first for $n \geqslant 5$, and its efficiency asymptotically approaches unity. Its chief advantage comes about when there is a possible outlier (discussed below).

Dixon in 1957 also considered linear estimators of the standard deviation for the normal distribution. These estimators are better than, although related to, the sample range. One such estimator is based upon the sum of that set of subranges $W_{(i)} = X_{(n-i+1)} - X_{(i)}$ that gives minimum variance. For a sample of size 10, for example, the unbiased estimator of the standard deviation based upon the range is $.325W_{(1)}$ and has an efficiency of 85.0 per cent relative to the unbiased complete sample standard deviation. Dixon's improved estimator is $.1968(W_{(1)} + W_{(2)})$, and its efficiency is 96.4 per cent.

Exponential distributions. The one-parameter exponential distribution, with density

$$f(x) = \frac{1}{\sigma} e^{-x/\sigma}, \qquad x > 0, \sigma > 0,$$

is most useful in life-testing situations, where each observation is automatically an ordered waiting time to failure or death. Benjamin Epstein and Milton Sobel in 1953 showed that the maximum likelihood estimator of σ based only on the first r order statistics is

$$\hat{\sigma}_{r,n} = \frac{X_{(1)} + X_{(2)} + \cdots + X_{(r)} + (n-r)X_{(r)}}{r}$$

and that $2r\hat{\sigma}_{r,n}/\sigma$ is distributed as chi-square with $2r$ degrees of freedom.

This estimator is a multiple of a simple weighted average of the first r uncensored observations. For other conditions of censoring, the best linear estimator σ^* described by Sarhan and Greenberg in 1957 may be used. Other simple estimators include those of Harter (1961) and Kulldorff (1963*a*).

For the two-parameter exponential distribution (with unknown left end point of range, α) Epstein and Sobel in 1954 considered best linear unbiased estimators α^* and σ^* based again upon the first r order statistics. These are

$$\alpha^* = \frac{(nr-1)X_{(1)} - X_{(2)} - X_{(3)} - \cdots - X_{(r-1)} - (n-r+1)X_{(r)}}{n(r-1)}$$

and

$$\sigma^* = \frac{-(n-1)X_{(1)} + X_{(2)} + X_{(3)} + \cdots + X_{(r-1)} + (n-r+1)X_{(r)}}{r-1}.$$

Still further simplified estimators for both the one-parameter and two-parameter exponential distributions have been considered; they use the optimal k out of n sample order statistics. Tables and instructions for the use of these can be found in Sarhan, Greenberg, and Ogawa (1963), Kulldorff (1963*b*), and Saleh and Ali (1966).

Large samples. In a large sample, unbiased estimators of location and scale with high asymptotic efficiency may be derived by selecting k suitably spaced order statistics, where k is considerably less than n. The spacing of the quantiles $\lambda_1, \lambda_2, \cdots, \lambda_k$ which produces an estimator with maximum efficiency is termed an optimum spacing.

The problem of optimum spacing of sample quantiles for purposes of estimation is related to asymptotically optimum grouping of observations for convenience in exposition or for purposes of contrast. Cox in 1957 considered optimum grouping for the normal distribution and obtained the same set of quantiles as Ogawa found in 1951 for optimum spacings. Kulldorff in 1958 derived optimal groupings for the exponential distribution, and the results were identical with those sample quantiles derived by Sarhan, Greenberg, and Ogawa (1963) for optimum spacings. The optimum spacing for estimation in the rectangular distribution, however, is obtained from the extreme observations, whereas optimal grouping requires equal frequencies.

Normal distribution. For the mean, Mosteller in 1946 considered the estimator

$$\mu^* = \frac{1}{k}\sum_{i=1}^{k} X_{(n_i)}, \qquad k = 1, 2, \text{ or } 3, \quad 1 \leqslant n_1 < \cdots < n_k \leqslant n.$$

The integers n_i are determined by k fixed numbers, $\lambda_1, \cdots, \lambda_k$, such that $0 < \lambda_1 < \cdots < \lambda_k < 1$, and $n_i = [n\lambda_i] + 1$. For $k = 3$, for example, Mosteller obtained $\lambda_1 = .1826$, $\lambda_2 = .5000$, and $\lambda_3 = .8174$, with asymptotic efficiency equal to 87.9 per cent. A different set of quantiles and a weighted average of them was selected by Ogawa in 1951 to minimize asymptotic variance. Tables for Ogawa's estimators are given in Sarhan and Greenberg (1962).

Estimation of both mean and standard deviation is a more complex problem; it has been solved only for symmetric spacing with $k = 2$, in which case the quantiles selected are $\lambda_1 = .134$ and $\lambda_2 = .866$.

Exponential distribution. For the one-parameter exponential distribution, the estimators of the form

$$\sigma^* = \sum_{i=1}^{k} b_i X_{(n_i)}$$

have been obtained with high maximum asymptotic relative efficiency for $k = 1(1)15$ by Sarhan, Greenberg, and Ogawa (1963). Kulldorff (1963*a*) has also done this, with greater arithmetic precision.

As an example, for $k = 3$,

$$\sigma^* = .4477X_{([.5295n]+1)} + .2266X_{([.8299n]+1)} + .0776X_{([.9655n]+1)}$$

has asymptotic relative efficiency .89.

Bivariate normal distribution. An estimator of the correlation coefficient, ρ, was devised by Mosteller in 1946 using the ranks of $2n < N$ observations. The procedure is to order the N observations on the x coordinate and to distribute the n largest and n smallest of them into an upper and a lower set based upon the y coordinate. The dividing line between the upper and lower sets is the median of the $2n$ observations on the y coordinate when the means and variances are unknown. Using the number of cases in the resultant four corners, an estimate of ρ can be obtained with the aid of a graph provided by Mosteller for varying levels of n/N. The optimal value of n/N is approximately 27 per cent, in which case the method has an efficiency slightly in excess of 50 per cent in comparison to the corresponding Pearson correlation coefficient. This method has merit where data are on punch cards and machines can be used for rapid sorting. By using the upper and lower 27 per cent of a sample, a similar adaptation can be made when fitting straight lines by regression, especially when both variables are subject to error (see Cureton 1966).

Outlying observations

An observation with an abnormally large residual (deviation from its estimated expectation) is called an outlier; it can arise either because of large inherent variability or because of a spurious measurement. A rule to reject outliers should be considered an insurance policy and not really a test of significance. The first attempt to develop a rejection criterion was suggested by C. S. Peirce in 1852 while he was studying observations from astronomy.

C. P. Winsor in 1941 proposed a procedure that now bears his name: a suspected outlier should not be rejected completely, but its original value should be replaced by the nearest value of an observation that is not suspect. For the normal distribution the symmetrically *Winsorized* mean is somewhat similar to the second Dixon estimator discussed above (the mean of all order statistics except the largest and smallest), but the former shows only a small loss of efficiency and is more stable than the latter (see Tukey 1962). The Dixon estimator was proposed for a different purpose but has been used when an outlying observation was suspect. In evaluating the utility of either of these procedures, or of any other, one must consider the probability of falsely rejecting a valid observation as well as the bias caused by retaining a spurious item. The usefulness of rejection criteria should be measured in terms of the residual error variance. [*These problems are discussed in more detail in* STATISTICAL ANALYSIS, SPECIAL PROBLEMS OF, *article on* OUTLIERS.] A few examples will be given here to illustrate rejection rules when it is suspected that the observation has an error in location and when the underlying distribution is normal with mean μ and variance σ^2.

Case of σ known. To test whether $X_{(n)}$ is an outlier when μ is known, use can be made of

$$B_0 = \frac{X_{(n)} - \mu}{\sigma},$$

whereas when μ is unknown, one can use

$$B_1 = \frac{X_{(n)} - \bar{X}}{\sigma}$$

or

$$B_2 = \frac{X_{(n)} - X_{(n-1)}}{\sigma}.$$

For detecting an outlier, the performance of B_1 is better than that of B_2 (and better than those of others not listed here), although B_2 is easier to compute. Tables for using B_1 can be found in Pearson and Hartley (vol. 1 [1954], 1966).

Case of σ unknown. If an independent estimator of σ is available from outside the sample in question, one can substitute it in B_1 to obtain a modified criterion. Tables of upper percentage points for this externally Studentized extreme deviate from the sample mean were prepared by David in 1956.

When no estimator of σ is available except from

the observed sample, two of the appropriate tests are

$$B_3 = \frac{X_{(n)} - \bar{X}}{s}, \qquad \text{where } s^2 = \frac{1}{n}\sum_{i=1}^{n}(X_i - \bar{X})^2,$$

and

$$r_{10} = \frac{X_{(n)} - X_{(n-1)}}{X_{(n)} - X_{(1)}}.$$

An example. Given five observations, 23, 29, 31, 44, and 63, is the largest observation a spurious one?

$$B_3 = \frac{63 - 38}{\sqrt{1016/5}} = \frac{25}{\sqrt{203.2}} = 1.75,$$

$$r_{10} = \frac{63 - 44}{63 - 23} = \frac{19}{40} = .475.$$

From the tables prepared by Frank E. Grubbs in 1950 for B_3 with $n = 5$, the 90th upper percentage point is 1.791, and the 95th upper percentage point is 1.869. Thus, the value of 63 gives a result above the 10 per cent level and would not be rejected by B_3 at ordinary levels. From the table prepared by Dixon in 1951 for r_{10}, the sample significance level is again seen to be between 10 per cent and 20 per cent, so 63 would again not usually be rejected, and the suspect value would be retained.

Multiple outliers. Tests can also be constructed for multiple outliers, and the efficiency of various procedures can be compared. Excellent reviews of this field were given by Anscombe (1960) and by Dixon in 1962.

Tests of significance

Counterparts to the standard tests of significance can be derived by substituting the median, midrange, or quasi midrange for the mean and by using the range or subranges in lieu of the sample standard deviation. Such tests will usually be lower in power, but in small samples the differences may be negligible.

Hypotheses on location. The test criterion for the difference between the location measure of a sample and a hypothetical population value based upon substitution of the range in the standard *t*-test, published by Joseph F. Daly in 1946 and by E. Lord in 1947, is

$$G_1 = \frac{\bar{X} - \mu}{X_{(n)} - X_{(1)}}.$$

An example. Walsh (1949) poses the problem of testing whether the mean ($\bar{x} = 1.05$) of the following sample of 10 observations differs significantly from the hypothesized population mean of 0. Ordered observations are −1.2, −1.1, −.2, .1, .7, 1.3, 1.8, 2.0, 3.4, and 3.7.

$$G_1 = \frac{1.05 - 0}{(3.7) - (-1.2)} = \frac{1.05}{4.9} = .214.$$

As tabulated by Lord, the critical values for a two-tailed test are 10 per cent = .186 and 5 per cent = .230, and therefore the hypothetical mean would not be rejected at the 5 per cent level.

Difference between means of two samples. Lord in 1947 devised the following test criterion for the difference between the means of two samples of size n from a normal population:

$$G_3 = \frac{|\bar{X} - \bar{Y}|}{X_{(n)} - X_{(1)} + Y_{(n)} - Y_{(1)}}.$$

An example. The problem is to test whether two samples, shown in Table 3, have a common population mean, assuming a normal distribution and equal but unknown variances. These data give $\bar{x} = 40.5$, $\bar{y} = 53.2$, and

Table 3

X-sample	Y-sample
27.6	43.3
35.5	48.7
45.0	53.6
46.7	56.5
47.6	63.9

$$G_3 = \frac{|40.5 - 53.2|}{(47.6 - 27.6) + (63.9 - 43.3)}$$

$$= \frac{12.7}{20.0 + 20.6} = .313.$$

The critical value for 5 per cent is .307; for 1 per cent it is .448. The null hypothesis of equality of the means is rejected at the 5 per cent level.

Equivalent tests were devised by J. Edward Jackson and Eleanor L. Ross in 1955 without the restriction that the two sample sizes be the same.

Tests on variances. One-sample, two-sample, and k-sample tests of variances and analysis of variance can be performed with the use of ranges and subranges. A good reference on this subject is Chapter 7 by David in Sarhan and Greenberg (1962).

BERNARD G. GREENBERG

BIBLIOGRAPHY

Those citations to the literature which are given in the text but are not specified below can be found in the bibliography of Wilks 1948 *or that of* Sarhan & Greenberg 1962. *The discussion in the latter monograph corresponds closely to the coverage in this article.*

ANSCOMBE, F. J. 1960 Rejection of Outliers. *Technometrics* 2:123–147.

BOTTS, RALPH R. 1957 "Extreme-value" Methods Simplified. *Agricultural Economics Research* 9:88–95.

CURETON, EDWARD E. 1957 The Upper and Lower Twenty-seven Per Cent Rule. *Psychometrika* 22:293–296.

CURETON, EDWARD E. 1966 Letter to the Editor. *American Statistician* 20, no. 3:49.

DANIELL, P. J. 1920 Observations Weighted According to Order. *American Journal of Mathematics* 42:222–236.

ENCKE, J. F. (1834) 1841 On the Method of Least Squares. Volume 2, pages 317–369, in *Scientific Memoirs: Selected From the Transactions of Foreign Academies of Science and Learned Societies, and From Foreign Journals*. Edited by Richard Taylor. London: Taylor. → First published in German in the *Astronomisches Jahrbuch*.

HARTER, H. LEON 1961 Estimating the Parameters of Negative Exponential Populations From One or Two Order Statistics. *Annals of Mathematical Statistics* 32:1078–1090.

KELLEY, TRUMAN L. 1939 The Selection of Upper and Lower Groups for the Validation of Test Items. *Journal of Educational Psychology* 30:17–24.

○KENDALL, MAURICE G. (1948) 1970 *Rank Correlation Methods*. 4th ed. London: Griffin; New York: Hafner.

KULLDORFF, GUNNAR 1963*a* On the Optimum Spacing of Sample Quantiles From an Exponential Distribution. Unpublished manuscript, Univ. of Lund, Department of Statistics.

KULLDORFF, GUNNAR 1963*b* Estimation of One or Two Parameters of the Exponential Distribution on the Basis of Suitably Chosen Order Statistics. *Annals of Mathematical Statistics* 34:1419–1431.

MOOD, ALEXANDER M.; and GRAYBILL, FRANKLIN A. (1950) 1963 *Introduction to the Theory of Statistics*. 2d ed. New York: McGraw-Hill.

NOETHER, GOTTFRIED E. 1949 Confidence Limits in the Non-parametric Case. *Journal of the American Statistical Association* 44:89–100.

OWEN, DONALD B. 1962 *Handbook of Statistical Tables*. Reading, Mass.: Addison-Wesley. → A list of addenda and errata is available from the author.

2PEARSON, E. S.; and HARTLEY, H. O. (editors) (1954) 1966 *Biometrika Tables for Statisticians*. Volume 1. 3d ed. Cambridge Univ. Press. → Reprinted with additions in 1970.

SALEH, A. K. MD. EHSANES; and ALI, MIR M. 1966 Asymptotic Optimum Quantiles for the Estimation of the Parameters of the Negative Exponential Distribution. *Annals of Mathematical Statistics* 37:143–151.

SARHAN, AHMED E.; and GREENBERG, BERNARD G. (editors) 1962 *Contributions to Order Statistics*. New York: Wiley.

SARHAN, AHMED E.; GREENBERG, BERNARD G.; and OGAWA, JUNJIRO 1963 Simplified Estimates for the Exponential Distribution. *Annals of Mathematical Statistics* 34:102–116.

SARNDAL, CARL E. 1962 *Information From Censored Samples*. Stockholm: Almqvist & Wiksell.

SARNDAL, CARL E. 1964 Estimation of the Parameters of the Gamma Distribution by Sample Quantiles. *Technometrics* 6:405–414.

TUKEY, JOHN W. 1962 The Future of Data Analysis. *Annals of Mathematical Statistics* 33:1–67, 812. → Page 812 is a correction.

WALSH, JOHN E. 1949 Applications of Some Significance Tests for the Median Which Are Valid Under Very General Conditions. *Journal of the American Statistical Association* 44:342–355.

WILKS, S. S. 1948 Order Statistics. American Mathematical Society, *Bulletin* 54:6–50.

Postscript

Research about order statistics has concentrated upon robustness, especially robustness of point estimation. A basic concept has been the use of linear combinations of order statistics to derive robust estimators of location and scale. [*See* ERRORS, *article on* EFFECTS OF ERRORS IN STATISTICAL ASSUMPTIONS.] Several excellent monographs on order statistics have appeared, of quality and breadth that show the subject has reached a mature stage and is by no means still a subtopic under the rubric of nonparametric statistics. Several monumental tables have appeared that make order statistics more widely applicable.

Robustness. Mosteller's 1946 article on some useful "inefficient" statistics anticipated the resurgence of interest in the robustness of order statistics. More recently, Tukey (1960) raised the question of robustness of efficiency in estimation of central tendency when the true distribution deviates only slightly from an assumed normal one. His interest in this field started with the Winsorized mean and trimmed means, discussed in the main article in connection with the rejection of outlying observations.

Huber (1964) studied in detail the estimation of a location parameter for contaminated normal distributions, using what he called a set of intermediaries (now sometimes referred to as Huber estimators) between the sample mean and sample median. He showed them to be asymptotically most robust.

The asymptotic theory of robust estimators formed by taking linear combinations of order statistics has been studied also by Gastwirth (1966), who discussed a procedure producing a robust estimator of the location parameter corresponding to a rank test in any symmetric unimodal distribution. Later, Chernoff, Gastwirth, and Johns (1967) searched for asymptotically best linear combinations of order statistics to estimate location and scale parameters for both censored and uncensored data. They suggested that the ultimate solution may require use of an "invariance principle" and compared their results to those of Govindarajulu (1965), finding some overlap between the two sets of estimators.

This growing interest in robustness of estimation led in 1970–1971 to a year-long seminar at Princeton University. From that seminar has come a highly important monograph by Andrews et al. (1972) that enumerates the properties of 68 robust estimators of location. Both large- and small-sample properties were studied by new asymptotic methods and by Monte Carlo sampling.

Huber (1972) gave an unusually clear review of the subject. He credits George E. P. Box with the first use of the term "robustness" in 1953, and discusses the first serious attempt to deal with non-normality by Simon Newcomb in 1886. He also considers the differing uses of "robustness." He treats three important ways of constructing estimators, namely, maximum likelihood type estimators (*M*-estimators), linear combinations of order statistics (*L*-estimates), and estimators derived from rank tests (*R*-estimates), and discusses the properties of each.

The history of robust estimation has inspired a good deal of interest. Stigler (1973) investigated the contributions of Simon Newcomb, Percy Daniell, and others to robust estimation and credits Legendre in 1805 with the notion of rejecting outliers as part of the use of least squares. Stigler also points out that Laplace in 1818 was the first to consider the distribution of the median and thus corrects an error in the main article, where it is stated that Encke made the first studies around 1834.

Newcomb, a mathematical astronomer, Stigler tells us, used an estimator in 1886 that gave less weight to the more deviant observations, thus providing a substitute for the sample mean when the parent distribution has tails more excessive than the normal. Stigler also treats in detail the contributions of Percy J. Daniell mentioned in the main article.

Further material on robustness continues to appear. For example, articles by Hampel (1974) and Johns (1974) are important in any discussion of robustness in estimation. Hampel studies the use of an influence curve and its properties to derive new estimators with specified robustness characteristics, and demonstrates relationships between Winsorized means, trimmed means, and Huber estimators. The influence curve is basically the first derivative of an estimator viewed as a functional, and is useful also in deriving asymptotic variances. Hampel also relates the influence curve to score-generating functions of rank tests and *R*-estimators, as well as the *M*-estimators. An interesting point in this paper is Hampel's advocacy of the median deviation (median of absolute values of deviations from median) as a robust estimator of scale. Johns deals with robustness in the sense of efficiency to estimate location for symmetric continuous distributions. The result is a series of estimators based upon linear combinations of order statistics with optimal coefficients estimated from the sample. The estimators are consistent and best asymptotically normal (BAN) with respect to the Cramér–Rao lower bound for variance.

Stigler (1974) makes further contributions; he considers weighted linear functions of order statistics for all types of distributions, and even includes grouped data. When the weights are based upon a smooth bounded function, the requirement that the distribution be continuous can be dropped. He illustrates by application to such estimators as the trimmed mean for location and Gini's mean difference for scale.

An annotated analytic bibliography on robustness studies prepared by Govindarajulu and Leslie (U.S. National Center for Health Statistics 1972) contains many papers on robustness involving order statistics.

Another interesting contribution to the literature on robustness is an introductory textbook that uses robust procedures but describes them by another term. The book, *Sturdy Statistics*, by Mosteller and Rourke, appeared in 1973. "Sturdy" is used in place of "robust" in part to avoid possibly confusing connotations; the authors focus upon the statistics themselves. The authors show how order statistics can be successfully explained and used in a basic course. For example, the median and other quantiles together with confidence limits for them, are treated in a distribution-free context. The work also shows how to use expected values of order statistics for the normal distribution and concludes with the use of a function of the sample range to estimate the population standard deviation.

Monographs. The monograph by David (1970) is a sequel to Sarhan and Greenberg's *Contributions to Order Statistics* (1962). Distribution theory, estimation, and short-cut procedures for estimation and testing are covered, along with the use of order statistics to treat outliers and slippage (where one population has "slipped" from others that are supposed to be identical). David also points out that the idea of using two quantiles with optimal spacing to estimate the mean in the normal distribution, specifically the 27th and 73d percentiles, was first suggested by Karl Pearson (1920). Pearson also studied the results of using the 7th and 93d percentiles to estimate optimally the standard deviation in the normal case. Pearson came to the 27th and 73d percentiles when he sought a pair of symmetric values to estimate the median with minimum variance. For estimating σ, he recommended the $\frac{1}{14}$ grade on either side, which leads approximately to the 7th and 93d percentiles. David indicates that in large samples, the values $\lambda_1 = 0.0694$ and $\lambda_2 = 1 - \lambda_1$ maximize efficiency. David gives the references for spacings in several other distributions, as well as for nearly optimal spacings in estimation problems. The monograph also contains an appendix guide to tables.

Volume 1 of Johnson and Kotz (1970) presents a remarkable richness of historical information and a comprehensive treatment of distributions, the relationships between them, and the problems of estimation. [*See* DISTRIBUTIONS, STATISTICAL.] A condensed account of the theory of order statistics is given at the outset, since many useful estimators are based upon order statistics. Chapter 13 gives extensive coverage of literature about the normal distribution. The last two chapters, on the Weibull distribution and extreme value distributions, are of special significance for order statistics and their applications.

Tables of order statistics. The two-volume (with a third, bibliographic, volume planned) set of tables by Harter (1969) is important for its detail and accuracy. Volume 1 deals with tests based upon the range and Studentized range for samples from a normally distributed population. Chapter 1 includes a discussion of the F' test proposed by Link (1950; 1952) in which the ratio of two sample ranges is used as a substitute for the traditional F ratio. The table of critical values for the F' test requires prior computation of three other tables, namely, the probability density function (pdf) of the range, and the pdf and the cumulative density function of the ratio of two ranges. These are given, as well as the power of the F' test in comparison with the power of the F test for selected sample sizes.

Chapter 2 concerns the Studentized range, in which the sample range is divided by an independent estimate of the standard deviation. This was first suggested by Student, and further considered by Newman (1939) for multiple comparisons of means. Keuls (1952) refined the procedures of Newman and opened a whole new field of multiple comparison tests. This chapter provides accurate critical values for several multiple comparison tests based on the range and the multiple range test. [*See* LINEAR HYPOTHESES, *article on* MULTIPLE COMPARISONS.]

Volume 2 deals with order statistics in the estimation of location and scale for various distributions (normal, rectangular, exponential, Weibull, extreme-value, and so on) including cases of doubly censored samples. The tables rely heavily on the use of quasi-ranges for samples from the normal distribution and the range for samples from the rectangular. Also tabulated are the expected values of order statistics of samples from various populations.

Another welcome addition is Volume 2 ([1966] 1972) of the highly successful *Biometrika Tables for Statisticians* by Pearson and Hartley. The volume includes tables for expected values, variances and covariances of normal order statistics, and for expected values of order statistics for the negative, gamma, and half normal distributions. Attention is given to the test statistic W proposed by Shapiro and Wilk (1965), a kind of internal Studentization for testing normality that is based upon the use of all the order statistics rather than just the range. Subsequent tables cover the two-sample Wilcoxon test, the Kruskal–Wallis test, Friedman rank test, and Kendall's concordance coefficient.

F. N. David et al. (1968) also present highly useful tables related to order statistics. The use of nonparametric tests has created widespread use of ordinal or ranked data. To improve rank order tests, the authors present detailed tables to replace ranks by normal centroids, medians, or normal scores. These last are expected values of normal order statistics, and the tables by David et al. extend to greater values than prior ones.

BERNARD G. GREENBERG

ADDITIONAL BIBLIOGRAPHY

ANDREWS, DAVID F. et al. 1972 *Robust Estimates of Location: Survey and Advances.* Princeton Univ. Press.

CHERNOFF, HERMAN; GASTWIRTH, JOSEPH L.; and JOHNS, M. V. JR. 1967 Asymptotic Distribution of Linear Combinations of Functions of Order Statistics With Applications to Estimation. *Annals of Mathematical Statistics* 38:52–72.

DAVID, F. N. et al. 1968 *Normal Centroids, Medians, and Scores for Ordinal Data.* Cambridge Univ. Press.

DAVID, HERBERT A. 1970 *Order Statistics.* New York: Wiley.

GASTWIRTH, JOSEPH L. 1966 On Robust Procedures. *Journal of the American Statistical Association* 61: 929–948.

GOVINDARAJULU, Z. 1965 Asymptotic Normality of Linear Functions of Order Statistics in One and Multisamples. Pages 1–26 in Z. Govindarajulu, *Asymptotic Normality of Linear Combinations of Functions of Order Statistics.* Part 2. Technical Report AF-AFOSR-741-65. Arlington, Va.: Office of Scientific Research, U.S. Air Force.

HAMPEL, FRANK R. 1974 The Influence Curve and Its Role in Robust Estimation. *Journal of the American Statistical Association* 69:383–393.

HARTER, H. LEON 1969 *Order Statistics and Their Use in Testing and Estimation.* 2 vols. Wright-Patterson Air Force Base, Ohio: Aerospace Research Laboratories, Office of Aerospace Research, U.S. Air Force. → Volume 1: *Tests Based on Range and Studentized Range of Samples From a Normal Population.* Volume 2: *Estimates Based on Order Statistics of Samples From Various Populations.*

HUBER, PETER J. 1964 Robust Estimation of a Location Parameter. *Annals of Mathematical Statistics* 35:73–101.

HUBER, PETER J. 1972 Robust Statistics: A Review. *Annals of Mathematical Statistics* 43:1041–1067. → The 1972 Wald lecture.

JOHNS, M. V. JR. 1974 Nonparametric Estimation of Location. *Journal of the American Statistical Association* 69:453–460.

JOHNSON, NORMAN L.; and KOTZ, SAMUEL 1970 *Continuous Univariate Distributions*. Volume 1. Boston: Houghton Mifflin. → Reissued in 1972 by Wiley.

KEULS, M. 1952 The Use of the "Studentized Range" in Connection With an Analysis of Variance. *Euphytica* 1:112–122.

LINK, RICHARD F. 1950 The Sampling Distribution of the Ratio of Two Ranges From Independent Samples. *Annals of Mathematical Statistics* 21:112–116.

LINK, RICHARD F. 1952 Correction to "The Sampling Distribution of the Ratio of Two Ranges From Independent Samples." *Annals of Mathematical Statistics* 23:298–299.

MOSTELLER, FREDERICK 1946 On Some Useful "Inefficient" Statistics. *Annals of Mathematical Statistics* 17:377–408.

MOSTELLER, FREDERICK; and ROURKE, ROBERT E. K. 1973 *Sturdy Statistics: Nonparametrics and Order Statistics*. Reading, Mass.: Addison-Wesley.

NEWMAN, D. 1939 The Distribution of Range in Samples From a Normal Population, Expressed in Terms of an Independent Estimate of Standard Deviation. *Biometrika* 31:20–30.

[2]PEARSON, E. S.; and HARTLEY, H. O. (editors) (1966) 1972 *Biometrika Tables for Statisticians*. Volume 2. 3d ed. Cambridge Univ. Press. → Both volumes are currently distributed by Dawson's of Pall Mall, an English firm in Folkestone, Kent.

PEARSON, KARL 1920 On the Probable Errors of Frequency Constants. Part 3. *Biometrika* 13:113–132.

SHAPIRO, S. S.; and WILK, M. B. 1965 An Analysis of Variance Test for Normality (Complete Samples). *Biometrika* 52:591–611.

STIGLER, STEPHEN M. 1973 Simon Newcomb, Percy Daniell, and the History of Robust Estimation 1885–1920. *Journal of the American Statistical Association* 68:872–879.

STIGLER, STEPHEN M. 1974 Linear Functions of Order Statistics With Smooth Weight Functions. *Annals of Statistics* 2:676–693.

TUKEY, JOHN W. 1960 A Survey of Sampling From Contaminated Distributions. Pages 448–485 in Ingram Olkin et al. (editors), *Contributions to Probability and Statistics: Essays in Honor of Harold Hotelling*. Stanford Univ. Press.

U.S. NATIONAL CENTER FOR HEALTH STASTISTICS 1972 *Annotated Bibliography on Robustness Studies of Statistical Procedures*. Prepared by A. Govindarajulu and R. T. Leslie. DHEW Publication No. (HSM) 72-1051. Vital and Health Statistics, Series 2: Data Evaluation and Methods Research, No. 51. Washington: Government Printing Office.

III
RUNS

A "run" is a sequence of events of one type occurring together in some ordered sequence of events. For example, if a coin is spun 20 times and the results are heads (H) or tails (T) in the following order:

$$\underline{HH}\,\overline{TTT}\,\underline{H}\,\overline{T}\,\underline{HHHH}\,\overline{TTT}\,\underline{H}\,\overline{TT}\,\underline{H}\,\overline{TT},$$

there are 10 runs in the sequence, as indicated by the lines shown. If the tossings occur at random, it is clearly more reasonable to expect the number of runs occurring in the above configuration than those in the configuration

$$HHHHHHHHHHTTTTTTTTTT,$$

which has the same number of heads and tails but only two runs. In dealing with runs as a statistical phenomenon, two general cases can be distinguished. In the first, the number of heads and the number of tails are fixed and interest is centered on the distribution of the number of runs, given the number of heads and tails. In the second, the number of heads and the number of tails themselves are random variables, with only the total number of spins fixed. The distribution of the number of runs in this case will be different.

A study of runs can assist one in deciding upon the randomness or nonrandomness of temporal arrangements of observations. Runs can also be used to form certain nonparametric or distribution-free tests of hypotheses that are usually tested in other ways. Examples of these uses are given below.

Simple runs. Interest in runs can be traced back to Abraham de Moivre's *Doctrine of Chances*, first published in 1781. Whitworth (1867) devotes some space to runs, and Karl Pearson (1897) discusses in an interesting manner the runs of colors and numbers occurring in roulette plays at Monte Carlo. His analysis is, in fact, faulty in certain respects, a point that was picked up by Mood (1940). (Some errors in Mood's own paper have been given by Krishna Iyer [1948].) In an unusual article on runs, Solterer (1941) examines the folklore which holds that accidents or tragedies occur in triplets by analyzing the dates of death of 597 Jesuit priests in the United States for the period 1900–1939. Using a simple run test, he shows that some grouping of deaths does, in fact, occur. Wallis and Moore (1941*a*; 1943) give further illustrations of the uses to which runs have been put. Many writers treat the following basic problem:

Given r_1 elements of one kind and r_2 elements of a second kind, with the $r = r_1 + r_2$ elements arranged on a line *at random*, what is the probability distribution of the total number of runs? The general answer is a bit complex in expression, but the first two moments of d, the number of runs, may be simply written out:

(1) Expected value of $d = E(d) = \frac{2r_1r_2}{r} + 1$,

and

Variance of d

(2) $= \text{var}(d) = \sigma^2(d) = \frac{2r_1r_2(2r_1r_2 - r)}{r^2(r-1)}$.

Furthermore, for large values of r_1 and r_2 the distribution of d tends to normality. This provides a ready and straightforward means of carrying out significance tests for the random arrangement of the r elements.

The power of the test will be good either when elements form rather too few groups—that is, when there is strong positive association of the elements —or when there are too many groups because of a disposition for the elements to alternate. The test may be applied to quantitative observations by dichotomizing them as above or below the sample median (or some other convenient quantile), as in the following example.

Example—lake levels. The data for this example are given in Table 1 and consist of the highest monthly mean level of Lake Michigan–Huron for each of the 96 years from 1860 to 1955 inclusive. Inspection of the figures shows that the median height was 581.3 feet. Each level below this figure is marked B, and each height equal to or above this figure is marked A. Thus, for these data, $r = 96$, $r_A = 49$, and $r_B = 47$. A dichotomization just a bit off the sample median (to avoid boundary problems) has been used. Although a dichotomization near the sample median (as opposed to some theoretical cutting point) was used to define the categories A and B, the null distribution of d is unaffected. The number of runs is counted and found to be 15. From (1) and (2) it is found that $E(d) = 49$ and $\sigma(d) = 4.87$. Testing the observed number of runs against the expected, the quantity

$$\frac{15 + 0.5 - 49}{4.87} = -6.87$$

is referred to the unit normal distribution. (The 0.5 above is the so-called continuity correction; see Wallis & Roberts 1956, pp. 372–375.) The probability of observing a value of −6.87 or smaller by chance, if the arrangement of A's and B's is random, is well below 0.00001, and hence it can be concluded that the observations are not randomly ordered. In other words, the lake has tended to be high for several years at a time and then low for several years. In practical terms, this means that an estimate of next year's highest level will usually be closer to this year's level than to the 96-year average.

Example—anthropological interval sift. The anthropological interval sift technique is described in a paper by Naroll and D'Andrade (1963). Traits often diffuse between neighboring cultures, forming clusters of neighbors with like traits. A sifting method seeks to sift out a cross-cultural sample, so that, ideally, from each geographical cluster of neighbors with like traits only one example will be considered in the cross-cultural sample. The object of the test is to see whether the sift has been successful in removing the correlation effects of neighboring societies with like traits.

The method is applied to narrow strips of the globe 600 nautical miles wide and thousands of miles long. From a random start, equal intervals are marked off along the length of the strip every so many miles. The first society encountered after each mark is included in the sample. The interval chosen must be large enough to ensure that neighboring members of a single diffusion patch are not included any more frequently than would be produced in a random geographic distribution. Everything depends upon the interval chosen, and a run test is used to decide whether the interval used is suitable.

For this purpose the 40 ordered societies obtained in the sample were dichotomized on each of four characteristics. For example, the first dichotomization was according to whether or not the society had a residence rule of a type theoretically associated with bilateral or bilineal kinship systems (bilocal, neolocal, uxorineolocal, or uxoribilocal). Although the values of the r_i in this example may seem to be random, it is reasonable to regard them as fixed and to apply the appropriate tests given the r_i. Hence, for each of the four characteristics looked at, the run test was applied, with the results shown in Table 2. Because of the relationships between the four characteristics, strong dependences exist among them; this is the reason for the identity, or near identity, of the results for characteristics 1 and 3, and for 2 and 4. Thus the

Table 2

Characteristic	d	r_1	r_2	Sample significance level*
1	9	34	6	0.137
2	19	18	22	0.337
3	9	34	6	0.137
4	19	22	18	0.337

* Probability of obtaining the observed number of runs or a more extreme result, if the two kinds of sign were arranged at random.

Source: Naroll & D'Andrade 1963, p. 1061. Reproduced by permission of the American Anthropological Association.

Table 1 — Lake Michigan–Huron, highest monthly mean level for each calendar year, 1860–1955

Year	Level in feet[a]	Two categories[b]	Change[c]
1860	583.3	A	
1861	583.5	A	+
1862	583.2	A	−
1863	582.6	A	−
1864	582.2	A	−
1865	582.1	A	−
1866	581.7	A	−
1867	582.2	A	+
1868	581.6	A	−
1869	582.1	A	+
1870	582.7	A	+
1871	582.8	A	+
1872	581.5	A	−
1873	582.2	A	+
1874	582.3	A	+
1875	582.1	A	−
1876	583.6	A	+
1877	582.7	A	−
1878	582.5	A	−
1879	581.5	A	−
1880	582.1	A	+
1881	582.2	A	+
1882	582.6	A	+
1883	583.3	A	+
1884	583.1	A	−
1885	583.3	A	+
1886	583.7	A	+
1887	582.9	A	−
1888	582.3	A	−
1889	581.8	A	−
1890	581.6	A	−
1891	580.9	B	−
1892	581.0	B	+
1893	581.3	A	+
1894	581.4	A	+
1895	580.2	B	−
1896	580.0	B	−
1897	580.85	B	+
1898	580.83	B	−
1899	581.1	B	+
1900	580.7	B	−
1901	581.1	B	+
1902	580.83	B	−
1903	580.82	B	−
1904	581.5	A	+
1905	581.6	A	+
1906	581.5	A	−
1907	581.6	A	+
1908	581.8	A	+
1909	581.1	B	−
1910	580.5	B	−
1911	580.0	B	−
1912	580.7	B	+
1913	581.3	A	+
1914	580.7	B	−
1915	580.0	B	−
1916	581.1	B	+
1917	581.87	A	+
1918	581.91	A	+
1919	581.3	A	−
1920	581.0	B	−
1921	580.5	B	−
1922	580.6	B	+
1923	579.8	B	−
1924	579.6	B	−
1925	578.49	B	−
1926	578.49	B	0
1927	579.6	B	+
1928	580.6	B	+
1929	582.3	A	+
1930	581.2	B	−
1931	579.1	B	−
1932	578.6	B	−
1933	578.7	B	+
1934	578.0	B	−
1935	578.6	B	+
1936	578.7	B	+
1937	578.6	B	−
1938	579.7	B	+
1939	580.0	B	+
1940	579.3	B	−
1941	579.0	B	−
1942	580.2	B	+
1943	581.5	A	+
1944	580.8	B	−
1945	581.00	B	+
1946	580.96	B	−
1947	581.1	B	+
1948	580.8	B	−
1949	579.7	B	−
1950	580.0	B	+
1951	581.6	A	+
1952	582.7	A	+
1953	582.1	A	−
1954	581.7	A	−
1955	581.5	A	−

a. Data for certain years are shown to two decimals to avoid ties.

b. *A*: 581.3 feet or more.
B: less than 581.3 feet.

c. This column will be used in a later example.

Source: Wallis & Roberts 1956, p. 566.

four tests are appreciably dependent in the statistical sense.

In all four cases the observed number of runs was slightly below the expected number. Application of the test described above showed, however, that there was no statistically significant evidence of overclustering and, hence, no statistically significant evidence of diffusion. Thus the sift had been successful in removing the statistically significant effects of diffusion. Of course, considerations of power would be necessary for a full analysis.

Runs with more than two categories. The foregoing can be extended to the situation where the r elements are subdivided into more than two types of element, say k types, so that there are r_1 elements of type 1, r_2 of type 2, and so on up to r_k of type k where $\sum_j r_j = r$. When the r elements are arranged at random, the first two moments of the number of runs, d, are now

$$(3) \qquad E(d) = r - \frac{1}{r}\sum_{j=1}^{k} r_j(r_j - 1),$$

and

$$(4) \qquad \begin{aligned} \text{var}(d) &= \sigma^2(d) \\ &= \frac{F_2(r-3)}{r(r-1)} + \frac{F_2^2}{r^2(r-1)} - \frac{2F_3}{r(r-1)}, \end{aligned}$$

where $F_2 = \sum_j r_j(r_j - 1)$, $F_3 = \sum_j r_j(r_j - 1)(r_j - 2)$. These formulas, provided that r and the r_j are all reasonably large, permit tests of significance for random groupings to be carried out.

Example—runs with multiple types. The example of runs with multiple types presented here is based on the work of E. S. Pearson and is quoted in David and Barton (1957). The data relate to the falls in the prices of shares on the London Stock Exchange during the Suez crisis period, November 6, 1956–December 8, 1956, inclusive. Five types of industrial activity were considered: A, insurance; B, breweries and distilleries; C, electrical and radio; D, motor and aircraft; E, oil. The closing prices as given in *The Times* (of London) for 18 businesses of each type were taken, and Table 3 shows, for each day, the type of industrial

Table 3

Nov. 6—A	Nov. 13—B	Nov. 20—E	Nov. 27—B	Dec. 4—C
Nov. 7—A	Nov. 14—C	Nov. 21—C	Nov. 28—E	Dec. 5—C
Nov. 8—D	Nov. 15—C	Nov. 22—E	Nov. 29—A	Dec. 6—D
Nov. 9—D	Nov. 16—C	Nov. 23—E	Nov. 30—E	Dec. 7—C
Nov. 10—A	Nov. 17—E	Nov. 24—E	Dec. 1—E	Dec. 8—B

activity for which the greatest number of the 18 showed a fall in price from the previous day. In the few cases where there were equal numbers for two types, that type which also showed the fewer rises in price was taken. For the data shown in Table 3, $r = 25$ with $r_A = 4$, $r_B = 3$, $r_C = 7$, $r_D = 3$, $r_E = 8$. There are 16 runs (d). From (3) and (4) above, the mean and standard deviation of d, given a random grouping of the letters, are $E(d) = 20.12$, $\sigma(d) = 1.24$. The probability of getting the observed value of d or a lower one by chance if the grouping is random is therefore found by referring the quantity

$$\frac{16 + 0.5 - 20.12}{1.24} = -2.91$$

to a unit normal distribution. The appropriate probability is 0.001, and it is therefore concluded that the test is picking out the fact that during the Suez crisis there was some persistence from day to day in the way in which different classes of shares were affected.

Runs in a circle. A modification of the type of run described above occurs when the line is bent into a circle so that there is no formal starting and ending point. If each configuration around the circle is equally probable, then it is possible to find the distribution of the total number of runs, d (see, for example, David & Barton 1962, p. 94, or Stevens 1939). The first two moments are

$$(5) \qquad E(d) = \frac{r_1 r_2}{r-1},$$

$$(6) \qquad \text{var}(d) = \sigma^2(d) = \frac{r_1 r_2 (r_1 r_2 - 3r_1 - 3r_2 + 1)}{(r-1)^2(r-2)}.$$

For small r, say, up to 20, the exact probabilities need to be enumerated to carry out significance tests (see, for example, Walsh 1962–1968), but for larger values of r the distribution can be assumed to be normal with the appropriate parameters. It should also be noted that in this analysis it is assumed that the circle cannot be turned over, as would be the case in considering the number of runs among the beads threaded on a bracelet.

Random totals. In the foregoing analyses it is assumed that the total composition of the sequence is known, which is to imply that the r_i are known quantities. The probability distribution and moments are thus conditional upon r_i being known. It is possible to take the argument a stage further and to consider the situation if the r_i are obtained as a result of some sort of sampling and are thus themselves random variables. For purposes of illustration, suppose that there are two kinds of elements and that the sampling is of a binomial form with the chance of an element being of the first kind as p and the chance of its being of a second kind as q (where $p + q = 1$), the elements being independent of one another. Then it can be shown that

$$(7) \qquad E(d) = 1 + 2(r-1)pq$$

and

$$(8) \qquad \begin{aligned} \text{var}(d) &= \sigma^2(d) \\ &= pq(4r - 6) - p^2q^2(12r - 20). \end{aligned}$$

A test based on the above moments might be used in a quality control situation where a previously estimated proportion, p, of articles are expected to have some characteristic. Consecutive items are examined and the number of runs, d, found in r items are counted in order to see whether the machine concerned is producing a random ordering of quality or not [*see* QUALITY CONTROL, STATISTICAL].

Runs up and down. Given a sequence of numbers, a sign, either + or −, can be attached to each number other than the first, the sign being + if the current number is above the previous number numerically or − if it is below. Using this sequence of +'s and −'s, one can count the number of runs (d'). If the original sequence of numbers was such that long sequences of roughly equal numbers were placed together, the number of runs as defined here would be large, since there would be a considerable element of chance as to whether two adjacent numbers were in ascending or descending order of magnitude. If there is, however, a gentle rise followed by a gentle fall in the original sequence, perhaps repeated several times, then the number of runs will be much smaller. Hence, the number of runs calculated in this manner is a measure of persistence of the trend. The probability distribution is a complicated one (details are given in David & Barton 1962, p. 154), but the first two moments of it can be evaluated fairly readily and give

$$E(d') = \frac{2n-1}{3} \tag{9}$$

and

$$\operatorname{var}(d') = \sigma^2(d') = \frac{16n-29}{90}, \tag{10}$$

where n is the total number of original observations, giving $n-1$ signs in the sequence. Again it may be shown that for large values of n, the distribution tends to normality.

Example—runs up and down. The data in Table 1 is used again, only in this example the aim is to see whether the directions of movement in water level tend to persist. It would be possible for high and low values to cluster together simply as a result of a few large changes, with changes of direction within the clusters varying as for independent observations.

To carry out a suitable test of the hypothesis that there is no persistence in the directions of the measurements, a sign was inserted against the maximum level for each year except the first. The sign is + if the level was higher than for the preceding year, − if it was lower, 0 if it was the same. (In this case the total number of runs, d', would be the same whether the 0 was considered as + or as −.) The longest movements in one direction were the two five-year declines from 1861 to 1866 and from 1886 to 1891. There were two four-year movements, the rise from 1879 to 1883 and the decline from 1922 to 1926.

The total number of runs, d', was 48. On the hypothesis that the + and − signs were arranged at random, the distribution of d' is approximately normal with moments from (9) and (10): $E(d') = 63.67$ and $\sigma(d') = 4.10$. Testing the observed against the expected value of d' the quantity

$$\frac{48 + 0.5 - 63.67}{4.10} = -3.7$$

is referred to a unit normal distribution. A one-tailed probability is required, since the alternative hypothesis is the one-sided one that there will be fewer runs than the null hypothesis indicates. The probability of a value of −3.7 or smaller arising by chance if there were no real persistence of movement in the same direction would be less than 0.001. Thus, the lake evidently has a tendency to move consecutively in the same direction more often than would be the case with independent observations. The clustering of high and low values, therefore, is not due (at least not exclusively) to a few large changes but in some part to cumulative movements up and down.

Test for two populations. Let $x_1, x_2, \cdots, x_{n_1}$ be an ordered sample from a population with probability density $f(x)$, and let $y_1, y_2, \cdots, y_{n_2}$ be a second ordered sample from the same population. Let the two samples be combined and arranged in order of magnitude; thus, for example, one might have

$$x_1 < y_1 < y_2 < x_2 < y_3 < x_3 < x_4 < x_5 < y_4 < \cdots.$$

To test the null hypothesis that the two samples come from the same population, let the two samples be combined as above, define a run as a sequence of letters of the same kind bounded by letters of the other kind, and count the total number of runs, d.

Thus, in the above example the first element forms a run, the second and third a run, the fourth a run, the fifth a run, the sixth to eighth a run, and so on. The first and last elements are bounded only on one side for this purpose so that if only the nine elements given here were involved, there would be six runs.

It is apparent that if the two samples are from the same population, the x's and y's will ordinarily be well mixed and d will be large. If the two populations are widely separated so that their ranges do not overlap at all, the value of d will be only two and, in general, differences between the populations will tend to produce low values of d. Thus, the two populations may have the same mean or median, but if the x population is relatively concentrated compared to the y population, there will be a long y run at each end of the combined sample and there will thus tend to be a low value of d.

The test is performed by observing the total number of runs in the combined sample, accepting the null hypothesis if d is greater than some specified number d_0, rejecting the null hypothesis otherwise. To do this, the first two moments of the null distribution of d, given in (1) and (2) above, may be used. This run test, originated by Wald and Wolfowitz (1940), is sensitive to differences in both shape and location between the two populations and is described in more detail in Mood (1950).

Example—two-sample test. This example of the two-sample test is taken from Moore (1959). Two random samples, each of 14 observations, were available, and it was desired to test whether they could be considered to have come from the same population or not.

Sample A: 9.1, 9.5, 10.0, 10.4, 10.7,
9.0, 9.5, 12.2, 11.8, 12.0,
10.2, 11.5, 11.6, 11.2

Sample B: 9.4, 11.3, 10.9, 9.4, 9.2,
8.5, 8.9, 11.4, 11.1, 9.3,
9.8, 9.6, 8.0, 9.7

Arranging these in numerically ascending order gave the following (where sample B is shown in italics):

8.0, *8.5*, *8.9*, 9.0, 9.1, *9.2*, *9.3*,
9.4, *9.4*, 9.5, 9.5, *9.6*, *9.7*, *9.8*,
10.0, 10.2, 10.4, 10.7, *10.9*, *11.1*, 11.2,
11.3, *11.4*, 11.5, 11.6, 11.8, 12.0, 12.2

A straightforward count shows that there are 10 runs; hence, $d = 10$. From (1) and (2), with $r_1 = r_2 = 14$ and $r = 28$, $E(d) = 15$ and var $(d) = 6.74$. Testing the null hypothesis that there is no difference between the two samples, the quantity

$$\frac{10 + 0.5 - 15}{\sqrt{6.74}} = -1.73$$

is referred to the unit normal distribution. The probability of that value, or a smaller value, arising by chance if there were no difference between the two samples is 0.045—that is, the difference is significant at the 5 per cent level but not at the 1 per cent level—and thus some doubt is thrown on the hypothesis that the two samples come from the same population. A standard two-sample t-test for samples from normal populations (see, for example, Moore & Edwards 1965, pp. 9–10) gives a similar indication.

Other run tests. Various other procedures based on runs have been proposed, a few of which are mentioned here. Length of runs as a criterion is discussed by Takashima (1955). The use of the longest run up or down (that is, numerically rising or falling) is discussed by Olmstead (1946), who gives a table to facilitate the use of his test. Mosteller (1941) gives a test on the use of long runs of observations above or below the median and discusses its use in the interpretation of quality control charts. Mann (1945) discusses a test for randomness based on the number of runs up and down. In testing goodness of fit, runs of signs of cell deviations from the null hypothesis have been studied (see, for example, Walsh 1962–1968, vol. 1, pp. 450–451, 463–465).

An extension of runlike concepts is used by Campbell, Kruskal, and Wallace (1966) to study the distribution of adjacencies between Negro and white students seating themselves in a classroom.

The power of run tests has been investigated by a number of writers. For example, David (1947) studies the power of grouping tests for randomness. Bateman (1948) investigates the power of the longest run test. Levene (1952) studies the power of tests for randomness based on runs up and down. The power of the two-sample tests above depends upon the kind of alternative that is being considered. In instances where the alternative can be precisely specified, there may be other tests, not primarily based on runs, that have better power.

Ties. Except in one of the above examples, where the course of action was obvious, the possibility of *ties*, that is, a pair of observations being precisely equal, has been ignored. Hence, observations could always be ordered uniquely without ties and the various statistics uniquely defined. However, in practice, observations are measured only to a few significant figures and ties will therefore sometimes occur. The simplest way to deal with this possibility in the situations discussed here is to assume that the tied observations are ordered at random. The appropriate tests are then carried out as before. Discussions of the problems of ties are given by Kendall (1948) and Kruskal (1952).

PETER G. MOORE

BIBLIOGRAPHY

BATEMAN, G. 1948 On the Power Function of the Longest Run as a Test for Randomness in a Sequence of Alternatives. *Biometrika* 35:97–112.

CAMPBELL, DONALD T.; KRUSKAL, WILLIAM H.; and WALLACE, WILLIAM P. 1966 Seating Aggregation as an Index of Attitude. *Sociometry* 29, no. 1:1–15.

DAVID, F. N. 1947 A Power Function for Tests of Randomness in a Sequence of Alternatives. *Biometrika* 34:335–339.

DAVID, F. N.; and BARTON, D. E. 1957 Multiple Runs. *Biometrika* 44:168–178.

DAVID, F. N.; and BARTON, D. E. 1962 *Combinatorial Chance*. New York: Hafner.

○FISHER, R. A. (1925) 1970 *Statistical Methods for Research Workers*. 14th ed., rev. & enl. New York: Hafner; Edinburgh: Oliver & Boyd.

○KENDALL, MAURICE G. (1948) 1970 *Rank Correlation Methods*. 4th ed. London: Griffin; New York: Hafner.

KRISHNA IYER, P. V. 1948 The Theory of Probability Distributions of Points on a Line. *Journal of the Indian Society of Agricultural Statistics* 1:173–195.

KRUSKAL, WILLIAM H. 1952 A Non-parametric Test for the Several Sample Problem. *Annals of Mathematical Statistics* 23:525–540.

LEVENE, HOWARD 1952 On the Power Function of Tests of Randomness Based on Runs Up and Down. *Annals of Mathematical Statistics* 23:34–56.

LUDWIG, OTTO 1956 Über die stochastische Theorie der Merkmalsiterationen. *Mitteilungsblatt für mathematische Statistik und ihre Anwendungsgebiete* 8:49–82.

MANN, HENRY B. 1945 On a Test for Randomness Based on Signs of Differences. *Annals of Mathematical Statistics* 16:193–199.

MOOD, ALEXANDER M. 1940 The Distribution Theory of Runs. *Annals of Mathematical Statistics* 11:367–392.

MOOD, ALEXANDER M. 1950 *Introduction to the Theory of Statistics*. New York: McGraw-Hill. → A second edition, by Mood and Franklin A. Graybill, was published in 1963.

MOORE, PETER G. 1959 Some Approximate Statistical Tests. *Operational Research Quarterly* 10:41–48.

[1]MOORE, PETER G.; and EDWARDS, D. E. 1965 *Standard Statistical Calculations*. London: Pitman.

MOSTELLER, FREDERICK 1941 Note on an Application of Runs to Quality Control Charts. *Annals of Mathematical Statistics* 12:228–232.

NAROLL, RAOUL; and D'ANDRADE, ROY G. 1963 Two Further Solutions to Galton's Problem. *American Anthropologist* New Series 65:1053–1067.

OLMSTEAD, P. S. 1946 Distribution of Sample Arrangements for Runs Up and Down. *Annals of Mathematical Statistics* 17:24–33.

PEARSON, KARL 1897 *The Chances of Death and Other Studies in Evolution*. Vol. 1. New York: Arnold.

SOLTERER, J. 1941 A Sequence of Historical Random Events: Do Jesuits Die in Three's? *Journal of the American Statistical Association* 36:477–484.

STEVENS, W. L. 1939 Distribution of Groups in a Sequence of Alternatives. *Annals of Eugenics* 9:10–17.

TAKASHIMA, MICHIO 1955 Tables for Testing Randomness by Means of Lengths of Runs. *Bulletin of Mathematical Statistics* 6:17–23.

WALD, ABRAHAM; and WOLFOWITZ, J. (1940) 1955 On a Test Whether Two Samples Are From the Same Population. Pages 120–135 in Abraham Wald, *Selected Papers in Statistics and Probability*. New York: McGraw-Hill. → First published in Volume 11 of the *Annals of Mathematical Statistics*.

WALLIS, W. ALLEN; and MOORE, GEOFFREY H. 1941a A Significance Test for Time Series Analysis. *Journal of the American Statistical Association* 36:401–409.

WALLIS, W. ALLEN; and MOORE, GEOFFREY H. 1941b *A Significance Test for Time Series*. Technical Paper No. 1. New York: National Bureau of Economic Research.

WALLIS, W. ALLEN; and MOORE, GEOFFREY H. 1943 Time Series Significance Tests Based on Signs of Differences. *Journal of the American Statistical Association* 38:153–164.

WALLIS, W. ALLEN; and ROBERTS, HARRY V. 1956 *Statistics: A New Approach*. Glencoe, Ill.: Free Press. → A revised and abridged paperback edition of the first section was published by Collier in 1962 as *The Nature of Statistics*.

○WALSH, JOHN E. 1962–1968 *Handbook of Nonparametric Statistics*. 3 vols. Princeton, N.J.: Van Nostrand. → Volume 1: *Investigation of Randomness, Moments, Percentiles, and Distributions*. Volume 2: *Results for Two and Several Sample Problems, Symmetry and Extremes*. Volume 3: *Analysis of Variance*.

WHITWORTH, WILLIAM ALLEN (1867) 1959 *Choice and Chance, With 1000 Exercises*. New York: Hafner. → The first and subsequent editions were published in Cambridge by Bell.

WHITWORTH, WILLIAM ALLEN (1897) 1965 *DCC Exercises: Including Hints for the Solution of All the Questions in* Choice and Chance. New York: Hafner. → A reprint of the first edition published in Cambridge by Bell.

ADDITIONAL BIBLIOGRAPHY

MARSHALL, CLIFFORD W. 1970 A Simple Derivation of the Mean and Variance of the Number of Runs in an Ordered Sample. *American Statistician* 24, no. 4: 27–28.

[1]MOORE, PETER G.; SHIRLEY, ERYLE A. C.; and EDWARDS, D. E. (1965) 1972 *Standard Statistical Calculations*. 2d ed. London: Pitman; New York: Wiley. → The first edition was written by Moore and Edwards.

IV
RANKING METHODS

A *ranking* is an ordering of individuals or objects according to some characteristic of interest. If there are n objects, it is natural to assign to them the *ranks* $1, 2, \cdots, n$, with 1 assigned to the object ranked highest. In this article it is, however, more convenient to employ the opposite and completely equivalent convention of assigning rank 1 to the lowest and rank n to the highest object. Ranking methods are concerned with statistics (*rank-order statistics*) constructed from the ranks, usually in random samples of observations. An ordinal scale of measurement clearly suffices for the calculation of such statistics, but ranking methods are also frequently used even when meaningfully numerical measurements are available. In cases when such measurements are available the ordered observations may be denoted by $x_{(1)} \leqslant x_{(2)} \leqslant \cdots \leqslant x_{(n)}$, where $x_{(i)}$ $(i = 1, 2, \cdots, n)$ is the ith *order statistic*. Only the rank i of $x_{(i)}$ occurs in rank-order statistics; use of $x_{(i)}$ itself leads to order statistics [*see* NONPARAMETRIC STATISTICS, *article on* ORDER STATISTICS].

Objects which are neighbors in the ranking always differ in rank by 1, although it may reasonably be objected that the actual difference between neighboring objects is often greater near the ends of a ranking than in the middle. This difficulty can be overcome by the use of *scores*, a score being a suitably chosen function of the corresponding rank. However, the objection turns out to have less force than it might appear to have, since in many important cases scoring has been found to be little more efficient than the simpler ranking.

It is clear that an unambiguous ranking is not possible when ties are present. Strictly speaking, ranking methods therefore require the assumption of an underlying continuous distribution so as to ensure zero probability for the occurrence of ties. In applications, ties will nevertheless insist on appearing. Commonly, each of the tied values is given the average rank of the tied group, a technique which leaves the sum of the ranks unchanged but reduces their variability. The effect is negligible if the number of ties is small, but in any case simple corrections can be applied to many of the procedures developed on the assumption of no ties.

Ranking methods are potentially useful in all fields of experimentation where measurements can be made on at least an ordinal scale. Much of the motivation for the development of the subject has in fact stemmed from the social sciences. To date, testing of hypotheses (for instance, identity of two or more populations) has been most fully explored. Point and interval estimation are sometimes possible, but usually the order statistics are needed in addition to the ranks. Ranks are also beginning to be used in sequential analysis and multiple-decision procedures. Ranking methods are not as flexible as parametric procedures developed on the assumption of normality, but they have a much greater range of validity, since they do not generally require knowledge of the underlying distribution other than its continuity (but see, for example, "Estimation," below). Most ranking methods therefore fall under the wider heading of nonparamétric statistics and play a major role in that subject.

When numerical measurements are replaced by their ranks, the question arises of how much "information" is thereby lost. Perhaps contrary to intuition, the loss in efficiency or power of the standard rank tests compared to the best corresponding parametric tests is usually quite small. The loss hinges on several factors: the particular test used, the nature of the alternative to the null hypothesis under test, and the sample size. Since the best parametric test in a given situation depends on the often uncertain form of the underlying distribution, one should also compare the performance of the two tests for other distributional forms which may reasonably occur. Whereas the significance level of the rank test remains quite unchanged, that of the parametric test may be seriously upset. Even when the parametric test is not too sensitive in this respect it can easily become inferior in power to the rank test. A more obvious point in favor of ranking methods is that they are much less affected by spurious or wild observations.

The following account begins with two-sample problems and then turns to several-sample situations and rank correlation. [*Nonparametric one-sample procedures, including ranking methods, are considered in* NONPARAMETRIC STATISTICS, *article on* THE FIELD.] The emphasis is on standard tests, but some estimation procedures are described, and reference is made to several multiple-decision procedures.

Two-sample problems

Let $x_1, x_2, \cdots, x_m$ and $y_1, y_2, \cdots, y_n$ be independent observations made respectively on the continuous random variables X with density $f(x)$ and Y with density $g(y)$. Consider the important question: Do the x's differ significantly from the y's; that is, do the observations throw serious doubt on the null hypothesis that $f = g$? The statistical test most appropriate for answering this question will depend on the alternative the experimenter has in mind. Often the y's are expected, on the whole, to be larger than the x's; that is, if $p = Pr(Y > X)$, the null hypothesis H_0: $f = g$, for which $p = \frac{1}{2}$, is tested against the alternative, H_1: $p > \frac{1}{2}$. A special case of this alternative occurs when $g(y)$ is merely $f(x)$ shifted to the right, so that the two populations differ in location only. Other alternatives of interest are $_1H$: $p < \frac{1}{2}$ and the two-sided alternative H_2 : $p \neq \frac{1}{2}$.

Note that H_0 specifies identity of f and g, not merely that the distributions have the same median (see Pratt 1964) or the same spread.

Wilcoxon's two-sample test. Wilcoxon's two-sample test enables one to test H_0 without knowledge of the common (under the null hypothesis) functional form of f and g. Arrange the $m + n$ observations (w_i) in combined ascending order of magnitude giving $w_1 \leqslant w_2 \leqslant \cdots \leqslant w_{m+n}$. The Wilcoxon (1945) statistic is the sum, R_x, of the ranks of those w's that came from $f(x)$. (For an equivalent counting procedure see Mann & Whitney 1947.) If H_0 is true, each observation has expected rank given by $[1 + 2 + \cdots + (m + n)]/(m + n) = \frac{1}{2}(m + n + 1)$, so that the expected value of R_x is $\frac{1}{2}m(m + n + 1)$. When H_1 is the alternative under consideration, values of R_x much smaller than this lead to the rejection of H_0 in favor of H_1. For a two-sided test, H_0 is rejected when R_x differs from $E(R_x)$ by too much in either direction. Actual significance points are given in many sources, for example, by Siegel and Tukey (1960) for $m \leqslant n \leqslant 20$. Outside this range a normal approximation, possibly with continuity correction, is adequate unless m and n differ greatly. Simply treat

$$z = [2R_x - m(m+n+1)]/\sqrt{\tfrac{1}{3}mn(m+n+1)}$$

as approximately a unit normal deviate.

An example. The amount of aggression attributed to characters in a film by nine members of each of two populations resulted in the following scores (Siegel & Tukey 1960):

x: 25 5 14 19 0 17 15 8 8

y: 12 16 6 13 13 3 10 10 11

These are combined into the w-series, with their origin noted as in Table 1. (The modified ranks will be discussed below.) Here $R_x = 1+3+5+6+13+14+16+17+18 = 93$ and $E(R_x) = \tfrac{1}{2}\cdot 9 \cdot 19 = 85.5$. Because R_x is this close to $E(R_x)$, there is evidently no reason for rejecting H_0. Tables show that for a two-sided test R_x would have to be as small as 62 or, by symmetry, as large as 109 to be statistically significant at the 5 per cent level.

Table 1 — Aggression scores and rankings

SCORE	SAMPLE	RANK	MODIFIED RANK
0	x	1	1
3	y	2	4
5	x	3	5
6	y	4	8
8	x	5	9
8	x	6	12
10	y	7	13
10	y	8	16
11	y	9	17
12	y	10	18
13	y	11	15
13	y	12	14
14	x	13	11
15	x	14	10
16	y	15	7
17	x	16	6
19	x	17	3
25	x	18	2

Test of relative spread. Although the foregoing example was used to illustrate Wilcoxon's test for differences in location, the alternative of interest was, in fact, that the two populations differ in *variability*. An appropriate test can be made by assigning in place of the usual ranks the modified ranks shown in the last row of Table 1, which are arranged so that low ranks are applied to extreme observations—both high and low—whereas high ranks are applied to the more central observations. Clearly, sufficiently low or high values of the sum R'_x of modified ranks will again lead to rejection of H_0. As before, there are three alternatives—say, H'_1, ${}_1H'$, H'_2, where H'_1 specifies that the x's are more spread out than the y's, etc. Precisely the same significance points apply as for the ordinary Wilcoxon test, since the ranks have merely been reallocated to produce a test statistic sensitive to the alternative under consideration. (Indeed, other alternatives, H'', can be tested by an assignment of ranks which tends to make R''_x small or large if H'' is true. As usual, the alternative should be specified prior to the experiment.)

Here $R'_x = 1+5+9+12+11+10+6+3+2 = 59$, which is significant at the 5 per cent (1 per cent) level against $H'_2(H'_1)$.

Unless the two populations have the same median or can be changed so that they do, interpretation of this test may be difficult.

Several-sample problems

One-way classification. Wilcoxon's procedure can be generalized to produce a test of the identity of several populations. Given the N observations x_{ij} ($i = 1, 2, \cdots, k$; $j = 1, 2, \cdots, n_i$; $\sum n_i = N$), arrange them in common ascending order. Let R_i denote the sum of the ranks of those n_i observations originating from the ith population. If $\bar{R}_i = R_i/n_i$, it is clear that large variations in these mean ranks will cast suspicion on the null hypothesis of equality of all k populations. A suitably standardized weighted sum-of-squares test statistic suggests itself. Kruskal and Wallis (1952) show that except for very small n_i,

$$(1) \qquad H = \frac{N-1}{N}\sum_{i=1}^{k}\frac{n_i[\bar{R}_i - \tfrac{1}{2}(N+1)]^2}{(N^2-1)/12} = \frac{12}{N(N+1)}\sum_{i=1}^{k}\frac{R_i^2}{n_i} - 3(N+1)$$

is distributed (under the null hypothesis of identity of the populations) approximately as chi-square with $k-1$ degrees of freedom. The last form of (1) is a convenient computing formula. In the central form it may be noted that $\tfrac{1}{2}(N+1)$ is the mean and $(N^2-1)/12$ the variance of a randomly chosen rank. In fact, the procedure is essentially a one-way analysis of variance performed on the ranks as variables.

Two-way classification. Friedman (1937) was concerned with data giving the standard deviations of expenditures on $n = 14$ different categories of products at $k = 7$ income levels. The problem was to determine whether the standard deviations differed significantly over income levels. The "problem of n rankings" considered by Kendall and Babington Smith (1939) can be handled in the same way and the procedure will be illustrated with such data. The $n = 4$ observers were asked to rank $k = 6$ objects, with the results shown in Table 2. An approximate test of equality among the objects is obtained by referring

$$\chi_r^2 = \frac{k-1}{k} \sum_{i=1}^{k} \frac{n[\bar{r}_i - \frac{1}{2}(k+1)]^2}{(k^2-1)/12}$$
$$= \frac{12}{nk(k+1)} \sum_{i=1}^{k} r_i^2 - 3n(k+1)$$

to tables of chi-square with $k - 1$ degrees of freedom. A measure of agreement between observers is given by the *coefficient of concordance*, denoted by $W = \chi_r^2/[n(k-1)]$, defined so as to have range (0,1), with $W = 1$ corresponding to complete agreement. (In the χ_r^2 expression $\bar{r}_i = r_i/n$.)

Table 2 — Results of four observers ranking six objects

		OBJECT					
		A	B	C	D	E	F
OBSERVER	P	5	4	1	6	3	2
	Q	2	3	1	5	6	4
	R	4	1	6	3	2	5
	S	4	3	2	5	1	6
Totals	r_i	15	11	10	19	12	17

In this example $\chi_r^2 = 32/7 = 4.57$, which is clearly not significant, and $W = 0.23$.

In cases where χ_r^2 is significantly large, that may be because one or more of the objects are "outliers" —that is, come from a population different from the bulk of the objects. Tables for the detection of such outliers have been provided by Doornbos and Prins (1958) and by Thompson and Willke (1963) [*see* STATISTICAL ANALYSIS, SPECIAL PROBLEMS OF, *article on* OUTLIERS].

Incomplete rankings. In a two-way classification, it is often impracticable to have each observer rank all the objects at one time. For example, if the objects are foods to be tasted, the block size is best restricted to 2 or 3, but each judge may be asked to perform several such tastings with intervening rest periods. By means of suitable balanced incomplete block designs it is still possible to treat this case by very similar methods [*see* EXPERIMENTAL DESIGN; *see also* Durbin 1951].

The method of paired comparisons. Formally, the method of paired comparisons is a special case of incomplete ranking where the block size is 2. Each of the observers expresses a preference for one of the objects in every pair he judges. Ordinarily all possible $n\binom{k}{2}$ pairwise comparisons are made, but fractional designs are also available.

The method has long interested psychologists (for "objects" read "stimuli") and can be traced back to G. T. Fechner. It received fresh impetus from the work of L. L. Thurstone (1927), who supposed, in effect, that a particular observer's response to object A_i can be characterized by a normal variable, Y_i, with true mean V_i. The probability that A_i is preferred to A_j ($A_i \rightarrow A_j$) in direct comparison is then $Pr(Y_i > Y_j)$. Thurstone distinguished five cases, of which the simplest (case V) assumes that (*a*) observer differences may be ignored and that (*b*) the Y_i are independent (actually, equal correlation suffices) normal variates with common variance. In this situation it is easy to estimate the V_i and to test their equality (see Mosteller 1951; Torgerson 1968).

Distributions other than normal have also been proposed for the Y_i. Postulating a sech² density for $Y_i - Y_j$, an assumption close to that of normality, Bradley and Terry (1952) arrive at the model

$$Pr(Y_i > Y_j) = \pi_i/(\pi_i + \pi_j), \qquad \pi_i \geqslant 0, \sum \pi_i = 1;$$

in other words, the odds on $A_i \rightarrow A_j$ are π_i to π_j. The π_i can be estimated by maximum likelihood. This approach can be extended to comparisons in triples and to multiple-choice situations (Luce 1959).

Implicit in these various models is the assumption that one-dimensional scaling of responses is appropriate. No such assumption is needed in the combinatorial approach of Kendall and Babington Smith (1940), who count the number of circular triads—that is, the number of times $A_i \rightarrow A_j$, $A_j \rightarrow A_k$, and yet $A_k \rightarrow A_i$. A sufficiently low count leads to rejection of the null hypothesis, H_0, of equality among the objects. The following procedure (David 1963) is completely equivalent for a single observer and provides a very simple general test. Let a_i be the total number of times that A_i is preferred to other objects. Then, under H_0,

$$T = \sum_{i=1}^{k} [a_i - \tfrac{1}{2}n(k-1)]^2/(4nk)$$

is distributed approximately as chi-square with $k - 1$ degrees of freedom. Large values of T lead to rejection of H_0, in which case it is possible to make more detailed statements about differences between the objects. The necessary multiple-decision procedures are analogous to those developed by Tukey and Scheffé for the separation of treatment means in the analysis of variance [*see* LINEAR HYPOTHESES, *article on* MULTIPLE COMPARISONS].

Rank correlation

For n pairs of observations (x_i, y_i) the product-moment coefficient of correlation

$$r = \frac{\sum(x_i - \bar{x})(y_i - \bar{y})}{\{\sum(x_i - \bar{x})^2 \sum(y_i - \bar{y})^2\}^{\frac{1}{2}}}$$

provides a convenient index $(-1 \leqslant r \leqslant 1)$ of the extent to which, as x increases, y increases $(r > 0)$ or decreases $(r < 0)$ linearly [*see* MULTIVARIATE ANALYSIS, *article on* CORRELATION METHODS].

Spearman's ρ. If r is applied to the ranks rather than to the observations, the result is Spearman's coefficient of rank correlation, more readily computed from

$$\rho = 1 - \frac{6\sum d_i^2}{n(n^2 - 1)},$$

where d_i is the difference in rank between y_i and x_i. Here $\rho = 1$ corresponds to complete agreement and $\rho = -1$ to complete reversal in the two sets of ranks.

An example. Consider the rankings of observers P and Q in Table 2. There, $d_i = 3, 1, 0, 1, -3, -2$, so that

$$\rho = 1 - \frac{6(24)}{6(6^2 - 1)} = 0.31,$$

indicating only slight agreement between P and Q. In fact, a value of $\sum d_i^2$ that is 4 or smaller or 66 or larger would be needed for rejection of the null hypothesis, H_0, of independent rankings at a 5 per cent level of significance (Owen 1962). For $n \geqslant 12$ an approximate test of H_0 can be made by treating $\rho\sqrt{n - 1}$ as unit normal.

Estimation. If the underlying distribution of (x, y) is bivariate normal, with correlation coefficient ρ_0, an estimate of ρ_0 which is unbiased in large samples is given by

$$r_0 = 2 \sin \tfrac{1}{6}\pi\rho = 2 \sin (30\rho)^\circ.$$

In the example, $r_0 = 0.33$.

Test of trend. Given the observations $y_1, y_2, \cdots, y_n$, an investigator may wonder whether some positional effect or trend is present. For example, if y_i is the reaction time of a subject to the ith presentation of a stimulus, fatigue could tend to increase the later observations. A possible test of randomness against such an alternative in this one-sample situation is to compare the ranking of the y_i with the natural ordering $1, 2, \cdots, n$, by means of ρ (with $d_i = \text{rank } y_i - i$). Clearly, a strong upward trend would lead to large values of ρ. Jonckheere (1954) treats the corresponding k-sample problem.

Other rank correlation coefficients. The coefficient ρ is a member of a class of correlation coefficients which includes r, Kendall's τ, and the Fisher–Yates coefficient, r_F, obtained by replacing x_i and y_i in r by their normal scores. The last three quantities are further discussed by Fieller and his associates (1957). In addition to the above applications, ρ, τ, and r_F may be used to test the equality of the true correlation coefficients in two bivariate normal populations and to investigate by means of partial rank correlation coefficients whether agreement between two rankings might be due to some extraneous factor (for example, Goodman 1959). For interpretations of various rank correlation coefficients see Kruskal (1958).

HERBERT A. DAVID

BIBLIOGRAPHY

○ *Of the books in this bibliography,* Kendall (1948) 1970 *corresponds most closely to the coverage of this article.* Owen 1962 *includes a very useful section on ranking methods and gives many of the tables needed to supplement in small samples the approximate tests discussed. The subject of paired comparisons is treated in* David 1963. Siegel 1956 *gives a helpful elementary account of nonparametric statistics for the behavioral sciences.*

BRADLEY, RALPH A.; and TERRY, M. E. 1952 Rank Analysis of Incomplete Block Designs. I: The Method of Paired Comparisons. *Biometrika* 39:324–345.

DAVID, HERBERT A. 1963 *The Method of Paired Comparisons.* London: Griffin; New York: Hafner.

DOORNBOS, R.; and PRINS, H. J. 1958 On Slippage Tests. III: Two Distribution-free Slippage Tests and Two Tables. *Indagationes mathematicae* 20:438–447.

DURBIN, J. 1951 Incomplete Blocks in Ranking Experiments. *British Journal of Psychology* 4:85–90.

FIELLER, E. C.; HARTLEY, H. O.; and PEARSON, E. S. 1957 Tests for Rank Correlation Coefficients. *Biometrika* 44:470–481.

FRIEDMAN, MILTON 1937 The Use of Ranks to Avoid the Assumption of Normality Implicit in the Analysis of Variance. *Journal of the American Statistical Association* 32:675–701.

GOODMAN, LEO A. 1959 Partial Test for Partial Taus. *Biometrika* 46:425–432.

JONCKHEERE, A. R. 1954 A Distribution-free *k*-sample Test Against Ordered Alternatives. *Biometrika* 41: 133–145.

○KENDALL, MAURICE G. (1948) 1970 *Rank Correlation Methods.* 4th ed. London: Griffin; New York: Hafner.

KENDALL, MAURICE G.; and SMITH, B. BABINGTON 1939 The Problem of *m* Rankings. *Annals of Mathematical Statistics* 10:275–287.

KENDALL, MAURICE G.; and SMITH, B. BABINGTON 1940 On the Method of Paired Comparisons. *Biometrika* 31: 324–345.

KRUSKAL, WILLIAM H. 1958 Ordinal Measures of Association. *Journal of the American Statistical Association* 53:814–861.

KRUSKAL, WILLIAM H.; and WALLIS, W. ALLEN 1952 Use of Ranks in One-criterion Variance Analysis. *Journal of the American Statistical Association* 47: 583–621.

LUCE, R. DUNCAN 1959 *Individual Choice Behavior: A Theoretical Analysis.* New York: Wiley.

MANN, HENRY B.; and WHITNEY, D. R. 1947 On a Test of Whether One of Two Random Variables Is Stochastically Larger Than the Other. *Annals of Mathematical Statistics* 18:50–60.

MOSTELLER, FREDERICK 1951 Remarks on the Method of Paired Comparisons. *Psychometrika* 16:3–9, 203–206, 207–218. → Part 1: The Least Squares Solution Assuming Equal Standard Deviations and Equal Correlations. Part 2: The Effect of an Aberrant Standard Deviation When Equal Standard Deviations and Equal Correlations Are Assumed. Part 3: A Test of Significance for Paired Comparisons When Equal Standard Deviations and Equal Correlations Are Assumed.

OWEN, DONALD B. 1962 *Handbook of Statistical Tables.* Reading, Mass.: Addison-Wesley. → A list of addenda and errata is available from the author.

PRATT, JOHN W. 1964 Robustness of Some Procedures for the Two-sample Location Problem. *Journal of the American Statistical Association* 59:665–680.

SAVAGE, I. RICHARD 1962 *Bibliography of Nonparametric Statistics.* Cambridge, Mass.: Harvard Univ. Press.

SIEGEL, SIDNEY 1956 *Nonparametric Statistics for the Behavioral Sciences.* New York: McGraw-Hill.

SIEGEL, SIDNEY; and TUKEY, JOHN W. 1960 A Nonparametric Sum of Ranks Procedure for Relative Spread in Unpaired Samples. *Journal of the American Statistical Association* 55:429–445.

THOMPSON, W. A. JR.; and WILLKE, T. A. 1963 On an Extreme Rank Sum Test for Outliers. *Biometrika* 50:375–383.

THURSTONE, L. L. 1927 A Law of Comparative Judgment. *Psychological Review* 34:273–286.

►TORGERSON, WARREN S. 1968 Scaling. Volume 14, pages 25–39 in *International Encyclopedia of the Social Sciences.* Edited by David L. Sills. New York: Macmillan and Free Press.

WILCOXON, FRANK 1945 Individual Comparisons by Ranking Methods. *Biometrics* 1:80–83.

Postscript

Several books on nonparametric statistics containing major sections on ranking methods have been published. Very readable elementary treatments by Conover (1971) and Hollander and Wolfe (1973) provide wide coverage and good bibliographies; the latter contains a particularly extensive set of tables. Another elementary book by Kraft and Van Eeden (1968) also features a large collection of tables. Lehmann (1975) treats randomization models at an elementary level but discusses the evaluation of power at an intermediate level. A more mathematical approach is taken in the concise accounts by Noether (1967) and Hájek (1969), whereas Gibbons (1971) gives a fairly detailed treatment at an intermediate level. The remarkable advanced book by Hájek and Sidák (1967) deals specifically with rank tests, and Puri and Sen (1971) provide an extensive advanced presentation of nonparametric methods in multivariate analysis.

HERBERT A. DAVID

[*See also the biography of* WILCOXON.]

ADDITIONAL BIBLIOGRAPHY

CONOVER, WILLIAM J. 1971 *Practical Nonparametric Statistics.* New York: Wiley.

GIBBONS, JEAN D. 1971 *Nonparametric Statistical Inference.* New York: McGraw-Hill.

HÁJEK, JAROSLAV 1969 *Nonparametric Statistics.* San Francisco: Holden-Day.

HÁJEK, JAROSLAV; and ŠIDÁK, ZBYNĚK 1967 *Theory of Rank Tests.* New York: Academic Press.

HOLLANDER, MYLES; and WOLFE, DOUGLAS A. 1973 *Nonparametric Statistical Methods.* New York: Wiley.

KRAFT, CHARLES H.; and VAN EEDEN, CONSTANCE 1968 *A Nonparametric Introduction to Statistics.* New York: Macmillan.

LEHMANN, E. L. 1975 *Nonparametrics: Statistical Methods Based on Ranks.* San Francisco: Holden-Day.

NOETHER, GOTTFRIED E. 1967 *Elements of Nonparametric Statistics.* New York: Wiley.

PURI, MADAN LAL; and SEN, PRANAB KUMAR 1971 *Nonparametric Methods in Multivariate Analysis.* New York: Wiley.

NONPROBABILITY SAMPLING

See under SAMPLE SURVEYS.

NONSAMPLING ERRORS

See under ERRORS.

NORMAL DISTRIBUTIONS

See DISTRIBUTIONS, STATISTICAL, *article on* SPECIAL CONTINUOUS DISTRIBUTIONS, *and the biographies of* GAUSS; LAPLACE; MOIVRE.

NORMALITY, TESTS OF

See GOODNESS OF FIT.

NUMERICAL ANALYSIS

See COMPUTATION.